冯顺山教授学术文集

A Collection of Academic Papers of Professor Shunshan Feng

《冯顺山教授学术文集》编辑组　编

U0433163

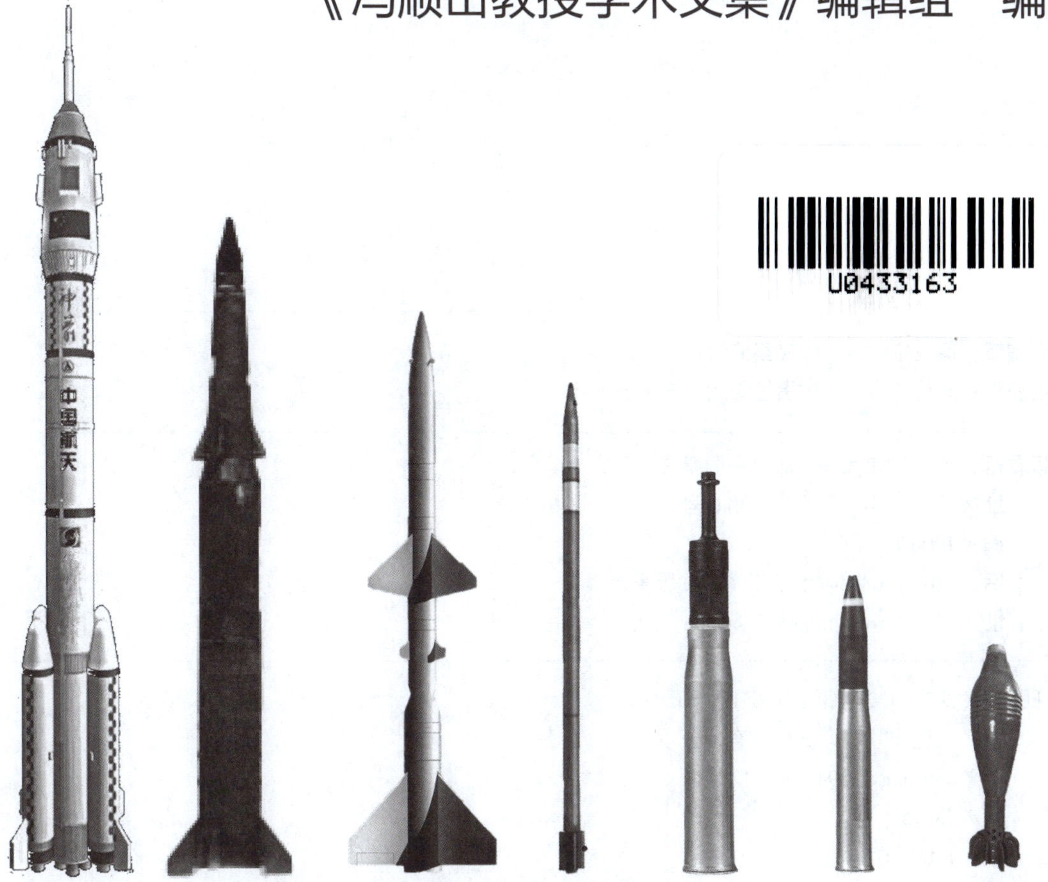

北京理工大学出版社
BEIJING INSTITUTE OF TECHNOLOGY PRESS

版权专有　侵权必究

图书在版编目(CIP)数据

冯顺山教授学术文集／《冯顺山教授学术文集》编辑组编. -- 北京：北京理工大学出版社，2023.2

ISBN 978-7-5763-2158-6

Ⅰ.①冯… Ⅱ.①冯… Ⅲ.①弹药-文集 Ⅳ.①TJ41-53

中国国家版本馆CIP数据核字(2023)第065439号

责任编辑：徐　宁　　**文案编辑**：国　珊
责任校对：周瑞红　　**责任印制**：李志强

出版发行／北京理工大学出版社有限责任公司
社　　址／北京市丰台区四合庄路6号
邮　　编／100070
电　　话／(010) 68944439（学术售后服务热线）
网　　址／http://www.bitpress.com.cn

版 印 次／2023年2月第1版第1次印刷
印　　刷／三河市华骏印务包装有限公司
开　　本／889 mm×1194 mm　1/16
印　　张／58.75
字　　数／1709千字
定　　价／215.00元

图书出现印装质量问题，请拨打售后服务热线，负责调换

冯顺山教授

冯顺山

Feng Shunshan，（1952. 8. 18—　　）

中国“毁伤技术与弹药工程”专家，教授/特聘教授，国防创新团队带头人。祖籍江苏省无锡市，1952 年出生于上海。1965 年就读于上海市长风中学（鲁迅中学），1969 年 5 月“知青上山下乡”赴黑龙江生产建设兵团六师六十团三连，1971 年 2 月入党。1973 年 9 月在北京工业学院“火箭战斗部技术”专业上学，1976 年 11 月毕业并留校任教。1982 年在职研究生毕业。

历任北京理工大学“战斗部与终点效应”实验室主任、力学工程系/机电工程系（八系）主任、机电工程学院院长、“爆炸科学与技术国家重点实验室”主任、教授、首席教授/责任教授、特聘教授、博士生导师、机械运载学部委员。主要社会兼职历任：兵器工业“弹箭技术”专业组副组长、原总装备部“枪炮弹箭”专业组专家、海军“导弹与舰炮技术”专业组专家、中国科协技术委员会第五届委员会委员、中国兵工学会理事、中国宇航学会理事、中国宇航学会无人飞行器分会“战斗部与毁伤技术”专业委员会第三届～第六届主任委员、中国兵工学会“爆炸与安全技术”专业委员会第二届～第四届主任委员、中国兵工学会“弹药技术”专业委员会第二届～第五届副主任委员、连续三届“国家安全生产专家”专业组副组长和专业组组长、国防科技奖/国家科技奖“兵器专业评审组”成员等。

从事“弹药战斗部技术”学科的教学、研究生培养、科学研究和学科建设工作，主要研究方向为：毁伤与弹药工程、爆炸技术及防护。20 世纪 80 年代率先在中国开展“杀伤战斗部威力参数精准测试技术”和多款新型战斗部研制，成果获得 1 项国家级 1 项部级科技奖。90 年代开展“高威力战斗部”“定向杀伤战斗部”和大型沉船“解体/清障”水下爆炸等技术研究，成果获得 4 项部级科技奖。21 世纪初原创“软毁伤电力目标”“无附带损伤拦截及毁伤”小型无人机、高效毁伤巡航导弹等技术，以及新概念弹药研制，成果获得 2 项国家级 7 项部级科技奖。2010 年以来原创“非致命封锁航母飞行甲板”“侵爆毁伤元”和“柔性防爆”等技术，以及有关新装备研制，成果获得 6 项部级科技奖励。参与组织制定军方“八五”～“十二五”弹药战斗部技术预先研究计划指南，提出科技前沿性发展的建议。对创建先进毁伤及弹药技术方法、前沿学术、创新型人才培养和一流学科建设做出了卓越贡献。

荣获“中国青年科技奖”、国务院特殊津贴、“全国高校先进科技工作者”、“国家有突出贡献中青年专家”、国防“511 人才”学术技术带头人、国防科技创新团队带头人和“北京市优秀教师”等十余项国家或部级荣誉奖励。2013 年作为中国工程院机械与运载学部院士有效提名人。

指导工学博士生 70 名，工学硕士生 39 名、博士后 7 名，发表学术论文 400 余篇，授权技术发明专利 110 余项，另受理 40 余项，出版著作教材和编制学术刊物合计 18 册。

《冯顺山教授学术文集》编辑组

2023 年元月

代 序

—授业恩师冯顺山教授

岁月不居，时节如流，七十之龄，忽焉已至。生命的积累，成就五味七彩的人生回忆，2022 年 8 月 18 日，授业恩师冯顺山教授 70 寿辰，汇锦《文集》以贺，回溯砥砺前行路，浅吟德艺双馨誉。承蒙众师弟师妹信任，为先生《文集》代序，欣然之余，诚恐有加。

1952 年 8 月 18 日，先生出生于上海，祖籍江苏无锡市。1969 年，先生作为“老三届”上海知青，“上山下乡”赴黑龙江生产建设兵团六师六十团三连，1973 年，推荐入学原北京工业学院（现北京理工大学）“火箭战斗部技术”专业学习，1976 年毕业并留校工作，1982 年在职硕士研究生毕业，1991 年破格晋升为教授，1993 年获批国务院学位委员会授权博士生导师。

先生是我国武器毁伤技术领域的著名专家，近 50 年职业生涯，笃志潜心“毁伤技术与弹药工程”学科方向的人才培养、科学研究、学科建设等工作，成就斐然，贡献卓著。

学科贡献，仰之弥高。先生自 1984 年担任“火箭战斗部技术”专业实验室主任伊始，历任力学工程系主任、机电工程学院院长、爆炸科学与技术国家重点实验室主任、毁伤技术与弹药工程学科方向首席教授，为专业/学科/实验室建设、实现“国内引领、国际一流”做出突出贡献。瞄准培养国防创新拔尖人才、建设大平台、承担大项目、产出大成果的发展目标，主持中关村校区戊区和靶道“战斗部与终点效应”实验室建设，开启支撑新时期人才培养和科研创新之路；主持“爆炸科学与技术”国家重点实验室申报、立项和建设验收，成为北京理工大学唯一国家重点 A 类实验室；主持“211”工程三期和“985”工程二期建设，实现办学和科研条件全面提质；主持国防科工委“战斗部基础条件”建设，建成河北怀来东花园“爆炸与毁伤”实验基地，为专业/学科强势发展和高水平科研创新提供关键支撑。特别是 1998 年以来，先生作为“毁伤技术与弹药工程”学科方向首席教授，高瞻远瞩，承前启后，开拓进取，围绕“一流师资队伍，一流课程体系，一流办学条件”建设，形成显著特色和优势，发挥无可替代的核心引领作用。近 15 年来，所在学科方向以第一完成单位和第一完成人获得国家技术发明二等奖 4 项、省部级科技奖一等奖 8 项，自主培养“四青”人才 3 人。2012 年弹药工程与爆炸技术本科专业获批首批国家卓越工程师培养计划专业，2019 年获批首批国家一流本科专业建设点，2019 年无人飞航工程系党支部获批首批全国高校样板支部，2021 年被授予首批国防科技工业先进集体。

学术造诣，钜儒宿学。先生是 2013 年中国工程院院士增选机械与运载学部有效候选人，中国青年科技奖获得者，国家有突出贡献中青年专家，国家教委高校先进科技工作者，国防科工委首批国防创新团队带头人，享受国务院政府特贴专家。曾任北京理工大学机械与运载学部第一届和第二届委员，“爆炸科学与技术”国家重点实验室学术委员会委员，原总装备部“枪炮弹箭”专业组专家，海军“导弹与舰炮技术”专业组专家，中国科协第五届委员会委员，中国兵工学会理事，中国宇航学会理事，中国宇航学会“战斗部与毁伤技术”专委会第三至第六届主任委员，中国兵工学会“爆炸与安全技术”专委会第二至第四届主任委员，中国兵工学会“弹药技术”专委会第二届至第五届副主任委员，国家科技奖评委，国防科技奖评委。

科学研究，创新垂范。先生是国防科技和武器装备创新的先行者，守正出奇，取得一系列创新性强、应用价值高、影响力大的原始创新成果。20世纪80年代，国内最先开展“破片杀伤战斗部威力场参数精准测试技术”研究，获国家科技奖1项；20世纪90年代，致力于“高威力战斗部技术”“定向杀伤战斗部技术”“大型沉船解体清障水下爆炸技术”等研究，获部级科技奖4项。21世纪前10年，着力“软毁伤电力设施”“无附带拦截毁伤小型无人机”“高效毁伤巡航导弹”等技术研究及装备研制，获国家科技奖2项、部级科技奖7项。近10年来，围绕“非致命封锁航母飞行甲板”“侵爆毁伤元”“柔性防爆”等技术研究及装备研制，获部级科技奖励6项。发表论文400余篇，授权发明专利110余项，出版著作、教材和编制学术刊物18册。

人才培养，教泽绵长。师者，匠心如光，止于至善，微以致远。先生造诣纵横理论创新、技术发明、工程应用和装备研制，力行国防科技创新实时融贯于人才培养，先后指导毕业博士生70名、硕士生39名、博士后7名，从业遍布高校、兵器、航天、船舶、航空、公安、军队等行业，传承与创新，一批毕业生业已成长为国防领域的中流砥柱。现任中央军委装备发展部副部长何玉彬将军，火箭军研究院导弹工程所所长张旭荣大校，陆军研究院炮兵防空兵所孙韬和徐立新大校，陆军试验训练基地王云峰大校，航天科技集团一院“长征十一号”运载火箭总师彭昆雅，航天科技集团一院14所党委书记李磊、国家技术发明二等奖获得者李国杰主任，航天科工集团四院红林公司总经理崔晓刚，兵器工业集团导航与控制所副所长周睿，获得多项省部级科技奖的中国工程物理研究院冯高鹏主任，兵器工业集团203所国家科技进步一等奖获得者罗健，公安部1所部级科技进步一等奖获得者刘春美，航天五院卫星应用总体部市场总监张晓东，航天科工集团四院41所点火专业首席专家王宇，中北大学学科方向带头人曹红松教授，北京理工大学蒋建伟教授、贵大勇教授、吴成教授、黄广炎教授等博士、博士后，不一而足，兼受教于先生门下。

老三届，黑龙江生产建设兵团，是先生人生起航的港湾，曾经的知青岁月，涵养了先生勤俭耐劳的品格；随后的大学和研究生求学经历，开启了先生科技人生的智慧，提供了先生扬帆起航的动力；从专业实验室主任、系主任、学院院长、国家重点实验室主任的职务履历，到破格晋升教授和博士生导师，从型号总师，到前沿国防科技创新和重大标志性科技成果取得，确立了先生的学术地位和行业影响力；从戊区、靶道，到西山实验中心和东花园实验基地建设，从国家重点实验室申报、建设和运行，到国家一流专业/学科形成，先生的付出和贡献，同感共识。

先生的成功、成就与贡献，我们无以复制，但教诲、鞭策、鼓励长存。

值此先生70寿辰，借代序《文集》之缘，携众师弟妹恭祝恩师：

福如东海，寿比南山，春晖永绽，如意安康。

敬撰

于北京理工大学

2023年元月7日

编者的话

21世纪初，我们就想编撰一部《冯顺山教授学术文集》，这个愿望终于在21世纪20年代初实现了。

冯顺山教授是上海市长风中学（鲁迅中学）“老三届知青”，1969年5月“上山下乡”赴黑龙江生产建设兵团，党龄51年、工龄53年。1976年留校任教，自此投身于我国国防科技和教育事业，从教46年。近半个世纪以来，冯顺山教授始终工作在一线将其全部才智和精力奉献给了我国兵器科学技术的发展、新型武器装备研发、人才培养和毁伤技术与弹药工程学科建设。他亲身经历并践行我国国防科技和一流学科建设蓬勃发展的过程，为弹药战斗部科技显著进步并向现代化发展做出了创新的重要贡献。

冯顺山教授喜欢挑战，注重原创、直面难题，亲力亲为、攻坚克难，成果高产。特别是针对重大重点军事需求，在目标软毁伤、无附带/低附带毁伤、非致命毁伤和高效杀伤等科技研究及有关新型弹药战斗部研制方面，奇思巧工，均取得原创性研究成果。有的起到立代建碑的作用，成为武器装备创新发展的典范，得到军委和军种首长的高度重视；多款新型弹药战斗部装备使我军具备某些特殊的新战力，助力非对称作战条件下“能打仗打胜仗”。不仅如此，冯顺山教授还在不同历史时期直接并深度参与我国国防科技工业和武器装备预研指南编制、战略发展建言献策等工作，他的许多建议和意见被高层主管部门采纳并付诸实施。

冯顺山教授在学科建设，尤其是先进专业实验基地建设方面，同样成果卓著，创建的“爆炸科学与技术”国家重点实验室和东花园“爆炸与毁伤”实验基地，成为学科可持续发展走向一流的重要科技研究平台和主体实验研究基地。冯顺山教授在创新型高层次人才培养方面，十分注重学生原创能力培养，他师德高尚，精于授业解惑，受到学生们敬爱，相当多的学生成为单位科技骨干或带头人。在毁伤与弹药工程学科（无人飞航工程系）建设方面，始终面向国防重点/重大需求，凝练前沿科技方向，守正出奇、进入一流、持续发展。教学科研和人才培养均取得多项标志性成果，有力支撑北京理工大学“兵器科学与技术”成为一流（A+）学科。他的许多新思想、新观念、新做法，长期成为学科建设和持续发展的启迪。

由于历史原因和篇幅所限，这部学术文集从冯顺山教授发表的400余篇论文中选录123篇，均是冯顺山教授及其指导的部分硕士生、博士生和博士后科研人员发表的学术论文。由于保密的原因，冯顺山教授的许多重要学术文件、研究报告、装备发展建议报告和学科发展报告无法编入文集，对于收录的文稿、科研项目和获奖项目名称，以及相关照片也做了保密性处理。但愿这部学术文集能够较好地反映冯顺山教授的学术思想和学术成就。

《冯顺山教授学术文集》编辑组

2023年元月

目　　录

第一部分　主要学术方向论文

第二部分　目标毁伤及杀爆战斗部技术

第三部分 短路软毁伤与导电纤维弹技术

第四部分　侵彻作用效应

第五部分　封锁类弹药技术

第六部分　爆炸冲击作用及防护

第七部分　其他弹药战斗部技术

第 一 部 分
主要学术方向论文

弹药战斗部技术的发展与创新

中国兵器工业集团公司　胡国强

北京理工大学　冯顺山

摘　要：本文论述了弹药战斗部在高新技术战争中的重要作用、三军武器装备对高效能先进战斗部的需求和国内外弹药战斗部的发展简况。说明了：①精确打击的实现涉及两个方面的问题，一是精确命中，二是高效能毁伤；②毁伤能力的提高将使精确打击起到事半功倍的效果；③新型目标、加固目标和新作战方式对毁伤能力提出了严重挑战；④新毁伤理念、方法、技术必须大力度超前发展，适应未来战争中多任务灵活打击的需要；⑤弹药战斗部已成为一种高新技术武器装备。阐述了2020年弹药战斗部重点发展方向与关键技术，指出了弹药战斗部领域存在着的主要问题，提出了“十五”期间成立总装弹药战斗部专业组等四条有助于弹药战斗部技术大跨度发展的策略和建议。

关键词：弹药；战斗部；毁伤

各军兵种中，不论是攻击型武器还是防御性武器，都需配置对目标起到毁伤作用的弹药或战斗部。战斗部是武器的有效载荷。现代化战争，尤其是高新技术战争，很大程度体现在武器装备的较量，尤其是高新技术武器装备的较量上。未来战争中，战斗人员面与面的对阵越来越少，基本不发生，利用各种射程的精确打击武器（如中远程炮弹、火箭弹、灵巧弹药、导弹、飞机等）攻击对方不同距离处的军事目标尤其是高价值军事目标，便可取得某种意义上的作战胜利，并产生重大的军事、政治和经济效果。1999年的科索沃战争表明，精确打击高价值目标（军事方面、经济方面以及政治方面的高价值目标）成为未来战争的一种十分重要的作战样式，引领作战思维的变革并走向现代化。

精确打击的实现涉及两个方面的问题，**一是精确命中，二是高效毁伤**，否则打击的任务不能完成。一个武器系统在实施打击任务时，需要高质量地完成以下五个阶段任务：一是定位目标或发现目标，二是由发射或动力装置接近目标，三是由制导系统命中目标，四是由战斗部毁伤目标，五是作战效果评估。其中，按预定要求毁伤目标是能否完成打击任务的最终标志。实战中，**要求尽可能用数量少的精确命中武器，在尽可能短的时间里，有效地毁伤军事目标，实现高效能打击。**

随着技术进步和毁伤需求，毁伤方式及能力得到快速发展，从以前的单一硬性毁伤（摧毁性的致命性的毁伤）到软性/非致命性毁伤（目标失效或失能后可以恢复）、威力可控毁伤、干扰性/诱饵性杀伤（失去或明显减弱攻击预定目标的能力）和爆炸产生电磁脉冲破坏电子器件或电路等多种多样的毁伤方式。**不同的毁伤方式用于对付不同类型的目标并适应不同作战思想，多种毁伤方式可产生灵活多模式的战术使用。**

毁伤能力由低向高发展，如破甲能力从几倍口径到10倍左右的口径，穿甲弹的长径比达30倍可击穿650～750 mm厚的装甲，云爆弹使冲击波威力的TNT（三硝基甲苯）当量增大到2～8倍，连续杆、离散杆等大破片技术可对目标产生显著切断效应，冲击反应破片可显著提升杀伤后效，子母弹可攻击多个目标或扩大攻击区域，串联战斗部对付加固（如披挂爆炸装甲等）的装甲目标、机场跑道和深层工事，干扰弹、诱饵弹使导弹攻击假目标或失去原来目标，等等。**毁伤能力的提高将使精确打击达到事半功倍的效果。**

但随着战场上新目标的不断产生、目标的加固技术、在对抗条件下生存能力提高以及战争思想的不断变革，给武器发展带来了严重的挑战，同时也给弹药战斗部技术或毁伤技术带来严重的挑战。只有创新，毁伤技术大跨度发展，才能应对挑战，并取得胜利。

1　三军武器装备和高新技术战争需要新型高效能弹药战斗部

随着国防科技和作战思想发展，军事目标的概念多样化，传统意义的有生力量、装甲车辆、防守阵地、火炮阵地、山头要塞、普通的飞机和军舰等虽仍是军事目标，但其重要性弱化。新目标层出不穷，典型的新目标有：巡航导弹，激光制导或末制导炸弹，高性能飞机，武装直升机，C^3I系统，雷达，航空母舰，高性能潜艇，先进的水中制导鱼雷，水中障碍，深层工事，发射井、电站和电网系统，空间轨道上的飞行器和数字化士兵等；还有那些关乎国民经济命脉的现代化设施（桥梁，交通枢纽，矿山，大型企业，通信网等）也都会成为新的攻击目标。**新毁伤方法、毁伤技术，必须大力度超前发展，**研究那些如何毁伤新目标以及潜在新目标的毁伤理念、机理、方法、技术，满足日益变化的现代化战争的需要。

毁伤与防护、易损性和生存力，是一对矛盾的相互促进发展的事物，打击能力的提高促进了防护能力的提高。高价值目标生存力提高技术日益受到重视，导弹智能化的反拦截技术、反干扰反诱饵技术，提高突防能力，装备、工事的抗打击加固，目标自身的被动和主动防御等，也都**对毁伤新方法新能力提出了挑战**。大威力、高效能和新概念的毁伤技术或弹药战斗部被强烈需求。高效能的弹药战斗部还可减少武器平台打击次数，这不但赢得作战时机，还可减少武器平台弹药库存，降低作战成本。

一弹多头（配置多种战斗部）、一头多种功能（瞬发、延时、自适应、定向杀伤、多毁伤等级）成为武器发展的一种重要趋势。战斗部类型也由原来的杀伤爆破型、杀爆燃型、破甲型、穿甲型等发展多种类型，并且不断增加。例如，破片聚焦、偏心引爆、壳体变形、起爆点适时选择、连续杆和离散杆等定向杀伤型战斗部，云雾爆轰战斗部，多自锻破片半穿甲型战斗部，分段杆式穿甲型战斗部，多级串联型战斗部，自适应、自瞄准型战斗部，子母型战斗部（包含开舱、抛撒子弹——定向抛、前抛、后抛、侧抛等），反地下深层目标战斗部，毫米波、红外、激光干扰/诱饵战斗部，烟雾隐身战斗部，碳纤维/导电纤维战斗部，红外照明战斗部、闪光致盲、发动机致熄、装备功能失效，人员失能等非致命杀伤战斗部，爆炸电磁脉冲战斗部，战场侦察战斗部，作战效果评估战斗部等。

战斗部不再是钢壳加炸药简单组合，先进战斗部的技术内涵向高新技术领域发展。新型材料（可燃金属、冲击反应材料、高性能合金、导电率优良的碳纤维/石墨/导电纤维，低易损性/钝感含能材料，特殊理化性能材料等）被大量应用；爆轰学、冲击动力学、弹道学、目标易损及软杀伤/非致命理论、运筹学等学科被综合应用于战斗部研究中；爆炸驱动控制、威力控制、起爆逻辑网络控制、系统优化、灵巧控制、安全可靠性、威力性能检测、毁伤评价和计算机仿真等技术深入渗透到弹药战斗部设计中。**弹药战斗部已成为一种高新技术的武器装备**。当自行开发一种特色新型武器时，首要问题是确定战斗部对目标的毁伤方式、毁伤等级和毁伤能力，然后再确定口径、推力、制导等问题。

总之，目标多样化及新目标的不断出现，作战思想或模式的变革，新理论、新技术的应用，使得**新型高效能弹药战斗部的需求空前强烈**。从事弹药战斗部或高效毁伤领域的科研人员，深感肩负历史使命，责任重大，创新前行。

2　国内外弹药战斗部的发展简况

2.1　美国弹药战斗部的发展计划

诸多国防科技先进国家中，美国在弹药战斗部方面属技术领先的国家。美国非常重视弹头技术进步和创新，认为毁伤能力的提高，不仅可实施多种打击，还可减少发射数量和库存量，可在军事上和经济上均取得显著的效益。这些思想可从美国1996年5月公布的“1996—2001美国国防技术领域计划”中得到体现。

计划中统筹考虑三军弹药战斗部的技术发展。子专题“常规武器（CW）”中，弹药、弹药组件、水雷、鱼雷是重点研究内容的组成部分。该子专题的**战略目标**中明确指出：“常规武器的特殊目标主要集中用来摧毁敌人人员、物质和基础设施的系统技术方面，但越来越强调通过非致命武器技术来使敌人丧失能力”。

美国将常规武器重点研究内容分为七个重点方向，“杀伤力与易损性”“弹头与炸药”是其中的两个方向，并设定“弹头与炸药”的**作战需求**为：

（1）新型/改进型空地导弹中的可瞄弹头，这种弹头可将杀伤概率提高到1，并且对导弹的需求减少20%～30%；

（2）杀伤力和抗现代对抗措施能力更强的自适应弹头，这种弹头可将弹药库存量需求降低30%～40%；

（3）穿甲能力提高300‰并且能摧毁硬度提高50%目标的穿甲武器；

（4）能够摧毁世界其他国家的ASW（反潜战）、ASUW（反水面舰船战）和SSTD（水面舰船鱼雷防御）威胁目标的多模鱼雷弹头，以及能杀伤已部署的和计划中的装甲的直接瞄准射击弹药，从而使战场交换比提高1～3倍。

“弹头与炸药”的研究主要目的定位于**提高武器的杀伤力、多任务灵活性、生存能力并降低费用**。其6年计划的重点研究目标是：

（1）摧毁先进的水下威胁；

（2）摧毁在储存、生产和部署中的大规模杀伤武器；

（3）摧毁先进的装甲和装甲防护系统；

（4）增大水下弹头和炸药的杀伤力与作战使用范围；

（5）摧毁加固的C^3I和采取了措施的地下目标；

（6）利用目标自适应弹头和先进炸药摧毁各种空中/地面威胁，利用超高速导弹对付时间关键目标。

从上面情况，可以看出弹药战斗部在美国武器研发中的重要地位和作用。美国发展毁伤技术有它自身的技术基础支撑和国家及军事需求。对于我军来说，不与之比年代、比品种，应根据近、中、长远作战需求发展特色的弹药战斗部技术，打出自己的毁伤优势。

2.2　国内弹药战斗部发展简况

国内弹药型号在“八五”“九五”有了一定的发展。我国的反装甲弹药，如破甲弹、穿甲弹、两级串联破甲战斗部，其威力性能达到国际较先进的水平。一些弹药、火箭弹、导弹战斗部型号研制成功，转入生产。

“八五”“九五”的预研取得了一批有价值有水平的成果，推动了“十五”型号研制或“十五”背景型号预研的发展。主要有以下几点：

（1）末敏弹技术，今年将完成演示验证；

（2）改性B炸药装药技术，可使装填TNT的弹威力提高30%～40%；

（3）红外隐身照明技术，毫米波干扰/诱饵技术，快速成烟烟幕干扰技术，箔条干扰技术，将转向型号上应用；

（4）爆炸开舱与抛撒子弹技术在型号上得到应用；

（5）一次引爆、二次引爆云爆弹技术转向型号应用；

（6）反机场跑道技术在型号上得到应用；

（7）软毁伤碳纤维战斗部技术取得了重要成果，转向武器化应用研究；

（8）毁伤反坦克导弹和反舰导弹的战斗部技术得到突破。

3　2020年弹药战斗部重点发展方向与关键技术

3.1　重点发展方向

（1）灵巧弹药；

（2）防空反导弹药及战斗部；
（3）反深层目标弹药及战斗部；
（4）反航空母舰、反潜、反鱼雷战斗部；
（5）电子对抗弹药及战斗部；
（6）软杀伤/非致命封锁弹药及子母战斗部；
（7）远程弹药；
（8）新型/新概念弹药及战斗部。

我国的弹药战斗部与美国相比，在型号种类方面约是它们的30%~35%，技术水平差距约为20年；预研课题数量方面差距可能更大，预研技术水平差距15~20年。

3.2 关键技术

3.2.1 目标易损性分析与毁伤准则

（1）巡航导弹、制导炸弹、武装直升机能被有效毁伤的方法和技术；
（2）深层工事（指挥、通信等高价值深层构筑物）、加固深层工事被摧毁的方法和技术；
（3）先进航空母舰、潜艇及鱼雷被摧毁的方法和技术；
（4）毫米波、红外、激光导引头的易干扰性或易诱饵性及失效准则，干扰诱饵方法与技术；
（5）装备、高价值目标、人员失效或失能的软毁伤或非致命杀伤方法和技术；
（6）易损性分析的计算机数值模拟和仿真技术；
（7）其他重要新目标在新型毁伤元作用下的易损性分析方法。

3.2.2 高威力或先进的战斗部技术

（1）不同弹道阶段毁伤导弹/制导炸弹的战斗部技术；
（2）摧毁武装直升机的技术；
（3）反深层目标、反航母的半穿甲战斗部技术；
（4）长杆状EEFP（延长爆炸成形弹丸）技术，长杆穿甲技术；
（5）多种方式的定向杀伤战斗部技术；
（6）子母战斗部技术；
（7）两级或多级反潜艇战斗部技术；
（8）复合干扰/诱饵剂（含水中应用），装填物抛撒与分散技术；
（9）电力、电网和武器装备（侧重电子装备）的软杀伤战斗部技术；
（10）人员失能的非致命战斗部技术。

3.2.3 新概念/新型弹药战斗部的关键技术

针对非对称战争对武器弹药的可能需求，自主创新研究新概念毁伤方式和方法，高效毁伤技术，推动新型战斗部发展。

4 存在的主要问题

4.1 应用基础、技术基础差

长期以来，我国弹药战斗部主要走仿制、仿研和少量从国外引进的道路。在仿制过程中，注重产品性能参数，对于设计思想、设计理论、应用科学、弹药系统总体设计技术、弹药体系构架，以及支撑性技术基础等重大问题很少予以有力的关注。以至于眼界狭窄、观念落后，弹药战斗部体系科学性较差，技术层次较低。由于严重缺乏应用基础研究、技术基础薄弱，因此自主发展高新技术弹药的综合能力较低，缺乏良性发展的学术及技术活力。

4.2 弹药战斗部预研薄弱

1. 预研力度小

弹药战斗部的预研力度较小，预研成果远不符三军发展先进武器的要求。弹药战斗部是一个较大系统，涉及的技术范围很广，研究中必须进行规模较大的破坏性试验，经费消耗较大，但其预研经费的力度仅相当于某些三级配套项目的经费，甚至还不如。导弹战斗部技术基本无预研，这是因为导弹及水中兵器的总体和主体在原航空、航天和船舶行业，但战斗部配套生产任务主要由兵器行业承担，不同的隶属关系和经费问题使得几方面都不设立预研课题，以至于导弹战斗部技术明显落后，严重影响导弹等武器装备研发的作战性能。

2. 缺乏目标易损性分析研究

由于长期仿制或仿研，只注重产品的结构和性能参数，很少进行目标易损性的预先研究工作。对于目标毁伤标准、杀伤准则、目标易损机理和毁伤机制，目标易损性数据和毁伤效能评价等方面研究成果很少，很不适应自行开发新弹药战斗部，不能在短期内形成我军特色毁伤能力需求的弹药战斗部。

3. 弹药战斗部系统总体技术弱

弹药战斗部通常是二级配套系统，对引信、传爆序列、装药/装填物、杀伤元素、威力、终点弹道、气动与稳定、毁伤概率、系统可靠性和使用安全性等须统筹考虑，提出技术性能及接口指标要求，并进行技术集成及功能优化。由于系统总体技术弱，没有很好发挥出应有的作用。

4.3 没有设立跨行业的弹药战斗部专业组

弹药战斗部为三军所需求，高效毁伤技术是跨兵种、跨行业的一种共性通用技术，但没有跨兵种、跨行业的专业组予以统筹考虑学术及技术发展、技术共享、预研经费的统筹使用和科研力量的综合利用等问题，科学策划/计划好近、中、长期的预研和新型弹药战斗部研发。

5 策略建议

（1）“十五”期间成立总装备部的弹药战斗部专业组，统筹考虑、安排弹药战斗部技术或毁伤技术的发展与应用，从而发挥好其在武器装备中的重要作用。

（2）大力加强毁伤前沿问题的应用基础研究，使重大的基础问题得到突破，支撑毁伤技术和型号应用的大跨度发展。建议成立毁伤科学与技术国防科技重点实验室。

（3）增强弹药战斗部“十五”预研力度，以便满足三军武器装备对其的重要需求。

（4）高度重视目标易损性分析的研究，使该项具有技术基础性的关键技术能对弹药战斗部发展起到有力的支撑作用。

2000 年 2 月在总装备部科技委年会上发表

非核终端毁伤技术

冯顺山　主笔

摘　要：针对我国武器装备发展战略，论证非核终端毁伤技术的重要地位和军事需求，以及终端毁伤能力严重低下等问题；阐述非核终端毁伤基础科研成果是支撑武器装备跨越发展的关键和原创技术之一，需得到应有重视；提出拟开展研究的重大基础性技术问题，以及高效毁伤科技发展的对策，明确提出建议：非核终端毁伤技术列为国防重大科技专项。

按：核与非核打击是世界新军事变革三大特征中的一个重要方面。战争的目的是对抗制胜，武器的目的是终端毁伤。武器对目标实施精确非核打击时需解决两大方面问题，一是打中目标，二是击毁目标。提高非核终端毁伤能力，是非核毁伤体系中的十分重要的方面，是武器非对称发展战略中应研究的重点，以便尽快形成非核威慑的特色。武器装备跨越或创新发展，终端毁伤方面的国防基础科研必须走在前面，需大力度超前发展，否则就会拖武器作战能力的后腿。非核终端毁伤技术研究是"强化基础"的重要组成部分。如何加强非核终端毁伤技术研究，尤其是加强基础和创新研究，是国防科工委实施基础能力发展战略的重要方面，美国、俄罗斯和欧洲等国已有这方面成功经验，收效十分显著。结合国情，我们应在管理、机制、计划、经费等方面采取有力措施，切实加强非核终端毁伤技术的研究工作。

非核终端毁伤技术是指不包括核爆炸和"生化武器"对目标毁伤的所有**终端常规毁伤技术**（以下简称"终端毁伤技术"），是研究**武器在目标处毁伤能力问题**，尤其是终端高效毁伤的科学与技术问题。在今后的高新技术局部战争中，如何毁伤和能否有效毁伤对抗方的高价值目标、坚固防护目标和新型目标以及提高武器终端毁伤效能等问题，都是十分重大的共性基础问题，如我军急需对付多种重要目标的多种类战斗部，一弹多头、一头多战斗部，现役及在研武器装备毁伤效能评估等方面遇到的终端毁伤问题都亟须解决。目前这些问题不能被较好地解决甚至是尚未找到办法解决，其根本原因是我国在非核终端毁伤技术领域的国防基础科研很薄弱，有些正处于起步研究阶段，解决问题的能力与需求差距很大。终端毁伤技术是推动武器装备跨越发展和创新的**重要原创技术之一**，这方面的能力就更为薄弱。

1　非核终端毁伤技术的重要地位和军事需求

1.1　重要地位

1.1.1　终端毁伤在国防科技发展中的重要地位

毁伤目标，是武器装备特有的功能，是使用陆、海、空、二炮武器装备进行作战的最终目的。终端毁伤是武器打击目标时的"**临门一脚**"，或是武器在弹道终端的"**打击功夫**"。在高新技术战争中，能否有效毁伤目标，是决定武器应用价值的根本，甚至关系到战争的胜负。

图1.1可说明终端毁伤能力在武器装备打击目标过程中的重要地位。

武器中担当**直接毁伤目标**任务的部件是战斗部，**战斗部是武器装备的有效载荷。终端毁伤，是武器到达目标处战斗部杀伤/摧毁目标的能力与效应；终端毁伤技术是研究军事目标在毁伤元作用下如何损伤的、战斗部威力及如何杀伤目标的系统理论、方法和手段；**终端毁伤技术是支撑以型号应用为背景的战斗部关键技术预研和先进战斗部装备研发、威力评价及正确使用的基础；也只有雄厚丰富的国防基础科研成果和先进系统的技术基础，才能满足主动与被动防御需求，才能满足武器装备对多种类高价值目标、先进目标、新型目标和未来目标**终端高效毁伤**的需求，才能形成有力的**非核终端毁伤威慑**。

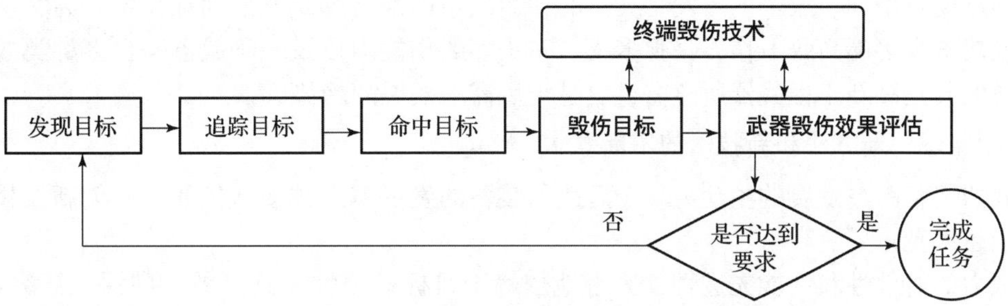

图 1.1　终端毁伤能力在武器装备打击目标过程中的重要地位

1.1.2　终端毁伤技术在战斗部预研和型号研制中的重要作用

武器装备虽应用的军种兵种不同、发射平台有差异，但终端毁伤某同类目标的毁伤技术是相同的。终端毁伤技术不仅是海、陆、空、二炮及各兵种武器战斗部装备的**共性或通用技术**，而且是需求变化最大、发展最活跃和种类需求最多的一类共性技术。其技术进步、发展和创新的成果，均可在三军武器的战斗部装备上得到应用。

图 1.2 表明非核终端毁伤技术在战斗部装备预研及型号研制中的重要作用。

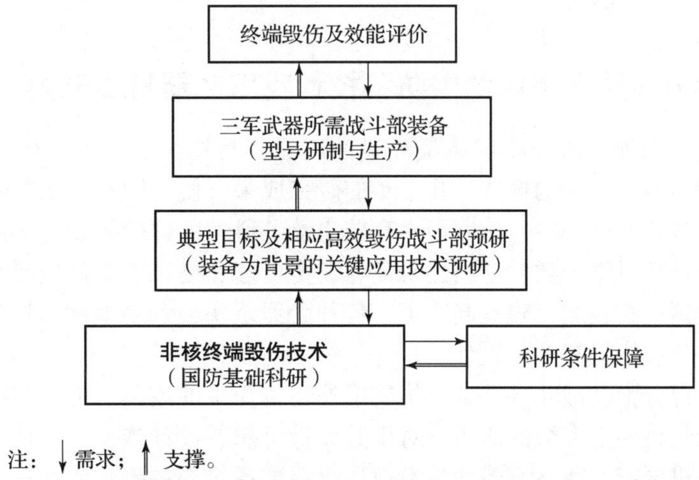

图 1.2　非核终端毁伤技术在战斗部装备预研及型号研制中的重要作用

非对称作战、自主创新和跨越发展的思想，对武器装备设计提出了很高的要求，然而对终端毁伤技术提出了更高的要求。例如，用什么方式毁伤/摧毁多种类高价值目标、如何分析重点目标易损性及用什么样的方法获取易损数据、毁伤机理及作用效应、战斗部实现大威力设计方法及理论、威力评价方法与技术，以及武器终端毁伤效能评估方法等基础性的问题严重地显现出来。科学地回答武器在目标处的**毁伤能力**问题，须开展系统深入的、有计划的、大力度的国防基础科研，从而建设好武器装备在近、中、长期均有着强烈需求的**终端毁伤技术平台**。

终端毁伤技术可为警用武器研发、民用爆破器材设计和应用技术提供有关科学技术的支撑，如石油、矿山和海洋矿产资源等的爆破及强冲击开采，还可在宇航等空间技术发展中起到重要作用。因而，它具有明显的军民两用技术特征。

1.2　军事需求

1.2.1　武器装备对提高终端毁伤能力的需求

战争表现的最显著特征之一是武器之间的对抗，也就是毁伤/摧毁对方目标能力的对抗。目标的进步，新目标的产生，尤其是高新技术目标的产生预示着战争中技术对抗的升级，还预示着打赢一场高技术局部战争的难度在增加。战争的爆发是突然的，而战争的准备却是长期而艰巨的。为实现武器对多目标灵活的战场攻击和终端高效毁伤，一弹多头、一头多毁伤能力以及一种武器平台多种类战斗部是今后重点武器装备尤其是高新工程武器所急需，也是今后需求发展的趋势。

1.2.2 新原理、新方法和新技术的不断需求

创新是武器装备跨越发展的主动力。今后武器装备的发展越来越多地依靠理论创新及技术创新，技术推动的作用越来越重要。

战斗部威力的有效性和性能先进性，紧密地依赖于目标易损性及终点效应的深入研究和毁伤方法、机理及技术的创新发展，否则战斗部威力与结构设计和技术无可靠的理论依据与试验数据。毁伤技术发展历史过程表明：聚能装药与杆式穿甲弹新技术使得破甲穿甲威力大幅度提高；云雾爆炸的应用，不仅冲击波的毁伤区域大大拓宽，并且产生窒息和温压效应；串联战斗部结构技术可有效用于对付爆炸反应装甲、硬目标及潜艇耐压层；子母战斗部技术使毁伤面积/效率大幅度提高；定向毁伤技术更有效地用于反导或反飞机；应用碳纤维短路毁伤机理、箔条及烟雾干扰机理、诱饵机理、光学致盲机理、爆炸电磁脉冲使电子器件失效机理和人员失能机理等的研究成果，开发出多种类的非致命毁伤的战斗部装备；等等。

国际上终端毁伤技术竞争激烈，显著的创新成果可能会导致某种可怕威力的武器出现。终端毁伤技术创新难度大、探索性强、涉及多学科和多技术领域，须经几代人的奋斗才会赶上世界先进水平，并力求取得竞争的优势地位。

2 终端毁伤能力问题严重制约国防科技和我军武器打击能力的发展

美国是终端毁伤技术和弹药战斗部装备最先进的国家，其军事实力不仅依赖于国家雄厚的经济实力，更重要地依赖于国防基础科研的优势。几十年的研究成果，使美国在终端毁伤技术领域始终处于最前沿。其现状和发展趋势主要体现在：建立了一套较完整、科学的评价体系，从武器系统的综合作战效能出发，在对战场目标进行科学分类、易损性研究、毁伤机理和终端毁伤效应研究的基础上，应用现代系统控制理论和数字虚拟现实技术，建立起适用于各种武器杀伤力和易损性的评估方法，已广泛用于武器系统综合作战效能评价和指导战斗部设计。

长期以来，我国大部分弹药战斗部装备走仿制国外装备为主的道路。由于照搬照抄的仿制方式，注重国外产品结构仿制及最终性能参数的研仿，对于其设计思想、设计依据、设计理论和相关基础研究基本不考虑。军事技术先进国家，将终端毁伤技术视作武器装备**核心关键技术之一**，保密要求很高，商业引进时通常列为**不转让技术**。例如，从国外引进诸多武器装备中（“双35”火炮及弹药、“苏27”飞机及配套弹药和某些导弹等），其战斗部技术均不转让，支撑战斗部研制的终端毁伤技术成果更是无法获得。

由于我国在终端毁伤技术方面弱势和空白，成为战斗部技术尤其是在**高效毁伤技术**发展中起到严重阻碍**作用**，以至于在不少武器毁伤能力设计方面科学性差或带有盲目性，甚至存在严重缺陷。例如，某些反机场跑道弹药设计威力不足，毁伤效能低下；反导武器毁伤导弹能力很低，甚至没有弄清楚武器在什么样的条件下才具备**有效毁伤**目标的能力；反水面水中目标时，不知道敌方的现代舰艇及水中武器抗毁伤能力多强，无法评估我方水中武器爆炸后对敌舰艇和水中兵器的毁伤能力；某些导弹对特定目标的毁伤等级说不清，不知应打几个波次才能达到预定的摧毁要求，因缺乏评估方法和数据故无法进行**毁伤效能**评价，以至于装备的订购量缺少依据；当自主开发或设计一种新战斗部时，发现根本无设计依据，几乎没有被打击目标的易损性数据，**毁伤准则**不清，无法设计战斗部；等等。另外，我国在战斗部或毁伤方面的专用实验及测试技术、专用制造技术和专用技术标准等的研究很薄弱，与国外先进水平相比差距很明显。只有通过长期的大力度的持续研究，才能动态缩小差距迎头赶上。

3 非核终端毁伤基础科研是支撑武器装备跨越发展的关键和原创技术之一

终端毁伤技术方面的基础问题，不仅是制约战斗部发展的关键问题之一，也是制约新型武器开发的

关键问题之一。美国国防部认为，要评估一个国家在未来军事冲突中的态势，首先要能精确地评估各种武器的**杀伤力**和**易损性**的能力（L/V）。然而更重要的是未来有哪些目标产生，如何毁伤，敌方会有什么样的新的终端毁伤手段，如何对抗才能不被动挨打。

美国研究了战场上几乎所有重要军事目标或可能成为重要军事目标的易损特性和对其毁伤机理，积累了丰富的实验数据，对各种目标毁伤模式研究较透彻，为爆炸、侵彻、杀伤干扰、非致命作用等毁伤机理建立了较完善的基础理论，并发展了 DYNA、AUTODYN 等大型的数值分析程序、易损性分析程序、武器对抗仿真程序和专家分析系统。在研究手法上已从实验研究发展到实验与理论建模、数值模拟及数字仿真相结合。对较重要的军事目标一般都建有相关毁伤准则。

苏联（俄罗斯）长期以来有相当数量的大学和专门科研机构研究目标易损性问题，只要是军事需要，几乎是不计成本地研究，建立了系统化的分析模型库和数据库，随时为了武器终端毁伤能力设计和毁伤技术创新提供基本数据。在目标材料动态特性和目标破坏机理方面的研究处于世界领先水平。

终端毁伤技术还是推动新型武器装备发展的**原创技术之一**，一种新的终端毁伤理念、方法和技术的产生，**导致新的武器装备产生**。例如，短路毁伤机理与作用效应研究成果，产生碳纤维/导电纤维弹（或战斗部）；云雾爆轰机理、爆炸作用效应与相应的战斗部结构技术研究成果，产生云爆弹或温压弹；对导引头干扰/诱骗机理及相应装填物研究成果，产生干扰弹和诱饵弹；对深层或超深层目标毁伤机理与毁伤效应研究成果，产生钻地弹或反深层目标战斗部；强电磁脉冲和强微波对电子器件作用的失效机理及小型高功率发生器技术的研究成果，产生电磁脉冲弹（战斗部）和微波弹；人员和装备多种类型的非致命失能机理与杀伤元素创新研究成果，产生各种类型的非致命杀伤战斗部；等等。

海湾战争、科索沃战争、阿富汗战争和第二次伊拉克战争中美国试用的一些终端毁伤技术创新的武器，诸如反深层目标弹药、碳纤维/石墨弹、高功率微波弹、云爆弹、温压弹、巨型炸弹、多模战斗部、多种类子母弹、“爱国者”导弹战斗部破片引爆“飞毛腿”导弹战斗部和高威力贫铀穿甲弹等，使得对目标的终端毁伤效能大幅度提高。终端毁伤技术新的研究成果还在美国的“NMD”“TMD”武器系统中得到了广泛应用。

4　非核终端毁伤技术需研究的重大基础问题

非核终端毁伤方面的基础科研是一个大体系，需通过丰富厚实的研究成果才能有力支撑武器装备的需求和发展，是需国家重点扶植的长远可持续发展的技术，型号预研不完全等同于基础科技能力增强；零散的一般力度的预研远达不到**加强基础**和**提高能力**的需求。经梳理，2020 年前需重点研究的基础性技术问题有以下九个方面。

4.1　目标易损性分析方法与评价技术

其主要研究目标分类及描述、防护特性、结构损伤等级与判据、功能失效等级与判据、在不同杀伤元素作用下的损伤模式和易损特性等的分析理论、方法和手段，研究目标的等效理论、方法与毁伤效应等效试验技术，研究真实目标易损性数据获取的实验方法、数据分析方法和毁伤程度的评价技术，为战斗部装备预研和研制提供共性基本方法和手段。

4.2　终端毁伤效能分析方法与评价技术

其主要研究武器装备在规定条件下达到规定使用指标能力的分析方法和试验评价技术，为装备研制和运用提供共性基本方法和手段。

4.3　终端毁伤机理与毁伤准则

终端毁伤机理主要研究战斗部毁伤目标所发生的现象及其规律性问题，在毁伤机理研究成果基础上，研究目标失效等级及在特定杀伤元作用下的毁伤准则。

4.4 大威力及控制威力技术

爆炸驱动及爆炸过程控制方法、杀伤元素形成及作用方向控制技术、新能源应用技术，以及研究战斗部采用什么样的结构技术、炸点控制技术等实现对目标最大破坏能力，达到高效毁伤目标的要求。

4.5 战斗部设计理论与设计方法

研究系统先进的战斗部设计理论、优化设计方法和数字化设计技术，研究战斗部、目标互相作用及战引最优配合问题。

4.6 终端毁伤的新概念、新原理和新技术途径

研究目标尤其是高价值目标被毁伤的新概念和新模式，研究/探索新的毁伤方法、原理和战斗部结构技术，研究大幅度提高终端毁伤能力的新技术途径。

4.7 对装填物、引信和火工等的新技术需求牵引

高效毁伤目标、新毁伤方式和新型战斗部结构对战斗部装填物、特种功能装置、引信及火工品等的新技术需求研究，从而牵引这些技术的发展，反之可起到技术推动作用。

4.8 终端毁伤方面的数字化平台技术

基础研究成果系统集成的总体技术、平台的数字化技术和升级接口技术；研究终端毁伤技术/战斗部技术/专用材料方面的数据库总体、数据规范及收集以及有效使用等问题。

4.9 毁伤方面专用的实验、制造和规范等技术基础

研究终端毁伤方面的先进实验方法、必需试验条件和重要参数的测试技术及测试手段；研究战斗部特种部件、重要结构件、装药及新装填物等的制造技术；研究战斗部设计、研制、试验、制造、验收、贮存和使用等方面的技术标准和规范。

5 对策及建议

5.1 将非核终端毁伤技术列为国防重大科技专项

基于终端毁伤技术的重要性、需求的急迫性和整体技术水平的落后状况，为满足打赢一场高技术条件下的局部战争对先进战斗部装备以及高效毁伤多种类高价值目标的需求，满足战斗部关键技术预研对毁伤方面基础研究成果强烈需求，满足2020年前武器装备发展战略对终端毁伤技术创新需求，以及国家安全的长远需求，须将非核终端毁伤技术列为国防基础科研重大关键技术领域并加大投入力度。

5.2 加强管理、健全组织

设立本重大专项的行政和技术管理机构，加强组织协调力度，形成有力、灵活、高效的顶层管理机构。设立本重大专项的项目组，集中兵器、航天、航空、船舶、有关高校、中科院、中物院的优秀专家队伍组成一个有机的整体，形成“国家队”。

5.3 统筹规划，加强基础

在相应基础条件建设中改变以往各单位零散、孤立地进行建设的局面，统筹考虑有关单位研究条件，提升重点或优势单位的研究能力，资源共享。统筹规划科研任务，加强基础研究，确保基础研究成果获得突破。并采取积极措施，提高设计、试制、制造、检测能力。

5.4 多学科交叉，适度竞争

通过多学科交叉，同时关注边缘学科、新学科的特点和研究成果，搭建创新研究队伍框架，建立创新研究工作的科学评价方法，不断加大本项目的创新力度。

5.5 自主发展为主，积极对外合作

坚持以我为主，形成一批自主知识产权的成果，但积极与友好国家合作，汲取和购买国外先进成熟的技术和软件，提高研究起点和效率。

5.6 加强非核终端毁伤国防基础科研条件保障

非核终端毁伤基础研究涉及爆炸、冲击、高速、大破坏和复杂环境及多制约条件下实施试验和数据获取的手段问题。由于试验过程的瞬时性、不可重复性、伴有严重破坏性和危险性，因此给研究带来很大难度，特别需要系统和先进的科研基础条件。国内非核终端毁伤基础研究的综合条件较差，远满足不了发展需求，应在整合国内各有关单位已有科研条件基础上，加大条件保障建设投入力度，有效支撑国防基础研究。

注：本文有删减

2004 年 4 月

爆炸冲击波与杀伤破片对飞机目标相关毁伤效应的研究[①]

冯顺山[②]，蒋浩征

（力学工程系）

提　要： 本文以小型导弹战斗部为研究对象，研究了破片杀伤作用和爆破作用对飞机目标的相关毁伤效应问题，并依据实验结果，应用数理统计方法给出了"杀爆相关毁伤准则"的工程计算式。计算式可用来评价和计算有关战斗部的威力效应，可供战斗部设计时应用。

An Investigation ON Fragmentb - LAST Interrelated damage effect on aircraft targets

Feng Shunshan，Jiang Haozheng

（Department of Dynamic Engineering）

Abstract： Fragment - blast interrelated damage effect og small sized missile warhead on aircraft targets is investigated. Based on the statistics from experimental results，an engineering approach for "Fragment - Blast Interrelated Damage Criterion" is presented. The approach may be adapted in estimating or calculating the damage potential，and of assistance in warhead design.

符号表

G_e——战斗部装药重量	（千克）
G_{be}——带壳战斗部爆炸后赋予冲击波能量的虚拟装药重量	（千克）
G_c——战斗部壳体重量	（千克）
q——破片重量	（千克）
α——战斗部装填系数	
$\bar{\alpha}$——破片速度衰减系数	（1/米）
V_0——破片初速	（米/秒）
V_R——破片在 R 处的存速	（米/秒）
R——破片或冲击波运动距离	（米）
Q——炸药爆热	（千卡/千克）
C_x——破片阻力系数	
ρ_0——当地空气密度	（千克·秒^2/米^4）
ϕ——破片形状系数	（米^2/公斤$^{2/3}$）
S——破片迎风面积的数学期望	（米^2）
r——等效圆破片面积的半径	（米）
N——破片数	

① 原文发表于《北京工业学院学报》1986（1）。

② 冯顺山：工学硕士，1980 年师从蒋浩征教授，研究方向"杀爆战斗部威力及毁伤效应"。

Ω_0——破片静态飞散角 （度）

v_J——破片密度 （块/厘米^2）

L——战斗部壳体产生破片的圆柱段长度 （米）

δ——战斗部壳体产生破片的圆柱段厚度 （米）

D——战斗部壳体外径 （米）

r_0——战斗部壳体内半径 （米）

r_m—在爆炸产物作用下壳体膨胀到破裂时内半径 （米）

ρ_q——战斗部壳体材料的密度 （千克/米^3）

σ_{sc}——静态飞散角的均方差 （度）

D_0——冲击波波阵面速度 （米/秒）

v_1——未扰动前的气体比容 （米^3/千克）

v_2——扰动后的气体比容 （米^3/千克）

P_1——未扰动前的气体压力 （公斤/厘米^2）

P_2——扰动后的气体压力 （公斤/厘米^2）

Δp_m——冲击波超压 （公斤/厘米^2）

$\bar{R}$——对比距离

t_+——冲击波正压区作用时间 （秒）

i_+——冲击波比冲量 （千克·秒/米^2）

χ_R——距离杀爆相关函数

$[R]$——未带孔靶板上的临界毁伤距离 （米）

$[R]_\chi$——带孔靶板上的临界毁伤距离 （米）

χ——下标，

a——下标，某一爆炸高度

s——下标，海平面

$\chi_{\Delta p}(r\cdot v_T)$——冲击波超压杀爆相关效应函数

$\chi_I(r\cdot v_T)$——冲击波比冲量杀爆相关效应函数

f_{a1}——在某一高度爆炸时冲击波压强修正系数

f_{a2}——在某一高度爆炸时冲击波冲量修正系数

f_{a3}——在某一高度爆炸时距离修正系数

K——战斗部选用炸药种类修正系数

σ_b——靶板（目标）材料的抗拉强度极限 （公斤/毫米^2）

a——与爆炸类型有关系数，一维平面爆轰 $a=1$，轴对称爆轰 $a=2$

b——与战斗部形状有关系数，轴对称 $b=2$

r——多方指数

h_m——靶板厚度 （米）

1　引言

常规导弹战斗部的作用通常是指杀伤、爆破和聚能三种，具有两种以上作用的战斗部称为复合作用战斗部或综合作用战斗部。综合作用战斗部对付不同的目标时，配上不同性能的引信，各作用的地位和相互间关系是不同的，终点效应也就不同。至今公开发表的战斗部作用及设计理论，仅分别考虑各个作用或是只考虑某单一作用的效应，均未考虑复合作用战斗部各作用之间的关系及其对同一局部目标的威力效应。譬如，破甲战斗部大部分具有一定厚度的金属外壳，但外壳形成的破片质量很小且仅能击中空域中的点，对于面目标和立体目标（如建筑物、地下设施等）其摧毁作用不很明显；而冲击波则涉及整个空域，因此对目标能有效发挥其威力。又如，非触发地—空和空—空导弹战斗部，一般均有杀伤、爆破两种作用，但因弹目距离较远，主要考虑破片杀伤效应。那些主要考虑单个作用的战斗部，在设计时自然主要关系某一作用的威力参数和其相应的结构参数。

综合作用战斗部对有些目标，必须考虑破片各作用之间的关系和各作用对同一目标的威力效应，否则就不能准确评价战斗部的威力效应。例如，触发起爆的小型综合作用战斗部对飞机上某一局部的毁伤效应，显然不只是一种作用产生的，而是几种作用共同产生且相互有关的效应，见图 1 和图 2。

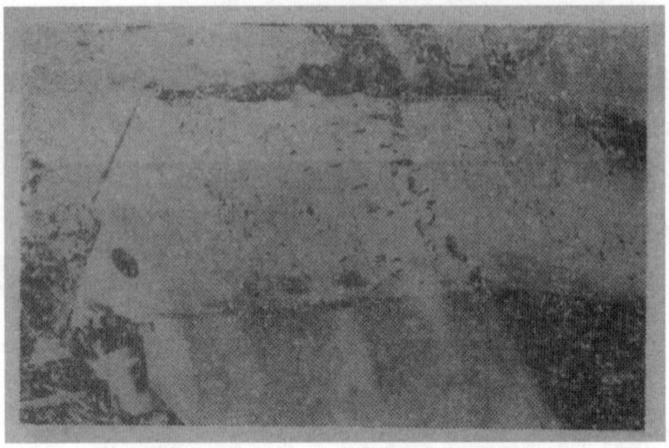

图 1　飞机垂直安定面被 1 号战斗部作用后的破片分布情况

图 2　图 1 的反面。在破片分布密度最大处目标毁坏（破裂与折断）

由上可知，综合作用战斗部其诸作用效应是否相关，取决于目标特性和弹目相互位置并在同一目标上至少有两种作用。

若综合作用战斗部的各个作用对目标的毁伤效应是相互有关，且不等于各个作用效应的代数和，本文称这种综合效应为几种作用的相关毁伤效应，简称“相关效应”。下面以 1 号小型超低空防空导弹战斗部为例，以飞机为目标分析各作用之间的关系。

1 号小型超低空防空导弹战斗部（下面简称“1 号战斗部”）具有杀、爆和聚三种作用。头部聚能罩产生的破片动能大，在 6 米远处能击穿 12 毫米厚的钢甲，但方向性很强，当战斗部舱前有其他舱段时这种作用对目标便不能有效发挥。因此，本文认为该战斗部可以忽略聚、爆两种作用对同一局部目标的相关毁伤效应，而着重研究杀、爆两作用的相关毁伤效应。

弹目距离很近时，战斗部侧向破片在飞散角范围内有相当高的破片分布密度，被破片飞散角覆盖的目标上形成一条破片孔带，带中破片分布密度最大处形成一条破片孔链，见图 1、图 3。如冲击波在目标上某局部特性（易损特性）已发生一定变化条件下继续或同时作用，所产生的最终毁伤效应与破片作用密切相关，并且不等于单个作用效应的代数和，见图 1、图 2。由杀伤、爆破两种作用引起的最终毁伤效应本文称为“杀爆相关毁伤效应”，简称“杀爆相关效应”。

图3　破片作用后在钢筒上形成破片孔链，图中两条线中所示

应用有关公式进一步论证上述看法。冲击波波阵面速度 D_0 的计算公式

$$D_0 = v_1[(P_2 - P_1)/(v_1 - v_2)^{\frac{1}{2}}]（米/秒） \tag{1}$$

冲击波超压 Δp_m 的计算公式[3]

$$\Delta p_m = \begin{cases} 20.06(\bar{R})^{-1} + 1.94(\bar{R})^{-2} - 0.04(\bar{R})^{-3} & 0.05 \leqslant \bar{R} \leqslant 0.5 \\ 0.67(\bar{R})^{-1} + 3.01(\bar{R})^{-2} - 4.31(\bar{R})^{-3} & 0.5 \leqslant \bar{R} \leqslant 7.09 \end{cases} \tag{2}$$

式中　$\bar{R}$——对比距离，$\bar{R} = R/\sqrt[3]{G_e}$；

G_e——装药重量，千克；

R——距离，米。

带壳战斗部装药爆炸后赋予冲击波能量当量[1]

$$G_{be} = KG_{be}\left[\frac{\alpha}{\alpha + 1 - a\alpha} + \frac{(a+1)(1-\alpha)}{a + 1 - a\alpha}\left(\frac{r_0}{r_m}\right)^{b(r-1)}\right]（公斤） \tag{3}$$

式中　K——炸药种类修正系数。

冲击波正压作用时间 t_+ 计算式[1]

$$t_+ = 1.35 \times 10^{-3} K^{1/6} R^{1/2} G_{be}{}^{1/6}（秒） \tag{4}$$

破片存速计算公式[1]

$$\left.\begin{aligned} V_R &= V_0 e^{-\bar{\alpha} R} \\ V_0 &= 80\sqrt{Q_V}/\sqrt{\frac{1}{\alpha} - \frac{1}{2}} \\ \bar{\alpha} &= \frac{1}{2} C_x \rho_0 g \phi q^{-1/3} \end{aligned}\right\} \tag{5}$$

式中　V_R——破片在 R 处的存速，米/秒；

V_0——破片初速，米/秒；

$\bar{\alpha}$——破片速度衰减系数，1/秒；

α——战斗部装填系数；

Q——炸药等容爆热，千卡/千克。

在静态地面条件下将 1 号战斗部有关数据代入上面公式进行计算，将计算结果绘制成曲线，见图4、图5、图6，这些曲线清楚表明：①冲击波波阵面速度大于破片速度（$D_0 > V_R$）的情况发生在较小的区域内，该区域中某一部分为冲击波绝对毁伤区，在没有被毁伤的地方则存在着“相关毁伤”，这是因为冲击波的压力不是瞬间消失而是有一定的作用时间，冲击波在目标上的反射和目标的动态响应使作用时间加长，破片在小于这作用时间内可飞达目标；②破片速度大于（等于）冲击波波阵面速度（$V_R \geqslant D_0$）的情况出现的区域远大于的 $D_0 > V_R$ 区域（且包围 $D_0 > V_R$ 的区域,）在这区域上杀爆相关效应无疑是存在的。战斗部设计者主要关心的是最大毁伤区域，故本文重点研究 $V_R > D_0$ 的情况。

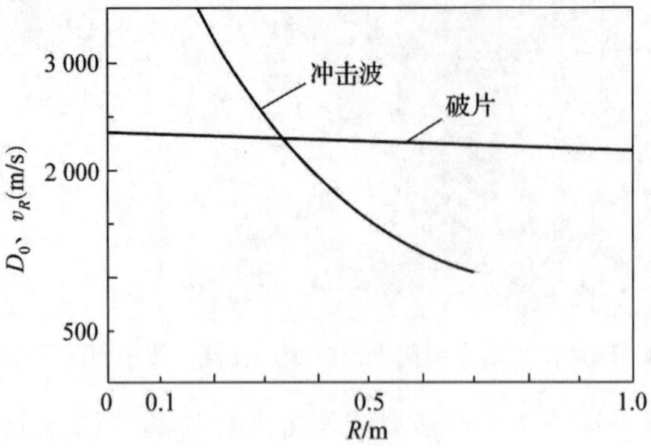

图4 D_0、$D_R \sim R$ 曲线

2 实验研究

如杀爆相关毁伤效应达到目标的极限毁伤程度，则相应的综合反映战斗部、目标有关参数的数学形式，本文称之为“杀爆相关毁伤准则”。

杀爆相关效应与冲击波单独作用的效应密切相关，因此，本文在冲击波局部毁伤准则的实验研究基础上，以圆柱形小型战斗部为研究对象考虑本问题。前已述及，本文主要考虑在破片已对目标作用条件下的相关效应。用大量有壳装药进行实验有许多难以克服的困难，且经济代价太高，为此本文提出了模拟破片打孔的实验方法。具体做法为：用模拟破片打孔的靶板置于裸装药柱产生的爆炸场中，观看其爆炸后的效应，测其数据。

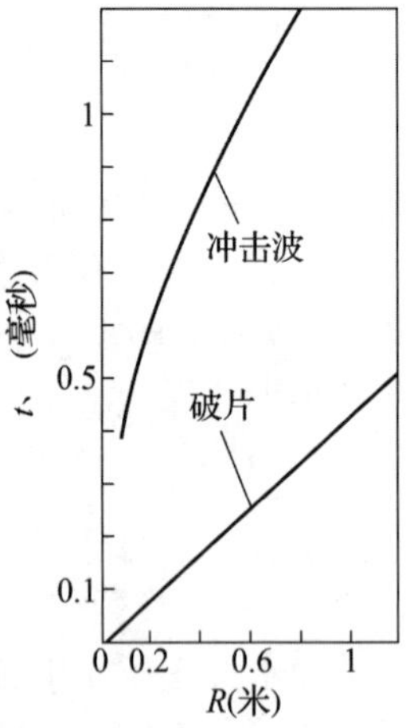

图5 t、$t_+ \sim R$ 曲线

图1和图3表明了在破片密度最大处形成一条破片孔链，破片孔链处自然成为目标上的最薄弱处，在冲击波作用下该处最易毁坏。理论上讲，破片密度最大处在飞散立体锥内的某一锥面上，实际上是在以该锥面为中心的某一区域内，实弹试验也证明了这一看法。目标上破片密度最大分布区域的长度取决于目标的形状、大小以及弹目交汇情况，本文主要确定“区域”（即破片孔链）的宽度。根据经验和一些试验的观察分析可确定：破片密度最大分布区域的宽度在理论分布锥面的 $\pm\sigma_{sc}$ 领域内。σ_{sc} 由式（6）确定

$$\sigma_{sc} = \frac{1}{A_\sigma}\Omega_0 \tag{6}$$

式中 σ_{sc}——静态飞散角的均方差，度；

Ω_0——破片静态飞散角，度；

A_σ——由实验确定的常数。

于是，“区域”的宽度 σ_J 由式（7）确定

$$\sigma_J = \frac{2 \times 180}{\pi} \cdot R_J\,\sigma_{sc} \tag{7}$$

式中 R_J——战斗部中心距 σ_J 中心的距离。

通常认为圆柱形战斗部破片在飞散角内服从正态分布，沿轴周向服从均匀分布。据正态分布特点，在 σ_J 区域内应占用全部破片（飞散角内）的68%。

与冲击波有关的主要参数是装药量 G_e、弹目距离 R、超压 Δp_m 和比冲量 i_+，与目标有关的主要参数是目标的等效厚度 h_m，强度 σ_b 等，综合这两方面的数学关系式资料* 中已给出。与杀伤破片有关的主要参数是破片数、贯穿孔的大小和 σ_J 区域内破片密度的分布特点。为使问题简化，根据有关试验提出下面

三点假设：①破片在σ_J内均匀分布，分布密度以v_J（块/厘米²）表示；②整体式战斗部采用破片大小（或重量）的数学期望值来描述破片性；③破片迎风面积的数学期望等效为圆面积，并由下式表示[1]

$$\left.\begin{aligned} S &= \phi q^{\frac{2}{3}} \\ r &= (S/\pi)^{\frac{1}{2}} \end{aligned}\right\} \tag{8}$$

式中 S——破片迎风面积的数学期望；

q——破片重量；

ϕ——破片形状系数；

r——等效圆破片面积的半径。

根据本文提出的假设和式（7）、式（8）导出v_J和r的计算式

$$v_J = \frac{101 \cdot N}{\pi^2 \cdot \Omega_0 \cdot R_J^2} \tag{9}$$

$$r = (\phi/\pi)^{\frac{1}{2}}[(1/N) \cdot \pi \cdot K_S \cdot L \cdot \rho_q \cdot \delta(D-\delta)]^{\frac{1}{3}} \tag{10}$$

式中 N——战斗部圆柱段产生破片的总数；

K_S——战斗部爆炸时壳体重量损失系数；

L、δ、D——壳体的长度、厚度和直径；

ρ_q——壳体材料的密度。

依据本文提出的观点和计算公式，设计和制作模拟破片打孔的靶板进行试验。

3 实验结果

靶板的破裂主要发生在模拟破片打孔的区域内，严重的可使靶板分成两块，除此之外有凹向变形和固靶螺钉孔处断裂，见图6、图7。

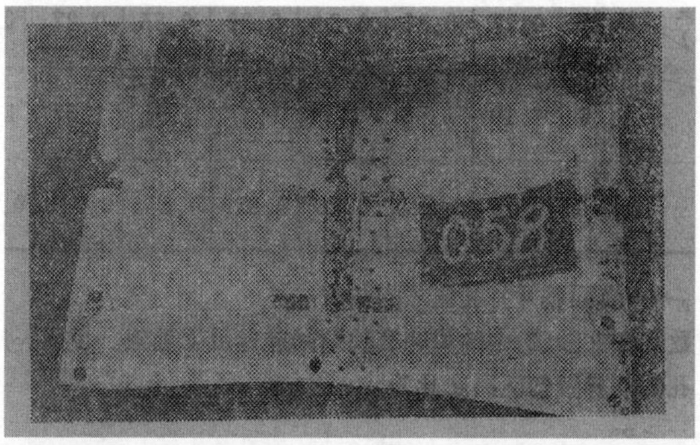

图6 模拟破片打孔靶板毁伤情况（$r=0.33$厘米，$v_J=0.27$块/厘米²，$[R]_x=0.58$米，靶板厚3×10^{-3}米）

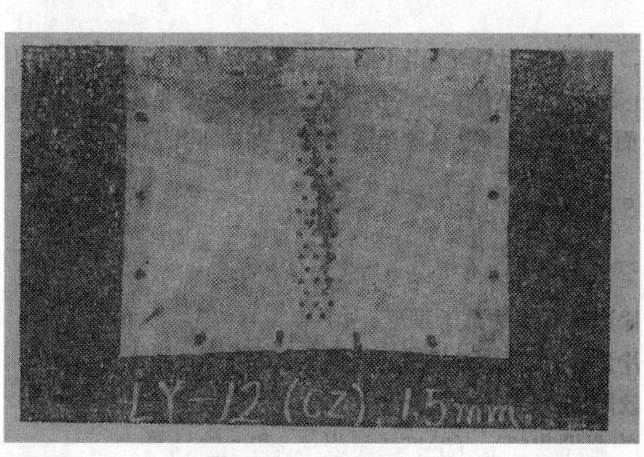

图7 模拟破片打孔靶板毁伤情况（$r=0.33$厘米，$v_J=0.39$块/厘米²，$[R]_x=1.05$米，靶板厚1.5×10^{-3}米）

确定模拟破片打孔靶板临界毁伤标准，即确定目标在杀爆相关作用下临界毁伤标准是一个新课题，

这需要深入研究目标在杀爆相关作用下的易损性问题。本文根据1号战斗部对飞机某些舱段的杀爆相关毁伤试验，以及冲击波对模拟破片打孔靶板的毁伤性试验分析，提出的鉴别靶板临界毁伤标准为：靶板上区域内连续断裂的长度必须大于（等于）该区域长度的2/3。符合该标准的靶板认为其应起的功用（譬如，作为受力件、气功效应件及保护件等）已基本丧失。图6、图7为符合临界毁伤标准的部分靶板。表1为符合临界毁伤标准的实验数据。

表1　实验数据（靶材均为 $LY-12$ 硬铝）

靶板厚 $h_m 10^{-3}$ 米	3	1.5	1.5	1.5	1.5	1.5	1.5	1.5	1.5	1.5
靶板上预制孔半径 r（厘米）	0.33	0.33	0.33	0.33	0.33	0.33	0.44	0.52	0.63	0.73
破片密度 v_J（块/厘米^2）	0.27	0.27	0.18	0.34	0.39	0.55	0.27	0.27	0.27	0.27
χ_R	1.20	1.19	1.13	1.4	1.47	1.65	1.37	1.43	1.48	1.70
断裂长/L_p	0.657	0.75	0.844	0.78	0.813	0.844	0.719	0.813	0.688	0.688

注：L_p——模拟破片打孔区域的长度。

根据表1的数据和文献［2］中给出的有关数据进行数据处理。设同一厚度的靶板其模拟打孔与不打孔的临界毁伤距离之比为

$$\frac{[R]_\chi}{[R]}=\chi_R \tag{11}$$

分析可知产生 χ_R 的原因是 r 和 v_J，则有

$$\chi_R=\chi(r\cdot v_J) \tag{12}$$

由式（11）导出

$$[R]_\chi=R(r,v_J,h_m,\sigma_b,G_e)=\chi(r,v_J)[R] \tag{13}$$

式中　$[R]_\chi$——经破片打孔的目标（靶板）临界毁伤的距离或范围，米；

$[R]$——未经破片打孔的目标（靶板）临界毁伤的距离或范围，米。

$\chi_R=\chi(r\cdot v_J)$——距离杀爆相关函数。

应用多元回归分析方法处理数据有

$$\chi_R=\chi(r\cdot v_J)=\begin{cases}1 & 0\leqslant 10rv_J\leqslant 0.34\\ 0.82+0.24v_J+4(r,v_J) & 0.34\leqslant 10rv_J\leqslant 2.2\end{cases} \tag{14}$$

其均方差为 $\sigma_{\chi R}=0.04$，回归相关系数 $|r|=0.95$，式中 r 与 v_J 的单位分别是"厘米"和"块/厘米^2"。相似于上面的处理过程有

$$[\Delta P_m]_x=\chi_{\Delta P}(r,v_J)[\Delta P_m] \tag{15}$$

$$[i_+]_x=\chi_1(r,v_J)[i_+] \tag{16}$$

$$\chi_{\Delta P}(r,v_J)=\begin{cases}1 & 0\leqslant 10rv_J\leqslant 0.2\\ 1.12-5.8(r,v_J) & 0.2\leqslant 10rv_J\leqslant 1\\ 4.83\times 10^{-2}(r,v_J)^{-1.1} & 1\leqslant 10rv_J\leqslant 2.2\end{cases} \tag{17}$$

$$\chi_1(r,v_J)=\begin{cases}1 & 0\leqslant 10rv_J\leqslant 0.2\\ 1.05-2.7(r,v_J) & 0.2\leqslant 10rv_J\leqslant 1\\ 2.65\times 10^{-1}(r,v_J)^{-0.48} & 1\leqslant 10rv_J\leqslant 2.2\end{cases} \tag{18}$$

式中　$[\Delta P_m]_\chi$、$[i_+]_\chi$——经破片打孔的目标（靶板）临界毁伤的入射超压（公斤/厘米^2）和比冲量（公斤·秒/米^2）；

$\chi_{\Delta P}(r,v_J)$、$\chi_1(r,v_J)$——超压、比冲量杀爆相关效应函数。

式（17）、式（18）二式的回归相关系数 $|r|$ 和剩余标准差 S 满足：$|r|>0.98$，$S<0.15$。由式（14）、式（17）、式（18）绘制的曲线见图8、图9。

$$[\Delta P_m]_{\chi a} = f_{a1} \cdot \chi_{\Delta P}(r, v_J) \cdot [\Delta P_m]_s \tag{19}$$

$$[i_+]_\chi = f_{a2} \cdot \chi_1(r, v_J)[i_+]_s \tag{20}$$

$$[R]_\chi = f_{a3} \cdot \chi_R(r, v_J) \cdot [R]_s \tag{21}$$

$$[G_e]_{\chi a} = \begin{cases} f_{a3} \cdot \chi_R^{-1}(r, v_J) \cdot [G_e]_s \\ (f_{a3})^2 \cdot \chi_R^{-2}(r, v_J) \cdot [G_e]_s \end{cases} \tag{22}$$

$$[h_m]_{\chi a} = \begin{cases} \dfrac{1}{0.245\sigma_b \cdot R}[A \cdot (f_{a3})^{-1} \cdot \chi_R(r, v_J) \cdot G_e^{\frac{2}{3}} - 16R] \\ \dfrac{1}{1.5\sigma_b \cdot R}[15 \cdot (f_{a3})^{-2} \cdot \chi_R^2(r, v_J) \cdot G_e + 247R^2] \end{cases} \tag{23}$$

式中，f_{a1}、f_{a2}、f_{a3}均称为高空修正函数，其具体表达式[2]为

$$\left.\begin{aligned} f_{a1} &= (P_a/P_s) \\ f_{a2} &= (G_a/G_s)^{\frac{2}{3}}(P_a/P_s)^{\frac{2}{3}}(T_s/T_a)^{\frac{1}{2}} \\ f_{a3} &= (G_a/G_s)^{\frac{1}{3}}(P_s/P_a)^{\frac{1}{3}} \end{aligned}\right\} \tag{24}$$

式（20）~式（23）中

$[G_e]_{\chi a}$——经破片打孔的目标（靶板）临界毁伤所需的炸药量（TNT 当量，千克）；

$[h_m]_{\chi a}$——可毁伤经破片打孔目标（靶板）的临界厚度，毫米；

脚标“χ”“a”“S”——杀爆相关，高空和海平面；

$[\Delta P_m]_s$、$[i_+]_s$、$[R]_s$、$[G_e]_s$——未打孔靶板临界毁伤值，由资料①给出。

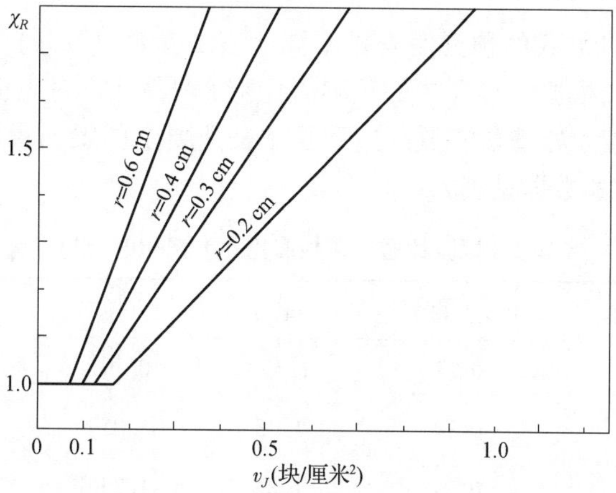

图 8　$\chi_R \sim v_J$ 曲线

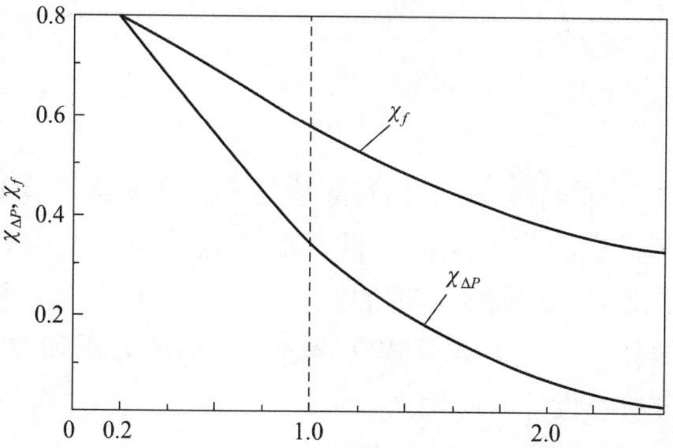

图 9　$\chi_{\Delta p}$，$\chi_1 \sim v_J$ 曲线

① 小药量爆炸冲击波对飞机目标的毁伤准则《爆炸与引爆技术》，北京工业学院 1985 年第二期

由式（19）、式（20）二式绘制的曲线见图 10、图 11。这些曲线清楚地表明：考虑了杀爆相关效应，战斗部的威力得到了较大的提高。图中横坐标 σ_b 表示靶板厚与靶材抗拉强度极限的乘积。

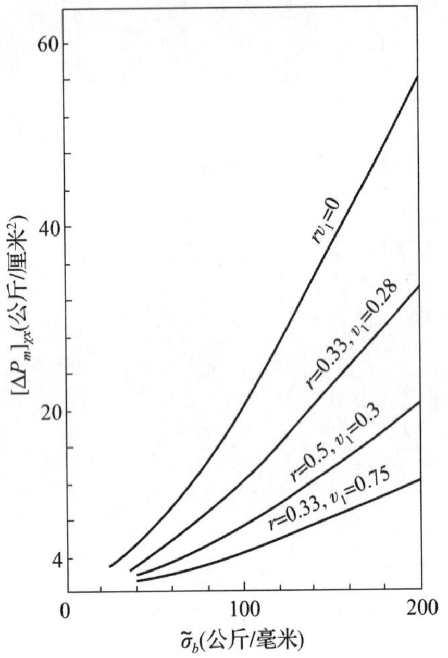

图 10　$[\Delta P_m]_{\chi s} \sim \sigma_b$ 曲线

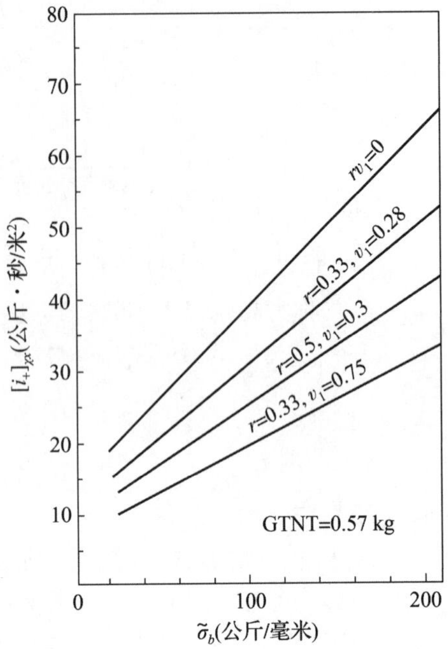

图 11　$[i_+]_{\chi s} \sim \sigma_b$ 曲线

用本文给出的“准则”计算式的预测值安排不同于表 1 参数（r，v）的试验，结果表明目标（靶板）毁伤程度达到了要求；又根据“准则”的预测值检验 1 号战斗部对飞机垂直安定面的毁伤情况。试验结果表明目标在破片密度最大处基本断裂，达到了本文所要求的毁伤程度，见图 1、图 2。这表明了“准则”的实际应用价值。试验数据见表 2。

表 2　试验数据（靶板厚均为 1.5×10^{-3} 米）

序号	r 厘米	v_J 块/厘米^2	χ_R	$[R]_\chi$ 米	实测 $[R]_\chi$	断裂长/L_p
1	0.475	0.27	1.334	0.791	0.97	0.75
2	0.4	0.39	1.45	1.059	1.05	0.781
3	0.4	0.55	1.83	1.336	1.34	0.5
飞机垂直安定面	0.33	0.276	1.25	0.35	0.338	基本断裂

4　几点讨论

本文的研究表明，相关效应问题在理论上和实际应用中都很重要。考虑相关效应，为进一步研究触发式或近炸式杀伤型战斗部对目标的作用问题、目标的终点效应以及毁伤概率提供了较符合实际的结果；能较准确地评价战斗部的威力；从而为合理的设计战斗部提供了理论依据。

本文提出的“杀爆相关毁伤准则”的计算式是通过一定理论分析和较大量试验得出的，与试验吻合较好，可供有关战斗部威力分析、设计时应用。

相关函数的值是有限的，不是随（r，v）的增大而任意增大，这主要因为冲击波压强随距离增大而迅速减弱，达到某一临界状态时已无毁伤能力，这说明 χ_R 大于某一值后相关效应各明显了。实验证明 $\chi_R>1.80$ 后计算式的预测值和试验的吻合度明显下降，见表 2 中序号 3，因此确定 $\chi_R\leq[\chi_R]=1.80$，从中导出 $10r\cdot v_J\leq2.2$。

本文对杀爆相关毁伤效应问题的研究是阶段性总结，还有待今后深入和完善。有些问题，诸如破片密度分布规律、动态条件杀爆相关效应的特点以及靶板的毁伤标准等要在今后进一步研究。

参加本文试验的还有汪绍飞等。

参 考 文 献

[1] 北京工业学院《爆炸及其作用》编写组. 爆炸及其作用：第八章和第十二章［M］. 北京：国防工业出版社，1979.
[2] PROCTOR J F. Internal blast damage mechanisms computer program［R］. AD－759002，1972.
[3] BAKER W F. Explosion in air［M］. Austin：University of Texas Press，1973.
[4] 中国科学院数学研究所统计组. 常用数理统计方法［M］. 北京：科学出版社，1974.

预制破片战斗部对地杀伤作用场计算模型[①]

蒋建伟[②]，冯顺山，何顺录，周兰庭，隋树元

（北京理工大学，力学工程系，冲击与波动实验室）

摘　要：本文建立了一种较实用的预制破片战斗部对地面目标杀伤作用场参数的计算模型。该模型考虑了战斗部破片初速及飞散方向角沿弹轴的分布；建立了弹体与地面坐标系；运用空间坐标变换实现了各坐标系参数的变换；进而推导了战斗部弹体坐标系下任一处破片着地的特征参数（着地速度、位置坐标以及破片密度）的关系式；确定了对地目标杀伤破片的杀伤区域形状和面积。本模型可用于战斗部方案论证中的方案选择以及产品的杀伤威力计算和评价

主题词：战斗部 破片初速 杀伤面积

一、引言

对付地面目标的导弹战斗部通常采用杀伤体制战斗部，最常见的结构形式是预制（包括刻槽、叠环和全预制）式破片结构。当战斗部在一定的作战条件（即炸高、落速、落角确定）下引爆后，战斗部将形成若干破片对地面目标进行杀伤，构成一个杀伤作用区。为了评定战斗部杀伤威力的大小，我国一般采用球形靶或扇形靶对战斗部进行地面静爆威力试验，尽管用试验结果来评价杀伤威力是比较客观的，但此方法需耗费大量的人力和物力，而且只有在加工产品后才能进行试验，对于处于研制阶段或方案论证阶段是无法进行试验的，再则静爆试验与实战条件还存在一定的差别。因此，建立一种较为适用的杀伤作用场计算与评价方法是很有必要的。目前，国内外普遍采用杀伤面积的概念（即所有使目标丧失战斗力概率权的面积之和）描述评判战斗部的杀伤成力，并有不少工程计算方法（1）（2），然而这些方法仍需要采用球形靶试验、破碎性试验以及破片速度的测量数据加之引入了一些假设条件（如破片速度沿弹轴不变、破片密度飞散正态分布等），故计算结果有时不能反映实际情况。这些方法在某些弹药（榴弹、迫击弹、航弹）的杀伤威力计算可以适用，但对预制破片式战斗部则不能适用，尤其是在战 斗部方案设计与论证阶段根本无法计算。

为此，本文建立了一整套实用的计算模型用以计算预制破片战斗部对地面目标的杀伤作用场。模型中采用了准确的战斗部破片初速及飞散方向角沿弹轴分布的公式；建立了弹体与地面坐标系；运用空间坐标变换实现了各坐标系参数的交换；进而推导了战斗部弹体坐标系下任一处破片密度）的关系式；运用能量和密度准则确定了对地目标杀伤破片的杀伤区域形状，此模型实际使用准确可靠，可较真实地反映预倒破片战斗部的杀伤威力。

二、计算模型

战斗部对地面目标的杀伤威力通常是以杀伤面积（即使目标失去战斗力概率作权的面积之和）这个量值加以衡量和评判的。很显然，战斗部威力的大小取决于战斗部自身条件和外部条件。战斗部的自身条件包括破片的初速与飞散方问角的分布，破片的质量空间分布等，战斗部外部条件主要是炸点参数（炸高、落速和落角）及目标特性（如目标分布特征、目标易损性），这些参数中炸高参数是可以确定的，但目标的特征参数则相当复杂，因为要对目标进行杀伤。从数学上讲，杀伤的概率应包括三部分，即目标的暴露概率；目标的命中概率；目标被命中而丧失战斗力的概率。通常也只考虑命中概率，命中

---

① 原文发表于《无人飞行器学会战斗部与毁伤效率专业委员会第一届学术会议论文集》1990。

② 蒋建伟：工学博士，2001 年师从冯顺山教授，研究中心爆管式子母战斗部抛撒技术及应用，现工作单位：北京理工大学。

概率则取决于命中目标的杀伤破片数量，也就是破片的密度分布问题。因此，本文建立杀伤作用场计算模型简化为求同时满足杀伤能量和破片密度要求的破片作用场形状和面积，这种做法比较直观。

以下分别建立各参数的计算模型。

2.1 破片初始威力参数计算模型

2.1.1 破片质量分布

预制破片战斗部爆炸后形成的破片质量分布不同于自然破片的质量分布，它的质量分布比较均匀，基本可假设为均匀分布（分布值为破片的平均质量）。关于单枚破片的平均质量q_f可由下式求得：

$$q_f = (1 - K_1)q_0 \tag{2.1}$$

式中：q_0——预制破片的理论质量

K_1——爆炸后破片的质量损失率（$K_2 = 10\% \sim 20\%$），一般取15%

2.1.2 破片初速与飞散角分布

战斗部破片初速和飞散方向角沿弹轴是变化的。图2.1为一战斗部结构示意图，设OX轴为弹轴线，O点为起爆点。

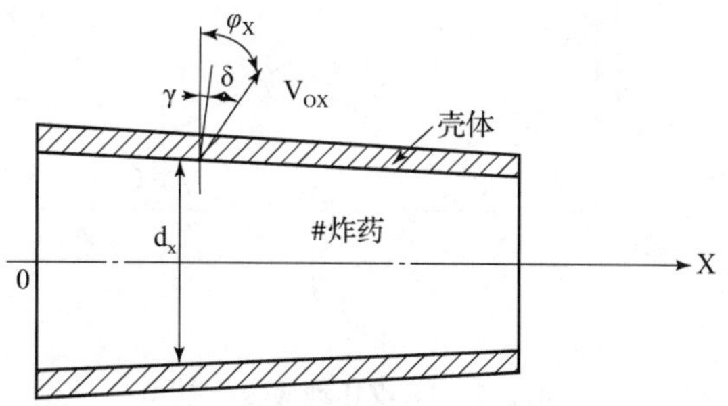

图2.1

据文献（3），可得预制破片战斗部轴向破片初速Vox对一端起爆

$$V_{0x} = K_2(1 - A\,e^{-\beta x/dx})(1 - C\,e^{-D(L-x)/dx})\sqrt{2E}\sqrt{\frac{\beta(x)}{1+0.5\beta(x)}} \tag{2.2}$$

中心起爆

$$V_{0x} = K_2(1 - A\,e^{-\beta|x|/dx})\sqrt{2E}\sqrt{\frac{\beta(x)}{1+0.5\beta(x)}} \tag{2.3}$$

式中：$2E$——格尼能量；

$\beta(x) = \frac{m(x)}{M(x)}$截面$x$处$\Delta x$厚炸药质量/壳体质量；

dx——截面x处的炸药柱直径；

K_2——能量利用系数，内刻槽结构$K_2 = 0.85 \sim 0.95$；

全预制结构$K_2 = 0.8 \sim 0.9$；整体式$K_2 = 1.0$

A、B、C、D——试验常数。A = 0.08，B = 1.11，C = 0.2，D = 3.03

战斗部的飞散方向角φ_x定义速度方向与赤道面的夹角，根据文献（4）有

$$\varphi_x = \delta + r = \sin^{-1}\left(\frac{v_{0x}}{2U} - \frac{v_{0x'}\tau}{2} - \frac{(v_{0x'}\tau)^5}{5}\right) + tg^{-1}\left(\frac{\Delta dx}{2\Delta x}\right) \tag{2.4}$$

式中：δ——壳体变形角；

r——赤道面与金属炸药界面法线夹角；

v_{0x}——在x截面破片初速；

$v_{0x'}$——在x截面破片初速梯度；

τ——加速度常数，按（5）可取 $\tau=0.4\dfrac{dx}{2\,v_{0x}}$；

U——爆轰波扫过炸药金属交界面的速度；

Δdx——在截面 x 处 Δx 轴向增量引起的炸药直径增量。

2.2 杀伤作用场计算模型

2.2.1 各坐标系的建立与诸参数的转换

设战斗部在距地面 H_0（定义起爆点至地面的垂直距离）的高空起爆，对应此时导弹的落速为 V_S，着角（弹轴与地面的夹角）为 θ，如图 2.2 所示，图中的 O_2yZH 为地面坐标系。

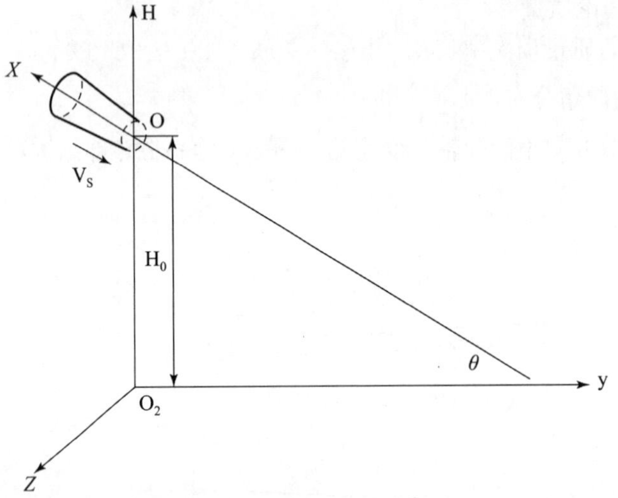

图 2.2 O_2yZH 地面坐标系

为了将战斗部弹体各部位的破片参数转换在地面坐标系下的量值，我们建立弹体柱坐标系 $OxR\delta$（如图 2.3），并引入弹体坐标系 $O_1\varepsilon\eta\xi$（如图 2.4）。

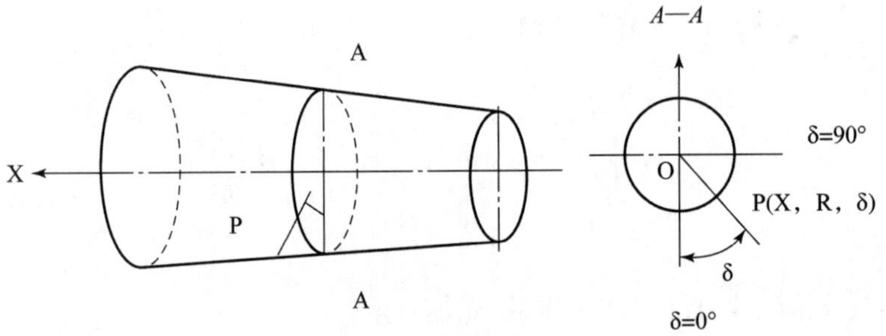

图 2.3 弹体柱坐标

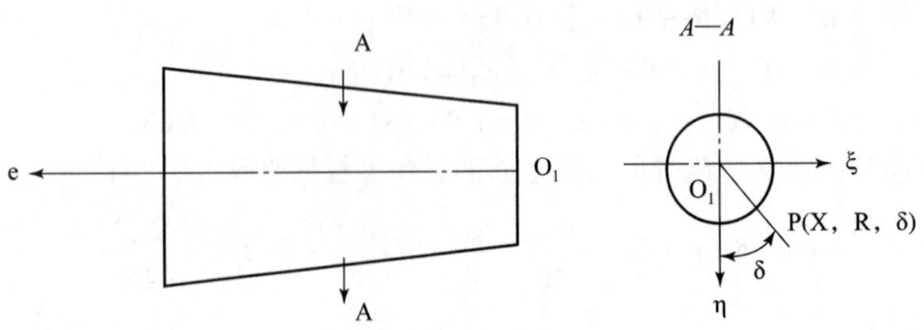

图 2.4 弹体坐标系

设 P（x，R，δ）为弹体柱坐标系只不过弹体上的任一点，此处的破片初速为 Vox，飞散方向角为 φ_x，则在该坐标系下的速度分量（Vx，V_R）为

$$v_x = v_0 x \sin \varphi x$$
$$v_R = v_{0x} \cos \varphi_x \tag{2.5}$$

在弹体坐标系中，P 处破片初速分量为（V_ε，V_η，V_ξ）按坐标变换有：

$$V_\varepsilon = v_x$$
$$V_\eta = V_R \cos \delta$$
$$v_\xi = V_R \sin \delta \tag{2.6}$$

设在地面坐标系中，P 处的空间坐标为（z，y，H）破片速度为（V_Z，V_y，V_H），按坐标变换法则有

$$\begin{pmatrix} z \\ y \\ H \end{pmatrix} = \begin{pmatrix} 0 & \sin \delta & 0 \\ \cos \theta & -\cos \delta \sin \theta & 0 \\ -\sin \theta & \cos \delta \cos \theta & 1 \end{pmatrix} \begin{pmatrix} x \\ R \\ H_0 \end{pmatrix} \tag{2.7}$$

$$\begin{pmatrix} v_z \\ v_y \\ v_H \end{pmatrix} = \begin{pmatrix} 0 & 0 & 1 \\ \cos \theta & -\sin \theta & 0 \\ \sin \theta & \cos \theta & 0 \end{pmatrix} \begin{pmatrix} V_\varepsilon \\ V_\eta \\ V_\xi \end{pmatrix} + \begin{pmatrix} 0 \\ v_S \cos \theta \\ v_S \sin \theta \end{pmatrix} \tag{2.8}$$

由式（2.5）（2.6）代入式（2.8）可得

$$V_z = V_{0x} \sin \varphi x$$
$$v_y = v_{ox} \cos \varphi_x \cos(\theta + \delta) + v_s \cos \theta$$
$$v_H = v_{ox} \cos \varphi_x \sin(\theta + \delta) + v_s \sin \theta \tag{2.9}$$

在地面坐标系下，P 处破片的合速度 V_0 为：

$$V_0 = \sqrt{V_z^2 + V_y^2 + V_H^2} \tag{2.10}$$

2.2.2　破片着地坐标及着地速度

由于破片初速很高，故弹道可近似直线。设破片速度 V_0 与地面坐标系的夹角 α、β、γ，则其方向余弦

$$\cos \alpha = v_y / v_0$$
$$\cos \beta = v_H / v_0$$
$$\cos \gamma = v_z / v_0 \tag{2.11}$$

破片自 H 高飞落地面，飞行距离 R 为：

$$R = H\, v_0 / v_H \tag{2.12}$$

破片在 O_2y 和 O_2Z 轴方向的飞行距离 y_s、z_s 为

$$y_s = R \cos \alpha = H \cdot v_y / v_H$$
$$z_S = R \cos \beta = H \cdot v_z / v_H \tag{2.13}$$

因此，P 处破片的着地坐标（y'，z'）为

$$y' = y + y_s = (x \cos \theta - R \sin \delta \sin \theta) + H\, v_y / v_H$$
$$z' = z + z_s = R \sin \delta + H\, v_z / v_H \tag{2.14}$$

按速度衰减公式可得，P 处破片落地速度 v_f 为

$$v_f = v_0 e^{-kR} = v_0 e^{-kHv_0/v_H} \tag{2.15}$$

式中：k——破片速度衰减系数。

2.2.3　破片着地密度

在战斗部弹体 P 处取一微元（轴向取 dx，径向取 $R_x d\delta$）PBCD（见图 2.5）

微元各点分别为：

P（x，Rx，δ），B（x + dx，Rx，δ）

C（x，Rx，$\delta + d\delta$），D（x + dx，Rx，$\delta + d\delta$）

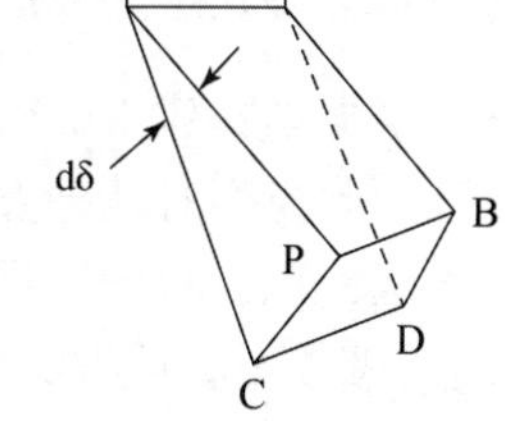

图 2.5

微元所含破片数量 dN 为

$$dN = \frac{R_x d\delta dx}{S_0} \tag{2.16}$$

式中：S_0——单枚破片理论面积；

Rx——弹体壳体中径。

2.2.4 杀伤作用场参数

按照前面的定义，战斗部爆炸后对地面的杀伤作用场是同时满足杀伤能量和破片密度要求的破片杀伤区域。设 A_1、A_2分别为满足杀伤能量和密度要求的面积，Ω_1、Ω_2分别为对于的区域，那么战斗部杀伤作用场面积 A 和区域 Ω 为

$$A = A_1 \cap A_2$$
$$\Omega = \Omega_1 \cap \Omega_z \tag{2.17}$$

2.2.4.1 按能量准则的杀伤区域参数

战斗部破片对地面目标杀伤，需具有一定的能量（动能或比动能），设［Es］为临界杀伤动能。那么杀伤破片必须满足

$$\frac{1}{2} q_f v_f^2 \geqslant [E_s] \tag{2.18}$$

将战斗部沿轴向划分成若干微圆环，对应每个微圆环轴向取 dx 由式（2.15）和式（2.18）可以获得圆环极限杀伤角

$$\{\delta_{x1}\} \quad (x = o, \ldots\ldots, L) \quad L\text{—战斗部长度}$$

将δ_{x1}代入式（2.14），得到微圆环极限杀伤坐标 $\{y'_{x1},\ z'_{x1}\}$ $(x = o, \ldots\ldots, L)$

对应的杀伤区微元面积 $d\,S_{x_1}$按式（2.18）积分得

$$d\,S_{x1} = \int_0^{dx_1} dS \tag{2.19}$$

落入 $d\,S_{x1}$面积内的破片数 $d\,N_{x1}$为：

$$d\,N_{x1} = \frac{2\pi\,R_x}{S_0} \cdot \frac{\delta\,x_1}{\pi} \cdot dx \tag{2.20}$$

极限杀伤坐标序列 $\{y'_x,\ z'_x\}$ $(x = o, \ldots\ldots, L)$ 构成了满足杀伤能量要求的杀伤区域Ω_1，对应杀伤面积 A_1和杀伤破片数 N_1

$$A_1 = 2\sum_{x=0}^{L} dS_{x_1}$$
$$N_1 = 2\sum_{x=0}^{L} dN_{x_1} \tag{2.21}$$

2.2.4.2 按破片密度准则的杀伤区域参数

由于战斗部破片沿弹轴径向分布均匀，因而破片着地的密度分布将呈现这样的规律，离地面坐标系 o，y 轴愈远，破片密度愈小，若密度小于一定限γ_0，我们假设就不起杀伤作用。即杀伤破片需满足

同样，可以按照$\gamma \geq \gamma_0$前面的方法，由式（2.16）~（2.19）及γ_0可以得到极限杀伤角 $\{\delta_{x_2}\}$ $(x = o, \ldots\ldots, L)$ 和极限杀伤坐标 $\{y'_{x_2},\ z'_{x_2}\}$，相应可获得杀伤面积 A_2和杀伤破片数 N_2。

三、计算实例与结论

根据本文建立的模型，用 FORTAN 语言编制了程序。结合某地地导弹的研制，我们在方案设计阶段，用此模型对几种方案（如纯圆柱形，锥形，锥形 + 圆柱形 + 锥形）进行了计算，得到了各种方案的杀伤区面积，对此结果，进行了优选，得到了采用锥形 + 圆柱形 + 锥形结构方案较好（即杀伤面积最大）。图 3.1 为此方案炸高 30 m、落角 45°、落速 520 m/s，按动能准则［E_J］ = 10 Kgm，密度准则γ_0 = 0.5 块/m^2的计算杀伤区域形状图，对应图中的杀伤面积 A = 221 82 m^2，杀伤破片数 N = 123 42（块）。

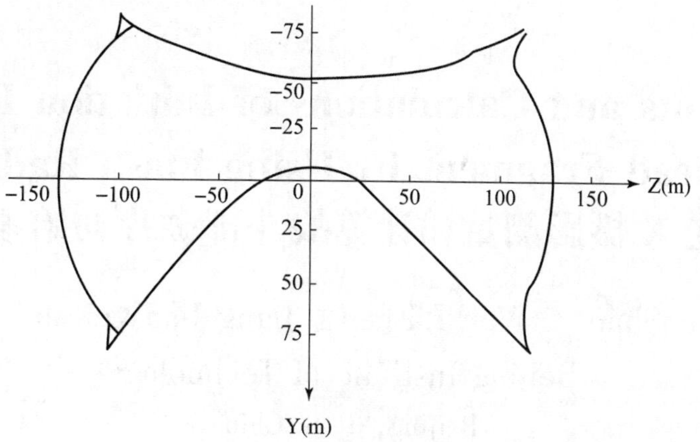

图 3.1　杀伤区形状图

实际使用结果表明，本文建立的模型可较准确地运用战斗部方案的优选以及杀伤威力的评判。

参考文献

［1］李景云，破片弹药杀伤面积的工程计算方法，1982

［2］于骐，杀伤弹药地面杀伤面积和杀伤半径的计算方法，兵工学报，弹箭分册，1984

［3］冯顺山等，战斗部破片初速轴向分布规律的实验研究，兵工学报，1987 年第 4 期

［4］Randers – Penerson，G “An Improved Equation for calculating Fragment Projection Angles” Proceeding of the second International symposium on Ballistics，1976

［5］Glenn Randers – Pchrson，P. R. Karpp，C. E. Anderson，Jr. H. J. Elichc，“Shortfrag users Guide” ADB046644，1980

Measurements and Calculations of Initiation Paramenters of Warhead Fragment by Using Flash Radiography
闪光 X 摄影测量和计算战斗部破片初始参数①

Feng ShunShan②, Wan LiZhen, Jiang JianWei and Li LyYin
Beijing Institute of Technology
Beijing, P. R. China

ABSTRACT: This paper describes the research of fragmentation warheads or pyrotechnical devices suited with metal casing by an experimental research method using flash radiography. The research includes casing expanding, casing rupturing, initial velocity distribution of fragments and dispersing direction distribution of fragments along the axis of symmetry. The experimental research method will be described and discussed in detail. Some radiographs of typical fragmentation warheads, and diagrams of initial fragments velocity and dispersing direction distribution of fragments along the axis of symmetry will be presented. Finally, initial fragment velocity distribution will be formulated by data processing and analyzing.

1 Introduction

Pyrotechnics as well as explosive is often charged in metal casing containers. Fragmentation warheads kill and wound targets mainly by the metal casing. No doubt, to realize und control the casing motion pattern at explosition will help the optimal design of fragmentation warheads and pyrotechnical devices as well as safety design of experiments. There are four primary parameters o describe the motion pattern: a) casing expanding. b) casing breaking or rupturing. c) initial fragment velocity distribution and d) dispersing direction distribution of fragments along the axis of symmetry. From those, the external and terminal ballistics of fragments can be outlined accurately.

Initial motion pattern of metal casing at explosition is a basic data to analyze the processes of blasting, blasting - combustion and explosition product expanding, and also to analyze and study the interaction between the metal casing and explosive. For those explosive devices, the function of which is mainly determined by their metal casing motion, the sum of all kinetic energy of fragments is proposed as a criterion, which practically applies to engineering design, to evaluate capabilities of the explosive devices.

By using flash radiography, the problems mentioned above can possibly be carried out. Some regular results can be obtained by checking the radiographs carefully, collecting data accurately and getting them into mathematic and theoretical analysis. This paper is concentrated on the study of initiation parameters of fragments.

2 Experimental Method

The experimental method can be briefly described as following take radiographs of moving casing or fragments at different time by two or more X - ray tubes, and then put them into automatic image processing or manual one.

① 原文发表于《"空间烟火工作组"第三届国际烟花大会和第十二届"国际烟火研讨会"》1987。
② 冯顺山：工学硕士，1980 年师从蒋浩征教授，研究方向"杀爆战斗部威力及毁伤效应"。

The principle arrangements of the experiments are shown in Figure 1.

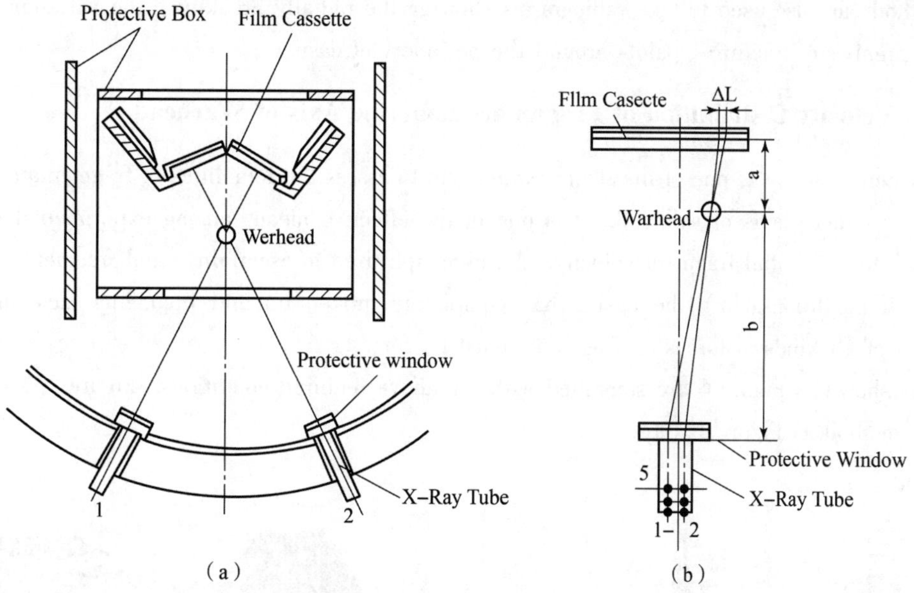

Figure 1　Experiment arrangements by flash radiography

Figure 1 (a) shows a method to take radiographs at different time respectively from different angles on two films. Figure 1 (b) shows a method to take radiograph at different time from one angle on one film. The latter could be used to trace the moving fragments. Because film in Figure 1 (b) has to be exposed twice, the experiment must be carefully arranged in order to make images on the film distinguishable.

Metal casing is broken down into a great number of fragments at high velocities after explosion. It is possible that fragments damage X - ray tubes, film cassettes and other facilities in the explosion experiment damage severely. So, effective protection which have no influences to the flash shooting on films should be taken. We have solved those problems and successed to carry out the experiment by flash radiography with explosive charges various in shape and weight ranging from tens grams to a thousand grams or more.

3　Measurements and Results

The four principle parameters of metal casing moving pattern, mentioned above, which dominate the motion of metal casing at explosion, will be respectively described as following.

3.1　Casing Expanding

The casing expanding pattern means the relation between the expanding rate of casing and time as well as the radial and axial position during the period of explosion products effectively acting on the casing. If the radiographs of casing expanding at different times after explosion be taken, for example see Figure 2 (a), (b), the data with regularity can be easily obtained.

3.2　Casing Breaking or Rupturing

Casing breaking or rupturing is here specified as patterns of radially and axially breaking or rupturing of casing, which can be applied to describe fragment size distribution, and to analyze mechanics of interacting between explosion products and metal casing. The method presented in Figure. 1 (b) can be used to take radiographs showing the occuring time of fracturing points, fracturing point distribution along the axis of warheads and the relation between this distribution and shapes of casing and explosive charge, when warheads are fixed vertically with the symmetry axis of X – ray tubes. If warheads are fixed parallel with the symmetry axis of X – ray tubes, the method can also used to take radiographs showing the radially breaking, the variation of thickness of casing and the number of fracturing paints around the periphery of casing.

3 Initial Velocity Distribution of Fragments along the Axis of Warhead

Initial fragment velocity is one of the major parameters to assess the lethalities of fragmentation warheads, it varies along the symmetry axis of warhead. It is one of the effective means among experimental methods to use flash radiography to test initial fragment velocity. A radiograph used to ascertain initial fragment velocity must be taken under such conditions: a) the casing has completely broken up and fragments have already formed; b) the resultant of all kinds of forces on fragments must be zero.

The curves shown in Figure 6 are staistified with the above required conditions. Figure 2. 3 and 4 are taken using the same method as Figure 1 (a).

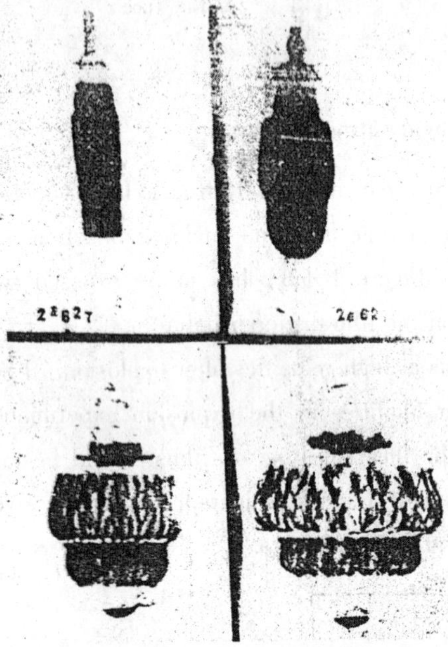

(a) (b)

Figure 2 Radiographs of casing expanding and breaking after explosion.

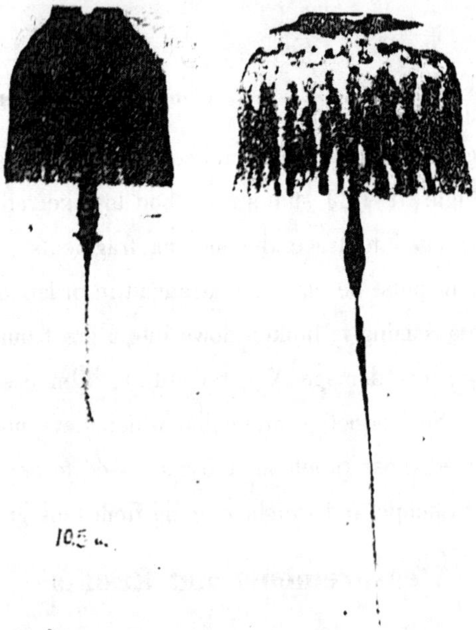

Figure 3 Radiographs of a shaped – charge at 10. 5 and 25 us after explosion.

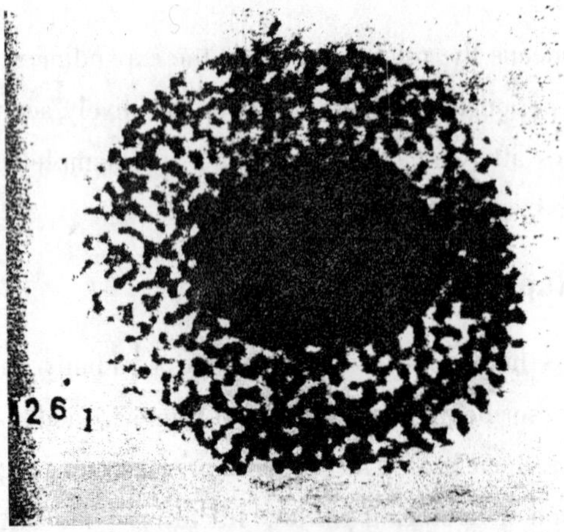

Figure 4 Radiograph of grenade filled with steel balls after explosion.

3.4　Fragment Dispersing Direction Distribution along the Axis of Warhead

Dispersing direction of fragments is another very important parameter of fragmentation warheads. If a piece of fragment does not move in expected directions and does not hit the targets as expected, it will be no use, no matter how much kinetic energy it has. By the way, when dispersing direction of fragments has been worked out, a more effective safety measure can be taken thereby. It is the distribution of fragment dispersing direction that determines the coverage and its pattern covered by the fragments on targets, the density distribution and the mass distribution of fragments in the coverage. The study of dispersing direction distribution of fragments can also contribute to the design of fragmentation warheads. When flash radiography is used to measure dispersing direction distribution of fragments, it is necessary to trace and shoot the fragments. The dispersing direction distribution pattern of fragments can be obtained by measuring radiographs, as long as the angle between the axis of warheads and the trajectories of moving fragments are known. The method illustrated in Figure 1 (b) makes it convenient to trace fragments. Figure 5 is taken by the method in Figure 1 (b) and 6 shows the dispersing direction distribution of fragments of a typical warhead.

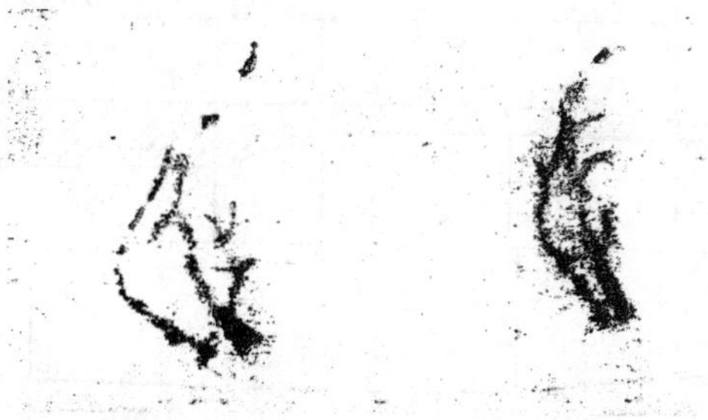

Figure 5　Radiography of fragments at two times after explosion showing the dispersing direction

While processing the data, pay attention to ΔL, which results in the distance S between two axis of X - ray tubes parallel installed, if the method illustrated in Figure 1b is applied. The distance L that a piece of fragment exactly travels is

$$L = \frac{b(l-S)}{a+b} - S$$

$$L = (l-k)(l-S) - S$$

Here, $k = a/(a+b)$, and then ΔL as

$$\Delta L = \frac{Sa}{b}$$

V_{oxl}: the initial velocity of fragments in Part i;

$\bar{V}_{oxl}$: the average initial velocity of fragments in Part i;

ϕ_{xl}: the dispersing direction angle of fragments in Part i;

$\bar{\phi}_{xl}$: the average dispersing direction angle of fragments in Part i;

$\bar{V}_{ox}$: the average velocity of all fragments;

$\bar{\phi}_{x}$: the average dispersing direction angle of all fragment;

Ω: the angle containing all fragments.

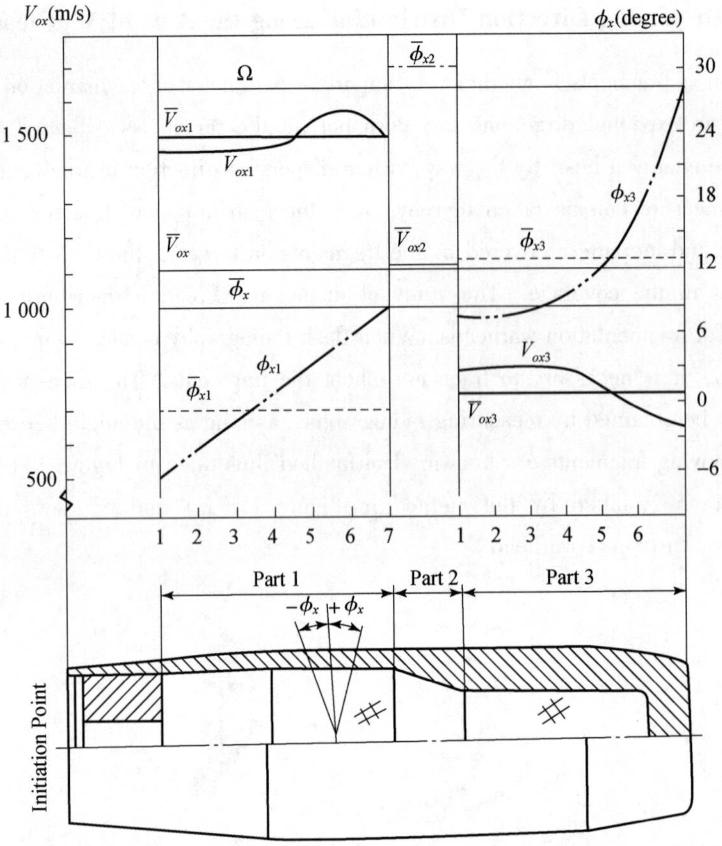

Figure 6 Distribution of initial velocity and dispersing direction of fragment from a typical warhead.

对空近炸战斗部的最优设计①

冯顺山②，蒋浩征

摘　要：本文以触发式或近炸式小型导弹战斗部为研究对象，建立了以破片杀伤飞机要害舱段中，典型舱段的比动能和总体设计对战斗部的9个约束函数，以及以“杀爆相关毁伤效应”为目标函数的工程优化设计数学模型。采用SUMT－DFP法求解。

关键词：战斗部；弹药；最优设计；终点效应

OPTIMUM DESIGN FOR ANTIAIRCRAFT MISSILE WARHEADS

Feng Shunshan, Jiang Haozheng

Abstract: This paper deals with the optimum design of mini－missile warheads of the proximity or contact type. Nine restraining functions including requirments of specific kinetic energy for fragments damaging the typical fatal cabins of an aircraft, and the requirement of overall design for the warhead are established. Mathematical model is set up for engineering optimization in which the interrelated fragmentation－blasting damage effect has been taken as the target function. The solution is worked out by using SUMT－DFP mothod.

Keywords: warhead; ammunition; optimum design; terminal effect

符号表

C_x——破片阻力系数

D——战斗部壳体外半径

G_e——装药重量

G_T——TNT炸药质量

G_b——有壳装药赋予冲击波能量当量

G_t——与炸药长度对应的圆柱段壳体质量

G_0——除引信外的战斗部总质量

G_R——战斗部壳体除G_Z外的质量

ΔG——优化设计计算质量与总体允许质量之差，简称剩余质量

G_a——位于空中某一高度爆炸时装药质量

G_J——将G_a换算成海平面爆炸时装药质量

$[h_m]$——战斗部毁伤靶板的极限厚度

K——有壳装药赋予冲击波能量当量修正系数

K_t——爆炸后壳体质量损失系数

K_Q——某种炸药与TNT炸药之爆热比值

L——炸药柱或战斗部壳体长度

ΔL——药柱中有其他物体时的装药质量修正量

---

① 原文发表于《兵工学报》1989（3）。

② 冯顺山：工学硕士，1980年师从蒋浩征教授，研究方向“杀爆战斗部威力及毁伤效应”。

N——与炸药长度对应圆柱段壳体产生的破片总数

P_S——海平面处大气压

P_a——某一高度大气压

Q_V——炸药的等容爆热

q——破片重量

R——战斗部中心距目标的距离

r_q——战斗部平均迎风面积等效圆面积半径

S——破片平均迎风面积

V_0——破片初速

V_R——破片存速

α——装填系数

$\bar{\alpha}$——破片衰减系数

δ——圆柱段壳体厚度

v_J——破片在σ_l区域内平均分布密度

ρ_0——当地空气密度

ρ_q——战斗部壳体材料的密度

ρ_ω——装药密度

σ_l——最大破片密度分布区域的宽度

σ_b——壳体材料的抗拉强度极限

ϕ——破片形状系数

χ——脚标，表示“杀爆相关”条件

Ω_0——破片静态飞散角

1 引言

目标函数是优化数学模型中的核心。通常一个实际工程问题中有若干个评价指标，以战斗部为例：重量、威力指标（一个或多个）、结构参数、毁伤概率等均可成为评价指标。在射击规律确定条件下，对空导弹战斗部的综合评价指标多采用坐标毁伤概率。

假如要害舱段间无损伤积累，并且认为毁伤任一要害舱段整个飞机即被毁伤。则由触发引信或近感引信启动的战斗部对飞机的毁伤概率，主要取决于导弹命中或靠近飞机要害舱段（发动机、驾驶舱、仪表舱，弹药舱和油箱等——本文称为第一类舱段）的概率。这类导弹命中要害舱段后，战斗部的破片足以毁伤该舱段。这是因为，近距离条件下，满足有效杀伤比动能的破片密度大大超过由非触发（近感）引信启动的战斗部期望达到的指标。所以，只要导弹命中要害舱段，即便不考虑爆破作用，也可认为战斗部杀伤破片对目标的坐标毁伤概率为1。

需要指出的是，飞机上的舱段均是缺一不可的。习惯上定义的飞机要害舱段是相对破片杀伤而言，那些所谓的非要害舱段仅仅被破片打些孔显然无伤大体，但在爆破作用下大块结构损伤却是有伤大体。所以对于杀、爆综合作用的触发或近炸战斗部来说，飞机上的所有舱段几乎都是要害舱段。如导弹命中（或靠近）非要害舱段（机翼、安定面、舵面、延长段和喷口等——本文称为第二类舱段），此时破片有效毁伤作用大大下降。而爆破作用的毁伤效果却突出。目前，在公开的文献资料中，还没有评定爆破作用对这些舱段的毁伤概率计算式，只是较粗略地规定某一舱段的破坏范围大于某一值时，评定毁伤该舱段的概率为1，否则为0。

1或0开关式地表示毁伤概率，导弹无论命中第一类舱段或第二类舱段，其坐标毁伤概率（如设为目标函数）对设计变量（威力参数和结构参数等）的变化会很不敏感，优化计算时难以寻找或找不到最

佳方案。

于是，进行由触发引信或近感引信启动的对空战斗部优化设计时，选择合适的目标函数是本文关心的首要问题。

小型触发引爆或近炸杀爆型战斗部优化设计时，其突出矛盾表现在：如何保证在有效毁伤第一类舱段前提下更有效地毁伤第二类舱段。

2　目标函数和设计变量

文献［1］的论述和结论表明，这种战斗部对第二类舱段的作用效应是“杀爆相关毁伤效应”。于是，本文提出：满足破片杀伤第一类舱段中典型舱段所需比动能，满足总体设计所要求的重量，壳体几何形状、强度和某些威力参数指标等前提下，以“杀爆相关毁伤效应”作为优化设计的目标函数。

设静态条件下，靶板是一块厚为 h_m 的有限圆弧形靶；并设圆柱形战斗部与弧形靶共轴，战斗部距靶板的距离为 R。弹与靶（目标）相互间的位置由图 2.1 所示。

将文献［1］中给出的“杀爆相关毁伤”准则中毁伤靶板极限厚度的数学关系式，写成一般形式

$$[h_m]_{\chi a}=h(r_q,v_J,\sigma_b,G_{Tbe},R,G_e,p_a) \tag{2.1}$$

将破片在目标上穿孔的面积等效成圆面积，该面积可用破片飞行时迎风面积的数学期望来表示[2]

$$r_q=(\phi,q) \tag{2.2}$$

战斗部结构特征、起爆方式以及弹目距离确定后，破片密度在目标上的分布规律也相应确定。杀爆相关效应研究中最关心目标上遭受破片密度最大的区域。此区域内的破片密度表达式写成函数关系式[1]

$$v_J=v(\Omega_0,\delta,q,R,K_s,\rho_a,L,D) \tag{2.3}$$

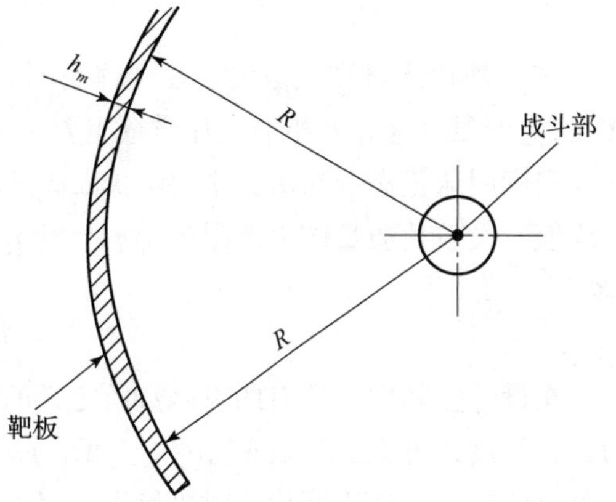

图 2.1　弹与靶的位置示意图

带壳条件下炸药赋予冲击波的能量当量为[2]

$$G_{Tbe}=G(\alpha,\delta,\rho_\omega,L,D,K,K_Q) \tag{2.4}$$

式中，$K_Q=Q_{iv}/Q_{TNT}$

将（2.2）、式（2.3）、式（2.4）代入式（2.1），有

$$[h_m]_{\chi a}=H(\alpha,q,\delta,\Omega_0,R,\rho_\omega,\rho_a,L,D,K,K_Q,K_s,G_a,R,p_a,\sigma_b) \tag{2.5}$$

式（2.5）便为目标函数的一般数学形式，它综合反映了战斗部的威力参数、结构参数、弹目距离、装药特性、目标特性、高度影响等对威力效应指标的影响，是一个多设计变量的单目标函数。

分析式（2.5）中设计变量可知，有些变量可事先确定，有的变量可作为离散变常数输入。例如，反映目标材料特性参数，战斗部离地面的距离和炸药种类等可视为已知参数，可事先确定的变量有 ρ_q，K，K_Q，p_a，σ_b，G_a 等；作为离散变常数的有 R，ρ_ω，K_s 等；战斗部外形尺寸通常是总体给定，则 D、L 也为已知变量。众多的设计变量中，可根据具体工程问题的需要选择优化设计变量，再将其余变量定为常量或变常量。

上面给出的目标函数是反映杀爆相关效应的目标函数。本文选择与杀爆效应密切相关的设计变量（α，q，δ，Ω_0）为优化设计变量，这 4 个设计变量基本决定圆柱形或圆锥形战斗部的威力参数和结构参数。

3　约束函数

根据已选定的设计变量，约束函数主要考虑以下几个方面。

（1）设战斗部总质量为 G_{0i}，导弹总体许可的最大质量为［G_0］（也可同时考虑许可最小质量），则质量约束函数由下式表示

$$G_{oi}=G(\delta,\rho_{\omega},\rho_{q},L,D,G_{B})\geqslant[G_{0}] \tag{3.1}$$

（2）破片杀伤要害舱段中典型舱段所需比动能约束函数。设破片贯穿典型舱段所需最小比动能为 $[e_Y]$，于是有

$$e_{Y}=e(\alpha,q,Q_{iv},\phi,C_{x},\rho_{0},R)\geqslant[e_{Y}] \tag{3.2}$$

（3）装填系数与壳体结构参数等之间的关系为

$$\phi(\alpha,\delta,D,L,\rho_{q},\rho_{\omega},G_{B})=0 \tag{3.3}$$

式（3.3）是一个等式约束。

（4）相关函数约束。文献［1］中的分析表明，相关函数 χ_R 不能任意大，有其恰当的边界值，因此有

$$1\leqslant\chi_{R}\leqslant\chi(\delta,q,\Omega_{0},R,D,L,\phi,\rho_{q},K_{s})\leqslant[\chi_{R}] \tag{3.4}$$

（5）壳体厚度。δ 的上限由式（3.1）制约，δ 的下限由总体要求的壳体允许强度、刚度等确定。如壳体材料已知，对应的壳体许可最小厚度为［δ］，便有

$$\delta\geqslant[\delta] \tag{3.5}$$

（6）破片质量和飞散角。在壳体材料、起爆方式确定条件下，可根据经验和必要的试验确定 q 与 Ω_0 的上下限。这儿所指的破片质量主要考虑破片最大分布密度区域内破片质量的分布情况。譬如，当破片密度服从正态分布时，主要考虑 $\pm\sigma$ 区域内破片的质量分布情况。本文取该区域内破片质量分布的平均值。设破片质量的上下限为 $[q_2]$、$[q_1]$，飞散角的上下限为 $[\Omega_2]$、$[\Omega_1]$，于是有

$$[q_{1}]\leqslant q\leqslant[q_{2}] \tag{3.6}$$

$$[\Omega_{1}]\leqslant\Omega_{0}\leqslant[\Omega_{2}] \tag{3.7}$$

在设计空间中，所有约束函数所定义的空间为可行域，它是一个曲面，在此曲面上寻找使目标函数为最大的设计方案 $X^{*}=(\alpha^{*},q^{*},\delta^{*},\Omega_{0}^{*})$。所有约束函数的边界值选择必须合理，有科学依据，同时要满足总体要求。在方案论证时可给出几组不同边界值进行比较。

4 最优化设计数学模型

由式（2.5）~式（3.5）式的具体表达式，构成工程应用的优化设计数学模型。极小化

$$\begin{aligned}G_{1}(X)&=H_{1a}(\alpha,q,\delta,\Omega_{0})=\min[h_{m}]_{1a}\\&=(65.31/\sigma_{b})-[1.62/(\sigma_{b}\cdot R)]\cdot(G_{s}/G_{a})^{1/3}\cdot(p_{s}/p_{a})^{1/3}\cdot\\&\quad[K\cdot\pi\cdot K_{Q}\cdot\rho_{\varpi}\cdot(L-\Delta L)]^{2/3}(D-2\delta)^{4/3}\cdot\\&\quad[0.82+404\cdot K_{s}\cdot L\cdot\rho_{q}\cdot\delta(D-\delta)\cdot(0.06+(\phi/\pi)^{1/2}\cdot q^{1/3}/(\pi\cdot\Omega_{0}\cdot q\cdot R^{2})]\cdot\\&\quad[(\alpha/(2-\alpha))+2(1-\alpha)\cdot(r_{0}/r_{m})^{2(\gamma-1)}/(2-\alpha)]^{2/3}\qquad 20\leqslant|h_{m}\cdot\sigma_{b}|\leqslant210\end{aligned} \tag{4.1}$$

$$\begin{aligned}G_{2}(X)&=H_{2a}(\alpha,q,\delta,\Omega_{0})=\min[h_{m}]_{2a}\\&=(-164.64/\sigma_{b})-[25/(\sigma_{b}\cdot R^{2})]\cdot(G_{s}/G_{a})^{2/3}\cdot(p_{a}/p_{s})^{2/3}\cdot\\&\quad[K\cdot\pi\cdot K_{Q}\cdot\rho_{\varpi}\cdot(L-\Delta L)]\cdot(D-2\delta)^{2}\cdot\\&\quad[0.82+404\cdot K_{s}\cdot L\cdot\rho_{q}\cdot\delta\cdot(D-\delta)\cdot0.06+(\phi/\pi)^{1/2}\cdot q^{1/3}/(\pi\cdot\Omega_{0}\cdot q\cdot R^{2})]^{2}\cdot\\&\quad[(\alpha/(2-\alpha))+2(1-\alpha)\cdot(r_{0}/r_{m})^{2(\gamma-1)}/(2-\alpha)]\qquad 210\leqslant|h_{m}\cdot\sigma_{b}|\leqslant750\end{aligned} \tag{4.2}$$

满足约束条件

$$G_{2}(X)=[G_{0}]-G_{B}-(\pi/4)\cdot[\rho_{\omega}\cdot(L-\Delta L)\cdot(D-2\delta)^{2}+4L\cdot\rho_{q}\cdot\delta(D-\delta)]\geqslant0 \tag{4.3}$$

$$G_{3}(X)=[6.4\times10^{2}\cdot\alpha\cdot Q_{iv}\cdot q^{1/3}\cdot\exp(-10^{5}\cdot C_{x}\cdot\rho_{0}\cdot\phi\cdot R\cdot q^{1/3})/(\phi\cdot q\cdot(2-\alpha))]-[e_{Y}]\geqslant0 \tag{4.4}$$

$$G_{4}(X)=D-2\delta-[\alpha\cdot(D^{2}\cdot L\cdot\rho_{q}+4\cdot G_{B}/\pi)/(\alpha\cdot L\cdot\rho_{q}+\rho_{\varpi}(L-\Delta L)\cdot(1-\alpha))]^{1/2}=0 \tag{4.5}$$

$$G_5(X)=[\chi_R]-0.82+404K_s\cdot L\cdot\rho_q\cdot\delta(\delta-D)\cdot(0.06+(\phi/\pi)^{1/2}\cdot q^{1/3})/(\pi\cdot\Omega_0\cdot q\cdot R^2)\geqslant 0 \tag{4.6}$$

$$G_6(X)=\delta-[\delta_1]\geqslant 0 \tag{4.7}$$

$$G_7(X)=q-[q_1]\geqslant 0 \tag{4.8}$$

$$G_8(X)=[q_2]-q\geqslant 0 \tag{4.9}$$

$$G_9(X)=\Omega_0-[\Omega_1]\geqslant 0 \tag{4.10}$$

$$G_{10}(X)=[\Omega_2]-\Omega_0\geqslant 0 \tag{4.11}$$

上面诸公式长采用的量纲：质量为 10^{-3} kg，长度为 10^{-2} m，角度为（π/180）rad，　$[h_m]$ 为 10^{-4} m。

5　计算方法与计算结果分析

目标函数和约束函数是四维欧氏空间可行域上的实值非线性函数，且有连续的二阶偏导数，本文采用 SUMT－DFP 方法求解。该方法是一种有代表性的参数制约法，它将有约束极小化问题转成无约束极小化问题的无穷序列，再采用求解无约束极值的方法求解。SUMT 法可求解式（5.1）所示的非线性规划问题。

极小化目标函数

$$\left.\begin{array}{l} f(X)\,X\in E^n \\ \text{满足于} \\ g_i(X)\geqslant 0 \quad i=1,2\cdots\cdots n \\ h_j(X)=0 \quad j=1,2\cdots\cdots p \end{array}\right\} \tag{5.1}$$

式（5.1）适用于内、外初始点，但最好给出内点，有助于加快计算机寻优过程。优化问题求解时，采用标准寻优子程序（MINI），使用者只需编制相应的主程序和函数段（FUNC）子程序。

本文用 FORTRAN 语言写成计算程序，主程序框图见图 5.1。以超低空导弹战斗部为实例进行了优化计算。

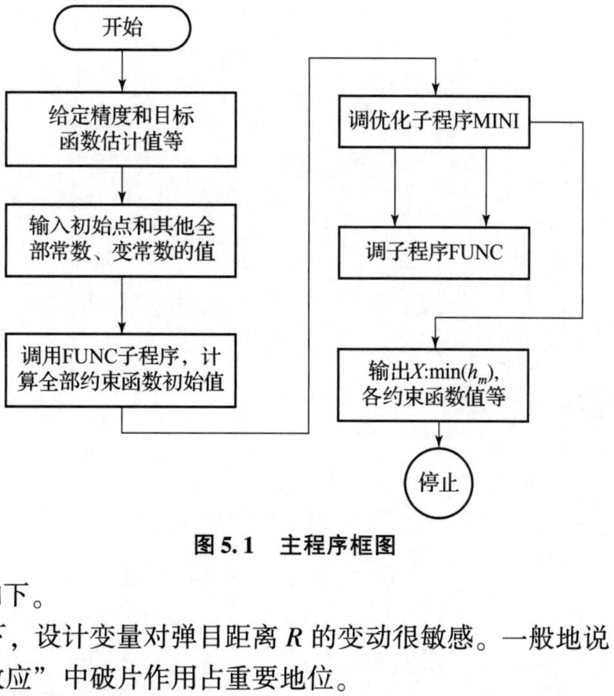

图 5.1　主程序框图

计算结果的基本特点如下。

(1) 等同的约束条件下，设计变量对弹目距离 R 的变动很敏感。一般地说，R 增大，壳体厚度随之增大，这说明“杀爆相关效应”中破片作用占重要地位。

(2) 等同的 R 和其他约束条件，设计变量对 $[q_1]$ 很敏感。$[q_1]$ 值减小，壳体厚度随之减小，这是因为 $[q_1]$ 减小后破片密度增加较快，也就是说，相关函数对破片密度的变化比破片质量的变化更为敏感。

(3) Ω_0 的优化计算值均趋于 $[\Omega_1]$，其目的在于提高破片飞散角内的破片分布密度。

(4) 随着 R 的变小，优化的结果希望战斗部的装填系数 α 增大，这表明要提高爆破威力。

6 对所建立的数学模型的讨论

(1) 数学模型基本反映了杀爆作用战斗部的威力参数和结构参数，在一定条件下，可定量预测破片杀伤作用和爆破作用之间的合理匹配关系。

(2) 数学模型对触发式或近炸杀爆战斗部的威力评价、方案论证、结构设计和改进提供有实际参考价值的预测值，以便提高效率，减少实验次数。

(3) 本文的工作是阶段性工作，还有待今后进一步完善。

参考文献

[1] 冯顺山，蒋浩征. 爆炸冲击波与杀伤破片对飞机目标相关毁伤效应的研究 [J]. 北京工业学院学报，1986 (1)：119－131.

[2] 北京工业学院《爆炸及其作用》编写组. 爆炸及其作用：下册 [M]. 北京：国防工业出版社，1979.

偏轴心起爆破片初速径向分布规律研究①

冯顺山[1]，蒋建伟[1②]，何顺录[1]，周申生[2]
（1. 北京理工大学；2. 航空航天二院）

摘　要：采用脉冲X光测试方法，研究了轴对称装药金属壳体在偏轴心起爆条件下，破片初速的径向分布规律，建立了破片初速径向分布的计算公式。

关键词：破片初速；爆炸驱动；破片式战斗部

ON THE PATTERN OF RADIAL DISTRIBUTION PATTERN OF INTIAL VELOCITY OF FRAGAMENTS UNDER ASYMMETRICAL INITIATION

Feng Shunshan，Jiang Jianwei，He Shunlu
(beijing Institute of Technology)
Zhou Shensheng
(Research Institute No. 2，Ministry of Aerospace Industry)

Abstract：Patterns of radial distribution pattern of initial velocities of fragments from a symmetrical charge structure subjected to asymmetrical initiation were studied using flash X - ray radiography. Formulae for the calculation of the distribution of the initial velocities of fragments along the radial direction were established.

Keywords：initial velocity，explosion driving，fragmentation warhead

轴对称装药结构的战斗部，起爆位置通常设置在轴线上。起爆后，轴向 X 横截面的微圆环壳体将被驱动，形成若干个具有相同初速，且尺寸均匀分布的破片。用于计算轴心起爆条件下破片初速的公式有多种，有代表性的是 Gurney 公式或修正的 Gurney 公式[1]。

工程应用中，有时也需采用偏轴心（下称偏心）起爆方法。这种起爆导致爆轰波的不对称，相应地作用在壳体上的冲量也呈现一定的分布规律。试验结果表明，壳体形成的破片的径向初速测试值与 Gurney 公式的计算值相差比较大。关于偏心起爆条件下破片初速径向分布规律的文献尚未见公开报道。鉴于偏心起爆过程涉及诸多复杂因素，用现有二维弹塑性流体模型进行数值模拟仍有一定困难。本文采用脉冲X光测试方法进行实验研究，对试验数据进行理论分析和数学处理，建立起表示偏心起爆破片初速径向分布与结构参数定量关系的数学模型。

1　实验研究

1.1　偏心起爆装药结构和有关说明

偏心起爆装药结构多种多样，为了建立更普遍的计算模型，本文以壳体厚度和装药直径线性变化的偏心线状起爆管的装药结构为研究对象。图1.1为装药结构示意图，图1.2为装药结构轴向 x 处横截面视图。

---

① 原文发表于《兵工学报》1993（增刊）。

② 蒋建伟：工学博士，2001年师从冯顺山教授，研究中心爆管式子母战斗部抛撒技术及应用，现工作单位：北京理工大学。

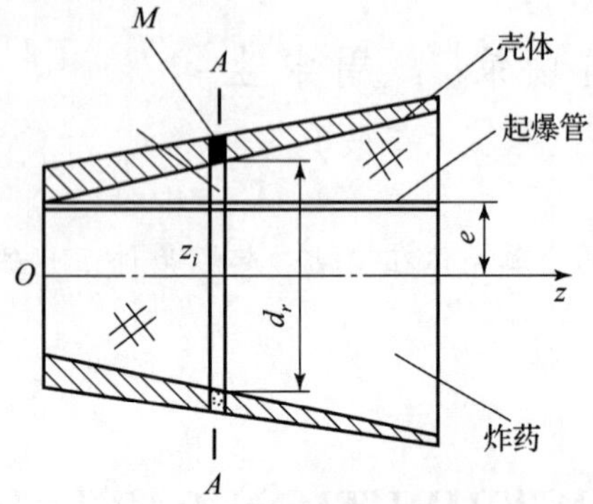

图 1.1 装药结构示意图

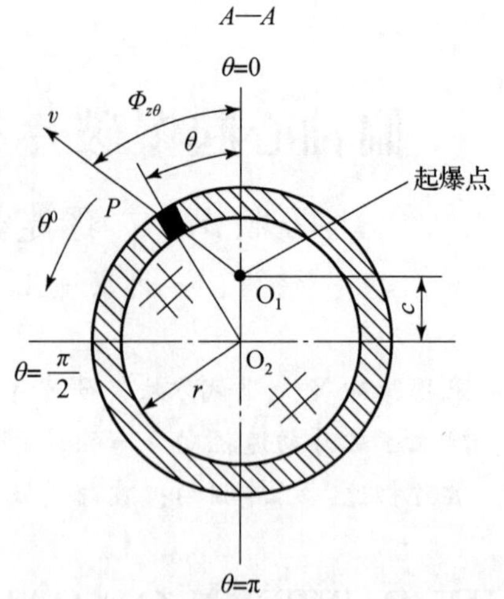

图 1.2 装药结构轴向 x 处横截面视图

图中 d_r 为轴向 x 横截面装药直径，装药半径为 $r=d_r/2$；m_x为轴向 x 横截面微圆环装药质量；M_x为微圆环壳体质量；β_x为装填比（m_x/M_x）；e_{yx}为偏心起爆点距对称轴心的偏心距；e_{yx}/r 为偏心系数；θ 为径向角，即图 1.2 中 $P_{x\theta}O_x$ 与 O_1O_x的夹角，定义 $\theta=0$ 为正反向，$\theta=\pi/2$ 为正中向，$\theta=\pi$ 为正正向，$0\leqslant\theta\leqslant\pi/2$ 为偏心起爆的反向，$\pi/2\leqslant\theta\leqslant\pi$ 为偏心起爆的正向；$\Phi_{x\theta}$为 x 横截面 θ 处的破片径向飞散角；$v_{x\theta}$为 x 横截面 θ 处的破片径向初速；定义 θ/π 为径向角系数。

1.2 试验研究方法

1.2.1 试验模型

对应图 1.1 中的不同横截面，其起爆点的 e_{yx}/r 都不相同，故壳体任一处的 $v_{x\theta}$可表示为（x，e_{yx}/r，θ/π）的函数。对图 1.1 的结构进行测试，以获得 $v_{x\theta}$的分布是很困难的。考虑到装药结构由若干个微圆环组成，故选取不同装填比的单个微圆环作为试验模型。图 1.3 为典型的试验模型。

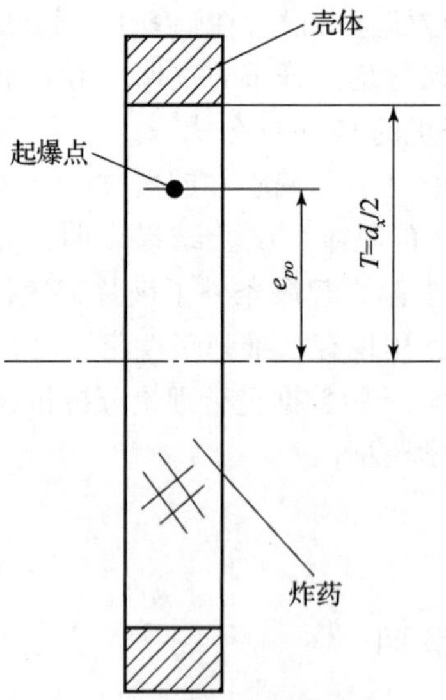

图 1.3 典型的试验模型

1.2.2 试验方法

采用 300 kV 脉冲 X 光摄影仪（曝光时间 30 ns）和射线管相互平行的摄影方法[2]，利用计算机图像处理系统对 X 光图像进行处理和数据分析。图 1.4 为试验布置图。

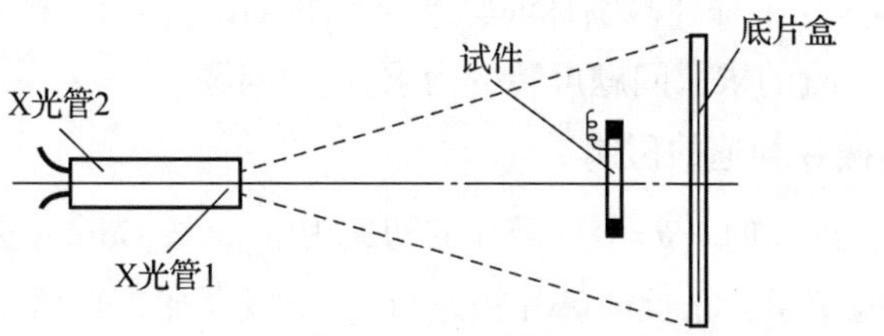

图 1.4　试验布置图

试验时，先用一个射线管拍摄试件的静止图像，后拍摄试件爆炸和达到初速后[1]的两个不同时刻的动态图像。三次拍摄都在同一张底片上成像，根据两个平行射线管的视角差、成像放大比和不同径向角处破片的位移，测得初速径向分布数据。

1.2.3　试验结果分析

对三组整体式和内刻槽对称圆柱形壳体试件分别进行试验，每组试件的结构参数都不同。对脉冲 X 光图像进行计算机处理和对测量结果进行分析，可得如下结论：

（1）偏心起爆条件下，$v_{x\theta}$随 θ 的增大而增大。在 $\theta=0$ 处的破片初速最小，如 $\beta_x=0.952$、$e_{yx}=1$ 时，此处破片初速仅是中心起爆破片初速的 0.75 倍；在 $\theta=\pi$ 处的破片初速最大，如 $\beta_x=0.952$、$e_{yx}=1$ 时，此处破片初速是中心起爆的 1.25 倍。β_x不同，比值相应变化（受 Gurney 公式函数关系制约）。

（2）在正反向（$\theta=0$）处，破片的初速与 e_{yx}/r 成反比；正正向（$\theta=\pi$）处，破片初速随 e_{yx}/r 成正比；但在正中向（$\theta=\pi/2$）处，不论 e_{yx}/r 如何变化，破片初速都很接近中心起爆的破片初速。

（3）偏心起爆下，不论 e_{yx}/r 如何变化，破片运动方向在垂直轴线的平面内的投影，即破片的径向飞散方向几乎与中心起爆的结果一样（即沿壳体法线方向飞散）。这表明偏心起爆不改变破片的径向飞散角。

（4）对整体式试件，破片的径向尺寸随 θ 的增大而减小，$\theta=0$ 时破片尺寸最大，$\theta>\pi/2$ 时破片尺寸不再有很明显的差异。对内刻槽壳体试件，径向破片尺寸基本相同。

2　数学模型的建立和分析

2.1　数学模型

2.1.1　基本假设

为了简化计算模型，做如下假设：①偏心起爆管的爆速远大于炸药的爆速；② $v_{x\theta}$主要与 e_{yx}/r 和 θ/π 有关；③破片运动方向在垂直轴线平面内的飞散方向沿该处壳体的法线方向飞散；④破片初速径向分布的数学表达形式取中心起爆破片初速轴向分布表达式[1]的形式，其内涵符合破片初速的径向分布规律。

依据上述假设，偏心起爆下 $v_{x\theta}$的表达式为

$$v_{x\theta}=F(x/d_x)\sqrt{2E}\sqrt{\frac{\beta_{x\theta}}{1+0.5\beta_{x\theta}}} \tag{2.1}$$

其中

$$F(x/d_x)=[1-Ae^{-1.11x/d_x}][1-Ce^{-3.03(L-x)/d_x}]$$

$$v_{0x}=[1-0.3615e^{-1.11x/d(x)}][1-0.1925e^{-3.030(L-x)/d(x)}]\sqrt{2E_g}\sqrt{\beta(x)/[1+0.5\beta(x)]}$$

式中　$F(x/d_x)$——装药结构两端部效应修正函数；

A、C——由试验确定的常数，对两端敞口的整体式壳体装药结构，当长径比大于 1.5 时，$A=0.36$、$C=0.19$；

$\sqrt{2E}$——炸药的 Gurney 能量；

$\beta_{x\theta}$——x 横截面 θ 处破片的等效装填比，$\beta_{x\theta}=f(e_{yx}/r,\theta/\pi)\cdot\beta_x$；

β_x——中心起爆条件下 x 坐标处微圆环的装填比，$\beta_x=\beta[x]$[1]。

从式（2.1）可以看出，$v_{x\theta}$的求取问题可转变为求$\beta_{x\theta}$的问题。

2.1.2 e_{yx}/r 对正反向破片初速的影响

当偏心系数为 e_{yx}/r 时，正反向（$\theta=0$）破片的初速为 v_{x0}，按 Gurney 公式分析，总存在一个相对于 β_x 的等效装填比β_{x0}。按试验结果，正反向破片初速与 e_{yx}/r 成反比，因而β_{x0}是 e_{yx}/r 的反函数。

设$f_1(e_{yx}/r)=\beta_{x0}/\beta_x$。那么，当 $e_{yx}/r=0$ 时，有 $f_1(e_{yx}/r)=1$；当 $e_{yx}/r=1$ 时，有 $f_1(e_{yx}/r)=\min$。构造函数

$$f_1(e_{yx}/r)=1-a_1(e_{yx}/r)^{a_2} \tag{2.2}$$

根据试验结果回归得常数 a_1，a_2，式（2.2）可写为

$$\beta_{x0}/\beta_x=\begin{cases}1-0.47(2e_{yx}/r)^{0.75} & 0\leqslant e_{yx}/r\leqslant 1/2\\ 1-0.53(e_{yx}/r)^{0.17} & 1/2\leqslant e_{yx}/r\leqslant 1\end{cases} \tag{2.3}$$

2.1.3 β_{x0}的径向分布

由试验结果知，v_{x0}随 θ 角增大而增大，因而可视β_{x0}随 θ 角增大而增大。图 1.2 中的 P_{x0}至 O_1点的距离不仅与有 e_{yx}/r 关，而且也是 θ 角的增函数。

设 $\lambda=|P_{x0}O_1|/r$。由几何关系得

$$\lambda=[4(e_{yx}/r)\sin^2(\theta/2)+(1-e_{yx}/r)^2]^{1/2} \tag{2.4}$$

同样，设$f_2(e_{yx}/r,\theta/\pi)=\beta_{x0}/\beta_x$。并构造函数

$$f_2(e_{yx}/r,\theta/\pi)=1+a_3(\theta/\pi)(e_{yx}/r)^{a_4}\cdot e^{a_5{}^{\lambda}} \tag{2.5}$$

由式（2.3）、式（2.4）即可得β_{x0}径向分布表达式

$$\beta_{x0}=\begin{cases}[1-0.47(e_{yx}/r)^{0.75}][1+2(\theta/\pi)(e_{yx}/r)e^{\lambda_1^{\lambda}}]\cdot\beta_x & \begin{matrix}0\leqslant\theta\leqslant\pi\\ 0\leqslant e_{yx}/r\leqslant 1/2\end{matrix}\\ [1-0.53(e_{yx}/r)^{0.17}][1+(\theta/\pi)(e_{yx}/r)^{0.1}e^{\lambda_2^{\lambda}}]\cdot\beta_x & \begin{matrix}0\leqslant\theta\leqslant\pi\\ 1/2\leqslant e_{yx}/r\leqslant 1\end{matrix}\end{cases} \tag{2.6}$$

将式（2.6）代入式（2.1），便可求得破片初速轴向与径向分布的数值 $v_{x\theta}$。

2.2 计算结果分析

2.2.1 偏心起爆 $v_{x\theta}$与中心起爆 v_x之比（$v_{x\theta}/v_x$）

将相同装药结构、不同起爆点（偏心、中心）的初速值进行比较，可清楚说明破片初速的径向分布特征。图2.1 展示了 $v_{x\theta}/v_x$随 θ 变化的计算曲线与实测值的对比情况。通过比较，它们之间相对误差值的平均值<2%。从图中还可看出，不论 e_{yx}/r 如何变化，在 $\theta=2$ 处，$v_{x\theta}/v_x$的值都接近于1。图中曲线充分表明，偏心起爆使大于 $\pi/2$ 区域内的破片初速得到不同程度的增加。

2.2.2 $\beta_{x\theta}/\beta_x$与 $E_{x\theta}/E_x$的分布特征

偏心起爆的等效装填比与中心起爆装填比的比值 $\beta_{x\theta}/\beta_x$反映了破片初速径向分布的内在规律。图 2.2 展示了本文数学模型的计算曲线。从曲线中可以发现，$\beta_{x\theta}/\beta_x$随 θ 的变化很显著，最大值约为 2；在 $\theta=\pi/2$（对应曲线 P 点）处，$\beta_{x\theta}/\beta_x=1$，充分说明了此处破片初速为什么很接近中心起爆的原因。

设 $E_{x\theta}$为偏心起爆下径向角（0，θ）内的破片动能和 E_x为中心起爆下（0，π）角内的破片动能之和。定义

$$f_3(\theta)=E_{x\theta}/E_x=\int_0^{\theta}v_{x\theta}{}^2d\theta/v_x^2\cdot\pi \tag{2.7}$$

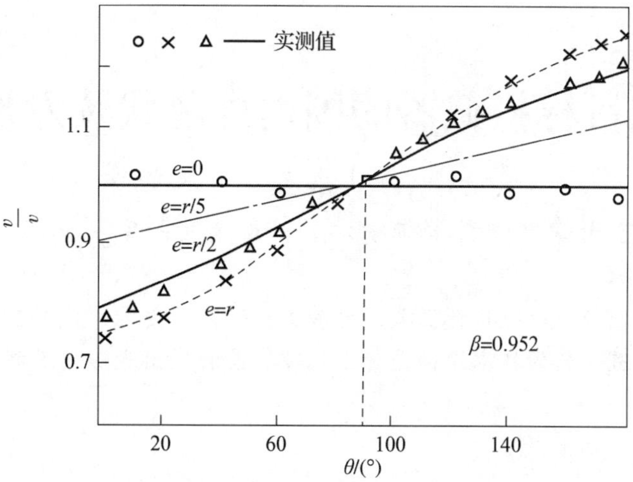

图 2.1　$v_{x\theta}/v_x-\theta$ 计算与实测分布曲线

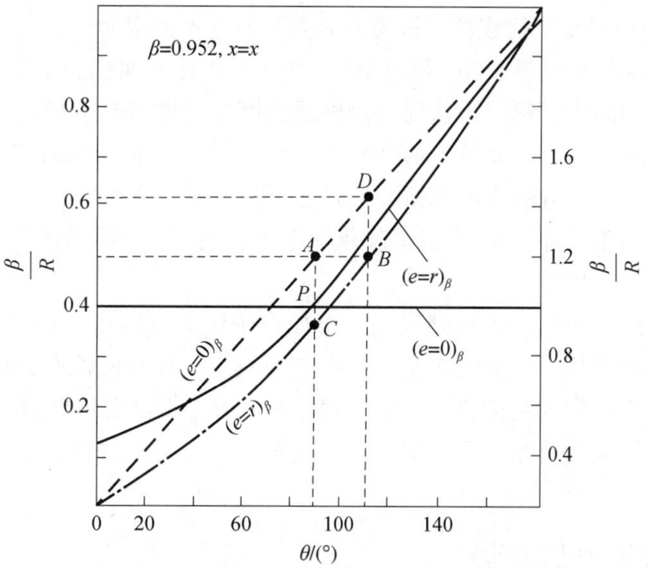

图 2.2　$\beta_{x\theta}/\beta_x-\theta$ 与 $E_{x\theta}/E_x-\theta$ 曲线

式（2.8）可用来分析破片杀伤动能在空间分布的情况，为实现破片在空间合理分配或更有效地利用破片动能提供技术途径。

图 2.2 中的 $E_{x\theta}/E_x-\theta$ 曲线表明，当 $e_{yx}/r=1$ 时，在 $0°\leqslant\theta\leqslant112°$ 内破片动能之和占壳体总动能的 1/2，余下的 68°内的破片动能之和也占壳体总能的 1/2。分析可知，如目标在这 68°角内，被毁伤的概率将会有明显的提高。曲线还表明，偏心起爆虽使破片径向初速呈现某种分布规律，但与中心起爆相比并不改变壳体形成全部破片的总动能，即对某一确定装药来说，无论起爆点如何变化，壳体被炸药驱动形成全部破片的总动能值为常数。这里所指的壳体不仅指整个壳体，也可是某一微圆环壳体。

3　结论

研究结果表明，偏轴心起爆使爆轰产物作用在壳体上的冲量不再呈现轴对称分布，从而使破片初速不仅具有轴向分布的特征[1]，还具有径向分布的特征；破片初速径向分布表达式（2.1）较客观地反映了破片初速与偏心系数 e_{yx}/r、径向角系数 θ/π 和 x 坐标的关系，其计算值与实验值吻合较好，可用于工程问题的计算。

反轨道目标拦截器的国内外现状及发展分析①

冯顺山，刘春美②，张旭荣

（北京理工大学爆炸科学与技术国家重点实验室，北京　100081）

摘　要：针对目前的国外反轨道目标拦截器的演变，研究了美国、俄罗斯的反轨道武器的状况，同时对俄罗斯和美国的反轨道武器的技术途径进行了比较。最后提出我国发展拦截器存在的问题，并指出相关的技术难点。

关键词：反轨道目标；拦截器

引言

1957 年苏联第一颗人造卫星发射成功，标志着人类开始进入太空时代，同时也拉开了太空军事化的序幕。在迄今为止世界各国发射的近 5 000 颗卫星中，2/3 是直接或间接用于军事目的的。这些卫星通过获取军事情报、监视敌方动向、了解气象状况、提供通信网络、发布导航信息等方式，配合和支持地面的军事行动，起到了扩大地面军事行动效果的“倍增器”作用。目前，美国和俄罗斯获取的大部分军事情报来自卫星的太空侦察，大部分远距离军事通信是通过卫星来实现的，许多武器投射和导航系统也都要依靠卫星来进行。除军用卫星外，美国和苏联/俄罗斯还发射了可用于军事目的的宇宙飞船、空间站、航天飞机、反卫星、反导弹武器等。

太空军事化将使未来战争由空中扩展到太空，正如空中优势曾在现代战争中起过特别重要的作用，太空优势也将显著地影响未来战争。世纪之交的几次局部战争，势必促使有关国家尤其是美、俄等发达国家发展航天武器装备，以夺取太空优势，进而达到控制太空、控制地球的目的。21 世纪，一个国家是否拥有太空这个制高点，将成为衡量一个国家军事力量是否强大的重要标志，而航天武器装备正是夺取太空这个制高点的利器。

1　国外研究现状及发展趋势

美、俄两国作为世界的两大军事强国，在拦截器尤其反卫星技术领域同样处于重要位置，它们在空间技术领域的现状和发展基本代表了当今世界反卫星领域的研究水平和现状。

1.1　美国

美国的太空控制战略从 20 世纪 50 年代苏联发射第一颗人造卫星前即开始萌发。此后，美国开展了一系列反卫星武器的研制、开发甚至部署活动。在研制反卫星武器的过程中，美国充分意识到在企图攻击摧毁敌方卫星系统的同时，敌方也极有可能攻击自己的卫星系统。为了保护自己的太空系统，美国一方面积极发展阻止敌方进入和利用太空的能力，另一方面努力提高己方和友方太空系统的生存能力。20 世纪 80 年代的星球大战（SDI）计划催生了比较成型的太空作战理论。1984 年，美国军方颁布了有太空作战内容的空军野战手册——AFM1 - 1。在 1991 年海湾战争中美国首次引入太空控制战略，在取得成功的同时也认识到己方太空系统可能会成为敌方打击的高价值目标。在以后的科索沃战争、阿富汗战争以及第二次海湾战争中，美国对太空系统的依赖程度越来越高，太空系统的脆弱性与其重要性一样得到了

---

① 原文发表于《2005 年弹药战斗部学术交流会论文集》。

② 刘春美：工学博士，2004 年师从冯顺山教授，研究等待型反空间轨道目标战斗部技术，现工作单位：公安部第一研究所。

越来越高度的重视。1997 年美国国家安全太空路线图制定小组公布了一系列太空控制技术计划，并详细绘制了全面的、体系化的关键技术全寿命周期实现路径图。1998 年出台了《美国太空司令部战略主导计划》，后又在此基础上制定了《2002 财年及以后的战略主导计划》。其中均包括专门的篇章系统论述美国的太空控制思想，从中不难看出美国太空控制思想在逐步丰富和发展。美国空军 1998 年 8 月 23 日正式颁发的太空作战条令文件 AFDD2－2 指出：太空控制是获得并维持太空优势的手段，通过太空控制可确保友军利用太空环境，同时使敌人不能使用太空；太空对抗所涉及的活动包括陆海空天部队、信息作战部队以及特种作战部队的所有相关活动。

AFDD2－2 明确表述了对太空控制直接有贡献的三大能力：一是太空态势感知能力，包括探测报告威胁和监视能力；二是进攻性太空对抗能力，即通过欺骗、打断、暂时取消效用、降效和摧毁等手段直接威胁敌人，不同程度地阻止敌人使用太空资源；三是防御性太空对抗能力，包括击败发动太空进攻的敌人、直接降低敌人威胁水平的主动防御和降低己方太空系统易损性，但不直接降低敌人威胁水平的被动防御。2000 年 2 月，美国国防科学委员会指出：美国必须把发展摧毁卫星的能力作为威慑战略的组成部分；实施进攻性空间控制，研制能使敌方卫星或其他航天系统失效或毁坏的武器。国防部 2000－2001 财年《空间技术指南》将空间控制技术作为重点支持对象。2001 年 1 月 11 日，美国政府公布了《国家安全太空管理与组织审议委员会的报告》并改组了美国太空军事力量的相应组织机构。2001 年 9 月，美国防部《四年一度防务评估报告》提出要控制空间即不让敌人利用之，并强调这是未来军事竞争的关键目标。

反卫星武器。美国于 1959 年率先进行了高空核爆炸试验，利用核爆炸效应能量进行反卫星；1964 年，美国首次部署雷神陆基反卫星核导弹；从 20 世纪 70 年代后期，美国重点转向研制动能和定向能非核反卫星武器，1978 年研制了一种带自动寻的两级固体反卫星导弹，其全长 5.4 m，直径约 0.5 m，重 1 179 kg，弹头部分装有小型拦截器。导弹由 F－15 战斗机携带到高空发射，在第 1、第 2 级助推器的推动下弹头相对速度达到约 13 km/s 时自动跟踪目标并与其相撞。1984—1985 年，美国用反卫星导弹进行了 5 次实弹跟踪目标与打靶试验，原计划再经过 7 次飞行试验后即可装备部队。后由于苏联在 1983 年采取单方面停止向空间发射反卫星武器的行动，美国国防部于 1988 年 3 月宣布终止这项历时 10 年的机载反卫星导弹计划。1997 年 10 月 17 日，美国陆军在新墨西哥州白沙导弹靶场进行了激光反卫星试验。试验结果表明激光照射使星上传感器达到饱和状态（暂时失效但不被破坏）。这次试验成功是美军激光反卫星武器的一个重要里程碑，标志着美国激光反卫星武器开始或即将拥有实战能力。

在美国空军最近出台的“转型飞行计划”（TFP）中，提出研制和部署空射反卫星导弹（ALASM）、地基激光器和改进型空天激光全球作战（EAGLE）中继镜等，在 2015 年以后部署。空射反卫星导弹是一种小型空射导弹，能够拦截近地轨道卫星；地基激光器能够穿透大气层向近地轨道卫星照射激光束，从而提供强大的、攻防兼备的空间控制能力；改进型空天激光全球作战中继镜利用机载、地基或者天基激光器与天基中继镜，投射不同能量的激光，从而扩大机载与陆基激光器的射程，实现大范围的杀伤。要求在 2010 年前具有反卫星通信能力，能阻止或中断敌方空间通信和预警信号。

轨道武器。过去美国开展过水星、双子星座和阿波罗计划，俄罗斯有东方号、上升号、进步号和联盟号。未来，按照美国总统布什 2004 年 1 月宣布的新的空间计划，在 2008 年前将研制和试验一种新型载人探测飞行器（CEV），并在 2014 年前完成首次载人任务。目前，NASA 正在进行方案论证和选择，近期可能采用一次性使用运载火箭发射，采用弹道式返回的方案，类似于过去的阿波罗飞船，但所采用的都是当今最先进的技术。

另外，根据美国空军“转型飞行计划”（TEP），美国将在 2015 年以后部署一种轨道转移飞行器（OTV），对美国空间资源进行在轨服务，显著提高这些空间资源的灵活性和作战能力，并对其提供有效的保护。而对敌方卫星，可以将其推离轨道或使其功能丧失。

空间飞行器－目前，美国国防高级研究计划局（DARPA）和美国空军正在联合执行“猎鹰”（从美国本土进行兵力应用和发射，FALCON）计划，旨在开发和验证可以使美国进行全球快速打击的近期目标和远期目标变为现实的一套通用技术，同时还将验证低成本、快速空间发射和运输的能力。“猎鹰”计划的近期目标（2010 年）是通用航空飞行器（CAV）/小型发射火箭（SLV）系统具备初始作战能力；远期目标（2025 年）是研制出高超声速巡航飞行器（HCV）。

总之，美国正在加大对轨道武器的研制力度，以进一步确立和巩固自己在轨道武器研制方面的优势地位。在未来，美国将重点发展空间操作飞行器、空间机动飞行器、轨道转移飞行器、通用航空飞行器等，其中将优先发展可通过空间或从空间进行对地攻击的轨道武器系统。

1.2 俄罗斯/苏联

苏联从 1964 年开始研制反卫星武器，在 1968—1971 年进行了有关接近、识别与推毁目标的飞行试验，1976—1982 年又进行了旨在提高实战能力的快速发射、拦截和新型制导技术的试验，从而已具备了反卫星作战的能力。苏联的反卫星武器是一种带有雷达或红外制导装置、轨道机动发动机和高能炸药破片杀伤战斗部的卫星，质量为 2.5 ~ 3 t，长 4.5 ~ 6 m，直径 1.5 m，由液体运载火箭从地面基地发射到与目标航天器同一平面的轨道，然后通过地面制导使它与目标交会，最后利用自身的制导装置接近目标到一定距离后爆炸，并摧毁目标。俄罗斯还研制了另外一种号称天雷的卫星，它更加先进，具有一定的轨道机动能力，可以在 5°~ 10°范围内改变轨道角，拦截不同高度敌方卫星。俄军在 1992 年承认其已经拥有和随时可以投入实战的这两种共轨式反卫星武器。但由于其发射阵地固定，因此只能攻击地球低轨道卫星。

苏联 1983 年后停止了反卫星武器的空间试验活动，但仍然是目前世界上唯一保持反卫星武器能力的国家。在此之后，苏联加紧研制天基反卫星导弹和激光武器。1981 年，苏联在宇宙 – 1276 号飞船上试验了光学制导的二级轻体反卫星导弹。1987 年，苏联成功研制了太空激光武器系统。

2001 年 6 月，俄罗斯把军事航天部队和军事导弹防御部队从战略火箭军中分离出来，并在其基础上组建新的军种 – 航天部队。新组建的航天部队将装备新型反卫星武器，作战高度为 5 000 km，可攻击敌方部署在地球低轨道上的侦察、导航、气象卫星和航天飞机。面对美国空间优势的压力，俄罗斯正在重新开始发展反卫星武器。

总之，反卫星武器已从以前的战略威慑作用转向战术应用，在未来战争中将作为一种战术性武器发挥作用。当前，美俄均已具有研制实战用反卫星武器的能力，但能力有限。今后将着重发展反卫星武器的精确制导技术、跟踪瞄准与传输技术、小型化技术等，以进一步提高反卫星武器的性能。为了争夺空间优势，保证国家安全，今后反卫星武器的竞赛将愈演愈烈。

2 美俄两国反轨道武器的技术途径对比

美俄两国目前发展的反卫星武器都是从本国的实际情况出发，尽可能发挥自己的优势，力图早些占领反卫星这一领域。俄罗斯利用大助推力火箭和卫星数量多的优势，美国则利用微电子技术先进的优势，都有一个先解决有无问题的愿望。

可行性。俄罗斯“宇宙号”反卫星由于采用追逐式的轨道交会和与目标相隔一段距离的爆炸摧毁方式，因此技术不太复杂，容易实现。相反，美国现在的 F – 15 机载反卫星系统方案，因采用直接上升的迎头碰撞方案，其制导精度要求相当高。因此这一技术的难度大，要求有高精度的末寻的导引头。

实用性。俄罗斯“宇宙号”反卫星系统虽然根据实验结果能够攻击低轨道上的卫星，但由于技术和设计上的缺陷，其实用性是有限的：发射条件受限，只有当目标卫星的轨道投影点接近反卫星发射场上空时，或者目标卫星只能在或接近发射台经度 11°范围飞过的瞬间才有可能实现截击，这一条件只有目标卫星轨道倾角大于发射场才能满足，这就意味着截击星必须平均等待 6 小时才能攻击所制定的目标，这样一天只有两次攻击机会，限制了系统快速反应能力；由于追逐式轨道交会，攻击时间长速度较慢，按照目前平均需要 3 小时的攻击速度，带来很多问题：截击星本身容易被发现、干扰和遭到破坏，并利于对方机动规避；由于采用笨重的大型液体火箭和笨重不易转移的固定发射阵地，不能机动发射和快速发射。

美国的 F－15 机载反卫星系统具有机动灵活、反应快、精度高等特点，不受固定阵地的限制，可以飞到任意高度发射导弹，使得作战时间大大缩短，且保证了一定的作战距离和作战高度。其主要是采用了高精度的末寻的导引头。

费效比美国准备用三四十亿美元来建立一套 F－15 机载反卫星系统。苏联的“宇宙号”反卫星系统看似简单，但要建立一套能实战的系统，费用并不比美国低。相比较，美国的 F－15 机载反卫星系统对低轨道卫星的威胁更大。

3　国内研究现状及发展趋势

我国现有的一系列军事、民事太空开发项目，为国内反卫星武器的发展提供了坚实的技术基础，如在轨卫星机动、在轨卫星控制、高能激光技术、卫星小型化技术等。其中卫星小型化技术既降低了发射费用，同时又可提高卫星在空中机动的灵活性，甚至在一定程度上还可满足“按需发射”的要求。

中国反卫星武器技术的研究始于 20 世纪 80 年代，其重点放在反卫星武器的可用技术上，包括动能杀伤器、高能激光、卫星干扰技术、卫星回收和跟踪技术、卫星小型化技术、卫星的地面控制、卫星保护技术、卫星通信、定向飞行、精确高度控制等。

3.1　掣肘之处

尽管我国在过去的 20 年里在航空航天技术领域也确实取得了巨大进展，但想要真正拿出一个实用的反卫星武器，尚有许多力不从心之处。

（1）较弱的跟踪能力。在太空跟踪技术领域，我国在很大程度上依赖别国支持，而战时是不可能得到这些支持的。除了国内的定位站，2 个国外的跟踪站和 4 个跟踪船，我国的跟踪系统尚不具备全球跟踪定位能力。

（2）有限的发射能力。虽然我国的发射能力一直在稳步提高，但离“按需发射”的要求还很远。在这种情况下打赢未来的天战和使用反卫星武器不可能。

（3）脆弱的发射基地。我国的发射、跟踪、定位设备都不是机动的，包括可能的地基激光武器发射站，这使得它们极易遭到攻击。

3.2　相关技术难点

发现目标并识别目标技术，即辨识敌方卫星技术；导弹的精确轨道控制技术，即如何将导弹送入预定轨道；红外制导末寻的导引头技术；稀薄空气动力学，空间轨道运动学和动力学问题；大气层外实现机动飞行必须依靠反推控制系统（推力矢量控制和小型的侧面燃烧火箭发动机控制），也是一个需要深入研究的新课题；高空姿态控制问题还需作较为详细和深入的研究；人工毁伤元素技术也有一定的难度，毁伤元素的尺寸分布、硬度、强度、接近速度，以及预警时间都需要进一步的研究。

4　总结

中国发展外空技术完全是为了和平利用外空，中国一向主张太空的非武器化，认为防止在太空进行军备竞赛符合世界各国利益。随着时间的推移，地面跟踪将会越来越精确，运载器的运载能力将会越来越大，制导也会越来越准确。因此，在不久的将来，也会有越来越多的国家将拥有自己的轨道目标拦截器。

一种聚焦式杀伤战斗部的设计方法①

冯顺山，黄广炎②，董永香

（北京理工大学爆炸科学与技术国家重点实验室，北京 100081）

摘 要：为研究聚焦式杀伤战斗部的设计方法，采用新的战斗部设计思想，以一端静态起爆条件下的轴对称聚焦式杀伤战斗部为例进行了设计研究，将圆弧形战斗部装药结构各微元段按一定角度旋转，得到了满足聚焦杀伤要求的战斗部装药曲线结构。计算表明，该聚焦式杀伤战斗部可控制破片的飞散方向并形成聚焦带，实现了对目标的剪切，毁伤。该设计方法对加强战斗部终点毁伤效应的工程设计有着重要意义。

关键词：爆炸力学；聚焦战斗部；战斗部设计；破片速度；破片飞散

A New Design Method of Fragment Focusing Warhead

Feng Shunshan，Huang Guangyan，Dong Yongxiang

（1 State Key Laboratory of Explosion Science and Technology，Beijing Institute of Technology，Beijing 100081）

Abstract：In order to study the design method of fragment focusing warhead，based on the new，design thought and design model，the design study was carried out，taking an example of axial symmetry fragment focusing warhead with static explosion at one side. The slight part of fragment，arc charging structure was circumrotated on certain angle. The warhead satisfying the desire，of focusing effect was achieved. The calculation result shows that the fragment focusing warhead，can control the dispersion direct ion of the fragment so as to form focus band，and the target can，be destroyed. The design method is of great value to the engineering design for strengthening the，terminal effects of warhead.

Keywords：mechanics of explosion；fragment focusing warhead；warhead design；fragment velocity；fragment dispersion

聚焦杀伤战斗部通过控制破片的飞散方向，使各破片联合作用形成聚焦带来实现剪切效应，从而对目标进行结构毁伤，其关键技术就是针对聚焦点，来有效控制破片的飞散方向。

聚焦式杀伤战斗部作为一种前沿的战斗部终点毁伤技术，受到国内外越来越多的科研学者的重视，为此进行了大量的实验及数值模拟研究，但至今大多数聚焦设计方案都没有提出确切的理论上的设计方法，只是根据目标杀伤要求，通过大量实验数据或者数值仿真效果，反推符合杀伤要求的聚焦战斗部装药结构，这种设计方法不仅耗费大量物力财力，而且运用范围也极其有限，效果并不明显。所以，建立一种经济有效的聚焦式杀伤战斗部设计方法很有意义。

本文以一端静态起爆条件下的轴对称聚焦式杀伤战斗部为例进行设计研究，建立设计模型，根据战斗部装药情况、壳体材料、端部情况及聚焦要求，通过计算圆弧形装药结构战斗部各微元段旋转角度 θ_x，得到满足杀伤要求的聚焦式战斗部的装药曲线结构。

---

① 原文发表于《弹道学报》2009（3）。

② 黄广炎：工学博士，2004 年师从冯顺山教授，研究封锁型弹药设计及效能分析方法，现工作单位：北京理工大学。

1 聚焦式杀伤战斗部设计思想及模型建立

1.1 聚焦设计思想

聚焦战斗部的设计主要是通过采取特殊的装药曲线结构和起爆方式来控制破片的飞散方向，使大部分破片向同一位置附近飞散，从而实现破片对目标的集中打击。不考虑起爆点爆轰驱动对破片飞散方向的影响[1]，图1所示装药结构曲线是以（x_f，y_f）为圆心且过（x_0，y_0）的圆弧战斗部，其破片将会向（x_f，y_f）聚焦。但由于不同起爆点爆轰波传播方向的影响，各破片将会偏离该处战斗部壳体法线并以一定角度 δ_x 飞散。如果忽略破片微元段壳体法线方向 φ_1 的变化，只要计算出 δ_x，然后将战斗部各微元壳体绕该段中心逆向旋转角度 $\theta_x=\delta_x$，就可以使该处破片朝聚焦点飞散。实际上，战斗部该处壳体旋转一定角度后，壳体法线与弹轴 x 正方向夹角 φ_1 将会变化，这样战斗部壳体旋转后该处破片飞散方向 $\delta'_x\neq\delta_x$。所以，要使战斗部产生最好的聚焦效果，旋转角度 θ_x 必须满足 $\delta'_x=\delta_x$。

1.2 聚焦战斗部设计模型建立

以一端起爆（起爆点为坐标原点）的轴对称聚焦式杀伤战斗部为研究对象建立设计模型，如图2所示，破片聚焦点坐标为（x_f，y_f）。假设聚焦战斗部壳体母线方程为

$$ax^2+2bx+cy^2+2dx+2y+f=0 \tag{1}$$

式中 a、b、c、d、e、f——系数。

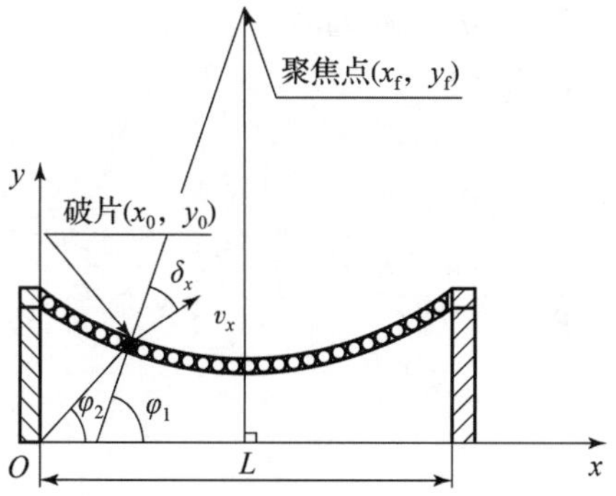

图1 聚焦战斗部设计图

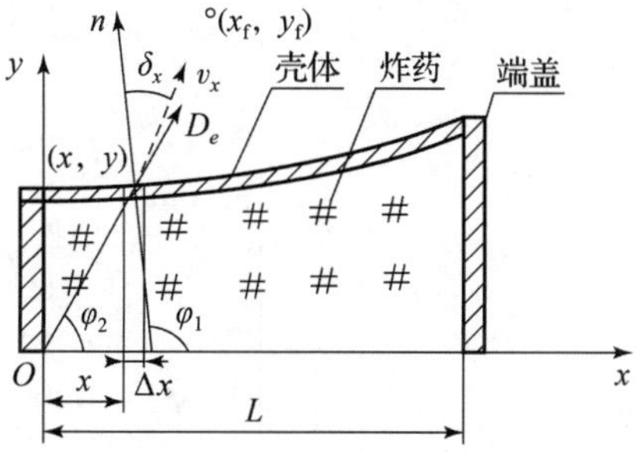

图2 聚焦战斗部示意图

可近似把战斗部看成由若干个微圆环组成，取微元段长度 Δx 进行分析。破片初速用格尼公式的修正式[2]计算：

$$v_x=k(1-A_1e^{-Bx/d(x)})(1-A_2e^{-C(L-x)/d(x)})\times\sqrt{2E_g}\sqrt{\beta(x)/(1+0.5\beta(x))} \tag{2}$$

式中 v_x——距起爆端 x 处壳体形成的破片初速；

B、C——实验确定的常数；

A_1、A_2——装药结构两端端部效应修正系数；

$\beta(x)$——炸药装填系数；

$d(x)$——装药直径；

k——炸药能量利用系数；

$\sqrt{2E_g}$——炸药的格尼常数（速度），取决于炸药的组分和特性。

破片飞散方向选用夏皮洛公式[3]计算：

$$\delta_x=\arctan\left[\frac{v_x}{2D_e}\cos\left(\frac{\pi}{2}-\varphi_1+\varphi_2\right)\right] \tag{3}$$

式中 δ_x——该处壳体破碎生成的破片速度矢量偏离壳体法线的偏角；

D_e——炸药爆速；

φ_1——距起爆端 x 处壳体微元法线与弹体对称轴（x 轴）的夹角；

φ_2——该处爆轰波阵面法线与弹体对称轴的夹角。

由图 2 易知，要使该战斗部杀伤破片在（x_f，y_f）处聚焦，必须满足：$\delta_x = \varphi_1 - \arctan\left(\dfrac{y_f - y}{x_f - x}\right)$，即

$$\varphi_1 - \arctan\left(\frac{y_f - y}{x_f - x}\right) = \frac{v_x}{2D_e}\cos\left(\frac{\pi}{2} - \varphi_1 + \varphi_2\right) \tag{4}$$

要满足式（4）所示破片飞散方向的要求，必须根据该处壳体初始飞散方向相应旋转一定角度 θ_x，使战斗部壳体旋转后该处破片飞散方向 $\delta'_x = \theta_x$。所以，根据聚焦杀伤要求以及相关起爆条件，在编制的杀伤战斗部破片初始威力参数计算程序的基础上，加入聚焦杀伤战斗部设计程序，在已知炸药成分、壳体材料、结构尺寸等参数条件下，就可以计算 θ_x。只要将图 2 所示战斗部各壳体微元都绕壳体法线逆时针旋转 θ_x（θ_x 为负时，表示顺时针旋转），再经过拟合计算得到旋转后的战斗部装药结构，就可设计出在（x_f，y_f）形成满足破片聚焦飞散的杀伤聚焦战斗部。

1.3 战斗部装药结构曲线各微元段偏转角度计算

把战斗部分成若干微元段进行分析，要使各微元段破片在爆炸驱动下向聚焦位置方向飞散，则必须把该微元段装药母线旋转角度 θ_x，使旋转后的 $\delta'_x = \theta_x$。在模型中通过计算转化，得到各微元段旋转角度为

$$\theta_x = \begin{cases} \arctan\dfrac{\dfrac{v_x}{2D_e}\cos\left(\dfrac{\pi}{2} - \varphi_1 + \varphi_2\right)}{1 - \dfrac{v_x}{2D_e}\cos\left(\dfrac{\pi}{2} - \varphi_1 + \varphi_2\right)} & \varphi_1 > \arctan\dfrac{y_f}{x_f} \\ \arctan\dfrac{\dfrac{v_x}{2D_e}\cos\left(\dfrac{\pi}{2} - \varphi_1 + \varphi_2\right)}{1 + \dfrac{v_x}{2D_e}\cos\left(\dfrac{\pi}{2} - \varphi_1 + \varphi_2\right)} & \varphi_1 < \arctan\dfrac{y_f}{x_f} \\ 0 \quad \varphi_1 = \arctan\dfrac{y_f}{x_f} \end{cases}$$

2 聚焦战斗部设计实例研究

任取一战斗部结构进行设计分析，战斗部长 L = 12 cm，假设要设计的聚焦战斗部的聚焦点坐标为（6，50），战斗部壳体中心（$L/2$）处壳体微元坐标为（6，2），装药为 B 炸药，一端起爆，两端堵塞，战斗部壳体由直径 0.4 cm 的预制钨合金球形破片组成。易求得初始圆弧形装药结构战斗部装药母线方程为

$$x^2 + y^2 - 12x - 100y + 232 = 0$$

要针对聚焦杀伤要求设计聚焦战斗部，必须使每个壳体微元都绕壳体法线旋转角度 θ_x。使用前面设计的模型和偏转角计算方法，用 VC ++ 编制计算程序，计算每段壳体微元所需旋转的角度 θ_x，如图 3 所示。

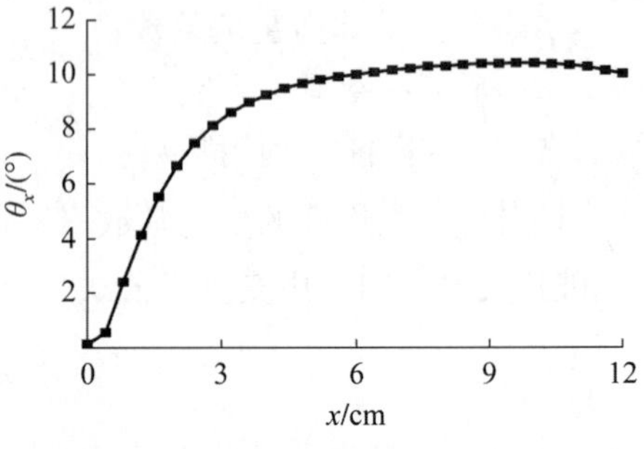

图 3　战斗部各微元段旋转角度 θ_x

把初始设计的凹圆弧形战斗部壳体微元都绕该段法线旋转角度 θ_x，得到新的战斗部各微元段装药直径随战斗部轴向的变化值，再用最小二乘法进行多项式拟合，就可以得到在（6，50）形成聚焦杀伤的聚焦杀伤战斗部。

受爆轰气体泄漏的影响，战斗部两端破片飞散方向的计算

值与实际存在一定误差，假如加厚战斗部两端的堵塞钢板厚度，考虑到初始设计的凹圆弧形战斗部大部分壳体微元需逆向旋转角度 $\theta_x \approx 10°$，所以可以把整个战斗部装药曲线绕曲线中心旋转 $\theta_x = 10°$，这样旋转后的战斗部将会在聚焦点处形成聚焦。

然而，旋转后有些微元部分对应的装药直径将发生变化，导致该战斗部微元段的炸药装填系数比 $\beta(x)$ 发生变化，这样该处微元破片初速会随之变化，从而 δ_x 也会变化，必须再考虑旋转后的战斗部破片飞散方向 δ_x 与旋转角度 θ_x 的关系。通过数值计算可知，绝大部分微元处壳体 $\delta_x - \theta_x < 1°$，如图 4 所示。这样，该段战斗部各破片偏离聚焦点的距离可忽略不计。

计算得到符合聚焦杀伤要求的战斗部装药结构曲线方程为

$$x^2 + y^2 + 4.67x - 98.54y + 129.06 = 0$$

3 破片聚焦效果的工程计算验证

假设距起爆端 x 处的破片飞散到聚焦高度时，其运动轨迹偏离聚焦点（x_f，y_f）的距离为 Δs_x。使用基于 VC ++ 编制的杀伤战斗部破片威力参数数值仿真的计算程序，来计算前面所设计的聚焦杀伤战斗部各破片对应运动轨迹在聚焦高度处偏离聚焦点的距离 Δs_x，如图 5 所示。

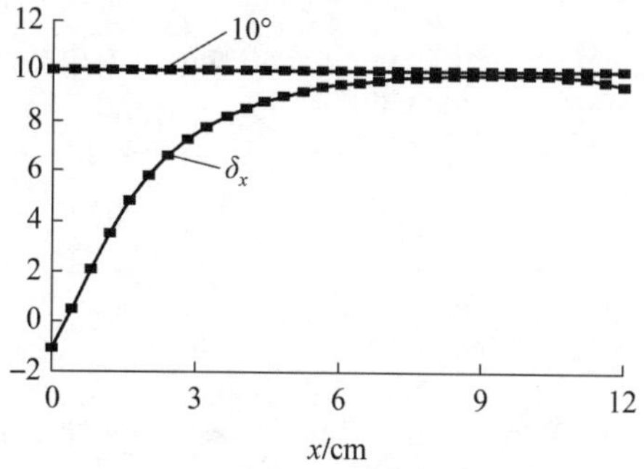

图 4 旋转 10°后战斗部各微元段的 δ_x

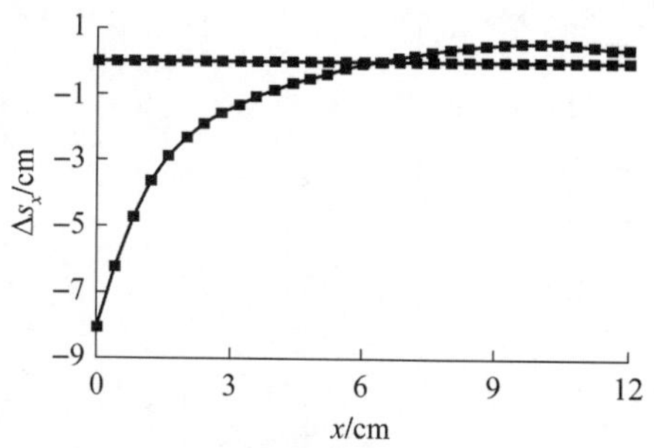

图 5 各破片偏离聚焦点的距离

由图 5 可知，战斗部大部分破片运动轨迹偏离聚焦点（x_f，y_f）的距离均满足 $\Delta s_x < 0.5$ cm。显然，在一端静态起爆情况下，战斗部破片在聚焦目标位置处形成聚焦杀伤。

如果聚焦战斗部打击的是高速运动的目标，要考虑各破片到达聚焦位置的同时性，假设各破片从爆轰驱动达到初速开始，运动到聚焦高度所用的时间为 t_x。同样在破片专家系统中编制计算程序，计算得到图 6 的数据曲线图。战斗部各破片达到聚焦位置的时间差距 $\Delta t < 0.1$ ms，所以，即使战斗部欲毁伤的目标在以 1 000 m/s 左右的速度运动的情况下，破片打击目标的聚焦宽度在 10 cm 以内，也能形成剪切作用。

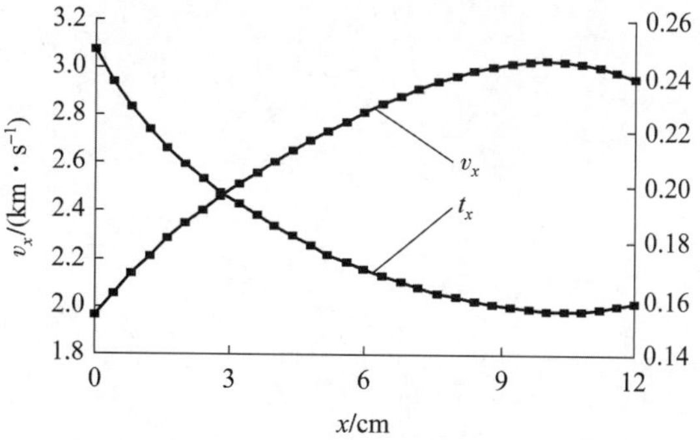

图 6 各破片的初始速度和到达聚焦位置的时间

4 结束语

在新的聚焦战斗部设计思想的基础上，建立了聚焦战斗部设计模型，提出了一种聚焦式杀伤战斗部的设计方法。对一端静态起爆条件下的聚焦战斗部设计进行了分析，根据战斗部装药情况、壳体材料、端部情况及聚焦要求，通过计算对应圆弧形装药结构战斗部各微元段旋转角度 θ_x，得到满足杀伤要求的聚焦式战斗部的装药曲线结构。该设计方法同样适合于其他起爆条件下的聚焦杀伤设计，对加强战斗部终点毁伤效应的工程设计有重要参考价值。

参 考 文 献

[1] 冯高鹏. 破片聚焦战斗部研究 [D]. 北京：北京理工大学，2005.
FENG G P. Study on the fragment focusing warhead [D]. Beijing：Beijing Institute of Technology，2005. (in Chinese).

[2] 冯顺山，崔秉贵. 战斗部破片初速轴向分布规律的实验研究 [J]. 兵工学报，1987 (4)：60 - 63.
FENG S S，CUI B G. An experimental investigation for the axial distribution of initial velocity of shells [J]. Acta Armamentarii，1987，4 (4)：60 - 63. (in Chinese).

[3] 隋树元，王树山. 终点效应学 [M]. 北京：国防工业出版社，2000.
SUI S Y，WANG S S. Terminal effects [M]. Beijing：National Defense Industry Press，2000. (in Chinese)

导电纤维弹关键技术分析[①]

冯顺山，张国伟[②]，何玉彬[③]，王 芳
（北京理工大学爆炸与安全国家重点实验室，北京 100081）

摘 要：导电纤维弹是以导电纤维作为对电力系统毁伤元素的软杀伤弹药。文中在分析国内外导电纤维战斗部研究状况基础上，对电力系统目标易损性及其毁伤机理、导电纤维丝束、气象条件对导电纤维丝束飘落的影响、运载平台以及子弹、丝团和引信匹配等关键技术进行了系统分析。研究结果对导电纤维弹设计有重要参考价值。

关键词：软杀伤；导电纤维；短路毁伤；战斗部

KEY Technology Analysis About Conductivity Fiber Bomb

Feng Shun shan, Zhang Guo wei, He Yu bin, Wang Fang
(Beijing Institute of Technology, Beijing 100081, China)

Abstract: Conductivity fiber bomb is a kind of soft wound ammunition used to destroy the electrical system by conductivity fiber. On the basis of analysis of conductivity fiber warhead studies abroad and domestic. The vulnerability and damage mechanism of electrical system, conductivity fiber, influence of the climatic conditions on the drop behavior of fiber, carrier platform and the match of submunition, fibers and fuze, and so on are discussed systematically. This results will provide a significant guide to design of conductivity fiber bomb.

Keywords: soft wound; conductivity fiber; short - cut damage; warhead

1 引言

导电纤维弹（俗称碳纤维弹）是采用导电纤维作为毁伤元素对电力系统进行短路毁伤的一种软杀伤弹药。它的战术使用可使敌方发电厂和高压变电站在某一段时间内供电中断，致使用电系统，如 C^3I 系统、雷达系统、后勤保障系统瘫痪或相当程度瘫痪，达到削弱敌作战能力、瓦解敌方战斗意志和赢得战场主动权的目的。

国内研究在弹用导电纤维材料技术、短路毁伤机理及毁伤电力设施效应、丝团的抛撒及展开技术和毁伤效能评价等方面均有突破。

2 电力系统目标及其易损性

2.1 电力系统目标

通常将用于生产、运输、分配和使用电能的发电机、变压器、输配电线路及各种用电设备联结在一起，再加上继电保护自动装置、调节自动化和通信设备等相应的辅助系统构成的统一整体称为电力系统。电力系统是战时重要的军事打击目标。

---

① 原文发表于《弹箭与制导学报》2004 年第 24 卷第 1 期。
② 张国伟：工学博士，2001 年师从冯顺山教授，研究电力系统易损特性及导电纤维子弹设计中若干问题，现工作单位：中北大学。
③ 何玉彬，兵器学科博士后，2000 年入站，合作导师冯顺山教授。

2.2 高压变电站短路毁伤易损性

高压变电站短路毁伤易损性是指，一个高压变电站对导电纤维丝束短路毁伤的敏感性。其包含三个重要方面，一是对单束导电纤维短路引弧的敏感性，即导电纤维丝束作用在高压电力设施上产生短路情况时，发生通流引弧的敏感性，这与导电纤维丝束的材料特性和电阻率等有关，也与电压大小有关；二是导电丝束通流引弧短路发生后，保护装置跳闸的敏感性，除了与整定时间有关外，主要与导电纤维丝束产生通流引弧的有效持续时间有关；三是保护装置重复合闸后，再次跳闸的敏感性，这与作用在高压电力设施上导电纤维丝束的量有关。

战斗部攻击电力系统，当导电纤维丝束飘落向高压电力设施时，二者开始发生相互作用，进而产生相应的电感效应、电热效应、电离效应（电弧放电效应）最终产生短路效应。短路故障形式有三相短路、两相短路、两相接地短路、单相接地短路等。短路故障可分为瞬时故障和永久故障。对于瞬时故障，电力系统通常采用重合闸功能予以解除；而对于一次甚至两次重合闸不成功的，则视为永久故障。

2.3 大电力系统短路毁伤易损性

大电力系统是电力工业发展的趋势，电力资源及电能可综合管理、优化配置合理使用。

大电力系统短路毁伤易损性是指，大电力系统对一个或多个变电站或高压开关站或重要高压设施受到导电纤维战斗部攻击失能后，产生联带效应造成自身全系统失能的敏感性。这与大电力系统自身的短路保护能力、系统运营管理能力、应急处理能力及用户载荷情况有关，当一个或几个高压变电站失能就能引起大区域电网甚至国家大电网失能，这表明电力系统很易损。

2.4 打击目标（对多个枢纽高压变电站打击）

区域大电力系统主要由多个发电厂、枢纽变电站等组成，在攻击时需要选择多个所属不同区域大电力网系统的枢纽变电站进行打击，同时使多个枢纽变电站、发电厂瘫痪，产生“多米诺”效应，最终使国家大电网甚至是几个国家联合的超大区域电网崩溃。

3 毁伤机理

3.1 导电纤维丝束短路机理

当单位长度质量很小的、表面镀金属层的导电纤维丝束落于目标上时，将造成大量的各种类型的短路效应。大电流作用下产生的高温，使导电纤维丝束表面的金属镀层被迅速蒸发，高温还使丝束自身及周围空气产生大量等离子体，形成金属气体及等离子构成的能发生小间隙放电的短路通道，继而产生电弧，使得短路效应高强度加剧并自持放电，直至保护装置跳闸电弧终止。

对于导电性能差的导电纤维，其通流引弧能力差，自持放电不能继续，不能产生使保护装置跳闸的短路效应；对于导电性能优的金属导线，则因高温将其熔断掉落，自持放电也不能继续。

3.2 短路封锁机理

短路封锁机理：一定数量的导电纤维丝束作用在高压电力设施上后，其潜在的短路毁伤能力使保护装置在一段时间内不能重合闸（每合闸一次便有某1～2根导电纤维丝束产生引弧短路效应，保护装置跳闸），供电能力丧失。

3.3 清除导电丝束机理

当用导电纤维战斗部对电力系统进行短路毁伤时，将有成千上万根导电纤维丝束作用于电网上，图1是南联盟电网电站被导电纤维弹攻击的情况。图2是南联盟电力系统工作人员清除电网上的导电纤维丝情况。

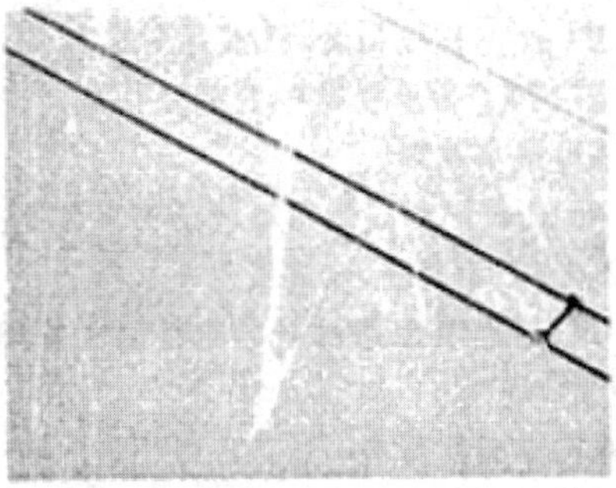

图 1 南联盟电力设施遭打击情况

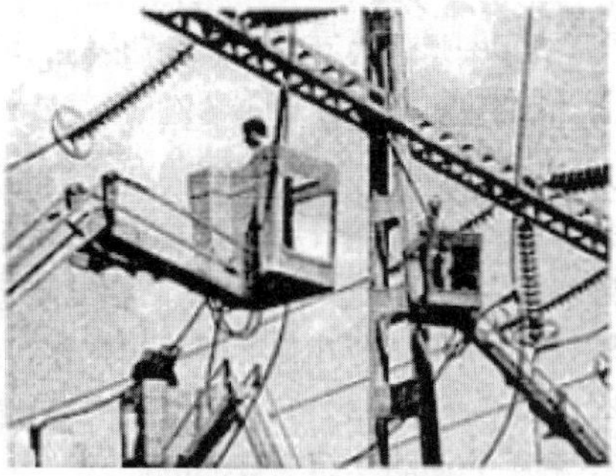

图 2 南联盟清除导电纤维情况

清除问题涉及两方面，从毁伤角度是增大清除难度，从防护角度是提高清除效能。增大清除难度体现在两个方面：一方面是攻击的导电纤维弹多，增加目标上导电纤维丝束的量；另一方面是采用难清除的导电纤维丝束，增加清除难度。后者涉及导电纤维丝束的材料、结构和性能设计问题，是提高清除难度的关键。应选择导电性优、由多根导电丝组合成丝束的、强度恰当、清除时易断但又不影响丝团展开和环境使用要求的、在目标上缠绕能力强的导电纤维丝束。只有当全部彻底地清除电力设施上的导电纤维丝束，同时还需清除邻近环境内的导电纤维丝束（风刮起可飘落在目标上），清除才真正有效。

3.4 电网大系统短路瘫痪机理

电网大系统中某一个或几个变电站或发电厂瘫痪后，那么该变电站和发电厂的负载则一般会由其他变电站和发电厂来承担，由于发电系统的总备用量有一定的比例，一般为 20%～25%，当新承担的负载超过其规定的供电能力时保护装置自动跳闸，发电机关闭；负载向其他高压变电站及发电厂转移时再次重复发生保护装置跳闸，发电机关机；若累积的负载转移可持续，则最终使整个电网大系统瘫痪。

4 相关技术问题

4.1 导电纤维丝束

导电纤维丝束可以是碳纤维，也可以是镀有金属层的玻璃纤维或其他具有导电性能的导电纤维，而镀有金属层的导电纤维可以是镀铜或镀铝等[3][4]。根据导电纤维丝束短路机理及清除导电纤维丝束机理设计其威力参数和其他应用参数，主要有：

（1）导电纤维丝束的电阻率，与每根丝镀金属层的厚度、连续性和丝的根数密切相关；

（2）电热性能即在高电压作用下的通流引弧能力，除上面一条外还与镀层材料及丝束材料密切相关；

（3）丝束抗拉强度，在满足导电丝束制备、丝团绕制和子弹承受勤务处理及作战时冲击条件下强度最小原则；

（4）在长贮和高低温环境条件下性能正常，要求不老化、不脆断，金属镀层不脱落。

4.2 风、雨等气象条件对导电纤维丝束飘落的影响

在抛撒高度基本一致的条件下，丝束的飘离距离随风速的增加而增加，同样的风速，高度越高，飘移距离随之增加。

当风速不大时，丝束的飘离距离较小，风速对丝束的飘离距离影响可以忽略或者可以通过修正距离的方法进行解决。由于从导电纤维战斗部中抛撒出来的导电纤维丝团数量多，因此，当风速过高时，不仅丝束的飘离距离明显增加，而且被抛撒出来的丝团在展开的过程中受大风影响，丝束之间互相缠绕且汇聚，形成一条较窄的带形覆盖区域，这使威力（丝束覆盖面积）减小。

在雨、雾等气象条件下，对丝团展开和飘落有影响，小雨、小雪与雾影响不大，中大雨雪将有明显影响，由于雨雪大量附着在导电纤维丝束上，严重影响了丝团展开和飘落过程，从而需修正导电纤维弹的落点，保证导电丝束对目标的有效作用。

4.3 子弹、丝团、引信的匹配

美国在科索沃战争中所采用的布撒器具有末制导、风力修正和定高抛撒子弹的能力，子弹可在低落速条件下抛撒丝团，并使丝团基本在自由下落条件下展开成丝束。能否在有利的高度抛撒子弹，子弹能否在最佳高度区内抛撒丝团，丝团能否承受弹道上的各种冲击环境，丝团能否在无初速自由下落条件下展开等是导电纤维子弹的关键技术问题。

这需要综合考虑导电纤维弹的下落速度、抛丝团高度和丝团展开问题，这些涉及定高引信或定高抛丝团技术问题、引战配合以及丝团设计问题。

4.4 运载平台

导电纤维弹可在多种武器上使用，如无人驾驶飞机、机载布撒器、导弹、火箭弹和炮弹。将导电纤维弹用于合适的武器上，取决于战术使用的有效性，需要具体问题进行具体分析。

5 结束语

尽管国内在导电纤维战斗部的研制方面有了较大的进展，但还存在着诸多关键技术和机理研究尚须突破，相信只要能够将问题重视起来，并加以认真分析努力攻克，完全可以在今后的一段时期内研制出性能优良的导电纤维战斗部，达到国际先进水平。

参考文献

[1] PRIDE R A, MCHATTON A D, MUSSELMAN K A. Electronic equipment vulnerability to fire released carbon fibers NASA TM－80219: [R]. 1980.

[2] BELL V L. The potential for damage from the accidental release of conductive carbon fibers from aircraft composites: [R]. 1980.

[3] 崔凯，刘强华．导电玻璃纤维的应用和发展 [J]. 玻璃纤维，1999 (1): 27.

[4] 肖友军．玻璃纤维表面化学镀 Cu 及其催化脱氢性能研究 [J] 表面技术，2000，29 (2): 11, 14.

爆炸事故原因的爆炸效应细观分析方法①

冯顺山

摘　要：爆炸事故在弹药战斗部研制或使用中意外发生，会带来方方面面的恶劣影响，分析事故发生的原因是处理事故和问题归零的关键依据。本文在分析爆炸事故分类、意外爆炸的基本起因基础上，提出爆炸事故原因的爆炸效应细观分析方法，即由炸药对壳体或紧密相联结构直接作用效应的微细观分析方法，辨识及确认或爆炸发生的最初起因，阐述了建立该方法的原理、诊断事故原因的原则和特点。通过两例弹药发生爆炸事故起因细观诊断的实践，表明该方法的科学性和实用性，为有关爆炸事故的直接原因分析提供一种有效分析手段，进而提出强化易爆产品设计安全性是将事故起因控制住的关键。

1　爆炸事故

爆炸作为一种特殊的强脉冲能量，具有很好的应用前景，但有益的应用必须以安全为前提。在现代生产中，诸如：把高压、高温或低温那种苛刻的极端状态引入生产过程中，生产、输送等作业的高速化，促进了所使用的测量仪表的自动化；人们对与爆炸有关的新物质、新用途、新系统认识的有限性；管理方面的不科学性和人为的失误或违章行为等；爆炸事故常有发生。爆炸事故是恶性事故，通常伴有人员伤亡和重大的经济损失。由于爆炸事故发生的突然性、高速传播或扩展的特点，现有的科学技术，对已发生的爆炸事故的控制能力有限，因此预防爆炸事故的发生，将爆炸起因限制在设计制造阶段和使用管理阶段，即抑制萌芽发生，具有十分重要的意义。

国家安全生产局前几年对全国重大事故隐患的普查情况表明：①重大事故（一次死亡 10 人以上，或直接经济损失 500 万元以上）隐患中，爆炸隐患占了 24%；②特别重大事故（一次死亡 50 人以上，或直接经济损失 1 000 万元以上）隐患中，爆炸隐患占 44.17%，可见爆炸事故的隐患问题很严重，减少重大、特别重大爆炸事故的发生，对国民经济的良性发展具有重要意义，其任务是艰巨的。

1.1　爆炸事故发生的基本情况

爆炸事故的发生可分为以下九种基本情况：①炮弹、火箭、导弹等弹药在发射时发生意外爆炸；②弹药、火炸药等易燃易爆危险品或装置在生产中的某个环节发生意外爆炸；③在运输过程中发生的意外爆炸；④在储存时发生意外爆炸；⑤在使用中发生意外爆炸；⑥易燃易爆危险品或装置在错误的管理和操作条件下发生爆炸；⑦研制新的含能材料或有关新系统时，由于认识方面问题，设计有缺陷，在与环境条件耦合概率发生爆炸；⑧人为地有意识地制造爆炸事故；⑨由于自然灾害产生爆炸事故。

1.2　爆炸事故的分类

根据事故统计，按照爆炸类型进行分类，可归纳出以下七种爆炸类型。

1. 安全装置失效型爆炸

易燃易爆系统中，安全装置失灵或环境条件超出安全装置或爆炸物质允许的极限值，使其安全保险功能失效，系统内的点火或引爆器件误动作，引燃或引爆系统，产生爆炸。

2. 需要有点火源的着火破坏型爆炸

装置、容器、管道、塔槽等内部的危险性物质，由点火源给以能量，引起着火、燃烧、分解等化学

---

① 原文发表于《第十届全国爆炸与安全技术交流会》，大会报告，2011 年。

反应，因此，压力急剧上升，发生伴随有容器破坏的爆炸。

3. 需要有点火源的泄漏着火型爆炸

容器内部的危险性物质，由于阀门打开、容器裂缝或微孔之类的破坏，泄漏到外部，与点火源接触而着火，引起爆炸或火灾。

4. 由于化学反应热蓄积的自然着火型爆炸

化学反应热的蓄积，使温度上升，其结果反应速度加快，使温度更加上升，当达到这种物质的着火温度时，就发生自然着火，引起火灾或爆炸。

5. 由于化学反应热蓄积的反应失控型爆炸

化学反应热的蓄积，使温度上升和反应速度加快，结果，该物质的蒸气压或分解气体的压力急剧上升，引起容器的爆炸性破坏。

6. 传热型蒸气爆炸

液体与其他高温物质接触时，发生快速传热，液体被加热，使之暂处于过热状态，而引起伴随急剧气化的蒸气爆炸。

7. 平衡破坏型蒸气爆炸

密闭容器内的液体，在高压下保持蒸气压平衡时，如果容器破坏，蒸气喷出，因内压急剧下降而失去平衡，使液体暂时处于不稳定的过热状态。由于急剧气化，残留的液体冲撞器壁，这种冲击压的作用使容器再次破坏，发生激烈的蒸气爆炸。

1.3 发生意外爆炸的基本起因

安全装置失灵或失效主要有以下四种情况：①引信中安全保险机构因有缺陷失灵，不能起到可靠保险作用；②隔火隔爆机构失灵或失效；③外界赋予的力量大于保险机构可承受的极限值，安全保险机构失效；④火工品或爆炸物安全性设计/制造有缺陷，与恶劣工作环境耦合，安全性能失效。

引起火灾和爆炸的点火源一般有下列八种：①电火花，②静电火花，③高温表面，④热辐射，⑤冲击和摩擦，⑥绝热压缩，⑦明火，⑧自然着火。

化学反应热蓄积引起的爆炸主要有以下八种情况：①分解引起的自然着火，②混合接触引起的自然着火，③吸水引起的自然着火，④氧化引起的自然火，⑤吸附引起的自然火，⑥聚合反应失控引起的爆炸，⑦合成反应失控引起的爆炸，⑧分解反应失控引起的爆炸。

过热液体蒸发引起的爆炸主要有以下五种情况：①传热引起的水蒸气爆炸，②传热引起的低温液化气的蒸气爆炸，③高压状态液体的蒸气爆炸，④火灾引起被加热的罐发生蒸气爆炸，⑤常温下液化气的蒸气爆炸。

2 爆炸事故原因的爆炸效应分析方法

2.1 爆炸事故原因的一般认识

一般来说，爆炸事故发生后，要进行事故原因的调查，其目的是希望今后不再发生由于同样原因而引起的爆炸事故。这同样也是为了找到防患于未然的措施。

爆炸事故原因的调查，必须备齐下述资料：

（1）工厂（爆炸现场）设施地形图和厂房（设施）的配置图；

（2）主要产品的生产量和生产工序图；

（3）平时和事故当天的操作状态；

（4）当天的气象状况；

（5）事故发生前的经过和事故概要；

（6）目击者提供的事项；

（7）当时者的服装和携带的工具；

（8）事故后的紧急处置和灾害情况；

（9）受到保护的爆炸现场，现场照片和示意图，伤亡人员的位置和伤亡情况；

（10）有关设备的设计图纸和配置图；

（11）原材料和产品的危险特性和试验结果；

（12）其他必要的事项。

爆炸事故发生后，为要调查原因，得出科学正确的结论，应当说不是轻而易举就能做到的，尤其是设计或制造方面的缺陷所致的原因。在事故调查中，如拘泥于先入为主的观点，恐怕有可能弄错原因，因此应尽可能避免这一点。考虑事故起因问题时，要忠实地立足于事实，对所有现象和证词都能做到完全合理的解释，要排除抽象理论，尽可能具体地对事故的原因和经过做出说明。在市场经济的今天，爆炸事故原因的确认问题涉及经济、法律责任和单位间利益等问题，使事故原因调查受到多方面的干扰，问题复杂化。但科技工作者应抱着科学的态度，实事求是，尽可能客观公正地找出事故原因。作者在辨识及诊断爆炸事故原因时采用的原则为：**物证为主，物证中现场已核实或通过鉴定的直接物证为主，直接物证中与事故原因密切相关的典型物证为主。**

爆炸发生后，还会引发二次或多次后续爆炸，之后还伴随燃烧和抢救，凌乱的爆炸现场和严重破坏装置、器件和环境，给事故原因调查带来困难。在实际问题中，拿出确切可信的证据说明爆炸事故产生的原因是一件难度很大的事情。

爆炸事故发生的原因一般有以下几种：①人为破坏，蓄意引发爆炸事故；②明显的违章、违规或错误操作产生爆炸事故；③自然灾害导致的爆炸事故，如地震、雷电等；④被外界明火源、飞行物或其他物体撞击产生燃烧爆炸；⑤原因不明的突发性爆炸事故。前面四种原因较清楚，容易弄清问题并说清事故的起因和发展过程，事故的处理也较容易。第五种事故原因很复杂，因素不确定，例如，安全设计或制造方面的缺陷、操作失误、管理方面的失误、化学反应自燃、外界因素（如天气、静电、磁场、电线老化、热辐射、动物作用等）、材料问题、器件或设备的质量或老化导致功能弱减失效等诸因素中的某一因素造成爆炸事故。正是由于造成爆炸事故原因的随机性、多样性和事故现场的破坏的严重性，诊断事故原因是一项复杂而艰巨的工作，通常需要多方面学科专家和专注精力查找事故原因。冷静的头脑、敏锐的洞察力、多学科的专业知识、正确的思维方法、丰富实践经验和科学的诊断方法，对于确认事故原因是非常重要的。

2.2　爆炸效应细观分析方法

由爆炸作用效应分析爆炸事故已是一种常用的分析方法，通常是根据爆炸的宏观威力和残体的运动特性及其碰撞效应去分析爆炸事故的特点，为事故原因分析提供某一方面的依据。例如，①通过分析残体的质量、材质、宏观形状、飞行距离、飞行高度、飞行方向、对其他物体的碰撞效应，分析残体的运动弹道和运动速度，进而求出其驱动所需的能量；②通过爆炸中心周围建筑物或结构件的破坏情况（板及柱变形弯曲等），估算冲击波强度，进而估算爆炸源的爆炸能量；③通过爆炸点处炸坑的直径、深度、土质、坑中爆炸残留物等数据，分析爆炸源（爆炸物质、爆炸量等）的情况；④通过人员目击、监控设备观测事故发生时的有关信息。这些都是有效分析方法，得以广泛应用。

本文针对弹药产品爆炸事故中的复杂难题，依据成功的关键在于细节的理念，提出爆炸事故原因诊断的方法为：**炸药对壳体或紧密相联结构件直接作用效应的细观分析方法**。其内涵及特征有以下几点。

（1）主要针对军用炸药类爆炸源对约束物作用的爆炸效应，通常是某一种爆炸装置，如弹药或战斗部或爆破器材爆炸作用效应。

（2）对那些紧贴炸药的壳体及结构件、或紧密邻近炸药的材料及结构件所形成的爆炸残体进行细观分析，即对那些直接受炸药及驱动作用的残体进行细观分析。

（3）此处所说的细观分析是指对残体外表、形状、破坏特征做精细分析。例如，收集到的一个小碎片或小零件上有爆炸直接作用效应的细微痕迹，精细观测表面凹凸部位与程度、变形部位与程度、破裂部位与程度，十分关注细微的效应情况，并区分是爆炸直接作的效应还是抛掷后与其他物体二次碰撞后产生的效应。

（4）残体每一部位的具体变形、破坏效应及特点，携带了炸药爆炸对其作用过程的信息，通过细微效应观测，以此来分析炸药爆轰/爆炸/爆燃全过程特征，从而分析并推测炸药发生爆炸时的条件特点，细微效应差异产生的机制，再诊断出产生这种条件的可能原因。

建立这一方法的原理为：

①炸药的爆轰情况与炸药起爆条件密切相关，即起爆驱动或作用能量与其完全爆轰、半爆、爆燃、反应波传播方向及波阵面反应参数密切有关；

②受爆炸直接作用的材料或结构件各部段宏观与细观效应和爆炸反应状态吻合，什么样的爆炸作用方向和作用力产生相应的破坏效应，如起爆点的方位不同且起爆能量不同，爆炸反应能作用效应或驱动效应有着微细差异或不同；又如爆炸反应程度及演化过程，与对应被驱动物的损伤微细效应特点不同；

③当爆炸反应物扩展到近10倍装药半径时，爆炸反应产物内的压力趋于均匀化，离炸药较远处的材料或结构件形成的爆炸残体，所携带的爆炸作用细节信息相对较少。

采用上面的分析方法，细微观测、精细推演，得到方向性看法，并结合关键细节的数字仿真研究或判断性对比实验，能诊断出最接近实际的结论。

由于爆炸物质、爆炸事故及产生事故原因的多样性，可以认为，分析爆炸事故原因的方法多种，不存在一种通用方法，不同方法适用不同问题，一事一议一诊，几种方法结合诊断事故原因准确度会更好；各种有效方法的集合形成了爆炸事故原因的分析方法。随着现代化发展，新问题随之出现，对爆炸事故原因的分析方法也将不断受到挑战。

3 两例爆炸事故原因分析情况简介

3.1 ××枪榴弹装配时发生意外爆炸事故的原因

3.1.1 枪榴弹及事故简介

枪榴弹为40 mm破甲杀伤枪榴弹，可用步枪发射（空包药筒发射）射程约200 m，可对付轻型装甲（爆炸聚能射流可击穿约240 mm厚的钢甲）和有生力量（几百枚杀伤破片可杀伤人员）。该弹在××厂总装后进行包装工序时（将全弹放入塑料包装筒）突然发生意外爆炸，造成6人死亡、7人受伤及车间部分毁坏的重大事故。事故发生后，停止生产，检查产品质量，经济损失重大。事故现场回收到该事故弹绝大部分残体，包含打入地内的聚能射流杵体。根据聚能射流的形状分析，认为该弹正常爆轰形成正常聚能射流。

通过分析各种可能引发事故的因素，并经过试验，排除了环境、榴弹内空包子弹、引信跌落失效等因素，认定事故弹引信传爆序列是正常动作后产生的爆炸。问题在于为什么在枪榴弹外包装时，引信会解保、传爆序列会正常工作。

3.1.2 枪榴弹引信简介

该引信是一个单保险机械触发型简易引信。由引信壳体、击针、传爆药柱、雷管、雷管座、保险钢球、保险杆、弹簧、保险杆定位火药、底螺和易燃薄片组成（结构图略）。其工作原理为：空包药筒的火药气体点燃易燃薄片并点燃保险杆定位火药，火药燃烧完毕后保险杆在弹簧的作用下到解保位置，随之保险钢球也到解保位置，此时，雷管座可轴向运动。弹碰到目标后，雷管座在惯性运动下前冲，其头部雷管与击针碰撞，雷管起爆并引爆传爆药柱及主装药。

3.1.3 雷管座、保险杆为典型分析对象

对回收的引信残体进行解剖，放大镜观测各零件在爆炸作用下的变形或破坏微细效应。发现如下问题：①雷管座上的钢珠孔（爆炸后已闭合）与钢珠在雷管座台上压痕位置不对应，相位差约90°，且钢球既不在保险位置又不在正常解除保险位置，而在非正常解除保险位置；②保险杆对应钢珠处呈光亮状

未受到损伤（其余表面被冲击和挤压熔蚀），也无钢珠挤压的痕迹，说明爆炸时钢珠不在保险位置。结论是，该弹包装时引信已解除了保险，包装筒轻冲击工作台（使弹尽快挤入筒内）使击针获得动能引爆雷管，这就是枪榴弹发生意外爆炸的直接原因。

根据上面的分析，进行了三发弹的对比试验，重复了事故弹的爆炸情况，回收到与事故相同性很好的引信残体，验证了上面结论。验证试验时还发现，若起爆点（在弹外用雷管起爆）位置不近似引信中的雷管位置，回收后的残体与事故弹的差异很大，不具有对比分析的意义。这说明起爆点对主装药爆轰特性及残体上的爆炸效应特征起支配作用。

3.2　××迫击炮钢珠杀伤爆破燃烧弹膛炸事故原因诊断

3.2.1　××迫击炮弹及事故简介

××迫击炮弹由尾翼、附加药包（发射用）、基本药管、尾管、弹体（外壳和预制钢珠套及金属锆环组成）、炸药和无线电引信组成。具有杀伤（钢珠预制破片和自然两种破片）、爆破和燃烧综合威力性能，主要对付有生力量、火炮阵地和运输车辆等目标。

事故弹采用迫炮进行动态爆炸完全性交验时发生膛炸事故，炮与弹形成大量碎块。事故已生产的批量迫弹产品军方拒收，造成报废等重大经济损失。通过分析，基本排除了由于发射药和引信方面产生的爆炸的原因。

3.2.2　迫弹战斗部装药结构简介

其结构特点为：由二层壳体组成，外层为钢壳（上下二段螺纹连接），内层是产生预制杀伤破片的钢珠层，铸塑的方法形成钢珠套。钢珠套由三段组成，上下为锥台状，中间为圆柱状，叠放在迫弹战斗部钢壳中（结构图略）。装药为注装 B 炸药。由于钢珠套的加工精度问题，轴向尺寸不足（间隙）采用毡垫找齐，满足装配尺寸要求。

3.2.3　战斗部外壳内壁上钢珠作用细观效应为典型分析对象

根据爆炸后回收的战斗部壳体残体及迫击炮残体尺寸，以及它们的飞行距离，表明战斗部内的装药虽非完全爆轰，但也达到相当程度的爆轰。精细分析大量钢珠作用于战斗部壳体每一段内壁上的撞击凹坑深浅特点、钢珠飞滑的痕迹及方向特点，发现与正常爆轰产生的钢珠飞散规律很不相同，细节分析表现出战斗部装药偏心起爆和多点偏心起爆的爆轰特征。根据这一分析结果，结合弹的装药结构特点，找到了战斗部装药产生非正常爆轰的直接原因，即迫击炮发射时，弹产生很大的驱动加速度，因钢珠套之间插入较厚的毡垫，上段钢珠套的惯性后坐位移与炸药柱的后坐位移不同步，在药柱表面处（尤其是钢珠套之间的联结部）产生强大的冲击摩擦力，摩擦热达到炸药的燃爆条件时，产生药柱的侧面起爆的意外爆炸事故。结论是，装药结构不符合发射安全要求，存在严重的设计和制造缺陷。

上述两个例子说明了爆炸效应细观分析方法在诊断事故原因时的具体应用情况，不同的事故特点不同灵活应用。通过找出事故原因，厂家改进了设计和制造工艺，加强了生产质量检查，去除了事故隐患。同时，吸取了教训，为弹药系统的安全及可靠性设计提供镜鉴，预防再次发生同类型爆炸事故。这两例事故也证明，弹药引信等易爆产品的安全性主要是设计出来的，应强化设计安全性和设计可靠性；然而制造也很重要，应高水平设计制造工艺，严保加工质量满足产品设计要求，保障使用安全。

参考文献

[1] 贝克．爆炸危险性及其评估 [M]．张国顺，等译．北京：群众出版社，1988.

[2] 北川徹三．爆炸事故分析 [M]．北京：化学工业出版社，1984.

[3] ×××枪榴弹爆炸事故分析报告（内部报告）

[4] ×××迫击炮弹爆炸事故分析报告（内部报告）

导电纤维丝团空中展开与飘落运动特性①

张广华②，冯顺山，葛　超，王　超，方国燚

（北京理工大学爆炸科学与技术国家重点实验室，北京　100081）

摘　要： 为深入了解导电纤维丝团空中展开与飘落的运动规律，建立了丝团空中运动的外弹道模型。以 MATLAB 数值计算软件为基础，基于四阶 RUNGEKUTTA 算法编写了相应的计算程序对模型进行求解。通过地面抛撒试验及靶场试验对丝团的空中展开及飘落情况进行了测试，数值计算结果与试验结果吻合程度较好，说明了模型的合理性。计算及试验结果对确定平台的抛丝团高度、丝团落点定位及进一步分析丝团对目标的毁伤效能提供了参考依据。

关键词： 导电纤维；丝团；软毁伤；电力系统；弹道模型

Spreading and Falling Law of Conductive Fiber Cluster Moving in the Air

Zhang Guanghua，Feng Shunshan，Ge chao，Wang Chao，Fang Guoyi

（State Key Laboratory of Explosion Science and Technology，Beijing Institute of Technology，Beijing 100081，China）

Abstract：To study spreading and falling law of the conductive fiber cluster，the trajectory model of the cluster was established. Based on the numerical calculation software of MATLAB，forth – order Runge – Kutta algorithm was used to write corresponding program，and the model was solved. Spreading and falling conditions of the clusters were tested by the ground scattering experiment and shooting range experiment，and the numerical calculation results are consistent with the experiment results，which proves that the model is reasonable. Numerical calculation results and test results provide references for determining scattering height of the shell，locating drop point of the cluster and further analyzing the damage efficiency to the target.

Keywords：conductive fiber；cluster；soft damage；power system；trajectory model

导电纤维丝团是导电纤维弹中的有效装填物，用于毁伤目标。丝团一般以子母弹的形式装填在母弹当中，根据母弹弹体结构的不同，装填丝团一般为几十个到上百个。每个丝团上都缠绕有几十米长的导电丝束，用于搭挂于目标上并引发短路效应。

丝团被抛出后，在运动惯性力、重力及气动阻力的联合作用下，一边向下运动，一边展开；展开后的丝束在空中飘移下落一段距离后以一定的概率搭挂于变电站或高压线等供电设施上，通过其短路引弧效应导致目标乃至整个区域内的电力系统瘫痪。

丝团空中展开及飘落的运动规律直接影响到展后的丝束能否准确地搭挂到被攻击的目标之上，并对其进行有效的毁伤，可以对导电纤维弹的终端弹道设计（抛丝团高度、速度等）提供依据。目前，国内外关于丝团及导电丝束空中运动规律的研究不多[1-3]，因此，本文建立了一种用于描述丝团空中运动的弹道模型，阐述其空中运动特性。

---

① 原文发表于《弹道学报》2014（1）。

② 张广华：工学博士，2011 年师从冯顺山教授，研究弹载导电丝束短路毁伤机理及弹道特性，现工作单位：西安 204 所。

1　丝团运动模型

1.1　假设条件

导电纤维丝团的结构如图 1 所示。

丝团从母弹中被抛出时的运动姿态及空中运动时受到的风力影响是不确定的，如果考虑所有因素，将会给研究带来较大的困难。因此，为了便于建立空中运动模型，特做如下假设：①丝团被抛出母弹时的运动姿态为水平运动；②丝束展开过程中只考虑丝轴绕丝束的旋转运动而忽略其他运动；③丝轴及丝束受到的风力及风向固定不变，风向为水平方向；④不考虑丝束在展开过程中的弯曲；⑤忽略阻力叶片绕丝束的转动。

1.2　丝团展开模型

以大地坐标系为基准坐标系，丝团展开过程示意图如图 2 所示。

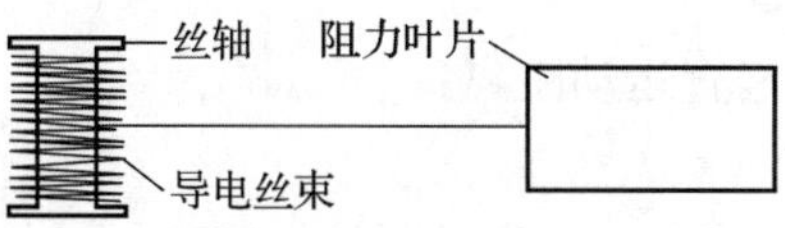

图 1　导电纤维丝团的结构

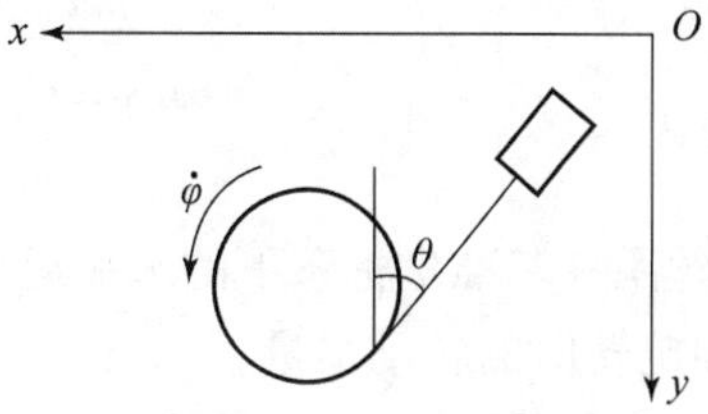

图 2　丝团展开过程示意图

丝束展开长度随时间的变化规律：

$$\dot{l} = R\dot{\varphi} \tag{1}$$

式中　l——丝束展开长度；

$\dot{l}$——丝束展开速度；

R——丝团的半径；

$\dot{\varphi}$——丝轴旋转角速度。

丝轴旋转时的角速度及角加速度的变化规律如下：

$$\dot{\varphi} = \frac{\dot{y} - \dot{y}_c}{R\cos\theta} \tag{2}$$

$$\ddot{\varphi} = \frac{\ddot{y} - \ddot{y}_c + \dot{\varphi}^2 R\sin\theta}{R\cos\theta} \tag{3}$$

式中　y——丝轴（丝轴连同未展开的丝束）在竖直方向的运动距离；

y_c——已展开丝束在竖直方向的运动距离；

θ——丝束与 y 轴间的夹角。

丝团运动过程中，随着丝束的不断展开，缠绕在丝轴上的丝团半径 R 也在不断变化，R 的变化规律为

$$-2\pi\frac{\mathrm{d}R}{\mathrm{d}t} = \frac{Rd}{a}\dot{\varphi} \tag{4}$$

式中　a——丝轴的长度，一般取为 1.5 ~ 2 cm；

d——单根丝束的直径，一般取为 0.2 ~ 0.5 mm。

1.3　未展开丝团的运动模型

未展开丝团包括丝轴及缠绕在丝轴上的未展开的丝束，其受力情况如图 3 所示。

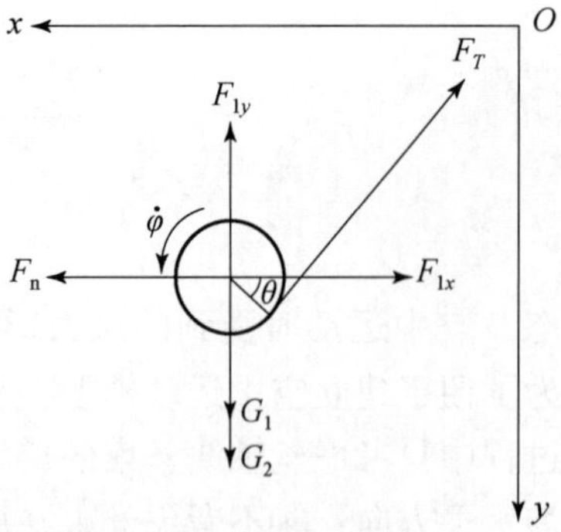

图 3　未展开丝团受力图

图 3 中，F_{1x}，F_{1y}分别为丝团受到的水平方向和竖直方向的空气阻力；G_1，G_2分别为丝轴及未展开丝束所受到的重力；F_T 为已展开丝束对丝团的拉力；F_{f1}为丝团受到的风动力。根据图 3 未展开丝团的受力情况，建立其运动方程：

$$m\ddot{x} = F_{f1} - F_T \sin\theta - F_{1x} \tag{5}$$

$$m\ddot{y} = G_1 + G_2 - F_T \cos\theta - F_{1y} \tag{6}$$

$$F_T R = J\ddot{\varphi} \tag{7}$$

式中　x，y——丝团在基准坐标系下的水平运动距离和竖直运动距离；

J——丝轴旋转时的转动惯量；

m——丝轴及未展开丝束的质量和，$m = m_z - lp_l$，其中，m_z为丝团初始质量，p_l为丝束线密度。

1.4　已展开丝束的运动模型

已展开丝束的受力情况如图 4 所示。

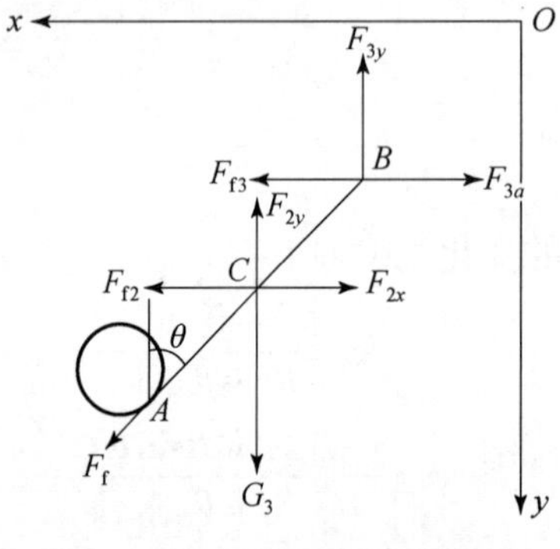

图 4　已展开丝束的受力情况

图 4 中，F_{2x}和 F_{2y}分别为已展开丝束受到的水平方向和竖直方向的空气阻力；F_{3x}和 F_{3y}分别为阻力叶片受到的水平方向和竖直方向的空气阻力；F_{f2}和 F_{f3}分别为已展开丝束及阻力叶片受到的风的作用力；G_3为已展开丝束受到的重力。A 点为已展开丝束与丝轴的交点，C 点为已展开丝束的中心点，B 点为阻力叶片与丝束的交点。

根据图 4 所示已展开丝束的受力分析，建立其运动方程：

$$m_c\ddot{x}_c = F_{f2} + F_{f3} + F_T\sin\theta - F_{2x} - F_{3x} \tag{8}$$

$$m_c\ddot{y}_c = G_3 + F_T\cos\theta - F_{2y} - F_{3y} \tag{9}$$

$$\begin{aligned} J_1\ddot{\theta} = {} & F_{f2}\frac{l}{2}\cos\theta + F_{f3}l\cos\theta + F_{2y}\frac{l}{2}\sin\theta + \\ & F_{3y}l\sin\theta - F_{2x}\frac{l}{2}\cos\theta - F_{3x}l\cos\theta - G_3\frac{l}{2}\sin\theta \end{aligned} \tag{10}$$

式中 x_c——已展开丝束的水平运动距离；

m_c——已展开丝束的质量，$m_c = lp_l$；

J_1——已展开丝束绕丝束与丝轴的接触点 A 转动时的转动惯量。

1.5 丝团展开及飘落运动方程组

联立式（1）~式(10)，辅以丝团运动过程中运动距离、速度之间的微分关系式，得到了基于假设条件下的丝团空中展开与飘落外弹道模型：

$$\begin{cases} \dfrac{\mathrm{d}x}{\mathrm{d}t} = \dot{x} \\ \dfrac{\mathrm{d}y}{\mathrm{d}t} = \dot{y} \\ \dfrac{\mathrm{d}x_c}{\mathrm{d}t} = \dot{x}_c \\ \dfrac{\mathrm{d}y_c}{\mathrm{d}t} = \dot{y}_c \\ \dfrac{\mathrm{d}\theta}{\mathrm{d}t} = \dot{\theta} \\ \dfrac{\mathrm{d}l}{\mathrm{d}t} = R\dot{\varphi} \\ \dfrac{\mathrm{d}R}{\mathrm{d}t} = -\dfrac{1}{2\pi}\dfrac{Rd}{a}\dot{\varphi} \\ \dfrac{\mathrm{d}\varphi}{\mathrm{d}t} = \dfrac{\ddot{y} - \ddot{y}_c}{R\cos\theta} \\ F_T = \dfrac{J\ddot{\varphi}}{R} \end{cases} \tag{11}$$

本文以 MATLAB 软件为基础，以四阶龙格－库塔算法为求解方法，编制了相应的程序对方程组（11）进行数值求解[4]。结合相关试验（地面抛撒试验、靶场飞行试验）结果及方程组（11）反推出了方程组（11）中的相关气动参数，并根据文献［5－6］提供的方法对丝团及丝束所受气动力进行计算；转动惯量根据文献［7］提供的方法计算。

2 试验研究

为了对本文所建立模型的合理性进行验证，进行了单个丝团的定向抛撒试验及靶场飞行试验。其中，抛撒试验主要用于研究丝团的空中展开规律，靶场飞行试验主要用于研究丝团的空中飘落规律。

2.1 试验丝团的技术状态

试验丝团的技术状态如表 1 所示，l_a和 l_b分别为阻力叶片的长度和宽度，l_{ss}为缠绕在丝轴上的丝束长度。

表 1 试验丝团的技术状态

mz/g	l_a/mm	l_b/mm	lss/m
5	20	10	20

2.2 地面抛撒试验

地面抛撒试验的现场布局如图 5 所示。

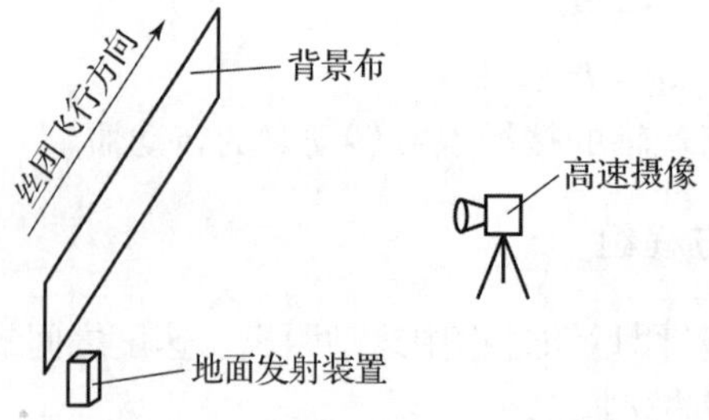

图 5　地面抛撒试验的现场布局

地面发射装置将丝团发射出去后，通过高速摄像设备记录其空中运动状态；将高速摄像设备拍摄的结果导入与之配套的后台软件上，并对其进行相应的后处理，即可得到丝团空中展开时的运动参数随时间的变化情况。

2.3　靶场试验

2012 年 6 月 28 日于某靶场进行了导电纤维丝团空中抛撒的飞行试验。

试验丝团装填于特定的抛撒装置中进行发射；抛撒装置运动到一定高度后将丝团从空中进行抛撒，并通过三坐标跟踪雷达测量得到了丝团空中运动时的各项参数。

3　结果分析

根据抛撒试验及靶场飞行试验的试验条件对方程组（11）中的各个变量赋予如表 2 所示的初始值。

表 2　方程组（11）变量初始值

试验	x/m	y/m	x_c /m	y_c /m	θ/rad	l/m	R/mm	F_T /N
地面抛撒试验	0	0	0	0	0	0.1	5	0
靶场试验	0	0	0	0	0	0.1	5	0

3.1　丝束展开规律

选取典型时刻的丝束展开长度值为分析对象，将数值计算结果与抛撒试验结果进行对比，对模型进行验证。因为地面抛撒试验的发射装置距离地面较低，试验时丝团在空中运动 0.48 s 后便已着地，所以只能对 0.48 s 之前的相关数据进行对比分析。试验及数值计算的结果如表 3 所示，l_e为试验值，l_s为计算值，Δl 为绝对误差，$\Delta l/l_e$为相对误差。

表 3　丝束展开长度随时间变化值

t/s	l_e /m	l_s /m	Δl/m	$(\Delta l/l_e)$/%
0	0	0	0	0
0.13	0.70	0.67	−0.03	−4.29
0.22	1.17	1.13	−0.04	−3.42
0.34	1.53	1.56	0.03	1.96
0.40	1.93	1.84	−0.09	−4.66
0.48	2.29	2.38	0.09	3.93

从表 3 可以看出，采用方程组（11）求解出的结果与实验值之间的绝对误差在 0.03 ~ 0.09 m，相对

误差在 1.96%~4.66%，二者吻合情况较好。

当抛撒速度为 38 m/s，风力条件为无风时，对方程组（11）进行求解，求解时间设定为 20 s，得出丝束展开长度 l 随时间 t 的变化情况如图 6 所示。

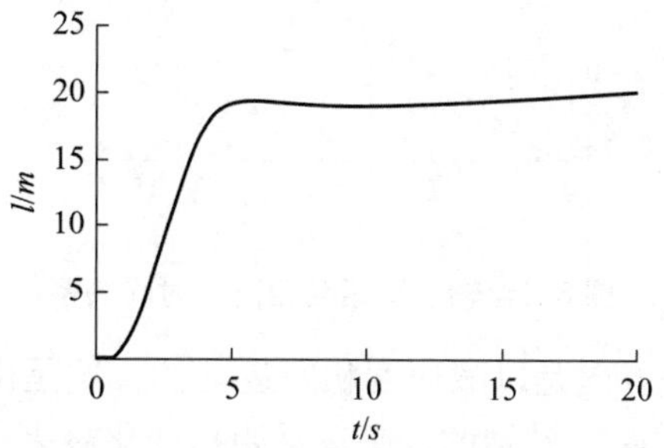

图 6　丝束展开长度－时间关系

从图 6 可以看出，运动初期，丝束以接近均匀速度展开，展开速度为 4.5~5 m/s，运动到 4.5 s 左右时完全展开。

3.2　丝团飘落规律

由于靶场试验时测试雷达距离抛撒装置较远，加之丝团体积很小，想要对丝团空中运动的全过程进行跟踪测试具有比较大的难度，所以只对其空中抛出位置及落地后的位置进行了测量，由此可以确定丝团在给定抛撒高度下的水平飘移距离。试验及数值计算的相关数据如表 4 所示，表中，y_p 为抛丝团高度；x_e，x_s 分别为丝团飘移距离试验值和计算值，Δx 为绝对误差，$\Delta x/x_e$ 为相对误差。

表 4　丝团漂移距离

y_p/m	x_e/m	x_s/m	Δx/m	$(\Delta x/x_e)$/%
14.4	60.3	57.6	-2.7	-4.48
46.3	69.8	73.7	3.9	5.59
84.9	89.9	84.5	-5.4	-6.00

从表 4 可以看出，丝团飘移距离的计算值与试验值之间的绝对误差在 2.7~5.4 m/s，相对误差在 4.48%~6%，二者吻合程度较好。

当抛撒速度为 139.9 m/s、风力条件为无风时，对方程组（11）进行求解，求解时间设定为 20 s，得出丝团的飘落规律如图 7、图 8 所示。丝团空中全展后的丝束长度有几十米，其运动状态为丝轴连带着已展开的丝束共同飘落。因此，丝轴的飘落状态可以反映出丝束的飘落状态，为了便于对丝团的飘落规律进行分析，下文中以丝轴的水平及竖直运动距离为研究对象，对其飘落规律进行研究。

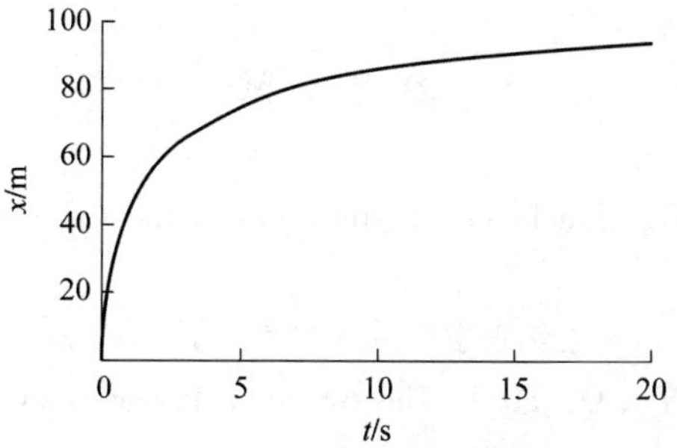

图 7　丝轴水平运动距离－时间关系

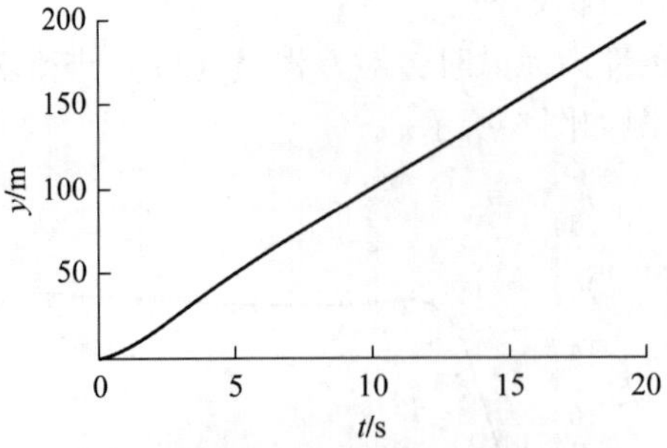

图8 丝轴竖直运动距离 – 时间关系

从图 7 可以看出，丝轴在水平方向是以减速度状态运动的，这是由空气阻力及已展开丝束对丝轴的拉力作用造成的；运动到 10 s 左右时，丝轴的水平运动距离变化越来越缓慢，说明在 10 s 左右时其水平方向的运动速度已经衰减到了接近于 0 的状态。在 20 s 的时间内，丝轴在水平方向共运动了 92. 3 m。

由图 8 可以看出，在运动初期 3 ~ 4 s 的时间内，丝轴在竖直方向是以加速状态运动的，这是由其自身重力大于竖直方向的空气阻力所造成的；随着竖直方向运动速度的不断增大，丝轴所受空气阻力也逐渐增大，最后，其自身重力与阻力达到平衡的状态。所以，在 4 s 过后，丝轴以接近匀速的状态下落，下落速度为 8. 5 ~ 9 m/s。在 20 s 的时间内，丝轴在竖直方向共下落了 195. 5 m。

根据图 7、图 8 所示内容，绘制丝轴的运动轨迹，图 9 所示内容与试验时观测到的丝轴运动轨迹吻合。

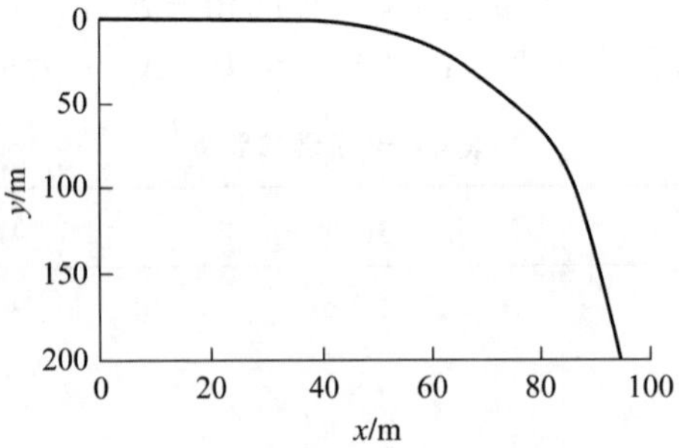

图9 丝轴空中运动轨迹

4 结论

（1）通过对导电纤维丝团空中展开与飘落过程中的受力情况进行分析，建立了基于一定假设条件下的丝团空中运动外弹道模型。此模型可以对丝团在不同抛撒条件（风速、抛撒丝团高度、抛撒速度等）下的展开及飘落情况进行预测。

（2）为了验证模型的合理性，进行了单个丝团的定向抛撒试验及靶场飞行试验，计算结果与试验结果吻合较好，表明本文所建立的丝团弹道模型合理。

参考文献

[1] COSTELLO M F, FROST G W. Simulation of projectiles connected by a flexible tether: ARL – CR – 456 [R]. 2000.

[2] 张国伟，冯顺山，孙学清，等．导电丝束空中展开规律 [J]．弹道学报，2006，18 (1)：48 – 50.
ZHANG G W, FENG S S, SUN X Q, et al. The spreading law of conductive fiber in the air [J]. Journal of ballistics, 2006, 18 (1): 48 – 50. (in Chinese)

贫铀材料冲击破碎和能量释放后效研究①

王成龙②，黄广炎，冯顺山

（北京理工大学爆炸科学与技术国家重点实验室，北京 10081）

摘　要：通过弹道枪发射，测试了同尺寸条件下 Al/PTFE、Al/W/PTFE 和两种贫铀配方材料，在 1 000 m/s的速度条件下，穿透 2 mmA3 钢板后对多层后效铝板的毁伤效果和在密闭容器内的能量释放效应。试验结果表明，贫铀配方材料兼具高能量与高密度特性，后效多层靶毁伤效果和能量释放效应显著；对比两种贫铀配方材料，贫铀复合材料和易碎贫铀合金材料，两者后效多层靶毁伤能力显著，能量释放效应相当，易碎贫铀合金材料密度高达 10.7 g/cm^3。同尺寸条件下单枚试样含能较高，且具有穿靶破碎效应，能量释放以燃烧形式为主；贫铀复合材料冲击过载和冲击温升敏感，具备自持反应效应，能量释放以燃爆形式为主。研究结果表明贫铀材料具有较好的研究价值和应用背景。

关键词：贫铀材料；后效毁伤；能量释放；冲击反应

1　引言

随着目前全球核武器与核电产业的迅猛发展，如何合理开发利用数量庞大的贫铀废料，已经成为亟须解决的问题。贫铀材料由于其高密度、高强度、高韧及自锐性等特点[1]，在工业和军事方面均有广泛的应用价值，尤其在武器方面，贫铀材料已经作为穿甲弹芯和复合装甲夹层被 20 多个国家装备应用。

冲击反应材料是近几年在弹药战斗部应用领域最前沿的技术之一，它既具有类似金属的力学强度。又含有与高能炸药相当的化学能，还具有与惰性材料类似的安全性，可以直接机械加工，只有在高速命中目标后才会发生剧烈爆炸反应[2]。典型基础金属/氟聚物配方为 Al/PTFE，其理论密度为 2.4 g/cm^3，为了提高材料密度有不少学者在其中加入一定量的钨粉，这样可以在一定程度上提高其密度，但是因为钨是一种惰性金属，钨粉的加入会明显降低材料的反应活性[3]，也有学者添加其他成分而显著改善其各项性能，但未见其公开相关配方[4]。易燃钨锆合金材料[5]和易碎钨合金材料也是弹药战斗部应用领域的重要研究方向，但是前者反应能量低，主体材料仍为惰性金属钨，后者仅能实现后效动能杀伤，并不具有反应后效。

贫铀材料是一种高密度的活性金属材料，可以替代冲击反应材料中的活性铝金属和高密度惰性钨金属，从而成为一种高密度、高活性冲击反应材料。同时在贫铀复合材料研究的基础上，为了提高和改善毁伤元的密度与侵彻性能，以实现不同的应用背景，本文提出了具有侵彻 - 破碎 - 燃烧效应的易碎贫铀合金材料。本文通过枪击试验的方法，对比测试了 Al/PTFE、A/W/PTFE、贫铀复合材料（简称 Du - F）和易碎贫铀合金（简称 Du - X）四种材料，同尺寸条件下以 1 000 m/s 的速度，穿透 2 mm 的 A3 钢板后，对后效多层靶的毁伤效果和在密闭容器内的能量释放效应。

2　贫铀材料特性与试样制备

本文研究主要对比了 Al/PTFE、A/W/PTFE、DU - F 和 DU - X 四种材料的冲击反应后效毁伤能力和冲击反应能量释放效应，其中 Al/PTFE、Al/W/PTFE 和 DU - F 为复合材料通过粉末混合压制烧结工艺，DU - X 为合金材料采用合金冶炼工艺制备杀伤元试样如图 1 所示，试验用材料性能参数如表 1 所示。

---

① 原文发表于《第十五届全国战斗部与毁伤技术学术交流会议文集》2017 年 11 月。

② 王成龙：工学博士，2013 年师从冯顺山教授，研究侵爆毁伤元冲击反应机理及作用后效，现工作单位：北京航天长征飞行器研究所（14 所）。

DU－F具有密度和能量的均衡结合，属于自持反应材料，可以在穿透防护层后进入与空气隔绝的目标内部（如飞行器导控舱段、燃油油箱、战斗部装药等）发生剧烈放热反应，从而有效毁伤目标。同时反应产物UF_6升华温度仅为56.4 ℃，因此在反应场中是以气态形式作为反应产物出现，从而进一步提高燃爆威力。

铀金属具有高强度、高韧及自锐性等特点，通过添加某金属元素（代号X）从而改变其拉压特性。该添加金属熔点低，在常温条件为四方晶系，具有很好的延展性，但是在高温（161 ℃）下，发生晶变，由四方晶系转变为斜方晶系，变为脆性材料。因此DU－X材料在对靶板的冲击过程中，产生冲击温升，结构中添加金属发生晶变，在侵彻靶板过程中，材料内部逐渐解体为碎片群，产生一种“瀑布”效应，穿透靶板后“瀑布”效应扩展为碎片云并伴随着铀金属的燃烧效应，对靶后目标形成大面积破坏引燃毁伤[6]。

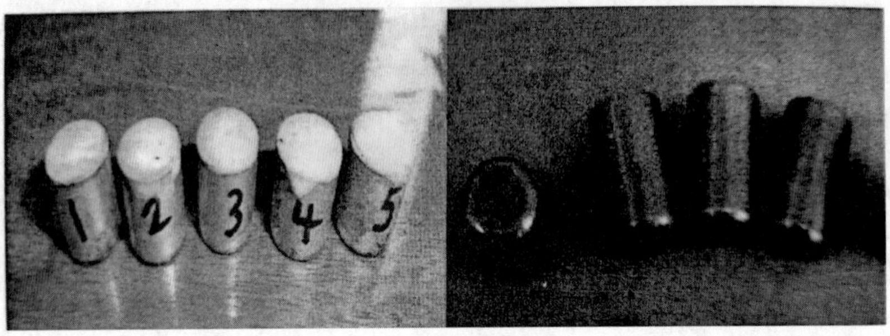

（a） （b）

图1 制备杀伤元试样

（a）包覆后的贫铀复合材料试样；（b）贫铀合金材料试样

表1 试验用材料性能参数

序号	材料类型	密度 g/cm³	尺寸¹ mm	质量¹ g	质量含能 kJ/g	单枚含能 kJ
1	Al/PTFE	2.21	10×15.6	2.71	8.63	23.4
2	Al/W/PTFE	5.68	10×15.3	6.83	1.78	12.2
3	DU－F²	5.60	10×15.3	6.73	3.25	21.9
4	DU－X²	10.7	10×15.5	13.0	4.79 燃烧能	64.6

注：1. 尺寸和质量取多次测量平均值；

2. 贫铀材料参数不含包覆成分。

3 试验方案

本文采用了两种试验方法综合研究反应材料冲击穿透板后，对多层后效靶的毁伤效应以及能量释放效应。

方案1：后效多层靶毁伤评价方案

试验布局如图2（a）所示，通过弹道枪发射材料试样，并通过高速摄影测量试样着靶速度，前靶板为2 mm厚A3钢板，靶后布置三层0.5 mm的LY－12硬铝板。试样撞击2 mm厚A3钢板后，诱发反应并发生破碎，在穿靶后形成伴随着反应现象的碎片云，通过高速摄影的火光现象可以判定后效反应的剧烈程度，通过后效铝板的破孔面积统计可以判定材料的穿靶破碎效应和后效毁伤能力[7]。

方案2：密闭容器压力测试方案

试验布局如图2（b）所示，通过弹道枪发射材料试样，并通过高速摄影测量试样着靶速度，试验本为密闭容器，前靶板为2 mm厚A3钢板，箱体侧表面固定3个压力传感器，底部撞击靶为30 mm厚A3钢板防止破片击穿。材料穿透靶板进入密闭容器内部发生剧烈反应，引起容器内产生压力变化，从而被壁面上的压力传感器所记录。材料穿靶后仍具有剩余速度，在密闭容器内呈现动态的反应域，因此在不同位置处设置了传感器，用来表征材料反应域的动态变化和能量分布特征[8]。

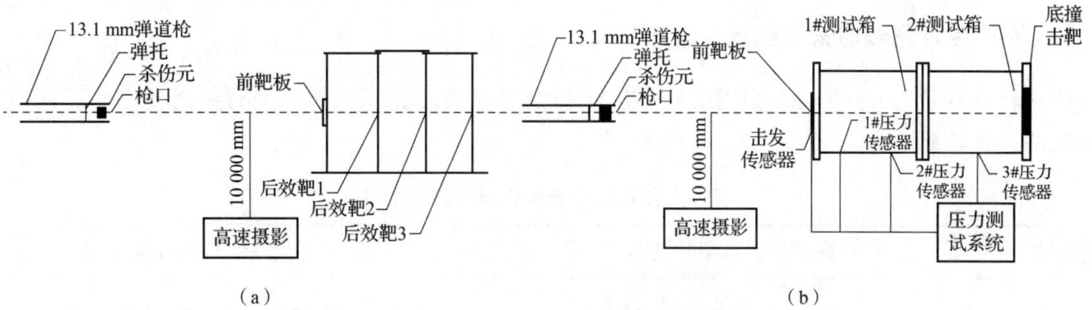

图 2　试验方案示意图

(a) 后效多层靶毁伤试验方案；(b) 密闭容器反应压力测试方案

4　试验结果与分析

4.1　前靶板穿孔效应

反应材料穿靶过程示意图如图 3 所示，由表 2 中各发生试验的前板穿孔大小可知，四种反应材料在 1 000 m/s 的速度条件下。对 2 mm 厚的 A3 钢板可造成 1.73～2 倍破片直径的穿孔效应。Al/PTFE、Al/W/PTFE 和 DU－F 都属于金属－聚合物类冲击反应材料，材料在撞击靶板时 PTFE 基体受到高速冲击过载发生冲击破碎和冲击分解反应 PTFE 基体和金属结合面也发生滑裂剥离现象材料头部压实墩粗；进入板后在继续经历压实墩粗以及压剪作用力、局部点火反应膨胀等作用，材料开始形成裂纹并逐渐扩展；在上述因素的共同作用下靶板穿孔加大最终达到接近 2 倍破片直径的穿孔效应（相同条件下的钢破片仅为 1.2～1.5 倍的穿孔直径）[9]。DU－X 材料属于合金材料，材料在侵彻过程中发生绝热剪切温升，X 金属发生高温晶变（161 ℃），DU－X 合金相解体材料破碎，U 金属在与靶板的高速摩擦过程中被引燃与空气发生燃烧现象：DU－X 合金在穿透靶板的过程中，不发生自持放热反应，同时 DU－X 合金材料密度较大（10.7 g/cm^3），考虑高密度材料对薄靶板的冲击响应问题，其穿孔效应仅为 1.73～1.8 倍的破片直径，略小于金属－聚合物类反应材料。

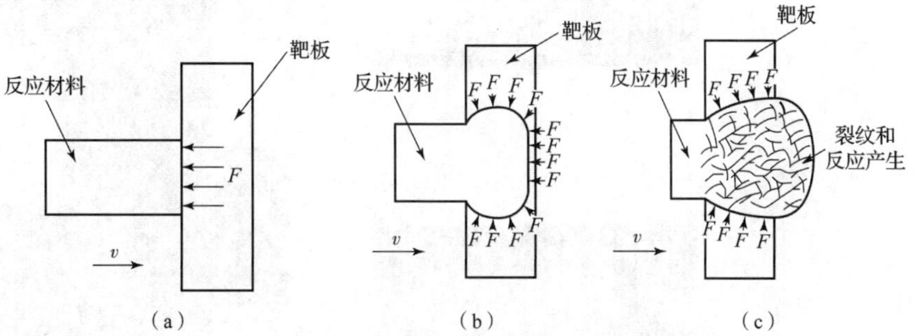

图 3　反应材料穿靶过程示意图

表 2　试验测试数据结果

材料类型	多层后效靶测试					密闭容器压力测试				
	速度 m/s	穿孔 mm	后效靶破孔面积 mm^2 第 1 层	第 2 层	第 3 层	速度 m/s	穿孔 mm	压力测试峰值 kPa 1#	2#	3#
Al/PTFE	1 075	Ø20	3 816	6 594	177	1 078	Ø19	162	192	83
Al/W/PTFE	1 009	Ø19	1 279	807	1 178	1 068	Ø18	67.5	60	64.5
DU－F	977	Ø18	11 775	12 325	4 475	984	Ø17.8	655	460	657
DU－X	965	Ø18	10 400	16 200	8 825	1 000	Ø17.3	242	155.5	222

4.2 多层后效靶毁伤试验结果

四种材料在穿透 2 mm 厚 A3 钢板后对多层靶板的破坏效应高速摄影典型时刻对比图和三层后效铝板破孔情况如表 3 所示。

表 3 多层后效靶毁伤试验结果照片

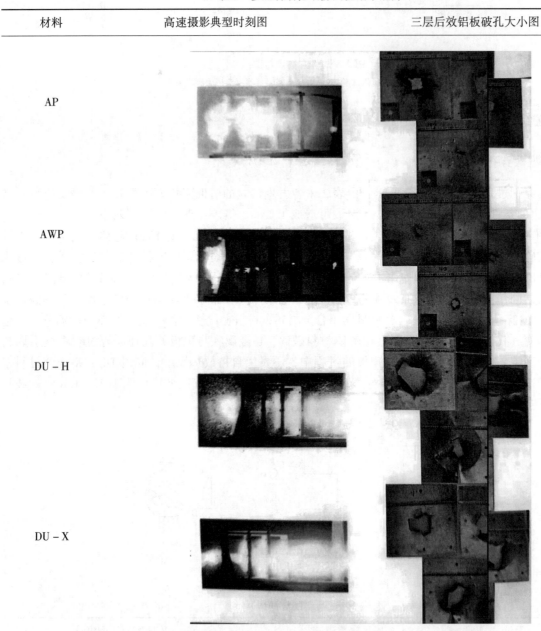

材料	高速摄影典型时刻图	三层后效铝板破孔大小图
AP		
AWP		
DU－H	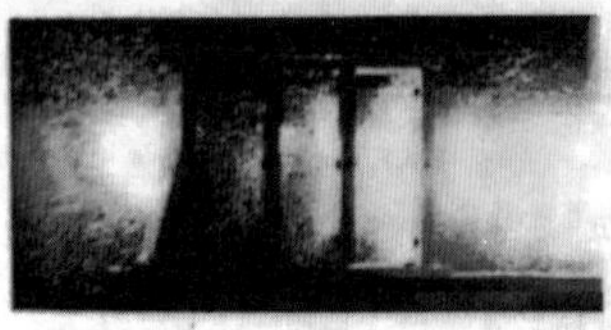	
DU－X	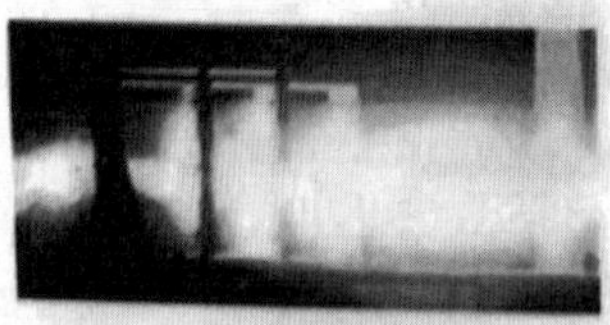	

Al/PTFE 材料（表中简称 AP）在多层靶板之间反应火光明显，但是因为其材料密度较低，其后效毁伤主要集中在第一层和第二层后效铝板上，第三层后效铝板仅有零星穿孔。

Al/W/PTFE 材料（表中简称 AWP），在加入一定惰性 W 金属之后，密度得到提升（从 2.21 g/cm^3 升高到 5.68 g/cm^3），但可反应的 Al/PTFE 成分显著减少，冲击敏感度明显降低。材料在撞击侵彻 2 mm 厚 A3 钢板时，靶前反应火光现象明显证明材料仍具有一定的反应能力，但是穿靶之后，材料仅有零星反应火光，大部分呈现为黑色团状粉末。通过观察三层后效铝板，各铝板破孔大小相当，主要为材料穿靶破碎后产生的碎片云团动能机械作用引起的贯穿效应[10]。

DU－F 材料，含活性高密度 U 金属成分，密度调节为 5.6 g/cm^3。DU－F 材料穿透 2 mm 厚 A3 钢板后，反应火光现象剧烈，并在完全贯穿三层硬铝板之后反应火光依然剧烈。通过对比三层硬铝板的破孔大小，大面积破孔主要集中在前两层，在穿透两层硬铝板后碎片云动能量减少，因此第三层破孔面积呈减小趋势。

DU－X 合金材料，其密度高达 10.7 g/cm^3，可以完全贯穿 2 mm 厚 A3 钢板和三层硬铝板，其仍具有大量剩余能量。从高速摄影典型时刻图，可以观察到反应火光非常剧烈，其主要是 U 金属成分与空气中氧气的剧烈燃烧现象。对比三层硬铝板的破孔大小，其破坏效应略优于 DU－F，且最大破孔为第二层硬铝板，这表明 DU－X 材料穿靶后的碎片云在穿透第一层硬铝板后仍然在纵向扩张，其后效破坏效应更加显著。

4.3 密闭容器压力测试结果

四种材料穿透 2 mm 厚 A3 钢板进入密闭容器内部发生反应，容器壁三个位置的传感器测得的反应压力峰值平滑曲线分别如图 4 所示。

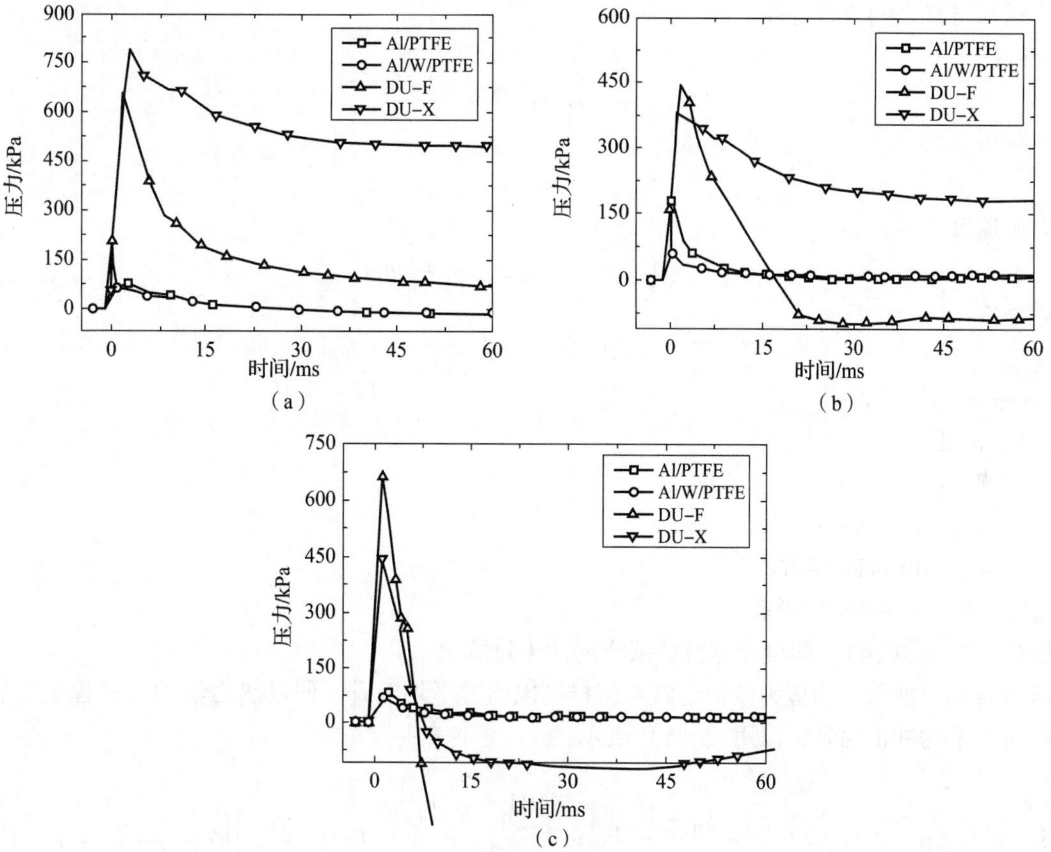

图 4 密闭容器压力测试结果

（a）1#传感器压力曲线；（b）2#传感器压力曲线；（c）3#传感器压力曲线

DU－X 合金材料密度高，在同尺寸条件下质量大，单枚试样含能远远高于其他材料，同时根据多层靶毁伤效果可知，材料在穿过靶板后易破碎形成碎片云。碎片云与空气接触发生燃烧反应释放能量，密闭容器内压力迅速上升，其压力曲线在四种材料中最高，如图 4（a）压力曲线。但是密闭容器内空气有限，随着反应场向密闭容器内运动，压力峰值出现下降（图 5），逐渐低于 DU－F 材料。DU－X 材料密度很高，穿过前靶板后仍具有很高的剩余动能，当材料运动到密闭容器末端与底部撞击板撞击，在密闭容器内会产生“二次扰动”使压力再次上升，如图 4（c）压力曲线，同时也可以看出 3#传感器的压力曲线特征明显有别于 1#传感器和 2#传感器的压力曲线。

DU－F 材料在1#传感器位置的压力峰值如图4（a），仅略低于DU－X材料，但明显高于Al/WPTFE材料和Al/PTEE材料。DU－F材料的单枚试样能量虽然仅为DU－X的1/3，但是DU－F可以由PTFE冲击分解提供氧化剂，发生自持反应并产生气体，其反应速度和反应威力高于DU－X材料的燃烧反应现象，因此DU－F材料的压力曲线在达到峰值后迅速下降。随着反应场向密闭容器内部运动，自持反应效果逐渐体现，其反应压力超过DU－X材料，如图4（b）和（c）所示，当材料运动到密闭容器末端与底部撞击板撞击，同样也出现“二次扰动”现象，但DU－F的“二次扰动”现象伴随着二次撞击反应，其压力上升更高。

Al/PTFE材料虽然质量含能比较高，但是材料密度低，同尺寸条件下单枚破片能量含量有限。Al/W/PTFE材料加入惰性W金属提高密度之后，材料能量密度明显下降，压力峰值曲线明显低于原Al/PTFE材料。

4.4 材料的冲击过载与冲击温升

试样撞击靶板产生的冲击过载可以根据一维应力波理论进行解析。当试样以速度 v_0 撞击靶板时，应力波在靶板和材料之间进行传播。

对于试样

$$P_f = \rho_f D_f u_f \tag{1}$$

对于靶板

$$P_t = \rho_t D_t u_t \tag{2}$$

应力波速度

$$D_f = a_f + b_f u_f D_t = a_t + b_t u_t \tag{3}$$

边界条件

$$P_f = P_t v_0 - u_f = u_t \tag{4}$$

式中 P——冲击波压力；

ρ——密度；

D——冲击波速度；

u——质点运动速度；

a 和 b——Hugoniot 参数；

下角标 f 和 t——试样和靶板。

根据式（1）~式(4)，即可计算材料试样的冲击过载 P。

根据Hugoniot曲线，当应力波传过材料试样可以引起平均温升。假设热力学动态过程是绝热的并忽略摩擦效应，平均冲击温升可以用式（5）表示：

$$T_1 = T_0 \exp\left[\frac{\gamma_0}{V_0}(V_0 - V_1)\right] + \frac{V_0 - V_1}{2C_v} P_1 + \frac{\exp\left(-\frac{\gamma_0}{V_0} V_1\right)}{2C_v} \int_{V_0}^{V_1} P \exp\left(\frac{\gamma_0}{V_0} V\right)\left[2 - \frac{\gamma_0}{V_0}(V_0 - V)\right] dV \tag{5}$$

式中 γ_0——Gruneisen 系数；

C_v——等压热熔；

P——冲击波压力；

V_0——试样的初始体积；

V_1——应力波传播后的试验体积。

材料为复合材料，其各项参数根据质量平均法进行了估算。

通过计算之后的四种材料冲击过载和冲击温升如图5中柱状图所示，结合不同位置处的测得压力峰值可以看出。DU－X合金材料的冲击过载和冲击温升都是最高的，其冲击温升达到569.65 K，已经达到其诱发晶变的温度（434.13 K），因此穿靶后破碎效果非常显著。DU－F材料的冲击过载和冲击温升都高于 Al/PTFE 材料，冲击过载是诱发材料反应和材料破碎的主要因素，冲击过载越高材料初始反应能量越高，穿靶后破碎效果越好，后效毁伤和能量释放效应越显著。冲击温升是材料进一步加剧反应的主要诱因，材料平均温度越高材料活化能越高，在冲击诱发反应之后，反应扩展更加迅速，能够更加迅速的达到反应峰值释放能量[11]。A1/W/PTFE 材料虽然冲击过载非常高，但是材料本身能量密度较低且冲击温度较低，材料更加惰性，因此反应并不充分，能量释放效应较低。

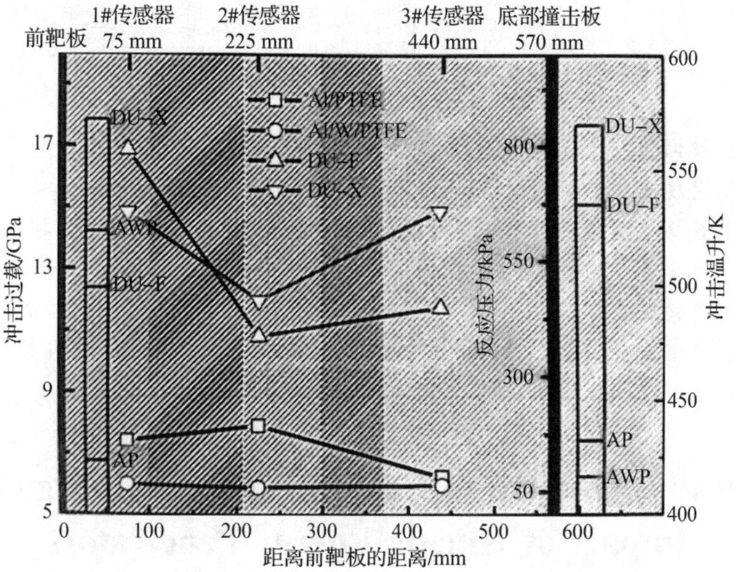

图 5　试样冲击过载、冲击波温升和不同测试位置压力峰值图

5　结论

通过后效多层靶毁伤测试和密闭容器压力测试，对比了 Al/PTFE、A1/W/PTFE、DU－F 和 DU－X 四种材料，在同尺寸 Ø10×15 mm，速度 1 000 m/s 的条件下，穿透 2 mm 厚 A3 钢板后的冲击破碎和能量释放效应。研究结果表明：

（1）贫铀材料（DU－F 和 DU－X）兼具高能量、高密度特性，后效多层靶毁伤效果和能量释放效应显著，具有较好的研究价值和应用前景。

（2）易碎 DU－X 合金材料，结合 X 金属特点，具备良好的穿破碎效应，虽然能量释放以燃烧形式为主，但是因为密度高（10.7 g/cm^3），其同尺寸单枚试样含能较高，能量释放效应同样显著。

（3）DU－F 材料冲击过载和冲击温升敏感，具备自持反应燃爆特性，在有限空气条件下，能量释放效应逐渐体现。

参考文献

［1］JACOBSEN M K，VELISAVLJEVIC N. High pressure elasticity and thermal properties of depleted uranium［J］. Journal of applied physics，2016，119（16）：165904.

［2］宋勇，潘名伟，冉秀忠，等．活性破片的研究进展及应用前景［C］//2014’（第六届）含能材料与钝感弹药技术学术研讨会论文集．中国四川成都，2014：3.

［3］WANG L，et al. Insensitive high－energy energetic structural material of tungsten－polytetrafluoroethylene－aluminum composites［J］. AIP advances，2015，5（11）：117142－117142－7.

［4］余庆波，王海福，等．缓冲材料对活性破片战斗部爆炸驱动影响分析［J］．北京理工大学学报，2013（2）：121－126.

动能侵彻体冲击带壳炸药装药的爆燃失效①

冯顺山[1]，赵宇峰[1②]，边江楠[2]，周彤[1]
（1. 北京理工大学爆炸科学与技术国家重点实验室，北京 100081；
2. 北京航空工程技术研究中心，北京 100076）

摘 要：针对动能侵彻体冲击下带壳炸药装药的失效问题，分析了炸药装药不同反应情况驱动壳体速度，提出用等效格尼速度或等效格尼能表征炸药装药失效等级，相应给出了求解等效格尼速度和等效格尼能的方法。基于等效格尼速度对带壳炸药装药失效等级进行了划分并得到判据，提出爆燃失效是一种有效的失效理念。结合实验、数值模拟和解析计算研究了动能侵彻体冲击下带壳B炸药的爆燃失效问题，分析了爆燃失效与正常爆轰失效的关系，结果显示当等效格尼速度约为正常爆轰反应条件下的1%时，可视为邻近炸药装药爆燃失效的下限值。

关键词：动能侵彻体；爆燃失效；带壳炸药装药；等效格尼速度；等效格尼能；等效破片初速

Deflagration Failure of Explosive Cased Charge Under Impact of Kinetic Energy Penetrators

Feng Shunshan[1], Zhao Yufeng[1], Bian Jiangnan[2], Zhou Tong[1]
(1. State Key Laboratory of Explosion Science and Technology, Beijing Institute of Technology, Beijing 100081, China;
2. Beijing Aeronautical Technology Research Center, Beijing 100076, China)

Abstract: Aiming at the failure problem of cased explosive charge under impact of kinetic energy penetrators, the velocity of case driven by explosive charge under different reaction rate was analyzed. The grade of characterizing explosive charge failure was proposed using equivalent gurney velocity or equivalent gurney energy. The method of solving equivalent gurney velocity and equivalent gurney energy was given accordingly. Based on equivalent gurney velocity, the failure grades of cased explosive charge were divided and the criterion was given, the idea of deflagration failure is proposed which is considered to be more effective to realize. The deflagration failure problem of cased composition B under impact of kinetic energy penetrators was studied based on experimental tests, numerical simulation and analytical calculation. The relations between deflagration failure and normal detonation failure were analyzed and the lower limit value of deflagration failure criterion was given. The results shows that it can be regarded as the lower limit of deflagration failure of explosive when equivalent gurney velocity reaches 1% of gurney velocity.

Key words: kinetic energy penetrators; deflagration failure; cased charge; equivalent gurney velocity; equivalent gurney energy; equivalent fragment initial velocity

CLC number: TJ55; O389 Document code: A DOI: 10.11943/CJEM2018215

1 引言

末端反导是一种有效的反导方式，其中毁伤目标战斗部使其解体失效是末端反导作战是否有效的关

---

① 原文发表于《含能材料》2019（4）。

② 赵宇峰：工学博士，2011年师从冯顺山教授，研究冲击反应破片对带壳炸药装药爆燃作用效应，现工作单位：中北大学。

键，否则无论是否命中预定目标，来袭导弹仍具有爆炸破坏能力，引发不可知的后果。所引出的核心问题是战斗部动能毁伤元冲击目标战斗部时，能否使其炸药装药发生有效反应，进而解体失效并丧失预定的毁伤能力。动能毁伤元撞击战斗部壳体产生的冲击波以及侵透壳体后对炸药的强剪切会使其发生反应，输出的能量驱动战斗部，壳体破裂程度及其破片的运动速度大小与炸药装药反应程度息息相关，从而决定了目标战斗部失效程度，最有效要求是目标战斗部在空中解体。通常认为炸药装药爆轰使战斗部解体失效是最理想的，但若目标战斗部采用钝感装药[1-3]，爆轰解体就难以实现，非爆轰解体失效问题得到关注。

关于带壳炸药装药在不同刺激条件下的反应情况，目前国内外关注重点仍放在炸药装药发生爆轰反应时的破片冲击失效临界阈值[4-8]，而其他失效等级的研究很少。美国2011年发布的最新非核弹药危险评估试验标准 MIL－STD－2105D 对装药在外界刺激下的反应等级进行了划分[9]；闵冬冬[10]与李德贵[11]基于单枚动能侵彻体冲击实验的方法对带壳炸药装药的失效等级进行了初步研究。MIL－STD－2105D 是基于弹药安全性的试验标准，将其应用在反导毁伤目标战斗部的研究中时，相应的失效等级的划分以及基本概念的阐述还需做出一定的调整；闵、李二人虽然对带壳炸药装药的失效等级进行了研究，但研究重点仍然是带壳炸药装药的爆轰解体失效。本研究关注非爆轰失效问题，提出了基于等效格尼速度和等效破片初速研究带壳炸药装药在动能侵彻体冲击条件下的爆燃失效等级及判据的思路，并通过实验与数值模拟相结合方法对其进行了计算和验证。

2　带壳炸药装药失效等级研究

2.1　炸药装药正常爆轰驱动壳体速度分析

带壳炸药装药失效等级与反应产物驱动壳体速度大小密切相关。本研究以炸药装药正常爆轰反应为基准，通过壳体破片运动速度的角度去分析炸药装药反应等级，故首先说明正常爆轰反应条件下的破片初速情况。

在带壳炸药装药正常爆轰驱动过程中，壳体膨胀破裂成一定数量的破片并运动飞散，破片获得的最大速度即为破片初速 v_0，其受炸药装药的种类、装药密度、起爆位置、爆轰波的传播方向，以及稀疏波作用等因素影响，目前可通过 AUTODYN 等数值模拟软件或工程计算方法近似求解。破片初速解析计算能量模型如下。

根据能量守恒定律可知，炸药装药正常爆轰所释放出的化学能 Q_e，一部分变为爆轰产物的比内能，一部分变为产物的动能，一部分变为壳体的动能，即

$$m_e Q_e = m_e E_{ie} + E_t + 0.5Mv_0^2 \tag{1}$$

式中　m_e——炸药装药质量，kg；

M——壳体质量，kg；

E_{ie}——爆炸产物的比内能，$J \cdot kg^{-1}$；

E_t——爆炸产物的动能，J；

v_0——破片初速，$m \cdot s^{-1}$。

式（1）在以下假设[12]基础上建立：爆轰波同时到达金属壳体表面，爆轰产物驱动壳体达到最大速度 v_0 时产物速度从起爆点至壳体内表面之间呈线性分布；不考虑起爆点对爆轰波形及传播方向的影响；不考虑壳体强度及其破裂时造成的能量损耗，即壳体同时破裂、同时被驱动达到最大速度 v_0。

对于圆柱形壳体 $E_t = m_e v_0^2/4$，则

$$2(Q_e - E_{ie}) = \left(\frac{1}{2} + \frac{M}{m_e}\right)v_0^2 \tag{2}$$

令 $E_g = Q_e - E_{ie}$，E_g 即为炸药装药格尼能，是炸药正常爆轰时释放的化学能与爆轰产物状态内能之差，$J \cdot kg^{-1}$。式（2）即可演变为求解壳体破片初速的公式，称为格尼公式：

$$v_0 = \sqrt{2E_g} \cdot \sqrt{\frac{\beta}{1+0.5\beta}} \tag{3}$$

式中 $\beta = m_e/M$——带壳炸药装药装填比；

$\sqrt{2E_g}$——炸药装药的格尼速度，具有速度的量纲，$m \cdot s^{-1}$。

对于装填密度为 1.68 $g \cdot cm^{-3}$ 的熔铸 B 炸药（质量比 RDX/TNT = 60/40），其格尼速度为2 697 $m \cdot s^{-1}$[13]。

格尼能 E_g 与格尼速度 $\sqrt{2E_g}$ 反映炸药正常爆轰驱动壳体的特性值，其数值由炸药装药的配方、制备工艺及其装填密度决定，可通过破片初速来反求。

2.2 等效格尼能与等效格尼速度

动能侵彻体冲击下带壳炸药装药有时会发生非正常爆轰，诸如弱爆轰、爆炸、爆燃、燃烧，炸药装药反应程度及驱动壳体的能力差异显著，此时通常所说的格尼能不适用于分析炸药装药发生非正常爆轰时驱动壳体的能力。炸药装药非正常爆轰条件下的能量释放，可将其视作为某种低能含能材料的正常爆轰所释放的能量，相比于原炸药装药的格尼能 Eg 明显弱化，特提出等效格尼能 E_{di} 和等效格尼速度 $\sqrt{2E_{di}}$ 的概念，进而定量描述其弱化程度。

等效格尼能与等效格尼速度的定义：炸药装药弱爆轰、爆炸或爆燃反应释放出的能量，等效为相同形状、质量、装填比的某种低能炸药装药正常爆轰的格尼能 E_{di}，称之为该炸药装药非正常爆轰条件下的等效格尼能；等效格尼能 E_{di} 对应的 $\sqrt{2E_{di}}$，称之为该炸药装药非正常爆轰条件下的等效格尼速度。

带壳炸药装药在非正常爆轰条件下输出的能量 E_{gF} 与正常爆轰条件下的格尼能 E_g 的比值 k_{Ei}^2 为

$$\frac{E_{gf}}{E_g} = \frac{Q_{ei} - E_{ti}}{Q_e - E_t} = k_{Ei}^2 \tag{4}$$

式中，脚标 F 表示非正常爆轰；k_{Ei}^2 由炸药装药反应速率或程度决定，也可由实验对比测定，取值范围 0～1，取值不同可表征炸药装药的反应等级；脚标 i 表示炸药装药反应等级，定性划分为 5 个等级，分别是 A（正常爆轰）、B（弱爆轰）、C（爆炸）、D（爆燃）、E（燃烧）。

基于等效格尼能的概念与式（4），可给出等效格尼能 E_{di} 的数学表达式：

$$E_{di} = E_{gr} = k_{Fi}^2 E_g \tag{5}$$

式中，脚标 d 表示等效条件。

基于等效格尼速度的概念与式（4）、式（5），可给出等效格尼速度 $\sqrt{2E_{di}}$ 的数学表达式：

$$\sqrt{2E_{di}} = \sqrt{2E_{gF}} = k_{Ei}\sqrt{2E_g} \tag{6}$$

式中，k_{Ei} 为等效格尼速度与格尼速度的比值，于是如式（7）：

$$0 \leqslant \sqrt{2E_{di}} \leqslant \sqrt{2E_g} \tag{7}$$

本研究拟由带壳炸药装药非正常爆轰反应条件下的等效格尼能或等效格尼速度的分布值指导分析目标战斗部炸药装药失效等级问题。

2.3 等效破片初速

由于炸药装药非正常爆轰反应的复杂性，理论上求解等效格尼能和等效格尼速度数值较为困难，较为简便有效途径可通过实验测量或数值模拟得到不同动能冲击条件下炸药装药反应驱动壳体破片的最大运动速度后求出等效格尼能或等效格尼速度。依据本研究提出的等效格尼能和等效格尼速度的定义，相应提出等效破片初速的概念，进而定量描述其弱化程度。

等效破片初速的定义：带壳炸药装药弱爆轰或爆炸或爆燃反应的等效格尼能所求得破片的最大速度，等效为相同形状、质量、装填比的某种低能炸药装药正常爆轰反应时破片初速，称之为原炸药装药非正常爆轰条件下的等效破片初速 v_{di} 在实验或数值模拟研究中，对比分析带壳炸药装药不同反应条件下驱动破片的速度，均为同坐标条件的破片速度，由下式表示：

$$v_{\mathrm{di}} = \sqrt{2E_{\mathrm{di}}} \cdot \sqrt{\frac{\beta}{1+0.5\beta}} \tag{8}$$

式（3）和式（8）仅能计算带壳炸药装药长径比足够大时破片最大初速。由于炸药装药爆轰非瞬时性和稀疏波的作用，实际破片初速值沿弹轴沿周向角呈分布状态。本课题组对不同起爆位置和装药结构条件下战斗部破片初速轴向分布及径向分布规律进行了大量的研究，通过脉冲 X 光摄影技术精准跟踪测量破片运动方向和速度变化历史，实现了任一破片初速的测试，提出了轴线起爆[14-15]（包含一端起爆、中心起爆和两端起爆）和偏轴心起爆[16-18]条件下求解一般结构形状（包含等壁厚和变壁厚）战斗部破片初速的工程计算式。轴对称等壁厚炸药装药结构，一端轴线起爆条件下破片初速的工程计算式为

$$v_{0x} = (1-0.3615e^{-1.111x/d})[1-0.1925e^{-3.030(L-x)/d}]\sqrt{2E_B}\sqrt{\frac{\beta}{1+0.5\beta}} \tag{9}$$

式中　v_{0x}——破片在 x 坐标处的初速值，随 x 坐标变化而变化，$\mathrm{m \cdot s^{-1}}$；

L——炸药装药的长度，m；

d——炸药装药的直径，m。

壳体破片初速值分布如图 1 所示，V_{0a}、V_{0b} 分别为壳体端部破片初速、V_{0c} 为破片最大初速，当炸药装药长径比足够大时，$V_{0c}=V_0$。

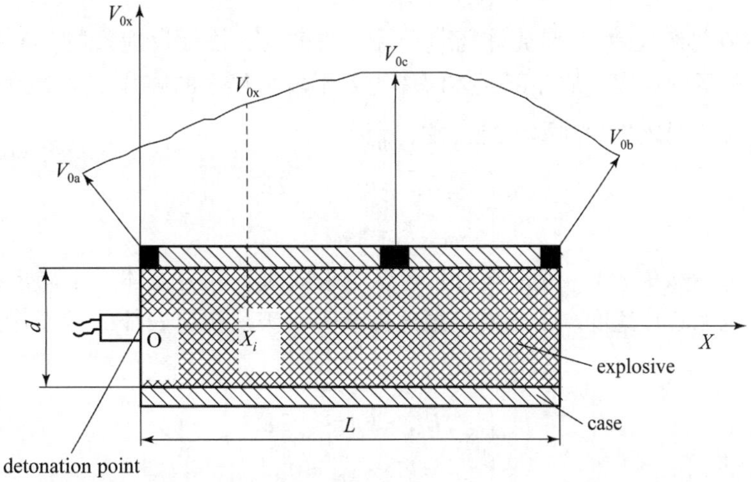

图 1　轴对称等壁厚炸药装药结构一端轴线起爆条件下破片初速值随弹轴分布示意图

故 x 坐标处的等效破片初速 v_{xdi} 的工程计算式（9）可变为

$$v_{xdi} = (1-0.3615e^{-1.111x/d})[1-0.1925e^{-3.030(L-x)/d}]\sqrt{2E_{di}}\sqrt{\frac{\beta}{1+0.5\beta}} \tag{10}$$

炸药装药非正常爆轰条件下，实测得某 x 坐标处的等效破片初速值，就可通过式（10）求得相同数值的等效格尼速度或等效格尼能；相反，由等效格尼速度通过式（10）亦可计算出不同 x 坐标处的等效破片初速。

2.4　带壳炸药装药失效等级

提出了带壳炸药装药的五个反应等级，对应指导战斗部解体失效等级定性划分为 A、B、C、D、E 五个等级，由等效格尼速度和等效破片初速表征的判据数学形式由式（11）所示：

$$\begin{aligned}\sqrt{2E_{si}} &= k_{Ei}\sqrt{2E_g} \\ V_{xdi} &= k_{Ei}V_{0x}\end{aligned} \tag{11}$$

式中　脚标 S——失效；

i——失效等级 A、B、C、D、E。

五个失效等级描述如下：

（1）正常爆轰失效（A 等级失效）。带壳炸药装药正常爆轰失效时，壳体正常破碎解体，由格尼速度或破片初速表征的带壳炸药装药正常爆轰失效判据 $\sqrt{2E_{SA}}$或 V_{xdi} 为

$$\begin{cases}\sqrt{2E_{SA}} = \sqrt{2E_g} \\ v_{xdA} = V_{0x}\end{cases} \tag{12}$$

（2）弱爆轰失效（B 等级失效）。带壳炸药装药弱爆轰失效时，弱爆轰反应产物压力很高对战斗部壳体仍具有很强的驱动能力，壳体显著破坏解体。由等效格尼速度或等效破片初速表征的带壳炸药装药弱爆轰失效判据 $\sqrt{2E_{SB}}$或 V_{xdB} 为

$$\begin{cases}\sqrt{2E_{SB}} = k_{CB} \cdot \sqrt{2E_B} \\ V_{xdB} = k_{CB} \cdot V_{0x}\end{cases} \tag{13}$$

（3）爆炸失效（C 等级失效）。带壳炸药装药爆炸失效时，爆炸产物对战斗部壳体仍具有较强的驱动能力，壳体仍可形成较多破片而解体，但尺寸和质量相对较大，壳体破片运动速度为每秒几百米量级，具有杀伤作用。由等效格尼速度或等效破片初速表征的带壳炸药装药爆炸失效判据 $\sqrt{2E_{SC}}$或 v_{xdC} 为

$$\begin{cases}\sqrt{2E_{SC}} = k_{EC} \cdot \sqrt{2E_{\mathrm{B}}} \\ v_{xdC} = k_{EC} \cdot v_{0x}\end{cases} \tag{14}$$

（4）爆炸失效（D 等级失效）。带壳炸药装药爆燃失效时，爆燃反应产物具有一定压力，可驱动壳体产生几米至上百米的运动速度，战斗部壳体解体为若干较大尺寸的破片。由等效格尼速度或等效破片初速表征的带壳炸药装药爆燃失效判据 $\sqrt{2E_{\mathrm{SD}}}$或 $v_{x\mathrm{DD}}$ 为

$$\begin{cases}\sqrt{2E_{\mathrm{SD}}} = k_{\mathrm{ED}} \cdot \sqrt{2E_g} \\ v_{x\mathrm{DD}} = k_{\mathrm{ED}} \cdot v_{0x}\end{cases} \tag{15}$$

（5）燃烧失效（E 等级失效）。带壳炸药装药燃烧失效时，装药壳体发生局部物理膨胀甚至有裂缝现象，但不形成破片也无破片速度。由等效格尼速度或等效破片初速表征的带壳炸药装药燃烧失效判据 $\sqrt{2E_{\mathrm{SE}}}$或 v_{xdE} 为

$$\begin{cases}\sqrt{2E_{\mathrm{SE}}} = k_{EE} \cdot \sqrt{2E_8} \\ v_{xdE} = k_{EE} \cdot v_{0x} = 0\end{cases} \tag{16}$$

综上所述，对于带壳炸药装药或战斗部被打击后发生正常爆轰、弱爆轰、爆炸和爆燃失效均可以使壳体解体产生破片并具有速度，虽四者失效程度差异很大，但都可达到解体效果，故只需使炸药装药爆燃失效即可达到使目标战斗部解体毁伤的目的。

3　带壳炸药装药爆燃失效研究

3.1　实验研究

实验系统由动能侵彻体、高速摄影相机、前置靶板、试件（带壳 B 炸药或带壳尼龙）、炸药装药反应效应鉴证靶、惯性块（模拟战斗部质量）组成。发射装置为 13.2 mm 口径的滑膛式弹道枪，由混合发射药对动能侵彻体加载，使其获得预定速度对试件作用。动能侵彻体的着靶速度 u_0 和试件壳体运动速度由高速摄影记录。终点弹道示意图如图 2 所示。

参试的动能侵彻体为平头圆柱形 45#钢破片，质量为 10 g、尺寸为 Φ12 mm × 12 mm；带壳炸药装药试件为圆柱形带壳 B 炸药，其质量比 RDX/TNT = 60/40、尺寸为 Φ38 mm × 20 mm（$L/d = 0.526$），外侧的 PVC 衬套厚度为 1.5 mm、金属壳体厚度为 2 mm，装填比 $\beta = 0.62$，装填密度为 1.68 g · cm^{-3}，前置靶板为 A3 钢材；带壳尼龙试件用于分析动能侵彻体冲击作用下压力径向扩张对壳体的径向膨胀速度的

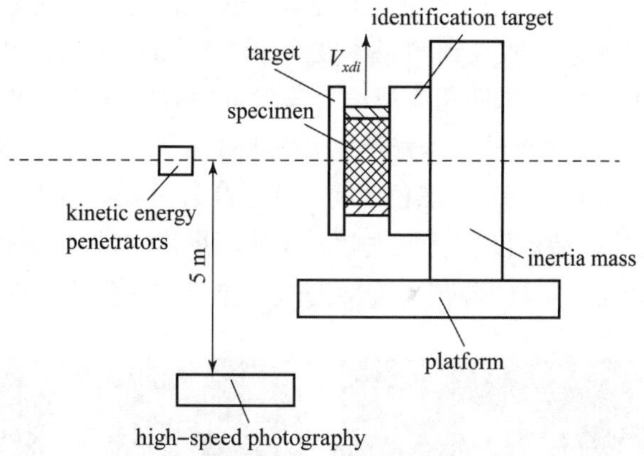

图 2　终点弹道示意图

贡献，其尺寸与 B 炸药相同，密度为 1.14 g · cm^{-3}。对比炸药装药爆燃失效条件下驱动破片的速度，均为壳体中部的破片速度，带壳 B 炸药照片如图 3 所示。

图 3　带壳 B 炸药照片

首先进行带壳尼龙试件的试验，测得无炸药装药条件下的壳体膨胀速度值 v_n。进行了动能侵彻体 1 171 m · s^{-1}的着靶速度冲击带壳尼龙试件试验，尼龙以及鉴证靶上动能侵彻体作用效应和壳体胀坏情况如图 4 所示，通过高速摄影测得壳片初速 v_n = 38 m · s^{-1}。故可判断带壳炸药装药爆燃失效时壳片初速应明显大于 v_n 才合理，炸药装药爆燃条件下试件驱动壳体真实速度应去除 v_n。

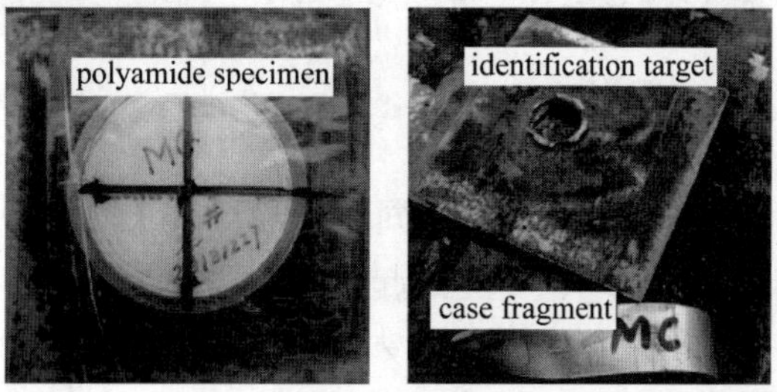

（a）　　　　（b）

图 4　动能侵彻体冲击带壳尼龙试件典型实验照片

（a）polyamide specimen；（b）identification target and case fragment

当动能侵彻体以 917 m·s^{-1}的着靶速度冲击带壳 B 炸药试件时，鉴证靶上除弹坑外无任何凹陷变形痕迹，实验平台无未反应炸药粉末，收集到的壳体破片如图 5（a）所示，单枚破片最大长度约为壳体周长的 50%，通过高速摄影测得 B 炸药试件壳片初速 v_B = 52 m·s^{-1}，判断试件在爆燃失效条件下等效破片初速 $v_{x dD} = v_B - v_n$ = 14 m·s^{-1}；当动能侵彻体着靶速度为 863 m·s^{-1}，实验平台有未反应炸药碎块如图 5（b）所示，前置靶板上附少量炸药装药反应黑色产物，通过高射摄影测得 B 炸药试件壳片初速 v_C = 37 m·s^{-1}，等同于 v_n，可判断带壳炸药装药发生燃烧失效，B 炸药燃烧对破片实际驱动速度为 0。对比分析 v_B 和 v_C 可知，动能侵彻体以 917 m·s^{-1}的着靶速度冲击试件接近使炸药装药反应的下限条件。

（a）

（b）

图 5　动能侵彻体冲击带壳 B 炸药典型实验照片

（a）u_0 = 917 m·s^{-1}；（b）u_0 = 863 m·s^{-1}

3.2　数值模拟研究

基于 AUTODYN 软件对动能侵彻体冲击带壳 B 炸药或带壳尼龙试件进行仿真，其仿真模型如图 6 所示。根据结构的对称性建立 1/2 二维有限元模型。动能侵彻体（45#钢破片）、前置靶板、壳体、尼龙、鉴证靶、惯性块均采用 Lagrange 算法，其中 45#钢破片、前置靶板、壳体的网格密度为 0.5 mm，鉴证靶和惯性块的网格密度为 1 mm，尼龙的网格密度为 0.1 mm，炸药装药采用 SPH 算法，网格密度为 0.1 mm，计算时采用 cm - g - μs 单位制。在试件壳体上设置 3 个 Gauges 观测壳片初速，3 个 Gauges 观测点位置 x/L 分别为 0.15、0.5、0.85。

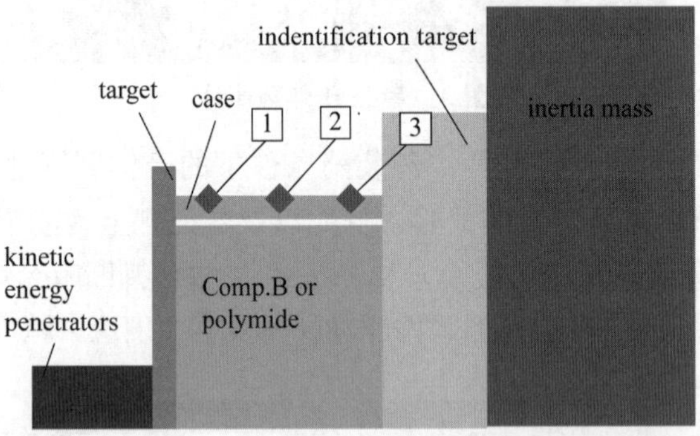

图 6　动能侵彻体冲击带壳 B 炸药（带壳尼龙）仿真模型

数值模拟中各试件材料、尺寸和密度与实验相同，45#钢和 A3 钢均采用 shock 状态方程和 Johnson - Cook 强度模型，材料参数取自文献［19］；B 炸药起爆状态方程采用 Lee - Tarver 三项点火增长模型方程、炸药反应产物和未反应物采用 JWL 状态方程，材料参数取自文献［20］；尼龙尺寸与 B 炸药相同，其材料参数取自 AUTODYN 材料数据库。

动能侵彻体冲击带壳尼龙、带壳 B 炸药时壳体 Gauges 观测点速度如表 1 所示，与实验结果对比可知，此时带壳 B 炸药发生爆燃失效，而尼龙径向膨胀能使壳片获得一定速度，两种条件下壳体三个观测点的平均速度差值即为 B 炸药爆燃对破片的实际驱动速度，得到爆燃失效条件下等效破片初速值为 14. 8 $m \cdot s^{-1}$。

表 1　数值模拟结果

	$u_0/(m \cdot s^{-1})$	velocity of Gauges/$(m \cdot s^{-1})$			velocity of case / $(m \cdot s^{-1})$
		Gauges 1	Gauges 2	Gauges 3	
cased polyamide	1 100	29. 8	38. 7	49. 5	39. 3
cased Composition B	910	43. 3	58. 7	60. 3	54. 1
actual driving speed of Composition B	—	13. 5	20. 0	10. 8	14. 8

Note: u_0 is impact velocity of kinetic energy penetrators.

3. 3　对比分析

基于实验和数值模拟得到的 B 炸药爆燃对破片实际驱动速度，通过式（10）可求出相应的等效格尼速度，对比如表 2 所示，再将实验得到的 B 炸药爆燃失效的格尼速度代入式（10），分别求出 x/L = 0. 15、0. 5、0. 85 处的破片速度后取平均值，其与实验结果对比如表 3 所示。基于上文分析以及式（9）可求得带壳 B 炸药试件正常爆轰时破片平均初速值为 1 216. 0 $m \cdot s^{-1}$，结合格尼速度可求得相应的 k_{ED} 的数值，如表 2 和表 3 所示。

表 2　数值模拟与实验结果对比

	$\sqrt{2E_{dD}}/m \cdot s^{-1}$	$v_{xdE}/m \cdot s^{-1}$	k_{ED}	
			$\sqrt{2E_{dD}}/\sqrt{2E_g}$	v_{xdD}/v_{0x}
simulation	32. 1	14. 8	0. 011 9	0. 012 2
experiment	30. 4	14	0. 011 3	0. 011 5
error	5. 6%	5. 7%	5. 3%	6. 1%

表 3　解析计算与实验结果对比

	$\sqrt{2E_{dD}}/m \cdot s^{-1}$	$v_{xdE}/m \cdot s^{-1}$	k_{ED}	
			$\sqrt{2E_{dD}}/\sqrt{2E_g}$	v_{xdD}/v_{0x}
calculate	29. 7	13. 7	0. 011 0	0. 011 3
experiment	30. 4	14	0. 011 3	0. 011 5
error	2. 3%	2. 1%	2. 7%	1. 7%

注：$\sqrt{2E_{dD}}$ and v_{xdD} are parameters of cased charge; k_{ED} is the ratio of to $\sqrt{2E_g}$ or v_{xdD} to v_{0x}

由上表的误差分析可知，数值模拟计算值与实验结果、式（10）计算值与实验结果基本一致，说明上述实验方法、数值模拟方法以及式（10）均可用于研究带壳炸药装药不同条件下的失效问题，由等效格尼速度和等效破片初速的途径可定量评价带壳炸药装药的反应/失效等级。

由 k_{ED} 的数值可知，当带壳炸药装药爆燃失效的等效格尼速度达到正常爆轰失效格尼速度的 1. 13% 时，炸药装药可实现爆燃反应且壳体结构解体失效。表 2 和表 3 数据分析表明，等效格尼速度或等效破片初速约为正常爆轰反应条件下的 1% 时，可视为邻近炸药装药爆燃失效的下限值，工程应用需考虑多种干扰因素，修正系数应大于 1。

4 结论

（1）提出由等效格尼速度和等效破片初速表征的带壳炸药装药五个失效等级的判据 $\sqrt{2E_{si}}$ 或 v_{xdi}。

（2）得到了 B 炸药爆燃失效等级下的等效格尼速度、等效破片初速以及 k_{ED} 值，认为等效格尼速度约为正常爆轰反应条件下的 1% 时，可视为临近炸药装药爆燃失效的下限值。

参考文献

[1] Gobalsecurity Org. Insensitive high explosives [EB/OL]. (2006 - 11 - 08). http://www.globalsecurity.

[2] SCOTT F. Melt pour loading of PAX - 21 into the 60mm M720E1mortar cartridge [C]//Insensitive Munitions and Energetic MaterialsTechnology Symposium. Arlington: NDIA, 2000: 91 - 99.

[3] STEVEN N, JOHN N, PAMELA F. Recent developments in reduced sensitivity melt pour explosives [C]//34th International Annual Conference of ICT, 2003.

[4] 田少康，李席，刘波，等．一种 RDX 基温压炸药的 JWL - Miller 状态方程研究 [J]. 含能材料，2017，25 (3): 226 - 231. TIAN S K, LI X, LIU B, et al. Study on JWL - Millerequation of state of RDX - based thermobaric explosive [J]. Chinese journal of energetic materials (Hanneng Cailiao), 2017, 25 (3): 226 - 231.

[5] 钱石川，甘强，任志伟，等．HNS - Ⅳ炸药一维冲击起爆判据的研究 [J]. 含能材料，2018，26 (6): 495 - 501. QIAN S C, GAN Q, REN Z W, et al. Study onone - dimensional shock initiation criterion of HNS - Ⅳ explosive [J]. Chinese journal of energetic materials (Hanneng Cailiao), 2018, 26 (6): 495 - 501.

[6] 王昕，蒋建伟，王树有，等．钨球对柱面带壳装药的冲击起爆数值模拟研究 [J]. 兵工学报，2017，38 (8): 1498 - 1505. WANG X, JIANG J W, WANG S Y, et al. Numericalsimulation on the initiation of cylindrical covered chargeimpacted by tungsten sphere fragment [J]. Acta armamentarii, 2017, 38 (8): 1498 - 1505.

[7] 刘鹏飞，智小琦，杨宝良，等．六棱钨柱冲击起爆带壳 B 炸药比动能阈值研究 [J]. 高压物理学报，2017，31 (5): 637 - 642. LIU P F, ZHI X Q, YANG B L, et al. Specific kineticenergy threshold of impacting initiation covered explosive B by six - prismed tungsten fragment [J]. Chinese journal of high pressure physics, 2017, 31 (5): 637 - 642.

[8] 路迎，王芳，卞晓兵，等．破片对复合壳体装药冲击起爆判据的研究 [J]. 兵工学报，2017 (增刊 1): 194 - 199. LU Y, WANG F, BIAN X B, et al. Shock initiation criterion of composite shell charges under impact of fragment [J]. Acta armamentarii, 2017 (S1): 194 - 199.

[9] Center NSW. Hazard assessment tests for non - nuclear munitions [S]. Military Standard No MIL - STD - 2105D, 2011.

[10] 闵冬冬．破片侵彻下战斗部装药失效机理及低易损性方法研究 [D]. 北京：北京理工大学，2014. MIN D D. The failure mechanism of warhead charge and the low vulnerability method by the fragment penetration [D]. Beijing: Beijing Institute of Technology, 2014.

[11] 李德贵．战斗部抗破片冲击引爆感度评价方法与实验研究 [D]. 北京：北京理工大学，2016. LI D G. The evaluation method and experimental research on anti - fragment penetration sensitivity of warhead charge [D]. Beijing: Beijing Institute of Technology, 2016.

[12] 张宝坪，张庆明，黄风雷．爆轰物理学 [M]. 北京：兵器工业出版社，1997: 305 - 311. ZHANG B P, ZHANG Q M, HUANG F L. Detonation physics [M]. Beijing: Weapon Industry Press, 1997: 305 - 311.

[13] 孙业斌．爆炸作用与装药设计 [M]. 北京：国防工业出版社，1987: 76 - 78. SUN Y B. Explosion effect and charge design [M]. Beijing: National Defense Industry Press, 1987: 76 - 78.

障碍型弹药对飞行甲板的封锁效能分析方法[①]

白洋[②]，李伟，冯顺山

（北京理工大学爆炸科学与技术国家重点实验室，北京　100081）

摘　要：基于对航母飞行甲板进行封锁作战的思想，开展了具有杀爆功能的障碍型子弹药对飞行甲板目标的封锁效能分析。对于封锁效能的分析归结于终点空间封锁概率。首先根据飞行甲板的功能和舰载机起飞作战流程，将甲板划分为保障起飞功能的3个特征功能区域：起飞区、调动区和停机区，并分别建立了其起飞功能失效的封锁判据，并且在飞行甲板坐标系中给出了封锁各功能区子弹药的有效障碍点区域确定方法；其次运用蒙特卡洛法模拟子弹药的落点，并利用矩阵仿真等方法给出了各个区域封锁概率的计算式；最后给出了具有杀爆功能的障碍型子弹药对航母飞行甲板的空间封锁概率的计算式。

关键词：航母飞行甲板；封锁效能；蒙特卡洛

The Analysis Method of Blockade Effectiveness on Barrier - Submunition Attacking the Fight Deck

Bai Yang, Li Wei, Feng Shunshan

(State Key Laboratory of Explosion Science and Technology, Beijing Institute of Technology, Beijing 100081, China)

Abstract: This paper carried out the research of the blockade effectiveness on barrier - submunition attacking the fight deck of aircraft carrier based on the ideas about blockading the fight deck of aircraft carrier. The analysis of blockade effectiveness attributed to the calculation of spatial blockade probability. This paper first divided the fight deck of aircraft carrier into three specific parts (launch area, dispatch area and parking area) which about the lanuch process according to the function of it and the operation process of CVW, and established the blockade criterion about the function of combat for each part and put forward the way to determine the function expression of each area for effective point of impact of submunition in the coordinate system of flight deck. Then, got the point of impact of submunitions by using the Monte - Carlo method and put forward the calculation method and computational formula about each area's blockade probability by the matrix simulation method and others, then put forward the computational formula about the spatial blockade probability of the fight deck. The calculation method about the blockade probability which this paper put forward could provide a reference to the terminal blockade, effectiveness for the weapon of blockade flight deck.

Key words: fight deck; blockade effectiveness; Monte - Carlo

引言

航空母舰作为海上最具价值的打击目标，如何对其进行有效打击使其丧失作战能力一直是各国武器专家研究的热点。从航母的作战模式分析，其主要依靠搭载的舰载机来完成作战任务。一旦舰载机无法正常起飞，航母就丧失了主要作战能力。保障航母舰载机起飞主要条件的是飞行甲板及其配套设施和人

① 原文发表于《北京理工大学学报》2015（35）。

② 白洋：工学博士，2012年师从冯顺山教授，研究弹药对典型目标的封锁方法和效能分析，现工作单位：航天科工四院四部。

员，通过采取使飞行甲板失效的方式来限制舰载机起飞是一种新的作战思想，这种不对航母造成致命性打击的“打得巧”作战思想受到了广泛重视。

引入对陆基机场的爆坑障碍封锁作战思想，利用子母弹技术对航母飞行甲板实施特殊障碍封锁作战，能够保证在不对航母造成毁灭性打击的情况下使之丧失作战能力。目前国内有些学者已经开展了相关的研究[1-4]，提出了例如串联嵌入子弹[2]和整体式嵌入子弹药[3]以及“铝热”弹[4]等。作者所在科技团队多年来对航母目标非致命和低附带损伤等研究方面取得了丰硕的成果，针对飞行甲板研发了具有杀爆功能的障碍型子弹药技术。本文基于此开展障碍子弹药封锁效能分析，为封锁威力设计提供依据。

依据航母甲板上与舰载机起飞作战相关部件，区别以往对航母甲板区域的划分，将甲板分为起飞区、停机区和调动区，根据各个区域作运行模式不 同，分别建立了不同的起飞功能封锁判据，提出障碍型子弹药对航母飞行甲板的终点空间封锁概率计算方法和给出了计算公式。

1　飞行甲板易封锁特性分析

1.1　飞行甲板保障作战特性分析

航空母舰主要是靠其上搭载的舰载机来完成各种作战目的，飞行甲板作为航母最重要的要害部位之一，一旦被封锁，在一定的时间内丧失舰载机起降功能，航母的作战能力将会大大降低，甚至丧失。根据舰载机作战流程[5]，可以归纳得出航母飞行甲板保障起飞作战功能为：

（1）舰载机提供挂弹，牵引至弹射区，弹射起飞等一系列服务，确保舰载机在具备足够的作战能力条件下安全起飞离开航母执行作战任务；

（2）无作战任务的舰载机提供停放的空间以及为即将执行作战任务的舰载机提供调度转移的服务。

对航母飞行甲板的封锁既是使之无法完成上述两种起飞作战功能。本文以典型航母为研究对象，将封锁航母飞行甲板的问题简化为最大限度地使舰载机无法有效弹射起飞。据此可将典型航母飞行甲板简化为 3 个区域：

（1）起飞区：主要包含 4 条起飞跑道和 4 个蒸汽弹射器，以及其他辅助起飞设施；

（2）停机区：美军典型航母飞行甲板上舰桥前可停放 26 架舰载机，舰桥左前侧可以停放 12 架，斜角甲板左舷后突出部位可停放 6~7 架，起飞行甲板停放舰载机数占舰载机总数的 50%[6]；

（3）调动区：指除去起飞区，航母飞行甲板上用于将舰载机牵引调度至弹射区的其他区域，其包括停机区。

1.2　飞行甲板起飞功能失效判据

由于航母飞行甲板 3 个区域的功能不同，分别建立不同的起飞功能失效判据。

（1）对陆基跑道封锁最主要的是确保跑道上不存在最小起降窗口（MLW）供飞机起飞降落 E，而对于飞行甲板而言，由于舰载机采用弹射起飞技术，这就决定了舰载机的起飞窗口是固定的，一旦舰载机的起飞窗口被损毁，则舰载机就无法起飞作战。

由于障碍型封锁子弹药不仅具有障碍舰载机起/降的能力，同时还具有杀爆能力，其对航母起飞区的有效封锁区域是结合子弹对舰载机的威慑杀伤半径 R 确定的，本文认为只要有两枚以上封锁子弹药落入指定区域，则该条起飞跑道被封锁。

（2）停机区：由于飞行甲板上一般停放了 50% 的舰载机，尼米兹级航母“华盛顿”号航母上一般停放了 45 架舰载机，这个数目基本决定了一个攻击波最多出动飞机的数目，对停机区的封锁实际上是对舰载机进行毁伤。本文做出如下假设：

①落入停机区的封锁子弹均满足对舰载机的最低杀伤要求（破片数目，破片密集度等）；

②子弹对舰载机造成杀伤的同时，封锁了子弹杀伤区域内的飞行甲板；

③不考虑封锁子弹药对舰载机的累积杀伤，不重复计算子弹毁伤重叠区域；

④一架舰载机遭到破片杀伤，不会引起其周围舰载机的毁伤。

本文基于以上 4 点假设，给出停机区的起飞功能封锁判据为毁伤面积：毁伤元对停机区毁伤的面积占停机区总面积的比例作为停机区的封锁概率。

（3）调动区：将毁伤陆基机场跑道从而使之不存在供飞机起飞的最小起降窗口的思想转化为毁伤航母飞行甲板的调动区，使舰载机无法在甲板上调动。本文引入最小调动窗（minimum dispatch window，MDW）的概念，MDW 指舰载机从停机区转移至起飞区的弹射器上所需要的最小区域。当落入调动区的子弹同时满足下述 3 点要求时，本文就认为调动区不存在使舰载机调动的 MDW，即调动区被封锁。

落入调动区的子弹数目要达到指定的数目 N，封锁子弹的数量等于调度区的面积与单枚子弹毁伤面积之比。

建立以航母纵向中心线为 X 轴，横向中心线为 Y 轴，两中心线交点为原点的航母坐标系，将落入调动区的子弹的横、纵坐标分别按从小达到排序，两相邻的横坐标和两相邻的纵坐标间距离均要小于 L_{max}，L_{max} 为两相邻横坐标和两相邻纵坐标之间允许的最大距离，其取决于子弹的毁伤半径以及舰载机的尺寸，本文规定 L_{max} 为 $2R$。

最小和最大的横坐标与调动区的左边界和右边界的距离均要小于 $L_{max}/2$；最小和最大的纵坐标与调动区的下边界和上边界的距离均要小于 $L_{max}/2$。

2 子弹药有效障碍点区域的确定

为了便于确定子弹药对于各个特征区域有效封锁的障碍点区域，将甲板上的建筑（舰桥等）简化为甲板的一部分。

2.1 起飞区有效封锁区域

以一条起飞跑道为例，为了简便计算，将封锁子弹的有效落点区域简化为矩形区域，同时结合航母飞行甲板轮廓曲线，得到要封锁该条起飞跑道子弹的有效落点区域，阴影区域由子弹毁伤半径 R 确定，如图 1 所示。

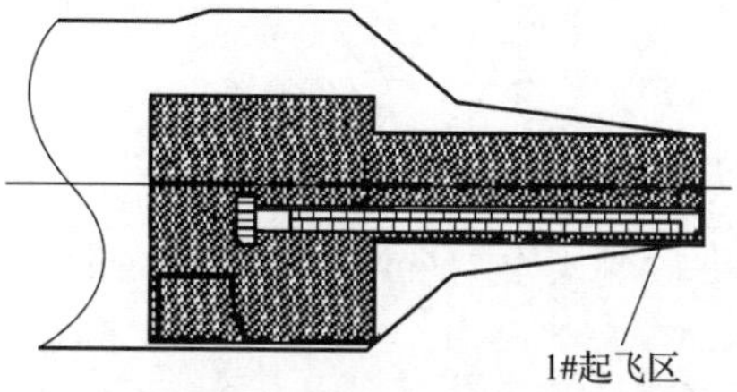

图 1 对于 1#起飞跑道子弹有效封锁落点区域

根据该区域在坐标系中的位置，可以建立封锁该条起飞跑道子弹药有效落点区域 A 时的函数表达式，同理可以分别得到封锁其余 3 条起飞跑道的子弹药的有效落点区域 A_{q2}、A_{q3}、A_{q4} 的函数表达式。由于各个起飞跑道的封锁区域存在重叠区，还要根据各个区域函数表达式得到重叠区域的函数表达式。

2.2 停机区有效封锁区域

将斜角甲板左舷后突出部位定义为 A 停机区，舰桥前以及左前侧区域定义为 B 停机区[7]，在考虑了子弹的毁伤半径之后，对于停机区的封锁，子弹有效障碍点区域如图 2 所示。

为了便于利用矩阵仿真[8]计算其封锁概率，在面积相等的条件下，将 A、B 停机区简化为矩形，分别得到 A、B 停机区在飞行甲板坐标系中的函数表达式，A_{tA}，B_{tB}。

2.3 调动区有效封锁区域

调动区是在除去 4 条起飞跑道后，最大限度包含航母飞行甲板的其余区域，其中为了简化调动区的函数表达式，将弹射起飞区简化为与飞行甲板轴线平行，如图 3 所示。

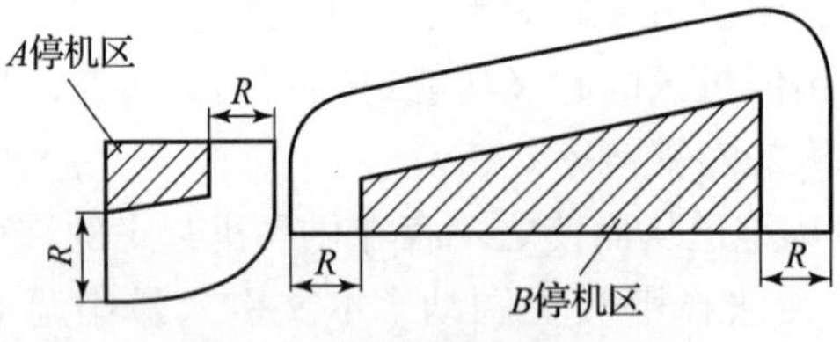

图2　对于停机区子弹有效封锁落点区域

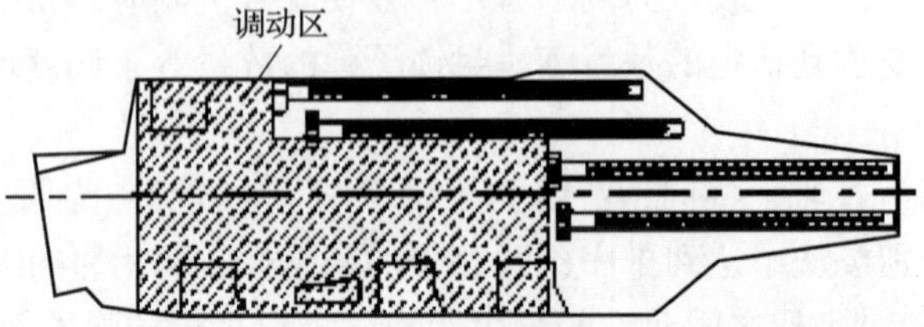

图3　调动区在航母飞行甲板上的分布图

在飞行甲板坐标系中确定调动区内子弹药的有效封锁障碍点区域函数表达式 A_d。

3　甲板起飞功能封锁概率的计算

本文基于蒙特卡洛法对航母飞行甲板的封锁概率进行计算。

假设发射 M 枚母弹，且母弹的落点服从以（X_m，Y_m，σ_1，σ_2，ρ）为参数的二维正态分布，其中（X_m，Y_m）为母弹瞄准点，σ_1，σ_2 分别为抛撒区域中心纵向、横向散布的均方差，ρ 为纵向和横向散布的相关系数。

以（0，0）作为瞄准点（飞行甲板纵向轴线和横向轴线的交点），且认为 $\sigma_1=\sigma_2=\sigma$，同时由于纵向和横向散布均方差相互独立所以 $\rho=0$，则母弹落点坐标（x，y）的概率密度函数为

$$f(x,y)=\frac{1}{2\pi\sigma^2}e^{-(x^2+y^2)/2\sigma^2} \tag{1}$$

根据圆概率偏差的定义可得

$$1-e^{-\frac{r^2CEP}{2\sigma^2}}=0.5 \tag{2}$$

由式（2）可得圆概率偏差和均方差的关系

$$\sigma=r_{CEP}/1.177\,4 \tag{3}$$

根据均方差的定义可得到

$$\begin{cases} X_i=X_m+Z_{1\sigma}=X_{mi}+Z_1\dfrac{r_{CEP}}{1.177\,4} \\ Y_i=Y_m+Z_{2\sigma}=Y_{mi}+Z_2\dfrac{r_{CEP}}{1.177\,4} \end{cases} \tag{4}$$

式中　X_i，Y_i——第 i 枚母弹的实际开舱抛撒点；

X_{mi}，Y_{mi}——第 z 枚母弹的瞄准点；

Z_1，Z_2——独立的标准正态分布随机数，且有

$$\begin{cases} Z_1=\sqrt{-2\ln r_1\cos(2\pi r_2)} \\ Z_2=\sqrt{-2\ln r_1\sin(2\pi r_2)} \end{cases} \tag{5}$$

式中　r_1，r_2——相互独立的在（0~1）区间服从均匀分布的随机数。

假设每枚母弹携带子弹数为 N，子弹在母弹抛撒圆域内服从均匀分布，以（X_i，Y_i）为圆心，抛撒半径为 R_y 产生圆域内均匀分布随机数作为子弹落点，则子弹抛撒落点坐标为

$$\begin{cases} X_{ij}=X_i+\gamma_1\cos t \\ Y_{ij}=Y_i+\gamma_2\sin t \end{cases} \tag{6}$$

式中　γ_1，t——（0，R_y），（0，2π）区间上服从均匀 分布的随机数；

x_{ij}，y_{ij}——第 i 枚母弹第 j 枚子弹的落点坐标。

根据大数定律：只有在相同条件下进行无数次的独立实验时，事件出现的频率才等于事件的概率．故只有模拟次数 n 达到一定次数时，其结果才能作为封锁概率。

3.1　起飞区封锁概率的计算

由于封锁子弹对 4 条起飞跑道的有效封锁区域存在重叠部分，根据概率知识可以得到起飞区的计算公式

$$\begin{aligned} P_{qf} &= P(A_{q1} \cup A_{q2} \cup A_{q3} \cup A_{q4}) \\ &= \sum_{j=1}^{4} P(A_{qj}) - \sum_{1 \leqslant i < j \leqslant 4} P(A_{qi}A_{qj}) + \\ &\quad \sum_{1 \leqslant i < j < k \leqslant 4} P(A_{qi}A_{qj}A_{Qk}) - P(A_{q1}A_{q2}A_{q3}A_{q4}) \end{aligned} \tag{7}$$

式中　$P(A_{qi}) = n_{qi}/n$，$(i=1, 2, 3, 4)$；

n_{qi}——n 次模拟中，第 i 条起飞跑道被封锁的次数；

$P(A_{qi})$——第 i 条起飞跑道的封锁概率。

3.2　停机区封锁概率的计算

对于停机区封锁概率的计算本文采用矩阵仿真法，其步骤为：

（1）对 A、B 停机区划分为若干个 1 m×1 m 的 小正方形网格，每个正方形的中心坐标代表该正方形的方位，同时分别建立两个大型矩阵 A_j 和 B_j，矩阵内的元素与网格数一一对应，每一个网格定义为一个矩阵元素，对矩阵元素进行初始化，赋值为 0；

（2）统计落入能够对停机区造成毁伤的有效区域内的子弹坐标；

（3）逐一确定每个小正方形的坐标与有效子弹坐标之间的距离，如果该距离小于子弹的毁伤半径，则该正方形被毁伤，将其初值 +1；

（4）分别统计 A_j 和 B_j 两个矩阵内非 0 的个数，即被毁伤的正方形个数，从而可以得到被毁伤的面积，然后除以其所对应停机区的面积即可得到该停机区的封锁概率。

由于 A、B 停机区在空间位置上相互独立，停机区的封锁概率为

$$P_{tj} = 1 - (1 - P_{tjA})(1 - P_{tjB}) \tag{8}$$

$$\begin{cases} P_{tjA} = \left. \sum\limits_{i=1}^{nA} P_{tjA/i} \right| n = \left. \sum\limits_{i=1}^{nA} \dfrac{A_{hsA/i}}{A_{tjA}} \right| n \\ P_{tjB} = \left. \sum\limits_{i=1}^{nB} P_{tjB/i} \right| n = \left. \sum\limits_{i=1}^{nB} \dfrac{A_{hsB/i}}{A_{tjB}} \right| n \end{cases} \tag{9}$$

式中　nA，nB——n 次模拟中命中 A、B 停机区的次数；

$A_{hsA/i}$，$A_{hsB/i,}$——一次命中 A、B 停机区的毁伤面积；

A_{tjA}，A_{tjB}——A、B 停机区的总面积。

3.3　调动区封锁概率的计算

根据调动区的封锁判据，调动区的封锁概率即不存在最小调动窗口（MDW）的概率为

$$P_{dd} = n_{dd}/n \tag{10}$$

式中　n_{dd}——n 次模拟中，不存在最小调度窗口的次数。

3.4　破片对靶板侵彻效果对比

综上可以得到航母飞行甲板封锁概率的计算表达式

$$P_{fs}=1-(1-P_{qf})(1-P_{tj})(1-P_{dd}) \tag{11}$$

为了体现出各个区域的重要性，区别各个区域作战性能的差异，在此本文引入区域权重系数 m。除了舰载直升机，其余固定翼飞机的起飞都必须利用起飞区的弹射器才能起飞，一旦起飞区被封锁，舰载机就无法作战，故本文定义起飞区的权重系数为 $m_q \leqslant 1$；停机区一般停放了约50%的舰载机，即使一轮打击将停机区内的飞机全部损毁，敌方仍可以从机库里调动飞机进行作战，故本文定义停机区的权重系数为 $m_{tj} \leqslant 0.5$；对于调动区，由于舰载机从停机区域转移至弹射器上都要经过调动区，一旦调动区被封锁，舰载机要完成起飞作战的任务将受到很大的限制，故本文定义调动区的权重系数 $m_{dd} \leqslant 0.75$。

考虑各区域的重要因子后，航母飞行甲板的封 锁概率

$$P_{fs}=1-(1-m_{qf}P_{qf})(1-m_{dd}P_{dd})(1-m_{tj}P_{tj}) \tag{12}$$

4 结束语

本文基于障碍型封锁子弹药对航母飞行甲板的封锁思想，展开了杀爆式障碍型子弹药对航母飞行甲板封锁效能的分析，提出了航母飞行甲板的空间封锁概率计算方法同时给出了计算公式。通过带入武器系统参数，如一波打击母弹数量，母弹瞄准点，母弹的 CEP，单枚母弹携带子弹数，子弹抛撒分布规律以及参数等；同时结合子弹药参数，如子弹对舰载机的毁伤半径等，利用本文所提出的方法，可以计算杀爆式障碍型封锁子弹药对航母飞行甲板的空间封锁概率。

参考文献

[1] 冯顺山，董永香．反航母侵立封子弹药技术总结报告（GF报告）[R]．北京：北京理工大学，2007－2012.
FENG S S, DONG Y X. The summary reports about the technology of submunitions attack aircraft carrier [R]. Beijing: Beijing Institute of Technology, 2007－2012. (in Chinese)

[2] 李真．串联嵌入弹的终点效应优化设计和数值模拟 [D]．南京：南京理工大学，2008.
LI Z. The terminal effects of tandem embedded projectile designing and numerical simulation [D]. Nanjing: Nanjing University of Science and Technology, 2008. (in Chinese)

[3] 张险峰，贾伟，陈万春．嵌入式封锁子弹药设计方法研究 [J]．北京理工大学学报，2012，32 (6)：561－564.
ZHANG X F, JIA W, CHEN W C. Structure design of embedded interdiction sub－munitions [J]. Transactions of Beijing Institute of Technology, 2012, 32 (6): 561－564. (in Chinese)

[4] 谭书怡，周中健，王树山．封锁航母飞行甲板弹作用原理试验研究 [C]//2005年弹药战斗部学术交流会．广州：[s. n.]，2005：186－189.
TAN S Y, ZHOU Z J, WANG S S. Research on the action principle experiment of interdiction munitions for the flight deck of aircraft carrier [C]//Proceedings of 2005 Academic Communication Meets of Munitions and Warhead. Guangzhou: [s. n.], 2005: 186－189. (in Chinese)

[5] 孙诗南．现代航空母舰 [M]．上海：上海科学普及出版社，2000.
SUN S N. Modern aircraft carrier [M]. Shanghai: Shanghai Popular Science Press, 2000. (in Chinese)

[6] 苏国华，舒健生，崔萍．多模弹药对机场跑道封锁时间的快速计算模型 [J]．火力与指挥控制，2011 (12)：64－66.
SU G H, SHU J S, CUI P. Fast calculation model of runways * blockage time based on multi－mold ammunition [J]. Fire control & command control, 2011 (12): 64－66. (in Chinese)

超近程反导武器系统探讨[①]

黎春林[②]，冯顺山

（北京理工大学爆炸灾害预防与控制国家重点实验室，北京　100081）

摘　要：简要介绍现代防空反导武器系统的发展概况，并分析研究了当前发展用于高价值目标防空反导的超近程反导武器系统的必要性，并对其使用特点进行了分析研究，提出了总体设想。

关键词：导弹防御；超近程；防空武器

Research on super close – in antimissile weapon system

Li Chunlin，Feng Shunshan

（BIT，National Key Lab of Prevention and Control Explosive Disasters，Beijing 100081，China）

Abstract：Briefly describes the development about the air defence and antimissile weapon systems. And analyzes the necessary to develop a super close – in antimissile weapon for the defence of high value objective.

And a tentative plan is given.

Keywords：Missile defence；Super close – in；Antiaircraft weapon

1　引言

当今，进攻性洲际弹道核导弹已成为国际间实现威慑力量平衡、确保互有把握摧毁的有力装备，战术导弹已成为地面机动发射平台、作战飞机、战舰乃至坦克的主要武器装备。在众多的导弹武器中，巡航导弹因其具有良好的通用性、超强的突防能力、远程精确打击能力和高效作战能力等特点，在近年发生的高技术现代局部战争中独领风骚，美、英等西方军事大国为了达到无风险作战的目的，大量使用了巡航导弹对高价值目标进行空中打击，使其成为现代战争中的重要空袭武器。

目前，导弹特别是战术巡航导弹的威胁日趋严重，对导弹的防御特别是高价值目标在未来战争中的生存问题已成为人们广为关注的焦点。为了在未来战争中夺取反空袭作战的胜利，大力发展反巡航导弹武器系统已是当务之急。

2　反导武器系统的发展概况

近年来，空袭兵器正在向隐身化、超距离打击和战术运用多样化方向发展，导弹的发展更是突飞猛进。而反导工作的研究也正随着新的作战需求不断推陈出新，出现了为数众多的反导武器系统。如美国的“爱国者”地对空导弹系统、英国的“海狼”舰载防空导弹系统、以色列的“巴拉克”舰载防空导弹系统等。

按防御区域分类，目前，导弹的防御可分为点防御和面防御，通常采用的反导方法有三种：①在导弹发射前，毁其于发射平台之上；②弹道拦截，主要采用分层拦截战术，由飞机空空导弹、面空导弹、电子对抗和点防御系统等，组成纵深梯次配置的拦截系统；③干扰导弹的目标标定、定位、导引和通信系统。反导的主要手段有：①电子对抗。采用电子干扰、电子假目标等方式进行对抗，一般用于远距离

① 原文发表于《现代防御技术》2003（Z1）。

② 黎春林：工学博士，2000年师从冯顺山教授，研究超近程反导武器系统及其战斗部技术，现工作单位：公安部一所。

（几十千米至几百千米）对抗；②假目标。采用金属箔条、红外假目标等，用于中近距离（1 km 至几十千米）；③地对空导弹。用于中近距离对抗；④防空截击机拦截。用于远距离拦截作战；⑤高射炮。用于低空（1～3 km）拦截等。

为应对新的反导作战形势的需求，反导武器系统的发展正趋向于以下几方面：①发展一体化的防空系统，寻求与反飞机、反战术弹道导弹共用探测器和拦截器，使未来的防空作战成为反飞机、反巡航导弹和反战术弹道导弹的一体化作战；②强调发展低成本防御系统，研制新型拦截装置，完善防御体系；③探索新原理、新技术在防空反导中的运用等；④重视开展国际合作，以减轻经济负担。

3　高价值目标的生存与发展超近程反导武器系统的必要性分析

近年来，发生的多场局部战争表明，高价值目标是战争中首先和主要被打击的对象，高价值目标的生存问题已成为当今防空反导问题研究的重要方向。

3.1　高价值目标

所谓高价值目标是指对战争胜负走向具有重要影响意义的目标群体，其中包括：①事关国家军事实力的重要组成部分，如航空母舰、大型战舰、军港、作战指挥中心、重要的通信中心及其网络、关键的军事交通中心及其网络、军用机场、导弹发射基地、飞机库、弹药库、油库、重要的防御阵地等；②事关国家政治实力的重要组成部分，如党政机关办公地、国家形象标志设施等；③事关国家经济实力的重要组成部分，如大型工厂、大型水库、大型发电站等；④事关国民生活安定的重要设施，如大中城市供电与供水网络系统、铁路与公路的重要地段及桥梁等，这些目标的价值是由其在战争中的地位与作用来确定的，并将随战争的发展发生变化。

3.2　高价值目标在战争中的生存评估

统观现有的防空反导系统，在以下几方面还存在明显的不足，在确保高价值目标生存方面还力不从心。

（1）早期探测导弹的能力差。在许多情况下，由于不可避免的气候影响、电子干扰、导弹的自身隐身性能的提高和其他工程因素，即使来袭的导弹以亚声速飞行，也难以在早期被发现。各种超声速巡航导弹的广泛使用更是大大缩短了导弹的飞行时间，降低了被发现的概率。同时，为能确定采用相应的对抗手段，必须对目标进行早期识别，鉴别来袭导弹的导引方式等，在今天的技术条件下实现是十分困难的。

（2）防导弹突防的能力有限。发达国家也非常重视导弹在作战中的生存问题，使其高度智能化，以增强导弹的突防能力，采取的主要措施有：

①采用先发现目标、隐蔽发射、超低空飞行、隐身技术、用红外或被动式雷达导引头、利用杂波干扰和人为干扰等措施提高导弹的隐蔽性；

②提高攻击速度，以快制胜，使敌防御系统没有采取有效防御措施的足够时间；

③提高机动性，使导弹在末段接近目标时跃起，然后再俯冲进行攻击或在距目标最后几千米时，导弹 10～15 的过载进行 S 形机动，进一步增大敌防御系统难度；

④采用全方位饱和攻击方式，利用敌防御系统同时跟踪目标的数量有限，发射多枚导弹从不同方位同时发起攻击；

⑤采用高空巡航、大攻角俯冲攻击，中高空巡航、低空突防，低空巡航、超低空突防等攻击方式等。这些措施的采用，使导弹的突袭方向、突袭时机变化多端，现有防御系统难以早期预警和实施有效的全方位防卫。

（3）现有防御系统的布防存在一定的缺陷。当前导弹的多层防御系统重点侧重于中远程拦截、近程摧毁，反导系统以战略纵深进行梯次布防，但毕竟防御成本大大高于进攻成本，系统数量有限，防御的空域有限。并且随着战争进程的发展不断消耗，导弹突防的概率随之增加。面对突防的导弹攻击，对高价值目标缺乏必要的点防御系统。

（4）现有反导系统防御的有效性差。由于来袭导弹的突然性和高速飞行，防御的有效时间短，应对仓促，在历次的武器性能试验中常有漏网之鱼。就是在海湾战争中，名噪一时的“爱国者”导弹对性能落后的“飞毛腿”导弹的拦截也并非万无一失。现有的毁伤拦截手段虽然有效，但在导弹发展日新月异的今天，新原理、新材料和新结构在导弹上得到了广泛应用，其对抗力已明显下降，拦截手段和毁伤方法急需创新。

3.3　超近程反导武器系统概念

超近程反导虽是一个新兴的反导概念，但在现代坦克防御中早已出现，是坦克主动防御反坦克导弹进攻的一种新技术概念，主要用于在几米范围内拦截来袭的反坦克炮弹（特别是空心装药战斗部）和导弹。早在1954年，Wales就为机械、磁学或光学炮管传感器系统注册了一项专利。在该方案中，利用起爆空心装药形成的线形高速射流，从侧面攻击来袭的空心装药战斗部，从而使得来袭炮弹或导弹早炸。俄罗斯很早就在T－55坦克上试验安装了DROZD系统，该系统有8个炮弹发射器，在炮塔上炮管的两侧各有4个。该系统发射弹药能在坦克前方7～8 m处爆炸并产生向外飞散的破片群，在1.5 m范围内破片密度能达到120枚/m^2，能够对付来袭速度在70～700 m/s的导弹或火箭。在1992年的迪拜展览会上俄罗斯首次展出了类似的ARENA主动防护系统。此外，德国、以色列等都在或多或少的技术基础上研制从可控制发射器中发射弹药对付来袭威胁的防御系统，所有这些都需要探测器、中心处理器和主动干扰装置。虽然不断有类似系统出现，但目前都还没有投入战场使用。

本文的超近程概念，我们将其定义为：导弹锁定目标后，对目标进行末端攻击时的弹目距离，一般在1 000 m以内。用于高价值目标防御的超近程反导防御系统指：布防在高价值目标周围，在覆盖高价值目标的几十米至几百米的空域内进行防御作战，根据获得的目标早期运动信息，预测其飞行轨迹，计算拦截点，采用快速反应的反导弹药系统，对突防的巡航导弹、激光制导炸弹和其他来袭目标进行毁伤。

3.4　我国发展超近程反导武器系统的必要性与可行性简析

我国防空反导体系的发展起步较晚，经多年的实践和射击试验表明，在超近程范围内，小口径弹药破片命中导弹后的杀伤效果不佳，产生的微小偏航作用不大，更不能直接引爆导弹，目前尚无有效的防御手段。为了应对各种突发事件和弥补近程防御体系的不足，为高价值目标（军事要地、党政机关等）在未来战争中的生存增加一道保护屏障，在现有对导弹的多层防御基础上研制超近程防御武器，对高价值目标提供超近程防御是十分必要的。

对于进入超近程范围内的导弹，因其已锁定目标，不再调姿，弹道固定，基本近似直线运动，故能被精确预测最佳拦截点（命中要害舱段的误差小于1 m），能充分发挥中口径战斗部的毁伤能力，使导弹解体或被直接引爆等，完成预定的作战任务。因此，对于高价值目标的固定空域防御来讲，采用提前探测、跟踪和锁定目标，定点拦截，利用中口径的快速反应弹药系统对导弹进行超近程拦截是可行的。

4　超近程反导武器系统组成与功能设置

4.1　超近程反导武器系统的组成

全武器系统主要由精确跟踪系统、指挥控制中心、武器平台与随动系统、发射装置和拦截弹药等组成，如图1所示。

（1）精确跟踪系统。该系统承担近距离的搜索、预警、快速跟踪和精确定位功能，主要包括精确跟踪雷达、红外探头、CCD探测装置以及必要的辅助跟踪设备。

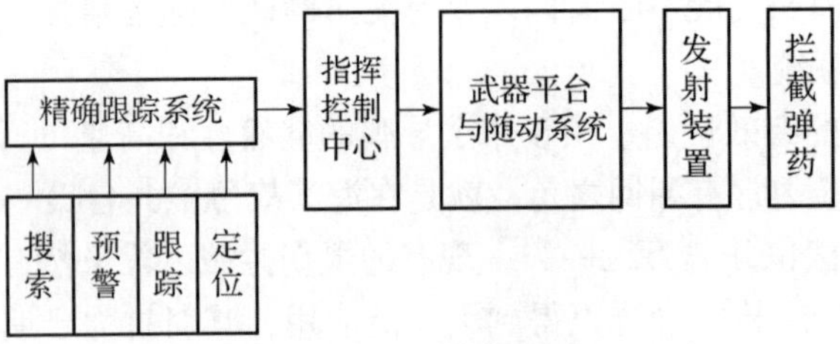

图1　武器系统结构图

（2）指挥控制中心。该中心为系统的信息处理、指挥控制部分，其硬件是高速处理数据的计算机。

（3）武器平台与随动系统。武器平台为发射装置和探测雷达等提供相对平稳的工作基台，随动系统精确执行指挥控制中心的控制指令，使发射装置快速转动，精确瞄准截击点。

（4）发射装置。采用多管集束弹药发射装置，对准截击点，平稳顺序发射。

（5）拦截弹药。拦截弹药是命中精度较高的弹药，具有对常规巡航导弹综合杀伤作用的高毁伤能力。

4.2　超近程反导武器系统的功能设置

超近程反导武器系统的功能设置为：

（1）该系统具有接收和处理上级防空反导系统提供的预警信息的能力；

（2）该系统具有独立的近距离快速探测、目标识别能力；

（3）该系统具有独立的指挥控制中心，能够快速对突发事件作出反应；

（4）该系统具有迅猛的反导弹药发射系统和随动系统；

（5）该系统机动灵活，便于野战布设；

（6）该系统的弹药毁伤目标的效能高，对来袭导弹具有毁灭拦截能力。

可按其中的最大数值进行归一化处理，由此就可求出最终目标的威胁等级评判结果。即

$$W_p = W'_p / \max(W'_p)$$

由此完成目标威胁等级的评判，从而对传感器资源和武器资源进行更加有效的分配或重分配，最终制定出正确的防御措施或进攻方案。

5　结束语

本文提出的超近程导弹防御系统对导弹的有效攻击距离定为近程防御系统的盲区，集目标探测技术、信息综合处理技术、伺服随动技术、多管发射技术、弹药毁伤技术等于一体，是一种新型防空反导武器装备。它的研制成功将部分弥补我国现代高技术战争中防空反导防御体系的不足，为高价值目标的防御增加一道新的防御措施。

参考文献

［1］张振鹏．国外反战术弹道导弹的现状与发展趋势［R］．北京：北京航空航天大学，1995.

［2］温德义．国外弹道导弹防御技术的发展及其对我国的威胁［R］．北京：中国国防科技信息中心，1995.

［3］瞿宝林．对巡航导弹防御的初步探讨［R］．北京：中国国防科技信息中心，1985.

第 二 部 分
目标毁伤及杀爆战斗部技术

爆炸冲击波毁伤效应的试验与计算①

蒋浩征，冯顺山②

摘　要：本文以SAM－7导弹战斗部为例，就爆炸冲击波对飞机目标的局部破坏作用进行试验研究；根据试验数据，用数理统计中的回归分析方法得出了工程计算的公式，可用来计算冲击波对飞机各要害部件或不同厚度的铝靶板的毁伤效应。

1　引言

在对付超低空入侵的飞机目标的防空武器中，单兵导弹占有重要地位。单兵导弹特点是全弹重量轻；机动性好。此外，由于采用红外导引，在目视条件或雷达等探测设备的监测下，命中精度较高。

为了充分发挥导弹有效载荷—战斗部的效率，目前单兵导弹皆采用综合作用型式（杀伤、爆破等）战斗部，当导弹以尾追或迎攻方式直接命中目标时，触发引信启动与传爆系列作用而引爆战斗部的爆药装药，壳体在爆轰产物作用下形成破片，在壳体破裂的同时，在空气中形成爆炸冲击波。由于这类型战斗部装填系数α较大（一般为0.4～0.5）而且在低空处爆炸，故爆炸冲击波是摧毁目标很重要的因素。而目前公开文献资料报道小装药量战斗部爆炸所形成的冲击波对飞机目标的局部毁伤效应是极其少的，因此很有必要研究小药量爆炸冲击波的毁伤效应。

2　战斗部爆炸所形成冲击波参数的计算

战斗部对飞机目标毁伤效应的理论计算或者试验研究，都必须知道离爆心不同距离处的冲击波的参数；通常采用冲击波的下列参数作为衡量毁伤效应的威力参数：冲击波起压ΔP_m心立击波正压作用时间t_+，冲击波比冲量i_+。

关于冲击波参数的计算方法，本文除采用国内较普遍采用的经验公式法外，还采用了数值计算法。现将SAM－7导弹战斗部裸装药与带壳装药两种情况的计算所采用公式与结果列举如下：SAM－7战斗部的主装药成分为纯黑铝（即A－Ⅸ－2），装药量为334 g，扩爆药成分为特屈儿，重量为34 g，其装药总重G_e为386 g。以TNT当量计，则G_e为570 g。

1. 经验公式法[1]

冲击波入射超压ΔP_m的计算公式采用有

Cagobckuu. M. A.　公式

$$\left.\begin{aligned}&\Delta P_m=10.7\frac{G_e}{R^3}-1(\mathrm{kg/cm^2})\left(\frac{R}{\sqrt[3]{G_e}}\leqslant 1\right)\\&\Delta P_m=0.76\frac{\sqrt[3]{G_e}}{R}+2.55\left(\frac{\sqrt[3]{G_e}}{R}\right)^2+6.5\left(\frac{\sqrt[3]{G_e}}{R}\right)^3\\&\qquad(\mathrm{kg/cm^2})\left(1\leqslant\frac{R}{\sqrt[3]{G_e}}\leqslant 15\right)\end{aligned}\right\}\quad(1)$$

---

① 原文发表于《爆炸与引爆技术》1983年第2期。

② 冯顺山：工学硕士，1980年师从蒋浩征教授，研究方向“杀爆战斗部威力及毁伤效应”。

Baker. W. 公式[2]

$$\left.\begin{aligned}&\Delta P_m=20.06\frac{\sqrt[3]{G_e}}{R}+1.94\left(\frac{\sqrt[3]{G_e}}{R}\right)^2-0.04\left(\frac{\sqrt[3]{G_e}}{R}\right)^3\\&\qquad\left(\frac{\text{kg}}{\text{cm}^2}\right)\left(0.05\leqslant\frac{R}{\sqrt[3]{G_e}}\leqslant 0.50\right)\\&\Delta P_m=0.67\frac{\sqrt[3]{G_e}}{R}+3.01\left(\frac{\sqrt[3]{G_e}}{R}\right)^2+4.31\left(\frac{\sqrt[3]{G_e}}{R}\right)^3\\&\qquad\left(\frac{\text{kg}}{\text{cm}^2}\right)\left(0.50\leqslant\frac{R}{\sqrt[3]{G_e}}\leqslant 70.9\right)\end{aligned}\right\}\tag{2}$$

式中 R——离战斗部爆心的距离，m；

G_e——战斗部装药重量（按 TNT 当量计，kg）。

SAM－7 战斗部爆炸后，离爆心不同距离处的冲击波入射超压 ΔP_m 的计算值见表 1。冲击波正压区作用时间 t_+ 及比冲量 i_+ 的计算公式采用文献［1］

$$t_+=1.35\times10^{-3}\sqrt{R}\sqrt[6]{G_e}(\text{s})\tag{3}$$

$$\left.\begin{aligned}&i_+=35\frac{G_e^{\frac{2}{3}}}{R}\left[\frac{\text{kg}-\text{s}}{\text{m}^2}\right]\left(\frac{R}{\sqrt[3]{G_e}}\geqslant 0.43\right)\\&i_+=15\frac{G_e}{R^2}\left[\frac{\text{kg}-\text{s}}{\text{m}^2}\right]\left(\frac{R}{\sqrt[3]{G_e}}\geqslant 0.43\right)\end{aligned}\right\}\tag{4}$$

表 1

		R/m	0.5	1.0	1.1	1.4	1.5	1.7	2.0	2.3	2.5	3.0	4.0	5.0
ΔP_m /(kg · cm^{-2})	裸装药	Cagobckuu 公式	40.7	6.54	5.16	2.9	2.47	1.86	1.31	0.98	0.83	0.58	0.35	0.24
		Baker 公式	29.1	5.1	4.1	2.35	2.02	1.54	1.1	0.84	0.71	0.51	0.31	0.22
	带壳装药	Cagobckuu 公式	20.9	3.6	2.84	1.67	1.4	1.1	0.8	0.61	0.53	0.38	0.28	0.17
		Baker 公式	15.4	2.9	2.3	1.39	1.2	0.9	0.68	0.53	0.45	0.33	0.15	0.15

SAM－7 战斗部爆炸后，距爆心不同距离处的冲击波 t_+ 的计算值见表 2。

表 2

	R/m	0.5	0.75	1.0	1.5	2.0	2.5	3.0	3.5	4.0	4.5	5.0
裸装药	t_+/10^{-3} s	0.87	1.07	1.23	1.51	1.74	1.94	2.13	2.3	2.5	2.6	2.75
	i_+/[(kg－s) · m^{-2}]	34.3	22.9	17.2	11.5	8.6	6.87	5.73	4.9	4.3	3.82	3.44
带壳装药	t_+/10^{-3} s	0.77	0.94	1.1	1.33	1.54	1.72	1.88	2.03	2.18	2.31	2.43
	i_+/[(kg－s) · m^{-2}]	21.1	14.1	1.05	7.02	5.3	4.2	3.5	3.01	2.63	2.34	2.1

2. 数值计算法

在参考 AD－759002[3]、文献［4］的基础上，建立爆炸冲击波参数计算的数学模型，编成了 BASIC

语言程序，在 MDR – Z_{80} 型微型计算机上进行数值计算。此法的优点是计算迅速，可同时计算得到6个冲击波参数，此外，还可计算冲击波波阵面内不同时间的超压值。

其计算原理是以裸状球形 TNT 炸药装药，重量为 0.454 kg（1 b）在海平面爆炸所测得的与计算的冲击波参数数据为基础，从装药表面到 2.342×10^3 m 的距离上，分为 108 个点的距离，列出冲击波参数数值。将各种爆炸条件（不同高度、带壳、战斗部各种装药形状与装药成分等）下的战斗部装药换算成裸状球形 TNT 装药在海平面爆炸所得到冲击波参数。对于任一给定的（不位于预定 108 点）距离 R 的 ΔP_m，t_+，i_+，可用插值公式求出。其公式[4]为

$$Y = Y_{j-1}\left(\frac{Y_j}{Y_{j-1}}\right)^F \tag{5}$$

$$F = \frac{1_n \dfrac{R_s}{R_{j-1}}}{1_n \dfrac{R_j}{R_{j-1}}} \tag{6}$$

式中　F——插值因子；

Y——所要计算的 ΔP_m，t_+，i_+ 值；

R_s——换算成海平面爆炸时的距离，m；

R_j、R_{j-1}——0.454 kg TNT 裸状球形装药在海平面爆炸时 108 距离结点中的两个距离（R_s 在其中间），m。

在空中爆炸的冲击波参数换算成海平面的爆炸的冲击波参数的换算公式为

$$R_s = R_H\left(\frac{G_{e.s}}{G_{e.H}}\right)^{\frac{1}{3}}\left(\frac{P_H}{P_s}\right)^{\frac{1}{3}} \tag{7}$$

$$\Delta P_{H.M} = \Delta P_{s.m}\left(\frac{P_H}{P_s}\right) \tag{8}$$

$$i_H = i_s\left(\frac{G_{e.s}}{G_{e.H}}\right)^{\frac{1}{3}}\left(\frac{P_H}{P_s}\right)^{\frac{2}{3}}\left(\frac{T_s}{T_H}\right)^{\frac{1}{2}} \tag{9}$$

式中　R——距离，m；

P——爆炸点的环境压力，kg/cm^2；

T——爆炸点的环境温度，K°；

s——下标，海平面状态；

H——下标，高空高度为 H 的状态。

在上述的计算中，对不同形状和不同成分、带壳的战斗部装药应乘相应的修正系数而换算成裸状球形 TNT 装药

$$G_{e.H} = K_1K_2K_3G_e \tag{10}$$

式中　G_e——战斗部原结构中的装药重量，kg；

K_1——装药形状的修正系数；

K_2——装药形状的修正系数；

K_3——带有壳体后的修正系数。

具体的K_1、K_2、K_3的修正数值见文献［4］。

由数值计算法所得出的带壳 SAM—7 战斗部爆炸后，在不同的高度 H 上，离爆心不同距离 R 处的冲击波入射超压 ΔP_m 和比冲量 i_+ 的部分数值见表 3。

表 3

距离 R/m	海平面 ΔP_m /(kg·cm^{-2})	海拔 50 m ΔP_m /(kg·cm^{-2})	海拔 50 m i_+ /[(kg-s)·cm^{-2}]	海拔 1 500 m ΔP_m /(kg·cm^{-2})	海拔 1 500 m i_+ /[(kg-s)·cm^{-2}]	海拔 2 500 m ΔP_m /(kg·cm^{-2})	海拔 2 500 m i_+ /[(kg-s)·cm^{-2}]
0.5	24.264	19.421	1.139	1.841	1.103	17.787	1.083
1.1	3.958	3.051	0.587	2.997	0.553	2.901	0.528
2.0	1.049	0.842	0.392	0.799	0.375	0.769	0.363
2.6	0.614	0.509	0.315	0.476	0.295	0.451	0.283
3.2	0.426	0.353	0.257	0.328	0.246	0.312	0.238

上述的冲击波参数的计算数值是为试验研究冲击波对飞机目标的局部作用时毁伤效应提供原始数据。

3 冲击波毁伤效应的试验研究

冲击波对目标的毁伤准则有超压破坏准则和冲量破坏准则。目前大多数公开文献资料报道的是大装药量冲击波对目标整体作用的数据，例如冲击波波阵面超压 ΔP_m 等于 1 kg/cm^2 时，各种类型飞机皆遭完全破坏，而小装药量的战斗部爆炸后所形成的冲击波作用于飞机目标时，即使冲击波波阵面超压 ΔP_m 等于 1 kg/cm^2 时，仅能使局部变形面达不到摧毁破坏，因此有必要试验研究冲击波局部毁伤效应。

用飞机上某些要害舱段进行大量试验虽有真实性，但因经济代价太高而不可能实现。考虑到现代军用飞机的结构为隔框和蒙皮组装而成，两者均为受力件，对于这种结构的强度可以换算成硬铝的等效强度。本试验采用一 定尺寸不同厚度的硬铝板为靶板，靶板边界固定，放置于裸装药柱爆炸产生的冲击波作用场中，观察其效应，测试其数据，并用飞机上尾翼进行对比试验。

对于一定厚度的铝板，必须承受到冲击波临界超压值 ΔP_{mb} 后才达到完全破坏。对于一定的装药量战斗部来说，根据冲击波参数的计算式（1）、式（2）或式（8），可以计算出冲击波超压值为 ΔP_{mb} 的距离。换句话说，在一定的装药量战斗部的爆炸作用下，对于一定厚度的靶板，存在着一个极限破坏距离；这就把冲击波超压确定靶板破坏准则问题变为确定极限破坏距离的问题。

试验采用 SAM-7 战斗部裸体装药 A-Ⅸ-2，重量 G_e 为 368 g，靶材采用 Ly-12 硬铝；在直径为 5 m 的爆炸筒内进行。实验装置见图 1。靶板用螺钉、压板进行周边固定，靶板尺寸主要采用 400×500 mm^2，厚度有 1 mm、1.5 mm、2.5 mm、3 mm。判别靶板完全损坏的原则为：显著的扯裂与变形；断裂的裂纹长度至少等于靶板的某一边界的长度，固定螺孔超过 2/3 以上拉断。

试验按优选法进行，首先在预测几个距离上，对两块不同厚（2、3 mm）铝板进行，从中找该铝板的极限破坏距离，将所得试验数据用无量纲之比得出极限距离经验公式。然后由经验式预测值安排不同厚度的铝板试验，并用飞机尾翼进行校核试验，对经验式修正而得出工程计算公式，图 2 是同一种厚度铝板在不同距离上实验结果。

表 4 列举了一些代表性的试验数据。

图 1　冲击波毁伤效应的实验装置

图 2 同厚度铝板在不同距离上的实验结果

表 4

序号	靶厚 b/mm	靶板面积 S /mm^2	炸距 R /m	靶板断裂纹总长 $\sum l_i$/mm	靶板凹陷的最大变形量 δ/mm	螺钉拉断数 n/个
1	3.0	350×500	0.5	420	210	9
2	3.0	400×500	0.5	930	250	9
3	2.5	400×500	0.6	400	130	14
4	2.5	400×500	0.6	330	140	13
5	1.5	400×500	0.73	340	110	11
6	1.0	400×500	0.92	380	160	11

4 冲击波毁伤效应计算公式

由上述试验数据经过处理可得知：冲击波超压 ΔP_m 与靶板厚 b、靶板强度 δ_b 之间存在着指数关系，见图 3。其关系式为

$$\Delta P_m = K(b\sigma_b)^{\frac{3}{2}} = 0.02(b\sigma_b)^{\frac{3}{2}} \tag{11}$$

式中 ΔP_m——冲击波入射超压，kg/cm^2；

b——靶板厚度，mm；

σ_b——靶板的静态抗拉强度极限，kg/mm^2，对 Ly－12 铝材，取 σ_b = 43 kg/mm^2；

K——系数，由试验确定为 0.02。

设

$$b\,\sigma_b = \overline{\sigma_b}(\text{kg/mm}) \tag{12}$$

式中 $\overline{\sigma_b}$——靶板的虚拟对比强度。

比冲量 i_+ 与靶板的虚拟对比强度 $\overline{\sigma_b}$ 之间关系曲线见图 4。其关系式为

$$i_+ = \begin{cases} 0.245\,\overline{\sigma_b} + 16(20 \leqslant \overline{\sigma_b} \leqslant 210) \\ 1.5\,\overline{\sigma_b} - 247(210 \leqslant \overline{\sigma_b} \leqslant 400) \end{cases} \tag{13}$$

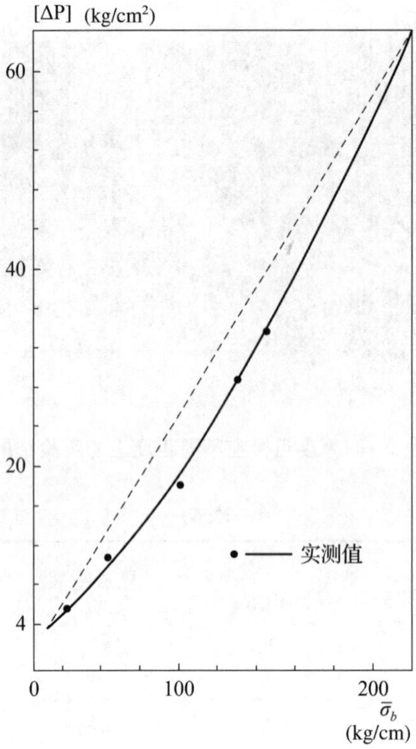

图 3

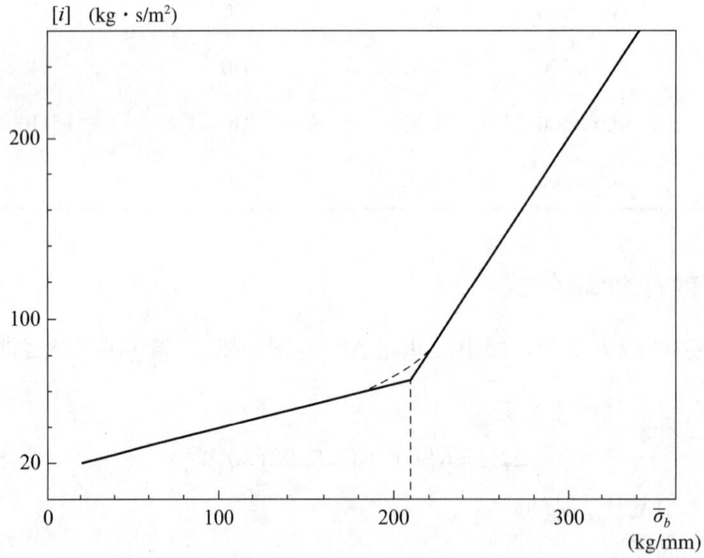

图 4

由式（4）、式（12）可得出极限破坏距离（R）与装药量 G_e、靶板虚拟对比强度$\overline{\sigma_b}$时关系曲线见图5，其关系式为

$$(R)=\begin{cases}\dfrac{35G^{\frac{2}{3}}}{0.245\,\overline{\sigma_b}+16}(20\leqslant\overline{\sigma_b}\leqslant 210)\\ \left(\dfrac{15G}{1.5\,\overline{\sigma_b}-247}\right)^{\frac{1}{2}}(210\leqslant\overline{\sigma_b}\leqslant 400)\end{cases}\tag{14}$$

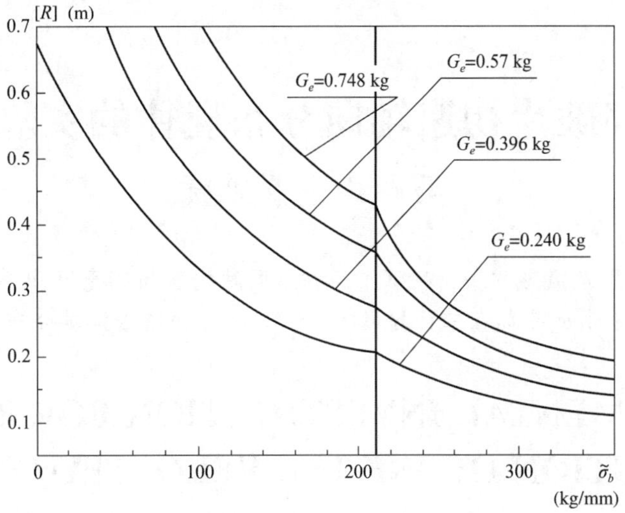

图 5

由式（12）计算得出的均方偏差（σ_b）为 0.022 m。

如在实际问题中给定距离（R）和靶板的虚拟强度，则所需战斗部的极限装药量（G_e）可由式（15）确定。

$$G_e=\begin{cases}\left[\dfrac{(R)}{35}(0.245\,\overline{\sigma_b}+16)\right]^{\frac{3}{2}}(20\leqslant\overline{\sigma_b}\leqslant 210)\\ \dfrac{(R)^2}{15}(1.5\,\overline{\sigma_b}-247)(210\leqslant\overline{\sigma_b}\leqslant 400)\end{cases} \tag{15}$$

上述式（11）、式（13）、式（14）、式（15）皆为由试验得出的冲击波对飞机目标局部破坏与毁伤的经验公式。

5　结束语

本文提出的爆炸冲击波对飞机目标的毁伤效应的计算式（9）、式（11）、式（12）、式（13）是在前人工作的基础上，经过一定的理论分析与较大量的试验结果而得出，用真实的飞机尾翼校正实验的数据吻合较好，可供战斗部设计时应用与参改。

在一定的装药 G_e 作用下，对一厚度 b 的靶板，在由试验确定的冲击波极限破坏距离 R 时所需超压 ΔP_m 和比冲量 i_+ 的数值；与在 G_e、R 一定时，由数值计算法和经验公式法中的 Baker 公式所计算得出的 ΔP_m 和 i_+ 基本相符。

应该指出：本文所得出公式是在特定的 SAM－7 装药量的条件下进行试验得到的，共适用范围经校核与分析可应用于装药量 G_e 为（$570\ g<G_e<5\times570\ g$），G_e 以 TNT 当量计。

由于试验仅限于一种硬铝材料，未考虑在近距离内装药形状对试验结果的影响；此外，鉴别靶板破坏的准则尚需进一步探讨。这些问题有待今后进一 步研究。

参加本次试验研究与计算工作的有崔秉贵、汪绍飞、魏晓涛、刘煜等同志和李立群、张贵江、施惠基、祝晓斌等同学。

参 考 文 献

[1] HENRYCH J. The dynamics of explosion and its use [M]. Amsterdam: Elsevier, 1979.
[2] BAKER W E. Explosions in air [M]. Austin: University of Texas Press, 1973.
[3] PROCTOR F J. Internal blast damage mechanism computer program: AD－759002 [R]. 1972.
[4] 殷海权，邬志兴. 炸药装药空爆时冲击波参数的数值计算 [J]. 兵工学报，1982 (2): 64－68.

战斗部破片初速轴向分布规律的实验研究①

冯顺山，崔秉贵

摘　要：本文用脉冲X光摄影方法，测试了几种不同结构装药钢壳战斗部的破片初速；经数据分析处理，获得一般装药结构杀伤战斗部在三种起爆方式下，破片初速沿弹轴分布的工程计算式。

ANEXPERIMENTAL INVESTIGATION FOR THE AXIAL DISTRIBUTION OF INITIAL VELOCITY OF SHELLS

Feng Shunshan，Cui Binggui

Abstract：Initial velocity of steel shells with various charge structures have been measured by using flash x－ray photography system. As the result of data processing，an engineering analysis for the distribution of initial velocities along the shell axis is presented. This approach applies to fragment warhead with general charge structures under three different initiation patterns.

1　实验方法

我们应用脉冲X光摄影技术，成功地实现了破片初速的测试。方法概述如下：①用X光摄影机拍摄战斗部爆炸后，壳体膨胀、破裂及飞散过程中多个不同时刻的图像。②在两张不同时刻的X光照片上，沿弹轴对应测量各破片移动的相对距离，由距离、时间和X光图像放大比例，便可准确地求得沿弹轴各破片在这一距离的平均速度值。

用于采集原始数据的X光照片必须是壳体已完全破裂形成破片，且破片所受合外力接近于零。因为：①壳体在爆轰产物或空气阻力作用下，明显加速或减速过程中所拍照片上的数据不是我们所要求的；②壳体破裂过程中，两块相邻的、未最后分离的破片，在运动时相互之间有明显的牵连作用（这种牵连作用主要与炸药种类、壳体材料和质量比有关）。试验表明：对于整体式钢质壳体弹药，爆炸后壳体膨胀至$3 \sim 3.5r_0$（装药半径）才满足上述要求。

一发轴对称试验弹，只要拍摄两张（或更多）不同时刻的X光照片，便可获得该弹在相应时刻内的一组（或多组）破片速度沿弹轴的分布值。用X光拍摄破片恰好达到初速时的照片是困难的，如果认为破片达到初速后短时间内破片初速不衰减（相应运动距离不大于$10\ r_0$），那么，由此期间拍摄的X光照片所测得的平均速度就可作为破片的初速。

被测试弹结构形状见图1.1，有三种结构：①壳体和装药均为圆柱形；②壳体（等厚壁）和装药均是圆锥形；③壳体外形是圆锥形（变壁厚）、装药为圆柱形。圆锥形壳体有四种锥度。每种弹测试三发。起爆方式为一端起爆。试验弹爆炸后的X光照片见图1.2。

从X光照片上，描绘同一发弹不同时刻的壳体破碎轮廓曲线即破片外缘曲线，测量对应于壳体不同位置的两轮廓曲线间的相对距离（图1.3），由两曲线间的时间差，便求得在这一距离上平均速度的分布规律。

①　原文发表于《兵工学报》1987年第4期。

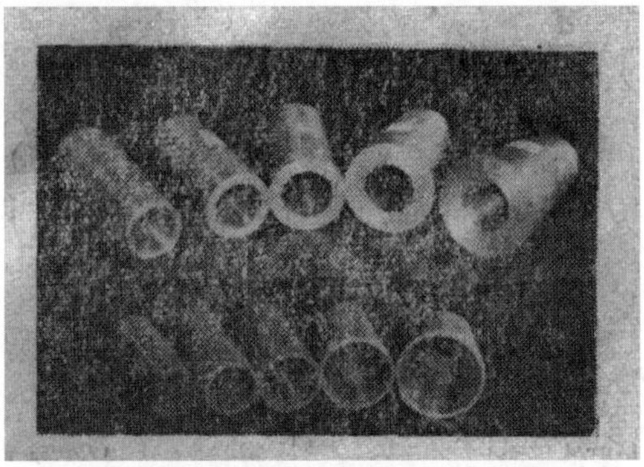

图 1.1 试验弹壳体照片，材料 45#钢，梯黑装药 50/50

图 1.2 圆柱形壳体爆炸后 40.5 μs 时的 X 光照片

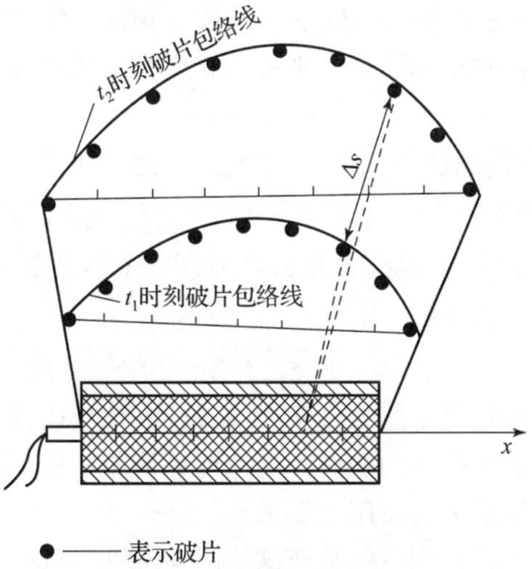

图 1.3 数据测量示意图

2 破片初速轴向分布规律

Gurney 方程只适于计算圆柱形战斗部破片初速，而战斗部有不同的结构形状，为计算非圆柱形（如锥形、腰鼓形、聚焦式）战斗部破片初速，必须建立一般结构形状战斗部破片初速轴向分布的工程公式。

本文处理问题的方法，是用 X 光摄影测试的结果来修正 Gurney 公式。本文认为一般装药结构战斗部破片初速轴向分布函数主要与破片位置 x 和对应的质量比 β 有关，即

$$v_{ox} = F(x/d, \beta) v_0 \tag{2.1}$$

分析试验数据可知，速度值发生在较大变化的位置在战斗部的两端，式（2.1）可写成

$$v_{ox} = F_1(x/d, \beta) \cdot F_2\left(\frac{L-x}{d}, \beta\right) v_0 \tag{2.2}$$

圆柱形装药结构的战斗部，其 β，d 均为常数，式（2.2）可写成

$$v_{ox} = F_1\left(\frac{x}{d}\right) \cdot F_2\left(\frac{L-x}{d}\right) v_0 \tag{2.3}$$

式中　$F_1(x/d)$——起爆端修正函数；

$F_2\left(\frac{L-x}{d}\right)$——非起爆端修正函数。

分析 X 光招牌你和从测得数据，可令

$$F_1(x/d)=1-A\mathrm{e}^{-Bx/d} \tag{2.4}$$

$$F_2\left(\frac{L-x}{d}\right)=1-C\mathrm{e}^{-D(L-x)/d} \tag{2.5}$$

式中，A、B、C、D 均为由试验数据确定的常数。

对于壳体和装药非圆柱形结构的战斗部，沿弹轴 β 为 x 的函数，即

$$\beta=\beta(x) \tag{2.6}$$

则 v_0 也不再是常量，是随 $\beta(x)$ 变化的：

$$v_0=v[\beta(x)] \tag{2.7}$$

如认为，沿弹轴微元形成的破片初速仍遵循 Gurney 公式的规律，式（2.7）可写成

$$v_0=\sqrt{2E_g}\sqrt{\beta(x)/[1+0.5\beta(x)]} \tag{2.8}$$

则初速 v_0 也是 x 的函数。

相对于圆柱形结构，非圆柱形轴对称装药结构，沿弹轴各截面的端效应是不同的。假定沿弹轴，变 β 与等 β（圆柱形壳体与装药）在 x/d 相同条件下，端效应相同。如锥形结构的起爆端与非起爆端仍用式（2.4）、式（2.5）、式（2.8）修正和计算，则式中 d 应为 x 的函数，即

$$d=d(x) \tag{2.9}$$

于是，一端起爆条件下，求解一般结构形状、两端敞口的杀伤战斗部破片初速的工程计算式为

$$v_{0x}=[1-A\mathrm{e}^{-Bx/d(x)}][1-C\mathrm{e}^{-B(L-x)/d(x)}]\sqrt{2E_g}\sqrt{\beta(x)/[1+0.5\beta(x)]} \tag{2.10}$$

由圆柱形装药结构的试验数据（9 发），根据方差最小原理，求得常数分别为：$A=0.361\,5$，$B=1.111$，$C=0.192\,5$，$D=3.030$，代入式（2.10），有

$$v_{0x}=[1-0.361\,5\mathrm{e}^{-1.11x/d(x)}][1-0.192\,5\mathrm{e}^{-3.030(L-x)/d(x)}]\sqrt{2E_g}\sqrt{\beta(x)/[1+0.5\beta(x)]} \tag{2.11}$$

只要知道战斗部的结构尺寸，就可求得 x 处的 $d(x)$、$\beta(x)$ 和该处破片初速。

对圆柱形装药结构、两种锥形装药结构的破片初速计算值与试验值的对比证明：式（2.11）的计算结果与试验值符合得很好，可以满足工程应用的要求。

由前面求得的修正函数，还可以得到一般装药结构战斗部两端起爆与中心起爆条件的破片初速轴向分布计算公式：

中心起爆：

$$v_{0x}=[1-0.192\,5\mathrm{e}^{-3.030x/d(x)}][1-0.192\,5\mathrm{e}^{-3.030(L-x)/d(x)}]\sqrt{2E_g}\sqrt{\beta(x)/[1+0.5\beta(x)]} \tag{2.12}$$

两端起爆：

$$v_{0x}=[1-0.361\,5\mathrm{e}^{-1.111x/d(x)}][1-0.361\,5\mathrm{e}^{-1.111(L-x)/d(x)}]\sqrt{2E_g}\sqrt{\beta(x)/[1+0.5\beta(x)]} \tag{2.13}$$

应用式（2.11）、式（2.12）、式（2.13）时，计算步长可根据弹丸结构的具体情况确定；整体式壳体，由壳体材料确定；预制或半预制式壳体，由破片尺寸确定。

3　结论

本文在分析和试验基础上给出了修正的计算式，计算值与试验值吻合得很好，它们适用于计算（预报）不同装药结构战斗部的破片初速轴向分布规律。

另外，本文试验中采用的是整体式、两端敞口、装药与端面平齐的试件；对于一端或两端有底盖，以及壳体长度大于或小于装药长度等情况下的初速分布规律，将在今后探讨。

小药量爆炸冲击波对飞机毁伤效应的研究[①]

冯顺山[②]，蒋浩征

摘　要：本文以小型导弹战斗部为例，就爆炸冲击波对飞机目标破坏作用和毁伤准则进行了试验研究，根据试验数据，用数理统计方法得出工程计算式，可用来计算冲击波对不同厚度铝板及飞机各舱段的毁伤效应。

1　引信

衡量爆炸冲击波威力效应常用冲击波对目标的毁伤准则（标准）来评定。反应冲击波毁伤准则的主要参数是超压 ΔP_m 和比冲量 i_+。确定爆炸冲击波的毁伤准则是一个较为复杂和困难的问题，这是因为：①有不同的爆炸源；②有各种各样不同特性的目标，且对冲击波的响应不同；③爆炸环境的不同。有关冲击波毁伤准则的实验和研究，前人已获得了许多有实际意义的研究成果，给出有关人、动物、建筑物、武器装备、飞机等在冲击波作用下使其毁伤的超压和比冲量准则。已公开发表的文献资料中给出的冲击波毁伤准则，大都是原子弹或大药量爆炸场对目标整体作用的实验数据和经验公式，据霍普金森爆炸相似理论可知，这些准则应用到小装药情况是不合适的。譬如，在空气中，冲击波波阵面的超压 ΔP_m 等于 1 kg/cm^2 时，各类飞机完全破坏；然而，在实验中看到诸如“SAM－7”这类小型战斗部（裸装）爆炸后即使冲击波波阵面超压 ΔP_m 约为 5 kg/cm^2，飞机遭受载荷的部位也仅能产生局部轻微变形，远达不到摧毁性的破坏程度。

地—空导弹和空—空导弹战斗部所配备的引信大多为非触发的，爆炸冲击波的传播距离增加而衰减较快，因此，通常不考虑爆炸冲击波对飞机目标的作用，主要考虑飞散破片对目标的杀伤作用。随着导弹制导技术的发展。有些导弹战斗部采用触发式（苏“SAM－7”，美“红眼睛”）和无线电近感式引信，使战斗部的杀伤威力有较大的提升。此类战斗部爆破作用占有重要地位，又是成为杀伤目标的主要作用。由于这些战斗部的装药是很有限的，爆炸冲击波只对目标的局部区域有效覆盖，在这种情况下，要准确计算战斗部对飞机上各舱段的毁伤效应和对目标的坐标毁伤概率，必须研究爆炸冲击波对目标的局部毁伤准则。

从事战斗部设计和威力效应研究者主要关心战斗部作用域目标上的威力参数、目标特性参数以及目标的最终毁伤效应。基于这种思想，本文以小型导弹战斗部为例，在地面就冲击波对飞机目标的局部毁伤作用和毁伤准则进行试验研究。根据试验数据处理得到工程计算公式，这些公式可用来计算冲击波对不同厚度铝板及飞机各舱段的毁伤效应。

2　试验方法

“局部”是一个相对的概念。对同一目标，部斗部有效范围不同，“局部”的含义不同，目标的毁坏情况也不相同。设某目标的一个投影面为长方体（图 1），其特征用 L 和 W 表示，图中阴影部分表示冲击波有效作用范围。图 1 中除（a）图外其余均为“局部”，相比较可看出（e）图中的阴影部分周围有约束边界，因此在该区域内形成毁坏的条件是最严格的，即典型的局部毁坏，本文研究的重点就是这种情况。

---

① 原文发表于《兵工学报弹箭分册》1987 年第 1 期。

② 冯顺山：工学硕士，1980 年师从蒋浩征教授，研究方向“杀爆战斗部威力及毁伤效应”。

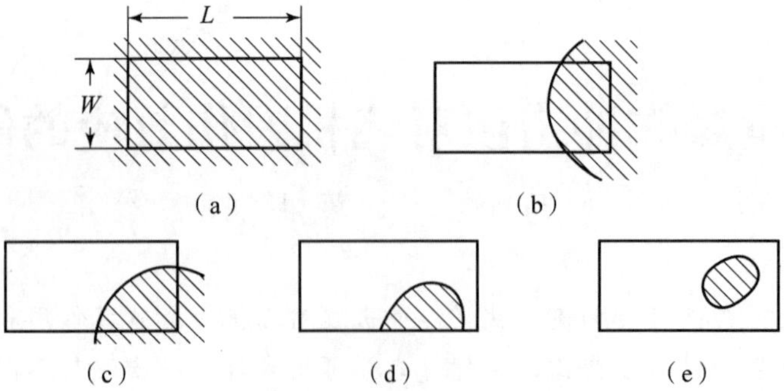

图1　局部作用示意图

制定毁伤准则的依据是，目标在爆炸冲击波作用场内的生存力研究。用飞机上某些舱段进行较大量试验，虽有其真实性却因经济代价太高而不能进行；又因飞机上各类舱段形状结构复杂，在空中爆炸载荷作用下的冲击响应具有多自由度，分析计算十分复杂；所以将问题化为单板单元处理，采用基本试验的观察方法。现代飞机的结构特点为蒙皮加隔框式，两者均为受力件，为简化问题，这种结构的强度可换算成硬铝板的等效强度。本实验采用一定尺寸的不同厚度的硬铝板作为靶板，靶板的边界固压在靶架上，置于裸装药柱产生的爆炸场中，观察其效应，测其数据，并用飞机尾翼进行对比试验。

对于一定厚度的铝板，必须承受到爆炸冲击波临界超压值［ΔP_m］或比冲量［i_+］后才会破坏。由爆炸冲击波参数计算公式可知，药量一定的战斗部在某个距离上可产生所要求的临界超压值［ΔP_m］或比冲量［i_+］，那么对一定厚度的靶板来说相应存在着一个临界毁伤距离。这样，就把冲击波毁伤靶板的准则（冲击波极限参数）问题变为首先确定临界毁伤距离［R］问题。

试验中主要采用长径比约为1：1的A—K—2裸装柱形药柱，重为368 g，装药密度是1.7 g/cm^3；靶材采用LY－12硬铝板，在直径为6 m的爆炸筒内进行试验。实验装置见图2，靶板用14个M10螺钉和压板进行周边固定。靶板尺寸主要是：400×500 mm^2；厚度主要有：1、1.5、2.5、3 mm。用压电晶体测压传感器进行超压测试，因传感器的强度问题并没有实测靶板表面处的入射超压；仅测试了距弹1 m以外的不同距离处的超压，其目的在于验证Baker超压计算公式的精度，考核其计算值与实测值的吻合情况（图2），以便用计算公式较精确获得靶板上入射冲击波超压的峰值。试验按优选法进行，以便能较快地找到某一厚度靶板的临界毁伤距离。

图2　试验中采用的靶架，弹悬挂在靶板上方其轴线与靶板平行

3　实验结果

靶板的毁伤特点为：①所有靶板呈凹形，只是炸距不同，凹的程度不同；②撕裂、断裂部分主要发生在与靶架固联的边界处，并且首先发生在边界的中心处和角上，然后沿边界扩延；③靶板上螺钉孔处发生断裂。靶板的毁坏规律基本相同于文献［4］中所讲的金属靶板的破坏规律。该文献的结论中指出靶架的作用起到应力提升器的作用，边界的中心处为应力最高点。

确定靶板（目标）临界毁伤标准是研究目标易损性的关键。从弹药工程角度，根据试验和对飞机舱段的易损性分析，本文提出判断靶板临界毁伤标准为：①显著的撕裂与变形，断裂裂纹长度总和 $\sum l_i$（不计螺钉孔的断裂长度）应满足关系式 $\frac{2}{3}\cdot\frac{L}{4}\leqslant\sum l_i\leqslant\frac{L}{4}$，$L$ 为靶板周长。②固定螺钉孔拉断数量不小于全部螺钉孔数的2/3。本文认为达到这两条标准，就可确认为靶板（目标）已处于临界毁伤状态；此时，它们已失去了其应起的基本作用（譬如作为受力件、气动效应件以及保护件等）。把符合标准的试验数据列于表1，表中的断裂总长不计螺钉孔处的断裂纹长度。确认为临界毁伤的部分靶板见图3、图4。

表1　符合标准的试验数据

序号	炸药 /g	靶板厚 /mm	靶板面积 /mm^2	炸距 /m	断裂总长 /mm	凹陷最大深度 /mm	螺钉孔拉断个数
1	368	3.0	350×500	0.50	420	210	9
2	368	3.0	400×500	0.50	460	250	9
3	368	2.5	400×500	0.60	400	130	14
4	368	2.5	400×500	0.60	330	140	13
5	368	1.5	400×500	0.73	340	110	11
6	368	1.0	400×500	0.92	380	160	11
7	368	1.0	400×500	0.92	360	150	10

图3　2.5 mm靶板的毁伤情况有Δ示意的为临界毁伤靶板

图4　3 mm 靶板放在不同炸距下的毁伤情况

处理数据时所用的基本公式[2,3]

$$\Delta p_m = \begin{cases} 20.06(\bar{R})^{-1} + 1.94(\bar{R})^{-2} - 0.04(\bar{R})^{-3} & 0.05 \leqslant \bar{R} \leqslant 0.5 \\ 0.67(\bar{R})^{-1} + 3.01(\bar{R})^{-2} - 4.31(\bar{R})^{-3} & 0.5 \leqslant \bar{R} \leqslant 7.09 \end{cases} \tag{1}$$

$$i_+ = \begin{cases} \dfrac{AG_e^{2/3}}{R} & 0.05 \leqslant \bar{R} \leqslant 0.5 \\ \dfrac{15G_e}{R^2} & 0.5 \leqslant \bar{R} \leqslant 7.09 \end{cases} \tag{2}$$

$$\bar{R} = R/\sqrt[3]{G_e}, \quad A = 34 \sim 36$$

式中　R——炸药中心距目标表面的距离，m；

G_e——战斗部装药（以 TNT 当量计算）重量，kg；

$\bar{R}$——对比距离。

处理数据时的最大允许偏差$|\sigma_R|$由式（3）表示

$$|\sigma_R| = \left|\frac{[R]_1 - [R]_2}{[R]_1}\right| \leqslant 0.05 \tag{3}$$

式中，下标 1、2 分别表示计算值与实测值。

由式（1）、式（3）二式，并采用数据处理与试验交叉进行信息互馈法[6]，得到临界超压［ΔP_m］$_s$与靶板厚度 h_m，靶板强度 σ_b之间的指数关系式

$$[\Delta P_m]_s = K(\sigma_b \cdot h_m)^{3/2} \quad 20 \leqslant \sigma_b \cdot h_m \leqslant 750 \tag{4}$$

式中　$[\Delta P_m]_s$——冲击波毁伤靶板的临界入射超压（kg/cm^2），下标 s 表示海平面；

σ_b——靶材的静态抗拉强度极限，kg/mm^2；

h_m——靶板的厚度，mm；

K——系数，由试验确定为 $K = 0.02$。

根据霍普金森爆炸冲击波比例定律[2]式（4）用于不同重量装药时应写成

$$[\Delta P_m]_s = 0.14\left(\frac{\rho_{\varpi i}}{\varpi_j}\right)^{\frac{1}{3}}(\sigma_b \cdot h_m)^{3/2} \tag{4$'$}$$

式中　ω_i——某种装药重量（TNT 当量）；

$\rho_{\varpi i}$——单位体积装药重量。

如令 $\tilde{\sigma}_b = \sigma_b \cdot h_m$，由式（1）、式（2）、式（4）和回归分析方法得到比冲量毁伤准则

$$[i_+]_S = \begin{cases} 0.245\tilde{\sigma}_b + 16 & 20 \leqslant \tilde{\sigma}_b \leqslant 210 \\ 1.5\tilde{\sigma}_b - 247 & 210 \leqslant \tilde{\sigma}_b \leqslant 750 \end{cases} \tag{5}$$

式中　$[i_+]_S$——冲击波毁伤靶板的临界入射比冲量（$kg \cdot s/m^2$）。

同样根据霍普金森爆炸冲击波比例定律和式（2）、式（5）得到目标距弹的临界毁伤距离（范围）$[R]$。

$$[R]_S = \begin{cases} \dfrac{AG_e^{2/3}}{0.245\tilde{\sigma}_b + 16} & 20 \leqslant \tilde{\sigma}_b \leqslant 210 \\ \left(\dfrac{15G_e}{1.5\tilde{\sigma}_b - 247}\right)^{\frac{1}{2}} & 210 \leqslant \tilde{\sigma}_b \leqslant 750 \end{cases} \tag{6}$$

如给定弹目距离 R_s 和 $\tilde{\sigma}_b$，则毁伤目标需要的临界装药重量 $[G_e]_S$ 由式（7）确定

$$[G_e]_S = \begin{cases} \left[\dfrac{R_S}{A}(0.245\tilde{\sigma}_b + 16)\right]^{\frac{2}{3}} & 20 \leqslant \tilde{\sigma}_b \leqslant 210 \\ \dfrac{R_S^2}{15}(1.5\tilde{\sigma}_b - 247) & 210 \leqslant \tilde{\sigma}_b \leqslant 750 \end{cases} \tag{7}$$

如给定 G_e、σ_b 及 R_s，则可以得到毁伤靶板的极限厚度 $[h_m]_S$ 为

$$[h_m]_S = \begin{cases} \dfrac{1}{0.245\sigma_b}\left(\dfrac{AG_e^{2/3}}{R_S} - 16\right) & 20 \leqslant \sigma_b[h_m]_S \leqslant 210 \\ \dfrac{1}{1.5\sigma_b}\left(\dfrac{15G_e}{R_S^2} + 247\right) & 210 \leqslant \sigma_b[h_m]_S \leqslant 750 \end{cases} \tag{8}$$

式（4）、式（5）、式（6）、式（7）、式（8）称为冲击波对目标局部毁伤准则的数学表达式。对于高空，各种装药密度和战斗部有金属外壳等条件下，计算对目标的毁伤效应，须将各式乘以各相应的修正函数。由式（4）′、式（6）绘制的曲线与实测值比较见图5、图6、图7。

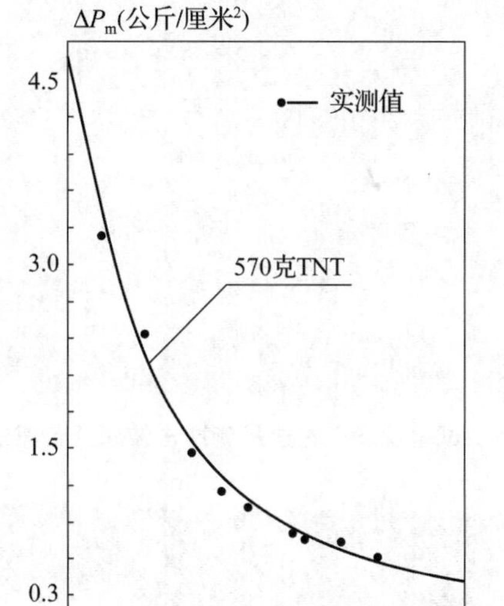

图5　ΔP_m—R 曲线与实测值比较

用表达式的预测值安排等效厚度分别为3.3、5.5 mm厚的飞机垂直舱面和垂直安定面的冲击波毁伤试验，试验结果表明目标的毁伤程度基本达到了要求见图8、图9、图10，从而说明了“准则”具有一定的实际应用价值。

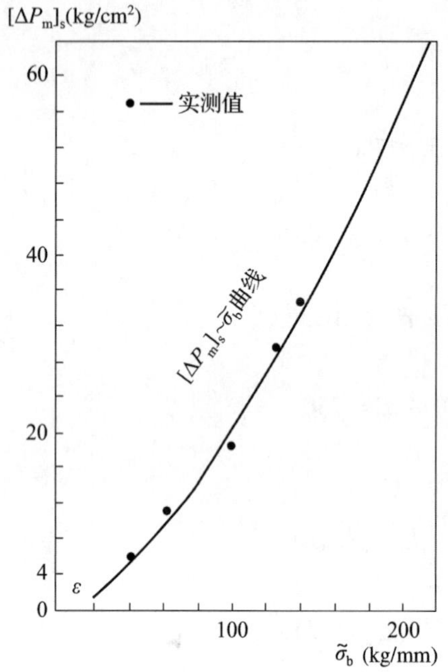

图 6　$[\Delta P_m]_s \sim \tilde{\sigma}_b$ 曲线与实测值比较

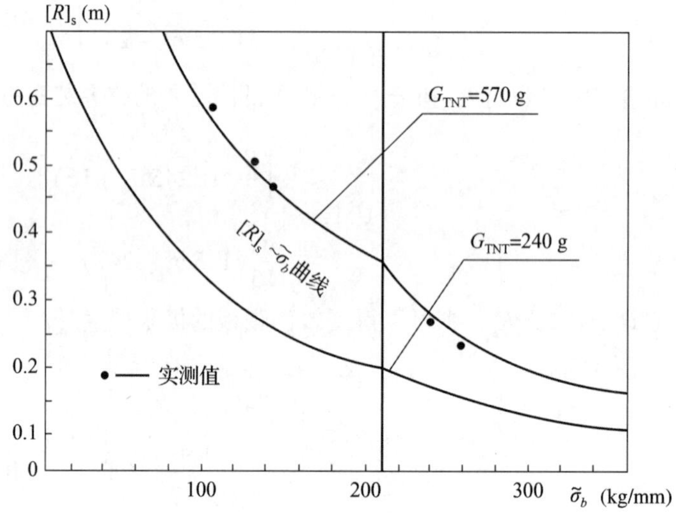

图 7　$[R]_s \sim \tilde{\sigma}_b$ 曲线与实测值比较

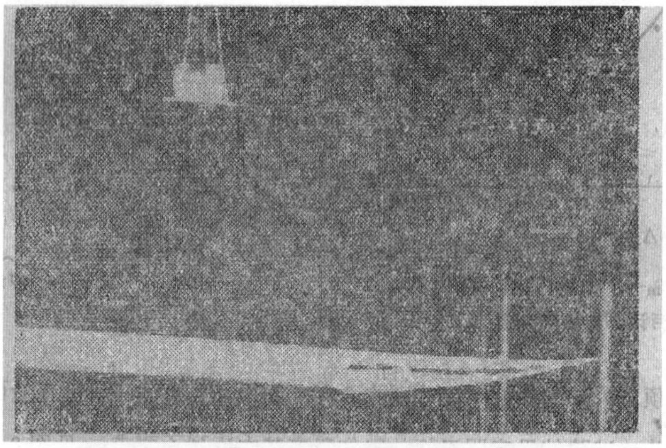

图 8　试验现场，右边是测冲击波超压传感器

图 9　飞机垂直舱面的毁伤情况，蒙皮从框架上撕裂，舱面在支撑处弯曲，炸心距舱面 0.47 m，炸口 1.0×0.5 m^2

图 10　飞机垂直安定面毁伤情况，可以看到主大梁基本断裂

4　结束语

本文给出的“准则”数学表达式是通过一定的理论分析与较大量试验结果得到的，与实验吻合较好，可供部斗部设计时应用。

依据爆炸比例定律“准则”表达式可适用于不同重量装药，至少可适用小于 100 kg 的装药情况。药柱长径比较大和药柱轴线垂直目标爆炸情况下，应用“准则”表达式要考虑加修正系数。

参考文献

[1]《爆炸及其作用》编写组．爆炸及其作用：第十二章［M］．北京：国防工业出版社，1979.
[2] BAKER W E. Explosions in air［M］. Austin: University of Texas Press, 1973.
[3] HENRYCH J: The dynamics of explosion and its use［M］. Amsterdam: Elsevier, 1979: 153.
[4] ROSS C A. Failure of metal and composite plates subjected to blast loadings, 1977.
[5] 中国科学院数学研究所统计组．常用数理统计原理［M］．北京：科学出版社，1974.
[6] 冯顺山，蒋浩征．超低空导弹战斗部的威力效应与最优设计．北京工业学院八系，1982.12

HIGH OBLIQUITY IMPACT OF WU SPHERE ON THIN PLATES①
钨球对薄靶板的大倾角撞击

Sun Tao②, Feng Shunshan, Jiang Jianwei, Sui Jianhui, Qi Baohai
Department of Mechanical Engineering of Beijing Institute of Technology Beijing 100081, P. R. China
Prof. Pei Sixing
Department of Mechanical Engineering of Taiyuan Institute of Machinery Taiyuan 030051, P. R. China

ABSTRACT: This paper presents a combined experimental – computational study of the impact) of Wu spheres on thin IY12C7. Aluminum and A3 steel plates at different oblique angles of 70°, 75° and 80°. The aluminum plates were nominally 2 mm thick, the steel plates were nominally 1 mm and 2 mm thick. The ricocheting phenomena were chiefly studied and the ricocheting velocities were determined according each oblique angle by experiment. The combination of velocities and materials resulted in plate perforation for some cases. Some plates were cut to allow an examination of their cross – section. 11e simulation were carried out with two methods, the first employing the rigid body model, the second with the AUTODYN – 3D code. The results of simulation agree quite well with experiments.

1 Introduction

Normal impact of projectile on targets is well studied in a wide range of targets thickness, projectile shapes and impact velocities. In contrast, oblique impact studies are more scarce and more difficult to characterize because of the occurrence of projectile and target deformation on most cases. The process of ricocheting differs in characteristic manner from simple reflection in dependence on whether plastic deformation of projectile and/or target and projectile break – up dominate [1]. Hence, the methods in these ricocheting studies were experimental techniques.

The dynamics of the impact of hard spheres against inclined thin mild steel and aluminium alloy plates were investigated in [2]. An analytical simulation was used by rigid body response of the sphere. The validation of this model for the case of ricocheting has been reported for special cases for velocities up to 1 000 m/s. Sun [3] also used the rigid body equations to analysis the harden Wu spheres impacted thin spaced layered aluminium targets with different oblique angles.

In this paper we applied the method of [3] in analysis the hard Wu sphere ricocheting on soft aluminium and A3 steel plates for the impact velocities under 1 000 m/s. While for the high striking velocity, the impact causes extensive projectile deformation and targets crater or perforation. the simulations were carried out with AUTODYN – 3D code. The experiments were supported by above numerical analysis.

---

① 原文发表于《第十六届国际弹道会议》1996 年。

② 孙韬：工学博士，1995 年师从冯顺山教授，研究反舰导弹结构解体毁伤效应，现工作单位：陆军研究院炮兵防空兵研究所。

2 Experimental Results

Projectiles are 4. 70 mm diameter tungsten alloy (93% – Wu – Ni – Fc) spheres. The aluminium plate is 2 mm thick while the A3 steel plates are of two kinds of thickness of l mm and 2 mm thick. Projectile is launched using a 12. 7 mm smooth bore powder gun. The impact velocities are measured and complemented to the surface of the plate with PM6680B high resolution programmable time/counter. The angles of obliquity for aluminium plate are 70R. 75° and 80°, for A3 steel plate are 70R and 80° (angle between line of fire and the normal of the target plate). Table l summarizes the static and dynamic material properties of projectile and target materials applied in the oblique impact experiments and numerical simulation as well. It contains the yielding stress σ_Y (or $\sigma_{0.2}$), elasticity module E, poisson number μ and failure strain ε_f.

Table l Material properties

Material	WU	LY12CZ	A3
ρ [g/cm^3]	18. 0	2. 70	7. 80
E [Mbar]	3. 02	0. 71	2. 16
μ	0. 30	0. 30	0. 30
σ_Y [MPa]	790	284	235
ε_f (%)	12. 0	280	45. 0

Figure 1 gives experimental results among the oblique angle and the ricocheting velocity. It denotes that all the impact of Wu spheres on the thin 1 mm or 2 mm A3 steel plate creates ricocheting for the oblique angle of 80° until the sphere break – up.

Figure 2, Figure 3 depict sone typical front sides and cross sections of craters.

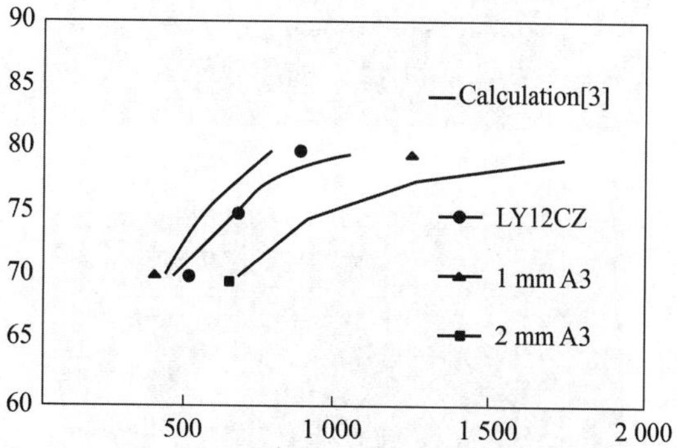

Figure 1 Oblique angle and ricocheting velocity

The experimental results show that the combination of velocities and materials results in plate perforation. This phenomenon also has been discovered in [4]. For the high oblique angle of 80°, ricocheting with perforation occurs when the impact velocity between 860 m/s and 920 m/s and up 920 m/s only creates perforation for LYl207. plates. While for 2 mm A3 targets, ricocheting with perforation happens when the striking velocity between 1 200 m/s and 1 570 m/s, above 1 570 m/s only ricocheting. It is estimated that at high impact velocity the sphere deforms seriously up to break – up. the rough surface of the crater in Figure 3 proves this at some extent.

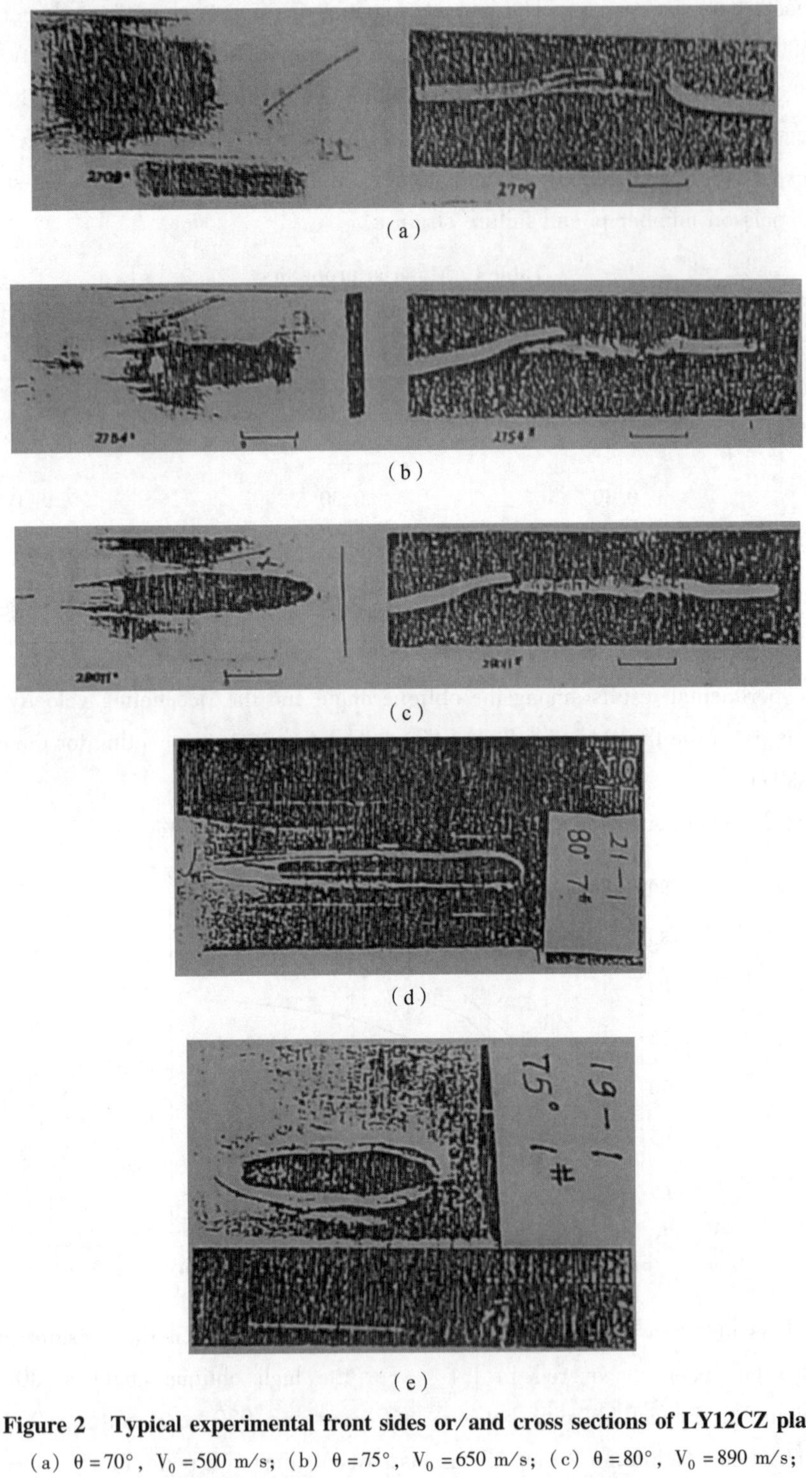

Figure 2 Typical experimental front sides or/and cross sections of LY12CZ plate

(a) $\theta=70°$, $V_0=500$ m/s; (b) $\theta=75°$, $V_0=650$ m/s; (c) $\theta=80°$, $V_0=890$ m/s; (d) $\theta=80°$, $V_0=1\ 010$ m/s; (e) $\theta=75°$, $V_0=1\ 720$ m/s

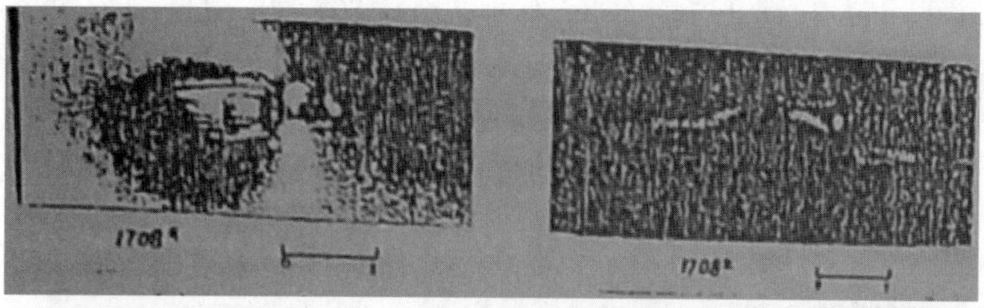

(a)

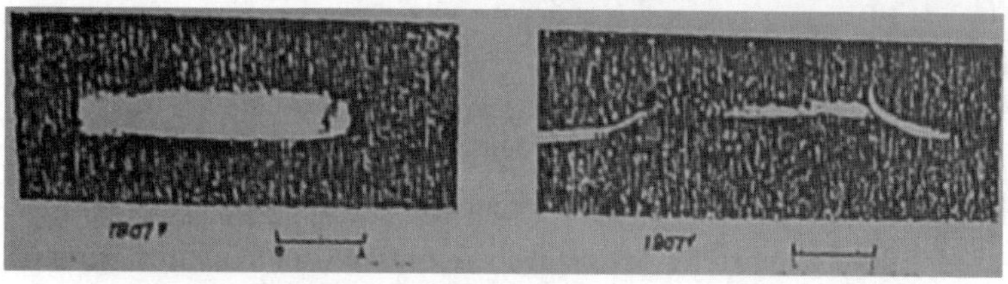

(b)

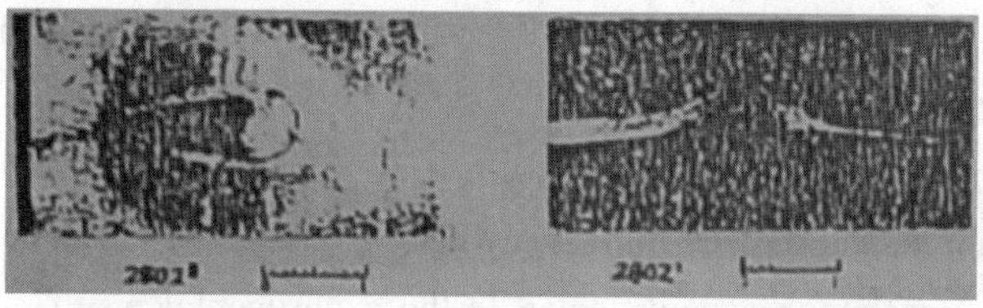

(c)

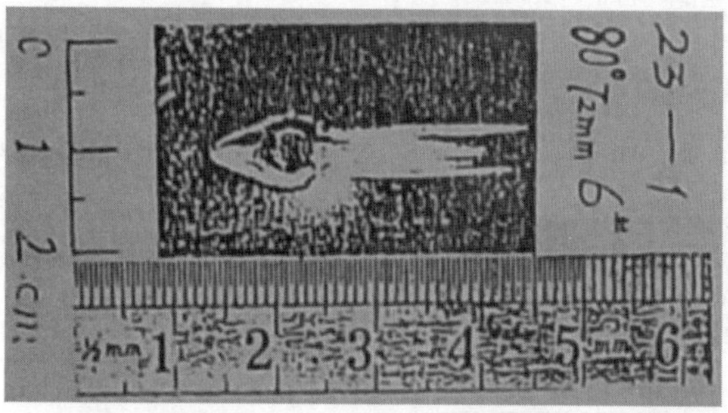

(d)

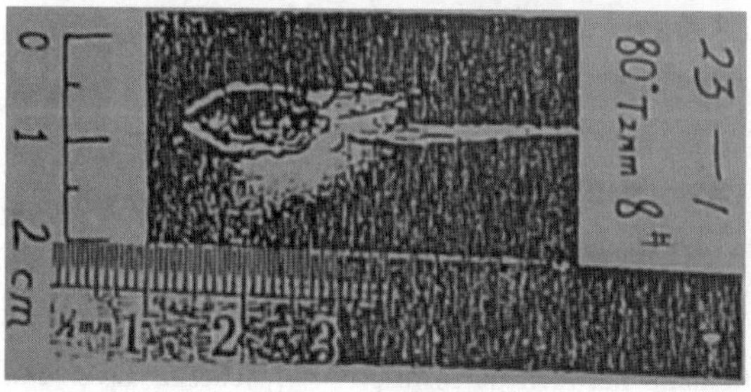

(e)

Figure 3　Typical experimental front sides and cross sections of A3 steel plate

(a) $\theta=70°$, $V_0=400$ m/s, 1 mm A3 steel plate; (b) $\theta=80°$, $V_0=1\ 050$ m/s, 1 mm A3 steel plate; (c) $\theta=80°$, $V_0=1\ 170$ m/s, 2 mm A3 steel plate; (d) $\theta=80°$, $V_0=1\ 440$ m/s, 2 mm A3 steel plate, front side; (e) $\theta=80°$, $V_0=1\ 750$ m/s, 2 mm A3 steel plate, front side

If the ricocheting takes place, the crater depth and length for Wu harden spheres on LY12CZ aluminum plates is greater than on A3 steel plates at the oblique angle of 80°. In contrary to the results obtained for the oblique angle of 70°, the crater length for 2mm aluminum and A3 steel material shows little dependence on impact velocity below 700 m/s. It should be noted that the crater length of Wu sphere on 2 mm A3 plate is nearly same or decreased with the impact angle increasing from 70° to 80°. In addition to this, exit direction of the ricocheting spheres is closely associated with the crater contours [1], Figure 2 and Figure 3 slows the obviously difference for two materials. Even though the yielding stress of two target materials differs few, the density of A3 steel is 3 times more than LY12CZ aluminum. For the same impact energy and direction, Wu spheres require more energy to create much large crater or ricocheting.

3 Numerical simulations

3.1 Rigid Body Model

Sun [3] has set up the two dimensional rigid body equations governing the translation motion of the center of the sphere. The calculation is carried out using the sane force coefficients mentioned in [3] either for LY12CZ or A3 targets in our paper. Only the number u, is changed from 0.34 for LY12CZ material to 0.40 acting on A3 steel to correspond with the experimental results. Value μ_0 determines the couple and lift force acting perpendicular to the velocity vector [4], the bigger the value lead to the bigger side direction effective forces, the crater length and depth qualitatively denote this.

Figure 1 shows the calculated results compared with experiments. The simulations for 2mm A3 plate with oblique angle of 80° shows that all the impact velocities will cause ricocheting phenomena on the surface of plate. We order the final calculated ricocheting velocity equal lo the experimental result making the sphere break - up.

Figure 4 compares the experimental perforation or ricocheting layers for spaced layered LY12CZ targets arranged same as [3] with computations for oblique angle of 70°. We can see that when the impact velocity increases the deviations between calculated and experiments ad up especially for impact velocity up to 1 000 m/s. Even though the sphere has been deformed, the capacity of perforation increases more than rigid body.

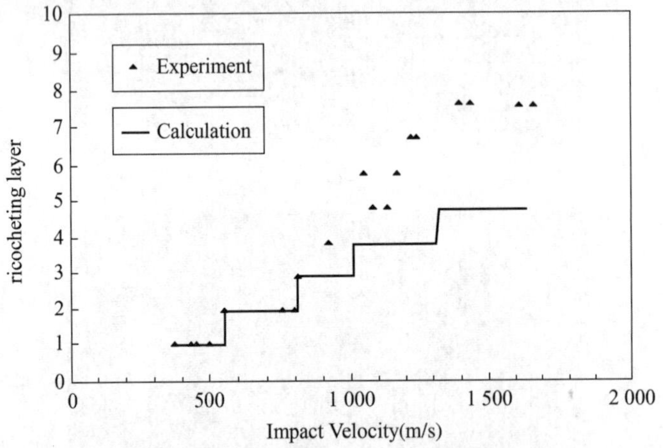

Figure 4 Relationship between perforation or ricocheting layer and impact velocity

3.2 AUTODYN Simulation

AUTODYN is used to simulate the impact problem. AUTODYN [5] includes a 3D Lagrange solver with eroding elements, which can handle material distortion and material erosion, is used in this study. We take advantage of the symmetry in the problem to reduce the size of the computational mesh: the problem is symmetrical with respect to a plane normal to the plate surface and containing the center of the sphere.

Figure 5 shows the initial configuration of the half symmetric idealization of the numerical example. The 3D calculation used 7 986 nodal points and 5 540 volume elements, for the sphere (726 nodes) and for the plate (7 260 nodes). The sphere and the plate are allowed to interact through a dynamically defined eroding impact surface. As elements reach a limiting value of geometric strain they are transformed (eroded) into free mass points. The eroded nodes can continue to internet with the existing structures with the full momentum conservation [4].

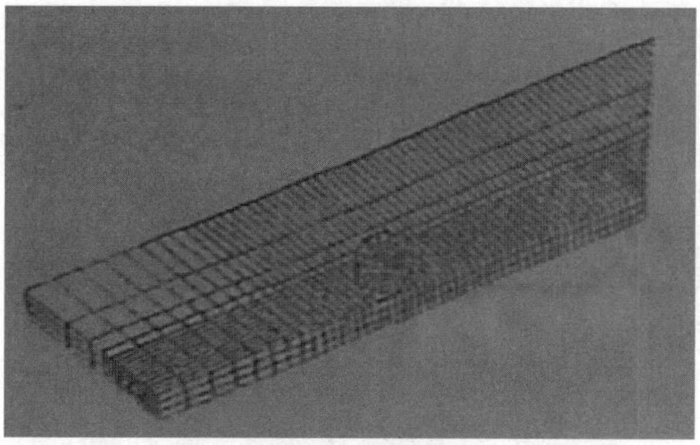

Figure 5 The initial configuration for sphere and plate

The materials are modeled by Shock Hugoniot linear equation, the main coefficients of the EOS for three materials are included in Table 2 and the elasto – plastic behavior is modeled by a von – Mises yield condition and for all materials we assume a failure mechanism based on maximum allowed effective plastic strain. Figure 6 shows the simulation results for Wu sphere striking 2 mm L. Y12C7Z. plates for selected times. Figure 7 gives some particular selected times for different oblique and striking velocity. Each figure shows a cross – section viewed for comparison with the photographs. The general shape agrees quite well with the photographs for different impacting cases.

(a)

(b)

(c)

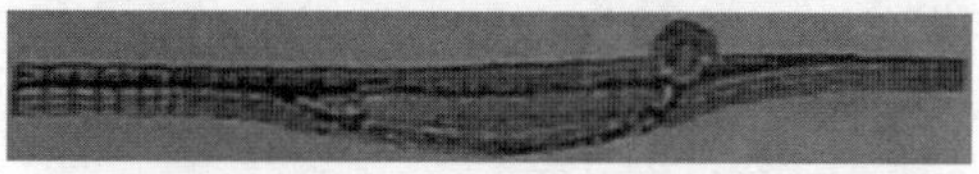

(d)

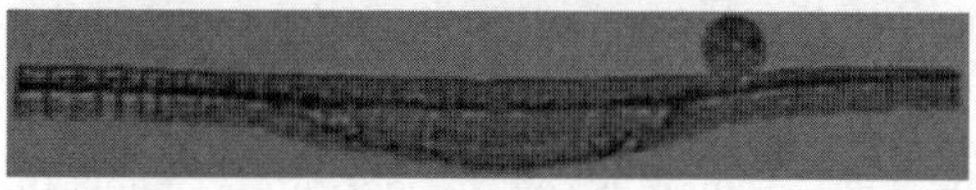

(e)

Figure 6 Numerical simulation results by AUTODYN for LY12CZ. θ =75, V_0 =650 m/s

(a) Grid plot for 20 μs; (b) Grid plot for 40 μs; (c) Grid plot for 60 μs; (d) Grid plot for 80 μs; (e) Grid plot for 100 μs

4 Conclusions

Experimental results of the high oblique impacting phenomena of Wu spheres on the thin 2 mm L. Y12C7Z. aluminium plates and 1 mm or 2 mm A3 steel plates are presented. The deformation and failure behavior caused the ricocheting or perforating behavior of the plates is dominated by the projectile and target material properties.

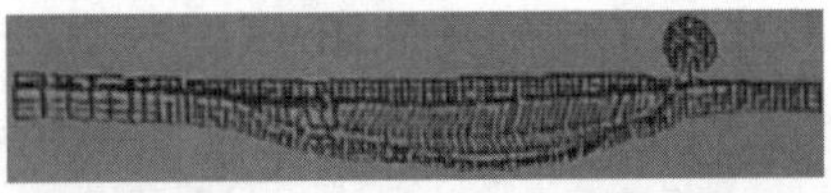

(a)

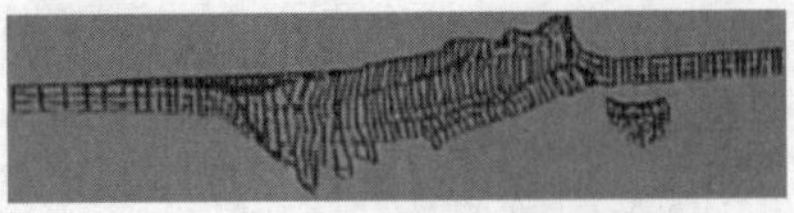

(b)

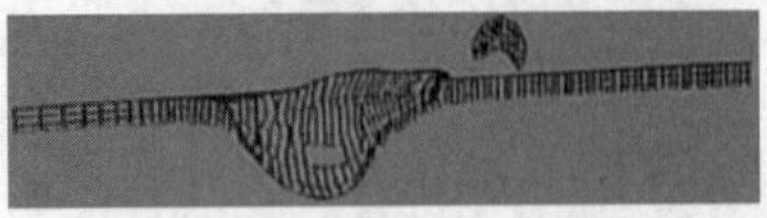

(c)

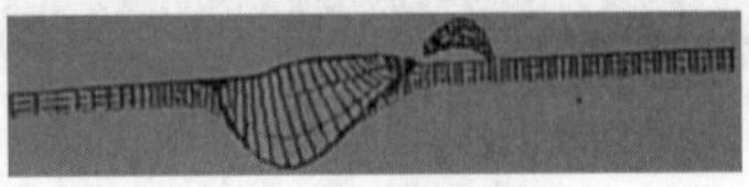

(d)

Figure 7 Numerical simulation results by AUTODYN

(a) Grid plot for 2 mm LY12CZ. $\theta=80°$, $V_0=890$ m/s, t=60 μs; (b) Grid plot for 2 mm LY12CZ. $\theta=75°$, $V_0=1\ 720$ m/s, t=30 μs; (c) Grid plot for 2 mm A3 steel. $\theta=80°$, $V_0=1\ 440$ m/s, t=22 μs; (d) Grid plot for 2 mm A3 steel. $\theta=80°$, $V_0=1\ 750$ m/s, t=16 μs

Two approaches arc employed in the numerical simulation: the rigid body model and the eroding Lagrange AUTODYN method. The rigid body model can be applied below some critical velocity for different materials while the AUTODYN – 3D code results in good simulations of the main features of the experiments. The comparison between calculations and the experimental result verifies the possibility to predict the impacting. behavior with high angle of obliquity.

Table 2 Values of EOS in AUTODYN simulation

Material	Wu	LY12CZ	A3
C (cm/μs)	0.40	0.538	0.497
S	1.268	1.337	0.604
τ	1.58	2.10	1.81

5 References

[1] SENF H, ROTHENHAUSLER H, SCHARPF F, et al. Experimental and numerical investigation of the ricocheting of projectiles from metallic Surfaces [C] //Proceedings of the 6th Symposium on Ballistics, Orlando, Florida, Oct. 1981.

[2] BACKMAN M E, FINNEGAN S A. Dynamics of the oblique impact and ricochet of nondeforming spheres against thin plates [M]. China Lake, CA: Naval Weapons Center, 1976.

[3] SUN T, PEI S X. The study of oblique impact of a deformable projectile at ductile spaced layered targets [C] //Proceedings of the 15th Symposium on Ballistics, Jerusalem, Israel, May 1995.

[4] KIVITY Y, MAYSELESS M, LUTTWAK G, et al. High obliquity impact of soft and hard spheres on thin plates [C] //Proceedings of 15th Symposium on Ballistics, Jerusalem, Israel, May 1995.

[5] Cowler M, et al. Numerical simulation of impact phenomena in an interactive computing environment [C] //International Conference on Impact Loading and Dynamic Behavior – IMPACT 87, Bremen, Germany, May 1987.

钨球对多层 LY12CZ 硬铝间隔靶侵彻分析①

孙韬[1②]，冯顺山[1]，裴思行[2]，张国伟[2]
（1. 北京理工大学，北京 100081；2. 华北工学院）

摘 要：提出一种基于损伤的球体冲击碰撞硬质目标的动态变形破裂模型，计算所得钨合金球体对 LY12CZ 硬铝间隔靶板正侵彻后球体最终变形状态、侵彻层数等与实验结果吻合较好。

关键词：侵彻；多层间隔靶；损伤

THE STUDY OF PENETRATION OF TUNGSTEN SPHERES AT LY12CZ SPACED LAYERED TARGETS

Sun Tao，Feng Shunshan
（Beijing Institute of Technology，Beijing，100081）
Pai Sixing，Zhang Guowei
（North China Institute of Technology）

Abstract：This paper gives the results of theoretical studies of deformable tungsten spheres impacting and penetrating the thin spaced layered LY12CZ plate targets. The damage has been taken into consideration in the failure analysis. A good agreement between observed and computed data relevant to penetration phenomena is obtained.

Key words：penetration；spaced layered targets；damage

多层 LY12CZ 硬铝间隔靶体系常用于等效和模拟某些飞行器结构的壳体厚度或相关力学特性。钨合金材料以其高强度、高密度等特性多被装填弹药战斗部进行中高空反导等。因此，钨合金材料破片对多层铝靶的侵彻研究具有重要意义。

一般而言，正撞击状态对材料的损伤是最严重的，侵彻效果也最好。相关的理论研究和试验结果很多，但对多层靶标的侵彻研究不多[1-4]。球形体的侵彻理论多假定弹体为刚性体，若球体碰撞产生大变形，一般采用动态有限元技术做冲击侵彻数值研究，或者不考虑球体形状变化而利用能量守恒、动量守恒原理等方法进行近似破坏分析。本文在 Recht 变形柱形弹体撞击靶板理论基础上[5]，对球体撞击靶体的变形给予近似分析，并应用材料损伤理论对钨合金球体冲击侵彻多层间隔铝靶的问题进行研究。

1 球体冲击塑性变形模型的建立

当具有圆球形表面的弹性体在其外表面受到均布动压力时，变形将从其外表面以球面波的方式向球心传播，其波速与半无限介质中的纵波波速相同。当外载不对称时，在球体内出现的波的形式类似于半无限介质中的纵波、横波及表面波[6]。为解决撞击过程中球体的塑性变形问题，采用基本假设：①撞击变形过程中无质量的损失，同时材料是不可压缩的；②忽略变形球体内横波与表面波对变形的影响；③变形是瞬时发生的，任一时刻的变形量均将小于球体初始半径。

① 原文发表于《兵工学报》1998 年第 2 期。

② 孙韬：工学博士，1995 年师从冯顺山教授，研究反舰导弹结构解体毁伤效应，现工作单位：陆军研究院炮兵防空兵研究所。

由假设①，设球体变形之后的几何形状是由两个椭球体各自一半组成，如图 1.1 所示。由此可得

$$\frac{4}{3}\pi R_0^3 = \frac{2}{3}\pi R^2(R_0 + h) \tag{1.1}$$

即

$$\pi R_0^2(R_0 + R_0) = A(R_0 + h) \tag{1.2}$$

式中 R_0——球体初始半径；

R，h——变形后椭球体的长半轴和最小短半轴，h 位于球体与靶标接触的一侧；

$A = \pi R^2$——球体变形后的最大截面积。

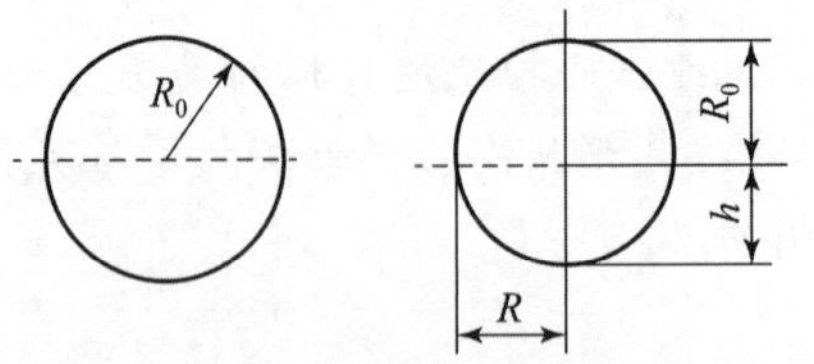

图 1.1 球体变形状态

考虑到直接求解变形球体中塑性波速 C_P 的困难性，由假设②不妨将波在球体内的传播简化为一维应变弹塑性波，近似取[7]

$$C_P = 0.80\sqrt{\frac{(1-\mu)E}{\rho_P(1+\mu)(1-2\mu)}} \tag{1.3}$$

式中 E、μ 及 ρ_P 分别为球体材料弹性模量、泊松系数和密度。

取 v_0 为弹体撞击靶板的入射速度，$v_0 < C_P$。设撞击开始时，弹体、靶体的接触面的运动速度为 v_1，接触应力为 $C_D \cdot \sigma_C$，C_D 为形状系数，得[3]

$$v_1 = \frac{v_0}{(1 + \rho_t C_H / \rho_P C_P)} \tag{1.4}$$

ρ_t，C_H 分别为靶板材料密度和膨胀压缩弹性波的传播速度。设在 dt 时间中，弹体中塑性区增长 dh，其实际增长速度为 $C_P - (v - v_1)$，则有[3,5]

$$\mathrm{d}h = [C_P - (v - v_1)]\mathrm{d}t = [C_P - v_L]\mathrm{d}t \tag{1.5}$$

式中 v——弹体运动速度；

v_L——弹体中塑性边界相对于运动靶体表面的速度。

$$v_L = v - v_1 \tag{1.6}$$

设 dh 这段材料原来都是未变形的弹体中的材料；A_0，A_1 分别为塑性区增长 dh 前后的球体最大截面积。由式（1.1）和式（1.5）得

$$A_0[R_0 + (h - C_P\mathrm{d}t)] = A_1 \cdot [R_0 + (h - (C_P - v_L)\mathrm{d}t)] \tag{1.7}$$

对照 Recht 柱形弹体撞击变形 dh 前后截面积的变化形式，可化简近似成

$$\left(\frac{A_0}{h} + 1\right)\frac{A_1}{A_0} - \frac{R_0}{h} = \frac{1}{1 - v_L/C_P} \tag{1.8}$$

对式（1.8）右边展开有

$$\frac{1}{1 - v_L/C_P} = 1 + \frac{v_L}{C_P} + \left(\frac{v_L}{C_P}\right)^2 + \cdots \tag{1.9}$$

忽略式（1.9）右边二阶以上高阶小量，代入式（1.8），并化简得

$$\frac{1}{1 - v_L/C_P} = 1 + \frac{v_L}{C_P} + \left(\frac{v_L}{C_P}\right)^2 + \cdots \frac{A_1}{A_0} = 1 + \frac{v_L}{C_P} \cdot \frac{1}{\dfrac{R_0}{h} + 1} \tag{1.10}$$

对球体而言，不同于柱形体的局部变形而呈现出整体变形形态。设作用在变形球体上的相关外力为动屈服流应力和某一与速度相关的函数 $f(v)$，即

$$\frac{\mathrm{d}}{\mathrm{d}t}\left[\frac{2}{3}\rho_P\cdot A(R_0+h)v\right]=-A[C_D\sigma_{yc}^D+f(v)] \tag{1.11}$$

式中　σ_{yc}^D——材料动屈服流应力。

由式（1.1）化简得

$$\mathrm{d}v=-\frac{1.5C_D\sigma_{yc}^D C_P\mathrm{d}t}{\rho_P C_P(R_0+h)}-\frac{1.5f(v)\mathrm{d}t}{\rho_P(R_0+h)} \tag{1.12}$$

近似取 $\mathrm{d}h=C_P\mathrm{d}t$，令 C_D 值为常数，积分式（1.11），利用起始条件 $t=0$，$v=v_0$，得

$$v=v_0-\frac{1.5C_D\sigma_{yc}^D}{\rho_P C_P}\ln\left(1+\frac{h}{R_0}\right)-F(v,t) \tag{1.13}$$

$$F(v,t)=\int_0^t\frac{1.5f(v)\mathrm{d}t}{\rho_P(R_0+h)} \tag{1.14}$$

将式（1.4）和式（1.6）代入式（1.13），得

$$v_L=\frac{v_0}{z^*}-\frac{1.5C_D\sigma_{yc}^D}{\rho_P C_P}\ln\left(1+\frac{h}{R_0}\right)-F(v,t) \tag{1.15}$$

其中，z^* 称为阻抗因子，$z^*=1+(\rho_P C_P/\rho_t C_H)$。由于在变形过程中 $v_L<v_0/z^*$，且撞击结束时 $v_L=0$[3,5]，当式（1.15）中 $\frac{v_0}{z^*}>\frac{1.5C_D\sigma_{yc}{}^D}{\rho_P C_P}$ 及球体撞击靶板产生变形的入射速度满足 $\sigma_{yt}+\frac{\rho_t v_0^2}{2}\geqslant\sigma_{yc}{}^D$；那么 $F(v,\ t)$ 在整个变形过程中大于零，其中 σ_{yt} 为靶板材料屈服流应力。这样，由式（1.10）和式（1.15）不妨取撞击靶板结束时最大截面积之比的最大值为

$$\frac{A_1}{S_0}=1+\frac{v_0}{z^*C_P(1+R_0/h)} \tag{1.16}$$

式中，S_0 为球体撞击靶板的初始最大截面积。这样，由式（1.1）和式（1.16）即可确定出球体撞击单一靶板而产生的变形的参量值 A_1 和 h。

2　侵彻模型及弹体冲击损伤模型

当弹体在侵彻靶板过程中伴随有变形时，动量守恒定律常用于粗略分析侵彻现象阶略去靶板得到的冲量，有

$$m_P v_0=(m_P+m_t)v_f \tag{2.1}$$

式中　m_p——弹体质量；

m_t——挤凿出的靶元质量；

v_f——靶元挤凿下来后的速度，也是弹体在击穿靶体后的速度，取 m_t 等于

$$m_t=\frac{\sigma_{yc}^D}{\sigma_{yt}+\rho_t v_0^2/2}\cdot\rho_t A_1\cdot h_1 \tag{2.2}$$

式中　h_t——靶板原始厚度。一般取弹体侵彻多层靶板的侵彻模型为[4]

$$m_P\frac{\mathrm{d}v}{\mathrm{d}t}=-P \tag{2.3}$$

式中，阻力 P 为 $P=C_0+C_1v+C_2v^2$，$C_0=C_DA\sigma_{tt}$ 为靶板塑性流阻力，$C_1v=6\pi\rho_t RC_S v$ 为摩阻或黏性系数 C_S 产生的黏性阻力，σ_{tt} 为靶板材料剪切屈服强度；$C_2v^2=C_DA\rho_t v^2/2$ 为压强阻力，球体形状发生变化时，C_D 值近似为 $C_D=C_{D0}+\frac{C_{DC}-C_{D0}}{R_C-R_0}(R-R_0)$，$R_C$ 为球体变形至破裂时的椭球长半轴，不妨取 C_{D0}、C_{DC} 的值分别为 0.5 和 1[3]。

当弹体冲击侵彻多层间隔靶时，弹体将受到一系列的冲击载荷作用，使材料的微细结构发生变化，引起微缺陷的成核、生长、扩展和汇合，导致材料宏观力学性能的劣化，最终形成宏观开裂等破坏形式。取 ω 为损伤变量，则[8]

$$\sigma_{yc}^{D}(\omega)=\frac{\sigma_{yc0}^{D}}{(1-\omega)},\ C_{P}(\omega)=\frac{C_{P0}}{(1-\omega)} \tag{2.4}$$

σ_{yc0}^{D}和 C_{P0}为无损材料动屈服流应力和塑性波速，C_{P0}由式（1.2）确定，σ_{yc0}^{D}值由实验确定。

ω 的演化规律可由细观力学方法或唯象学方法确定[8]，本文将采用唯象的方法建立适当的损伤模型以便于工程应用。设球体侵彻第 L 层靶共用 N 个步骤或计算周期，$\xi_{i,L}$为弹体最大截面半径在第 i 个计算周期从 $R_{i-1,L}$增至 $R_{i,L}$时的参照量，且满足

$$\xi_{i,L}=\ln\frac{R_{i,L}}{R_{i-1,L}} \tag{2.5}$$

则穿透 L 层后总的 ξ_L 为

$$\xi_L=\sum_{i=1}^{N}\ln\frac{R_{i,L}}{R_{i-1,L}}=\ln\frac{R_{N,L}}{R_{0,L}} \tag{2.6}$$

式中，$R_{0,L}$为初始撞击 L 层时弹体最大截面半径；$R_{N,L}$为穿透后的最大截面半径。定义 ω 为

$$\omega=\omega_C\cdot\ln\frac{R_M}{R_0}/\ln\frac{R_C}{R_0} \tag{2.7}$$

式中，R_M 为穿透第 M 层靶后弹体最大截面半径；R_C 由实验确定；ω_C 为材料损伤临界值，一般性研究中，可取 $\omega_C=0.5$[8]。

3 实验结果与计算

选用质量为 1 g、直径 4.7 mm 的 93% W－Ni－Fe 钨合金球体，放置 20 层 LY12CZ 硬铝靶板，靶板厚度均为 2 mm，每两层的间隔为 30 mm。表 3.1 给出了钨球材料和铝板材料的相关力学性能。采用 12.7 mm滑膛枪发射弹丸，入射初速由 PM6680B 高分辨电子计时仪测量。回收未破裂钨球，测量后表明钨球质量并未损伤，其变形几何状态基本符合上述假定，见图 3.1，图中（a）为初始状态；图 3.1（b）、（c）为撞击后状态。

表 3.1 材料力学性能

材料	ρ/kg · m^{-3}	E/GPa	σ_y/MPa	μ	δ
钨合金	18 000	302	790	0.30	0.08
LY12CZ	2 700	70	284	0.30	0.19

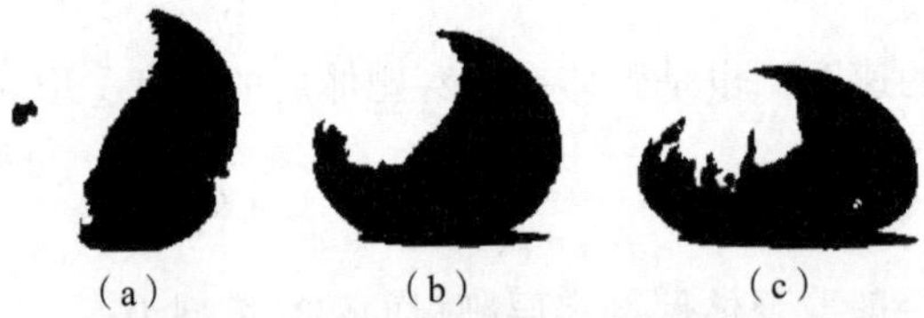

图 3.1 实验所得钨球撞击变化状态

试验得到钨球临界破裂直径为 5～5.1 mm，取式（2.3）中 C_S 值为 0.06，σ_{yc0}^{D}值简单地取表 3.1 中的值，计算所得钨球侵彻间隔铝靶层数 M 及球体变形最大截面直径 D_M同入射速度 v_0的关系见图 3.2、图 3.3。可见，计算结果与实验有相当的一致性。实验得出钨球在入射速度大于 1 200 m/s 时破裂，计算对应的破裂速度阈值为 1 210 m/s。钨球变形计算值与实验偏差的原因在于动态加载下，材料动应力值一般大于静态加载下的数值，这样由公式得出的球体撞击靶板产生变形的入射速度要增大。设钨合金材料动态加载时屈服应力比静态值增大 1.5 倍，则球体开始变形速度由静态值的 612 m/s 增大到 817 m/s，图 3.3表明了这一趋势。此外球体的变形恢复没有考虑，撞击后内部出现的其他波及波的相互作用等都对变形有影响。值得注意的是，倘若将球体的临界破裂直径提高到 5.5 mm，其他参量不变，计算显示球体的最大侵彻层数比原先增加 3 层，弹丸的破裂速度也由 1 210 m/s 增至 1 450 m/s。

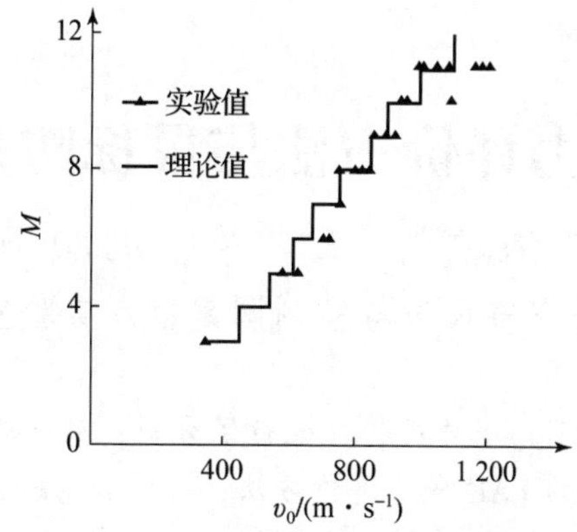

图 3.2 穿透层数计算与实验比较

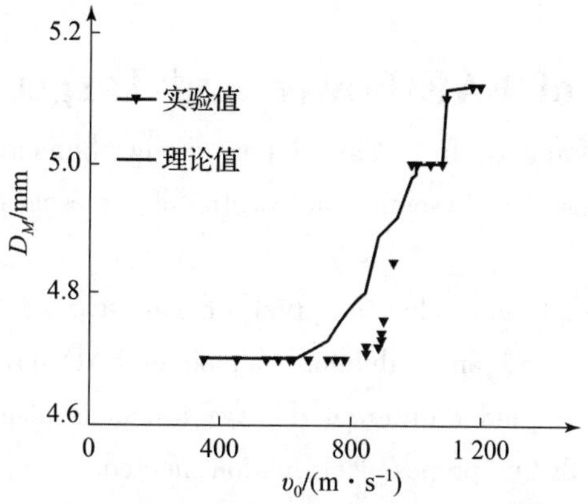

图 3.3 球体变形最大截面直径计算与实验比较

4 结论

本文在相关假设和 Recht 柱形弹体变形模型的基础上，给出了球形弹体撞击靶板过程中变形量的近似表达式，并结合损伤理论建立了变形球体冲击侵彻多层间隔靶的数学模型，计算所得侵彻层数、变形参量及破裂速度等与实验结果吻合较好，可应用于材料特性和侵彻分析等研究中。

参考文献

[1] BACKMAN M E, GOLDSMITH W. The mechanics of penetration of projectiles into targets [J]. International journal of engineering science, 1978, 16: 1-99.
[2] ZUAKAS J A. High velocity impact dynamics [M]. San Francisco, CA: John Wiley & Sons, Inc., 1990.
[3] 钱伟长. 穿甲力学 [M]. 北京: 国防工业出版社, 1984.
[4] SUN T, PEI S X. The study of oblique impact of a deformable projectile at ductile spaced layered targets [C] // Proceedings of 15th International Symposium on Ballistics. Jerusalem, Israel: 1995, 3 (2): 387-394.
[5] RECHT R F. Tarlor ballistics impact modelling applied to deformation and mass loss deformation [J]. International journal of engineering science, 1978, 16: 809-827.
[6] GREY R M, ERINGEN A C. The elastic sphere under dynamic and impact loads. OWR Tech Rep, 1955 (8).
[7] 马晓青. 冲击动力学 [M]. 北京: 北京理工大学出版社, 1992.
[8] GURRAN D R, SEAMAN L, SHOCKEY D A. Dynamic failure of solids [J]. Physics reports, 1987, 147 (5-6): 253-388.

FAE威力评价方法与目标防护分析[①]

王海福[②]，王　芳，冯顺山
（北京理工大学爆炸灾害预防与控制国家重点实验室，北京　100081）

摘　要：对燃料空气炸药（FAE）爆炸场特性进行了分析；从冲击波超压－冲量毁伤准则出发，提出一种以靶板毁伤效应为评价依据的FAE威力评价方法，并结合易损性等效原理，对不同毁伤等级下目标防护问题进行了分析的。

关键词：燃料空气炸药；威力评价；毁伤效应；目标易损性

Evaluation Method of FAE Power and Target Protection Analysis

Wang Haifu，Wang Fang，Feng Shunshan
（National Key Lab. of Prevention and Control of Explosion Disasters，BIT）

Abstract：Characteristics of explosion fields from fuel air explosive（FAE）are analyzed. Based on shock overpressure－impulse damage criterion，an evaluation method of FAE power based on damage effects of the target is developed. Target protection under different damage orders is also discussed by means of combining vulnerability equivalent principle with this proposed evaluation method.

Key words：fuel air explosive（FAE）；evaluation of power；damage effect；target vulnerability

燃料空气炸药（FAE）爆轰过程所需氧气取自爆炸现场的空气，从而大大提高了常规武器战斗部的装药利用效率。另外，由于FAE实施分布爆炸，其云雾区爆轰压力较低，但超压及冲量作用范围却相对较大[1-3]。因此，建立科学、有效的FAE威力评价方法，对FAE武器性能定位、战术使用以及目标防护具有重要意义。

关于FAE爆炸威力问题，曾分别提出和使用过爆炸潜能、冲击波能量以及冲击波参数等评价方法[4]，但这些方法本身存在某些不足，评价结果不尽相同，甚至相差很大，从而在某种程度上降低了评价结果的有效性和应用价值。本文在分析FAE爆炸场特性及冲击波毁伤准则基础上，提出了一种基于靶板毁伤效应的FAE威力评价方法，并从易损性等效的角度对目标防护问题进行了分析。

1　FAE爆炸场特性分析

FAE实施分布爆炸，与常规凝聚相高能炸药相比，云雾区爆轰压力较低，但超压及冲量作用范围相对较大，其爆炸场特性典型曲线如图1所示，图中的ΔP代表超压，R为距爆心的距离。从图1可以看出，尽管FAE云雾区爆压较TNT装药低，但FAE爆炸场具有较大的云雾爆轰区（$R=R_0$），而且空气冲击波超压随传播距离衰减速率较TNT爆炸场缓慢，当距爆心距离超过某一范围后（$R\geq R_c$），FAE爆炸场超压却大于TNT装药。

---

① 原文发表于《中国安全科学学报》1998年第5期。

② 王海福：工学博士，1993年师从冯顺山教授，研究战斗部威力及效应评价，现工作单位：北京理工大学。

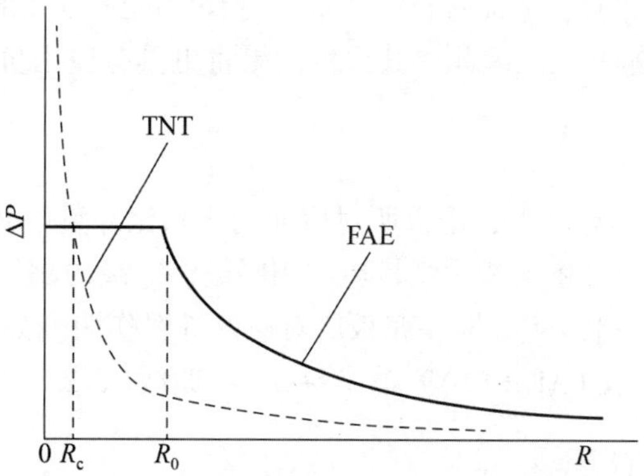

图1　FAE及TNT装药爆炸场超压－距离关系

表1给出了典型FAE燃料与空气按化学当量配比条件下的爆轰参数计算结果。从表1可以看出，气态和液态燃料FAE云雾区C－J爆轰压力仅为2 MPa左右，含铝固态燃料FAE云雾区C－J爆轰压力也只能达到4 MPa左右，而装药密度在1.58 g/cm^{3}左右的TNT炸药，C－J爆轰压力可高达20 GPa左右，两者相差近万倍量级。

表1　典型FAE燃料在理想条件下的爆轰参数计算结果

燃料或装药	爆轰速度/(m·s^{-1})	C－J压力/MPa
环氧乙烷	1 816	1.93
0.5铝粉+0.5环氧丙烷（固态燃料）	2 231	4.32
TNT装药	6 945	19 293

关于FAE爆炸场冲量问题，可通过环氧乙烷和TNT装药各1 kg爆炸作用于刚壁上总冲量来加以分析。根据经典爆轰理论，总冲量可用式（1）表述：

$$I=\frac{8}{27}mD \tag{1}$$

式中，m、D分别表示装药质量和爆速。根据理想的爆轰配比条件，1 kg环氧乙烷完全爆轰约需7.85 kg空气来提供约1.57 kg的氧气，即1 kg环氧乙烷相当于8.85 kg环氧乙烷－空气混合物，取TNT装药爆速$D_T=7\ 000$ m/s，环氧乙烷爆速$D_F=2\ 000$ m/s，则环氧乙烷FAE爆炸作用于刚壁的总冲量为TNT装药的2.53倍。

2　FAE威力评价方法

2.1　评价准则

冲击波对目标毁伤作用通常用峰值超压（ΔP）、正压作用时间（t_+）和比冲量（i）三个参数来度量。因此，其毁伤准则有[5,6]：超压准则、冲量准则和超压－冲量准则（P－I准则）。

超压准则认为，只有当作用于目标的超压达到或超过某一临界值Δp_c时，才能对目标造成一定程度的毁伤。显然，超压准则忽略了冲击波的正压作用时间因数。研究表明，只有当t_+与目标自振周期T间满足$t_+\geqslant 10T$时，峰值超压才对目标毁伤起决定作用。

冲量准则认为，只有当作用于目标的比冲量达到或超过某一临界值i_c时，才能对目标造成一定程度的毁伤，冲量准则的不足在于忽略了目标毁伤存在临界超压的事实。研究表明，只有当t_+与T间满足$t_+\leqslant T/4$时，比冲量才对目标毁伤起决定作用。

P－I准则认为，冲击波对目标毁伤效应由Δp和i共同决定，只有当两者同时达到或超过某一临界

值 Δp_c 和 i_c，即 $\Delta p \geqslant \Delta p_c$ 及 $i \geqslant i_c$ 时，才能对目标造成一定程度的毁伤。显然，较之于超压准则和冲量准则，P-I 准则更具科学性和普遍意义。因此，其评价结果将更具有应用价值。

2.2 评价方法

根据 P-I 准则，在某一毁伤等级下，靶板所对应的超压和比冲量值并不唯一，而是由一系列超压和比冲量组合（Δp，i）所构成[5,6]。将那些对靶板造成相同毁伤等级的超压与比冲量组合所构成的曲线称之为靶板等毁伤曲线，在各毁伤等级下，同一靶板所对应的等毁伤曲线如图 2 中曲线 1、2、3、4 和曲线 5 所示，曲线 7 和曲线 8 分别代表 FAE 和 TNT 爆炸场 $\Delta p \sim i$ 曲线。

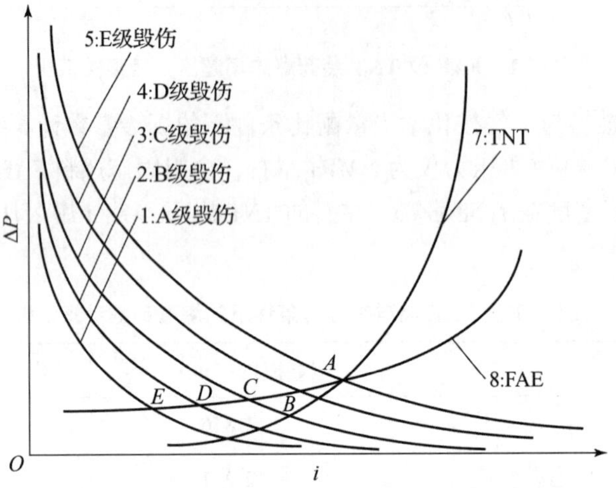

图 2　Δp，i 准则法评估 FAE 爆炸威力原理图

从图 2 可以看出，如果通过靶板毁伤效应试验，获得不同厚度 δ（δ 表征易损性）靶板，在不同毁伤等级 D 下，所对应的等毁伤曲线和 FAE 爆炸场中 $\Delta p \sim i$ 曲线。联立关系式（2）和式（3），可得到靶板在 FAE 爆炸场中遭受不同等级毁伤时所对应的 Δp 和 i 值，即如图 2 中 A、B、C、D 和 E 点所对应的坐标值。因此，根据球形 TNT 装药近地爆炸冲击波参数经验计算公式[2,3]总能找到一个装药量为 m^T 的 TNT 爆炸场（例如，对应于 A 点的等效 TNT 装药量爆炸场 $\Delta p \sim i$ 曲线，如图 2 中曲线 7 所示），当靶板离该 TNT 爆炸场中心距离与 FAE 相同时，两者对靶板所造成的毁伤等级相同。因此，根据等距离等毁伤效应等威力原则，可以认为 m^T 即为该试验 FAE 在距爆炸场中心距离为 R 处的等效 TNT 威力当量。

$$\Delta p = f_1(i, W, D) \tag{2}$$

$$\Delta p = f_2(i) \tag{3}$$

$$\Delta p = 0.103\frac{\sqrt[3]{m^T}}{R} + 0.403\left[\frac{\sqrt[3]{m^T}}{R}\right]^2 + 1.27\left[\frac{\sqrt[3]{m^T}}{R}\right]^3 \quad \left(1 \leqslant \frac{R}{\sqrt[3]{m^T}} \leqslant 15\right) \tag{4}$$

$$i = 333\frac{\sqrt[3]{m^T}}{R} \quad (R > 12r_0) \tag{5}$$

式中　Δp——冲击波超压，MPa；

i——比冲量，(Pa · s)；

R——距爆心距离，m；

r_0——球形 TNT 装药半径，m。

3　目标防护分析

靶板是一般意义上的目标。根据靶板毁伤效应，一方面可对 FAE 爆炸威力问题做出全面、科学的评价，另一方面还可对具体目标在特定毁伤等级下的防护问题做出分析。靶板与具体目标间的显著差异在于易损性的不同，因此，对具体目标进行防护分析的关键在于建立靶板与具体目标间的易损性等效关系。对目标易损性用构件特性 T_{obj} 来表征，靶板易损性仍以其厚度 δ 表征，具体目标与靶板间的易损性等效关系为

$$W=f_3(T_{obj}) \tag{6}$$

式（6）物理意义表明，对于构件特性 T_{obj} 的具体目标，其易损性可等效为厚度 δ 的某种标准靶板。假设测得 FAE 爆炸场中超压~距离（$\Delta p \sim R$）和比冲量~距离（$i \sim R$）关系分别为

$$\Delta p=f_4(R) \tag{7}$$

$$i=f_5(R) \tag{8}$$

联立关系式（7）和式（8）可给出式（3）的具体形式，其物理意义如图 2 中曲线 8 所示。根据等毁伤曲线的物理意义，如果，距 FAE 爆炸场中某一距离为 R_F 处的（Δp，i）组合处于厚度为 δ 标准靶板某一等毁伤曲线的右上方，则在该毁伤等级下 FAE 爆炸场对处于半径为 R_F 范围内的构件特性为 T_{obj} 目标均有效。因此，从目标防护的角度看，要使该目标所遭受的毁伤程度轻于该毁伤等级，目标距 FAE 爆炸场中心的距离应大于 R_F。

由此可见，基于该评价方法，并结合靶板与特定目标间的易损性等效关系，可对 FAE 爆炸场中的具体目标在特定毁伤等级下的防护问题作出全面分析与评价。

4 结束语

燃料空气炸药（FAE）实施分布爆炸，与常规凝聚相高能炸药相比，其云雾区爆轰压力相对较低，但超压和冲量作用范围却较大。针对 FAE 这一显著特征，笔者从经典爆轰理论的角度，对 FAE 爆炸场参数进行了分析；在此基础上，从冲击波超压 - 冲量毁伤准则出发，并提出了一种基于靶板毁伤效应的 FAE 威力评价方法。该评价方法显著优点在于，一方面，采用了等距离等毁伤效应等威力评价原则，因而，评价结果更具科学性和实用价值；另一方面，根据该评价方法，并结合靶板与具体目标间的易损性等效关系，还可对 FAE 爆炸场中的具体目标在特定毁伤等级下的防护问题作出全面分析与评价。

（收稿日期：1997 年 6 月；作者地址：北京白石桥路 7 号；北京理工大学爆炸灾害预防、控制国家重点实验室；邮编：100081）

参 考 文 献

[1] SEDGWICK R T，KRATA H R. Fuel air explosives：a parametric investigation：AD - A159 177 /5/HDM [R]. 1979.

[2] 许会林，汪家华. 燃料空气炸药 [M]. 北京：国防工业出版社，1980.

[3] 孙业斌，惠明军，曹欣茂. 军用混合炸药 [M]. 北京：兵器工业出版社，1995.

[4] 惠君明. 燃料空气炸药威力评判的讨论 [J]. 兵工学报 · 火化工分册，1995 (2)：50 - 54.

[5] ABRAHAMSON G R，LINDBERG H E. Peak load - impulse characterization of critical pulse loads in structural dynamics [M] //Dynamic response of structure. New York：Permagon Press，1972.

[6] SEWELL R G S. Blast damage criterion [R]. NAYWEPS Report 8469，1964.

燃料空气炸药武器化应用条件①

王海福②，冯顺山

（北京理工大学爆炸灾害预防与控制国家重点实验室，北京　100081）

摘　要：针对燃料空气炸药（FAE）战斗部在常规武器系统中的应用问题，从 FAE 爆炸场特性、FAE 战斗部作用原理以及武器系统发射过载的角度出发，对 FAE 战斗部爆炸威力优势区最小云雾直径、所对付的目标特性以及所应用的弹种进行了分析。

关键词：燃料空气炸药；FAE 战斗部；冲击波效应

An Analysis of Restricted Application of Fuel Air Explosive to Conventional Weapons

Wang Haifu，Feng Shunshan

（Department of Engineering Safety Beijing Institute of echnology Beijing 100081）

Abstract：Based on the characteristics of FAE parameters in explosion fields，the explosion principle of FAE warhend and the launching load of conventional weapons，an analysis of the minimum diameter of FAE clouds with higher explosion power compared with TNT explosive 、the characteristics of targets suitable for being attacked by FAE warhead，and types of conventional weapons suitable for transplanting FAE warhead is carried out.

Key words：fuel air explosive；FAE warhead；shock effect

燃料空气炸药（fuel air explosive，FAE）是近三四十年发展起来的一种新型爆炸能源，其显著特征在于：FAE 爆轰过程所需的氧气取自爆炸现场的空气中，因而可大大提高装药效率；FAE 实施分布爆炸，因而其云雾区爆轰压力较低，但超压作用范围和比冲量却较大。FAE 武器化应用开创了常规武器提高威力的新途径，使常规弹药战斗部技术发生了重大革新。通过对 FAE 武器化应用条件分析，为 FAE 弹药发展方向、型号研制及其战术使用提供参考。

1　FAE 爆炸场特性

表 1 给出了两种常用 FAE 燃料与空气按化学当量配比的爆轰参数计算结果。从表 1 可以看出，FAE 爆轰参数并不理想，云雾区爆轰压力仅为 2 MPa 左右，爆轰速度在 1 800 m/s 左右，而装药密度在 1.60 g/cm^{3}左右的 TNT 炸药，其爆轰速度在 7 000 m/s 左右，C－J 爆轰压力可高达 20 MPa 左右，两者间爆轰压力相差近万倍量级。

表 1　几种常用 FAE 燃料与空气按化学当量配比的爆轰参数计算结果

燃料或炸药	爆轰速度/(m·s^{-1})	C－J 爆轰压力/MPa
环氧乙烷	1 816	1.93

① 原文发表于《弹箭技术》1998 年第 10 期。

② 王海福：工学博士，1993 年师从冯顺山教授，研究战斗部威力及效应评价，现工作单位：北京理工大学。

续表

燃料或炸药	爆轰速度/(m·s^{-1})	C－J 爆轰压力/MPa
环氧丙烷	1 833	2.05
TNT 装药	6 945	19 293

有关 FAE 与相同装药量 TNT 炸药的爆炸场超压随距离变化规律如图 1 所示。根据表 1 计算结果并结合图 1 可以看出，尽管 FAE 云雾区爆轰压力较 TNT 装药小得多，但 FAE 爆炸场具有较大的云雾爆轰区，而且空气冲击波超压随传播距离的衰减速率较 TNT 炸药缓慢，当距爆心的距离超过某一范围后，FAE 爆炸场超压却大于 TNT 装药。

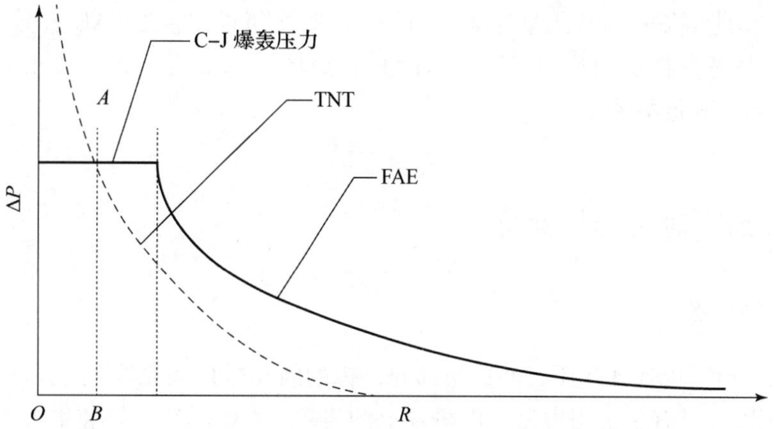

图 1　FAE 及 TNT 装药爆炸场超压随距离变化规律

有关 FAE 爆炸场冲量问题，可通过环氧乙烷、环氧丙烷燃料以及 TNT 装药各 1 kg 爆炸后作用于刚壁上的总冲量大小加以分析，根据经典爆轰理论，总冲量 I 可为

$$I = \frac{8}{27}mD \tag{1}$$

式中　m、D——装药量和爆速。

计算结果列于表 2，从表 2 可以看出，FAE 爆炸作用于刚壁的总冲量较相同装药量的 TNT 炸药大得多。

表 2　FAE 与 TNT 装药爆炸作用于刚壁总冲量对比

燃料	完全爆轰所需氧气量/kg	FAE 量/kg	与 TNT 爆炸总冲量之比
环氧乙烷（1 kg）	1.57	8.85	2.31
环氧丙烷（1 kg）	3.14	11.9	3.14

2　FAE 战斗部威力优势区及最小云雾半径

从图 1 可以看出，当距爆心距离超过某一范围（图 1 中 AB 线所对应的距离后），FAE 爆炸场超压开始大于 TNT 装药，这表明，FAE 爆炸场存在一个显著发挥威力效能的威力优势区及最小云雾半径。根据爆炸相似律及大量试验结果，球形 TNT 装药在普通地面爆炸时，空气冲击波超压（取前三项）可表达为

$$\Delta P = 0.103\frac{\sqrt[3]{W}}{R} + 0.403\left(\frac{\sqrt[3]{W}}{R}\right)^2 + 1.27\left(\frac{\sqrt[3]{W}}{R}\right)^3 \quad (2)$$

$$\left(1 \leqslant \bar{r} = \frac{R}{\sqrt[3]{W}} \leqslant 15\right)$$

式中 R——距爆心距离，m；

W——TNT 药量，kg；

ΔP——冲击波超压，MPa。

根据关系式（2），当 $\bar{r}=1$ 时，$\Delta P = 1.78$ MPa，这表明，对于装药量为 W(kg) 的球形 TNT 炸药，当距爆心距离为 $\sqrt[3]{W}$（m）时，其爆炸场超压已小于 FAE 云雾区的爆轰压力。考虑到 FAE 实际爆轰压力要受云团中燃料浓度分布特性、环境及气候等因素的影响，可以认为这一超压值基本上与 FAE 云雾边缘处的爆轰压力相当。由此可见，与装药量为 W(kg) 的 TNT 炸药相比，当 FAE 云雾半径 R 超过 $\sqrt[3]{W}$(m) 后，其爆炸场将开始显著发挥威力优势。因此，FAE 爆炸场发挥威力优势的最小云雾半径 R_{cm}(m) 与 TNT 装药量 W(kg) 间应满足关系

$$R_{cm} \geqslant \sqrt[3]{W} \quad (3)$$

3 FAE 战斗部所对付的目标特性

3.1 从目标易损性看

表 3、表 4 分别给出了爆炸场超压、冲量与部分目标的毁伤效应关系[4,5]，从表 3、表 4 及图 1 可以看出，FAE 战斗部用以对付有生集团力量、停机坪上的飞机、C3I 阵地、轻装甲或无装甲车辆、土木建筑工事、舰面设施以及雷场等“软或中软”性质的目标，可显著发挥威力效能，但用以对付以坦克为代表的“硬或坚固”目标，其效能并不显著。

表 3 爆炸场冲击波超压、冲量与部分目标毁伤效应关系

超压/kPa	冲量/(kPa · s)	目标毁伤效应
0.1 ~ 5	0.01 ~ 0.015	门窗玻璃安全无损
8 ~ 10	0.016 ~ 0.02	门窗玻璃局部破坏
15 ~ 20	0.05 ~ 0.10	门窗玻璃全部破坏
25 ~ 40	0.10 ~ 0.30	门窗、隔板被破坏；干砌砖墙、铁皮烟囱被摧毁
45 ~ 70	0.30 ~ 0.60	轻型结构严重破坏；输电线、铁塔倒塌；大树连根拔起
75 ~ 100	0.50 ~ 1.00	砖瓦结构房屋全部破坏；钢结构建筑物严重破坏；行进中的汽车破坏

表 4 爆炸场冲击波超压与动物毁伤效应关系

超压/kPa	<9.81	9.81 ~ 19.61	19.61 ~ 39.23	39.23 ~ 58.84	>58.84
伤亡等级	无伤	轻伤	中伤	重伤	死亡

3.2 从目标运动特性看

二次引爆型 FAE 战斗部从燃料抛撒引信（一次引信）启动以引爆中心抛撒药，到云雾起爆引信（二次引信）启动以引爆 FAE 云团并对目标实施毁伤作用过程，存在一定的时间延迟，因此，二次引爆

型 FAE 战斗部对目标的毁伤效能除与目标本身的易损性有关外，在某种程度上还要受目标运动速度的影响。表 5 给出了国外已用于装备部队的两种典型二次引爆型 FAE 武器设计参数。

表 5　两种典型二次引爆型 FAE 武器战斗部设计参数

武器型号	燃料类型	子炸弹装药量/kg	云雾尺寸/m	起爆延迟时间/ms
CBU－55B	环氧乙烷	33	2.5×15	125
SLUFAE	环氧丙烷	38	3.5×16.5	150

为便于分析问题，并不失一般性，假设 FAE 云团的形状特征为类圆柱体，高径比取 $H:D=1:5$，而且不同燃料量 FAE 云雾形状满足几何相似放大原则，云雾区外空气冲击波峰值超压随传播距离衰减满足指数规律，则 FAE 战斗部所装填的燃料量 Q(kg) 与云雾半径 R_c(m) 间的关系以及 FAE 爆炸场中空气冲击波峰值超压 ΔP(MPa) 随传播距离 X(m) 的衰减关系可分别表述为

$$R_c = 2.42\sqrt[3]{Q} \tag{4}$$

$$\Delta P = P_c \exp[-\alpha(X - R_c)] \quad (X \geqslant R_c) \tag{5}$$

式中　P_c——云雾区爆轰压力，MPa；

α——云雾区外空气冲击波衰减系数。

假设 FAE 战斗部精确命中目标，并以超压准则为目标毁伤依据，则二次引爆型 FAE 战斗部对运动目标实施有效毁伤的目标最大运动速度 V_m(m/s) 应满足

$$V_m \leqslant \frac{2.42\alpha\sqrt[3]{Q} + \ln P_c - \ln P_{cr}}{\alpha t_d} \tag{6}$$

式中　t_d(s)——引爆延迟时间；

P_{cr}——目标临界毁伤超压，其他参数意义同上。

根据 FAE 爆炸场冲击波峰值超压试验结果，得到衰减系数 $\alpha \approx 0.15$，因此，利用式（6）并结合表3和表4，即可就二次引爆型 FAE 战斗部对各类目标造成不同毁伤等级时的目标最大运动速度进行分析和确定。表6 给了 CBU－55B 子炸弹对两类典型目标造成严重毁伤时所对应的目标最大运动速度分析结果。

表 6　CBU－55B 子炸弹对典型目标造成严重毁伤的目标最大运动速度分析结果

目标	动物	钢结构建筑物或行进汽车
目标毁伤程度	重伤	严重破坏
目标运动速度 V_m	≤900 km/h≈0.75M（马赫数）	≤740 km/h≈0.6M

从表6 可以看出，二次引爆型 FAE 武器对地面运动目标（速度一般在 120 km/h 以下）的毁伤效能基本不受其运动速度的影响，对于速度在 0.6 M（马赫数）以上的空中高速运动目标，必须对二次引爆型 FAE 武器引战配合进行必要的修正以弥补因起爆延迟所造成偏差，否则，二次引爆型 FAE 战斗部对空中高速运动目标的毁伤效能将显著下降，采用一次引爆型 FAE 战斗部可有效避免上述不足。

4　FAE 战斗部所应用的弹种

FAE 战斗部基本作用原理为：对于二次引爆型 FAE 战斗部，先由一次引信引爆中心抛撒药柱，并利用中心抛撒药爆轰所产生的高温、高压产物气体迅速将装填在战斗部内的燃料抛撒出去，燃料与现场空气混合后形成可爆性 FAE 云团，经适当延时后二次引信（云雾起爆引信）启动将 FAE 云团引爆，并利用云雾区爆轰波和云雾区外的空气冲击波对目标实施毁伤作用。一次引爆型 FAE 战斗部的主要区别在于通过燃料配方技术使燃料在抛撒的同时实现爆轰，从而减少了二次引信的起爆问题。由此可见，无论一

次或二次引爆型 FAE 战斗部，都存在一个燃料抛撒问题，因此，FAE 战斗部壳体强度约束条件较常规凝聚相装药战斗部要弱得多，否则无法实现燃料的正常抛撒。这表明常规武器系统的发射过载在很大程度上决定着 FAE 战斗部的可移植性。

对于 FAE 航空炸弹，由于实施机载投放，因此，基本无须考虑发射过载问题；对于 FAE 火箭弹和导弹系统，由于弹药发射初速很低，在飞行过程中依靠发动机提供源动力使弹药逐渐加速，因此，其发射过载也较小，一般仅为几至十几，这一发射过载对 FAE 战斗部不会造成显著影响。对于炮弹和枪榴弹系统，由于弹丸在炮管极有限的距离内依靠药筒中火药爆燃的高压迅速加速到所需的出膛初速，因此，发射过载及弹体近受载荷均很高。

表 7 给出了几种常用火炮系统的发射过载数据。从表 7 可以看出，炮弹系统的发射过载较航空炸弹、火箭弹或导弹大近万倍量级，在这样高的发射过载作用下，要确保 FAE 战斗部壳体强度既能满足发射过程中不发生塑性变形或破裂的要求，同时又能实现燃料正常抛撒，就当前 FAE 战斗部结构设计及其作用原理而言，技术上存在相当大的难度，需进一步从 FAE 战斗部结构设计和开舱技术等方面做系统、深入的研究。

表 7　几种常用火炮系统的发射过载数据

火炮系统	小口径高炮	线膛火炮	迫击炮	无座力炮
发射过载	40 000	10 000 ~ 20 000	4 000 ~ 10 000	5 000 ~ 10 000

5　结论

（1）FAE 云雾区爆轰压力较 TNT 装药小近万倍，但云雾区外空气冲击波峰值超压随传播距离的衰减速率较 TNT 装药缓慢，FAE 战斗部显著发挥威力优势的最小云雾半径 R_{cm}（m）与 TNT 装药量 W(kg) 间应满足 $R_{cm} \geqslant \sqrt[3]{W}$。

（2）FAE 战斗部（包括一次和二次引爆型）所对付的目标，除应具有“软或中软”性质外，二次引爆型 FAE 战斗部对目标运动速度还有一定的要求，当目标运动速度超过云雾半径 R_{cm}（m）与 TNT 装药量 W(kg) 间应满足 $R_{cm} \geqslant 3W$。

（3）FAE 战斗部（包括一次和二次引爆型）所对付的目标，除应具有“软或中软”性质外，二次引爆型 FAE 战斗部对目标运动速度还有一定的要求，当目标运动速度超过战斗部具有良好的可移植性；对于炮弹、枪榴弹等具有很高发射过载的常规武器系统，就当前 FAE 战斗部结构设计及其作用原理而言，实现这种移植在技术上存在相当大的难度，需进一步从战斗部结构设计和开舱技术方面作系统、深度的研究。

参 考 文 献

[1] 许会林，汪家华．燃料空气炸药［M］．北京：国防工业出版社，1980.
[2] SDEGWICK R T，KRATZ H R. Fuel air explosives：a parametric investigation：AD - A159 177/5/HDM［R］．1979.
[3] 北京工业学院八系．爆炸及其作用：下册［M］．北京：国防工业出版社，1979.
[4] SEWELL R G S. Blast damage criterion［R］．NAYWEPS Report 8469，1964.
[5] 李铮．空气冲击波作用下的安全距离［J］．爆炸与冲击，1990，10（2）：135.
[6] 孙业斌，惠君明，曹欣茂．军用混合炸药［M］．北京：国防工业出版社，1995.
[7] 魏惠之，等．弹丸设计理论［M］．北京：国防工业出版社，1985.

基于靶板毁伤效应的燃料空气炸药威力评价方法探讨[①]

王海福[②]，王　芳，冯顺山

（北京理工大学爆炸灾害预防与控制国家重点实验室，北京　100081）

摘　要：从冲击波超压－冲量毁伤准则出发，提出一种以靶板毁伤效应评价燃料空气炸药威力的方法，并就实施评价方法的技术途径进行了探讨。

关键词：毁伤效应；燃料空气炸药；威力评价

Evaluation Method of FAE Power Based on its Damage Effects upon the Target

Wang Haifu，Wang Fang，Feng Shunshan

（National Key Lab of Prevention and Control of Explosion Disasters，

BIT，Beijing 100081，China）

Abstract：Based on shock overpressure－impulse damage criterion，a method is developed to evaluate FAE power according to its damage effects upon the targets，the related operation and technique are also discussed.

Key words：damage effect；evaluation of power；fuel air explosive（FAE）

1　引言

燃料空气炸药（FAE）显著特征在于：一方面，其爆轰过程所需氧气取自爆炸现场的空气，因而可大大提高战斗部装药的使用效率；另一方面，FAE 实施分布爆炸，其云雾区爆轰压力较低，但超压及冲量作用范围却相对较大[1,2]。建立全面、科学的 FAE 爆炸威力评价方法，对 FAE 武器化、性能定位及其战术使用具有重要意义。

有关 FAE 爆炸威力问题，众多同行分别提出和使用过爆炸潜能、冲击波能量以及冲击波参数（超压和比冲量）等评价方法[3,4]，但上述方法由于本身存在某些局限和不足，评价结果也不尽相同，甚至相差很大，因而在某种程度上降低了评价结果的有效性和应用价值。本文在分析冲击波毁伤准则的基础上，从冲击波超压－冲量毁伤准则出发，提出一种以靶板毁伤效应评价 FAE 爆炸威力的方法，并就实施该评价方法的技术途径进行了探讨。

2　FAE 爆炸威力评价准则及方法

2.1　评价准则

冲击波对目标毁伤作用通常用峰值超压（Δp）、正压作用时间（t_+）和比冲量（i）3 个参数来度量，因此，相应的冲击波毁伤准则有[5,6]：超压准则、冲量准则和超压－冲量准则（P－I 准则）。

超压准则认为，只有当作用于目标的超压达到或超过某一临界值 Δp_c 时，才能对目标造成一定程度

---

① 原文发表于《含能材料》1999 年第 1 期。

② 王海福：工学博士，1993 年师从冯顺山教授，研究战斗部威力及效应评价，现工作单位：北京理工大学。

的毁伤。显然，超压准则忽略了冲击波正压作用时间。研究表明，只有当 $t_+ \geqslant 10T$ 时（式中 T 为目标自振周期），峰值超压才对目标起决定毁伤作用。

冲量准则认为，只有当作用于目标的比冲量达到或超过某一临界值i_c时，才能对目标造成一定程度的毁伤。冲量准则的不足之处在于它忽略了目标毁伤存在临界超压作用的事实。研究表明，只有当$t_+ \leqslant T/4$ 时，比冲量才对目标毁伤起决定作用。

P－I 准则认为，冲击波对目标毁伤效应由 Δp 和 i 共同决定，只有当两者同时达到或超过某一临界值 Δp_c和i_c，即 $\Delta p \geqslant p_c$及 $i \geqslant i_c$时，才能对目标造成一定程度的毁伤作用。显然，与超压准则和冲量准则相比，P－I 准则更具科学性和普遍意义。因此，采用 P－I 准则对 FAE 爆炸威力进行评价更具应用价值。

2.2 评价方法

根据 P－I 准则，靶板在某一毁伤等级下所对应的超压和比冲量值并不唯一，而是由一系列超压和比冲量组合（Δp，i）所构成。将那些对靶板造成相同毁伤等级的超压与比冲量组合所构成的曲线称之为靶板等毁伤曲线，同一靶板在各毁伤等级下所对应的等毁伤曲线如图 1 中曲线 a，b，c，d 和 e 所示，曲线 f 和 g 分别代表 FAE 和 TNT 爆炸场 $\Delta p - i$ 曲线。

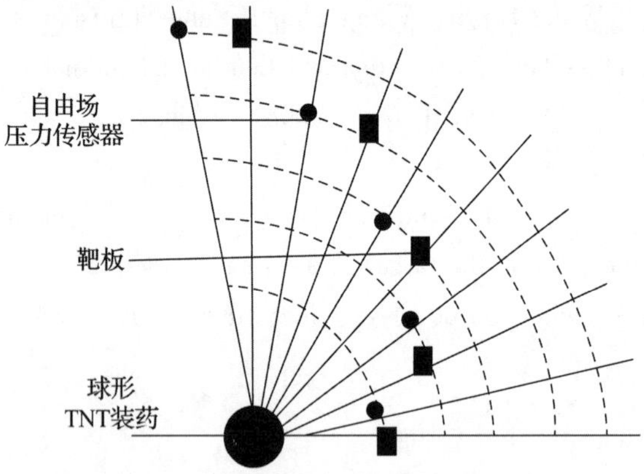

图 1　P－I 准则法评价 FAE 爆炸威力原理图

通过靶板毁伤效应试验获得不同厚度 δ（靶板易损性用其厚度 δ 来表征）靶板在各毁伤等级（D）下所对应的等毁伤曲线以及 FAE 爆炸场中 $\Delta p - i$ 曲线，其函数关系为

$$\Delta p = F_1(i, \delta, D) \tag{1}$$

$$\Delta p = F_2(i) \tag{2}$$

从图 1 可以看出，通过联立式（1）和式（2），可得到靶板在 FAE 爆炸场中遭受不同等级毁伤时所对应的 Δp 和 i 值（图 1 中 A，B，C，D 和 E 点所对应的坐标值）。因此，借助球形 TNT 装药近地爆炸冲击波参数经验计算式（3）和式（4）[5]，总能找到一个装药量为m_T的 TNT 爆炸场（对应于 A 点的等效 TNT 装药量爆炸场 $\Delta p - i$ 曲线如图 1 中曲线 g 所示），当靶板离该 TNT 爆炸场中心的距离与 FAE 相同时，两者对靶板所造成的毁伤等级相同。因此，根据等距离等毁伤效应等威力原则，可以认为m_T即为该试验 FAE（燃料量为m_F）在距爆炸场中心距离为 R 处的等效 TNT 威力当量。

$$\Delta p = 0.103\,\frac{\sqrt[3]{m_T}}{R} + 0.403\left(\frac{\sqrt[3]{m_T}}{R}\right)^2 + 1.27\left(\frac{\sqrt[3]{m_T}}{R}\right)^3 \quad \left(1 \leqslant \frac{R}{\sqrt[3]{m_T}} \leqslant 15\right) \tag{3}$$

$$i = 340\,\frac{m_T^{2/3}}{R} \quad (R > 12r_0) \tag{4}$$

式中　m_T——球形 TNT 装药质量，kg；

Δp——冲击波超压，MPa；

i——比冲量，Pa · s；

R——测点距爆心的距离，m；

r_0——球形 TNT 装药半径，m。

3　评价技术

根据上述评价方法，在对 FAE 爆炸威力进行具体评价时，需着重解决两方面的问题：一是获得不同厚度（δ）靶板在各毁伤等级（D）下所对应的等毁伤曲线，二是获得 FAE 爆炸场 $\Delta p-i$ 曲线。

3.1　靶板等毁伤曲线的确定

等毁伤曲线反映靶板（或目标）本身的固有特性，对于厚度一定的靶板，等毁伤曲线只取决于毁伤等级，与爆炸场特性无关。这表明，靶板等毁伤曲线既可借助 FAE 爆炸场，也可借助 TNT 爆炸场毁伤效应试验来获得。但从试验可操作性和结果可靠性看，采用 TNT 爆炸场更为适宜。因此，实验中可采用球形 TNT 装药作为爆炸源，通过在距爆心不同距离处分别设置试验靶板和自由场压力传感器并根据回收靶板残骸以及传感器所获信号，可对靶板毁伤等级及其所对应的 Δp 和 i 值进行确定。实验中有关靶板和传感器布置如图 2 所示。

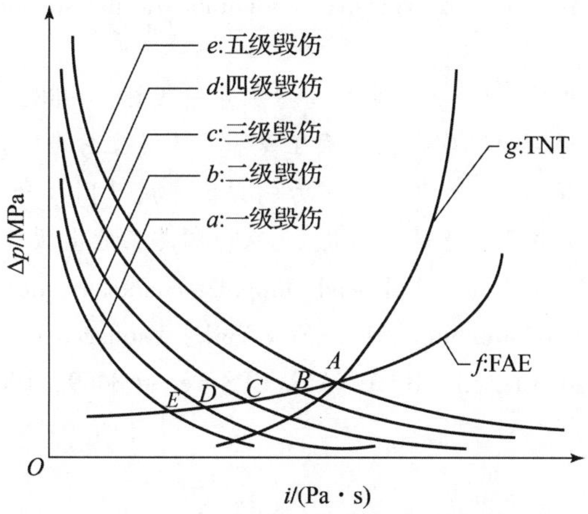

图 2　靶板等毁伤曲线实验测试示意图

由于准确测量比冲量参数较为困难，先借助球形 TNT 装药近地爆炸冲击波参数经验计算式（3）和式（4）对实验结果进行修正，然后采用 Abrahamson 和 Lindberg 理论等毁伤曲线的无量纲函数形式[6,7]：

$$\left(\frac{i}{i_0}\right)^2=\frac{3(\Delta p/p_0)}{4(\Delta p/p_0-1)}\quad(1<p/p_0<2)\tag{5}$$

$$\left(\frac{i}{i_0}\right)^2=\frac{3(\Delta p/p_0)}{3(\Delta p/p_0-2)}\quad(p/p_0\geqslant 2)\tag{6}$$

对修正后实验数据进行拟合以进一步减小实验误差，从而确立关系式（1）的具体形式。式中，p_0 和 i_0 分别为靶板在准静态条件下的临界毁伤超压和比冲量。

3.2　FAE 爆炸场超压与比冲量关系的确定

如图 2 所示，通过某一厚度（δ_0）靶板在 FAE 爆炸场中毁伤效应以及传感器所获信号对靶板毁伤等级及其所对应的 Δp 和 R 间关系确定，即

$$\Delta p=f(\delta_0,R,D)\tag{7}$$

将关系式（7）与式（1）联立，可求得各实测 Δp 相对应的 i 值，进而确立关系式（2）的具体形式。

最后，通过联立式（1）~式(4)，可对 FAE（燃料量为 m_F）爆炸中距爆心不同距离处的等效 TNT 威力当量做出评价，在冲击波 P－I 毁伤准则和靶板毁伤效应基础上建立 FAE 爆炸威力的评价方法。

4 结束语

本文从冲击波超压 - 冲量毁伤准则出发，提出一种以靶板毁伤效应为依据的 FAE 威力评价方法，并就其实施途径进行了探讨。该方法的显著优点在于：一方面，由于采用了等距离等毁伤效应等威力评价原则，因而评价结果更具科学性和实用价值；另一方面，由于靶板等毁伤曲线的测定在 TNT 爆炸场中进行，而 FAE 爆炸场中超压 - 比冲量关系，只需借助某一厚度靶板在 FAE 爆炸场中的毁伤效应测试结果即可获得，因此，评价技术途径具有较好的可操作性。从应用的角度看，一方面，利用该方法不仅可就 FAE 对特定易损性目标的毁伤效应做出评价，而且还可进一步就 FAE 武器所对付的目标特性做出分析和评判；另一方面，借助该评价方法，FAE 战斗部设计者还可根据目标特性，从优化 FAE 爆炸场超压和比冲量组合关系（Δp，i）的角度来进一步提高 FAE 武器的威力性能。

参考文献

[1] SEDGWICK R T，KRATA H R. Fuel air explosives：a parametric investigation [R]. AD - A159177/5/HDM，1979.

[2] 许会林，汪家华. 燃料空气炸药 [M]. 北京：国防工业出版社，1980.

[3] 惠君明. 燃料空气炸药威力评判的讨论 [J]. 兵工学报 · 火化工分册，1995 (2)：50 - 54.

[4] 惠君明，刘荣海，彭金华，等. 燃料空气炸药威力的评价方法 [J]. 含能材料，1996 (3)：123 - 128.

[5] 北京工业学院八系. 爆炸及其作用：下册 [M]. 北京：国防工业出版社，1979.

[6] ABRAHAMSON G R，LINDBERG H E. Peak load - impulse characterization of critical pulse loads in structural dymanics，in dynamic response of structure [M]. New York：Permagon Press，1972.

[7] SEWELL R G S. Blast damage criterion [R]. NAYWEPS Report 8469，1964.

FAE 爆轰参数计算与性能设计探讨[①]

贵大勇[②]，冯顺山，刘吉平，俞为民

（北京理工大学，北京市　100081）

摘　要：采用 VLW 状态方程及其程序计算环氧丙烷/铝粉燃料的 FAE 爆轰参数，研究了分散浓度（氧平衡）对 FAE 战斗部威力性能的影响。结果表明，按照微负氧设计，有利于提高 FAE 爆轰威力和燃料能量利用率；铝粉配比的增加能提高 FAE 爆轰参数，但配比过高不利于燃料分散，导致 FAE 氧平衡不好，反而影响 FAE 战斗部的威力性能。

关键词：燃料空气炸药（FAE）；爆轰参数；分散；性能设计

Calculation of Detonation Parameters and Performance Designs of FAE

Gui Dayong，Feng Shunshan，Liu Jiping，Yu Weimin

（Beijing Institute of Technology · Beijing 100081）

Abstract：In this paper，the detonation parameters of FAE for propylene oxide/aluminum powder were calculated with VLW EOS and its code，and the effects of dispersion concentration of the fuels in the air（oxygen balance）DD the performance of FAE warhead were studied，The results shows that the design according to slight oxygen deficiency is favourable for the improvement of detonation properties of FAE and the energy utilization ratio of fuels，and the increase of ratio of aluminum powder in the fuel is beneficial to detonation properties of FAE，but too high ratio of aluminum powders bad for the dissemination of the fuels and results in the decrease of the performance of FAE warhead instead

Key words：fuel air explosive（FAE）；detonation parameter；dispersion；performance design

1　引言

燃料空气炸药（FAE）因具有能量高、分布爆炸等传统弹药所不具备的特点而受到国内外的广泛重视。与凝聚炸药不同，燃料空气炸药要利用空气中的氧参与反应，因此其组成、密度等不仅与燃料本身有关，还与抛撒、分散的状况和周围空气环境密切相关。所以，FAE 的抛撒与分散的技术水平直接影响到云雾区质量，从而影响到 FAE 的起爆与威力效果。

近年来，针对 FAE 战斗部冲击波威力较低的问题，有许多人通过使用高能燃料的途径进行研究，并对 FAE 云雾形成规律进行试验和数值模拟，取得了一定进展[1-3]。但对于燃料抛撒与分散对 FAE 战斗部威力性能的影响，进而指导 FAE 战斗部的设计，没有较深入的研究报道。

本文针对以环氧类化合物（如环氧乙烷（EO）、环氧丙烷（PO）等）及其添加金属粉（如铝粉、镁粉等）为燃料的 FAE 战斗部，通过计算 FAE 的爆轰参数研究燃料在不同配比时 FAE 的爆轰特性，并研究 FAE 云雾区不同分散状况下的爆轰性能，目的在于为 FAE 战斗部设计提供指导。

---

①　原文发表于《弹箭与制导学报》2000 年第 1 期。

②　贵大勇：工学博士，1997 年师从冯顺山教授，研究固液燃料混合 FAE 战斗部设计，现工作单位：深圳大学。

2 FAE 爆轰参数的计算

2.1 计算方法

在计算炸药爆轰参数的方法中，应用最广的 BKW 状态方程已有五六十年的发展历史，其计算程序比较适用于计算凝聚类炸药，但对于计算液体和粉状燃料存在严重缺陷。基于维里（Virial）理论和相似论建立起来的 VLW 状态方程，近年已发展为从常压到几十万大气压广阔压力范围均适用的通用状态方程，其炸药密度范围适用于从 0.005 g/cm^3 到 3.5 g/cm^3，且克服了 BKW 状态方程的爆温计算值较实验值偏低的问题[4,5]。因此，对于燃料空气炸药爆轰参数，本文采用 VLW 爆轰产物状态方程及其程序进行计算。VLW 状态方程表达式如下：

$$\frac{PV}{RT} = 1 + B \cdot \left(\frac{b_0}{v}\right) + \frac{B^*}{T^{*1/4}} \sum_{N-3}^{M} (n-2)^{-n} \left(\frac{b_0}{v}\right)^{n-1} \tag{1}$$

式中 B^*——无量纲第二维里系数。专采用了 Lennard - Jones 势时，则

$$B^* = \sum_{j=0}^{\infty} \left[-\frac{2^{j+\frac{1}{2}}}{4j!} \Gamma\left(\frac{j}{2} - \frac{1}{4}\right) T^{*-(2j+b_0)} \right. \tag{2}$$

式中 $T^* = \frac{KT}{\varepsilon}$，$b_0 = \frac{2}{3}\pi N \sigma^3$。

这里：ε，σ 为 Lennard—Jones 势参数。

2.2 不同配比的 FAE 爆轰性能的比较

本文以环氧丙烷液体燃料及其与铝粉构成的液固混合燃料为例，采用上述 VLW 状态方程及其程序计算 FAE 的爆轰参数。为比较不同配比条件下 FAE 的爆轰性能，计算中取各 FAE 的氧平衡条件一致，氧的量按正好使 FAE 组成中的 H 全部氧化成 H_2O、AL 全部氧化成 AL_2O_3、C 全部氧化成 CO 进行计算确定，此时氧平衡（O，H）为微负氧平衡。表 1 是环氧丙烷与铝粉在不同配比时计算得到的爆轰性能参数。将其中爆压对铝粉含量作图，如图 1 所示。

表 1 不同配比 PO/AL 粉燃料空气炸药爆轰参数计算值

燃料	配比/%		爆压	爆速	爆温	爆热
	PO	AL				
P0	100	0	2.308 0	1 839	2 712	374.7
P9	90	10	2.458 3	1 890	2 951	405.2
P8	80	20	2.605 4	1 934	3 189	429.7
P7	70	30	2.732 5	1 970	3 418	448.0
P6	60	40	2.841 8	2 002	3 653	463.6
P5	50	50	2.950 1	2 034	3 922	480.4
P4	40	60	3.076 1	2 071	4 266	503.0
P3	30	70	3.237 7	2 106	4 673	537.4

计算结果表明，随燃料中铝粉比例增加，燃料空气炸药的爆压、爆速、爆温等呈直线升高，表明在理想抛撒和分散，且燃料较充分反应的条件下，增加燃料中铝粉的含量，能显著提高燃料空气炸药的威力性能。

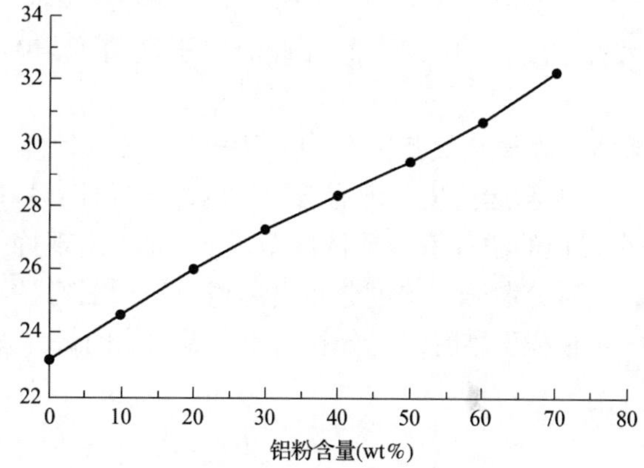

图1　FAE 爆压对铝粉含量的关系曲线

2.3　燃料分散浓度对 FAE 爆轰性能的影响

以上是在燃料抛撒为微负氧的假设条件下计算所得。然而，对于实际 FAE 战斗部，其分散燃料的浓度在中心区及边沿较低，在整个云雾区范围内呈一定分布，这样构成不均匀的氧平衡环境。根据经验及数值模拟一般如图 2 所示[3]。

如图 2 所示，对应于不同的爆炸分散半径处，云雾处于不同的氧平衡，可细分为五个区域（*a*，*b*，*c*，*d*，*e*）。在正氧平衡范围，*e* 区对应浓度极低，甚至不能起爆的区域；在负氧平衡范围，*a* 区对应于极高浓度区域（根据抛撒情况也可能不存在），甚至超出了爆炸极限；*b* 区对应负氧平衡区域（对于 PO，−0.191 2≤0，B < −0.119 6）；*c* 区对应微负氧平衡区域（对于 PO，0.119 6≤O，B <0）；*c*，*d* 交界处为零氧平衡点；*d* 区对应正氧平衡区域（对于 PO，0 < O B≤0.072 55）。而且，对于不同的装药条件，形成的 FAE 云雾浓度或氧平衡分布状态也将有所不同。

为了研究不同分散状况对 FAE 爆轰性能的影响，采用 VLW 程序计算不同浓度或不同氧平衡的 FAE 爆轰参数，结果如图 3 ~ 图 5 所示。

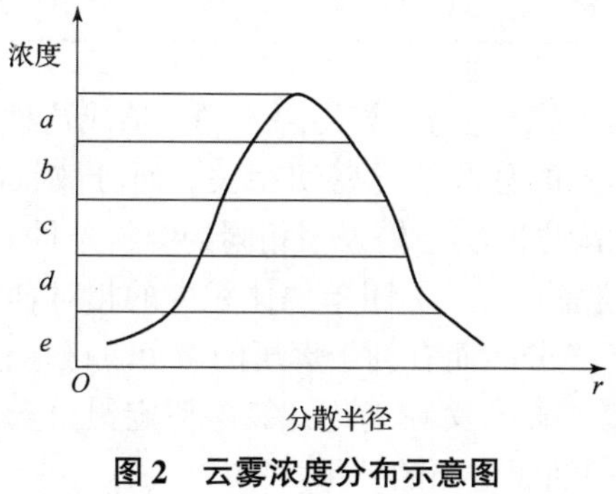

图2　云雾浓度分布示意图

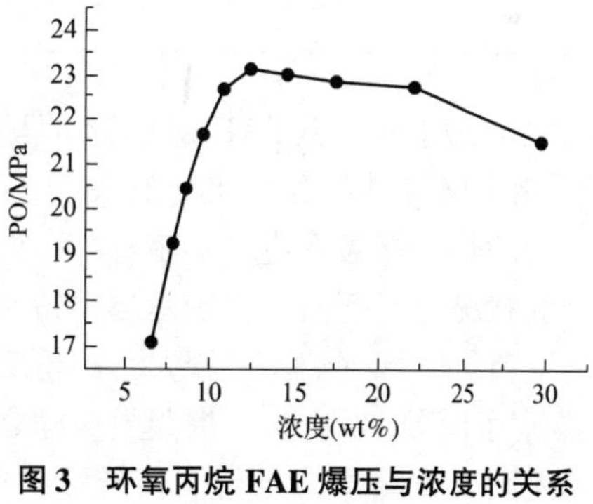

图3　环氧丙烷 FAE 爆压与浓度的关系

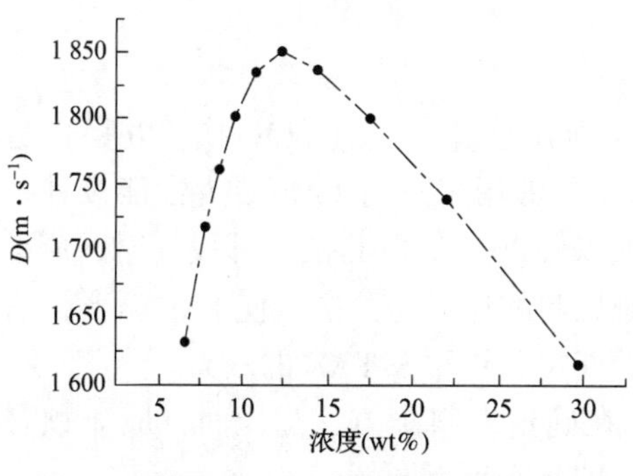

图4　环氧丙烷 FAE 爆速与浓度的关系

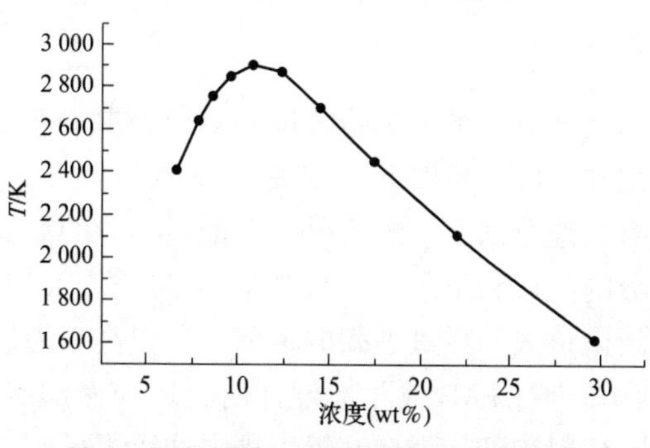

图5　环氧丙烷 FAE 爆温与浓度的关系

可以看出环氧丙烷在空气中的质量分散浓度（wt%）达到12.35%时（此时氧平衡值为0.068 06），FAE爆轰压力和爆速达到最大；浓度在10.78%时（此时氧平衡值为－0.029 69），FAE的爆温处于最高值。

不同氧平衡下的计算结果表明在微负氧范围（c区）内，FAE比严重负氧平衡、正氧平衡甚至零氧平衡的FAE具有较高的爆压、爆速和爆温值，即有着更好的爆轰性能；同时说明燃料的分散浓度能够显著影响FAE的爆轰，存在一个最佳的浓度值（或浓度范围），此时出现峰值爆压或爆温。因此，从战斗部设计的角度出发，应使燃料抛撒分散开，浓度分布更理想，使得云雾浓度有更大的范围处于c区，尽量减小a和e存在区域，这样才有利于燃料的充分利用和爆轰性能的最佳发挥。

3 与实验结果的对比

对不同配比的燃料的计算结果与文献实验数据进行比较，如表2所示。由表2可以看出，对于单一环氧丙烷燃料，本文计算超压值与实验结果吻合较好；对于含铝粉的液固混合燃料，其超压变化规律在铝粉比例较低时计算结果与实验结果是一致的。

表2 计算数据与实验数据的比较

燃料代号		PO	P88	P76	P70	P59	P53
配比/%	PO	100	88	76	70	59	53
	Al粉	0	12	24	30	41	47
文献［6］实验数据	ΔP_{max}/MPa	2.500 7	2.451 7	3.187 2	3.265 6	3.098 9	1.274 9
	PO(wt%)	6.69	5.50		5.02	3.73	
文献［7］实验数据	ΔP_{max}/MPa	2.109					
	PO(wt%)	9.62					
本文计算值	ΔP_{max}/MPa	2.074 5	2.386 4	2.554 9	2.631 2	2.751 3	2.816 3
	PO(wt%)	9.62	13.21	11.85	11.14	9.74	8.93

当铝粉比例大于30%时，其FAE超压反而呈下降趋势，这与计算结果不同。造成这种差别的原因在于FAE的爆轰性能不仅与燃料配方组分有关，还与燃料的分散状况密切相关。而分散状况直接影响到FAE中燃料组分的燃烧与爆轰反应。文献［6］中不同配方的实验均采用相同的结构条件和一致的参数。不同的配方其分散状况有所不同，特别是当铝粉比例较高时，云雾团单位体积内的相对含量（wt%）明显降低，造成严重负氧区域（a区）较大，铝粉反应不充分，而有利于爆压的微负氧区域（c区）较小，整个燃料的能量利用率低，所以实际爆轰效应随之降低。而本文在理论计算中假定铝粉全部参与反应且完全燃烧，因此得到FAE爆轰参数计算值随铝粉含量增加而增加的结果。

4 讨论

根据本文计算结果与实验数据的对比及其分析，本文所用计算方法（VLW状态方程）适用于燃料能充分分散和反应的条件下FAE爆轰参数的计算与评价，能够用来进行FAE威力性能设计。对于燃料组分不能有效地全部反应的情况，其爆轰参数的计算需加以修正。在FAE战斗部设计时，应综合考虑高能燃料组分的配比和抛撒，铝粉含量不可过高，以及控制燃料的抛撒与分散，使FAE云雾区有尽可能宽的区域处于微负氧和零氧平衡的范围，以提高高能燃料的能量利用率及FAE的威力。

此外，在进行FAE战斗部设计时，还应考虑到：①在确定燃料配方、配比时，应兼顾有效作用范围；②此时燃料浓度是否在其爆轰极限的可靠范围之内，即是否可靠起爆。

由本文2.2节可看出，在相同的微负氧平衡条件下，可计算得到相同质量环氧丙烷燃料及其与铝粉构成的不同比例混合燃料的云雾体积。随着燃料中铝粉配比的增加，虽然爆压增加，但云雾体积却直线下降，即作用范围呈直线减小。

同样，可计算得到在相同质量时不同分散浓度下单一环氧丙烷燃料的云雾体积随分散浓度的升高整体呈指数形式下降。因此，选择较低的分散浓度作为依据来进行 FAE 战斗部的设计，有利于提高 FAE 的威力效应。对于单一环氧丙烷燃料，选择 9%~14% 的分散浓度比较合适。

对表 1 中各配方计算燃料在空气中的浓度，并与相应爆炸极限比较，如表 3 所示。

表 3 不同配比混合燃料中 PO 在空气中的浓度

燃料	PO	P9	P8	P7	P6	P5	P4	P3
铝粉配比(wt%)	1	10	20	30	40	50	60	70
PO(wt%)	14.47	13.43	12.32	11.14	9.88	8.52	7.07	5.51
PO($g \cdot m^{-3}$)	217.9	203.3	187.6	170.6	152.3	132.4	110.6	86.91

由表 3 中的数据可以看出，这几种含铝混合燃料及单一环氧丙烷燃料按照微负氧设计时，其较敏感组分（PO）的浓度均处在环氧丙烷的爆炸极限（45~360 g/m^3）之内。因此，按照微负氧设计，满足 FAE 可靠起爆条件。当然，含铝粉比例不同，其浓度分布不同，能满足起爆的区域大小也将随之变化。综合考虑 FAE 爆轰参数、有效作用范围、起爆可靠性和铝粉的分散性，并参照实验数据，液固混合燃料中铝粉的配比按 30%~40% 设计是可行的方案。

5 结束语

（1）采用 VLW 状态方程及其程序计算环氧丙烷及其与铝粉在不同配比时混合燃料的 FAE 爆轰参数，并与实验结果进行了对比，表明本文所用计算方法适用于燃料能较充分分散及反应的条件下 FAE 爆轰参数的计算与评价，能够用来进行 FAE 威力性能设计。

（2）将 FAE 战斗部云雾按浓度分布或不同的氧平衡划分成五个区域，能够清楚地研究燃料抛撒状况对 FAE 威力的影响，指导 FAE 战斗部的设计。在微负氧范围内，FAE 比严重负氧、零氧及正氧平衡的 FAE 具有较高的爆压、爆速和爆温值；表明燃料的分散浓度或氧平衡状态能够显著影响 FAE 的爆轰，且存在一个最佳的浓度值或浓度范围。对于环氧丙烷燃料，选择分散浓度 9%~14% 比较合适。

（3）在相同微负氧假设条件下，增加燃料中铝粉的含量，能显著提高燃料空气炸药的爆轰性能参数。综合考虑 FAE 爆轰参数、有效作用范围、起爆可靠性和铝粉的分散性，并参照实验数据，液固混合燃料中铝粉的配比按 30%~40% 设计是可行的方案。

参 考 文 献

[1] 杨立中，张春云，张正才，等. 高能 FAE 燃料的选择 [J]. 南京理工大学学报，1998，22 (1)：19-22.

[2] 蒲加顺，白春华，梁慧敏，等. 多元混台燃料分散爆轰研究 [J] 火炸药学报，1998，21 (1)：1-5.

[3] BING S，DIMITRI G，RONAID P，et al. The numerical modeling of an explosive dissemination system [C] //Proceedings of the 20th International pyrotechnics Seminar，1 994：1061-1081.

[4] 吴雄. 应用 VLW 状态方程计算炸药的爆轰参数 [J]. 兵工学报，1985 (3)：11-20.

[5] 张熙和，云主惠. 爆炸化学 [M]. 北京：国防工业出版社，1989.

[6] 惠君明. 计算机在高威力 FAE 配方设计中的应用 [C] //1987 年兵工学会第二届年会暨信息发布会，1987.

[7] SEDGWICK T R，KRATZ H R. Fuel air explosives：a parametric investigation [C] //Proceedings of the 10th Symposium on Explosives and Pyrotechnics，San Francisco，1979.

反导战斗部破片作用场分析①

黎春林②，冯顺山

（北京理工大学，北京市 100081）

摘 要：针对在反导战斗部工程设计过程中，为了合理控制破片空间作用场，对破片空间作用场的计算方法进行了理论推导，得出了具有普遍意义的反导战斗部破片空间作用场的理论分析模型。

关键词：战斗部；破片；作用场；反导

Analysis on Antimissile Warhead Fragment Impact Field

Li Chunlin, Feng Shunshan

Abstract: It is the kernel matter during the antimissile warhead engineering design how to control the fragment space impact field reasonably. This paper briefly analyses the calculating method of the fragment space impact field. The general theory analytic model of the antimissile warhead fragment impact field is given.

Key words: warhead; fragment; impact field; antimissile

1 引言

众所周知，破片式杀伤战斗部在防空反导中对付各类空中目标时有较好的适应性，利用大威力破片对导弹进行解体毁伤是常用的反导方法之一。破片式战斗部对目标的毁伤原理是利用战斗部爆炸后形成高速破片杀伤场，靠高速破片的动能对目标产生贯穿、引燃或引爆作用，达到毁伤目标的目的。在反导战斗部的设计过程中，提高单位面积内的有效毁伤破片密度，是提高对导弹的毁伤率的有效途径，通常采用定向、汇聚等多种破片结构形式，以期达到控制破片在空中的散布，形成高密度破片带。

破片式杀伤战斗部通过调整破片的飞散角、单枚破片质量、破片数和破片初速等手段来适应不同的作战需要，其中调整破片飞散角，提高有效作用空间内破片的密集度是重要手段之一。由于破片的形成与飞散是一个十分复杂的过程，影响破片的作用场的因素很多，传统上是以实验研究和统计规律为基础指导弹药设计。本文在对破片速度轴向分布规律的研究基础上，对汇聚型破片的空间作用场的计算模型进行理论推导，提出了具有普遍意义的分析模型，对不同时刻破片空间作用场的计算具有指导意义。

2 模型的简化

常见的破片式杀伤战斗部，其破片的分布形式基本上都是轴对称的旋转空间，空间问题可以简化为平面问题进行研究。汇聚型破片战斗部是常见的反导战斗部结构之一，其主要结构特点是对不同位置的破片提前预置一个破片飞散方向角，当战斗部爆炸时，使破片按预定的方向汇聚，以达到提高预定空间内破片密度的目的。选择一端起爆的轴对称战斗部作为研究对象，假定爆炸冲击波近似看作球面波，冲击波作用场做一维传播。如图1所示，以起爆点为圆心，弹轴为 x 轴，建立平面坐标系，战斗部壳体由函数 $y=f(x)$（假定破片的尺寸为无穷小，$f(x)$ 为连续函数且一阶可导）确定，任取壳体上一点 $A(x, y)$ 处的微元 dl 作为研究对象，微元内破片的初速大小为 x 的函数，记为 $V(x)$。

---

① 原文发表于《弹箭与制导学报》2003年第3期。

② 黎春林：工学博士，2000年师从冯顺山教授，研究超近程反导武器系统及其战斗部技术，现工作单位：公安部一所。

2.1　破片飞散方向的计算

图2所示壳体上任一点 $A(x, y)$ 处，研究对象微元 $\mathrm{d}l$(AB 段) 成破片过程。爆轰波波阵面（传播速度为 D）由 $A-A_1$，运动到 $B-B_1$，前进距离为 $D\cdot\mathrm{d}t$，在 A 点，壳体向外运动速度（即破片速度）由 0 增至 $V(x)$，壳体转体 δ 角，假定 AB 段只改变方向，不改变长度[1]，在等腰三角形 ABC 中，根据正弦定理可得

$$\frac{AC}{\sin\delta}=\frac{AB}{\sin\left(\frac{\pi-\delta}{2}\right)}$$

$$AC=\frac{1}{2}V(x)\,\mathrm{d}t \tag{1}$$

$$AB=D\times\frac{\mathrm{d}t}{\cos\theta} \tag{2}$$

式中　θ 是爆轰波波阵面与弹体内壁的法线方向夹角，根据几何关系可得 θ 的表达式为

$$\theta=\arctan\frac{y}{x}-\arctan\frac{\mathrm{d}y}{\mathrm{d}x} \tag{3}$$

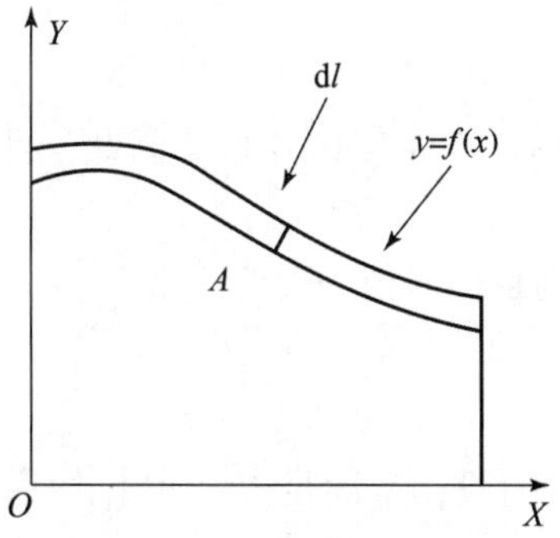

图1　战斗部壳体示意图

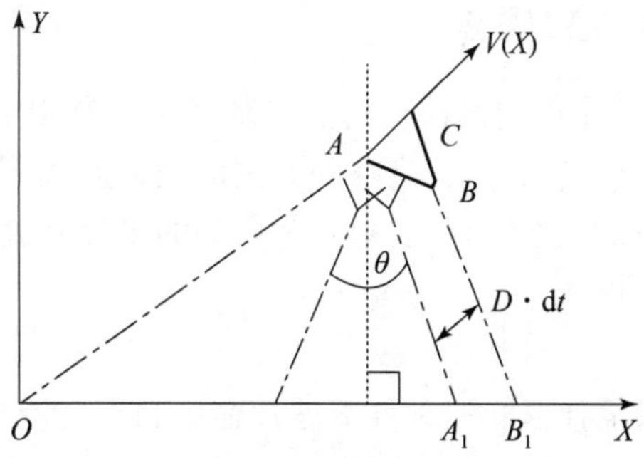

图2　微元几何关系

将式（2）和式（3）代入式（1），同时考虑到占值较小，可取 $\sin(\delta/2)=\delta/2$，得破片转体角 δ 的表达式为

$$\delta=\frac{V(x)}{2D}\cos\left(\arctan\frac{y}{x}-\arctan\frac{\mathrm{d}y}{\mathrm{d}x}\right) \tag{4}$$

2.2　破片初速的计算

为了获得有普遍意义的精确计算破片初速的表达式，前人进行了大量有成效的工作，经过大量的理论与实验研究，总结了一些可供工程设计和计算的公式[2]。考虑到起爆位置和断面的影响，本文采用文献［3］得到的求解任意壳体形状的战斗部破片初速轴向分布的一般表达式：

$$V(x)=[1-A\mathrm{e}^{-Bx/d(x)}][1-C\mathrm{e}^{-D(L-X)/d(x)}]\sqrt{2Eg}\sqrt{\frac{\beta(x)}{1+\frac{1}{2}\beta(x)}} \tag{5}$$

式中　x——起爆点到计算部分的距离；

$d(x)$——装药对应 x 处的外径；

$\beta(x)$——x 点处对应的装填系数；

L——壳体的有效长度。

A，B，C，D 为由试验数据确定的常数，一端起爆条件下，$A=0.361\,5$，$B=1.111$，$C=0.192\,5$，$D=3.03$。

由式（5）可知，只要知道对应 x 处的 $d(x)$ 和 $\beta(x)$，便可求得该处的破片初速。

2.3 破片空间作用场的描述

假定破片形成后的运动为直线运动，则由几何关系可得壳体上任一点 A(x, y) 处的破片形成后在给定坐标系内任一时刻的位置坐标计算式为

$$\begin{cases} X(t) = x + \int_0^t V(t)\sin\left(\dfrac{\pi}{2} + \arctan\dfrac{dy}{dx} - \delta\right)dt \\ Y(t) = y + \int_0^t V(t)\sin\left(\dfrac{\pi}{2} + \arctan\dfrac{dy}{dx} - \delta\right)dt \end{cases} \tag{6}$$

式中 $V(t)$——破片在空间任一时刻的存速，由式 (7)[1] 给出：

$$V(t) = V(x)e^{-aR(t)} \tag{7}$$

其中 a——速度衰减系数；

$R(t)$——破片 t 时刻的行程。

$$R(t) = \sqrt{[X(t)-x]^2 + [Y(t)-y]^2} \tag{8}$$

由计算机进行叠代计算，将式 (4)、式 (5) 代入式 (6) 和式 (7)，便可求出破片在空间任一时刻的作用场。

3 目标函数确立

汇聚破片战斗部的设计目的是控制破片在空中的散布，形成高密度破片带。因此，目标函数选定为求设计汇聚点 (x_0, y_0) 一定空间 S 内的破片数 N 最大值。

取空间 S 以汇聚点 (x_0, y_0) 为中心的半径为 R 的圆，则目标函数为

$$N = \max\sum_{t=t_1}^{t_2} N_i \tag{9}$$

式中 t_1, t_2 满足弹目交汇条件下破片命中目标的有效作用时间段；N_i 满足式 (10) 约束条件的有效破片数。

$$\min\sqrt{[x(t)-x_0]^2 + [y(t)-y_0]^2} \leqslant R, \quad t_1 \leqslant t \leqslant t_2 \tag{10}$$

4 计算与分析

根据以上模型，可以计算出由给定曲线 $y=f(x)$ 形成由战斗部壳体在爆炸过程中产生的破片在任一时间内的空间作用场，以便于对战斗部壳体形状进行优化设计。图 3 是某一给定曲线函数确定的战斗部壳体线段 AC，爆炸后形成的破片连线在不同时刻 (t_i) 对应的空间位置图 (A_iC_i)。

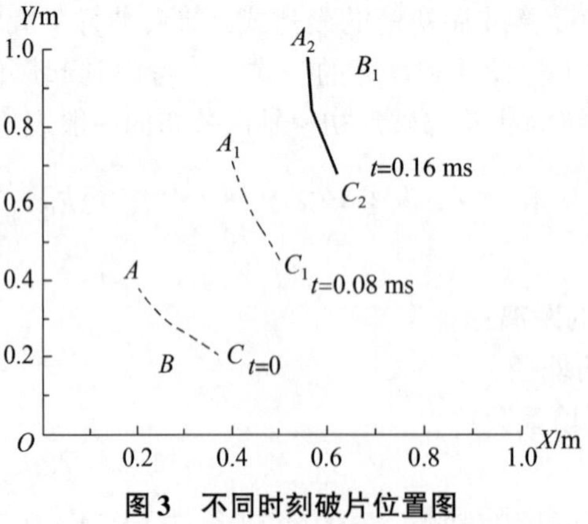

图 3　不同时刻破片位置图

图 3 显示出弹药壳体曲线 AC 在爆轰波作用下，在 2 个不同时间产生的破片的连线 (A_iC_i) 的几何形状和位置，图中，O 为起爆点，OX 为弹体轴线，虚线为 AA_1 点形成破片运动方向，虚线为 C 点形成破

片的运动方向，虚线 BB_1 为壳体曲线 AC 形成破片的几何汇交中心线。由图示表明，破片在逐渐向以虚线所示的中心线 BB_1 汇聚，破片在特定方向上的密度在加大，但同时由于不同位置对应的战斗部装填系数不同造成速度差影响，破片连线（A_iC_i）在不断地拉长，相邻破片之间的距离在增加破片到达汇聚点存在一个时间差。

5 结论

（1）反导战斗部汇聚型破片结构设计目的是获得在弹目交汇过程中，在预定空间内的杀伤破片密度的最大化。本文在力求模型具有普遍代表性的前提下，推导出由任一曲线确定的战斗部壳体形成破片的空间飞散方向计算模型，并根据给定的壳体形状函数表达式，求出任意点（x，y）的一阶导数 dy/dx 和 $\beta(x)$，便可计算出战斗部爆炸后任一时刻破片的作用场，考虑到战斗部的轴对称性，可将计算结果延伸到三维空间。

（2）汇聚型破片战斗部结构设计中，在考虑破片飞散方向的汇聚过程时，要注意沿轴向破片速度变化的影响，随着破片飞行时间的增长，虽然破片在某一给定方向上有汇聚效应，破片的方向密度在增加，但随着破片飞行时间的增加，单个破片之间的相对距离在拉大，破片到达汇聚区存在一个时间差，对于较高速运动的导弹目标来说，若弹目距离较长，则单位时间内作用在目标上的破片密度有可能降低，达不到汇聚型破片的设计目的，必须合理控制破片的速度差。

（3）该计算模型可以为汇聚型战斗部壳体的几何形状设计提供计算值，结合相应的弹目交汇条件、破片的毁伤判据、破碎模型[4]等，可用于对汇聚型战斗部壳体的优化计算和对导弹的毁伤效能进行分析。

参考文献

[1] 爆炸及其作用编写组. 爆炸及其作用：下［M］. 北京：国防工业出版社，1979.

[2] CHARRON Y J. Estimation of velocity distribution of fragmenting warheads using a modified Gurney method［R］. AD－A 74759，1979.

[3] 冯顺山，崔秉贵. 战斗部破片初速轴向分布规律的实验研究［J］. 兵工学报，1987（4）：60－63.

[4] LLOYD R M. Convention warhead system physics and engineering design［M］. Reston，VA：AIAA，1998.

空间轨道目标终端毁伤方式研究①

刘春美[1,2]②，冯顺山[1]，张旭荣[1]，董永香[1]

（1. 北京理工大学“爆炸科学与技术”国家重点实验室，北京 100081；

2. 中北大学机械工程与自动化学院，山西太原 030051）

摘 要：对目前空间轨道目标终端毁伤方式进行了分析，研究了硬毁伤与软毁伤方式对目标的毁伤手段，美国动能武器中反卫星的动能拦截弹技术及其杀伤增强装置，俄罗斯的反卫星能力。分析了目前直接碰撞技术的缺点，提出了非直接碰撞的等待式威力场且可实现多次打击能力的观点。

关键词：终端毁伤方式；动能拦截弹；KKV；非直接碰撞

The Research on Terminal Lethality Modes to Spacecraft

Liu Chunmei[1 2], Feng Shunshan[1], Zhang Xurong[1], Dong Yongxiang[1]

(1. The State Key Laboratory of Explosion Science and Technology,

Beijing Institute of Technology, Beijing 100081, China;

2. Mechanical Engineering & Automation School, North University of China, Taiyuan 030051. China)

Abstract: Terminal lethality mode to spacecraft at present was analyzed. The lethality ways on hard lethality and soft lethality were studied. Then the technology of kinetic energy intercept pellet and its enhanced device to anti－satellite was investigated especially in America and Russia. At last the shortcoming was analyzed on Kinetic Kill Vehicle. And it was proposed that the point of undirected impact, waiting power field and realizing multi－hit would come true

Keywords: terminal lethality mode; kinetic energy intercept projectile; kinetic kill vehicle; undirected impact

0 引言

迄今为止，在世界各国发射的空间飞行器中，2/3 以上是直接或间接用于军事目的的。一些卫星通过获取军事情报、监视敌方动向、了解气象状况、提供通信网络、发布导航信息等方式，配合和支持地面的军事行动，起到了扩大地面军串行一动效果的“倍增器”作用[1]。除军用卫星外，美国和苏联/俄罗斯还发射了可用于军事目的的宇宙飞船、空间站、航天飞机及反卫星、反导弹武器等。太空军事化将使未来战争由空中扩展到太空，正如空中优势曾在现代战争中起过特别重要的作用，太空优势也将显著地影响未来战争。世纪之交的几次局部战争，势必促使有关国家尤其是美、俄等发达国家发展航天武器装备，以夺取太空优势，进而达到控制太空、控制地球的目的。

空间飞行器在军车上的广泛应用，构成了对敌方军事力散的威胁，必然带来空间防御问题，导致了反卫星武器的研制，尤其是动能反卫星武器。美国已经演示验证了反卫星和反弹道导弹，在飞行试验中证实其技术可行性，有的动能反卫星武器甚至已进入工程研制发展阶段，在 21 世纪的最初 10 年将具备

① 原文发表于《中北大学学报（自然科学版）》2006 年第 27 卷增刊。

② 刘春美：工学博士后，2007 年师从冯顺山教授，研究等待式拦截幕形成及对飞行器作用机理与应用，现工作单位：公安部第一研究所。

作战能力[2,3]，依靠动能毁伤空间轨道目标技术已受到世界各军事强国的大力关注。

本文主要对动能反轨道目标终端毁伤技术的最新进展及发展趋势进行研究，并对毁伤方式进行讨论，提出一些自己的看法。

1 终端毁伤方式研究

总结各国在反卫星技术的研究成果，可以把对轨道目标的终端毁伤方式分为硬毁伤与软毁伤两类[4-7]。

1.1 硬毁伤方式

硬毁伤方式是通过攻击性武器毁伤空间飞行器从而达到摧毁目的，包括动能武器、定向能武器等，现有的动能武器利用武器本体与目标碰撞，可以一举摧毁敌方卫星或航天器，或利用增强装置中发出的破片，通过打击航天器内部局部的设施，使之永久性失效。

动能武器硬毁伤方式采用的手段：一是使用反卫星武器毁伤，如使用截击卫星、动能拦截弹直接碰撞目标造成目标的完全粉碎性失效；二是使用核能或化学能毁伤，如使用携带核装药战斗部或高能炸药的反卫星天雷、反卫星导弹等，在目标附近引爆、摧毁目标，美国动能武器中的天基反卫星导弹是种小型的动能武器，可以部署在航天站或者专用航天平台上对敌方的卫星和天基武器系统发起攻击；反卫星导弹对轨道高度低于1 000 km的航天器有较强的攻击力。

定向能武器有高能激光武器、粒子束武器与微波武器，通过发射高能激光束、粒子束、微波束直接照射目标进行破坏。利用定向能杀伤手段摧毁空间目标，具有速度快、攻击空域广的特点，但技术难度较大。

1.2 软毁伤方式

软毁伤方式是通过破坏敌方空间飞行器的传感器、通信设备、通信链路、太阳能电池板来毁伤目标，使目标暂时失效、工作性能降低或者彻底瘫痪。软毁伤方式对抗的手段主要依靠电子干扰设备、干扰剂、射束能武器来干扰星上的各个系统等。

（1）由于所有的空间飞行器都与地面接收站、接收机或空间的其他飞行器之间存在信息交换，有信息交换就有通信链路，通过电子干扰武器发送功率、频率、调制方式匹配的电磁信号，可以有效干扰卫星的通信链路，或者通过电子阻塞使卫星的上载信息或下载信息获取能力得到暂时有效抑制，或者通过电子欺骗给敌方发送错误的信息，达到反空间轨道目标的目的。

（2）低功率的激光束能武器瞄准的目标是侦察卫星的光学系统和传感器，或者卫尾的姿态传感器之类的薄弱部件，通过射束能武器的光电效应、热效应来达到致盲卫星、干扰卫星、压制卫星的作用。

（3）在敌方卫星的轨道上或卫星的监视视场内释放金属碎片与颗粒、气溶胶、特种材料等干扰物，使敌方卫星的电池板、光学器件或者天线等暂时失去效能或者功能退化，如对敌方卫星大阳能电池板喷涂一层油漆使其失能，则卫星上供电系统就会受到严重影响，导致星载计算机系统缺电不能正常工作，或通信系统因缺电而无法与地面联络；对于卫星上的天线进行频率干扰，使其无法接收与发送信息；对卫星上的探测用的光学镜头及红外摄像头喷涂特殊材料，使其无法“看到”或侦察到目标，或者在轨道上释放导电纤维丝团，一旦丝团缠绕到卫星上，就会造成天线或电池短路，毁坏整个敌方卫星系统。

（4）另外还可以用寄居卫星，这是一种体积极小、能寄附在敌方卫星上的小型卫星，能在战时根据己方相应的指令对敌方卫星进行干扰或摧毁。

当然有时硬毁伤与软毁伤没有明显的区别，对卫星的毁伤程度依赖于所使用武器的状况，如用激光武器或微波武器攻击卫星上的电子系统，功率稍大就会造成硬毁伤使电子系统完全破坏乃至永久性破坏，功率较低时可能仅仅使其传感器等临时失效，经过一段时间后可恢复使用。

2 国外动能武器终端毁伤方式的现状及发展趋势

2.1 动能武器概念

动能武器属于硬毁伤方式，是利用高速飞行器的巨大动能，以直接碰撞的方式摧毁目标，也可以通过弹头携带的高能爆破装置在目标附近爆炸产生密集的金属碎片或霰弹击毁目标。动能武器中的动能拦截弹由动能杀伤拦截器（kinetic kill vehicle，KKV）和推进系统两部分组成，推进系统用于把 KKV 发射到预定空域并将其加快到足够高的速度，KKV 能够精确自主寻的，通过直接碰撞摧毁目标。依据部署的位置，动能武器可分为地基、海基、空基和天基等类型。动能拦截弹系统，主要用于反导弹和反卫星[8]，是目前在技术上比较成熟、重点发展的动能武器。目前发展这种技术的国家有美国、英国、法国、俄罗斯和以色列等国。

2.2 美国发展的反卫星动能拦截弹

自 20 世纪 80 年代以来，美国在发展动能武器，特别是动能拦截弹的技术上取得了重大进展，不仅初步演示验证了用动能拦截弹拦截弹道导弹和反卫星，证实其技术上具有可行性，而且在实现动能拦截弹的轻小型化方面取得了重大突破，从而为发展反卫星武器系统开辟了一种新的途径。美国在动能武器的发展方面处于世界绝对领先地位。

用于反卫星的动能拦截弹，如美国空军研制并试验的由飞机携带在空中发射的微型寻的拦截器（MHV）和由美国陆军研制的地基动能反卫星（KEASAT）拦截弹等。

美国大部分动能拦截器本身没有战斗部、引信等结构，减轻了弹的重量，从而也大大降低了拦截器的质量；同时由于利用弹本身的巨大动能毁伤目标，因而杀伤威力大。当然其必须有极高的制导精度作为前提。在制导精度有限的情况下，其大都采用了杀伤增强装置。例如，智能卵石，为保证制导精度达到足以使拦截器本体直接碰撞卫星的杀伤效果，设有伞型杀伤增强装置。这种装置具有折叠式的可径向展开的伞状结构，金属伞展开后迎着近于法线方向与卫星目标相撞，可增大碰撞面积。采用穿透与冲击两级杀伤机理相互补充，以提高杀伤概率，冲击机构是一个增强聚薄膜板，穿透机构则是分布在薄膜上的小球。这个可膨胀的聚酯薄膜与高密度小球组合使用，可达到穿透和压碎卫星结构、碰撞毁掉关键附属部件的目的。

外大气层拦截器（ERIS）：通过拦截器本体直接碰撞杀伤目标。ERIS 前端是红外导引头，为扩大拦截器杀伤机构的横剖面，采用了中心可展开的杀伤增强机构，是一个搀和金属粉的塑料充气网，展开呈分角形，展开直径 0.94 ~ 3 m，重 5.8 kg，可能会将其用作地基直接上升式反卫星武器。

随着技术的不断发展，美国研制的各种动能拦截弹战斗部技术已经出现了轻小型化、智能化、通用化的趋势。

2.3 俄罗斯/苏联发展的反卫星武器

苏联从 1964 年开始研制反卫星武器，在 1968—1971 年进行了有关接近、识别与摧毁目标的飞行试验，1976—1982 年又进行了旨在提高实战能力的快速发射、拦截和新型制导技术的试验，从而具备了反卫星作战的能力。苏联的反卫星武器是一种带有雷达或红外制导装膛、轨道机动发动机和高能炸药破片杀伤战斗部的卫星，质量为 2.5 ~ 3 t，长 4.5 ~ 6 m，直径 1.5 m，由液体运载火箭从地面基地发射到与目标航天器同一平面的轨道，然后通过地面制导使它与目标交会，最后利用自身的制导装置接近目标到一定距离后爆炸，并摧毁目标。共轨式反卫星拦截器是一种以卫星反卫星的地基拦截器，也称作“拦截卫星”或“杀伤卫星”。其战斗部的主要技术指标如下：质量 363 kg，采用近炸引信和高能炸药，在接近目标数十米处爆炸，通过产生的大批破片式毁伤元击毁卫星；破片为菱形，质量 0.5 ~ 2 g，破片初速 305 ~ 610 m/s；当靠近目标至一定距离，相对速度达到每秒几十米到几百米时，战斗部引爆，其破片摧毁卫星。

俄罗斯还研制了另外一种号称天雷的卫星，它更加先进，具有一定的轨道机动能力，可以在 5°~10° 范围内改变轨道角，拦截不同高度敌方卫星。俄军在 1992 年已经承认其拥有和随时可以投入实战的这两种共轨式反卫星武器。但由于其发射阵地固定，因此只能攻击地球低轨道卫星。除了已部署的共轨式反卫星拦截器和正在研制的定向能反卫星武器外，俄罗斯/苏联还有以下几种具有反卫星能力的武器："橡皮套鞋"反导系统具有拦截美国低轨道卫星的能力，航天飞机和航天站可作为天基反卫星武器平台支援反卫星作战，射频武器具有对卫星进行电子干扰的能力，机载反卫星武器，太空雷反卫星武器。

上述研究现状表明，动能式轨道拦截的空间技术已经得到长足的发展，但随着探测技术与智能控制技术的发展，空间领域的武器逐渐向增强打击精度和效能、增加探测功能、研制增强杀伤威力场的方向发展，空间轨道发展特种威力场、实现多次打击技术也必然是未来的研究方向之一。

3　技术分析及结论

多年来有些专家学者希望用直接碰撞途径 KKV 打击空间轨道目标，笔者认为它存在以下缺点：

（1）目前的动能拦截弹均采用直接碰撞硬毁伤方式，实质上这是一种理想的毁伤方式，需要极高的打击制导精度。实际上，未来的空间轨道目标可能具有智能规避能力，如同导弹打飞机，由于人员操纵飞机具有高度机动性，而单发炮弹的脱靶率很高，导致需要多发炮弹连续打击以确保打击精度，这还是在声速下发生的，若目标在空间轨道中超高速运动，飞行速度达到几千米每秒，甚至十几千米每秒，交汇速度则更高，稍有差错则稍纵即逝，对抗双方不容易形成直接碰撞，直接命中概率很低直接影响打击精度。

（2）目前的动能拦截弹均采用自杀式的直接碰撞毁伤，只能进行一次性打击，属于极低效费比的方式。

（3）碰撞后造成双方硬毁伤粉碎性破坏，且容易形成更多的空间垃圾。

随着科技的发展，终端毁伤方式中的直接碰撞难度在增加，对抗双方不易形成直接碰撞，从而对非直接碰撞提出挑战。笔者认为：①可以发展一种威力半径可调的等待式的杀伤场；②在目标交汇前形成一种均匀分布且慢速扩张的威力场；③甚至可多次实现这样的威力场以打击多个目标。

显然这样的非直接碰撞更有效，毁伤概率只是均匀分布的概率问题，打击精度高；一次打击后没有达到预定效果，可实现多次打击多个目标，提高效费比；打击后只是局部破坏航天器结构使之完全失效，即发生局部硬毁伤而不会产生大量空间碎片，对己方卫星不会造成损伤，也不会造成毁伤能力的浪费。

参考文献

[1] 林聪镕，徐飞．世界航天武器装备［M］．长沙：国防科技大学出版社，2001：1－2.

[2] 刘绍球，高淑霞，高永浩，等．反卫星武器研制状况的分析研究［J］．系统工程与电子技术，1994（1）：16－21.

[3] 张纯学．动能武器及其发展［J］．飞航导弹，2004（8）：21－24.

[4] 熊伟，贾鑫反卫星方式的技术与性能分析［J］．电子对抗技术，2004，19（5）：36－40.

[5] WRIGHT D，GREGO L，GRONLUND L. The physics of space security：a reference manual［M］．Cambridge，MA：American Academy of Arts and Sciences，2005：117－138.

[6] MISAWA M. Model satellite in stiffness design of deployed satellite antennas［R］．AIAA－2001－1229，2001：1－7.

[7] 王敏．反卫星光电传感器激光对抗技术［J］．光电对抗与无源干扰，1999（2）：19－22.

[8] STEGMAIER J，GRANNAN M. Kinetic energy anti－satellite weapon system［R］．AIAA 92－1355，1992：451－460.

破片对空间目标毁伤效应的数值仿真研究①

李　磊②，冯顺山，董永香

（北京理工大学爆炸科学与技术国家重点实验室，北京　100081）

摘　要：在信息战争中，卫星等空间目标对战争胜负起着至关重要的作用，空间武器化是未来战争发展的趋势，研究动能破片对空间目标防护结构的毁伤能力具有现实意义。文中应用非线性动力学分析软件 AUTODYN－2D 的光滑质点流体动力学（SPH）方法对破片超高速碰撞多层靶板进行了研究，得到了破片的形状、质量、冲击速度和角度对破片毁伤能力的影响规律，为高效毁伤空间目标的破片设计提供了依据。

关键词：破片；空间目标；毁伤效应；数值仿真

Numerical Simulation of Damage Effect of Fragment on Space Target

LI Lei②，Feng Shunshan，Dong Yongxiang

Abstract：Satellites play a vital role in modern information war and space weapon will be fully used in future war. It is significant to research damage ability of kinetic energy fragment to protect structure of space target. Interactive Non－Linear Dynamic Analysis Software AUTODYN－2D is employed to research hypervelocity impact of fragment on multi－sheet bumpers. Effects of fragment shape，mass，impact velocity and angle to its damage ability and the way to design fragment which damage space target efficiently are available.

Keywords：fragment；space target；damage effect；numerical simulation

1　引言

从毁伤的角度出发，利用 AUTODYN－2D 数值仿真软件的 SPH 方法[4]研究了球形、柱形和杆形破片对给定的空间目标的毁伤能力的差别，最终为高效毁伤空间目标的破片设计提供依据。

2　空间目标防护结构简化模型

有效毁伤空间目标必须有效突破其防护结构。航天器的 Whipple 防护方案，其主要思想是采用两层板防护结构，前板使超高速碎片破碎甚至液化、汽化，经过板间距，碎片云扩散，这样就达到了使碎片能量发散的目的，从而减轻了对后板或舱壁的破坏。

多层防护结构的防护性能要远比 Whipple 防护结构好，但是不同的防护结构的性能是和防护间距相关的。对于大多数航天器，由于外形尺寸的限制，防护间距一般不超 110 mm，所以绝大多数航天器采用充填式防护结构。在空间目标普遍采用多层防护结构的背景下，把防护结构近似简化成三层 6061－T6 铝靶来研究破片的毁伤效果，三层靶板厚度分别为 1.5 mm、3 mm 和 4.5 mm，间距为 80 mm 和 40 mm，具体布置图见图 1。

---

① 原文发表于《弹箭与制导学报》2005 年第 1 期。

② 李磊：工学博士，2003 年师从冯顺山教授，研究空间目标易损性及毁伤效应，现工作单位：北京航天长征飞行器研究所（14 所）。

3　数值模拟材料模型和状态方程

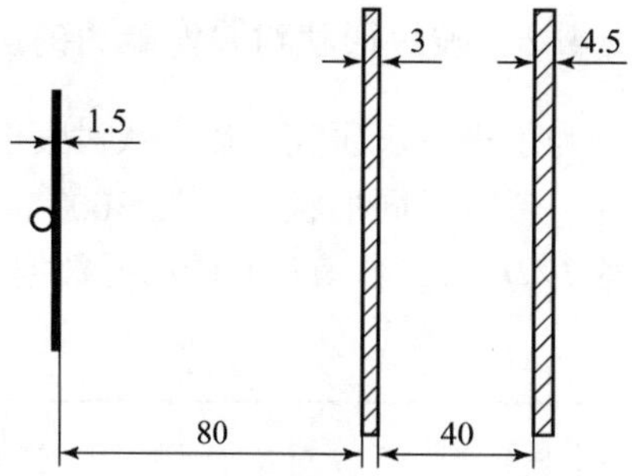

图 1　靶板布置示意图

破片在超高速碰撞下将会发生液化甚至汽化等相变，一旦破片发生冲击相变，其毁伤能力将会受到很大的影响，因此在数值计算时，采用具有描述相变情况能力的状态方程。文中对破片采用更能适用于熔化和汽化情况的 Tillotson 状态方程，而对靶板采用超高速数值模拟中广泛采用的 Mie – Grüneisen 状态方程，材料强度采用 Steinberg – Guinan 模型，采用最大主应力失效准则，当材料的最大拉应力或者最大剪应力超出了所定义的失效应力就认为材料失效。2024 – T3 Al 和 6061 – T6 Al 的最大拉应力分别定义为 2.5 GPa 和 2.6 GPa[2]。其他的材料参数采用 AUTODYN 提供的材料库中的数据，同时在强度模型中引入热软化指数来近似模拟液化对材料强度的影响。

4　模拟结果以及分析

4.1　数值仿真结果与实验对比

Piekutowski 对非球形锌材料破片超高速碰撞所产生的碎片云进行了大量的实验研究[3]，得到了不同形状破片超高速碰撞后碎片云分布的照片，具体实验数据如表 1 所示，其中 R 为球形破片半径和柱形、杆形破片端面半径，L 为柱形、杆形破片轴向长度。

表 1　实验数据

项目	球形破片	柱形破片	杆形破片
几何形状/mm	$R=2.883$	$R=2.528$ $L=5.055$	$R=1.999$ $L=14.148$
撞击速度/(km · s^{-1})	4.980	5.055	4.970

文中采用铝材料破片和靶板，在破片的形状和碰撞速度与文献［3］实验相同的条件下进行了超高速碰撞的数值仿真研究，并与实验进行了对比，虽然所用材料不同，结果有所差异，但是碎片云的分布大体相同，比较结果见图 2。

(a)

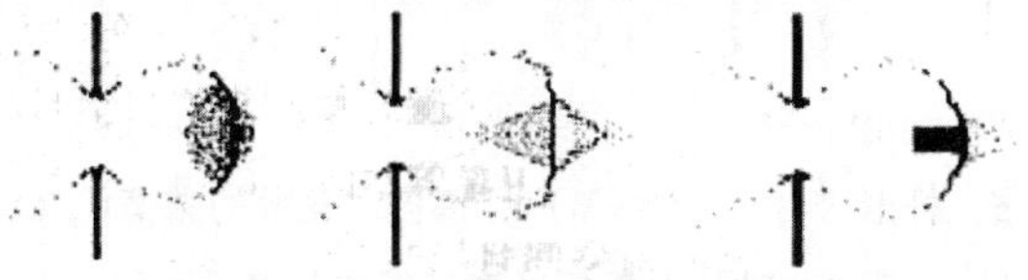

(b)

图 2　不同形状破片超高速碰撞的数值仿真结果与实验照片对比

(a) 不同形状锌材料破片超高速碰撞靶板的实验照片；(b) 不同形状铝材料破片超高速碰撞靶板的数值仿真结果

从实验和数值仿真中都可以看出，在碰撞速度近似相等的条件下，球形破片和柱形破片贯穿第一层靶板后就发生了破碎，形成了碎片云并在两靶间距中扩散，而杆形破片并没有完全破碎，仍有完整的部分存在，因此可以预测杆形破片的毁伤能力要比球形破片和柱形破片强。

4.2 破片形状对毁伤能力的影响

为了进一步研究破片形状对毁伤能力的影响，对相同质量不同形状的破片的毁伤能力进行数值仿真研究，考察不同形状破片在5 000 m/s的速度下对多层靶板的毁伤能力，其中，柱形破片$L/D=1$，杆形破片$L/D=3$。数值仿真的初始数据如表2所示。

表2 数值仿真的初始数据

项目	球形破片	柱形破片	杆形破片
几何形状/mm	$R=3.40$	$R=2.88$ $L=5.76$	$R=2.00$ $L=112.00$
撞击速度/($km\cdot s^{-1}$)	5.00	5.00	5.00

图3是球形破片碰撞多层靶板的数值仿真结果，18.7 μs时碎片云到达第二层靶板并穿孔，大部分碎片云被挡在第一靶板、第二靶板中间；39.4 μs时到达第三层靶板，但是由于剩余速度不够，未能穿透第三靶板。图4是长径比为1的柱形破片碰撞多层靶板的数值仿真结果，16.7 μs时碎片云到达第二靶板，碎片云最终未能穿过第二靶板。图5是长径比为3的杆形破片碰撞多层靶板的数值仿真结果，16.7 μs时碎片云到达第二层靶板，破片前部形成了碎片云，后部仍保持完整并与第二靶板超高速碰撞，再次形成碎片云，于29.1 μs时到达第三靶板，最终穿过第三靶板。从数值仿真结果中很容易分辨不同形状破片的毁伤能力的高低，杆形破片的毁伤能力远比球形和柱形破片好，柱形破片在正碰撞时毁伤能力最低，对靶板的穿孔情况见表3。

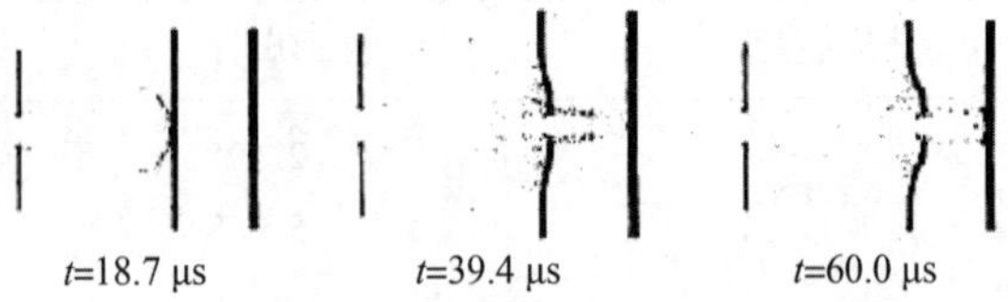

图3 球形破片碰撞多层靶板的数值仿真结果

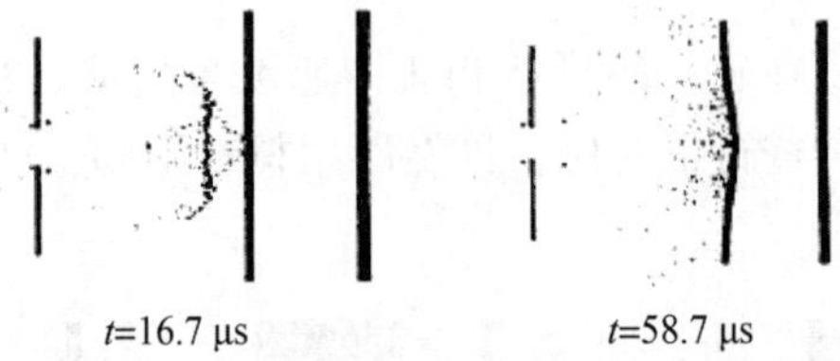

图4 柱形破片（$L/D=1$）碰撞多层靶板的数值仿真结果

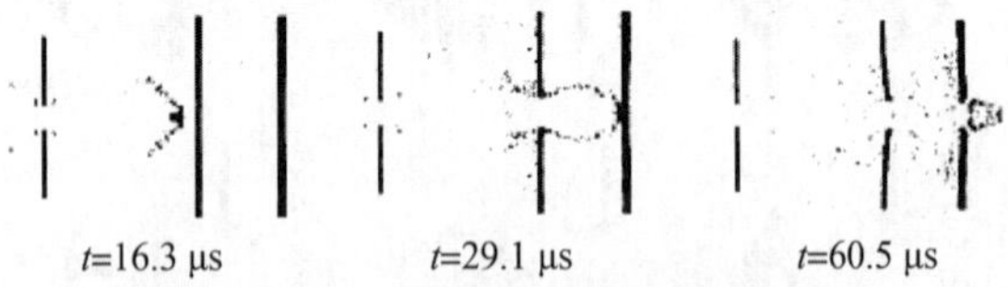

图5 杆形破片（$L/D=3$）碰撞多层靶板的数值仿真结果

表3 破片对靶板超高速撞击的穿孔直径 单位：mm

靶板	球形破片	柱形破片	杆形破片
第一靶板	15.8	14.6	10.2
第二靶板	37.6	0	14.4
第三靶板	0	0	12.2

4.3 杆形破片在不同碰撞角度下的毁伤能力

从4.2节可以看出，在相同质量下，杆形破片的毁伤能力远大于其他两种形状的破片，但是破片在与目标碰撞时不一定是正碰撞，其碰撞角将会对破片的毁伤能力产生影响，采用与4.2节中相同的杆形破片，针对其轴线与速度方向夹角为10°、30°、60°三种情况进行数值仿真研究。

从数值仿真的结果（图6和图7）可以看出，杆形破片在碰撞角为10°和30°时，碎片云很容易穿透第三层靶板，在碰撞角为60°、时间为59.8 μs时，碎片云都被第三层靶板拦截，但是第三层靶板已经出现局部大变形和裂纹。

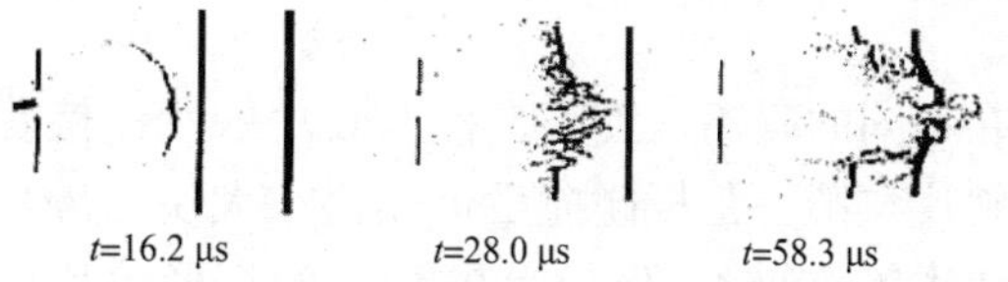

图6　杆形破片在碰撞角为10°时的仿真结果

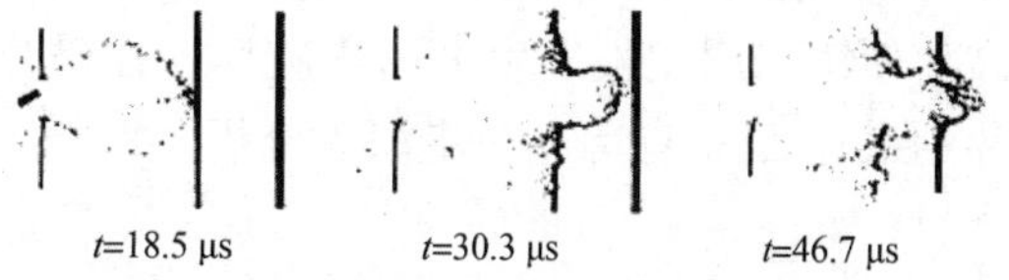

图7　杆形破片在碰撞角为30°时的仿真结果

4.4 碰撞速度与临界质量的关系

文中破片的临界质量定义为在给定速度和正碰撞条件下破片刚好穿透三层靶板时的质量，利用数值仿真研究这一虚拟实验工具，研究三种不同形状的破片在不同的正碰撞速度下其临界质量的变化规律，数值仿真结果见图8，从图9可以看出，球形破片在5 000 m/s时的临界质量大于3 000 m/s时的临界质量，原因可能是碰撞速度为5 000 m/s时球形破片发生了碎化，侵彻能力降低，碰撞速度大于5 000 m/s时，临界质量随着速度的增加而减小，杆形破片在超高速碰撞下前端破碎，但是后端仍然保持完整，因此在相同的碰撞速度下其临界质量远远小于球形破片和柱形破片，但是当碰撞速度大于5 000 m/s以后，由于强烈的碰撞，其完整的部分越来越少，对后板的侵彻能力降低，因此临界质量出现增加的趋势。

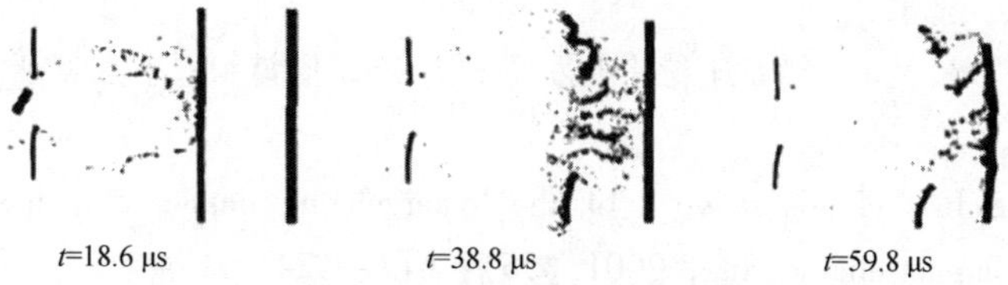

图8　杆形破片在碰撞角为60°时的仿真结果

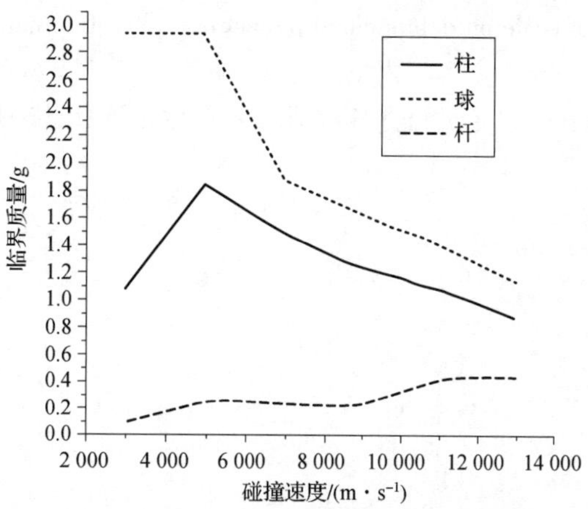

图9　碰撞速度与临界质量之间的关系

5 讨论

影响破片形成碎片云主要有以下两个因素[6]：一是破片的动能，二是破片的特征尺寸与靶板厚度比，破片与靶板碰撞瞬间，冲击波分别在破片和靶板中反向传播，在各自的自由端反射成拉伸波，拉伸波相互作用即在该处产生拉伸断裂，在数值仿真中，柱形破片轴向尺寸与板厚比不是很大，在柱形破片内发生拉伸波的相互作用而导致层裂破碎。若破片的特征尺寸与靶板厚度比很大，破片中传播的冲击波还没有来得及在自由端反射就被靶板中反射的拉伸波追赶上，由于追赶卸载，破片将不会发生拉伸破碎，因此数值计算中的杆形破片后部并未发生破碎，破片前端破碎是由于材料承受不住相当强大的压应力而导致的。

从数值仿真中可以看出，在不同的碰撞角度下，杆形破片依然保持着较强劲的毁伤能力，其原因在于在某一临界碰撞角度以前，破片与第一靶板碰撞后仍然存在大块的碎片，这造成了能量集中，对后续靶板有很强的毁伤能力，碰撞角小于60°时，质量为0.42 g的杆形破片可以穿透第三层靶板，但是相同碰撞速度下，球形破片和柱形破片的临界穿透质量分别为1.846 g和2.198 g。从数值仿真中还可以看出，在一定碰撞角度下，杆形破片使第二层靶板发生大面积破坏，在破片场中可以增大其他破片直接碰撞第三层靶板的概率，杆形破片的最佳 L/D 有待进一步深入分析。

6 结束语

只要采用合理的算法、状态方程和材料模型，数值仿真就可以比较好地反映超高速碰撞的物理过程，从数值仿真中可以看出，在正碰撞速度为5 000 m/s的条件下，相同质量不同形状的破片对多层靶板的毁伤能力有很大的差别，杆形破片由于在碰撞后后部仍保持完整，对后板的毁伤能力远大于球形破片和柱形破片，而且在一定碰撞角度下，杆形破片仍然保持着较强的毁伤能力，因此设计高效毁伤空间目标的破片时，杆形破片是比较好的选择。

参考文献

[1] 张伟，庞宝君，邹经湘，等. 航天器微流星和空间碎片防护方案［J］. 哈尔滨工业大学学报，1999（2）：18－22.

[2] 张伟，曲焱喆，贾斌，等. 弹丸超高速撞击防护屏碎片云数值模拟［J］. 高压物理学报，2004，18（1）：47－52.

[3] PIEKUTOWSKI A J. Debris clouds reduced by the hypervelocity impact of non－spherical projectiles［J］. International journal of impact engineering，2001（26）：613－624.

[4] Autodyn users manual revision 4.3［Z］. Century Dynamics，2003.

[5] PIEKUTOWSKI A J. Effects of scale on debris cloud projectiles［J］. International journal of impact engineering，1997（20）：639－690.

[6] 张庆明，黄风雷. 超高速碰撞动力学引论［M］. 北京：科学出版社，2000.

地地战役战术导弹武器系统作战效能分析①

黄龙华[1②]，何黎明[2]，冯顺山[1]

(1. 北京理工大学爆炸科学与技术国家重点实验室，北京 100081；
2. 南昌陆军学院，南昌 330103)

摘 要：针对地地战役战术导弹武器系统作战效能评价过程的复杂性，提出了一种基于模糊层次分析法理论的数学评估模型，对该型导弹系统的综合评价指标体系进行了介绍，并且通过编程实现了综合评判过程的智能化，比传统的评判方法具有更高的精度和可操作性，从而为该型导弹武器系统的作战效能分析提供了一种新方法，具有一定的参考价值。

关键词：地地战役战术导弹；武器系统；作战效能；模糊综合评价

The Fuzzy Synthetic Evaluation of the Operational Effectiveness of ground - to - ground Campaign Tactical Missile

Huang Longhua[1], He Liming[2], Feng Shunshan[1]

(1 State Key Laboratory of Explosion Science and Technology, Beijing Institute of Technology, Beijing 100081, China; 2 Army)

Abstract: To reduce the complexity of operational effectiveness evaluation for ground - to - ground campaign tactical missile, a two - stage fuzzy combined evaluation model was given and corresponding evaluation process was described in this paper. In addition, the evaluation method which overcame the shortcoming of the old method, can be programmed easily and applied widely due to its simple, normative arithmetic formulas.

Keywords: ground - to - ground campaign tactical missile; weapon system; operational effectiveness; fuzzy synthetic evaluation

1 引言

文中提出基于模糊层次分析法对地地战役战术导弹武器系统作战效能进行评估的模型。建立了以“火力突击能力、战场生存能力、机动能力、电子对抗能力”为基础的作战效能评价的多层指标体系，有效地提高了评判过程的科学性和准确度。

2 地地战役战术导弹武器系统作战效能因素指标体系

根据地地战役战术导弹武器系统所担负的任务和特点，重点分析了影响作战效能的主要因素，建立了如图 1 所示的地地战役战术导弹武器系统作战效能因素指标体系。

该指标体系[1]将地地战役战术导弹武器系统作战效能从火力突击能力、战场生存能力、机动能力和电子对抗能力四个方面进行全面评价。文中的重点是强调该型导弹系统的作战效能集中体现在火力突击能力和战场生存能力两个方面。其中火力突击能力从以下四个方面进行评估：射击精度，命中与毁伤，导弹突防能力和射击反应能力，命中与毁伤和导弹突防能力应考虑到对不同目标进行突击时所选用的不

---

① 原文发表于《弹箭与制导学报》2005 年第 12 期。

② 黄龙华：工学博士，2003 年师从冯顺山教授，研究弹药终端毁伤效能评估方法，现工作单位：华东交通大学。

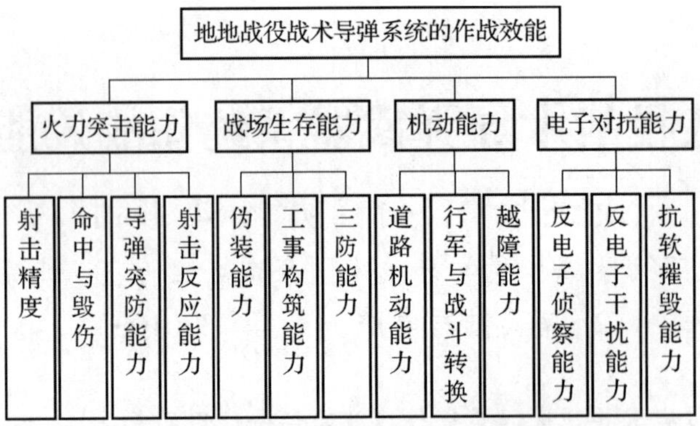

图1　地地战役战术导弹武器系统作战效能指标体系

同导弹类型。战场生存能力则从伪装能力、工事构筑能力和三防能力等方面进行评估。机动能力则从该武器系统的道路机动能力、行军与战斗转换和越障能力进行评估。电子对抗能力评估则主要集中在反电子侦察能力、反电子干扰能力和抗软摧毁能力。

由于各影响因素的综合性、模糊性的存在，对该型导弹作战效能的分析无法完全采用定量的方法。因此运用模糊层次分析法更能贴近实际情况，从而得到更为准确的评判结论。

3　作战效能的模糊层次分析方法

模糊综合评价是对受多种因素影响的事物做出全面评价的一种十分有效的多因素评价方法。在对地地战役战术导弹武器系统作战效能进行综合评价的过程中，由于涉及的因素很多，且先前建立的评价指标体系是递阶层次结构的，因此，采用多级模糊综合评判法[2]，可以更科学合理地进行效能分析。即先按最低层次的各个因素进行综合评价，然后一层一层依次往上进行评价，直到最顶层，从而得出评价的结果。其主要有以下步骤。

3.1　建立因素集

因素集是影响评价对象（作战效能）的各种因素所组成的集合。将因素集 $u=\{u_1,u_2,\cdots,u_n\}$ 分成若干组，即 $u=\{u_1,u_2,\cdots,u_k\}$，并以此作为第一级因素集。设 $u=\{u_1,u_2,\cdots,u_{ji}\}$ 为第二级因素集。图2所示为多层指标体系结构。

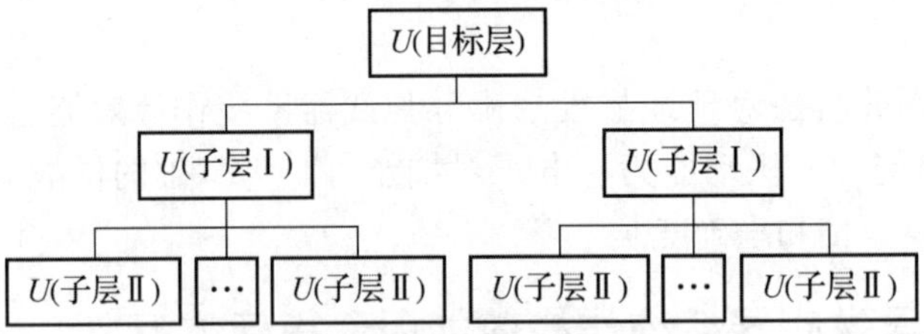

图2　多层指标体系结构

3.2　建立备选集

备选集是评价者对作战效能可以做出的各种总的评价所组成的集合，常采用专家评估法来确定备选集中各元素的取值，记作 $v=\{v_1,v_2,\cdots,v_n\}$，式中 $v_j(j=1,2,3,\cdots,m)$ 代表各种可能的评判结果。备选集评测结果一般为：好，较好，一般，较差，差。

3.3　确定权重集

权重反映了各个因素在综合评价过程中所占有的地位或作用。u_i 中各因素相对 u_i 的权重分配为 $A_i=$

$(a_1, a_2, \cdots, a_{ik})$，同时应满足：$a_{i1} + a_{i2} + \cdots + a_{ik} = 1$。

3.4 进行模糊综合评价

先对第二集因素集 $u = \{u_1, u_2, \cdots, u_{ji}\}$ 的 j 个因素进行单因素评判。所构成的单因素评价矩阵为

$$\boldsymbol{R}_i = \begin{bmatrix} r_{11}^{(i)} & r_{12}^{(i)} & \cdots & r_{1m}^{(i)} \\ r_{21}^{(i)} & r_{22}^{(i)} & \cdots & r_{2m}^{(i)} \\ \cdots & \cdots & \cdots & \cdots \\ r_{j1}^{(i)} & \cdots & \cdots & r_{jm}^{(i)} \end{bmatrix}$$

设 $u_i = \{u_{1i}, u_{2i}, \cdots, u_{ji}\}$ 的权重为 $A_i = (a_1, a_2, \cdots, a_{ik})$

求得单因素综合评价矩阵：

$$\boldsymbol{B}_i = A_i \cdot \boldsymbol{R}_i = (b_1, b_2, b_3, \cdots, b_j)$$

在对单层模糊综合评判时，可以采用以下四种方法[3]：

（1）模糊变换法。通过模糊变换的方式即“最大最小”原则获得评价结果向量，计算公式为

$$b_j = \bigvee_{i=1}^{n} (a_i \wedge r_{ij}) \quad ,j = 1,2,\cdots,m$$

（2）以“乘”代替“取小”，计算公式为

$$b_j = \bigvee_{i=1}^{n} (a_i r_{ij}) \quad ,j = 1,2,\cdots,m$$

（3）以“加”代替“取大”，计算公式为

$$b_j = \sum_{i=1}^{n} (a_i \wedge r_{ij}) \quad ,j = 1,2,\cdots,m$$

（4）加权平均，计算公式为

$$b_j = \sum_{i=1}^{n} a_i \wedge r_{ij} \quad ,j = 1,2,\cdots,m$$

其中，许多文献资料也说明了，由于在加权平均算法中按普通矩阵乘法计算权向量与评价矩阵的乘积，它的评价结果向量中包括了所有共同作用，是一种真正将影响因素“综合”考虑的评价方法，因此它的结果更符合客观情况。文中也就是采用加权平均算法，并对其他的算法进行了比较。

再按第一因素集 $u = \{u_1, u_2, \cdots, u_k\}$ 做综合评价。

设 $u = \{u_1, u_2, \cdots, u_k\}$ 的权重为 $A_i = (a_1, a_2, \cdots, a_{ik})$，则总评价矩阵为

$$\boldsymbol{R} = \begin{bmatrix} \underline{B_1} \\ \underline{B_2} \\ \vdots \\ \underline{B_K} \end{bmatrix}$$

综合评价矩阵 $\underline{\boldsymbol{B}} = \underline{A} \circ \underline{\boldsymbol{R}}$

3.5 综合评判结果的计算

为了充分利用综合评价提供的信息，将评判等级与相应的分数相结合，计算出综合评判结果 V'。

$$V' = \left(\sum_{j=1}^{n} b_j \cdot v_j' \right) \Big/ \sum_{j=1}^{n} b_j$$

其中：$v' = (v_1', v_2', v_3', v_4') = (1, 0.88, 0.76, 0.68, 0.6)$。

4 实例分析

由专家对某型地地战役战术导弹武器系统进行性能评价，每个性能分为五个级别，分别是：好、较

好、一般、较差、差，得到表1所示的评价集。并将其评价过程在用 Visual Basic 6.0 语言编写的程序[3]中实现。

表1　地地战役战术导弹武器系统作战效能指标评价集

级别	火力突击能力（u_1）				战场生存能力（u_2）		
	射击精度（u_{11}）	命中与毁伤（u_{12}）	导弹突防能力（u_{11}）	射击反应能力（u_{10}）	伪装能力（u_{11}）	工事构筑能力（u_{12}）	三防能力（u_{11}）
1（好）	0.38	0.20	0.42	0.25	0.18	0	0.20
2（较好）	0.46	0.35	0.28	0.45	0.32	0.24	0.18
3（一般）	0.14	0.22	0.15	0.14	0.28	0.46	0.40
4（较差）	0.20	0.15	0.15	0.08	0.12	0.10	0.22
5（差）	0	0.08	0	0.08	0.10	0.20	0

级别	机动能力（u_2）			电子对抗能力（u_2）		
	道路机动能力（u_{11}）	行军与战斗转换（u_{12}）	越障能力（u_{11}）	反电子侦察能力（u_{11}）	反电子干扰能力（u_{12}）	抗软摧毁能力（u_{11}）
1（好）	0.10	0.08	0.10	0.20	0.28	0.24
2（较好）	0.15	0.20	0.10	0.32	0.45	0.42
3（一般）	0.30	0.42	0.35	0.18	0.14	0.16
4（较差）	0.25	0.15	0.25	0.18	0.12	0.08
5（差）	0.20	0.15	0.10	0.12	0	0.10

对各种因素确定其权重。为更科学地确定权重数，可依据专家的评价集，利用程序所提供的方法，反演算出最佳的权重。

因素类权重为

$A=(a_1,a_2,a_3,a_{ik})=(0.42,0.26,0.18,0.14)$

$A_1=(a_{11},a_{12},,a_{13},a_{14})=(0.34,0.26,0.22,0.18)$

$A_2=(a_{21,}a_{22,},a_{23})=(0.46,0.32,0.22)$

$A_3=(a_{31,}a_{32,},a_{33})=(0.32,0.46,0.22)$

$A_4=(a_{41,}a_{42,},a_{43})=(0.35,0.19,0.46)$

将上述数据输入程序中，并进行第一层次的模糊综合评判计算，得到结果向量：

$\underline{B}_1=(0.319,0.39,0.163,0.154,0.035)$

$\underline{B}_2=(0.127,0.264,0.364,0.135,0.11)$

$\underline{B}_3=(0.091,0.162,0.366,0.204,0.155)$

$\underline{B}_4=(0.234,0.393,0.163,0.123,0.09)$

将上述结果向量建立一个新的判断矩阵，和前面给出的因素类权重 A 输入程序内进行第二层次的模糊综合评判计算，得到

$\underline{B}=(0.226,0.317,0.252,0.154,0.217)$

所以

$$V' = (0.226, 0.317, 0.252, 0.154, 0.217) \cdot \begin{bmatrix} 1 \\ 0.88 \\ 0.76 \\ 0.68 \\ 0.6 \end{bmatrix} = 0.992$$

从综合评判的结果来看，专家对该型地地战役战术导弹武器系统的作战效能给予了比较高的评价。当有多个型号的导弹系统可供选择时，可使用该模糊综合评判程序对其分别进行综合评判，选择一种作战效能最高的型号。

5　结束语

文中通过编程的方式实现了地地战役战术导弹武器系统作战效能的多层次模糊综合评判，较好地解决了影响因素不容易量化的问题，提高评判结果的准确度。

参考文献

[1] 张廷良，陈立新. 地地弹道式战术导弹效能分析 [M]. 北京：国防工业出版社，2001.
[2] 杨纶标，高英仪. 模糊数学原理及运用 [M]. 广州：华南理工大学出版社，2001.
[3] 李鸿吉. 模糊数学基础及实用算法 [M]. 北京：科学出版社，2005.
[4] 甄涛，王平均. 地地导弹武器作战效能评估方法 [M]. 北京：国防工业出版社，2005.

动能反空间目标战斗部的研制现状及若干问题分析[①]

李　磊[②]，冯顺山，董永香，何玉彬

（北京理工大学爆炸科学与技术国家重点实验室，北京　100081）

摘　要：在信息战争中，卫星等空间目标对战争的胜负起着至关重要的作用，空间武器化是未来战争发展趋势，发展反空间目标战斗部技术具有战略意义。在分析美国和苏联动能反卫星武器的研制状况的基础上，提出了其存在的问题，并就动能反空间目标战斗部设计中的若干关键问题进行了讨论，对高效毁伤空间目标战斗部技术的发展提出了一些建议。

关键词：反空间目标战斗部；空间目标；动能破片；超高速

引言

空间的地理位置得天独厚，在政治、经济、军事、外交等方面都具有非常重要的应用价值。在军事上，航天系统可提供通信广播、侦察监视、导航定位、导弹预警、气象保障、地形测绘、核爆炸探测和搜索救援等作战支援。在未来信息化战争中，航天系统是实施远程精确打击的必要手段，侦察监视等卫星系统可实时或近实时地为武器装备提供目标信息和打击效果评估，导航定位卫星可提高武器弹药的投掷与命中精度，航天系统在未来信息化战争中将发挥至关重要的作用，军事卫星系统已经成为整个战略系统中必不可少的部分。为了削弱对方的战争能力，摧毁敌方在太空中运行的各类卫星等空间目标将是打击敌人的有效手段。因此各军事大国争相发展航天技术，制定空间武器发展战略，力争控制空间，空间军事化和武器化的趋势不可避免。可以预见，随着航天技术的发展，以夺取“制天权”为目的、以军用航天器为主战力量、以军用航天器之间的攻击与防御为主要作战形式的“天战”将会是未来战争的发展趋势。

毁伤空间轨道目标战斗部技术受到了世界上各军事强国和我国有关部门的大力关注，本文主要对动能反空间轨道目标武器的发展以及趋势进行分析，进而对这种战斗部技术的一些关键问题进行讨论，提出自己的看法，以促进该技术的发展，为我国在未来战争中掌握主动权建立技术保障。

1　动能反卫星武器的国外研制状况

美国和苏联（俄罗斯）是世界上拥有卫星最多的两个国家，也是世界上最早发展反卫星武器的两个国家，这两个国家反卫星技术的发展已经有40多年的历史，已经研制了共轨式、破片杀伤式和激光式等多种反卫星武器，因此美苏两国的反卫星技术的发展和现状代表了当今世界在反卫星领域的研究水平和现状。

1.1　苏联反卫星武器的研制情况

从1968年到1983年的15年间，苏联在共轨式反卫星武器方面投入大量的力量进行研究，共进行了20多次试验，总成功率为60%以上。拦截最高高度为2 000 km左右，从地面发射到拦截卫星的时间在1小时之内。共轨式反卫星拦截器是一种以卫星反卫星的地基拦截器，也称作“拦截卫星”或“杀手卫

① 原文发表于《弹药战斗部学术交流会论文集》2005年第3期。

② 李磊：工学博士，2003年师从冯顺山教授，研究空间目标易损性及毁伤效应，现工作单位：北京航天长征飞行器研究所（14所）。

星”。战斗部质量363 kg，采用近炸引信引爆高能炸药，在接近目标数十米内爆炸，通过产生的大量碎片击毁卫星，破片呈菱形，质量0.5~2 g，破片速度305~610 m/s。拦截目标卫星时间1小时左右（第一圈轨道内拦截）到3.8小时（第二圈轨道内拦截）。其作战方式主要有两种：第一种是共轨道攻击，即拦截卫星进入与目标卫星轨道平面和高度相近的轨道，通过自身机动，在绕地球飞行若干圈后，逐渐接近目标，当靠近目标到一定距离、相对速度达到每秒几十米到几百米时，战斗部引爆，其碎片摧毁卫星。第二种是快速上升攻击，即把拦截器射入与目标卫星的轨道平面相同而高度较低的轨道，然后通过机动快速上升接近目标，可在第一圈轨道内就把目标卫星摧毁，显然，这种拦截方式的作战效率比第一种更高。苏联反卫星轨道拦截示意图如图1所示。

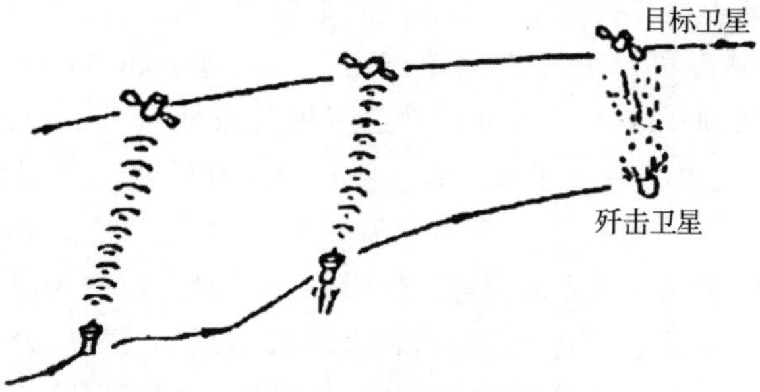

图1 苏联反卫星轨道拦截示意图

除了已部署的共轨式反卫星拦截器等武器外，苏联还有以下几种具有反卫星能力的武器。

（1）“橡皮套鞋”反导系统具有拦截美国低轨道卫星的能力；

（2）航天飞机和航天站可作为天基反卫星武器平台支援反卫星作战；

（3）射频武器具有对卫星进行电子干扰的能力；

（4）机载反卫星武器；

（5）太空雷反卫星武器；

（6）动能反卫星武器。

1.2 美国反卫星武器的研制情况

美国比较重视动能反卫星武器的研究，1981年，美国空军完成了机载反卫星导弹的地面试验；1984年开始进行空中发射的飞行试验，至1985年共进行了5次；1985年9月13日，首次成功地用反卫星导弹击毁一颗在500多千米高轨道上的军用实验卫星。如图2所示这种反卫星武器是利用性能优越的战斗机快速爬升时的能量，使导弹能突破发射时所受到的大气层的阻力，然后助推火箭点火迎向对方的卫星，依靠激光回转仪导引最后撞击目标。该导弹的战斗部为动能碰撞杀伤。

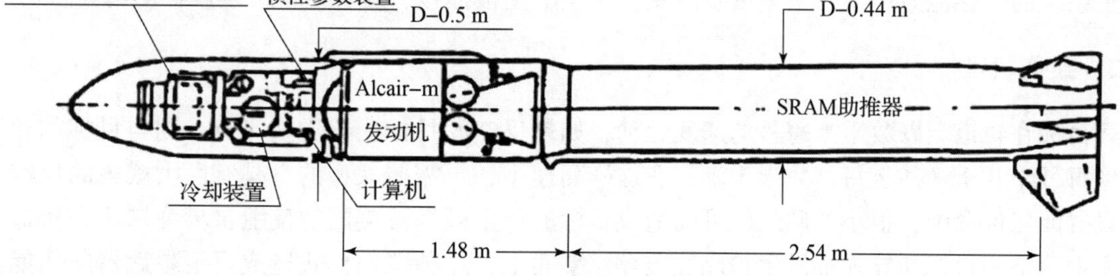

图2 反卫星导弹总体结构图

美国的另一种成熟的动能反卫星武器是“智能卵石”反卫星拦截弹。该拦截弹从航天器上发射，以高速动能杀伤拦截器直接碰撞目标。拦截弹具有体积小、重量轻的特点，可以大量布置到空间，高1 m，直径10~13 cm，推力器从接到指令到最大机动响应为毫秒级，加速度达到10 g，机动速度达到600 m/s以上，可以满足碰撞杀伤卫星的要求。为保证制导精度达到足以使拦截器本体直接碰撞卫星的杀伤效果，设有伞型杀伤增强装置。这种装置是折叠式的可径向展开的伞状结构，金属伞展开后迎着近于法线方向与卫星目标相撞，可增大碰撞面积。冲击机构是一个增几强 W51 脂薄膜板，穿透机构则是分布在薄膜上的小球。这个可膨胀的聚酯薄膜与高密度小球组合使用，可达到穿透和压碎卫星结构、碰撞毁掉关键附属部件的目的。

2　反空间目标战斗部设计中的若干问题分析

2.1　空间目标及其易损性

空间目标主要指人造地球卫星和各种空间站和空间平台等。

人造地球卫星按照轨道高度可以分为近地轨道卫星（1 000 km 以下）、中高轨道卫星（1 000～10 000 km）、高轨道卫星（36 000 km）和大椭圆轨道卫星（近地点 300～500 km，远地点 40 000 km 以上）；按照功能主要可以分为气象卫星、导航卫星、侦察卫星和军用武器卫星等。典型的卫星由不同功能的若干份系统组成，一般可以分为有效载荷和保障系统两大类。有效载荷随着卫星的任务不同而不同，如侦察卫星的照相机和科学卫星的望远镜等。保障系统一般可以分为以下分系统：①结构系统，用于支撑和固定卫星上的各种仪器设备，卫星结构材料大多采用铝、镁等轻合金材料以及钛合金材料，碳纤维复合材料和金属基复合材料也得到了使用。②热控制系统，用来保障卫星上各种仪器设备在复杂的环境中处于允许的温度范围内。③电源系统、控制系统、数据管理系统和测控系统等分系统。

空间站和各种空间平台属于载人航天器，结构要比卫星复杂得多。其主要组成系统包括有效载荷、结构系统、环境控制和生命保障系统、电源系统、数据管理系统、通信和测控系统、热控制系统和推进系统等。

空间环境非常复杂，存在大量的空间碎片和微流星，这些碎片对航天器构成了非常严重的威胁。但是各种航天器要发射到距离地面很远的位置，因此质量不可能很大，其防护结构非常薄弱。例如，各种人造地球卫星仅对舱外裸露的散热器件和散热管道等部件根据需要进行局部防护，而载人航天器一般采用多层防护靶防护，防护结构要确保经得住二次碎片冲击，这种防护结构的防护性能相对好一些。空间目标在轨道上超高速运动，在这样的高速度下，空间目标就显得十分脆弱，很小的金属破片就可以造成卫星的破坏。比如破片对卫星造成很小的穿孔，即使没有破坏其内部的关键部件，也会对目标造成严重的毁伤，因为穿孔破坏了卫星的保障系统。空间目标内部结构十分复杂，受到破坏后很容易丧失原有的功能。总之，与地面目标相比，空间目标是非常易于损坏的。

随着空间武器化的趋势，各军事强国都对卫星系统进行改进，增强其防卫能力。比如：提高卫星的轨道机动能力，使卫星具备规避打击能力；采用分布式小卫星，小卫星分布在同一轨道的不同位置，或分布在不同的轨道上，反卫星武器很难将其全部摧毁，即使部分损坏，其余仍能继续工作，从而提高了整个卫星系统的生存能力；或者采用后备卫星，一旦卫星被摧毁，后备卫星可以候补使用等。

2.2　毁伤机理

空间目标在轨道上以数千米每秒的速度运动，如果反空间目标武器逆轨飞行，则与目标的相对速度可以达到每秒十几千米甚至每秒数十千米。在这样高速下，不需要大型的、具有巨大威力的反空间目标武器，只要能保证命中，很小的破片就可以造成目标的严重破坏。美国的反卫星导弹属于直接命中杀伤机理，命中后空间目标和导弹都发生粉碎性破坏，事实上，这种毁伤模式造成反卫星武器毁伤能力的浪费，因为小破片对卫星的穿孔破坏可以与导弹直接命中卫星具有相同的毁伤效果，即卫星失去原有的功能。而且，导弹与目标的粉碎性破坏会产生大量的空间碎片，这些碎片在太空中扩散，也会对己方的航天器造成严重的威胁。美国的“智能卵石”反卫星拦截器的战斗部是伞型杀伤增强装置，接近目标时迅速展开，采用穿透与冲击两级杀伤机理相互补充，以提高杀伤概率。但是这种杀伤机构在对目标进行破坏的同时，不同程度地会对自己造成损伤，而且这种反卫星拦截器不具备多次打击能力。苏联的“杀手卫星”是采用突然变轨接近目标、引爆高能炸药驱动破片的方式来毁伤空间目标。这种攻击属于自杀性攻击，在摧毁对方卫星的同时也对自己造成严重的损伤。

若反空间目标武器与空间目标处于尾追条件，则动能破片相对目标的速度就比较低，远没有达到超高速作用的速度。但是在相对低速的条件下，反空间目标武器就可以有时间来瞄准目标，或者发射的弹丸有足够的时间来启动制导系统。因此在这种条件下，可以在反空间目标武器上安装定向动能抛射器来发射破片射击目标。

2.3　动能破片

迎击条件下，动能破片对空间目标的作用属于超高速碰撞范围，目前国内外对于超高速碰撞的研究主要集中于空间目标如何防护微流星和空间碎片的超高速撞击。主要的防护方法就是采用 Whipple 防护结构，即多层靶板防护。该结构由前板、后板和一定的间隙组成，前板使高速破片充分粉碎甚至熔化或汽化，这些小碎片形成碎片云并随着时间延长而逐渐扩散，这样就造成了破片能量的耗散，大大减轻了其对后板的破坏。破片的形状对碎片云的形状有很大的影响，如果破片超高速碰撞靶板以后所产生的碎片相对集中或者存在未粉碎的大碎片，则这种破片由于碰撞后仍保持能量集中而具备良好的毁伤能力。破片质量也是其毁伤能力的决定性因素之一，但是一种航天器的有效载荷不可能很大，破片的数量又要满足空间上的分布密度，因此每个破片的质量在保证有效毁伤空间目标的前提下越小越好，这就提出了一个临界质量的问题。所谓临界质量，就是在一定的碰撞条件下，破片能穿透空间目标防护结构的最小质量。因此对破片的威力性能进行优化，确定破片的最佳外形和破片临界穿透质量，可以为高效毁伤空间目标的破片设计提供依据。

2.4　数值仿真技术

在逆轨道拦截条件下，动能破片与空间目标的碰撞速度一般超过实验室所能得到的速度，以及地面实验条件的局限和实验的高成本等，图 3 为破片贯穿多层防护结构的数值仿真，这些都使得数值仿真技术成为研究超高速碰撞毁伤效应的有力工具。常用的数值计算方法有拉格朗日（Lagrange）方法、欧拉（Euler）方法、拉格朗日和欧拉混合（ALE）方法以及光滑质点流体动力学（SPH）方法。拉格朗日方法和欧拉方法不能很好地模拟超高速碰撞的情况，而采用 SPH 方法和合适的材料模型可以得到与试验非常接近的结果。通过数值仿真可以弥补和解决试验条件达不到的超高速碰撞的设计问题，可以进行大量的计算机虚拟试验，得到有规律的结果，也可以展示超高速碰撞的瞬时情况、认识超高速作用机理等。因此，开展超高速碰撞的数值仿真研究是设计优化反空间目标动能破片的有效途径。

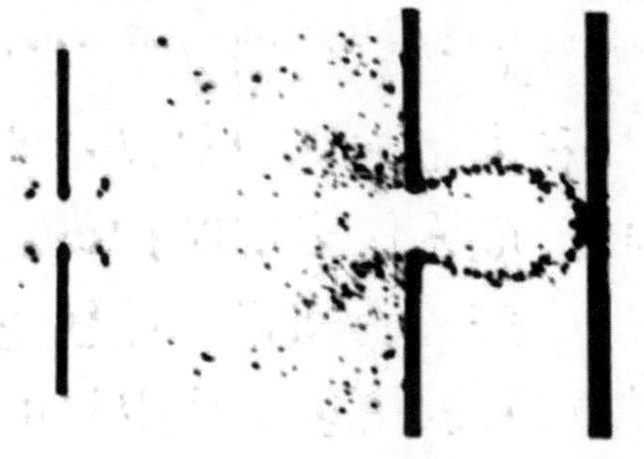

图 3　破片贯穿多层防护结构的数值仿真

2.5　特殊动能杀伤场

有时出于对政治、军事和外交等多方面的考虑，对空间作战提出了特殊的威力场要求。大多数空间目标并不具有作战功能，无须摧毁也能达到毁伤效果，比如抛撒蔽物质使空间目标丧失通信功能，或者抛撒导电物质使其部件发生短路等。一旦卫星等空间目标发生故障，很难进行维修，因此这种特种杀伤场可以起到与动能摧毁同样的作战效果，而且空间目标丧失使用功能后并不会形成大量的空间碎片，由此可见，特殊动能杀伤场也是反空间目标的一种有效途径。

3 结论

（1）从上述分析可以看出，美俄等军事强国早在40多年前就开展了动能反空间目标武器的研制，并且达到了实战水平。随着科技的发展，空间作战不可避免，大力发展反空间目标战斗部技术具有战略意义。

（2）先进的军用卫星系统为了提高安全性和可靠性，或采用分布式小卫星系统，或有后备卫星，一旦卫星被毁坏，系统的整体功能并不会受到大的破坏。因此，反空间目标战斗部应该具备高效毁伤、多次打击的能力。一次打击没有达到预定效果，可以再次打击。

（3）空间目标在轨道以超高速运动，如果拦截器迎面拦截，则相对速度更高。在这样的高速度下，直接命中的概率比较低。如果拦截器在目标前散布开均匀分布的破片，等待对方目标与破片碰撞，这样既提高命中概率，拦截器自身也不会受到损伤。因此均匀分布的等待式破片威力场的碰撞杀伤是毁伤空间目标的新途径。

（4）达到一定毁伤效果的条件下，单枚破片的质量越小越好，因此对动能破片的威力性能进行优化设计具有现实意义。

（5）由于空间作战条件特殊，在地面很难进行试验模拟，计算机数值仿真技术是设计高效毁伤空间目标战斗部的有力工具。

（6）由于空间目标远离地面，出现故障便难于维修，因此能导致空间目标出现故障的特殊威力场具备和动能摧毁相同的作战效果，这种威力场也是反空间目标的一种新途径。

参考文献

[1] 刘绍球，高淑霞，高永浩，等. 反卫星武器研制现状的分析研究［J］. 系统工程与电子技术，1994(1)：38.

[2] 楮桂柏. 航天技术概论［M］. 北京：中国宇航出版社，2002.

[3] 曲广吉，韩增尧. 破片超高速撞击动力学建模与数值仿真技术［J］. 中国空间科学技术，2002，22(5)：26－30.

[4] 张庆明，黄风雷. 超高速碰撞动力学引论［M］. 北京：科学出版社，2000.

[5] HAYHURST C J，CLEGG R A. Cylindrically symmetric SPH simulations of hypervelocity impacts on thin plates［J］. International journal of impact engineering，1997，20：337－348.

基于战斗部微圆柱分析的破片飞散特性研究①

黄广炎[1②]，刘沛清[1]，冯顺山[2]

(1. 北京航空航天大学航空科学与工程学院，北京 100083；

2. 北京理工大学爆炸科学与技术国家重点试验室，北京 100081)

摘　要：基于战斗部的微圆柱方法分析了一端起爆情况下爆轰波对破片微元的驱动飞散机理。通过联合同样基于战斗部微圆柱的破片初速端部修正方法，得到了一种基于微圆柱的破片飞散方向沿战斗部轴线分布的计算方法。并基于扇形靶板及经纬度试验测试分析方法，开展了战斗部破片飞散方向的试验研究。验证了该基于微圆柱的战斗部破片飞散方向计算方法的准确性。破片飞散特性研究对于深入开展战斗部破片飞散方向的控制与应用研究提供了重要的参考依据。

关键词：战斗部；破片飞散特性；破片初速；微圆柱；试验研究

Research on Dispersion Characteristic of Fragment Based on Micro - column Analyses for Warhead

Huang Guangyan[1], Liu Peiqing[1], Feng Shunshan[2]

(1. School of Aeronautics Science and Engineering, Beihang University, Beijing 100083, China;

2. State Key Laboratory of Explosion Science and Technology, Beijing Institute of Technology, Beijing 100081, China)

Abstract: The dispersion mechanism of warhead fragment impelled by shockwave under one - end initiation was studied based on the analysis method of micro - column. Combined with an amendable method for the fragment initial velocity, a calculation method for the distribution of dispersed direction of fragment along the warhead axis was built based on the micro - column. The experiment research on the fragment dispersed direction was carried out with a sector target and longitude - latitude analysis method. The accuracy of the calculation method for the fragment dispersed direction was verified. The investigation on dispersion characteristic of fragment pro - vides an important basis for the control and application of the fragment dispersed direction.

Keywords: warhead; fragment dispersion characteristic; fragment velocity; micro - column; experiment research

0　引言

战斗部爆炸后，破片在空间的飞散分布是确定破片杀伤作用场必须研究的一个重要问题，尤其是在未来封锁类弹药战斗部的非致命毁伤、大威慑范围研究趋势下，对战斗部破片飞散方向的控制方法与预测计算研究将越来越重要。

破片飞散方向的影响因素较为复杂，主要包括战斗部的装药结构、端部约束、引爆方式、壳体和炸药材料等，国内外很多顶级专家学者都对该课题展开了科学研究。泰勒提出了最早的确定破片初始飞散方向的方程[1]，但因其所假设的破片没有加速度且速度均匀，爆炸气体密度均匀且遵循定常流动过程，所以泰勒公式在战斗部两端附近的计算不能准确地说明破片飞散角。在泰勒之后，很多人进行了大量的

---

① 原文发表于《兵工学报》2010 年第 4 期。

② 黄广炎：工学博士，2004 年师从冯顺山教授，研究封锁型弹药设计及效能分析方法，现工作单位：北京理工大学。

理论和试验工作，对泰勒公式进行了修正，其中：兰德－皮尔森考虑了沿着金属表面紧挨着两点[2]，每一点都按照一定的方程做加速运动，具有较好的计算准确度。夏皮洛将泰勒理论具体加以应用[3]，提出桶形战斗部的破片飞散方向计算公式，该方法只适合起爆系统在战斗部中心线上、战斗部外壳对称的情况。在国内，20 世纪八九十年代，赵振荣提出了对破片初始速度求导的飞散方向计算方法，但这种方法要求初速表达式是理想化的；周培基也曾提出泰勒公式的修正，取得了不错的效果。近年来，冯顺山、钱立新等在破片飞散方向的理论分析、试验研究或数值仿真方面开展了科学的研究工作，进一步提高了破片飞散方向控制技术在战斗部研究中的应用[4-5]，但在战斗部破片飞散方向机理及试验研究方面近年工作较少。在前人研究工作的基础上，基于微圆柱分析方法，以一端起爆杀伤战斗部为研究对象，开展相关破片飞散方向的机理及试验研究。

1 战斗部破片爆炸驱动飞散机理

战斗部壳体破片在爆轰产物气体驱动下飞散，若瞬时爆轰条件下，爆轰波传播方向与破片壳体初始位置垂直，如球形装药，则壳体破片将沿其初始位置的法线方向抛射。对于圆柱形、桶形战斗部，若将战斗部看成由若干个微圆柱组成，每个微圆柱对应一个破片微圆壳（由多个破片微元径向排布组成）和微圆柱段装药，假如每个微圆柱都有一个起爆点，且战斗部各微圆柱起爆点同时起爆，则战斗部各破片微元都将沿该微元段壳体法线飞散，如图 1 所示。

实际上，战斗部起爆点数目有限，一般是一端起爆或中心起爆，甚至是偏心起爆，而且装药不可能瞬时爆轰，即使是预制破片战斗部或刻槽战斗部，其在破片飞散前战斗部壳体都有膨胀变形，故每个破片微元大多会偏离壳体法线方向。

以一端起爆战斗部为例讨论，如图 2 所示，同样可把战斗部看成由若干个微圆柱组成。参考泰勒公式，假设破片为瞬时加速，爆炸气体为线性，相邻破片微圆柱互不干扰。

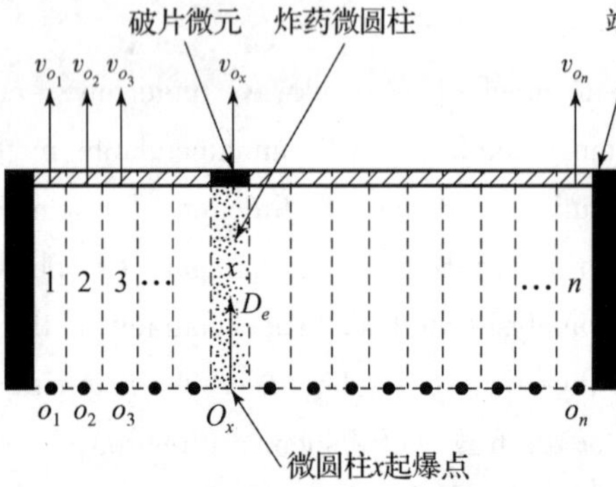

图 1 战斗部微圆柱破片微元垂直飞散

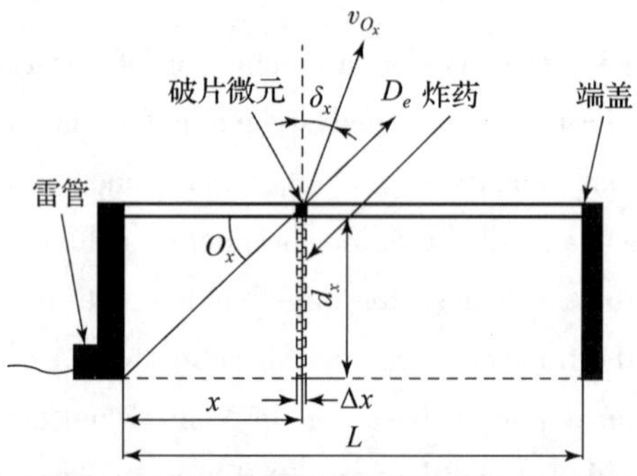

图 2 一端起爆条件下破片微元飞散方向

战斗部任意破片微元对应起爆点均有特定位置。当爆轰波从壳体表面掠射时，该微圆柱段 Δx 壳体破片飞散方向将偏离该段壳体法线方向一定角度 δx。将微圆柱段 Δx 爆轰波传播与破片微元段飞散关系局部放大研究，如图 3 所示：x 为微圆柱段 Δx 壳体破片质心；ab 为 Δx 壳体微元爆轰驱动前初始位置；cb 为 Δx 壳体微元爆轰波掠射后的飞散位置，其间爆轰波波阵面由 aa' 传播到 bb'，传播时间为 Δt。

在等腰三角形 xbx' 中，由正弦定理有

$$\frac{xx'}{xb} = \sin\alpha/\sin[(\pi-\alpha)/2] = 2\sin\frac{\alpha}{2} \tag{1}$$

假设该微圆柱段破片微元抛射速度为 v_{0x}，在 Δx 时间内的平均速度为 v_{0xm}，则有

$$\begin{cases} xx' = v_{0xm}\Delta t = v_{0x}\Delta t/2 \\ xb = D_e\Delta t/2\cos\theta_x \\ \delta_x = \alpha/2 \end{cases} \tag{2}$$

将式（2）代入式（1）有

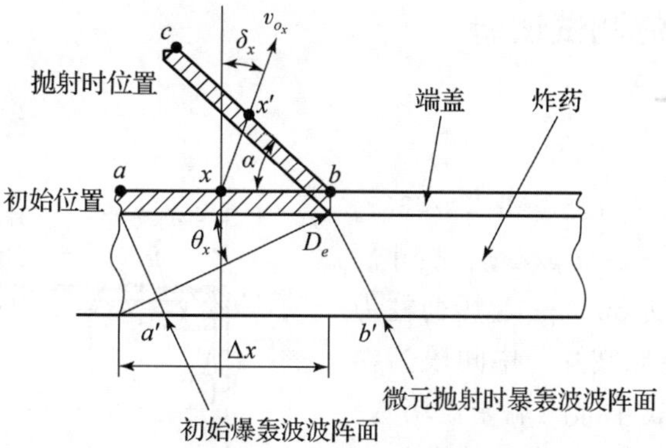

图 3　Δx 微圆柱段破片微元驱动飞散示意图

$$\delta_x = \arcsin\left(\frac{v_{0x}}{2De}\cos\theta_x\right) \tag{3}$$

式中　δ_x——Δx 微圆柱段破片微元抛射方向与壳体法线方向的夹角；

θ_x——爆轰波入射角；

D_e——炸药爆速；

v_{0x}——破片微元初速。

对于战斗部破片初速的分析，半个世纪以来，格尼公式在欧美国家一直被认为是计算破片初速较有效的公式。其能反映炸药对金属的加速能力，但其假设瞬时爆轰、壳体等壁厚、各破片微元的初速都相等与实际并不相符。事实上，战斗部各破片微元初速沿着战斗部轴向具有变化梯度，战斗部引爆方式是各式各样的，战斗部壳体也不是无限长的圆柱体，受轴向稀疏波的影响，作用于壳体各微元的冲量不相等，且炸药的爆轰总是以有限速度进行，反应区后的稀疏波使一部分能量损耗，故战斗部两端部的破片初速要比中间段的破片初速低。

同样以一端起爆战斗部为例，为了得到更接近实际的结果，把战斗部微元近似看成由许多微圆柱组成来讨论，如图 2 所示。认为一般装药结构战斗部破片初速轴向分布函数主要与破片位置 x 和对应的质量比 β_x 有关，即

$$v_{0x} = H(x/d_x, \beta_x)v_0 \tag{4}$$

破片速度值发生较大变化的位置在战斗部的两端：

$$v_{0x} = H_1(x/d_x, \beta_x)H_2\left(\frac{L-x}{d_x}, \beta_x\right)v_0 \tag{5}$$

式中　$H_1(x/d_x)$——起爆端修正系数；

$H_2\left(\frac{L-x}{d_x}\right)$——非起爆端修正系数。

通过用 X 光摄影测试的结果来修正格尼公式，分析 X 光照片和从测试得到数据，认为

$$H_1(x/d_x) = 1 - A\mathrm{e}^{-Bx/d_x} \tag{6}$$

$$H_2\left(\frac{L-x}{d_x}\right) = 1 - C\mathrm{e}^{-D(L-x)/d_x} \tag{7}$$

式中　A、B、C、D——由试验数据确定的常数。

假设沿弹轴各微元环形成的破片仍遵循格尼公式的规律：

$$v_0 = \sqrt{2E_g}\sqrt{\beta_x/[1+0.5\beta_x]} \tag{8}$$

于是，一端起爆条件下，适用各种结构形状的杀伤战斗部的破片初速的工程计算式为

$$v_0 = [1 - A_1\mathrm{e}^{-Bx/d_x}][1 - A_2\mathrm{e}^{-C(L-x)/d_x}] \cdot \sqrt{2E_g}\sqrt{\beta_x/[1+0.5\beta_x]} \tag{9}$$

将式（9）代入式（3），可得基于战斗部微圆柱的破片微元飞散方向计算公式为

$$\delta_x = \arcsin\{[1 - A_1\mathrm{e}^{-Bx/d_x}][1 - A_2\mathrm{e}^{-C(L-x)/d_x}] \cdot \sqrt{2E_g}\sqrt{\beta_x/[1+0.5\beta_x]}\cos\theta_x/2D_e\} \tag{10}$$

2 破片飞散方向静爆测试试验

2.1 试验战斗部结构

试验用杀伤战斗部壳体材料为35CrMnSiA/42Cr，炸药装药为8701，装药密度$\rho=1.72\ g/cm^3$，战斗部长度为140 mm，装药总质量为500 g，起爆药柱为A5炸药。战斗部两端内置钨球破片，中间段为内刻槽壳体破片，一端起爆，战斗部两端端盖堵塞，战斗部结构如图4所示。沿战斗部轴向共计排布18个典型破片（钨球破片和刻槽破片），将战斗部看成由18个微圆柱组成，在战斗部表面标记母线，在该标记母线上，18个编号典型破片质心位置呈直线排布。

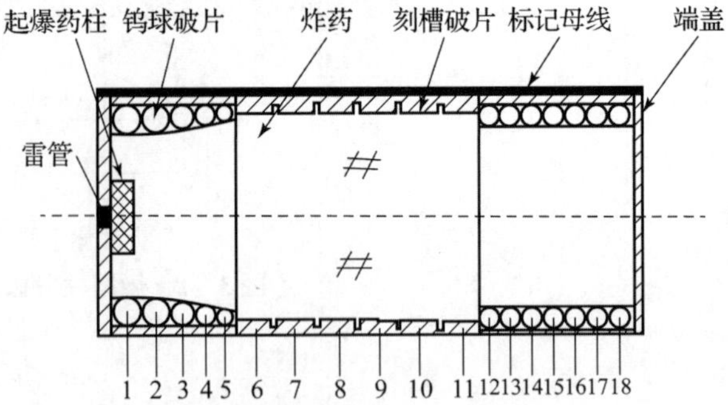

图4 试验用战斗部结构示意图

2.2 靶板拦截设计

根据经验值估算，该试验战斗部破片飞散方向与战斗部轴线夹角应在70°~120°。但在试验中，为保证收集到同一母线上的所有编号破片，可根据战斗部位置布置130°左右的扇形拦截靶，在距战斗部中心6 m处共布置10块1.5 m×2 m×0.004 m的A3钢板$L_0\sim L_9$。每块靶板对应战斗部中心的圆心角为14°，战斗部非起爆端指向逆时针方向第2块靶下边缘处（0°方向），如图5所示。战斗部表面标记母线位置对准靶板水平中心位置处，战斗部距地面高度为0.75 m，如图6所示。

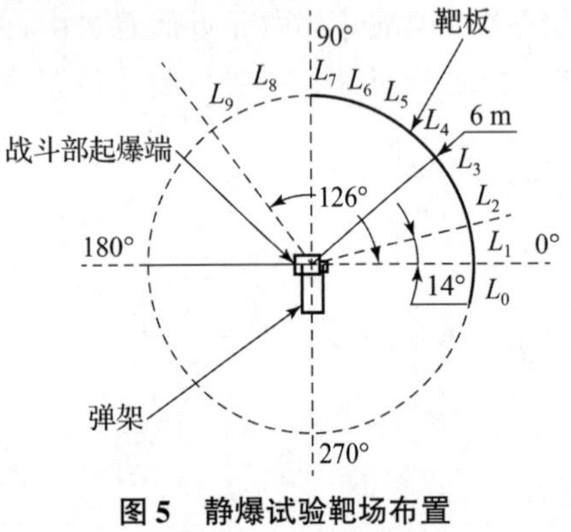

图5 静爆试验靶场布置

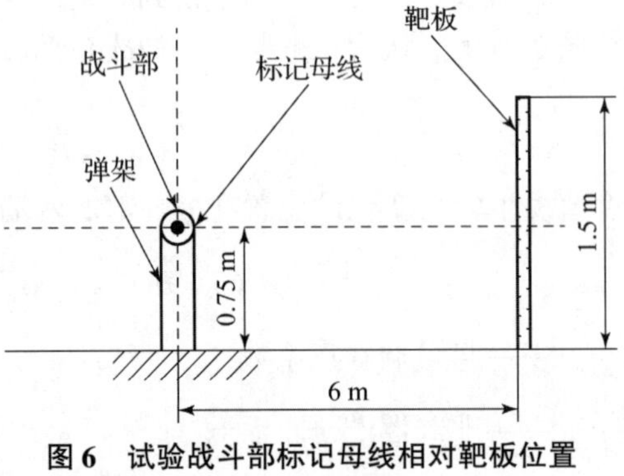

图6 试验战斗部标记母线相对靶板位置

为了方便记录战斗部静爆后各编号破片的分散分布，在各靶板上进行经纬度划分。以图5所示逆时针方向第2块钢板L_1下边缘为零经度线，相对战斗部中心每隔2°逆时针方向画经度线，在与战斗部标记母线同一水平高度的靶板位置画直线作为零纬度线，相对战斗部中心每隔1°画纬度线，横向线表示纬度，纵向线表示经度，横向每格为2°，纵向每格为1°。

3 试验结论与分析

3.1 靶板对战斗部标记母线上各编号破片的拦截

沿战斗部轴线方向的标记母线上共计排布18个编号破片：6个刻槽破片和12个钨球破片。将战斗部看成由18个微圆柱组成，每个微圆柱由装药微圆柱和若干个同一编号破片构成的微圆筒组成，战斗部起爆中心线、标记母线和靶板零纬度线处于同一水平位置，故统计各靶板零纬度线附近的刻槽破片和钨球破片的经纬度值，即可得到该战斗部破片飞散方向沿轴线的散布统计。战斗部标记母线上的编号破片及非起爆端部端盖自然破片对 6 m 处 A3 钢板击的侵彻情况如图 7 所示。

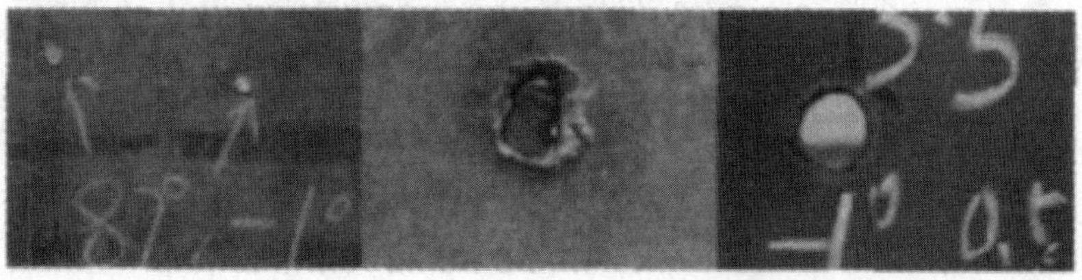

图 7　靶板对战斗部钨球破片的拦截

3.2　破片飞散方向试验测试与计算数据对比分析

据图 4 所示试验战斗部标记母线上的破片编号以及钨球破片、刻槽破片在战斗部静爆后对扇形靶板的侵彻分布的数据统计，得到战斗部各编号破片沿战斗部轴线飞散方向分布规律，如图 8 所示。图中破片飞散方向的计算数据为根据试验战斗部结构参数，使用式（10）计算得到。

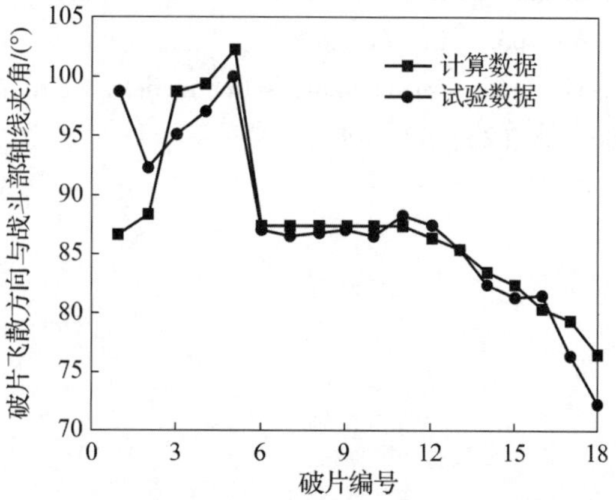

图 8　战斗部破片飞散方向的试验测试与计算数据对比

对比图 8 的试验数据与计算数据可知，两者在战斗部 88% 长度部分的飞散数据较为接近。试验所测战斗部起爆端编号 1 和 2 的钨球破片飞散方向与理论计算值均略有差异，试验测试得到的破片飞散方向为 98°和 92°，而根据式（10）计算得到的数值为 86.6°和 88.2°，这主要是因式（10）的破片飞散方向计算所假设的破片加速为瞬时、爆炸气体为线性分布以及相邻破片微圆柱互不干扰，与实际并不相符。在起爆端破片飞散时，实际战斗部结构已开始膨胀变形，起爆端邻近破片微圆柱段将对起爆端破片有约束，爆轰气体有泄漏，存在爆炸稀疏波的影响，故起爆端附近的破片飞散方向计算值与试验值有偏差，必须加以一定的修正。

在战斗部刻槽段的半预制破片及非起爆端的钨球破片飞散方向的试验测试值与计算值基本一致，说明式（10）对于战斗部破片飞散方向的计算可靠性较高，尤其是编号为 3 ~ 5 的钨球破片。因为战斗部在该位置，装药结构（破片壳体与炸药接触面）有突变，见图 4，该段壳体法线方向与战斗部轴线夹角大于 90°，造成破片飞散方向与战斗部轴线夹角分别为 95.2°、97.5°、100.2°，而计算值则分别为 98.6°、99.3°、102.2°，虽然存在一定误差，但基本反映了该微圆柱段的破片飞散方向的突变特性。

对于战斗部非起爆端尾部的编号破片 17、18，试验值为 76.3°、72.3°，计算值为 79.2 °、76.4 °，存在 3°左右的误差，主要是因战斗部尾部端盖造成的爆轰气体泄漏影响，假如在增加尾部端盖厚度的情况下，计算值与试验值的误差将进一步减小，可以认为两者基本符合。

4　结论

基于战斗部微圆柱分析方法，研究了一端起爆情况下爆轰波对破片微元的驱动飞散机理，并联合破片初速对起爆点及端盖约束修正的计算方法，得到了基于微圆柱分析的战斗部破片飞散方向计算公式。同时，还开展了基于微圆柱的战斗部破片飞散方向静爆试验测试研究。通过对战斗部标记母线上各编号破片在靶板上拦截位置的经纬度值统计分析，得到了试验战斗部各编号破片的飞散方向；通过与文中建立的基于微圆柱的破片飞散方向计算公式得到的计算数据对比分析，验证了该计算方法的准确性，对战斗部破片飞散方向的控制方法与预测计算研究提供了重要的参考依据。

参考文献

[1] WALTERS W P, AUKAS J A. Fundamentals of shaped charges [M]. New York: Wiley, 1989.

[2] RANDERS P G. An improved equation for calculating fragment projection angle [C] // Proceedings of the 2nd International Symposium on Ballistics, Daytona Beach, FL, 1976: 9 - 11.

[3] LIOYD R M. Conventional warhead systems physics and engineering design [M]. Virginia: American Institute of Astronautics and Aeronautics Inc, 1998.

[4] 冯顺山，崔秉贵. 战斗部破片轴向分布规律的试验研究 [J]. 兵工学报，1987，4 (4): 60 - 63.
FENG S S, CUI B G. Experimental research on fragments dispersion of forward - firing warhead [J]. Acta Armamentarii, 1987, 4 (4): 60 - 63. (in Chinese)

[5] QIAN L X, LIU T, ZHANG S Q, et al. Fragment shot - line model for air - defense warhead [J]. Propellants, explosives, pyrotechnics, 2000, 25 (2): 92 - 98.

战斗部破片对目标打击迹线的计算方法①

黄广炎[1②,2]，冯顺山[2]，刘沛清[1]

（1. 北京航空航天大学航空科学与工程学院，北京　100083；

2. 北京理工大学爆炸科学与技术国家重点试验室，北京　100081）

摘　要：提出了破片对目标打击迹线概念，通过建立预制或半预制破片战斗部对目标杀伤作用场参数的计算模型，以及在空间坐标系下破片各威力特性参数的相互转换，得到了一种破片打击迹线的计算方法，并基于VC和MATLAB联合编程实现了战斗部破片打击迹线作用场的分布计算。通过破片对目标打击迹线可定量分析战斗部所有破片的整个空间分布，并直观描述每个破片的飞散方向、能量衰减及对目标打击能力变化情况，从而可精确地评价破片对目标不同要害位置的打击效果。该杀伤作用场参数计算模型和破片对目标打击迹线计算方法可用于战斗部起爆姿态选择和对不同目标杀伤效果的威力评价，为战斗部方案设计和战斗部终点毁伤效应分析计算提供了新的参考依据和技术途径。

关键词：爆炸力学；打击迹线；MATLAB；破片；战斗部；VC

A visual C ~ (++) and Matlab - based computational method for shot—lines of warhead fragments to a target

Huang Guangyan[1,2], Feng Shunshan[2], Liu Peiqing[1]

(1. School of Aeronautic Science and Engineering, Beijing University of Aeronautic & Astronautic, Beijing 100083, China;

2. State Key Laboratory of Explosion Science and Technology, Beijing Institute of Technology, Beijing 100081, China)

Abstract: The existent shot—line concept was extended based on the visual C ~ (++) and Matlab codes. A computational method was developed by setting up the computational model of the damage field parameters, which consisting of the initial velocity, dispersion direction, velocity attenuation, and so on, for the preformed or half—preformed fragment warhead attacking the target, and the trans—formation of the fragment power parameters in the space coordinate. And a new computational method was achieved to predict the shot—line field distribution of the fragment warhead by using the combined program with visual C ~ (++) and MATLAB. A computational example was given by applying the achieved computational method to determine the shot - line field distribution of the fragment warhead under different explosion conditions attacking the plane target. The computational results described distinctly the damage field distribution of the warhead fragments to the plane and reflected visually the trajectories and energy attenuations of all the warhead fragments. The achieved computational method was validated by the comparison between the computational data with the known experimental results. This research can provide a new reference for the warhead design and terminal damage effect analysis.

Keywords: mechanics of explosion; shot - line; MATLAB; fragment; warhead; VC

---

① 原文发表于《爆炸与冲击》2010年第7期。

② 黄广炎：工学博士，2004年师从冯顺山教授，研究封锁型弹药设计及效能分析方法，现工作单位：北京理工大学。

1 引言

破片是杀伤或杀爆战斗部毁伤的重要元素，战斗部爆炸后的破片初速、飞行轨迹、能量衰减等杀伤特性参数非常复杂。以往基本上根据实验判断毁伤效果，但攻击目标的差异、战斗部装药结构、起爆条件等相关因素的变化都必须要用不同的实验方案，耗费人力、物力巨大。计算机仿真技术的发展，带来了战斗部毁伤领域新的发展契机。

在系统直观描述破片飞行弹道及杀伤能力方面，钱立新等[1]采用破片场仿真方法对大面积杀爆战斗部作用过程进行了威力评定；钱立新等[2]基于防空战斗部提出了破片打击迹线的概念，研究了可描述杀伤、聚焦、多P和连续杆战斗部杀伤威力的打击迹线计算模型；蒋建伟等[3]为模拟战斗部破片的宏观可控性和微观随机性，建立了描述广义杀伤战斗部起爆及破片飞散的射击迹线仿真模型；冯顺山等[4]对战斗部破片初速的轴线分布规律进行了实验研究；蒋建伟等[5]建立了破片杀伤威力参数的计算模型。

本文中，基于VC和MATLAB，进一步提出破片对目标打击迹线的概念，探讨杀伤战斗部破片对典型目标打击迹线的计算模型，形成新的战斗部对特定目标的终点毁伤效应评估概念。

2 破片对目标打击迹线概念

破片对目标打击迹线与描述单个弹丸的运动轨迹及侵彻行为的射击线类似，破片对目标打击迹线是一条带有变化质量、速度和飞散方向的线段，该线段起点为战斗部起爆瞬间破片原始所处壳体微元的位置，终点为该破片飞行一段距离后能量衰减至临界有效杀伤特定目标的位置，用以表述破片的运动轨迹和对欲打击目标的有效毁伤范围。因此，战斗部破片对目标打击迹线随打击目标的不同而不同。

战斗部起爆后对目标的杀伤作用场可用多条破片对目标打击迹线描述，每条破片对目标打击迹线分别对应不同战斗部壳体位置破片的运动轨迹和对特定目标打击能力变化情况。即使是同一战斗部，如所打击的目标对象发生变化，战斗部的破片对目标打击迹线分布情况也将不同。

在已知战斗部装药结构、起爆姿态条件下，可以得到战斗部起爆驱动破片的离散初始威力特性参数，包括每个破片的质量、几何外形、三维空间坐标值、速度和飞散方向。以这些初始条件为基础，根据空气中外弹道方法计算所有破片的整个空气运行轨迹以及能量衰减情况，选择战斗部欲打击的目标并确定破片对该特定目标的毁伤准则，即可得到战斗部破片对目标打击迹线作用场的分布情况，从而精确评价破片对目标不同位置的打击效果。

3 破片杀伤作用场威力参数计算模型

破片对目标打击迹线的计算，需建立战斗部破片杀伤作用场各威力特性参数的计算模型。破片对目标的杀伤威力性能参数主要包括质量、初速、飞散方向、速度（能量）衰减等。取战斗部装药结构如图1所示（预制破片壳体、轴对称装药、一端起爆），分别讨论破片各威力参数。

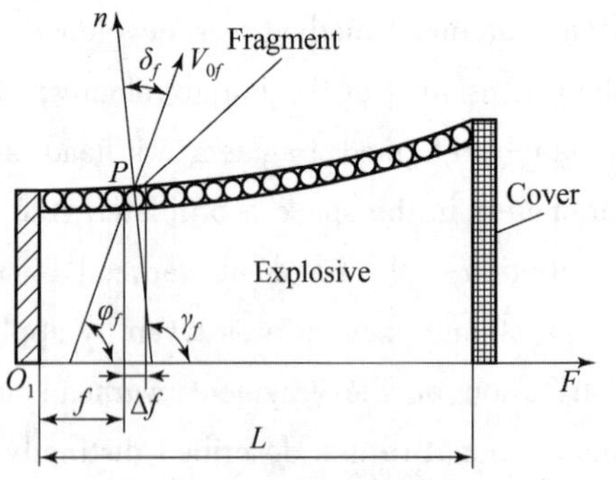

图1 战斗部装药结构图

3.1 破片初速

采用对战斗部破片初速格尼公式修正的计算方法[4]，该方法适合各种装药结构、不同起爆方式下的战斗部破片初速计算，是一种广义破片初速计算方法。

认为一般装药结构战斗部破片初速轴向分布函数主要与破片位置f和对应的质量比β有关，即

$$v_{0f}=H(f/d,\beta)v_0 \tag{1}$$

破片速度值发生较大变化的位置在战斗部的两端：

$$v_{0f}=H_1(f/d,\beta)H_2\left(\frac{L-f}{d},\beta\right)v_0 \tag{2}$$

式中　$H_1(f/d)$——起爆端修正系数；

$H_2\left(\frac{L-f}{d}\right)$——非起爆端修正系数。通过 X 射线测试结果修正格尼公式。

$$H_1(f/d)=1-A\mathrm{e}^{-Bf/d} \tag{3}$$

$$H_2\left(\frac{L-f}{d}\right)=1-C\mathrm{e}^{-D(L-f)/d} \tag{4}$$

式中　A、B、C、D——由实验数据确定的常数。

假设沿弹轴各微元环形成的破片仍遵循格尼公式：

$$v_0=\sqrt{2E_g}\sqrt{\frac{\beta(f)}{1+0.5\beta(f)}} \tag{5}$$

于是一端起爆条件下，适用各种结构形状的杀伤战斗部的破片初速

$$v_{0f}=[1-A_1\mathrm{e}^{-Bf/d(f)}][1-A_2\mathrm{e}^{-C(L-f)/d(f)}]\sqrt{2E_g}\sqrt{\frac{\beta(f)}{1+0.5\beta(f)}} \tag{6}$$

3.2　破片飞散方向

使用兰德 - 皮尔森对泰勒公式的修正计算方法，考虑了沿着金属表面紧挨着的两点，每一点都按照下面的方程做加速运动[5]：

$$v_t=v(1-\mathrm{e}^{\frac{T-\tau}{\tau}}) \tag{7}$$

v、T、τ 都随初始位置变化，得到破片飞散计算的修正公式为

$$\delta_f=\arcsin\left[\frac{v_{0f}}{2u}-\frac{v'_{0f}\tau}{2}-\frac{(v'_{0f}\tau)^2}{5}\right] \tag{8}$$

式中　δ_f——破片飞散方向与壳体法线方向的夹角；

v'_{0f}——破片初速轴向梯度；

u——爆轰波扫过炸药壳体交界面的速度；

$u=D_e/\cos\theta$，θ_f——爆轰波传播方向与破片微元处壳体装药接触界面的夹角；

τ——破片加速常数，$r=0.4d(f)/(2V\mathrm{or})$，$d(,)$ 为距起爆端场处的炸药直径。

战斗部各破片飞散方向角为

$$\varphi_f=\gamma_f-\delta_f$$

式中　γ_f——炸药、破片接触界面的法线与装药轴线的夹角；

φ_f——破片飞散方向与战斗部轴线的夹角。

3.3　子弹破片速度衰减

破片在运动过程中因受空气阻力速度不断衰减，当速度减小到某一值时将不再满足杀伤作用的要求，丧失杀伤能力。假设破片飞行距离为 s 时，破片速度衰减为[6]

$$vfs=v_{0f}\mathrm{e}^{-\alpha s}\approx v_{0f}\mathrm{e}^{-\frac{C_D\rho\varphi m-1/3}{2}s} \tag{9}$$

式中　C_D——常数，对球形破片 $C_D=0.97$；

ρ——当地空气密度；

m——破片质量。

3.4　杀伤作用场坐标系建立以及参数转换

假设战斗部起爆作战几何空间如图 2 所示，战斗部距地面高度为 H，速度为 v_d，着角为 θ。$Oxyz$ 为地面坐标系，O_1F 为战斗部轴线，O_1 为起爆点，P 为战斗部任一破片，破片 P 相对战斗部的初速为 v_{0f}，相对地面坐标系的破片初速为 v_{d0f}，飞散方向与战斗部轴夹角为 φ_f。P' 为在地面的着点。假如到地面时，

破片初速仍满足对某目标的杀伤要求，则 PP' 为破片 P 的有效射击迹线。为便于将各破片参数转换成地面坐标系下的数值，引入柱坐标系 $O_1F\eta\varepsilon$ 和体坐标系 $O_1F\eta\zeta$，如图 3 所示。

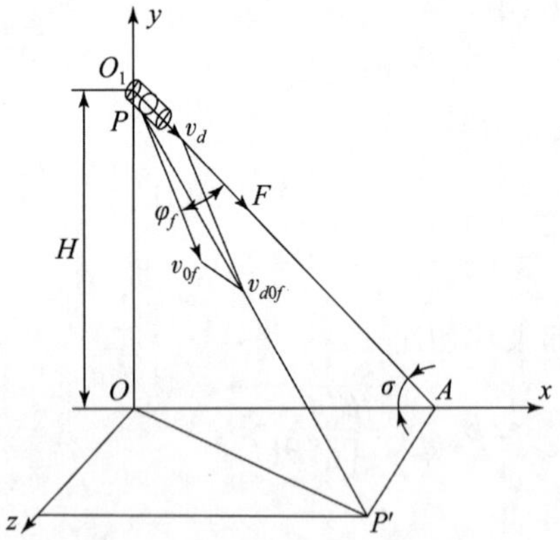

图 2　战斗部起爆作战几何空间

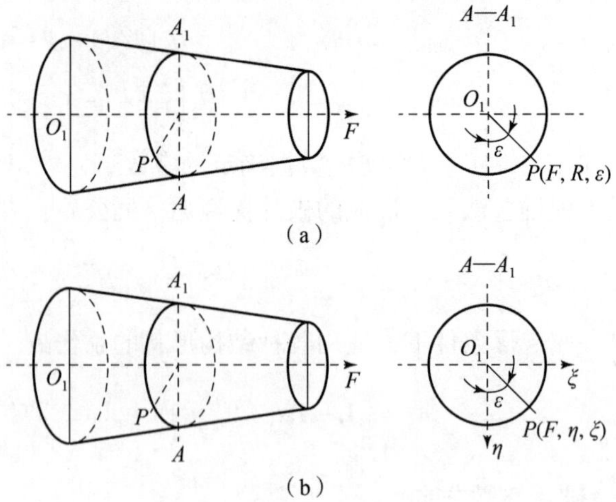

图 3　战斗部坐标系

(a) 柱坐标系；(b) 体坐标系

破片 P 相对地面坐标系的初始坐标

$$\begin{pmatrix} x_P \\ y_P \\ z_P \end{pmatrix} = \begin{pmatrix} \cos\sigma & -\cos\varepsilon\sin\sigma & 0 \\ -\sin\sigma & -\cos\varepsilon\cos\sigma & 1 \\ 0 & \sin\varepsilon & 0 \end{pmatrix} \begin{pmatrix} f_P \\ R_P \\ H \end{pmatrix} \tag{10}$$

破片 P 相对地面坐标系的初始速度

$$\begin{aligned} \begin{pmatrix} v_{Px} \\ v_{Py} \\ v_{Pz} \end{pmatrix} &= \begin{pmatrix} \cos\sigma & -\sin\sigma & 0 \\ \sin\sigma & \cos\sigma & 0 \\ 0 & 0 & 1 \end{pmatrix} \begin{pmatrix} v_{Pf} \\ v_{P\eta} \\ v_{P\varepsilon} \end{pmatrix} + \begin{pmatrix} v_d\cos\sigma \\ v_d\sin\sigma \\ 0 \end{pmatrix} \\ &= \begin{pmatrix} \cos\sigma & -\sin\sigma & 0 \\ \sin\sigma & \cos\sigma & 0 \\ 0 & 0 & 1 \end{pmatrix} \begin{pmatrix} v_{0f}\sin\varphi_f \\ v_{0f}\cos\varphi_f\cos\varepsilon \\ v_{0f}\cos\varphi_f\sin\varepsilon \end{pmatrix} + \begin{pmatrix} v_d\cos\sigma \\ v_d\sin\sigma \\ 0 \end{pmatrix} \end{aligned} \tag{11}$$

求解可得

$$\begin{cases} x_P = f_P\cos\sigma - R_P\cos\varepsilon\sin\sigma \\ y_P = -f_P - R_P\cos\varepsilon\sin\sigma + H \\ z_P = R_P\sin\varepsilon \end{cases} \tag{12}$$

$$\begin{cases} v_{Px} = v_{0f}\sin\varphi_f\cos\sigma - v_{0f}\cos\varphi_f\ \cos\varepsilon\sin\sigma + v_d\cos\sigma \\ v_{Py} = v_{0f}\sin\varphi_f\sin\sigma + v_{0f}\cos\varphi_f\ \cos\varepsilon\ \cos\sigma + v_d\sin\sigma \\ v_{Pz} = v_{0f}\cos\varphi_f\sin\varepsilon \end{cases} \tag{13}$$

4　破片对目标打击迹线计算

将破片杀伤作用场各威力特性参数的计算模型用 VC 和 MATLAB 联合编程，可针对不同装药结构和起爆方式的战斗部对目标的有效打击迹线作用场分布进行数值计算，计算步骤详见图 4

4.1　破片打击迹线作用场分布计算实例

计算图 5 所示杀伤战斗部在 6 种不同起爆姿态下对飞机目标打击迹线分布，计算中破片对飞机目标的杀伤标准使用 800 J/cm^2 的比动能杀伤准则。杀伤战斗部壳体材料为 35CrMnSiA/42Cr，炸药装药为

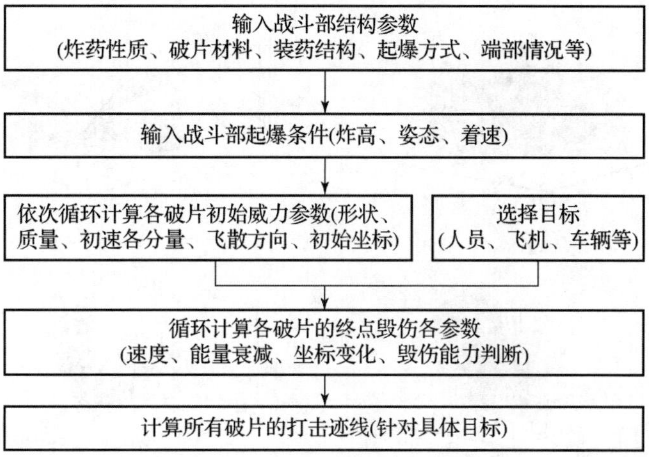

图 4　计算步骤

8 701，装药密度 $p = 1.72\ g/cm^3$，战斗部长度为 140 mm，装药总质量为 500 g，起爆药柱为 A5 炸药，战斗部两端内置钨球破片，中间段为内刻槽壳体破片，一端起爆，战斗部两端端盖堵塞。沿战斗部轴向排布 18 个典型破片（钨球破片和刻槽破片），在战斗部表面标记母线，在该标记母线上，各典型破片质心位置呈直线排布[7]。

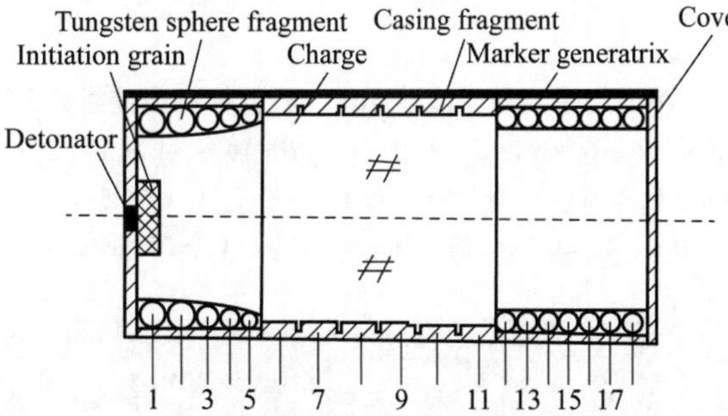

图 5　杀伤战斗部结构示意图

计算结果如图 6 所示，图中每根细线分别代表相应战斗部壳体位置破片在比动能满足对飞机的毁伤要求下的飞行弹道。计算结果比较清晰地表述了战斗部在不同起爆姿态下破片对飞机目标的杀伤作用场分布，直观地体现了战斗部所有破片的飞行轨迹和能量衰减情况。在作战空间，通过判断飞机各要害舱段所处置区域破片打击迹线穿过的情况，就可以精确评估破片对飞机目标的毁伤效果，从而可对应调整战斗部最佳起爆姿态，达到对目标的最佳毁伤效果。

4.2　与实验数据的对比

杀伤战斗部静爆实验[7]见图 7。在距离杀伤战斗部心 6 m 处共布置 10 块长 1.5 m、高 2 mm、厚 4 mm的 A3 钢板（等效飞机目标），形成 130°左右的扇形拦截靶。战斗部非起爆端指向逆时针方向第 2 块靶下边缘，战斗部表面标记母线位置对准靶板水平中心位置。战斗部距离地面高度 0.75 mm，以逆时针方向第 2 块钢板下边缘为零经度线，相对战斗部中心每隔 2°逆时针方向画经度线，在与战斗部标记母线同一水平高度的靶板位置画直线作为零纬度线，相对战斗部中心每隔 1°画纬度线。

破片对靶板打击迹线分布如图 8 所示。图中，ξ 为模拟目标靶板经度，ζ 为模拟目标靶板纬度。由图可知，计算的杀伤战斗部破片打击迹线与靶板位置交汇点坐标数据和实验中对靶板上的破片穿孔位置的统计数据基本吻合，较真实地体现了杀伤战斗部对目标的杀伤效果，验证了本文中所建立的破片杀伤作用场各威力参数计算模型和破片对目标打击迹线计算方法的正确性。

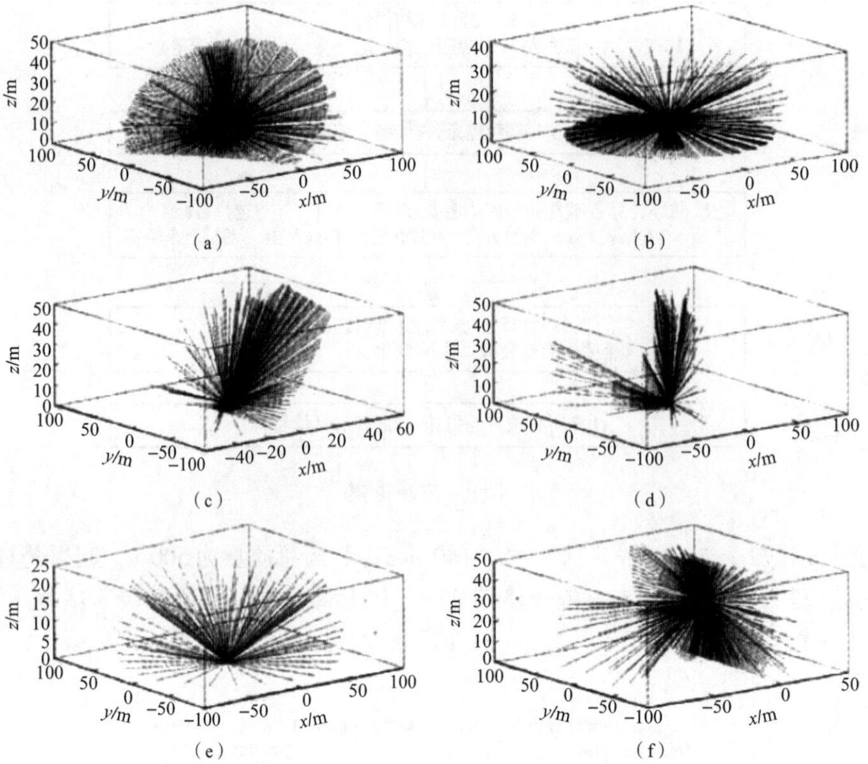

图 6　不同起爆姿态下战斗部对飞机目标打击迹线分布图

(a) $H=10$ m, $V_d=60$ m/s, $\sigma=200°$; (b) $H=10$ m, $V_d=100$ m/s, $\sigma=0°$;
(c) $H=3$ m, $V_d=70$ m/s, $\sigma=50°$; (d) $H=3$ m, $V_d=65.5$ m/s, $\sigma=271°$;
(e) $H=0$ m, $V_d=0$, $\sigma=0°$; (f) $H=30$ m, $V_d=200$ m/s, $\sigma=284°$

图 7　靶场实验布置

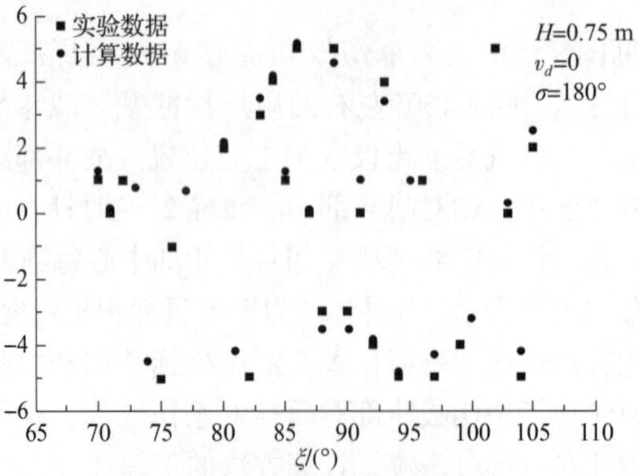

图 8　破片对靶板打击迹线分布

5　结论

为了完善破片对目标打击迹线评估方法这种新的战斗部终点毁伤效应评估概念，在提出破片对目标打击迹线概念的基础上，建立系统的破片杀伤作用场各威力参数计算模型和破片对目标打击迹线计算方法，基于 VC 和 MATLAB 联合编程，对某战斗部在不同起爆姿态下对飞机目标打击迹线散布情况进行了计算，计算结果直观地反映了战斗部破片对目标的毁伤能力，为战斗部方案设计和战斗部终点毁伤效应分析提供了新的参考依据和技术途径。

参 考 文 献

[1] QIAN L X，LIU T，ZHANG S Q，et a1. Fragment shot—line model for air—defence warhead [J]. Propellants，explosives，pyrotechnics，2000，25 (2)：92 - 98.

[2] 钱立新，屈明，刘彤. 再入式大面积杀爆战斗部威力评定方法研究 [J]. 弹道学报，2001，13 (3)：42 - 46.
QIAN L X，QU M，LIU T. Lethality assessment study of reentry large - scale blast - fragmentation warhead [J]. Journal of Ballistics，2001，13 (3)：42 - 46.

[3] 蒋建伟，卢永刚，钱立新. 射击迹线技术在战斗部破片场仿真中的应用 [J]. 弹箭与制导学报，2001，21 (1)：29 - 34.
JIANG J W，LU Y G，QIAN L X. Application of shot line model in simulation of fragment warhead [J]. Journal of projectiles，rockets，missiles and guidance，2001，21 (1)：29 - 34.

[4] 冯顺山，崔秉贵. 战斗部破片初速轴向分布规律的实验研究 [J]. 兵工学报，1987，11 (4)：61 - 63.
FENG S S，CUI B G. An experimental investigation for the axial distribution of initial velocity of shells [J]. Acta Armamentarii，1987，11 (4)：61 - 63.

[5] 蒋建伟，冯顺山，何顺录. 一种用于评价和计算杀伤威力的数学模型 [J]. 北京理工大学学报，1992，12 (2)：90 - 94.
JIANG J W，FENG S S，HE S L. A model for evaluating and calculating the lethal power [J]. Transactions of Beijing Institute of Technology，1992，12 (2)：90 - 94.

[6] 隋树元，王树山. 终点效应学 [M]. 北京：国防工业出版社，2000.

[7] 黄广炎，冯顺山. 基于战斗部微圆柱分析的破片飞散特性研究 [C]//第九届全国冲击动力学学术会议论文集，2009：165 - 169.

巡航导弹目标易损特性及其等毁伤曲线研究[①]

陈赟[1②]，冯顺山[1]，房玉军[2]

（1. 北京理工大学爆炸科学与技术国家重点实验室，北京　100081；
2. 中国兵器装备研究院，北京　10089）

摘　要：巡航导弹目标的易损特性的研究对于反导系统的战斗部设计有着重要的指导意义。首先定义了巡航导弹在近程条件下的有效毁伤形式，确定了巡航导弹的易损部件。其次对其毁伤等级进行有效划分，建立了近程防御巡航导弹条件下的毁伤判据。最后根据超压－冲量毁伤准则建立了巡航导弹在不同毁伤等级下的等毁伤曲线的数学表达式。

关键词：目标易损性；毁伤判据；等毁伤曲线

1　引言

在历次现代战争中，巡航导弹在实施远程饱和式攻击和空中纵深目标精确打击方面表现出色，由于巡航导弹具有超低空飞行，远程攻击，精确制导，使其突防概率极高，加之具有战略和战术双重作战任务和能力，迅速成为当今武器发展的重点。各国在大力发展巡航导弹的同时，加紧对巡航导弹的近程拦截方法的研究，需要对巡航导弹的易损特性及毁伤等级进行研究。本文定义了巡航导弹在近程条件下的有效毁伤形式：爆炸解体毁伤和结构解体毁伤，并确定了巡航导弹的易损部件。对其毁伤等级进行有效划分，获得了近程防御巡航导弹条件下的毁伤判据。根据超压－冲量毁伤准则建立了巡航导弹在不同毁伤等级下的等毁伤曲线的数学表达式。本文的研究成果可为巡航导弹的毁伤系统提供设计依据。

2　巡航导弹的易损特性分析

在近程反导的条件下，有效的毁伤形式是应使巡航导弹发生严重的解体破坏，丧失终点毁伤攻击能力。目标解体破坏有两种形式：爆炸解体和结构解体。对于巡航导弹来说，成型装药形成的高速侵彻体和杀爆相关毁伤可以分别实现这两种解体破坏。

爆炸解体破坏：引爆导弹的战斗部舱。目标导弹战斗部爆炸解体毁伤，是最高毁伤等级的解体毁伤。

结构解体破坏：在近程/超近程范围内，反导弹药使导弹出现如下任何一种形式的破坏，称之为结构解体破坏：

（1）导弹战斗部壳体破裂：壳体破裂成若干块，装药破碎散落，战斗部整体不具备对目标实施有效毁伤；

（2）导弹气动外形严重改变而产生严重的气动失稳：导弹在某两个舱体之间的连接部位断开，导弹发生显著的弯曲且气动件严重破损；

（3）导弹同时出现（1）和（2）两种破坏形式。

以“战斧”巡航导弹为例，不同用途的“战斧”巡航导弹，在弹体设计上采用了模块化设计思想，以 BGM－109C 为例，导弹的结构图如图 1 所示。针对近程反导毁伤要求，要害舱段为战斗部舱、整个导弹弹体结构，其重点是中舱与前后舱的连接接口处，相关毁伤的目的是在“接口”处发生

---

① 原文发表于《战斗部与毁伤技术专业委员会第十二届学术交流会文集》2011。

② 陈赟：工学博士，2005 年师从冯顺山教授，研究超近程反导方法与威力效应，现工作单位：中国兵器装备研究院。

断裂。

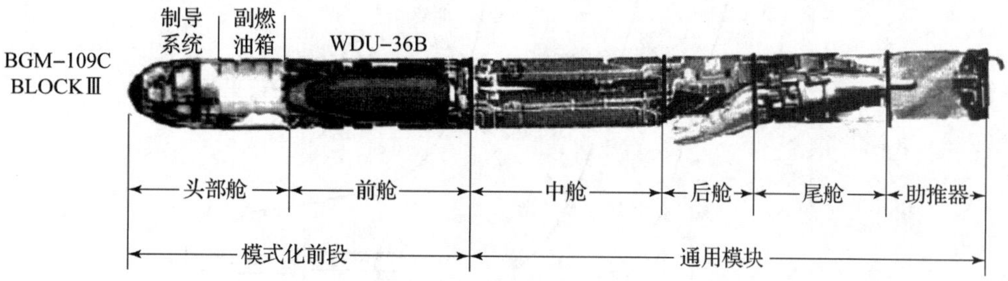

图 1　BGM-109C 巡航导弹结构图

通常情况下，巡航导弹的毁伤等级分为[1]：

K 级：巡航导弹被击中后彻底性毁伤，在空中结构解体或爆炸解体；

C 级：巡航导弹被击中后，部分功能丧失而不能完成其预定任务，如偏航、或航程减少等。

在近程条件下，原有的毁伤等级不能有效对毁伤程度进行描述，将 K 级毁伤进一步划分为：

KK1 级：巡航导弹被击中后，立即发生爆炸解体，或战斗部结构解体；

KK2 级：巡航导弹被击中后，其战斗部虽未发生爆炸，但发生弹体严重的结构解体效应，至少分成两段；

KKC 级：巡航导弹被击中后，其战斗部既未发生爆炸，也未断裂成两段，仅发生严重的结构损坏效应。

将每个毁伤等级进一步细分为若干档，如表 1 所示。

表 1　巡航导弹目标的毁伤等级

毁伤等级	毁伤程度分档	毁伤效果
KK1	KK11	目标战斗部被引爆
	KK12	战斗部壳体解体
KK2	KK21	导弹目标结构解体
	KK22	导弹目标飞行一段时间后，在气动力的作用下结构解体
KKC	KKC	导弹发生严重的结构损坏效应

3　目标结构解体毁伤判据

KK11 级毁伤由导弹的战斗部被直接引爆来实现，本文主要研究杀爆相关效应对目标的 KK12、KK2、KKC 级结构解体毁伤的判据。由于杀爆相关毁伤作用可以简化为破片作用基础上的冲击波的作用，因此杀爆相关毁伤效应可等效为目标易损特性发生变化单冲击波的毁伤效应，符合超压－冲量毁伤准则。

超压－冲量毁伤准则认为毁伤效应是由超压和冲量共同决定的，导致目标造成同等级别的毁伤的超压和冲量的值并不唯一，而是由一系列超压和冲量的组合所构成的，冲击波对目标毁伤取决于超压和冲量两者的组合效应，相同毁伤等级的超压和冲量的组合构成该毁伤等级的等毁伤曲线，利用相对变形量心和相对裂纹长度，来定量地描述目标受损反映为冲击波对巡航导弹造成结构破坏。巡航导弹结构等毁伤曲线如图 2 所示。

对于裂纹扩展问题，在断裂力学中应用较为广泛的是动态断裂准则，即 K 准则，该准则认为当材料应力强度因子 K_1 达到某一临界值 K_{1c}时，裂纹失稳扩展[2-3]。

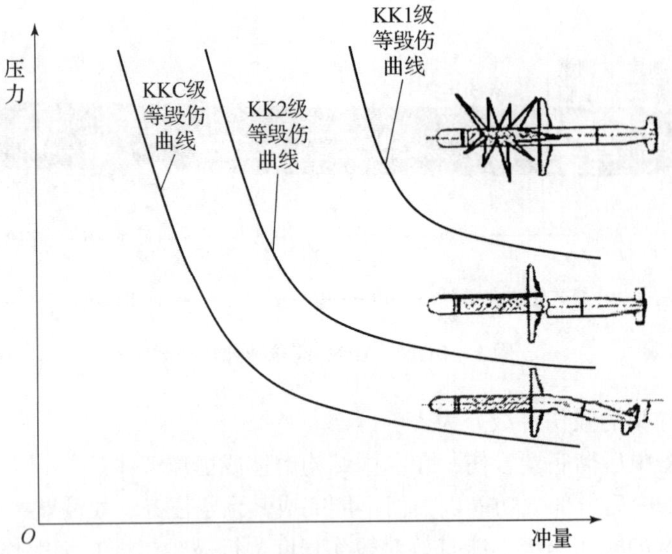

图2　巡航导弹结构等毁伤曲线

$$K_1 \geqslant K_{1c} \tag{1}$$

式中　K_{1c}——材料临界断裂强度因子。

对于圆柱薄壳，假设长度为$2L$，半径为R，均匀厚度为h_s，含有$2a$的横向裂纹，忽略破片孔对裂纹的影响，壳体单位长度作用拉伸载荷N_0，其弯曲的应力强度因子为[4]

$$K_1 = F_b(\lambda a)\frac{N_0}{h_s}\sqrt{\pi a} \tag{2}$$

$$\lambda a = \frac{[12(1-\nu^2)]^{\frac{1}{4}}}{\sqrt{R_s h_s}}$$

其中：

$$N_0 = \sigma \cdot A = \frac{M}{W}\left[\frac{1}{4}\pi(D^2 - d^2)\right] \tag{3}$$

$$W = \frac{I}{\frac{D}{2}} = \frac{\frac{\pi}{64}(D^4 - d^4)}{\frac{D}{2}} \tag{4}$$

式中　ν——泊松比；

$F_b(\lambda a)$——通过查表得到；

W——抗弯截面系数；

σ——单向拉伸正应力；

A——截面面积；

D——圆柱壳外径；

d——圆柱壳内径。

将各参数值代入上式，并考虑到圆柱靶在冲击波的作用下产生的变形对结果的影响，对$2a$进行估算，γ的临界值为0.4，即当弹体上产生的相对裂纹长度$\gamma = 0.4$时，在导弹气动力的作用下，将导致结构断裂。

文献［5］从弹药工程角度，根据试验和对飞机舱段的易损性分析，提出了判断目标的临界毁伤判据：当连续断裂的长度大于（等于）该区域长度的2/3，认为其应起的功能已基本丧失。对于本文来说，战斗部结构上功能的丧失意味着战斗部的结构解体，弹体功能的丧失意味着导弹的结构解体。

目标在近程条件下的KK级毁伤判据如表2所示。

表 2　目标在近程条件下的 KK 级毁伤判据

毁伤等级	划分准则
KK12	战斗部壳体相对裂纹长度：$\gamma \geqslant 0.667$
KK21	导弹上相对裂纹长度：$\gamma \geqslant 0.667$
KK22	导弹上相对裂纹长度：$0.4 \leqslant \gamma \leqslant 0.667$
KKC	导弹的相对变形量：$\varphi_c \geqslant 0.5$

4　等毁伤曲线

采用冲击波威力等效靶对目标的毁伤程度进行研究。采用爆炸模拟试验与数值仿真结合的方法来获得目标的等效靶板：冲击波对于圆柱靶的试验采用爆炸试验来进行，对等效靶板的作用采用数值仿真的方法来进行，并获得靶板不同的变形情况，通过仿真和试验，获得了目标在受冲击波作用、破片与冲击波相关作用两种情况下，各毁伤等级对应的等效靶板的变形挠度，见表 3、表 4。

表 3　巡航导弹的毁伤等级与冲击波威力等效靶板变形量的关系之一

	毁伤级别	等效靶板临界变形量
冲击波单独作用时	KK21	83 mm
	KKC	75.1 mm
	K	60.3 mm

表 4　巡航导弹的毁伤等级与冲击波威力等效靶板变形量的关系之二

	毁伤级别	等效靶板临界变形量
冲击波单独作用时	KK21	60.3 mm
	KK22	48 mm
	KKC	36 mm
	K	25 mm

通过表 3、表 4 可以看出，当目标只受到冲击波作用和杀爆相关毁伤破片孔密度较小时，两者的 KK21 级毁伤对应的变形挠度相同，均为 83 mm，但 KKC 级毁伤的毁伤变形挠度有较大差异，这说明即使破片孔密度较小，也能够大幅度提高目标的毁伤效果；当破片与冲击波相关作用时，在原来能够造成目标 KKC 级甚至于 C 级毁伤等级的条件下，可以提高到 KK21 级毁伤，大大提高了对目标毁伤等级。

超压 - 冲量破坏准则的模型为[6]

$$(\Delta P - P^*)(i_+ - i^*) = DN \tag{5}$$

$$i^* = \frac{\pi h}{4x}\sqrt{2\rho\sigma_\gamma w_0} \tag{6}$$

$$P^* = \frac{\pi^4}{64x^2}\sigma_\gamma h w_0 \tag{7}$$

其中　ρ——靶板的密度；

σ_γ——屈服强度；

h——靶板厚度；

x——边长的一半；

w_0——挠度。

通过大量仿真与计算，拟合出这几种情况下的等毁伤曲线，具体参数如表 5 所示，KK22 级毁伤下单冲击波作用和杀爆相关作用的等毁伤曲线如图 3 所示。

表 5　等毁伤 P－I 曲线参数等效靶板挠度

作用情况	等效靶板挠度 w/mm	P^*/MPa	i^*/(Pa·s)	$\dot{D}N$
冲击波单独作用时	83	1.48	1 248	575
	75.1	1.34	1 127	543
	60.3	1.07	902	509
破片与冲击波相关作用时	60.3	1.07	902	509
	48	0.85	721	325
	36	0.64	541	98
	25	0.45	375	225

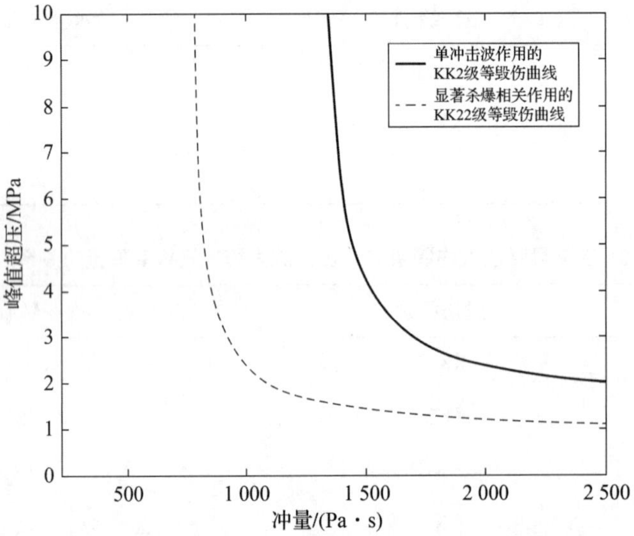

图 3　KK2 级毁伤下单冲击波作用和杀爆相关作用的等毁伤曲线

5　结论

本文针对巡航导弹的近程拦截的特点，定义了近程反导条件下的巡航导弹的解体毁伤的概念，并进一步划分了目标的 K 级毁伤，根据动态断裂准则，计算得出各毁伤等级的临界条件，建立近程防御巡航导弹条件下的毁伤判据。根据超压－冲量毁伤准则建立了巡航导弹在不同毁伤等级下的等毁伤曲线的数学表达式，相同的毁伤等级下，存在杀爆相关效应所需要的超压和冲量的数值都要小于单冲击波作用下的超压和冲量的数值，因此，若同样质量的装药对目标作用，达到相同的毁伤等级，杀爆相关作用比单纯冲击波作用的距离要大得多，即杀爆相关作用提高了毁伤威力，或者是为了获得相同的威力可以减小毁伤元的装药的质量，这对于战斗部的设计，提供了重要依据。本文的研究成果可为巡航导弹反导系统的战斗部的设计提供设计依据。

参考文献

[1] 李向荣．巡航导弹目标易损性与毁伤机理研究［D］．北京：北京理工大学，2006.

Axial Distribution of Fragment Velocities from Cylindrical Casing Under Explosive Loading①

爆炸载荷下圆柱形壳体破片速度的轴向分布

Huang Guangyan②, Li Wei, Feng Shunshan

State Key Laboratory of Explosion Science and Technology, Beijing Institute of Technology, Beijing 100081, P. R. China

The initial velocity distribution of fragments from cylindrical casing, which detonates at one end of explosive charge, is the key issue in the field of explosion technology and its protecting. Most of the formulae available can predict the average velocity or the velocities of fragments at the middle part of the cylindrical casing. However, due to the influence of rarefaction waves, often referred to as 'edge effects', which are generated at the two ends of the charge, most methods cannot accurately predict the velocities of fragments near the two ends of the cylindrical casing. The Fragment Velocities dispersal along the axis of a cylindrical casing, which was detonated at one end, was investigated experimentally in this paper, and the initial velocity distribution was obtained by a flash - radiograph technique. Thus, a formula for Fragment Velocities dispersing along the axis of the cylindrical casing, which is based on the Gurney formula, was proposed by the theoretical analysis and data fitting with the edge effects from the two ends of the charge. This formula has higher accuracy and wider applicability in the study of the Fragment Velocities dispersing along the axis of cylindrical casing under interior blast loading.

1 Introduction

The cylindrical casing filled with explosive charge as a typical symmetrical structure is highly representative in the field of explosive technology and its applications, especially in the design of conventional warheads. The initiation point is usually located at one end along the center - line of the charge. The cylindrical casing is driven by the highly pressurized products, then expands and eventually ruptures into a large number of fragments with different velocities and angles. It is of importance to calculate the initial velocities and the subsequent velocity distribution of fragments for design in explosion techniques or efficiency analysis in explosion applications. The dynamic issue of fragmentation of metal casing has been of interest to the researchers in the field of explosion driven and its applications. Some formulae to calculate the initial Fragment Velocities had been developed for decades. *Gurney* [1] proposed a typical formula based on the energy distribution to calculate the initial Fragment Velocities of cylindrical casing, which can be expressed as:

$$v_0 = \sqrt{2E} \cdot \sqrt{\beta/(1+0.5\beta)} \tag{1}$$

where $\sqrt{2E}$ is Gurney constant of the explosive, v_0 is the Gurney velocity, and b is the ratio of the mass of charge to that of metal casing.

Gurney assumed that all parts of the metal casing rupture at the same stress and with the same initial fragment velocities along the axis of the cylindrical casing. However, Gurney formula gives different results with the experimental data for the rarefaction wave generated at the ends of the cylindrical charge. The rarefac - tion wave at

---

① 原文发表于《International Journal of Impact Engineering》75 (2015) 20 - 27。

② 黄广炎：工学博士，2004 年师从冯顺山教授，研究封锁型弹药设计及效能分析方法，现工作单位：北京理工大学。

the ends and the fraction energy of the metal casing are not considered in Gurney formula. Some researchers investi - gated the rarefaction wave at the ends and developed some correction formulae based on the Gurney formula.

Based on the experimental data, *T. Zulkouski* [2] developed a correction formula $C_f(x)$ by adopting an exponential form to describe the influence of rarefaction wave at the ends on the initial Fragment Velocities, which can be expressed as:

$$C_f(x) = (1 - e^{-2.3617x/d})(1 - 0.288e^{-4.603(L/x)/d}) \tag{2}$$

where x is the distance to the detonation end along the axis of cylindrical casing. d is the diameter of the charge. L is the length of the casing.

The formula to predict the initial Fragment Velocities consid - ering the rarefaction wave at the ends can be obtained by multi - plying Equation (1) with Equation (2). *T. Zulkouski* concluded that t he initial energy of fragment increases with the increasing of the ratio of length to diameter. The optimized design of warheads is obtained when the ratio beyond 1.4. *Randers - Pehrson* [3] estimated the initial Fragment Velocities by recalculating b based on the geometric equivalence, which results from the fact that not all charge along the axis is fully used because rarefaction wave takes away the energy from the explosion products, that is to say, prac - tical b is changing along the axis and is calculated by geometric equivalence. *Y. J. Charron* [4] and *E. Hennequin* [5] proposed the correction formula for b according to the geometric equivalence mentioned by *Randers - Pehrson* [3]. And the correction formula based on the Gurney's formula [1] for calculating the initial frag - ment velocities along the axis is shown as:

$$v(x) = \sqrt{2E}\left[\frac{F(x) \cdot \beta}{1 + 0.5F(x) \cdot \beta}\right] \tag{3}$$

Where $F(x) = 1 - \left[1 - \min\left(\frac{x}{2R}, 1.0, L - x/R\right)\right]$ is the correction formula for β.

A formula is proposed to calculate the relationship of the expanding velocity of the casing with the expanding radius [6]. The casing begins to rupture when the expanding radius is several times the initial radius of the casing. The fragment velocities in this moment are regarded as the initial fragment velocities. This method is based on a lot of experiment data and experiences, but the trend of initial fragment velocities dispersing can be obtained:

$$v(x) = \frac{v_d}{16}\sqrt{\frac{\beta}{2+\beta}\left[1 - \left(\frac{r_0}{r}\right)^4\right]}$$
$$\left[1 + 6\alpha(1-\alpha) + \frac{3}{2}\alpha \cdot \ln\frac{3-2\alpha}{\alpha} + 6\alpha(1-\alpha)(2\alpha - 1) \cdot \ln\frac{3-2\alpha}{2(1-\alpha)}\right] \tag{4}$$

Where $\alpha = x/l$, x is the distance from the detonation end, l is the length of the casing, v_d is the detonation velocity of the charge, r_0 is the initial radius of the casing and r is the expanding radius in one moment.

In terms of blast impulse on the casing, the rupture of the casing and the explosion products escaping, Hutchinson [7 - 9] showed a new thought and the calculation formula is derived based on im - pulse/momentum analysis.

The Equations mentioned above are proposed inconsideration of rarefaction wave at the ends, but the calculated results had a big difference with experimental data [10 - 13]. The initial fragment velocities of the detonation end from Equation (2) and those of the two ends from Equation (3) are all zero, which is unmatched with the experimental phenomenon. The initial fragment velocities at the two ends from Equation (4) are not zero but much smaller than the experimental data. Thus, a more accurate formula should be proposed for predicting the initial fragment velocities.

2 Experimental

The initial fragment velocities dispersing along the axis of cylindrical casing under interior blast loading was tested by using the flash – radiograph in the experiments of this paper. For safety rea – sons, a set of protection – devices had been designed including protection – box, protection – plates, film – protection devices and support, the schematic view of test set – up is shown in Figure 1.

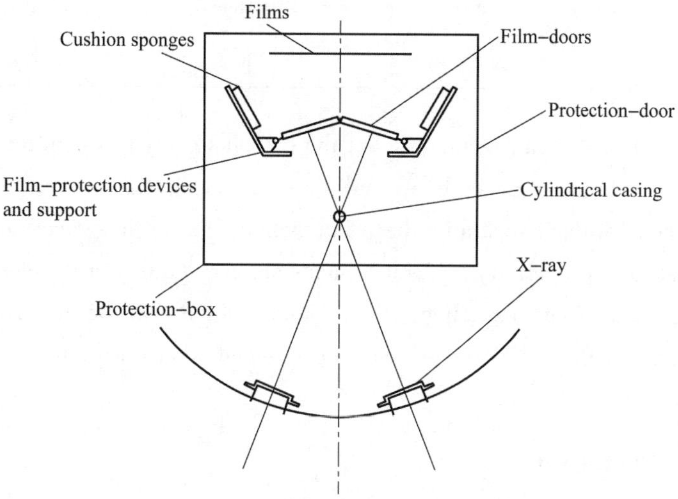

Figure 1 Schematic view of test set – up.

The cylindrical casing was hanged up on the roof protection – plate within protection – box. The axis of casing was perpendicular to the ground and located at the intersection generated by two beams of flash – radiograph which was symmetric about the center line (the dash line of Figure 1) of protection – box. The protection – plates were hung up around the protection – box, which had a good effect on preventing the fragments escaping from protecting box damaging people and other devices. The film – protection devices were behind the cylinder casing and included film – doors, film doors supports, hinges and cushion sponges. The film – doors were the key component of film – protection devices. The film – doors betweenthe films and the cylindrical casing can prevent fragments damaging the films and that can make the films be normal exposed. The film – doors supports made the film – doors rotating by the hinges and impacting on the cushion sponge when fragments impacting the film – doors, which transformed the kinetic energy of fragments into the film – doors and made this energy absorbed by cushion sponge finally. So the film – doors have a better protective effect onthe films.

The AISI 1045 steel was adopted in the experiments as the ma – terial of metal casing. The explosive filled in the casing was TNT50/ RDX50 and was detonated at one end. The parameters of cylindrical casing are listed in Table 1.

For a special cylindrical casing, the fragment velocities dispersing along the axis of it can be obtained by only two photos which were taken at different time. It is very difficult to take the photos exactly when the fragments reach the initial velocity. If the assumption that the fragment velocities does not decrease in an interval just a while after the fragments reach the initial velocity [14] is adopted, the average velocity of fragments in this interval can be regarded as the initial velocity of fragment. The method of measuring the initial velocity of fragment is described as follows: Firstly, a static radiograph was taken as the benchmark photo before the explosive was detonated. Secondly, multiple timings were selected, corresponding to the casing expanding, casing fracture and fragment dispersing for the images to be captured by the flash – radiograph. Two typical photos in different times are selected to compare with the benchmark photo and their axis must coincide. The relative displacements of fragments in different po – sitions X_i (These positions had been determined in the experiment and will be described below) along the axis of cylindrical casing were measured respectively. So the initial velocity of fragment along the axis of cylindrical casing can be calculated through the relative displacements and the interval of photos.

Table 1 The parameters in No. 1 test.

Specimen	Length of casing L/mm	Exterior radius of casing D/mm	Interior radius of casing d/mm	Density of casing ρ_m/g cm^{-3}	Density of explosive P_c/g cm^{-3}
No. 1	77.30	29.68	23.60	7.85	1.70

For a specimen, a lot of photos can be taken, but the two photos used as data for analyzing must have some properties including the casing fracturing completely and the resultant drive to the frag – ment almost zero. That is because the adjacent fragments without separation have interaction force in the process of casing's frac – turing. In addition, after adjacent fragments separating, if frag – ments are accelerating or decelerating apparently under the effect of detonation products and air resistance, the resultant force will be not zero. And W. J. Stronge's experiments indicated that the above requirement could be satisfied when the radius of casing expands to 3e3.5 times its the original radius [13].

3 Results and Discussion

3.1 Data Analysis of X – ray Photos

Due to a distance existing between negative and casing and the light source of flash – radiograph treated as a point, the picture on the photo is larger than the actual casing. The magnification ratio has to be calculated for data analysis to be quantitative. Based on the benchmark photo, the axial and radial magnification ratios are calculated 1.285 and 1.267, respectively. Using charges with the parameters listed in Table 1, three tests were conducted. Two pic – tures have been selected from one of the tests and can be compared with each other, as shown in Figs. 2 and 3. The a, b, c, d, e and f are mark points on the fragment dispersing profi le from detonation end in turn and these points are one – to – one correlation on Figs. 2 and 3. The non – uniformity of fragments dispersing along the axis can be seen and is in good agreement with previous experiments [10 – 13]. Through the examination of the experimental radiographs, it can be found that the fragment radial displacement gradient of point a to c is the biggest and the fragment radial displacement of that is the smallest along the axis from the detonation end. The majority of fracture points occur in the Region 1, as shown in Figs. 2 and 3, which makes the size of fragment smallest in this region. However, at the non – detonation end, the fragment radial displacement gradient of points d to e is smaller while the actual radial displacement is bigger than at points a to c. In this region 2, the fracture planes are less circumferential than in region 1 and the fragments are narrower. The region 3 corresponding to point f (non – detonation end) is filled with smaller fragments. The fragment radial displacement gradient is the smallest and the actual radial displacement is greatest along axis between points c and d. The least number of fracture planes occur in Region 4 in which the narrowest fragments are found.

The above observation can be explained as follows: due to the detonation and rarefaction wave both propagating firstly from the detonation end (Point a), from point a to c, the closer to detonation end, the weaker the effect of driving force on fragments. In addition, the rarefaction wave decreases quickly with propagating in the explosion productions, which results in a big difference of driving capability and that of radial displacement gradient from point a to c. Thus, the large shear deformation which exceeds the dynamic shear strength occurs in a lot of places of Region 1, which leads to a lot of axial fractures in this region. The small and dense fragment can be seen in Region 1 of Figs. 2 and 3. Because of the rarefaction wave and the escape of explosion productions in Region 1, the acceleration of fragments in this region is lower than other region and the displacement of fragment is smaller.

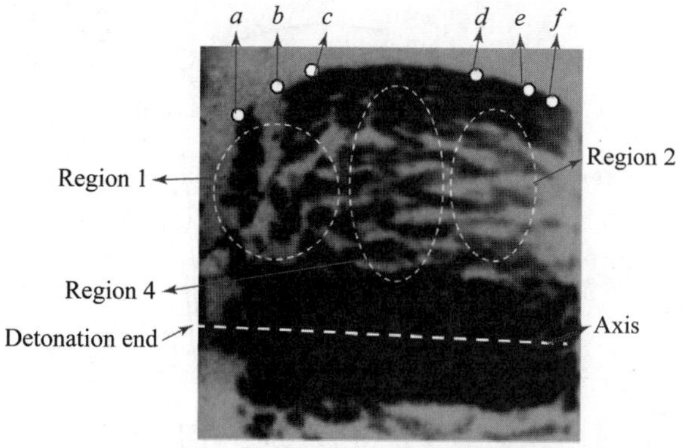

Figure 2 The flash - radiograph photo of No. 1cylindrical casing at 21. 6 μs

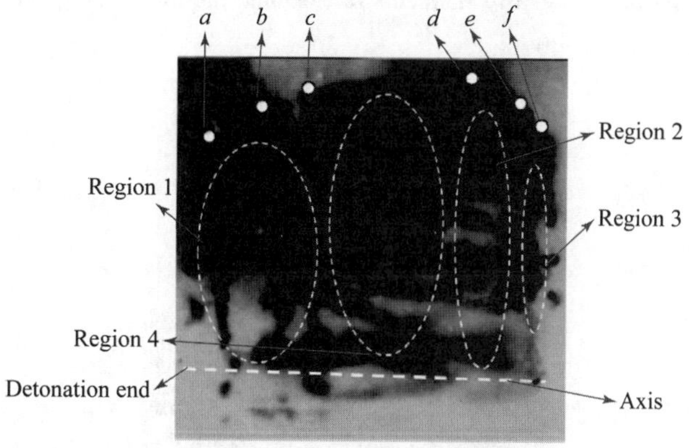

Figure 3 The flash - radiograph photo of No. 1cylindrical casing at 44. 2 μs

The rarefaction from two ends have a little effect on middle part of cylindrical casing, so the radial displacement gradient along axis is the smallest, which may not generate the shear stress exceeding the dynamic shear strength of casing material and generates a little of fractures in axial direction in Region 4. The region from point d to e is driven by detonation wave firstly and is influenced by rare - faction wave until the detonation wave reaches the non - detonation end. So the radial displacement gradient of fragments from point d to e is smaller than that of fragment from point a to e, which leads to a little of the shear failure and fracture point. And major frag - ments are narrow as shown in Figs. 2 and 3. In Region 3 of Figure 3, some debris was generated at the detonation end due to a great change of displacement gradient caused by a sharp rarefaction wave from non - detonation end.

At the detonation end, due to the decrease of radial drive towards the end, the expansion of the casing is more spherical than cylin - drical. This means that the casing material experiences dynamic strains in two senses, both around the axis (circumferential) and along the axis (axial). It therefore breaks into the polygonal pieces characteristic of spherical dynamic fracture. From point c onwards, the radial drive is more uniform with length along the axis, and the casing fractures under circumferential strain only and breaks into the long, slender fragments originally described by Mott (1943).

3.2 Data Measurement

The fragment profiles of two moments shown in Figs. 2 and 3 are drawn in Figure 4.

The dashed line is the fragment profile at time t_1 and the cor - responding measurement has been made on the radiograph in Figure 5. The solid line is the fragment profile at time t_2 and the cor - responding measurement has been made on the radiograph in Figure 6.

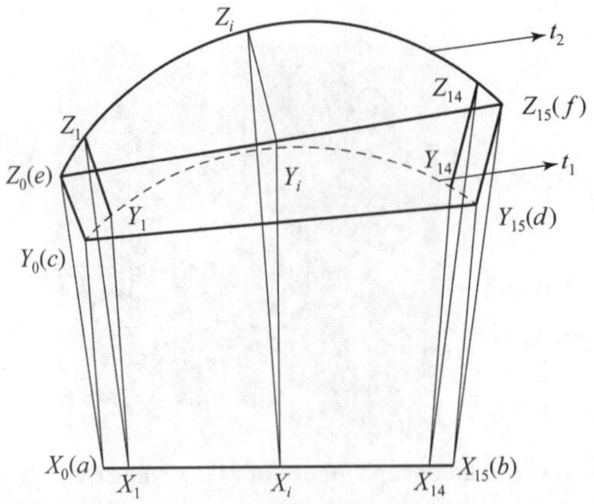

Figure 4 The profiles of two moments to calculate the initial Fragment Velocities.

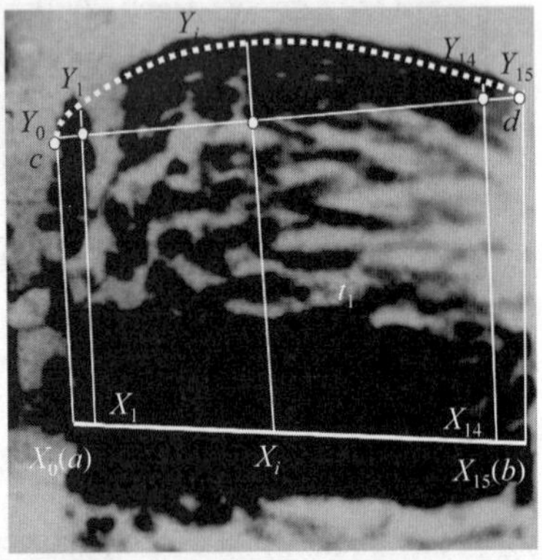

Figure 5 The flash – radiograph photo of No. 1 cylindrical casing at 21. 6 ms for measurement.

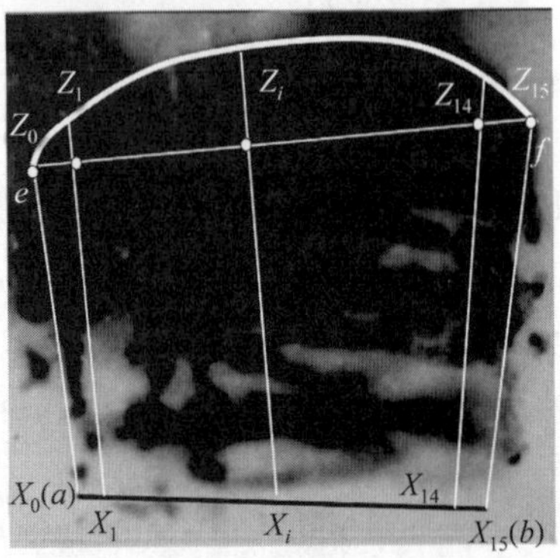

Figure 6 The flash – radiograph photo of No. 1 cylindrical casing at 42. 2 ms for measurement.

The initial velocity measurement method is as follows: Firstly, the axial line a – b, the dashed curve c – d above the end points of the fragment profile at time t_1 (from Figure 5) and the solid curve e – f from time t_2 (Figure 6) were each divided into 15 intervals uniformly. The equal division points of a – b are named as X_i. The points X_i are connected with the corresponding equal division points of c – d, respectively, to extend this to the fragment profile at time t_1 and create the points Y_i. The distance of X_iY_i is the dynamic displace – ment of the corresponding fragment as shown in Figure 5. In the same way, the points X_i are connected with the equal division points e and f, respectively, to extend to the fragment profile at time t_2 and create the points Z_i. The distance of X_iZ_i is the dynamic displace – ment of the corresponding fragment as shown in Figure 6. The length of solid lines between points Y_i and points Z_i are the displacements in the interval $t_2 - t_1$. From these, the average velocity of fragments along the axis can be obtained based on the axial and radial magnification ratios of 1. 285 and

1.267. The average velocity listed in Table 2 is the mean value of experiments in three groups.

3.3 Comparison with the Calculation Result

Substituting the parameters from Table 2 into Equations (1) – (4), respectively, the calculated results and the experiments in this paper are compared in Figure 7 and the extent to which their pre – dictions deviate from the above experimental data is shown as Figure 8.

In Figure 7, the abscissa represents the coordinate along the axis of the charge and the original point is the detonation point. The fragment velocities at middle part of casing is the biggest and that at two end of casing is the smallest for the calculated results of all three Equations, which indicates that the effect of rarefaction wave on fragment velocities had been considered in all these Equations. The calculated results of Equation (2) agree best with the experi – mental results at the non – detonation end, while there is a big dif – ference between all the predictions and the data at the detonation end. Equations (2) and (3) are obviously wrong at the detonation end because they predict the fragment velocities at this end being zero. However, the calculated results of Equations (3) and (4) have a good agreement with experiments at which the maximum velocity occurs but have a big different at else parts of casing, especially, since the predicted fragment velocities of (3) at the two ends of casing are both zero.

Figure 8 shows the errors between the calculated results of Equations (2) – (4) and experimental results. It is found that the errors of the three Equations at the middle part of casing are all relatively small, which indicates that the effect of rarefaction wave on frag – ment velocities had been considered in all these Equations. But there are big errors at the end of cylindrical casing. It is concluded that these three Equations cannot predict accurately the fragment velocities dispersing along the axis of cylindrical casing, especially at the ends of casing. So a new formula needs to be established to predict accurately the fragment velocities dispersing along the axis of cylindrical casing.

Table 2 The measured distances from flash – radiograph photos and the average Fragment Velocities.

No. i	X_iY_i /mm	X_iZ_i /mm	Y_iZ_i /mm	V/ms^{-1}
0	47.1	65.7	21.6	954
1	51	70.5	25.5	1 129
2	53.4	73.1	26.7	1 182
3	55.2	74.6	27.6	1 222
4	56.4	75.4	28.7	1 270
5	56.9	76.3	29.5	1 309
6	57.2	76.9	30.6	1 354
7	56.9	77.3	31.5	1 396
8	56.1	77.8	32.3	1 432
9	55.2	78.1	32.7	1 445
10	54.6	78.3	33.2	1 469
11	53.8	78.5	33.7	1 493
12	53.1	77.7	33.1	1 463
13	52.3	76.2	31.6	1 398

continued

No. i	X_iY_i /mm	X_iZ_i /mm	Y_iZ_i /mm	V/ms^1
14	51.6	74.5	30.4	1 347
15	50.6	72.9	27.3	1 208

4 Model

4.1 Establishing of the Formula

According to the comparison between the three Equations and experiments, it is found that the calculated results from Equation (2) agree well with the experimental data from all cylinders except at the detonation end. So the analysis formula in this paper is established on the base of Equation (2). The flash - radiograph show that the fragment velocities are relative small and the fragment velocities gradient along axis are relatively big in the two ends and around them. But an opposite phenomenon occurs in the middle part. The reason of the phenomenon mentioned above has been explained. So two terms are inserted in Gurney formula for correction:

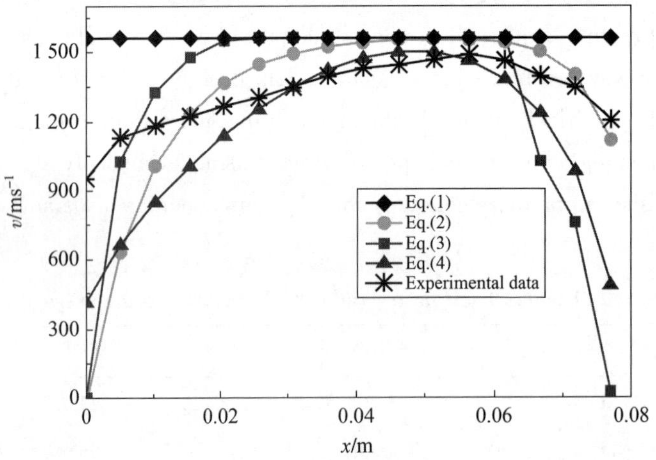

Figure 7 Axial distribution of Fragment Velocities by Eqs. (1) - (4) and experiments

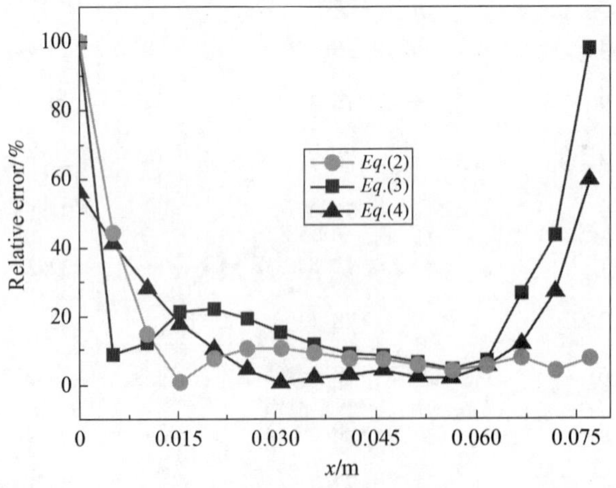

Figure 8 Relative error of Eqs. (2) - (4) along the axis of the cylindrical casing

$$v_{0x} = F_1(x/d) \cdot F_2[(L-x)/d]v_0 \tag{5}$$

Where, F_1 (x/d) is the corrected term at detonation end; F_2 $[(L x)/d]$ is the corrected term at non – detonation end; L is the length of cyLindrical casing; v_0 is the initial velocity from Gurney formula and the E_g is the Gurney energy determined by the type of explosive. The Gurney energy E_g should be investigated firstly:

For calculating the Gurney energy E_g, *Hardesty* and *Kennedy* [15] developed a formula based on the explosion heat of different explosives. *Danel* and *kazandjian* [16] corrected the relationship between the detonation velocity of a charge and its Gurney energy at infinite expansion. *Mohammad. Hossein* and *Keshavarz* [17] analyzed the formula mentioned above and developed a formula based on the constituent elements of chemical formula of explosive instead of the explosion heat:

$$\sqrt{2E} = 0.404 + 1.020\rho_0 - 0.021c + 0.184b/d + 0.303d/a \tag{6}$$

Where, c is the nitrogen atom number in chemical formula, b/d is the ratio of the atom number of hydrogen with oxygen, d/a is the ratio of the atom number of oxygen with carbon, r_0 is the density of explosive. The coefficient in the Equation (6) is fitted by the experiment data. This formula is close to the essential nature of explosive due to its being based on the chemical formula of explosive. In addition, it is not necessary to assume the compo – nent of explosion products and the explosion heat and the calculated results agree with the experimental data. So the Equation (6) is adopted in this paper for calculating the Gurney energy.

According to the flash – radiograph photos and the experimental data measured, it is found that the initial fragment velocities decreases exponentially close to two ends of cylinder. The Equation (2) had considered about this instead of the detonation end and the part near it and assumed that the explosive at the detonation end has no effect of driving cylinder casing, this is to say, the initial fragment velocities at the detonation end is zero, which is obvi – ously different from experimental data. This paper developed a new formula considering the effect of the explosive at the deto – nation end to correct the Equation (2):

$$F_1(x/d) = 1 - Ae^{-Bx/d} \tag{7}$$

$$F_2[(L-x)/d] = 1 - Ce^{-D(L-x)/d} \tag{8}$$

Where, A, B, C and D are the correction coefficients determined by experimental data. So the Equations (7) and (8) are substituted in Equation (5):

$$v_{0x} = (1 - Ae^{-Bx/d})(1 - Ce^{-D(L-x)/d}) \cdot \sqrt{2E} \cdot \sqrt{\beta/(1+0.5\beta)} \tag{9}$$

The correction formula$C_f(x)$ is zero in the Equation (2) at the detonation end, which leads to the initial fragment velocities equal to 0. But the correction formula $(1-A)(1-C\,e^{-DL/d})$ in Equation (9) is obviously not equal to 0. If the correction coefficients are suitable, the Equation (9) can accurately predict the initial fragment veloc – ities along the axis.

4.2 Determining the correction coefficients

The correction coefficients A, B, C and D can be determined by the initial fragment velocities from experimental data. The Equa – tion (9) is very difficult to determine by the method of linear fit. In addition, the difference between the maximum initial fragment velocities (1 493 m/s) and the Gurney velocity (1 566 m/s) is 4.7%, which indicates that the rarefaction wave from detonation and non – detonation ends has little effect on the maximum initial fragment velocities, that is to say, the initial fragment velocities from detonation end to the position where the maximum initial fragment velocities occurs is almost determined by the detonation end instead of non – detonation end. The initial fragment velocities from non – detonation end to the position where the maximum initial fragment velocities occurs is in a similar way. So the correction coefficient A and B can be determined by the initial fragment velocities from detonation end to the position where the maximum initial fragment velocities occurs and the correction coefficient C and D can be determined by the initial fragment velocities from non –

detonation end to the position where the maximum initial fragment velocities occurs.

Firstly, based on the analysis above, the Equation (8) can be assumed as 1 for determining the correction coefficient A and B, that is to say, the effect on the non - detonation end is neglected. So the Equation (9) can be expressed as:

$$v_{0x} = (1 - Ae^{-Bx/d}) \cdot \sqrt{2E} \cdot \sqrt{\beta/(1+0.5\beta)} \tag{10}$$

The correction coefficient A is determined by the initial frag - ment velocitiesv_0 at the detonation end using the experimental data and is calculated as 0. 361.

$$A = 1 - v_0/[\sqrt{2E} \cdot \sqrt{\beta/(1+0.5\beta)}] \tag{11}$$

The correction coefficient A is substituted into Equation (10) and the coefficient B is obtained by fitting the initial fragment velocities along the axis using the least square method. The correction coef - ficient C and D are obtained by the same way and are equal to 0. 192 and 3. 03.

The correction formula at the detonation end:

$$F_1(x/d) = 1 - 0.361e^{-1.111x/d} \tag{12}$$

The correction formula at the non - detonation end:

$$F_2[(L-x)/d] = 1 - 0.192e^{-3.03(L-x)/d} \tag{13}$$

Atx = L (77 mm), the value of the correction formula at the detonation end:

$$F_1(x/d) = 1 - 0.361e^{-1.111\times 77/23.5} \approx 0.990\,5 \to 1 \tag{14}$$

Atx = 0, the value of the correction formula at the nondetonation end:

$$F_2[0] = 1 - 0.192e^{-3.03\times 77/23.5} \approx 0.999\,990\,6 \to 1 \tag{15}$$

It can be concluded that the advantage of the correction formula using exponential form includes correcting efficiently at bothends almost without any effect on the other end. Thus, the correction formula is reasonable and a unified formula to calculating initial fragment velocities along axis of cylinder casing is obtained. The correction coefficient A, B, C and D are substituted into Equation (9) and the unified formula is obtained as:

$$v_{0x} = (1 - 0.361e^{-1.111x/d}) \cdot (1 - 0.192e^{-3.03(L-x)/d}) \cdot \sqrt{2E} \cdot \sqrt{\beta/(1+0.5\beta)} \tag{16}$$

5 Verification

For validating the general applicability of Equation (16), another, more heavily, cased cylinder experiment was fired as No. 2, with structural parameters listed in Table 3. This charge had a Gurney beta of 0. 148, compared to 0. 372 for the first set of charges.

The same experimental method as section 2 was adopted for No. 2 test. Figure 9 shows the flash radiographs at two different times after No. 2 test cylinder casing was detonated. The same measure - ment method as section 3. 2 was adopted for the Figure 9 to obtain the initial fragment velocities along axis of No. 2 test.

The structural parameters listed in Table 3 are substituted into Equation (16) and the calculated curve is shown as Figure 10. In addition, the experimental data is compared with the calculated curve. It is clearly shown that: the calculated curve remains in good agreement with the further experimental data. However, it is found that all experimental data is below the calculated curve. This could be because the formula established in this paper neglects the fracture energy of casing and assumes that the fracture energy is much smaller than explosion energy (this may be less true for the thicker casing). So the Equation (16) has a general application in calculating the initial fragment velocities along the axis of cylindrical casing.

To further demonstrate the accuracy of Equation (16) for a wider range of test conditions, a test data of an explosive - filled cylindrical casing ($\frac{L}{D} = 2$, $\beta = 0.429$, $\sqrt{2E} = 2\,840$ m/s) in Ref. [16] was choose to be comparison, as shown in Figure 11. It also evidenced that Equation (16) has a good agreement with fragment velocities distribution at two end parts of explosive – filled cylindrical sample in Ref. [16]. Thus, the new expression can predict the fragment velocities well for most simple conditions of explosively loaded cylinders.

6 Discussion

The Equations (2) e (4) are compared with Equation (16) in Figure 12 and the structural parameters listed in Table 2 are adopted in these Equations. In addition, the error between every Equation and experimental data is shown in Figure 13. As shown in Figure 12, the Equation (16) has the best agreement and the smallest error with experimental data of all Equations mentioned above. The maximum relative error of Equation (16) is 6.7% and the average value of the error is 3.86%, which indicates that the Equation (16) can predict the initial fragment velocities along the axis of a finite – length cylindrical casing to an acceptable degree of accuracy.

It can be shown in Figure 13 that the initial fragment velocities of non – detonation end is 26.6% (254 m/s) bigger than that of detonation end. This can be seen in Figs. 2 and 3: the degree of fracture at the detonation end is much more serious than that at the non – detonation, which leads to the gas escape at the detonation end much more serious than that at the non – detonation. So the driving force is decreasing sharply at the detonation [17]. The phenomenon mentioned above can be explained as follow: the detonation begins with the rarefaction wave from the detonation end. So the effect of rarefaction on the detonation is much more than that on the non – detonation. Due to the serious decrease of rarefaction wave in the process of detonation wave propagating over the explosive and the effect of rarefaction on middle part of casing much little than two ends, the gradient of the radial pressure along the axis of cylindrical casing is the biggest at the detonation end, which causes big tensile stress and shear stress in this region [18 – 21]. Thus, the serious fracture can be seen at the detonation end, while the fraction at the non – detonation is smaller and that at the middle part is littlest of all casing.

7 Conclusions

The influence of the rarefaction wave, which is generated at the two ends of the cylindrical casing, on the initial fragment velocities along the cylinder axis was studied in this paper. Some tests of detonating on one end of the cylindrical casing, which is filled with explosives charge, had been conducted, and the velocity distribution of fragments was obtained by a flash – radiograph technique. The correction formula $F_1(x/d)$ for detonation end and $F_2[(Lx)/d]$ for non – detonation end is proposed based on Equation (2). Furthermore, the correction coefficients A, B, C and D were all fitted by the experimental data. Therefore, Equation (16), which can accurately calculate the initial fragment velocities along the axis of cylindrical casing, was obtained by considering the influences of rarefaction wave from the two ends. By comparing the calculated results from Equation (16) and experimental data, it was found that the maximum relative error is 6.7% and the average relative error is less than 3.86%, which indicates that Equation (16) has better general feasibility to predict the initial fragment velocities dispersing along the axis of casing.

The new expression can predict the fragment velocities well for most simple conditions of explosively loaded cylinders. But because of the complexity of explosively loaded shells and variability of detonation conditions, to predict a much wider range of test conditions, we need to investigate further on the influence of different end cover and length – diameter ratio of explosively loaded cylinders, correction expression of explosively loaded circular truncated cone and non – central detonation conditions. All of these will be investigated based on these analytical models in our subsequent research.

Acknowledgments

This work was partially sponsored by Natural Science Foundation of China under Grant No. 11102023,

No. 11172071, and Foundation of State KeyLaboratory of Explosion Science and Technology of China under Grant No. YBKT15 - 02. The authors wish to express their appreciation to personnel at Intensive Loading Effects Research Group of State Key Laboratory of Explosion Science and Technology, where the experimental data reported on this paper were collected.

References

[1] GURNEY R W. The initial velocities of fragments from bombs, shells and grenades: ATI36218 [R]. Report No 405. Aberdeen Proving Ground, Maryland: Ballistic Research Lab, 1943.

[2] ZULKOSKI T. Development of optimum theoretical warhead design criteria: Report No AD - B015617 [R]. China Lake, California, USA: Naval Weapons Center, 1976: 39 - 44.

[3] RANDERS - PEHRSON G. An improved equation for calculating fragment projection angles [C] //Proceedings of International Symposium on Ballistics, Daytona Beach, U. S. A, 1976, 9 - 11 March. Daytona Beach, FL. Washington, D. C.: American Defense Preparedness Association, 1976.

[4] CHARRON Y J. Estimation of velocity distribution of fragmenting warheads using a modified Gurney method: ADA074759 [R] . Ohio, U. S. A: Air Force Institute of Technology, Wright - Patterson AFB Oh School of Engineering, 1979: 113 - 6. ADA074759: [R] .

[5] HENNEQUIN E. Influence of the edge effects on the initial velocity of fragments from a Warhead [C] // Proceedings of 9th International Symposium on Ballistics, Shriv - enham, U. K, 1986.

[6] Beijing Institute of Technology《explosion and its effects》Compile Group. explosion and its effects [M]. Beijing: National Defense Industry Press, 1973.

[7] HUTCHINSON M D. The escape of blast from fragmenting munitions casings [J] . International journal of impact engineering, 2009, 36: 185 - 92.

[8] HUTCHINSON M D. Replacing the equations of Fano and Fisher for cased charge blast equivalence - I ductile casings [J] . Propellants, Explosives, Pyrotechnics, 2011, 36: 310 - 331.

[9] HUTCHINSON M D. Replacing the equations of Fano and Fisher for cased charge blast impulse - II Fracture strain method [J] . Propellants, Explosives, Pyrotechnics, 2012, 37: 605 - 608.

[10] MARTINEAU R L, ANDERSON C A, SMITH F W. Expansion of cylindrical shells subjected to internal explosive detonations [J] . Experimental mechanics, 2000, 40: 219 - 225.

[11] CHHABILDAS L C, THORNHILL T F, REINHART W D, et al. Fracture resistant properties of AerMet steels [J]. International journal of impact engineering, 2001, 26: 77 - 91.

[12] JAANSALU K M, DUNNING M R. Fragment velocities from thermobaric explosives in metal cylinders [J]. Propellants, Explosives, Pyrotechnics, 2007, 32: 80 - 86.

[13] STRONGE W J, MA X Q, ZHAO L T. Fragmentation of explosively expanded steel cylinders [J]. International journal of mechanical sciences, 1989, 31: 811 - 823.

[14] FENG S S, CUI B G. An experimental investigation for the axial distribution of initial velocity of shells [J]. Acta armamentarii, 1987, 4: 60 - 63.

[15] HARDESTY D R, et al. Thermochemical estimation of explosive energy output [J]. Combust and flame, 1977, 28: 45 - 59.

[16] KÖNIG P J. A correction for ejection angles of fragments from cylindrical wareheads [J]. Propellants, Explosives, Pyrotechnics, 1987, 12: 154 - 157.

[17] DANEL J F, KAZANDJIAN L. A few remarks about the Gurney energy of condensed explosives [J]. Propellants, Explosives, Pyrotechnics, 2004, 29: 314 - 316.

[18] KESHAVARZ M H. New method for prediction of the Gurney energy of high explosives [J]. Propellants, Explosives, Pyrotechnics, 2008, 33: 316 - 320.

Effect of Eccentric Edge Initiation on the Fragment Velocity Distribution of a Cylindrical Casing Filled with Charge①

偏心起爆对圆柱形装药壳体破片速度分布的影响

Li Wei②, Huang Guangyan, Feng Shunshan

State Key Laboratory of Explosion Science and Technology, Beijing Institute of Technology, Beijing 100081, P. R. China

Cylindrical casing filled with charge under eccentric point initiation at one end is commonly used in warheads and can occur in some explosion accidents. The fragment velocity distribution of the casing is an important parameter in warhead design, structure protection and safety. A formula composed of three terms, including the influence formula of eccentric initiation, rarefaction wave and incident angle of the detonation wave was proposed in the present work to determine the fragment velocity distributions of a cylindrical casing under eccentric point initiation at one end. The influence of the incident angle of detonation wave is proposed as the key factor. A numerical simulation model was proposed, validated and used to determine the influence of the incident angle of detonation wave. Its corresponding formula was established and validated with the established numerical simulation model. The rest of two factors can be calculated with previously reported formulas. The formula proposed in the present work was further validated with the established numerical model. The results indicate that the calculation formula can accurately predict the fragment velocity distributions along the axis and circumference of a cylindrical casing under eccentric point initiation at one end.

1 Introduction

Cylindrical casing filled with charge is the most common structure applied in explosion security protection, explosion control technique and warhead design. Fragment velocity distribution is one of most important parameters of the casing. Many experiments and numerical simulation model about cylindrical casing filled with charge under axial edge initiation have been reported. However, due to the accident stimuli or poor explosion control technique, the initiation point at the end could be off the axis of the cylindrical casing filled with charge, leading to eccentric edge initiation. The direction detonation wave propagation and the energy distribution of explosion product on the cylindrical casing will change under eccentric edge initiation. In addition, according to Shapiro formula, the direction fragment flying is affected by the initiation point of charge and the direction detonation wave propagation [1], hence, the fragment velocity distribution is significantly different from that under axial initiation. Therefore, it is necessary to predict the fragment velocity distribution of cylindrical casing filled with charge under eccentric edge initiation.

The fragment velocity distribution along the axis of the cylindrical casing under central point initiation at the near end has been extensively studied theoretically, experimentally and numerically in the past decades (Figure 1). Assuming the gas density is uniform, Gurney proposed that the initial fragment velocity was a function of the

---

① 原文发表于《International Journal of Impact Engineering》80 (2015) 107 - 115。

② 李伟：工学博士，2012 年师从冯顺山教授，研究内置冲击反应材料弹丸对装药结构的侵爆机理，现工作单位：中国工程物理研究院。

charge to metal ratio (mass ratio of explosive charge to casing) [2]. A constant, Gurney energy, was proposed in his formula and was determined experimentally. However, the influence of rarefaction wave from the ends of cylindrical casing was not considered in his formula. Zulkouski [3] reported a correction formula to describe the influence of rarefaction wave. However, the velocity at the near end cannot be determined with this formula. Karpp and Predebon [4] validated Gurney formula and further confirmed that the maximum fragment velocity can be easily determined with the formula but the influence of rarefaction wave was neglected. They used a finite - difference code, HEMP, to measure fragment velocities and proposed that gas leakage caused by casing breakup also could affect the fragment velocity distribution. Charron [5] modified Gurney formula by multiplying the mass ratio of charge o casing by a term (Eq. (16)) that describes the effect of rarefaction wave at two ends. Eq. (16) is based on the conclusion given by Pehrson [6] that the maximum propagation distances of the rarefaction waves from the near (Figure 1) and from the far end are 2R and R (R: radius of charge), respectively. Zlatkis [7] investigated the edge effect of fragment dispersion and used a peripheral frame to minimize end effects. A calculation formula derived from impulse/momentum analysis was reported to determine the blast impulse on the casing, the rupture of the casing and the explosion product escaping [8 - 11]. Hornberg [12] conducted some cylinder test and investigated material deformation caused by charges, in addition, he calculated and compared the Gurney velocity of many kinds of charges. Martineau [13] and Singh [14] reported a relationship between the velocity and time of a shell subjected to internal high - explosive detonations. Jaansalu [15] measured the fragment velocity of a metal cylinder casing using flash X - ray imaging technique and discussed the influences of casing thickness, casing material and explosion on the fragment velocity. Feng [16] considered rarefaction wave at two ends as the key factor on the fragment velocity distribution along the axis of a cylindrical casing under central point initiation at one end and conducted a series of experiments to measure the fragment velocity distribution with pulsed X - ray. According to Feng's experiments, Feng and Huang [17] added some tests and proposed a highly accurate semiempirical formula based on the experimental data obtained by his and Feng's experiments.

Due to the high velocity and large amount of fragments generated by the altered initiation point on the charge, eccentric initiation has attracted considerable attentions. Waggener [18] added a term to Gurney formula to determine the fragment velocity of a cylindrical casing under eccentric line initiation. The term repre - sents the maximum enhanced velocity in the target direction, which equals to 1.25. Later, Wang [19] demonstrated that the velocity gain should be 1.1 instead of 1.25. Held [20] added modified term, $2\cos(\theta/2)$, to Gurney formula to calculate the velocities of a single eccentric initiation line on a cylindrical casing. The modified term is multiplied by the mass ratio (β) of charge to casing in Gurney formula, where q is the circumference of the point on the casing. The velocity calculated from the modified formula is consistent with the experimental results but not applicable to $\theta = 180°$. Feng [21] calculated of the velocities of a cylindrical casing under eccentric initiation using a similar formula but with a different modified term that was obtained from experimental data. The calculation formula can be used to accurately predict the fragment velocity distribution along the circumference of cylindrical ring. However, the modified term has ten fitting parameters that could be changed with the altering of ratio b. Wang [22] added a term $k = 0.9\beta^2 - 0.23\beta + 0.14$ to the formula. Thus, the formula can be applied to the casings with different β values. All the formulas mentioned above are suitable to calculate the fragment velocity distribution along the circumference of a cylindrical ring under eccentric point initiation and under eccentric line initiation. However, the eccentric point initiation at the near end (Figure 1) is usually adopted in cylindrical casings filled with charge for most of warhead and sometimes occurs in some devices. To assess the potential effect of the munition and its hazard in a credible accident, it is necessary to predict the fragment velocity distribu - tions along the axis and circumference of cylindrical casings under eccentric point initiation at the near end. The formulas mentioned above cannot meet this need.

In the present work, a formula was established to predict the fragment velocity distributions along the axis and circumference of cylindrical casings under eccentric point initiation at the near end. In this formula, the effect of the incident angle of the detonation wave was proposed as a key parameter to determine the velocity distribution. The influence formula of incident angle of the detonation wave was obtained by a novel scheme of numerical simulations which had been firstly validated by experimental data. The results obtained from the formula are consistent with the data obtained with other established numerical simulation models, indicating that the formula can accurately predict the fragment velocity distributions along the axis and circumference of a cylin - drical casing under eccentric point initiation at the near end.

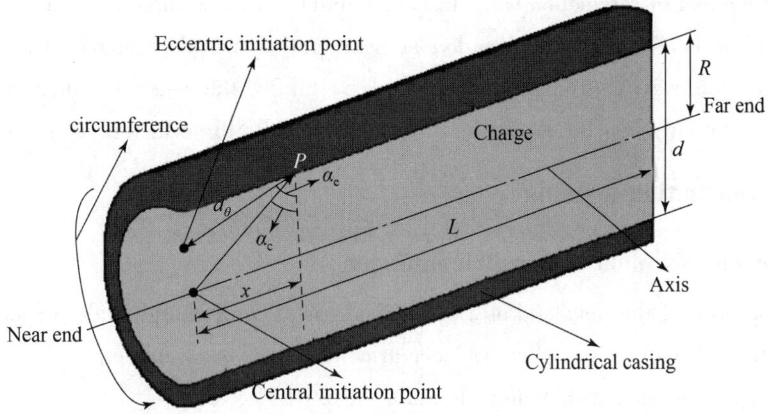

Figure 1 Cutaway schematic diagram of a cylindrical casing with charge under eccentric point initiation at near end.

2 Calculating Formula

2.1 Description of Problem

As shown in Figure 1, the rarefaction wave from initiation at near and far ends, the eccentric initiation and the incident angle of detonation wave at each position in the casing caused by eccentric point initiation at the near end can affect the fragment velocity distribution. Thus, all these three factors should be considered in determining the initial fragment velocity distributions along axis and circumference of a cylindrical casing under eccentric point initiation at near end. Feng's formula included the rarefaction wave from initiation at near end and can be used to calculate the initial fragment velocity distribution along axis of a cylindrical casing under central point initiation at near end [17] (Figure 1). Wang's formula included the effect of eccentric initiation on circumferential velocity distribution and thus can be used to calculate the initial fragment velocity distribution along circumference of a cylindrical casing under eccentric line initiation [22]. However, the incident angle of detonation wave at every position of casing under eccentric point initiation at near end was not considered in Wang's formula. Although Feng's formula included the incident angle of detonation wave under the point initiation at near end, the incident angle of detonation wave under eccentric point initiation (ae) is different from that under central point initiation (ac) (Figure 1). Therefore, Feng's formula cannot be directly used to determine the influence of the incident angle of detonation wave of a cylindrical casing under eccentric point initiation at near end. In full, the formulas for the three factors need to be re - established.

2.2 Assumption and solution

Both eccentric and central initiation points are located at the near end and the ratio of length to diameter of cylindrical charge is usually higher than 2. Hence the strength, at any point P, of the end effect, even from the near end, will be little affected by the position of the detonation point and that, if L/D > 2, the two end effects will act in isolation from each other. Thus, it can be assumed that the difference between the rarefaction waves in cylindrical casings under eccentric and central point initiation at the near end is negligible. Therefore, all three factors, including the effect of eccentric initiation on circumferential velocity distribution, the effect of rarefaction waves from the initiation at near and far ends on velocity distribution and the effect of incident angle of deto - nation wave at every position (such as point *P* in Figure 1) of the casing on velocity distribution, should be taken into account. The effect of eccentric initiation on circumferential velocity distribution can be determined with Wang's formula [22]. The remaining two factors can be calculated by terms for the rarefaction wave and incident angle of detonation wave in Feng's formula with some modification [17]. Based on the assumption mentioned above, the term for the effect of rarefaction wave in Feng's formula can be directly used to determine the effect of rarefaction wave under the eccentric point initiation at near end of cylinder casing. The term for the effect of incident angle of detonation wave needs to be corrected by taking the eccentric initiation point position into account. In full, a new formula can be established by combining the three factors as described below.

2.3 Establish calculation formula

2.3.1 The influence formula of eccentric initiation

Figure 2 is the top view of the near end of a cylindrical casing with charge under eccentric point initiation at the near end. As mentioned above, the effect of eccentric initiation on circumferential velocity distribution of a cylindrical casing can be calculated with Wang's formula (Eq. (1)) [22].

where $v(a, \theta)$ is the circumferential velocity; E is Gurney energy, a is the eccentric coefficient (Figure 2), R is the radius of charge, θ is the circumferential angle (Figure 2); β is the mass ratio of charge to metal casing, $k = 0.90\beta^2 - 0.23\beta + 0.14$ [22].

2.3.2 The influence formula of rarefaction wave

Neglecting, for now, the effect of eccentric initiation, Feng's formula [17] (Eq. (3)) is composed of the terms for the influences of the rarefaction wave (h(x/d)) and incident angle of detonation wave $g(\alpha_C)$ [16]. $g(\alpha_C)$ is the incident angle of detonation wave under axial point initiation (Figure 1):

where d is the diameter of charge (Figure 1).

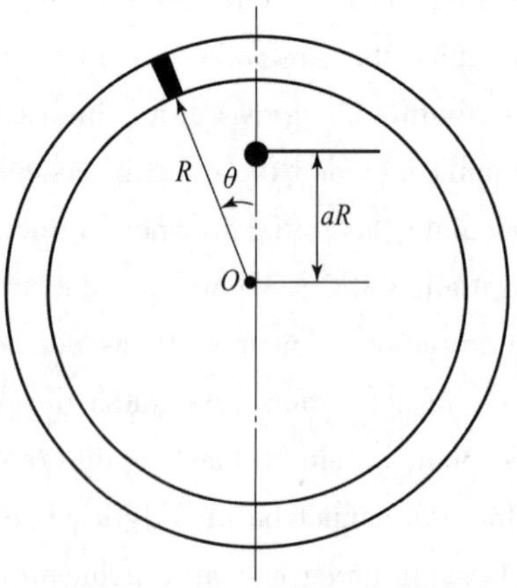

Figure 2 The top view of the near end of a cylindrical casing with charge under eccentric point initiation at the near end.

where x is the axial distance between the near end and a special position P, L is the length of the cylindrical casing (Figure 1).

The third term in Eq. (3) is Gurney formula and the first two terms are a correction formula f(x/d) (Eq. (4)). The influence of rarefaction wave h (x/d) can be calculated with Eq. (5).

2.3.3 The Influence Formula of Incident Angle of Detonation Wave

As shown in Figure 1, the incident angle of the detonation wave under eccentric point initiation (α_c) is different from that under central point initiation (ac). Thus, Eq. (2) cannot directly be used to calculate the incident angle of detonation wave under eccentric point initiation. A specific position P in the casing can be determined by its distance from the near end x and the circumferential angle q. Thus, the incident angle, (α_e), of the detonation wave under eccentric point initiation at the near end can be expressed as Eq. (6).
where dq is the distance between eccentric initiation point and the position P. It can be calculated with Eq. (7).

In Eq. (7), d_θ equals to zero with $a = 1$ and $\theta = 0$, which is not applicable to Eq. (6). However, the eccentric initiation point cannot be at the interior surface of cylindrical casing in practice. There is always a distance between eccentric ignition point and the interior surface of cylindrical casing and a cannot be 1. Therefore, Eq. (7) is reasonable and can be used in determining the influence of the incident angle of detonation wave under eccentric point initiation. The influence of the incident angle of detonation wave under eccentric point initiation, $g(\alpha_e)$, can be obtained by substituting the α_C in the formula of $g(\alpha_C)$ with α_e.

2.3.4 Discussions

Based on discussions above, the velocity distribution v (x, q) of fragments in a cylindrical casing under eccentric point initiation at the near end can be obtained by combining Eq. (1) – (7) as one equation (Eq. (8)). In Eq. (8), the first term is an expression for the influence of the rarefaction wave, the second term is an expression for the influence of the eccentric initiation and the third term is an expression for the influence of the incident angle of detonation wave under eccentric point initiation. It can be seen that the in – fluence formulas of the incident angles of detonation waves ($g(\alpha_C)$ and $g(\alpha_e)$) are the challenge. Since $g(\alpha_C)$ is caused by the incident angle of detonation wave, it is the ratio of the velocities at two situations, where one incident angle is zero for any position in the cylindrical casing and another one changes with the altering of position in the cylindrical casing. To determine the velocities at these two situations, a numerical simulation model was proposed and validated with experimental data.

3 Numerical Model Validation

The numerical simulation is carried out in a finite difference – engineering package AUTODYN. This software has been widely used to analyze the nonlinear dynamic impacts and explosions. It allows the application of different algorithms such as Euler – Lagrangian, Arbitrary Lagrangian Euler (ALE) and SPH to solve the fluid – structure problems [23]. In the present work, the SPH method was adopted to simulate the explosive fragmentation process where the casing material is plastically deformed and eventually ruptured by the inner charge [24]. To obtain a accurate and reliable numerical model in AUTODYN, Feng's work under central point initiation at the near end [16, 17] and under eccentric initiation [21] were used to calibrate the model.

3.1 Numerical Model I: Cylindrical Casing under Central Point Initiation

3.1.1 Description of the Experiment: Cylindrical Casing under Central Point Initiation

The experiments in Feng's work were conducted to investigate the fragment velocity distribution along the axis of a cylindrical casing filled with Composition B (COMP – B) under central point initiation. The metal casing was fabricated with AISI 1045 steel and the parameters of the cylindrical casing and charge are listed in Table 1.

3.1.2 Numerical model of Feng's cylindrical casing experiment

The symmetry of the cylindrical casing, charge and initiation point allows the software to model only a quarter of the casing (Figure 3). Both the cylindrical casing and COMP – B are discretized by a set of particles that are assigned with a mass interaction among themselves without direct connectivity [24]. The SPH method requires a large number of uniformly distributed particles to provide reasonably accurate outputs. Thus, the size of particles should be as small as possible. However, extremely small size of particles makes the numerical process time - consuming. The minimal fragment in AUTODYN is composed of at least two particles. In the present work, a particle radius of 0.4 mm was chosen based on Figure 6, which resulted in 195, 456 particles (Figure 3). These particles were enough to provide accurate outputs and the process efficiency was improved by the parallel processing.

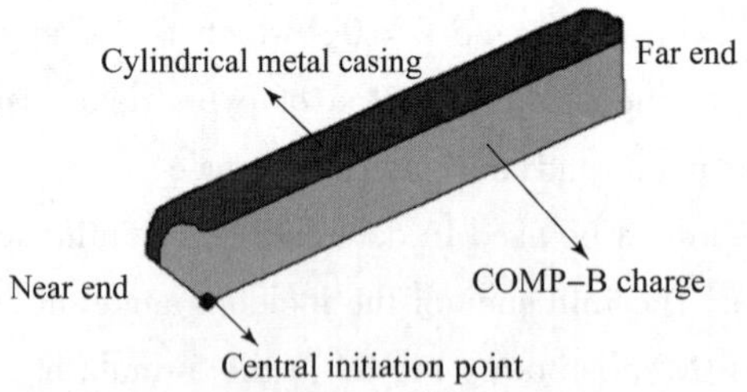

Figure 3 The numerical model of a quarter of cylindrical AISI 1045 steel casing and COMP - B charge under central point initiation.

3.1.3 Material Model

The Johnson and Cook (JC) model (Eq. (9)) is the most widely used to describe materials subjected to large deformation, high strain rate and thermo softening [25]. The model is based on Von - Mises plasticity, where the strength of yielded stress depends on the state of equivalent plastic strain, strain rate and temperature. In the present work, this model was selected to model the material behavior of the AISI 1045 steel cylindrical casing.

where A, B, C, m and n are the constants of the material, ε_{ep} is the equivalent plastic strain, $\dot{\varepsilon}^* = \dot{\varepsilon}_p \dot{\varepsilon}_0$ is the relative plastic strain rate at a reference strain rate $\dot{\varepsilon}_0 = 1\ s^{-1}$. The homologous temperature is defined as $T^* = (T - T_r)(T_m - T_r)$, where T is the current temperature, T_r is the room temperature and T_m is the melting temperature of the material.

The parameters of the material used to construct the Johnsone Cook relation are listed in Table 2 [26]. The principal strain failure model and stochastic failure model based on the Mott distribution were adopted to simulate the formation of natural fragments. The Mott stochastic failure model was used to characterize the failure strain of the casing material with a random distribution. Each cell in the numerical model has a slightly different strain where the cell fails [27 - 29]. Therefore, the material contains weak spots where the failure is expected [30].

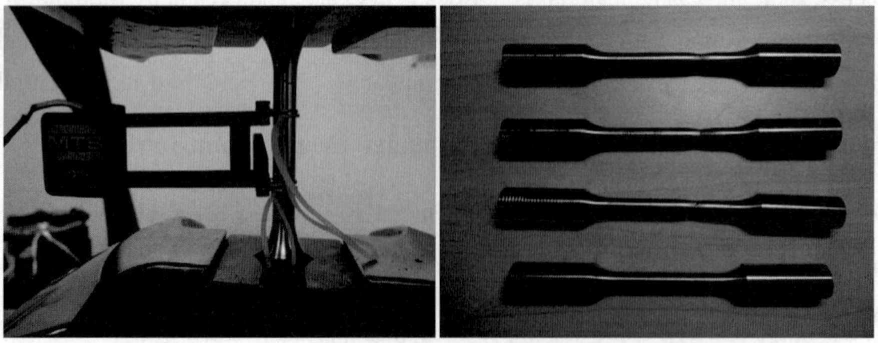

Figure 4 Standard tensile tests of AISI 1045 specimens.

In AUTODYN, the principal tensile failure strain can be determined with the standard tensile test in a typical principal strain failure model. The Mott stochastic failure model [29] is expressed as:

where (1 - p) is the probability that there is no facture for the strain values less thanε. C and γ are constants,

which are inputs for AUTODYN. The parameter g is directly related to the fracture strain deviation and can be estimated by the function of the mechanical characteristics of materials (Eq. (11)).
where *PF* is the true stress at the fracture and $\in_F$ is the plastic strain at fracture. Assuming the stressestrain curve for high strains fits the strain – hardening law, the parameter *P2* is the proportionality coefficient of the law:

Table 2 Johnsonecook parameters of AISI 1045 steel.

Material	A(Mpa)	B(Mpa)	n	C	m	Tm(K)
AISI 1045	507	320	0.28	0.064	1.06	1 793

where P_b is the blast pressure, E is the internal energy per initial volume, V is the initial relative volume, ω, C_1, C_2, r_1 and r_2 are the constants of the material. The properties and parameters of the material are listed in Table 3 [23].

3.1.4 Results and Discussions

The experimental results and the numerical outputs of the fragment distributions at the almost same moments (52.6ms and 53.1 ms) are shown in Figure 5. From Figure 6, it is can be seen that the velocity of fragments increases with particle size decreasing, while, the velocities of fragments have no big difference between particle size 0.3 mm and 0.4 mm. In addition, it is clear that the numerical fragment velocity distributions (particle size 0.3 mm and 0.4 mm) and experimental fragment velocity distributions are in good agreement. Hence, 0.4 mm radii particles were enough to provide accurate results and the process efficiency. These indicate that the numerical model is able to give reliable predictions of the fragment velocity distribution of a cylindrical casing under central point initiation at the near end.

3.2 Numerical Model II: Cylindrical Ring Under Eccentric Initiation

3.2.1 Description of the Experiment: Cylindrical Ring under Eccentric Initiation

The cylindrical ring experiments in Feng's work were modeled to investigate the fragment velocity distribution along the circumference of a cylindrical ring filled with COMP – B under eccentric point initiation. The metal ring is fabricated with AISI 1045 steel. The parameters of the ring and charge are listed in Table 4.

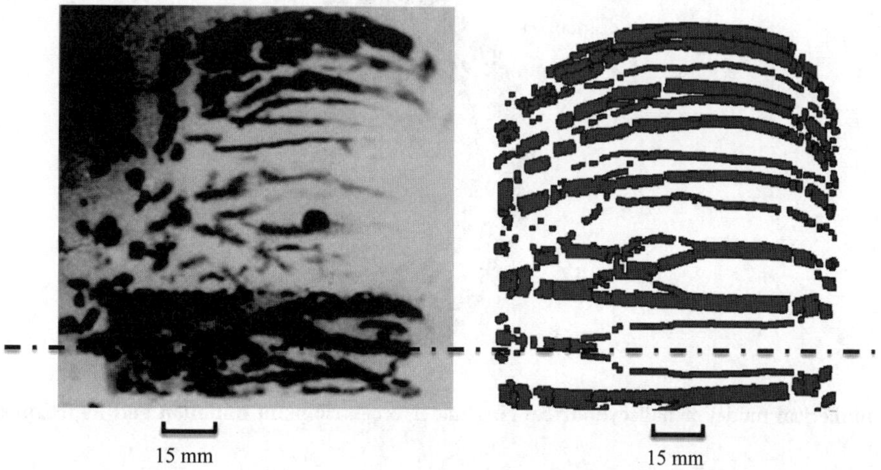

Figure 5 The X – ray photo (52.6 ms) (left) and numerical simulation picture (53.1 ms)(right) of a cylindrical casing under central point initiation at the near end.

3.2.2 Numerical Simulation of Feng's Cylindrical Ring Experiment and the Results

The symmetry of the cylindrical casing, charge and initiation point allows the model to simulate half of the casing (Figure 7). The same numerical model and material as these in section 3. 1 were adopted to simulate the fragment velocity distribution along the circumference of cylinder ring in AUTODYN. It can be clearly seen from Figure 7 that the fragment distribution along the circumference of the cylindrical ring is not uniform. The fragment distribution near the eccentric initiation point is denser while that far away from the eccentric initiation point is sparser. Figure 8 shows the good agreement between the experiment and numerical simulation outputs. These indicate that the numerical model is able to give reliable predictions of velocity distribution on a cylindrical ring under eccentric initiation.

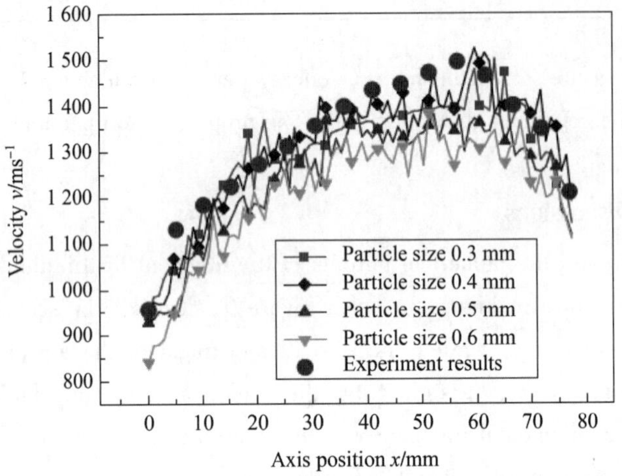

Figure 6 The agreement between the experimental and numerically simulated velocity distribution with results along the axis of the cylindrical casing.

Table 4 Dimensions of the cylindrical ring and density of explosive.

Length of casing (L, mm)	Exterior diameter of casing (D, mm)	Interior diameter of casing (d, mm)	Density of casing (ρ_m, g · cm^{-3})	Density of explosive (P_e, g · cm^{-3})
13. 23	49. 09	44. 07	7. 85	1. 707

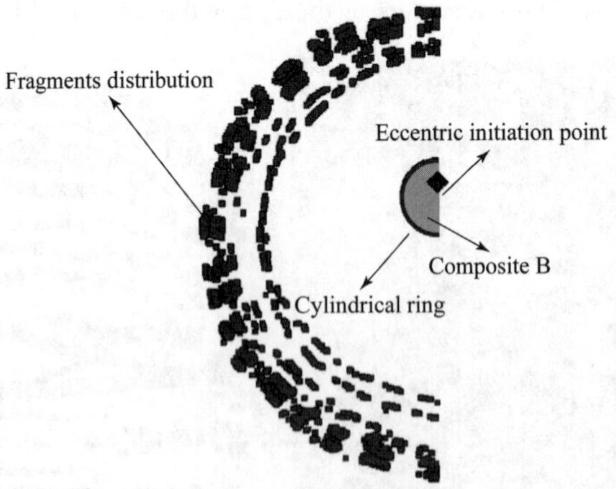

Figure 7 The numerical model of half cylindrical ring under eccentric point initiation and its fragment distribution.

4 The Influence Formula of Incident angle of Detonation Wave

The validated numerical model was then used to simulate the effect of the incident angle of detonation wave on the velocity of fragments of a cylindrical casing. As discussed in section 2. 3. 4, the influence formula of the incident angle of detonation wave g(ac) is the ratio of the fragment velocities at two situations where one incident

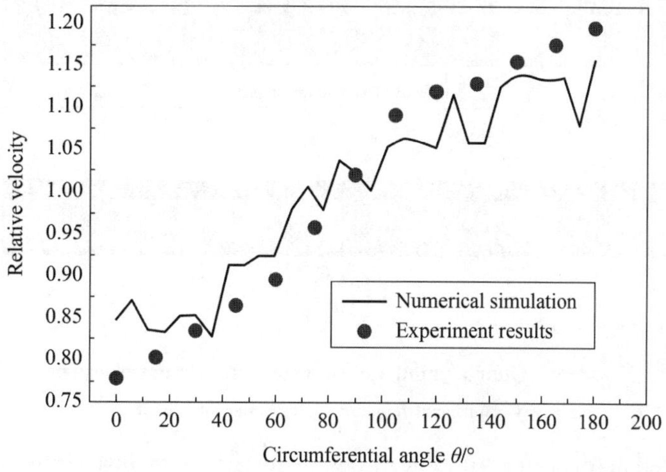

Figure 8 The good agreement between the experimental and numerically simulated fragment velocity distributions along the circumference of cylindrical ring.

angle is zero for any position in the cylindrical casing and another one is changed with the altering of the position in the cylindrical casing. To achieve these two situations, axial line initiation of cylindrical casing was used to obtain the zero incident angle of detonation wave and the central point initiation at near end was adopted to obtain the changeable incident angle (Figure 12).

4.1 Numerical Simulation

Table 5 Three cases (cylindrical casing with charge) with different diameter.

Cases no.	Effective length of casing L_f/mm	Exterior diameter of casing D/mm	Interior diameter of casing d/mm	Length extension L_e/mm
1	77.30	31.36	28.32	28.32
2	77.30	26.64	23.60	23.60
3	77.30	22.70	19.66	19.66

The effects of rarefaction wave and eccentric initiation on the fragment velocities could lead to different experimental and numerical g(ac). Therefore, these two factors must be eliminated with a suitable numerical simulation. The effect of eccentric initiation can be eliminated by setting axial line initiation and central point initiation. The effect of rarefaction wave can be eliminated by extending the length of the cylindrical casing from its two ends. Charron [5] proposed that the maximum propagation distance of the rarefaction wave from the near end of a cylindrical casing under central point initiation at the near end is 2R. Here, R is the interior radius of casing, e. g. the radius of the charge. The maximum propagation distance of the rarefaction wave from the far end is R, indicating that the rarefaction wave has no influence on other part of charge. Thus a formula based on Gurney formula was developed:

where β is the mass ratio of charge to metal casing, F(x) is a correction term:

where x is the axial distance between the near end and a specific position P(Figure 1), L is the length of the cylindrical casing.

To assure the effect of rarefaction wave is eliminated completely, the extended lengths of both ends (Le) are set as 3R (1.5d) (Figure 9).

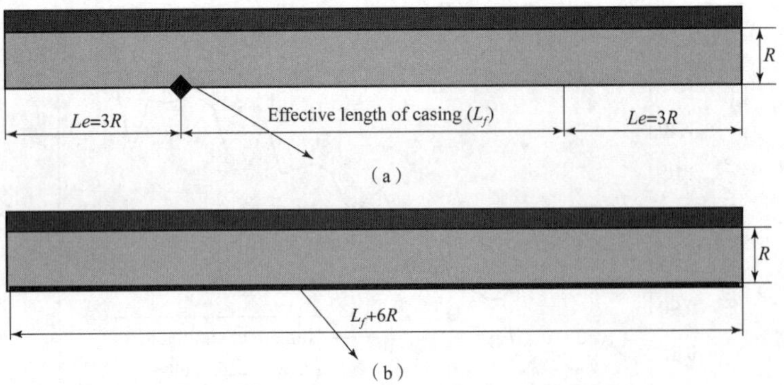

Figure 9 Quarter model of an extended cylindrical casing:
(a) Central point initiation, (b) Axial line initiation.

Three cases with different diameters (Table 5) were designed to investigate whether the diameter is a influent factor of g(ac). Their numerical model is shown in Figure 10.

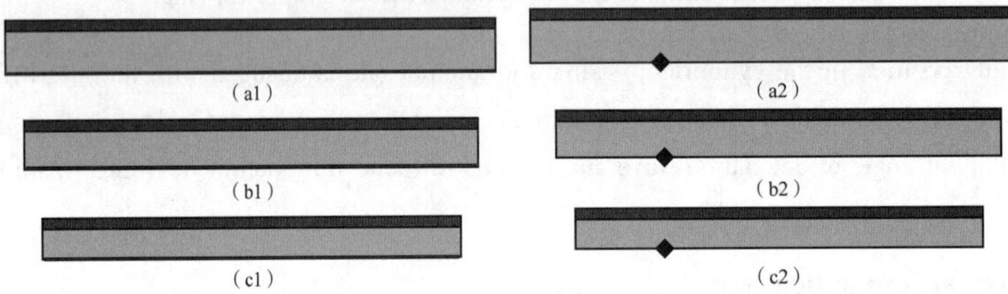

Figure 10 Quarter model of three cases (cylindrical casing and charge) with different diameters in Table 5. (a1) Case No. 1 under axial line initiation, (a2) Case No. 1 under central point initiation, (b1) Case No. 2 under axial line initiation, (b2) Case No. 2 under central point initiation, (c1) Case No. 3 under axial line initiation, (c2) Case No. 3 under central point initiation.

4.2 Results and Formula of $g(\alpha_c)$

It can be clearly seen from Figure 11, that the fragment velocities of all three cases under axial line initiation are a constant in the effective length range L_f, indicating that the rarefaction waves have no influence on the fragment velocity and the incident angle of detonation wave is the only variable under axial line initiation and central point initiation (Figure 12). The fragment velocity distributions along the axis of the cylindrical casing within effective length range under axial line initiation and central point initiation are defined as $V_l(2x/d)$ and $V_p(2x/d)$, respectively. It is clear that x equals 0 at the central initiation point. Therefore, $g(\alpha_c)$ can be calculated by inserting Eq. (2) in Eq. (17).

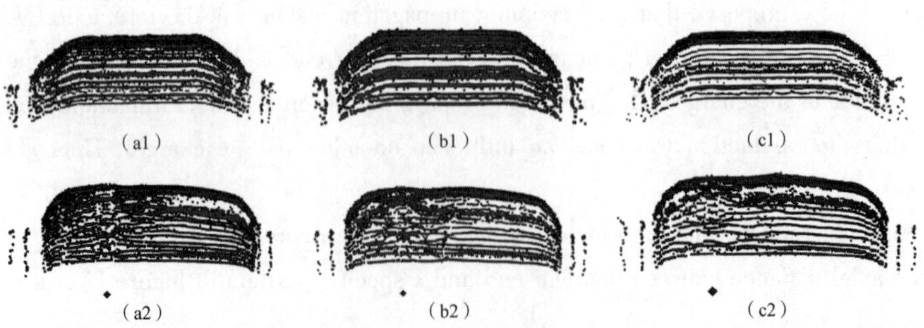

Figure 11 The fragment distributions of the extended cylindrical casings under axial line initiation and central point initiation: (a1) Case No.1 under axial line initiation, (a2) Case No. 1 under central point initiation, (b1) Case No. 2 under axial line initiation, (b2) Case No. 2 under central point initiation, (c1) Case No. 3 under axial line initiation, (c2) Case No.3 under central point initiation.

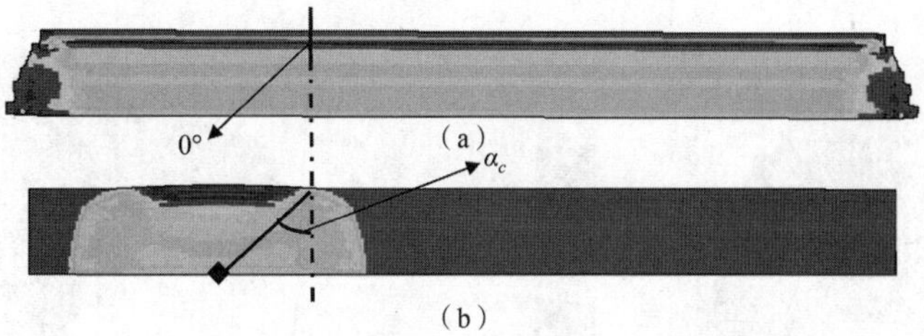

Figure 12 The incident angle of detonation waves under axial line initiation (a) and central point initiation (b).

As shown in Figure 13, the three cases with different interior diameters show similar relationship between $g(\alpha_c)$ and incident angle of detonation wave, indicating the diameter of charge has no influence on $g(\alpha_c)$. Therefore, $g(\alpha_c)$ is just the function of the incident angle of detonation wave (α_c). The fitted curve was obtained by of the average of $g(\alpha_c)$ – incident angle of detonation wave of the three cases (Figure 13) and the curve follows Eq. (18).

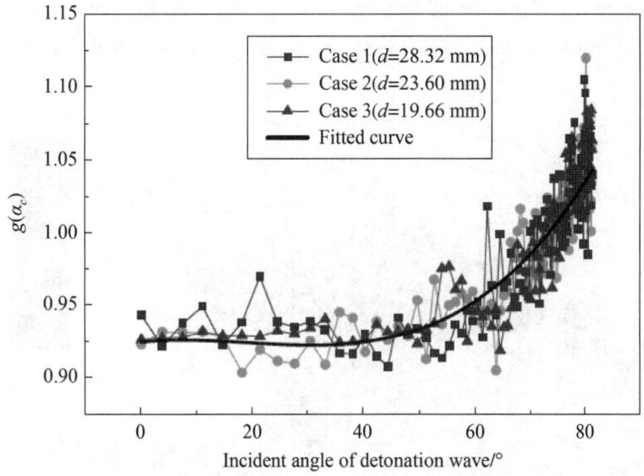

Figure 13 The relationship between g(ac) and incident angle of detonation wave

As discussed in section2. 3. 3, $g(\alpha_e)$ can be obtained by substituting α_c with α_e in $g(\alpha_c)$ based on Eq. (7) for d_θ, the distance between the eccentric initiation points and the position *P*. The calculating formula for fragment velocity distributions along the axis and circumference of a cylindrical casing under eccentric point initiation at the near end is completed by inserting the formulas of $g(\alpha_c)$ and $g(\alpha_e)$ in Eq. (18).

5 Validation of the Formula Proposed in the Present Paper

The validated numerical model was used to verify whether Eq. (8) can accurately predict the fragment velocity distributions along the axis and circumference of a cylindrical casing under eccentric point initiation at the near end. The cylindrical casing and charge with parameters listed in Table 1 was used in both numerical simulation and calculation with Eq. (8). The eccentric initiation point was set 1/2R off the axis at the near end, resulting in an eccentric coefficient of 0.5 (Figure 1). The fragment velocity distributions obtained with numerical simulation and Eq. (8) are shown in Figure 14 (a) and (b), respectively.

Figure 15 shows the good agreement between the calculation results with Eq. (8) and numerical simulation outputs. The outputs were obtained with 2 604 data points and the relative deviations of most data points obtained with Eq. (8) from these obtained with numerical simulation are less than 5%. The number of data points with deviation values higher than 10% is less than 5. These outputs indicate that Eq. (8) can be used to accurately predict the fragment velocity distributions along the axis and circumference of a cylindrical casing under eccentric point initiation at the near end.

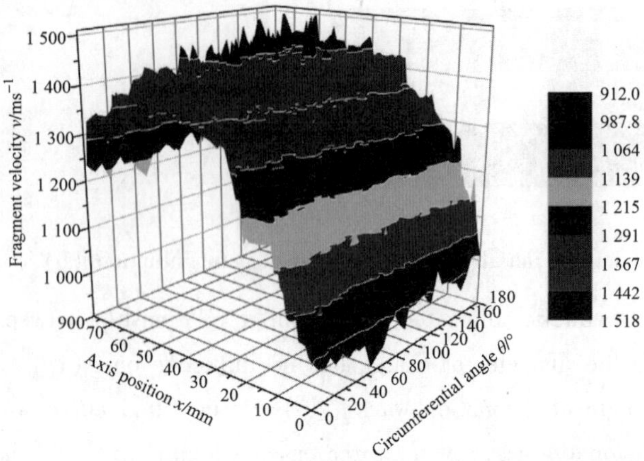

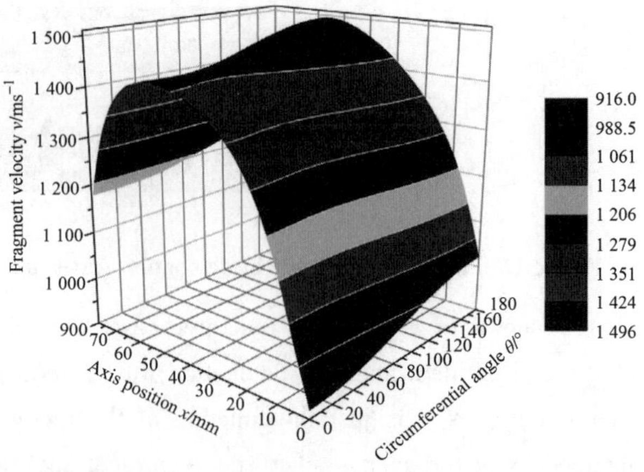

Figure 14 The fragments velocity distributions of the cylindrical casing under eccentric point initiation at the near end.

(a) Numerical simulation outputs; (b) Calculation results with Eq. (8).

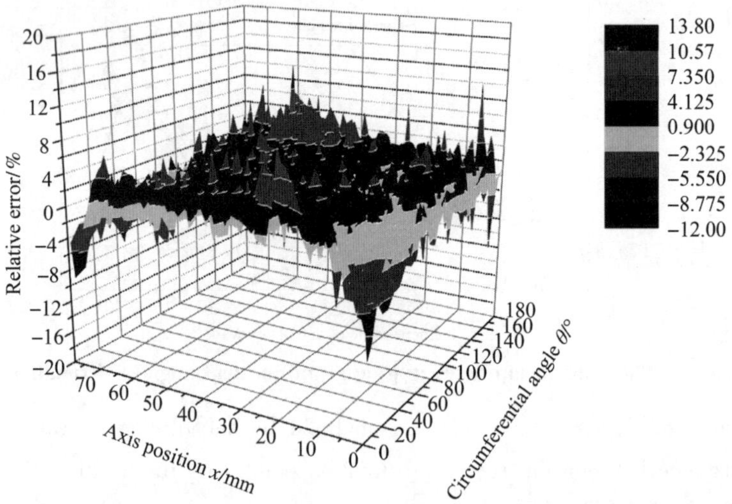

Figure 15 The relative deviation of the fragments velocity distribution obtained with Eq. (8) from that obtained with numerical simulation.

6 Conclusion

A formula was proposed to predict the fragment velocity dis - tributions along the axis and circumference of a cylindrical casing under eccentric point initiation at one end. The formula is composed of three influence factors, e. g. the rarefaction wave from initiation at near and far end, eccentric initiation and the incident angle of the detonation wave. Their calculation formulas were proposed and discussed. Among them, the first two factors can be determined with Wang's formula and by decoupling Feng's formula, prospectively. A novel numerical simulation model was proposed, validated against flash X - ray images and data from experiment and used to simulate the influence of incident angle of detonation wave. A new formula was proposed to determine the influence formula of the incident angle of detonation wave. The formula proposed in the present work was further validated with the established numerical simulation model. The results indicate that the fragment velocity distribution obtained with the formula is in good agreement with that obtained with numerical simulation model. Therefore, this formula could guide the theoretical study and optimal design of warheads.

Acknowledgments

This work was partially sponsored by Foundation of State Key Laboratory of Explosion Science and Technology of China under Grant No. YBKT15 –06 and Natural Science Foundation of China under Grant No. 11102023. The authors wish to express their appreciation to personnel at Intensive Loading Effects Research Group of State Key Laboratory of Explosion Science and Technology, where the experimental data reported on this paper were collected.

References

[1] SUI S Y, WANG S S. Terminal effects [M]. Beijing: National Defense Industry Press, 2000: 35 –40.

[2] GURNEY R W. The initial velocities of fragments from bombs, shells and grenades [R]. BRL Report No 405. Aberdeen Proving Ground MD, USA: Ballistic Research Lab, 1943: 50.

[3] ZULKOSKL T. Development of optimum theoretical warhead design criteria: AD – B015617 [R]. China Lake, CA: Naval Weapons Center, 1976: 6 –8.

[4] KARPP R R, PREDEBON W W. Calculations of fragment velocities from naturally fragmenting munitions [R]. BRL Report No. 2509, AD – B007 377. Aberdeen Proving Ground MD, USA: Ballistic Research Lab, 1975: 18 –24.

[5] CHARRON Y J. Estimation of velocity distribution of fragmenting warheads using a modified gurney method: ADA074759 [R]. Air Force Institute of Technology Wright – Patterson AFB, OH School of Engineering, 1979: 27 –29.

[6] RANDERS – PEHRSON G. An improved equation for calculating fragment projection angles [C] //Proceedings of 2nd International Symposium on Ballistics, Daytona Beach, U. S. A, 1976: 9 –11.

[7] ZLATKIS A, KORIN N, GOFMAN E. Edge effects on fragments dispersion [C] //23th International Symposium of Ballistics, Tarragona, Spain, April 16 –20, 2007: 177 –184.

[8] HUTCHINSON M D. The escape of blast from fragmenting munitions casings [J]. International journal of impact Engineering, 2009, 36: 185 –192.

[9] HUTCHINSON M D. With – fracture Gurney model to estimate both fragment and blast impulses [J]. Central European journal of energetic materials, 2010, 7: 175 –180.

[10] HUTCHINSON M D. Replacing the equations of Fano and Fisher for cased charge blast equivalence – I ductile casings [J]. Propellants, explosives, pyrotechnics, 2011, 36: 310 –313.

[11] HUTCHINSON M D. Replacing the equations of Fano and Fisher for cased charge blast impulse – II fracture strain method [J]. Propellants, explosives, pyrotechnics, 2012, 37: 605 –608.

[12] HORNBERG H, VOLK F. The cylinder test in the context of physical detonation measurement method [J]. Propellants, explosives, pyrotechnics, 1989, 14: 199 –211.

[13] MARTINEAU R L, ANDERSON C A, SMITH F W. Expansion of cylindrical shells subjected to internal explosive detonations [J]. Experimental mechanics, 2000, 40: 219 –225.

[14] SINGH [M], SUNEJA H R, BOLA M S, et al. Dynamic tensile deformation and fracture of metal cylinders at high strain rates [J]. International journal of impact Engineering, 2002, 27: 939 –954.

[15] JAANSALU K M, DUNNING M R, ANDREWS W S. Fragment velocities from thermobaric explosives in metal cylinders [J]. Propellants, explosives, pyrotechnics, 2007, 32: 80 –86.

Fragment Velocity Distribution of Cylindrical Rings Under Eccentric Point Initiation[①]

偏心点起爆作用下圆柱环的破片速度分布

Huang Guangyan[②], Li Wei, Feng Shunshan

State Key Laboratory of Explosion Science and Technology,

Beijing Institute of Technology, Beijing 100081, P. R. China

Abstract: A design mode, in which a casing is filled with a charge initiated off – centre (eccentric point initiation), is common in the field of explosion and structural protection. The fragment velocity distribution along the circumference of the casing is distinctly non – uniform due to the difference in blast loading around the circumference of the casing. To quantify the fragment velocity distribution, two kinds of cylindrical rings with different structural parameters were adopted. Static explosive experiments with three eccentric coefficients (0, 0.5, 1.0) were conducted with pulsed X – ray diagnostics. Using coefficients derived from experimental data and calculating the effects of both the eccentricity of initiation and angle around the circumference, an angle – de – pendent ratio β_θ of charge mass to casing mass has been derived as a mean to modify the fragment velocity formula of Gurney for this application. The derived formula was shown to correctly predict the fragment velocity distribution around the circumference of the cylindrical ring. The calculated velocity distributions show good agreement with the experimental results.

Keywords: Fragment velocity; Gurney; Explosive load; Eccentric point initiation; Cylindrical ring

1 Introduction

Most axially symmetrical casings filled with charge are given axial initiation at one end. The fragment velocity and angles around the circumference of any section across the casing axis are uniform and lead to an evenly – distributed kinetic energy field of fragments around the axial of the casing. However, in some applications of explosive munitions [1, 2], eccentric initiation is adopted instead of axial initiation. Taking a section of cylindrical casing (regarded as a ring) for discussion, the initiation point will in this case be off the center of the circle. Around the circumference, the ring is under an axially asymmetric energetic loading, generated by the eccentric initiation, which causes that the impulse exerted on the ring is asymmetrically distributed. Furthermore, the fragment velocity and the kinetic energy around the circumference of the ring are variable. Relevant explosion research is therefore directed to obtain the property of fragments velocity distribution around the circumference of cylindrical ring under eccentric initiation.

The investigation of the fragment velocity of cylindrical casing under axial initiation has been studied for decades. There exist many formulae to calculate the fragment velocity of cylindrical casing under axial initiation, such as the typical Gurney formulae [3] and the corrected Gurney formulae [4 – 7]. However, it is more difficult and complex to predict the fragments velocity distribution under eccentric initiation. Waggener described

① 原文发表于《Propellants Explos. Pyrotech》2015, 40, 215 – 220。

② 黄广炎：工学博士，2004 年师从冯顺山教授，研究封锁型弹药设计及效能分析方法，现工作单位：北京理工大学。

the asymmetrically initiated warheads and the mechanism on increasing the velocities of the warhead fragments in the target direction, in addition, he estimated velocities from the warhead by adding an additional coefficient 1.25 in Gurney's formula which takes the enhanced velocity in the target direction into account [8, 9]. Hennequin obtained the enhanced coefficients on fragments velocity at the target direction by conducting the experiments where cylindrical casings with different weight ratios of the charge to the casing were initiated at an eccentric point [10]. Huang investigated the radial velocities for a double line initiation at different distance from the initiation points to the fragment layers [11]. Held used a modified form of Huang's formula [11] to calculate the velocities for single eccentric line initiation warheads. The prediction of Held's formulae gave a good agreement with the experimental data in target direction, while discrepancies in other directions occurred. Wang [12] proposed a formula to predict the radial distribution of the fragment velocities of the velocity enhanced warhead by using a correctional function. His formula calculates the velocity gains with azimuth angle and gave an accurate prediction comparable with experiment data.

Due to various eccentric initiation warheads and too many factors involved into the process of eccentric initiation, for the purpose of establishing a general calculating model, warheads with a thickness of casing and diameter of charge that vary in a linear fashion are treated as research objects in this paper. Pulsed X-rays were used to provide velocity diagnostics on the experimental tests. However, it was impossible to obtain the fragment velocity around azimuth direction in every section of a warhead due to the sight obstruction of fragments at other axial sections. For an axisymmetric warhead, it can be treated as the combination of many infinitesimal rings with the same axial line. Hence, a single ring is treated as the model to be investigated by experiments.

2 Experimental Section

2.1 Methods

The cylindrical ring filled with charge was adopted as the typical experimental model in Figure 1. The radius and mass of the charge are r and m. The mass of cylindrical ring is M. $\beta = m/M$ is the mass ratio of the charge to cylindrical ring. e_p is the distance from the eccentric initiation point to the axis of cylindrical ring. e_p/r is defined as the eccentric coefficient. θ is the azimuth angle, which is represented by the angle between OP_θ and OO_p. θ/π is defined as the azimuth angle coefficient. ϕ_θ is the scattering angle, which is represented by the angle between O_pP_θ and OO_p. v_θ is the fragment velocity at the azimuth angle θ. This fragment velocity v_θ can be expressed as a function of the two independent variables, e_p/r and θ/π: $v_\theta = f(e_p/r, \theta/\pi)$.

A 300 kV pulsed X-ray with two parallel X-ray tubes was used in the test, as shown in Figure 2. X-ray 0 took a static photograph of specimen (the cylindrical ring filled with charge) at time t_0. X-rays 1 and 2 took two photographs at different times t_1 and t_2 in sequence. The pictures were compiled in one photograph, as shown in Figure 3.

The fragment velocities can be calculated based on the interval time $\Delta t = t_2 - t_1$ and the corresponding displacement Δx in every azimuth angle. Due to the distance L_1 from X-ray tubes to specimen and the distance L_2 from specimen to the negative plate, an amplification coefficient m, which is the size ratio of the image in photo to the real object, exists. To obtain the displacement Δx in every azimuth angle between the interval time Δt, the fragment profile at time t_1 should be moved the distance ΔL, which is caused by the distance between two parallel X-ray tubes in order to coincide the circle center of specimen O_1 at time t_1 with O_2 at time t_2, as shown in Figure 3. The off-set distance ΔL was obtained from the image of a fiducial steel sphere. The ray from O_2 intersects with the moved fragments profile in P1 and the fragments profile at time t_2 in P_2. Hence, Dx is the distance P_1P_2 and the average fragment velocity is:

$$v(\theta) = \frac{\Delta x}{\Delta t} \tag{1}$$

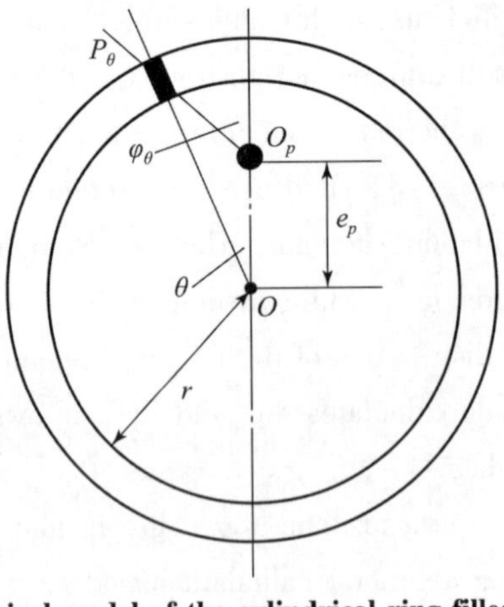

Figure 1 Typical model of the cylindrical ring filled with charge.

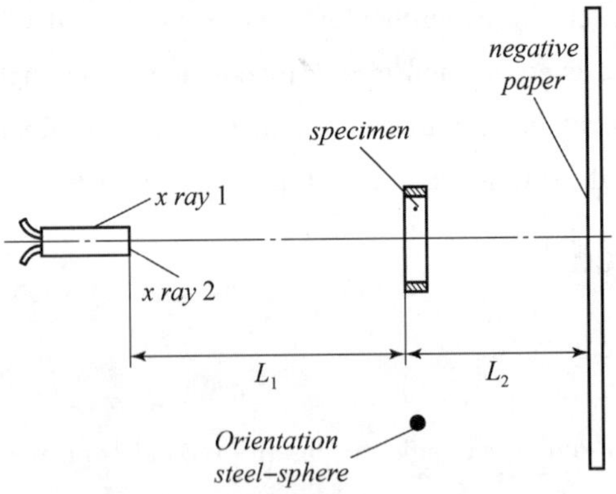

Figure 2 Schematic view of test set - up

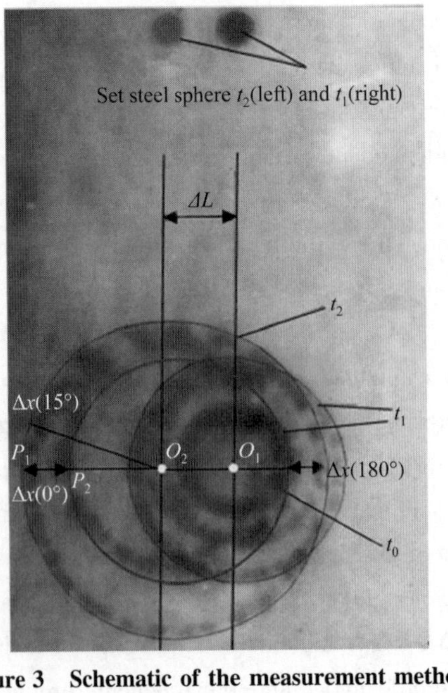

Figure 3 Schematic of the measurement method.

The points P_1 and P_2 are changing with the angle q of the ray from point O_2 (the azimuth angle).

Firstly, two kinds of specimens with different structural parameters were used for the cylindrical rings with charge, which are listed in Table 1. AISI 1045 steel was adopted in the experiments as the material of metal casing. The explosive filled in the casing was TNT50/RDX50. The densities of AISI 1045 and TNT50/RDX50 are 7. 85 g cm^{-3} and 1. 68 g cm^{-3}. The eccentric coefficients are selected as 0, 0. 5 and 1. 0 for these two kinds of specimen. For every eccentric coefficient, three tests were conducted and the average values treated as the fragment velocities.

Table 1 Two kinds of specimens with different structural parameters.

Specimen	No.	Length of ring L [mm]	Diameter of charge d [mm]	Thickness of ring d [mm]	Mass ratio of charge to ring β	Eccentric coefficient epx/r
1#	1 - 1	13. 23	44. 07	2. 51	0. 961 5	0
	1 - 2					0. 5
	1 - 3					1
2#	2 - 1	15. 78	53. 69	3. 04	0. 930 9	0
	2 - 2					0. 5
	2 - 3					1

2. 2 Experimental Results

According to the measurement methods mentioned above, the enhancement of the fragment velocity with varying azimuth angle of the two kinds of specimens is shown in Figure 4. The fragments velocity enhancement coefficient is defined as the ratio of the fragments velocity under eccentric initiation to that under axial initiation.

In Figure 4, it can be seen that there is a greater change of the fragments velocity enhancement coefficient with azimuth angle with the 1 # specimen compared to the 2 # specimen. The 1 # specimen has a smaller enhancement co - efficient in the azimuth angle region 0 - 908and a bigger enhancement coefficient in 90 - 1808 than the 2 # specimen. Furthermore, it shows that there is little difference between the fragments velocity enhancement coefficients of 1# and 2# under the eccentric initiation $e_p = 0.5$, but a larger difference under the eccentric initiation $e_p = 1$. This can be explained as follows: While the length of 2# specimen is almost equal to that of 1# specimen, the diameter of the 2# specimen is 21. 8% greater than that of 1#, which allows a greater influence of the rarefaction wave generated from the ends of charge [13]. Hence, the influence of eccentric initiation will be weakened and the fragments velocity enhancement coefficient of 2# will be closer to 1 than that of 1#. For the eccentric initiation coefficient 0. 5, the difference of enhancement coefficient between specimens 1#

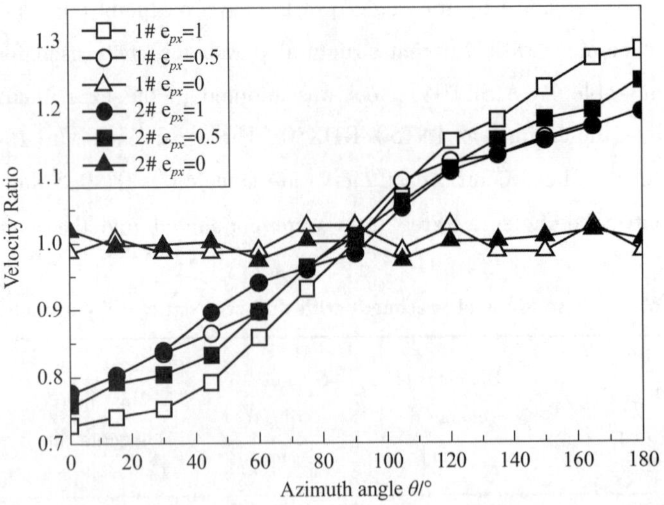

Figure 4 Change of the fragment velocity enhanced coefficient with azimuth angle of specimens 1# and 2#.

and 2# is very small, due both to the weaker eccentric initiation effect and to the influence of the rarefaction wave. Hence, the enhanced effect of 2# is closer to axial initiation with respect to 1#.

3 Calculated Model for the Distribution of the Fragment Velocity Along the Azimuth

To simplify the calculational model, an assumption was made that the fragments velocity along the azimuth direction v_θ is just a function of the eccentric initiation coefficient e_p/r and azimuth angle coefficient θ/π. In addition, it was assumed that the scattering angle of fragments would be the same as the normal direction of cylindrical ring. Hence, based on the fragments velocity calculating formulae of axisymmetric casing filled with charge under axial initiation, v_θ can be corrected and expressed as: [14]:

$$v(\theta) = \sqrt{2E\beta(\theta)/[1+0.5\beta(\theta)]} \tag{2}$$

where, $\sqrt{2E}$ is Gurney constant of the explosive, $\beta_\theta = f(e_p/r, \theta/\pi)$, β_θ is the equivalent mass ratio of charge to cylindrical ring at the azimuth angle θ and β is the actual mass ratio of charge to cylindrical ring under axial initiation.

From Equation (2), it can be seen that the calculation of v_θ is transformed into that of β_θ. v_0 is defined as the fragment velocity at $\theta = 0$ and there must be a β_0 equivalent to β at $\theta = 0$. According to experimental results, the fragments velocity decrease with increasing ep/r in the azimuth angle region of 0 ~ 90°. Hence, β_0 is an inverse function of e_p/r. If $f_1(e_p/r) = b_0/b$, then $f_1(e_p/r)$ will be equal to 1 at $e_p/r = 0$ and to minimum value at $e_p/r = 1$. $f_1(e_p/r)$ can be constructed as:

$$f_1(e_p/r) = 1 - a_1(e_p/r)^{a_2} \tag{3}$$

where, a_1, a_2 are the constants obtained by regressing the experimental data. Based on these, Equation (3) can be expressed as:

$$f_1(e_p/r) = \beta_0/\beta = \begin{cases} 1 - 0.47(2e_p/r)^{0.75} & 0 \leqslant e_p/r \leqslant 0.5 \\ 1 - 0.53(2e_p/r)^{0.17} & 0.5 \leqslant e_p/r \leqslant 1 \end{cases} \tag{4}$$

From the experimental results, v_θ is increasing with increase of θ. It can concluded that β_θ is increasing with the increasing of θ. The distance $|P_\theta O_1|$ from point P_θ to point O_1 is a function of e_p/r and θ, as shown in Figure 1.

λ is defined as the distance coefficient $|P_\theta O_1|/r$ and can be expressed as:

$$\lambda = [4(e_p/r)\sin^2(\theta/2) + (1 - e_p/r)^2]^{0.5} \tag{5}$$

In the same way, $f_2(e_p/r,\ \theta/\pi)$ is defined as β_θ/β_0 and can be constructed as:

$$f_2(e_p/r, \theta/\pi) = \beta_\theta/\beta_0 = 1 + a_3(\theta/\pi)(e_p/r)^{a_4}e^{a_5\lambda} \tag{6}$$

where a_3, a_4 and a_5 are constants obtained by regressing the experimental data. Equation (6) can expressed as:

$$f_2(e_p/r, \theta/\pi) = \begin{cases} 1 + 2(\theta/\pi)(e_p/r)e^{K_1\lambda} & \begin{matrix} 0 \leqslant \theta \leqslant \pi \\ 0 \leqslant e_p/r \leqslant 0.5 \end{matrix} \\ 1 + (\theta/\pi)(e_p/r)^{0.1}e^{K_2\lambda} & \begin{matrix} 0 \leqslant \theta \leqslant \pi \\ 0.5 \leqslant e_p/r \leqslant 1 \end{matrix} \end{cases} \tag{7}$$

where, K_1 and K_2 are 0.6 and 0.74. Combining Equation (4) and (7), the function of β_θ can be expressed as:

$$\beta_\theta/\beta = f(e_p/r, \theta/\pi) = \begin{cases} [1 - 0.47(2e_p/r)^{0.75}][1 + 2(\theta/\pi)(e_p/r)e^{K_1\lambda}] & \begin{matrix} 0 \leqslant \theta \leqslant \pi \\ 0 \leqslant e_p/r \leqslant 0.5 \end{matrix} \\ [1 - 0.53(2e_p/r)^{0.17}][1 + (\theta/\pi)(e_p/r)^{0.1}e^{K_2\lambda}] & \begin{matrix} 0 \leqslant \theta \leqslant \pi \\ 0.5 \leqslant e_p/r \leqslant 1 \end{matrix} \end{cases} \tag{8}$$

An expression for v_θ can now be obtained by substituting Equation (5) and (8) into Equation (2).

4 Comparison of Calculated and Experimental Results

4.1 Azimuthal Distribution of the Fragment Velocity

To demonstrate the applicability of the above formulae, a comparison of calculated and experimental results is shown in Figure 5 and Figure 6.

From Figure 5, for specimen 1#, and for $e_p/r = 0.5$, it can be seen that a good agreement between calculated results and experimental data exists, except a slight deviation in the azimuth angle region of 110° – 130°. Here the calculated results show that the enhancement coefficient of the fragment velocity increases smoothly with increasing azimuth angle in this region, while the experimental data maintain near – constant higher values. For $e_p/r = 1$, good agreement between the calculated results and the experimental data exists, except a slight deviation in the azimuth angle region of 30° – 50°.

From Figure 5 and Figure 6, the fragment velocity enhancement coefficient for specimen 2# is closer to 1.0 than that of 1#, which indicates that the bigger the diameter of charge, the longer the time of detonation and expansion and the more uniform the phase of detonation product. Furthermore, the difference between the calculated curves and the experimental data of 2# is less than that of 1#.

4.2 Azimuthal Distribution of the Kinetic Energy of the Fragments

E_θ is defined as the sum of the kinetic energies of the fragments in the azimuth angle range from 0 to θ, E_π is defined as the sum of the kinetic energies of the fragments in the azimuth angle range from 0 to π. The ratio of E_θ to E_π can be expressed as:

$$\frac{E_\theta}{E_\pi} = \frac{\int_0^\theta v_\theta^2 d\theta}{\int_0^\pi v_\theta^2 d\theta} = \frac{\int_0^\theta \left(\frac{v_\theta}{v}\right)^2 d\theta}{\int_0^\pi \left(\frac{v_\theta}{v}\right)^2 d\theta} = \frac{T_\theta}{T_\pi} \tag{9}$$

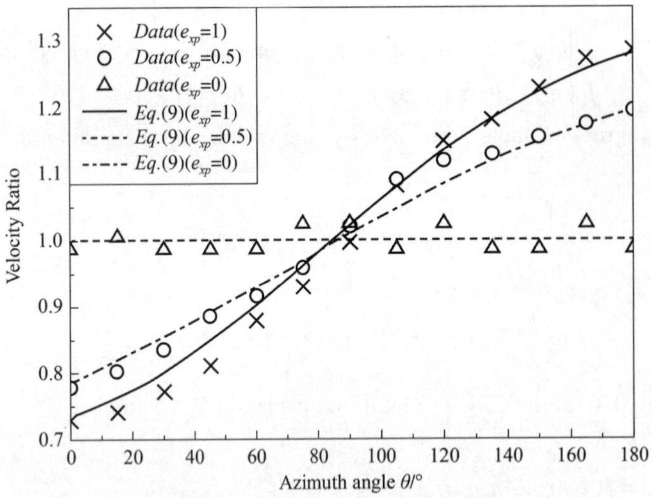

Figure 5 Comparison of calculated and experimental fragment velocity enhanced coefficients changing with azimuth angle 1#.

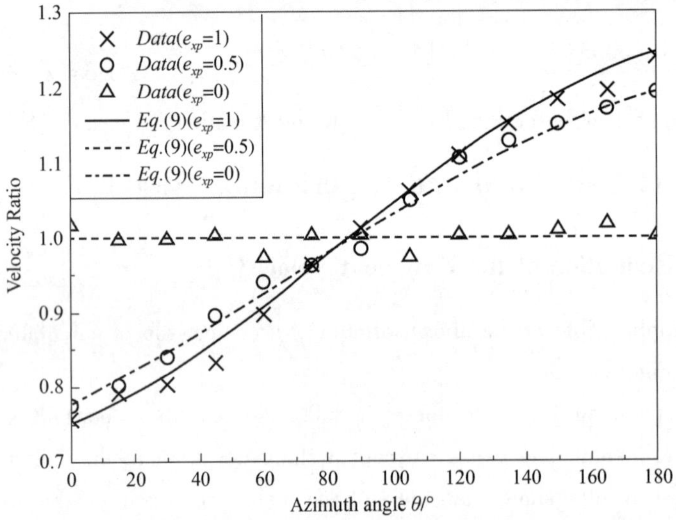

Figure 6 Comparison of calculated and experimental fragment velocity enhanced coefficients changing with azimuth angle 2#.

where, v is the fragments velocity of cylindrical ring under axial initiation. T_θ is defined as:

$$T_\theta = \int_0^\theta \left(\frac{v_\theta}{v}\right)^2 d\theta \tag{10}$$

By substituting Equations (2), (5) and (8) into Equation (10), T_θ can be obtained as:

$$T_\theta = \int_0^\theta \left(\frac{v_\theta}{v}\right)^2 d\theta = \int_0^\theta \frac{\beta_\theta/(1+0.5\beta_\theta)}{\beta/(1+0.5\beta)} d\theta = \frac{1}{\beta/(1+0.5\beta)}\int_0^\theta \frac{a+b\theta}{1+0.5(a+b\theta)} d\theta \tag{11}$$

where, a and b can be obtained from Equations (5) and (8) by:

$$a = \begin{cases} 1-0.47(2e_p/r)^{0.75} & 0 \leqslant e_p/r \leqslant 0.5 \\ 1-0.53(2e_p/r)^{0.17} & 0.5 \leqslant e_p/r \leqslant 1 \end{cases} \tag{12}$$

and

$$b = \begin{cases} \dfrac{2}{\pi}[1-0.47(2e_p/r)^{0.75}](e_p/r)e^{K_1\lambda} & 0 \leqslant e_p/r \leqslant 0.5 \\ \dfrac{1}{\pi}[1-0.53(2e_p/r)^{0.17}](e_p/r)^{0.1}e^{K_2\lambda} & 0.5 \leqslant e_p/r \leqslant 1 \end{cases} \tag{13}$$

By substituting Equation (11) into Equation (9), the azimuth distribution of the kinetic energies of the fragments can be obtained. The azimuth distributions of specimens 1# and 2# are shown in Figure 7 and Figure 8. In addition, the experimental points for kinetic energy were derived from the experimental velocity data. As shown in Figure 7 and Figure 8, there is a good agreement between calculated curves and experimental data.

As shown in Figure 7 and Figure 8, there is a good agreement between calculated curves and experimental data.

It can be seen that the fraction of the total fragment kinetic energy over azimuth angle from 0 to 90° is 0.5 under axial initiation. This is because the shock wave and detonation product exert on the cylindrical ring, uniformly. However, for $e_p/r = 0.5$, to find the mid-point in the kinetic energy distribution, the azimuth angle range has to be from 0 to 106° and from 0 to 107° for 1# and 2#, respectively. For $e_p/r = 1$, the required azimuth angle ranges are from 0 to 112° and from 0 to 113° for 1 and 2#, respectively. This indicates that the detonation energy of charge is in homogeneously distributed around the azimuth direction of cylindrical ring. With increasing eccentric coefficient e_p/r, the azimuth angle range near the initiation point with 50% energy ratio is increasing, while that far from the initiation point is decreasing. Hence, the part of ring far from the initiation point receives more detonation energy and the fragment velocity at 1808 is as its maximum value. The energy ratio curves are concave upward under eccentric initiation, i. e. there is pronounced enhancement of the fragments velocity at the part of ring farthest from the initiation point, as shown in Figure 7 and Figure 8.

In addition, it can be seen from Figure 7 and Figure 8 that the energy ratio curve of $e_p/r = 0.5$ runs below that of $e_p/r = 0$ and above that of $e_p/r = 1$. The change of e_p/r from 0 to 0.5 is same as that from 0.5 to 1, but the difference between energy ratio curves from $e_p/r = 0.5$ to 1 is much less than that from $e_p/r = 0$ to 0.5. This indicates that the effect of gradually increasing eccentricity of initiation in enhancing velocity at points far from the initiation point diminishes as e_p/r increases.

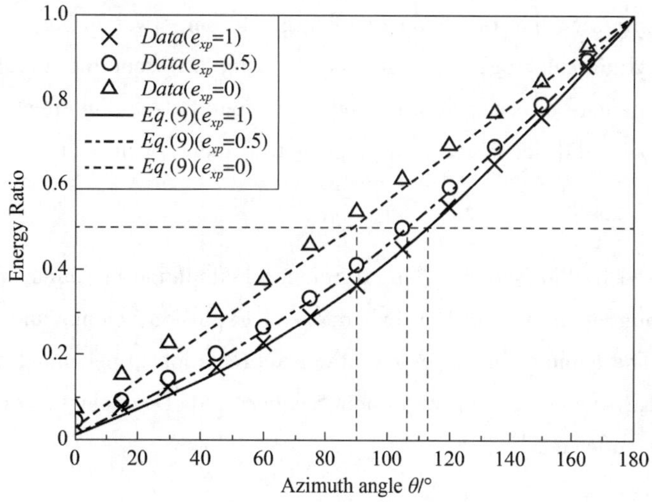

Figure 7 Azimuth distribution of the fragments with energy ratio 1#.

5 Conclusion

Linked to an experimental investigation, the new formulae for the azimuth distribution v_θ of the fragment velocities of cylindrical rings under eccentric initiation are proposed. Two kinds of typical specimens have been tested under eccentric initiation ($e_p/r = 0$, 0.5 and 1) and the experimental data of the azimuth distributions of the fragments velocities were recorded by means of 300 kV pulsed X-ray sets with two parallel X-ray tubes.

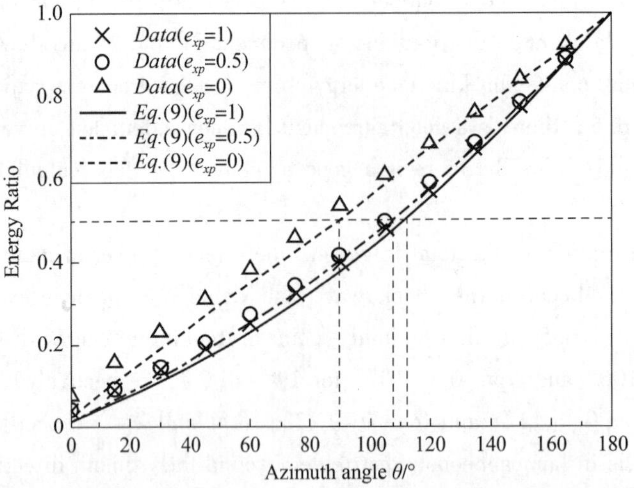

Figure 8 Azimuth distribution of the fragments with energy ratio 2#.

Based on a modification of Gurney's formula, new formulae for vq the fragment velocity as a function of azimuth angle q and eccentric initiation coefficient e_p/r are proposed. Firstly, the calculation of v_θ was translated into that of an equivalent mass ratio β_θ of charge to cylindrical ring at every azimuth angle θ. Secondly, the mass ratio β_0 was obtained by considering the effect of eccentric initiation coefficient e_p/r on the mass ratio b at azimuth angle θ; Finally, the equivalent mass ratio β_θ was obtained by considering the effect of azimuth angle q on the relationship between β_θ and β_0. Above all, the formulae of v_θ can be obtained. Comparison with the experimental data, the calculated results of the formulae of v_θ has a good agreement, which indicates that the formulae of v_θ can predict the fragments velocity azimuth distribution of cylindrical ring under eccentric initiation.

In addition, the calculated fragments kinetic energy azimuth distribution has been compared with experimental data. It could be seen that the detonation energy of charge was in homogeneously distributed around the azimuth direction of cylindrical ring; the parts of the ring far from the initiation point receiving more detonation energy with the increase of eccentric initiation coefficient. Also, the effective enhancement on the part farthest from the initiation point did not increase in proportion to increase in e_p/r.

Acknowledgments

This work was sponsored by the National Natural Science Foundation of China under Grant Nos. 11102023 and 11172071, and the Foundation of State Key Laboratory of Explosion Science and Technology of China under Grant No. YBKT15－02. The authors wish to express their appreciation to personnel at Intensive Loading Effects Research Group of State Key Laboratory of Explosion Science and Technology, where the experimental data reported on this paper were collected.

References

[1] FENG S S. New technology to enhance damage efficiency of missile fragmentation warhead [R]. Internal Report GFYJ 93, Beijing Institute of Technology, Beijing, P. R. China, 1994.

[2] LLOYD R M. Physics of direct hit and near miss warhead technology: Progress in Astronautics and Aeronautics, Volume 194 [M]. Reston: American Institute of Aeronautics and Astronautics, 2001.

[3] GURNEY R W, The initial velocities of fragments from bombs, shells and grenades [R]. BRL Report No 405, Ballistic Research Laboratory, Aberdeen Proving Ground, MD, USA, 1943.

杀爆战斗部对地面目标毁伤威力的评估方法及应用①

付建平[1②]，郭光全[2]，冯顺山[3]，陈智刚[1]，赵太勇[1]，侯秀成[1]

（1. 中北大学地下目标毁伤技术国防重点学科实验室，山西太原 030051；
2. 晋西工业集团有限责任公司，山西太原 030051；
3. 北京理工大学机电学院，北京 100081）

摘 要：研究一种实现杀爆战斗部威力场数值仿真的综合性能分析系统。按照战斗部仿真的功能要求，设计了系统总体框架、接口和模块功能。该系统实现了基于 LS－DYNA 计算结果的动态破片场生成，用射击迹线仿真模型确定了破片的飞行弹道。以某战斗部为例进行了算例分析，并与试验结果进行了对比。分析结果表明，仿真与试验结果基本吻合。该系统可为战斗部总体结构设计和威力评估提供帮助。

关键词：兵器科学与技术；杀爆战斗部；数值模拟；威力评估；计算机应用

Damage Assessment Method and Application of Blast－fragmentation Warhead against Ground Target

Fu Jianping[1], Guo Guangquan[2], Feng Shunshan[3], Chen Zhigang[1],
Zhao Taiyong[1], Hou Xiucheng[1]

(1. National Defense Key Laboratory of Underground Damage Technology, North University of China, Taiyuan 030051, Shanxi, China;
2. Jinxi Industries Group Co., Ltd., Taiyuan 030051, Shanxi, China;
3. School of Mechatronical Engineering, Beijing Institute of Technology, Beijing 100081, China)

Abstract: A comprehensive performance analysis system for numerical simulation of blast－fragmentation warhead power field is researched. In accordance with the functional requirements of warhead simulation, the system architecture, module function and interfaces are designed. In the system, the dynamic frag－ment fields are formed based on LS－DYNA simulation results, and the fragment trajectories are established by using the shot－line model. The simulated results of this system are compared with the test results by taking a blast－fragmentation warhead for an example. The result shows that the numerically simulated re－sults are in agreement with the test results. This system can be very helpful to the overall structure design and lethality assessment of warhead.

Keywords: ordnance science and technology; blast－fragmentation warhead; numerical simulation; damage assessment; computer application

0 引言

杀爆战斗部是一种最常见的战斗部类型，战斗部装药爆炸产生大量破片和爆炸冲击波，对车辆、人员、导弹、雷达等目标产生破坏。随着计算机技术的快速发展，战斗部威力评估的可视化逐渐成为一种

① 原文发表于《兵工学报》2015.06，37。

② 付建平：工学博士，2013 年师从冯顺山教授，研究基于新型毁伤元战斗部设计方法与威力效应，现工作单位：中北大学。

潮流。国外开发的多款战斗部威力评估软件，如美国的 BRL – CAD、荷兰的 TAR – VAC、瑞典的 AVAL 以及德国的 Split – X，已在战斗部设计中得到广泛应用，推动了战斗部设计水平的进步[1]。自 20 世纪 90 年代末以来，国内许多学者[2-8]对威力场模型描述方法、可视化技术、数据结构、仿真系统等开展了多种尝试和探索，极大地促进了战斗部威力评估软件开发。

本文吸取了国外战斗部威力评估软件的设计思想，采用成熟的理论分析模型，集成了 True – Grid 建模工具、LS – DYNA 求解工具、LS – PrePost 后处理以及自编数据处理程序，开发了杀爆战斗部综合性能分析系统。

1 杀爆战斗部综合性能分析系统设计

1.1 系统总体框架

杀爆战斗部综合性能分析系统实现了基于 LS – DYNA 计算结果的静、动态破片场生成，使用射击迹线模型确定了破片的飞行弹道，并集成了破片飞散角、杀伤面积、杀伤概率以及冲击波超压计算等程序，实现了参数化建模、仿真计算、威力性能评估的快速化、一体化。在设计系统框架时，按照不同的功能进行模块划分，每个模块都提供相应的输入、输出接口。在每个模块内部，根据功能需要再将其分解为输入模块、处理模块和输出模块等子模块。

杀爆战斗部综合性能分析系统的总体框架如图 1 所示。该系统主要由结构设计模块、理论分析模块、数值仿真模块、数据库管理模块和效能评估模块 5 个主要模块组成。

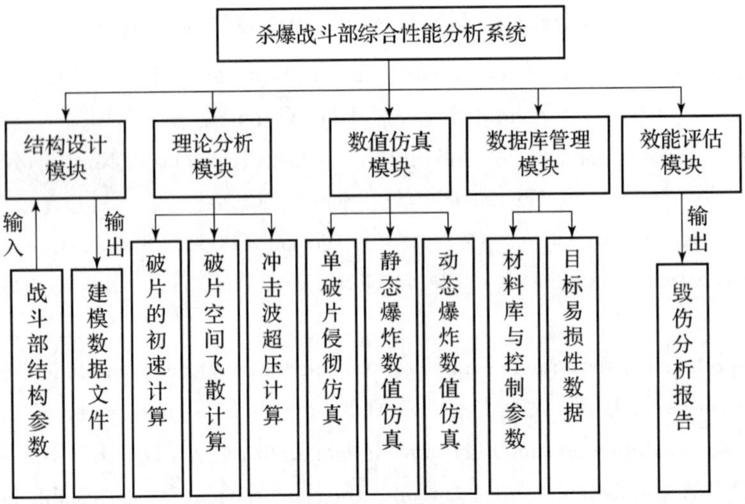

图 1 系统模块图

各主要功能模块通过一定格式的数据文件进行数据传递和调用，软件各模块相对独立利于提高系统的安全性和维护性，同时也能满足数据处理的实时性要求。

1.2 破片信息统计方法

1.2.1 质心点跟踪法

质心点跟踪法只适用于预制破片，半预制破片及自然破片无法使用。在建立破片有限元模型时，将每个破片的质心点添加到同一个跟踪节点集合，在输出的 nodout 文件中储存每个时刻节点集中每个破片质心点的信息。

在 nodout 文件中，结果数据按照时间步长分块存储，读取最后一个时间步长的数据作为破片的初始状态。文件中每个数据块的各列分别对应破片节点编号、破片沿 3 个坐标轴的位移分量、速度分量、加速度分量和位置坐标分量。

该方法的优点是跟踪点设置和破片信息提取十分简单，并且非常高效，尤其是对预制破片数量特别多的大型战斗部最适用。缺点是适用范围太窄，只能统计预制破片信息，而且在建模时要保证质心点正好位于网格节点上。

1.2.2 破片识别法

为了能够识别和跟踪自然破片与半预制破片，对该系统提出了一种新的方法。这种方法可以广泛地用于所有破片的信息统计。数据处理时，根据两个破片无共节点的特征，识别出组成每个破片的所有单元和节点，从而得到每个破片的质心位置、速度、破片质量等信息，为后续的威力计算提供数据。此方法的破片信息是通过对破片中每个节点和单元信息的矢量叠加得到，准确度高。

该方法的优点是适用范围广，对预制破片、自然破片和半预制破片都可以进行破片信息统计。

1.3 破片分布密度及杀伤面积计算

以战斗部在离地面约 0.75 m 处爆炸、人体靶是立姿为例，则炸点与目标连线基本上是一条水平线。此时，对目标的暴露面积应计算它在垂直于地面平面内的投影面积。设 σ 为目标的密度（单位地平面上的目标个数），则每个目标所占的面积为 $1/\sigma\ \mathrm{m}^2$。由于目标均匀分布，所以每个目标所占的地面为一正方形，边长为 $\sqrt{1/\sigma}$ m。相当于在底面边长等于 $\sqrt{1/\sigma}$ m 的正方形、高为 1.5 m 的立体空间内，有一个尺寸为 0.25 m×0.5 m×1.5 m 的人体模拟靶。假设上述两个体积相对于炸点的方位角都是等概率的，则可以得到它们在垂直于地面平面内的平均投影面积分别为 $\frac{4\times\sqrt{1/\sigma}}{\pi}\times 1.5\ \mathrm{m}^2$ 和 $\frac{2\times(0.25+0.5)}{\pi}\times 1.5\ \mathrm{m}^2$。

因此，在 $\frac{4\times\sqrt{1/\sigma}}{\pi}\times 1.5\ \mathrm{m}^2$ 的面积内的一个目标，被一个有效杀伤破片击中的概率为

$$\frac{\dfrac{2\times(0.25+0.5)\times 1.5}{\pi}}{\dfrac{4\times\sqrt{1/\sigma}}{\pi}\times 1.5}=\frac{0.75}{2\times\sqrt{1/\sigma}} \tag{1}$$

以 ρ 表示不同距离上有效杀伤破片在垂直于地面平面内的密度，则一个目标至少被一个破片击中的概率为

$$P_k(x,y)=1-\left(1-\frac{0.75}{2\times\sqrt{1/\sigma}}\right)^{\rho\frac{4\sqrt{1/\sigma}}{\pi}\times 1.5} \tag{2}$$

在一般情况下，$\frac{0.75}{2\times\sqrt{1/\sigma}}$远小于 1，所以

$$P_k(x,y)=1-\mathrm{e}^{(1-\frac{0.75}{2\sqrt{1/\sigma}})^{\rho\frac{4\sqrt{1/\sigma}}{\pi}\times 1.5}}=1-\mathrm{e}^{-\frac{2.25\rho}{\pi}} \tag{3}$$

式中 ρ——$(x,\ y)$ 处有效杀伤破片在垂直于地面平面内的密度，块/m^2。

杀伤面积能用数学方法计算。假设地面上围绕（x，y）点的某一微元面积 $\mathrm{d}x\mathrm{d}y$ 内目标密度为 $\sigma(x,y)$，则该面积内的目标数为 $\sigma\mathrm{d}x\mathrm{d}y$。预期杀伤数 E_c 可表示为

$$E_c=\int_{-\infty}^{+\infty}\int_{-\infty}^{+\infty}\sigma(x,y)P_k(x,y)\,\mathrm{d}x\mathrm{d}y \tag{4}$$

再进一步假设，目标在地面上均匀分布，则 $\sigma(x,\ y)$ 可简单地表示为一个常数 σ，那么式（4）就写成

$$A_L=\frac{E_c}{\sigma}=\int_{-\infty}^{+\infty}\int_{-\infty}^{+\infty}P_k(x,y)\,\mathrm{d}x\mathrm{d}y \tag{5}$$

式中：A_L 具有面积量纲，被称为杀伤面积，也称为平均效率面积。杀伤面积并不是炸点附近地面上一块真实的面积，它是一个加权面积，杀伤概率是它的权。杀伤面积标准是比较各种战斗部对给定目标综合杀伤威力的一种评定标准。

1.4 使用流程

根据使用者输入的战斗部结构参数，系统生成战斗部模型，并理论分析其杀伤参数。调用 LS - DYNA 求解器对战斗部进行仿真计算。然后，基于 LS - DYNA 的仿真计算结果，结合弹药杀伤威力计算方法，用效能评估模块对战斗部的静、动态威力进行计算和评估。由于数值计算生成的结果只是战斗部爆炸后破片在很小范围内的飞散状态，并且当破片在空气中飞散时，还要考虑空气阻力对破片速度的影响。因此，效能评估模块把数值计算结果作为破片初始状态，用射击迹线模型[9,10]计算后生成静态和动态破片威力场。

进一步地分析可以得到战斗部对人员、轻型装甲目标的毁伤情况，还可以计算出杀伤半径、杀伤面积和毁伤概率等威力性能指标。

如果存储在数据库中的材料库、计算控制参数和目标易损性数据不全面，则需使用数据库管理模块添加、删除、查询、修改等操作后，再使用效能评估模块。

2 应用算例

2.1 静态爆炸仿真

以某柱形预制破片战斗部为例进行仿真分析。战斗部主要由装药、壳体（内衬和外壳）、端盖和预制破片等组成。其端盖材料为 35CrMnSi 钢，壳体材料为 LY12，破片由小钨球、大钨球和预刻槽钢环组成，破片总数为 15 000 余枚，破片总重 50.1 kg，起爆方式为单端中心点起爆。图 2 为用该系统建立的某杀爆战斗部结构有限元模型。

用 LS - DYNA 对上述模型进行数值模拟，并对数值模拟结果进行破片场分析。图 3 为战斗部破片场静态仿真结果。计算得到在 300 μs 时刻破片的平均速度为 2 057 m/s。

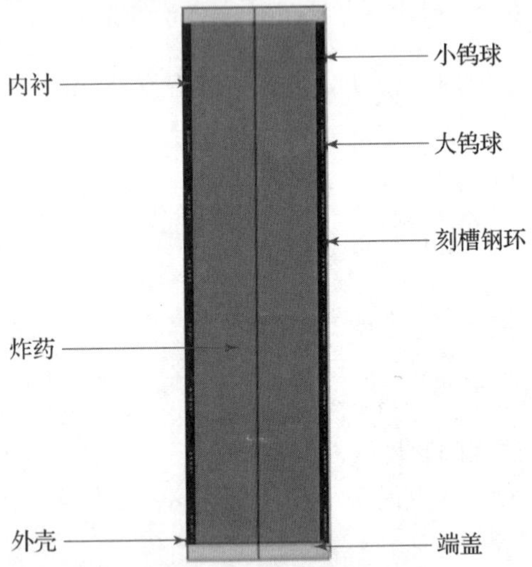

图 2 战斗部结构有限元模型

图 3 战斗部破片场静态仿真结果

通常把杀伤概率 $P_k=0.9$ 的等概率线包围的杀伤面积称为密集杀伤面积。对静态仿真数据处理后，得到战斗部静爆时杀伤概率为 0.9 的密集杀伤面积为 20 969.8 m^2。通过系统的理论分析模块计算出距爆炸中心 30 m 处的冲击波超压值为 0.025 MPa。

2.2 动态爆炸仿真

预制破片弹的动态杀伤场主要取决于其末端弹道条件，如炸高、落速、落角，破片静态飞散角对其动态杀伤场的分布也有较大影响。图 4 为动态计算时战斗部着地示意图。图 4 中，v_c 为弹丸的落速，θ_c

为弹丸的落角，H 为炸高，战斗部是关于 Oxz 平面对称的。图 5 为战斗部在 $\theta_c = 65°$、$v_c = 500$ m/s、$H = 21$ m 时动态爆炸仿真后的地面破片场。

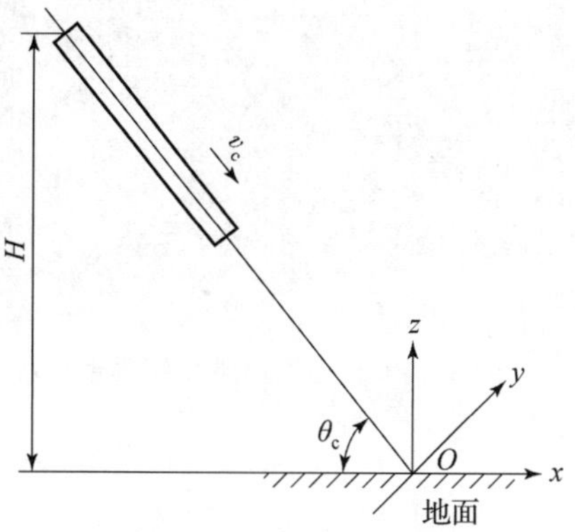

图 4　动态仿真示意图

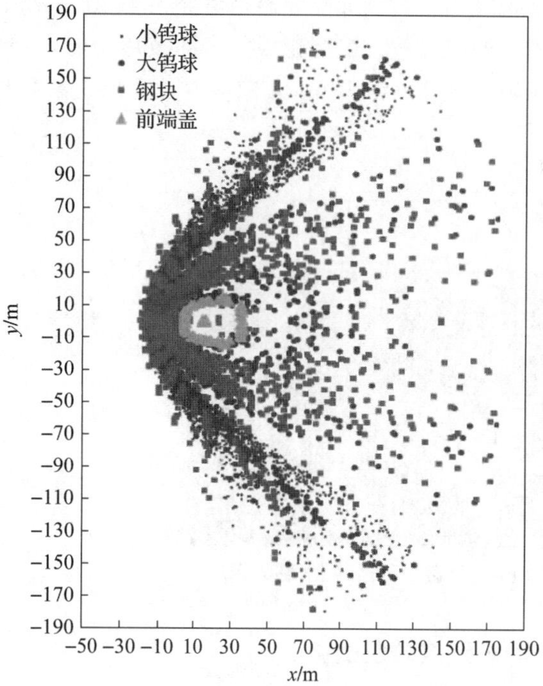

图 5　动态破片场地面分布

从图 5 中可以发现，在飞行中近炸时，由于落角和牵连速度影响，破片场的空间分布与静态条件下的分布发生明显变化，破片场由同心圆变为后密集、前疏散的发散状分布，两侧破片与静爆时相比变化较小。

3　试验结果与分析

3.1　静爆试验

试验时，爆心离地面高度为 1.5 m，战斗部底朝上，底端雷管起爆，松木靶宽 6 m、高 3 m、厚 25 mm，装甲钢宽 1.2 m、高 3.6 m。图 6 为试验现场布置图。

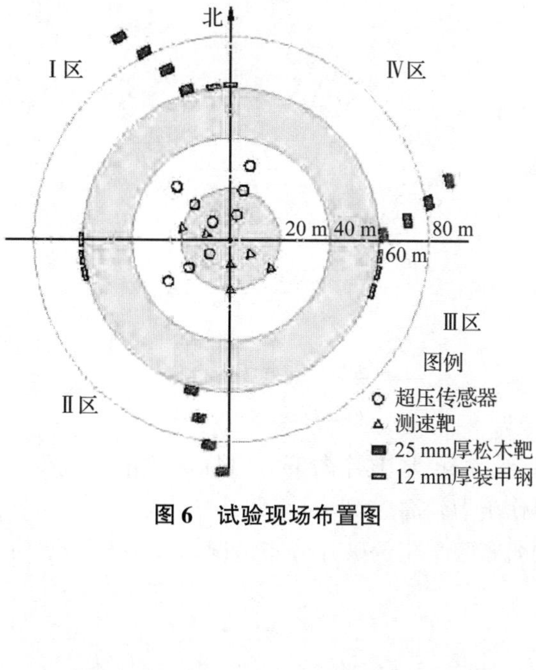

图 6　试验现场布置图

试验测得 30 m 处冲击波超压值为 0. 028 MPa，测得距爆心约 10 m 处破片群速度结果为：破片速度范围为 2 022 ~ 2 302 m/s，平均速度为 2 156 m/s。

试验结果表明，大钨球能穿透距爆炸中心 60 m 处 12 mm 厚装甲钢靶板，穿透率为 87. 5% 。统计出各个松木靶板的破片数，击穿靶板的全部破片和嵌入靶板破片的半数之和为杀伤破片。根据不同距离靶板上的杀伤破片数 N，绘出 $N-R$ 曲线，计算出扇形靶试验密集杀伤半径为 88. 2 m。穿靶情况见图 7 和图 8。

图 7　80 m 处穿透 25 mm 松木靶的情况

图 8　60 m 处均质装甲靶穿透情况

3. 2　动态终点效应试验

结合飞行试验，对战斗部进行了动态终点效应试验，在预计落区布设了扇形靶、厚钢板、帆布、模拟轻型装甲车等效应物。试验弹准确在目标区上空起爆，破片向地面飞散。战斗爆炸过程见图 9。

（a）

（b）

图 9　破片飞散过程

（a）试验；（b）仿真

从图 9 可知，战斗部起爆后，能控制破片向下方目标区飞散，减少了向上方飞行预制破片的数量。形成的打击区域集中，终点毁伤威力较高。

按扇形木靶和水平布设的帆布收集靶的破片分布情况以及破片沿弹轴左右对称分布的特点，绘制了对人员的杀伤区域分布图，如图 10 所示。密集杀伤区域为射向前 54 m，后 32 m，左右各 89 m，密集杀伤面积约为 10 700 m^2。区域形状大致为长轴垂直射向、短轴平行射向的“元宝型”形状。根据地面布设的钢靶板和地面破片打击痕迹作出的大钨球破片密集打击区域如图 11 所示。区域面积约为 3 500 m^2。

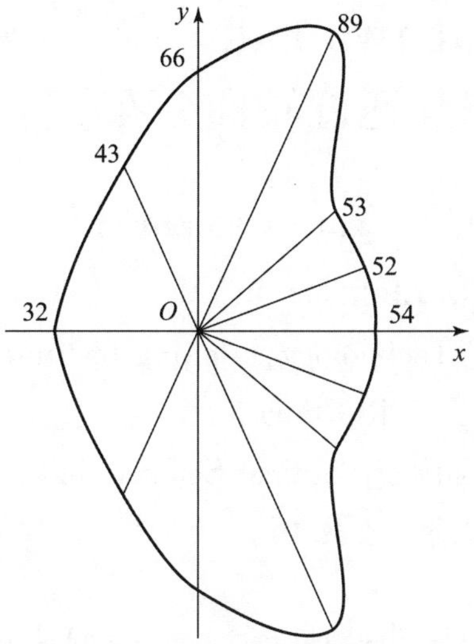

图 10　对人员的密集杀伤区域

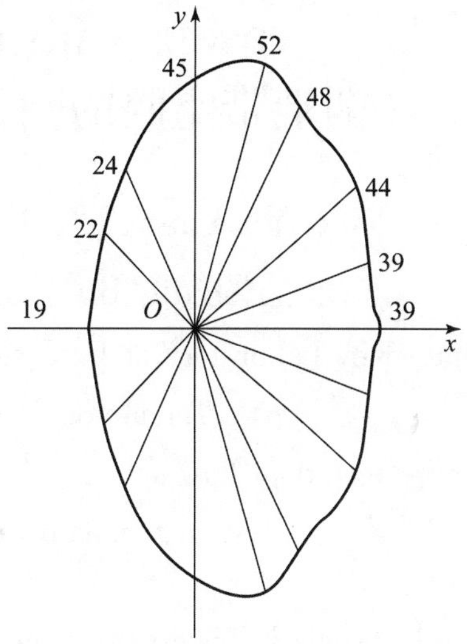

图 11　大钨球密集飞散区域

由图 10、图 11 与图 5 的动态仿真结果对比可知，仿真计算结果与试验情况基本吻合，但是分布区域偏小。表 1 列出了战斗部的主要计算和试验结果。从表 1 可以发现，计算值都略微偏小，但是误差不大，可以满足工程应用的要求。计算中使用的材料模型和参数有待进一步改进。

表 1　战斗部仿真计算和试验结果

结果类型	破片平均速度/($m \cdot s^{-1}$)	30 m 冲击波超压/MPa	动态杀伤面积/m^2
计算结果	2 057	0.025	9 634
试验结果	2 156	0.028 8	10 700

4　结论

本文详细介绍了杀爆战斗部综合性能分析系统的总体框架、接口设计、功能模块组成以及相关的核心算法。以某杀爆战斗部为例，进行了仿真分析。通过与战斗部试验结果对比可知，综合性能分析系统具有较好的计算精度，计算结果与试验结果基本吻合，可作为设计人员进行相关工作的工具使用。

参考文献

[1] 杨云斌，屈明，钱立新．破片战斗部威力仿真方法与仿真软件研究 [J]．计算机仿真，2007，24 (10)：14－19.

[2] 陈留涛，丁刚毅，金乾坤．战斗部威力场仿真系统设计与实现 [J]．兵工学报，2009，30 (10)：1297－1302.

[3] 胡玉涛，蒋邦海，卢芳云，等．基于 LS－DYNA 计算结果的破片战斗部虚拟打击仿真 [J]．弹道学报，2012，24 (1)：27－31.

[4] 孙传杰，余春祥，李会敏，等．防空战斗部总体结构设计及威力评估软件系统开发 [J]．弹箭与制导学报，2013，33 (2)：47－48，56.

Axial distribution of fragment velocities from cylindrical casing with air parts at two ends①

两端带空腔的圆柱形壳体破片速度轴向分布

Gao Yueguang[a]②, Feng Shunshan[a], Yan Xiaomin[b],

Zhang Bo[a], Xiao Xiang[a], Zhou Tong[a]

a State Key Laboratory of Explosion Science and Technology, Beijing Institute of Technology, Beijing 100081, P. R. China

b Anti－terrorism Law school, Northwest University of Political Science and Law, Xi'an, Shaanxi Province, 710063, P. R. China

Abstract: In practical applications of cylindrical explosive－filled casing, non－explosive materials or structures need to be filled at two ends, which makes the ends of the casing have no corresponding explosive charge drive. If the non－explosive materials or structures are regarded as air, it is called cylindrical casing with air parts at two ends. The casing of air parts at two ends will be accelerated by the shock waves and detonation products generated from explosive charge, which cannot be predicted by Gurney formula because the mass ratio of explosive charge to metal casing is 0. In the present work, according to theoretical analysis and simulation, it was found that the existence of air parts at two ends would have no effect on the velocity distribution of fragments generated from the casing with explosive charge. In addition, the fragment velocities generated from air parts decreased exponentially with the increase of the distance from two ends, and the fragment velocities generated from air part at non－detonation end would also present an exponential decay with the increase of the length of air part at non－detonation end. A new formula was proposed to predict the axial distribution of fragment velocities of cylindrical casing with different length of air parts at two ends, which was detonated at one end. The proposed formula was further validated by experimentally verified numerical simulations, and the results showed that the average relative error was no more than 6.28%, RMSE was no more than 60.23 m/s, and R^2_{adj} was more than 0.958, which indicated that the new formula could predict the axial distribution of fragment velocities very well. This work could provide reference for the theoretical analysis and the design of similar warhead.

Keywords: Fragment velocity; Gurney equation; cylindrical casing; numerical simulation.

1 Introduction

Considerable attention has been paid to the dynamic responses of the cylindrical casing filled with an explosive charge in the field of warhead design, structure protection, and anti－terrorism technology. After the explosive charge is detonated, the cylindrical casing is pressurized from within, then expands and eventually breaks into a large number of fragments with different velocities and angles when the internal gas pressure is blow the yield stress of the casing metal [1]. Among them, the fragment velocity distribution has received more

① 原文发表于《International Journal of Impact Engineering》140（2020）103535。

② 高月光：工学博士，2017 年师从冯顺山教授，研究半侵彻弹药战斗部对甲板目标威力效应，现工作单位：在校生。

attention because of the rapid development of irregular warheads and the higher requirements of variant explosives in the field of public security, and various experimental, numerical, and theoretical methods have been conducted for studying fragment velocity.

There have been many studies on fragment velocity distribution of explosive – filled casing. In addition, some formulas to calculate the fragment velocity have been proposed and validated by experimental results. Based on the conservation of energy, Gurney [2] proposed a typical formula to estimate the fragment velocity. He assumed that the explosive was detonated instantaneously, and the potential energy of that before detonation was completely converted into the kinetic energies of the detonation product gases and metal casing after detonation. In his theoretical model, the fragment velocity is a function of the mass ratio of the explosive charge to the metal casing, which can be expressed as:

$$v_0 = \sqrt{2E}\sqrt{\frac{1}{1/\beta + 0.5}} \tag{1}$$

where v_0 is Gurney velocity, $\sqrt{2E}$ is the Gurney constant of explosive, and β is the mass ratio. The β can be expressed as C/M, where C and M are the mass of explosive and metal casing, respectively. Although this model was based on a series of simplifying assumptions, some studies (e. g., [3]) have shown that the calculation results of Gurney's formula are in good agreement with the experimental maximum fragment velocity of warheads.

In Gurney's model, the explosive is detonated simultaneously and all fragments have same velocities along the axis of the cylindrical casing. Thus, the fragment velocity axial distribution is free from the effect of rarefaction waves generated at two ends of the cylindrical casing. It is consistent with most of the experimental results in the middle of cylindrical casing with big length – diameter ratio, but the error is large near the end. The dynamic response of the explosive – filled casing detonated at one end was recently studied numerically and experimentally [4 – 11]. König [12] took the effect of axial rarefaction waves on the axial distribution of fragment velocities and their direction into account and then proposed a modified formula to estimate the velocity and ejection angle of fragments. Anderson et al. [13] further explained how axial rarefaction waves affected the velocity and direction of fragments by carrying out numerical studies. Li et al. [14] carried out numerical simulations to study the effect of axial rarefaction waves and incident angle of detonation waves and eccentric initiation. The results indicated that the fragment velocity axial distribution of cylindrical casing was mainly affected by axial rarefaction waves and incident angle of detonation waves. Kong et al. [15] conducted numerical researches on cylindrical casings with and without end caps, respectively. It indicated that axial rarefaction waves caused the fragment velocity distribution along the axis to be inconsistent.

Based on previous researches, some researchers investigated the effect of axial rarefaction waves at two ends and developed some correctional formulas based on Gurney formula. According to a kind of geometric equivalence method, Hennequin [16], Randers – Pehrson [17], and Charron [18] proposed a correctional formula, which was shown in Eq. (2). The formula can calculate the fragment velocity axial distribution:

$$v_x = \sqrt{2E}\sqrt{\frac{F(x)\beta}{1 + 0.5F(x)\beta}} \tag{2}$$

where the correctional formula can be expressed as $F(x) = 1 - \min\{x/2R, 1, (L-x)/R\}$, x is the axial distance from the detonation point, and R is the radius of explosive.

Zulkouski [19] added a correctional formula in front of the Gurney formula by conducting a large number of experiments. The correctional formula, $C_f(x)$, is an exponential function that describes the influence of rarefaction waves generated at two ends, which can be expressed as:

$$C_f(x) = (1 - e^{-2.3617x/d})(1 - 0.288e^{-4.603(L-x)/d}) \tag{3}$$

where x is the axial distance from the detonation point, d is the diameter of explosive, and L is the length of the cylindrical casing. The correctional formula implies that the effect of axial rarefaction waves is only related to the diameter of explosive and the distance from the detonation end. However, the calculation results obtained with Eq. (3) are unmatched with experimental phenomenon because fragment velocity at the detonation end is zero.

Huang et al. [20] experimentally investigated fragment velocity distribution along the axis of the cylindrical casing obtained by a flash - radiograph technique. Based on experimental results, the formula proposed by Zulkouski [19] was improved. The modified formula verified by another experimental test and Grisaro's experimental data, which can be expressed as:

$$v_{0x} = (1 - 0.361e^{-2.3617x/d})(1 - 0.192e^{-3.03(L-x)/d})\sqrt{2E} \cdot \sqrt{\beta/(1+0.5\beta)} \quad (4)$$

where v_{0x} is the fragment velocity at a certain position on the cylindrical casing.

After many experiments, a formula [21] was proposed to calculate the relationship between the expanding velocity of the casing and the expanding radius. It assumed that the casing ruptured when the radius of the casing expanded several times larger than the initial radius, and the fragment velocity at this moment was considered as the initial fragment velocity. The formula was shown as follows:

$$v_0(x) = \frac{v_d}{16}\sqrt{\frac{\beta}{2+\beta}\left[1-\left(\frac{r_0}{r}\right)^4\right]} \cdot \left[1 + 6\alpha(1-\alpha) + \frac{3}{2}\alpha \cdot \ln\frac{3-2\alpha}{\alpha} + 6\alpha(1-\alpha)(2\alpha-1) \cdot \ln\frac{3-2\alpha}{2(1-\alpha)}\right] \quad (5)$$

where α is the ratio of axial distance from the detonation point to length of charge, v_d is the detonation velocity of the explosive, r_0 is the initial radius of casing, and r is the expanding radius in one moment.

Guo et al. [22] conducted X - ray flash - radiograph experiments to study the velocity axial distribution of fragments from two kinds of non - cylindrical casings. According to the "extended hard core", a formula was proposed to calculate the axial distribution of fragment velocities from non - cylindrical symmetry explosive - filled casing. The experiment results showed good consistency with calculation results obtained by Eq. (6). It can also predict the velocity distribution of fragment from cylindrical casing.

$$v(x) = \sqrt{2E}/\sqrt{\frac{1}{\beta(x)(1-k^2(x))} - \frac{k(x)}{3(k(x)+1)} + \frac{1}{2}} \quad (6)$$

where x is the axial distance from the detonation point, $\beta(x) = C(x)/M(x)$ is the filled ratio at a certain position x, $C(x)$ and $M(x)$ are the mass of explosive and metal casing, respectively, $k(x) = r_i(x)/R(x)$ is the invalid radius ratio, and $r_i(x)$ and $R(x)$ are radius of invalid explosive and explosive, respectively.

An et al. [23] investigated the fragment velocity distribution along the axis of cylindrical casings packed with hollow - charge. Based on the X - ray flash - radiograph experimental results, a formula was proposed for predicting the axial distribution of fragment velocities from cylindrical casings with hollow - charge detonated at one end, which was further validated by experimental data and numerical results. The formula can be expressed as:

$$v(a,x) = (1 - 0.361e^{-1.11x/d - 0.31 \cdot a}) \cdot (1 - 0.192e^{-3.30(L-x)/d + 0.68 \cdot a}) \cdot \sqrt{2E} \cdot \sqrt{\frac{1}{1/(\beta(a) \cdot e^{-(0.31\beta_0^2 - 0.79\beta_0 + 2.29) \cdot a^{2.4}}) + 0.5}} \quad (7)$$

As experiments are usually time - consuming, expensive and sometimes impractical, Numerical simulation method is validated to be able to simulate dynamic fracture under dynamic loading [24]. Many efficient methods for dynamic fracture have been proposed in many application scenarios, including finite element method [25], meshfree method such as SPH method [26 - 28], phase field method [29], a hybrid continuum discrete - element method [30], and peridynamics [31], just to name a few. A simplified model was proposed to predict impulsive loads on submerged structures to account for fluid - structure interaction, which showed good agreement with calculations of complete fluid - structure model [32]. Rabczuk [26] used MLSPH method to study the dynamic failure of concrete structures, and proposed a viscous damage model which can capture well the so -

called strain rate effect for high dynamic loading. An explicit phase field method was proposed for dynamic brittle fracture [29], which can avoid the numerical difficulty in convergence and the calculation of anisotropic stiffness tensor in the implicit phase field method. Ren [28] derived the dual - support SPH by means of variational principle and proposed the implicit SPH formulation, which can solve many problems solved by FEM.

In the above calculation model, the explosive charges are all axially continuous, the fragment velocity distribution of cylindrical casing is only affected by the effect of rarefaction waves generated at two ends. In some cylindrical casings, the non - explosive material or structure is needed to be filled therein. As a result, the explosive charge is axially discontinuous, the propagation of detonation waves is also discontinuous, and the driving process of metal casing is diverse from continuous charge. In the present work, the non - explosive material or structure is regarded as air, which is called the cylindrical casing with air parts at two ends. Due to the combined effects of axial rarefaction waves, the shock wave and detonation products generated by explosive charges, existing axial distribution formulas of fragment velocities from cylindrical casing cannot predict that from air parts since the mass ratio of the explosive charge to the metal casing in the air part is 0. In the present work, after theoretical analysis and numerical simulations, a new formula was proposed to predict the axial distribution of fragmentation velocities from cylindrical casing with air parts at two ends, which was detonated at the center of one end. This formula was based on the Gurney formula and used a uniform form of the correctional function to characterize the velocity axial distributions of fragments from explosive charge part and air parts at two ends. The new formula was further validated by experimental verified numerical simulations, which showed a good agreement with each other. Hence, the proposed formula can be used for predicting the axial distribution of fragment velocities from cylindrical casing with air parts at two ends. The structure of the paper is shown in the flowchart (Figure 1).

2 Theoretical Analysis

Figure 2 shows the axial sectional view of a cylindrical casing with air parts at two ends, which includes three parts of air part A, air part B, and explosive charge C. The corresponding casing is also divided into three parts of casing A, B, and C. Explosive charge is detonated at the center of one end. The diameters of explosive and metal casing are respectively d and D. The diameters of casing from explosive charge and air parts are the same. The lengths of air parts A and B are L_A and L_B, $L_A = ad$, $L_B = bd$. a and b are the ratio of the length of air parts A and B to the diameter of explosive charge, respectively. The length of explosive charge is L_C, and the total length of casing is L.

The shock wave and detonation products generated after the explosion of explosive charge would rapidly diffuse in the circumferential and axial directions. The metal casing of explosive charge would expand, deform, and break into pieces due to the circumferential upward extrusion. At the same time, axial rarefaction waves would also propagate into the shock wave and detonation products along with the axial motion of the shock wave and detonation products, which rapidly attenuated the pressure of the shock wave and detonation products. During the axial outward diffusion process of the shock wave and detonation products, the casing A and B will also be accelerated by them, resulting in expansion deformation and even rupturing into fragments. Known from the analysis of the above, the energy of fragments generated from air parts A and B comes from the explosive charge C. Therefore, it can be considered to add the correctional function on the basis of the Gurney formula of the explosive charge C to predict the fragment velocities from air parts A and B, which was accelerated by the shock wave and detonation products generated by explosive charge C. Based on the Gurney formula of the same mass ratio of explosive charge to the metal casing, a new formula can be obtained to predict the distribution of fragment velocities generated by the cylindrical casing with air parts at two ends. We assume that the pressure of the shock wave and detonation products decreases as the distance from the end of the explosive charge increases. Therefore,

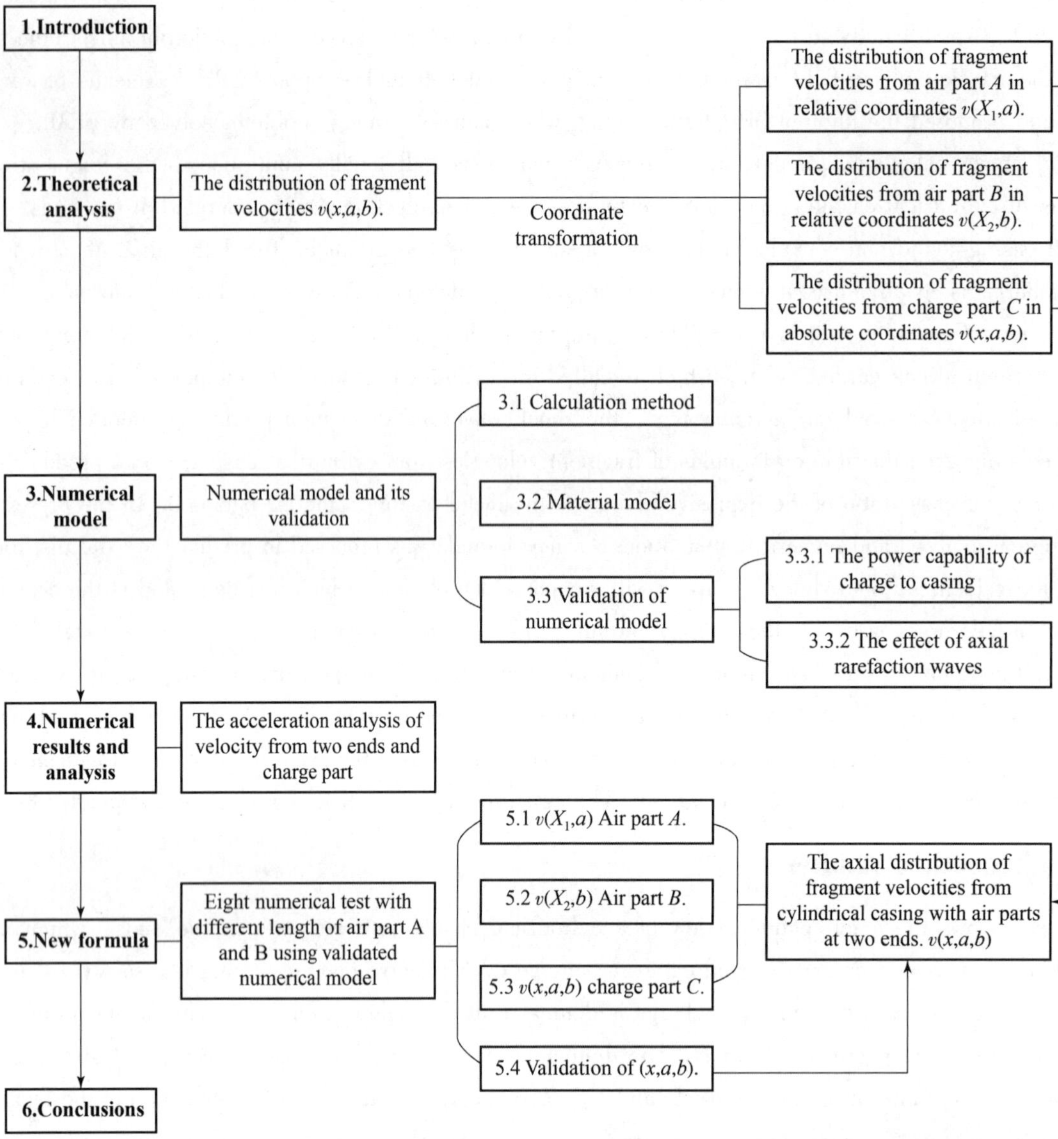

Figure 1 The structure of the paper.

the correctional function is a function of the length of air parts. The shock wave and detonation products diffuse through the air part, meanwhile the axial rarefaction waves continue to depressurize the pressure. As a result, fragment velocity close to the end of the explosive charge will be lower than that calculated by Gurney formula, which has been widely studied in previous researches.

Based on the above analysis, the axial distribution of fragment velocities generated by air part A, air part B, and explosive charge C can be obtained by adding correctional function to Gurney formula under the condition of the same mass ratio of explosive charge to metal casing. Therefore, the axial distribution of fragment velocities is a function of the axial position and the length of air parts A and B, which can be expressed as:

$$v(x,a,b) = \sqrt{2E}\sqrt{\frac{1}{1/\beta_0 f(x,a,b) + 0.5}} \tag{8}$$

where $v(x, a, b)$ is the axial distribution of fragment velocities from cylindrical casing with air parts at two ends, x is the axial distance from the detonation end, and the direction is from the detonation end to the non-detonation end, β_0 is the mass ratio of explosive charge to metal casing of explosive charge part C, $f(x, a, b)$ is the correctional function of the axial position and the length of air parts A and B.

Previous study [18] indicated that the maximum distances of axial rarefaction waves were $2r_c$ and r_c from detonation end and non - detonation end, respectively, where r_c is the radius of the explosive charge. Hence, the length of the explosive charge needed to be large enough to eliminate the effects of axial rarefaction waves from the end of explosive charge on the maximum velocity of fragments. It is also considering the practical application of cylindrical casing is often a slender casing with large ratio of length to diameter. In the meantime, in order to achieve the greatest possible damage effects, the length of air parts at two ends is not infinite. Therefore, in view of practical application situation, at the same time reducing the cost and improving the efficiency of investigation, it is assumed that the length of the explosive charge is greater than twice the charge diameter, the length of the air part is less than 1.5 times of the charge diameter, $L_A < 1.5d$, $L_B < 1.5d$, $L_C > 2d$.

Based on the above assumption, axial rarefaction waves generated from detonation end and non - detonation end would not affect each other, which means air part A and air part B would have no influence on each other and the fragment velocity from them can be solved independently. Hence, the relative coordinates of air parts A and B was established, and the axial distribution of fragment velocities from air parts A and B would be discussed in their own relative coordinates. The axial distribution of fragment velocities from explosive charge C can be produced under the initial coordinates (i. e. absolute coordinates). As shown in Figure 2, the absolute coordinate is set as x, the origin is at the detonation end, and the direction is from the detonation end to the non - detonation end. The relative coordinate of air part A is set as X_1, which is in the opposite direction as the absolute coordinates, and the origin is at the detonation end. The axial distribution of fragment velocities from air part A in relative coordinates is denoted as $v(X_1, a)$, and the relative correctional function is $f(X_1, a)$. The relative coordinate of air part B is X_2, the origin is at the non - detonation end, the direction is same as the absolute coordinates. The axial distribution of fragment velocities from air part B in relative coordinates is named as $v(X_2, a)$, and the relative correctional function is $f(X_2, a)$. The velocity distribution of fragment generated by explosive charge C can still be described by Eq. (8). In the end, the axial distributions of fragment velocities from air parts A and B in relative coordinates are shown as follows:

$$v(X_1,a) = \sqrt{2E}\sqrt{\frac{1}{1/\beta_0 f(X_1,a) + 0.5}} \tag{9}$$

$$v(X_2,b) = \sqrt{2E}\sqrt{\frac{1}{1/\beta_0 f(X_2,b) + 0.5}} \tag{10}$$

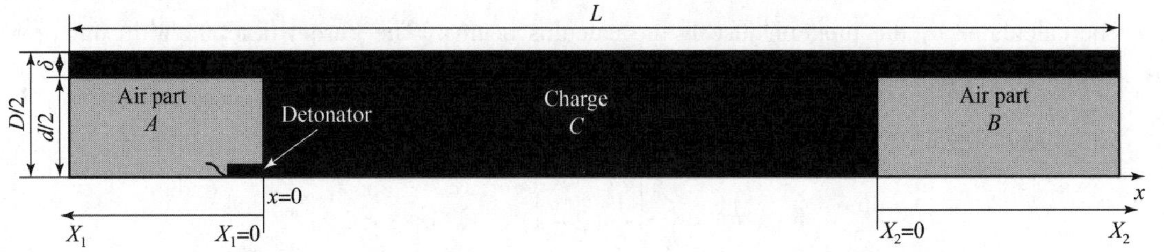

Figure 2 The axial sectional view of a cylindrical casing with air parts at two ends.

Since the convected velocity of the relative coordinates is 0, the axial distribution of fragment velocities from air part A and B, after obtaining in their own relative coordinates, can be obtained through the coordinate change in absolute coordinates. For the axial distribution of fragment velocities from air part A in absolute coordinates $(-L_A < x \leqslant 0)$, it would be acquired as long as $X_1 = -x$ is substituted into Eq. (9). In the same way, the axial distribution of fragment velocities from air part B in absolute coordinates $(L_C < x \leqslant (L_B + L_C))$ can be solved by substituting $X_2 = x - (L_B + L_C)$ into Eq. (10).

3 Numerical Model

3.1 Calculation Method

The process of casing expansion, rupture and breakage caused by the shock wave and detonation products generated from explosive charge is a very complicated process, in which many complex physical phenomena are difficult to be simulated by traditional calculation methods. Large deformation and failure of materials cause large errors in the traditional finite element method (FEM). The smoothed particle hydrodynamics (SPH) method in AUTODYN is a gridless method [15], which could be avoid of the limitation of FEM and Euler mesh and material boundary problem, especially suitable for solving the instantaneous and dynamic large deformation problems such as explosion, and high – speed collision [33 – 36]. Many previous studies have confirmed that the SPH method can be used to simulate dynamic fracture and fragmentation, to obtain fragment velocity and even calculate the distribution of fragment size [9, 11, 14, 15, 33, 37 – 39]. Hence, in the present study, the SPH method was used to simulate the expansion and fragmentation of cylindrical casing.

In SPH method, the material is represented by a finite set of particles using a smoothing procedure in which the value of $f(x)$ is replaced by $\langle f(x) \rangle$ at a point x in the domain Ω. Integral representation of a function $f(x)$ is an exact one:

$$f(x) = \int_{\Omega} f(x')\delta(x - x')dx' \tag{11}$$

where $f(x)$ is a function of a position vector x, Ω is the function domain, which can be a volume that contains the x, and where $f(x)$ is defined and continuous, and $\delta(x - x')$ is a Dirac function, which has the properties:

$$\delta(x - x') = \begin{cases} 1 \rightarrow x = x' \\ 0 \rightarrow x \neq x' \end{cases} \tag{12}$$

By replacing the Dirac function with a smoothing function $W(x - x', h)$, the integral representation of $f(x)$ is approximated by $\langle f(x) \rangle$:

$$\langle f(x) \rangle = \int_{\Omega} f(x')W(x - x', h)dx' \tag{13}$$

where W is the smoothing kernel function, or smoothing function, or kernel function, h is the smoothing length, which can be calculated by the program just as the calculus begins. The kernel function W in our program is a cubic B – spline kernel function:

$$W(s,h) = \frac{\alpha}{h^n}\begin{cases} \frac{2}{3} - s^2 + \frac{1}{2}s^3 & 0 \leqslant s < 1 \\ \frac{1}{6}(2 - s)^3 & 1 \leqslant s < 2 \\ 0 & s \geqslant 2 \end{cases} \tag{14}$$

where the constant α has the values 1, $15/7\pi$ or $3/2\pi$ in function of the space dimension (1D, 2D or 3D). The kernel function has to satisfy some conditions, the three most important of which are as follows:

$$\int_{\Omega} (x - x')dx' = 1 \tag{15}$$

$$\lim_{h \to 0} W(x - x', h) = \delta(x - x') \tag{16}$$

$$W(x - x', h) = 0 \quad when \quad |x - x'| > kh \tag{17}$$

where k is a constant related to the kernel function for point at x, which defines the effective non – zero area of the kernel function. Derivatives of $\langle f(x) \rangle$ are integrated using the kernel function as:

$$\nabla\langle f(x)\rangle = \int_{\Omega} f(x')\ \nabla W(x - x', h)\,dx' \tag{18}$$

The volume weight for each particle can be calculated by m^j/ρ^j, and the discrete forms of mass momentum, and energy conservation equation can be expressed respectively as:

$$\frac{d\rho^i}{dt} = \rho^i \sum_{j=1}^{N} \frac{m^j}{\rho^j}(v_\alpha^j - v_\alpha^i)\frac{\partial W^{ij}}{\partial x_\alpha^i} \tag{19}$$

$$\frac{dv_\alpha^i}{dt} = -\sum_{j=1}^{N} m^j\left(\frac{\sigma_{\alpha\beta}^i}{\rho^{i2}} + \frac{\sigma_{\alpha\beta}^j}{\rho^{j2}}\right)\frac{\partial W^{ij}}{\partial x_\beta^i} \tag{20}$$

$$\frac{dE^i}{dt} = -\frac{\sigma_{\alpha\beta}^i}{\rho^{i2}}\sum_{j=1}^{N} m^j(v_\alpha^i - v_\alpha^j)\frac{\partial W^{ij}}{\partial x_\beta^i} \tag{21}$$

where $W^{ij} = W(x^i - x^j, h)$, t is the time, x is the spatial coordinate, ρ is the material density, E is the specific internal energy, v_α is the velocity component, $\sigma_{\alpha\beta}$ is the stress tensor component, and the subscript $\alpha(\alpha = 1, 2, 3)$ and $\beta(\beta = 1, 2, 3)$ are the component indices. Simulation would be carried out by calculating Eq. (19) ~ (21) combined with the material model and the initial and boundary conditions.

Considering the symmetry of the cylindrical casing, the simulation model only needs a quarter of it. The metal casing and explosive charge were divided into a large number of particles of a certain quality, there is no direct connection relation between particles. The predecessors' research results showed that the calculation precision of the SPH method is closely connected with particle size. The smaller the particles tend to the higher calculation precision, but also means consuming too much computing resources. Li [14] carried out numerical simulation with different discretization and found out that a particle with a radius of 0.4 mm can calculate the fragment velocity with sufficient accuracy compared with the experimental results. In order to ensure sufficient calculation accuracy while taking into account the efficiency, the diameter of the particle was set as 0.4 mm.

3.2 Material Models

The accuracy of material model is very considerable for the correctness of simulation results. In this paper, the metal casing is made of AISI 1045 steel and the explosive charge is COMP – B. These two materials are widely used to study the velocity distribution of fragments, and relevant simulation and tests can provide references.

Johnson – cook model is widely used to describe the material subjected to large deformation, high strain rate and thermal softening effects. The Johnson – cook model was taken as the material model of AISI 1045 steel to describe the dynamic fracture process of the casing:

$$\sigma = (A + B\varepsilon_{ep}^n)(1 + C\ln\dot{\varepsilon}^*)(1 - T^{*m}) \tag{22}$$

where A, B and C are the yield strength, the strain strengthening coefficient and the strain rate strengthening coefficient, respectively, n and m are the strain strengthening exponent and soften exponent, ε_{ep} is equivalent plastic strain, $\dot{\varepsilon}^*$ is the relative plastic strain rate, and T^* is the homologous temperature, which can be expressed as $T^* = (T - T_r)/(T_m - T_r)$, where T_r is room temperature and T_m is the melting temperature.

The formation of fragments was simulated by the principal strain failure model and stochastic failure model. The Mott stochastic failure model, which was widely used to simulate the fracture and fragmentation process of metal casing, was used to describe the fragmentation of the cylindrical casing, which can be expressed as:

$$dp = (1 - p)Ce^{\gamma\varepsilon}d\varepsilon \tag{23}$$

where $(1 - p)$ is the no fracture probability at the strains less than ε, C and γ are constants. According to Li's standard tensile test of AISI 1045 steel [14], the value of γ is 53.8. The properties and parameters of the metal casing are listed in Table 1.

Table 1 Johnson – Cook parameters of AISI 1045 steel.

Material	A(Mpa)	B(Mpa)	n	c	m	T_m(K)	γ
AISI 1045	507	320	0.28	0.064	1.06	1793	53.8

Standard JWL equation of state (EOS) was adopted to simulate the dynamic response of COMP – B in the present work. JWL model is the most common model to describe the explosion propagation driven by detonation wave, which is defined as a function of relative volume and internal energy per initial volume, and it can be expressed as:

$$P_b = C_1\left(1 - \frac{\omega}{r_1 \nu}\right)e^{-r_1 \nu} + C_2\left(1 - \frac{\omega}{r_2 \nu}\right)e^{-r_2 \nu} + \frac{\omega E}{\nu} \quad (24)$$

where P_b is the blast pressure, V is the initial relative volume, E is the initial energy per initial volume, and ω, C_1, C_2, r_1 and r_2 are constants. The parameters of the explosive charge are listed in Table 2.

Table 2 Parameters of explosive model.

Material	ρ_0/g · cm^{-3}	D/$m \cdot s^{-1}$	P_{CJ}/GPa	E_0/KJ · m^3	C_1/GPa	C_2/GPa	R_1	R_2	ω
COMP – B	1.717	7 980	29	8.5×10^6	542	7.68	4.2	1.1	0.24

3.3 Validation of Numerical Models

The simulation results are reliable as long as the simulation model has high accuracy. In this paper, it is necessary to verify the power capability of the explosive charge to the meatal casing, that is, the maximum velocity of fragments, and the influence of axial rarefaction waves on the fragment velocity. Hence, the numerical model was validated by previous experimental results.

3.3.1 Validation of the Model for the Power Capability of the Charge to the Casing

The power capability of the charge to the casing in our numerical simulation was first validated. According to the experiment design came from Wang [41], a validation model was established in Figure 3. The explosive charge was TNT (1.6 g/cm^3, detonation pressure 18.56GPa, detonation velocity 6.93 km/s), and the metal casing was AISI 1045 steel (7.85 g/cm^3). The parameters of the specimen were listed in Table 3. In their experiment, the charge was initiated at the center of one end, and the whole acceleration process of cylindrical casings under internal explosive loading was recorded by an arrayed Doppler Photonic System (DPS). The final velocity measured by DPS was 1.25 km/s, and the velocity in simulation was 1.18 km/s. The relative error was 6%, which meant that the numerical model can accurately simulate the power capability of the charge to the casing.

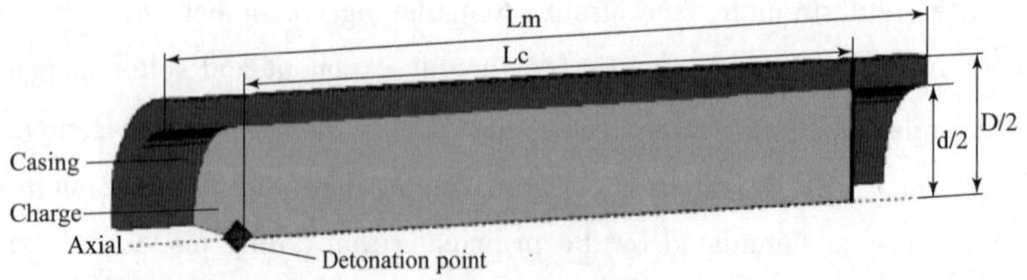

Figure 3 Numerical validation model for the power capability of the charge to the casing.

Table 3 Parameters of the model for the power capability of the charge to the casing

Length of the charge Lc/m	Length of the casing Lm/mm	Inner diameter d/mm	Outer diameter D/mm	Number of particles
200	240	50	61.2	2,349,696

3.3.2　Validation of the Model for Axial Rarefaction Waves

The simulation model for the effect of axial rarefaction waves was then evaluated. The first cylindrical casing came from Huang's first test [20], where the axial distribution of fragment velocities was obtained by a flash – radiograph technique. The explosive charge and metal casing were made from COMP – B and AISI 1045 steel, respectively. The charge was point initiated at the center of one end. The parameters are listed in Table 4, and the numerical model is shown as Figure 4 The particle radius was set to the same (0.4 mm). The simulation results were compared with the experimental results in Figure 5, which showed good agreement with each other.

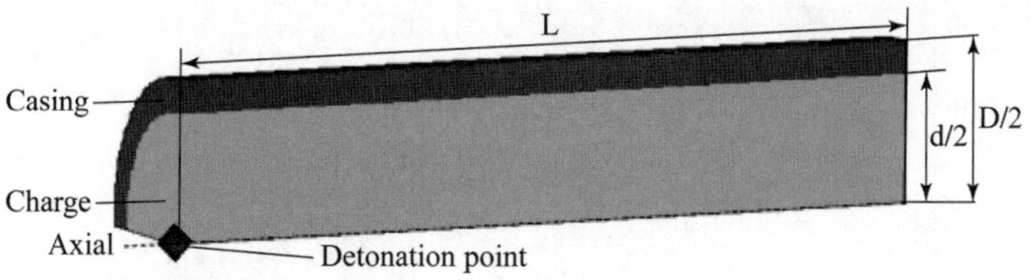

Figure 4　numerical validation model for axial rarefaction waves.

The Second cylindrical casing came from Huang's second test [20]. The parameters are listed in Table 4. Figure 6 shows the comparison of the velocity distribution of fragments between simulation and experiment, and it can also be clearly seen that the simulation results are consistent with the experimental results, which suggests that the simulation model is accurate for determining the effect of axial rarefaction waves.

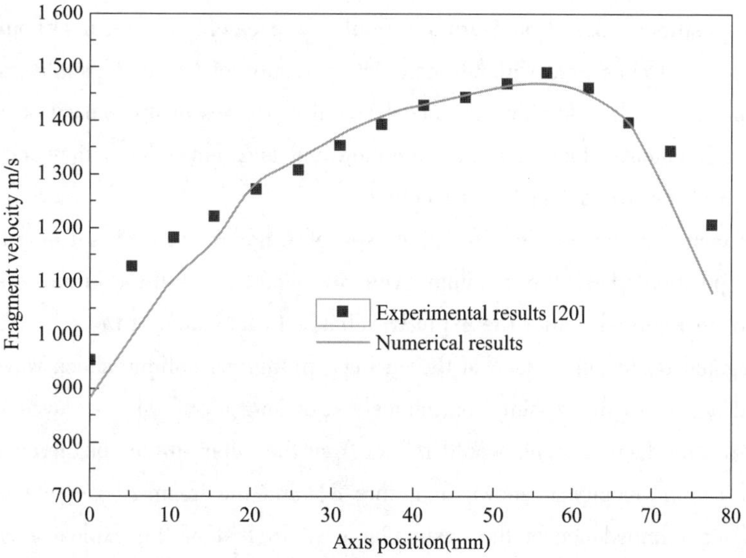

Figure 5　Comparison of the fragment velocities from simulation and Huang's first test.

Table 4　Parameters of the numerical validation model for axial rarefaction waves.

No	Length L/mm	Inner diameter d/mm	Outer diameter D/mm	Number of particles
1	77.3	23.6	29.68	212,200
2	77.15	23.56	36.95	211,789

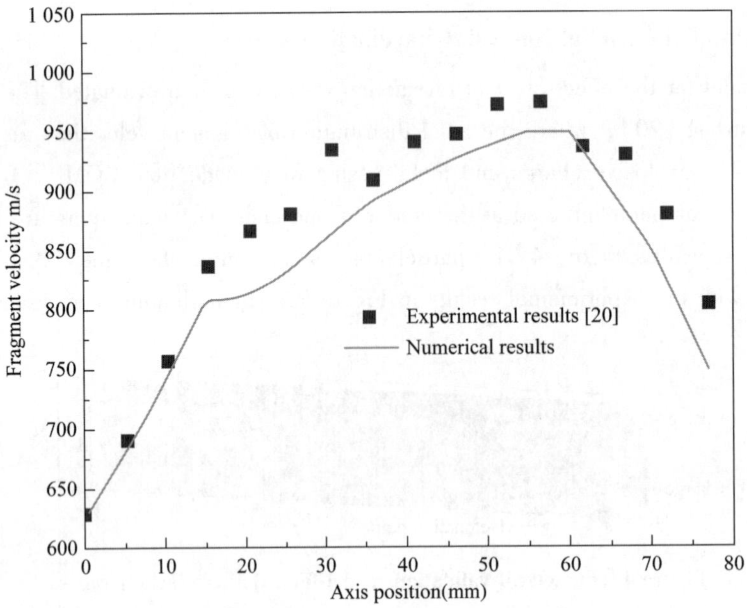

Figure 6 Comparison of the fragment velocities from simulation and Huang's second test.

4 Numerical Results and Analysis

Figure 7 shows the acceleration curves of fragment velocities from detonation end, non – detonation end, and the position where the maximum fragment velocity occurs, and the mass ratio of explosive charge and metal casing is same in three positions. Based on Gurney formula, the calculation results should be same. However, the simulation results were not consistent with the calculation results of Gurney formula due to the neglect of the effect of axial rarefaction waves. In addition, the acceleration processes of three position were also distinct. The acceleration processes of fragment velocities from detonation end and non – detonation end also represented those from the starting position of air part *A* and *B*, respectively.

In order to better compare the acceleration processes of fragment velocities from detonation end, non – detonation end and the position where the maximum velocity occurs, the three curves were moved to the same start position. As shown in Figure 8, after the explosive charge *C* detonated at the center of one end, the sliding detonation waves propagated on the inner face of the casing, producing oblique shock waves in the metal casing. The shock waves would accelerate the casing continuously. As long as the shock waves propagated to the outer surface of the casing, the rarefaction waves would reflect from the outer surface of the casing. At this point, the metal casing *C* showed a high velocity as shown in Figure 8, and the fragment velocity jumped from point A to point D. Due to the impact impedance of the casing larger than that of the explosive charge, reflected shock waves inside the metal casing *C* had continued to spread in the detonation products, meanwhile the rarefaction waves reflected from the outer surface of casing also followed the reflected shock waves to propagate into the detonation products. The reflected shock waves converged to the axis of symmetry and the divergence of those continued outward reflection and attenuation. Behind the forward shock wave, the shock waves were reflected and propagated between the inner and outer surfaces of the metal casing *C* (Figure 9). Due to the combined effects of the shock waves and detonation products, the metal casing *C* continued to accelerate, and the velocity curve between D and E rose with the fluctuations, which was consistent with the experimental results of Wang et al. [40]. The detonation products drove the metal casing *C* to continue accelerating to the maximum velocity after the shock waves' effect.

In the vicinity of the detonation end, after the explosive charge was detonated in the center, the casing *A* began to accelerate just like metal casing C, and the shock waves arrived to the inner surface of casing A near the detonation end. However, because the shock waves here were affected by axial rarefaction waves, their pressures were not as high as those of the shock waves in explosive charge C, and the dispersion direction of detonation products in air part A was opposite to that of the shock waves. Therefore, although a lot of detonation products flied outward from the detonation end, only little kinetic energy was taken away with them. As a result, velocity curve from detonation end only jumped from point A to point B. Then axial rarefaction waves continued to influence the acceleration effect of the shock waves and detonation products on the casing A. In the end, the slope of the velocity curve became smaller and the finial fragment velocity significantly decreased.

The explosive charge detonated from the detonation end, and propagated along the axial direction. When explosive charge at the non - detonation end was detonated, the detonation products generated by the intermediate explosive charge continued to propagate forward and replenished energy for the shock waves because the dispersion direction of the detonation products is consistent with the propagation direction of the shock waves. Therefore, although axial rarefaction waves had strong attenuation effect on the fragment velocities at the non - detonation end, the pressure attenuation of the shock waves was not as fast as that at the detonation end. After fragment velocity accelerated from point A to point C (Figure 8), the pressure of shock waves declined, and axial rarefaction waves would have a significant impact on the acceleration effect of the shock waves and detonation products at this time.

According to one - dimensional dispersion theory of detonation products, the dispersion direction of the detonation products was in contrast to the propagation direction of detonation wave at the detonation end, but they were same at the non - detonation end. As a result, the ratio of kinetic energy of fragments at the non - detonation end and detonation end was $E_b/E_a = 16/11$ [21], and the ratio of that in simulation was 1.49, the relative error was 3%, which once again proved the reliability of our simulation model.

The fragment velocity curves of gauge points at the same distance from the detonation end and non - detonation end were shown in Figure 10, $X_1 = 3$ meant the gauge point was at a distance of 3mm from detonation end on casing A, X_2 represented the gauge point was on casing B.

As mentioned above, the casing A near the detonation end and casing B near the non - detonation end cannot reach the ideal Gurney velocity due to the effect of axial rarefaction waves. However, in the vicinity of the non - detonation end, the attenuation effect of axial rarefaction waves was weakened by the accumulated detonation products from explosive charge C. The detonation products continued to diffuse into air parts on both sides, and axial rarefaction waves continued to weaken the pressure of shock waves and detonation products at the same time. With the increase of the distances from the detonation end and non - detonation end, the acceleration effect of the shock waves on the casing was rapidly attenuated by the axial rarefaction waves. As shown in Figure 10, when the distances from the detonation end and the non - detonation end was less than 6mm, the fragment velocity generated by casing B from the non - detonation end was bigger than that generated by casing A at the same distance from the detonation end. However, when the distances from the detonation end and the non - detonation end were more than 6mm, the casing A and B at the same distances from the detonation end and non - detonation end would have the same fragment velocities. Therefore, compared with the cumulative acceleration effect of the shock waves and detonation products, the influence of axial rarefaction waves on the shock waves and detonation products began to occupy the leading role, which made the fragment velocity only related to the distance from explosive charge ends. In addition, detonation products continued to diffuse outward and became sparse, and the acceleration effect of them on the metal casing of air parts also decayed. After a certain distance, the casing eventually formed a same velocity, which indicated that the casing did not break in the axial direction and a whole fragment was formed (Figure 11).

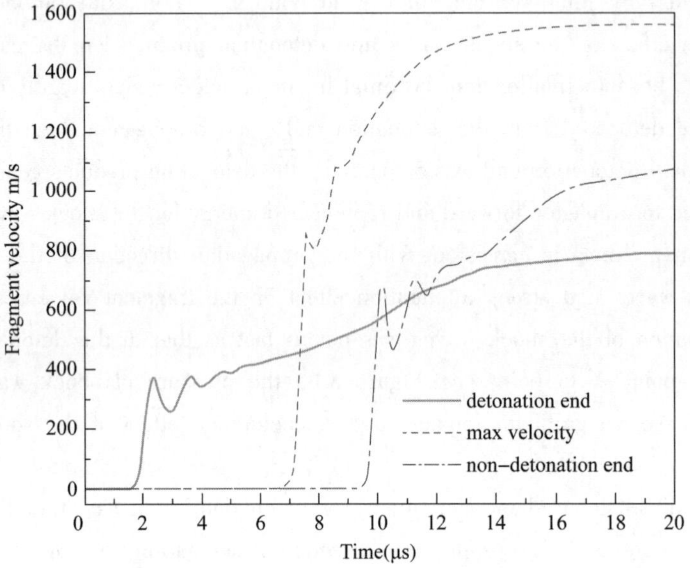

Figure 7　Velocity – time curves of the detonation end, non – detonation end, and the position where max velocity occurs.

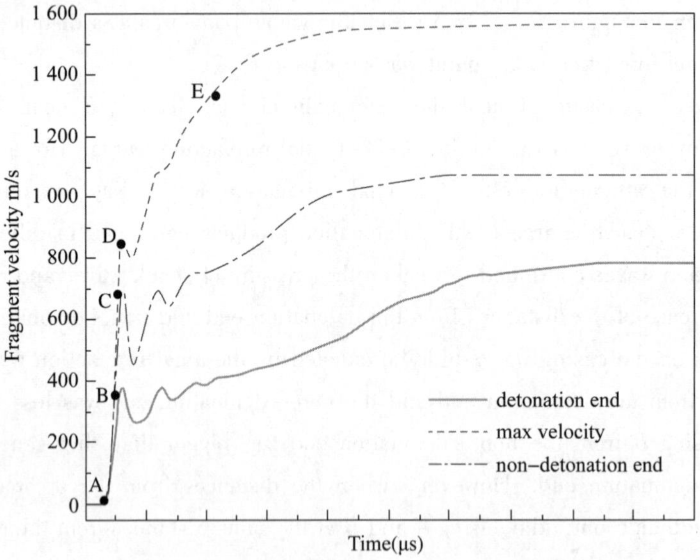

Figure 8　Velocity curves of three position（Begin at the same start position）.

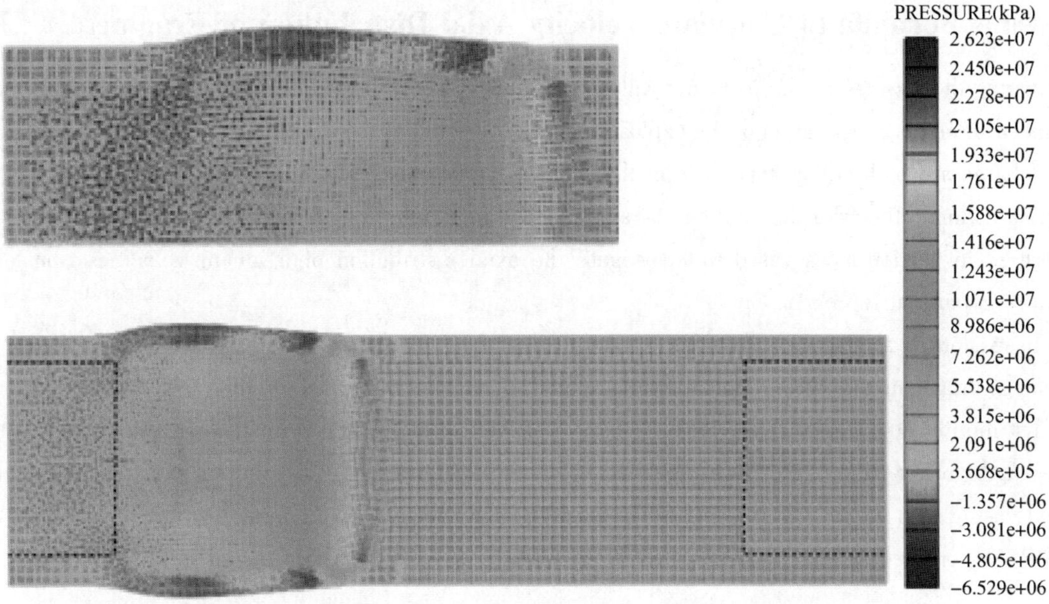

Figure 9　Reflection of shock waves between the inner and outer surface of the casing.

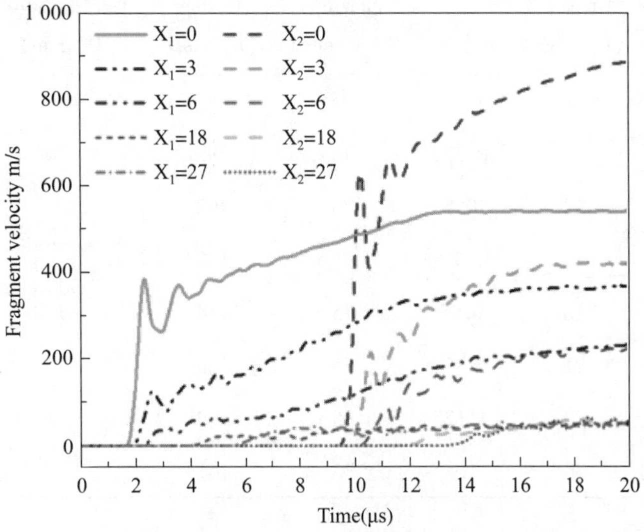

Figure 10　The acceleration curves of gauge points at the same distance from detonation end and non – detonation end.

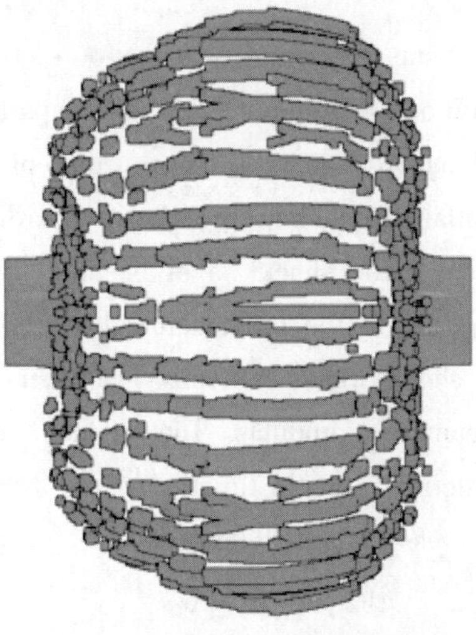

Figure 11　The distribution of fragments at 50 μs.

5 New Formula to Calculate Velocity Axial Distribution of Fragment

No matter the casings of air parts or explosive charge, they all produce the destruction fragments with a certain speed after the acceleration of the explosive charge. Therefore, the fragments generated from air parts and explosive charge are both vital components of the damage performance of the whole cylindrical casing, and it is necessary to estimate the axial distribution of fragment velocities from the whole cylindrical casing. The simulation model verified in section 3 was used to investigate the axial distribution of fragment velocities from cylindrical casing with air parts at two ends.

As discussion in section 2, two relative coordinates about air parts A and B were established, the axial distribution of fragment velocities from air parts A and B can be discussed in their own coordinates, and the velocity distribution of fragment from charge C can be solved in absolute coordinates. Eight numerical tests with different lengths of air parts A and B were performed to investigate the axial distribution of fragment velocities from cylindrical casing with air parts at two ends. The metal casing and explosive charge were made by AISI 1045 steel and COMP - B, respectively. The parameters of the eight numerical tests were listed in Table 5.

Table 5 Parameters of ten cylindrical casings with air parts at two ends.

No.	Air part A (L_A, mm)	a	Air part B (L_B, mm)	b	charge C (L_C, mm)	Casing (L, mm)	Out diameter (D, mm)	Inner diameter (d, mm)	Number of particles
1	9	0.375	6	0.25	75	90	30	24	220 623
2	12	0.5	9	0.375	75	96			226 517
3	15	0.625	12	0.5	75	102			232 411
4	18	0.75	15	0.625	75	108			238 305
5	21	0.875	18	0.75	75	114			244 200
6	24	1	21	0.875	75	120			250 094
7	27	1.125	27	1.125	75	129			258 935
8	30	1.25	36	1.5	75	141			270 724

5.1 Fragment Velocities Distribution from Air Part A in Relative Coordinates

By setting gauge points at different distances from the detonation end, we obtained the velocity distribution of fragments from air part A under the condition of different lengths of air part A, and plotted it in Figure 12. As can be seen from Figure 12, although the lengths of air part A were different, the fragment velocities generated from them still presented the same exponential attenuation in relative coordinates. This also meant that the axial rarefaction waves had a great attenuation of the shock waves because the deflection direction of the detonation products at the detonation end was opposite to the propagation direction of the detonation wave. After a short distance from the detonation end, the shock waves basically lost their acceleration effect on the casing. The casing was mainly accelerated by the detonation products. The fragment velocity was hardly affected by the length of air part A. Hence, the correction function is just a function of the axial position, which is named as $g(X_1)$, that is, $f(X_1, a) = g(X_1)$, and the expression is as follows:

$$g(X_1) = G_1 + G_2 e^{-X_1/G_3} \tag{25}$$

where G_1, G_2, and G_3 are constants, which can be obtained by fitting the simulation results using the least square method. The results showed $G_1 = 0.001$, $G_2 = 0.397$, and $G_3 = 2.25$, respectively ($R^2 = 0.95$). Hence, the correctional function of the axial distribution of fragment velocities from air part A in relative coordinates is shown as follows:

$$f(X_1, a) = 0.397 * e^{-X_1/2.25} + 0.001 \tag{26}$$

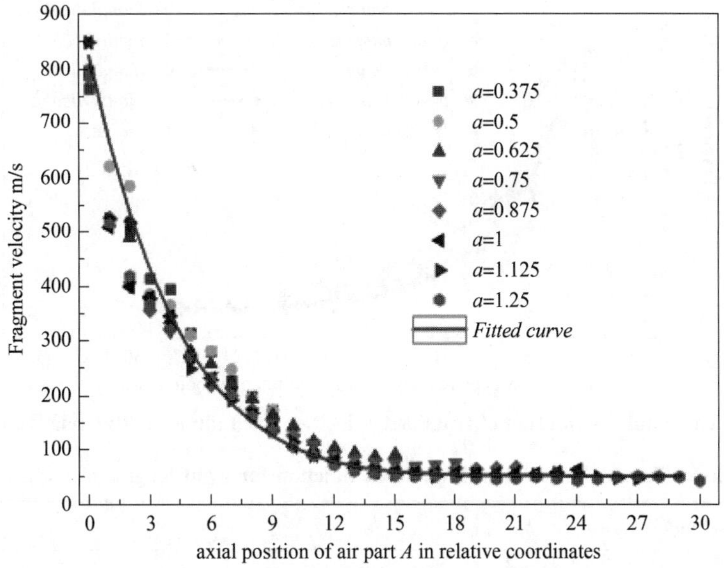

Figure 12 the axial distribution of fragment velocities from air part *A* in relative coordinates.

5.2 Fragment Velocities Distribution from Air Part *B* in Relative Coordinates

The scatters of different colors represented the fragment velocities in relative coordinates under the conditions of different lengths of air part B (Figure 13). The axial position represented the distance from the non – detonation end. It can be clearly found that with the increase of the distance from the non – detonation end, the fragment velocity gradually declined exponentially, but the attenuation rules of fragment velocity were not the same. Therefore, the correction functions under different lengths of air part B were all supposed to be exponential functions, which denoted as $h(X_2, b)$, that is, $f(X_2. b) = h(X_2, b)$:

$$h(X_2, b) = h_{1i} + h_{2i} e^{-X_2/h_{3i}} \tag{27}$$

where h_{1i}, h_{2i} and h_{3i} are constants, $i = 1, 2, 3 \cdots 8$, which represents the number of numerical tests. After fitting the eight groups of simulation data of fragment velocity distribution from air part B, the results were listed in Table 6, and the relationship between the constants and parameter b was shown in Figure 14. h_2 and h_3 were almost constant ($h_2 = 0.173$, $h_3 = 2.919$), while h_1 showed the characteristic of exponential decay with the increase of parameter b, which can be described as follows:

$$h_1 = n + q e^{-b/k} \tag{28}$$

where n, q, and k are constants, and they are obtained by the simulation results using the least square method, that is, $n = 0.001$, $q = 0.254$, and $k = 0.112$ ($R^2 = 0.98$). Substituting Eq. (28) into Eq. (27), the correction function of the axial distribution of fragment velocities from air part B in relative coordinates was obtained, which can be expressed as:

$$f(X_2, b) = 0.173 * e^{-X_2/2.919} + 0.254 e^{-b/0.112} + 0.001 \tag{29}$$

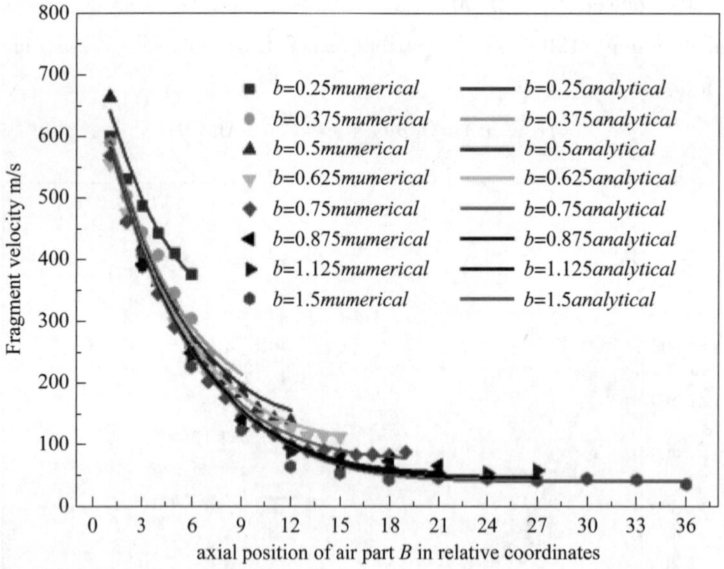

Figure 13 The axial distribution of fragment velocities from air part *B* in relative coordinates.

Table 6 Parameters of the correctional function for eight lengths of air part *B*.

Length of Air part*B*, L_b(mm)	b	h_1	h_2	h_3	R^2
6	0. 25	0. 029	0. 173	2. 919	0. 93
9	0. 375	0. 009	0. 175	2. 92	0. 99
12	0. 5	0. 006	0. 172	2. 916	0. 94
15	0. 625	0. 004	0. 170	2. 918	0. 99
18	0. 75	0. 002	0. 171	2. 922	0. 99
21	0. 875	0. 001	0. 174	2. 923	0. 98
27	1. 125	0. 000 8	0. 173	2. 918	0. 98
36	1. 5	0. 000 6	0. 173	2. 921	0. 99

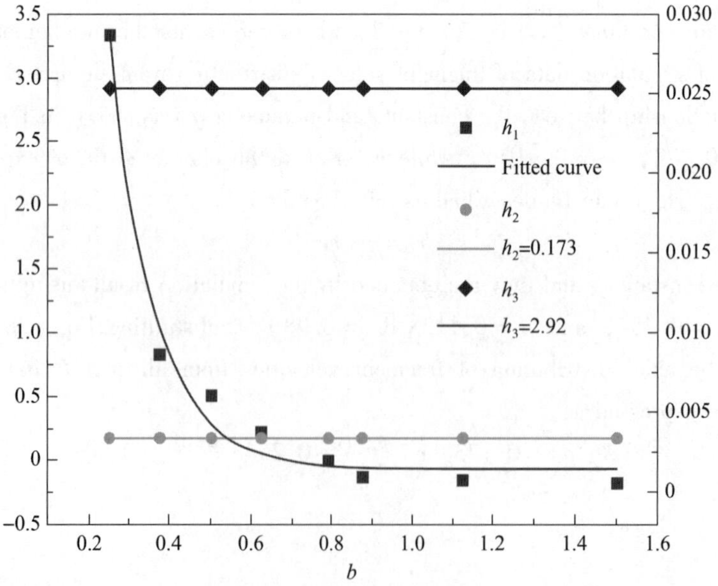

Figure 14 The relationship between coefficients (h_1, h_2, and h_3) and *b*.

5.3 Fragment Velocities Distribution From Charge Part *C* in Absolute Coordinates

The axial distribution of fragment velocities from explosive charge C was shown in Figure 15, where $a=0$ and $b=0$ represented the usual cylindrical casing without air parts at two ends. As shown in Figure 15, it can be found that the fragment velocity distribution generated from explosive charge C with different lengths of air parts at two ends had no difference with that from the cylindrical casing without air parts at two ends. Although the casing C was still connected to the casing A and B, this did not affect the fragment velocity distribution of the casing C, which indicated that the attenuation effect of the axial rarefaction waves on the shock waves and detonation products was the main reason that the fragment velocity near the end of explosive charge C was smaller than the Gurney velocity, and the length of the casing had nothing to do with it. Therefore, the axial distribution of fragment velocities from cylindrical casing with different lengths of air parts at two ends can be replaced by that from cylindrical casing without air parts at two ends. The correctional function of the axial distribution of fragment velocities from explosive charge C could be explored on basis of Eq. (4), which had a high precision compared with experimental data. Eq. (4) was on the basis of the Gurney formula by correction function to revise the effect of axial rarefaction waves on the fragment velocity. But its correction function expression was not suitable for ours. Hence, the correction function in our formula needed to be rebuilt on the basis of Eq. (4). The modified model of Eq. (4) can be expressed as:

$$v_{0x} = \sqrt{2E} \cdot \sqrt{\frac{1}{1/\beta f(x) + 0.5}} \tag{30}$$

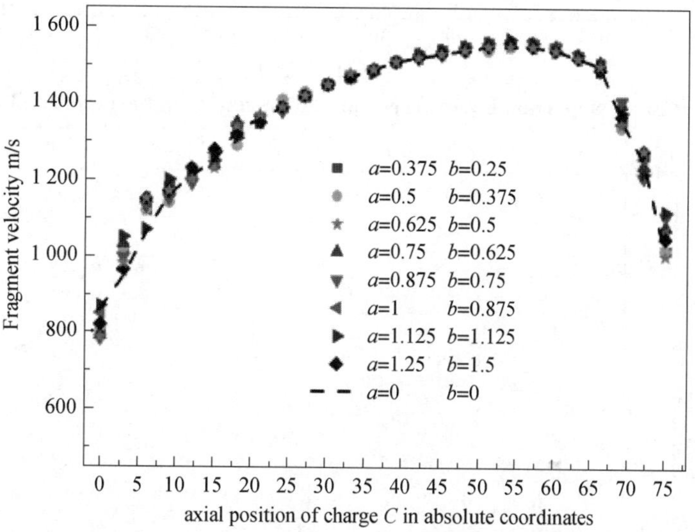

Figure 15　The axial distribution of fragment velocities from explosive charge *C* in absolute coordinates.

where $f(x)$ is the correctional function used to describe the effect of the axial rarefaction waves from two ends, which can be expressed as:

$$f(x) = (1 - A_1 e^{-B_1 x/d})(1 - C_1 e^{-D_1(L-x)/d}) \tag{31}$$

where A_1, B_1, C_1 and D_1 are coefficients, L is the length of the cylindrical casing, and d is the diameter of the explosive charge.

The first and second terms of the correctional function were used to correct the effect of the axial rarefaction waves from the detonation end and non-detonation end, respectively, the coefficients A_1, B_1, C_1 and D_1 can be determined by the experiment data of Huang's study [20]. Therefore, using the same solution in Huang's study [20], the coefficient A_1, B_1, C_1 and D_1 were obtained. The value of them were 0.689, 0.953, 0.385 and 2.3, respectively. Figure 16 shows the comparison between the calculation results of the Eq. (4) and Eq.

(30) and the experiment data. In addition, the error between equations and experimental data was shown in Figure 17. As shown in Figure 16 and Figure 17, both the Eq. (30) and Eq. (4) have good agreement and small error with experimental data. The maximum relative error of Eq. (30) was 4.6% and the average value of the error was 4.54%, the maximum relative error of Eq. (4) was 6.7% and the average value of the error was 3.86%, which indicated that the Eq. (30) can predict the fragment velocities along the axis of cylindrical casing without air parts at two ends. Hence, the correctional function of the axial distribution of fragment velocities from explosive charge C can be expressed as:

$$f(x,a,b) = (1 - 0.689e^{-0.953x/d})(1 - 0.385e^{-2.3(L_C - x)/d}) \tag{32}$$

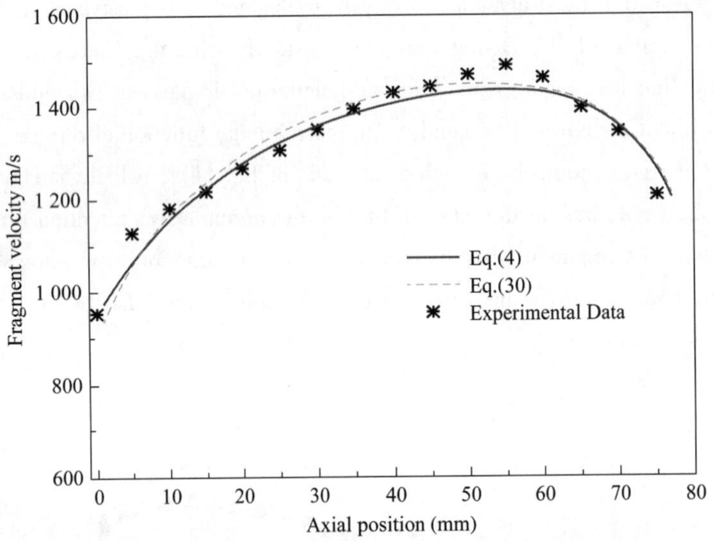

Figure 16　Comparison between the calculation results and experimental data.

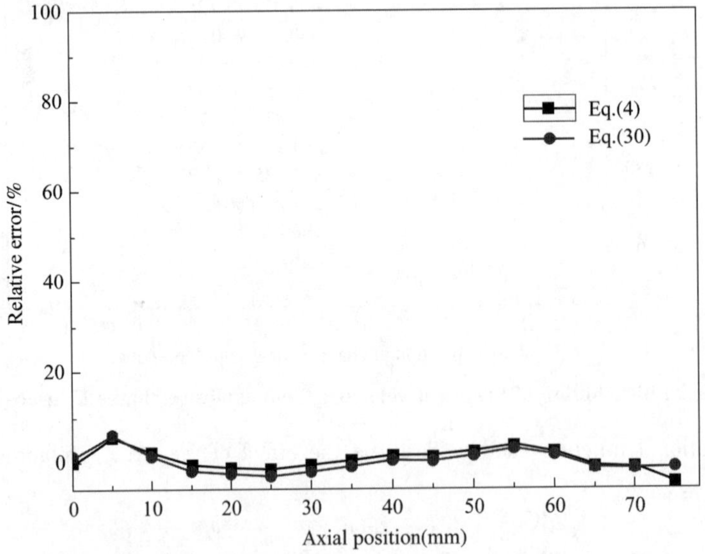

Figure 17　Comparison of relative errors of two equations along the axis of cylindrical casing.

By substituting Eq. (32) into Eq. (8), and transforming the axial position of Eq. (9) and Eq. (10) in their own relative coordinates into the absolute coordinates, the axial distribution of fragment velocities from cylindrical casing with air parts at two ends can be expressed as follows:

$$v(x,a,b)=\begin{cases}\sqrt{\dfrac{2E}{1/\beta_0(0.397e^{x/2.25}+0.001)+0.5}} & -L_A<x\leqslant 0\\ \sqrt{\dfrac{2E}{1/\beta_0((1-0.912e^{-1.73x/d})(1-0.385e^{-2.3(L_C-x)/d}))+0.5}} & 0<x\leqslant L_C\\ \sqrt{\dfrac{2E}{1/\beta_0(0.173*e^{-(x-L_C)/2.919}+0.254e^{-b/0.112}+0.001)+0.5}} & L_C<x\leqslant L_B+L_C\end{cases} \quad (33)$$

5.4　Validation of the Proposed Formula

Eight additional cylindrical casings with different length of air parts at two ends and β_0 were numerically carried out to further demonstrate the reliability of Eq. (33) (Table 7), and they had metal casing of AISI 1045 steel and explosive charge of COMP – B.

The axial distribution of fragment velocities of eight validation tests was showed in Figure 18 and Figure 19, where scatters were numerical results and lines were calculation results (Eq. (33)). It can be seen that the calculation results from new formula matched the experimentally verified numerical simulation results. The relative error was used to assess the global error between the predicated value and true value, root – mean – square – error (RMSE) is a measure of accuracy, to compare forecasting errors of the predicated value and true value, and adjusted R^2 was used to estimate the fitting accuracy of the fitting formula. The average relative error, RMSE, and R^2_{adj} were shown in Table 8. The average relative errors were no more than 6.28%, The RMSE of the first group was no more than 60.23m/s, and the second group was no more than 46.23m/s, which meant that the error between the predicated value of Eq. (33) and the simulation results was very small. R^2_{adj} (more than 0.958) indicated that the proposed formula had a high fitting precision relative to the simulation results. The results mentioned above indicated the proposed formula was accurate enough to predict the axial distribution of fragment velocities from cylindrical casing with air parts at two ends.

Table 7　Comparison of calculated and numerical results of the axial distribution of fragment velocities from cylindrical casing with air parts at two ends.

No.	Air part A (L_A, mm)	a	Air part B (L_B, mm)	b	charge C (L_C, mm)	Casing (L, mm)	Out diameter (D, mm)	Inner diameter (d, mm)	Number of particle
V1 – 1	6	0.2	12	0.4	75	93	34	30	268,488
V1 – 2	12	0.4	18	0.6		105			277,560
V1 – 3	15	0.5	24	0.8		114			284,364
V1 – 4	24	0.8	30	1		129			295,704
V2 – 1	10	0.25	5	0.125	100	115	46	40	611,323
V2 – 1	15	0.375	10	0.25		125			629,873
V2 – 3	25	0.625	20	0.5		145			642,873
V2 – 4	30	0.75	40	1		170			665,409

Table 8　The results of the validation simulations.

No.	Average relative error	RMSE	R^2_{adj}
V1 – 1	3.67%	60.23	0.965

continued

No.	Average relative error	RMSE	R^2_{adj}
V1 - 2	1. 98%	51. 32	0. 968
V1 - 3	1. 68%	51. 43	0. 992
V1 - 4	6. 28%	57. 58	0. 958
V2 - 1	0. 5%	43. 36	0. 989
V2 - 1	0. 63%	43. 89	0. 993
V2 - 3	1. 07%	46. 23	0. 995
V2 - 4	1%	39. 66	0. 991

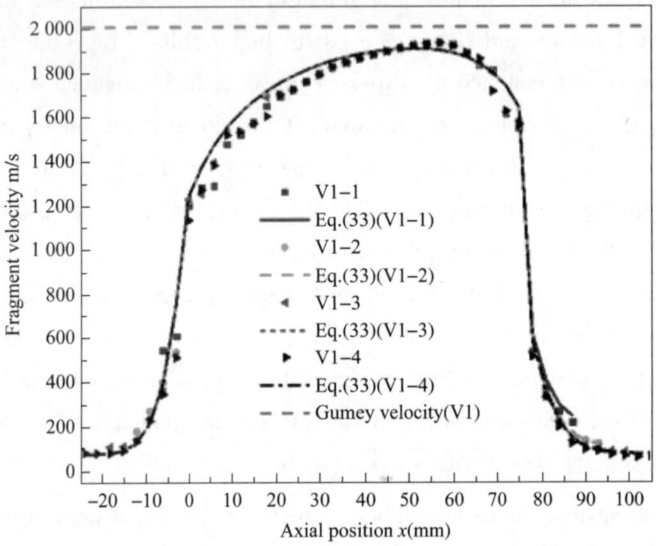

Figure 18 Comparison between the calculations and numerical results (the first group).

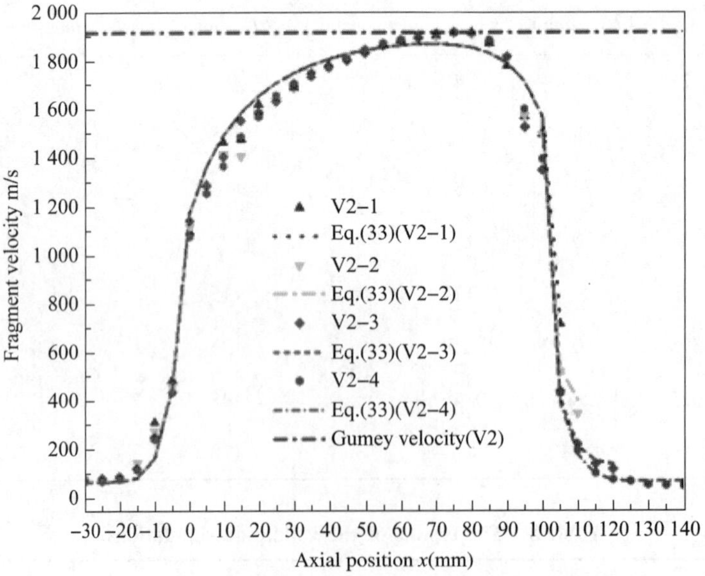

Figure 19 Comparison between the calculation and numerical results (the second group).

6 Conclusions

The axial distribution of fragment velocities from cylindrical casing with air parts at two ends was investigated by experimentally verified numerical simulation. The casings from air parts at two ends would be accelerated and ruptured, the fragment velocities from them cannot be calculated by formulas proposed before. The effects of the shock waves and detonation products from explosive charge on casings of air parts at two ends were explored. It can be found that the effect of the axial rarefaction waves from air part *A* on the shock waves and detonation products was greater than that from air part *B* since the dispersion direction of detonation products was opposite to the propagation direction of detonation waves. Based on Gurney formula, two correctional functions in relative coordinates were proposed, and the axial distributions of fragment velocities from air parts at two ends in absolute coordinates were obtained through the coordinate change since the convected velocity of the relative coordinates was 0. The correctional function of air part *A* was only related to the axial position, but that of air part *B* was not only a function of the axial position but also a function of the length of air part *B*. The coefficients in correctional function were obtained by fitting simulation results. The axial distribution of fragment velocities from charge *C* was almost unaffected by the length of air parts at two ends. In order to ensure the unity of form, a new formula was established based on the previous formula. In the end, a modified formula was proposed to predict the axial distribution of fragment velocities from cylindrical casing with air parts at two ends. In addition, the proposed formula was further verified by the experimentally verified numerical simulation. The results showed that the average relative error was no more than 6.28%, RMSE was no more than 60.23 m/s, and R_{adj}^2 was more than 0.958, which indicated that the calculations of the proposed formula had good agreement with numerical results. Therefore, the proposed formula can be used to predict the axial distribution of fragment velocities from cylindrical casing with air parts at two ends, and will be helpful to the theoretical analysis and optimal design of novel warheads.

Acknowledgments

The authors are very grateful for the support received from State Key laboratory of Explosion Science and Technology of China.

References

[1] TAYLOR G. The fragmentation of tubular bombs, Advisory Council on Scientific Research and Technical Development, 5 (1963) 202 – 320.

[2] GURNEY R W. The initial velocities of fragments from bombs, shell and grenades [R]//Ballistic Research Lab. Aberdeen Proving Ground, Maryland, 1943.

[3] BOLA M, MADAN A, SINGH M. Expansion of metallic cylinders under explosive loading [J]. Defence science journal, 1992, 42: 157 – 163. https://doi.org/10.14429/dsj.42.4375.

[4] REN G, GUO Z, FAN C, et al. Dynamic shear fracture of an explosively – driven metal cylindrical shell [J]. International journal of impact engineering, 2016, 95: 35 – 39. https://doi.org/10.1016/j.ijimpeng.2016.04.012.

[5] LI Y, LI Y H, WEN Y Q. Radial distribution of fragment velocity of asymmetrically initiated warhead [J]. International journal of impact engineering, 2017, 99: 39 – 47. https://doi.org/10.1016/j.ijimpeng.2016.09.007.

[6] FENG S S, CUI B G. An experimental investigation for the axial distribution of initial velocity of shells [J]. Acta armamentarii, 1987, 4: 60 – 63.

不同端盖厚度的圆柱形装药壳体破片初速分布[①]

高月光[②]，冯顺山，刘云辉，黄　岐

（北京理工大学爆炸科学与技术国家重点实验室，北京　100081）

摘　要：端盖厚度是杀伤战斗部威力精准设计时需考虑的重要因素之一，为此开展了一端中心起爆且带不同厚度端盖的圆柱形装药壳体破片初速分布特性的研究。基于理论分析和SPH数值仿真方法，建立不同厚度端盖的圆柱形装药壳体数值模型，进而提出破片初速轴向分布理论计算公式。结果表明：随着端盖厚度的增加，轴向稀疏波作用递减，壳体两端破片速度相应增加；爆轰产物加速效应是起爆端破片增速的主因，爆轰波加速效应是非起爆端破片增速的主因；相对起爆端，非起爆端破片加速更为显著。得到的公式更适用于工程中不同端盖厚度和材质的条件，误差显著减小，可为有关杀伤战斗部破片参数精准设计提供参考。

关键词：兵器科学与技术；破片初速；格尼公式；轴向稀疏波；数值模拟

Velocity distribution of fragment from cylindrical casing with end caps of different thicknesses at two ends

Gao Yueguang，Feng Shunshan，Liu Yunhui，Huang Qi

（State Key Lab of Explosion Science and Technology，Beijing Institute of Technology，Beijing 100081，China）

Abstract：The thickness of end cap is one of the important factors to be considered in the accurate design of antipersonnel warhead，the velocity distribution of fragments from cylindrical casing with end caps of different thickness at two ends was explored，which was detonated at the center of one end. Based on theoretical analysis and SPH numerical simulation，the numerical models of cylindrical casing with different thickness end caps were established，and then the theoretical formula of axial distribution of the initial fragment velocity was proposed. The results showed that the effect of axial rarefaction waves decreased，and the fragment velocity near two ends accordingly increased with the increase of the thicknesses of two end caps. The acceleration effect of detonation product is the main cause of fragments near the detonation end，while the acceleration effect of detonation wave is the main cause of fragments near the non－detonation end. Compared with the fragment near the detonation end，the acceleration of the fragment near the non－detonation end was more severe. The proposed formula was more suitable for end caps of different thicknesses and materials in engineering and can the relative error is significantly reduced，which can provide reference for the accurate design of related antipersonnel warhead.

Keywords：ordnance science and technology；initial fragment velocity；gurney equation；axial rarefaction waves；numerical simulation

0　引言

破片型战斗部利用炸药爆炸后产生具有一定动能的破片来毁伤目标，破片初速是研究和评估战斗部

---

① 原文发表于《兵工学报》2022。

② 高月光：工学博士，2017年师从冯顺山教授，研究半侵彻弹药战斗部对甲板目标威力效应，现工作单位：在校生。

威力和防护结构性能的重要指标[1,2]。战斗部通常是圆柱形装药壳体，采用一端起爆的方式。许多研究人员通过试验、数值仿真和理论分析等方法对破片初速进行了研究，并获得了可以解释破片形成和预测破片初速的理论和计算公式，研究成果为武器设计、公共安全、反恐等领域提供必要的参考[3]。

在计算圆柱形装药壳体破片初速的理论和公式中，格尼公式因其简单、精度高而广泛应用。基于一些简单的假设和能量守恒规律，格尼[4]提出了计算破片初始速度的经典公式，其中破片初速与炸药和壳体质量比相关，公式如下：

$$v_0 = \sqrt{2E}\sqrt{\frac{1}{1/\beta + 0.5}} \tag{1}$$

式中 v_0——破片初始速度；

$\sqrt{2E}$——装药的格尼速度；

β——装药与壳体质量比。

虽然计算的结果略高于破片真实速度，但相对误差均在 10% 以内，说明格尼公式具有较高的精度[5]。虽然格尼公式具有很好的精度，且广泛应用于破片初速的计算中，但是格尼公式只能计算破片的最大速度，真实中圆柱形装药壳体并非无限长，端部由于存在轴向稀疏波，破片初速会降低，因此，越来越多的研究人员开始对破片的加速过程进行深入研究，并获得了更精确的结果[6-8]。考虑到轴向稀疏波对速度的衰减作用，研究人员在格尼公式的基础上进行了修正，建立了新的计算公式，更加准确地预测破片初速的分布。黄广炎等[9]采用 X 光试验装置获得精准的破片初速，并提出了新的改进公式，解决了端部破片初速为 0 的问题，计算公式具有很高的精度，公式如下所示：

$$\begin{aligned} v_{0x} = {} & (1 - 0.361\mathrm{e}^{-2.3617x/d})(1 - 0.192\mathrm{e}^{-3.03(L-x)/d}) \cdot \\ & \sqrt{2E} \cdot \sqrt{\beta/(1 + 0.5\beta)} \end{aligned} \tag{2}$$

孔祥韶等[10]对圆柱形装药壳体的动态过程进行了研究，发现轴向稀疏波会造成靠近端部破片初速的降低。郭志威等[11]研究了轴向稀疏波对破片初速的影响，发现轴向稀疏波对破片加速过程有显著影响。此外，他们还进行了 X 光试验，研究了不同数量端盖的圆柱形装药壳体破片初速情况，结果表明端盖的存在能够减弱稀疏波对破片初速的影响。高月光等[12]对两端带有空气段的圆柱形装药壳体进行了数值研究，发现轴向稀疏波与炸药同时作用于空气段壳体，使其破裂成具有一定速度的破片，并得到了两端带有空气段的圆柱形装药壳体破片初速的轴向分布规律。

在上述理论和计算模型中，可以发现轴向稀疏波通过影响破片加速过程来降低破片初速，而端盖的存在会削弱轴向稀疏波作用。然而，少有研究量化端盖对轴向稀疏波和破片初速的影响。在实际应用中，战斗部大部分为两端带不同厚度端盖的圆柱形装药壳体，端盖的存在提高了破片的速度，增加了有效破片数，破片初速的合理预测对战斗部毁伤效能分析及结构设计具有十分重要的意义，可克服因只采用格尼速度计算破片初速造成对战斗毁伤性能的过高评估。因此，本文研究的两端有不同厚度端盖的圆柱形装药壳体破片初速分布具有更大的实际应用价值。本文在理论分析的基础上，采用数值仿真方法对一端中心起爆且带不同厚度端盖的圆柱形装药壳体进行研究，分析其加速过程，在格尼公式的基础上建立的破片初速分布公式，给出其适用范围，并通过试验和数值仿真进一步验证了所提出的公式的正确性。

1 理论分析

图 1 为两端带有不同厚度端盖的圆柱形装药壳体横截面示意图，装药为一端中心起爆。h_1 和 h_2 为起爆端和非起爆端端盖厚度，d 和 D 分别代表装药和壳体的直径，L 为壳体长度。在实际应用中，圆柱形装药壳体通常是薄壁壳体，且两端端盖的厚度也不会是无穷大的。因此端盖厚度可以通过壳体厚度进行表征。

建立起如图 1 所示的坐标系，坐标原点为起爆点，方向由起爆端指向非起爆端。在真实战斗部中，壳体通常为细长壳体，且轴向稀疏波对起爆端和非起爆端破片初速的影响距离分别为 $2R$ 和 R，R 是装药半径[11]。因此，本文讨论中的圆柱形装药壳体均假设长度大于两倍装药直径来避免两端稀疏波的相互叠加影响。

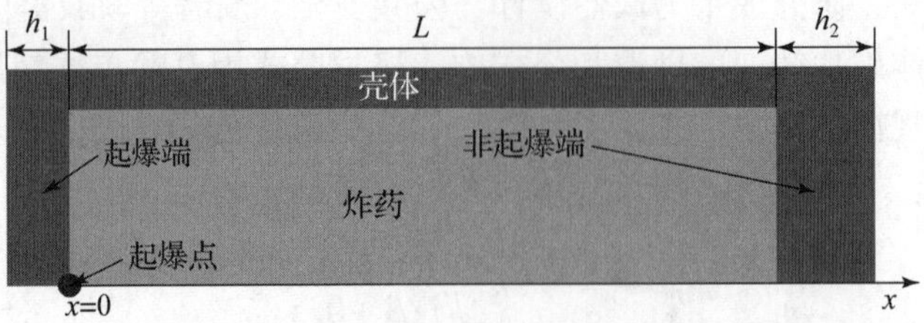

图 1　两端带有不同厚度端盖的圆柱形装药壳体横截面示意图

下面推导端盖破片速度与圆柱壳体破片速度，依据动量守恒定理，得到端盖和壳体的动量守恒方程为

$$P(t)\,\mathrm{d}S\cdot\mathrm{d}t=\rho\delta\mathrm{d}S\cdot\mathrm{d}V \tag{3}$$

其中　ρ——圆柱壳体或端盖的密度；

δ——圆柱壳体或端盖的厚度（端盖厚度记为 h）；

S——爆轰产物与金属壳体的接触面积，积分可得

$$V=\frac{\int P(t)\,\mathrm{d}t}{\rho\delta} \tag{4}$$

则圆柱壳体和端盖的速度分别为

$$V_c=\frac{\int P(t)\,\mathrm{d}t}{\rho_c h} \tag{5}$$

$$V_e=\frac{\int P(t)\,\mathrm{d}t}{\rho_e\delta} \tag{6}$$

$$\frac{V_c}{V_e}=\frac{\rho_e h}{\rho_c\delta}=\mathrm{k} \tag{7}$$

其中下标 e 代表端盖，下标 c 代表圆柱壳体，由上述公式可得，圆柱壳体破片与端盖破片速度比是关于壳体密度与厚度乘积的比。因此，提出端盖厚度无量纲参数 k_1 和 k_2 为起爆端和非起爆端端盖厚度和密度乘积与壳体厚度和密度乘积的比值（$h_1=k_1\rho_c\delta/\rho_{e1}$，$h_2=k_2\rho_c\delta/\rho_{e2}$），$\rho_{e1}$ 和 ρ_{e2} 是起爆端和非起爆端端盖厚度密度。当两者材料相同时，其速度之比仅为端盖与壳体厚度的函数。

格尼公式是基于装药与壳体的动量守恒和能量守恒得到的，本质是关于装药与壳体质量比的函数。端盖本质上也是属于壳体，其产生的破片也占用了炸药部分能量，因此在求解端盖厚度对圆柱壳体破片速度的影响时，采用无量纲参数 k_1 和 k_2 可以很好地表征端盖部分壳体质量对于圆柱壳体破片速度影响。端盖厚度的增加，本质相当于增加了不同轴向位置处的质量比 β，进而改变了破片速度，能够反应无量纲中因变量与自变量的因果关系，反映现象的本质。轴向位置通过与装药直径相比得到无量纲参数本质也是反映轴向位置处质量比的变化，进而改变轴向位置处的破片速度。因此，得到新的不同端盖厚度圆柱形装药壳体破片速度轴向分布公式为

$$v_x=\sqrt{2E}\sqrt{\frac{1}{1/\beta f(x/d,k_1,k_2)+0.5}} \tag{8}$$

2　数值模拟

2.1　仿真方法

光滑粒子流体动力学（SPH）方法是求解强动载条件下材料动力学响应的一种先进仿真方法，其能

够有效避免材料大变形时网格畸变问题，这是拉格朗日方法所不能解决的。Randles 等[13]采用 SPH 方法，通过光滑粒子和广义边界条件等改进方法，对圆柱形装药壳体的破碎进行了分析，仿真结果与实验结果吻合较好。SPH 是一种无网格方法，广泛用于模拟材料瞬时动态大变形。SPH 方法可以避免欧拉网格和材料边界问题。在过去的几年中，许多研究人员利用 SPH 方法对材料动态冲击进行了模拟，仿真结果均与实验结果吻合较好[14-17]。

SPH 方法是基于插值理论，利用插值函数将连续动力学转化为积分方程。对于函数 $f(x)$ 在参数 h 定义的核函数 W 的半径内的核估计 $\langle f(x)\rangle$，可以通过对相邻粒子积分函数 $f(x)$ 得到

$$\langle f(x)\rangle = \int_{\Omega} f(x')W(x-x',h)\,\mathrm{d}x' \tag{9}$$

对于核函数 W，其满足两个条件：①它是偶函数；②它是标准函数 $\int_{\Omega} W(x-x',h)\,\mathrm{d}x = 1$。$h$ 是该函数的宽度，当 h 接近 0 时 $\langle f(x)\rangle$ 等于 $f(x)$。如果将质量 dm 用 $\rho dx'$ 代替，那么积分 $\langle f(x)\rangle$ 可以表示积分区域内所有粒子的总和，依据粒子的质量、速度、动能守恒方程可进行相应的求解：

$$\frac{\mathrm{d}\rho^i}{\mathrm{d}t} = \rho^i \sum_{j=1}^{N} \frac{m^j}{\rho^j}(v_\alpha^j - v_\alpha^i)\frac{\partial W^{ij}}{\partial x_\alpha^i} \tag{10}$$

$$\frac{\mathrm{d}v_\alpha^i}{\mathrm{d}t} = -\sum_{j=1}^{N} m^j\left(\frac{\sigma_{\alpha\beta}^i}{\rho^{i2}} + \frac{\sigma_{\alpha\beta}^j}{\rho^{j2}}\right)\frac{\partial W^{ij}}{\partial x_\beta^i} \tag{11}$$

$$\frac{\mathrm{d}E^i}{\mathrm{d}t} = -\frac{\sigma_{\alpha\beta}^i}{\rho^{i2}}\sum_{j=1}^{N} m^j(v_\alpha^i - v_\alpha^j)\frac{\partial W^{ij}}{\partial x_\beta^i} \tag{12}$$

因为圆柱形装药壳体是对称的，所以只需要对它的 1/4 进行建模。考虑到 SPH 方法为无网格方法，模型中的材料是由大量的粒子构成的。粒子是独立的，它们之间的关系由上述方程确定。实验证明，SPH 方法的准确性与粒子的大小有关。粒子越小，模拟的准确性越高，但也意味着将耗费更多的时间。李伟等[17]采用 SPH 方法对不同直径的粒子模型进行了仿真，结果表明，直径为 0.4 mm 的粒子即可获得足够精度的结果，且该直径粒子在多篇关于破片速度仿真的文章中均获得比较理想的结果[11,12]。考虑到计算精度和时间的平衡，我们将模拟中的粒子直径设置为相同的直径（0.4 mm）。

2.2 材料模型

本文中的壳体和炸药材料分别为 AISI 1045 钢和 B 炸药，这是研究壳体爆炸和破片初速的最常用材料。AISI 1045 钢爆炸后会发生膨胀破裂，其过程可以采用 Johnson - Cook(JC) 模型来描述：

$$\sigma = (A + B\varepsilon_{\mathrm{ep}}^n)(1 + C\ln\dot{\varepsilon}^*)(1 - T^{*m}) \tag{13}$$

其中，A、B、C、m、和 n 都是常数，$\varepsilon_{\mathrm{ep}}$ 代表等效塑性应变，$\dot{\varepsilon}^*$ 是无量纲塑性应变率，T^* 是无量纲温度满足 $T^* = (T - T_r)/(T_m - T_r)$。壳体材料 JC 模型参数如表 1[18]所示。

表 1 1045 钢 JC 参数[18]

A/MPa	B/MPa	n	c	m	T_m/K	γ
507	320	0.28	0.064	1.06	1 793	53.8

炸药爆炸过程可以用 JWL 模型模拟：

$$P_b = C_1\left(1 - \frac{\omega}{r_1\nu}\right)\mathrm{e}^{-r_1\nu} + C_2\left(1 - \frac{\omega}{r_2\nu}\right)\mathrm{e}^{-r_2\nu} + \frac{\omega E}{\nu} \tag{14}$$

式中 P_b——炸药的爆压；

E——单位体积炸药内能；

ω、C_1、C_2、r_1 和 r_2——材料常数，炸药 JWL 参数如表 2[19]所示。

表 2　B 炸药 JWL 参数[19]

$D/(\mathrm{m\cdot s^{-1}})$	P_{CJ}/GPa	$E_0/(\mathrm{kJ\cdot m^3})$	C_1/GPa	C_2/GPa	R_1	R_2	ω
7 980	29	8.5×10^6	542	7.68	4.2	1.1	0.24

2.3　仿真模型验证

仿真模型只有具有足够的精度，才能用于模拟材料的真实性能。根据之前的分析，轴向稀疏波和端盖是影响破片初速的两个主要因素。因此，只要能验证这两个因素对破片初速的影响，就可以验证数值模型的精确性。轴向稀疏波对破片初速的影响可以采用黄广炎等[9]的试验结果进行验证，端盖对破片初速的影响可以采用郭志威等[11]的试验结果进行验证。黄广炎等的 X 光试验中的圆柱形装药壳体两端没有端盖，即端盖厚度为 0。郭志威等的 X 光试验中两个圆柱形装药壳体中一个仅非起爆端带有端盖，另一个两端均带有端盖，端盖的厚度与壳体厚度相等。由于他们均采用 X 光设备获得的破片初速，因此试验结果具有很高的精度。三个试验的圆柱形装药壳体的参数如表 3 所示，其样本示意图如图 2 所示。依据他们的试验建立相应的数值仿真模型。

表 3　黄广炎[9]和郭志威[11]试验中样本参数

No.	d/mm	D/mm	L/mm	δ/mm	h_1/mm	h_2/mm
1	24	30	77	3	0	0
2					0	3
3					3	3

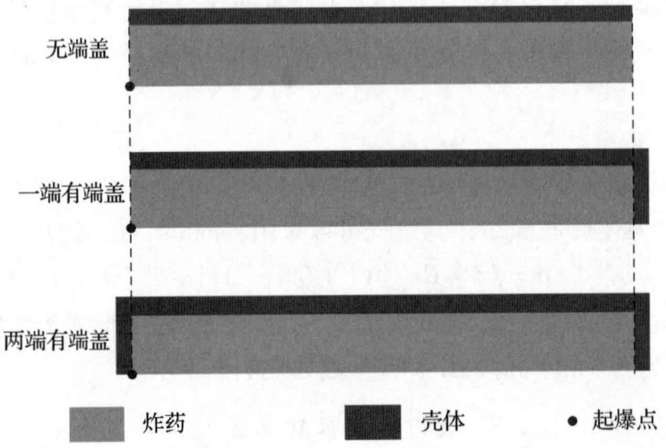

图 2　试验样本示意图

数值仿真结果与试验结果对比如图 3 所示，其中 E 为试验结果，S 为模拟结果。可以发现端部没有端盖的圆柱形装药壳体破片初速的仿真结果中，轴向稀疏波使得端部附近破片的速度明显减小，其破片初速的轴向分布与黄广炎等的试验结果具有很好的吻合性，结果表明仿真模型是能够精准模拟轴向稀疏波对破片初速的影响。同时与郭志威等的端部带有不同数量端盖的圆柱形装药壳体的试验结果对比表明，端盖的存在确实能够有效减弱轴向稀疏波对破片初速的削弱作用，从而提高端部破片初速，仿真得到的破片初速分布与试验获得的破片初速分布十分吻合，说明仿真模型是能够有效模拟端盖对于破片初速的影响，从而获得精确的破片初速分布。

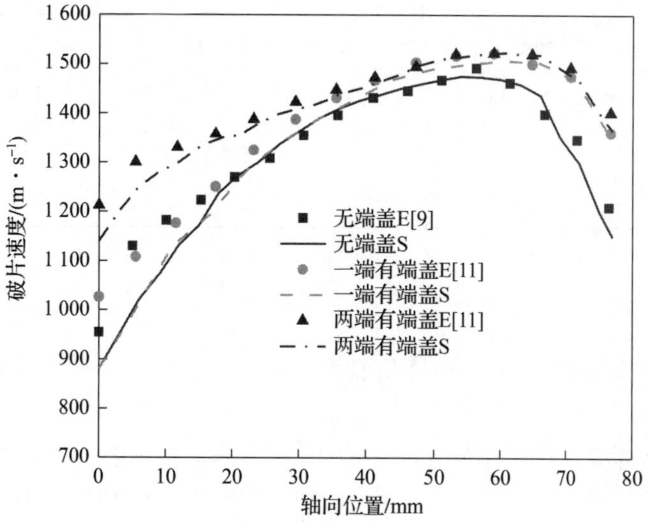

图 3　试验结果与仿真结果比较图

综上所述，所建立的仿真模型是能够精确地模拟出端盖和稀疏波对破片初速的影响，因此可以用来进一步求解两端带有不同厚度端盖的圆柱形装药壳体破片初速的轴向分布。

2.4　仿真样本模型

通过前一节中对仿真模型的校核，结果表明其能够精确模拟端盖和轴向稀疏波对圆柱形装药壳体破片初速的影响，因此，我们利用该仿真模型，设置两端不同厚度的圆柱形装药壳体，求得两端不同厚度端盖的圆柱形装药壳体破片初速分布。建立的数值仿真模型如图 4 所示，两端带有不同厚度端盖的圆柱形装药壳体参数如表 4 所示。参数 k_1 和 k_2 为起爆端和非起爆端端盖厚度与壳体厚度的比值，其比值为 0 代表该端部没有端盖，仿真模型的粒子直径设置为相同的 0.4 mm，每个仿真模型的粒子数见表 4。

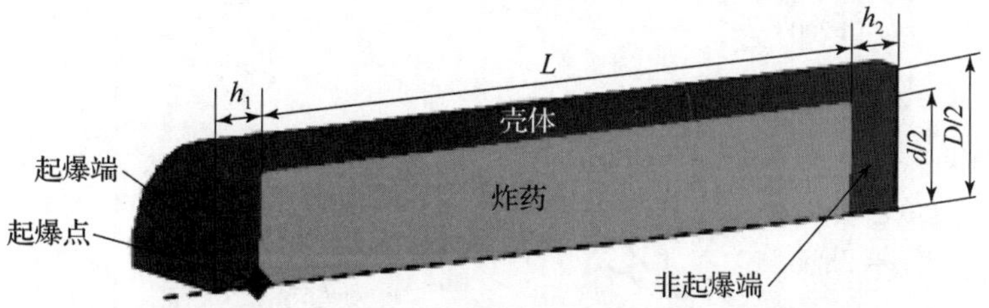

图 4　两端带不同厚度端盖的圆柱形装药壳体仿真模型

表 4　10 个带有不同厚度端盖的圆柱形装药壳体参数

No.	d/mm	D/mm	L/mm	δ/mm	k_1	k_2	粒子数
1	24	30	60	3	0	0	211 392
2					0	1	142 350
3					0.5	0.5	142 350
4					0.7	0.7	220 640
5					1	1	224 604
6					1.5	1.5	231 210
7					2	2	237 816

续表

No.	d/mm	D/mm	L/mm	δ/mm	k_1	k_2	粒子数
8					2.5	2.5	244 422
9					3	3	251 028
10					3.5	3.5	257 634

3　结果与分析

由于壳体的长度足够长（$L/d>2$），两端的轴向稀疏波不会相互影响。因此，两端端盖厚度对破片初速的影响可以单独分析。

图5为起爆端破片的加速过程，k_1表示起爆端端盖的厚度，$k_1=0$表示起爆端无端盖。如图5所示，对于起爆端破片，只要存在端盖（无论厚度多少），端盖会降低轴向稀疏波对初始冲击波的影响，使得破片在初始冲击波的加速作用下达到几乎相同的速度点A，而没有端盖的壳体在初始冲击波的加速作用下由于轴向稀疏波的作用，速度只能达到a点。随后的爆轰波继续加速起爆端壳体，使其速度从A点达到B点，可以发现，由于端盖的存在，壳体速度从b点增加到了B点，这是因为端盖减弱了轴向稀疏波对初始冲击波后爆轰波的削弱作用，但这种速度的增加并随端盖厚度而改变。此后，爆轰产物将破片初速从B点增加到C点。可以看出，起爆端端盖的存在显著提高了爆轰产物对破片的加速作用，且这种加速作用随着起爆端端盖厚度的增加而增强，破片初速从c点增加到了C点，其原因是起爆端端盖显著减少了爆轰产物的轴向分散和轴向稀疏波对爆轰产物的削弱作用。此外，起爆端端盖的存在也缩短了破片的加速过程，使破片能更快地达到最大速度，如起爆端有端盖的圆柱形装药壳体产生的破片初速达到最大速度点C的时间比无端盖的圆柱形装药壳体产生的破片初速达到最大速度点c的时间提前了15%（2.5 μs）。

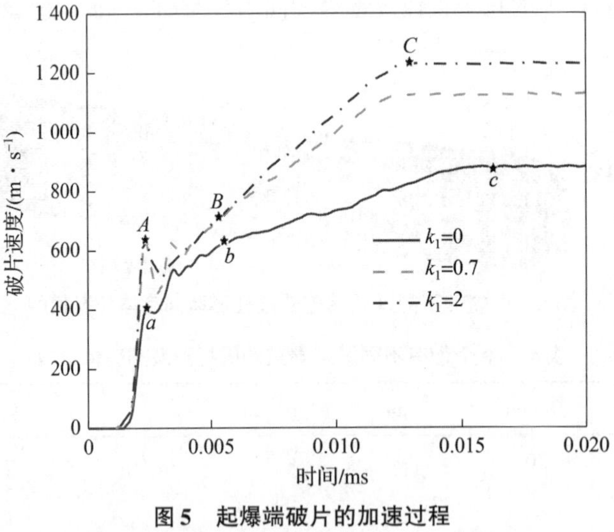

图5　起爆端破片的加速过程

图6所示为非起爆端破片的加速过程。同样的原因，由于非起爆端端盖的存在，壳体速度从点d增加到点D，但这种初始冲击波对壳体的加速效果并没有随着非起爆端端盖厚度的改变而改变。但不同于起爆端壳体加速过程的是，非起爆端端盖厚度的增加改变了后续爆轰波对壳体的加速作用，为此，破片初速能够从点e提高到了点E。对于非起爆端破片，非起爆端端盖的存在并没有增强爆轰产物对破片的加速作用，非起爆端带有端盖的破片初速从点E增加到点F的速度增加量与无端盖条件下破片初速从点e增加到点f的速度增加量接近，这是因为爆轰产物的运动方向是从起爆端指向非起爆端，这使得轴向稀疏波作用被后续爆轰产物所减弱，非起爆端爆轰产物对破片的加速作用受到轴向稀疏波的影响相对起爆

端破片较弱，因此，后续爆轰产物加速作用受端盖厚度影响程度较小，不同厚度端盖对非起爆端破片的加速效应主要体现在爆轰波的加速作用显著增加。

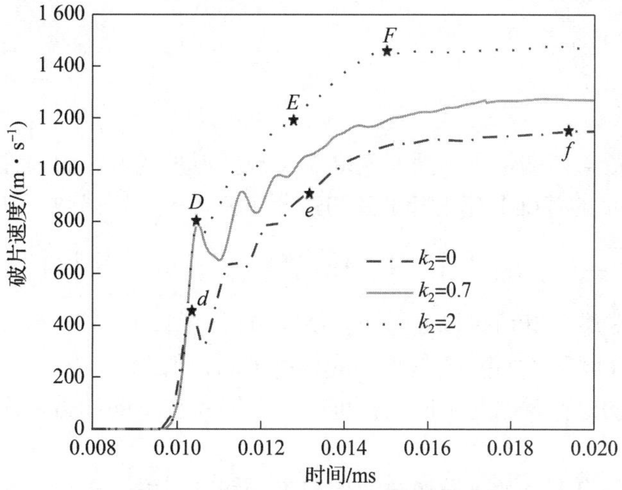

图 6　靠近非起爆端破片加速过程图

从图 5 和图 6 可以看出，起爆端端盖主要通过增加爆轰产物对破片的加速效应使得靠近起爆端破片初速相应增加，而非起爆端端盖主要通过增加爆轰波对破片的加速效应使得非起爆端破片初速相应增加。

两端带有不同厚度端盖的圆柱形装药壳体破片初速分布仿真结果如图 7 中散点所示，记为 S。其中爆轰产物不断地从起爆端向非起爆端运动，使得爆轰产物对非起爆端的破片的加速作用大于对起爆端的破片的加速作用，同时端盖又增加了爆轰波对非起爆端破片的加速效应；而对起爆端破片，端盖虽增加了爆轰产物的加速作用，但爆轰波加速作用并未显著增加。因此，相对起爆端破片，非起爆端破片增速更加明显，也更加接近格尼速度。

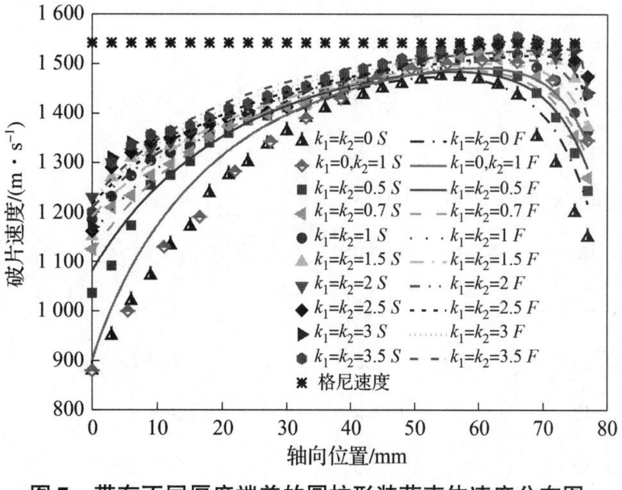

图 7　带有不同厚度端盖的圆柱形装药壳体速度分布图

4　破片初速分布公式

4.1　修正函数

如第 3 节所述，两端端盖的存在改变了破片的加速度过程，导致两端破片初速以不同程度的增加。黄广炎等[9]提出的公式能够精准地预测两端没有端盖的圆柱形装药壳体破片初速的轴向分布，本文中破片初速分布将在黄广炎等人提出的公式基础上建立[9]，包含无端盖情况（$h_1=h_2=0$），并将其按照本文修正函数形式进行修正[12]，修正函数表示如下：

$$f\left(\frac{x}{d},0,0\right)=(1-0.69\mathrm{e}^{-0.95x/d})\cdot(1-0.39\mathrm{e}^{-2.3(L-x)/d}) \tag{15}$$

式中 x——离起爆点的轴向距离；

d——装药直径；

L——装药长度。

由于新公式可以完全覆盖两端带有不同厚度端盖（包括厚度为0）的圆柱形装药壳体破片初速轴向分布，因此可以在式（15）的基础上建立起破片初速分布公式的修正函数，其表达式为

$$f\left(\frac{x}{d},k_1,k_2\right)=(1-A\mathrm{e}^{-Bx/d})\cdot(1-C\mathrm{e}^{-D(L-x)/d}) \tag{16}$$

式中 A、B、C 和 D——常数，其值可以通过仿真结果拟合得到；

k_1 和 k_2——起爆端和非起爆端端盖厚度与壳体厚度的比值。

不同端盖厚度条件下的这些常数是不同的，可以先求解其在不同端盖厚度条件下的值，再分析其与端盖厚度的关系。

由于圆柱形装药壳体满足两端的轴向稀疏波作用不会相互影响。因此，基于黄广炎等[9]的理论，从起爆端到最大破片初速位置之间的破片初速由起爆端端盖厚度决定，根据这部分破片初速分布可以计算出常数 A 和系数 B。同样，从非起爆端到最大破片初速位置之间的破片初速由非起爆端端盖厚度决定，通过这部分破片初速分布可以得到常数 C 和 D。

根据上述分析，在计算常数 A 和系数 B 时，假设式（16）右侧括号为1，即破片初速仅由起爆端的端盖厚度决定。常数 A 可以由起爆端处破片初速得到，系数 B 可以通过拟合这部分破片初速分布计算得到。常数 C 和 D 可以用类似的方法得到。由此得到不同端盖厚度条件下的系数如表5所示，系数与 k_1、k_2 的关系如图8所示，显然，在不同的端盖厚度条件下，系数发生了变化。

表5　9个仿真模型得到的修正函数参数

k_1/k_1	A	B	C	D
0	0.689	0.385	0.953	2.3
0.5	0.592	0.35	0.9	2.54
0.7	0.512	0.3	0.881	3.177
1	0.453	0.25	0.808	3.606
1.5	0.445	0.239	0.81	8.165
2	0.42	0.17	0.792	7.586
2.5	0.43	0.15	0.82	11.78
3	0.42	0.146	0.914	21.9
3.5	0.42	0.15	0.993	39.96

常数 A、B、C 和 D 与 k_1、k_2 之间存在明显的函数关系，由于两端破片初速不相互影响，常数 A 和 B 只与起爆端端盖厚度有关，即只与 k_1 有关，与 k_2 无关，常数 A 和 B 可以由 k_1 来计算，常数 C 和 D 只与非起爆端端盖厚度有关，即只与 k_2 有关，与 k_1 无关，常数 C 和 D 可以由 k_2 来计算。由图8中系数 A、B、C 和 D 与 k_1、k_2 之间的关系也可以得到，系数 A、B 与比值 k_1 的关系分别为指数关系和二次函数关系，系数 C、D 与比值 k_2 的关系可以用指数函数表示。其关系如下：

$$A=a+m_1\mathrm{e}^{-k_1/n_1} \tag{17}$$

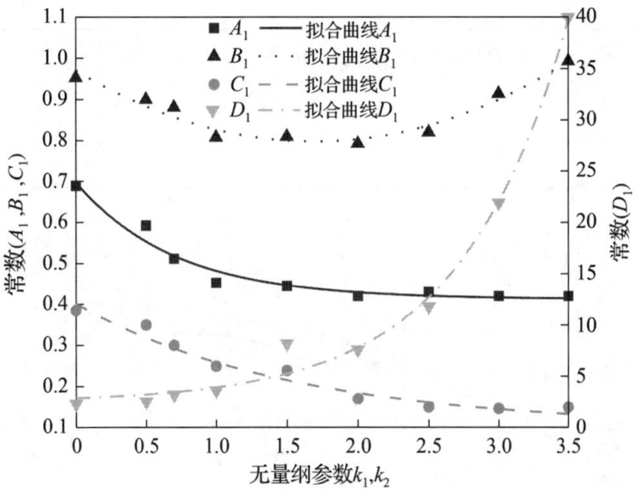

图 8　参数与 k_1 和 k_2 的关系

$$B = b + fk_1 + gk_1^2 \tag{18}$$

$$C = c + m_2 e^{-k_2/n_2} \tag{19}$$

$$D = d + m_3 e^{-k_2/n_3} \tag{20}$$

其中，a、m_1、n_1、b、f、g、c、m_2、n_2、d、m_3和 n_3均为常数，可以通过拟合图 8 中的常数 A、B、C 和 D 得到。拟合得到 $a = 0.41$，$m_1 = 0.29$，$n_1 = 0.71$（$R^2 = 0.95$），$b = 0.97$，$f = -0.2$，$g = 0.06$（$R^2 = 0.946$），$c = 0.10$，$m_2 = 0.3$，$n_2 = 1.54$（$R^2 = 0.96$），$d = 2.46$，$m_3 = 0.4$，$n_3 = 0.77$（$R^2 = 0.989$）。将式（17）~式（20）代入式（16）中，即可得到修正函数：

$$\begin{aligned} f\left(\frac{x}{d},k_1,k_2\right) = & (1-(0.41+0.29e^{-\frac{k_1}{0.71}}) \cdot \\ & e^{-(0.97-0.2k_1+0.06k_1^2)\frac{x}{d}}) \cdot \qquad 0 \leqslant k_1 \leqslant 10 \\ & (1-(0.1+0.3e^{-\frac{k_2}{1.54}}) \cdot \\ & e^{-(2.46+0.4e^{\frac{k_2}{0.77}})\frac{(L-x)}{d}}) \qquad 0 \leqslant k_2 \leqslant 5 \end{aligned} \tag{21}$$

将式（21）代入式（8），即可获得计算两端带不同厚度端盖的圆柱形装药壳体破片初速分布公式，由式（8）计算的结果与仿真结果的对比如图 7 中所示，其中散点代表仿真值，记为 S，曲线代表计算值，记为 F。可以发现不同厚度端盖条件下公式的计算结果与仿真结果均具有较好的吻合性，相对误差最大不超过 4.6%，因此，提出的破片初速公式可以用来计算两端带有不同厚度端盖的圆柱形装药壳体的破片初速轴向分布，并且包括常见两端没有端盖的情况（端盖厚度为 0），具有更广泛的适应性。

4.2　公式验证

为了验证提出的破片初速公式的准确性，采用前人的试验结果[20]进行验证。试验中的壳体和炸药材料为 1045 钢和 B 炸药。试验中壳体具有三种不同厚度的端盖，参数见表 6，式（8）的计算结果与试验结果对比如图 9 所示。可以发现每种端盖厚度条件下公式计算结果均与实验结果吻合较好，所提公式及黄广炎等的公式与试验结果的误差比较如图 10 所示。可以发现式（8）最大相对误差仅为 6.8%，平均相对误差仅为 1.5%，相对文献［9］公式误差显著减小，表明所提公式具有足够的精度。

表 6　试验样本参数表

No.	d/mm	D/mm	L/mm	δ/mm	k_1	k_2
1	25	31	75	3	1	1

续表

No.	d/mm	D/mm	L/mm	δ/mm	k_1	k_2
2					2	2
3					3	3

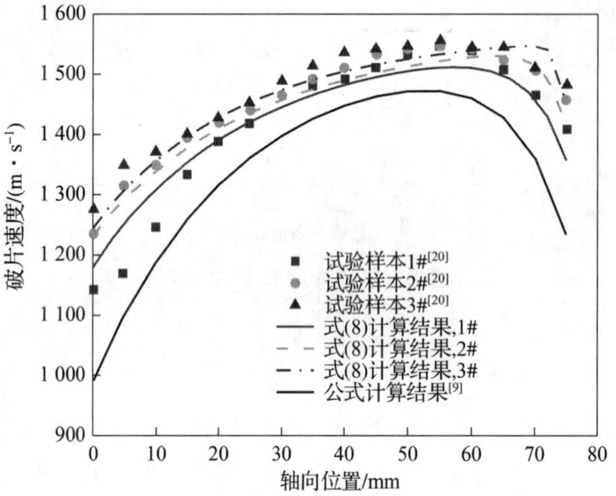

图 9　试验结果与计算结果对比图

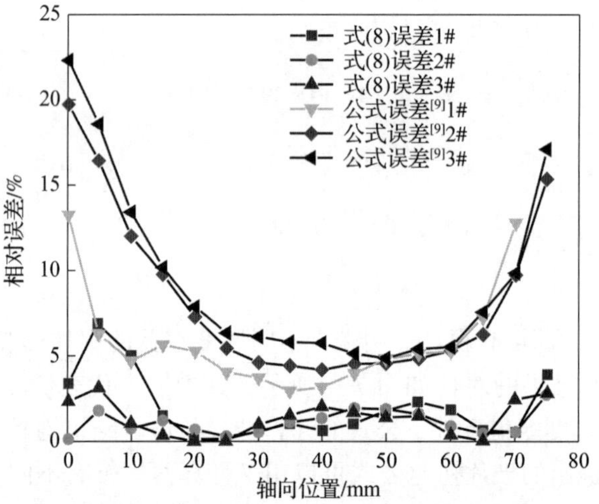

图 10　计算结果相对试验结果的相对误差图

为验证所提公式适应于不同材质的炸药和材料，采用前人的两项试验结果进行验证。其中试验 No. 9302 中炸药和壳体材料为 Octol 和 1020 钢，壳体为预制破片，长径比为 2，$\beta = 0.93$，两端无端盖；No. 9457 中炸药为 Octol，壳体为 1020 预制破片，内含有 1.6 mm 厚的 6061 - T4 铝合金衬套，端盖为 2024 - T3 铝合金，厚度 6.35 mm，具体试验参数如表 7 所示，详情见文献［21］中表 I。试验结果与式（8）计算结果比较如图 11 所示。由图可以发现，对于不同炸药和壳体材质，式（8）均能获得精确的预测结果，相对试验 No. 9302，式（8）平均相对误差仅为 3.13%，产生误差一部分原因是壳体采用预制破片后壳体会提前断裂导致爆轰产物提前泄露造成破片速度低于整体型壳体产生的自然破片速度；相对试验 No. 9457，预制破片内侧放置了 1.6 mm 厚的6061 - T4 铝合金衬套，使得爆轰产物不至于过早泄露，因此式（8）得到的计算结果与试验结果相比平均误差仅为 0.46%，且试验证明了式（8）可适应于端盖与壳体不同材质，证明了公式中所提无量纲参数 k_1 和 k_2 的合理性。

表 7　试验参数表

No.	d/mm	D/mm	L/mm	δ/mm	k_1	k_2
9320	126	140.8	252.08	7.38	0	0
9457	125.98	148.86	251.96	5.08	0.308	0.308

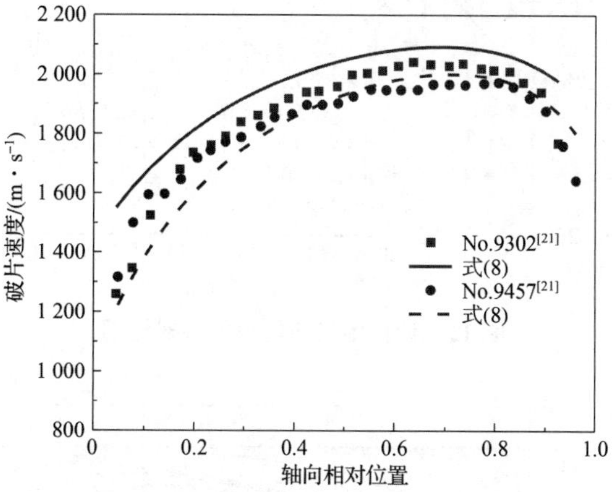

图 11　试验结果与公式计算结果比较图

同时，为了验证两端端盖厚度不会影响另一端端盖附近的破片初速分布，我们采用上述经过试验验证的仿真模型，设计了两端带有不同厚度端盖的圆柱形装药壳体模型来进行验证。壳体和装药材料与先前的一致。仿真模型参数如表 8 所示。仿真结果和式（8）的计算结果对比如图 12 所示，可以发现虽然两端端盖厚度不同，但是公式计算结果仍然与仿真结果具有很好的一致性，说明无论两端的端盖厚度是相同还是不同，一端端盖的厚度不会影响另一端破片初速分布。

表 8　验证仿真参数表

No.	d/mm	D/mm	L/mm	δ/mm	k_1	k_2	粒子数
V1－1	30	34	75	2	0	0	264 792
V1－2					0.5	1	274 704
V1－3					1	1.5	281 312
V1－4					2.5	2	294 528
V1－5					3	2.5	301 136

以上试验结果和仿真结果的比较表明，所提的破片初速公式能够较准确地预测端盖厚度对圆柱形装药壳体破片初速的影响。

4.3　讨论

图 13 是不同端盖厚度条件下式（8）计算得到的破片初速分布，其中壳体和装药材料与上述材料一致，尺寸为：$L=90$ mm，$d=36$ mm，$D=44$ mm。

由图 13 可以发现，随着端盖厚度的增加，轴向稀疏波对两端破片的影响逐渐减小，两端破片初速逐渐接近格尼速度；当非起爆端端盖厚度是壳体厚度 5 倍以上（$k_2 \geqslant 5$）时，轴向稀疏波的影响基本被端盖作用抵消，非起爆端破片初速基本达到格尼速度；当起爆端端盖厚度是壳体厚度 10 倍以上（$k_1 \geqslant 10$）时，轴向稀疏波对起爆端破片的影响范围相对无端盖时的影响范围缩减到 10% 以内（$<0.2R$），破片基本达到格尼速度；端点处（$x=0$，$x=L$）破片初速偏低是由于其同时处于端盖上，端盖的等效装填比小，因此速度会偏低，但仅仅是一点位置处速度偏小，对速度分布的影响可忽略。

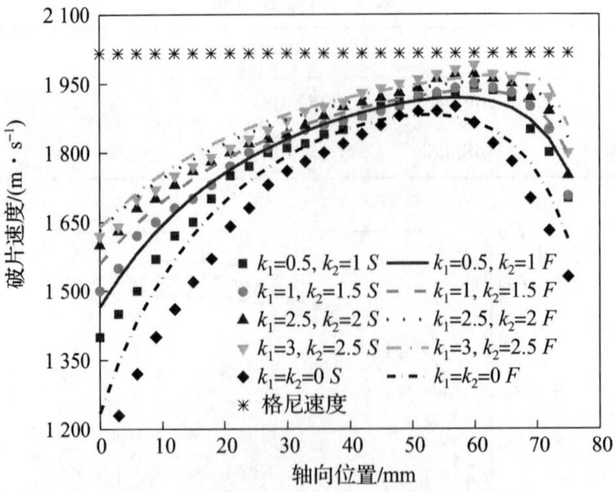

图 12　计算结果与仿真结果对比图

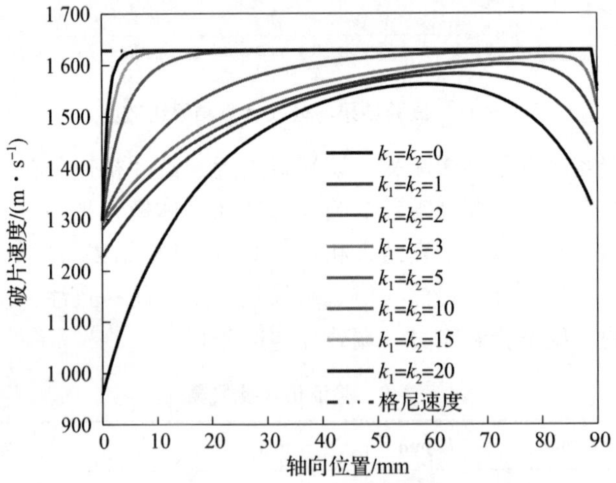

图 13　不同端盖厚度圆柱形装药壳体破片初速分布

因此式（8）的实际应用范围：$0\leqslant k_1\leqslant 10$，$0\leqslant k_2\leqslant 5$，如式（21）所示，端盖厚度超过此范围后可直接采用格尼公式计算。同时实际应用中战斗部两端端盖厚度为有限厚度，因此式（8）对于战斗部有效破片数的计算及毁伤效能分析具有十分重要的实际应用价值。同时，两端端盖厚度的关系也间接验证了起爆端轴向稀疏波的影响范围是非起爆端轴向稀疏波影响范围的 2 倍[12]。

5　结论

针对不同厚度端盖对圆柱形装药壳体破片初速的影响，采用理论和仿真计算的方法进行了量化分析，获得如下结论。

（1）建立带不同厚度端盖的圆柱形装药壳体仿真模型，并基于试验数据，验证了模型的合理性。

（2）仿真分析得到端盖对破片加速过程影响规律：随着端盖厚度的增加，轴向稀疏波作用递减，两端破片速度相应增加，相对起爆端破片，非起爆端破片加速更明显；冲击波加速能量的增加是造成起爆端破片初速增加的主要因素，非起爆端破片初速增加则更多归功于爆轰波加速能量的增强。

（3）建立了可预测不同厚度端盖的圆柱形装药壳体破片初速分布理论公式，并通过仿真确定了关键参数，公式适应范围为：$0\leqslant k_1\leqslant 10$，$0\leqslant k_2\leqslant 5$，当超出上限值，轴向稀疏波对端部破片初速的影响基本忽略，即两端破片基本达到格尼速度。试验对比结果表明公式可以适应不同材料的壳体和炸药，相对前人公式误差显著减小，其对杀伤战斗部提高有效破片数和毁伤效能评估具有实际的指导价值。

子母弹反机场跑道封锁概率快速计算方法研究①

刘云辉[1②]，冯　源[2]，高月光[1]，冯顺山[1]

（1. 北京理工大学爆炸科学与技术国家重点实验室，北京　100081；

2. 中国兵器科学研究院，北京　100089）

摘　要：传统子母弹反机场跑道封锁概率计算方法复杂、耗时长，不能实现毁伤效果侦察评估弹真实战场条件下的实时效能量化计算要求。为此，提出基于神经网络算法的子母弹反机场跑道封锁概率快速计算方法，该方法利用BP神经网络算法对原有计算模型进行拟合，获得可实现实时计算的封锁概率计算模型。仿真结果表明，该方法不仅可继承所学习的传统蒙特卡洛模型的优点，还可克服其准度、精度与耗时性之间的矛盾，评估所需时间从传统蒙特卡洛法的秒级缩短至毫秒级，实现战场毁伤效能实时计算要求，具备实用性和通用性。研究成果可为各类毁伤效果侦察弹对机场跑道封锁效能实时评估计算提供可行方法。

关键词：兵器科学与技术；神经网络；蒙特卡洛法；机场跑道

Research on fast calculation method of the probability of airstrips blockaded by Cluster Munitions

Liu Yunhui[1], Feng Yuan[2], Gao Yueguang[1], Feng Shunshan[1]

(1. State Key Laboratory of Explosion Science and technology, Beijing University of technology, Beijing 100081;

2. China Academy of ordnance Sciences, Beijing 100089)

Abstract: The traditional calculation method of the probability of airstrips blockaded by Cluster Munitions is complex and time - consuming, which can not meet the requirements of real - time effectiveness quantitative calculation of damage effect reconnaissance and evaluation missile under real battlefield conditions. Therefore, a fast calculation method for the blockade probability of cluster munitions against airport runways based on neural network algorithm is proposed. This method uses BP neural network algorithm to fit the original calculation model to obtain a blockade probability calculation model that can be calculated in real time. The simulation results show that this method can not only inherit the advantages of the traditional Monte Carlo model, but also overcome the contradiction between its accuracy, accuracy and time - consuming. The evaluation time is reduced from the second level of the traditional Monte Carlo method to the millisecond level, which meets the real - time calculation requirements of battlefield damage effectiveness, and has practicability and universality. The research results can provide a feasible method for real - time evaluation and calculation of the effectiveness of various damage effect reconnaissance bombs on airport runways.

Key words: ordnance science and technology; neural networks; Monte - Carlo method; runway

---

① 原文发表于《火力与指挥控制》2022。

② 刘云辉：工学硕士，2019年师从冯顺山教授，研究障碍弹对飞行甲板封锁效能评估方法，现工作单位：江西洪都航空工业集团有限责任公司。

0 引言

机场跑道作为现代战争中夺取制空权的关键，其作用日益彰显。近几年的多次局部战争表明，若能在战争中对敌方机场跑道进行有效封锁，则可在后续作战中掌握主动权。而若能及时并准确评估对机场跑道的封锁程度，分析弹药对其的封锁概率，对掌控战场态势、制定后续作战方案具有显著的军事意义。

在子母弹对机场跑道封锁概率计算模型方面，国内已有大量学者对此开展了研究，现有的模型通常采用蒙特卡洛法，通过像素仿真法[1]、空间遍历法[2]、区域搜索法[3-4]或随机抽样法[5]等方法对最小升降窗口进行搜索，进而计算出子母弹对机场跑道的封锁概率。像素仿真法、空间遍历法的计算准度受到所取的像素点大小、遍历步长的限制，要求取具有一定准度、精度的结果需要大量的时间与计算机资源。区域搜索法基于子弹药落点搜索最小起降窗口，其计算准度与评估速度相比，像素仿真法、空间遍历法得到了大大的提高，但是其只能搜索平行于跑道的直起降窗口，对于斜起降窗口则力有未逮，评估结果相比实际值偏大。随机抽样法基于蒙特卡洛法原理在跑道上随机抽取大数量的矩形来判定是否有完整的最小起降窗口，可有效应对斜起降状况，在取样次数达到一定数量的前提下，相比之前的方法可大大提高结果准确度，但是，蒙特卡洛法的复用使得基于随机抽样法建立的封锁概率计算模型的计算时间大大增加。

综上，基于蒙特卡洛方法的封锁概率计算模型准确度已逐步提高，并可考虑到斜起降窗口的情况，但是这些计算模型一直未脱离蒙特卡洛法的制约，求取精度与评估耗时的矛盾性一直存在，其达到秒级、分级的评估时间，已无法适应时代的发展。随着毁伤效果侦察评估弹药的提出与研制，研究更快速的封锁概率计算方法已成为无法回避甚至必须解决的问题。

考虑到神经网络具有可以逼近任何非线性映射关系的能力[6]。在效能评估领域，已有众多使用案例。如在鱼雷作战效能[7]、潜射反舰导弹作战效能[8]、电子战效能评估[9]、炮光集成武器系统作战效能[10]等方面已有应用；在基传统于蒙特卡洛方法计算模型的基础上，采用神经网络算法对封锁概率计算模型进行拟合，在此基础上可实现概率的快速计算，使得评估时长缩短至到毫秒级，同时也可兼顾蒙特卡洛方法结果的优点。

1 子母弹反机场跑道封锁概率计算模型

1.1 子弹药落点散布计算模型

以机场正中心为坐标原点建立坐标系，机场跑道长、宽分别为 L、B。假设共对机场发射 N 枚母弹，第 i 枚母弹预定抛撒坐标地面投影为（x_i，y_i），在应对一条机场跑道的情况下，追求最大封锁概率，母弹预定瞄准坐标一般会设定为机场跑道中轴线，即 $y_i=0$。母弹实际抛撒坐标（x_i'，y_i'）（$i=1, 2, 3, \cdots, N$）为

$$\begin{cases} o=0.849\,3\cdot \mathrm{CEP} \\ x_i'=o\cdot r_1+x_i \\ y_i'=o\cdot r_2 \end{cases} \tag{1}$$

式中 CEP——子母弹平台圆概率偏差；

r_1、r_2——服从标准正态分布 $N(0, 1)$ 生成的随机数。

在上述基础上，假设每枚母弹装填 n 枚子弹药，考虑子母弹抛撒半径 R 及抛撒中心盲区半径 R_0，则第 i 枚母弹的第 j（$j=1, 2, 3, \cdots, n$）枚子弹药的落点坐标（x_{ij}，y_{ij}）为

$$\begin{cases} x_{ij}=x_i'+\sqrt{t_1}\cdot\cos(t_2) \\ y_{ij}=y_i'+\sqrt{t_1}\cdot\sin(t_2) \end{cases} \tag{2}$$

式中　t_1——服从均匀分布 U(R_0^2，R^2) 生成的随机数；

t_2——服从均匀分布 U(0，2π) 生成的随机数。

1.2　子母弹对跑道封锁概率计算模型

基于蒙特卡洛法，在随机生成的子弹药落点基础上，考虑子弹药的毁伤半径 R_e，在子母弹的打击下机场跑道是否存在可供飞机起降的最小起降窗口。若机场存在完好的最小起降窗口，则说明此次仿真计算中封锁失败；若机场在子母弹打击下不存在完好的最小起降窗口，则说明此次仿真中，封锁成功。

假设在仿真中共进行 M 次蒙特卡洛计算，其中有 M_b 次计算中跑道被完全封锁，则子母弹对机场跑道的封锁概率 P_b 为

$$P_b = \lim_{M\to\infty}\frac{M_b}{M} \tag{3}$$

在子母弹抛撒参数，CEP 性能、机场尺寸、起降窗口尺寸等参数不变情况下，封锁概率可视为母弹预定抛撒坐标的函数，即

$$P_b = f(x_1, x_2, \cdots, x_N) \tag{4}$$

式中　x_i——第 i 枚母弹的预定抛撒横坐标；

f——封锁概率非线性函数。

2　基于神经网络的算法研究

2.1　BP 神经网络理论

BP(back propagation) 神经网络是一种基于误差反向传播算法的神经网络，由一层输入层、一层或多层隐藏层和一层输出层组成的多层网络结构。图 1 所示为其基本结构。由于 BP 神经网络具有可以拟合任意非线性映射关系的能力，因此可以用来对子母弹反机场跑道封锁概率与子母战斗部预定瞄准坐标之间的非线性关系进行非线性拟合。

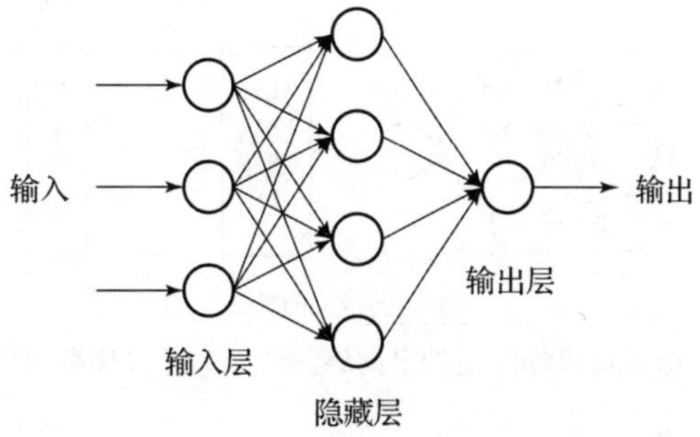

图 1　BP 神经网络结构

网络中每个神经元以上一层的所有神经元的输出作为输入，相应的输出为

$$h_i(j) = \begin{cases} x_j & ,i = 1 \\ g\left[b_i(j) + \sum_{k=1}^{N_{i-1}} w_i(j,k)\cdot h_{i-1}(k)\right] & ,i > 1 \end{cases} \tag{5}$$

式中　$h_i(j)$——第 i 层第 j 个神经元的输出，以输入层为第一层，输出层为最后一层；

$b_i(j)$——该神经元的偏置系数；

N_{i-1}——上一层神经元的数量；

$w_i(j, k)$——该神经元的权重系数；

g——神经元的激活函数，取 S 型函数，即

$$g(x)=\frac{1}{1+\mathrm{e}^{-x}} \tag{6}$$

在网络进行训练学习时，根据网络输出的预测值 P_{out} 与输入样本值 P 的误差来不断调整每一个神经元中的权重系数和偏差系数。由于只有一个输出，故误差计算公式为

$$L=(P_{\text{out}}-P)^2 \tag{7}$$

每次学习后，权重系数 w 以及偏差系数 b 的调整为

$$w_i(j,k)^{\text{new}}=w_i(j,k)-\eta\cdot\frac{\partial L}{\partial w_i(j,k)},\quad i>1 \tag{8}$$

$$b_i(j)^{\text{new}}=b_i(j)-\eta\cdot\frac{\partial L}{\partial b_i(j)},\quad i>1 \tag{9}$$

式中　η——学习率，一般取 0.01~0.5。

经过多次迭代学习训练，反复调整权重系数以及偏差系数，整体误差渐次降低，直至达到预定误差 ε，即可获得对目标非线性函数有较好拟合性能的神经网络模型。

2.2　算法思想

整个算法分为两部分：

（1）随机生成一定数量的 $(x_1,x_2,\cdots,x_N)$（x_i~ U($-L/2,L/2$)，$i=1$，2，…，N），输入子母弹对机场跑道封锁概率计算模型模拟 N 枚母弹打击机场跑道，对其封锁概率进行评估计算，形成初始数据集。

（2）将形成的数据集输入神经网络中，构建合适的神经网络结构中进行迭代优化计算，获得最优权重系数以及偏差系数，以获得对封锁概率非线性函数具有较好拟合性能的神经网络模型。

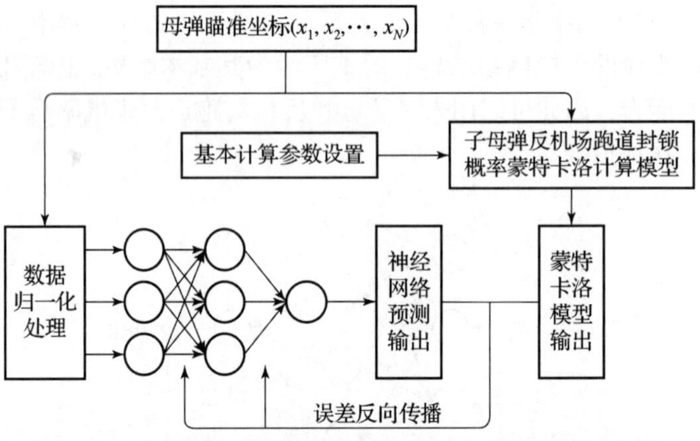

图 2　子母弹反机场跑道神经网络封锁概率计算模型训练示意图

3　案例计算与结果分析

3.1　计算条件

（1）机场跑道：长 $L=2\,400$ m、宽 $B=60$ m；

（2）飞机最小起降窗口：长 $L_1=800$ m、宽 $B_1=20$ m；

（3）最小起降窗口搜索方法：随机抽样法，抽样次数为 10 000 次；

（4）子母战斗部参数：子弹药数量 $n=150$ 枚，抛撒半径 $R=120$ m，抛撒中心盲区半径 $R_0=15$ m，子弹药毁伤半径 $R_e=1$ m，CEP=100 m；

（5）蒙特卡洛模拟仿真计算次数：$M=10\,000$ 次；

（6）子母弹数量 N：2 枚；

（7）数据集数量：20 000 组；

（8）预定误差 ε：1×10^{-6}。

3.2　计算结果与分析

任选数组母弹瞄准坐标（x_1，x_2），分别使用空间遍历法、区域搜索法、随机抽样法以及神经网络法，在相同的计算机运行环境条件下，在 MATLAB 中编程对封锁概率进行计算，并计算了神经网络法对比随机抽样法的相对误差。其中，空间遍历法各方向的遍历步长设定为 0.25 m，随机抽样法抽样次数为 10 000 次，其各方法封锁概率计算结果如表 1 所示，相对误差为神经网络法结果与随机抽样法结果的相对误差。

表 1　仿真计算结果精准度分析

序号	母弹瞄准坐标（x_1，x_2）	空间遍历法 P_k	区域搜索法 P_q	随机抽样法 P_s	神经网络法 P_{out}	相对误差
1	(−615, 656)	0.25%	0.24%	0.19%	0.19%	0.00%
2	(−337, 233)	4.44%	4.40%	4.21%	4.06%	0.15%
3	(−323, 353)	16.40%	16.25%	15.41%	15.48%	0.07%
4	(−329, 466)	22.00%	21.85%	20.50%	20.68%	0.18%
5	(−400, 400)	30.61%	30.41%	28.61%	28.63%	0.02%

空间遍历法由于遍历步长的选取，相比区域搜索法会存在一定的漏判，其结果相对区域搜索法应偏大，而区域搜索法由于无法搜索斜跑道，其结果相对随机抽样法结果偏大，随机抽样法所得封锁概率结果具有最高的准确度。分析表中数据可发现，其结果与上述理论分析一致。由于神经网络学习对象为基于随机抽样法的蒙特卡洛模型，其计算结果与随机抽样法结果最大相对误差为 0.18%。结果相对随机抽样法结果的精度符合工程应用要求。各方法封锁概率计算所需时间结果如表 2 所示。

表 2　仿真计算耗时结果分析

序号	母弹瞄准坐标（x_1，x_2）	空间遍历法 t_k	区域搜索法 t_q	随机抽样法 t_s	神经网络法 t_{out}
1	(−615, 656)	20.91 h	6.32 s	5.95 min	1.81 ms
2	(−337, 233)	23.31 h	7.13 s	1.06 h	1.79 ms
3	(−323, 353)	20.83 h	8.69 s	3.52 h	1.93 ms
4	(−329, 466)	20.56 h	9.55 s	4.91 h	2.02 ms
5	(−400, 400)	33.59 h	10.30 s	6.61 h	1.84 ms

空间遍历法计算所需时间达到小时级，区域搜索法基于子弹药落点的跑道搜索原理大大提高了计算所需时间，但也达到了秒级，计算结果相对快速，但由于无法考虑斜起降情况而使得结果准确性有待提高，随机抽样法计算结果具有最高的准确性，但是其计算所需时间达到分级、小时级，并且封锁概率越大，所需时间越长。神经网络法在继承随机抽样法的准确度的情况下，其评估耗时大大缩短，达到了毫秒级。

4　结论

针对传统基于蒙特卡洛法的毁伤效果侦察评估弹对子母弹反机场跑道封锁概率模型计算速度慢，无法实现战场实时评估问题，提出使用神经网络对传统蒙特卡洛模型进行学习和拟合，建立了基于神经网络的子母弹反机场跑道封锁概率计算模型，并对2枚母弹条件下的封锁概率进行了仿真计算，仿真结果表明：该方法能较好地拟合基于蒙特卡洛方法的子母弹反机场跑道封锁概率计算模型，学习训练好的神经网络模型能够比较准确地给出反机场跑道封锁概率结果，计算精度可满足工程应用要求，同时评估时间相比于传统蒙特卡洛法得到了大大缩短，从秒级缩短至毫秒级，实现战场毁伤效能实时计算要求，具备实用性和通用性。本方法可为各类毁伤效果侦察弹对机场跑道封锁效能实时评估计算提供可行方法。

参考文献

[1] 王凤泰，唐雪梅．用像素模拟仿真法计算子母弹头的毁伤效率［J］．现代防御技术，2000，28（5）：29－35.

[2] 杨云斌，李小笠．机场跑道目标易损性分析方法研究［J］．弹箭与制导学报，2010，30（2）：141－144.

[3] 舒健生，陈永胜．对现有跑道失效率模拟模型的改进［J］．火力与指挥控制，2004，29（2）：99－102.

[4] 寇保华，杨涛，张晓今，等．末修子母弹对机场跑道封锁概率的计算［J］．弹道学报，2005，17（4）：22－26，49.

[5] 黄寒砚，王正明，袁震宇，等．跑道失效率的计算模型与计算精度分析［J］．系统仿真学报，2007，19（12）：2661－2664.

[6] DING S，SU C，YU J. An optimizing BP neural network algorithm based on genetic algorithm［J］. Artificial intelligence review，2011，36（2）：153－162.

[7] 贾跃，赵学涛，林贤杰，等．基于BP神经网络的鱼雷作战效能模糊综合评估模型及其仿真［J］．兵工学报，2009，30（9）：1232－1235.

[8] 刘吉军，李刚，王宗亮．基于神经网络模型的潜射反舰导弹作战效能分析［J］．四川兵工学报，2013（1）：47－49.

[9] 高彬，郭庆丰．BP神经网络在电子战效能评估中的应用［J］．电光与控制，2007，14（1）：69－71.

[10] 刘国强，陈维义，程晗，等．基于BP神经网络的炮光集成武器系统作战效能评估与预测［J］．海军工程大学学报，2019（3）：55－59.

Fragment characteristics of cylinder with discontinuous charge①

非连续装药圆柱壳体破片特征

Gao Yueguang②, Feng Shunshan, Xiao Xiang, Zhang Bo, Huang Qi

State Key Laboratory of Explosion Science and Technology,

Beijing Institute of Technology, Beijing 100081, PR China

Abstract: In real applications of warheads, not all charges are continuous. For specific purposes, such as power control and intermittent initiation, non - explosive materials are filled into the casing, which caused the charge to be discontinuous. In addition, detonators and control devices are often used with non - explosive materials. The effect of axial discontinuity of the charge on the fragment was not considered. In this study, we assumed that the non - explosive material was air, and divided the cylinder axially into charge and air parts axially. Existing formulae can neither accurately forecast fragment velocity from the charge part, nor predict that from the air part because of the non - existence of explosives. A cylinder with a discontinuous charge was investigated using a numerical method, and two charge parts were detonated from the air part because the initiator was usually in the non - explosive material to save space. It was found that the explosive products from the two charges were superimposed in the middle of the air part, forming a superimposed high - pressure zone, which accelerated the air part and the charge part casings near the air part. The influence distance of the superimposed high - pressure zone on the charge parts is 2R, where R is the charge radius. A formula was established to predict the fragment velocity distribution. The simulation results were verified by X - ray radiography, where the circumferential half casing of each specimen was semi - preformed and the other half was natural. It was found that the project angle and axial density of the fragments were related to the axial - driven energy gradient but not to the shear stress exceeding the dynamic shear strength. The air part of the semi - preformed shell overlapped and formed a dense zone, whereas that of the natural shell broke at three abrupt axial - driven energy gradients and formed two fragments with greater mass and kinetic energy. The experimental data showed that the average relative error of the proposed formula was less than 4. 5% , indicating that the proposed formula can predict the fragment velocity distribution from such a cylinder with sufficient accuracy. This study provides a basis and reference for the study of cylinders with discontinuous charges.

1 Introduction

The rupture and fracture of the exploded cylinder were analyzed to assess the expected effect of the warhead and the resistance of the structure. The cylinder expands and breaks into various fragments at high velocities when an explosive is initiated [1, 2]. Considering the terminal effect of fragments, their characteristics have been widely studied through experimental, numerical, and theoretical method, and the results provide a basis for further research in several fields such as public security, anti - terrorism and weapon design.

Various theories have been proposed to characterize the fragmentation of cylinders. In addition, a large

① 原文发表于《International Journal of Impact Engineering》173 (2023) 104479。

② 高月光：工学博士，2017 年师从冯顺山教授，研究半侵彻弹药战斗部对甲板目标威力效应，现工作单位：在校生。

number of formulae have been established to predict fragment velocity, among which the Gurney formula is widely used for its simplicity and high accuracy in most circumstances. Based on a few simple implicit assumptions and the conservation of energy, Gurney [3] developed a universal formula for calculating the initial fragment velocity, which was related to the energy of a particular explosive and the ratio of the explosive mass to the casing mass.

$$v_0 = \sqrt{2E}\sqrt{\frac{1}{1/\beta + 0.5}} \tag{1}$$

where v_0 is the initial fragment velocity, $2E$ is the Gurney characteristic velocity for a given explosive, and β is the mass ratio of the explosive to metal casing. Although the Gurney velocities slightly exceeded the final velocities with different mass ratios and explosives in the study, the relative errors were all within 10 percent [4], which meant that the Gurney formula had high accuracy.

Because Gurney assumed that the cylinder was infinitely long and ignored the rarefaction waves at the two ends, the Gurney formula only calculated the maximum velocity of the fragments. The calculation results were consistent with most of the fragment velocities. However, the error is large near the end due to the rarefaction waves from the two ends. Several studies have been investigating the dynamic response of a cylinder initiated at one end [5 - 12]. Kong et al. [13] investigated the dynamic process of cylinders where the casing had or did not have end caps, which also showed that the fragment velocity decreased at the two ends due to the rarefaction waves.

In addition, the rarefaction waves from the two ends were analyzed in detail, and aimproved Gurney formulae were established. Charron [14] used the geometric equivalence method and removed a part of the explosive from the two ends, which was assumed to be not involved in the reaction. Then the local mass ratio with the removed explosive was incorporated into the Gurney formula, and the calculation results demonstrated good agreement with the experiment, particularly at two ends. After conducting a large number of experiments, Zulkouski [15] proposed a correction function in the form of an exponential function to correct the Gurney formula, which can describe the variation in fragment velocities along the axial position caused by rarefaction waves. However, the fatal flaw of the formula was that the fragment velocity at the detonation end was obviously incorrect since the calculation result was 0. Hence, Huang et al. [16] performed an X - ray experiment and proposed an improved formula, that can solve the problem where the fragment velocity at the detonation end is zero. In addition, the calculation results were highly precise.

$$v_{0x} = (1 - 0.361e^{-2.3617x/d})(1 - 0.192e^{-3.03(L-x)/d})\sqrt{2E} \cdot \sqrt{\beta/(1 + 0.5\beta)} \tag{2}$$

Based on several experimental data, a formula [17] was established, which can calculate the fragment velocity using the expanding radius of the casing. The results indicated that the fragment velocity was stable after the casing expanded to its maximum critical radius.

In the calculation model mentioned above, the rarefaction waves were only generated from two ends because the explosive charge was axially continuous, and the mass ratio at any axial position was not zero. However, in some real applications of exploded cylinders, the explosive charge must be axially discontinuous when the non - explosive material or structure is filled into the casing, which caused the acceleration process of the metal casing to bea different from that of a cylinder with an axially continuous explosive. In the present research background, the non - explosive material was considered as air, and the cylinder can be referred to as a cylinder with a discontinuous charge. The cylinder is divided axially into air and charge parts. The casing of the air part would not stand still; on the contrary, it broke into fragments with different velocities and masses because of the acceleration effect from the charge part on both sides, and eventually broke into a large number of fragments. The existing formulae can neither accurately predict fragment velocities from charge parts, nor predict those from the air part since the casing of the air part had no explosive response. Based on the Gurney formula, the velocity of the air part casing would be 0 because the mass ratio (β) in the air part was 0, which is opposite to the fact. In our previous work [18], a cylinder with air parts at the two ends was analyzed. However, the position of the air component can significantly affect the dynamic process of the casing, which makes the previous model inappropriate for the present condition. Hence, a model of a cylinder with discontinuous charge was established, and two charge parts were detonated instantaneously from the air part since the initiator was usually in the non - explosive material to save space. Based on a theoretical analysis, we used numerical simulations to explore and analyze the fragmentation of the casing. A new formula was established and demonstrated using an X - ray experiment. In addition, the circumferential half casing of each specimen was semi - preformed and the other half was natural, which was used to explore the distribution of fragments. Discontinuous charge was studied for the first time in this study. These results can provide a basis for the design of warheads.

2 Theoretical analysis

2.1 Simplifications and assumptions

Figure 1 shows the cross - section of a cylinder with a discontinuous charge. Because the two charge parts are initiated instantaneously from the air part, the two charge parts can be considered to be centrally initiated simultaneously. The explosive charge was separated into two parts, denoted as charge parts A and B, and the air part was named as air part C. L, L_A, L_B, and L_c are the lengths of the casing, charge part A, charge part B and air part C, respectively. The corresponding casings of the charge part A, charge part B and air part C were casings A, B, and C. d and D are the inner and outer diameters of the casing, respectively. The fragment velocities from air part C were affected by the effect of rarefaction waves and the acceleration effect from the two charge parts, which were all related to the diameter of explosive and axial position. To achieve dimensionless parameters, the length of the air part C must be correlated with the diameter of the explosive. Therefore, a new parameter a was defined, which is the ratio of the air part length to the explosive diameter ($a = L_C/d$).

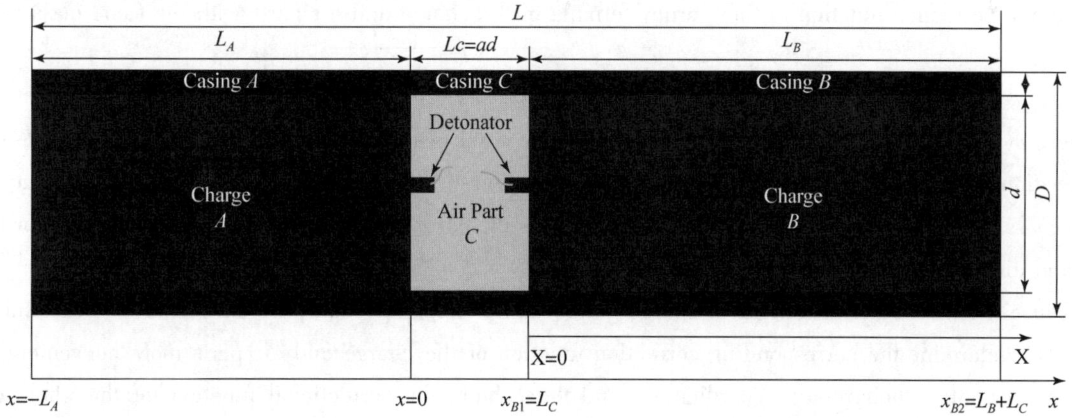

Figure 1 Cross - section of cylinder with discontinuous charge.

2.2 Limitations

In the practical warhead of a cylinder filled with explosives, considering that the main effect of the air part involves placing the non - explosive material or structure, the miniaturization of non - explosive material or structure implies a small length - diameter ratio. Meanwhile, to ensure the maximum terminal effect of fragments, and form sufficiently large number of effective fragments, the length of the air part needs to be controlled within a reasonable scope. Therefore, the length of the air part discussed in this paper did not exceed twice of the explosive diameter ($L_C \leqslant 2d$). The influence distance of rarefaction waves is 2R , and the influence distance from the non - detonation end is R, where R is the radius of the explosive [14]. In addition, the length of the charged parts needs to be long enough to guarantee the terminal effect of fragments, and the real warhead is usually a slender cylinder, implying that the length - diameter ratio is quite large. Therefore, the lengths of the charge parts *A* and *B* were assumed to be no less than two times the charge diameter ($L_A \geqslant 2d$, $L_B \geqslant 2d$). Hence, limitations of this study are as follows:

1) $L_C \leqslant 2d$;

2) $L_A \geqslant 2d$, $L_B \geqslant 2d$.

2.3 Establish the new formula

In addition to the effect of rarefaction waves from the end, the explosive charge near the air part was affected by the acceleration effect of the explosive products from another charge part. Hence, the acceleration history of fragments from the charge part near the air part would be obviously different, and it would also change under different lengths of air part *C*. Owing to the axial diffusion of the explosive energy on both sides, the casing of the air part expands and ruptures, forming fragments with a certain velocity and mass. However, the fragment velocities from the air part cannot be obtained using the Gurney formula and any improved form because the casing does not have a corresponding explosive. The acceleration energy of air part casing came from the charge parts *A* and *B*. Theoretically, the fragment velocity from air part *C* can be calculated by adding the corresponding correctional function to the Gurney formula of the charge part. In addition, the fragment velocity from air part *C* was not only related to the axial position, but also to the length of air part *C* because of the influence of the uneven axial energy distribution and the effect of rarefaction waves.

In conclusion, regardless of the fragments from the charge parts or air part, the fragment velocities were all affected by the axial position and length of the air part. The acceleration energy of all fragments originated from the explosive with the same mass ratio. To pursue simplicity and unity of the calculation form, we can add the corresponding correction function to the Gurney formula of the charge and air parts with the same mass ratio:

$$v(x,a) = \sqrt{2E}\sqrt{\frac{1}{1/\beta_0 \cdot f(x,a) + 0.5}} \tag{3}$$

where $v(x,a)$ denotes the fragment velocity. x is the axial position in absolute coordinates, and its origin is the detonation end of charge part *A* (Figure 1). The direction of the axis is from charge part *A* to charge part *B*, β_0 is the mass ratio of charge parts *A* and *B*; and $f(x,a)$ is the correctional function, which is suitable for both the charge and air parts.

Because the detonation conditions of charge parts *A* and *B* were the same, we only needed to determined one of them. To determine the corresponding correction function of the charge and air parts more conveniently, it is necessary to establish their relative coordinates, and then obtain the correctional function for the whole cylinder under the absolute coordinates by coordinate transformation. The correctional function of air part *C* can be solved in absolute coordinates, and the correction function of the charge part can be explored in relative coordinates, which is denoted as X, as shown in Figure 1. The charge part would consider charge part B as the research object. The relative correctional function is $f(X,a)$, and the relative fragment velocity from charge part B can be expressed as:

$$v(X,a) = \sqrt{2E}\sqrt{\frac{1}{1/\beta_0 \cdot f(X,a) + 0.5}} \tag{4}$$

As discussed above, after the fragment velocity from charge part *B* in the relative coordinates is obtained, the fragment velocity from charge part *A* can also be obtained. Finally, the fragment velocity from charge parts *A* and

B in absolute coordinates can be acquired by coordinate transformation:

$$x = \begin{cases} -X & -L_A \leqslant x < 0 \\ x & 0 \leqslant x \leqslant L_C \\ X + L_C & L_C < x \leqslant L_C + L_B \end{cases} \tag{5}$$

3 Numerical simulation

3.1 Method

Smoothed Particle Hydrodynamics (SPH) is a attractive method for solving the dynamic response of a material under high dynamic loading conditions, which is difficult to handled using Lagrangian finite elements because severe mesh distortions occur when the material has a large deformation. Based on the interpolation theory, the SPH method transforms continuous dynamics into an integral equation using an interpolation function. For the kernel estimation of a function $f(x)$ within the radius of the kernel function W defined by parameter h, the kernel estimate $\langle f(x) \rangle$ of function $f(x)$ at a certain position can be acquired by integrating the adjacent particles:

$$\langle f(x) \rangle = \int_{\Omega} f(x') W(x - x', h) dx' \tag{6}$$

The kernel function W satisfies the following two conditions: (i) it is an even function, $W(x - x', h) = W(x' - x, h)$, and (ii) it is a wrap function, $\int_{\Omega} W(x - x', h) dx = 1$. h is the width of the kernel function and $\langle f(x) \rangle$ is equal to $f(x)$ when h approaches zero.

Belytschko et al. [33] previously demonstrated that Eulerian kernels exhibit spurious stretching instability. The tensile instability caused by the Eulerian kernel, can lead to an artificial fracture. Lagrangian kernel functions can remove instabilities but are only applicable to moderate deformations [34]. Rabczuk and Belytschko [35] explored a method for transitioning from Lagrangian to Eulerian kernels when the nodes were cracked. The SPH processor in AUTODYN – 3D is a Lagrangian meshless method with a Lagrangian kernel. In addition, a pre – processor for automatically packing SPH particles into arbitrary; explosive burn logic; a smoothing length that can vary in space and time depending on local density changes [36]. In the past few years, several researchers have carried out simulations of dynamic impacts using the SPH method, which were consistent with the experimental results [9, 13, 26, 28 – 32].

3.2 Material model and parameters

As the most commonly used materials for studying the fragmentation of materials, AISI 1045 steel and COMP – B were for the metal casing and explosive charge, respectively. AISI 1045 steel expands and fractures after the explosive detonates, which can be described by the Johnson – Cook (JC) model:

$$\sigma = (A + B\varepsilon_{ep}^{n})(1 + C\ln \dot{\varepsilon}^{*})(1 - T^{*m}) \tag{7}$$

where A, B, C, m and n are constants; ε_{ep} represents the equivalent plastic strain, $\dot{\varepsilon}^{*}$ is the relative plastic strain rate; and T^{*} is the homologous temperature, which can be obtained as $T^{*} = (T - T_r)/(T_m - T_r)$. Table 1 lists the of metal casing parameters.

Table 1 Parameters of AISI 1045 steel [37].

$\rho_0/g \cdot cm^{-3}$	A(MPa)	B(MPa)	n	c	m	T_m(K)	γ
7.85	507	320	0.28	0.064	1.06	1 793	53.8

The JWL model is a widely used model to simulate the detonation and explosion of explosive charges, which can be adopted to describe the dynamic response of COMP – B, and is described as:

$$P_b = C_1\left(1-\frac{\omega}{r_1\nu}\right)e^{-r_1\nu}+C_2\left(1-\frac{\omega}{r_2\nu}\right)e^{-r_2\nu}+\frac{\omega E}{\nu} \tag{8}$$

where P_b is the blast pressure of the explosive charge; E is the initial energy per initial volume; ω, C_1, C_2, r_1 and r_2 are constants; and V is the initial relative volume. The properties and parameters of the explosive charges are listed in Table 2.

Table 2 Parameters of COMP – B.

ρ_0/g · cm^{-3}	D/m · s^{-1}	P_{CJ}/GPa	E_0/KJ · m^3	C_1/GPa	C_2/GPa	R_1	R_2	ω
1.717	7 980	29	8.5×10^6	542	7.68	4.2	1.1	0.24

Because the specimen was symmetrical, only a quarter of it required to be modelled. It has been proven that the accuracy of the SPH method depends on the particle size. The accuracy of the simulation tended to be higher for smaller particles, which also indicated a time – consuming process. To determine the optimal dilation parameter h, the optimal parameters were determined by comparing the simulation results with the test results for different particle diameters. The experiments referred to previous experimental results of our laboratory [16], and the specific parameters can be found in the original literature [16]. The materials and models of the explosive and shell were the same as those described above. The comparison results are shown in Figure 2, and it can be observed that the fragment velocity decreased with increasing particle diameter. When the particle diameter is 0.3 mm and 0.4 mm, there is basically no significant difference in the fragment velocity, and it is in good agreement with the test results, which also coincides with the previous simulation results [31]. In order to save computer resources, a particle diameter was selected as 0.4 mm, which can reliably predict the fragment velocity distribution from the exploded casing.

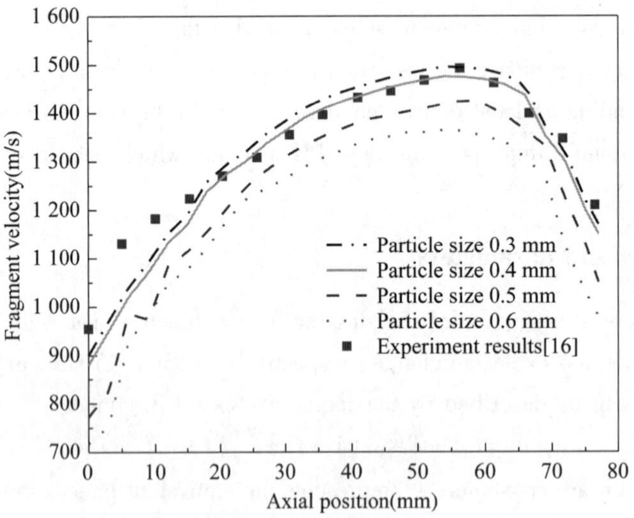

Figure 2 Influence of the dilation parameter.

The 1/4 numerical model is shown in Figure 3, where the gauge points were located on the casing surface to record the fragment velocity (points 1 – 25), and the distance of each point was 3 mm. Ten numerical specimens were established, as listed in Table 3.

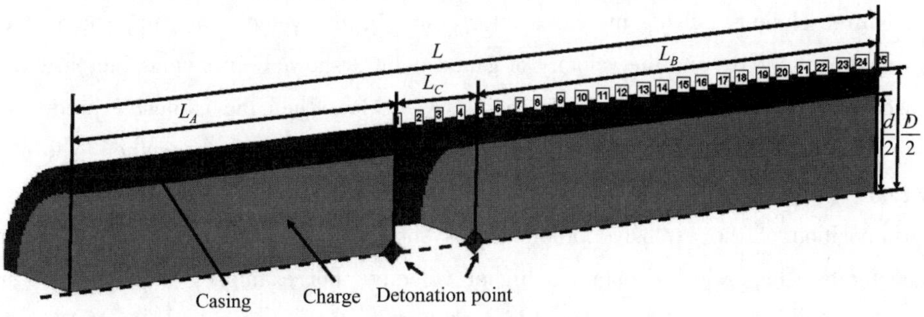

Figure 3　Numerical model of cylinder with discontinuous charge under two central point initiations.

Table 3　Parameters of ten cylinder with discontinuous charge.

No.	L_A (mm)	L_B (mm)	L_C (mm)	L (mm)	D (mm)	d (mm)	a	Number of particles
1	48	60	6	114	30	24	0.25	302, 805
2	48	60	9	117	30	24	0.375	305, 750
3	48	60	12	120	30	24	0.5	308, 696
4	48	60	15	123	30	24	0.625	311, 641
5	48	60	18	126	30	24	0.75	314, 586
6	48	60	21	129	30	24	0.875	317, 531
7	48	60	30	138	30	24	1.25	326, 336
8	48	60	36	144	30	24	1.5	332, 257
9	48	60	42	150	30	24	1.75	338, 147
10	48	60	48	156	30	24	2	344 038

3.3　Results and discussion

The line numbers represent the gauge points (Figure 4 – Figure 8), and the distribution and order of the gauge points are similar to those in Figure 3. Because the length of the air part had a huge influence on the acceleration propagation, we analyzed the acceleration history of fragments generated from the charge and air parts with different lengths of the air part. After analyzing the simulation results in detail, the influence of air parts with different lengths can be divided into two obvious groups, and Cases No. 3 and No. 6 can represent those two groups. Figure 4, Figure 5 and Figure 6 belong to Case No. 3, and Figure 7 and Figure 8 belong to Case No. 6,

Because two detonation points were initiated simultaneously, the fragment velocities of air part *C* exhibited symmetry, and only half of them were shown in Figure 5 and Figure 7 for Cases No. 3 and No. 6, respectively. After two central detonation points are simultaneously initiated, detonation waves and explosive products were generated from the detonation end of charge parts *A* and *B*. While the detonation waves continued to detonate the explosive, the detonation products began to diffuse into the air and attenuate due to the influence of the rarefaction waves (Figure 4a). Figure 5 shows that the acceleration process of casing *C* and the velocity of the gauge points increased sharply to points A, B and C due to detonation waves. Owing to the influence of rarefaction waves, the first peak velocity decreases and the arrival duration increases as the distance from the explosive increases. Subsequently, the velocity increased in a pulsating manner since the detonation waves and rarefaction waves would alternately dominate. As time increased, the detonation waves became weaker due to the rarefaction waves, and the farther away from the explosive charge, this phenomenon would occur sooner. For example, the velocity of gauge point 1 had three obvious pulsing increases after point A, the velocity of gauge point 2 had two obvious pulsating increases after point B, and the velocity of gauge point 3 showed an obvious increase after point C. The same rule was used for Case No. 6 in Figure 7 (points A – D). When the explosive products generated from charge parts *A* and *B* met in the middle of air part *C*, they were superimposed together to form a superimposed high – pressure zone (Figure 4b). The width of the superimposed high – pressure zone also increased with continuous superimposition of the explosive products, as shown in Figure 4b – d. The superimposed high – pressure zone accelerated the casing to obtain a higher velocity. For example, gauge point 3 accelerated to a higher velocity after point c where a superimposed high – pressure zone occurred (Figure 5). As the width of the superimposed high – pressure zone increased, the velocities of gauge points 2 and 1 accelerated in succession to points b and c (Figure 5), and the velocities of gauge point 1 – 3 almost reached the same velocity.

After the superimposed high – pressure zone arrives at charge parts *A* and *B*, it also accelerates the casing, as can be seen from points d – i in Figure 6 (Figure 4d – f). However, the pressure in the superimposed high – pressure zone gradually decreased as the width increased, and the accelerating effect gradually decreased. The acceleration from points d – i to the maximum velocity decreased with increasing distance from the superimposed high – pressure zone. The superimposed high – pressure zone weakened the effect of the rarefaction waves, which accelerated the fragment velocity. Therefore, the influence distance of the superimposed high – pressure zone should be the same as that of the rarefaction waves [14]; that is, the influence distance of the superimposed high – pressure zone was 2R for the detonation end, where R is the radius of charge.

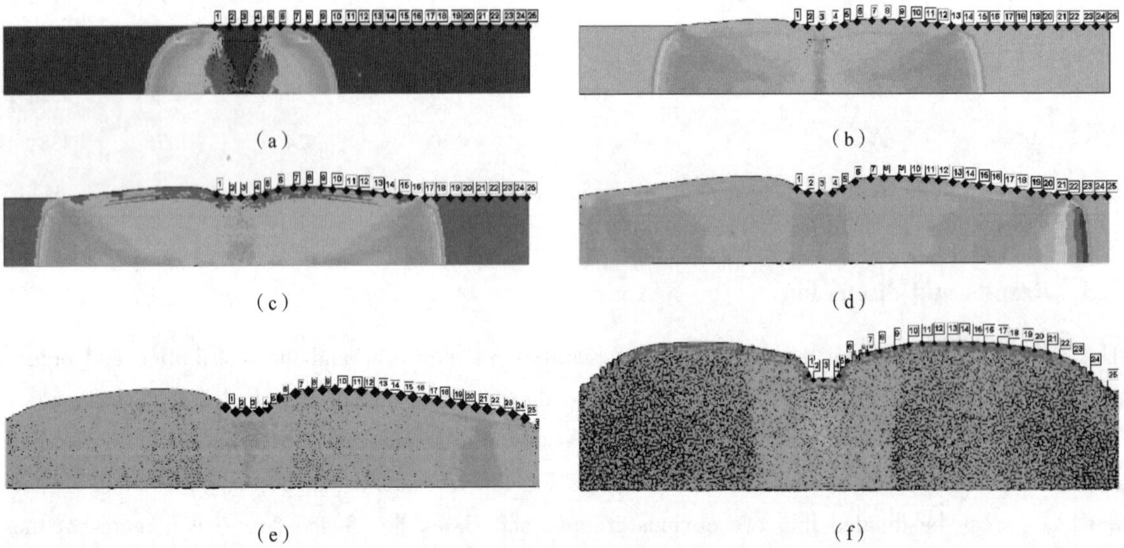

Figure 4 Acceleration process of cylinder (Case No. 3).

(a) 0.002 ms; (b) 0.004 ms; (c) 0.005 ms; (d) 0.007 ms; (e) 0.01 ms; (f) 0.02 ms

In addition, the superimposed high – pressure zone was affected by the rarefaction waves. The pressure and width of the superimposed high – pressure zone decreased with an increase in the length of the air part *C*. The acceleration curves of gauge points 1 – 3 after point a/b/c had almost the same slope in Case No. 3 (Figure 5). However, the slope of the acceleration curve of gauge points 1 – 4 increased as the distance to the detonation end decreased (Figure 7), indicating that the rarefaction waves weakened the accelerating effect of the superimposed high – pressure zone; and the further away from the middle of air part *C*, the greater the weakening effect. When the length of the air part *C* reached a certain value ($L_C > 0.75d$, as discussed in the next section), the superimposed high – pressure zone would be weakened by the rarefaction waves before reaching the charge parts *A* and *B*, and it would hardly accelerate the casing of the charge parts (Figure 8). Specifically, the superimposed high – pressure zone would only affect the fragment velocities from the air part *C*.

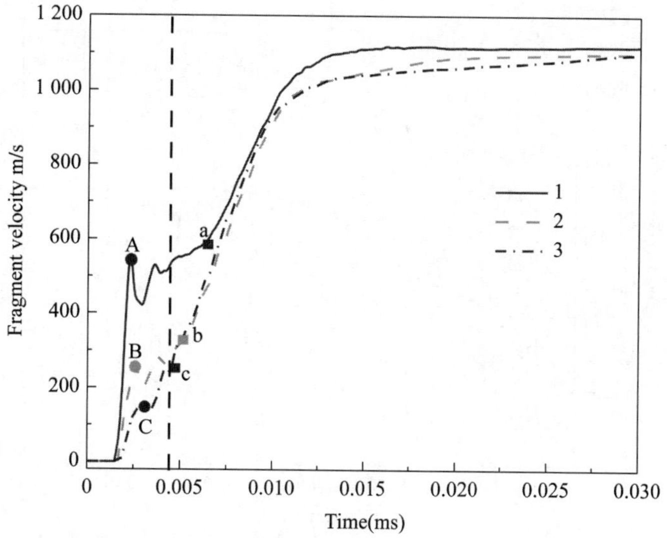

Figure 5 Acceleration process of gauge points on casing *C* (air part, Case No. 3).

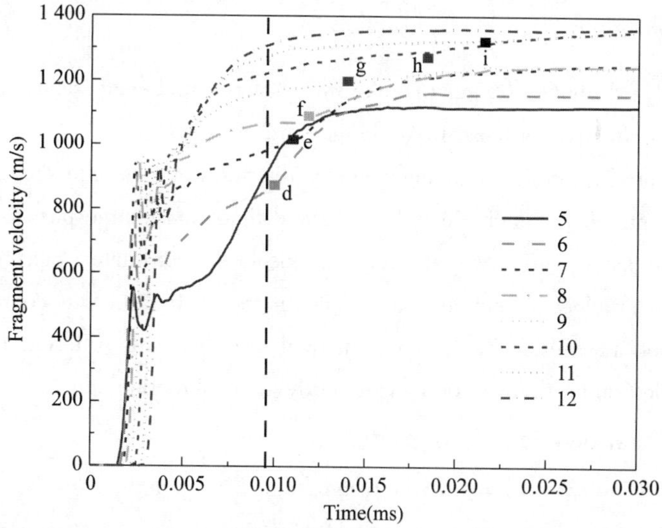

Figure 6 Acceleration process of gauge points on casing *B* (charge part, No. 3).

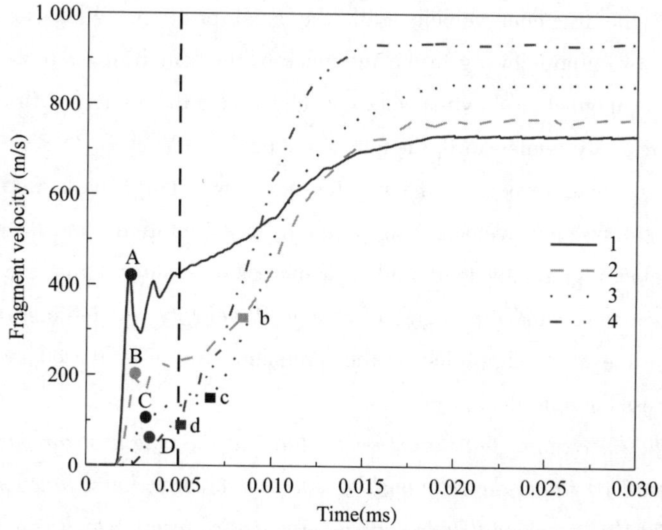

Figure 7 Acceleration process of gauge points on casing *C* (air part, Case No. 6).

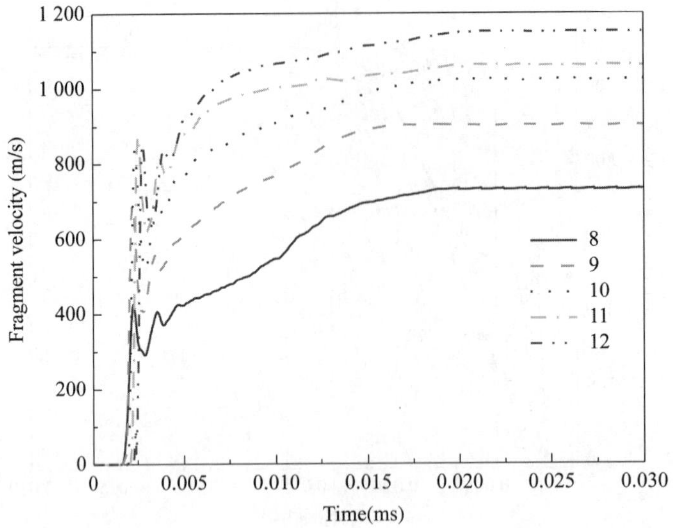

Figure 8 Acceleration process of gauge points on casing *B* (charge part, No. 6).

3.4 Estimating formula

As discussed and analyzed above, casing *C* broke into a large number of fragments with different velocities, which can be considered as effective as those from charge parts *A* and *B*.

As discussed in Section 2, for the fragment velocity distribution from the charge parts, we only need to determine that from charge part *B* in relative coordinates, and that from charge part *A* can also be obtained since the initial condition is the same, and two of them in absolute coordinates can be obtained by coordinate transformation. The fragment velocity from air part *C* is discussed in absolute coordinates without coordinate transformation. As discussed in Section 3.1, ten numerical samples with different lengths of air part *C* were established, which could determine the accuracy of the correctional function.

3.4.1 The correctional function of air part *C*

In Figure 9, $a = 0.25S$ indicates $L_C = 0.25d = 6$ mm, and S a numerical result, and F is the fitted curve of the simulation results. As discussed above, the fragment velocity from air part *C* was not only related to the length of air part *C* but also to the axial position. The axial position is standardized in Figure 9, which can better demonstrate the change in the fragment velocity with the axial position. When analyzing the effect of axial position, we divided it into two groups owing to the influence of the length of air part *C*.

When $0 < a \leqslant 0.75$, the fragment velocities were constant under the same length of air part *C*, which meant that the fragment velocity was only related to the length of air part *C*. When $0.75 < a \leqslant 2$, the fragment velocities were symmetrically distributed with respect to the middle since two charge parts were initiated simultaneously. Consequently, only half of the fragment velocity was required for exploration, and the other half was obtained by symmetry. The fragment velocity gradually increased with increasing distance from the detonation end due to the superimposed high-pressure zone, and the degree of change in velocity was more severe with an increase in the length of air part *C* because the weakening effect of the rarefaction waves increased and the accelerating effect of the superimposed high-pressure zone decreased.

When $0 < a \leqslant 0.75$, *the correctional function is only related to the length of air part* C, *which is denoted as* $g(a)$, $f(x,a) = g(a)$. *Figure* 10 *shows the relationship between the correctional function and the length of the air part* C. *It can be found that the correctional function was exponential form, which can be expressed as*:

$$g(a) = G_1 + G_2 e^{-a/G_3} \tag{9}$$

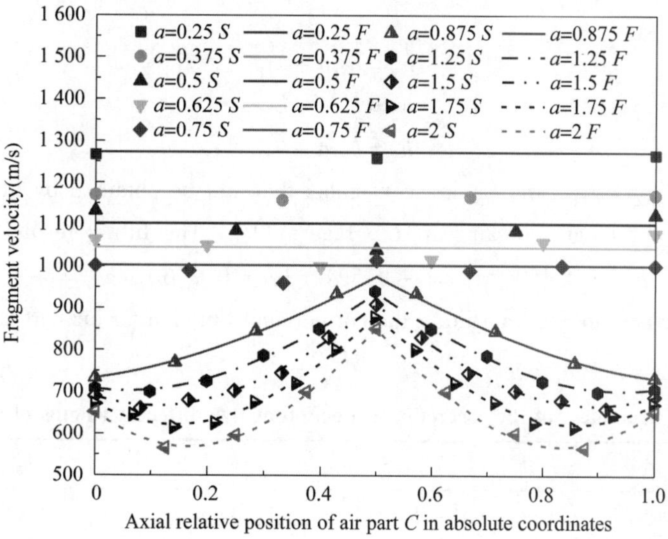

Figure 9　Fragment velocity from air part *C* in absolute coordinates.

where G_1, G_2 *and* G_3 *are constants. After assessing the fitting accuracy of the simulation results in Figure* 13, *we can obtain appropriate values, which give* $G_1 = 0.3$, $G_2 = 0.7$ *and* $G_3 = 0.35$ ($R^2 = 0.998$). *It can be found that the data points were in good agreement with the fitted curve.*

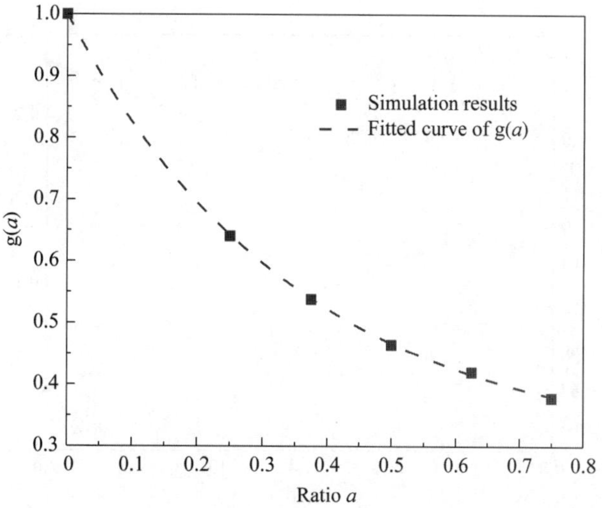

Figure 10　Relationship between correctional function and the ratio *a* ($0 < a \leqslant 0.75$).

When $0.75 < a \leqslant 2$, the correctional function was not only associated with the length of air part C, but also with the axial position. However, it presents quadratic relations of different curvatures under different length of air part C, which is denoted as $h(x,a)$ ($f(x,a) = h(x,a)$).

$$h(x,a) = \begin{cases} h_1\left(\dfrac{x}{L_C}\right)^2 + h_2\left(\dfrac{x}{L_C}\right) + h_3, & 0 \leqslant x < \dfrac{L_C}{2} \\ h_1\left(\dfrac{L_C - x}{L_C}\right)^2 + h_2\left(\dfrac{L_C - x}{L_C}\right) + h_3, & \dfrac{L_C}{2} \leqslant x < L_C \end{cases} \tag{10}$$

where h_1, h_2 and h_3 are coefficients. The two parts of the correctional function were symmetrical. The coefficients for different lengths of air part C can be obtained by fitting the numerical results; the fitting results are listed in Table 4. The relationship between the three coefficients and the ratio a is shown in Figure 11. Obviously, the coefficients h_1, h_2 and h_3 are not constants, and their relationships with the ratio a are all linear forms, which can be expressed as:

$$h_1 = h_{11}a + h_{12} \tag{11}$$

$$h_2 = h_{21}a + h_{22} \tag{12}$$

$$h_3 = h_{31}a + h_{32} \tag{13}$$

where h_{11}, h_{12}, h_{21}, h_{22}, h_{31}, and h_{32} are constants, that can be obtained by fitting the coefficients of the five cylinders under different lengths of air part C (Figure 11). The fitting results give $h_{11} = 0.807$, $h_{12} = -0.251$, ($R^2 = 0.992$), $h_{21} = 0.489$, $h_{22} = -0.524$ ($R^2 = 0.996$), $h_{31} = -0.038$, $h_{32} = 0.23$ ($R^2 = 0.995$). After substituting Eq. to Eq. into Eq. , the correctional function for air part C can be obtained ($0.75 \leqslant a < 2$).

Table 4 Parameters of the correctional function of five different lengths of air part C.

L_C(mm)	a	h_1	h_2	h_3	R^2
21	0.875	0.455	0.096	0.197	0.958
30	1.25	0.758	-0.087	0.183	0.986
36	1.5	0.96	-0.21	0.173	0.95
42	1.75	1.161	-0.332	0.164	0.965
48	2	1.363	-0.454	0.154	0.968

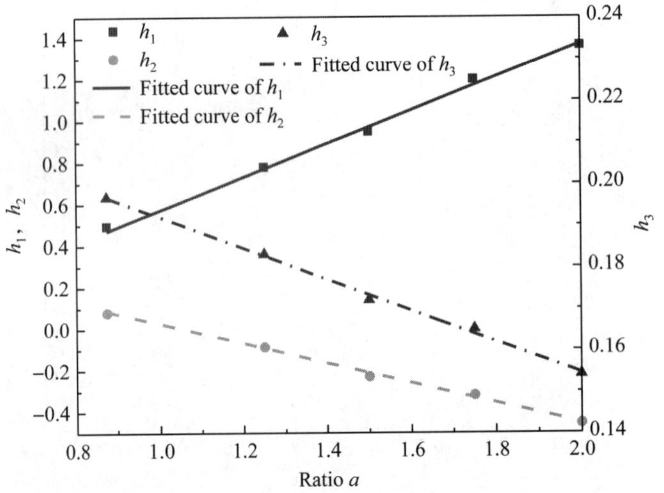

Figure 11 Relationship between the coefficients (h_1, h_2 and h_3) and the ratio a.

In conclusion, the correctional function $f(x, a)$ from air part C in absolute coordinates can be summarized as follows:

When $0 < a \leqslant 0.75$:

$$f(x,a) = 0.3 + 0.7e^{-a/0.35} \tag{14}$$

When $0.75 < a \leqslant 2$:

$$f(x,a) = \begin{cases} (0.807a - 0.251)\left(\dfrac{x}{L_C}\right)^2 + (0.524 - 0.489a)\left(\dfrac{x}{L_C}\right) + 0.23 - 0.038a, & 0 \leqslant x < \dfrac{L_C}{2} \\ (0.807a - 0.251)\left(\dfrac{L_C - x}{L_C}\right)^2 + (0.524 - 0.489a)\left(\dfrac{L_C - x}{L_C}\right) + 0.23 - 0.038a, & \dfrac{L_C}{2} \leqslant x < L_C \end{cases} \tag{15}$$

3.4.2 Correctional function of explosive charge B

Figure 12 shows the fragment velocities from explosive charge B in relative coordinates under different lengths of air part C, where $a=0$ indicates the cylindrical casing detonated at one end with no air part. $a=0F$ [18] represents the calculation results of our previous study [18]. S represents the simulation result. It showed that the fragment velocities from charge part B near the air part C were relatively higher due to the acceleration effect of the superimposed high - pressure zone. The acceleration effect gradually decreased with increase in the length of air part C, and almost disappeared when the length of air part C exceeded $0.75d$. Since the acceleration range of the superimposed high - pressure zone was only 2R, the fragment velocity from another part of casing B was almost the same as that from the cylinder detonated at one end without air part. Hence, most of the fragment velocities from charge part B could be obtained by the previous formula (Eq.), which was proposed by Huang, and can predict the fragment velocity from cylinder with high accuracy. However, the correctional function is not suitable in the present situation. In our previous study [18], we obtained a new formula by modifying Eq. (2), and the correctional function has the same format as Eq. .

$$f(X)=(1-0.689e^{-0.953X/d})(1-0.385e^{-2.3(L_B-X)/d}) \tag{16}$$

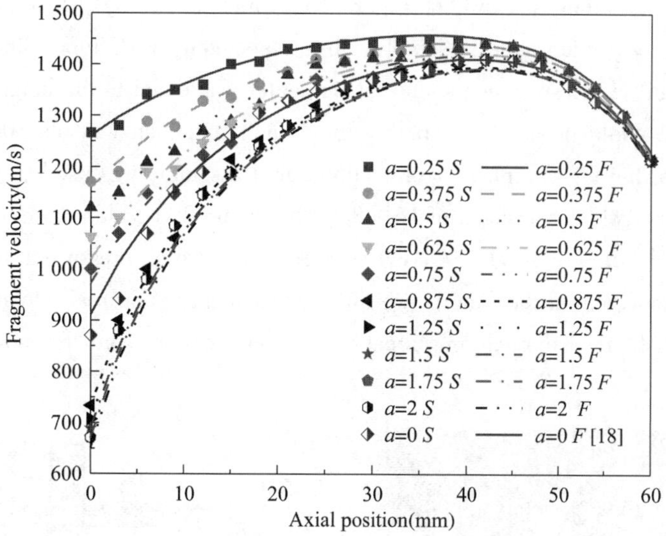

Figure 12 Fragment velocity distribution from charge part *B* in relative coordinate.

Based on the solution process in Eq. , the first bracket represents the influence of rarefaction waves at the detonation end, and the second bracket represents the influence of rarefaction waves at the non - detonation end. The air part did not affect the non - detonation end, therefore, the contents in the second bracket remained unchanged, and the correctional function for charge part B can be expressed as follows:

$$f(X,a)=(1-Ae^{-BX/d})(1-0.385e^{-2.3(L_B-X)/d}) \tag{17}$$

where A and B are the constants. A was obtained from the fragment velocity at the detonation end of the air part.

$$A=\begin{cases}1-g(a)=0.703(1-e^{-a/0.352}) & 0\leqslant a\leqslant 0.75\\ 1-h_3=0.038a+0.77 & 0.75<a\leqslant 2\end{cases} \tag{18}$$

B in Eq. was obtained by fitting the fragment velocity between the detonation end and the position where the maximum velocity appeared. For the sake of simplicity, B was assumed the same as that in Eq. , the correction function for charge part B was obtained as follows:

When $0<a\leqslant 0.75$:

$$f(X,a)=(1-0.703\cdot(1-e^{-a/0.352})\cdot e^{-0.953X/d})(1-0.385e^{-2.3(L_B-X)/d}) \tag{19}$$

when $0.75<a<2$:

$$f(X,a)=(1-(0.038a+0.77)\cdot e^{-0.953X/d})(1-0.385e^{-2.3(L_B-X)/d}) \tag{20}$$

After Substituting Eq. and Eq. into Eq. , the fragment velocity from charge part B can be obtained, as shown by the lines in Figure 12. It can be found that the calculation results were consistent with the simulation results even if B was assumed as constant. The fragment velocity from two charge parts in absolute coordinates can be obtained by the coordinate transformation of Eq.. Finally, the fragment velocity distribution from the cylinder with a discontinuous charge can be obtained.

4 X – ray experiments

4.1 Configuration and analysis method

Through an analysis of the simulation results, a modified formula for the velocity distribution of fragments from a cylinder with a discontinuous charge was obtained. X – ray radiography was performed to verify the accuracy of the proposed formula. Previous studies have shown that X – ray radiography can accurately measure fragment velocity distributions [5, 16, 38, 39].

The experimental configuration is shown in Figure 13, which is mainly primarily composed of an X – ray tube, protection structure, protection panel assembly, film, specimen, PVC tube, and thin plate. A PVC tube was used to adjust the height of the specimen such that its center was equal to the height of the X – ray tube, and a complete fragment profile could be obtained on the film. Each specimen was placed vertically on a PVC tube with a detonator parallel to the X – ray tube. A protection panel assembly was used to protect the film and record the distribution of fragments, which consisted of 16 (8 +4 +4) mm 2A12 aluminum alloy and 5 mm wood. Two X – ray tubes were arranged side – by – side, X – ray were released at high pressure at two different moments, and the fragment profile was displayed on the film at two different moments. The real moving distance of the fragment in a certain period can be obtained through amplification ratio conversion, and the moving speed of the fragment can then be obtained.

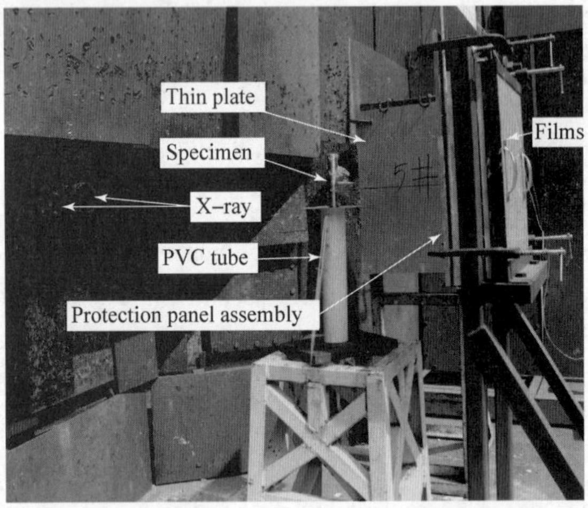

(a)

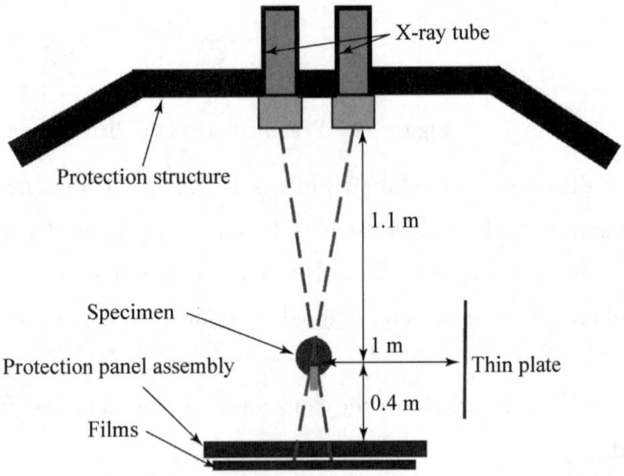

(b)

Figure 13 X – ray experimental configuration.

(a) Layout of test site; (b) Schematic diagram of test device

The schematic of test specimens is shown in Figure 14a, and the real specimens was in Figure 14 (b), which were obtained by rotating the casing by 90° in Figure 14 (a). Detonator and booster methods were adopted to achieve simultaneous initiation of both charge parts. A hole was punched in the center of one side of the air shell, which was perpendicular to the film. The detonator passed through the hole and was connected to the booster with its center aligned with the booster center, which ensured that the detonator initiated the booster and then simultaneously detonated two charge parts. Because the detonator and hole were perpendicular to the film, the fragment velocity parallel to the film was unaffected. The diameter of the detonator is 6. 9 mm. The booster was made of RDX 8701 explosive, with a diameter of 10 mm and length equal to that of the air part. Simultaneously, to obtain the velocity and distribution of fragments, and to prevent the velocity from being measured owing to the overlapping of fragments, half of the shell was cut at a depth of 2 mm, which was divided into nine equal parts in the circumferential direction, and the axial spacing was 5 mm. The other half was a natural shell for comparison. Three specimens were used, and the parameters are listed in Table 5.

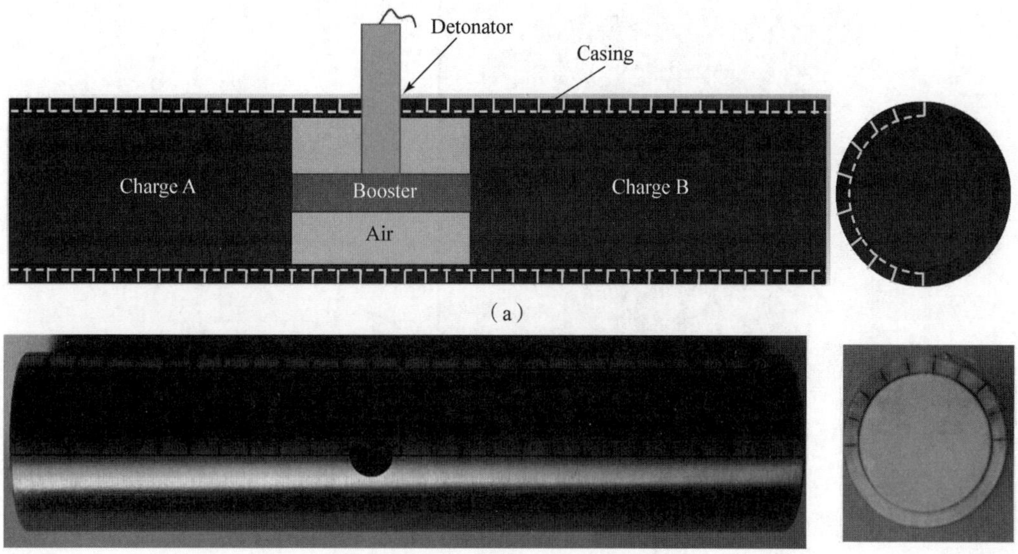

(a)

(b)

Figure 14 Schematic of test specimens.

(a) Schematic of the specimen's structure; (b) Real specimens (rotating 90°)

Table 5 Parameters of three test specimens.

No.	L_A(mm)	L_B(mm)	L_C(mm)	L(mm)	D(mm)	d(mm)	a	t_0	t_1
1#	48	60	12	120	30. 05	23. 95	0. 5	30. 7	50. 65
2#	48	60	24	132	29. 95	23. 85	1	20. 8	35. 6
3#	48	60	48	156	29. 97	23. 89	2	25. 7	40. 75

The fragment velocity processing is shown in Figure 15. The fragments obtained at these two moments were superimposed onto one film. The first moment is t_0, and the second moment is t_1. The two moments were determined through simulation to ensure that the fragment reached its maximum speed. The moments of the three specimens are listed in Table 5. The symmetry axes of the two moments were determined using fragment contours. As the fragment reached its maximum velocity at the millisecond level, the size of the PVC tube barely changed. The diameter of the PVC tube was 75 mm, and the diameter of the steel ball was 8mm, resulting in a large error. Therefore, the amplification ratio can be determined based on the PVC tube size. After obtaining the fragment contour of the two moments, the symmetry axis of the two moments were aligned with each other. The distance of the fragment from the corresponding position was measured using CAD software. The radiographic images were 2 – dimensional projection of the 3 – dimensional distribution of the fragments. Therefore, the distance travelled by a fragment between the two – timed images was a projection of the actual flight path and the calculated velocity was a component of the actual velocity. Only the measurements of the case fragments that were projected at right angles to the X – ray beam were the correct determination of their velocity.

If the fragment contour was simply equalized, the measurement error of the fragment velocity would increase because of the difference in the axial moving distance of the fragments at different axial positions. To measure the fragment velocity more accurately, the fragments of the air and charge parts were divided into equal sections with an interval of 6 mm, as shown in Figure 15 (b). The moving distance Δl_i of each point can be measured using the CAD software, and the actual moving distance can be obtained by multiplying the amplification ratio (k). The fragment velocity distribution was obtained using $v_i = k\Delta l_i/(t_1 - t_0)$. The fragment velocities on both sides were measured twice to reduce the measurement error.

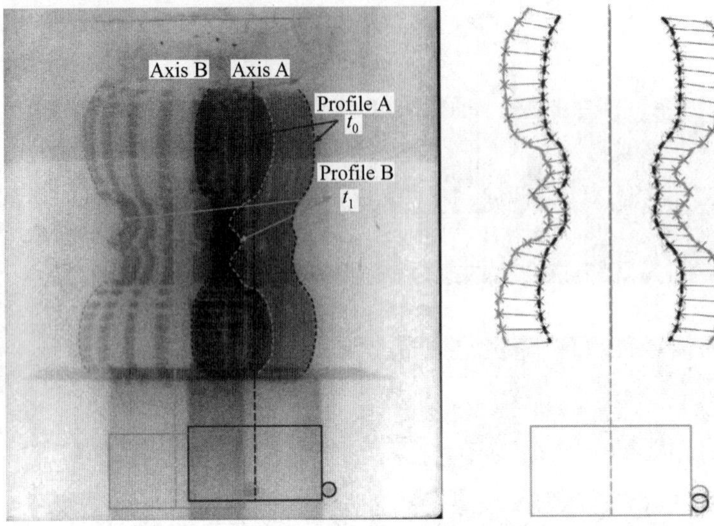

(a) (b)

Figure 15 Schematic of fragment speed processing.

(a) X-ray film; (b) Image alignment processing

The fragments of the air part can be accelerated by a booster, which should be subtracted. According to energy conservation, the kinetic energy of the air part casing per unit length was equal to the energy of the equivalent COMP-B and booster per unit length. The equivalent COMP-B represents the energy from the two charge parts. By converting the energy of the equivalent COMP-B and booster per unit length into the fragment kinetic energy per unit length, the fragment velocity of the air part without the influence of the booster (v_{0B}) can be obtained, and the formula is as follows:

$$E_m = E_B + E_R \tag{21}$$

$$E_m = \frac{1}{2}mv_{0e}^2 \tag{22}$$

$$E_B = \frac{1}{2}mv_{0B}^2 \tag{23}$$

$$E_R = \frac{1}{2}mv_{0R}^2 \tag{24}$$

$$v_{0R} = \sqrt{2E_R}\sqrt{\frac{1}{1/\beta_R + 0.5}} \tag{25}$$

$$v_{0B} = \sqrt{v_{0e}^2 - v_{0R}^2} \tag{26}$$

where E_m, E_B, and E_R are the kinetic energy of the casing, equivalent COMP-B, and RDX-8701 per unit length, respectively. m is the casing mass per unit length. v_{0e} was the experimental velocity of the fragments. v_{0R} was the maximum velocity of the casing accelerated by the booster per unit length. v_{0B} is the maximum velocity of the casing per unit length accelerated by the equivalent COMP-B. β_R is the mass ratio of the casing to RDX-8701. $\sqrt{2E_R}$ is the Gurney velocity of RDX-8701, which can be obtained using the following formula [40]:

$$\sqrt{2E_R} = 0.52 + 0.28D_e \tag{27}$$

where D_e is the detonation velocity of RDX - 8701 (8800 m/s). Hence, $\sqrt{2E_R}$ was equal to 2984 m/s.

4.2 Results and discussion

4.2.1 Comparison with numerical simulation

Figure 16 shows the comparison of the casing contours in the experiment and simulation, which shows that they had good consistency. The simulation accurately reproduced the deformation of the shell in the test.

In the simulation, the end of the casing easily formed small fragments, which can be clearly observed in 3# specimen. In addition, the acceleration effect of the superimposed high - pressure zone on the air part casing can be clearly observed; especially 2# and 3# specimens presented the same symmetric quadratic distribution as in the simulation.

The difference between the shell of the air part in 1# specimen and that in the simulation was that it broke into two obvious large fragments in the axial direction, which did not occur in the simulation. The shell of the charge part near the detonation end also broke into various smaller fragments, which was reflected in the simulation and test results.

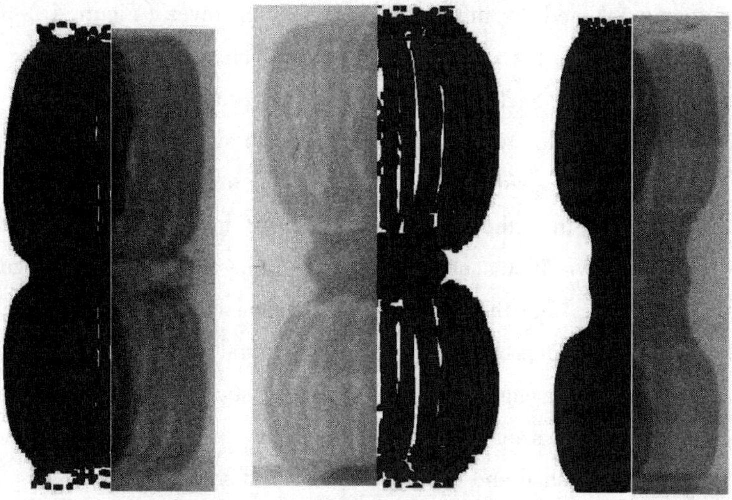

(a)　　(b)　　(c)

Figure 16　Comparison of shell contour in simulation and experiment.

(a) 1#; (b) 2#; (c) 3#

4.2.2 Distribution of fragments

The first layer of the protection panel assembly under the impact of fragments is shown in Figure 17, where the black dotted line represents the center height of the air part. It was assumed that only one fragment made a single crater and that a fragment did not break - up along its flight path to the plate. As half of the casing was semi - preformed, the natural casing and semi - preformed casing were symmetrical and parallel to the X - ray tube, the impact craters formed by the semi - preformed fragments and natural fragments were symmetrically distributed on the protection panel assembly. Natural fragments formed impact craters in the green box. The yellow box is a regular arrangement of impact craters formed by semi - preformed fragments. The distribution of impact craters caused by semi - preformed fragments represented the direction and distribution of fragments because the rupture of semi - preformed fragments after an explosion is controllable. Because the initial conditions of the two charge parts were the same, only one of them was analyzed. The distribution of the fragments was shown in Figure 18, where each region corresponds to that shown in Figure 17. Owing to the influence of the rarefaction waves from the non - detonation end, the fragment in Region I had a larger project angle, did not impact the protection panel and formed impact craters.

In terms of the distribution density of fragments from charge parts, the impact craters showed that the distribution of semi - preformed fragments can be divided into three regions, namely Region II, Region III and Region IIV. The number of preformed fragments and the width of each region in a single row are listed in Table 6. In Regions II to IV, the fragment density in Region III was the largest, followed by Region IV, and Region II was the smallest. This was similar to the distribution of fragments from the charge part in Figure 18, which also can be found in a previous study, where the axial fracture of natural fragments was owing to the fact that the radial shear force exceeded the dynamic shear strength [16]. However, the fracture of semi - preformed fragments has unrelated to the dynamic shear strength. Therefore, the explanation for the most serious rupture of the natural shell at the end of the previous article was not suitable for semi - preformed fragments. It was believed that the change in the fragment distribution density was not owing to the radial shear force but because of the change in the axial - driven energy gradient. The minimum fragment density in Region II was owing to the effect of rarefaction waves gradually increasing from the non - detonation end, which led to an axial decrease in the driven energy and an increase in the project angle and axial distance of the impact craters. Region IV was also affected by rarefaction waves, but the fragment density was higher than that in region II, because the acceleration of detonation products formed by another charge part weakened the influence of rarefaction waves (Figure 4e - f). In region III, there was no influence of rarefaction waves, the axial distance of the impact craters was the smallest, the fragment density was the largest, and the project angle of the fragments was small.

Owing to the acceleration effect of the two charger parts on both sides, the detonation products accelerated the casing of the air part. However, as the detonation products spread in the air part, their energy density declined axially, which caused the fragments from the ends of the air part to converge to the middle, resulting in the overlapping of fragments in Region V. It was obvious that the project angle of the fragments increased with an increase in the length of the air part, and the overlapping of fragments in Region V would be more obvious, as shown in Figure 18. The distribution of impact craters formed by natural fragments on the protection panel was not as obvious as that of semi - preformed fragments because of the stochastic nature of the fracture processes; however, the results were consistent with those of semi - preformed fragments, which can be obtained from Figure 18. The obvious difference between the natural shell and the semi - preformed shell is the fracture of the air shell. As shown by the red circle in Figure 17, the natural shell of the air part was broken into two obvious fragments in the axial direction, and the kinetic energy of the fragments was significantly larger, resulting in the penetration of the 12mm 2A12 Aluminum alloy panel. The fracture position was located at the axial middle of the air part and the junction of the air and charge part, and there was an evident axial - driven energy gradient. The axial - driven energy gradient at the junction of the air and charge parts was caused by the existence of explosives. The axial - driven energy gradient in the middle of the air part is the energy peak formed by the superposition of the detonation products from the two charge parts. Therefore, the air part shell was axially broken into two fragments, which were unaffected by the length of the air part. Table 6 Number of preformed fragments and width of each region in a single row.

No.	Region Ⅱ		Region Ⅲ		Region Ⅳ		Region Ⅴ	
	Width/cm	Number of fragments	Width/cm	Number of fragments	Width/cm	Number of fragments	Width/cm	Number of fragments
1#	2.8	2	3.4	4	4.2	3	2.2	3
2#	4.8	2	4.8	5	1.7	1	5.1	6
3#	8	2	6	6	3.3	3	5.3	9

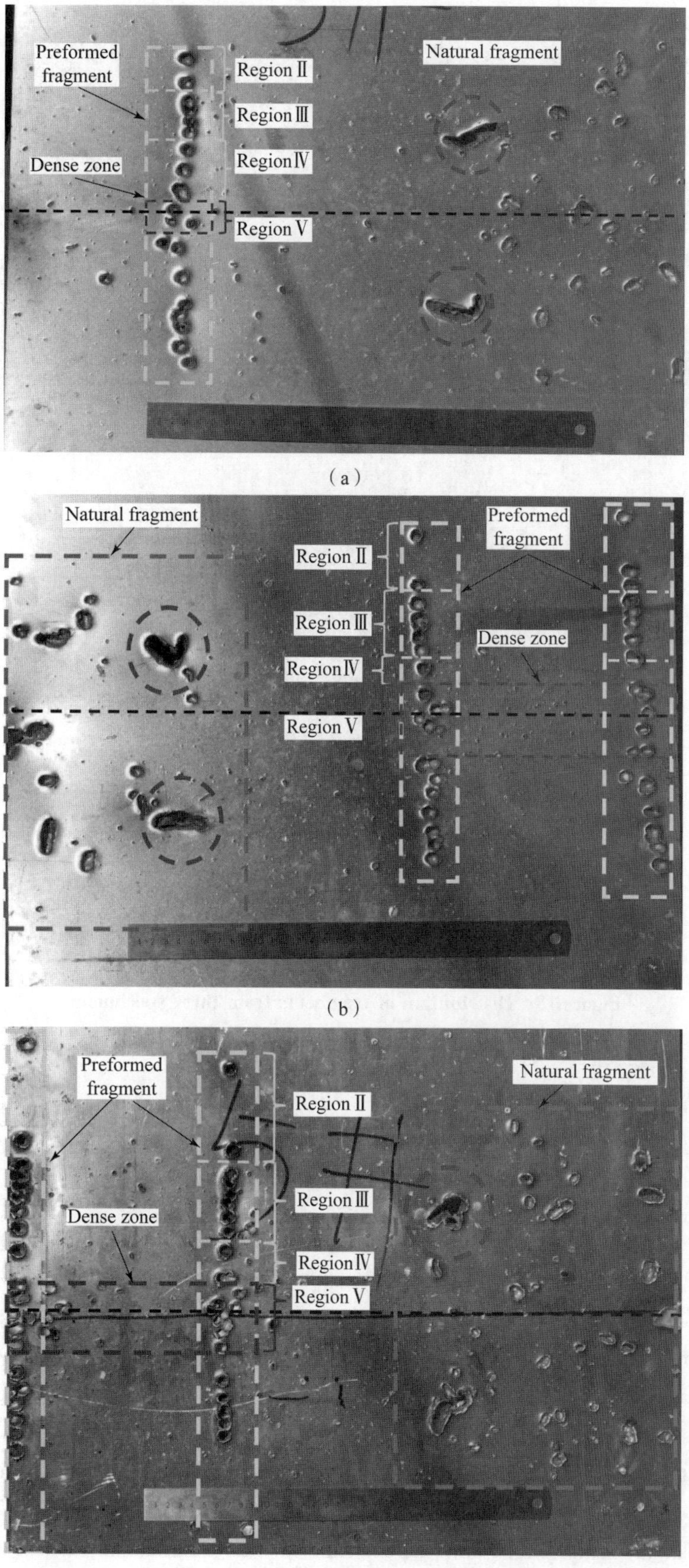

Figure 17　Distribution of impact craters on the protection panel.

(a) 1# specimen; (b) 2# specimen; (c) 3# specimen

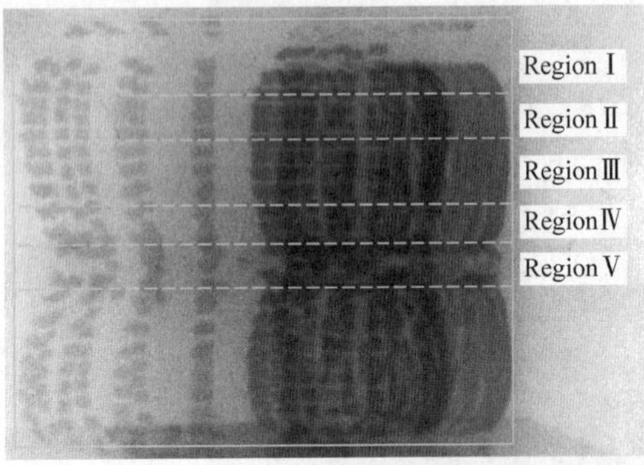

(a)

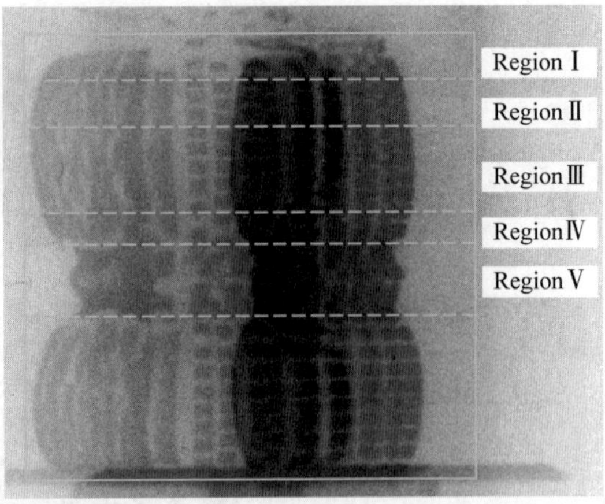

(b)

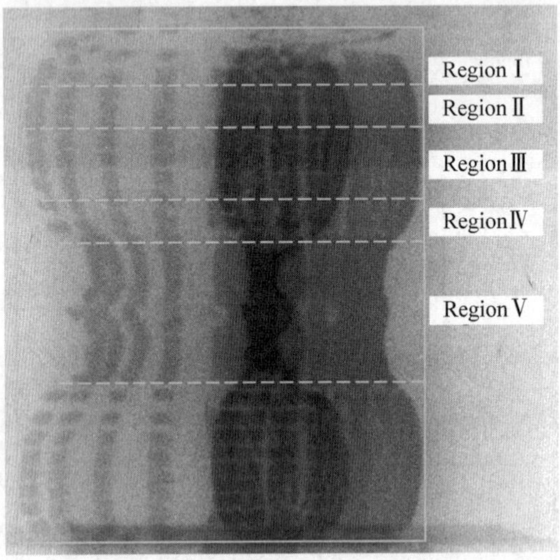

(c)

Figure 18 Distribution of fragments from three specimens.

(a) 1#; (b) 2#; (c) 3#

4.3 Validation of the proposed formula

The analysis method described in the previous section was used to measure the left and right sides of the fragment contour of each specimen several times, and the fragment velocity was obtained and is shown as scatters in Figure 18 to 20. *E* represents the experimental data, and *F* represents the calculation results of the proposed formula. Figure 18 shows the fragment velocity distribution from the air part in absolute coordinate. Figure 19 shows the fragment velocity distribution of the two charge parts for the three specimens, where it began from the detonation end. Figure 20 shows the velocity distribution of the fragment from the entire casing in absolute coordinates. The average standard deviations of the fragment velocities for the three specimens were 17 m/s, 34 m/s, and 23 m/s, respectively, which indicated that the accuracy of the analysis method meets the requirements, and that the linear cutting treatment of the shell does not significantly change the fragment velocity.

Figure 18 indicates that the fragment velocity from the air part with different lengths can be divided into two groups, and the calculation results are in good agreement with the experimental data. Figure 19 shows that the fragment velocity from both the charge parts of each specimen was consistent after coordinate normalization processing, which also indicated that the length of the charge part had no influence on the velocity distribution of the fragment because the influence distance of rarefaction waves on the fragment velocity from the detonation end and non - detonation end was limited. The fragment velocity distribution for the entire casing in Figure 20 shows that the calculation results of the proposed formula are in good agreement with the experimental data in absolute coordinates. Figure 21 shows that the average relative error for the three specimens was less than 4.5% , indicating that the proposed formula can accurately predict the fragment velocity distribution for cylinder with discontinuous charge with sufficient accuracy.

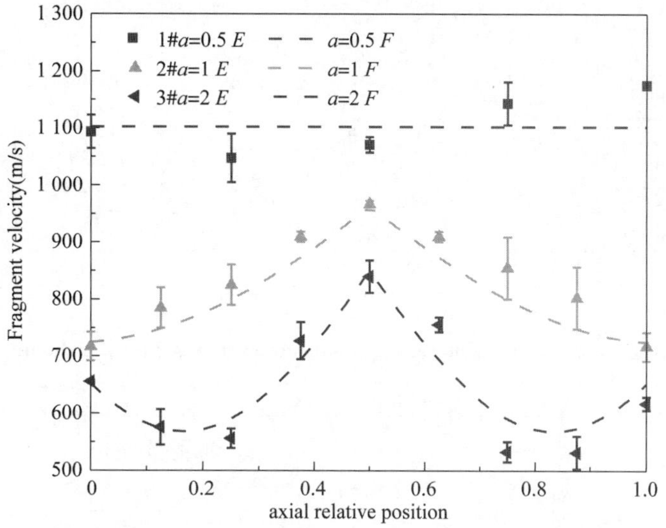

Figure 19 Fragment velocity distribution form air part in absolute coordinate.

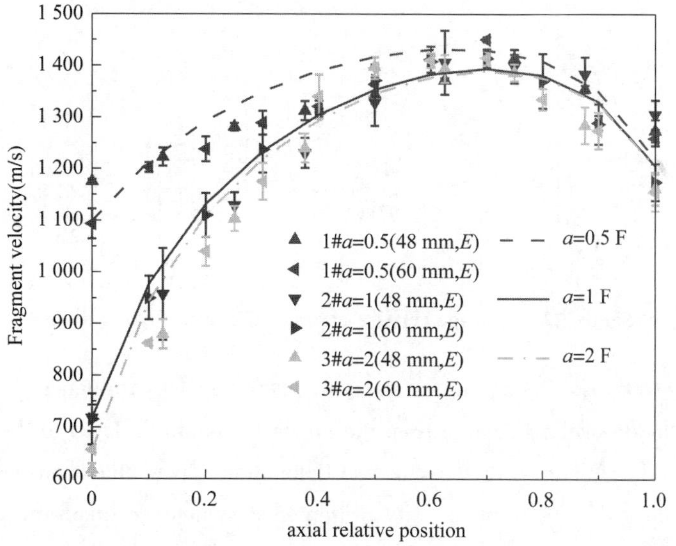

Figure 20 Fragment velocity distribution from charge part in relative coordinate.

5 Conclusions

A cylinder with a discontinuous charge was first proposed and analyzed using numerical simulations and X - ray radiography. The casing of the air part broke into a large number of fragments owing to the acceleration effect of detonation waves and explosive products from the two charges on both sides.

The acceleration process of the fragment was analyzed in detail, and the ratio a was proposed and used in the correctional function to show the influence of the length of the air part. The fragment velocity was distributed in axial symmetry because the two detonation points were initiated simultaneously. The detonation waves and explosive products diffused into the air , and superimposed in the middle of the air part to form a superimposed

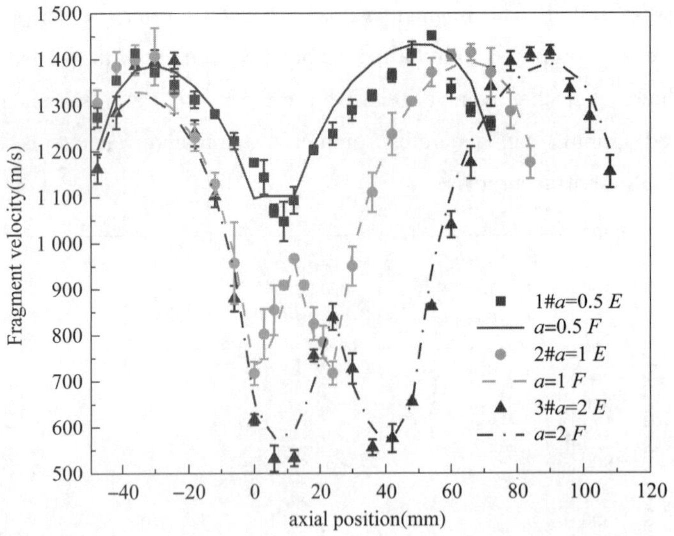

Figure 21 Fragment velocity distribution in absolute coordinate.

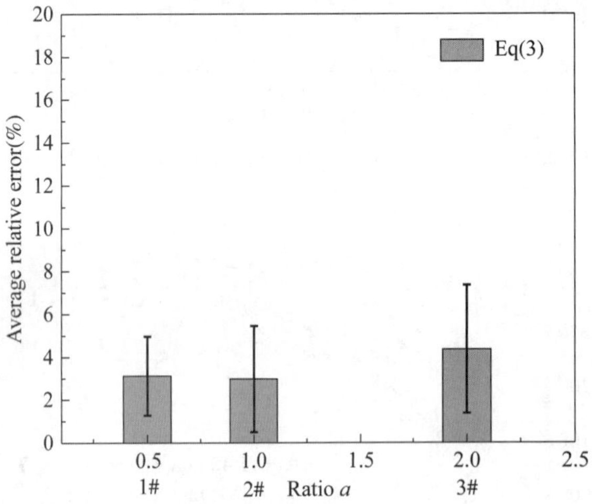

Figure 22 Average relative error of the proposed formula.

high – pressure zone. When $0 < a \leq 0.75$, the rarefaction waves had less influence on the superimposed high – pressure zone, and the velocity of the fragment from the air part was only related to the length of the air part and unrelated to the axial position. When $0.75 < a < 2$, the detonation waves and explosive products were weakened by the rarefaction waves, and the fragment velocity presented a symmetric quadratic distribution. Owing to the acceleration of the superimposed high – pressure zone, fragments from two charge parts close to the air part were accelerated to a higher velocity than those from the detonation end of a cylinder detonated at one end. The influence distance of the superimposed high – pressure zone on the charge part is 2R, where R is the charge radius.

In the experiment, the circumferential half of each specimen was semi – preformed and the other half was natural. It was found that the project angle and axial density of the fragments were related to the axial driving energy gradient but not to the shear stress exceeding the dynamic shear strength. The air part of the semi – preformed shell overlapped and formed a dense zone, while that of the natural shell axially formed two fragments with greater mass and kinetic energy owing to the axial – driven energy gradient occurring at the middle of the air part and the junction of the air and charge parts. The shell contour in the simulation was in good agreement with that in the experiment. A formula was established for predicting the fragment velocity distribution based on the simulation results, as demonstrated by the experimental data. The results revealed that the average relative error of the proposed formula was less than 4.5%, indicating that the proposed formula can predict the fragment velocity distribution from such cylinder with sufficient accuracy. The research presented in this study can provide a theoretical foundation for the design of novel warheads.

Acknowledgements

The authors would like to thank my colleague for their assistance and Prof. Feng for his guidance.

References

[1] G. Taylor, The fragmentation of tubular bombs, Advisory Council on Scientific Research and Technical Development, 5 (1963) 202 - 320.

[2] W. J. Stronge, X. Ma, L. Zhao, Fragmentation of explosively expanded steel cylinders, International Journal of Mechanical Sciences, 31 (1989) 811 - 823.

[3] R. W. Gurney, The initial velocities of fragments from bombs, shell and grenades, in, Army Ballistic Research Lab Aberdeen Proving Ground Md, 1943.

[4] M. Bola, A. Madan, M. Singh, Expansion of metallic cylinders under explosive loading, Defence Science Journal, 42 (1992) 157 - 163.

[5] S. Feng, B. Cui, An experimental investigation for the axial distribution of initial velocity of shells, Acta Armamentarii, 4 (1987) 60 - 63.

[6] Q. Lixin, L. Tong, Z. Shouqi, Y. Yunbin, Fragment Shot - Line Model for Air - Defence Warhead, Propellants, Explosives, Pyrotechnics, 25 (2000) 92 - 98.

[7] T. Hiroe, K. Fujiwara, H. Hata, H. Takahashi, Deformation and fragmentation behaviour of exploded metal cylinders and the effects of wall materials, configuration, explosive energy and initiated locations, International Journal of Impact Engineering, 35 (2008) 1578 - 1586.

[8] D. Goto, R. Becker, T. Orzechowski, H. Springer, A. Sunwoo, C. Syn, Investigation of the fracture and fragmentation of explosively driven rings and cylinders, International Journal of Impact Engineering, 35 (2008) 1547 - 1556.

[9] I. Cullis, P. Dunsmore, A. Harrison, I. Lewtas, R. Townsley, Numerical simulation of the natural fragmentation of explosively loaded thick walled cylinders, Defence Technology, 10 (2014) 198 - 210.

[10] G. Ren, Z. Guo, C. Fan, T. Tang, H. Hu, Dynamic shear fracture of an explosively - driven metal cylindrical shell, International Journal of Impact Engineering, 95 (2016) 35 - 39.

[11] Y. Li, Y. - h. Li, Y. - q. Wen, Radial distribution of fragment velocity of asymmetrically initiated warhead, International journal of impact engineering, 99 (2017) 39 - 47.

[12] Y. Gao, S. Feng, B. Zhang, T. Zhou, Effect of the length - diameter ratio on the initial fragment velocity of cylindrical casing, in: IOP Conference Series: Materials Science and Engineering, IOP Publishing, 2019, pp. 012020.

[13] X. Kong, W. Wu, J. Li, F. Liu, P. Chen, Y. Li, A numerical investigation on explosive fragmentation of metal casing using Smoothed Particle Hydrodynamic method, Materials & Design, 51 (2013) 729 - 741.

[14] Y. J. Charron, Estimation of velocity distribution of fragmenting warheads using a modified Gurney method, in, Air Force Inst of Tech Wright - Patterson AFB OH School of Engineering, 1979.

[15] T. Zulkoski, Development of optimum theoretical warhead design criteria, in, Naval Weapons Center China Lake CA, 1976.

Fragment characteristics from a cylindrical casing constrained at one end①
一端全约束圆柱形装药壳体破片特征分布

Gao Yueguang[a]② , Feng Shunshan[a] , Xiao Xiang[a] , Feng Yuan[b] , Huang Qi[a]

[a]State Key Laboratory of Explosion Science and Technology,

Beijing Institute of Technology, Beijing 100081, PR China

[b]China Research and Development Academy of Machinery Equipment, Beijing, 100089, China

Abstract: One end is constrained in a partially penetrated warhead, resulting in a significantly different acceleration process from that of an unconstrained warhead. However, existing formulas and models are only applied to a cylindrical casing in an unconstrained state. To study the fragment characteristics of a cylinder constrained at one end, X-ray radiography was performed in which the charge was detonated at the center of the other end, and the acceleration process of fragments was analyzed in detail by using numerical simulation. The results indicated that the size and kinetic energy of the fragments near the constrained end increased as the radial pressure gradient decreased axially. More importantly, the fragment velocity near the constrained end exceeded the maximum velocity calculated using the Gurney equation, which was found for the first time. A new formula was proposed to predict the fragment velocity distribution from a cylindrical casing constrained at one end; the average relative error was less than 3.3%. In addition, the simulation results were in good agreement with the experimental results. This indicates that the reflected shock wave generated from the constrained end made the fragment reach the Gurney velocity faster; however, the continuous acceleration of detonation products with higher pressure caused by the reflected shock wave was the final reason for the fragment velocity exceeding the theoretical velocity. The fragment velocity was affected by a combination of rarefaction and reflected shock waves. As the length-to-diameter ratio exceeded 2, the increasing number of fragment velocities exceeded the theoretical velocity. The results of this study can provide important guidance for warhead structure design and terminal effect evaluation.

1 Introduction

The rupture and movement of the exploded cylinder were analyzed to assess the terminal effect of the warhead and resistance of the structure [1, 2]. The casing expands and breaks into several fragments with different velocities and directions after the charge is detonated [3-8]. When evaluating the terminal effect of fragments, fragment velocity has been widely considered through experimental, numerical, and theoretical method, and the results provide a basis for further research in various fields such as public security, anti-terrorist, and weapon design.

Several theories have been proposed regarding the dynamics of material [4, 9-13], particularly the fragmentation of cylinders [1, 14-18]. In addition, a large number of formulae have been established to predict

① 原文发表于《International Journal of Mechanical Science》，2023 年 1 月 27 接收。

② 高月光：工学博士，2017 年师从冯顺山教授，研究半侵彻弹药战斗部对甲板目标威力效应，现工作单位：在校生。

the fragment velocity, among which the Gurney formula is widely used owing to its simplicity and high accuracy in most circumstances. Based on a few simple implicit assumptions and the conservation of energy, Gurney [19] developed a universal formula for calculating the initial fragment velocity, which is related to the energy of a particular explosive and the ratio of the explosive mass to the casing mass.

$$v_0 = \sqrt{2E}\sqrt{\frac{1}{1/\beta + 0.5}} \tag{1}$$

where v_0 is the initial fragment velocity, $\sqrt{2E}$ is the Gurney constant of a given explosive, and β is the mass ratio of the explosive to metal casing. Although the Gurney velocities slightly exceeded the final velocities with different mass ratios and explosives in the study, the relative errors were all within 10 percent [20], which indicating high accuracy of the Gurney formula.

Because Gurney assumed that the cylinder was infinitely long and ignored the rarefaction waves at the two ends, the Gurney formula only calculated the maximum velocity of the fragments. The calculation results were consistent with most fragment velocities. However, the error is large near the end owing to the rarefaction waves from the two ends. Several studies have investigated the dynamic response of a cylinder initiated at one end [21 – 28]. Kong et al. [29] investigated the dynamic process of cylinders where the casing had or did not have end caps, which also showed that the fragment velocity decreased at the two ends owing to rarefaction waves.

In addition, the rarefaction waves from the two ends were analyzed in detail, and improved Gurney formulae were established. Charron [30] used the geometric equivalence method and removed a part of the explosive from the two ends, which was assumed to be not involved in the reaction. Then the local mass ratio with the removed explosive was incorporated into the Gurney formula, and the calculation results showed good agreement with the experiment, particularly at the two ends. After conducting a large number of experiments, Zulkouski [31] proposed a correction function in the form of an exponential function to correct the Gurney formula, which can describe the variation in fragment velocities along the axial position caused by rarefaction waves. However, the fatal flaw of the formula was that the fragment velocity at the detonation end was obviously incorrect because the calculation result was zero. Hence, Huang et al. [32] conducted an X – ray experiment and proposed an improved formula that can solve the problem where the fragment velocity at the detonation end is zero. Additionally, the calculation results exhibited a high degree of precision.

$$v_{0x} = (1 - 0.361e^{-1.111x/d})(1 - 0.192e^{-3.03(L-x)/d})\sqrt{2E}\cdot\sqrt{\beta/(1+0.5\beta)} \tag{2}$$

Gao et al. [33] conducted a simulation study on a cylindrical charge shell with air parts at the two ends. Although no corresponding charge was observed for the air parts, the air part casing still generated fragments with certain velocities owing to the shock wave and detonation products generated by the middle charge. They proposed a new correctional function for the Gurney formula, which was used to correct the mass ratio at each axial position and obtain the axial velocity distribution of the fragments.

$$v(x,a,b) = \sqrt{2E}\sqrt{\frac{1}{1/\beta f(x,a,b) + 0.5}} \tag{3}$$

where a and b are the ratio of the air part length to the charge diameter. The correctional function of Huang 's formula [32] for the fragment velocity from the charge part was revised as:

$$v_{0x} = \sqrt{2E}\sqrt{\frac{1}{1/[\beta(1-0.689e^{-0.953x/d})(1-0.385e^{-2.3(L-x)/d})] + 0.5}} \tag{4}$$

Gao et al. [34] investigated a cylindrical casing with end caps of different thicknesses, and obtained a formula to predict the fragment velocity distribution with sufficient accuracy. Liao [35] also investigated the effect of the end cap on the fragment velocity, and a new formula was established that was similar to Huang's formula [32]. Based on various experiments, a formula [36] was established to calculate the fragment velocity using the expanding radius of the casing. The results indicated that the fragment velocity was stable after the casing expanded to its maximum critical radius. In addition, by changing the position of the detonation point, the fragment velocity under eccentric initiation condition can be obtained [37 – 42], and the velocity enhancement effect under single and double eccentric initiation conditions can be obtained through detonation wave theory [27, 43]. Considering the influence of eccentric initiation, rarefaction waves, and incident angle, Li [44] obtained an all – direction fragment velocity distribution.

In a previous study, the cylindrical charge casing was in an unconstrained state, and the shock wave and explosive products at the two ends were free to propagate after the explosive was initiated. For a partially penetrated warhead, the projectile can be optimized to achieve the embedding effect and delay the explosion on the target surface [45, 46]. The physical model of such a partially penetrated projectile is a cylindrical casing constrained at one end. However, the constrained end would make the acceleration process of the fragments significantly different from that of a cylinder without a constrained end. The spatial distribution of the fragments is also significantly different [47 – 50]. More importantly, the partial fragment velocities of the cylinder constrained at one end exceeded the theoretical Gurney velocity, which had never been observed.

In this study, X – ray radiography was performed to investigate fragment distribution and velocity from a cylindrical casing constrained at one end. The acceleration process of the fragment velocities was analyzed using the smoothed particle hydrodynamics (SPH) method in AUTODYN. The research results can provide a theoretical basis for the design and evaluation of related warheads.

2 X – ray experiments

In this section, the fragment characteristics of a cylindrical casing constrained at one end are obtained by X – ray radiography. The X – ray experimental configuration and the installation of the sample are described in Section 2. 1. Subsequently, the fragment velocity processing method is characterized in Section 2. 2. Ultimately, the velocity distribution and spatial distribution characteristics of the fragments are characterized in Section 2. 3.

2. 1 Configuration

The velocity distribution of the fragments was acquired by X – ray radiography, which has been used in various research studies to obtain the fragment velocity with high accuracy. The experimental configuration is shown in Figure 1; it is primarily composed of two X – ray tubes, a protection structure, protection panels, a film, a specimen, and a fixed plate. The fixed plate was constrained to the correct height to ensure that the casing outline could be fully displayed in the film. A through hole was opened in the middle of the fixed plate, and the solid body of the cylindrical casing was inserted into it. The diameter of the hole was 24 mm, which is equal to the diameter of the solid body and less than the outer diameter of the shell (30 mm). The length of the solid body was 40 mm, and the solid body of the casing remained stationary because of the fixed plate after the charge exploded from the center of one end. The cylindrical casing was detonated at the center of one end (Figure 1b). The protection panels were used to protect the film and record the distribution of fragments, which consisted of 16 (8 + 4 + 4) mm of 2A12 aluminum alloy and 5 mm of wood. The two X – ray tubes were arranged side – by – side. The X – ray was released at a high pressure at two different moments, and the fragment profile was displayed on the film at two different moments. Through amplification ratio conversion, the real moving distance of the fragment in a certain period can be obtained, and the moving speed of the fragment can then be determined.

A schematic of the test specimens is shown in Figure 2, where d and D are the charge and outer diameters of the shell, respectively, L is the charge length; and h is the height between the lower end face of the charge and the fixed plate, which is expressed by the dimensionless parameter k_2 multiplied by the charge diameter. x is the

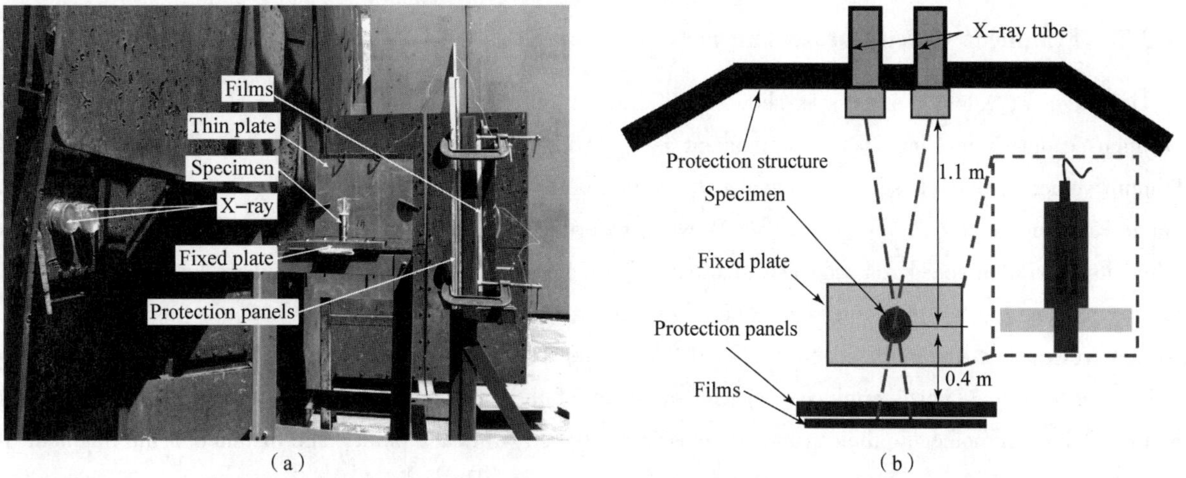

Figure 1 X - ray experimental configuration.

(a) Layout of test site; (b) Schematic of test device. The X - ray test device included two X - ray tubes, a fixed plate, protection panels, films, a thin plate, and a specimen. After the sample was detonated, the fragment distributions at two different moments were obtained on the same film, and the fragment spatial distribution characteristics were obtained on the protection panel.

axial position coordinate of the fragment velocities at the detonation end. The end cap of the non - detonation end between the charge and fixed plate was named the "head cap." A solid body was inserted into the fixed plate to constrain the casing. AISI 1045 steel and Q345B steel were used as the casing and the fixed plate, respectively. The explosive charge was composed of COMP - B. The thickness of the fixed plate was 20 mm, which prevented the casing from moving. The fixed plate measured 300 × 400 mm. The parameters of these two specimens are listed in Table 1.

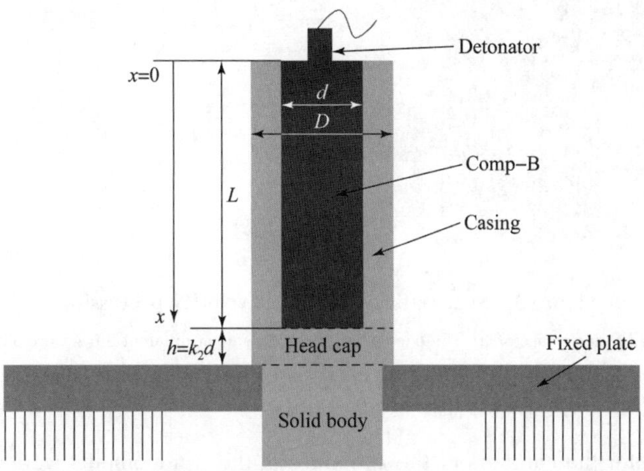

Figure 2 Schematic of test specimens. The specimen was composed of a charge (COMP - B), head cap, solid body, and cylindrical casing, which was detonated at the center of the end by a detonator. The solid body was inserted into the fixed plate's through - hole. *D* and *d* are the diameters of casing and charge, respectively; *L* is the length of the charge; *h* is the height of the head cap; k_2 is the ratio of the head cap to the explosive diameter; *x* is the axial position.

Table 1 Test specimen parameters.

No.	L(mm)	D(mm)	d(mm)	k_2	t_0(μs)	t_1(μs)
1	70	30	24	1	30.65	50.7
2				0.25	30.7	50.65

L: the length of the charge; D: outer diameter of the casing; d: diameter of the charge; k_2: Ratio of head cap to explosive diameter; t_0 and t_1 are the first and second moments of the X - ray tube emission.

2.2 Fragment velocity processing method

The fragment velocity process is shown in Figure 3. The fragments obtained at these two moments were superimposed onto one film. The first moment is t_0, which is the time when the fragment reaches a stable maximum velocity, and the second moment is t_1, which is the appropriate time after the first moment to obtain the average fragment velocity. The two moments were determined through simulation to ensure that the fragment reached its maximum speed, as shown in Figure 3. The moments of the three specimens are listed in Table 1. The symmetry axes of the two moments were determined using the fragment contours. When the fragment reached its maximum velocity at the millisecond level, the diameter of the solid body did not change. Therefore, the amplification ratio can be determined from the diameter of the casing solid body. After obtaining the fragment contours of the two moments, their symmetry axes were aligned with each other. The distance of the fragment from the corresponding position was measured using the CAD software. The radiographic images were a 2 - dimensional projection of the 3 - dimensional distribution of the fragments. Therefore, the distance traveled by a fragment between the two images was a projection of the actual flight path, and the calculated velocity was a component of the actual velocity. Only measurements of the case fragments projected at right angles to the X - ray beam allowed for the correct determination of their velocity.

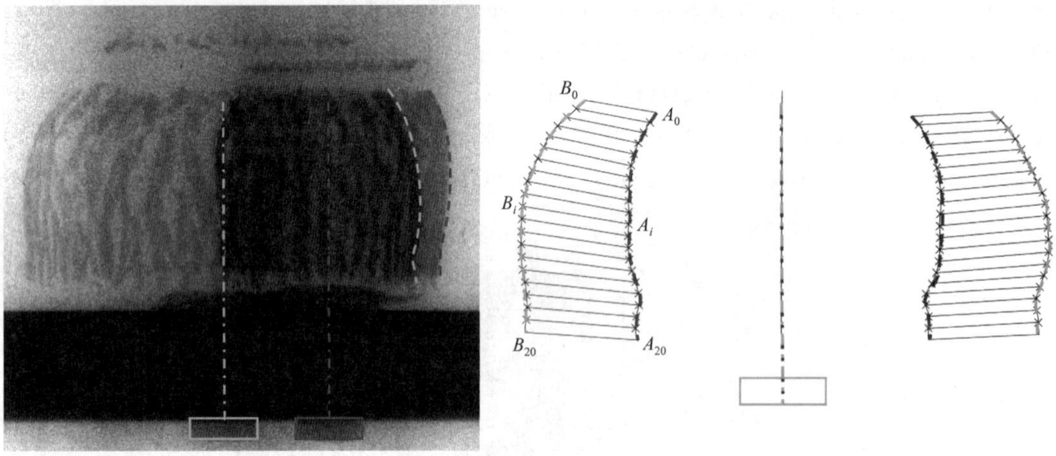

(a) (b)

Figure 3 Schematic of fragment velocity processing.

(a) X - ray film. (b) Image alignment processing. The contours of the fragments at different times were obtained from the film, and the contours were extracted and divided equally by the CAD software to obtain the moving distances of the casing at different axial positions. The magnification ratio was determined by the diameter of the solid body at the bottom.

In the CAD software, after the contours of the two sides of the two moments were described, each was divided equally. To correspond the fragments at each axial position as accurately as possible, the profile was divided into 20 equal parts. The specific length of each point was determined and named Δl_i. The size of the solid body of the shell was measured to obtain the amplification ratio (k), and the actual distances of each point was obtained ($k\Delta l_i$). The real fragment velocity distribution can be obtained using $v_i = k\Delta l_i/(t_1 - t_0)$. The fragment velocities on both sides were measured twice to reduce the measurement error.

2.3 Experimental results and discussion

After the charge was detonated, the shell broke into a large number of fragments, and the fragments impacted the protection panel at a distance of 0.4 m and formed impact craters on it. Therefore, the size and distribution of the impact craters can represent the fragments [47, 48]. The impact crater on the protection panel is shown in Figure 4, where the dotted black line represents the height of the upper surface of the fixed plate. It can be clearly observed that the fragments can be divided into three regions, from the detonation end to the constrained end. Region 1 was closer to the detonation end, and the fragments were small fragments characterized by a large axial distribution range, that is, a large project angle. A large project angle also led to a significant decrease in the distribution density of fragments on the protection panel, and the spatial distribution density continued to decrease with moving distance. Owing to the size limitation of the protection panel, fragments with large project angles were beyond the range of the protection panel and would not impact the protection panel. Therefore, some fragments did not affect the protection panel. The fragment size and kinetic energy were small, as were the impact craters. Region 2 corresponded to the middle part of the shell; the fragments were of medium size, and the axial distribution range was smaller than that of Region 1. The fragments have a larger mass and larger kinetic energy, and the size of the impact craters was larger than that of Region 2. Region 3 is close to the fixed plate. The size of the fragments was significantly larger than those of the other fragments, and the number of fragments was smaller. However, the fragments have more kinetic energy, forming a larger impact crater. The widths of Region 2 of specimens 1# and 2# were 8.8 and 7.8 mm, respectively, and the width of Region 3 was 4.5 mm. The reason for the smaller size of the fragments near the detonation end was that the radial pressure gradient along the axial direction produced large tensile and shear stresses in this region [5, 51 - 53]. However, the constrained end caused the radial pressure gradient to be significantly reduced, resulting in a larger fragment than that in the other parts.

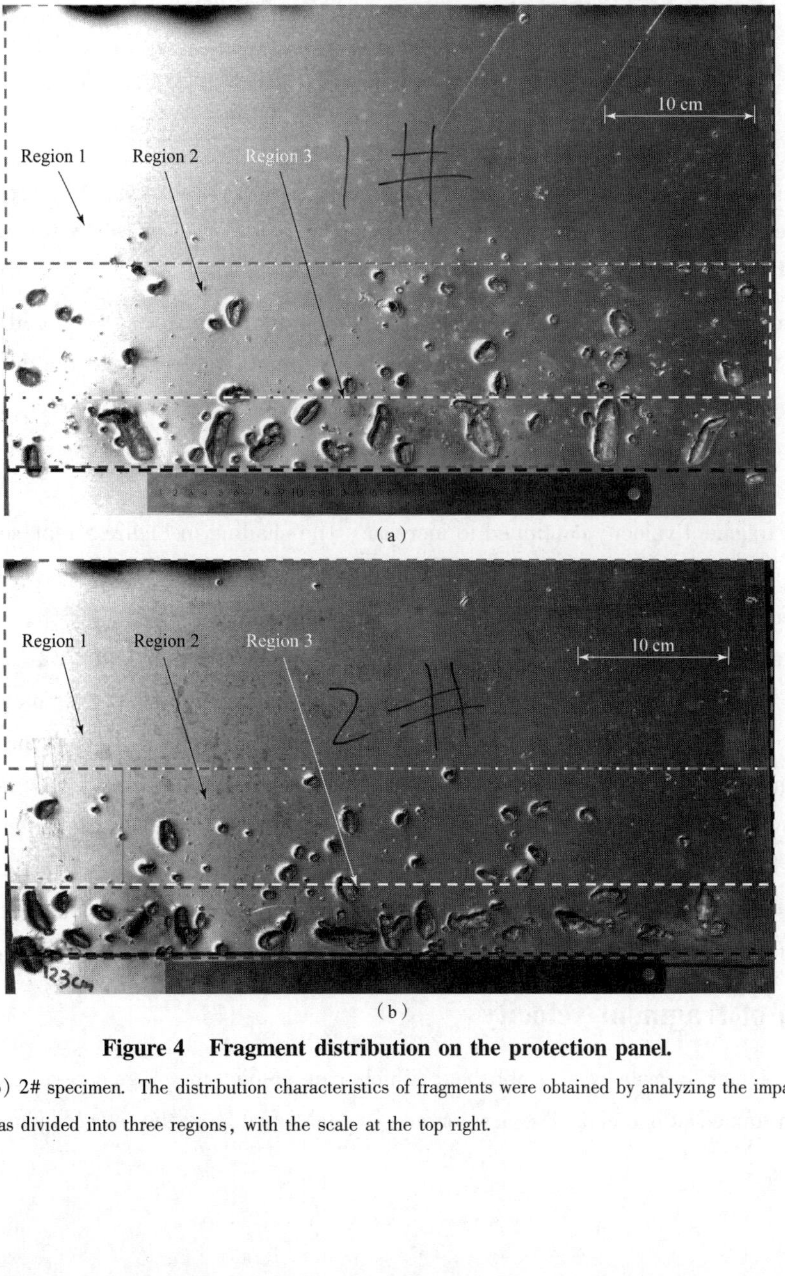

Figure 4 Fragment distribution on the protection panel.

(a) 1# specimen. (b) 2# specimen. The distribution characteristics of fragments were obtained by analyzing the impact craters on the protection panel. The impact crater was divided into three regions, with the scale at the top right.

Based on the fragment velocity processing method described in Section 2.1, the fragment velocity distributions of the two specimens were obtained, which can be seen as scattered in Figure 5. The theoretical fragment velocity calculated based on the Gurney equation was 1 542 m/s. The standard deviation reflects the degree of dispersion in the data set. When repeated measurements were performed, the standard deviation of the measurement values represented the precision of the measurement. The average standard deviation of the fragment velocities for the two specimens was 42 m/s, which did not exceed 2.5% of the maximum fragment velocity, indicating high accuracy of the fragment velocity processing method.

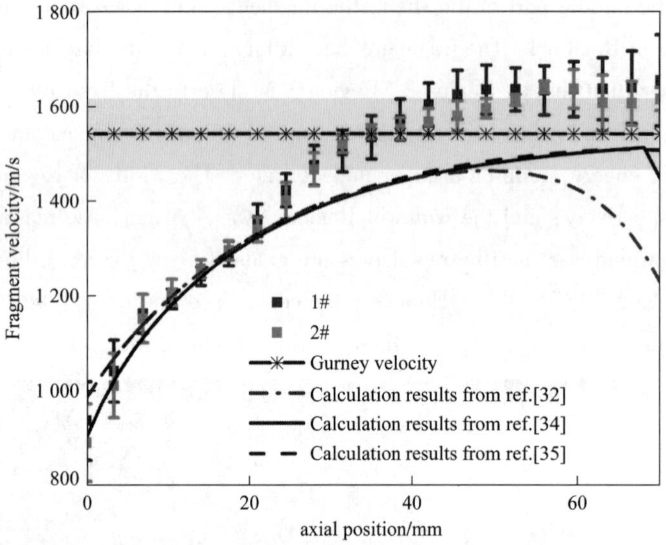

Figure 5 Comparison between the experimental data and calculation results. 1# and 2# represented the experimental data from two specimens. Gurney velocity represented the theoretical velocity calculated by Gurney equation. The other three lines represented calculation results from previous formulae.

The fragment velocity near the detonation end was small, approximately 59% of the theoretical velocity calculated using the Gurney equation, which was affected by the rarefaction waves at the detonation end. With an increase in the axial position, the influence of rarefaction waves gradually decreased, the fragment velocity increased gradually, and reached the theoretical value at approximately 1.5 times the radius of charge from the detonation end.

Since then, the fragment velocity continued to increase. The shading in Figure 5 represents a tolerance region with an admissible error of ±5% for the Gurney velocity. It was found that the fragment velocity was stable after exceeding the theoretical velocity by 5%, and that the maximum did not exceed 7.3%. The fragment velocity distribution was not affected by the height of the head-end cap. In addition, this was also different from the fragment velocity distribution of a cylindrical shell with end caps, because the end caps only reduced the effect of rarefaction waves, making the fragment velocity closer to the theoretical velocity. Even if the end cap at the non-detonation end is equivalent to an infinite thickness, the calculation results from Gao [34], Liao [35] and Huang [32] are shown as lines in Figure 5. It can be found that the existing formulas cannot predict the fragment velocity distribution from the cylindrical casing constrained at one end. In particular, the calculated fragment velocities at the constrained end were all lower than the theoretical velocity of the Gurney equation, which was quite different from the experimental results. Therefore, there is an urgent need to develop new formulas.

3 Formula of fragment velocity

In this section, a new formula was established, which can predict the fragment velocity distribution from a cylindrical casing constrained at one end. The new formula is established, and the relevant parameters were confirmed in Section 3. 1. The accuracy of the formula is analyzed by comparing them with the test results in Section 3. 2.

3. 1 Establishing the formula

To predict the fragment velocity distribution from a cylindrical casing constrained at one end, a new formula is established by adding a correctional function based on the Gurney equation:

$$v_{0x} = \sqrt{2E}\sqrt{\frac{1}{1/\beta f(x) + 0.5}} \tag{5}$$

where $f(x)$ is the correctional function. The fragment velocity near the detonation end was less than the theoretical velocity because of the influence of the rarefaction waves. The maximum influence distances of the rarefaction waves at the detonation and non - detonation ends were 2R and R, respectively, where R is the radius of the charge. To avoid the interaction of rarefaction waves at the two ends, and because the actual warhead usually has a large length - diameter ratio, the casing length was assumed to be greater than two times the charge diameter. Previous studies have also proven that the detonation and non - detonation ends do not interact with each other and can be studied independently [32, 33]. Therefore, the correctional function considered the influence of detonation and constrained ends. Because the status of the detonation end in this study is the same as that in our previous study [33], the effect of rarefaction waves at the detonation end in the correctional function can take the same form as in the first parenthesis in Eq. . The trends of the correctional function with the axial position were calculated based on the experimental data of the two specimens, which are shown as scatters in Figure 6. It can be observed that the correctional function is in an exponential form, and the influence of the constrained end can be in an exponential form similar to that of the detonation end. Hence, the correctional function can be expressed as:

$$f(x) = (1 - 0.689e^{-0.953x/d})(A - Be^{-x/d}) \tag{6}$$

where A and B are constants obtained by fitting the experimental data. The results show that A = 1. 2 and B = 0. 2 ($R^2 = 0.982$). The fitted curve of the correctional function is shown as a line in Figure 6, which is in good agreement with the experimental results. After substituting Eq. into Eq. , the velocity distribution of fragments from a cylindrical casing constrained at one end can be obtained as:

$$v_{0x} = \sqrt{2E}\sqrt{\frac{1}{1/[\beta(1 - 0.689e^{-0.953x/d})(1.2 - 0.2e^{-x/d})] + 0.5}} \tag{7}$$

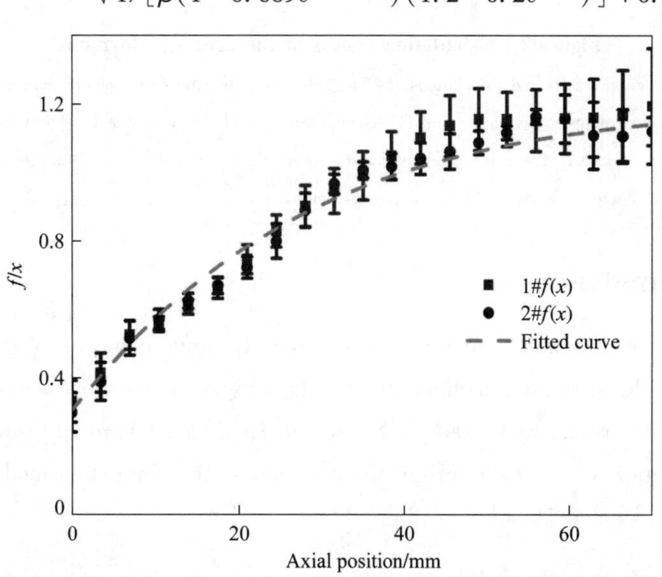

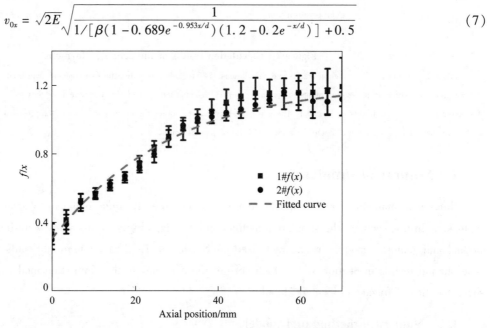

Figure 6 Correctional functions for two specimens. 1#$f(x)$ and 2#$f(x)$ represented the correction functions from experiment. The fitted curve was the best fit to all the data.

3.2 Validation

Through the above derivation, a new formula was obtained to predict the fragment velocity distribution from a cylindrical casing constrained at one end, and a comparison between the calculation of the new formula and the test results is shown in Figure 7 (a). In calculating the fragment velocity at the detonation end, the new formula followed the same trend as the previous formula (Figure 5), demonstrating once again that the influence of rarefaction waves was dominant at the detonation end and the effect of the constrained ends could be ignored. Second, the most obvious improvement of the new formula was that it could accurately predict the fragment velocities at the constrained end, because it exceeded the theoretical velocity of the Gurney equation. Finally, the results of the new formula exhibit the same trend as the experimental results, indicating that the new formula can calculate the effect of the detonation and constrained ends on the fragment velocity with sufficient accuracy.

The relative error reflects the reliability of predictions. The average relative error of the calculated results for the new formula are shown in Figure 7 (b). It can be found that the average relative error of the new formula did not exceed 3.3% for the two specimens. This indicates that the new formula is highly reliable and can be used to predict the fragment velocity distribution from a cylindrical casing constrained at one end.

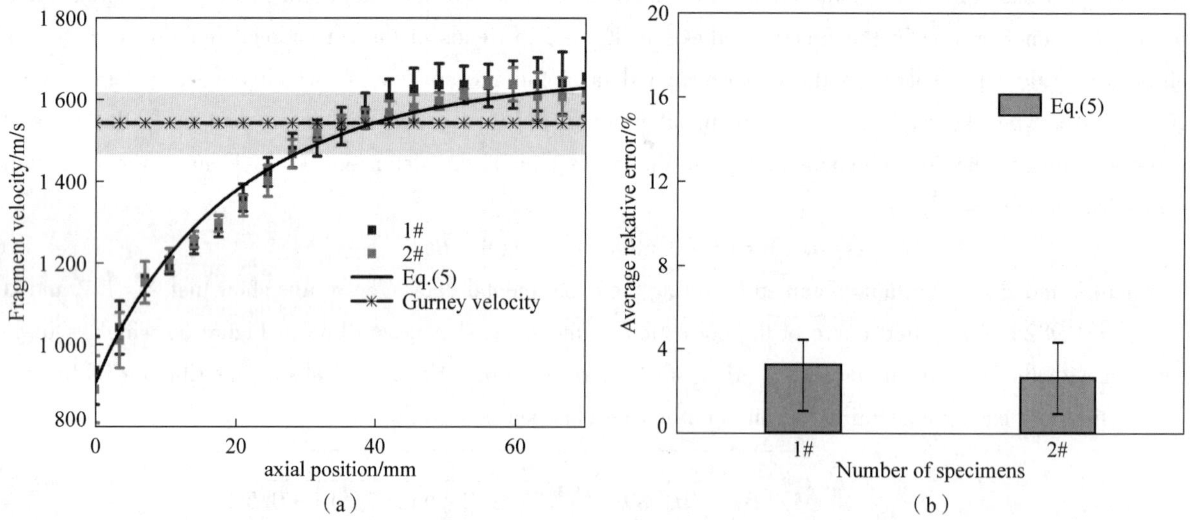

Figure 7 Calculation results of the proposed formula.

(a) Comparison of the experimental results and calculations. 1# and 2# represented the experimental data from two specimens. The line of Eq. (5) was the calculation result of the proposed formula. Gurney velocity represented the theoretical velocity calculated by the Gurney equation. (b) Average relative error of the new formula. The average relative error between the calculated value of the proposed formula and all measured values for the two specimens was used to evaluate the accuracy of the proposed formula.

4 Numerical simulation

Numerical simulation is a powerful tool to analyze the dynamic behavior of materials in the process of explosion. In this section, the simulation method and its adaptability to this study were introduced. The numerical method and material model are characterized in Section 4.1. The simulation results are compared with the experimental results in Section 4.2, which verifies the accuracy of the simulation model and describes the spatial characteristics of fragments in more detail.

4.1 Numerical method and model

This section uses the SPH method to simulate shell fragmentation and charge explosion. The theory and advantages of SPH method are introduced in Section 4. 1. 1. The size and material parameters of the model are described in Section 4. 1. 2.

4. 1. 1 SPH method

Smoothed particle hydrodynamics (SPH) is an appealing method for solving the dynamic response of materials under high dynamic loading conditions, which is challenging for Lagrangian finite elements owing to severe mesh distortions when the material has large deformations. The SPH method is a meshfree method widely used to simulate instantaneous and dynamic large deformations [29, 54 - 57]. The SPH method can avoid Euler grid and material boundary issues. Recently, various researchers have used the SPH method to characterize dynamic impacts in latest decades, with good agreement with experimental results [58 - 62].

The SPH method, which is based on interpolation theory, uses an interpolation function to convert the continuous dynamics into integral equations. The kernel estimate $\langle f(x)\rangle$ of function $f(x)$ at a specific position can be obtained by integrating the adjacent particles for the kernel estimation of a function $f(x)$ within the radius of the kernel function W defined by the parameter h:

$$\langle f(x)\rangle = \int_{\Omega} f(x')W(x - x',h)dx' \tag{8}$$

The kernel function W performs under two conditions: (i) it is an even function, $W(x - x',h) = W(x' - x,h)$, and (ii) it is a wrap function, $\int_{\Omega} W(x - x',h)dx = 1$. Because h is the width of the kernel function, $\langle f(x)\rangle$ equals $f(x)$ as h approaches zero.

The foundation of the SPH method can be established when the volume weight of each particle is replaced by m^j/ρ^j, which can estimate material parameters such as strain rates, velocity, and accelerations. The conservation equations are as follows:

$$\frac{\mathrm{d}\rho^i}{\mathrm{d}t} = \rho^i \sum_{j=1}^{N} \frac{m^j}{\rho^j}(v_\alpha^j - v_\alpha^i)\frac{\partial W^{ij}}{\partial x_\alpha^i} \tag{9}$$

$$\frac{\mathrm{d}v_\alpha^i}{\mathrm{d}t} = -\sum_{j=1}^{N} m^j\left(\frac{\sigma_{\alpha\beta}^i}{\rho^{i2}} + \frac{\sigma_{\alpha\beta}^j}{\rho^{j2}}\right)\frac{\partial W^{ij}}{\partial x_\beta^i} \tag{10}$$

$$\frac{\mathrm{d}E^i}{\mathrm{d}t} = -\frac{\sigma_{\alpha\beta}^i}{\rho^{i2}}\sum_{j=1}^{N} m^j(v_\alpha^i - v_\alpha^j)\frac{\partial W^{ij}}{\partial x_\beta^i} \tag{11}$$

4. 1. 2 Model size and material parameters

As this specimen was symmetrical, only a fourth of it had to be modeled. Given that the SPH method is gridless, the material in the model comprises a mass of particles. The particles were autonomous, and their relationship was determined using the aforementioned equations. It has been established that the accuracy of the SPH method is directly proportional to particle size. With smaller particles, the simulation accuracy was higher; however, it was also a time - consuming process. Li [44] used the SPH method to simulate different particle diameters and discovered that a particle with a diameter of 0. 4 mm was sufficient to achieve reliable data. In our simulation, the particle diameter was set to the same diameter to achieve a satisfactory balance between calculation accuracy and efficiency (0. 4 mm).

The numerical model for Specimen 1# is shown in Figure 8. The cylinder parameters are listed in Table 1. The velocity gauges on the casing surface recorded the velocity history of the fragments as they moved with the casing. The distance between the velocity gauges was set as 2 mm. The size of the fixed plate was set to 30 mm × 30 mm × 10 mm to reduce to computing time, and the bottom was fixed with a fixed boundary to simulate the fixed condition.

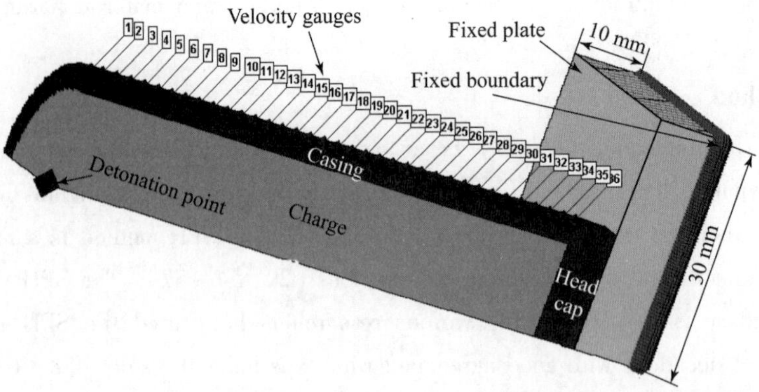

Figure 8 3D numerical model. It is a quarter – model. The velocity gauges on the casing were used to obtain the velocity distribution of fragments, and the axial spacing is 2 mm. The model was made of a casing, a head cap, and a charge. The charge was detonated by the detonation point at the center of the end. The casing, head cap, and fixed plate were automatically connected in AUTODYN – 3D. The fixed boundary simulated the fixed condition of the fixed plate.

The JWL is the most commonly used model to simulate the explosion of a charge, and it was adopted to describe the explosion of COMP – B:

$$P_b = C_1\left(1 - \frac{\omega}{r_1 \nu}\right)e^{-r_1\nu} + C_2\left(1 - \frac{\omega}{r_2\nu}\right)e^{-r_2\nu} + \frac{\omega E}{\nu} \tag{12}$$

where P_b is the blast pressure of explosive charge, E is the initial energy per initial volume. ω, C_1, C_2, r_1, and r_2 are constants, and V is the initial relative volume. The properties and parameters of the explosive charge were listed in Table 2.

Table 2 Parameters of COMP – B.

$\rho_0/(g \cdot cm^{-3})$	$D/(m \cdot s^{-1})$	P_{CJ}/GPa	$E_0/(kJ \cdot m^3)$	C_1/GPa	C_2/GPa	R_1	R_2
1.717	7 980	29	8.5×10^6	542	7.68	4.2	1.1

AISI 1045 steel is the most commonly used shell material in the study of the dynamic response after the explosion, and its parameters have long been recognized [33, 44]. The fracture and rupture of the casing can be simulated using the Johnson – Cook (JC) model:

$$\sigma = (A + B\varepsilon_{ep}^n)(1 + C\ln \dot{\varepsilon}^*)(1 - T^{*m}) \tag{13}$$

where A, B, C, m, and n are all constants, ε_{ep} represents the equivalent plastic strain, $\dot{\varepsilon}^*$ is the relative plastic strain rate; and T^* is the homologous temperature, which can be obtained as $T^* = (T - T_r)/(T_m - T_r)$. The parameters of the AISI 1045 steel are listed in Table 3 [63]. The casing would fracture when it expanded to a certain extent; therefore, the failure and erosion criteria must be added in the simulation. Based on previous studies, the plastic strain failure criterion in this study was adopted with a value of 0.23 [44], and the erosion criterion was the geometric strain with a value of 2.0 [64]. The JC model was also adopted for the constitutive model of the fixed plate, as listed in Table 3 [65].

The Shock EOS is a Mie – Gruneisen EOS based on the shock Hugoniot of dynamic fracture. This form of EOS is widely used and adequately represents most materials, and can be expressed as:

$$p = p_H + \Gamma_0 \rho_0 (e - e_H) \tag{14}$$

where it is assumed that $\Gamma_0\rho_0$ = constant and

$$p_H = \frac{\rho_0 c_0^2 \mu(1 + \mu)}{1 - (s - 1)\mu^2} \tag{15}$$

$$e_H = 0.5\left(\frac{p_H}{\rho_0}\right)\left(\frac{\mu}{1 + \mu}\right) \tag{16}$$

where Γ_0 is the Gruneisen parameter, $\mu = \left(\frac{\rho}{\rho_0}\right) - 1$, ρ and ρ_0 are the current and initial densities, respectively; and c_0 is the bulk sound speed [61, 66]. This EOS form is limited in that it is applicable only for a limited range of velocities because it does not allow for material phase changes such as melting or vaporization. The 'Steel 1006' material model from the Autodyn library has a similar yield stress (350 MPa) to that of the steel used in our experiment [64]. Hence, the default AUTODYN Shock EOS material constants of 'Steel 1006' were used for both metal materials in the simulation, as listed in Table 4 [67].

Table 3 Parameters of JC model.

Material	ρ_0/g · cm^{-3}	A(MPa)	B(MPa)	n	c	m
AISI 1045	7.85	507	320	0.28	0.064	1.06
Q345B	7.85	370	405	0.374	0.065	1.02

Table 4 Parameters of EOS.

c_0(m/s)	s	Γ_0	μ
4 569	1.49	2.17	81.8

4.2 Comparison with experimental results

A comparison of the fragment distribution between the test and simulation results is shown in Figure 10. The left side shows the fragment distribution obtained at two moments in the test, and the right side showed the simulation result at the first moment. The non - uniformity of fragments diffused along the axis is recognizable, which is in accordance with previous experiments [4, 68 - 70]. When the test and simulation results were compared, it was discovered that the fragmentation of the shell at different positions was similar. The distribution of the fragments in Figure 9 corresponds well with that of the protection panel in Figure 4. The difference lies in the fact that only part of the fragments in Region 1 of Figure 9 impacted Region 1 of the protection panel in Figure 4, and the rest of the fragments flew out of the range of the protection panel owing to excessive project angle. According to the fracture of the shell at the second moment on the left, the fragments can be divided into three regions, and each region corresponds to the region shown in Figure 4. The shell near Region 1 is subjected to the effect of rarefaction waves, resulting in an increase in the radial displacement gradient. The shell was broken into smaller fragments. With an increase in the distance from the detonation end, the effect of the axial rarefaction waves is weakened, the shear fracture of the shell is reduced, and the fragment size increases, as can be seen from the size of the impact crater in Region 2 of Figure 4, which was firstly described by Mott. Because the shell in Region 3 has almost the same velocity, the radial displacement gradient of the shell is extremely small, which causes the shell to break into larger fragments, as shown in Figure 4.

The stress concentration at the connection between the head cap and shell led to a fracture. Simultaneously, the head cap also broke into small fragments in Region 4 under the effect of the shock wave and detonation product. The fragments from Region 4 of 1# and 2# specimens impact the fixed plate and formed shallow impact craters, as shown by the impact craters within the red and yellow circles in Figure 10b, respectively. Fragments in Region 4 ricocheted after impacting the fixed plate, and the ricochet fragments impacted different regions on the protection panel in Figure 4. The ricochet characteristics of fragments can be found in another study by our team [71].

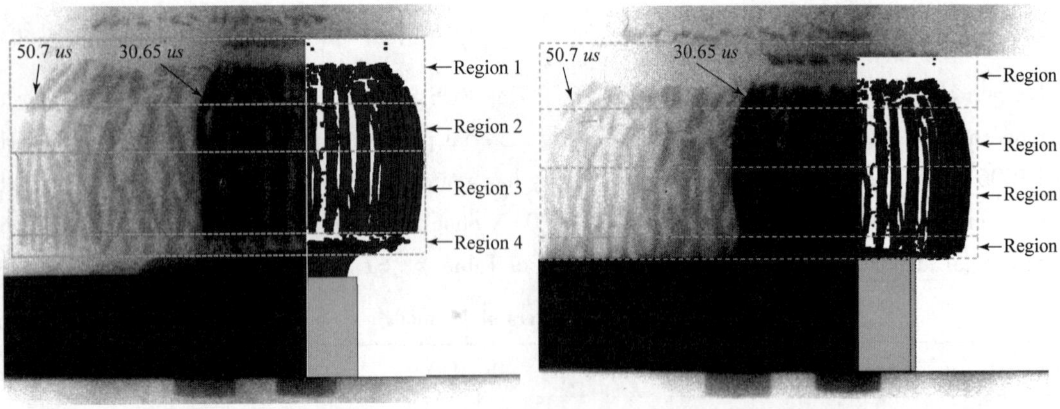

(a) (b)

Figure 9 Comparison of the fragment distribution between test and simulation at the same time.

(a) 1# specimen, (b) 2# specimen. The gray figure on the left is the test result, and the colored figure on the right is the simulation result. 30.65 μs and 50.7 μs represent two moments in the X-ray. The fragment velocity distribution at the first moment (30.65 μs) is compared. The fragments were divided into four regions according to their fragmentation and spatial distribution.

The residual body of the head cap after the explosion is shown in Figure 10. The height of the head cap of specimen 1# was twice the diameter of the charge, and the residual part of the end cap was more likely to show a tensile fracture, whereas the height of the head cap of specimen 2# was only a quarter of the diameter of the charge. The shock wave and detonation products first interact with the center of the head cap, causing localized higher temperature and steel discoloration. Owing to the restriction of the fixed plate, the head cap forms a strong shear band in the radial direction, forming a flat shear surface.

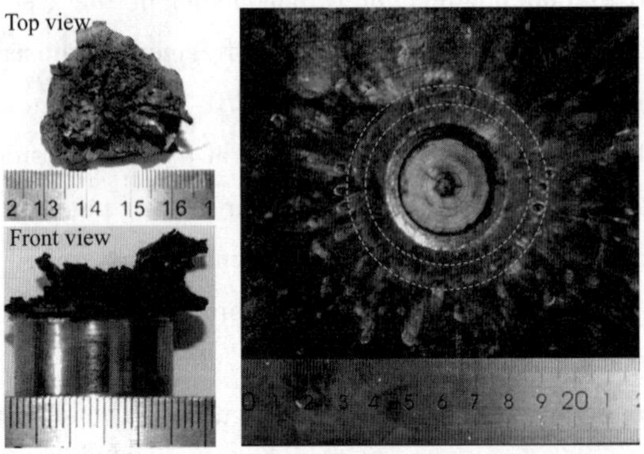

(a) (b)

Figure 10 Residual part of the head cap.

(a) 1# specimen head cap residual part. The left was the top view, the right was the front view, and the bottom scale had a minimum scale of 1 mm. (b) 2# specimen head cap residual part. The impact craters in the red circle came from Region 4 of 1#, and the impact craters in the yellow circle came from Region 4 of 2#.

In the simulation, the distribution of the fragment velocities was obtained through gauges and compared with the test and theoretical formula, as shown in Figure 11 (a). The simulation results were found to be in good agreement with the test results, and the simulation results near the detonation end were slightly lower than the test results, indicating that the influence of rarefaction waves was stronger in the simulation, which may be due to neglecting of the effect of air in the simulation. The fragment velocity near the fixed plate was consistent. In addition, it can be found that the height of the head cap did not affect the fragment velocity. Figure 11 (b) shows that the average relative errors of the simulation results of the two specimens were 3.8% and 3%, respectively, indicating that the simulation model had good consistency with the experiment. Consequently, our simulation model can be used to simulate casing fractures and ruptures.

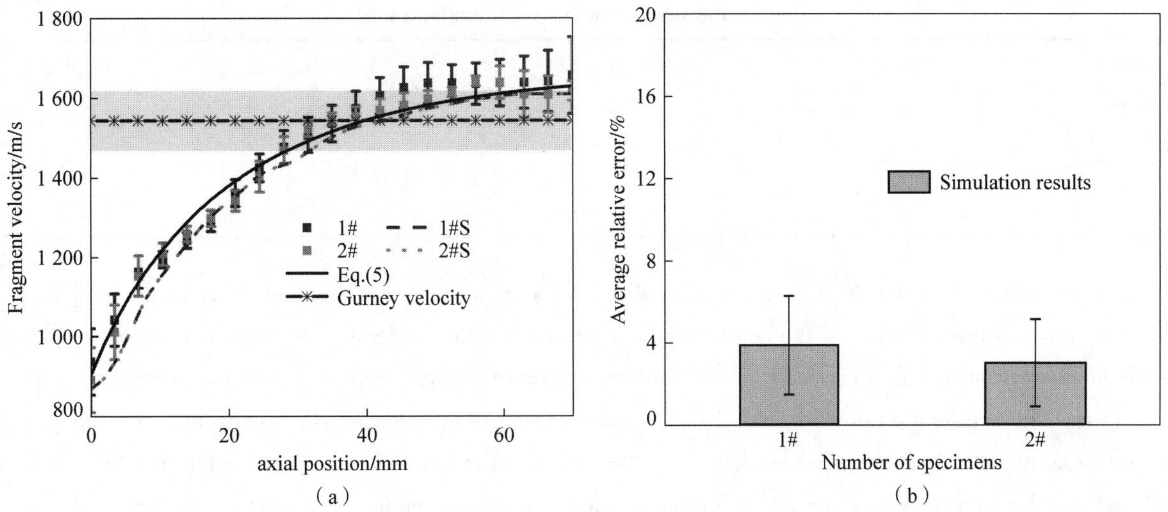

Figure 11　Simulation results.

(a) Comparison of the experimental data and simulation results. 1# and 2# represented the test data points of different specimens. 1#S and 2#S represented the simulation results of the corresponding test specimens. Eq. (5) shows the calculation results from the proposed formula. Gurney velocity was the theoretical value from Gurney equation. (b) Average relative error of the simulation results. It showed the average relative error of all velocity gauges for each model from the simulation.

5　Discussion

In this section, the above verified simulation model is used to analyze the relevant parameters. The reason why the fragment velocities exceeded the theoretical velocities is obtained in Section 5. 1. The effect of shell length on fragment velocity distribution is discussed in Section 5. 2.

5. 1　Acceleration process of fragment

In the previous section, the authenticity of the simulation model was confirmed. This section uses the simulation model to analyze the influence of the constrained end and explain why the fragment velocities near the constrained end were greater than the theoretical velocities predicted by the Gurney equation. The above simulation model was used to establish the model shown in Figure 12. The difference was that model No. 3 had no head cap on the non – detonation end, model No. 4 had an unconstrained end on the non – detonation end, and model No. 5 had a constrained end on the non – detonation end, which was the same as that of specimen 1#. The model parameters are listed in Table 5.

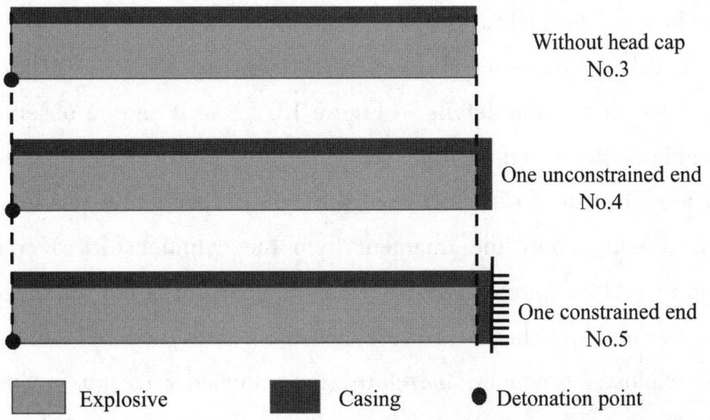

Figure 12　Three types of simulation models. Each simulation model consists of explosives and casing, which are detonated by the detonation point at the center of one end. The difference between the three models was the non – detonation end. Model No. 3 had no head cap, model No. 4 had an unconstrained end, and model No. 5 had a constrained end.

Table 5　Parameters of three simulation model.

No.	L(mm)	D(mm)	d(mm)	h
3	70	30	24	0
4				6
5				6

The simulation of model Nos. 3 and 4 can also be found and verified in our previous study [72]. To objectively analyze the influence of the constrained end on the fragment velocity, the maximum fragment velocities of each model are shown in Figure 13 (a), and their axial positions were selected to satisfy generality and randomness. The acceleration processing of the maximum velocity from specimen No. 5 (specimen 1#) adopted data from the gauge point, where the axial distance from the detonation end was 32 mm. For a more intuitive comparison of the two acceleration processes, they were moved to the same starting point, as illustrated in Figure 13 (b).

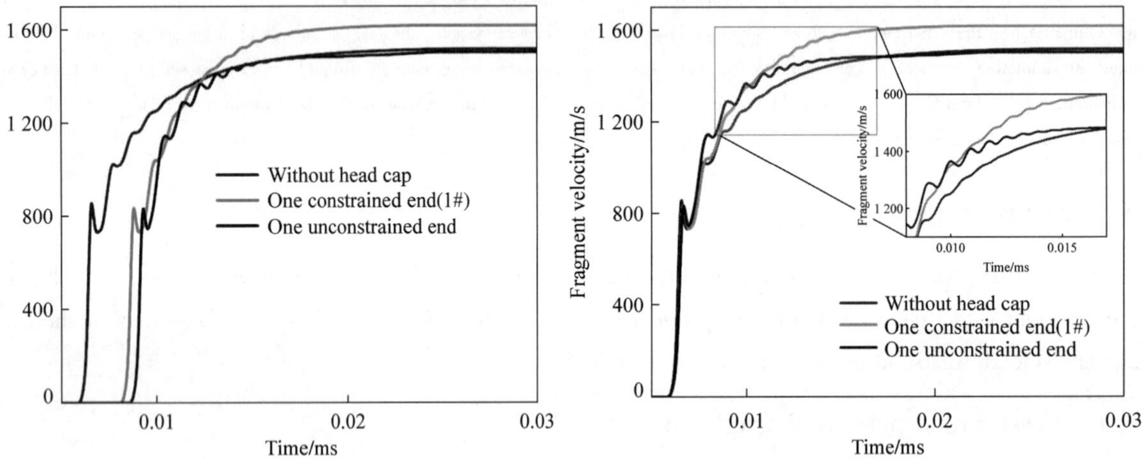

(a)　　　　　　　　(b)

Figure 13　Comparison of the acceleration processing of the maximum velocities.

(a) Original acceleration process, (b) Move to the same start position. The red line represented the acceleration curve of the maximum velocity generated from the shell without head end cap (No. 3 in Table 5), The green line was from the shell with an unconstrained end at the non – detonation end (No. 4 in Table 5), and the blue line was from the shell with a constrained end at the non – detonation end (No. 5 in Table 5), with the same parameters as the 1# specimen.

Previous experiments have proven that the acceleration process of cylindrical shells can be divided into two stages: the initial acceleration stage of the shock wave and the further acceleration of detonation products [73, 74]. The above figure can be used to analyze the acceleration difference between the two parts to determine the real reason why the fragment velocity exceeded the theoretical velocity.

By comparing the larger version of the details in Figure 13 (b), it can be observed that the existence of the end cap accelerated the fragment after a certain time. However, the final velocity of the fragment from the cylinder with an unconstrained end was consistent with that from the cylinder without a head cap, which was equal to the theoretical Gurney velocity. Finally, only the fragment from the cylinder with a constrained end continued to accelerate and exceeded the theoretical Gurney velocity. The comparison of the three acceleration processes can be explained as follows: when the incident shock wave reached the constrained end, the wave impedance of the end was greater than that of the explosive products; therefore, the reflected wave was a compression wave, which was called a weak reflected shock wave. The reflected shock wave accelerated the shell from the non – detonation end, which accounted for the sudden increase in the velocity of the fragments in simulation models 4 and 5.

The movement direction of the detonation products corresponds to the non – detonation end. The swelling of the detonation products caused the connection between the unconstrained end cap and cylindrical shell to first produce stress concentration and fracture. The detonation products dispersed outward from the fracture, and rarefaction waves entered simultaneously. In addition, with the continuous movement of the unconstrained end cap, the pressure and density of the detonation products continued to be attenuated and sparse, and the final fragment velocity did not increase or exceed the theoretical Gurney velocity. However, the pressure of the detonation product was not rapidly attenuated because of the immobility of the constrained end, and the detonation product continued to accelerate the fragment and exceeded the theoretical Gurney velocity.

Therefore, the reflected shock wave changed the accelerating process of the fragment, which made the fragment quickly reach its maximum velocity, and increased the pressure of the detonation products. However, the flying away of the unconstrained end cap caused rapid attenuation of the detonation product pressure, which did not accelerate the fragment to reach its maximum velocity. In contrast, the constrained end causes the detonation product with a higher pressure to remain for a longer period of time, allowing the fragment to continue to accelerate and exceed the theoretical Gurney velocity.

The principle of action of the reflected shock wave is illustrated Figure 14. The arrival time of the reflected shock wave at different axial positions can be calculated using the one - dimensional isentropic equation:

$$t = 1.42(1.7L - x)/D_e \tag{17}$$

The derivation process can be found in the Appendix. For example, the axial position of the fragment from a cylinder with constrained end in Figure 13 was 62 mm, and the time for the reflected shock wave to reach that position can be calculated, which was 0.01 ms. It can be found that the velocity increase also started at 0.01 ms (Figure 13), which again proved that the acceleration was caused by the reflected shock wave.

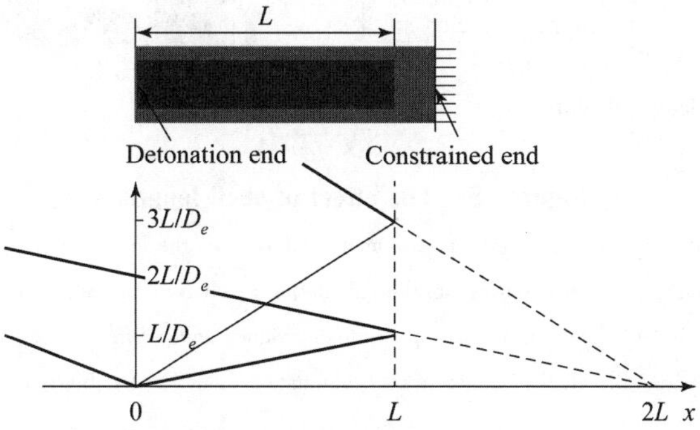

Figure 14 Schematic of reflected shock wave. L is the length of charge. D_e is the detonation velocity of explosive.

5.2 Effect of shell length

Figure 15 (a) shows the fragment velocity distributions from the cylindrical casing constrained at one end under different length - to - diameter ratios, where S and F represent the simulation results and calculated values, respectively. First, the calculated value of the proposed formula was in good agreement with the simulation results, which also proves that the proposed formula can predict the fragment velocity with sufficient accuracy. Second, it can be found that with an increase in the length - to - diameter ratio, an increasing number of fragment velocities exceed the theoretical velocities owing to the detonation products with higher pressure caused by the reflected shock wave from the constrained end. In addition, by summarizing the axial positions of the fragment velocity less than the theoretical velocity under different length - diameter ratios, it can be concluded that the rarefaction waves at the detonation end were the reason for the decrease in the fragment velocity and that their influence distance was twice the charge diameter. When the charge aspect ratio was reduced to 2, the effect of the reflected shock wave was completely offset by the rarefaction waves from the detonation end, and the fragment velocity no longer exceeded the theoretical velocity, which is very similar to the previous literature finding that rarefaction waves have significant influence when the length - to - diameter ratio is below 1.5 [75]. When the length - diameter ratio was less than 2, the effect of the rarefaction waves from the detonation end took over, and the reflected shock wave from the constrained end did not cause an increase in the detonation product pressure; thus, it did not increase the fragment velocities. The variation in the average relative error between the calculated and simulated values as a function of the length - to - diameter ratio is shown in Figure 15 (b). When the length - to - diameter ratio was less than 2, owing to the disappearance of the reflected shock wave effect, the average relative error of the proposed formula increased, but it was still less than 4%. When the length - to - diameter ratio was greater than 2, the average relative error decreased even more. Considering the fact that the warhead usually has a large length - to - diameter ratio, which is usually greater than 2, the accuracy of Eq. satisfies the following requirements.

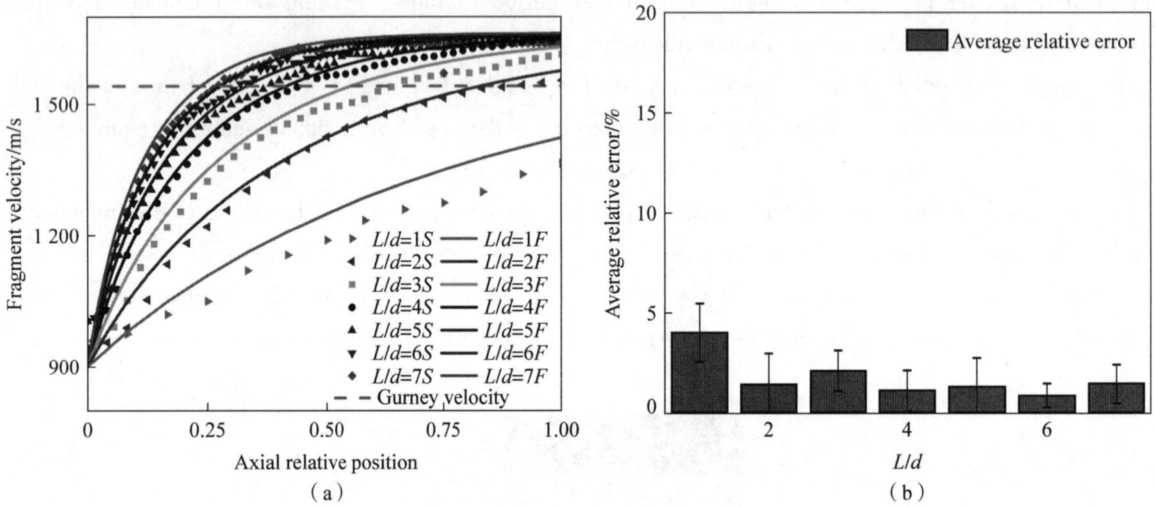

Figure 15 The effect of shell length.

(a) Fragment velocity distributions under different length - to - diameter ratios. L is the length of charge; d is the diameter of charge; L/d is the length - to - diameter ratio; S is the simulation result; F represents the calculation result from the proposed formula; Gurney velocity is the theoretical value calculated by the Gurney equation. (b) Average relative error of calculated values under different length - to - diameter ratio. The average relative error and standard deviation were calculated by all velocity gauges of each simulation model under different length - to - diameter ratios.

6 Conclusions

Aiming at the fragment characteristics of a cylindrical casing constrained at one end, experiment and numerical simulations were performed in this study. Two specimens were tested using X - ray radiography and the velocity and distribution of the fragments were obtained. The experimental results indicated that the constrained end increased the fragment velocity to exceed the theoretical Gurney velocity. Four regions were identified in this study. Owing to the large radial displacement gradient, the number of fragments in Region 1 was the largest, whereas their size was the smallest. In Region 2, the fragment size increased owing to the decrease in the radial displacement gradient. The radial displacement gradient in Region 3 was the smallest, and the fragment size and kinetic energy were the largest. Region 4 came from the head cap, where small fragments were formed at the connection between the head cap and cylindrical shell, and a ring of impact craters was formed on the fixed plate.

Based on the experimental data, a new formula is established to predict the fragment velocity distribution from a cylindrical casing constrained at one end. Compared with the experimental data, the average relative errors of the new formula were less than 3.3%, indicating high accuracy of the new formula. Through the analysis of the fragment acceleration process in the simulation, it was found that the reflected shock wave generated from the constrained end made the fragment reach the Gurney velocity faster, but the continuous acceleration of detonation products with higher pressure caused by the reflected shock wave was the final reason why the fragment velocity exceeded the theoretical velocity. A formula for calculating the onset time of the reflected shock wave at different axial positions was established using theoretical calculation.

The influence of the length – to – diameter ratio was also analyzed. The results showed that the fragment velocity was affected by the rarefaction wave at the detonation end and reflected shock wave at the constrained end. The proposed formula can predict the fragment velocity from a cylinder constrained at one end with different length – to – diameter ratios. When the length – to – diameter ratio was less than 2, the effect of the rarefaction waves from the detonation end took over, and the reflected shock wave from the constrained end did not cause an increase in the detonation product pressure; thus, it did not increase the fragment velocities. As the length – to – diameter ratio exceeded 2, the reflected shock wave dominated, and an increasing number of fragments had a velocity exceeding the theoretical velocity. The results of this study provide theoretical guidance for the design and evaluation of related warheads.

Acknowledgements

The authors would like to thank my colleague for their assistance and Prof. Feng for his guidance.

Appendix

The theoretical model used in this study is shown in figure. The explosive was detonated at one end, and the constrained end was regarded as a rigid wall. After the explosive exploded, a detonation wave formed. When it arrived at the right rigid wall, it formed a reflected left – shock wave. Because the entropy production of the detonation product is small, it can be regarded as an isentropic flow. According to the one – dimensional isentropic flow theory of gas and assuming that the isentropic ratio γ is equal to three, the one – dimensional isentropic equation of the detonation products can be expressed as follows:

$$\begin{cases} \frac{\partial}{\partial t}(u+c)+(u+c)\frac{\partial}{\partial x}(u+c)=0 \\ \frac{\partial}{\partial t}(u-c)+(u-c)\frac{\partial}{\partial x}(u-c)=0 \end{cases} \tag{A1}$$

Consider the left – end initiation as an example: before the detonation wave reaches the right – end rigid wall, that is, $t=\frac{L}{D_e}$, D_e is the detonation velocity of the explosive. A luster wave is a simple right – going wave zone that can be solved using a unique solution of Eq. (A1):

$$\begin{cases} x=(u+c)t+F_1(u) \\ u-c=const. \end{cases} \tag{A2}$$

where the value of $F_1(u)$, and the constant can be obtained based on the initial conditions. When the time is zero, x is 0, and thus, $F_1(u)$ is 0. The coefficients of the explosive products after the detonation waves are the C – J coefficients, where $u_j=1/4\,D_e$, $c_j=3/4\,D_e$. Hence, the solution can be obtained as:

$$\begin{cases} x=(u+c)t \\ u-c=-\frac{1}{2}D_e \end{cases} \tag{A3}$$

Hence, the right – going simple wave state parameters are as follows:

$$\begin{cases} u=\frac{x}{2t}-\frac{D_e}{4} \\ c=\frac{x}{2t}+\frac{D_e}{4} \end{cases} \tag{A4}$$

When $t > \frac{L}{D}$, the right - going simple wave reaches the right rigid wall, and the reflected first left - going shock wave propagates in the right - going simple wave zone and form a composite wave field, which can be described as:

$$\begin{cases} x = (u+c)t \\ x = (u-c)t + F_2(u-c) \end{cases} \tag{A5}$$

When $t = \frac{L}{D}$, x is L and u is equal to zero at the wall. For the first right - going wave, $u + c = u_j + c_j = D_e$. Hence, the speed of sound C at the rigid wall is equal to D_e. $F_2(u-c)$ can be obtained, which is equal to $2L$. The composite wave field can be expressed as:

$$\begin{cases} x = (u+c)t \\ x = (u-c)t + 2L \end{cases} \tag{A6}$$

Because u is equal to zero at the rigid wall, the speed of sound at the rigid wall can be obtained based on the above equation:

$$c = \frac{L}{t} \tag{A7}$$

According to the acoustic approximation of a weak shock wave, the velocity of the reflected shock wave is:

$$D'_e = \frac{1}{2}[(u_1 - c_1) + (u_2 - c_2)] \tag{A8}$$

When γ is 3, $u + c = u_0 + c_0$, where u_0 and c_0 are the parameters of the explosive products after the detonation wave arrives, as shown in Eq. (A4). Hence, the propagation equation of the reflected shock wave is:

$$\frac{dx}{dt} = D'_e = u - c = \frac{x}{2t} - \frac{D_e}{4} - \frac{L}{t} \tag{A9}$$

Integrating Eq. (A9) yields

$$x = -\frac{D_e t}{2} - \frac{\sqrt{LD_e t}}{2} + 2L \tag{A10}$$

Eq. (A10) can be approximated by the following linear equation:

$$x = 1.7L - D_e t/1.42 \tag{A11}$$

Thus, the initial acceleration time of the reflected shock wave to the fragment at each axial position is:

$$t = 1.42(1.7L - x)/D_e \tag{A12}$$

References

[1] H. Y. Grisaro, A. N. Dancygier, Characteristics of combined blast and fragments loading, International Journal of Impact Engineering, 116 (2018) 51 - 64.

[2] R. R. Karpp, W. W. Predebon, Calculations of Fragment Velocities from Naturally Fragmenting Munitions, in, ARMY BALLISTIC RESEARCH LAB ABERDEEN PROVING GROUND MD, 1975.

[3] G. Taylor, The fragmentation of tubular bombs, Advisory Council on Scientific Research and Technical Development, 5 (1963) 202 - 320.

[4] W. J. Stronge, X. Ma, L. Zhao, Fragmentation of explosively expanded steel cylinders, International Journal of Mechanical Sciences, 31 (1989) 811 - 823.

[5] G. I. Taylor, Analysis of the explosion of a long cylindrical bomb detonated at one end, The Scientific Papers of Sir Geoffrey Ingram Taylor, 3 (1963).

第 三 部 分
短路软毁伤与导电纤维弹技术

导电丝束镀层金属引发相间电弧的机理研究①

梁永直②，冯顺山
（北京理工大学爆炸科学与技术国家重点实验室，北京　100081）

摘　要：文中从电磁感应、气体放电及电弧理论出发，根据武器上应用的导电丝束作用于高压导线相间所引发放电电弧的特性，提出了导致电弧产生的四种物理作用效应的观点，并建立了描述其特征的数学模型；分析了镀层金属引发电弧的初始状态、转化趋势及其转化的临界判据，对导电丝束强制引弧的物理过程及其作用机理做了初步的探讨和研究。

关键词：导电丝束；高压导线；金属蒸汽态电弧；气体态电弧；强制引弧

Study on the Mechanism of Arc Discharge of Conductivity Fiber Between Wires of High – voltage

LIANG Yong zhi，FENG Shun shan
（Beijing Institute of Technology 100081，China）

Abstract：In this paper，based on the characteristics of arc discharge between wires of high – voltage，a new kind of viewpoint about the mechanism of arc discharge and its corresponding maths model are presented. The critical criterion of arc convertion from metallic to gaseous phase is obtained by analysis the arc discharge condition and its changing trend. The conception of compulsing arc is also discussed in this paper.

Keywords：conductivity fiber；high – voltage；metallic arc；gaseous arc；compulsing arc

1　引言

现代社会生活人们高度依赖电力，电力系统的运行安全与稳定直接关系到国计民生、社会经济生活以及军事国防安全。导电纤维（碳纤维）战斗部对电力系统实施攻击（强制引弧短路），可以造成短路毁伤效应，使得变电站、发电厂等电力设施瘫痪，在相当一段时间内失能，导致工业和社会运营、防空作战系统以及通信、运输等系统被迫处于瘫痪状态。在武器上应用的导电丝束是一种非致命杀伤元素，其强制引弧能力是衡量导电丝束性能的重要指标，其研究成果对导电丝束设计和制造提供依据。因此，为了获得不同导电丝束对高压线路的引弧特性，从导电丝束毁伤高压线路的角度出发，对导电丝束的强制引弧物理过程及其作用机理做了初步的探讨和研究。

2　导电丝束作用于高压导线初期时金属蒸汽的形成及其数学模型

2.1　导电丝束作用的物理过程

导电丝束在飘落、搭接到高压导线的过程中，首先受到来自导线周围环境强电磁场的作用而历经电

① 原文发表于《弹箭与制导学报》2004 年第 24 卷第 2 期。
② 梁永直：工学博士，2000 年师从冯顺山教授，研究导电丝束对高压电网短路毁伤机理及效应，现工作单位：加拿大科技公司。

感效应，在此过程中，丝束表面镀层金属在强电场的作用下引发强场致发射而使其蒸发、汽化成金属蒸汽；同时，丝束中电流的 I^2R 能量使得丝束本身产生急剧高温，也使得丝束表面的金属镀层产生热游离效应而迅速蒸发、汽化，二者效应作用于丝束及其表面镀层金属使其成为金属蒸汽而充斥高压导线相间，前者为金属蒸汽的初始来源，后者为主要来源。金属蒸汽的介入破坏了高压导线相间气体绝缘介质的安全距离，并且在强电场作用下带电质点的有序移动加速破坏了相间的绝缘条件，其过程可由带电质点的电流密度 J 描述，如式（6）和式（7）所示。在相间间隙充满金属蒸汽的条件下，引发空气间隙击穿放电，形成金属蒸汽态电弧。随后，导电丝束在高温、强光、强电场综合作用下，完全汽化并且使周围空气产生强烈电离，形成金属离子、电子和等离子体构成的电弧短路通道[6-8]。图 1 所示为导电丝束介入高压导线初始瞬间表面金属镀层汽化的示意图。

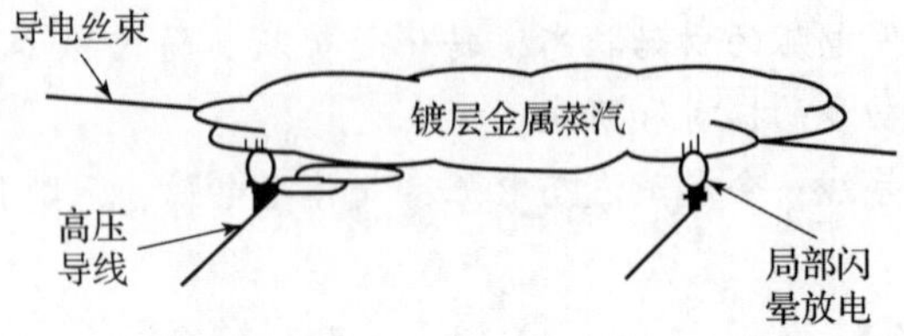

图 1　导电丝束介入高压导线初始瞬间表面金属镀层汽化

2.2　电感效应作用的数学模型描述

导电丝束作用于高压导线的电感效应，就其本质而言，服从于 Faraday 电磁感应定律，即发电、用电是机械能通过切割磁力线与电能的相互转换，基于实验的 Faraday 电磁感应定律给出了在一封闭回路上的由纯切割磁场和非切割磁通变化综合感应的总电动势[2]，即

$$\begin{aligned} E(t) &= \oint E_t(t)\cdot \mathrm{d}l = \oint E_c(t)\cdot \mathrm{d}l + \oint E_n(t)\cdot \mathrm{d}l \\ &= -\frac{\partial}{\partial t}\iint[\boldsymbol{B}_c(t)+\boldsymbol{B}_n(t)]\cdot \mathrm{d}s = -\frac{\partial}{\partial t}\iint \boldsymbol{B}_t(t)\cdot \mathrm{d}s \end{aligned} \tag{1}$$

式中　$\boldsymbol{E}_t(t)$——总感生电场强度；

$\boldsymbol{E}_c(t)$——切割磁场感生电场强度；

$\boldsymbol{E}_n(t)$——非切割磁场感生电场强度；

$\boldsymbol{B}_c(t)$——切割磁场强度；

$\boldsymbol{B}_n(t)$——非切割磁场强度。

但在实际的作用过程中，导电丝束的搭接可能是错综复杂的闭合或非闭合回路，上述任一种感生电动势可发生在导体微元上，沿一闭合回路上的总感生电动势只是沿回路各个微元上感生电动势的积分，因此，获得其感生电动势的微分形式，可解决导电丝束在各种复杂条件下的非闭合导体段上感生电动势的计算问题。

（1）纯切割磁场产生的动生电动势微分形式。

$$\mathrm{d}\boldsymbol{\varepsilon}_c(t) = [U_{dl\cdot b}(t)\times \boldsymbol{B}(t)]\cdot \mathrm{d}l = E_c(t)\cdot \mathrm{d}l \tag{2}$$

式中　$U_{dl\cdot b}(t)$——微元 dl 与磁场 $\boldsymbol{B}(t)$ 的相对速度 $U_{dl\cdot b} = U_{de} - U_b$。

（2）非切割磁场产生的感生电动势微分形式。

$$\begin{aligned} \mathrm{d}\varepsilon_n(t) &= \int_L\left[\iint \frac{\boldsymbol{r}_{sdE}\times \boldsymbol{B}'(t)(\boldsymbol{n}'_b\cdot \mathrm{d}s)\mathrm{d}L}{4\pi r_{sdE}^2}\right]\cdot \mathrm{d}l \\ &= \int_L \frac{\boldsymbol{r}_{sdE}\times \boldsymbol{\Phi}'(t)\mathrm{d}L}{4\pi r^2}\cdot \mathrm{d}l \end{aligned} \tag{3}$$

式中　$\boldsymbol{\Phi}'(t)\mathrm{d}L$——磁通量变化率微元，即 $\boldsymbol{\Phi}'(t) = \frac{\partial}{\partial t}\iint B(t)\cdot \mathrm{d}s$。

综合上述，Faraday 电磁感应定律可表示为

$$\begin{aligned}\varepsilon_{ab}(t) &= \varepsilon_c(t) + \varepsilon_n(t) \\ &= \oint\left[U_{de\cdot b}(t) \times \boldsymbol{B}(t) + \int_L \frac{\boldsymbol{r} \times \Phi'(t)\,\mathrm{d}l}{4\pi r^2}\right] \cdot \mathrm{d}l\end{aligned} \tag{4}$$

（3）非闭合线段的感生电动势计算。

$$\begin{aligned}\varepsilon_{ab}(t) &= \int_a^b \left[\boldsymbol{E}_c(t) + \boldsymbol{E}_n(t)\right] \cdot \mathbf{d}\boldsymbol{l} \\ &= \int_a^b \left[U_{dl\cdot b}(t) \times \boldsymbol{B}(t) + \int_L \frac{\boldsymbol{r} \times \Phi'(t)\,\mathrm{d}l}{4\pi r^2}\right] \cdot \mathrm{d}l\end{aligned} \tag{5}$$

2.3　强场致发射效应的数学模型描述

当导电丝束镀层金属内电子能量超过金属的逸出功时，电子会从金属表面逸出，脱离金属的势垒束缚而形成传导电流。当外加电场作用时，电子由无限远处逐渐接近金属表面时，由于金属内部镜像电荷的吸引，电子的位能随位置的变化而逐渐降低，进而陷入基态，如图 2 中曲线 1 所示[1]。

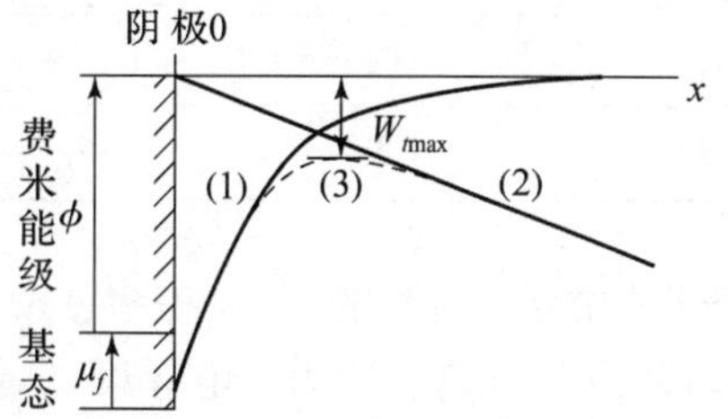

图 2　导电丝束镀层金属表面电子在电场中位能变化曲线

当导电丝束介入高压电场时，距离表面镀层金属 X 处的电子按曲线 2 所示的能量 $W_f = eEx$，随 x 轴的增加而进入较低能级，二者叠加得曲线 3，其描述了导电丝束介入高压电场时表面镀层金属内电子的能量与电子逃离金属表面的距离之间的关系。其中 $W_{t\max}$ 为在外加电场作用下金属本身对电子构成势垒的逸出功 Φ 的降低值。由 Schottky 理论[1]，其场致发射效应可用电流密度 J 表示为

$$J = AT^2 \exp\left\{-\frac{e}{KT}\left[\Phi - \left(\frac{eE}{4\pi\varepsilon_0}\right)^{0.5}\right]\right\}(\mathrm{A/cm^2}) \tag{6}$$

$$W_{t\max} = -e\,(eE/4\pi\varepsilon_0)^{0.5} \tag{7}$$

式中，k，T，E，Φ 分别表示玻尔兹曼常数、导线温度、导线电极表面电场强度、金属的逸出功及杜斯曼常数 $A \approx 120.4(\mathrm{A/cm^2 \cdot K^2})$。

2.4　电热效应的数学模型描述

历经电感效应后，丝束本身有短路电流通过并产生热量，由于时间很短（由实验证实，通常为 ms 级以下），该热量来不及向周围介质扩散，从而将电能转换成热能 Q 累积到丝束内，有

$$Q = \int_0^t RI^2(t)\,\mathrm{d}t = \int_0^t I(t)\varepsilon_{ab}(t)\,\mathrm{d}t \tag{8}$$

其中，$\varepsilon_{ab}(t)$ 由本文 2.2 小节式（2）~式（5）确定。

又因为单位长度丝束的累积热量可表示为 $Q = \int_0^t q_s\rho_s C_s \frac{\mathrm{d}T(t)}{\mathrm{d}t}\mathrm{d}t$，将其代入则有丝束的最大累积热载荷为

$$Q_{\max} = \int_0^t RI^2(t)\,\mathrm{d}t = \int_0^t q_s\rho_s C_s \frac{\mathrm{d}T(t)}{\mathrm{d}t}\mathrm{d}t \tag{9}$$

式中，q_s，ρ_s，C_s，$T(t)$ 分别表示丝束截面积、丝束的密度、丝束材料的比热和温度。

2.5　热游离效应的数学模型描述

当气体介质的温度升高时，气体分子的运动碰撞也随之加剧，热游离效应占据主导作用，它可用表

征热游离强弱的游离度 x 来描述。其主要影响因素有介质温度、介质的游离电位和气体压力。图 3 所示为不同介质气体在不同温度 T 时的游离度 x[3]。表 1 给出了不同气体的游离电位值[4]。

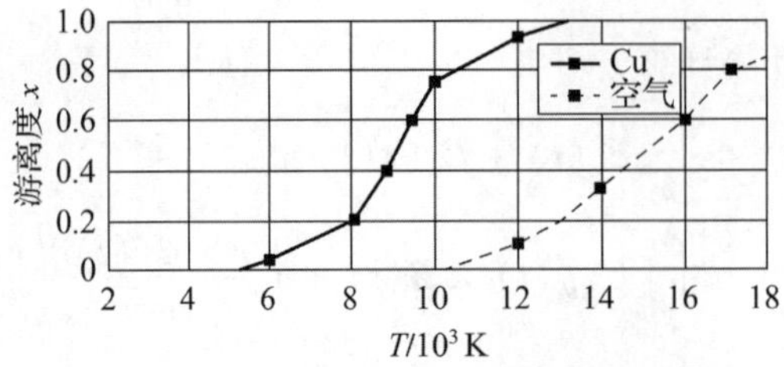

图 3　不同介质气体在不同温度时的游离度

表 1　气体介质的游离电位　/V

气体介质	H	H2	O	O_2	N_2	He
游离电位	13.6	15.4	13.6	12.5	15.8	24.5
气体介质	Al	Cu	Fe	W	Hg	
游离电位	6.0	7.72	7.86	7.89	10.4	

由图 3 和表 1 可以看出，金属蒸汽的游离电位比空气小得多，Al 蒸汽原子大约在 4 000 K 时就有了明显的游离，而空气要在 10 000 K 时才出现游离。这为导电丝束金属镀层材料的选择提供了理论依据。

气体压力增大，则游离程度减小。由于金属蒸汽的游离电位小于空气，因而更易于游离。金属蒸汽电弧中带电质点的游离状况可以用 Saha 理论[4]描述为

$$\frac{x^2}{1-x^2}p = 3.20\times10^{-5}T^{2.5}\exp\left(-\frac{eU_y}{KT}\right) \tag{10}$$

式中，K，p，T，U_y，x 分别为玻尔兹曼常数（$K=1.37\times10^{-23}$ J/K）、气体压力（kPa，1 atm = 101.3 kPa）、介质温度（K）、介质的游离电位（V）及游离度（$0\leqslant x\leqslant1$）。

此时，在介质为金属蒸汽电弧条件下，可推得带电质点的游离度为

$$x = \frac{5.66\times10^{-3}T^{1.25}\exp(-583\,9U_y/T)}{[3.20\times10^{-5}T^{2.5}\exp(-1.17\times10^4U_y/T)]^{0.5}} \tag{11}$$

3　金属蒸汽电弧向气体电弧的扩散及其临界判据

随着金属蒸汽电弧通道的形成，短路电流的急剧上升，当电流达几安培及其以上时（实际情况可达数百上千安培），I^2R 能量产生剧烈高温导致周围环境气压升高，金属蒸汽扩散而密度降低；当温度达到空气介质游离电位的临界温度 T_m 时，金属电弧周围空气气体分子开始强烈电离，使得金属蒸汽态电弧向气体态电弧的过渡。金属蒸汽态电弧的持续时间与周围环境气压、电弧电流、外加磁场有关。

假设这一内部扩散过程在理想气体中进行，其规律遵循 Stefan Maxwell 公式，则由电弧动态能量平衡原理，即电弧的发热量等于弧柱含热量的增加与散热损耗之和，得

$$I_hU_h = \frac{\mathrm{d}Q}{\mathrm{d}t} + P_s \tag{12}$$

而 $Q = q_h\rho C_T(T_m - T_0)$，代入式（12）整理后得，金属蒸汽电弧向气体电弧转化的临界判据为

$$T_m \geqslant \frac{1}{q_h\rho C_T}\int_0^t (I_hU_h - P_s)\,\mathrm{d}t \tag{13}$$

式中，q_h，ρ，C_T，I_h，P_s 分别为金属电弧的弧截面、浓度、比热、弧电流、弧电压和散热损耗。

实验证实[5]，在电弧电流恒定时，周围环境气压升高，则电弧总的持续时间变大，而金属蒸汽态电弧的持续时间缩短；在小电流状况下（3 A 以下）时，电弧为以金属蒸汽态为主。随着电流的加大，全

部燃弧时间中金属蒸汽态电弧所占比例降低。其原因为 I^2R 能量使得周围气体分子的电离及参与电弧内部电流的传输过程较早加入；外加磁场能够促进电弧运动，实际上是促进了周围环境介质分子的电离过程，以及促进了电离形成离子和电子参与电弧电流的传输过程，缩短了金属蒸汽态电弧的持续时间。

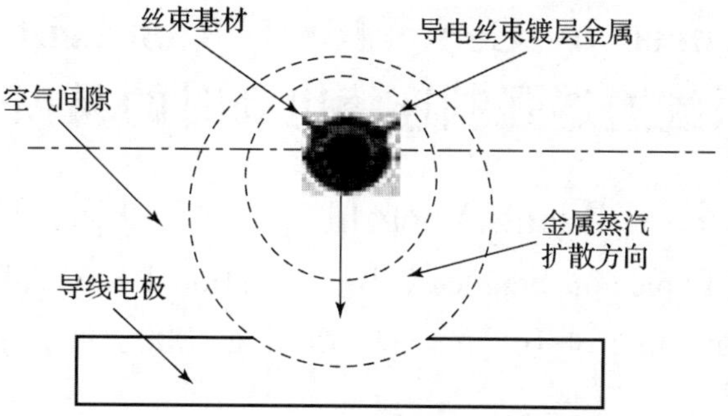

图 4　导电丝束金属蒸汽扩散模型

4　结论

本文首次从理论上提出了军事上使用的导电丝束作用于高压导线相间而产生的强制引弧的概念，分析了金属镀层引发电弧的初始状态、转化趋势及其转化的临界判据，提出了金属镀层导致电弧产生的四种物理作用效应的观点，并建立了描述其相应特征的数学模型；对其物理过程及其作用机理进行了理论探讨和定性研究。

研究了金属蒸汽态电弧的形成机理，认为在导电丝束介入高压导线相间的初期，导电丝束金属镀层在强电场作用下的场致发射效应是金属蒸汽形成的初始来源，随后丝束历经的强电场电感效应而产生的 I^2R 能量引发的热游离是金属蒸汽形成的主要原因。

分析了金属蒸汽态电弧向气体态电弧扩散的影响因素，给出了其扩散模型，建立了金属蒸汽态电弧向气体态电弧转换的临界判据。

参考文献

[1] 中野义映. 高电压技术 [M]. 北京：科学出版社，2004.
[2] 冯顺山，等. 导电纤维战斗部若干问题分析 [C] //无人驾驶飞行器学会战斗部与毁伤效率专委会2003年会论文集，2003.
[3] 徐国政，张节容，钱家骊，等. 高压断路器原理和应用 [M]. 北京：清华大学出版社，2000.
[4] 林莘. 现代高压电器技术 [M]. 北京：机械工业出版社，2002.
[5] 荣命哲，贾文慧，王瑞军. 低压电器电极间发生的电弧放电及其特性 [J]. 低压电器，1998 (3)：9 - 14.
[6] Carbon fiber risk analysis：NASA CP - 2074 [S]. MARCH，1979.
[7] Carbon/graphite fiber risk analysis and assessment study：NASA - CR - 159212 [R]. Jan，1980.
[8] High performance fibers：ADA274693 [R]. 1994.

Insecure Analysis on Electric Power System when Experienced by Compulsively Consecutive Arc①
电力系统遭遇强制连续电弧时的易损性分析

Feng Shunshan, Liang Yongzhi②, He Yubin, Hu Haojiang

(Explosion Science& Tech. National Key Lab,

Beijing Institute of Technology, Beijing 100081, P. R. China)

Abstract: In this paper. based on the characteristics of arc discharge which is caused by conductive fibers applied on weapons through the impact between high - voltage (HV) wires, a new concept of compulsively consecutive are CCA) is presented firstly. A mathematical description of four types of effects is given and further to the critical criterion of are conversion from metallic to gaseous phase is also obtained. A numbers of proposals are set forth by analyzing the insecure factors of electric power system and their developing trend after effects of CCA, based on the blocking mechanism, removal mechanism and power system resumption mechanism of conductive fibers.

Keywords: compulsively consecutive arc (CCA); conductive fibers; metallic arc; gaseous arc; insecure factors

1 Introduction

During the daily operation of the power system, disturbances of some improper and unsafe factors of an kinds by nature or human beings are frequently experienced. Such as short circuits either transient or short duration acted upon the high—voltage (HV) power facilities on casual cases by the effect of conducting materials or a thundering strike, leading to the corona. flashing, arcing and discharging of the IIV transmission lines and substation facilities and as a consequence. disaster of short circuits happened. Eastern interconnect of blackout on August 14th is a typical case [5]. Under current circumstances and international background of regional wars and anti - terrorism activities, unsafe factors of the cases increase to deliberately ruin the HV Power system, causing the severe damages. Especially at the early stage of the Gulf War and Kosovo War in 1990's, the soft wound attacks by the US force to strike the power system of Iraq and Yugoslavia where ammunition involved conductive fibers (carbon fibers) with metallic coated were employed, causing the severe short circuits damaging effects. As such, majority of critical substations and power stations experienced pow supply failures and the whole national power grid were collapsed.

Conductive fibers on weapons is a kind of soft wound element. It can cause compulsively consecutive arc (CCA) to HV power system. bringing the severe short circuit damages [1, 2]. To take effective mitigation measures and to minimize the loss, it is necessary to study the physical process: functional mechanism of CCA and its effects when applied on the weapons acted upon the HV. large capacity transmission lines and further to analyze its impact and strategy thereof associated with the power grid and eventually provide a theoretical evidence and technical background to ensure the safe operation of power system.

---

① 原文发表于《安全科学与技术国际会议》(2004)。

② 梁永直：工学博士，2000 年师从冯顺山教授，研究导电丝束对高压电网短路毁伤机理及效应，现工作单位：加拿大科技公司。

2　CCA Effects of Conductive Fibers and its Mathematical Description.

2.1　Concept of CCA

When some conductive fibers or conductive material acts on phase to phase or phase to ground of the HV power facilities, short circuit arc occurs and will he sustained till unit trips. Such event happens and even by several reclosing will be in vain. In general. CCA, as its name goes. is a phenomenon. a specially designed and manufactured conductive materials (including any conductive fibers, powder of any forms) but acted on the power facilities to destroy the insulation condition, causing the effects of short circuit arc and discharge and then unit trips. To be further, CCA is not singular one hut frequent occurrence to make the reclosing of unit loss its function while a "permanent trouble" is formed (loss power supply for a period of time, which has to be cleared). The jeopardy by the CCA effects differs from the one experienced by nature. The arc when happened tinder normal condition. it is transient, poor and short duration. Sometimes it does not to trip the units or just trip once to clear the short circuit. However, the CCA effect, especially for the key power facilities, the short circuit failure will be possible to extend the whole power system, causing the power grid interference, disturbance, malfunction and local disconnection from grid and further to the large blackout. Such occurrence is beyond control by the system protection. It brings adverse effects on the power system safe operation. Therefore, careful study and more attention are needed to study and analyze.

2.2　Physical Process of CCA Effects

The text describes the physical process of the function of conductive fiber when acted on the HV transmission lines.

The lab study demonstrates that the indication of the physical function of the conductive fiber shows that far field attracts and near fields expels. The reason is that the effect of electric induction, which makes the attractive force between the conductive fiber and conductor is greater than the expelling force governed by the principle of Faraday. The far field expels, which means that the two objects when situated closer. the free electrons have moved, making the surfaces of two objects bear the same polarity [2].

When the fiber coated with metal drops to the HV transmission lines, the magnetic inductive effects occurs at the surrounding electro - magnetic forces. During the process, the coated metal attached to conductive fiber begins to emit electrons, evaporate and then turn to metallic gas. Meanwhile, the I^2R energy in the fiber makes the fiber itself extremely overheated, leading to the effect of free float at coated metal and have it rapidly evaporated. Such effect when acted on the coated metal turns it into the metallic gas, which spreads over the HV transmission lines. The former is the initial source of metallic gas while the latter is the main source. The intervention of metallic gas causes the breakage of the insulation distance between HV wires. Under the force of strong electric filed, the movement of electrons breaks rapidly the phase to phase insulation, its process is indicated by the current density as J, as shown in equation (5). When the phase to phase clearance is filled with metallic gas, the discharge happens at air clearance, resulting in the formation of metallic arc. As a consequence, the electrons from the base material of conductive fiber and the surrounding air are created under such high temperature and strong field. The conductive substance has increased and the arcing short circuit path formed even have stronger the short circuit capability.

2.3　Mathematical Description of CCA Effects

a) Electro - induction effect

The effect of electro – induction when acted up on HV transmission lines by conductive fiber, as it name goes, is subject to the electron – induction principle of Faraday. I. e. the power generation is converted from the mechanical energy into electric energy. The principle of Faraday is shown as follow [2]:

$$E_t(t) = \int_c \vec{E}(t)\vec{d}l + \int_n \vec{E}(t)\vec{d}l = -\frac{\partial}{\partial t}\iint [B_c(t) + B_n(t)]\vec{d}s \tag{1}$$

in which, E is total magnetic inductive field strength; E is cutting magnetic inductive filed strength; E is non – cutting magnetic inductive filed strength; B) is cutting magnetic field strength; B) is non – cutting magnetic field strength;

However, the actual process is that the conductive fiber may be dropped onto the complicated closed or non – closed circuit, either of the inductive electromotive force of which may occur at its conductor micro – unit, the total inductive electromotive force generated along with a closed circuit is only the integral of various circuit micro – units. Therefore, the differential form generated by the inductive electromotive force can resolve these issues to calculate the sectional induction electromotive force of conductive fiber under various complicated non – closed conductors.

①Differential format of inductive electromotive force generated by pure – cut of magnetic fields

$$d\varepsilon_c(t) = [\vec{v}_{dl\cdot b}(t) \times \vec{B}(t)] \cdot \vec{d}l = \vec{E}_c(t) \cdot dl \tag{2}$$

in which, $\vec{v}_{dl\cdot b}(t)$ is relative velocity at dl vs $\vec{B}(t)$, $\vec{v}_{dl\cdot b} = \vec{v}_{de} - \vec{v}_b$.

②Differential format of inductive electromotive force generated by non – cut of magnetic fields

$$d\varepsilon_n(t) = \int_L \left[\iint \frac{\vec{r}_{sdE} \times \vec{B}'(t)(\vec{n}_{b'} \cdot \overrightarrow{ds}) \cdot dL}{4\pi r_{sdE}^2}\right] \cdot \vec{d}l = \int_L \frac{[r \times \phi'(t)\vec{d}L]}{4\pi r^2} \cdot dl \tag{3}$$

in which, $\phi'(t)\vec{d}L$ is magnetic flex ratio, namely $\phi'(t) = \frac{\partial}{\partial t}\iint B(t) \cdot ds$.

③Calculation of induction electromotive force at non – closed conductor section

$$\varepsilon_{ab}(t) = \int_a^b [\vec{E}_c(t) + \vec{E}_n(t)] \cdot \vec{d}l = \int_a^b [\vec{v}_{dl\cdot b}(t) \times \vec{B}(t) + \int_L \frac{[r \times \phi'(t)\vec{d}L]}{4\pi r^2} \cdot dl \tag{4}$$

b) Strong filed to emission effect

When the energy of the coated metal's electrons surpasses their bounding force, the electrons will escape from their metal surface, then the current is being formed. According to the theory of Schottky [6], its emitting effect is indicated by the current density as J

$$J = AT^2 \exp\left\{-\frac{e}{kT}\left[\phi - \left(\frac{eE}{4\pi\varepsilon_0}\right)^{0.5}\right]\right\}(\mathrm{A/cm^2}) \tag{5}$$

in which, k, T, E, ϕ respective to Borman constant, conductor temperature, conductor polarity surface filed strength. metal escaping force and Dusman constant $A \approx 120.4\ (\mathrm{A/cm^2K^2})$.

c) Electro – thermal effect

Subject to the electro – induction effect, the short circuit current flows at conductive fiber and heat is then generated. As the duration is quite short, (lab test proves that it is less than ms grade), Such generated heat spreads to the surrounding media, turning the electric energy into the thermal energy Q, which is accumulated onto the fiber. The maximum thermal load is:

$$Q_{\max} = \int_0^t RI2(t)\,\mathrm{d}t = \int_0^t q_s \rho_s C_s \frac{\mathrm{d}T(t)}{\mathrm{d}t}\mathrm{d}t \tag{6}$$

in Which, q_s, ρ_s, C_s, $T(t)$ respective to fiber section area, fiber density, fiber material specific heat and temperature.

d) Effect of thermal dissociation

When the temperature of gaseous media increases, the movement and collision of gaseous molecule becomes stronger and the effect of thermal dissociation plays the leading role. It can be described as x. The main affecting factors include media temperature, dissociative potentials and gaseous pressure The dissociative potential of metallic gas is much less than that of the air. The Al when at 4 000 K appears to have the dissociative indication but the same for air is around 10 000 K. This provide theoretical evidence during the choice of metallic coated material. When the gaseous pressure is greater, the less dissociative it appears. The dissociative situation of electrons in the metallic gas may be described by Saha [4] theory as:

$$\frac{x_2}{1-x_2}p = 3.20\times10^{-5}T^{2.5}\exp\left(-\frac{eU_y}{KT}\right) \tag{7}$$

in which, K, p, T, U_y, X respective to Borman constant ($K=1.37\times10^{-23}$ J/K) gaseous pressure (kpa, 1 atm = 101.3 kPa)、media temperature (K)、media dissociative potentials (V) and its dissociative degree $0\leqslant x\leqslant1$. Subject to the metallic gas media, the dissociative degree is:

$$x = \frac{5.66\times10^{-3}T^{1.25}\exp\left(\dfrac{-5839U_y}{T}\right)}{\left[3.20\times10^{-5}T^{2.5}\exp\left(\dfrac{-1.17\times10^{-4}U_y}{T}\right)\right]^{0.5}} \tag{8}$$

2.4 Conversion and Critical Criterion of the Arc Form During the CCA Process

With the formation of metallic arc path, the short circuit current increases drastically. When the current reaches to certain ampere and above (the actual current may reach to hundreds of amps). The high temperature generated by I^2R energy will lead to the increase of ambient atmosphere and dispersion of metallic vapor and the decease of its density. When the temperature reaches to the air media dissociative potential T_m, the surrounding air located around the metallic arc begin to electrocute, making the metallic vapor turn into the arc as shown in Figure 1 [2]. Its duration related to the ambient atmosphere, arc current and external applied magnetic fields.

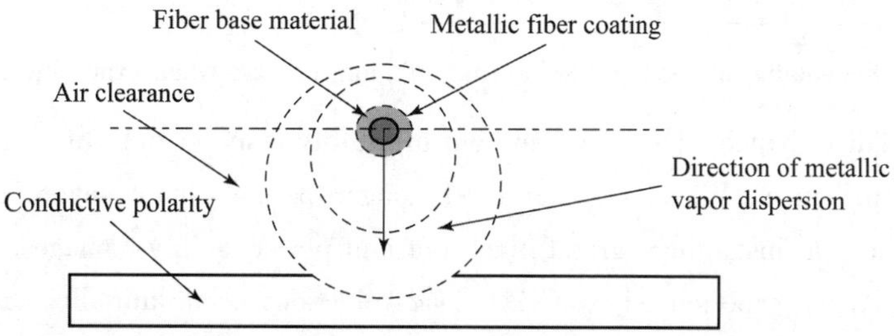

Figure 1　Model of metallic fiber vapor dispersion

Assuming that such internal dispersion is processed in ideal gas, the formula of Stefan Maxwell will apply. i. e. the theory of arc energy balance, the heating value generated by arc is equal to the addition of heat arch column plus the heat loss.

$$I_hU_h = \frac{dQ}{dt} + P_s \tag{9}$$

Where$Q = q_h\rho C_T(T_m - T_0)$ the critical criterion is:

$$Tm \geqslant \frac{1}{q_h\rho C_T}\int_0^t (I_hU_h - P_s)\,dt \tag{10}$$

in which, q_h, C_T, I_h, U_h, P_s respective to metallic are sectional area, concentration, specific heat, arc current. arc voltage and heat loss.

3 Insecure Analysis on Electric Power System After CCA

3.1 Insecure Analysis on Power System when Experienced by the CCA Effects

When the HV power system is experienced by CCA, the acting points selected to destroy are those like super - high HV transmission lines or key substations. Though the protection is in function when short circuit occurs, i. e. the relay to detect the trouble and the faulty sections will be isolated, followed by 1 ~ 3 time reclosing. Should it be in vain is deemed to the permanent trouble and reclosing is not permitted then.

Firstly, short circuit caused by CCA is a severe, repeated and also complicated failure. Conductive fiber is tapped onto the conductors. causing all kinds of short circuit failure forms (three phases, two phases, two phases to ground. single phase to ground); Secondly, the fiber may be tapped to the transformers, HV insulators or metal hardwares to cause the severe burning and permanent damage. Under such circumstance, troubles of muti - type are formed. Such as fibers interference the system protection function (HV circuit breakers), making it malfunction or in operable, which in turns affects the downstream facilities and cause the chain reactions of the power system. Even the protection could be in normal function, it is sure case that the substation would loss its function [6]. Schematics of the troubles and its affecting process when experienced by CCA is shown at Figure 2. At a case when transformer in trouble, the unsafe factors of the power system shall include:

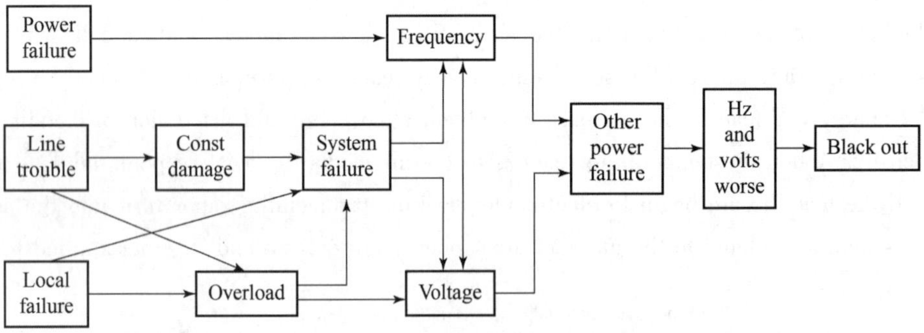

Figure 2 Schematics of the troubles and its affecting process when experienced by CCA

a) Sustained system disturbance: The power system instability resulted from the breakup of the power angle may lead to the grid disturbance. When retained for a longer period, more severe consequence will occur, including the further to the grid instability, grand black out and power facility damages.

b) Load Rejection: When experienced by CCA, load rejections occur initially starting form the local area due to the overload of certain components, especially for the looped power grid, leading to the grand black out. Shutdown the units and load rejection are the main means to minimize the consequences. However, the actions by putting out of service of certain components lead to overload of adjacent power facilities. leading to the shutdown chain reaction.

c) Frequency collapse: When experienced by CCA, the disconnection of some local systems may cause the serious shortage of active power, which will lead to the drops of frequency and further to grid failure. Reduction of load at low frequency is an effective measure to prevent it from going into worse. However, the setting at grid low frequency is coordinated with the generator. Low frequency protection to assure that the generator remains in service to the extent and time that the frequency drop will not result in the activation of generator low frequency protection. As such the chain action occurs with the continued drops of frequency and eventually leading to the grand black out.

d) Voltage collapse: When experienced by CCA. the voltage drops due to the short circuit, the large - sized generator or the main transmission lines at power receiving end will trip, the demand for reactive power increases, but the reactive power at the capacitors in parallel will decrease, which leads to the voltage collapse. The effective measure to avoid further voltage drop is to reject the load. But such device shall prevent improper action when voltage drops resulted from the short circuit.

3.2 The Blocking Mechanism to the Power System by Conductive Fibers

The conductive fibers lead to the open – air power system loss its function at expected period. In general, a numbers of conductive fibers are required to act upon the power facilities. To have a HV transformer loss its function only requires a couple of conductive fibers (including the process of reclosing). Upon failure of a power system, the effects will continue as a numbers of fibers remained or bound unto the power facilities. They have the potential CCA capacity, which will make the auto protection devices at a time period loss the function of reclosing (at each time of reclosing, the conductive fiber will cause the CCA effects of short circuit and the activation of auto protection device). A normal operated power system is in such a way being blocked off. Should more of the same occur at critical power system, the regional power grid or state power grid will be blocked off.

3.3 Removal of Metallic Fibers and Power Resumption Mechanism

When the conductive fibers are applied during the battle to attack the power system, thousands of fibers will act upon the power grid. The removal of the conductive fibers involves two aspects: to maximize the removal difficulties in terms of short circuit wound point. To enhance the removal effectiveness in terms of prevention point. As for the former. it involves two aspects: the first is the attacks by more fibers to increase the amount of conductive fibers on the targets. The second is to use the hard to remove fibers to increase its removal difficulties. The latter involves the design of the fiber regarding to its material, structure and performance. which is the key for the removal. Not only the conductive fibers on the power facilities be completed removed, but also the same on the adjacent fibers on the target flown by wind be removed as well upon completion of which, it is then deemed to be a complete removal.

During our experiment on a real 110 kv substation and simulated test on 330 kv transformer, we found that the removal technology is quite important. As the conductive fibers consisted of dozen of pieces, diameter of each one is $<20\ \mu$m. During removal process, the tiny pieces are in form of outspread and still remain on the insulators or power poles. The short circuit happens during reclosing.

4 Some Proposals to Enhance the Power System Security Under the Condition of CCA

a) To minimize the loss resulted from this event, some preventive measures in terms of technology and security on the HV power system shall be considered.

b) As to blocking mechanism, removal mechanism when power system experienced by CCA, proper devices shall be furnished to clear the troubles in a more effective, soonest and scientific manner in order to resume the normal operation of power system. Meanwhile, a specific study should be established on the fiber clearing technology and the technical way to check the effects and final assessment upon completion of the fiber removal.

5 Conclusions

This article clearly introduced the concept of the CCA when being employed on the weapons acted upon the HV electric power system. It described the CCA process of conductive fibers and set up the dispersion model and the critical criterion of arc conversion from metallic to gaseous phase. It analyzed the unsafe factor of electric power system and their developing trend after effects of CCA. Based on the blocking mechanism, fiber removal and power system resumption mechanism of conductive fibers, and eventually gave certain proposals.

References

[1] FENG S S. Key technology analysis about conductive fiber bomb [J]. Journal of projectiles, rockets, missiles and guidance, 2004, 24 (1): 40-42.

[2] LIANG Y Z, FENG S S. Study on the mechanism of are discharge of conductive fiber between wires of high-voltage [J]. Journal of projectiles, rockets, missiles and guidance, 2004, 24 (2): 64-66

[3] Carbon/graphite fiber risk analysis and assessment study [R]. NASA-CR 159212, Jan. 1980.

[4] High performance fibers. ADA274693, 1994.

[5] U. S. -Canada Power System Outage Task Force. Interim Report: Causes of the August 14th Blackout in the U. S. and Canada. Nov. 2003

[6] YUAN J X. Security and Stability Control of Power System, 1996: 289.

导电丝束空中展开规律①

张国伟[1, 3②]，冯顺山[2]，孙学清[1]，梁永直[2]

（1. 中北大学机械电子工程系，太原 030051；2. 北京理工大学爆炸与安全国家重点实验室，北京 100081；3. 北京航空航天大学航空科学与工程学院，北京 100083）

摘　要：针对长导电丝束缠绕成的丝团在空中展开及飘落过程中受到风速、风向及丝团抛撒初速等各种因素的影响，对该丝团的空中展开及飘落过程进行了受力分析，建立了描述丝束展开规律的数学模型，其计算结果与试验结果对比，两者吻合较好。研究结果对确定丝团的抛撒高度及丝束飘落问题处理具有重要的参考价值。

关键词：导电纤维；丝团；毁伤；电力系统

The Spreading Law of Conductive Fiber in the Air

ZHANG Guo－wei[1,3]，FENG Shun－shan[2]，SUN Xue－qing[1]，LIANG Yong－zhi[2]

（1. Department of Mechanic and Electronic Engineering，Taiyuan 030051，China；2. State Key Laboratory of Prevention and Control of Exploration Disaster，Beijing Institute of Technology，Beijing 100081，China；3. Aeronautical Science and Engineering College，Beihang University，Beijing 100083，China）

Abstract：The fiber rows circled by long conductive fiber are affected by wind speed，direction and the original casting speed of fiber rows in their spreading and falling process. In this paper，we carried on some mechanical analysis for these fiber rows in their spreading and falling process. A mathematical model describing the spread of the fiber rows was set up. The computation results are accordant with that of the tests. The research results are important in confirming the casting height of fiber rows and in dealing with the fall of the fiber.

Keywords：conductive fiber；fiber rows；damage；electric power system

在1991年的海湾战争中，美国在“战斧”巡航导弹上首次使用导电纤维子弹破坏伊拉克的电力系统。随后，在1999年的科索沃战争中，美国再一次使用了经过改进的导电纤维子弹，并取得了极佳的作战效果。通过布撒器或其他运载工具将导电纤维子弹运送到目标上空一定的高度，子弹被抛出后，降至预定高度，再将导电纤维丝团抛撒出来。丝团在重力、风力及空气阻力等的作用下，边下落、边展开，最后展开的导电纤维丝束落于电网上，造成电力系统的短路毁伤而导致局部甚至大面积停电事故。

缠绕于丝团上的导电纤维丝束在落于电网前的展开长度是否满足要求，在风的作用下导电纤维丝束能否可靠落于目标上等都直接关系到短路毁伤效能。关于柔性长丝束丝团展开规律的研究未见公开报道[1]，因此，对柔性长丝束丝团空中展开规律进行研究是十分必要的。

1　假设条件

导电丝束缠绕在丝轴上，丝束一端有阻力叶片，丝团结构[2]如图1所示。丝团在空中被抛撒出来后，其姿态是随机的。另外，风也是一个不定值。在研究导电纤维丝束空中运动规律时若全面考虑其影响因素，必将为研究带来很大的困难。为便于对导电纤维丝束空中运动规律进行研究，特作如下假设：

---

① 原文发表于《弹道学报》2006年第18卷第1期。

② 张国伟：工学博士，2001年师从冯顺山教授，研究电力系统易损特性及导电纤维子弹设计中若干问题，现工作单位：中北大学。

①丝团在空中抛撒出后，速度为水平方向；②导电纤维丝束在展开下落过程中受到的风力是均匀稳定的，即风速和风向均不变；③导电纤维丝束在展开下落过程中所受到的阻力仅与导电纤维丝束上的阻力叶片、丝团运动速度及风速有关；④导电纤维丝束在展开下落过程中，仅考虑丝轴绕轴线方向的转动；⑤不考虑导电纤维丝束在展开过程中的弯曲。

2 数学模型

在上述假设条件下，建立如图2所示的直角坐标系。由于受丝团重力及阻力叶片阻力的作用，丝团在展开下落过程中必然在阻力叶片的下方。图2表示在某一瞬时丝团及阻力叶片的相对位置及导电纤维丝束展开情况。图中，C_1 为丝团的圆心（质心）；C_2 为已展开导电纤维丝束的质心；A 为展开导电纤维丝束与丝团（未展开导电纤维丝束）的切点；B 为展开导电纤维丝束的端点；φ 为丝团的转角；θ 为丝束的转角；l 为已展开导电纤维丝束的长度。

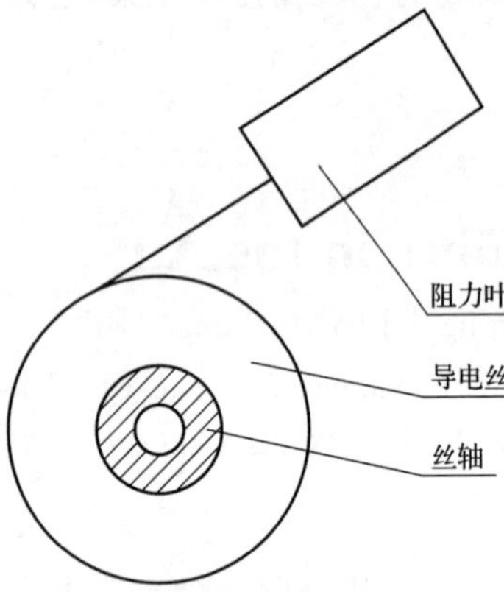

图1 丝团结构

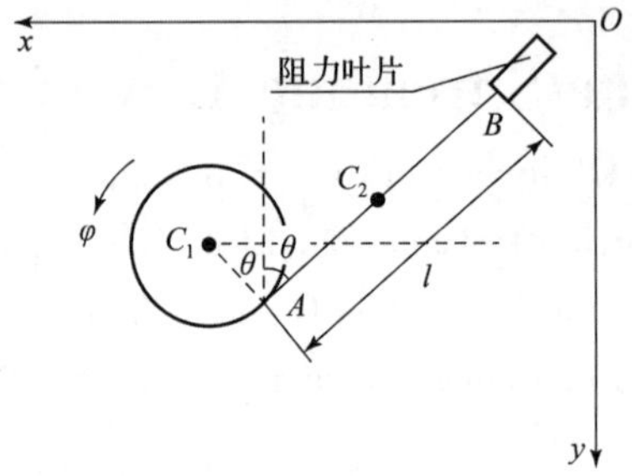

图2 某一瞬时丝团的位置及导电纤维丝束展开情况

2.1 丝团受力分析

丝团在下落过程中受到的力有丝轴重力 P、未展开导电纤维丝束的重力、丝团在展开下落过程中所受到的风力 f_2 及导电纤维丝束受到的沿导电纤维丝束方向的总拉力 T，见图3。

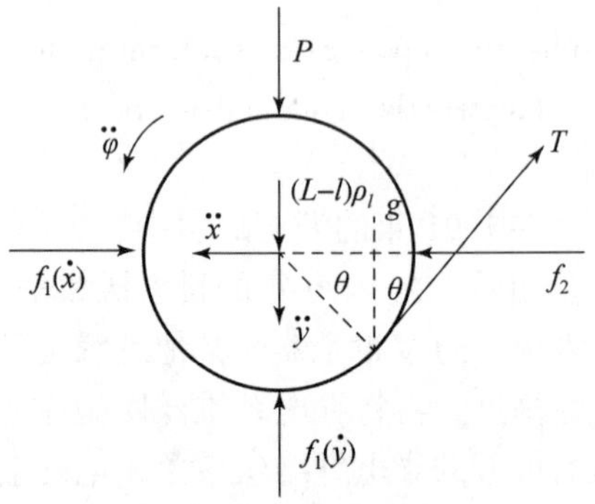

图3 丝团展开过程受力图

2.2 微分方程组的建立

分析丝团受力并根据牛顿第二定律，可以得到

$$m\ddot{x} = f_2 - f_1(\dot{x}) - T\sin\theta \tag{1}$$

$$m\ddot{y} = p + (L-l)\rho_l g - T\cos\theta \tag{2}$$

$$\frac{1}{2}\cdot\frac{P}{g}r^2\ddot{\varphi} + \frac{1}{2}(L-l)\rho_l(R^2-r^2)\ddot{\varphi} = TR \tag{3}$$

$$m = \frac{P}{g} + (L-l)\rho_l \tag{4}$$

假设风的作用方向如图3所示，则式（1）应改写为

$$m\ddot{x} = f_2 - f_1(\dot{x}) - T\sin\theta$$

上述式中，m 为丝团质量（含丝轴和未展开导电纤维丝束两部分）；L 为丝团上所缠绕的导电纤维丝束的总长；R 为不包含已展开导电纤维丝束部分的丝团半径；r 为丝轴半径；ρ_t 为导电纤维丝束的线密度；T 为导电纤维丝束受到的总拉力（沿导电纤维丝束方向）；$f_1(\dot{x})$、$f_1(\dot{y})$ 为丝团在展开下落过程所受到的空气阻力，见图3。

$$f_2 = \frac{1}{2}C_f\rho_2 v_B^2 S \tag{5}$$

式中　C_f——阻力叶片的阻力系数；

ρ_2——空气密度；

v_B——阻力叶片的运动速度，即图 4 中的 B 点速度；

S——阻力叶片的迎风面积。

由图 2 可以得到

$$\begin{cases} xc_2 = x - R\cos\theta - \dfrac{l}{2}\sin\theta \\ yc_2 = y + R\sin\theta - \dfrac{l}{2}\cos\theta \end{cases} \tag{6}$$

式中，xc_2、yc_2 分别为已展开导电纤维丝束中点的飘移距离和下落高度。

图 4 中的 C_2 点满足关系式：

$$mc_2\ddot{x}c_2 = T\sin\theta - f_1(\dot{x}) \tag{7}$$

$$mc_2\ddot{y}c_2 = T\cos\theta + mc_2 g - f_1(\dot{y}) \tag{8}$$

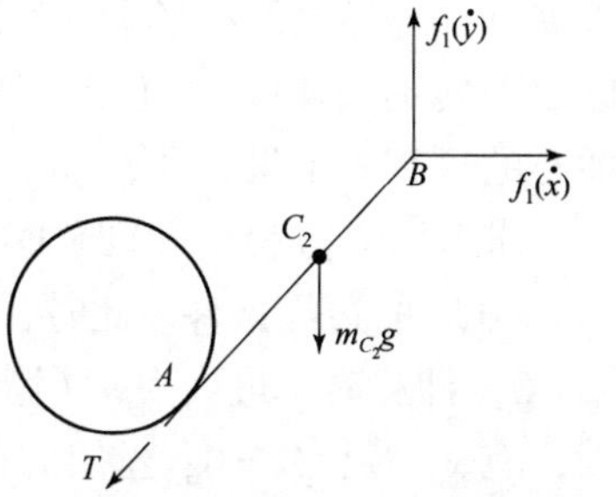

图 4　已展开导电纤维丝束的受力图

式中　mc_2——已展开导电纤维丝束的质量。

$$\frac{1}{12}mc_2 l^2\ddot{\theta} = -f_1(\dot{x})\frac{l}{2}\cos\theta + f_1(\dot{y})\frac{l}{2}\sin\theta \tag{9}$$

式中，

$$mc_2 = l\rho_l \tag{10}$$

丝团下落过程中，导电纤维丝束不断展开，缠绕在丝团上的导电纤维丝束不断减少，使缠绕导电纤维丝束的丝团直径不断减小，即 R 为一变量：

$$-2\pi\frac{\mathrm{d}R}{\mathrm{d}t} = \frac{a}{d}R\varphi d \tag{11}$$

式中　d——导电纤维丝束直径；

a——丝轴宽度。

在丝团下落展开过程中，风对其影响最大。图 3 表示的是风的作用方向与丝团的运动方向同在一个二维的平面内。由式（1）～式（11）共 12 个方程构成联立方程组，即

$$\begin{cases} m\ddot{x} = f_2 - f_1(\dot{x}) - T\sin\theta \\ m\ddot{y} = P + (L-l)\rho_l g - T\cos\theta \\ \dfrac{1}{2}\cdot\dfrac{P}{g}r^2\ddot{\varphi} + \dfrac{1}{2}(L-l)\rho_l(R^2 - r^2)\ddot{\varphi} = TR \\ m = \dfrac{P}{g} + (L-l)\rho_l \\ f_2 = \dfrac{1}{2}C_f\rho_2 v_B^2 S \\ mc_2\ddot{x}c_2 = T\sin\theta - f_1(\dot{x}) \\ mc_2\ddot{y}c_2 = T\cos\theta + mc_2 g - f_1(\dot{y}) \\ xc_2 = x - R\cos\theta - \dfrac{l}{2}\sin\theta \\ yc_2 = y + R\sin\theta - \dfrac{l}{2}\cos\theta \\ \dfrac{1}{12}mc_2 l^2\ddot{\theta} = -f_1(\dot{x})\dfrac{l}{2}\cos\theta + f_1(\dot{y})\dfrac{l}{2}\sin\theta \\ mc_2 = l\rho_l \\ -2\pi\dfrac{\mathrm{d}R}{\mathrm{d}t} = \dfrac{a}{d}R\varphi d \end{cases} \tag{12}$$

3 边界条件

在式（12）中共有12个变量，分别是：已展开导电纤维丝束的飘移距离 xc_2 及下落高度 yc_2、展开导电纤维丝束的长度 l、沿导电纤维丝束拉长方向上的总拉力 T、导电纤维丝束的转角 θ 及角速度 $\dot{\theta}$、含未展开部分导电纤维丝束的丝团质量 m、丝团半径 R（随导电纤维丝束的展开不断减小）、丝团的转动角度 φ 及转动角速度 $\dot{\varphi}$、已展开导电纤维丝束的质量 mc_2、导电纤维丝束受到的空气阻力 f_1。在使用中笔者关心的只有3个量，即 l、xc_2、yc_2 及其与时间的关系。

求解联立方程组（12）时，已知量为 L、ρ_1 及 P。同时还需要确定在丝团被抛撒出来的瞬间，即时间 $t=0$ 时的边界条件。此时，边界条件包括：丝团初始坐标 x_0（空中抛撒丝团的位置点 y_0，取 $x_0=0$，$y_0=0$，计算时 y 最大为 h（抛撒丝团的高度）。丝团被抛出瞬间的速度值通过试验获得，根据假设条件①，有 $\dot{x}_0=v$，$\dot{y}_0=0$；丝团的初始角速度 $\varphi_0=0$，初始转角 $\varphi_0=0$；导电纤维丝束的初始角速度 $\theta_0=0$，初始转角 $\theta_0=\dfrac{\pi}{2}$；导电纤维丝束的初始展开长度 $l_0=0$。

4 计算结果分析

在确定了边界条件后，即可得到联立方程组（12）的数值解。实际应用中，风不确定且风速不稳定[3]。在风速为2 m/s、丝团初始抛撒高度为90 m、丝团初始抛出速度为15 m/s、导电纤维丝束长度为30 m的初始条件及假设风的方向与丝团抛撒方向一致的条件下得到的计算结果见图5～图7。

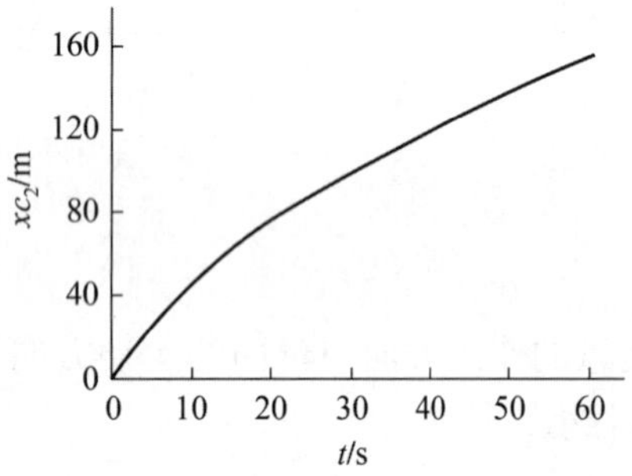

图5 导电纤维丝束飘移距离与下落时间关系曲线

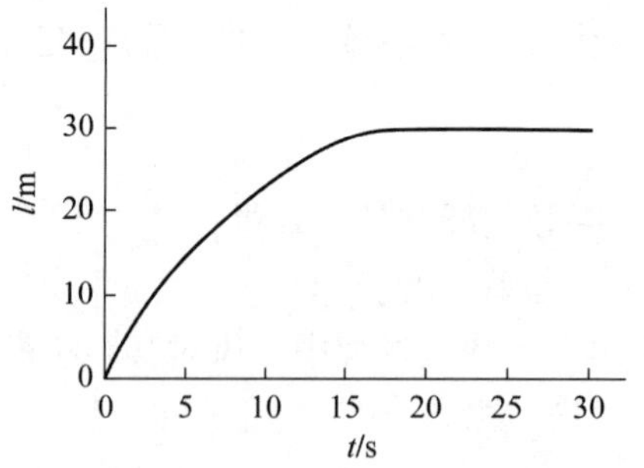

图6 导电纤维丝束展开长度与下落时间关系曲线

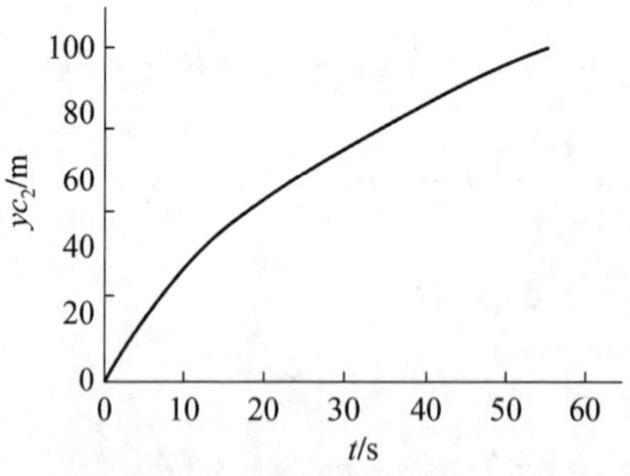

图7 导电纤维丝束下落高度与下落时间关系曲线

分析图5～图7可以得出：

（1）导电纤维丝束的飘移距离随时间增加而加大，但随时间增加，飘移距离增加速率逐渐降低，最后以接近风的速度匀速飘移；

（2）导电纤维丝束的展开长度开始时展开速率较快，然后逐渐变缓，直至最后全部展开。由于计算时将导电纤维丝束的长度限定为30 m，而在40 m左右的高度就可完全展开，故图6中曲线在30 m处斜率产生突变；

（3）导电纤维丝束下落高度的变化趋势为开始快，然后逐渐变慢，最后基本保持匀速下落，其原因是：在开始下落时，丝团上的丝束展开少，在垂直方向上力以丝团重力为主，下落加速度大，体现在曲线上即曲线的斜率大；而随着丝团的下落，导电纤维丝束展开长度增加，在垂直方向上的重力逐渐减小，且受到已展开部分的牵连，下落加速度降低，直至最后缠绕在丝团上的导电纤维丝束全部展开，带有丝轴的导电纤维丝束以接近恒定的速度下落。

为验证计算结果的准确性，笔者进行了相应的试验验证。试验条件：风速为 1.8 m/s、丝团抛撒高度为 90 m，初始抛撒速度为 18 m/s。测得的试验结果：导电纤维丝束的飘移距离为 120 m，下落时间为 42 s，导电纤维丝束全部展开。

5　结论

试验结果与计算结果的对比表明，计算结果真实可信，为实战中确定丝团的抛撒高度、子弹落点的选择提供了可靠依据。

参考文献

[1] COSTELLO M F, FROST G W. Simulation of two projectiles connected by a flexible tether: ARL - CR - 456 [R]. 2000.

[2] 张国伟. 电力系统易损特性及导电纤维子弹设计中若干问题研究 [D]. 北京：北京理工大学，2004.

[3] 温克钢. 风和雨 [M]. 北京：中国青年出版社，1963.

导电纤维丝束在高压导线相间引弧试验研究①

王　芳②，梁永直，冯顺山

（北京理工大学爆炸科学与技术国家重点实验室，北京　100081）

摘　要：通过高压小容量和高压大容量条件下的引弧短路试验，对导电纤维丝束引发高压导线相间短路电弧的内在规律进行了研究。由高压小容量和高压大容量引弧短路试验结果表明，导电纤维丝束引发高压导线相间短路电弧的机理主要是导电纤维丝束引发了气体间隙弧光放电电弧，形成自持放电现象所致。

关键词：导电纤维丝束；高压导线；强制引弧

Experimental Research on the Electrical Arc Generated by the Conductive Fiber between the Conducting Wires with High Voltage

Wang Fang, Liang Yongzhi, Feng Shunshan

(Beijing Institute of Technology, Beijing 100081, China)

Abstract: Under the condition of small capacity source and large capacity source with high voltage, the experiments on the electrical arc generated by the conductive fiber between wires with high voltage are conducted. Based on the experimental research, the instinct mechanism of the electrical arc is caused by the self - sustaining electrical discharge of air generated by the conductive fiber between the conducting wire.

Keywords: conductive fiber; conducting wire with high voltage; electric arc

1　引言

导电纤维弹是一种装填导电纤维丝束（镀金属的碳纤维丝束或玻璃纤维丝束）的软杀伤弹药，主要利用导电纤维丝束使电力系统输变电设施和高压线路产生引弧短路效应，使电力系统在一定时间内失去输变电和供电能力，乃至电力系统全线崩溃、瘫痪，从而达到威慑敌方、赢得战场主动的目的。导电纤维丝束作为一种新型毁伤元，研究其对高压线路的短路毁伤机理具有十分重要的研究意义。文中通过高压小容量和高压大容量条件下的引弧短路试验，对导电纤维丝束引发高压导线相间短路电弧的内在规律进行了研究[1,2]，为导电纤维丝束对高压电力设施的短路毁伤机理研究提供了科学依据。

2　高压小容量条件下的引弧试验研究

2.1　试验装置

为了能够对不同类型导电纤维丝束引弧短路性能做一客观的定性定量描述，试验选择在高电压、小容量电源装置上进行导电纤维丝束搭接相 - 相间高压线路的引弧短路试验，试验装置如下。

---

① 原文发表于《弹箭与制导学报》2006 年第 26 卷第 2 期。

② 王芳：工学博士，1996 年师从冯顺山教授，研究导电丝束空中展开及毁伤特性，现工作单位：北京理工大学。

2.1.1　工频通流试验装置（见图1）

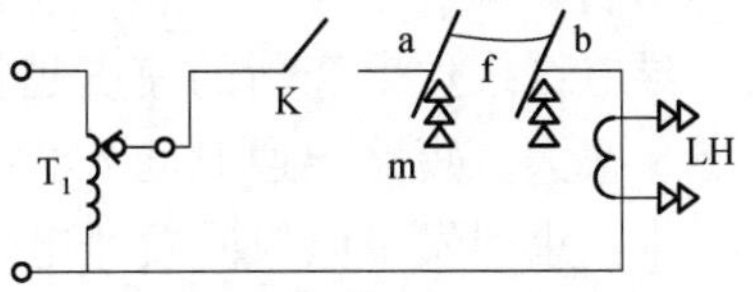

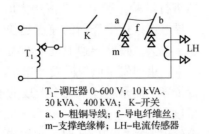

T_1–调压器 0~600 V；10 kVA、30 kVA、400 kVA；K–开关
a、b–粗铜导线；f–导电纤维丝；
m–支撑绝缘棒；LH–电流传感器

图1　导电纤维丝束引发小开距空气间隙短路电弧通流试验电路

2.1.2　单根导电纤维丝束在短空气间隙下引弧试验装置（见图2）

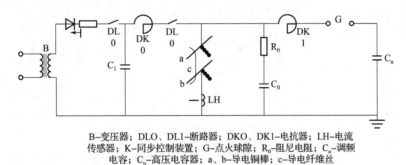

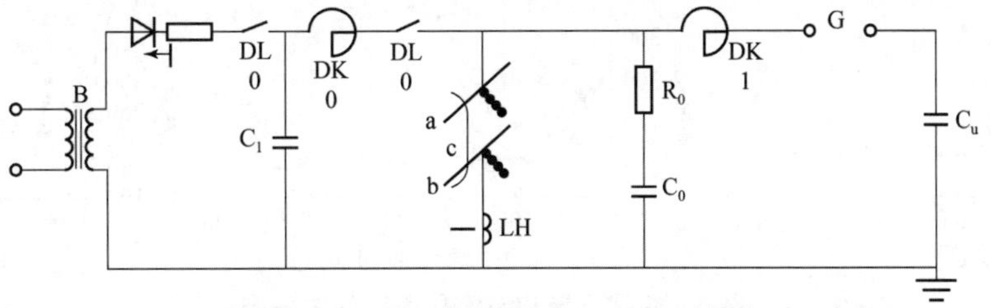

B–变压器；DLO、DL1–断路器；DKO、DK1–电抗器；LH–电流传感器；K–同步控制装置；G–点火球隙；R_0–阻尼电阻；C_0–调频电容；C_u–高压电容器；a、b–导电铜棒；c–导电纤维丝

图2　威尔合成回路试验电路图

2.2.1　导电纤维丝束引发小开距空气间隙短路电弧通流试验

分别选用1#镀铜碳纤维丝束、2#镀铝玻璃纤维和3#镀铜玻璃纤维三种导电纤维丝束来进行短路电弧通流试验研究，具体方法为：利用图1所示试验电路，将被试导电纤维丝束的搭接长度固定，逐步升高对导电纤维丝束所施加的电压，记录每次试验系统短路电流波形。根据电流波形做出其短路电流平均值和电流导通时间，得到短路电流 I 与电流导通时间 t 的关系曲线。

2.2.2　单根导电纤维丝束在小开距空气间隙下引弧试验

分别采用三种单根导电纤维丝束，即1#镀铜碳纤维丝束、2#镀铝玻璃纤维和3#镀铜玻璃纤维来进行短路电弧通流试验研究，获取其短路电弧电流波形图，以评价其在小开距空气间隙下的引弧能力。具体方法为：试验电路采用图2所示威尔合成回路，合成回路是采用两个独立电源（即低电压大电流的电流源和高电压小电流的电压源）来代替直接试验时的一个高电压大电流电源，两个电源先后加在被试装置上进行试验。由于每个电源容量较小，从而可用较小的设备获得较大的试验容量。

2.3　试验结果与分析

2.3.1　导电纤维丝束引发小开距空气间隙短路电弧通流试验分析

三种导电纤维丝束的短路电流 I 与电流导通时间 t 的关系曲线如图3所示。从图3中可以看出，不同材料的导电纤维丝束有基本相同的通流特性。

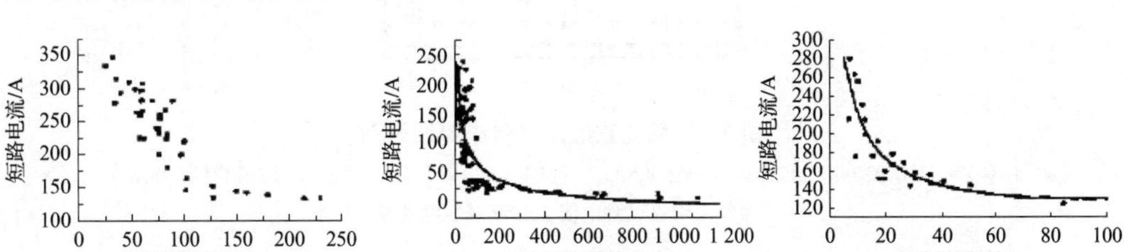

图3　短路电流和电流导通时间关系图

试验表明，通过导电纤维丝束的电流 I 越大，电流导通时间 t 越短。试验过程中大部分导电纤维丝束都是被直接烧断的，由于使用调压器进行试验，所以当电流增大时，电源所提供的容量也随之增加。从通流特性曲线可以看出，电流越大，导电纤维丝束烧断的速度越快。但在试验中也发现，在电流 I 和电压 U 很大时，偶尔会出现短路后电流 I 很大、短路导通时间 t 也很长的情况。这种情况同时伴有爆裂声，根据试验现象判断这是由于导电纤维丝束引发了电极间短路电弧。根据电弧理论，一旦出现弧光放电，电流不再是通过导电纤维丝束进行传导，而是通过空气中形成的电弧通道传导电流，而表面上看起来是导电纤维丝束的通流能力突然变得很大。

在此试验中，继续升高对导电纤维丝束所施加的电压时，短路电流平均值 I 和电流导通时间 t 的关系如图 4 所示。

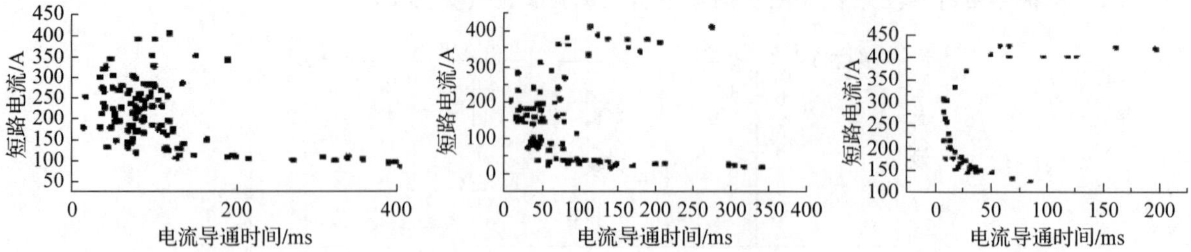

图 4　短路电流和电流导通时间关系图

由图 4 可以看出，导电纤维丝束导致小开距间隙电弧放电的显著标志是导电纤维丝束导通电流明显增大，而且导通持续时间增长；同时，还可以看出：不同材料的导电纤维丝束的特性基本相似。

2.3.2　单根导电纤维丝束在小开距空气间隙下引弧试验分析

图 5 为 1#、2#和 3#单根导电纤维丝束的引弧短路通流试验电流波形图。可以看到，当选用单根带金属镀层纤维丝束来进行试验时，不同金属镀层纤维丝束引发的系统短路电流大小接近，进一步印证引弧后短路电流的大小主要由系统容量决定。

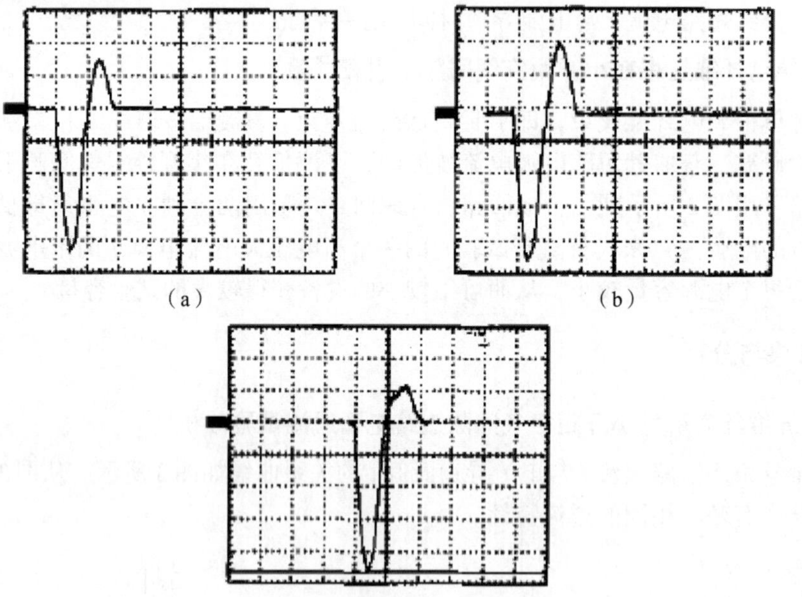

（a）　（b）

（c）

图 5　引弧短路通流试验电流波形图

（a）1#丝束引弧试验电流，波形峰值电流 8 500 A；（b）2#丝束引弧试验电流，波形峰值电流 8 650 A；（c）3#丝束引弧试验电流，波形峰值电流 9 140 A

2.3.3　试验结论

在所施电压较低的情况下，电流导通时间 t 随短路电流 I 的增大而缩短；而当电压达到一定值时，电流导通时间 t 则随短路电流 I 的增大而突增。同时，伴随明亮的弧光放电电弧和巨大响声，其原因为：导电纤维丝束引发了气体间隙弧光放电电弧，形成自持放电现象所致。

由图4可知，所施加电压到达一定引弧电场强度后，三种不同材料的导电纤维丝束所能引起的系统导通电流 I 大小基本相同，这说明此时的短路电流 I 已不是由导电纤维丝束本身的导通能力来决定，而是由系统容量来决定。

采用威尔合成回路对小开距空气间隙（试验间距为4~20 cm）的短路试验表明：单根导电纤维丝束可以引发空气电弧，造成很大的短路电流，且不同镀层导电纤维丝束所能引起的系统导通电流大小基本相同，引弧后短路电流的大小与威尔合成回路的内部阻抗及系统容量有关，而与导电纤维丝束本身的关系不大。

由于电力系统的外绝缘基本属于长空气间隙，一般110 kV及其以上电压等级电力系统的外绝缘，相间绝缘距离应大于2.5 m。上述在短空气间隙下导电纤维丝束引发空气间隙短路电弧的结论能否推延到长空气间隙的真实电力系统中去，还需要通过试验研究来加以证实。

3　高压大容量条件下的引弧试验研究

3.1　试验参数设定与试验方法

3.1.1　试验参数设定

第一种试验参数为电压系统110 kV，最大相间距离为2.5 m，第二种试验参数为电压系统161 kV，相间距离为4.5 m。按照电力系统自保护的整定时间为110 kV下一般不大于0.12 s，要求试验系统的继电保护装置在检测到短路故障后0.12 s时，将试验系统的保护断路器跳闸，分断短路故障。

3.1.2　试验方法

将一束导电纤维丝用夹子事先悬挂于110 kV试验母线的上方约2 m的位置。当发电机运行状态达到额定工作条件，变压器输出电压达到110 kV时，由主控室发出释放指令，两只继电器同时通电而夹子打开，则导电纤维丝束从空中自由降落于110 kV试验母线上，形成短路电弧。

3.2　试验结果及结论

试验采用高速摄像机对试验过程进行了录像，截取的照片如图6所示。

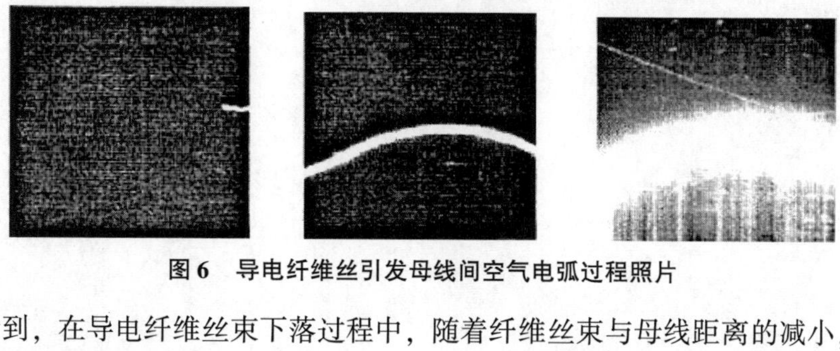

图6　导电纤维丝引发母线间空气电弧过程照片

由图6可以看到，在导电纤维丝束下落过程中，随着纤维丝束与母线距离的减小，间隙中电场强度逐渐增大，开始沿着导电纤维丝束发生放电现象。

根据流柱理论，空气间隙达到一定电场强度后，在电场作用下，在空气间隙中形成大量电子和离子，使得空气间隙导电，间隙被击穿而形成电弧，电流随之剧增，而电弧压降又随电流增大而下降。这是因为导电纤维丝束在开始的短时间内通过很大的电流，使导电纤维丝束瞬间具有很高的温度，在很短时间内，导电纤维丝束已被烧断烧毁，其金属外层由于高温成为金属蒸汽，而金属蒸汽比空气更容易游离，金属蒸汽的游离使得间隙导电，形成电弧通道，同时由于高温作用空气游离也很剧烈，所以电弧会沿导电纤维丝束引弧。导电纤维丝束引发短路电弧后，即形成自持放电现象而不会熄灭，直至被试验系统的高压断路器强制分断。

导电纤维丝束短路 110 kV 或 161 kV 电力系统的等效试验，证实了高压小容量条件下的试验结论同样适用于高压大容量、长空气间隙的真实电力系统。

4 结论

高压小容量条件下的引弧试验表明，在所施电压较低的情况下，电流导通时间 t 随短路电流 I 的增大而缩短；而当电压达到一定值时，电流导通时间 t 则随短路电流 I 的增大而突增。其原因为：导电纤维丝束引发了气体间隙弧光放电电弧，形成自持放电现象所致。继续提高所施加电压到一定场强后，三种不同材料的导电纤维丝束所能引起的导通电流 I 大小基本相同，这说明此时的短路电流 I 已不是由导电纤维丝束本身的导通能力来决定，而是由系统容量来决定。

高压大容量条件下的引弧试验证实了高压小容量条件下的试验结论同样适用于高压大容量、长空气间隙的真实电力系统。

参 考 文 献

[1] 梁永直，冯顺山．导电丝束镀层金属引发相间电弧机理研究 [J]．弹箭与制导学报，2004，24 (2)：64－66.

[2] 冯顺山，梁永直，胡浩江．电力系统在导电纤维弹攻击下短路毁伤研究与安全防护分析 [C]//2004 第八届全国爆炸与安全技术学术会议论文，2004.

超音速伞－弹系统三维有风弹道计算方法研究①

李国杰②，冯顺山，曹红松，王云峰

（北京理工大学爆炸科学与技术国家重点实验室，北京　100081））

摘　要：通过分析摩擦层的大气运动特点，建立了中低空风场随高度的分布函数，分析了伞－弹系统从超声速初始状态减速到稳态过程中的受力环境，并推导了其三维动力学运动方程；以自由下落悬浮式侦察弹为实例。利用相关软件对在三种典型工况下的伞－弹系统有风弹道及无风弹道进行了仿真，并与该系统的实测弹道进行了对比，仿真结果具有较高的可信度，从而验证了该计算方法的可行性与实用性。可为同类问题的弹道设计提供参考。

关键词：降落伞；风场；有风弹道；无风弹道；数值仿真

Calculation Method Study on 3D Windiness Trajectory of Supersonic Parachute－projectile Systems

Li Guojie, Feng Shunshan, Cao Hongsong, Wang Yunfeng

(State Key Laboratory of Explosive Science and Technology, Beijing Institute of Technology, Beijing 100081, China)

Abstract: By analyzing the movement feature of the atmosphere in friction layer. a distributing function adapt to the altitude of the mid－low－altitude wind field is build. The force condition of the parachute－projectile system which decelerates from the initial state of supersonic to the stable speed state is analyzed. and its dynamics equation of motion of 3D is deduced. Taking example for suspended reconnaissance missile which fall freely, using the interrelated program the windiness trajectory and no wind trajectory of the parachute－projectile system under three representative work conditions are simulated, and the results are contrasted with the measured trajectory. The simulated results are creditable, thereby the feasibility and practicability of this calculation method is validated, and it could give some reference to the trajectory design of the same problems.

Keywords: parachute; wind field; windiness trajectory; no wind trajectory; numerical simulation

1　引言

在绝大多数情况下风场对射弹外弹道的影响不可以忽略，风的影响往往是决定性的，尤其对于需要定点投放或需要一定滞空时间的伞－弹系统更是如此，因此对有风弹道计算方法的研究至关重要。在外弹道学中把近地面层中的风称为低空风，而将近地面层以上的风称为高空风，鉴于大气风场的复杂性和实际的应用价值，文中仅有针对性的对超声速伞－弹系统在中低空大气即摩擦层（0～2 km）中的减速有风弹道的分析方法进行了研究[1]。

2　风场模型的建立

大气风场的性质复杂而多变[1]。由于地形、经纬度、高度、温度、空气密度、大气压强等因素的影

---

①　原文发表于《弹箭与制导学报》第27卷第5期。

②　李国杰：工学博士，2002年师从冯顺山教授，研究自主攻击子弹系统总体及对跑道毁伤，现工作单位：北京航天长征飞行器研究所（14所）。

响，再加上诸多的随机因素，大气风场尤其是摩擦层中的风速和风向随时间的变化很迅速[2]。文中用平均风代替实际风，且不考虑阵风（或突风）影响。所谓平均风是指变化着的风在一段时间内的平均值，而瞬时风与此平均风之差称为阵风，如图 1 所示。虽然瞬时风的垂直分布规律性较差，带有很大的偶然性，但如考察大尺度平均风随高度的变化，还是可以找到一定规律的。

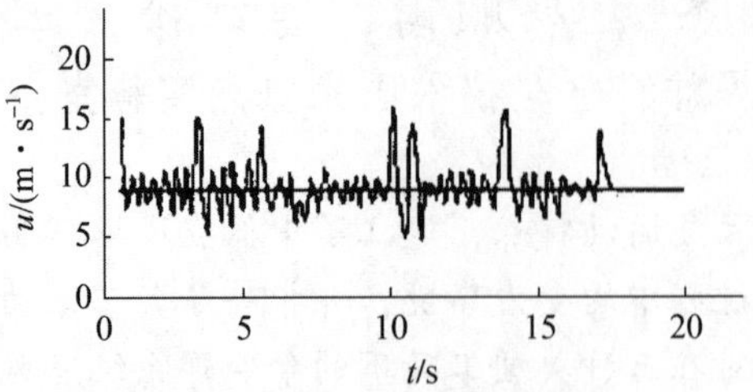

图 1　平均风与阵风示意图

摩擦层中风场的特点是因为摩擦而造成了非地转运动，实际观测也表明近地面附近的空气运动有明显跨过等压线向低压方向偏转的现象，最后的运动状态是达到气压梯度力、科氏力和摩擦力三力平衡。

2.1　近地面层风场模型（0 ~ 100 m）

除贴地层（厚度不到 2 m）外近地面层湍流摩擦力比分子黏性力大得多，因为本层很薄，其各属性如动量、热量和水汽等的垂直湍流输送通量可以认为不随高度改变，又叫常值通量层。同样，在平均情况下，其风向也几乎不随高度变化。本层风速的垂直切变较大，湍流运动较激烈。根据对本层风速观测资料的分析，可以确定其平均风速随高度呈对数分布或指数分布，这也得到了流体力学的理论解释。

假定近地面层中的空气微团按绝热过程运动，不抵抗阿基米德浮力做功，即中性平衡状态，即可最终推导出著名的风速随高度分布的对数定律：

$$\bar{u} = \bar{u}_1(\ln((h+z_0)/z_0)/\ln((h_1+z_0)/z_0)) \tag{1}$$

式中　$(h_1,\ \bar{u}_1)$——某一组（高度，风速）实测值；

z_0——地面粗糙参数（可视为风速为零的高度，m），这里取荒地地形条件参数 $z_0 = 0.025$。风向角下节给出。

2.2　上部边界层风场模型（100 ~ 2 000 m）

上部边界层又称爱克曼层，本层中湍流摩擦力与地转偏向力及气压梯度力同等重要，湍流运动影响逐渐减小，由于湍流黏性力的作用，风向和风速大小都随高度变化。为方便计算，建立坐标系时 z 轴沿地转风（自由大气大尺度运动中气压梯度力与水平科氏力接近平衡时的空气运动）方向。将摩擦力代入大气运动方程组，假设条件为：①地转风为常值；②理论上 $h\to\infty$ 时，风速和风向均趋近于地转风，但实际测量结果表明，平均而言，在 $h = 1\ 000$ m 时，实际风接近于地转风，即 $h_{\mathrm{H}} = 1\ 000$ m；③边界条件：$h = 0$ 时，$u = 0$。最终推导出风速沿高度分布为

$$\begin{cases} u_x = u_g(1 - \mathrm{e}^{-ah}\cos ah) \\ u_z = u_g \mathrm{e}^{-ah}\sin ah \end{cases} \tag{2}$$

合成大小及偏转角：

$$|\bar{u}| = u_g\sqrt{1 - 2e^{-ah}\cos ah + \mathrm{e}^{-2ah}} \tag{3}$$

$$\tan\varphi = (\mathrm{e}^{-ah}\sin ah)/(1 - \mathrm{e}^{-ah}\cos ah) \tag{4}$$

式中：地转风 u_g 可由式（1）100 m 处风速代入式（2）计算得出；$a = \sqrt{f/2k_1}$，由摩擦层高度公式 $h_{\mathrm{H}} = \pi/a = \pi\sqrt{2k_1/f}$，代入 $h_H \approx 1$ km、$f \approx 10^{-4}\ \mathrm{s}^{-1}$，可求得 $k_1 \approx 5\times10^4\ \mathrm{cm}^2\cdot\mathrm{s}^{-1}$，则 $a = 3.142\times10^{-3}\ \mathrm{m}^{-1}$；以 u_x、u_z 为横纵坐标可得到著名的“爱克曼螺线”。随着高度的增加，风向向右偏转（顺时针偏转），由式

（4）代入 100 m 高度可求得近地面层风向与地转风夹角近似为 0.636 5（36.5°）。

3　伞－弹系统运动方程的建立[3]

3.1　基本假设

伞－弹系统的运动是一个非常复杂的过程，其受力状态、环境参数、系统外形与姿态及随机参数等都很难精确地建立数学模型。假设：①引入大尺度随高度变化的平均风场，且风向与地面水平，不考虑阵风与垂直风影响；②伞－弹系统姿态轴线总与来流方向一致，即瞬时调整到与相对运动 u 一致；③暂不考虑阻力伞的拉直与充气阶段；④把伞－弹系统作为一个集中在质心处的质点处理；⑤不考虑系统升力；⑥忽略地球的科氏加速度和曲率的影响；⑦视地球重力加速度为常数 $g=9.81\ m/s^2$。

3.2　运动方程的建立

为了计算方便建立地坐标系时，x 轴方向与高空地转风方向一致，y 轴垂直向上，坐标原点 O 为开伞点在地面上的投影。Ox、Oz、Oy 构成右手一系。系统对地速度（绝对）ω、系统相对空气速度（相对速度）v、风速（牵连速度）u 三者的关系为：$\omega=v+u$，即 $v=\omega-u$，u 由式（1）、式（2）确定。如图 2 所示。

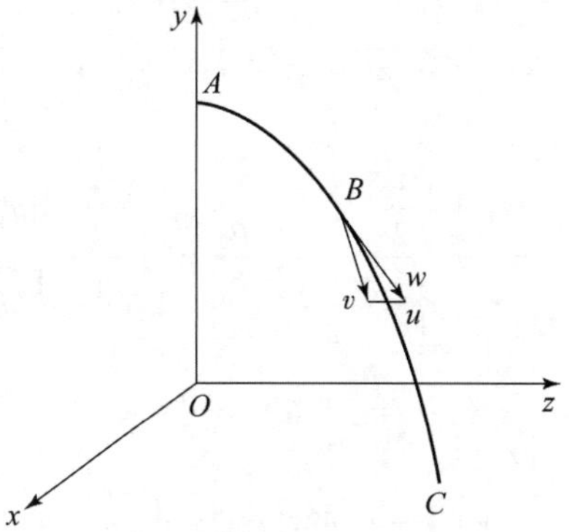

图 2　坐标轨迹示意图

由假设知 $u_y=0$，则相对速度在坐标轴上的分解为

$$v_y=\omega_y, v_x=\omega_x-u_x, v_z=\omega_z-u_z \tag{5}$$

设系统质量为 m，系统气动阻力为 F_d，空气密度为 $\rho_{(h)}$ 则气动阻力 F_d 各分量为

$$\begin{aligned} F_{dx} &= -(1/2)\rho_{(h)}v_x|v|(CA) \\ F_{dy} &= -(1/2)\rho_{(h)}v_y|v|(CA) \\ F_{dz} &= -(1/2)\rho_{(h)}v_z|v|(CA) \end{aligned} \tag{6}$$

式中：$\rho_{(h)}$ 由式（5）求得，系统阻力特征函数（CA）。由风洞试验或理论模拟得到。根据牛顿第二定律：

$$\begin{gathered} m\frac{d\omega_x}{dt}=F_{dx},\ m\frac{d\omega_y}{dt}=F_{dy}-mg \\ m\frac{d\omega_z}{dt}=F_{dz},\ X=\int_0^t\omega_x(t)dt \\ Y=\int_0^t\omega_y(t)dt,\ Z=\int_0^t\omega_z(t)dt \end{gathered} \tag{7}$$

剩余高度

$$h=h_0-Y \tag{8}$$

4　实例弹道解算

侦察评估子弹是典型的中低空减速伞－弹系统，它需要较低的稳定落速，滞空时间较长，因此通过解算有风弹道确定其落点区域对于实现子弹的战术任务至关重要。

4.1　问题描述

子弹在 $h=1\ 500$ m 高空以 $2Ma$ 的速度开伞减速减旋，终态稳定落速 $w=15\sim17$ m/s；弹径 $D=0.069$ m，截面积 $A=0.003\ 739\ \text{m}^2$；伞绳长度 $L=1.7$ m，成型幅半流带条伞阻力特征 $C_uA_u=0.16$，伞衣展开面积（表面积）约为 $A_u=0.4\ \text{m}^2$，伞衣充满球半径 0.24 m，最大迎风面积 $A_d=0.181\ \text{m}^2$；伞－弹系统质量 $m=2.26$ kg。

4.2　阻力系数的求解

4.2.1　标准大气与激波前后流场参数计算

标准大气 T_0，p_0，ρ_0，则在任意 Hm 高空处，有

$$
\begin{gathered}
T=288.15-0.006\ 5h, p_h/p_0=(T_h/T_0)^{5.255\ 88}\\
\rho_h/\rho_0=(T_h/T_0)^{4.255\ 88}, v_s=\sqrt{\gamma RT}
\end{gathered}
\tag{9}
$$

式中：空气比热比 γ，空气气体常数 R，若设初始参数 T_1、p_1、ρ_1，声速 u_{s1}，初始速度 $v_1==2v_{s1}$，则激波过后有

$$
\begin{gathered}
Ma_2^2=\frac{1+\frac{\gamma-1}{2}Ma_1^2}{\gamma Ma_1^2-\frac{\gamma-1}{2}},\ \frac{\rho_2}{\rho_1}=\frac{\frac{\gamma+1}{2}Ma_1^2}{1+\frac{\gamma-1}{2}Ma_1^2}\\
\frac{p_2}{p_1}=\frac{2\gamma}{\gamma+1}Ma_1^2-\frac{\gamma-1}{\gamma+1}\\
\frac{T_2}{T_1}=\frac{2+(\gamma-1)Ma_1^2}{(\gamma+1)Ma_1^2}\left(\frac{2\gamma}{\gamma+1}Ma_1^2-\frac{\gamma-1}{\gamma+1}\right)
\end{gathered}
\tag{10}
$$

4.2.2　高速绕流下阻力系数

因为 $L/D=1.7/0.069=24.64$，由相关公式知伞前流速最小点的速度也在弹前来流速度的 0.95 倍以上，因此伞前来流在初步计算时可简化的认为与弹前未扰动的自由气流相同。则可设总阻力包括子弹压差阻力、子弹激波阻力、伞衣激波阻力和伞衣压差阻力，即 $F_d=F_{pm}+F_{um}+F_{uu}+F_{pu}$，令 $F_d=(1/2)\rho_1v_1^2CA$ 推导得

$$
C=C_m\rho_2v_2^2/\rho_2v_1^2+(p_2-p_1)(A+A_d)/(0.5\rho_1v_1^2A)+C_u\rho_2v_2^2A_u/\rho_1v_1^2A \tag{11}
$$

式中：$C_m=0.847$ 为弹体低速阻力系数；其他参数可由式（9）、式（10）求得。得拟合曲线如图 3 所示，拟合方程为

$$
\begin{gathered}
C=-401.166\ 36+1\ 145.797\ 58Ma-\\
1\ 189.660\ 18Ma^2+651.162\ 58Ma^3-\\
183.538\ 941Ma^4+21.051\ 44Ma^5\\
(1.0Ma-2.0Ma)
\end{gathered}
\tag{12}
$$

4.2.3　低速绕流下阻力系数

亚声速时系统阻力系数视为恒定值，由 $F_d=F_{pm}+F_{pu}$ 并忽略子弹尾流影响可推导得

$$
C=F_d/(\rho_1v_1^2A/2)=C_m+C_uA_u/A=43.639 \tag{13}
$$

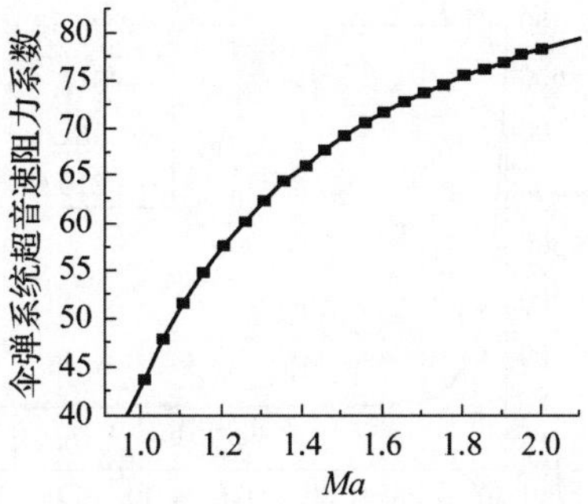

图 3　超声速下系统阻力系数随马赫数的变化曲线

4.3　初始条件

设初始轨迹角（与水平面夹角）$\beta=-30°$；偏航角（水平面投影与 z 轴夹角）为 $\alpha=0°$、36.5°、216.5°，分别表示与地转风风向相同、与地面风风向相同、与地面风风向相反；初始 $|w_0|=669$ m/s，$w_x=|w|\cos\beta\cos\alpha$，$w_y=|w|\sin\beta$，$w_z=|w|\cos\beta\sin\alpha$；起始点坐标（$x_0$，$z_0$，$y_0$）为（0，0，1 500）；$m=2.26$ kg；地面实测风速（y_1，u_1），取（2，3）。

4.4　仿真结果

根据上述已知条件，将运动方程离散化求解，得到了三种初始偏航角下的有风和无风三维弹道，如图 4 所示。为更清楚地表示各相对关系，分别绘制了三个二维图，如图 5～图 7 所示。图 8 则给出了弹道风对子弹终态落速的影响。

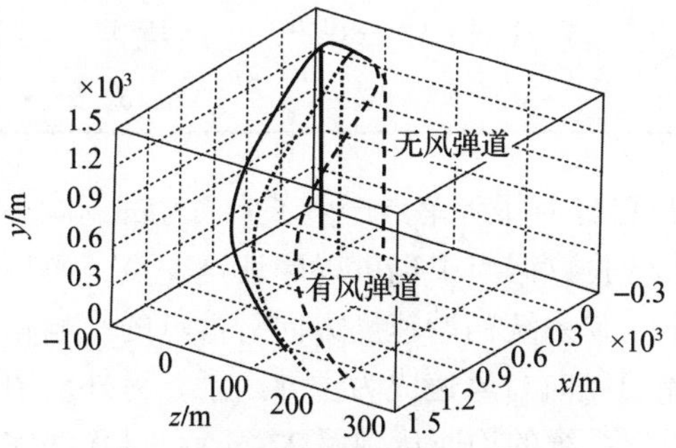

图 4　有风和无风弹道三维坐标图

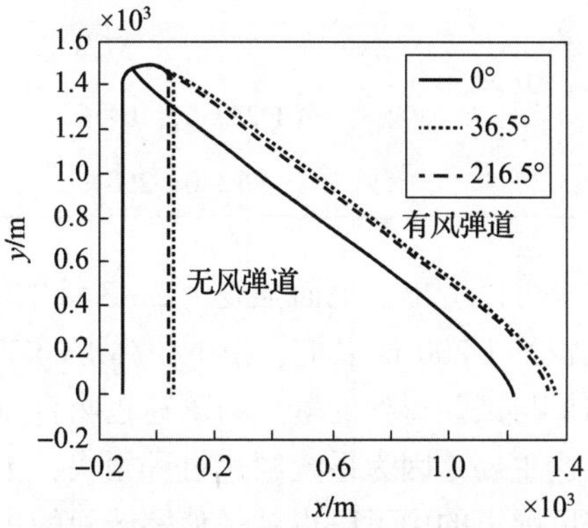

图 5　有风和无风弹道纵向二维坐标图

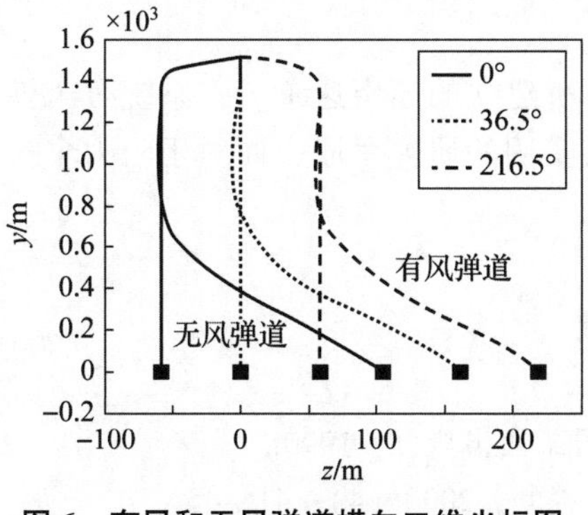

图 6　有风和无风弹道横向二维坐标图

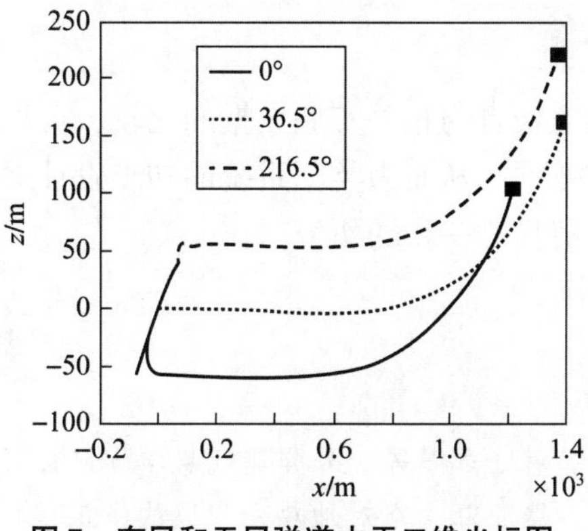

图 7　有风和无风弹道水平二维坐标图

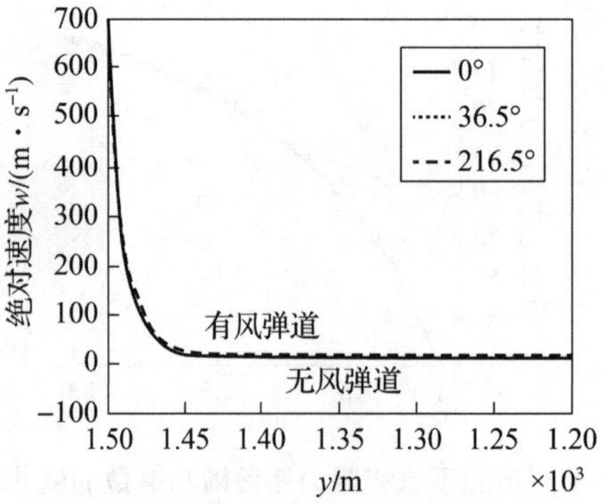

图 8　有风和无风弹道主段落速坐标图

4.5　结果分析

以上得到了有风弹道与无风弹道的仿真结果，现将所得数据与相关实验数据列表 1 如下。其中共同数据有轨迹角 -30°、总用时 95.7 s。试验条件为地面顺风 3 m/s。

表 1　仿真与试验结果列表

项目	条件		落点(x,z)/m	w_x/(m·s^{-1})	w_z/(m·s^{-1})	w_y/(m·s^{-1})	w/(m·s^{-1})	轨迹角/(°)
仿真	0°	无风	(97.2, 0)	0	0	-14.9	14.9	-89.0
		有风	(1 394.6, 161.5)	3.43	2.53	-14.9	15.5	-72.7
	36.5°	无风	(78.1, 57.8)	0	0	-14.9	14.9	-89.0
		有风	(1 375.1, 220.0)	3.45	2.543	-14.9	15.6	-72.7
	216.5°	无风	(-78.1, -57.8)	0	0	-14.9	14.9	-89.0
		有风	(1 220.6, 104.6)	3.27	2.41	-14.9	15.52	-72.9
试验	36.5°	有风	(1 340, 250)	—	—	—	16	—

由表 1 可以看出，地面强度为 2～3 级的轻微弹道风已经对子弹落点产生了实质性的影响，往往会使落点顺风飘移 1 200 m 以上，中下段弹道向左弯曲，从试验数据当中也可以印证这一点，而且相同条件下仿真结果与试验得到的落点与落速也吻合得很好。伞-弹系统初始速度方向对落点的影响相对成为次要因素，这主要是因为系统减速过程很快，于是弹道风的风向与风速成为主导因素。另外从图 8 可以看到，在有风弹道的中下段也就是侦察子弹的主要作用段，系统的均速达到了 22 m/s，已无法完整地实现其作战任务，因此作战应用时一定要考虑弹道风是否满足其应用条件。

5　小结

通过数值仿真与相关实验数据的比较，可以说明文中所建立的超声速伞-弹系统有风弹道的计算方法是行之有效的，从而为类似系统的初步设计提供了比较实用的研究手段，而对于阻力伞拉直与充满过程的加入有待作进一步的研究。

参考文献

[1] 汤晓云，韩子鹏，等．外弹道气象学 [M]．北京：兵器工业出版社，1990.
[2] 陈晓风，张正超．空对地攻击中风的修正 [J]．光电与控制，2000 (4)：41-45.
[3]《降落伞技术导论》编写组．降落伞技术导论 [M]．北京：国防工业出版社，1977.

弹用导电纤维丝束设计准则研究①

王　芳②，冯顺山

（北京理工大学爆炸科学与技术国家重点实验室，北京　100081）

摘　要：为解决弹用导电纤维丝束的设计问题，从提高导电纤维丝束的引弧能力和满足导电纤维子弹的使用要求出发，对导电纤维丝束的设计准则进行了研究。根据上述准则对导电纤维丝束进行了设计与制备，通过小容量条件下的引弧放电实验表明，所设计的导电纤维丝束具有很好的强制引弧能力，可高效短路毁伤高压露天电网。

关键词：导电纤维丝束；短路引弧；高压电网

1　引言

导电纤维子弹是一种装填导电纤维丝束（镀金属的碳纤维丝束或玻璃纤维丝束）的软杀伤弹药，主要利用导电纤维丝束使电力系统输变电设施和高压线路产生引弧短路效应，使电力系统在一定时间内失去输变电和供电能力，乃至电力系统全线崩溃、瘫痪。因此，弹用导电纤维丝束性能的好坏直接影响到导电纤维子弹对高压电力系统的毁伤效果。本文通过导电纤维丝束对高压导线相间的引弧机理研究，同时考虑导电纤维子弹的使用要求，对弹用导电纤维丝束在设计过程中所必须遵循的设计准则进行了研究，研究结果可为弹用导电纤维丝束的设计及制备提供科学依据。

2　导电纤维丝束设计准则研究

2.1　丝束材料分层设计准则

由导电纤维丝束在高压导线相间的引弧机理表明[1]，在高压电场作用下，丝束表面镀层金属产生电感效应、强场致发射效应及热游离效应，导致镀层金属被蒸发、气化和扩散，产生的大量金属蒸汽充斥于高压导线相间，使得空气间隙击穿放电，形成金属蒸汽态电弧；随后，丝束基材被加热而汽化，周围空气介质在高温、强电场综合作用下产生强烈电离，产生自由电子和离子，使得高压导线相间导电物质雪崩般增加，形成短路能力更强的电弧短路通道。导电纤维丝束引发电弧的作用过程如图 1 所示。

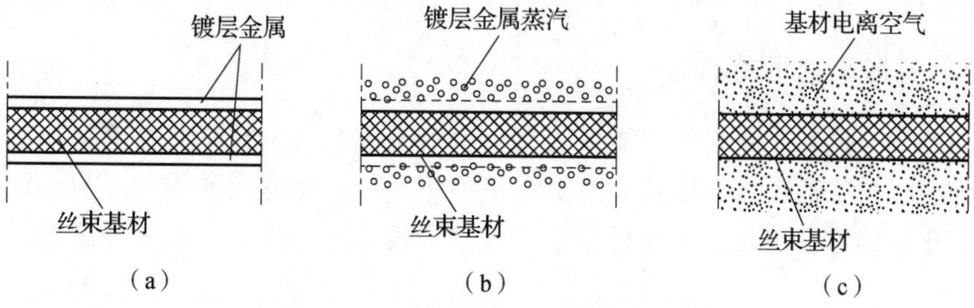

图 1　导电纤维丝束短路引弧机理示意图

（a）导电纤维丝束结构示意图；（b）镀层金属蒸汽示意图；（c）丝束基材电离空气示意图

为了使导电纤维丝束能够在导线相间引发可自持放电的电弧，需要从两方面加以考虑：①丝束本身

① 原文发表于《战斗部与毁伤效率专业委员会第十届学术年会论文集》2007。

② 王芳：工学博士，1996 年师从冯顺山教授，研究导电丝束空中展开及毁伤特性，现工作单位：北京理工大学。

需提供足够多的气体态带电粒子进入空气间隙参与粒子的定向移动；②丝束还需有加热电离周围空气的能力，从而使带电粒子持续不断地进入空气间隙。

因此将导电纤维丝束进行分层设计，即在导电纤维丝束的基材表面镀上一层金属，镀层金属在蒸发汽化后形成足够多的带电粒子进入空气间隙，起到引发电弧的作用；导电纤维丝束基材在电热效应作用下，起到加热电离周围空气，使其不断产生带电粒子的作用。在此需进一步说明的是：丝束提供的气体态带电粒子，可以是丝束本身或其镀层金属汽化而成，也可以是电离周围空气而来，但在此过程中丝束本身不可液化而滴落，所以丝束本身的材料、结构、直径等性能参数需满足一定的要求，方可引发间隙放电并进入自持阶段。

丝束镀层金属可以是铝、铜、镍等具有良好导电性能的金属，而基材则宜选用玻璃纤维、碳纤维、聚丙烯腈纤维或其他高分子纤维材料等。通过试验并综合考虑各种因素证实，镀铝玻璃纤维丝束作为弹用导电纤维丝束具有优良的综合性能。

2.2 表面电阻率准则

由导电纤维丝束的引弧机理可知，在空气间隙被金属蒸汽击穿导通的瞬间，初始短路电流产生的电能将全部被转化为热能累积于导电纤维丝束内，使得丝束基材加热周围空气而导致其电离，产生更多的导电粒子。因此，导电纤维丝束获得的累积热量为

$$Q = I^2 \rho L t \tag{1}$$

式中 I——短路电流的瞬时值，A；

ρ——单根导电纤维丝束镀层金属的电阻率，Ω · mm²/m；

L——搭接于导线之间的导电纤维丝束的长度，m；

t——通流时间，s。

导电纤维丝束引发空气间隙击穿而产生电弧时，应使 Q 不小于导电纤维丝束镀层金属和丝束基材被汽化所需要的最小热量 $Q_{\min}$，即 $Q \geqslant Q_{\min}$。通常在高压电网系统中短路电流 I 值一般在几万 A，因而，式（1）中 Q 的大小主要取决于 I 值，考虑到 I 值的大小与导电纤维丝束的电阻 R 成反比关系，得到导电纤维丝束引发空气间隙击穿而产生持续短路电弧所需满足的判据为

$$\rho \leqslant \frac{1}{I^2 L t} Q_{\min} \tag{2}$$

由式（2）可知，若要使导电纤维丝束产生足够的热量，需要产生较大的短路电流，也即要求弹用导电纤维丝束材料必须具有较小的表面电阻。较小的表面电阻也是保证导电纤维丝束具备镀层均匀性、连续性和良好导电性的一个重要指标。试验研究[2]证实：一般情况下，可取单位长度导电纤维丝束的表面电阻 $\rho \leqslant 30$ Ω/m。

2.3 镀层金属的最小逸出功准则

金属的逸出功反映金属离子获得外部能量脱离金属表面成为自由粒子所需的最小能量。虽然逸出功小的金属易于成为导电粒子而引弧，但并非最适宜于弹用导电纤维丝束的镀层金属。因此，考虑制造成本、工艺性及抗氧化性等综合因素，在满足制备经济性、长储性条件下，选用金属逸出功最小镀层金属的准则。

2.4 丝束基材的绝缘导热准则

为使导电纤维丝束引弧时在电网自动重合闸动作后，不至于全部蒸发汽化而消失，要求丝束基材在此过程中具有自我保护的能力，即具有潜在的强制引弧能力，从而继续封锁电网系统。为此，要求丝束基材具有绝缘性能，同时在引弧过程中，还需具有导热性能以便电离空气。因此，基材适宜选用绝缘导热材料，如玻璃纤维、聚丙烯腈纤维等，一旦导电纤维丝束表面的金属镀层被局部破坏，导电纤维丝束

就会变成绝缘丝束，失去导电能力，从而对自身形成一种保护机制。

2.5　丝束的可缠绕准则及强度准则

丝束的可缠绕准则是指导电纤维丝束要具有很好的柔性，在缠绕成丝束团时不断，经研究表明，纤维直径越细，缠绕丝束团时越不容易被折断。也就是说，降低纤维丝的直径，可有效提高纤维丝的柔性。一般情况下玻璃纤维丝的直径可以小至十几微米范围，完全满足要求。

丝束的强度准则，是指导电纤维丝束在缠绕成丝束团的过程中，需要承受一定的拉力和张紧力，而在导电纤维子弹整个作用过程中，需要承受很大的冲击过载和空气阻力。因此在设计导电纤维丝束时，必须考虑在上述条件下导电纤维丝束不断。

2.6　丝束的清除时效准则

导电纤维丝束的清除时效准则，是指在满足丝束团缠绕力要求的前提下，使丝束拉伸强度最小的原则。遵循这一原则设计出的导电纤维丝束既具有一定的拉伸强度，便于缠绕成丝团并被抛撒分散开，而且在抛撒和搭接电网时不会被拉断，但是在清除时一扯即断，极大地增加了清除难度，具有很好的清除时效。

3　导电纤维丝束短路引弧能力等效试验研究

根据上述设计准则对导电纤维丝束进行了设计与制备，并按照输出电压 35～220 kV 和相间距 1.6～10 m 条件下对导电纤维丝束进行了大量的小容量等效引弧放电试验研究，对所设计的导电纤维丝束的引弧短路能力进行了考核。试验装置原理图如图 2 所示，典型试验结果如表 1 和图 3 所示。

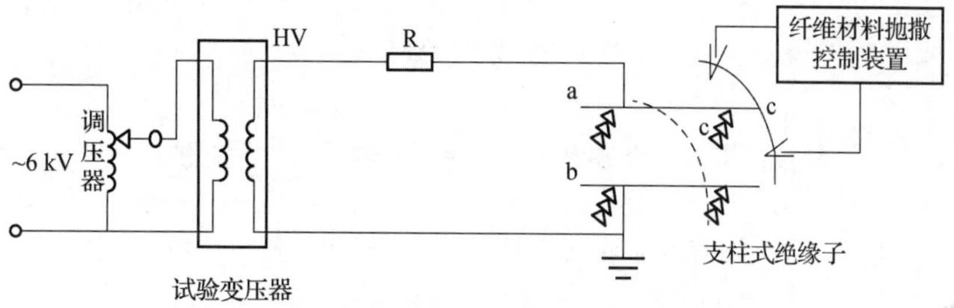

图 2　导电纤维丝束引弧短路试验装置原理图

表 1　不同电压等级、相间距引弧试验数据

电压/kV	相间距/m	引弧时间/s	引弧后丝束状况
66	2.5	0.333	汽化
66	3.5	0.250	汽化
110	4.0	0.333	汽化
110	5.7	0.250	汽化
161	4.5	0.583	汽化
161	5.7	0.375	汽化
220	7.0	0.292	汽化
220	7.2	0.292	汽化

试验结果表明：所设计的导电纤维丝束作用在相间时即可引发短路电弧，引弧成功率达 100%，引弧时间均大于保护装置设定的跳闸时间，丝束全部“汽化或离子化”，无碳状残留物，可高效短路毁伤

35～550 kV 的高压露天电网。

（a）　　　　　　　　　　　　（b）

图 3　高压小容量条件下导电纤维丝束引弧放电照片

（a）161 kV、相间距为 5.7 m；（b）220 kV、相间距为 7.0 m

4　结论

根据对导电纤维丝束在高压导线相间的引弧机理研究，从提高导电纤维丝束的引弧能力和满足导电纤维子弹的使用要求出发，对导电纤维丝束的设计提出了如下几种设计准则：分层设计准则，表面电阻率准则，最小逸出功准则，绝缘导热准则，可缠绕准则及强度准则，清除时效准则等。根据上述准则对导电纤维丝束进行了设计与制备，通过小容量条件下的引弧放电实验表明，所设计的导电纤维丝束具有很好的强制引弧能力，可高效短路毁伤高压露天电网。

参 考 文 献

[1] 梁永直，冯顺山．导电纤维丝束镀层金属引发相间电弧机理研究［J］．弹箭与制导学报，2004，24（2）：64－66.

[2] 王芳，梁永直，冯顺山．导电纤维丝束在高压导线相间引弧试验研究［J］．弹箭与制导学报，2006（2）：221－223.

超音速伞弹流场特性数值分析[①]

完颜振海[②]，冯顺山，董永香，周 彤
（北京理工大学爆炸科学与技术国家重点实验室，北京 100081））

摘 要：伞弹系统超声速流场特性直接关系到伞弹在超声速条件下的气动力性能，对流场特性进行分析和研究能够为伞弹设计和使用提供参考依据。通过建立分块结构化计算网格，基于有限体积法和SST湍流模型对飘带伞和平头子弹组成的伞弹系统进行超声速数值模拟，采用数值纹影法显示流场分布。结果表明，伞弹超声速流场结构复杂，包含多种激波形态和流态。通过研究得到：伞弹流场相互耦合，伞使得弹体尾流驻点距离减小；伞弹流场类型可分为开式流动和闭式流动两类，临界点为喉部位置，流场类型对伞的阻力系数影响很大；随着马赫数的增加，弹体头部激波脱体距离以及弹体和伞的尾流驻点距离都不断减小，变化幅度也减小。

关键词：流体力学；超声速伞弹；流场；SST湍流模型

Numerical Analysis of Flow Field Characteristics of Supersonic Submunition with Parachute

WANYAN Zhen – hai, FENG Shun – shan, DONG Yong – xiang, ZHOU Tong
(State Key Laboratory of Explosive Science and Technology, Beijing Institute of Technology, Beijing 100081, China)

Abstract: The flow field of the supersonic submunition with parachute has a big influence on its aerodynamic performance. The analysis of its characteristics and law of variation can provide reference basis for its design and usage. CFD simulations of flat – faced submunition with disk – streamer parachute were carried out based on multi – block structured grids. Finite volume method and SST turbulence model were used in the simulation. The flow fields were displayed by numerical schlieren method. The results show that the flow field of the submunition with parachute is complicated and has several shock – wave and flow patterns. It is concluded that the flow field of the parachute influences counterpart of submunition and decreases its wake stagnation distance. The type of flow field can be divided into open type and closed type by the neck station of the submunition wake, and has a big influence on the drag coefficient of the parachute. As the Mach number increases, both the detached distance of the submunition's head shock and wake stagnation distances of the parachute and submunition decrease, as well as the decline amplitudes.

Keywords: fluid mechanics; supersonic submunition with parachute; flow field; SST turbulence model

0 引言

子母战斗部在现代弹药中得到广泛应用，用来扩大毁伤范围和增强对目标的毁伤效能。考虑落速、着姿及重点效应等诸多因素，一些子弹采用钝头外形。超声速抛撒的钝头子弹的下落过程中一般经历超声速、跨声速、亚声速三个阶段，在跨声速阶段容易失去动态稳定性[1]，解决方法之一是使用超声速

---

① 原文发表于《兵工学报》2011年第5期。

② 完颜振海：工学博士，2006年师从冯顺山教授，研究封锁型导弹子母弹气动及弹道特性，现工作单位：创生科技（河北雄安）有限公司。

伞。常用的超声速伞一般为柔性带条伞，而在前体超声速尾流的影响下，带条伞可能会因为充气不足而导致开伞困难，开伞后也会发生伞衣喘振和摆振现象[2,3]。为避免柔性伞开伞问题和伞衣喘振现象，本文提出采用刚性伞帽代替柔性伞衣的方法，并选用能够承受超声速条件下较大开伞动载的飘带作为伞绳。考虑装填因素，飘带伞一般为等（亚）口径伞，飘带折叠后安放在弹尾，从而提高空间利用率。

超声速伞弹的流场特性决定其气动力性能，从而对子弹的超声速运动状态影响显著，因此有必要对其进行分析和研究。超声速伞弹流场结构复杂，以往的研究大多采用预测的方法。Henke[4]基于外形和自由流条件预测轴对称物体超音速边界层和尾流特性，但该方法不适合用于本文研究的飘带连接的伞弹情况。Noreen 等[5]提出了一种计算超声速物伞系统流场特性的方法，但该方法基于无黏性假设，并且需要事先通过理论或试验方法获得尾流场分布，在实际应用中受到限制。

通过风洞试验能够得到超声速伞弹的气动力大小，但是还无法细致地得到伞弹流场分布情况，因此本文在试验验证基础上采用计算流体力学方法研究伞弹流场分布。数值模拟已经在降落伞的研究中得到了广泛应用[6]，然而目前在超声速条件下对伞弹结合进行的分析和研究鲜有见到。本文针对飘带伞平头弹的气动布局特点，通过建立分块结构化计算网格，基于 SST 湍流模型采用有限体积法模拟了伞弹的超声速流场，对流场的结构进行了分析，并研究了流场随弹伞间距及马赫数的变化规律。

1　计算模型

1.1　模型尺寸

对应于风洞试验，本文对飘带处于拉直状态下的伞弹进行建模，模型尺寸如图 1 所示，弹体为平头凹槽型，弹体直径为 d，弹长 $L=2.5d$，弹伞间距为 $L_{mp}=5.6d$，弹伞直径比 $d/d_p=1.2$，伞帽为钢质材料，飘带材料为双层聚酰胺纤维，数量为 4。

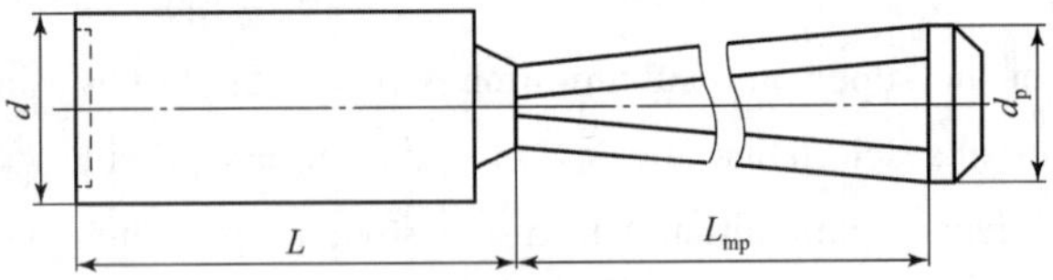

图 1　模型尺寸

1.2　计算网格

为了进行对比分析，分别建立了单独弹体和伞弹的计算网格。单独弹体采用结构化 H 型网格，流向、径向、周向网格点数分别为 110、61、61，网格总数为 409 310。伞弹采用分块结构化 H 型 6 面体网格，共分 24 块，物面网格和流场网格分别如图 2 和图 3 所示，流向、径向、周向网格点数分别为 200、61、61，网格总数为 744 200。弹体和伞的物面网格以及伞弹附近网格分别如图 2 和图 3 所示。

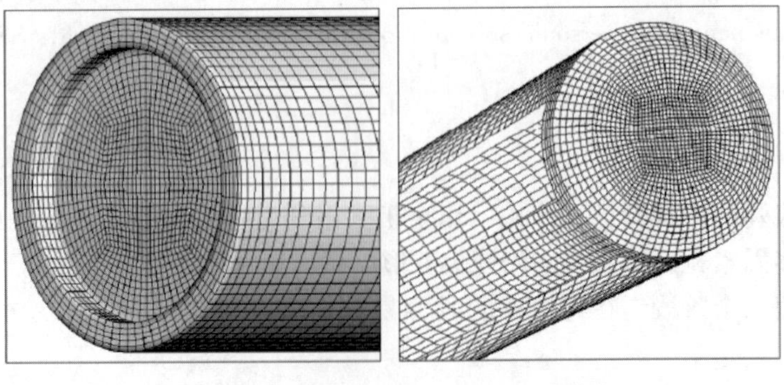

（a）　　　　　　　　（b）

图 2　弹体和飘带伞物面网格图

（a）弹体；（b）飘带伞

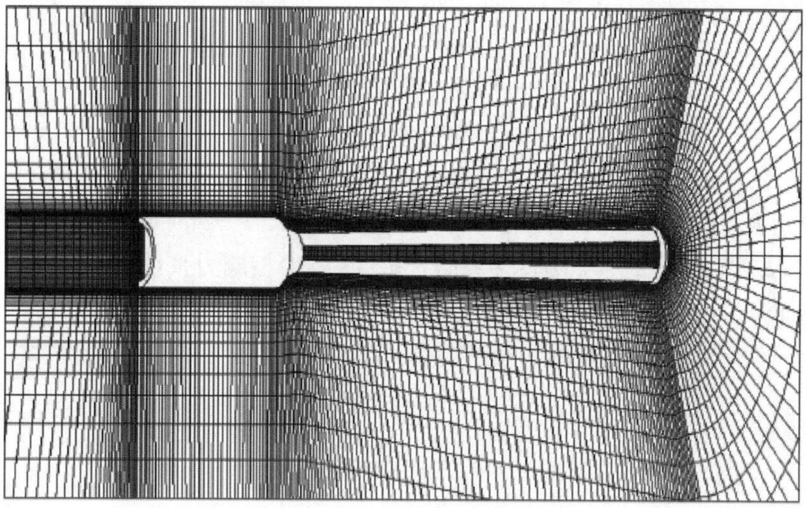

图 3　伞弹附近网格

在实际求解过程中，使用网格细化技术进行局部加密，以捕捉厚度很薄的激波层。图 4 给出了马赫数为 1.5 时，弹体头部附近加密前后的网格对比图，最终网格数为：单独弹体 666 327，伞弹 1 028 367。

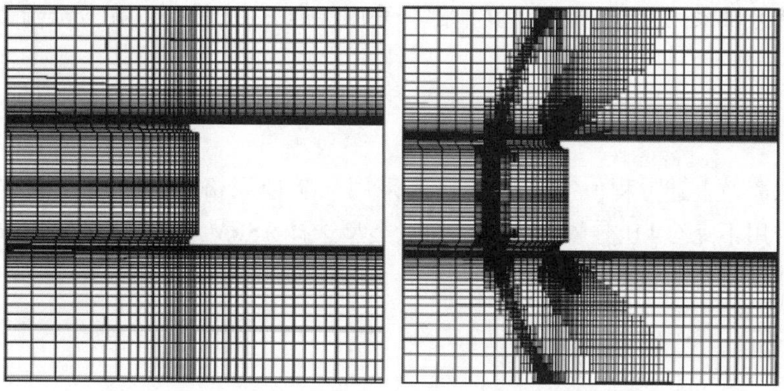

（a）　　　　　　　　　　　　　　（b）

图 4　马赫数为 1.5 时弹体头部附近网格

（a）原模型；（b）基于压力梯度 2 次加密后

1.3　边界条件

由于飘带的透气性较差，在模型中将其近似作为壁面处理。物面采用非滑移壁面条件，流场采用超声速入口出口边界条件，入口边界条件为温度 281 K，静压 88 792 Pa，马赫数计算范围为 1.5 ~ 3.0，对应雷诺数范围为 $2.18\times10^6 \sim 4.36\times10^6$。

2　计算方法

2.1　控制方程

对流场中任意控制体的体积 V 和矢量面积微元 $\mathrm{d}\boldsymbol{S}$ 应用积分形式的 Navier – Stokes 方程，不考虑体积力和外部热源，其在笛卡儿坐标系下的向量形式为

$$\frac{\partial}{\partial t}\int_V \boldsymbol{Q}\mathrm{d}V + \oint_S \boldsymbol{F}\mathrm{d}\boldsymbol{S} = \oint_S \boldsymbol{G}\mathrm{d}\boldsymbol{S} \tag{1}$$

向量 $\boldsymbol{Q}$、$\boldsymbol{F}$、$\boldsymbol{G}$ 的定义为

$$Q = \begin{bmatrix} \rho \\ \rho u \\ \rho v \\ \rho w \\ E_t \end{bmatrix}, \quad F = \begin{bmatrix} \rho \boldsymbol{v} \\ \rho u \boldsymbol{v} + p\boldsymbol{i} \\ \rho v \boldsymbol{v} + p\boldsymbol{j} \\ \rho w \boldsymbol{v} + pk \\ (E_t + p)\boldsymbol{v} \end{bmatrix}, \quad Q = \begin{bmatrix} 0 \\ \boldsymbol{\tau}_x \\ \boldsymbol{\tau}_y \\ \boldsymbol{\tau}_z \\ \boldsymbol{\tau} \cdot \boldsymbol{v} + k \nabla T \end{bmatrix}$$

式中，ρ、v 和 p 分别为控制体体积内流体的密度、速度和压强，$v = u\boldsymbol{i} + v\boldsymbol{j} + w\boldsymbol{k}$；$E_t$、$T$ 分别为单位体积的总能和静温，$E_t = \rho(e + |\boldsymbol{v}|^2/2)$；$k$ 为传热系数；$\boldsymbol{\tau}$ 为黏性摩擦应力张量。

对于量热气体而言，还满足理想气体状态方程和内能关系式：

$$p = \rho RT \tag{2}$$

$$e = h - p/\rho \tag{3}$$

式中　R——气体常数；

　　h——比焓。

2.2　湍流模型

对于可压缩流动，对 Navier – Stokes 方程采用 Favre 平均，将变量分为统计平均项和波动项两部分，由于引入雷诺应力项，需要湍流模型封闭方程。本文采用 Mentor[7] 提出的 SST 模型，该模型将计算区域分为两部分，在近壁区域采用 $k-\omega$ 模型，其他部分采用标准 $k-\omega$ 模型，在分流流动模拟方面能够给出较为准确的结果。

2.3　求解方法

使用有限体积法离散上述方程，空间离散方式采用二阶迎风格式，时间离散方式采用欧拉隐式格式，离散后的方程采用不完全 LU 分解法计算。通量类型选择 ASUM + 格式，此种格式能够提供激波非连续性的精确求解并能保持各标量的正性[8]。

3　结果与分析

3.1　数值方法有效性验证

本文研究的伞弹在沈阳空气动力研究所 FL – 1 风洞进行了试验。表 1 给出了分别由数值计算和风洞试验得到的单独弹体和伞弹的阻力系数 C_D 的对比情况，从中可以看出数值计算与试验结果很接近，相对误差在 4% 以内，从而验证了方法有效性。

表 1　数值计算与风洞试验结果对比

弹种	马赫数	阻力系数 C_D		相对误差/%
		数值计算	风洞试验	
单独弹体	1.5	1.546 7	1.489 4	3.85
	2.0	1.765 2	1.701 0	3.77
伞弹	1.5	1.993 0	1.922 9	3.64
	2.0	2.111 4	2.050 1	2.99

3.2　流场特性分析

流场采用数值纹影法显示，单独弹体马赫数为 2.0 时的零攻角稳态流场分布如图 5 所示，平头凹槽形的头部产生脱体弓形激波，顶点处激波与来流方向垂直，其类型为正激波。弓形激波分为强激波和弱激波两部分，激波后的流场也分为亚声速流动和超声速流动两个区域，图中的点画线为二者的分界线即声速线。由于弹体中段母线与来流保持一致，其对气流阻滞作用大大减弱，从而在头部凸出部分产生膨胀波。头部激波作用下向外偏转的气流通过膨胀波后向弹体表面偏转，同时由于过度膨胀在弹体中段前沿再附着后形成再附压缩波，气流经过再附压缩波后方向基本与自由流方向一致。弹体中段后边缘产生膨胀波，同时弹体边界层分离后形成自由剪切层，其内部形成回流区，其结束位置称为驻点。由于弹体的轴对称性，剪切层逐渐收缩直到达到极限位置生成再压缩波。驻点后的尾迹区，主要由气流边界层汇集而成，随着距离增加，流向速度逐渐增大，在喉部位置处达到声速。

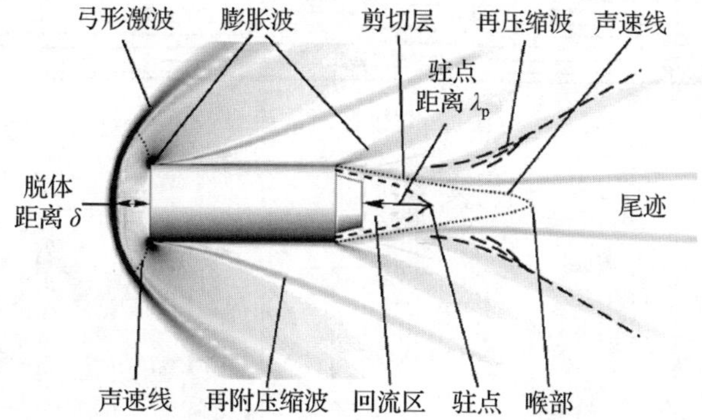

图 5　马赫数为 2.0 时单独弹体流场的数值纹影

马赫数为 2.5 时伞弹流场的数值纹影如图 6 所示，从图中可以看出飘带伞位于弹体的尾流场中。弹体的尾流可以分为三个区域，即图中的回流区、收缩区和尾迹区，其分界点分别为弹体的尾流驻点和喉部位置[4]。通过建立不含飘带的弹伞模型对比分析可以得到，有飘带时弹体的尾流驻点和喉部位置均比无飘带时靠前。

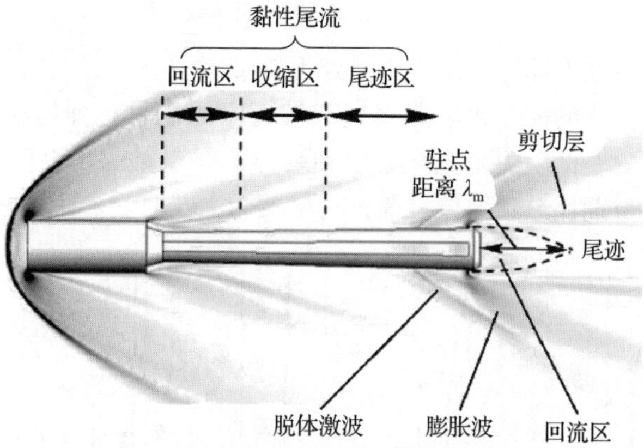

图 6　马赫数为 2.5 时伞弹流场的数值纹影

图 7 给出了伞弹附近的流线图，由于飘带的存在，伞帽被分为非连接飘带区和连接飘带区两部分。在非连接区，伞帽前端产生脱体激波，但由于受弹体尾流影响其激波并不完整，头部部分区域缺失，与弹体头部类似，伞帽前部边缘区域形成膨胀波；但与弹体不同的是，由于伞帽内凹且宽度较短，分离运动强烈，因此剪切层收缩效应不明显，没有出现再压缩波。对于连接区而言，仅发生膨胀波和剪切层分离。与弹体类似，伞的后端亦形成底部回流区和尾迹。

为了显示横向流场的结构，图 8 给出了马赫数为 1.5 时距离弹体尾部 $3d$ 处伞弹流场的横截面压强 p 与自由流压强 p_∞ 之比的等值线图及马赫数等值线。从图中可以看出，飘带主要改变了在其内部的流场，在其影响下，有飘带区形成低压区，无飘带区产生高压区，因此同样的径向距离处无飘带区马赫数较高。

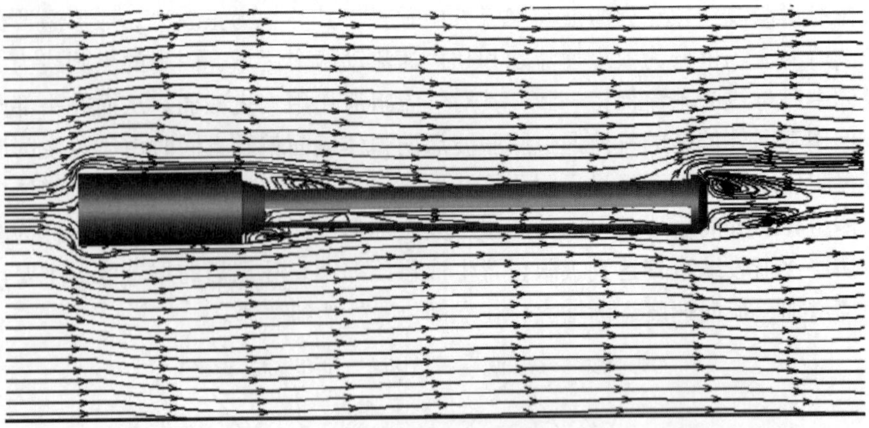

图7　伞弹附近的流线

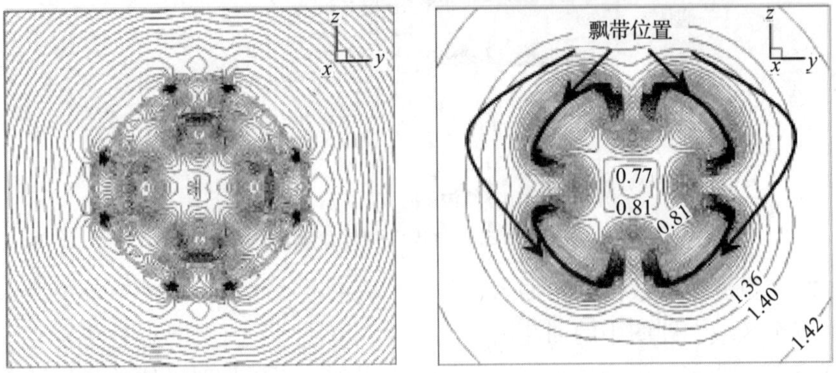

（a）　　　　（b）

图8　马赫数为1.5时距离弹体尾部3*d*处的等值线

（a）压力系数；（b）马赫数

图9给出了伞弹弹体和单独弹体尾部不同距离处无飘带区中心位置纵向±0.6*d*范围内的压强比分布曲线。从图可以看出，伞弹弹体的压强比在距离底部较近区域比单独弹体小，较远区域比单独弹体大。近区域的压降导致伞弹弹体的底部阻力系数大于单独弹体，马赫数为1.5时增加5%左右；远区域的压升导致伞弹弹体的尾流驻点位置比单独弹体靠前，马赫数为1.5时驻点距离λ_m减小8%左右。

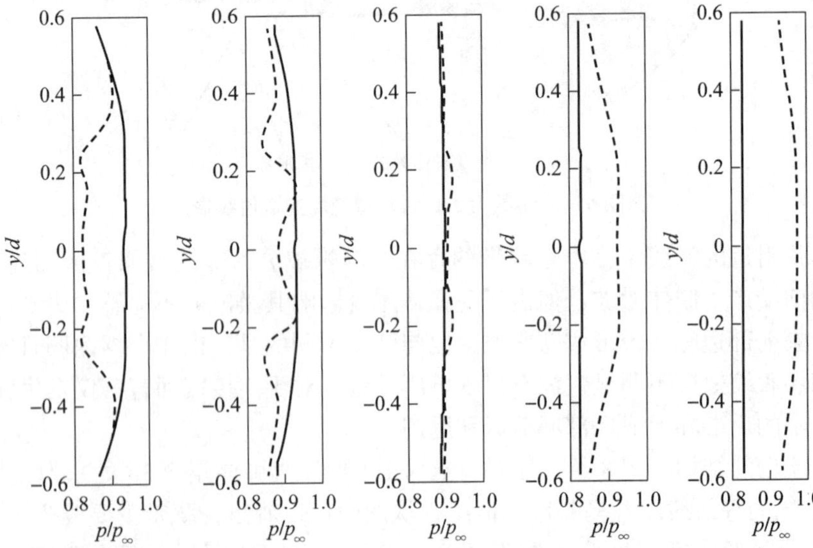

（a）　（b）　（c）　（d）　（e）

图9　马赫数为1.5时弹体尾部压强比分布图（虚线代表伞弹，实线代表单独弹体）

（a）$x/d=1$；（b）$x/d=2$；（c）$x/d=3$；（d）$x/d=4$；（e）$x/d=5$

3.3　流场特性随弹伞间距的变化

文献［9］指出，超声速前后体存在两种流场类型：一种为开式流动，另一种是闭式流动。当流动为开式时，前体尾部和后体前端的滞止区连通，气流流经前体后端时跳过间隔；当流动为闭式时，后体前端形成激波，并且前体尾部和后体前端的滞止区域不再连通。Charwat 等[10]通过试验发现前后体存在一个临界距离，小于此临界距离流动为开式，大于临界距离则为闭式，但没有给出相应的判断依据。

本文通过变化弹伞间距的研究发现，临界距离与弹体尾流喉部位置有关，这是因为当弹伞间距小于弹体喉部位置距离时，伞的前端为亚声速流，因此扰动可以向前传播，从而造成弹体尾部与伞前端滞止区连通，形成开式流动；而当弹伞间距大于弹体喉部位置距离时，伞的前端存在超声速流，从而形成头部激波，由于扰动无法穿过激波，因此形成闭式流动。例如，马赫数为 2.5 时弹体喉部位置距离约为 2.29 倍弹径，因此当弹伞间距小于此值时伞弹流场类型为开式流动，而大于此值时应为闭式流动。图 10（a）和图 10（b）分别给出了弹伞间距为 $2d$ 和 $3d$ 时伞弹的流场，从图中可以看出，两种流场分别为开式流动和闭式流动，符合上述结论。同时经过计算发现，伞的阻力系数与伞弹的流场类型密切相关，以图 10 给出的两种情形为例，图 10（b）中伞的阻力系数是图 10（a）中伞的阻力系数的 4 倍。由于飘带会对弹体的喉部位置产生影响，因此也会改变临界距离的大小，通过对不含飘带的伞弹计算得到，马赫数为 2.5 时伞弹的临界距离约为 2.57 倍的弹径，相比带有飘带的伞弹而言临界距离变大。

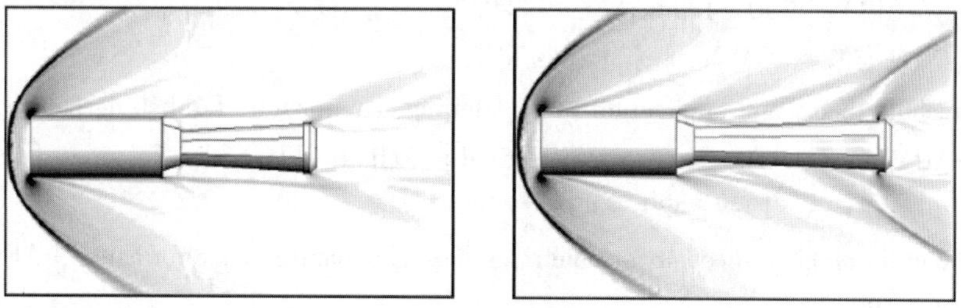

（a）　　　　　　（b）

图 10　伞弹流场

（a）马赫数为 2.5，$L_{mp}=2d$；（b）马赫数为 2.5，$L_{mp}=3d$

3.4　流场特性随马赫数的变化

对比不同马赫数下伞弹流场分布可以得出，随着马赫数的增加，弹体头部弓形激波强度越来越大，正激波点位置越来越靠近头部，强激波曲率半径和弱激波的激波角越来越小。通过前面分析，可以选取以下流场特性参数研究流场随马赫数变化，分别为：弹体头部激波脱体距离 δ，弹体尾流驻点距离 λ_m，伞的尾流驻点距离 λ_p。在选定的弹伞间距下，这些参数与弹径的比值随马赫数的变化如表 2 所示。从表中可以看出，随着马赫数的增加，激波脱体距离越来越小；无论是弹体还是伞，尾流驻点距离都不断减小。同时可从表中得出，流场特性参数的变化幅度也随着马赫数的增加不断减小。

表 2　流场参数随马赫数变化

参数	马赫数			
	1.5	2.0	2.5	3.0
δ/d	0.63	0.43	0.36	0.31
λ_m/d	1.14	0.72	0.52	0.50
λ_p/d	1.87	1.27	1.17	1.14

4 结论

本文通过数值模拟得到了亚口径伞弹超声速流场，通过分析可以得出，伞弹流场结构复杂，包含弓形激波、斜激波、膨胀波、再压缩波多种激波形态，以及自由流、剪切流、回流等各种流态。同时，可以得出以下结论。

（1）伞弹流场相互耦合，不仅弹体尾流对伞产生影响，同时飘带伞改变了弹体的尾流场分布，使得无飘带区弹底近区压强减小，远区压强增大，并在飘带附近形成低压区，弹体的尾流驻点向底部靠近。

（2）随着弹伞间距的不同，超音速伞弹的流场类型可以分为开式流动和闭式流动两种，其临界距离为弹体尾流的喉部驻点距离，闭式流动伞的阻力系数比开式流动大得多。

（3）随着马赫数的增加，弹体头部弓形激波脱体距离、强激波的曲率半径、弱激波的激波角都不断减小，弹体和伞的尾流驻点距离也越来越小，相对变化幅度亦减小。

本文研究可以为伞弹设计和应用提供参考，同时也为同类问题的研究提供参考。

参考文献

[1] MARKO W J. Dynamic stability of high – drag planetary entry vehicles at transonic speeds [J]. Journal of spacecraft, 1969, 6 (12): 1390 – 1396.

[2] 王利荣. 降落伞理论与应用 [M]. 北京: 宇航出版社, 1997.
WANG L R. Parachute theory and application [M]. Beijing: Astronautic Publishing House, 1997. (in Chinese)

[3] HIRAKI K, HINADA M. Flow measurement around rigid parachute – like bodies in supersonic free stream: AIAA – 97 – 1529 [R]. US: AIAA, 1997: 429 – 438.

[4] HENKE C W. Analysis of high speed axisymmetric wakes and parasonic performance: AFFDL – TR – 67 – 192 [R]. US: AFFDL, 1968.

[5] NOREEN R A, RUST L W JR, RAO P P. Analysis of the supersonic flow field about a forebody – decelerator combination: AFFDL – TR – 71 – 35 [R]. US: AFFDL, 1972.

[6] 余莉, 明晓. 降落伞的计算机仿真研究 [J]. 航天返回与遥感, 2005, 26 (4): 6 – 9.
YU L, MING X. Investigation of parachute's emulator and computer simulation [J]. Spacecraft recovery & remote sensing, 2005, 26 (4): 6 – 9. (in Chinese)

[7] MENTOR F R. Two – equation eddy – viscosity turbulence models for aerodynamic flows [J]. AIAA journal, 1994, 32 (8): 1598 – 1605.

[8] LIOU M S, STEFFEN C J. A new flux splitting scheme [J]. Journal of computational physics, 1993, 107 (1): 23 – 39.

[9] LIN T C, FEDELE J B, BAKER R L, et al. Engineering model for re – entry vehicle turbulent wakes [J]. Journal of spacecraft, 1980, 17 (2): 123 – 128.

[10] CHARWAT A F, ROOS J N, DEWEG F C, et al. An investigation of separated flows—part I: the pressure field [J]. Journal of aeronautical sciences, 1961, 28: 457 – 470.

平头弹超声速尾流对飘带伞气动特性的影响①

冯顺山，完颜振海②

（北京理工大学爆炸科学与技术国家重点实验室，北京　100081）

摘　要：本文提出了一种用于超声速抛撒平头子弹药的飘带伞结构的新型超声速伞，研究弹体超声速尾流对伞气动特性的影响。在风洞试验结果验证的基础上，通过建立多套分块结构化网格模型，采用有限体积法和SST湍流模型对单独弹和伞弹分别进行超声速数值模拟，用数值纹影法显示弹体尾流场并进行了分析，得到了平头弹超声速尾流对飘带伞气动特性影响的变化规律。结果表明，弹体尾流对伞气动特性的影响随马赫数的增加而变大，随弹伞间距与弹径比和伞弹径比的增加而变小，随飘带宽度与弹径比的增加先基本保持不变后变大。

关键词：平头弹；超声速尾流；飘带伞；气动特性

Influence of Flat – Faced Submunition'S Supersonic Wake Flow on the Aerodynamic Characteristics of Disk – Streamer Parachute

FENG Shun – shan, WANYAN Zhen – hai

(State Key Laboratory of Explosion Science and Technology, Beijing Institute of Technology, Bering 100081, China)

Abstract: A new disk—streamer supersonic parachute and its application to supersonic dispersed flat—faced submunitions are introduced in this research. The influences of submunition's wake on the parachute are investigated to provide references for the design and use of the parachute. Based on the validation of wind tunnel experiment, multiple multi—block grid models were built. Supersonic numerical simulations of the parachute, the submunition and their combination were conducted using finite volume method and SST turbulence model. The wake flow field of the submunition was given and analyzed by using numerical schlieren method. The influences of flat— faced submunition's wake flow on the aerodynamic characteristics of the parachute were acquired. which could be reflected by the parachute drag coefficient. The results show that the influence increases as the Mach number increases, namely the parachute drag coefficient decreases as the Mach number increases. Also。the influence decreases as ratio of the distance between submunition and parachute to submunition's diameter increase as well as the ratio of parachute diameter to submunition's diameter increase. As for the influence of streamer width, it basically stays the same at first then increases as the streamer width increases.

Key words: flat – faced submunition; supersonic wake; disk – streamer parachute; aerodynamic characteristics

子母战斗部在现代弹药中得到广泛应用，用来提高弹药作战威力、扩大毁伤范围、增强突防能力。考虑战斗部结构、稳定性、落速、触地姿态等诸多因素，很多子弹药采用钝头外形。超声速钝头体在减速到跨声速和亚声速时容易失去动态稳定性，产生大攻角状态下的大幅度摆动，甚至发生翻转，从而影响亚声速伞的打开[1-2]，解决方法之一是使用超声速伞。一般的超声速伞绝大多数为带条伞，而在前体

① 原文发表于《北京理工大学学报》2011年第6期。

② 完颜振海：工学博士，2006年师从冯顺山教授，研究封锁型导弹子母弹气动及弹道特性，现工作单位：创生科技（河北雄安）有限公司。

超声速尾流的影响下，带条伞会因为充气不足而导致开伞困难，而且开伞后会发生喘振和摆振，严重影响伞及弹体的稳定性，也可能造成伞衣和伞绳的结构疲劳破坏[3-4]剖。因此，笔者提出一种新型的飘带伞，伞衣为刚性材料，通过飘带与弹体相连。刚性伞帽避免了伞衣喘振现象，开伞过程仅包括抛伞和飘带拉直两个阶段，避免了传统降落伞充气阶段这一复杂的物理过程[5]，增强了伞的可靠性。同时，飘带具有一定的宽度，抗拉强度大，能够承受较大的开伞动载。考虑装填等因素，在应用于子弹药时飘带伞一般为等口径伞或亚口径伞，飘带折叠后贴在弹尾，节约空间从而提高装填效率。

对于这种新型伞的设计使用，需要对其气动特性主要是阻力特性进行分析和研究。对于钝头体超声速尾流的分析目前尚无统一的理论分析方法，一般通过风洞实验或飞行实验进行测量[5-6]。计算流体力学在弹药领域上得到了广泛的应用[7]，文中在风洞实验结果验证的基础上，通过建立多套分块结构化网格模型，对单独弹和伞弹进行超声速数值模拟，得出了伞的阻力特性随马赫数、弹伞间距与弹径比、伞弹径比、飘带宽度与弹径比的变化规律。

1　数值计算方法

1.1　控制方程

对流场中任意控制体的体积 V 和矢量面积微元 d$\boldsymbol{S}$ 应用积分形式的 N－S 方程，不考虑体积力和外部热源，其在笛卡儿坐标系下的向量形式为

$$\frac{\partial}{\partial t}\int_{v}\boldsymbol{Q}\mathrm{d}V+\oint_{S}\boldsymbol{F}\cdot\mathrm{d}\boldsymbol{S}=\oint_{S}\boldsymbol{G}\cdot\mathrm{d}\boldsymbol{S} \tag{1}$$

向量 $\boldsymbol{Q}$，$\boldsymbol{F}$，$\boldsymbol{G}$ 的定义为

$$\boldsymbol{Q}=\begin{bmatrix}\rho\\ \rho u\\ \rho v\\ \rho\omega\\ E_{\mathrm{t}}\end{bmatrix},\mathbf{F}=\begin{bmatrix}\rho\boldsymbol{\nu}\\ \rho u\boldsymbol{\nu}+p\boldsymbol{i}\\ \rho v\boldsymbol{\nu}+p\boldsymbol{j}\\ \rho\omega\boldsymbol{\nu}+p\boldsymbol{k}\\ (E_{\mathrm{t}}+p)\boldsymbol{\nu}\end{bmatrix},\boldsymbol{G}=\begin{bmatrix}0\\ \boldsymbol{\tau}_x\\ \boldsymbol{\tau}_y\\ \boldsymbol{\tau}_z\\ \boldsymbol{\tau}\cdot v+k\,\nabla T\end{bmatrix}$$

式中：ρ，$\boldsymbol{\nu}$，E_{t}，T，p 分别为控制体体积内流体的密度、速度、总能、温度和压强，$\boldsymbol{\nu}=u\boldsymbol{i}+v\boldsymbol{j}+\omega\boldsymbol{k}$，$E_{\mathrm{t}}=\rho(e+|v|^2/2)$，$k$ 为传热系数；$\boldsymbol{\tau}$ 为黏性摩擦应力张量。

对于量热气体而言，还满足理想气体状态方程和内能关系式：

$$p=\rho RT \tag{2}$$

$$e=h-p/\rho \tag{3}$$

式中　R——摩尔气体常数；

　　h——比焓。

1.2　湍流模型

对于可压缩流动，对 N－S 方程采用 Favre 平均，将流动变量分为统计平均项和波动项两部分，由于引入雷诺应力项，需要湍流模型封闭方程。文中采用 Mentor[8]提出的 SST 模型，其将计算区域分为两部分，在近壁区域采用 $k-\omega$ 模型，其他部分采用标准 $k-e$ 模型，在分离流动方面模拟结果较好。

1.3　求解方法与计算平台

使用有限体积法离散上述方程，空间离散方式采用二阶迎风格式，时间离散方式采用欧拉隐式格式，离散后的方程采用不完全 LU 分解法计算。通量类型选择 ASUM＋格式，此种格式提供激波非连续性的精确求解并能保持各标量的正性[9]。

采用 2 台 Dell PowerEdge T610 服务器并行计算，单台服务器配置为 4 个 4 核 Intel E5520 Xeon 处理

器，8 ×4 G 内存。

2　数值计算模型

2.1　模型尺寸

对飘带处于拉直状态下的伞弹进行建模，模型尺寸如图 1 所示，弹体为平头凹槽型，弹体直径为 D，弹长径比为 $L/D=2.5$。弹伞间距、伞径和飘带宽度分别为与弹径的比值分别为 L_{mp}，D_p 和 W，与弹径的比值分别为 $l=L_{mp}/D$，$d=D_p/D$ 和 $\omega=W/D$。风洞试验模型为 $l=5.6$，$d=0.83$ 和 $\omega=0.29$，同时也是笔者研究的基准模型。

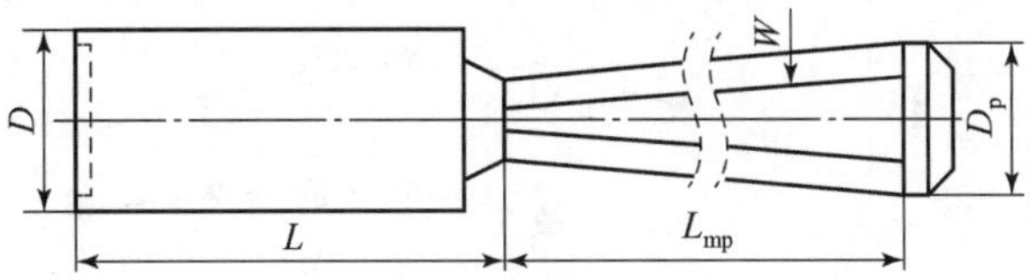

图 1　模型尺寸

2.2　计算网格及边界条件

网格质量对于计算结果非常重要，文中全部采用结构化网格。为了进行对比分析，分别建立了单独弹体和伞弹的计算网格。

单独弹体采用结构化 H 型网格，其物面网格如图 2 所示，流向、径向、周向网格点数分别为 110、61、61，网格总数为 409 310。

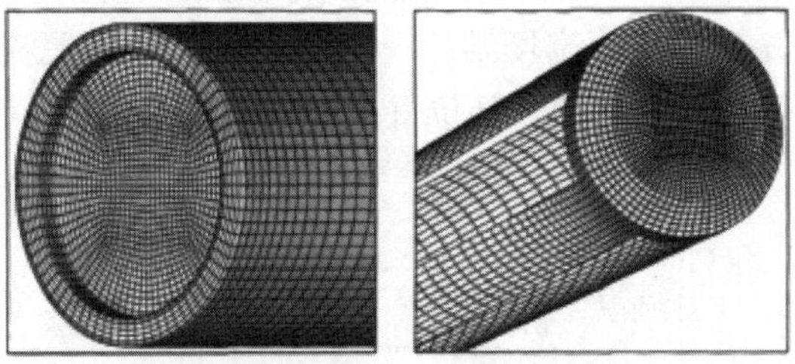

（a）　　　　（b）

图 2　弹体和飘带伞物面网格图

（a）弹体；（b）飘带伞

伞弹采用分块结构化 H 型六面体网格，共分 24 块，网格如图 3 所示。由于需要对不同弹伞间距、伞弹径比和飘带宽度进行建模，因此共计建了 22 套网格，其中流向网格点数在 200 ~ 250 之间，径向、周向网格点数分别为 61、61，对应的网格总数范围为 744 200 ~ 930 250。

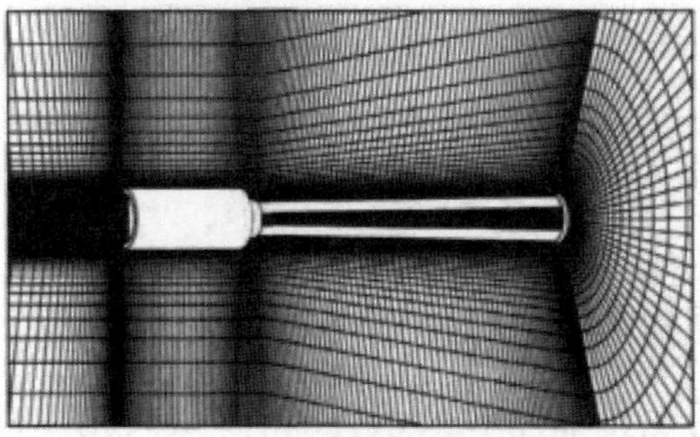

图 3　伞弹附近网格图

采用超声速入口出口边界条件，入口边界条件为温度 281K，静压 88 792 Pa，马赫数计算范围为 1.5 ~ 3.0，对应雷诺数范围为 $2.18\times10^6\sim4.36\times10^6$。

3 结果与分析

3.1 方法验证及弹体尾流场分布

采用风洞试验结果对数值方法进行验证，选取弹体的横截面作为参考面积，单独弹体和伞弹的阻力系数随马赫数的变化如图4所示。从图中可以看出，数值计算的结果与风洞试验比较吻合，相对误差不超过4%。

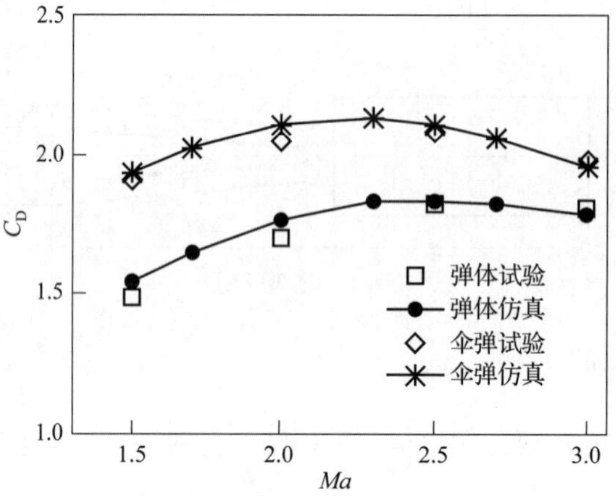

图4 数值计算与风洞试验结果对比图

使用数值纹影法显示马赫数为2时单独弹体的流场结构如图5所示，弹体尾部区域包含膨胀波、剪切层分离、底部汇流、再压缩波等复杂的流场结构。总体上说，弹体尾流可以分为三个部分：回流区、收缩区和尾迹区[4]，其分界点为尾流驻点和喉部位置，即分离流线和声速线的极限位置。

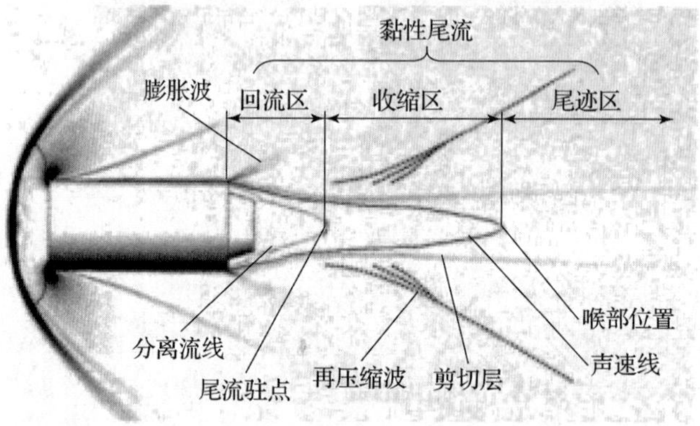

图5 单独弹体的超声速尾流场结构

3.2 弹体尾流对于伞阻力特性的影响

3.2.1 阻力系数随马赫数的变化

伞的阻力系数 C_D 随马赫数的变化如图6所示，随着马赫数的增加，伞的阻力系数不断减小，表明弹体尾流对伞的气动特性的影响越来越大。当马赫数由1.5增加到3.0时，伞的阻力系数减少了60%。

3.2.2　阻力系数随弹伞间距与弹径比的变化

在基准模型的基础上，保持其他参数不变，改变弹伞间距，以分析尾流影响随距离的变化，其结果如图 7 所示。由图可知，随着弹伞间距的增加，伞的阻力系数 C_d 逐渐变大。按照变化趋势可分为三个区域，分别是 $l<3$，$3\leqslant l\leqslant 5$ 和 $l>5$，对应于伞位于弹体尾流的回流区、收缩区和尾迹区。阻力系数在三部分均呈近似线性增长，以收缩区增加速度最快。从图中也可以看出弹体尾流影响的距离很远，弹伞间距在 20D 之外伞的阻力系数仍在增加。

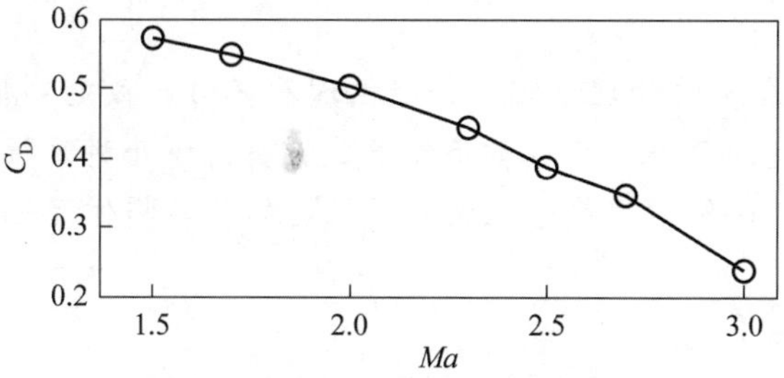

图 6　伞阻力系数随马赫数的变化

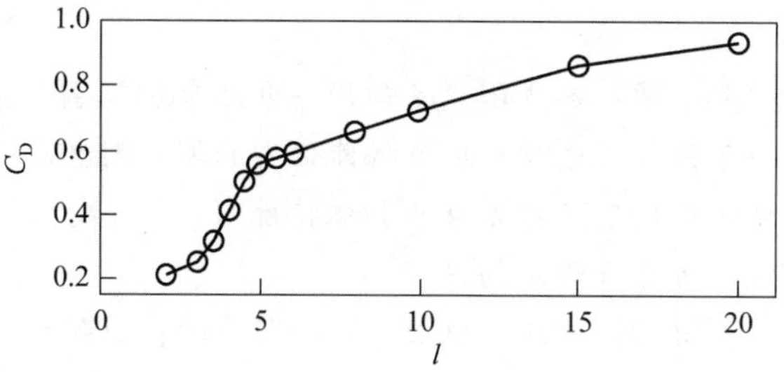

图 7　Ma＝1.5 时伞阻力系数随 l 变化

3.2.3　阻力系数随伞弹径比的变化

同样保持其他参数不变，改变伞的直径，研究尾流影响随伞弹径比的变化，其结果如图 8 所示。伞的阻力系数 C_D 随伞弹径比的增加而增大，即弹体尾流对较小直径的伞的气动特性影响较大，伞弹径比为 0.5 时的阻力系数仅为 1.0 时的 21.2%。

3.2.4　阻力系数随飘带宽度与弹径比的变化

伞的阻力系数随飘带宽度与弹径比的变化如图 9 所示，当 $\omega\leqslant 0.3$ 时，阻力系数基本保持不变，飘带宽度继续增大时，阻力系数开始减小。

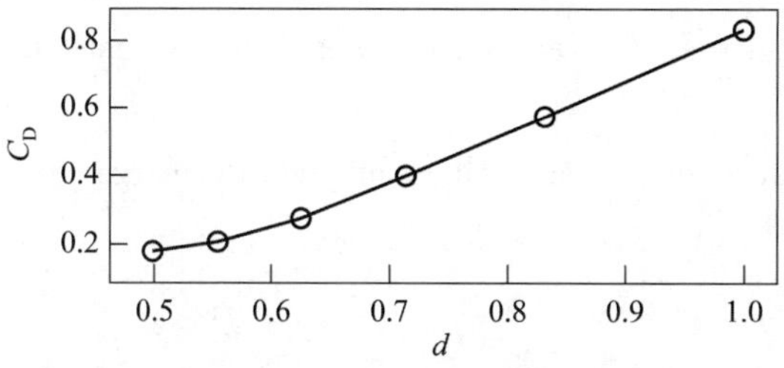

图 8　*Ma*＝1.5 时伞阻力系数随 d 变化

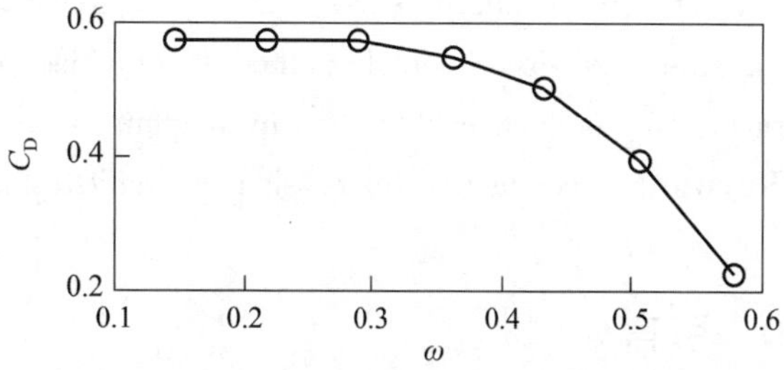

图 9　*Ma*＝1.5 时伞阻力系数随飘带宽度变化

4　结论

结合数值仿真与风洞试验对平头弹超声速尾流对亚口径飘带伞的气动特性的影响进行了分析和研究，从中可以得出以下结论：①平头弹超声速尾流对亚口径伞气动特性的影响与马赫数、弹伞间距、伞径大小以及飘带宽度有关；②随着马赫数增加，弹体尾流对伞的阻力特性影响越来越大，马赫数从 1.5 增加到 3.0 时，伞的阻力系数减少 60%；③增大弹伞间距可以减小弹体尾流的影响，伞位于弹体尾流不同区域时，其阻力系数增加速率不同，位于收缩区时增加最快；④增加伞的直径同样可以减小弹体尾流的影响，马赫数为 1.5 时等口径伞的阻力系数为半口径伞 4 倍多；⑤飘带宽度增加时，伞的阻力系数先是保持基本不变后减小，马赫数为 1.5 时，其临界飘带宽度约为 0.3 倍弹径。

相关研究结果可以为超声速伞弹设计提供指导，也可用于伞弹的外弹道计算。

弹用导电丝束引弧机理及试验研究①

梁永直[1②]，冯顺山[2]，张广华[2]，王超[2]
（1. 太原理工大学，太原 030024；
2. 北京理工大学爆炸科学与技术国家重点实验室，北京 100081）

摘 要：本文阐述了弹用导电丝束相间引弧发生、发展的物理过程、作用机理及其自持放电判据，给出了弹用导电丝束引弧性能优劣的影响因素与条件，通过高压、小容量条件下试验获得了几种改性碳基丝束和玻璃纤维丝束的引弧特性 U－d 曲线，为测试、评价弹用导电丝束引弧性能的优劣提供了一种有效、可靠及经济的方法。

关键词：弹用导电丝束；短路试验；引弧机理；自持放电

Arc Mechanism and Experimental Research of Conductive Fiber Used in Weapon

LIANG Yongzhi[1], FENG Shunshan[2], ZHANG Guanghua[2], WANG Chao[2]
(1 Taiyuan University of Technology, Taiyuan 030024, China; 2 State Key Laboratory of Explosion Science and Technology, Beijing Institute of Technology, Beijing 100081, China)

Abstract: The causes of arc discharge, physical procedures of development, mechanisms of effect and self－contained discharge criteria caused by conductive fiber between high—voltage wires were discussed in this paper. The influencing factors of arc—discharge performance with conductive fiber used in ammunition was proposed. The arc peculiarity curve (U—d) of some kinds of improved fibers with carbon—based and fiberglass were obtained by experimental method. It provides an effective and economic method to evaluate the arc performance of conductive fiber used in weapon.

Keywords: conductive fiber; short—circuit testing; arc mechanism; self—contained discharge

0 引言

弹用导电丝束是一种非致命性毁伤元素，它对高压电力设施具有强制连续引弧的特点，短路危害性很大。在电力系统的日常运行中，也经常会有一些偶然的、人为的导电材料对高压电力设施产生短路作用，引起高压线路、变电站等设施起晕、闪络、放电引弧、短路故障，直至大停电等危害，“北美 8.14 大停电”就是一个典型的事例[1]。为了获得不同导电丝束对高压线路的引弧特性，文中从导电丝束毁伤高压线路的角度出发，对导电丝束的引弧物理过程及其作用机理进行了研究，并通过在高压小容量条件下的试验对理论结果进行了试验验证。

---

① 本文发表于《弹箭与制导学报》2012 年第 6 期。

② 梁永直：工学博士，2000 年师从冯顺山教授，研究导电丝束对高压电网短路毁伤机理及效应，现工作单位：加拿大科技公司。

1　导电丝束对高压线路引弧放电机理

1.1　导电丝束对高压线路物理作用过程

当导电丝束自由飘落接近到高压线路上空时，高压线路周围强电场作用使得导电丝束被由弱至强的电磁场力吸引到导线的强电场区，丝束的介入破坏了导线相间气体绝缘介质的安全距离，引发空气间隙击穿放电；同时，导电丝束在强电场、大电流作用下产生高温，使丝束表面的金属镀层迅速蒸发、汽化，高温强电场还使丝束自身及周围空气产生大量等离子体，形成金属离子和等离子体构成的电弧短路通道[2]。在此过程中，贡献到空气间隙中的带电质点来源有：

（1）碰撞电离。在强电场作用下，高速运动的质点与气体中的中性原子或分子碰撞时，使得中性质点中的电子游离出来成为正离子和电子，其电离判据为

$$eEx \geqslant W_i \tag{1}$$

式中　e——质点电荷量；

E——电场强度；

W_i——电离位能，

$W_i = eU_i$，U_i——气体的电离电位；

x——质子碰撞前的行程。

（2）表面游离。导电丝束金属镀层获得能量后表面电子游离而产生的带电质点为其首要部分；不同的金属有不同的逸出功，而且其值的大小与其表面状况，如粗糙度、氧化程度等有密切相关。

（3）热电离。丝束引弧后产生高温，使得镀层金属蒸汽发生热电离。对一般气体。当温度 $T \geqslant$ 7 000～8 000 K 时，而对金属蒸汽则温度 $T \geqslant$ 3 000～4 000 K 时，在气体中就可发生热电离而产生足够多的自由电子，形成导电通道。

（4）光电离。丝束引弧后产生的强光辐射可引起气体分子发生光电离作用。光电离判据为

当频率为 v 的光子，满足

$$hv \geqslant W_i \text{ 或} \leqslant hc/W_i \tag{2}$$

则会使气体分子发生电离。其中：h 为普朗克恒量，$h = 6.63 \times 10^{-34}$ J·s；c 为光速，$c = 3 \times 10^8$ m/s；W_i 为电离能；λ 为辐射光波长，m。

1.2　高压导线在导电丝束作用下的自持放电引弧判据

在引弧放电过程中，气体中的带电质点除在电场作用下做定向运动，消失于电极上而形成外回路的电流外，还可能因扩散和复合而使其带电质点在放电空间消失。扩散是带电质点从高浓度区域向低浓度区域的移动从而使浓度变得均匀的过程，而复合则是带异号电荷的质点相遇还原为中性质点的过程，同时以光辐射的形式将电离时获得的能量释放出来，这种光辐射又可导致间隙中的其他原子或分子光电离[5]。因此，扩散和复合并不一定意味着对放电过程的削弱，对于导电丝束来说，正是由于扩散和复合而使得带电质点充满整个空气间隙，放电从电场中最强的那个点开始并且完成由非自持放电向自持放电机制的转变。

1.2.1　导电丝束作用下高压导线相间放电的电子崩理论

由碰撞电离的电子崩理论可知，非自持放电向自持放电转变的机制与气体压强 P 和间隙距离 d 的乘积 pd 有关[6]。要达到自持放电的条件，必须在间隙内初始电子崩消失前产生新的二次电子来取代外电离因素产生的初始电子。

图 1 所示为计算相间间隙中二次电子数的增长示意图。假设初始电子数为 n_0，到达 x 距离处时电子数已达到 n，这 n 个电子在如距离中会产生 dn 个新电子，其中：

$$\mathrm{d}n = \alpha n \mathrm{d}x \tag{3}$$

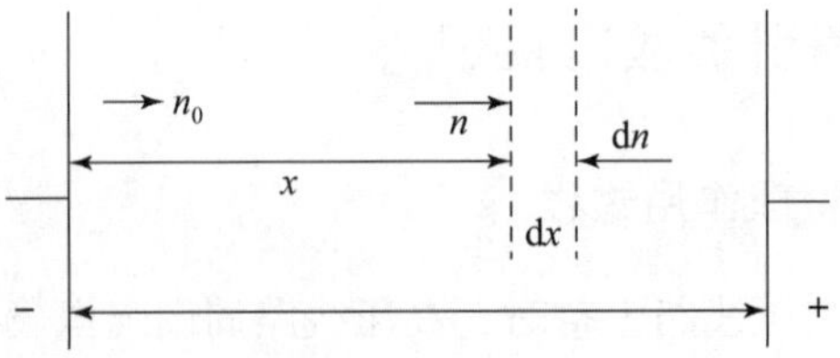

图1　计算相间间隙中电子数增长图

其中，α 为电子碰撞电离系数，表示一个电子沿电场方向运动 1 cm 行程中所完成的碰撞电离平均次数。对式（3）分离变量并积分，得到初始电子从阴极运动到阳极的过程中新增加电子数为

$$\Delta n = n - n_0 = n_0(e^{\alpha d} - 1) \tag{4}$$

其中，d 为板间距。

由式（1）已知，碰撞电离判据为 $x \geqslant U_i/E$。粒子的自由行程长度等于或大于某一距离 x 的概率为 $e^{-x/\lambda}$，其中 λ 为粒子的平均自由行程长度。因此，由碰撞引起电离的概率为 $e^{-U_i/E\lambda}$。则电子碰撞电离系数 α 可表示为

$$\alpha = \frac{1}{\lambda}e^{-U_i/E\lambda} \tag{5}$$

对于相间某种特定气体，有关系：$\lambda \propto T/p$ 存在，即 λ 与温度 T 成正比而与气压 p 成反比。

因此，当气温 T 一定时，式（4）可改写为

$$\alpha = Ap e^{-Bp/E} \tag{6}$$

其中，A 为与气体性质有关的常数；$B = AU_i$。

1.2.2　短间隙高压导线间的自持放电判据

导电丝束引弧试验现象表明，当 pd 值小于 26 kPa · cm（短间隙）时，电弧放电充满整个相间空间，如图 2 所示，符合 TOWNSEND 理论。

图2　短间隙放电电弧充满整个相间空间

当一个初始电子到达阳极时，电子崩中正离子数为（$e^{\alpha d} - 1$）个，这些正离子到达阴极时将产生 $\gamma(e^{\alpha d} - 1)$ 个二次电子，如果二次电子数等于 1，则放电就可以在无外电离因素的情况下维持下去[7]。故导电丝束引弧短间隙自持放电的判据为

$$\gamma(e^{\alpha d} - 1) = 1 \tag{7}$$

因为 $e^{\alpha d} \gg 1$，所以式（7）可写为

$$\gamma e^{\alpha d} = 1 \tag{8}$$

将式（6）代入式（8）整理，可得击穿电压 U_b 表达式为

$$U_b = Bpd/\ln(Apd/\ln(1/\gamma)) \tag{9}$$

1.2.3　长间隙高压导线间的自持放电判据

由导电丝束引弧试验可知，当 pd 值大于 26 kPa · cm（长间隙）时，电弧放电呈现细通道、曲折形，如图 3 所示，符合流注理论，即一旦流注产生，放电就可以由本身产生的电离而自行维持，流注的形成

条件就是自持放电的条件。故导电丝束长间隙引弧自持放电的判据为

$$e^{\alpha d} = 常数,或\ \alpha d = 常数 \tag{10}$$

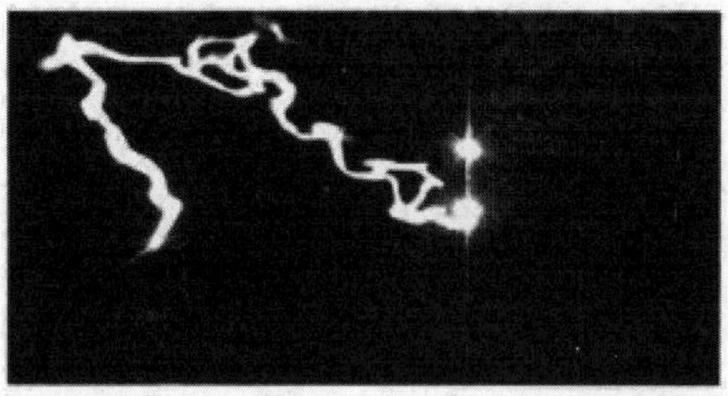

图3 长间隙放电呈现的曲折形、细通道电弧特性

虽然流注理论与 TOWNSEND 理论自持放电判据具有相同的形式，但二者维持自持放电机理是不同的。

式（10）所表述的为理论自持放电判据，可为提高丝束引弧能力及丝束的制备提供理论依据。而弹用导电丝束引弧可否形成自持放电，不仅与丝束本身的材质、表面镀层金属材料及均匀性、初始通流能力、抗烧蚀性和抗熔断性等有关，而且与电源系统的容量、绝缘介质类型及周围环境的气压、温度、湿度等有关。因此，要获得能满足武器上使用的某种导电丝束自持放电判据，需要通过大量试验得到该种导电丝束在特定环境、容量条件下的测量值，并加以分析而理论化。

弹用导电丝束对高压线路的强制连续引弧效应作为一个全新的课题，有许多理论和试验方面的工作需要逐步完善，文中试验即是在上述理论指导下研制出不同类型的1#、2#、3#、4#、5#导电丝束，并通过在高压小容量条件下的准等效原则进行引弧试验，从而对各种不同类型的导电丝束性能做出测试、评价，为弹用导电丝束的制备提供第一手资料。

2 试验装置和试验方法

2.1 试验装置

为了能够对不同类型导电丝束引弧性能做一客观的定性定量描述，文中选择在高电压、小容量电源装置上进行导电丝束搭接相－相间高压线路的短路引弧试验，以便对不同类型导电丝束进行筛选和检验；同时，研究导电丝束的导电率、基体材质特性、镀层金属与引弧性能之间的关系。为此，利用一套额定电压 345 kV、容量 345 kV · A 的工频试验装置进行了高压小容量的引弧短路试验，模拟的相间距离为 20～400 mm，其试验装置电路图如图 4 所示。

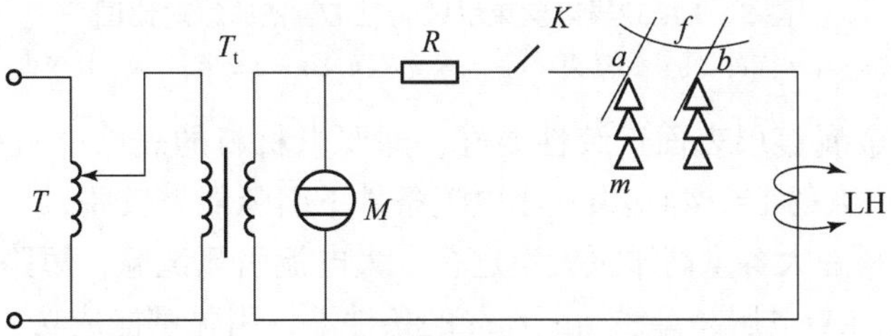

图4 导电丝束引发相间短路试验装置电路图

T－调压器；T_r－变压器；M－电压表；R－保护电阻；K－开关；

a，b－粗铜棒；f－导电丝束；m－支撑绝缘棒；LH－电流传感器

2.2 试验方法

对于1#～5#导电丝束，逐渐增大其绝缘间隙距离 d，分别测量不同间隙下引弧击穿电压 U 值，记录电弧持续时间 t，并观察放电电弧形态及特性，绘制丝束引弧特性 $U-d$ 曲线。对于每一间隙施加 5 次工

频电压，求其平均值作为该丝束在该间隙下的引弧击穿电压。

3　试验结果分析及结论

（1）通过试验观察了导电丝束长、短间隙放电电弧特性，验证了导电丝束短间隙放电电弧特性符合TOWNSEND 理论；而长间隙放电电弧特性符合流注理论，即流注形成条件就是自持放电的条件。

（2）由 1#～5#丝束引弧特性 $U-d$ 曲线（图5）可知，丝束的引弧性能与其镀层金属材质、表面均匀性、氧化程度、每束根数以及丝束基体材料有关。2#、3#丝束所用材料完全相同，但每束根数不同，在小开距条件下（≤10 cm），1k 根/束的 3#丝束性能要优于 30 根/束的 2#丝束。由此表明：在相同条件下，引弧性能与镀层金属蒸汽的密度有关；在大开距条件下，此情形表现不明显。5#丝束的镀层金属逸出功小、抗氧化性要优于其他几种材质，故其引弧综合性能也优于其他丝束，与理论分析相一致。

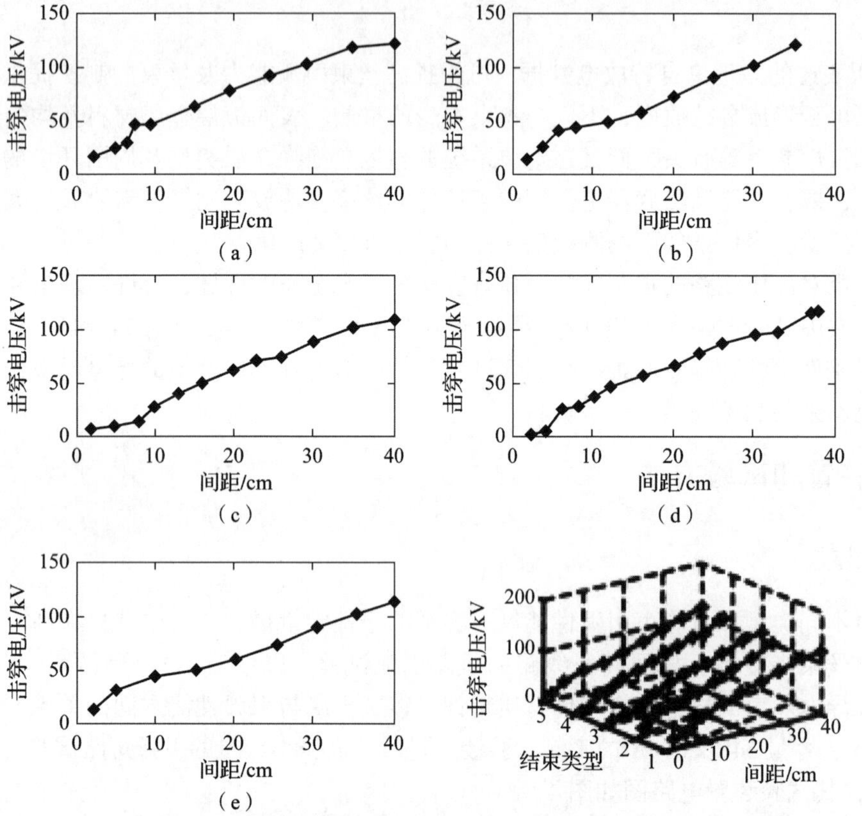

图5　1#～5#导电丝束引弧特性 $U-d$ 曲线对比图

（a）1#丝束；（b）2#丝束；（c）3#丝束；（d）4#丝束；（e）5#丝束；

（3）试验发现，改性金属镀层表面连续性要好，即要求材料的表面电阻率越小越好（一般应小于60 Ω/m），而表面电阻率很大的 1#、4#丝束在小电流条件下引弧能力较差。

（4）上述几种丝束材料在大容量高压试验室进行了大电流引弧试验，短路电流为 500 A 和 1 000 A，所有被试的导电丝束在 161 kV 电压、母线间距 4 m 的条件下，引弧都能成功，说明导电丝束短路真实电力设施所需的条件极易满足；同时也表明引弧试验装置的容量很大时，不能有效地对导电丝束进行筛选和检验，而小电流引弧试验装置的系统容量很小，对丝束引弧成功的条件相对苛刻，可以较好地分辨筛选出丝束的引弧性能优劣，满足后者一定满足前者。因此，高压小容量条件下的引弧试验是评价导电丝束引弧性能优劣的一种有效、可靠及经济的方法。

刚性伞稳定式子弹药的气动特性分析①

周　彤②，冯顺山，张晓东，方　晶

（北京理工大学爆炸科学与技术国家重点实验室，北京　100081）

摘　要： 本文结合数值仿真和柔性材料气动力修正算法，对刚性伞稳定式子弹的气动力特性进行了分析。仿真基于多面体网格模型，并采用有限体积法和SST（Shear－Stress Transport）湍流模型。飘带伞弹和丝绳伞弹的气动特性计算结果表明，由于飘带的摆振效应，亚声速时飘带伞弹的阻力系数大于丝绳伞弹，而超声速时由于飘带对弹底回流的阻滞，飘带伞弹的阻力系数小于丝绳伞弹。通过与试验数据的对比，证明本文所建立的分析方法是合理可行的。

关键词： 刚性伞；柔性索；子弹药；数值仿真；多面体网格

Analysis on Aerodynamic Characteristics of Rigid－parachute Stabilized Submunition

ZHOU Tong，FENG Shun－shan，ZHANG Xiao－dong，FANG Jin

（State Key Laboratory of Explosion Science and Technology，BIT，Beijing，100081，China）

Abstract： On the basis of numerical simulation and correction computation for flexible materials，the aerodynamic characteristics of rigid－parachute stabilized submunition was analyzed. Polyhedron meshes were adopted；finite volume methods and SST turbulence model were applied for calculation. The results show that the drag coefficient of ribbon－parachute submunition is higher than rope－parachute submunition in subsonic flows because of shimmy of the ribbons. However，the ribbons resist some parts of reflux under the projectile，lead to a decline of drag coefficient in supersonic flows. The calculation results have a good agreement with experiment data，which shows the rationality and practicability of the present analytical methods.

Keywords： rigid－parachute；flexible streamer；submunition；polyhedron mesh；numerical simulation

子弹药的稳定方式通常有两种：一种是柔性体稳定（如柔性飘带、降落伞等），另一种是刚性体稳定（如刚性尾翼等）。本文研究的是一种由刚性体和柔性体组合的新型稳定装置，主要由刚性伞和柔性索组成。传统的降落伞充气阶段复杂，特别是在低速条件下伞衣展开时间长，阻力难以达到预定要求，子弹容易产生明显的摇摆。刚性伞则避免了伞衣喘振现象，开伞过程仅包括抛伞和柔性索拉直两个阶段，伞衣阻力面积恒定，即使在低速条件下也能提供稳定的阻力；数值计算模型易于建立，有利于提高气动力系数和外弹道的计算精度。

柔性索可分为飘带和丝绳两类。国内外进行过一些有关带单独飘带的子弹药气动特性研究，Auman等[1,2]对采用飘带式稳定的M42子弹进行了飞行测试和风洞试验；张维全[3,4]对飘带式子弹的飘带气动力特性进行了研究；唐良锐[5]对飘带的高亚声速特性进行了风洞试验研究。完颜振海等[6,7]对飘带连接刚性伞稳定装置的超声速气动特性进行了研究。采用丝绳连接刚性伞帽的稳定装置国内外尚没有研究，丝绳相比飘带而言，由于直径相对于弹径而言非常小，对流场的扰动几乎可以忽略不计，主要通过刚性

① 原文发表于《弹道学报》2013年第4期。

② 周彤：工学博士，2009年师从冯顺山教授，研究封锁类子弹药气动特性及其弹道性能，现工作单位：北京理工大学。

伞帽提供稳定力矩。因此本文通过数值仿真和柔性材料的气动力修正算法，分别建立两种不同柔性索的刚性伞稳定式子弹的气动特性分析方法，比较其气动力系数和流场结构，为刚性伞稳定式子弹的设计使用及其外弹道的设计分析提供重要的参考依据。

1　数值分析

1.1　数值计算模型

两种刚性伞稳定式子弹药的计算模型如图 1 所示，弹头为尖拱形，稳定装置分别由 4 根飘带和 4 根丝绳与刚性伞组成，简化模型中将飘带和丝绳作为刚体处理。

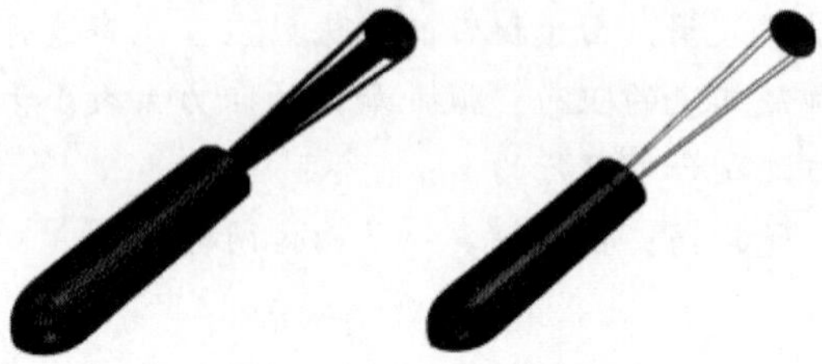

（a）　（b）

图 1　两种刚性伞稳定式子弹药的计算模型

（a）飘带伞弹；（b）丝绳伞弹

弹体直径为 D，弹长为 L，伞径为 D_p，弹体长径比 $L/D=4$，伞弹径比 $D_p/D=1$；飘带和丝绳的长度均为 L_p，$L_p/D=5$；飘带宽度为 W，飘带宽度与弹径比 $W/D=0.28$。

计算域划分采用蜂窝状多面体网格技术，蜂窝状多面体网格和四面体、六面体网格相比有更多的相邻单元，一个典型的多面体网格平均有 14 个面，梯度的计算和当地的流动状况预测更为准确，由于需要的网格数相对较少，约为四面体网格的 1/4，解算速度也会大大增快。

近壁面采用结构化加密网格，以更好地捕捉边界层的复杂流动。加密程度满足高 y^+ 壁面处理要求，壁面到第一层网格单元的形心的距离 δy 的预测值满足式（1）的要求：

$$\frac{\delta y}{y^+}=8.6L\,Re_L^{-\frac{13}{14}},y^+\leqslant 1 \tag{1}$$

式中，y^+ 为壁面函数，其定义为 $y^+=y\rho u_\tau/\mu$，y 为到壁面的距离，u_τ 为壁面摩擦速度，ρ、μ 分别为流体的密度和黏性，Re_L 为以含能涡速度和含能涡尺度为特征量的雷诺数。

飘带伞弹的网格总数约为 16 万，丝绳伞弹的网格总数约为 18 万。

入口和出口均采用自由来流的边界条件，弹体和飘带伞的边界条件为无滑移壁面，马赫数计算范围为 0.3 ~ 2.5。

1.2　数值计算方法

本文采用雷诺平均 Navier - Stokes（RANS）方程的数值模拟方法，定义速度矢量为 $\boldsymbol{u}=(\boldsymbol{u}_i,\ \boldsymbol{u}_j)$，空间坐标向量为 $\boldsymbol{x}$，时间为 t，在给定平均运动的边界条件和初始条件下数值求解雷诺方程[8]：

$$\begin{cases}\dfrac{\partial\langle u_i\rangle}{\partial t}+\langle u_j\rangle\dfrac{\partial\langle u_i\rangle}{\partial x_j}=-\dfrac{1}{\rho}\dfrac{\partial\langle p\rangle}{\partial x_i}+\\ \nu\dfrac{\partial^2\langle u_i\rangle}{\partial x_j\partial x_j}-\dfrac{\partial\langle u_i'u_j'\rangle}{\partial x_j}+\langle f_i\rangle\\ \dfrac{\partial\langle u_i\rangle}{\partial x_i}=0\end{cases} \tag{2}$$

式中，〈·〉代表平均量。u_i'，u_j' 为脉动速度；ν 为运动黏性系数；p 为压强；ρ 为密度；f_i 为单位质量的体积力。

初始条件:

$$\langle u_i \rangle (\boldsymbol{x},0) = V_i(\boldsymbol{x}) \tag{3}$$

边界条件:

$$\langle u_i \rangle |_{\Sigma} = U_i(\boldsymbol{x},t), \langle p \rangle (\boldsymbol{x}_0) = p_0 \tag{4}$$

式中，$V_i(\boldsymbol{x})$，$U_i(\boldsymbol{x},t)$ 和 p_0 是已知函数或常数；$\sum$ 是流动的已知边界；$\boldsymbol{x}_0$ 是流场中给定点的坐标。

雷诺方程中的 $\langle u_i' u_j' \rangle$ 是未知量，必须附加封闭方程才能数值求解雷诺方程。本文采用剪切应力运输模型（SST）[9]，SST 是双方程模型，在近壁面处采用 Wilcox $k-\omega$ 模式，在边界层边缘和自由剪切层采用 $k-\varepsilon$ 模式，二者之间采用混合函数过渡，SST 模型结合了 $k-\omega$ 和 $k-\varepsilon$ 湍流模型的优点，在模拟边界层以及预测分离流动方面具有更高的精度和可信度。

SST 模型的湍动能 k 的输运方程为

$$\begin{aligned} &\frac{\partial \rho k}{\partial t} + \frac{\partial}{\partial x_j}[\rho u_j k - (\mu + \sigma_k \mu_t)\frac{\partial k}{\partial x_j}] \\ &= \tau_{ij} S_{ij} - \beta^* \rho \omega k \end{aligned} \tag{5}$$

湍流比耗散率 ω 方程为

$$\begin{aligned} &\frac{\partial \rho \omega}{\partial t} + \frac{\partial}{\partial x_j}\left[\rho u_j \omega - (\mu + \sigma_\omega \mu_t)\frac{\partial \omega}{\partial x_j}\right] \\ &= P_\omega - \beta \rho \omega^2 + 2(1 - F_1)\frac{\rho \sigma_\omega}{\omega} \cdot \frac{\partial k}{\partial x_j} \cdot \frac{\partial \omega}{\partial x_j} \end{aligned} \tag{6}$$

其中，σ_k 和 σ_ω 分别为 k 和 ω 的湍流普朗特数；β，β^* 为常数；τ_{ij} 为壁面剪切应力；S_{ij} 为平均应变率；μ_t 为涡黏性系数；P_ω 为耗散率的生成项；F_1 为混合函数。

网格模型的离散化采用有限体积法，空间离散方式采用二阶迎风格式，对连续动量和能量控制方程采用耦合解算技术联立求解。亚跨声速段耦合无黏通量采用 Roe's FDS 格式[10, 11]，在超声速段采用 AUSM + 格式[12, 13]，这两种格式都是基于迎风差分的概念，与 Roe's FDS 格式相比，AUSM + 格式能够准确地捕捉冲击和间断接触，提供激波非连续性的精确求解并保持各标量的正性，更适用于超声速可压缩流动。状态方程选择理想气体，动态黏性项采用 Sutherland 定律：

$$\frac{\mu}{\mu_0} = \left(\frac{T}{T_0}\right)^{\frac{3}{2}} \left(\frac{T_0 + S}{T + S}\right) \tag{7}$$

式中　T_0 和 μ_0——参考温度和黏性；

　　　S——Sutherland 常数。

2　柔性材料阻力系数修正

上一节介绍的简化模型将飘带作为刚体来处理，在超声速的条件下，飘带基本上处于紧绷拉直状态，这种简化模型可以认为与实际情况相差不大[6, 7]。而在亚声速的条件下，飘带容易产生摆振、弯曲、变形等复杂的物理现象，并且其对流场的扰动作用不可忽略，因此有必要对亚声速条件下飘带的阻力系数进行修正。丝绳的直径相比弹径来说非常小，对流场的扰动几乎可以忽略不计，所以本文不对其气动力进行修正。

在数值计算结果的基础上，对飘带伞稳定装置的阻力系数进行修正计算。飘带对阻力的影响主要有表面摩擦阻力以及飘带摆振消耗气流能量产生的阻力，按照平头旋成体的阻力系数经验公式估算刚性伞的阻力系数，等于伞帽的迎面阻力系数与底部阻力系之和。因此稳定装置的阻力系数为

$$C_{xw} = C_{xfr} + C_{xwr} + C_{xpp} + C_{xbp} \tag{8}$$

式中　C_{xfr}——飘带的摩擦阻力系数；

　　　C_{xwr}——飘带的波动阻力系数；

　　　C_{xpp}——伞帽的迎面阻力系数；

C_{xbp}——伞帽的底部阻力系数。

假设飘带表面的附面层全部呈现紊流状态，不计飘带质量对气动力参数的影响：

$$C_{xfr} = n_1 \cdot 2\,(C_f)_{Ma=0} \cdot \eta_M \cdot \frac{S_r}{S_B} \tag{9}$$

$$C_{xwr} = n_2 \cdot C_{xfr} \tag{10}$$

$$\eta_M = \frac{1}{\sqrt{1+0.12Ma_\infty^2}} \tag{11}$$

式中 C_f——等效平板摩阻系数；

S_r——飘带的面积；

S_B——弹体的横截面积；

η_M——压缩性修正系数；

n_1——飘带数量；

n_2——战旗阻力修正系数，$n_2 = 9.0$。

按照平头旋成体的阻力系数曲线进行拟合处理[14]：

$$C_{xpp} = (0.42Ma_\infty^2 - 0.04Ma_\infty + 0.68)\frac{S_p}{S_B} \tag{12}$$

$$C_{xbp} = -C_{pb}\frac{S_p}{S_B} \tag{13}$$

$$C_{pb} = \begin{cases} -0.122(Ma<0.6) \\ 0.8Ma_\infty^2 - 1.11Ma_\infty + 0.5(0.6 \leqslant Ma_\infty \leqslant 1.0) \end{cases} \tag{14}$$

式中 S_p——刚性伞帽的面积。

3 结果与对比

3.1 计算方法验证

采用飘带伞稳定装置的平头弹的风洞试验由课题组在沈阳空气动力研究所 FL－1 风洞进行。如图 2 所示，亚声速时数值计算结果和试验数据相差较大，原因可能是由于在亚声速时飘带伞的摆动幅度较大，而刚体简化模型中假设飘带伞为静止状态。而经过气动力修正的计算结果与试验结果的相对误差小于 5.5%，因此认为本文所建立的分析方法可以满足计算要求。

3.2 伞弹零攻角气动特性

两种刚性伞稳定式子弹药在零攻角时的阻力系数修正计算结果如图 3 所示，由于飘带的摆振效应，飘带伞弹在亚声速时阻力系数大于丝绳伞弹，而在超声速区，其阻力系数小于丝绳伞弹。

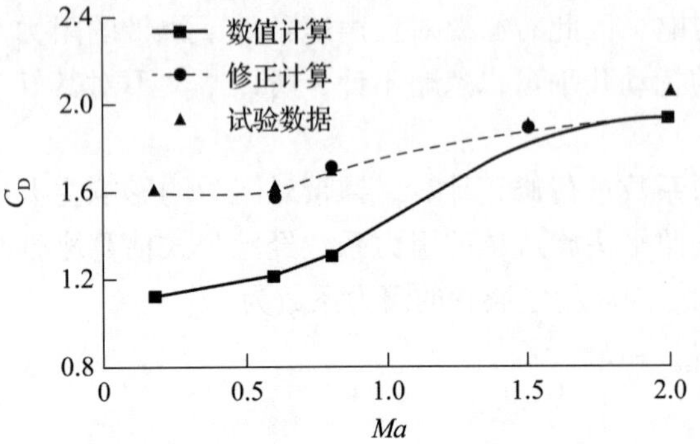

图 2 计算结果与实验数据对比

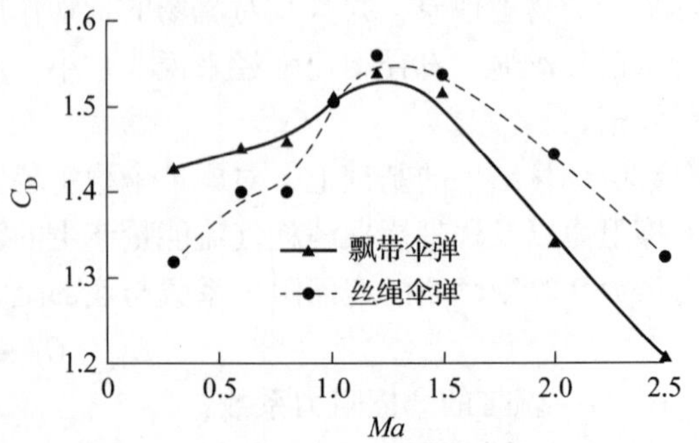

图 3 零攻角时两种刚性伞稳定式子弹药的阻力系数对比

对比两种伞弹在 $Ma=1.5$ 时的流场压力等值线，如图 4 所示，浅色代表压力大的区域。可以看出，飘带伞弹在靠近弹体底部的区域由于尾流被飘带干扰，部分回流受到阻挡，迎面作用于刚性伞帽上的高压气流范围小于丝绳伞弹，飘带环绕的区域内气流的方向基本上平行于飘带的方向，刚性伞帽背风面形成两组尾涡。而丝绳伞弹的弹底流场类似于单独弹体的底部流场，气流沿着弹底形成两组漩涡，周围的气流向轴线收缩，然后作用于刚性伞帽上，向外扩张，刚性伞帽背风面的负压区大于飘带伞弹。因此在超声速时，若不计飘带的摆振效应，丝绳伞弹的阻力系数大于飘带伞弹。

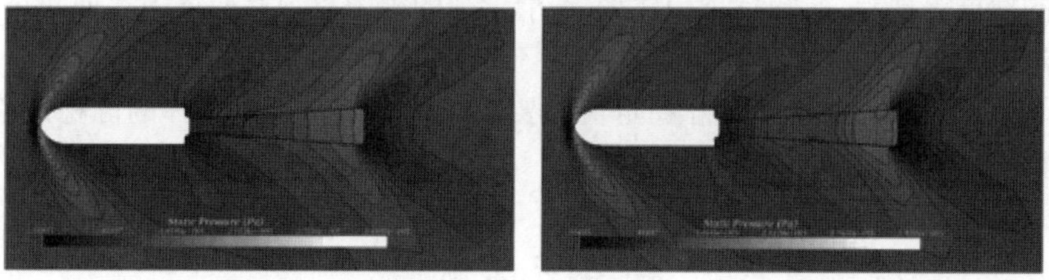

(a)　　　　　　　　　　(b)

图 4　子弹药在 $Ma=1.5$ 时的流场压力等值线

(a) 飘带伞弹；(b) 丝绳伞弹

3.3　伞弹有攻角超声速气动特性

在超声速的条件下，本文所建立的仿真模型能够较好地模拟刚性伞稳定式子弹药的气动特性，因此超声速段的数值计算结果可以作为分析实际流场的依据。

当自由来流的 $Ma=1.5$ 时，两种刚性伞稳定式子弹药的法向力系数以及俯仰力矩如图 5、图 6 所示，其中俯仰力矩系数的计算以弹体的中心为参考点。两种刚性伞稳定式子弹药的俯仰力矩系数均随着攻角的增大而负向增大，表明这两种弹是静稳定的，丝绳伞弹的静稳定性稍优于飘带伞弹。

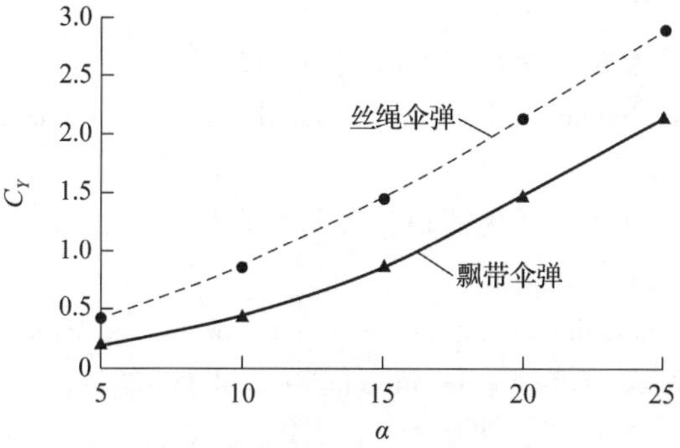

图 5　$Ma=1.5$ 时子弹药法向力系数随攻角的变化规律

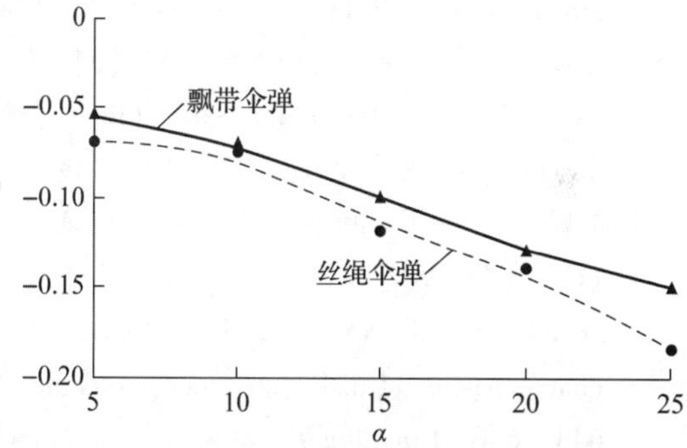

图 6　$Ma=1.5$ 时子弹药俯仰力矩随攻角的变化规律

两种刚性伞稳定式子弹药的流场压力等值线如图 7 所示，飘带对弹底回流的阻挡更加明显，作用于刚性伞帽上的气流压力小于丝绳伞弹，因此飘带伞弹的刚性伞帽受到的合力在弹轴法向上的分力小于丝绳伞弹，也就是说此时飘带伞提供的恢复力矩小于丝绳伞。

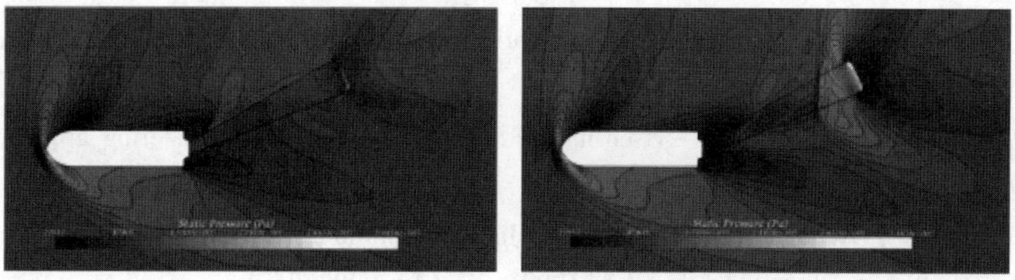

(a)　　　　　　　　　　(b)

图 7　$\alpha=25°$ 时子弹药流场压力等值线

(a) 飘带伞弹；(b) 丝绳伞弹

4 结论

本文结合数值仿真和气动力修正算法对刚性伞稳定式子弹药的气动特性和流场结构进行了分析和研究，并对比了计算结果与试验数据，可以得出以下结论：①本文建立的数值模型与修正算法能够满足刚性伞稳定式子弹药的气动力系数的计算要求，为这种类型的子弹药的弹道设计与分析提供了参考依据；②对比两种刚性伞稳定式子弹药，飘带伞弹在亚声速时阻力系数大于丝绳伞弹，这是因为飘带在亚声速时存在较为明显的摆振等现象，其波动阻力使得整个伞弹的阻力系数有所增大；而超声速时丝绳伞弹的阻力系数较大，这是因为绷直的具有一定宽度的飘带干扰了弹底的回流，减小了刚性伞帽上的迎面压力。③两种刚性伞稳定式子弹药均满足静稳定性的要求，超声速时丝绳伞弹的静稳定性优于飘带伞弹，在使用中可以根据实际要求选用不同的柔性索。

参考文献

[1] AUMAN L M, WILKS B. Application of Fabric Ribbons for drag and stabilization [C]. //18th AIAA Aerodynamic Decelerator Systems Technology Conference and Seminar. Munich, Germany. 2005: 1 - 9.

[2] AUMAN L M, DAHLKE C W. Aerodynamic characteristics of ribbon stabilized grenades: AIAA, 2000 - 0270 [R]. 2000

[3] 张维全. 柔性飘带稳定式子弹的飘带气动力设计 [J]. 兵工学报, 1989 (1): 13 - 21.
ZHANG W Q. Aerodynamic design of ribbons for ribbon stabilized projectile [J]. Acta armamentarii, 1989 (1): 13 - 21. (in Chinese)

[4] 张维全. 飘带式子弹气动力工程计算方法 [J]. 兵工学报, 1991 (4): 75 - 82.
ZHANG W Q. Engineering calculation method for aerodynamic force of ribbon stabilized projectile [J]. Acta armamentarii, 1991, (4): 75 - 82. (in Chinese)

[5] 唐良锐. 柔性飘带气动特性初探 [J]. 气动实验与测量控制, 1996, 10 (1): 9 - 13.
TANG L R. Elementary exploration of flexible ribbon aerodynamics [J]. Aerodynamic experiment and measurement & control, 1996, 10 (1): 9 - 13. (in Chinese)

[6] 冯顺山, 完颜振海. 平头弹超音速尾流对飘带伞气动特性的影响 [J]. 北京理工大学学报, 2011, 31 (6): 643 - 646.
FENG S S, WANYAN Z H. Influence of flat - faced submunition's supersonic wake flow to the aerodynamic characteristics of disk - streamer parachute [J]. Transactions of Beijing Institute of Technology, 2011, 31 (6): 643 - 646. (in Chinese)

[7] 完颜振海, 冯顺山, 董永香, 等. 超音速伞弹流场特性数值分析 [J]. 兵工学报, 2011, 32 (5): 520 - 525.
WANYAN Z H, FENG S S, DONG Y X, et al. Numerical analysis of flow field characteristics of supersonic submunition with parachute [J]. Acta Armamentarii, 2011, 32 (5): 520 - 525. (in Chinese)

[8] 张兆顺, 崔桂香, 许春晓. 湍流大涡数值模拟的理论与应用 [M]. 北京: 清华大学出版社, 2008.
ZHANG Z S, CUI G X, XU C X. Method and application of large - eddy simulation for turbulence [M]. Beijing: Tsinghua University Press, 2008. (in Chinese)

[9] MENTER F R. Two - equation eddy - viscosity turbulence modeling for engineering applications [J]. AIAA journal, 1994, 32 (8): 1598 - 1605.

[10] WEISS J M, MARUSZEWSKI J P, SMITH W A. Implicit solution of preconditioned Navier - Stokes equations using algebraic multigrid [J]. AIAA journal, 1999, 37 (1): 29 - 36.

[11] WEISS J M, SMITH W A. Preconditioning applied to variable and constant density flows [J]. AIAA journal, 1995, 33 (11): 2050 - 2057.

导电纤维丝团在严酷弹道环境下的性能研究①

张广华[1②]，冯顺山[1]，胡松涛[2]

（1. 北京理工大学爆炸科学与技术国家重点实验室，北京 100081；

2. 安徽红星机电科技股份有限公司，合肥 231135）

摘　要：为研究强冲击过载、高速旋转及高抛撒速度等严酷弹道环境对导电纤维丝团的工作性能影响，对两种典型结构丝团进行了理论分析与试验研究。结果表明，高过载对丝束强度影响较大，丝轴两端具有挡轮结构的丝团抗过载性能要远远高于无挡轮结构的丝团，使得丝团能够适应各种强过载冲击。通过理论分析得到了母弹或子弹旋转运动条件下丝团与子弹发生相对转动时的判据及丝束的损伤判据，还得到了子弹抛撒丝团时的丝束损伤判据。研究结果可为丝团适应严酷弹道环境的结构设计提供依据。

关键词：兵器科学与技术；导电纤维丝团；弹道过载；短路毁伤

Performance Study on Conductive Fiber Cluster Under Severe Ballistic Environment

Zhang Guanghua[1], Feng Shunshan[1], Hu Songtao[2]

(1 State Key Laboratory of Explosion Science and Technology, BIT, Beijing 100081, China

2Anhui Hongxing Mechanical and Electrical Technology Co. Ltd, Hefei 231135, China)

Abstract: To study influence of shock overload, high speed rotation and dispersing velocity on working performance of conductive fiber cluster, theoretical and experimental study was researched on two typical fiber clusters. The results show that high overload has great effect on fiber, fiber cluster with protective structure on the shaft has greater anti - overload performance than the no protective one, which help it bear various high overloads. Relative rotating criterion between bullet and fiber cluster and damage criterion of fiber under spinning condition were obtained through theoretical analysis, so was damage criterion of fiber on spreading. The results provide references to fiber cluster structure design that adapted to severe ballistic environment.

Keywords: ordnance science and technology; conductive fiber cluster; ballistic overload; circuit damage

0　引言

导电纤维丝团（以下简称“丝团”）是导电纤维弹中的有效装填物，属于软毁伤元，作用于高压电力系统后通过产生的强制性连续引弧效应对电力系统进行短路毁伤[1]。丝团通常装填在导电纤维子弹中，若干个子弹装填在航弹、火箭弹、导弹、榴弹或迫击炮弹等子母战斗部中[2]。根据子弹弹体结构的不同，装填丝团的个数从几十个到上百个不等，每个丝团上都缠绕有几十米长的导电丝束，用于引发对目标的短路毁伤效应[3]。

丝团在飞行弹道上会受到发射、母弹抛撒及子弹抛撒等多次弹道过载的影响，对于采用落地反抛式的导电纤维弹，丝团还会受到子弹落地冲击过载的影响。对于采用旋转稳定的弹丸，母弹或子弹的高速

① 原文发表于《中国兵工学会第十一届爆炸与安全技术学术年会论文集》2014。

② 张广华：工学博士，2011 年师从冯顺山教授，研究弹载导电丝束短路毁伤机理及弹道特性，现工作单位：西安 204 所。

旋转会导致丝团产生较大的离心过载，并导致子弹内的丝团与子弹壳体发生相对旋转，从而影响到丝团从子弹中抛出时的抛撒姿态、抛撒速度等参量，并进一步影响到丝团的空中展开效果、落点散布及落地后的有效包络面积。此外，丝团从子弹内抛撒出去并展开的过程中，高抛撒速度会对丝束结构产生很大的影响，严重时会导致丝束断裂。

目前，国内关于丝团在严酷弹道环境下的性能研究不多[2]，国外更是未见报道。因此，开展此项研究是很有必要的，本文即对导电纤维丝团在严酷弹道环境下的相关性能进行了理论分析与试验研究。

1 丝团结构及弹中静态分布

1.1 丝团结构

丝团主要由丝轴、丝束及辅助件构成。丝轴材料为经过性能优化后的一种工程塑料，具有较高的结构强度及抗冲击性能；丝束材料是一种表层镀有金属的柔性纤维[4]，长度一般为 20 ~ 40 m，单位长度质量小于 0.04 g/m。丝束按照特定的排列规则缠绕在丝轴上，并通过外部辅助件对其进行包裹，根据丝束缠绕方式的差异，可分为 M 型丝团和 Z 型丝团。图 1 所示为丝团包装件及展开结构示意图。

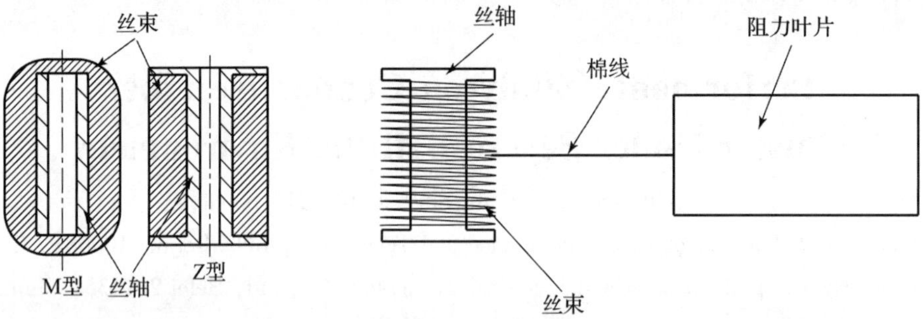

图 1 丝团包装件及展开结构示意图

1.2 丝团在子弹中的静态分布

丝团通常装填在导电纤维子弹（又称丝筒）中，装填数量根据子弹大小为几十个到百余个不等。图 2 所示为丝团在子弹内的装填状态。通过图 2 可以看出，丝团沿中心管方向（轴线方向）装填若干层，每层又分为内、外两圈。

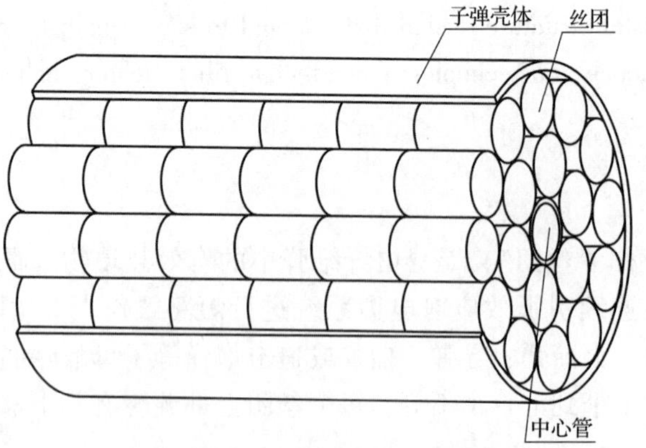

图 2 丝团在子弹内的装填状态

综合图 1、图 2 可以看出，M 型丝团在弹道过载作用下产生的破坏主要包括以下两方面：一方面发射及子弹落地过程中产生的轴向过载对丝轴及丝束产生的结构损伤；另一方面母弹或子弹旋转导致丝团与子弹壳体和中心管发生相对转动并对丝束结构造成损伤。Z 型丝团在弹道过载作用下产生的破坏主要包括：轴向过载对丝轴结构的损伤及丝团与子弹壳体和中心管相对转动时对丝束造成的损伤。

2　轴向冲击过载对丝团的影响

无后坐力炮及迫击炮发射的弹药过载在5 000 g左右，榴弹炮发射的弹药过载大多在10 000 g以上，中大口径榴弹及加农炮榴弹的发射过载可以达到30 000 g。当子弹采用落地反抛式抛撒丝团时，落地瞬时所产生的过载峰值在13 000 g以上[5]，落在岩石等硬质材质时的过载峰值在30 000 g以上。丝团，尤其是丝束在发射和终点弹道所受轴向过载会对其结构完整性产生很大的影响，甚至发生损伤，失去正常工作的功能。

2.1　轴向过载对丝轴的影响

丝轴的主要作用是制造丝团时支撑丝束的缠绕及排布状态，工作时保证丝束绕丝轴快速有效地展开。因此，要求丝轴必须具有一定的结构强度，这样才能保证在高过载条件下丝轴及丝束仍然能保持初始缠绕状态的完整性并使丝束有效展开。

丝轴在轴向过载下所受压力 σ_z 为

$$\sigma_z = \frac{m_z \cdot a_s}{S_z} \tag{1}$$

式中，m_z 为丝轴质量；a_s 为弹道过载；S_z 为丝轴受力面的特征面积。由此得出丝轴的抗压强度 σ_{bc} 的设计准则：

$$\sigma_{bc} > \lambda_x \sigma_z \tag{2}$$

式中，λ_x 为安全系数，一般取1.2～1.4。根据以往飞行试验所测结果，丝团飞行过程中所受弹道过载均小于30 000 g，取 $a_s = 30\ 000$ g，$m_z = 5$ g，$S_z = 15.7\ \text{mm}^2$，$\lambda_x = 1.4$，根据式（2）可知丝轴在30 000 g过载下应满足如下强度准则：$\sigma_{bc} > 13.38$ MPa。丝轴材料为经过性能优化的工程塑料，绝大多数工程塑料的抗压强度都远远大于13.38 MPa，由此可以得出：轴向过载虽然会对丝轴产生一定的冲击效应，但是不会造成其结构性的损伤。这与飞行试验现场观察到的现象是一致的，回收到的丝轴全部都保持结构的完整性。

2.2　轴向过载对丝束的影响

丝束是以很细的纤维材料为基体制作而成，静态拉伸试验结果显示，丝束在静态条件下的极限抗拉强力 $F_i \leqslant 10$ N，在高过载作用下易发生结构损伤。

为了研究丝束在高过载下的结构损伤，对其进行了马歇特锤击试验[6]，试验装置如图3所示，图3（a）为试验机本体，图3（b）为安装在击锤上的辅助件，辅助件内装有丝团。

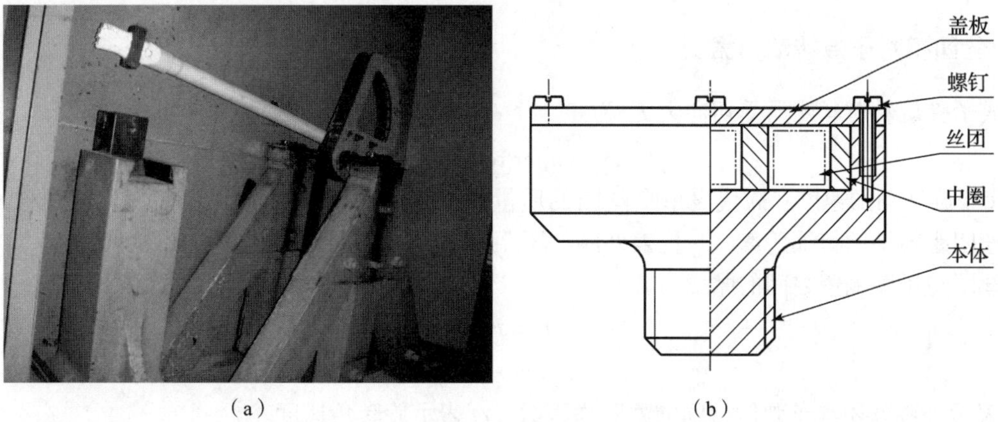

（a）　　（b）

图3　锤击试验机

（a）试验机本体；（b）安装在击锤上的辅助件

图3所示试验机分别对丝团进行了2 700～35 000 g的锤击试验，主要检测经过高过载冲击后的丝束结构是否完整，毛丝是否存在断、乱的现象，并对试验后的丝团进行了地面动态展开试验，结果如表1所示。

表1　锤击及地面动态展开试验结果

过载/g	锤击试验		动态展开试验	
	M型丝团	Z型丝团	M型丝团	Z型丝团
2 700	完好	完好	完全展开	完全展开
4 700	完好	完好	完全展开	完全展开
9 200	损伤	完好	展开过程中丝束断裂	完全展开
14 700		完好		完全展开
19 000		完好		完全展开
24 000		完好		完全展开
29 000		完好		完全展开
35 000		完好		完全展开

从表1可以看出，Z型丝团抗过载性能要远远高于M型丝团，这主要是由于M型丝团中的丝束直接包裹在丝轴外部，因此，与丝轴两端部位接触的丝束在高过载冲击下会产生应力集中，导致丝束结构极易发生损伤；而Z型丝团由于丝轴两端有挡轮保护，有效地防止了冲击过程中的应力集中现象，因此大大提高了丝束的抗过载性能。通过动态展开试验结果还可以看出，丝束结构的完整性直接影响到丝团的动态展开效果。

3　母弹或子弹旋转对丝团的影响

对于旋转稳定弹丸，母弹或子弹的旋转可能会导致丝团在子弹内部相对于子弹发生转动，影响丝团被抛撒出子弹时的姿态，并进一步影响丝团的空中展开与落点散布。

从图2可以看出，对于每层丝团，每个内圈的丝团都卡在了两个外圈丝团之间，同时，发射时的高过载会使各层丝团之间紧密连接。因此，做出以下假设：若丝团相对于子弹转动，应该是所有丝团以一个整体相对于子弹转动。

3.1　丝团相对子弹转动判据

母弹或子弹旋转时丝团所受切向力 F_t 为

$$F_t = n_i m r_i \alpha + n_o m r_o \alpha \tag{3}$$

式中，α 代表旋转角加速度；m 代表单个丝团的质量；n_i 代表内圈丝团个数，n_o 代表外圈丝团个数；r_i 代表内圈丝团距转轴的特征距离，r_o 代表外圈丝团距转轴的特征距离；r_i 及 r_o 如图4所示。

外圈丝团与子弹侧壁之间的摩擦力 f_b：

$$f_b = F_n \mu_{os} \tag{4}$$

$$F_n = n_i m\omega^2 r_i + n_o m\omega^2 r_o \tag{5}$$

式中，μ_{os} 表示外圈丝团与子弹侧壁间的静摩擦系数；ω 表示旋转角速度。

底层丝团与子弹端盖间的摩擦力 f_d：

$$f_d = (n_i + n_o) m a_s \mu_{ds} \tag{6}$$

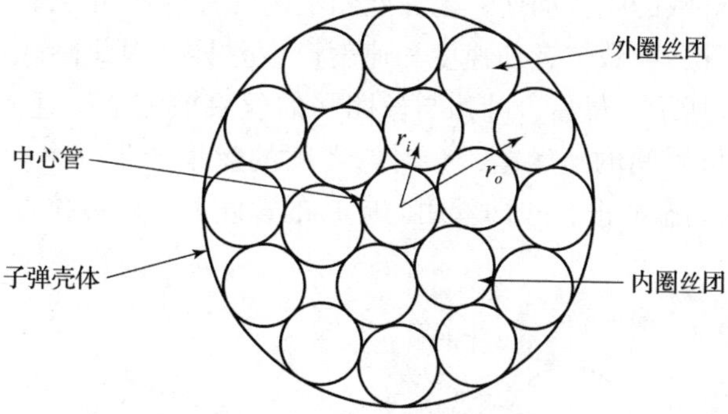

图 4　丝团装填示意图

为了便于丝团的装填和抛撒，装填在子弹内的丝团与子弹壳体和中心管之间都留有一些缝隙，发射过程中由于母弹或子弹高速旋转产生的离心力和轴向过载的影响，会导致内圈丝团与中心管之间、顶层丝团与子弹端盖之间处于“虚接触”状态，即只接触而不存在作用力。因此，丝团相对子弹发生旋转运动时在这两处位置产生的作用力可以忽略。

综上所述，得到母弹或子弹旋转时丝团所受总摩擦力 f_z：

$$f_z = f_b + f_d \tag{7}$$

当丝团所受切向力 F_t 大于摩擦力 f_z 时，相对子弹发生转动。结合式（3）~ 式（7），得到丝团相对子弹发生转动时的判据如下：

$$\alpha > \frac{\omega^2 \mu_{os}(n_i r_i + n_o r_o) + a_s \mu_{ds}(n_i + n_o)}{n_i r_i + n_o r_o} \tag{8}$$

3.2　丝束损伤判据

母弹或子弹旋转过程中，外圈丝团与子弹侧壁接触部位会产生摩擦力。当摩擦力方向与丝束缠绕方向相同时，摩擦力便会对缠绕在丝轴上的丝束起到拉力的作用，对丝束结构产生破坏。

当丝团与子弹相对静止时，摩擦力为 f_{os}，当两者相对旋转时，摩擦力为 f_{od}，表达式如下：

$$f_{os} = m\omega^2 r_o \mu_{os} \tag{9}$$

$$f_{od} = m\omega^2 r_o \mu_{od} \tag{10}$$

式中，μ_{od}代表丝团与子弹侧壁间的动摩擦系数。为了保证丝团在任意运动状态都不产生结构损伤，需满足以下条件：

$$F_i > (f_{os}, f_{od})_{\max} \tag{11}$$

由于物体间的静摩擦系数 μ_{os} 要大于动摩擦系数 μ_{od}，因此，式（11）可简化为

$$F_i > f_{os} \tag{12}$$

结合式（9）、式（12），得到丝束在母弹或子弹旋转条件下的损伤判据如下：

$$\omega \geqslant \sqrt{\frac{F_i}{m r_o \mu_{os}}} \tag{13}$$

4　丝团抛撒对丝束的影响

图 5 所示为丝团从子弹中抛撒出去后的运动示意图。

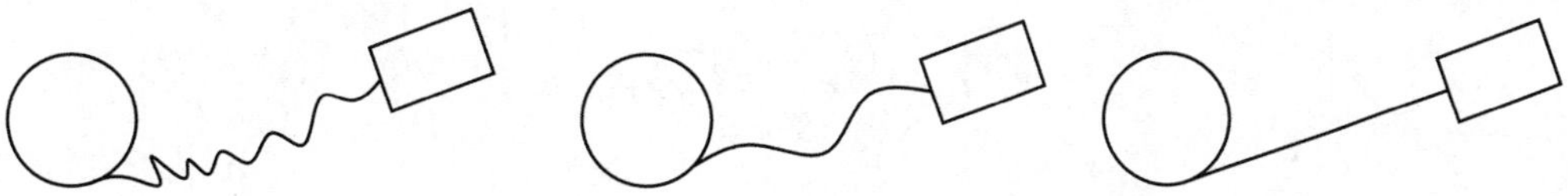

（a）　　（b）　　（c）

图 5　丝团抛撒示意图

抛撒初始，高抛撒速度导致的强扰动使包裹在丝团上的叶片与丝团分离，如图5（a）所示；由于丝团和叶片所受空气阻力的差异，导致二者加速度和速度产生差异，因此使得二者之间的丝束在运动过程中渐渐被拉直，如图5（b）所示；最后，叶片与丝团间的丝束被完全拉直，如图5（c）所示。在丝束被拉直的瞬间，由于丝团与叶片间的速度差，会对二者间的丝束产生一个动态冲击，该过程可等效为：丝团以速度 v_d 对静止丝束的动态冲击，v_d 即丝团与叶片的速度差。丝束拉直过程中，丝团和叶片的受力如图6所示。

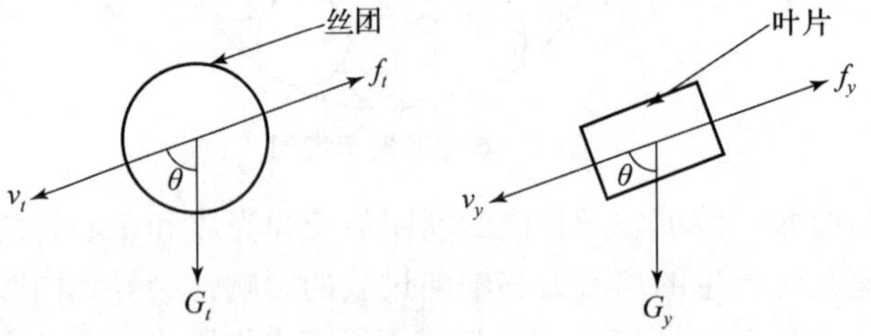

图6　丝团及叶片受力图

图中，v_t、v_y 分别代表丝团和叶片的飞行速度，G_t、G_y 分别代表丝团和叶片的重量，f_t、f_y 分别代表丝团和叶片所受空气阻力：

$$f_t = \frac{1}{2} c_t \rho s_t v_t^2 \tag{14}$$

$$f_y = \frac{1}{2} c_y \rho s_y v_y^2 \tag{15}$$

式中，c_t、c_y 分别代表丝团和叶片的阻力系数；ρ 代表空气密度；s_t、s_y 分别代表丝团和叶片的迎风面积。通过图6所示，得到丝团和叶片的运动方程如下：

$$m_t \frac{\mathrm{d}v_t}{\mathrm{d}t} = G_t \cos\theta - \frac{1}{2} c_t \rho s_t v_t^2 \tag{16}$$

$$m_y \frac{\mathrm{d}v_y}{\mathrm{d}t} = G_y \cos\theta - \frac{1}{2} c_y \rho s_y v_y^2 \tag{17}$$

对式（16）、式（17）进行积分，得到丝团和叶片的速度表达式如下：

$$v_t(t) = \sqrt{\frac{c_t \rho_t s_t g \cos\theta}{2m_t}} \cdot \tanh\left[t\sqrt{\frac{c_t \rho_t s_t g \cos\theta}{2m_t}} + a\tanh\left(v_p \sqrt{\frac{c_t \rho_t s_t}{2m_t g \cos\theta}} \right) \right] \tag{18}$$

$$v_y(t) = \sqrt{\frac{c_y \rho_y s_y g \cos\theta}{2m_y}} \cdot \tanh\left[t\sqrt{\frac{c_y \rho_y s_y g \cos\theta}{2m_y}} + a\tanh\left(v_p \sqrt{\frac{c_y \rho_y s_y}{2m_y g \cos\theta}} \right) \right] \tag{19}$$

式中，v_p 代表丝团的抛撒速度。

丝束所受动态冲击的动荷因数 K_d 为[7]

$$K_d = \sqrt{\frac{v_d^2}{g\delta_{st}}} = \frac{F_d}{G_t} \tag{20}$$

式中，δ_{st} 为静荷位移，即丝团重力 G_t 以静载荷方式作用于丝束时沿冲击方向产生的位移，F_d 为丝束拉直瞬间所产生的动态载荷，v_d 为丝团与叶片的速度差：$v_d = v_{ta} - v_{ya}$，v_{ta}、v_{ya} 分别为丝束拉直瞬间丝团和叶片的瞬时速度，通过式（18）、式（19）得出。当动载荷 F_d 大于丝束的抗拉强力 F_i 时，丝束断裂，由此得到丝团抛撒时的丝束断裂损伤判据：

$$G_t \sqrt{\frac{(v_{ta} - v_{ya})^2}{g\delta_{st}}} \geqslant F_i \tag{21}$$

5　结论

本文对导电纤维丝团在严酷弹道环境下的工作性能进行了理论分析与试验研究，结果显示，弹道过载对丝束结构影响较大，Z 型丝团的抗过载能力要远高于 M 型丝团，在高过载弹道环境下，应优先选用 Z 型丝团。通过理论分析探讨了母弹或子弹旋转对丝团的影响，得到了丝团与子弹发生相对转动时的转动判据及丝束的损伤判据。通过理论分析得到了子弹抛撒丝团过程中的丝束损伤判据。研究结果可为严酷弹道环境下的丝团结构设计提供依据。

参考文献

[1] 梁永直，冯顺山，张广华．弹用导电丝束引弧机理及试验研究 [J]．弹箭与制导学报，2012，32 (6)：179 - 182.

LIANG Y Z, FENG S S, ZHANG G H. Arc mechanism and experimental research of conductive fiber used in weapon [J]. Journal of projectile, rockets, missiles and guidance, 2012, 32 (6): 179 - 182.

[2] 胡松涛．弹道过载对导电纤维丝团性能影响研究 [J]．弹箭与制导学报，2014，34 (1)：79 - 82.

HU S T. Influence on property of conductive fiber under the ballistic overload conditions [J]. Journal of projectile, rockets, missiles and guidance, 2014, 34 (1): 79 - 82.

[3] 张广华，冯顺山，葛超．导电纤维丝团空中展开与飘落运动特性 [J]．弹道学报，2014，26 (1)：35 - 39.

ZHANG G H, FENG S S, GE C. Spreading and falling law of conductive fiber cluster moving in the air [J]. Journal of ballistics, 2014, 26 (1): 35 - 39.

[4] 梁永直．导电丝束对高压电网短路毁伤机理及效应研究 [D]．北京，北京理工大学：2004.

LIANG Y Z. Research of mechanism and effects on high voltage power grid when experienced short damage by conductive fibers [D]. Beijing: Beijing Institute of Technology, 2004.

[5] 张国伟．电力系统易损特性及导电纤维子弹设计中若干问题研究 [D]．北京：北京理工大学，2004.

ZHANG G W. The study of the electrical system vulnerability and some problem in the design of the conductive fiber submunition [D]. Beijing: Beijing Institute of Technology, 2004.

[6] 刘伟钦．火工品制造 [M]．北京：中国电力出版社，2007：115 - 127.

LIU W Q. Initiating explosive devices manufacturing [M]. Beijing: China Electric Power Press, 2007: 115 - 127.

[7] 梅凤翔，周际平，水小平．工程力学：下册 [M]．北京：高等教育出版社，2003：258 - 264.

MEI F X, ZHOU J P, SHUI X P. Engineering mechanics: Volume 2 [M]. Beijing: Higher Education Press, 2003: 258 - 264.

Computational Investigation of Wind Tunnel Wall Effects on Buffeting Flow and Lock – in for an Airfoil at High Angle of Attack①

风洞壁面效应对大攻角翼型的颤振与锁定影响数值研究

Zhou Tong[a] ②, Earl Dowell [b] , Feng Shunshan[a]

[a] School of Mechatronical Engineering, Beijing Institute of Technology, Beijing, P. R. China

[b] Department of Mechanical Engineering and Material Science, Duke University, Durham, NC, USA

Abstract: Wind tunnel walls may significantly affect the aerodynamic characteristics of an airfoil if the distance between walls is not large enough and the experimental data may be influenced by the confinement effect. The purpose of this paper is to investigate the wall effects on a NACA 0012 airfoil at high angle of attack (the flow over the airfoil is assumed to be fully separated). The effects on both a static airfoil and also an oscillating airfoil (pitching motion) are studied. Reynolds number varies from 10^5 to 3×10^5. For a static airfoil, there is a dominant frequency of the flow oscillations called the Strouhal or buffet frequency. Strouhal number of an airfoil in confined flow is affected by blockage ratio and angle of attack. A new wall correction formula for the Strouhal number has been determined. For an oscillating airfoil, lock – in is determined by a combination of excitation frequency and amplitude of the oscillating airfoil. The reason of difference between computational and experimental lock – in regions are explained. Lock – in region decreases with increase of blockage ratio. Reynolds number has slight impact on lock – in region, both in unconfined flow and confined flow.

Keywords: Wall effects; Buffeting; Lock – in; CFD; NACA 0012 airfoil

Nomenclature

c	airfoil section chord, meter
C_l	lift coefficient
C_m	moment coefficient
C_{ps}	pressure coefficient at separation
D	cross – stream length scale of object, meter
D_w	wake width, meter
f_e	exciting frequency of pitching motion, Hz
f, f_n	natural shedding frequency/ buffeting frequency, Hz
h	wind tunnel width of side walls, meter
k	blockage ratio of wind tunnel, $k = D/h$ or $k = c\sin\alpha/h$
Re	Reynolds number
St	ordinary Strouhal number, $St = fc\sin\alpha/U_\infty$

① 原文发表于《Aerospace Science and Technology》2019（95 卷，文章 105492）。

② 周彤：工学博士，2009 年师从冯顺山教授，研究封锁类子弹药气动特性及其弹道性能，现工作单位：北京理工大学。

St_c, St_c^*, St_c^{**}	corrected Strouhal number
t	physical time, second
t_n	non - dimensional time, $t_n = tU_\infty / c$
U_∞	free stream velocity, m/s
α, α_m	nominal angle of attack, mean angle of attack, degree
θ	amplitude of pitching motion, degree
ω	angular velocity of pitching motion, rad/s
ξ	correction factor related to wind tunnel wall effects

1 Introduction

Nonlinear phenomena of an airfoil/wing have generated great interest due to the risk of unstable or structure failure. High angle of attack, transonic flow and elastic structure are three of the most important sources of nonlinearity[1-4]. Two - dimensional wind tunnel experiments have been carried out to investigate the buffeting flow over a static airfoil and an oscillating airfoil with pitching motion[5,6]. However, when the distance between the wind tunnel walls is not large enough compared to the characteristic length of the object investigated, the wall effects cannot be neglected because they may significantly affect the aerodynamic characteristics of the object. Test data become less reliable due to notable blockage effects. Ota *et al.* [7] classified wall effects into two categories. The first is the confinement effect. That is, the tunnel walls prohibit the extension of streamlines around the body, resulting an increase local flow velocity. The second is a displacement effect due to the boundary layers developing on the walls. The displacement effect is assumed to be smaller than the confinement effect since the boundary layer is thin at high Reynolds number. Extensive investigations have been carried out on wind tunnel wall effects for decades by many investigators, both experimentally and computationally. In this paper, the main concern is how the wall affects the buffeting and lock in phenomenon of an airfoil at high angle of attack. Buffeting is a fluid instability caused by separation in the flow around a bluff body. In the wake, it is possible to detect a dominant vortex - shedding frequency, which corresponds to the instability of the separated boundary layers. Strouhal number is often used to describe the oscillating flow mechanisms. The original form of Strouhal number is $St = fD/U_\infty$, in which f is the vortex shedding frequency in cycles per unit time, D is the cross - stream length scale of the object, and U_∞ is the freestream velocity. When a body is oscillating at a certain frequency and amplitude, it can synchronize the buffeting frequency to the motion frequency if the two frequencies are sufficiently close to each other. This phenomenon is known as a lock - in [8-10]. Specifically, there are two kinds of lock - in in separated flow. One is for uncoupled problems (considering the structural motions is prescribed and unaffected by the flow), as in the present study. This mechanism of lock - in is attributed to the nonlinear aerodynamic resonance. Lock - in only occurs when the exciting frequency fe is close to natural buffet frequency fn and the maximum vibration amplitude is often acquired at $fe = fn$, like in Tang et al. 's experiment [6]. The other mechanism is for coupled problems (considering the structural motion and fluid motion to be self - excited and become unstable leading to a limit cycle oscillation). Gao et al. [11,12] attribute the latter lock - in mechanism to a linear coupled - mode flutter (the coupling between one structural mode and one fluid mode). In this case, the lock - in region did not display a symmetrical distribution against the buffet frequency and it is often located on the side of higher frequency [13].

Malavasi and Blois [14] measured the flow structure around an elongated rectangular cylinder near a solid wall by using a 2D PIV technique. The flow patterns of time - averaged flow fields highlight the significant distortions due the solid surface. Malavasi and Zappa [15] analyzed the changes in force coefficients and in the vortex shedding Strouhal number when the cylinder is placed at various distances from the bottom wall and different values of attack angle. Both force coefficients and shedding frequencies are significantly affected by the presence of the boundary if the distance to the boundary is smaller than three times the dimension of the cylinder cross section perpendicular to the flow.

Okajimaet al. [16] studied the blockage effects on a rectangular cylinder with a computational model. Direct Navier – Stokes simulation (DNS) was used for laminar flow at Reynolds number lower than 10^3 and a k – epsilon turbulence model for flow at Reynolds number around 4×10^3. The simulation results show that for the stationary airfoil cases, the lift and drag forces, and vortex shedding Strouhal numbers all increase with the increase of blockage ratios. Duraisamy *et al.* [17] studied wall tunnel effects on subsonic airfoil flows using various approaches including classical theory, computational simulations and experimental data. They also found an augmentation of the lift magnitude for steady flow over an airfoil.

It is known that the wall effects onthe flow around bluff bodies are greater than on streamlined bodies without flow separation. Some theoretical and empirical correction formulas for the wall effects on bluff bodies have been published. Maskell [18] developed a theory of blockage constraint on the flow past a bluff body, using an approximate relation describing the momentum balance in the flow outside the wake and two empirical auxiliary relations. After that, several correction formulae [19 – 21] have been proposed based on Maskell's method.

Ota*et al.* [7] proposed a correction formula of the two – dimensional wall effects not only for the mean forces, but also the fluctuating forces and the Strouhal number. The latter formula is

$$St_c = St(1 - k\xi) \tag{1}$$

St_c is the corrected Strouhal number and k is the blockage ratio (cross – stream length scale of object D divided by wind wall distance h). ξ is an empirical correction factor. Ota *et al.* suggested $\xi = 0.82$ for an inclined plate.

Ota's formula was applied to correct the Strouhal number for an inclined flat plate by Chen and Fang [22]. A correction factor of 1.21 was adopted based on Ota's correction formula and the least – squares method. The Strouhal numbers measured with different blockage ratios were corrected to yield a single value. However, the corrected Strouhal number varied from 0.135 to 0.15 for angle of attack from 65° to 90°, while the Strouhal number for unconfined flow is 0.136. Besem *et al.* [5] also used Ota's empirical formula to correct the Strouhal number of an airfoil at high angle of attack in confined flow. The parameter ξ is optimized by a least squares method and the minimum of the root mean square is found for $\xi = 1.6$. At $Re = 3\times10^5$, the corrected Strouhal number is 0.143 ± 0.003 for angle of attack from 25° to 70°. The universal Strouhal number for unconfined flow is 0.16.

Yeung [23,24] investigated the self – similarity of confined flow past a bluff body, including a flat plate, a cone, a circular cylinder and a sphere. The pressure at separation and the form drag are found as functions of blockage ratio. He derives a linear relationship between the drag and the base pressure observed empirically. He suggested several correction equations for the Strouhal number of a flat plate. One of the simplest equations is

$$St_c = St(1 - k)^{\xi} \tag{2}$$

In two other two equations, St_c is related to D_w and U_S. D_w is the wake width. U_s equals to $\sqrt{1 - c_{ps}}$, while c_{ps} is pressure coefficient at separation. In practice, the wake width concept is vague and the estimation of U_s is not easy.

Thus in the literature there are two simplified calculation procedures to correct the Strouhal number, which are $St_c = St(1 - k\xi)$ and $St_c = St(1 - k)^{\xi}$. The corrected Strouhal number depends on the blockage ratio k and a constant value ξ.

Okajimaet al. [25] studied the blockage effects on the force of a rectangular cylinder which is forced to oscillate in a uniform flow at Reynolds number of 4×10^3. It was found that the mean values of drag forces of an oscillating rectangular cylinder and the Strouhal numbers depend greatly on blockage ratios, while the phase differences between lift and oscillatory displacement are much less dependent on the blockage ratios. Lian [26] investigated the blockage effects on the aerodynamics of pitching wing using a computational approach. Simulations show that the pitching motion can mitigate the blockage effect. Besides, the sidewalls can lead to strong span – wise variation in the flow field. Chen and Fang[27] found that the lock in regime was narrower with an increase of the blockage ratio for a flat plate normal to the freestream.

In previous work, several researchers [6,28] have noted the analogy between the classic Von Karman Vortex Street behind a bluff body [29 – 34] and the flow oscillations that occur for flow around a static NACA 0012 airfoil at sufficiently large angle of attack. The Strouhal number for the airfoil is calculated by $St = fc\sin\alpha/U_\infty$, where α is angle of attack and c is chord length. For a pitching airfoil, lock – in is found for certain combinations of airfoil oscillation frequencies and amplitudes when the frequency of the airfoil motion is sufficiently close to the buffeting frequency. The flow oscillations are entrained by the airfoil motion and the flow responds only at the airfoil motion frequency. In experimental studies [5,6], the distance between wind tunnel walls is not large enough compared to the cross – stream dimension of the airfoil to avoid wind tunnel wall effects. The current work focuses on the effects of wind tunnel walls on a static airfoil and a pitching airfoil, respectively. The angle of attack of the airfoil varies from 30° to 90°, and the wake is fully separated. The Reynolds numbers studied are from 10^5 to 3×10^5. The goal of this paper is to explore the wall effects on a NACA 0012 airfoil at high angle of attack and provide a better understanding of the available experimental data for the airfoil.

2 Computational Methods and Validation

2.1 Computational Model

The computational model for the current study follows the wind tunnel experiment of a NACA 0012 airfoil, which was conducted in the Duke University low speed wind tunnel [5,6]. The quarter chord point of the airfoil is fixed in the central line of test section. The airfoil has a chord length $c = 0.2553$m and the dimension of wind tunnel test section is $0.7 \times 0.53\text{m}^2$. The blockage ratio is defined by $k = c\sin\alpha/h$, where h is the distance between two side walls. A sketch of the experimental set – up is show in Figure 1. The distance between the side walls is $h = 0.7$ m. The airfoil is mounted by spanning the tunnel which minimizes the tip effects and enables two – dimensional flow over the airfoil. All the micro – pressure sensors are placed in a mid – span rib. Angle of attack of the airfoil varies from 30° to 90°, which gives a corresponding blockage ratio from 0.182 to 0.365.

With the development of CFD, unsteady Reynolds – Averaged Navier – Stokes solver (RANS) is recognized as being able to give similar results with experiment for buffeting flow[35,36]. In Lian's research [26], the blockage effects on the aerodynamics of a flat wing have been investigated using a computational approach. He compared the force from the flow field of the 2D and 3D wing. Neither the lift nor drag coefficients showed a significant difference between the 2D and 3D results. In this paper, we adopted a 2D calculation domain and a polyhedral mesh. The dimension of the first computational cell near the wall boundary is about 7.5×10^{-5} chord lengths of the airfoil. The unsteady flow field around a static airfoil is simulated by the Star CCM + CFD code, using a 2D Reynolds – Averaged Navier – Stokes solver. The method uses second – order temporal discretization. The second – order upwind discretization scheme is used for evaluating face values for convection and diffusion fluxes. The inviscid fluxes are evaluated by using the Weiss – Smith preconditioned Roe's flux – difference splitting scheme [37,38]. Velocity of free stream varies from 5.75 m/s to 17.24 m/s. As a result, Reynolds number based on chord length of airfoil is from 10^5 to 3×10^5. The air flow is assumed to be incompressible and the density of air was set to a constant value. Boundary conditions include velocity – inlet, pressure – outlet and solid wall. The wall $y+$ value is smaller than 1.

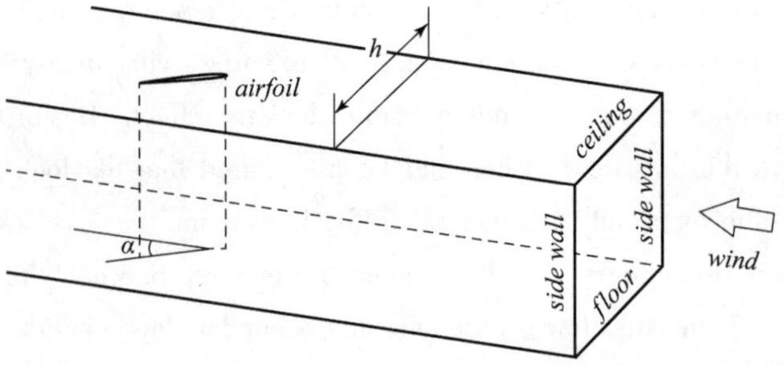

Figure 1 A sketch of experimental set - up

Great care has been taken in the computational simulation procedure, including optimization of grid density, selection of turbulence models and time step. More details can be found in the reference [28]. Some of the conclusions are as follows:

1) Very high CFD mesh resolution of the boundary layer is not necessary for the airfoil at high angle of attack. A number of cells of the order of 10^4 (including 350 points around the airfoil surface, and 5 - 10 points in boundary layer) is sufficient to predict the fundamental flow shedding physics.

2) The simulation results show that the Standard Spalart - Allmaras (SA) turbulence model and the Shear - stress transport (SST) k - omega turbulence model provide similar solutions for the time history of the lift coefficient. Both of them are capable of predicting the buffeting flow due to flow separation at high angle of attack. The SA turbulence model is adopted for further study in this paper.

3) Simulation results are numerically stable when the time - step is 10^{-3} seconds or less.

2.2 Validation

Computational simulation results for the static airfoil in confined flow were compared with the experimental results [5,6]. Figure 2(a) shows the buffeting frequency versus flow velocity. It can be seen that the computed buffeting frequencies have good agreement with the experimental buffeting frequencies. Figure 2(b) shows the Strouhal number versus angle of attack at $Re = 3 \times 10^5$. Strouhal number of the wake is given by $St = f c \sin\alpha / U_\infty$. As shown in Figure 2(b), Strouhal number increases with increase of angle of attack. However, the blockage ratio also changes at different angle of attack, which makes the Strouhal number a function of a combination of angle of attack and blockage ratio. More work needs to be done to separate the effects of these two factors.

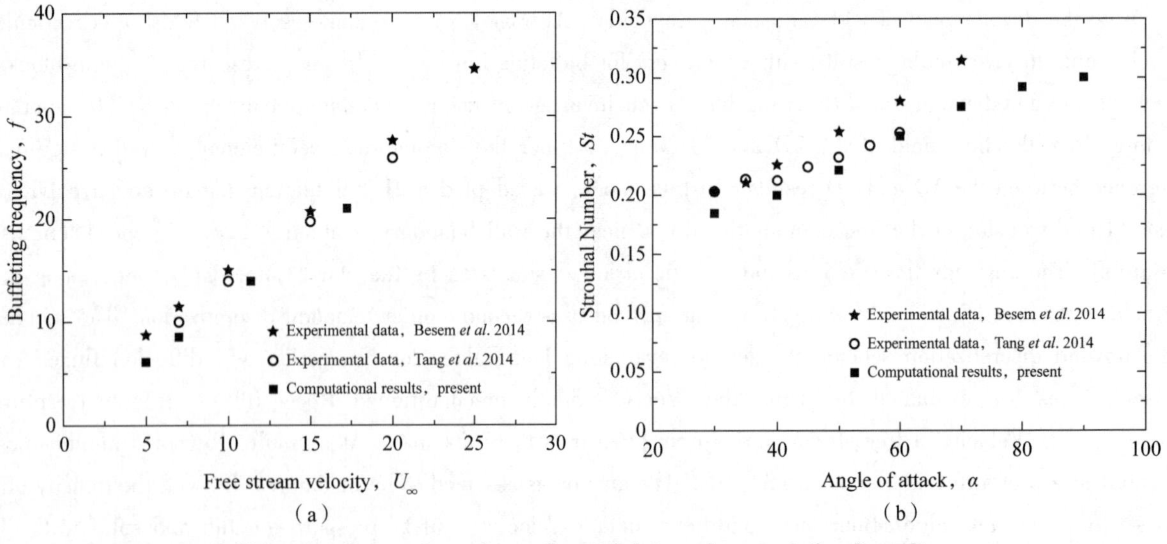

Figure 2 Comparison of experimental data and computational results

(a) Buffeting frequency vs. Flow velocity; (b) Strouhal number vs. Angle of attack

3　Wall Effects on Static Airfoil

3.1　Airfoil in Unconfined Flow

In order to establish a reference condition for the buffeting flow in the wind tunnel experiment, the aerodynamic characteristics of an NACA 0012 airfoil in unconfined flow have been studied first. The wall boundaries were expanded to seven times the chord length away from the central line and boundary conditions assume flow symmetry.

Figure 3(a) shows the Strouhal number in unconfined flow as it varies with Reynolds number. Computational results in this paper are compared with Besem *et al.*'s CFD results which were obtained by using a nonlinear frequency domain, Harmonic Balance code [5]. The present results are in quite good agreement with Besem *et al.*'s CFD results. As seen in the figure, the Strouhal number is independent of Reynolds number from 0.87×10^5 to 3.5×10^5.

The Strouhal number versus angle of attack is shown in Figure 3(b). For an airfoil in unconfined flow, a universal Strouhal number (0.162 ± 0.018) can be obtained, as one would expect. Besem *et al.* also found a Strouhal number around 0.16 at several angles of attack. Thus the change in Strouhal number due to confined flow is clear.

The Strouhal number reaches a maximum around angle of attack of 50°, where the lift coefficient also has a maximum [28]. The airfoil has a sharp trailing edge, so the separation point at the trailing edge is assumed to be fixed. However, the leading edge has a curved shape. The separation point at the leading edge may move under different flow conditions, unlike for a flat plate with sharp edges.

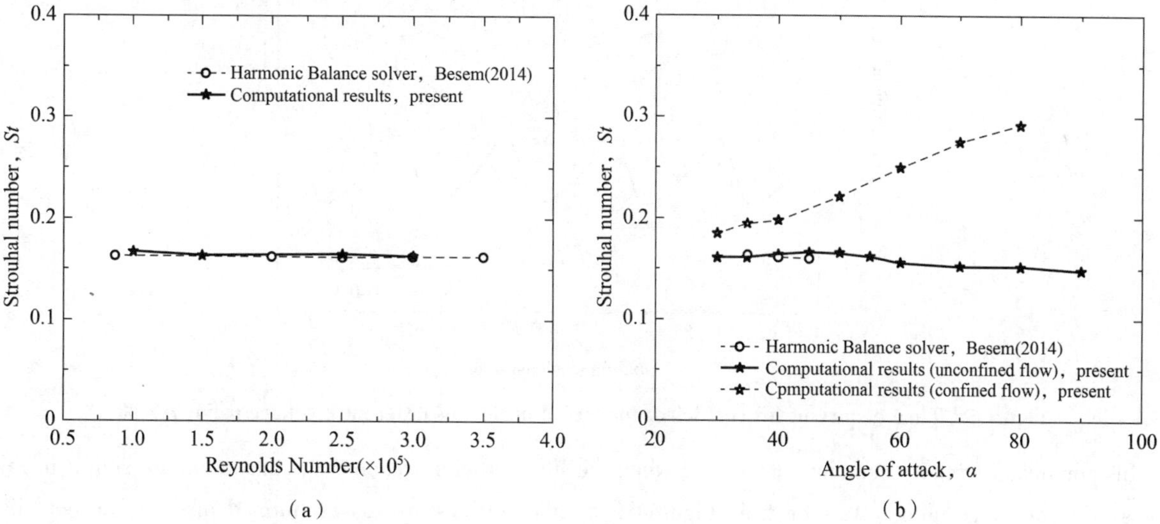

Figure 3　Strouhal number of flow over the airfoil in unconfined flow and confined flow

(a) Strouhal number vs. Reynolds number; (b) Strouhal number vs. Angle of attack

3.2　Effect of Blockage Ratio

In this section, the angle of attack is fixed and the wall distance h is changed to obtain a different blockage ratio. Effect of blockage ratio on Strouhal number for an airfoil in confined flow is shown in Figure 4. The Strouhal number increases almost linearly with increase of blockage ratio.

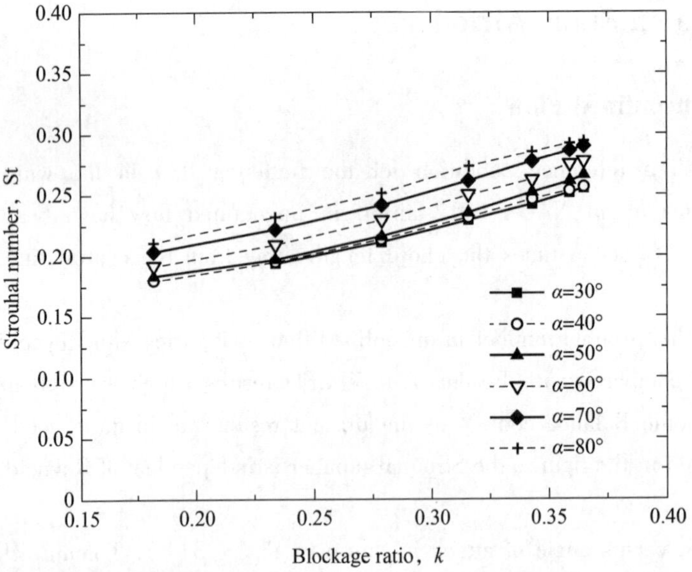

Figure 4 Effect of blockage ratio on Strouhal number for airfoil in confined flow

Figure 5 presents the time history of the lift coefficient for an airfoil for different blockage ratios. The angle of attack is 40° and Reynolds number is 3×10^5. The blockage ratio ranges from 0 to 0. 365. The average value of lift coefficient increases with increase of blockage ratio.

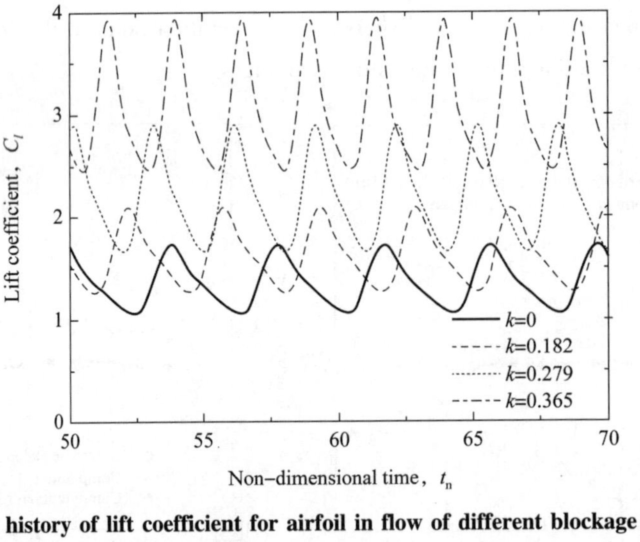

Figure 5 Time history of lift coefficient for airfoil in flow of different blockage ratio, $\alpha=40°$

Instantaneous vorticity contours at the time when the lift coefficient reaches the maximum are shown in Figure 6 for several blockage ratios. As seen from Figure 6(a), the vortices are loosely formed in the wake both in the streamwise and lateral directions, which represents a small value of vortex shedding Strouhal number. When the blockage ratio increases to a certain degree as in Figure 6(b) -6(d), the formation of vortices is affected by the wall constraint. With the increase of blockage ratio, the vortices are forced to form in a narrower space. Also the formation of vortices in the lateral direction is constrained and the wall constraint weakens the vortices in the far wake.

The Strouhal number in confined flow can be described by two simple equations mentioned in Section 1. The constant parameter ξ is optimized by a least square method. As seen in Figure 7, the correction equations are $St_c = St(1-k)^{1.2}$ and $St_c = St(1-1.15k)$, respectively. These two equations give quite similar results. The corrected Strouhal number is independent of the blockage ratio. However, it still changes with angle of attack.

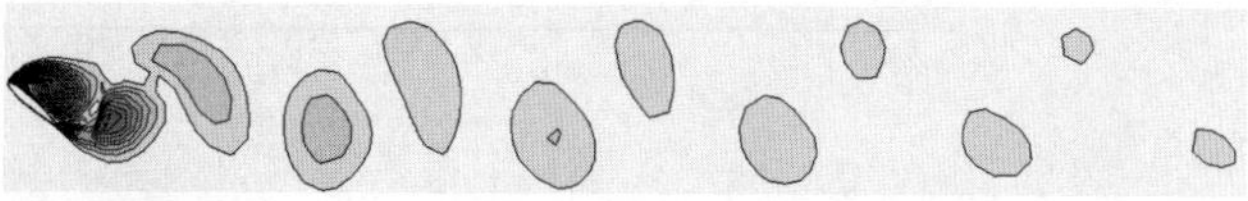

(a)

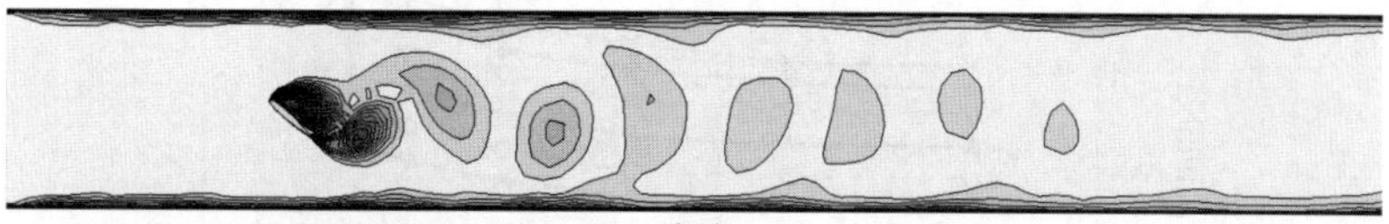

(b)

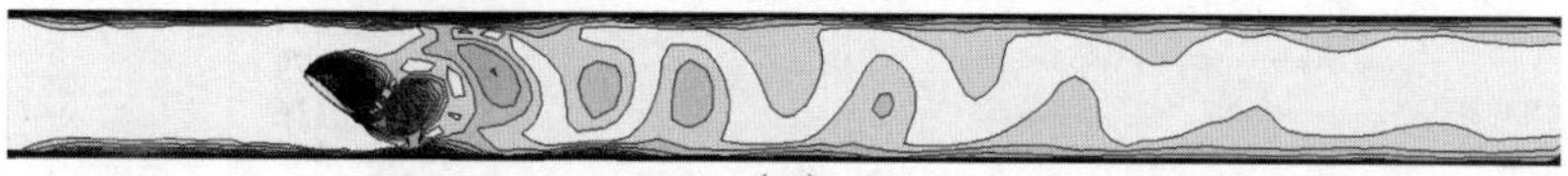

(c)

(d)

Figure 6　Instantaneous vorticity contours of the flow for different blockage ratios

(a) $k=0$; (b) $k=0.182$; (c) $k=0.279$; (d) $k=0.365$

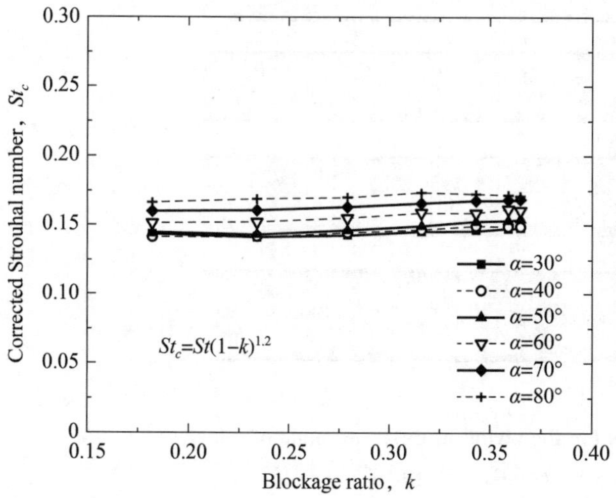

(a)

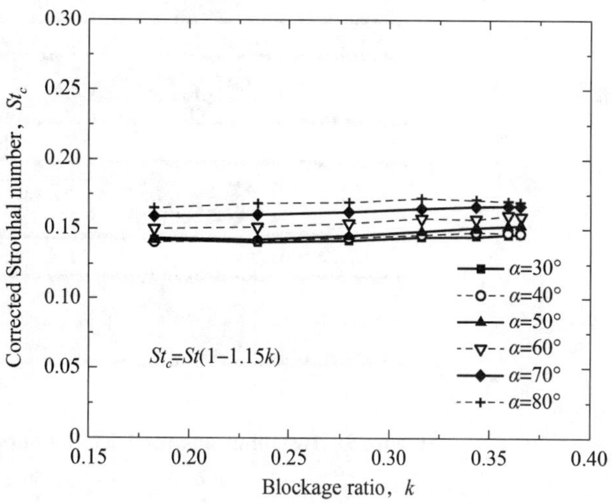

(b)

Figure 7　Corrected Strouhal number in confined flow

(a) Corrected Strouhal number, $St_c=St(1-k)^{1.2}$; (b) Corrected Strouhal number, $St_c=St(1-1.15k)$

3.3　Effect of Angle of Attack

Effect of angle of attack on Strouhal number for airfoil in confined flow is shown in Figure 8. The x - axis represents $\sin\alpha$. It can be seen that Strouhal number increases with increase of angle of attack, especially at extremely high angle of attack.

Instantaneous vorticity contours at the time when the lift coefficient reaches a maximum are shown in Figure 9 for various angles of attack. The blockage ratio of flow is kept constant at 0.365. The restriction of the walls on the formation of vortices is significant for such a high blockage ratio. It is well known that body shape has a great influence on the wake. For an airfoil in unconfined flow at different angles of attack, the dominant frequencies of the flow oscillations, as well as the separation point and position of vortices, are quite different from those for confined flow.

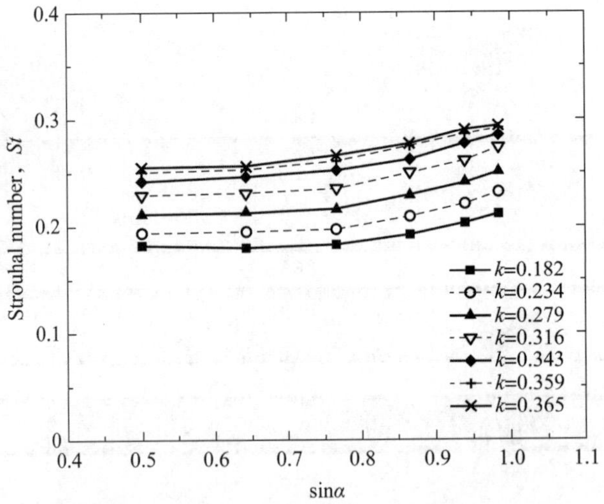

Figure 8　Effect of angle of attack on Strouhal number for airfoil in confined flow

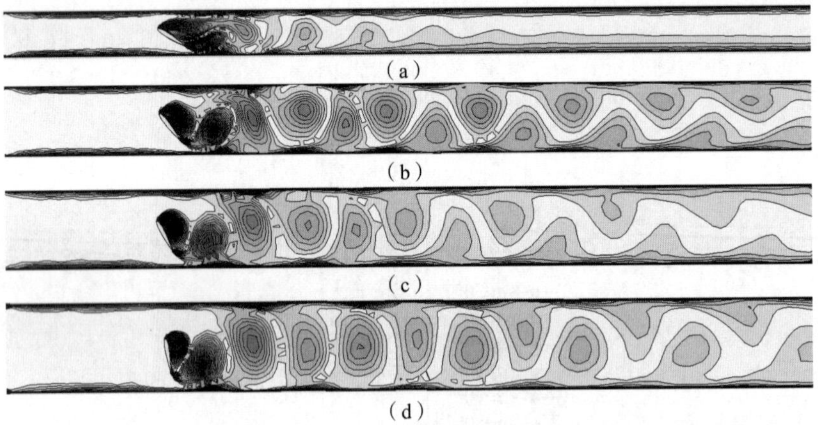

Figure 9　Instantaneous vorticity contours for the airfoil at different angle of attack

(a) $\alpha=30°$; (b) $\alpha=50°$; (c) $\alpha=60°$; (d) $\alpha=80°$

As seen in Figure 10, the lift coefficient of the airfoil in confined flow increases significantly, as does the buffeting frequency. A high angle of attack is likely to reinforce this trend, while the buffeting frequency is relatively low. It is noteworthy that at relatively high angles of attack (such as 50° and 60°), the time response is non - sinusoidal both in unconfined flow and confined flow. Streamlines around the airfoil at T_1 and T_2 in unconfined flow (as labeled in Figure 10) are shown in Figure 11. For $\alpha = 40°$, the time histories were sinusoidal. The vortices shed from the leading edge and trailing edge alternate, like in the classic von Karman vortex street. As seen from Figure 11 (a), the trailing - edge vortex reached its maximum and began to shed, which led to the minimum of the lift coefficient. The reattachment point moved to near the center of the suction - side surface. In case of $\alpha=60°$, (as seen in Figure 11 (b)), the tailing - vortex fused with the leading - edge vortex during the process of growing, which led to a temporary rise of lift coefficient.

Figure 12 shows the position of the stagnation point on the windward side for an airfoil in unconfined flow and confined flow. It can be seen that the stagnation point moves toward the trailing edge when the flow is confined. Displacement of stagnation points for airfoil in unconfined flow and confined flow increases as the angle of attack increases. While the stagnation point is moving toward to the trailing edge, the fluid mass passing the leading edge is increasing. This seems to give rise to a higher local velocity for vortices shedding from the leading edge.

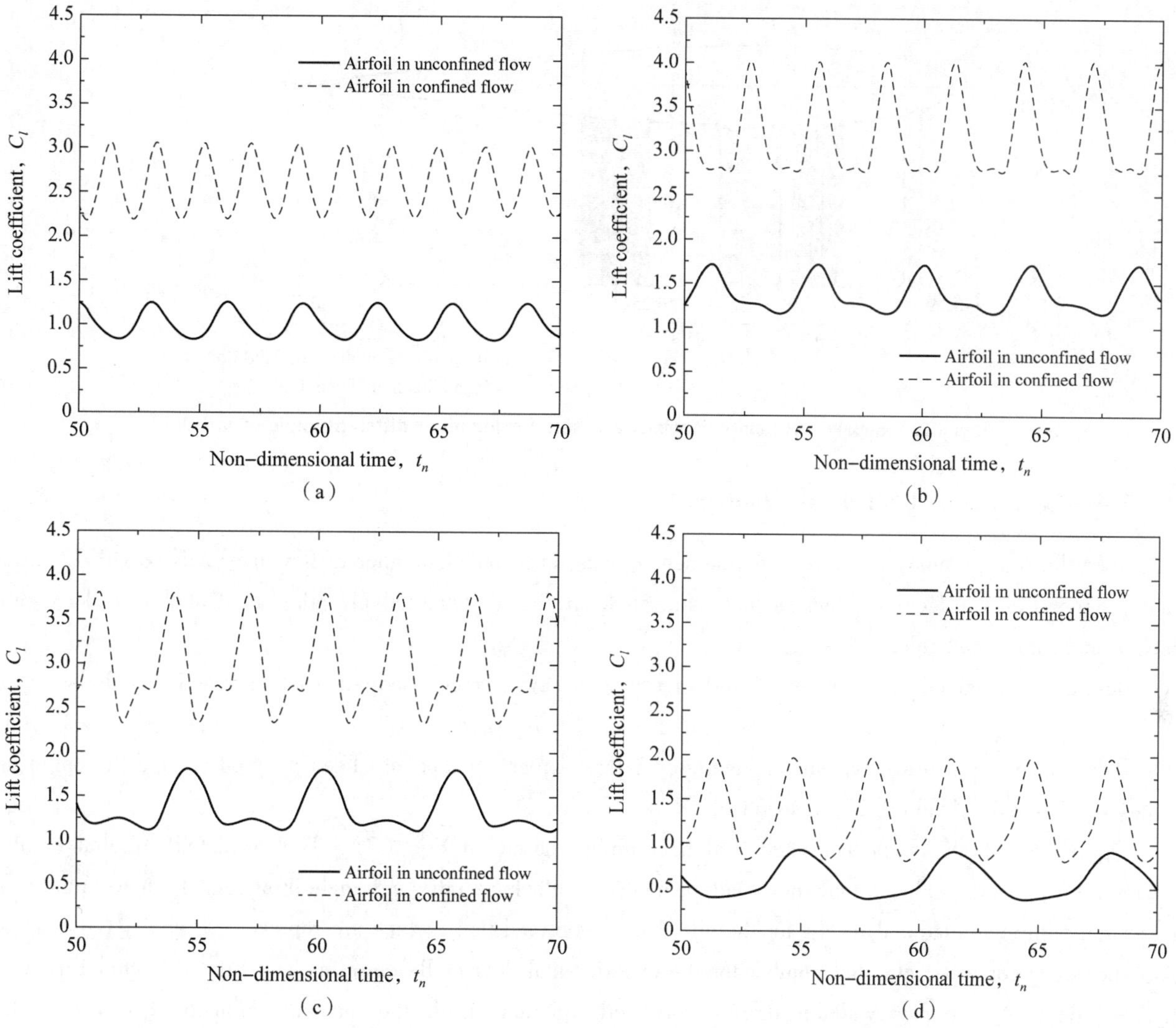

Figure 10　Lift coefficient for airfoil in confined flow and unconfined flow

(a) $\alpha=30°$; (b) $\alpha=50°$; (c) $\alpha=60°$; (d) $\alpha=80°$

In summary, angle of attack has important effect on both time histories and the responses to wall confinement. It must be considered for investigating the wall effects on an airfoil at high angle of attack.

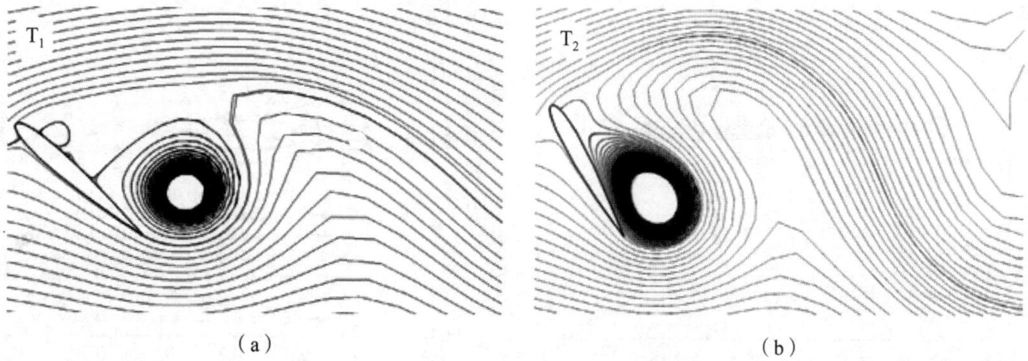

Figure 11　Streamlines around the airfoil at T_1 and T_2

(a) T_1 ($\alpha=40°$); (b) T_2 ($\alpha=60°$)

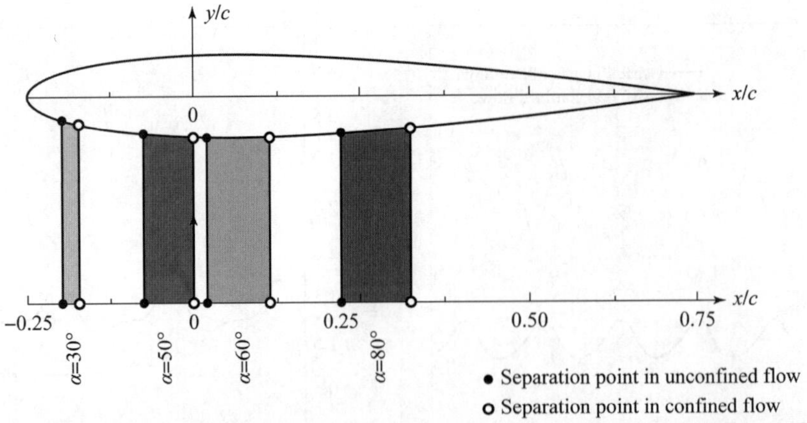

Figure 12　Separation points at windward side for airfoil at different angle of attack

3.4　Correction of Strouhal Number

According to the analysis above, confinement effects on Strouhal number of flow over a NACA 0012 airfoil mainly depend on two factors: blockage ratio and angle of attack. Qualitatively, the effect of the blockage ratio plays a more important role.

Based on regular pattern of Figure 4 and Figure 8, a new correction equation is expressed as follows.

$$St_c = St(1 - ak)(1.7 - \sin\alpha^3)^b \tag{3}$$

Here, a and b are two constant parameters. A least square method has been adopted to give the optimized values which are 1.15 and 0.28, respectively.

Figure 13 (a) shows the corrected Strouhal number based on Eq. (3). The corrected Strouhal number collapses to the same scale, which shows independence of blockage ratio and angle of attack. It shows quite good agreement with computational results for the airfoil in unconfined flow. which are represented by $k=0$. Figure 13 (b) shows the corrected Strouhal number for the experimental data of Besem *et al* [5]. Eq. (1) and Eq. (2) are also adopted to compare with the present corrected equation, while the constant parameter ξ is 1.6 for the former and 1.5 for the latter. Based on Eq. (1) and Eq. (2), the corrected Strouhal number collapses to approximately 0.14 and 0.15, respectively. The present Eq. (3) gives similar corrected results to computational data in unconfined flow, especially when $\alpha \geqslant 40°$.

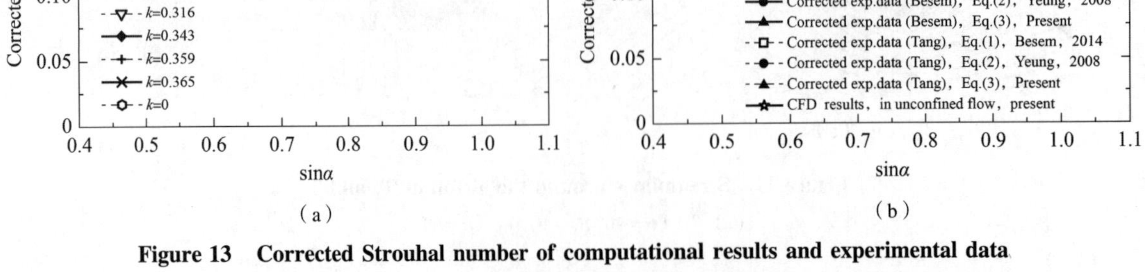

(a)　　　　(b)

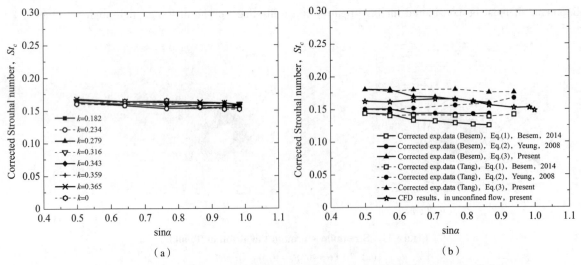

Figure 13　Corrected Strouhal number of computational results and experimental data

(a) Corrected Strouhal number of computational results; (b) Corrected Strouhal number of experimental data

4 Wall Effects on Oscillating Airfoil

4.1 Comparison with Experimental Data

When a body is oscillating at a certain frequency and amplitude, it can synchronize the buffeting frequency to the airfoil motion frequency if the two frequencies are sufficiently close to each other. This phenomenon is known as lock - in. In a previous study, the lock - in phenomenon for certain combinations of a NACA 0012 airfoil pitching motion frequency and amplitude has been found both experimentally and computationally. The lift coefficient peak amplitude due to pitch excitation for the lock - in case is much larger than for the unlock - in case, which may be a contributor to structural failure. As a result, the lock - in phenomenon is of great concern for the study of an oscillating airfoil. The boundary of the lock - in region is described by a combination of frequency and amplitude of the pitch motion, which indicates the critical condition of lock - in.

In the present study, the airfoil was forced to oscillate periodically in pitch at the quarter chord point around a given mean angle of attack α_m. The angular velocity of pitching motion at a given time t can be expressed as

$$\omega = 2\pi f_e \theta \cos(2\pi f_e t) \tag{4}$$

The oscillation amplitude θ varies from 1° to 3°. Now the Reynolds number is 10^5, as same as Besem's experiment [5]. The mean angle of attack is 40° and the blockage ratio is 0. 234. Figure 14 shows the comparison of computational lock - in region and experimental lock - in region. For completeness computational results are shown for both confined and unconfined flow. Later the differences in the computational results will be discussed in more depth. It can be seen that the experimental lock - in region is wider than that obtained from computation.

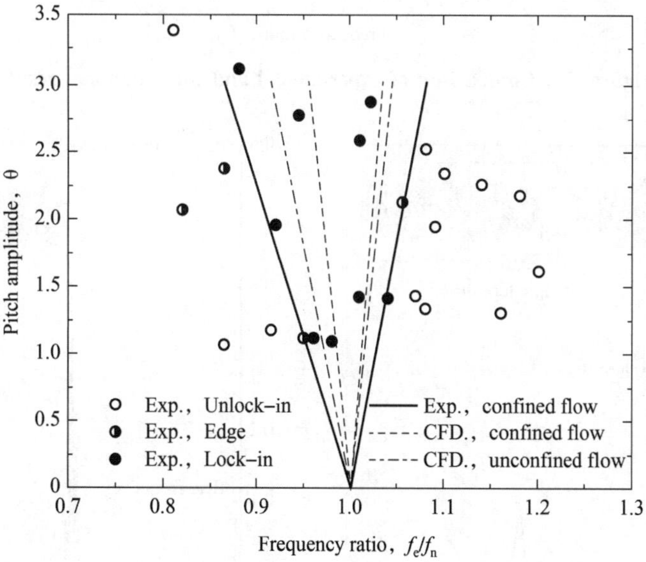

Figure 14 Comparison of experimental and computational results

Due to the background wind tunnel noise, the time history of lift coefficient appears random to a certain extent. Separating the excitation frequency and the buffet frequency is a challenge. Also, when the excitation frequency is quite close to buffeting frequency, more sample points are needed to resolve the two different frequencies in the FFT (Fast Fourier Transformation). For instance, there are nearly 4 periods in 20 seconds whenfe = 0. 97 fn and θ = 1°. The beat frequency is roughly equivalent to 2 Hz. In Besem et al. 's experiment, about 1500 samples were taken at a sampling rate of 256 Hz. As a result, the sample length is at least 5. 8 seconds. We compared the FFT analysis results of fe = 0. 97 fn when sample length is 5. 8 seconds and 10. 2 seconds, respectively. As seen from Figure 15 (a), if the sample length is 5. 8 seconds, the FFT analysis indicates a lock - in, which means the beat frequency has been ignored. On the contrary, when the sample length is 10. 2 (two complete cycles), the FFT analysis indicates unlock - in, as seen in Figure 15 (b).

Beat frequency as a function of frequency ratio is shown in Figure 16. It is found that the nearer the excitation frequency is to the buffeting frequency, the lower is the beating frequency. Basically, sample length needs to be larger than two periods to identify the beat frequency. If we analyze a sample of 5. 8 seconds, a beat frequency lower than 0. 345 will be ignored, as seen in the gray area in Figure 17. In this case, the lock - in region will be expanded obviously. This may explain why experimental lock - in region is wider than computational lock - in region, at least in part.

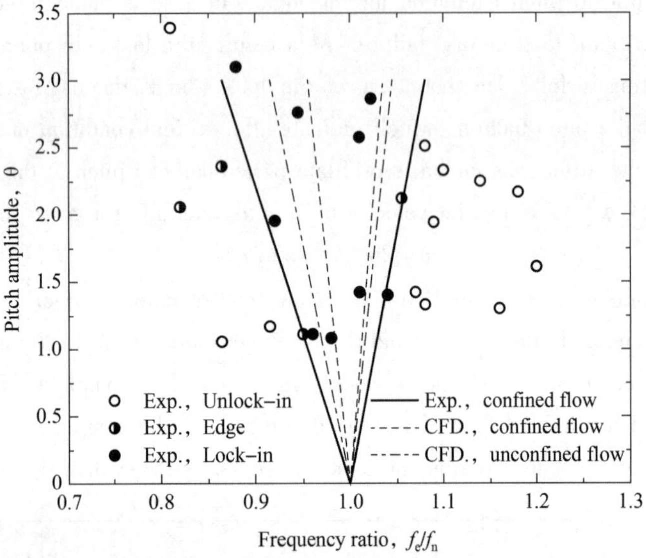

Figure 15 Comparison of experimental and computational results

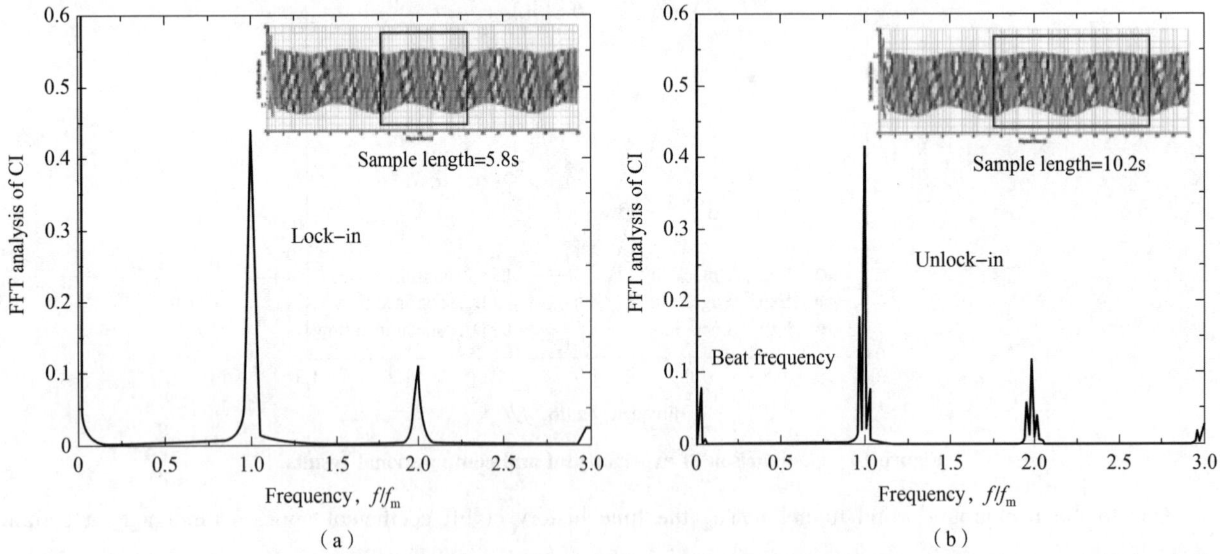

Figure 16 FFT analysis results of different sample length

(a) Sample length 5. 8 s; (b) Sample length 10. 2 s

4. 2 Effects of blockage ratio

Effect of blockage ratio on lock - in region of airfoil at angle of attack $\alpha = 40°$ is shown in Figure 18 (a) Reynolds number is 10^5. In this figure, the solid line represents $k = 0$, which means the airfoil is in unconfined

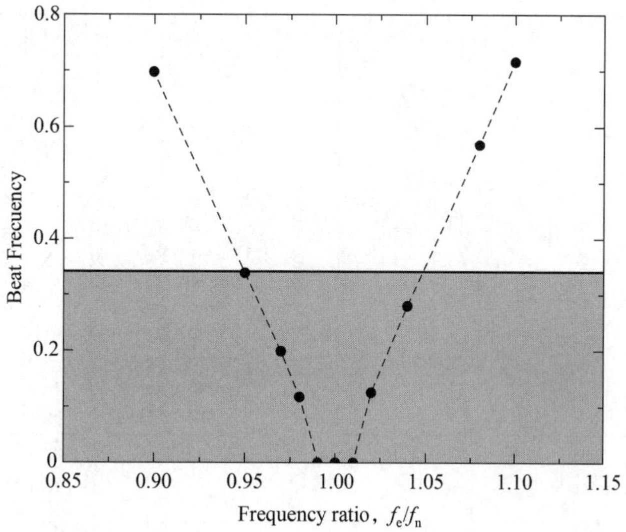

Figure 17　Beat frequency as a function of frequency ratio

flow. The dash – dot – line represents $k = 0.234$, which means the airfoil is in confined flow and the wall distance is the same as experiment. With increase of blockage ratio, the lock – in region of airfoil decreases. A similar conclusion has been found by Chen and Fang [27] for a flat plate in a uniform stream. Both findings suggest that wall effect results in a decrease of lock – in region. And the lock – in region is not symmetric about the natural shedding frequency both in confined flow and unconfined flow.

As discussed in the Section 3.2, vortices in wake of a static airfoil are forced to form faster in confined flow, which leads to a higher Strouhal number. When it comes to the oscillatory cases, the natural frequency of the wake is more difficult to synchronize to the exciting frequency for high Strouhal numbers, as the energy transfer is faster and there is less time for response. This interpretation might also explain why the lock – in region is not symmetric. In the left region of lock – in, the Strouhal number is smaller than that in the right region, which indicates that lock – in is more likely to occur.

Lock – in region at different Reynolds number in unconfined flow ($k = 0$) and confined flow ($k = 0.234$) are shown in Figure 18(b). As seen in the figure, Reynolds number has slight impact on lock – in region in the range of 10^5 to 3×10^5, both for unconfined flow and confined flow. In another word, the lock – in region almost remains the same if only velocity of inlet flow increases. In confined flow, the wall effect will result in an increase of local flow velocity, as one would expect. This is clarified as confinement effect according to Ota *et al.* [7]. According to the present discussion, it may be assumed that wall effects are more than just an increase of local flow velocity for an oscillating airfoil.

5　Conclusions

Wall effects on a static airfoil and an oscillating airfoil at high angle of attack have been computationally studied. Two – dimensional flow over a NACA 0012 airfoil is simulated by using the Star CCM + code. The simulation results have been validated by comparison with experimental data from a Duke University wind tunnel study. Effects of both blockage ratio and angle of attack are investigated separately and together, which provide a better understanding of the experimental data.

A dominant vortex shedding frequency can be obtained for a static airfoil at high angle of attack ($\alpha > 30°$). A universal Strouhal number scaling (0.162 ± 0.018) is obtained for an airfoil in unconfined flow at angles of attack ranging from 30° to 90°. As one can expect, walls close to the airfoil will affect the formation of vortices.

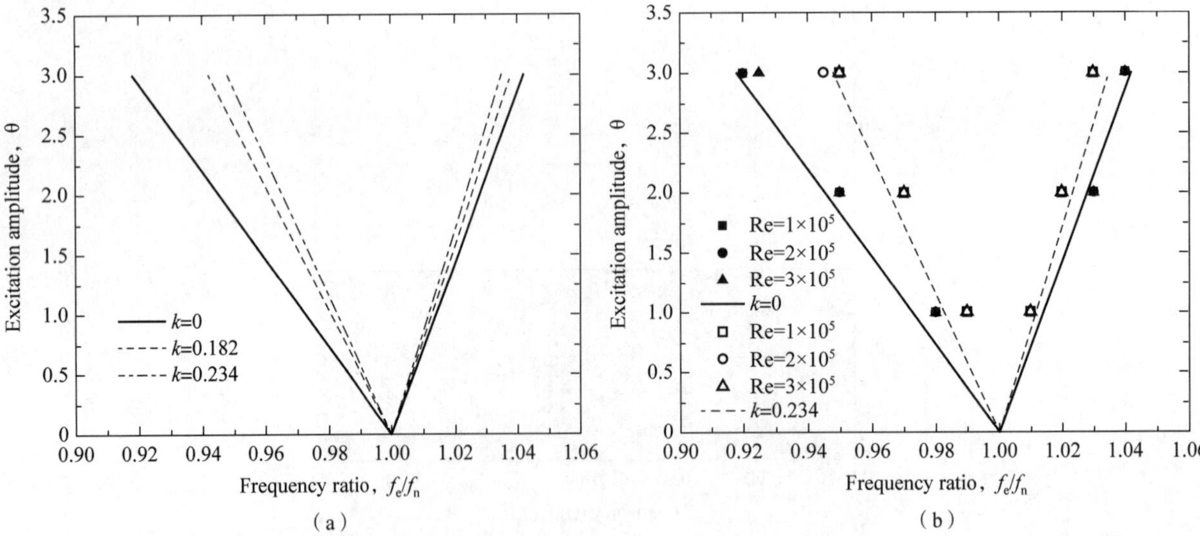

Figure 18 Lock – in region under different conditions

(a) Effect of blockage ratio; (b) Effect of Reynolds number

With an increase of blockage ratio, the vortices are forced to form in a narrower space and it corresponds to a higher Strouhal number. The previous researchers believed that the wall effect on Strouhal number can be corrected by a simple equation only related to the blockage ratio. In this paper, we find that Strouhal number depends on the angle of attack even when the blockage ratio is held constant. Taking into account both blockage ratio and angle of attack, a new correction formula for the Strouhal number in confined flow has been proposed. It has wider application compared with the existing formulas.

Wall effects on an oscillating airfoil (pitching motion) are also studied. Lock – in is determined by combination of the excitation frequency and amplitude when the buffeting frequency synchronizes to the excitation frequency. Computational lock – in region of airfoil in confined flow is narrower than the experimental lock – in region. The sample length of the flow time history is considered to be the main cause of difference between experimental and computational lock – in regions. Wall effects result in a decrease of lock – in region. Furthermore, lock – in region decreases with increase of blockage ratio. Both for unconfined flow and confined flow, Reynolds number in the range of 10^5 to 3×10^5 has slight impact on lock – in region.

Declaration of Competing Interest

None declared.

Reference

[1] NIU J, LEI J, LU T. Numerical research on the effect of variable droop leading edge on oscillating naca 0012 airfoil dynamic stall [J]. Aerosp. Sci. Technol, 2018, 72: 476 – 485. https://doi.org/10.1016/j.ast.2017.11.030.

[2] RAVEH D, DOWELL E. Frequency lock – in phenomenon for oscillating airfoils in buffeting flows [J]. J. Fluids Struct., 2011, 27 (1): 89 – 104. https://doi.org/10.1016/j.jfluidstructs.2010.10.001.

[3] GAO C, ZHANG W, YE Z. Reduction of transonic buffet onset for a wing with activated elasticity, Aerosp. Sci. Technol., 2018, 77: 670 – 676. https://doi.org/10.1016/j.ast.2018.03.047.

[4] LIU J, LUO K, SUN H, et al. Dynamic response of vortex breakdown flows to a pitching double – delta wing [J]. Aerosp. Sci. Technol., 2018, 72: 564 – 577. https://doi.org/10.1016/j.ast.2017.10.008.

Experimental Study on the Flapping Dynamics for a Single Flexible Filament in a Vertical Soap Film①

单根柔性丝线在竖直皂膜水洞中的摆振动力学特性试验研究

Feng Shunshan, Chen Chaonan②, Huang Qi, Zhou Tong

School of Mechatronical Engineering, Beijing Institute of Technology, 5 South Zhongguancun Street, Haidian District, Beijing 100081, PR China

Abstract: A flexible filament with one end fixed in a flowing soap film may show sustained flapping. For a certain filament, this flapping occurs only when the flow velocity exceeds a threshold value, which is called critical velocity. A previous study with silk filament demonstrated the following three stable dynamical states: stretched – straight, flapping, and bistable. In the present study, experiments involving filaments made of various materials (spandex, polyester, silk), with different diameters and lengths were tested in the vertical soap film. The effects of some representative parameters including length, diameter, and bending stiffness of the filament, inflow velocity on the amplitude, and frequency of the filament flapping were examined to reveal the flapping dynamics of the filament. Additionally, the critical velocity was specifically determined. The results in two – dimensional fluids showed a discrepancy with previous 3D studies. This discrepancy was mainly because filaments were not entirely immersed in the soap film. Modified models were raised based on previous 3D theories and concurred with our experimental results as well as previous works. The underlying mechanism of flag – in – wind problems was able to be revealed by the two – dimensional experiments and modified models in this study.

Keywords: flexible filament; Soap film; Flapping dynamics; Two – dimensional fluids; Critical velocity

1 Introduction

Interactions between flexible plate – like bodies and their surroundings are extremely common in daily life. Examples include the fluttering of a flag in the wind (Zhang et al., 2000), passive oscillating of fishes, flapping of paper in the printing industry (Watanabe et al., 2002a), and the falling of leaves. The flag – in – wind problem has attracted increasing research interest because it involves difficult but significant fluid – structure interaction (FSI) dynamics (Zhu and Peskin, 2003).

The first experimental study on the flag – in – wind problem was conducted by Taneda (1968) in the mid – 1960s. In his experiment, several fabrics (silk, muslin, flannel, and canvas) and shapes (rectangles and triangles) were used for a flag and put in a wind tunnel with a splitter plate placed in the wake of a flag. The wave mode, wave velocity, frequency, and drag were determined for flags with Reynolds numbers ranging from 103 to 3 × 105. A stable state was reported for the flag when the flow velocity is slow while an increase in the velocity led to regular oscillations. The results also indicated that the flag exhibited irregular flapping when the speed was continuously increased. The results were important in terms of future studies. However, the critical bifurcation

① 原文发表于《Journal of Fluids and Structures》2019 (86)。

② 陈超南：工学博士，2015 年师从冯顺山教授，研究母弹爆炸抛撒子弹药气动弹道特性，现工作单位：比亚迪股份有限公司（深圳）。

was neglected. Additionally, a few findings including the fact that the onset of flapping was not significantly affected by the flag mass are not in accordance with our current understanding. Subsequently, Huang (1995) used an aluminum foil with a thickness of 0. 05 mm as a test plate in an axial flow. The aim of the experiments was to confirm the improved theory. Only the length of the plate was varied in the experiment to examine its effect on the flutter boundary. This was followed by several experiments that focused on the dynamics of flexible plates as flag - in - wind problems. Watanabe et al. (2002a) conducted experiments on paper flutter in a wind tunnel, and they used sheet paper and web paper with different materials, sizes, and tension to examine their effects on paper flutter. The results indicated that the primary factors that influence sheet flutter included bending stiffness and mass ratio while web flutter was additionally influenced by a tension parameter. The critical bifurcation for flutter was obtained in the form of dimensionless inflow speed and dimensionless mass ratio, and their results anticipated many of further studies. Subsequently (Eloy et al., 2011) considered the effect of plate aspect ratio on the instability threshold and argued that a hysteresis loop is potentially caused by undesired two - dimensional plate deflections and that it is difficult to avoid 3D effects for plates at a high aspect ratio. (Chen et al. 2014) explored the transition route from a static steady to a chaotic state by fixing dimensionless mass ratio at unity and increasing the dimensionless velocity. (Malher et al. 2017) investigated the influence of a hysteretic damper on airfoil flutter instability. They performed a parametric study of aeroelastic as well as hysteresis model parameters to form a complete picture of the bifurcation scenario and highlight the capacity of the hysteretic damper in precluding the occurrence of stall.

Shelley et al. (2005) first examined the 3D flag - in - wind problem in a water tunnel, and this is relevant to obtain an understanding of the dynamics of towed bodies and effects of body inertia. Their soft plate was composed of Mylar sheet glued by long copper strips, and the findings indicated that body inertia played an important role in overcoming the stabilizing effects of finite rigidity and fluid drag.

As a result of the symmetry, the problem was simplified as a two - dimensional problem while Zhang et al. (2000) first investigated the flapping dynamics of a silk filament placed in a flowing soap film as a quasi - two - dimensional water tunnel. Their results indicated that three stable dynamical states, namely stretched - straight, flapping, and bistable, exist in the filament that was controlled by the length of the filament. When the length of filament exceeded a specific critical length L1, the filament continued to flap in the flow while the filament did not flap if the length was shorter than another specific critical length L2. When the filament length ranged between L1 and L2, a bistable state existed in which the filament can either flap or not based on its initial condition and external perturbations. Abderrahmane and Paidoussis (2011) found that the flapping of filament is quasi - periodic while the oscillation is chaotic at higher flow velocity conditions. They also reported a new phenomenon wherein the filament behavior continuously switches between the stretched - straight and flapping states without any external perturbation, and this is associated with a subcritical Hopf bifurcation. Bandi et al. (2013) used a rigid ring attached to an elastic filament to investigate its dynamics in a flowing soap film. The findings indicated two stable states, namely stretched - straight state and spontaneous oscillation state, and a critical bifurcation existed between the same. A few parameters (flow speed, length of the supporting string, ring mass, and ring radius) were examined with respect to their effects on the bifurcations. Orchini et al. (2015) observed that a pendulum suspended in a fast - flowing soap film exhibited sustained oscillations when the flow velocity exceeds a threshold value along with geometrical constraints, and the instability was suppressed by attaching a sufficiently long filament to the rear of the pendulum. However, the bifurcation condition was not discussed in their study. Another significant research about flow - structure interactions in soap films was conducted by Lacis et al. (2014), who investigated dynamics of a rigid plate attached to a cylinder in soap films. Their results were expected to explain the mechanism to use the existing energy in the flow for organisms' locomotion.

Theoretical studies included a solution proposed by Theodorsen (1935), who investigated the stability of airfoils under the self – oscillation and internal mechanism. He established fundamental theories for the fluttering of airfoils based on potential flow and the Kutta condition. Huang (1995) employed Theodorsen's classical solution for the fluid loading as a problem related to oronasal snoring, and the stability of the plate is investigated through an initial value problem. Huang's theory proposed two dimensionless parameters namely mass ratio of the thickness of a layer to the thickness of the plate and the velocity ratio of the flow to a reference value that controls the critical condition for instability. Experiments were executed and the results indicated the data was in close agreement with the predictions. Watanabe et al. (2002b) focused on the phenomenon of paper flutter. Based on their experiment, they developed two different methods of analysis. The first method is a flutter simulation by using a Navier – Stokes code to determine the unsteady lift force, amplitude of flutter, and airflow around a paper sheet. The second method analysis was based on a potential – flow analysis of an oscillating thin airfoil via an eigenvalue analysis to determine stability. They proposed two dimensionless parameters, namely dimensionless speed and dimensionless mass ratio. The results indicated that the flutter mode of paper successively changed from the fourth order to the third order to the second order with an increase in the dimensionless mass ratio. Moreover, they calculated a critical curve for the flutter of paper. Shelley et al. (2005) solved the problem by using an inviscid incompressible fluid. The results also focused on the wave number and were more applicable for the instability of a flag. However, Argentina and Mahadevan (2005) improved a model based on Theodorsen's fundamental theory by considering various physical mechanisms including finite length, small albeit finite bending stiffness of the flag, unsteadiness of the flow, added mass effect, and vortex shedding from the trailing edge. The critical speed for the onset of flapping as well as the frequency of flapping was predicted through the theory while the underlying mechanisms for nonlinear bistability were not explained by the theory. Alben and Shelley (2008) studied this flag – in – wind problem through a model for an inextensible flexible sheet in an inviscid 2D flow with a free vortex sheet, and they used two control parameters, namely the dimensionless mass of the flag and dimensionless rigidity. They numerically solved the fully – nonlinear dynamics and observed a transition from a power spectrum dominated by discrete frequencies to an apparently continuous spectrum of frequencies.

The previous 3D experimental results exhibited good agreement with the theories while the two – dimensional experimental results did not (Shelley et al., 2005). The filaments appeared much more stable than the predictions of two – dimensional models. This discrepancy resulted from the filament's gravity that stabilizes it against flapping (Eloy et al., 2011; Shelley et al., 2005). However, only a few experimental studies focused on the flapping dynamics of the filament in the soap film as well as the critical velocity for different kinds of filaments for a 2D flag – in – wind problem. This is potentially due to the instability of the soap film and the narrow range of inflow velocity that requires an appropriate bending stiffness of the filament to obtain sufficient data for instability analysis. In our experiments, glycerin is added into the solution to reduce water evaporation, and the filaments with low elastic modulus (ranges from about 15 MPa to 5 000 MPa) are selected as they are easy to flap.

In this study, the flapping dynamics of a single flexible in a soap film were examined. The experimental results were compared with the previous experiments and theories. The filaments seemed more stable than the predictions of theoretical models, unlike previous common opinion that this discrepancy resulted from the gravity that stabilizes the filament against flapping. We found another reason (perhaps the principal reason) for two – dimensional cases that the phenomenon was because the diameter of the filament was much bigger than the thickness of the soap film. Thus, the destabilizing pressure forces were expected to be modified when applying 3D theories in the two – dimensional flag – inwind problem. Additionally, a correction factor was proposed and the experimental results showed good agreement with modified models. The three – dimensional flag – in – wind problem could be examined by the two – dimensional experiments and modified models in the study.

2 Experimental setup and methods

The onset of the experiment is presented in Figure 1 It involves a gravity driven device that consists an overflow cup, supporting frame composed of nylon threads (with a diameter of 1 mm), and a reservoir. The device is not novel and was used in several previous studies (Zhang et al., 2000; Gharib and Derango, 1989; Belzale and Gharlb, 1997; Kellay et al., 1995). Typically, the film thickness is in the range of several microns, and this is one ten-thousands of the film width (Rutgers et al., 2001). Therefore, the film is considered as a 2D model in the experiment. The flowing area is divided into the following three parts: contraction, parallel and expansion. The flow is assumed to reach its stable velocity in the parallel part such that the filament in which an end fixed through a thin syringe is placed in the part. Particle Image Velocimetry (PIV) method was applied to measure the velocity of flows. We used a high-speed camera at a rate of 2000 images per second to record the movement of the filament. A sodium lamp was chosen as the only light source since its wavelength is similar to the thickness of film such that the optical light reflecting on upper and lower surfaces forms interference fringes that allow the visualization of the flow field.

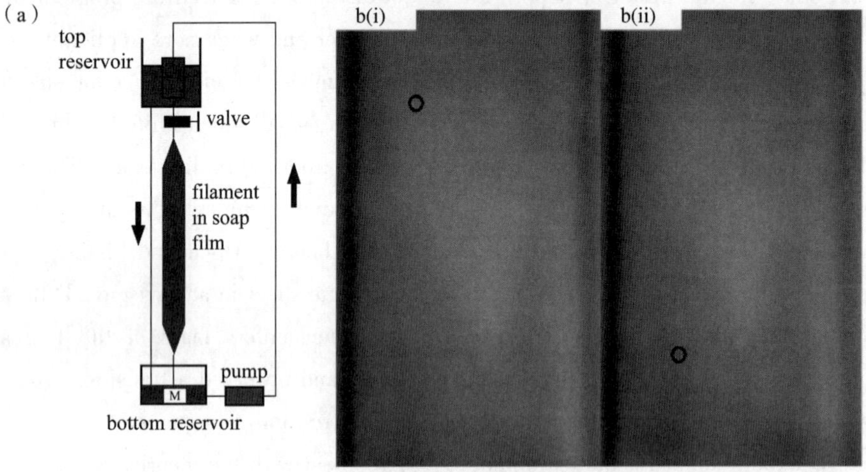

Figure 1 Experimental details

(a) Schematic diagram of experimental setup; (b) Inflow speed measurement method

Table 1 Important properties of the filaments used in the experiments.

Serial number	Industrial specifications	Material	Diameter (mm)	Linear density ($\times 10^{-4}$ g/cm)	Bending stiffness ($\times 10^{-12}$ N·m^2)
F1	20D	Spandex	0.046 1	0.222 5	3.07
F2	40D	Spandex	0.0701	0.432 9	16.99
F3	70D	Spandex	0.096 2	0.825 6	52.79
F4	140D	Spandex	0.127 2	1.598 2	201.68
P1	25D	Polyester	0.051 4	0.259 3	1.05×10^3
P2	35D	Polyester	0.061 6	0.370 4	2.16×10^3
S1	/	Silk	0.104 5	0.986 3	243.78
Zhang et al. (2000)	/	Silk	0.15	2.0	100

The filaments we used were made of spandex, polyester and silk with several different diameters. The diameter of the filament was measured by using a scanning electron microscope as shown typically for spandex filaments in Figure 2. As shown in Figure 3, a cantilever method was applied to obtain the bending stiffness of the filament that was estimated as follows (Dadeppo and Schmidt, 1971):

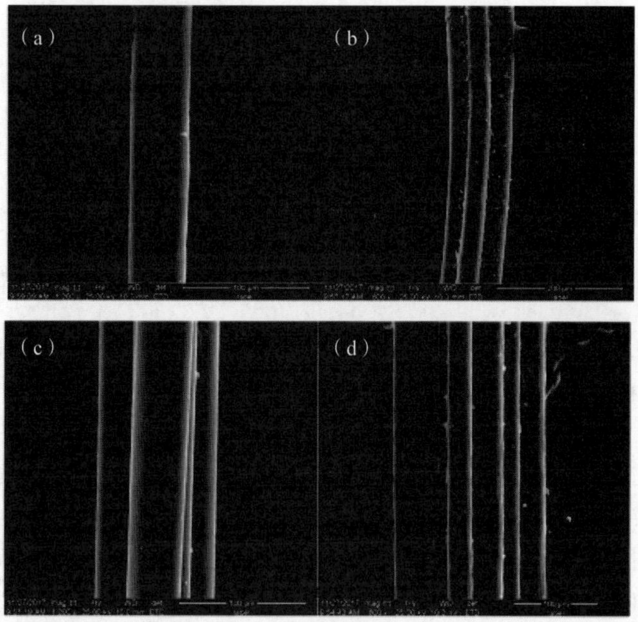

Figure 2 Typical scanning electron microscopy (SEM) movie frames for an individual filament, where (a), (b), (c), and (d) are filaments F1, F2, F3, and F4, respectively.

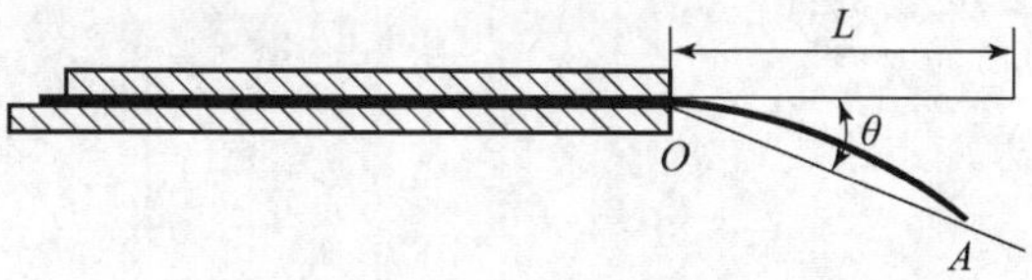

Figure 3 Principle of the cantilever method.

$$B = m_L g L^3 \cos\left(\frac{\theta}{2}\right) / 8 \tan\theta \tag{1}$$

where m_L denotes the linear density of the filament, g denotes the acceleration due to gravity, and θ denotes the inclination to the horizontal of the straight line OA.

A few important properties of the filaments are listed in Table 1. Zhang's experiment (Zhang et al., 2000) is listed at the same time as a comparison.

The thickness of the soap film in parallel part hp was obtained as follows:

$$h_p = Q/(L_p \times U) \tag{2}$$

where Q denotes the quantity of flow per second, L_p denotes the width of film in parallel part, and U denotes inflow velocity. Figure 4 shows the results of film thickness h_p as functions of inflow velocity.

As shown in Figure 4, the film thickness in parallel part ranges from 2 μm to 4 μm, and is typically far less than the diameter of the filaments.

3 Results and discussion

Three stable states were successfully reproduced in the experiment as indicated by Zhang et al. (2000). Figure 5 shows the typical stretch - straight and flapping states. Figure 6 shows the frequency relative to the flow velocity at a certain length and diameter. Flapping occurred suddenly at a specific velocity (U_{c2}) when the flow

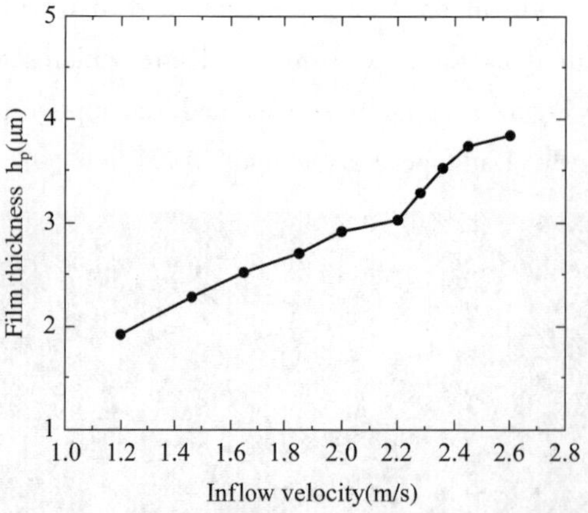

Figure 4 Film thickness in parallel part in relative to the inflow velocity.

velocity gradually increased. However, it did not cease when the velocity was subsequently lowered past the onset velocity until a lower velocity (U_{c1}, $U_{c1} < U_{c2}$). This hysteretic behavior is observed in almost all the four cases of spandex for every length. The range of the hysteresis loop increases with an increase in the bending stiffness (Figure 6 (a)) and increase in the filament length (Figure 6 (b)). This is qualitatively in agreement with the results of Watanabe et al. (2002a), wherein they observed the presence of a hysteretic behavior for sheet flutter and the absence of the same for web flutter, and no hysteretic corresponded to a weak effect in any manner. In our study, we consider U_{c1} as the real critical velocity in the subsequent analysis for hysteretic behavior that can be regarded as bistable phenomenon.

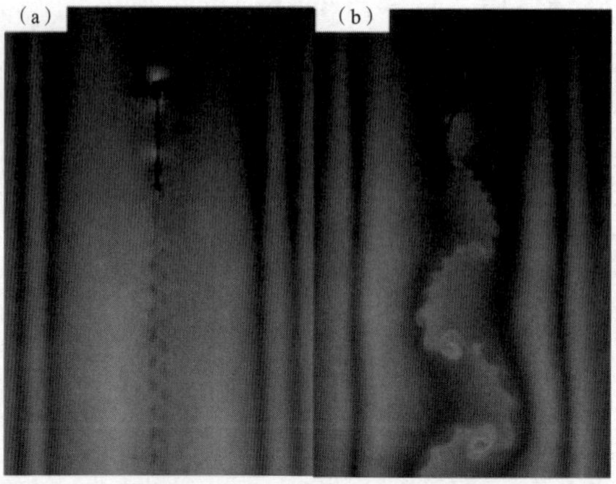

Figure 5 Experimental states

(a) Typical stretch - straight state. Filament *F*4, Length $L = 2.0$ cm, inflow velocity $U = 1.6$ m/s; (b) Typical flapping state.

Figure 7 shows the experimental measurements of the flapping frequency and amplitude as functions of inflow velocity. As Zhang's experiments showed, with respect to a specific length of velocity, the amplitude exhibited quasi - periodic behavior while the frequency increased with increases in the inflow velocity. The Strouhal number for filament was defined as $St = fA/U$ and was examined in the experiment, where f and A denote frequency and amplitude of the filament respectively. Figure 8 shows the experimental results for the changes in the St number with the inflow velocity. As shown in Figure 8, flapping frequency increased with increases in the velocity, while the Strouhal number remained consistent at approximately 0.2 - 0.3. A previous study (Taylor et al., 2003) about flying and swimming animals revealed that the Strouhal number in the cruise was maintained at approximately 0.2 - 0.4 for high power efficiency, and this is similar to the underlying mechanism between passive flapping of a filament and active flapping of an organism.

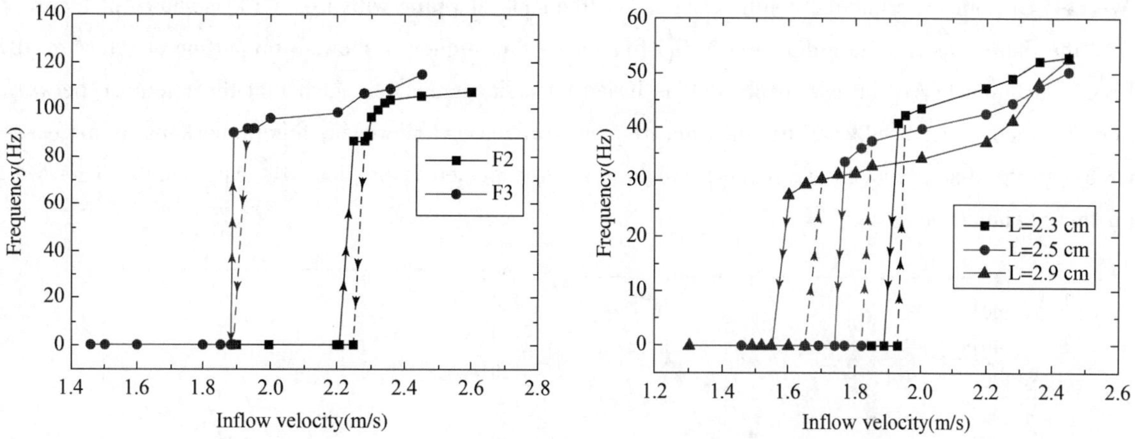

(a)　　　　　　　　　　(b)

Figure 6　Flapping frequency as function of inflow velocity

(a) Condition of a specific filament length $L=1.1$ cm; (b) Condition of a specific filament diameter $d=0.134\ 1$ mm (F4).

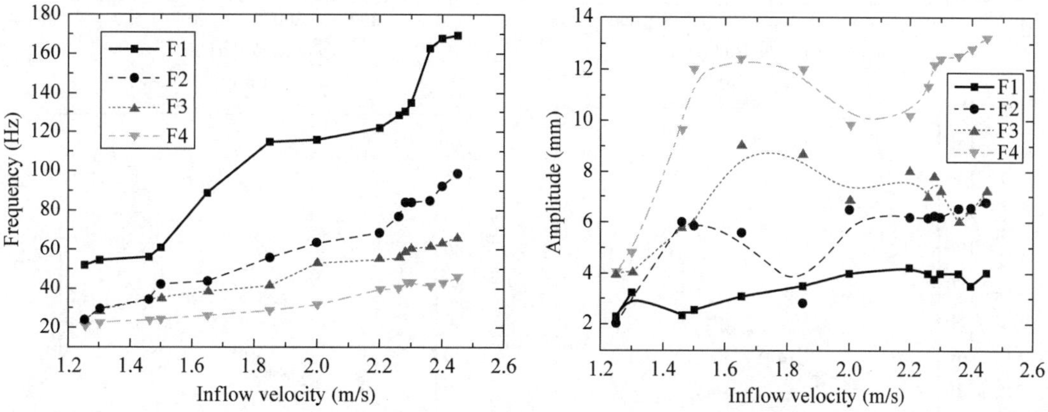

(a)　　　　　　　　　　(b)

Figure 7　The flapping frequency and amplitude of the filament as functions of inflow velocity. Filament length $L=3.5$ cm.

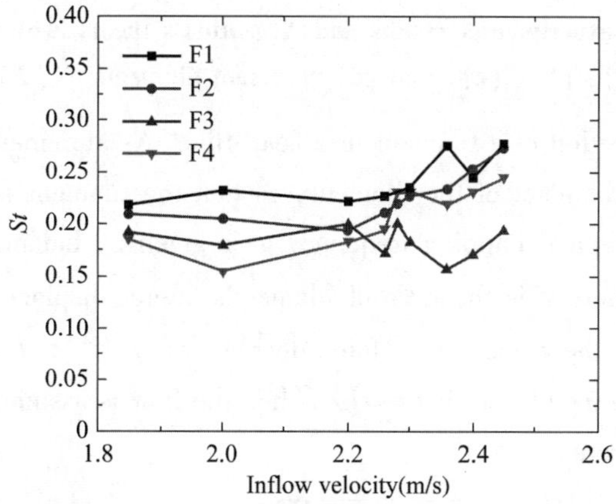

Figure 8　The Strouhal number of the filament as function as inflow velocity. Filament length $L=4$ cm.

In Argentina's theory, the frequency ω yields the following expression:

$$\omega \sim \left(\frac{\rho_f U^2}{\rho_s dL}\right)^{1/2} \tag{3}$$

where ρ_f denotes the film density, U denotes the inflow velocity, and ρ_s, d, L denote the density, diameter, and length of the filament respectively.

We compare our experimental results of spandex filaments' flapping with Eq. (3) as shown in Figure 9. As shown in the figure, for a certain diameter of the filaments, the frequency shows a proportion of (ρ_f U^2/ρsdL) 1/2. This is attributed to Argentina's theory that is based on a linear analysis such that the tolerance between real situation and theory increases when the flapping frequency increases. However, this proportionality increases with an increase in the diameter of the filaments, which seems unexpected in previous 3D experiments. Then we try to explain this phenomenon.

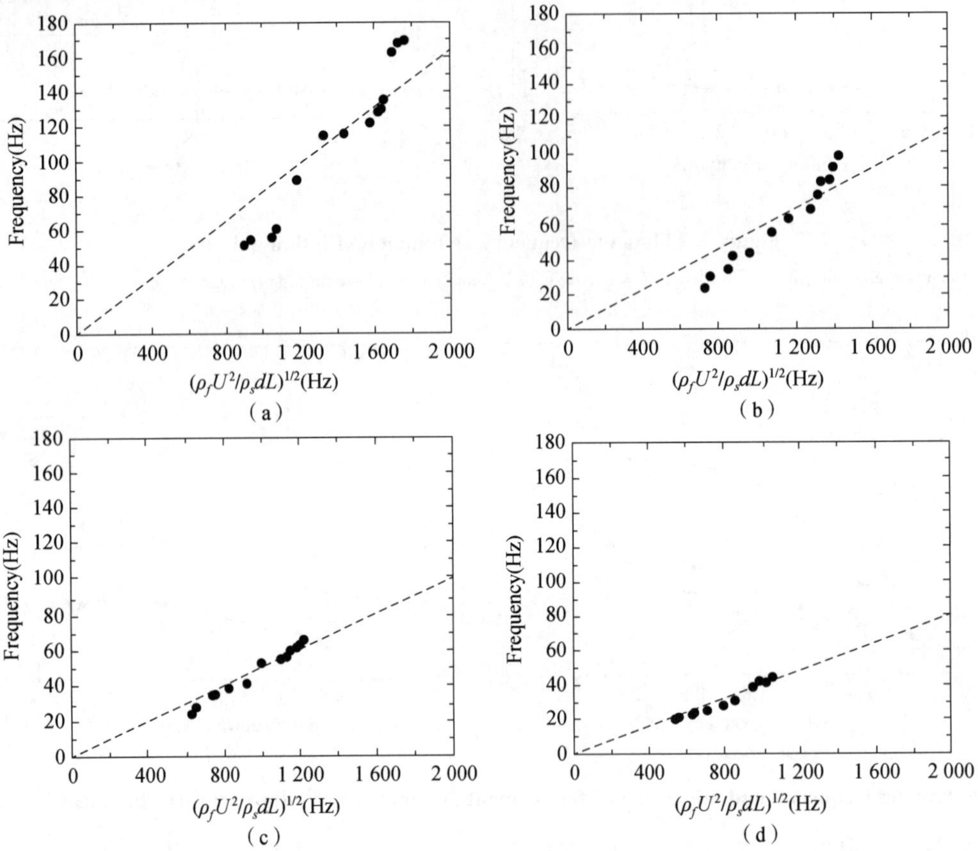

Figure 9 Comparison of the experimental results and Argentina's theory with respect to the frequency. Filament length L = 3. 3 cm. Additionally, (a), (b), (c), and (d) represent filaments $F1$, $F2$, $F3$ and $F4$, respectively.

Figure 10 shows the cross section of a filament in a soap film. As aforementioned, the film thickness of the parallel part is far less than the diameter of the filament, so that the filament is not completely immersed in the film. In Argentina's theory, the typical flapping frequency ω is given by balancing plate inertia $\rho_s d\omega^2\tau$ with the aerodynamic force$\rho_f U^2(\tau/L)$, where τ is the ratio of filament's lateral displacement Y to the filament length L. Here in two - dimensional fluids, the aerodynamic force decreases to $\rho_f U^2(\tau/L)\cdot\delta$. A schematic description of aerodynamic force is shown in Figure 11, so that $\delta = h/d$when the flow is assumed to be impressible and inviscid. Then the flapping frequency scales as

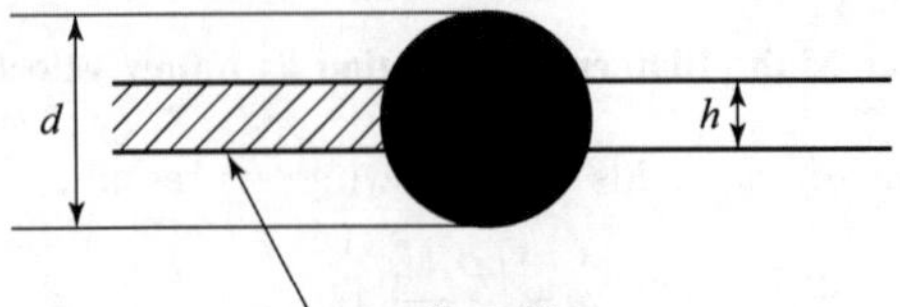

Figure 10 Cross section of a filament in a soap film.

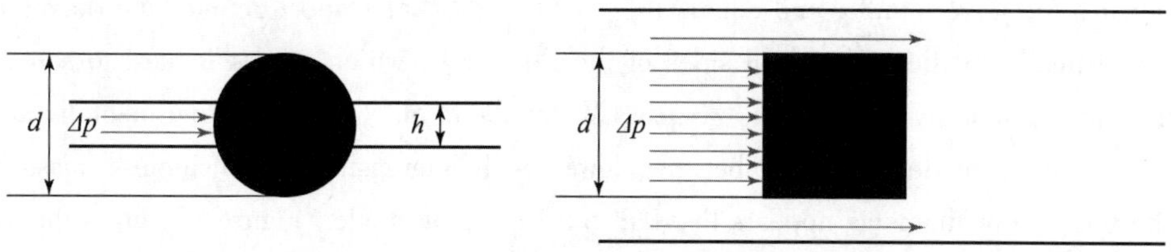

Figure 11　Force diagram of filament in 2D fluids and flag in 3D fluids.

$$\omega \sim \left(\frac{\rho_f U^2 h_p}{\rho_s d^2 L}\right)^{1/2} \tag{4}$$

Figure 12 shows the experimental results as well as previous studies. As shown in the figure, the experimental results agree well with the modified models. However, the filament appears flapping more swiftly than the predictions especially for the small diameter of the filament. The main reason is that the Reynolds number is approximately 200 for thin filaments, where the assumption of inviscid flow seems unreasonable. The actual aerodynamic force is greater than predicted value, as is the flapping frequency. With respect to the critical bifurcation, the flag motion of small lateral deflection z (x, t) in 3D fluids is driven by the linearized Euler – Bernoulli equation as follows (Eloy et al., 2011):

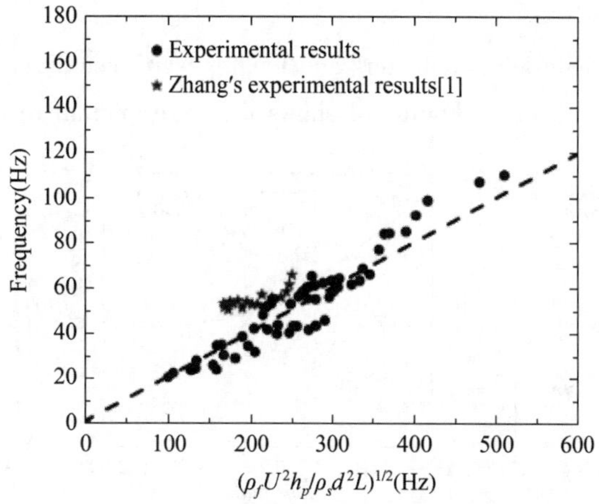

Figure 12　Experimental results of flapping frequency.

$$m_L \partial_t^2 z + B\partial_x^4 z = \Delta p(x,t) \cdot H \tag{5}$$

where m_L is linear density of the flag, B is the bending stiffness of the flag, $\Delta p(x, t)$ is the pressure difference between the front and back of the flag, and H is width of the flag. Assume $S = m_L/\rho_f dL$ and $U^* = U\sqrt{\rho_f dL^3/B}$ as dimensionless mass ratio and velocity respectively. Coupling this model to the fluid dynamics yields the dispersion relation linking ω and wave number k of perturbation of the form $z(x,t) = \eta_0(y)e^{i(\omega t + kx)}$.

$$\frac{1}{2}(-S\omega^2 + U^{*-2}k^4)k = (\omega + k)^2 \tag{6}$$

As previously mentioned, the aerodynamic force for filaments in soap film is expected to be multiplied by δ, so that the right side of Eq. (5) is rewritten as $\Delta p(x, t) \cdot H \cdot \sigma$. Afterwards the dispersion equation is coming as follows:

$$\frac{1}{2}(-S\omega^2 + U^{*-2}k^4)k = (\omega + k)^2 \cdot \sigma \tag{7}$$

where $dk = k^2(2 + kS) - 2SU^{*2}$. The system is neutrally stable if $dk \geqslant 0$ and unstable if $dk < 0$. The critical dimensionless velocity for flapping is as follows:

$$U_c^* = k\sqrt{\frac{2\sigma(U) + kS}{2\sigma(U)S}} \tag{8}$$

where k denotes the wave number and equals $(i-1/2)\ \pi$, $(i \geqslant 1)$ and i denotes the degree of the flapping mode. There are critical velocities U on both sides of the Eq. (8), which makes it hard to solve the equation. Here, we simplify the formula by assuming $h_p \equiv 3\ \mu\text{m}$. Watanabe et al. (2002a) experimentally and theoretically confirmed that the flutter mode of paper became lower with increasing dimensionless mass ratio. In our experiments, the flapping of filaments appears to be in third flapping mode. Figure 13 shows the vibration image of the filament processed by high - pass filtering.

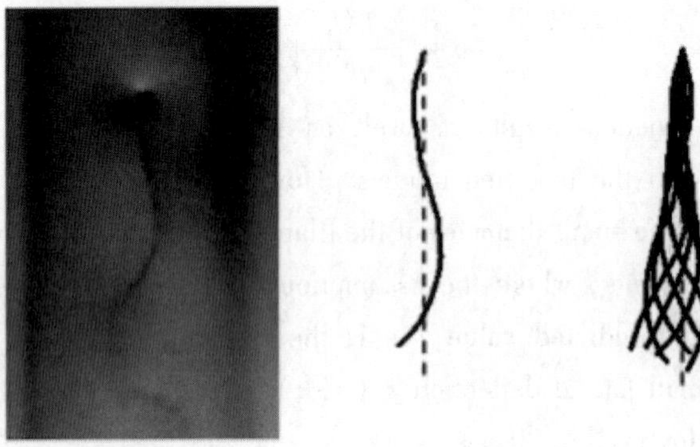

Figure 13 Vibration mode of filaments. Filament F3, length L = 2.8 cm, inflow velocity U = 2.2 m/s.

All filaments with different materials, diameters and lengths were tested via the experiment, while the inflow velocity ranges from 1.2 m/s to 2.6 m/s. Figure 14 shows the experimental results for all filaments. The results

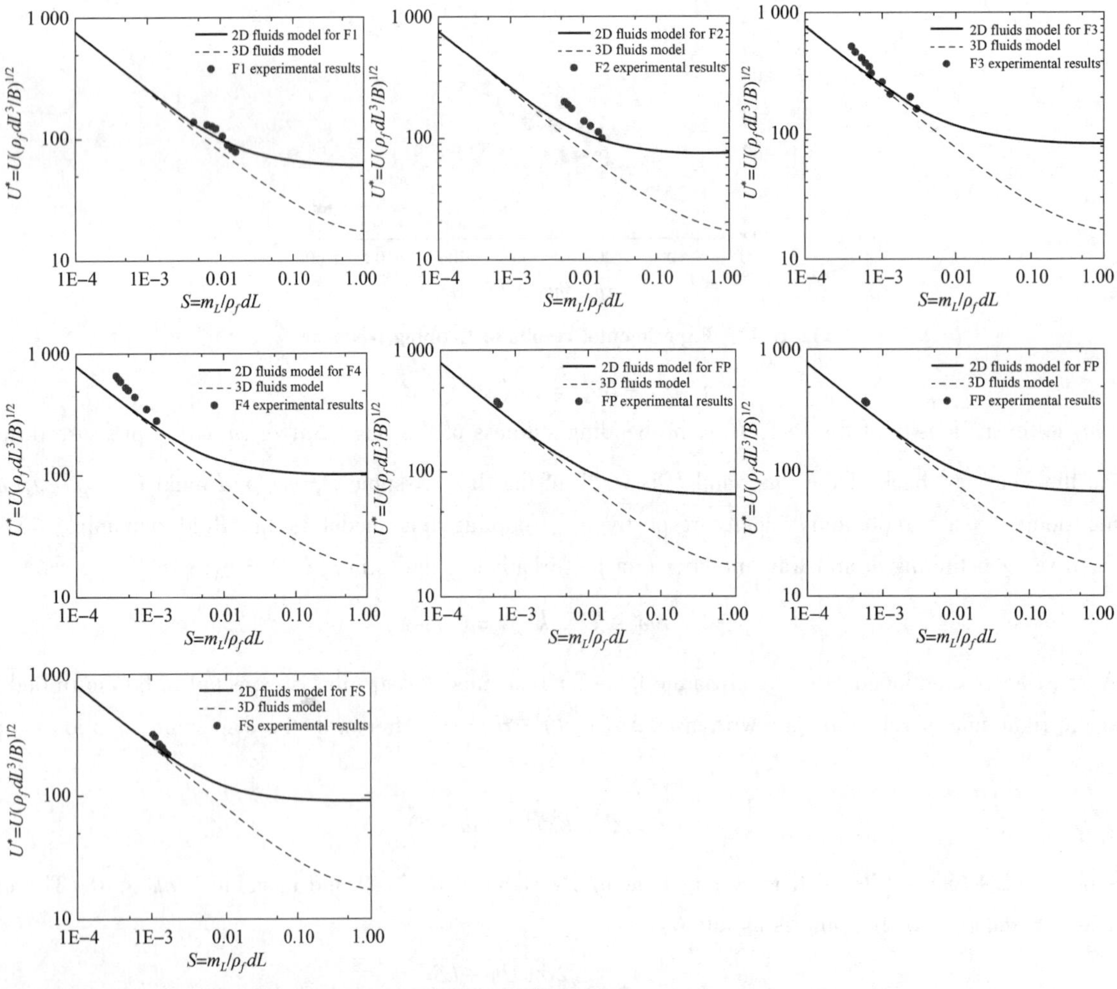

Figure 14 Critical curve of dimensionless velocity in relative to the dimensionless mass for all filaments.

showed good agreement with the modified model, although filaments still appeared to be more stable than the predictions. Figure 15 shows Zhang's experimental result in conjunction with predictions. Shelly predicted a critical length of ~ 0.9 cm while our modified model had a prediction of 1.4 cm. The experimental result was approximate 2.3 cm; our modified model reduced the discrepancy between experimental results and theory models while this discrepancy still existed (the gravity still played an important role in it). Additionally, Shelley did not consider the boundary conditions of the filament and especially vortex shedding from the trailing edge, which was a defect for predicting the critical velocity. This calls for more investigation of theoretical models and additional experiments.

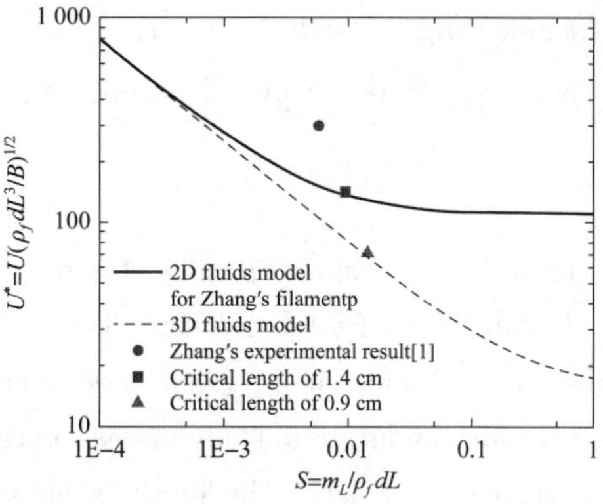

Figure 15　Comparison of Zhang's experimental result (Zhang et al., 2000) with prediction models.

4　Summary and Conclusion

In the present study, a filament with one end fixed was used in a flowing soap film. The results indicated a stronger hysteretic behavior in the critical bifurcation with higher bending stiffness and longer filaments. Subsequently, the dynamics of a single filament were examined in a flowing soap film with different filaments' lengths, diameters, bending stiffness, and inflow velocities to understand the underlying mechanism of the flapping of filaments. The results revealed that the amplitude exhibits a quasi – period while the frequency exhibits monotonicity with respected to the changes in the other parameters.

Moreover, the new phenomenon found in the paper was that the results of two – dimensional flag – in – wind problems showed a discrepancy compared to earlier work of 3D researches, especially for the flapping frequency and critical velocity above which the filament began to flap. This discrepancy was mainly because filaments were not entirely immersed in the fluids. Therefore, the aerodynamic force was expected to be lower than that in 3D conditions. With regard to flapping frequency, we amended Argentina et al.'s model and it exhibited acceptable agreement with our experimental results as well as Zhang et al.'s. As for critical velocity, a modified model based on Shelley et al.'s theory was proposed at the same time. It was found that the predicted critical velocities of the modified model were closer to the experimental results than those of the original model, while there still existed a difference because of the gravity that stabilized the filaments in the vertical soap film.

Acknowledgments

This work was supported by the National Natural Science Foundation of China (No 11702025).

References

[1] ABDERRAHMANE H A, PAIDOUSSIS M P, 2011. Flapping dynamics of a flexible filament. Phys. Rev. Lett., 84: 066604.

Study on the Application Scheme of Aerodynamic Coefficient Identification Based on the Differential Evolution Algorithm①

基于差分进化算法的气动参数辨识应用方案研究

Liu Pengfei[a ②], Cao Hongsong[a], Feng Shunshan[b], Liu Hengzhu[a], Du Ye[a]

a College of Mechatronics Engineering, North University of China, Taiyuan 030051, *China*;

b State Key Laboratory of Explosion Science and Technology, Beijing Institute of Technology, Beijing 100081, China;

Abstract: The present paper studies the application schemes of aerodynamic coefficient identification based on the differential evolution (DE) algorithm from free flight data. Two identification schemes utilizing the DE algorithm are proposed, including the whole discrete point (WDP) scheme and the differential interval linear (DIL) scheme. Several comparative tests are conducted to study the performances of the two schemes by using the noisy simulated flight data and the measured radar data. The results show satisfactory performance in estimating the aerodynamic coefficients for both two schemes. The WDP scheme provides more accurate and robust coefficients estimates in exchange of more than two times greater computational effort compared with the DIL scheme.

Keywords: aerodynamic coefficients; free flight data; identification; differential evolution algorithm; parameter estimation.

Nomenclature

Projectile Dynamic Model		Differential Evolution	
(I_A, I_C, I_C)	projectile inertia components ($kg\cdot m^2$)	$\boldsymbol{H}$	state vector of the ballistic model
(p, q, r)	projectile angular velocity vector components (*rad/s*)	ε	deviations of the measured state from the calculated state
(u, v, w)	projectile linear velocity vector components (*m/s*)	O	object function
(x, y, z)	projectile location vector components (*m*)	n	number of the measured points
(α, β)	projectile aerodynamic angles of attack (*rad*)	Ma	Mach number
$(\gamma, \varphi_2, \varphi_a)$	projectile Euler angles (*rad*)	$[Ma_1, Ma_2]$	Mach number interval
(X, Y, Z)	aerodynamic force vector components (N)	$DE/m/d/c$	*m*, vector to be perturbed *d*, number of difference vectors *c*, type of crossover

① 原文发表于《Mathematical Problems in Engineering》2020 (08)。

② 刘鹏飞：工学博士，2014 年师从冯顺山教授，研究弹道修正弹气动参数辨识与导引方法，现工作单位：中北大学。

(L, M, N)	aerodynamic moment vector components (N · m)	NP	number of individuals
ρ	atmospheric density (kg/m^3)	G	number of generations
d	projectile reference diameter (m)	k, k_{max}	parameters within the vectors
l	projectile reference length (m)	V	mutant vector
S	projectile reference area (m^2)	U	target vector
CG	projectilecenter of gravity location from the nose (m)	F	weight factor
m	projectile mass (kg)	CR	crossover rate
V_{CG}	magnitude of the mass center velocity (m/s)	Subscripts	
V_r	Projectile radial velocity (m/s)	E	earth frame
C_D	drag coefficient	D	aerodynamic drag force vector components
C_{D0}	zero - yaw drag coefficient	C, M	calculated and measured states of the projectile
C_{D2}	yaw drag coefficient ($1/ra\ d^2$)	L, U	lower and upper limits of the search space
$C_{l\alpha}$	lift force coefficient derivative with respect to the total yaw angle ($1/rad$)		
C_{ypa}	Magnus force aerodynamic coefficient ($1/ra\ d^2$)		
$C_{m\alpha}$	pitching moment aerodynamic coefficient ($1/rad$)		
C_{npa}	Magnus moment aerodynamic coefficient ($1/ra\ d^2$)		
C_{lDD}	roll moment aerodynamic coefficient due to fin cant ($1/rad$)		
C_{lp}	roll rate damping moment aerodynamic coefficient ($1/rad$)		
C_{mq}	pitch rate damping moment aerodynamic coefficient ($1/rad$)		

1 Introduction

The identification of aerodynamic coefficients, based on free flight data, is one of the important technologies in the field of projectile aerodynamics [1 - 5]. Compared with wind tunnel tests or numerical methods, aerodynamic coefficient identification can use actual flight data and full - scale projectiles. This could ensure higher fidelity and enable it to become a critical approach for determining the aerodynamic coefficients, especially in the verification phase of aerodynamic configuration design, ballistic flight control design and fire table generation [1, 2, 6-11].

In recent years, with the improvements in computer performance, modern intelligence optimization algorithms have been developed and applied to identify aerodynamic coefficients. Compared with the traditional gradient based identification method [12-21], intelligence optimization algorithms have extreme superiority in obtaining the global optimum and handling discontinuous nonconvex objectives [22]. Due to the great value in academics and engineering, a lot of work has been devoted to the research of intelligence optimization algorithms for aerodynamic coefficient identification. The genetic algorithm (GA) was used to estimate projectile aerodynamic coefficients from free flight range data [21]. The adaptive GA was used for nonlinear aerodynamic parameter identification of an axis-symmetrical tactical missile [23]. The partial swarm optimization (PSO) algorithm was applied to the aerodynamic coefficient identification of a spinning symmetric projectile [24]. The artificial bee colony (ABC) algorithm was applied to identify the high angle of attack aerodynamic parameters [25].

The differential evolution (DE) algorithm proposed by Storn and Price is a differential operator based evolutionary optimization technique [26-28]. This algorithm requires very few control parameters and adopts real-number-coding to add diversity to the difference distributions [27-29]. These unique features make it accurate, robust, easy to use, and converge fast compared with previous intelligence optimization algorithms. The DE algorithm has already been successfully used in the fields of aerodynamic design, mechanical design, and digital filter design [26, 30-32].

In the aerodynamic coefficient identification from the whole trajectory data, it is well known that a key process is to split the measurements into multiple small intervals, and then, the coefficients can be identified in the intervals. In the previous research [1, 2, 12, 13, 15-19, 21, 23-25], the coefficients in an interval are usually modeled as constants. In the work of Qi, the drag coefficients in each interval are modeled as a cubic spline [17]. The identification results are in good agreement with the measured data. However, due to the limited measurements of the small interval and the constant assumption of the coefficient, the unbalanced noise distribution would lead to overfitting of the data, and the coupling effect between the data intervals would be ignored. These drawbacks reduced the accuracy and robustness.

In thisarticle, two identification schemes based on the DE algorithm are proposed to improve the performance of identifying the aerodynamic coefficients from trajectory data, including the whole discrete point (WDP) scheme and the differential interval linear (DIL) scheme. Several comparative tests are conducted, and the characteristics of these two schemes with regards to the computational time, accuracy, and robustness are discussed.

2 Projectile Flight Dynamic Model

In this study, the system model of aerodynamic coefficient identification is a standard six-degree-of-freedom (6DOF) model typically used in flight dynamic modeling of projectiles. Figure 1 gives two standard reference frames used to develop this model. The 6DOF consist of the inertial position components of the projectile mass center (x, y, z) from an inertial frame and the standard aerospace sequence Euler angles (γ, ϕ_2, ϕ_a).

The kinematic and dynamic differential equations for describing the motion of the projectile in flight are given as follows [1, 33]:

$$\begin{bmatrix} \dot{x} \\ \dot{y} \\ \dot{z} \end{bmatrix} = \begin{bmatrix} \cos\varphi_2\cos\varphi_a & -\sin\varphi_a & -\sin\varphi_2\cos\varphi_a \\ \cos\varphi_2\sin\varphi_a & \cos\varphi_a & -\sin\varphi_2\sin\varphi_a \\ \sin\varphi_2 & 0 & \cos\varphi_2 \end{bmatrix} \begin{bmatrix} u \\ v \\ w \end{bmatrix} \tag{1}$$

$$\begin{bmatrix} \dot{\gamma} \\ \dot{\varphi}_2 \\ \dot{\varphi}_a \end{bmatrix} = \begin{bmatrix} 1 & 0 & -\tan\varphi_2 \\ 0 & -1 & 0 \\ 0 & 0 & 1/\cos\varphi_2 \end{bmatrix} \begin{bmatrix} p \\ q \\ r \end{bmatrix} \tag{2}$$

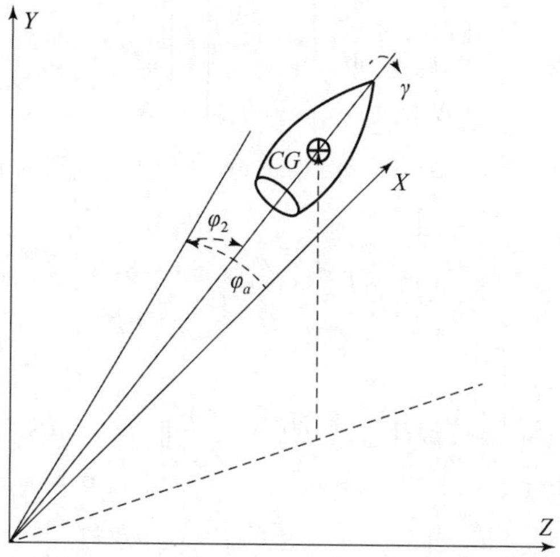

Figure 1　Schematic of the projectile's position coordinates.

$$\begin{bmatrix}\dot{u}\\ \dot{v}\\ \dot{w}\end{bmatrix}=\begin{bmatrix}X/m\\ Y/m\\ Z/m\end{bmatrix}+\begin{bmatrix}0 & -r & q\\ r & 0 & -r\tan\varphi_2\\ -q & r\tan\varphi_2 & 0\end{bmatrix}\begin{bmatrix}u\\ v\\ w\end{bmatrix} \tag{3}$$

$$\begin{bmatrix}\dot{p}\\ \dot{q}\\ \dot{r}\end{bmatrix}=\begin{bmatrix}I_A & 0 & 0\\ 0 & I_C & 0\\ 0 & 0 & I_C\end{bmatrix}^{-1}\left(\begin{bmatrix}L\\ M\\ N\end{bmatrix}+\begin{bmatrix}0 & -r & q\\ r & 0 & -r\tan\varphi_2\\ -q & r\tan\varphi_2 & 0\end{bmatrix}\begin{bmatrix}I_A p\\ I_C q\\ I_C r\end{bmatrix}\right) \tag{4}$$

The forces acting on the projectile in equation (3) consists of gravity (G) and the aerodynamic force. The aerodynamic force is split into a standard aerodynamic force (A) and Magnus aerodynamic force (M). Gliding flight is assumed in this study. The combination of forces is expressed in equation (5).

$$\begin{bmatrix}X\\ Y\\ Z\end{bmatrix}=\begin{bmatrix}X_G\\ Y_G\\ Z_G\end{bmatrix}+\begin{bmatrix}X_A\\ Y_A\\ Z_A\end{bmatrix}+\begin{bmatrix}X_M\\ Y_M\\ Z_M\end{bmatrix} \tag{5}$$

$$\begin{bmatrix}X_G\\ Y_G\\ Z_G\end{bmatrix}=-mg\begin{bmatrix}\cos\varphi_2\sin\varphi_a\\ \cos\varphi_a\\ -\sin\varphi_2\sin\varphi_a\end{bmatrix} \tag{6}$$

$$\begin{bmatrix}X_A\\ Y_A\\ Z_A\end{bmatrix}=\frac{1}{2}\rho V_{CG}^2 S\left(-C_D\begin{bmatrix}-\cos\alpha\cos\beta\\ \sin\alpha\\ \cos\alpha\cos\beta\end{bmatrix}+C_{l\alpha}\begin{bmatrix}\sin^2\beta+\cos^2\beta\sin^2\alpha\\ -\sin\beta\cos\alpha\cos\beta\\ -\cos^2\beta\sin\alpha\cos\alpha\end{bmatrix}\right) \tag{7}$$

$$\begin{bmatrix}X_M\\ Y_M\\ Z_M\end{bmatrix}=\frac{1}{2}\rho V_{CG}^2 S\, C_{yp\alpha}\left(\frac{pd}{V_{CG}}\right)\begin{bmatrix}0\\ -\sin\alpha\cos\beta\\ \sin\beta\end{bmatrix} \tag{8}$$

Inequation (7), the drag coefficient C_D is estimated by considering the zero - yaw drag coefficient C_{D0} and the yaw drag coefficient C_{D2} [2].

$$C_D=C_{D0}+C_{D2}=C_{D0}+(C_{l\alpha}+0.5C_{D0})(\alpha^2+\beta^2). \tag{9}$$

It is known that C_D has the most significant effect on the trajectory, and C_{D0} represents the main part of C_D[34].

The moment acting on the projectile in equation (4) consists of the moment due to the A, the moment due to the M and the unsteady aerodynamic moments (UA), as shown in equation (10).

$$\begin{bmatrix} L \\ M \\ N \end{bmatrix} = \begin{bmatrix} L_A \\ M_A \\ N_A \end{bmatrix} + \begin{bmatrix} L_M \\ M_M \\ N_N \end{bmatrix} + \begin{bmatrix} L_{UA} \\ M_{UA} \\ N_{UA} \end{bmatrix} \tag{10}$$

The moments inequation (10) are expressed in equations (11), (12), and (13).

$$\begin{bmatrix} L_A \\ M_A \\ N_A \end{bmatrix} = \frac{1}{2}\rho S l V_{CG}^2 C_{m\alpha} \begin{bmatrix} 0 \\ \sin\alpha\cos\beta \\ -\sin\beta \end{bmatrix} \tag{11}$$

$$\begin{bmatrix} L_M \\ M_M \\ N_N \end{bmatrix} = \frac{1}{2}\rho S l V_{CG}^2 (-C_{np\alpha}) \left(\frac{pd}{V_{CG}}\right) \begin{bmatrix} 0 \\ -\sin\beta \\ -\sin\alpha\cos\beta \end{bmatrix} \tag{12}$$

$$\begin{bmatrix} L_{UA} \\ M_{UA} \\ N_{UA} \end{bmatrix} = \frac{1}{2}\rho S l V_{CG}^2 \begin{bmatrix} C_{lDD} + C_{lp} pd/V_{CG} \\ C_{mq} qd/V_{CG} \\ C_{mq} rd/V_{CG} \end{bmatrix} \tag{13}$$

The aerodynamic angles of attack α and β are then defined by the following:

$$\tan\alpha = w/u, \tag{14}$$

$$\tan\beta = v/\sqrt{u^2 + w^2}. \tag{15}$$

The above equations are the major equations of the projectile flight dynamic model. More details regarding the projectile flight model can be found in [33]. All aerodynamic coefficients (C_{D0}, C_{D2}, $C_{l\alpha}$, $C_{yp\alpha}$, $C_{m\alpha}$, $C_{np\alpha}$, C_{lDD}, C_{lp} and C_{mq}) depend upon the local Mach number and are obtained during simulation by linear interpolation. To solve this model, the fourth - order Runge - Kutta algorithm is employed in this article. In addition, all calculations are performed with a real record of meteorological data.

3 DE Algorithm Based Aerodynamic Coefficient Identification

As mentioned above, the aerodynamic coefficient identification problem can be posed as an optimization problem; the basic task is to seek the set of coefficients (θ) that best fit the measured data, namely, to minimize the cost function $O(\theta)$ at different Mach numbers. In this article, the maximum likelihood criterion is chosen to formulate the object cost function, and thus, $O(\theta)$ is defined as:

$$O(\theta) = \sum_{i=1}^{n} [\varepsilon^T(i) R^{-1} \varepsilon(i) + \ln|R|] \tag{16}$$

where θ is the vector of unknown coefficients and $\varepsilon(i)$ is the deviation of the measured state vector from the calculated state vector, which is defined as:

$$\varepsilon(i) = \boldsymbol{H}_M - \boldsymbol{H}_C \tag{17}$$

where $\boldsymbol{H}_M$ and $\boldsymbol{H}_C$ are the measured and calculated state vectors, respectively, which are obtained from the 6 - DOF dynamic simulation with the identified coefficients in one step. R is the covariance matrix of measurement noise. As the statistical property of the measurement noise is unknown, the optimum estimation $\hat{R}$ usually replaces R:

$$R \approx \hat{R} = \frac{1}{n} \sum_{i=1}^{n} \varepsilon(i)\, \varepsilon^T(i) \tag{18}$$

The detailed implementation steps for the identification of aerodynamic coefficients using the DE algorithm can be described as follows.

Step1: Initialization operation: Generating NP ($i = 1 \cdots NP$) individuals randomly in the whole search space. The initial population is given by the following:

$$\theta_k^{i,G} = \theta_L^{i,G} + r \times (\theta_L^{i,G} - \theta_U^{i,G}) \tag{19}$$

where r is a uniformly distributed random number in the interval [0, 1] and $\theta_k^{i,G}$ is the kth component of the individual θ^i in the G generation. $\theta_L^{i,G}$ and $\theta_U^{i,G}$ are the lower and upper bounds of $\theta_k^{i,G}$, respectively.

Step2: Evaluate the fitness of each individual in the population according to equation (16).

Step3: Create a new population as follows.

a. Mutation operation: A mutant vector is generated according to the following:

$$V_k^{i,G} = \theta_k^{r1,G} + F(\theta_k^{r2,G} - \theta_k^{r3,G}) \tag{20}$$

b. Crossover operation: A new trail vector is generated by mixing the mutant vector and the target vector.

$$U_k^{i,G} = \begin{cases} V_k^{i,G}, & if\, r^k \leqslant CR^{i,G} or\ k = k_{rand} \\ \theta_k^{i,G}, & otherwise \end{cases} \tag{21}$$

wherer^k is a random number taken from the uniform distribution [0, 1].

c. Selection operation: Selection determines whether the target or the trial vector survives to the next iteration. For each mutant vector and its corresponding target vector in the current population, the object function values are compared using a greedy selection scheme. The selection operation is described as follows:

$$\theta_k^{i,G} = \begin{cases} U_k^{i,G+1}, & if\ O(U_k^{i,G+1}) < O(U_k^{i,G}) \\ \theta_k^{i,G}, & otherwise \end{cases} \tag{22}$$

Step4: Repeat steps 2 – 3 until the termination criteria are satisfied. In this article, the termination criteria are defined as follows:

$$G > G_{\max}. \tag{23}$$

The flowchart of the implemented DE algorithm is shown in Figure 2.

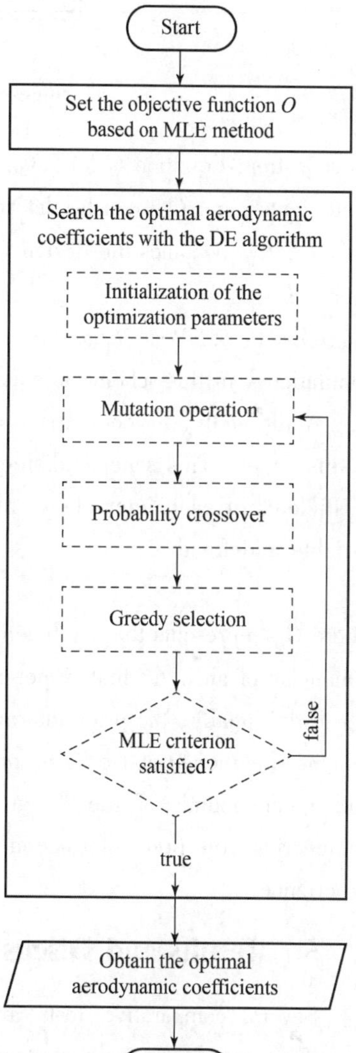

Figure 2　Flowchart of the implemented DE algorithm.

4　Identification Schemes

The first scheme, which is the whole discrete point (WDP) scheme, as show in Figure 3, treats the measured trajectory as a whole. The trajectory would not be divided into several intervals, and all of the coefficients corresponding to different Mach numbers in an aerodynamic coefficients table would be identified simultaneously. In this scheme, the design variable is defined as:

$$\theta = [C_{M_1}, C_{M_2}, \cdots, C_{M_j}, \cdots C_{M_{J-1}}, C_{M_J}]^T, \tag{24}$$

where C_{M_j} represents the coefficient corresponding to the $j-th$ Mach number in the coefficients table, $j = 1, 2, \cdots, J$, and J is the number of Mach numbers. The dimension of an individual in population k is equal to the number of Mach numbers J.

In the second scheme, which is the differential interval linear (DIL) scheme, as show in Figure 3, the whole trajectory would be divided into several intervals. The relationship between the coefficients and the Mach numbers in an interval is treated as a linear function, where the slope does not have to be zero. A constant factor K_{M_n} in each interval is introduced in this scheme, which is defined as:

$$K_{M_n} = \frac{C_{real}}{C_{prior}}, \tag{25}$$

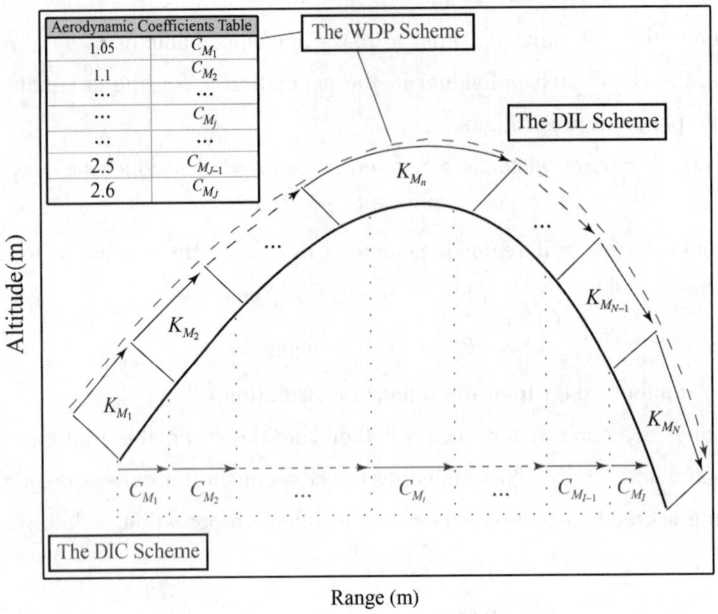

Figure 3 Schematic diagram of the identification schemes.

As seen from Equation (25), K_{M_n} represents the gap between the prior value and the real value of aerodynamic coefficients, and C_{real} can be determined by identifying K_{M_n} in each interval successively from the free flight data. Namely, K_{M_n} becomes the design variable in this optimization problem, which is defined as:

$$\theta = K_{M_n} \tag{26}$$

where $n = 1, 2, \cdots, N$, and N is the number of intervals. It is easy to find that the dimension of an individual in population k in this scheme is equal to the number of the correction factor in each interval.

Additionally, for comparison research, the conventional differential interval constant (DIC) scheme is used in this paper. This scheme divides the whole trajectory into several intervals as in the DIL scheme; yet, the coefficients in each interval are modeled as constants and identified successively. In this scheme, the design variable is defined as:

$$\theta = C_{M_i}, \tag{27}$$

where C_{M_i} represents the coefficient in the *ith* interval, $i = 1, 2, \cdots, I$, and I is the number of intervals. The dimension of an individual in population k in this scheme is equal to the number of coefficients in each interval.

Additionally, the prior information, namely, the general trend or characteristics, is required for all of the schemes mentioned above. This prior information could be used to divide the Mach number interval, to set lower and upper bounds of the design variable, to obtain the feasibility of identification, etc. In the practical engineering, the prior information could be obtained from wind tunnel tests, numerical methods, or engineering experience.

5 Results and Discussion

Several comparative tests are conducted in two different scenarios to study the performances of the identification schemes presented in this paper. A finned stabilized projectile is selected as the test bed, and the physical characteristics of this projectile are listed in Table 1. In order simplify the problem, the present study focuses on the identification of the drag coefficient, and methods presented in this paper could be easily used for other aerodynamic coefficients. Other aerodynamic coefficients are input as known values.

Table 1　The projectile's physical properties.

m (kg)	I_A ($kg-m^2$)	I_C ($kg-m^2$)	CG from nose (m)	d (m)
45. 340	0. 115	31. 627	1. 190	0. 122

In the first scenario, the simulated measurement data is generated from a 6 – DOF projectile dynamic simulation. To model the random noise, which inevitably exists in the trajectory data measured by radar, the Gaussian noise is added into the simulated position and velocity. Three standard deviations are used equal to 0. 2, 1. 0, and 5. 0 times the nominal standard deviation are considered, and the nominal standard deviation σ is 3 m and 0. 01 m/s for the position and velocity, respectively. The identification is conducted 50 times for each scheme, and the final identification results are the average values of the 50 identification results. The meteorological data and the initial launch conditions used in this scenario are chosen to be the same as the second scenario, which will be described in the next paragraph.

In the second scenario, phased array radar data from a free flight test is used as the measurement data to identify the aerodynamic drag coefficient. In this test, available measurements are the position data and radial velocity data, and the meteorological data were recorded with both ground and balloon – carried instruments throughout the test events. In Figure 4, the recorded data for the pressure, temperature, wind speed, and wind direction as a function of the altitude are shown. For the present projectile, the meteorological data below 12 km is mainly used. In Table 2, the initial launch conditions of the test are listed.

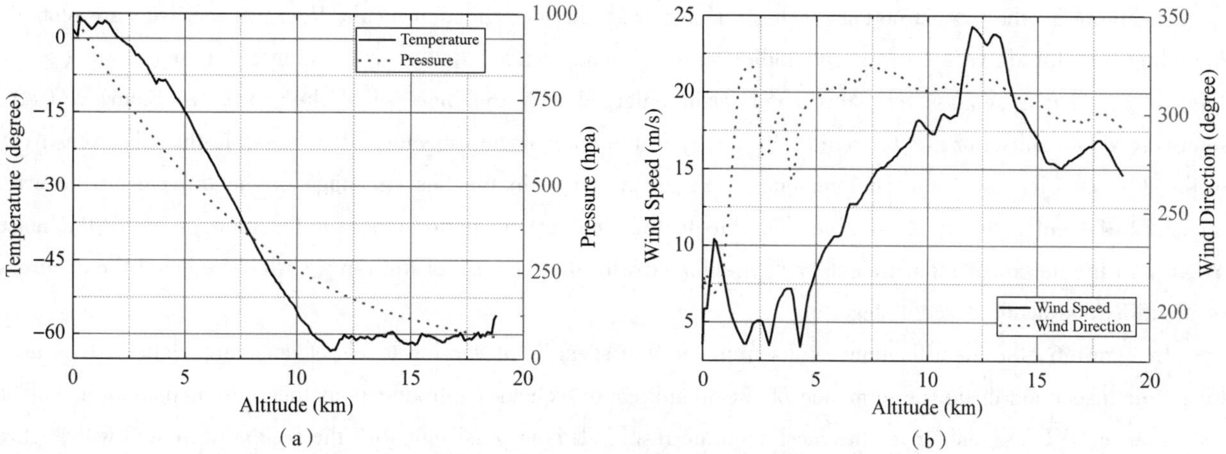

Figure 4　Meteorological data versus altitude

(a) temperature and pressure versus altitude; (b) wind speed and wind direction versus altitude

Table 2　Initial conditions for the two scenarios.

State	First Scenario/Second Scenario
Launching elevation angle (degrees)	42. 7
Launching azimuth angle (degrees)	358. 953
Velocity (m/s)	41. 317
Spin – rate (rad/s)	25. 133

Table 3 describes the initialization of the parameters implemented in the DE algorithm. The main control parameters that need to be set are the number of individuals (NP), the weight factor (F) and the crossover rate (CR). Every control parameter is tuned through appropriate tradeoff studies for the optimal process. Detailed information about the principle of tuning the control parameters can be found in [27, 28, 30, 35]. Additionally, to improve the efficiency of identification processing, the multiprocessing technique is used in the program.

Table 3 Parameter initialization for the three schemes.

Parameters	Identification Schemes		
	DIL	WDP	DIC
Number of individuals (*NP*)	80	160	80
Weight factor (*F*)		0.8	
Crossover rate (*CR*)		0.9	
Maximum number of generation (G_{max})		300	
Mutation strategy		DE/current - to - best/1/bin	
Selection strategy		best - parent - child	
Termination criteria		G_{max}	
Number of measurement data in interval	25	All	25
Number of parallel processes		15	

In Figure 5, the reconstructions of the trajectory with the identification results from two scenarios are plotted, including the altitude versus range and radial velocity versus range. In the first scenario (from (a) to (c) in Figure 5), as can be observed, the match error enlarged with the increase of the noise. In Figure 5 (c), relatively significant error can be found, especially in the apex of the trajectory. It is possible that high sensitivity of the drag coefficient in transonic region. In Figure 5 (d), in the descent stage of the trajectory, the error appears to be enlarged. It is possible that due to the limit of the performance of the radar system, the noise existing in the measured radar trajectory data enlarged with the increase of the range, which reduced the accuracy of identification at the descent stage.

In Figure 6, the identification results from the first scenario at varying levels of noise are plotted. It is used for closer inspection of the performance of the identification schemes introduced in the present paper. It can be seen that the WDP scheme gives the most accurate result. This is consistent with the results presented with Figure 5. As can be observed from Figure 5, the reconstruction of the trajectory of the WDP schemes matches the true data very well whether in the apex region or the descent stage of the two scenarios. This is because more measurements are considered in the WDP scheme, which enable the sampled noise to approach a normal distribution, and then leads to a more consistent nominal cost. This advantage can avoid the overfitting of data and increase the accuracy of the identification. Additionally, because the WDP scheme identifies all of the coefficients at different Mach numbers simultaneously, the error generated in the early stage of the trajectory would not be accumulated and spread to the later stage. In contrast, since the DIC scheme and the DIL scheme identified the coefficients along the trajectory interval successively, the coupling error would gradually accumulate and reduce the identification accuracy. Additionally, in each interval of the aerodynamic coefficients table, the coefficients are obtained by linear interpolation. It is actually that the relationship between the coefficients and Mach number is treated as a linear function, and the WDP scheme identifies both the slope and intercept. Namely, the WDP scheme can use the first - order information of the model of aerodynamic coefficients.

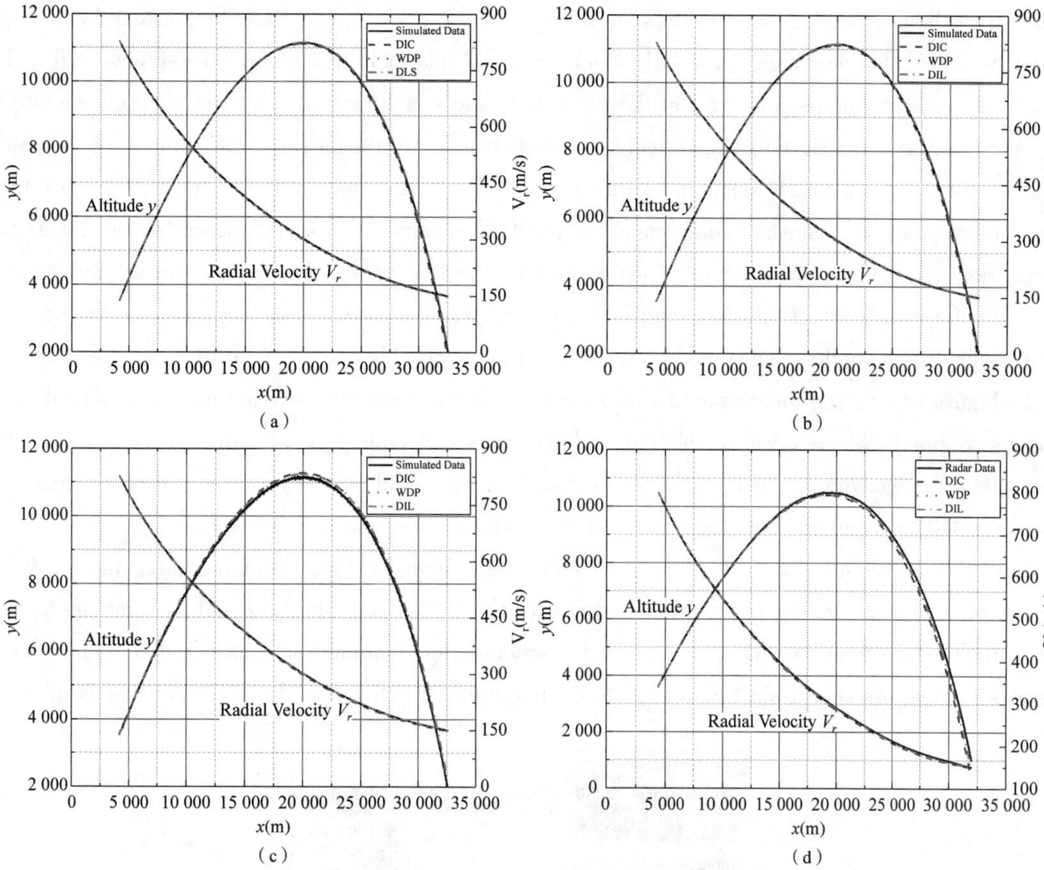

Figure 5　The trajectory matching for different schemes from the two scenarios

(a) $\sigma_{noise}=0.2\sigma$; (b) $\sigma_{noise}=1.0\sigma$; (c) $\sigma_{noise}=5.0\sigma$; (d) radar data

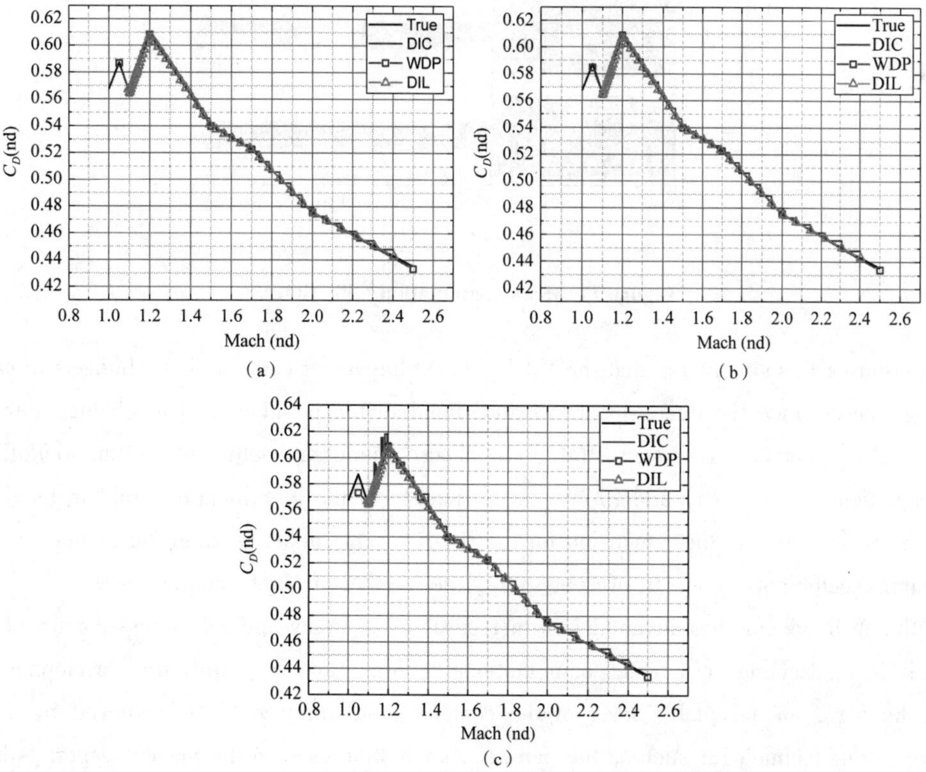

Figure 6　Identification results from simulated radar data for three schemes

(a) $\sigma_{noise}=0.2\sigma$; (b) $\sigma_{noise}=1.0\sigma$; (c) $\sigma_{noise}=5.0\sigma$

For the DIL scheme, it is intended to improve the identification accuracy. However, as can be observed from Figure 5 and Figure 6, the accuracy of the DIL scheme is much closer to that of the conventional DIC scheme. It is found that the sums of the absolute error of the range from the second scenario for the DIL scheme and the DIC scheme are 1. 5E +5 and 2. 15E +5, respectively, whereas it is 3. 1E +4 for the WDP scheme. It is possible that the prior information is not accurate enough. Although the DIL scheme models the coefficients in an interval with a linear function instead of a constant function as in the DIC scheme, the specific form of the linear function, namely, the prior information, may not be close enough to the real value. Moreover, the introduced correction factor would simultaneously affect the slope and intercept, which would lose some accuracy. In the future work, it is suggested to choose both the slope and intercept as design variables to be identified in the WDP scheme. Moreover, in Figure 6(c), significant fluctuation from the results of the DIC scheme can be observed, especially in the transonic region. This is because the DIC scheme uses constants to model the aerodynamic coefficients, which means that the converging order is zero - order. In the field of optimization, it also means that the robustness of the DIC scheme is worse than those of the WDP scheme and the DIL scheme.

In Figure 7, the computation CPU times using different schemes with simulated data and radar data are plotted. In all of the tests, the computation CPU times for the DIC scheme and the DIL scheme are very close to each other, which varied from 75—81 s. For the WDP scheme, the computation time is much longer than those of the other two schemes. As seen from Figure 7, the computation CPU time for the WDP scheme is up to 211 s.

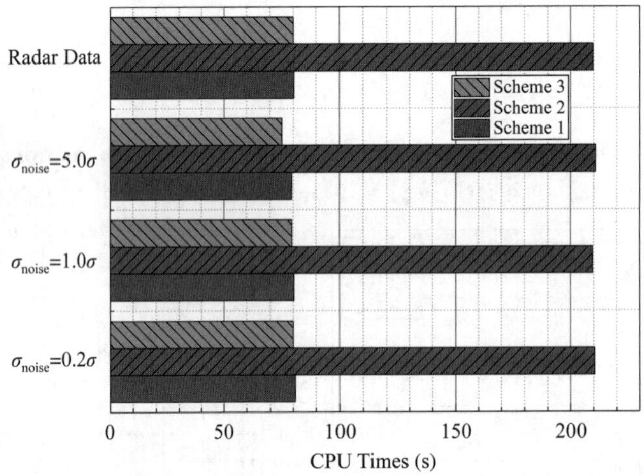

Figure 7 Mean computation CPU times.

A direct reason for this situation is that the WDP scheme has larger computational budgets in each iteration of the optimization process since the dimension of an individual in the population is much higher than in the DIC scheme and the DIL scheme. In present study, we choose the drag coefficient as the identified parameter. According to Equations (24), (26) and (27), the dimension of the identification problem for the DIC scheme and the DIL scheme is one, and the dimension for the WDP scheme depends upon the number of Mach numbers in the aerodynamic coefficients table. In present study, the number of Mach numbers is 9.

Although the WDP scheme has advantages with regards to accuracy and robustness compared with the other two schemes, it is in exchange of greater computational effort. However, with the development of computer science and technology, an acceptable level of the computational time could be achieved by utilizing high - performance computing technology, such as the parallel computation used in the present paper. Additionally, one can improved the efficiency of the identification by using multiple schemes. For example, in the different Mach number regions, one can apply the different schemes depending upon the degree of fluctuation of the aerodynamic coefficients. For the WDP scheme, one can divide the trajectory into two or three intervals to improve the efficiency while maintaining the accuracy in cases, where the number of Mach numbers in the aerodynamic coefficients table is much more than 9, as in the present paper.

6 Conclusions

Two identification schemes based on DE algorithm - based aerodynamic coefficient identification are proposed. Several comparative tests are conducted by identifying the aerodynamic coefficients from the noisy simulated flight data and radar data. The results show that both schemes presented in this paper can accurately estimate the aerodynamic coefficients. The WDP scheme and the DIL scheme improve the robustness of the identification based on the DE algorithm. Since the WDP scheme treats the trajectory as a whole, which effectively reduces the coupling error, the WDP scheme gives relatively better identification results compared with the DIL scheme and the DIC scheme. However, the computation times are more than two times greater compared with the other two schemes.

For application of the DE algorithm - based projectile aerodynamic coefficient identification in practical engineering, it is suggested that future work should focus on the following objectives. First, the fusion of multiple schemes or a new scheme might be developed to improve the overall identification performance with limited measurements. Second, the efficiency of the DE algorithm should be further improved, especially in dealing with multidimensional problems.

Data Availatility: The data used to support the finding of this study are available from the corresponding author upon request.

Conflicts of Interest: The authors declare no conflict of interest.

References

[1] MCCOY R L. Modern exterior ballistics: the launch and flight dynamics of symmetric projectiles, Rev. 2nd ed, Schiffer Pub, Atglen, PA, 2012.

[2] H. Zipeng. Exterior ballistics of projectile and rockets [M]. Beijing: Beijing Institute of Technology Press, 2008.

[3] ILIFF K W. Parameter estimation for flight vehicles [J]. Journal of guidance, control, and dynamics, 1989, 12: 609 - 622. https://doi.org/10.2514/3.20454.

[4] KLEIN V, MORELLI E A. Aircraft system identification: theory and practice, American Institute of Aeronautics and Astronautics, Reston, VA, 2006. https://doi.org/10.2514/4.861505.

[5] JATEGAONKAR R V. Flight Vehicle System Identification: A Time Domain Methodology, American Institute of Aeronautics and Astronautics, Reston, VA, 2006. https://doi.org/10.2514/4.866852.

[6] WERNERT P, THEODOULIS S. Modelling and stability analysis for a class of 155 mm spin - stabilized projectiles with course correction fuse (CCF) [C] //AIAA Atmospheric Flight Mechanics Conference, American Institute of Aeronautics and Astronautics, Portland, Oregon, 2011. https://doi.org/10.2514/6.2011 - 6269.

[7] OH S Y, KIM S C, LEE D K, et al., Magnus and spin - damping measurements of a spinning projectile using design of experiments [J]. Journal of spacecraft and rockets, 2010, 47: 974 - 980.

[8] J. Lijin, T. J. S. Jothi. Aerodynamic characteristics of an ogive - nose spinning projectile, Sādhanā. 43 (2018) 63. https://doi.org/10.1007/s12046 - 018 - 0857 - 3.

[9] COSTELLO M, SAHU J. Using computational fluid dynamic/rigid body dynamic results to generate aerodynamic models for projectile flight simulation, Proceedings of the Institution of Mechanical Engineers, Part G: Journal of Aerospace Engineering. 222 (2008) 1067 - 1079. https://doi.org/10.1243/09544100JAERO304.

第 四 部 分
侵彻作用效应

弹体对带有预制孔混凝土靶的侵彻分析计算①

徐立新[1②]，李大勇[2]，冯顺山[1]

（1. 北京理工大学，北京 100081；2. 95972部队，甘肃酒泉 735018）

摘 要：通过弹体对带有预测孔混凝土靶板侵彻的几何关系的分析，推导出了弹体对混凝土靶板侵彻的数学模型。这一结果可供串联战斗部的设计参考。

关键词：侵彻；预制孔洞；土盘；弹体；混凝土靶板

Engineering Analytical Method of The Penetration into Semi – Infinite Concrete Target with Prefabricated Hole

Xu Lixing[1]，Li Dayong[2]，Feng Shunshan[1]

（1. Beijing Institute of Technology，Beijing 100081，China；2. 95972，Gangsu，735018，China）

Abstract：On the Basis of engineering analytical method in this paper，though analysis of the spatial geometric relations of the perforation，formulate the mathematic mode on penetration problem.

Keywords：penetration；prefabricated hole；soil – disc；missile；concrete slab

1 引言

弹体对带有预制孔洞的混凝土靶板的侵彻是一个非常复杂的固体动力学问题，尤其是对于聚能战斗部先期在混凝土靶板侵彻出的孔洞，其内部的靶板材料及孔洞的几何形状和尺寸、材料的性质和变形，应变率，热性质，摩擦效应，裂纹传播，材料的断裂和破碎的机理都不清楚。这些影响弹体侵彻的多种因素非常复杂而又难以处理，长期以来，主要依靠大量试验研究。弹体沿着预制孔洞的混凝土靶板侵彻的过程中，靶板的抗侵彻能力随弹体侵彻深度的变化规律，弹体速度随侵彻的深度的变化规律尚未见诸报道和介绍。文中从孔洞的孔壁扩张机理出发，对于侵彻问题做适当的假设和简化，采用分析方法，提出了预测带有预制孔洞的混凝土靶板在中低速弹体冲击下的侵彻。该模型可供反坚固硬目标的串联战斗部设计参考。

2 基本假定和基本方程

以正侵彻讨论弹体对带有预制孔洞的混凝土靶板侵彻（图1）问题，给出以下假定：

（1）假定弹体以 $V_0 \leq 500$ m/s 的初速沿带有预制孔洞的混凝土靶板进行侵彻，混凝土靶板被认为是塑性介质；

（2）假定预制孔洞的形状是规则的圆柱形，内壁光滑；

（3）不考虑混凝土内部的裂缝对侵彻的影响；

---

① 原文发表于《弹箭与制导》2002年第3期。

② 徐立新：工学博士，1999年师从冯顺山教授，研究串联战斗部对混凝土目标侵彻效应，现工作单位：陆军研究院炮兵防空兵研究所。

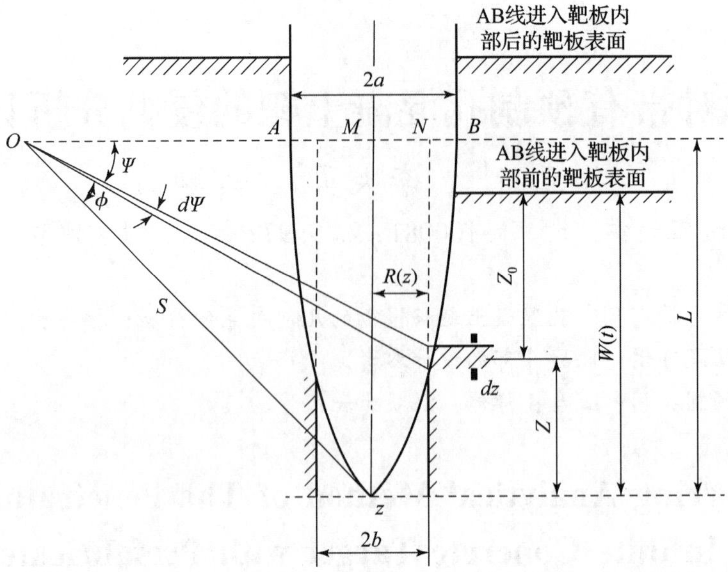

图1　弹体侵彻靶板的几何关系示意图

（4）设靶体的质点运动始终保持在平面内，平行于自由表面。弹体头部为卵形，其在侵彻过程中，推动孔洞周围的混凝土使其在弹体横截面上做径向运动。

（5）把混凝土看成由一系列厚度为 ΔZ，相互平行的土盘组成，忽略土盘间的耦合作用，则土盘内的混凝土是在弹体表面作用下，做一维径向运动。

（6）弹体为圆柱形，头部为卵形，其形成的弹坑为圆柱形，则塑性介质的运动呈柱对称形式；

（7）由于弹体侵彻速度较低，弹体不易变形，故忽略弹体的塑性变形，将其视为刚体。忽略土盘的塑性变形，其塑性本构服从理想塑性变形的 Tresca 准则。

（8）混凝土密度的变化仅与坐标有关，而与时间无关。

（9）由于这些混凝土材料弹性变形的恢复能力较小，假定靶板材料满足浮动锁应变模型。

在基本假设下，土盘的径向运动可简化为一维的。由动量定理和质量守恒定律可得

运动方程：
$$\rho_0 ru = -(r+u)\frac{\partial \sigma_r}{\partial r} - (\sigma_r - \sigma_\theta)\frac{\partial (r+u)}{\partial r} \tag{1}$$

连续方程：
$$\frac{\rho_0}{\rho} r = (r+u)\frac{\partial (r+u)}{\partial r} \tag{2}$$

屈服准则：
$$\sigma_r - \sigma_\theta = f(\sigma) \equiv \tau_0 + \mu\sigma \tag{3}$$

式中　r——拉格朗日坐标；

u——径向位移；

ρ_0 和 ρ——靶的初始和瞬时密度；

σ_r 和 σ_θ——土盘中的径向真应力和周向真应力（以压为正）；

u——质点径向位移；

τ_0 和 μ——材料常数。

靶板材料的本构方程为

$$p = K(1-\rho/\rho_0) = K\eta \tag{4}$$

$$p = (\sigma_r + \sigma_\theta + \sigma_\varphi)/3; \sigma_\theta = \sigma_\varphi \tag{5}$$

$$\tau = \lambda p + \tau_0; \tau_0 = [(3-\lambda)/3]Y \tag{6}$$

其中，p 为压力；η 为体积应变；K 为模式；V 为泊松系数；Y 为混凝土单轴压缩强度。

弹丸以初速 V_0 沿孔径为 b 的孔洞侵彻，在 $t=0$ 时刻，侵彻从一半径为 b 的圆形孔穴开始，则土盘内边界 $r=b$ 处，径向应力 $\sigma_r(b,t)=p(t)$，径向位移 $u(b,t)=R(\psi_0,t)$，则初始条件为

$$\psi_0 = \arccos\left(\frac{s-a+b}{s}\right) \tag{7}$$

因弹丸侵彻速度不大，可设波前密度 $\rho=\rho^*$ 为一不变的密度，忽略弹性应变，消去 σ_θ，式（2）可积分为

$$(r+u)^2=(1-\eta^*)r^2+R^2(t) \tag{8}$$

$$\eta^*=1-\rho_0/\rho_* \tag{9}$$

此处 $R(t)$ 是孔道腔壁的运动，因在波前 $r=r^*$ 处 $u=0$，则波前位置为

$$r^*=R/(\eta^*)^{\frac{1}{2}} \tag{10}$$

其中，“*”表示波阵面（波前）上的值。

过波前，质量守恒定律和动量守恒定律为

$$\dot{u}^*=\dot{r}^*\eta^*,\dot{\sigma}_r=\rho_0\eta^*(\dot{r}^*)^2 \tag{11}$$

将式（8）对 t 微分，则在波前上有

$$r^*\dot{u}^*=R\dot{R} \tag{12}$$

由以上方程（8）、式（11）、式（12）可得波前传播速度与径向应力

$$u^*=\dot{R}(\eta^*)^{\frac{1}{2}},\sigma_r{}^*=\rho\dot{R}^2 \tag{13}$$

据文献［1］可知：

$$(r+u)^v\sigma_r(r)=p(t)-\rho_0\int(r+u)^{v-1}r\frac{\partial^2u}{\partial t^2}\mathrm{d}r-(r^*)^v\sigma_r{}^*-\left(\frac{\tau_0}{\mu}\right)[(r^*)^v-(r+u)^v] \tag{14}$$

利用方程（8）和式（13），可将式（14）转化为

$$\sigma_r(r)=p(t)-\frac{\rho_0(\dot{R}+R\dot{R})}{(r+u)^v}\int_r^{r^*}r[(1-\eta^*)r^2+R^2]^{(\frac{v}{2}-1)}\mathrm{d}r\frac{\rho_0(R\dot{R})^2}{(r+u)^v}\int_r^{r^*}r[(1-\eta^*)r^2+R^2]^{(\frac{v}{2}-1)}$$
$$\mathrm{d}r-\rho_0R^2[\eta^*+(1-\eta^*)\xi^2]^{-\frac{v}{2}}-\frac{\tau_0}{\mu}\{[\eta^*+(1-\eta^*)\xi^2]^{-\frac{v}{2}}-1]\} \tag{15}$$

积分后得

$$\sigma_r(r)=p(t)-\frac{\rho_0(\dot{R}+R\ddot{R})}{v(1-\eta^*)^v}\{[\eta^*+(1-\eta^*)\xi^2]^{-\frac{v}{2}}-1\}$$
$$-\frac{\rho_0\dot{R}^2\eta^*}{(2-v)(1-\eta^*)}\{[\eta^*+(1-\eta^*)\xi^2]^{-\frac{v}{2}}-1-[\eta^*+(1-\eta^*)\xi^2]-1\} \tag{16}$$
$$-\rho_0R^2[\eta^*+(1-\eta^*)\xi^2]^{-\frac{v}{2}}-\frac{\tau_0}{\mu}\{[\eta^*+(1-\eta^*)\xi^2]^{-\frac{v}{2}}\}$$

其中：

$$\xi=r/r^*,r^*=R/(\eta^*)^{\frac{1}{2}} \tag{17}$$

由图 1 可知，对于卵形弹头，长为 L 的弹头，其圆弧曲率半径为 S，角度为 φ，则有

$$R(\psi,t)=S(\cos\psi-\cos\varphi) \tag{18}$$

在弹体侵彻过程中，对某一固定土盘，由于土盘深度保持不变，可求得

$$\dot{R}(\psi,t)=\dot{W}\tan\psi \tag{19}$$

$$\ddot{R}(\psi,t)=\ddot{W}\tan\psi-\frac{\dot{W}}{S\cos^2\psi} \tag{20}$$

当 $\psi=\psi_0$ 时，$\dot{W}=V_0$，对 $\mu=0$ 的特殊情况，则有

$$\sigma_r(r)=p(t)-\rho_0R^2\left\{1-\frac{\eta^*(1-\xi^2)}{2[\eta^*+(1-\eta^*)\xi^2]}\right\}+$$
$$\left[\frac{\rho_0(R^2+R\ddot{R})}{2(1-\eta^*)}+\frac{\tau_0}{2}\right]\ln[\eta^*+(1-\eta^*)\xi^2]+\frac{\tau_0}{2}\ln[\eta^*+(1-\eta^*)\xi^2] \tag{21}$$

弹丸卵形部的径向应力可由式（16）、式（17）和式（21）确定，并计入式（18）、式（19）和式（20）及$\xi=0$，若求出的径向应力为负值，则表明弹丸与混凝土靶体已经分离。

现在考虑弹丸卵形头部，讨论卵形弹丸头部的径向应力：

$$dF_n = 2\pi S^2[\cos(\psi) - (S-a+b)/S]\sigma_n d\psi \tag{22}$$

作用在弹头的表面切向力：

$$dF_\tau = \mu dF_n \tag{23}$$

式中，a 和 b 分别为弹体和预制孔洞的半径。

$$Z_0(\psi,t) = L - S\sin\psi \tag{24}$$

$$Z(\psi,t) = W(t) - Z_0 \tag{25}$$

式中，W、Z_0、Z 分别为侵彻深度、土盘深度、盘尖距。

通过对弹表元素状态分析可知：

$$\sigma_Z = \sigma_r \frac{\tan(\psi+\alpha)}{\tan\psi} \tag{26}$$

弹体所受的垂直阻力：

$$P_Z = -\int_{\psi_0}^{\phi} 2\pi RS\sin\psi\sigma_Z d\psi = -\int_{\psi_0}^{\phi} 2\pi RS\sin\psi \frac{\tan(\psi+\alpha)}{\tan\psi}\sigma_r d\psi \tag{27}$$

这里，$\psi_0 = \sin^{-1}\left(\frac{s-a+b}{s}\right)$

质量为 M 的弹体的运动方程为

$$M\ddot{W} = P_Z \tag{28}$$

3 计算过程

通过以上的分析和讨论，根据计算精度要求，把靶体分为 ΔZ 土盘薄层，由前一时刻的弹丸速度 $\dot{W}$ 和加速度 $\ddot{W}$，根据式（19）、式（20）和式（21）计算出各土盘的 R、$\dot{R}$ 和 $\ddot{R}$；再计算出各土盘的 $\dot{P}$；由式（28）计算出弹体所受的垂直 P_Z；用弹体的运动方程（29）求出弹体的新加速度 $\ddot{W}$；以一定的时间步长积分求出新时刻的弹丸速度 $\dot{W}$ 和侵彻深度 W。对以上步骤重复，直到达到最大侵彻深度，即 $\dot{W}=0$。

参考文献

[1] 徐立新．对二次侵彻靶板试验的研究和探讨［C］//中国宇航无人飞行器学会战斗部与毁伤效率专业委员会第七届学术年会论文集，2001.
[2] 尹放林．弹体侵彻深度计算公式对比研究［J］．爆炸与冲击，2000，20（1）：79－82.
[3] 王晓虹．应变软化岩土材料内扩孔问题解析解［J］．工程力学，1999，16（5）：71－76.
[4] 李永池．混凝土靶抗贯穿的一种新工程分析方法［J］．爆炸与冲击：2000，20（1）：13－18.

动能弹对预侵彻破坏的混凝土的侵彻计算模型[①]

蒋建伟[②]，何　川，冯顺山，门建兵

（北京理工大学爆炸灾害预防、控制国家重点实验室，北京　100081）

摘　要：本文针对动能弹侵彻已被侵彻破坏过的混凝土目标，将侵彻过程分为两个阶段，对建立在空穴膨胀理论基础上的经典的Forrestal关于弹丸侵彻混凝土分析模型进行改进，考虑了预侵彻破坏的混凝土中的穿孔对再次侵彻的影响，同时还考虑预毁伤留下的弹坑和混凝土的强度弱化对侵彻过程的影响，从而建立该问题的理论分析模型。最后通过计算实例，得到侵彻深度和着速的关系曲线。理论计算结果和实验结果吻合较好。证明了模型的可用性。分别讨论了侵彻深度和相对孔径与混凝土的强度弱化程度的关系。

关键词：预侵彻破坏；混凝土；侵彻；目标强度弱化

1　引言

弹丸对混凝土的侵彻问题一直是一个热点，很多研究就集中在弹丸对无预侵彻破坏的混凝土的侵彻，包括一系列的经验公式、理论计算模型和数值模拟[1,2]。而对于侵彻已受到预侵彻破坏的混凝土，如两级串联战斗部，由前级聚能射流开孔，后级随进动能弹继续侵彻混凝土，该问题是一个全新的研究课题。本文旨在改进基于空穴膨胀理论的基础上的Forrestal分析模型，分析混凝土已有的侵彻孔对侵彻过程的影响，并考虑预侵彻破坏对混凝土强度的弱化效果，从而为该问题建立一个较为完善的分析计算模型。

2　计算模型

图1为一直径为d、着速为v_0的动能弹对预侵彻破坏的孔径为d_1、深度为P_1的混凝土的侵彻模型示意图，也是本文的基本假设，侵彻过程分为图1所示的两个阶段，侵深分别为P_1和P_2。

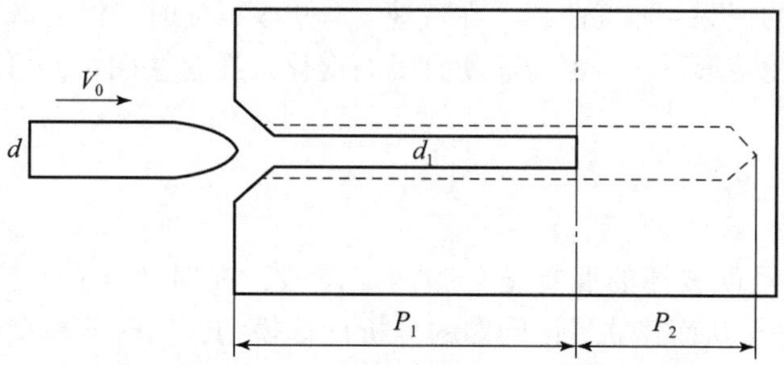

图1　动能弹侵彻预侵彻破坏混凝土的模型示意图

利用Forrestal计算模型，将第一侵彻阶段分为开坑和穿孔两个阶段，代表开坑和穿孔的临界点的k值是对文献［3］中的数值模拟和文献［4］中的实验所得结果进行修正，并考虑预侵彻破坏对弹坑形成的影响所得。

在第一阶段的穿孔阶段，根据弹体所受的轴向阻力要应用空穴膨胀理论来计算。根据空穴膨胀理

① 原文发表于《爆炸与冲击》2003（增刊）。

② 蒋建伟：工学博士，2001年师从冯顺山教授，研究中心爆管式子母战斗部抛撒技术及应用，现工作单位：北京理工大学。

论，当弹头部侵入混凝土时，作用在弹丸上的阻力只与与靶板材料接触部分的表面有关，由于这一阶段侵彻的有预破坏孔的混凝土，因此，弹头部表面有一部分没有与混凝土接触，计算阻力时，对弹头部表面的积分区间是 $\phi_1 \sim \pi/2$，而不是 $\phi_0 \sim \pi/2$[5]，通过计算，阻力的表达式可以简单表示为

$$F = \alpha + \beta v^2 \tag{1}$$

另外，在理论模型中，由于侵彻目标时受到预侵彻破坏的混凝土，考虑混凝土强度弱化，分别体现在和混凝土强度相关的两个参数 S 和 f_e，根据弱化的关系式得出

$$S_2 = 0.609\,5S \text{ 或 } f_{c2} = 0.609\,5f_e \tag{2}$$

即在理论模型中如用 S_2 代替 S，或用 f_{c2} 代替 f_c，就考虑了混凝土在预侵彻破坏时强度弱化对侵彻的影响。

于是通过以上改进的理论分析模型，就可以推导出到第一侵彻阶段和第二侵彻阶段的临界点的速度计算公式

$$v_1 = \sqrt{((\alpha + \beta v_1'^2)\exp(-(2\beta(P_1 - kd)/m)) - \alpha)/\beta} \tag{3}$$

式中，v_1' 表示第一侵彻阶段的开坑结束时弹丸的速度。

利用 Forrestal 分析模型，就可推导出第二阶段侵彻深度的计算公式

$$P_2 = \frac{2m}{\pi d^2 \rho N}\ln\left(1 + \frac{N\rho v_1^2}{Sf_c}\right) \tag{4}$$

从而完成总的侵彻深度的计算。

3 计算实例及讨论

采用一实验模型[8]作为计算讨论模型，该模型为动能弹对被聚能射流侵彻破坏过的混凝土侵彻，其中弹丸直径 125 mm，弹全长 336.6 mm，圆柱部为 175 mm，弹丸的 CRH 值为 1.366 4，质量 17.63 kg。混凝土靶的密度 2 400 kg/m³，初始抗压强度 35 MPa，靶板厚为 3 × 1.2 m。采用聚能装药进行预侵彻破坏，预破坏孔直径 100 mm，预破坏孔深度 940 mm，弹丸着速分别为 741 mm、814 mm、906 m/s，侵彻深度 P 分别为 1.96 mm、2.20 mm、2.30 m。

在上述实验条件下，采用本文的计算模型，得出侵彻深度和着速的曲线关系图，可以看出，当考虑强度弱化时（$S_2 = 0.609\,5S$），侵彻深度和实验数据吻合较好。当不考虑强度弱化时，其侵彻深度相对要小，且和实验数据误差较大。

为了解相对孔径和强度弱化程度对侵彻深度的影响，利用计算模型专门就这些参数进行讨论。画出了不同的相对孔径时对应的侵深 - 速度曲线，在着速 $v_0 = 906$ m/s 时，侵彻深度 - 相对孔径的函数关系曲线，以及不同的强度弱化情形下，侵深 - 速度曲线的变化。在这些图中，可以很清楚地看到各个参数对侵彻深度的影响。

4 结论

将动能弹侵彻带有预侵彻破坏的混凝土分为两个阶段，在对建立在空穴膨胀理论基础上的经典 Forrestal 分析模型进行改进，从而建立了该问题的分析计算模型，同时该模型中考虑了预破坏孔和混凝土的强度弱化对侵彻过程的影响。通过一个实验模型进行了计算讨论，理论计算结果和实验结果较好地吻合，可以进行工程应用。同时得到以下一些参数对侵彻的影响：着速是影响侵彻深度的重要因素，当 $v_0 < 300$ m/s 时，即使相对孔径 q 很大，侵彻深度增加也很小；而当 $v_0 > 500$ m/s 时，侵彻深度的增加非常明显。相对孔径 q 也是一个影响侵彻深度的重要因素，当 $q < 0.3$ 侵彻深度变化很小，为了取得很好的侵彻效果，应该取 $0.4 < q < 0.8$。混凝土的强度弱化越厉害，侵彻深度就越大，并且随着着速的增大，侵彻深度增加越快。

地下硬目标毁伤分析[①]

王云峰[②]，冯顺山，董永香，王　玥
（北京理工大学爆炸科学与技术国家重点实验室，北京　100081）

摘　要：本文分类研究了深层目标，讨论分析了目标易损面积及纯易损面积、结构毁伤、功能毁伤的概念，指出了目标易损面积与命中条件下的毁伤概率的相关性；构造了可以描述目标易损性的数学函数，理论上可通过对$f(\mu)$的分析，了解目标易损性变化规律，方便描述目标的毁伤过程，预测目标的毁伤状况。

关键词：深层目标；易损性；毁伤

Damage Analysis of Deep Buried Hard Target

Wang Yunfeng，Feng Shunshan，Dong Yongxiang，Wang Yue
（State Key Laboratory of Explosion Science and Technology，Beijing Institute of Technology，Beijing 100081，China）

Abstract：Deep buried targets are classified. The concept of vulnerable area，pure vulnerable area，structural damage and function damage of the target are discussed and researched. The relativity of target vulnerable area and damage probability under hit condition is pointed out. Mathematical function which is used to describe the vulnerability is built. Theoretically，through the analysis to function $f(\mu)$ we can reach the goal of understanding to variational law of the vulnerability，description of the damage process to target and forecast to damage condition of target.

Keywords：deep buried hard target；vulnerability；damage

1　引言

地下硬目标价值极[1]。随着武器系统制导精度、战斗部毁伤威力的不断提高，毁伤深层硬目标的能力大幅度提高，而新的防护技术又给防护工程体系提供了能够对抗随之出现的新型武器系统的可能性。毁伤元的毁伤能力是防护工程安全设计、加固改造参考的重要依据之一，分析目标构成材料特性、目标结构类型、毁伤元的毁伤特性及威力、弹目交汇条件等是硬目标毁伤分析中的基本问题，文中从总体上对地下硬目标毁伤问题进行了初步分析，供地下硬目标毁伤研究参考。

2　地下硬目标及其分类

2.1　地下硬目标

按不同埋藏的深度以及常规战斗部可实现的毁伤能力，可把地下硬目标分为深层目标和超深层目标；目标功能不同，防护形式、防护层厚度等有所区别[2]，例如，单层C^3I设施、穹顶式地面C^3I设施、飞机掩体、单层地下指挥所、地对地导弹战备掩体、地上多层建筑物下面的掩体等类目标，其防护层等效 30 MPa 混凝土厚度为 1.8 ~ 6.4 m 单体目标一般由顶部防护层、侧墙、底板、通风口、入口、防护门、

---

① 原文发表于《弹箭与制导》2006 年第 4 期。

② 王云峰：工学博士，2002 年师从冯顺山教授，研究深层目标易损性及爆炸毁伤效应，现工作单位：吉林白城 31 基地。

外部通信、内部人员及设备（计算机、大屏幕指显、通信设备等）等部分组成。

2.2 硬目标的分类

硬目标的分类比较复杂，没有统一的分类方法，大体如下。

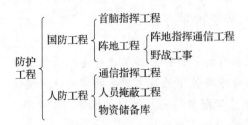

（1）按功能分类。

- 防护工程
  - 国防工程
    - 首脑指挥工程
    - 阵地工程
      - 阵地指挥通信工程
      - 野战工事
  - 人防工程
    - 通信指挥工程
    - 人员掩蔽工程
    - 物资储备库

（2）按构筑方式分类。

- 防护工程
  - 国防工程
    - 永备工事
    - 野战工事
  - 人防工事
    - 掘开式人防工程
    - 坑道式人防工程
    - 地道式人防工程

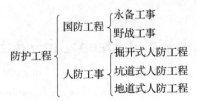

（3）按常规战斗部侵彻能力（30 MPa 混凝土）分类。

- 防护工程
  - 地面硬目标(＜1.5 m)
  - 浅层硬目标(3~6 m)
  - 深层硬目标(6~12 m)
  - 超深层硬目标(＞12 m)

（4）按直接承载炸弹当量分类。

- 防护工程
  - 战略意义的重要指挥所(2 000 lb)
  - 一般弹药库(1 000 lb)
  - 永备工事(500 lb)

3 毁伤分析

结构毁伤（或称物理毁伤）：指由于毁伤元的作用，使目标整体内在联系或组合方式发生了影响完成任务的改变或失效破坏。功能毁伤：在毁伤元的作用下，目标完成任务的能力受损或者内部人员和设备的运转效率降低，即功能毁伤。

（1）易损面积与目标毁伤：假设目标被弹面上有 e 个不同防护水平的区域，每个区域的面积为 M_i，$i=1$，2，3，…e，且 M_i 上的防护水平相同，如果毁伤元命中 M_i 的条件下 M_i 被毁伤的条件概率为 P_i（不考虑 M_i 上的差异），那么令

$$A_k = \sum_{i=1}^{e} P_i M_i$$

则称 A_k 为目标易损面积（部位），它是目标被弹面被其上面的毁伤概率加权后的面积总和。P_i 的量化与目标结构、坚固程度、弹目交汇条件、毁伤单元的毁伤威力等因素相关，表示了目标被弹面抗打击能力，它是一个相对的概念，如果 M_i 上的 $P_i=1$ 时，则该面处处易损。很容易理解：如果战斗部的侵彻深度能力超过防护层厚度，那么防护层（整个面积上）对该侵彻能力来说处处易损。

（2）纯毁伤面积与目标毁伤：如果 M_i 被毁伤的条件下目标，即毁伤的条件概率为 P_i^*，再令

$$A^* = \sum_{i=1}^{e} P_i^* M_i$$

定义 A^* 为目标的纯毁伤面积，它是与目标结构相关的量，体现了目标要害部位的关重特性，P_i^* 也代表

了 M_i 被毁伤后关联目标毁伤的权重，显然 A^* 表达了目标的易毁属性。通风口、外部通信系统、防护门等部位就具有这类性质。

（3）目标毁伤概率：目标的纯易毁面积与目标总被弹面积的比值，表达式为

$$P_y = A^* / \sum_{i=1}^{e} M_i = \sum_{i=1}^{e} P_i^* M_i / \sum_{i=1}^{e} M_i$$

4　目标毁伤过程的数学描述

目标完成任务的能力：目标完成任务过程中潜在能力的发挥水平。初始状态（各方面状态完好）时目标完成任务的能力为100%，如果侵爆作用后导致了防护层结构破坏，必然使目标完成任务能力受损，用 η 度量这个过程中目标完成任务能力下降水平（以下简称为下降水平，$0<\eta\leqslant 1$），假设战斗部在某个侵彻深度 x_0 处爆炸时，对应目标毁伤的最低级别（轻微毁伤），η 的“下降水平”为 η_0，如果装药量不变，侵爆深度增加，η 的“下降水平”必然增加（这里考虑的是目标随侵彻爆炸位置加深过程表现出来的宏观毁伤统计结果）。设防护层厚度为 d，侵爆深度变量为 x，令 $\mu=x/d$，见图1。以 μ 为横轴、η 为纵轴建立（μ，η）的直角坐标系，直到某一个深度 x_i，当 $\mu_i=x_i/d$ 时，防护层将被爆炸贯穿，防护层一旦被贯穿，爆炸产物、介质碎块等进入目标内部，通常认为此时目标被摧毁（或归结为摧毁性毁伤级别），此时 η 的“下降水平”约为1。显然 η 是侵彻过程 μ 的函数，令：$\eta=f(\mu)$，把 η 定义为目标完成任务能力损失水平函数。

从物理意义上可知，损失水平函数曲线 $f(\mu)$ 是一条 S 形曲线，如图2，具有如下性质：

（1）$\mu\to x_i/d$ 时，$\lim\limits_{\mu\to x_i/d} f(\mu)\to 1$。极限值处损失率 η 接近1，此时侵彻不一定要穿透防护层，符合目标毁伤物理事实。

（2）$f(\mu\geqslant x_i/d)=1$。说明只要战斗部侵彻到防护层某一深度 x_i，在 $\mu\geqslant x_i/d$ 以后，目标完成任务能力全部丧失，爆炸作用会对目标造成摧毁性毁伤。

（3）$\mu=\mu_0$，$f(\mu_0)=\eta_0$。η_0 可从最低的毁伤级别算起，表示了毁伤的最低级别，侵彻深度小于处对目标造成的毁伤不足以评价，符合毁伤分级思想。

（4）$u\to-\infty$ 时，$f(\mu)\to 0$。表明爆点在 $\mu_0\leqslant x_0/d$ 以外，侵彻深度越小，对目标毁伤作用越小，$\mu\to-\infty$ 在工程上没有意义，只是表示数学上的严密性而已。

（5）$f(\mu)$ 为单调增函数，曲线一定有拐点，如果没有拐点 $f(\mu)\in[0,1]$ 是不可能的。

综上有

$f(\mu)\in(0,1)$，$\mu\in(-\infty,x_i/d)$，$\lim\limits_{\mu\to x_i/d} f(\mu)\to 1$，$f(\mu\geqslant x_i/d)=1$。

经过研究，数学函数 $f(\mu)=\dfrac{1}{a+c\mathrm{e}^{-b\mu}}$（$a>0$，$b>0$，$c>0$）满足以上（1）、（2）、（3）、（4）、（5）特征条件，能描绘出图2表示的曲线。

该函数的极值由系数 a 决定，当 $a=1$ 时，函数 $f(\mu)$ 的极值为1，系数 c 和 b 决定 $f(\mu)$ 取得极限值的位置及曲线形状，适当选择 b 和 c 可以使函数 $f(\mu)$ 在 $\mu=x_i/d$ 附近处接近极值，而且在 $\mu>x_i/d$ 以后的值都非常接近1。

当然，想得出 $f(\mu)$ 并不是一件容易的事情。试验能得到目标的毁伤水平，由毁伤水平可以确定目标完成任务能力的变化水平，尽管毁伤水平与完成任务能力变化水平并不一定是简单对应关系，只要有样本量，理论上总可以得到（η_j，x_j/d），$j=1$，2，$\cdots i$，拟合试验数据确定系数 b 和 c，就能得出目标完成任务能力损失水平函数。

函数 $f(\mu)$ 有非常重要的意义，只要确定了函数 $f(\mu)$，对 μ 求导得：$\mathrm{d}\eta/\mathrm{d}\mu=f'(\mu)$，$\mathrm{d}\eta/\mathrm{d}\mu$ 表示了目标在侵彻爆炸单元作用下完成任务能力损失水平变化的变化速率，即真正数学意义上表述的目标对侵爆作用（以侵爆深度表现）的易损性，进一步研究还可以分析目标毁伤分级的问题、预测毁伤水平等。

5 人员和设备的冲击波毁伤

在没有建立专门的地下硬目标毁伤判别标准以前，可参考洞室、水电工程、隧道、矿山巷道等的爆破振动、破坏等评价准则。比如，衡量爆炸对洞室安全的判别方法有质点峰值振动速度法，等效比距离法[3]、安全距离估算法[4]等，行业不同、国别不同，判别标准存在较大差异。超压值为0.02~0.1 MPa时，人员从轻微挫伤到大部分死亡[5]；超压值为0.039 5~0.058 8 MPa时，重1 t的设备从离开基础到翻倒破坏，0.27~0.34 MPa，25 cm钢筋混凝土墙强烈变形、形成大裂缝、混凝土脱落，大于0.34 MPa倒塌；0.048~0.055 MPa，24~36.5 cm砖墙充分破坏；0.034~0.041 MPa裸露电线折断[6]；0.004 9~0.009 8 MPa装配玻璃破坏，大于0.2 MPa汽车玻璃破碎[7]。按经验公式计算（冲击波片超压曲线见图3）：10 kg左右装药在地面爆炸，距爆点11 m左右处的超压为0.046 MPa，从以上数据推断，10 kg装药的战斗部在目标内爆炸，能够毁伤200~400 m^2 目标内的人员和设备。

6 结论

地下硬目标毁伤研究具有十分重要的战略意义，非常迫切，但由于地下硬目标大多涉及核心秘密，资料不公开，工程浩大以及惊人的耗费，以至于不太可能进行等比例毁伤验证研究，可见地下目标毁伤研究存在相当大的困难。分析了目标对侵爆作用易损性的数学意义，构造的 $\eta=f(\mu)$ 能够表述目标毁伤实质，数学描述与物理事实完全统一，表达了研究的一种方法和概念，理论上可通过对 $f(\mu)$ 的分析，了解目标易损性变化规律，描述目标的毁伤过程，方便毁伤级别的划分，预测目标的毁伤状况，可供地下目标毁伤研究参考。

参考文献

[1] 王振宇，李清献，任辉启，等. 科索沃战争及其启示［J］. 防护工程，2000，1.

[2] SPECIAL FEATURE Adding new punch to cruise missiles Specialized warheads can defeat hard and buried targets JANES INTERNATIONAL DEFENSE［J］. 1998，41（1）.

[3] 甄胜利，霍永基. 水工结构爆破振动效应研究及安全分析［J］. 工程爆破，2003，9（4）.

[4] 陶颂霖. 爆破工程［M］. 北京：冶金工业出版社，1979.

[5] 随树元，王树山. 终点效应学［M］. 北京：国防工业出版社，2000.

[6] 哈努卡耶夫. 矿岩爆破物理过程［M］. 刘殿中，译，李国乔，校. 北京：冶金工业出版社，1980.

[7] 张志毅，王中黔，主编，冯叔瑜主审. 交通土建工程爆破工程师手册［M］. 北京：人民交通出版社，2002.

舰舷结构的能量法等效[①]

吴俊斌[1②]，冯顺山[2]，董永香[2]

(1. 北京机电工程研究所，北京 100074；

2. 北京理工大学爆炸科学与技术国家重点实验室，北京 100081)

摘 要：舰船装甲结构主要由面板和加强筋组成，在战斗部的设计中一般将舰船装甲结构简化为均质等厚度的靶板，这就需要在两者之间建立等效关系，而能量法等效是众多等效方法中的一种。在研究半穿甲战斗部的靶板等效中，通过选择舰舷结构中加强筋交叉点作为研究对象，选择某弹丸作为侵彻体，考虑了冲塞阻力、靶板变形吸能以及冲塞剪切做功等方面的影响，最终得出了给定靶板的等效厚度。

关键词：舰舷结构；能量法；等效；侵彻；弹丸

引言

军舰是列入海军编制，用于完成战斗任务和保障任务的战斗舰艇和特种舰艇。装甲防护是指舰船为防护各种战术武器攻击而设置的厚度大于 12 mm 的板材结构。以航空母舰为例，其装甲防护是生存能力的主要保障条件之一，是整体防御体系中的最后一道防线。航母舰体装甲结构保护的部位是飞行甲板、弹药舱、主机舱、舰机舱、指挥控制室、机库和油舱等部位。

对于装甲结构，均采用面板加加强筋的结构，既减轻了结构重量，又提高了结构强度和刚度。从以往的弹丸对钢靶板的侵彻试验来看，加强筋主要在以下几个方面影响靶板：加强筋对靶板的变形和破坏形式起重要作用，且作用机理复杂；在一定的速度范围内随弹丸着靶速度的降低，靶板的塑性变形区域增大，吸收弹丸的能量也随之增加，其侵彻过程中速度降落的幅度也在增加；对于塑性较差的靶板，筋与板的耦合效应对弹丸弹道参数的影响比靶板厚度增加对弹道参数的影响大。

1 能量法等效

1.1 等效原则

能量法等效的原则是：若某均质靶板与实际舰舷结构吸收弹丸动能的能力相同，也就是说，当某厚度的均质靶板和实际舰舷结构在同一弹丸以相同状态侵彻时，弹道极限速度相同，或者相同着靶速度的剩余速度相同，则认为两者等效。

1.2 基本假设

本次讨论有两个基本假设：假设弹丸为钝头弹，其侵彻靶板时形成冲塞；靶板及弹丸的弹性变形吸收的能量忽略不计。

1.3 等效靶厚度分析

弹丸侵彻靶板前后，有能量守恒方程：

$$\frac{1}{2}\mu_1 \pi R^2 v_0^2 = \frac{1}{2}(\mu_1 \pi R^2 + \mu_2 \pi R^2) v_f^2 + E_1 + E_2 \tag{1}$$

---

① 原文发表于《2005 年弹药战斗部学术交流会论文集》2005。

② 吴俊斌：工学硕士，2004 年师从冯顺山教授，研究舰舷结构等效靶，现工作单位：航天三院三部。

其中，μ_1、μ_2分别为单位截面积弹丸和靶板质量，

R 为弹丸和冲塞半径（此处认为二者半径相同），

v_0为弹丸着靶速度，

v_f为弹丸和冲塞的剩余速度，

E_1为靶板塑性变形吸收的能量，

E_2为侵彻过程中剪切力做的功。

设某一时刻弹丸和靶板的共同速度为v_s，由动量守恒定律有

$$\mu_1 \pi R^2 v_0 = (\mu_1 \pi R^2 + \mu_2 \pi R^2) v_s$$

整理得

$$v_s = \frac{1}{1 + \dfrac{\mu_2}{\mu_1}} v_0 \tag{2}$$

又有

$$E_1 = \frac{1}{2}\mu_1 \pi R^2 v_0^2 - \frac{1}{2}(\mu_1 \pi R^2 + \mu_2 \pi R^2) v_s^2 \tag{3}$$

将（2）式代入（3）式，得到

$$E_1 = \frac{1}{2}\frac{\mu_1\mu_2}{\mu_1 + \mu_2}\pi R^2 v_0^2 \tag{4}$$

对于等厚度均质靶板，在弹丸侵彻过程中，剪切阻力

$$F = 2\pi R(H - x) Y_{sd} \tag{5}$$

式中，H 为靶板厚度，

X 为弹丸在靶板中的运动距离，

Y_{sd}为靶板材料动态剪切强度。

剪切力 F 所做的功为 $dA = Fdx$

$$A = \int_0^h Fdx = \int_0^h 2\pi R(H - x)\, Y_{sd} = 2\pi RHh\, Y_{sd} - \pi R\, h^2\, Y_{sd} \tag{6}$$

式中，h 是弹丸在靶板中的侵彻深度，若穿透靶板，则 $h = H$，则（6）式可变为

$$E_2 = A|_{h=H} = \pi R\, H^2 Y_{sd} \tag{7}$$

在弹丸侵彻靶板时，弹丸的能量损失为

$$\Delta E = \frac{1}{2}m_p(v_0^2 - v_f^2) = \frac{1}{2}m_t v_f^2 + E_1 + E_2 \tag{8}$$

式中，m_p、m_t分别为弹丸和靶板冲塞的质量，

v_0为弹丸着靶速度，

v_f为弹丸和冲塞的剩余速度，

E_1为靶板塑性变形吸收的能量，

E_2为侵彻过程中剪切力做的功。

将（4）、（7）式代入（8）式，得到弹丸侵彻均质靶板时的能量损失为

$$\Delta E = \frac{1}{2}m_t v_f^2 + \frac{1}{2}\frac{m_p m_t}{m_p + m_t} v_0^2 + \pi R\, H^2 Y_{sd} \tag{9}$$

对于非均质靶板（结构靶），侵彻过程中剪切力所做的功E_2不同于均质靶板，其值为 $E_2 = \int_0^h Fdx$，F 为侵彻深度 x 的函数，不同结构 F 不同。

$$\Delta E' = \frac{1}{2} m_t' v_f^{2\prime} + \frac{1}{2}\frac{m_p' m_t'}{m_p' + m_t'} v_0^{2\prime} + \int_0^{H'} F dx \tag{10}$$

对比（9）、（10）两式，令 $\Delta E' = \Delta E$，可以得到等效均质靶厚度 H。

2　等效分析

2.1　舰舫结构

据某文献[4]报道，某舰船目标由 19. 05 mm 厚的钢板、背面用 T 型梁和角钢交叉焊接构成（结构见图 1）。对于靶标最严重的情况是靶标的交叉加强部分，把该处确定为 31. 75 mm 厚的均质钢板等效，在仿真计算中使用这个等效靶标。

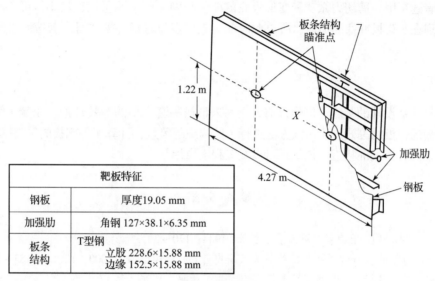

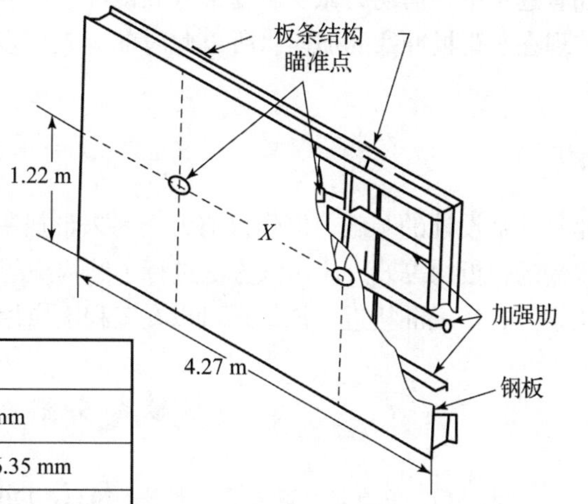

靶板特征	
钢板	厚度19.05 mm
加强肋	角钢 127×38.1×6.35 mm
板条结构	T型钢 立股 228.6×15.88 mm 边缘 152.5×15.88 mm

图 1　靶板结构

2.2　等效靶计算

本次仅计算弹头侵彻结构靶中加强筋交叉处。弹丸侵彻该处时剪力所做的功

$$E_2' = \int_0^h 2\pi R(h - x) Y_{sd} dx +$$

$$2\int_0^s 2R arcsin\left(\frac{1}{2R}\right)(s - x) Y_{sd} dx + 2\int_0^t 2R arcsin\left(\frac{w}{2R}\right)(t - x) Y_{sd} dx +$$

$$2\int_0^{s_1} 2R arcsin\left(\frac{t_1}{2R}\right)(s_1 - x) Y_{sd} dx + 2\int_0^{t_1} R arcsin\left(\frac{w_1}{R}\right)(t_1 - x) Y_{sd} dx$$

积分后得，

$$E_2' = R Y_{sd}\left(\begin{array}{c}\pi h^2 + 2arcsin\left(\frac{t}{2R}\right)s^2 + 2arcsin\left(\frac{w}{2R}\right)t^2 \\ + 2arcsin\left(\frac{t_1}{2R}\right){s_1}^2 + arcsin\left(\frac{w_1}{R}\right){t_1}^2\end{array}\right)$$

其中，板厚 $h = 19.05$ mm

“T” 型加强筋　厚 $t = 15.88$ mm

立缘高 $s = 228.6$ mm

边缘宽 $w = 152.5$ mm

角钢　厚$t_1=6.35$ mm

立缘高$s_1=127.0$ mm

边缘宽$w_1=38.1$ mm

弹丸半径　$R=200$ mm

弹丸质量　$m_p=300$ kg

靶板材料动态剪切强度　$Y_{sd}\approx40$ kg/mm^2

在战斗部侵彻试验中，某战斗部以 288.4 m/s 的速度侵彻 32 mm 厚均质船用钢靶板，其出靶速度为 164.0 m/s，因为该试验的靶板与本次研究的靶板（均质靶）厚度接近，根据等效原则，结构靶与均质靶消耗战斗部的能量相同，所以本次试验数据可以用于本计算。

将上述数据代入（10）、（9）式，得出等效靶厚度为 30.86 mm。

若弹丸是超音速穿甲，剪切力做功和变形吸收的能量与弹丸的总能量相比较小，可以忽略，靶板对弹丸的阻力主要体现在对靶板冲塞的加速，所以此时的等效可以仅考虑冲塞质量。若按此思路该靶板可以等效为 30.35 mm。

3 结束语

靶板等效是战斗部设计的基础，但到目前为止尚未形成一套完整的等效理论。能量法等效可以用于亚音速钝头弹所侵彻的靶板的等效。应用该方法进行了结构靶的等效，计算所得靶板的等效厚度 30.86 mm 和报道[4]的等效靶厚度 31.75 mm 接近，该结果可以在工程上应用。

参 考 文 献

[1] 钱伟长. 穿甲力学 [M]. 北京：国防工业出版社，1984：170－393.

[2] 金建明，唐平. 舰舷结构与均质靶板的等效方法研究 [J]. 弹道学报，2000，12（1）：83－87.

[3] 张庆明等. 破片贯穿目标等效靶的极限速度 [J]. 兵工学报. 1996. 第 17 卷第 1 期. 21～25.

[4] R. Hassett, J. C. S. Yang, J. Richardson and H. Walpert. STUDY OF PENETRATION FORCES FOR SUPERSONIC WARHEAD DESIGNS. 227～236.

小型侵爆战斗部毁伤多层靶板威力分析与仿真①

李国杰②，冯顺山，曹红松，王云峰

（北京理工大学爆炸科学与技术国家重点实验室，北京　100081）

摘　要：通过分析弹靶结构，优选可靠的分析方法，对小型半穿甲侵爆战斗部毁伤多层靶板目标的威力进行了理论计算，并利用 AUTODYN－2D 计算软件，建立了最优化数学模型，对侵彻与爆炸成坑过程进行了分段数值仿真，并与相关实弹数据进行比较，各组数据均显示该种结构的侵爆战斗部在特定弹目交汇条件下，弹体强度满足要求不会产生明显变形，峰值过载不超过装药应力安全限，而且爆破中心位置理想，毁伤效果满足战术指标要求，仿真结果具有较高的可信度，可为同类战斗部的方案设计提供参考。

关键词：战斗部；数值仿真；多层靶板；混凝土

Analysis and Numerical Simulation on Damage effect of Minitype Penetrate－Blast Warhead to Multi－layer Target

Li Guojie，Feng Shunshan，Cao Hongsong，Wang Yunfeng

（State Key Laboratory of Explosion Science and Technology，Beijing Institute of Technology，Beijing 100081，China）

Abstract：By analyzing the structure of the warhead and target，credible analysis method was chosen，and the damage effect of the semi－armor piercing and penetrate－blast minitype warhead destroying the multi—layer target Was analyzed in theory. Using the computing software AUTODYN－2D，the optimization mathematic model Was built，and the damage process of penetrating and blasting Was simulated，and then the results were contrasted with the data of some correlated live ammunitions. Each group of data shows that，under the specific warhead/target initial conditions，the body intensity of the penetrate—blast warhead of that structure meets the specification and would not transform obviously，and the peak value of the overload would not exceed the safety limit of its charge stress，and furthermore the position of explosion center is perfect，and the damage effect meets the specification too. The simulated results are creditable，they could give some reference to the scheme design of the like warhead.

Keywords：warhead；numerical simulation；multi－1ayer target；concrete

引言

随着科学技术的发展和新一代航空武器装备的使用，现代战争的主要作战手段已经从大规模装甲兵团推进转到了远程精确打击和大规模空袭上，这就必须取得战场制空权，因此如何快速高效地将敌对方的机场或空军基地封锁或摧毁以达到削弱其空中作战和保障能力的目的便成了各国战略考虑的重中之重，而跑道又以其重要性和不可隐蔽性成为被打击的首选目标。

本文以打击机场跑道弹药设计为应用背景对小型半穿甲侵爆战斗部毁伤多层靶板目标的威力进行了理论计算，并利用 AUTODYN－2D 软件对此战斗部毁伤机场跑道的过程进行了数值仿真，通过结果的对比分析对此种战斗部方案做出了可行性论证。

---

① 原文发表于《系统仿真学报》2007 年第 20 期。

② 李国杰：工学博士，2002 年师从冯顺山教授，研究自主攻击子弹系统总体及对跑道毁伤，现工作单位：北京航天长征飞行器研究所（14 所）。

1 理论分析

1.1 机场跑道结构分析

跑道一般分为标准四层结构，由表及里，第一层为混凝土道面，30 号砼，厚度 0.4 m；第二层为砂找平层，厚 3 ~ 5 cm，其作用是较准确地保证道面的厚度和标高以及减少摩擦使道面板可较自由伸缩；第三层为碎石基础，由经过严格碾压的石子组成，厚约 0.6 m；第四层为达到“最大密实”状态的压实土基层，厚度不小于 0.5 m。结构如图 1 所示。

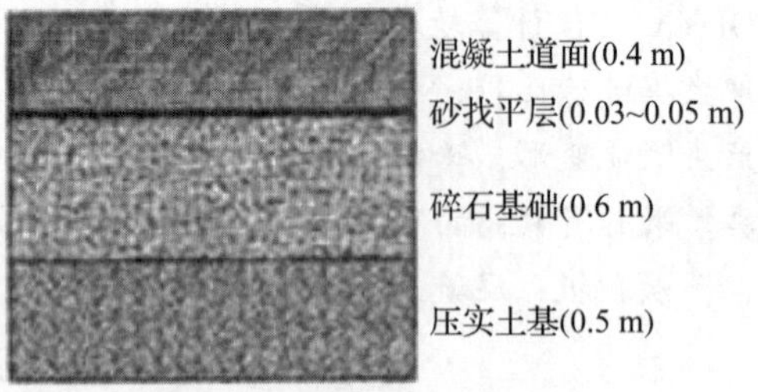

图 1 机场跑道的典型结构

1.2 战斗部结构分析

为了获得跑道毁伤的最佳效果，战斗部必须侵彻到一定深度再爆炸，一般应达到碎石基础层。可以达到这种目的的弹药基本有两种形式，一种是普遍采用的是动能半穿甲侵爆战斗部，另外一种是前级聚能开坑后级随进爆破两级串联复合战斗部。由于子弹总体参数的限制，而且灵巧子弹药结构又不宜过于复杂，因此在满足基本毁伤效能的基础上优先选择了动能半穿甲侵爆战斗部，具体结构如图 2 所示。

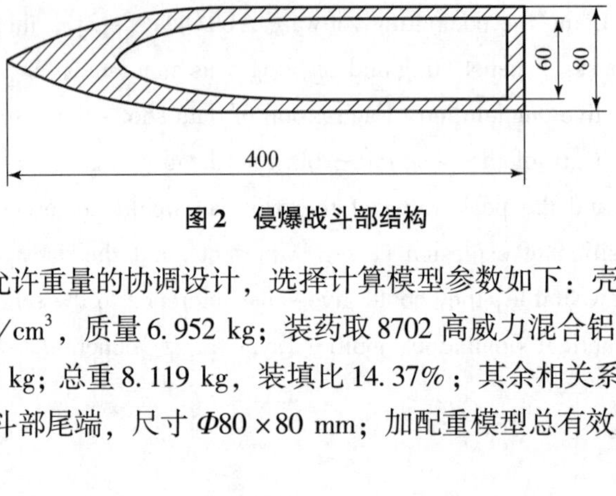

图 2 侵爆战斗部结构

经过战斗部限定体积和允许重量的协调设计，选择计算模型参数如下：壳体取 S – 7STELL 材料，屈服强度 1 539 MPa，密度 7.75 g/cm^3，质量 6.952 kg；装药取 8702 高威力混合铝炸药，密度 1.794 g/cm^3，质量 1.167 kg，TNT 当量 1.832 kg；总重 8.119 kg，装填比 14.37%；其余相关系统结构配重：质量 3.116 kg，按密度 7.75 g/cm^3 附加到战斗部尾端，尺寸 $\Phi 80 \times 80$ mm；加配重模型总有效质量：$m = 11.235$ kg。

1.3 威力评估

1.3.1 侵彻深度计算[2]

目前计算弹体对混凝土材料的侵彻深度公式有 40 多种，但因为每种公式的建立基础都不尽相同，如有纯经验的、以量纲分析为基础的及以空腔膨胀理论为基础的等，所以各自都有相应的适用范围和应用条件，因此选用公式时要针对具体情况优化选择。这里选用非常具有代表性的 Young 公式，它以庞大的试验数据量为基础，考虑的因素较周全，适用范围较广，且引入了截面密度的概念，因此比较而言具有相当的准确性，而对于解决侵彻多层介质靶板的问题，它也给出了出色的解决途径。

应用 Young 公式计算战斗部对混凝土分层结构的侵彻：

$$X = 0.000\,018\,1 SNK\,(m/A)^{0.7}(V - 30.48) \quad V > 61\ \text{m/s} \tag{1}$$

$$V_{ex} = V\left[1 - (T_n/\cos\theta - L/K_3)/X_n\right]^{0.5} \text{对厚混凝土} \tag{2}$$

式中：$S=0.245\ 78K_c(11-P)(t_cT_c)^{-0.06}(\sigma_c)^{-0.3}$，对于卵形弹头形状影响系数 N：$N=0.18(R_T/D-0.25)^{0.5}+0.56$，对于混凝土和岩石质量比例换算系数 K：$K=0.450\ 37\ m^{0.15}$，对于非冻土质量比例换算系数 K：$K=0.274\ 41m^{0.4}$。

其中 X——侵彻深度；

m——弹体总有效质量；

A——子弹横截面积；

V——子弹初速；

V_{ex}——单层穿透速度；

T_n——第 n 层靶板厚度；

θ——弹着角（与靶面法线方向夹角）；

L——子弹头部长度；

K_3——层面介质影响系数；

X_n——在第 n 层介质中 无界状态下的侵彻距离；

K_c——目标宽度影响系数；

P——混凝土中按体积计算的含钢百分率,%；

t_c——混凝土凝固时间，年，若 $t_c>1$，取 $t_c=1$；

T_c——目标厚度，以弹体直径为单位，若 $T_c>6$ 则取 $T_c=6$；

σ_c——材料抗压强度；

R_T——弹头部曲率半径；

D——弹体直径；另外攻角 β。初始条件见表 1。

表 1 侵彻深度计算已知条件列表

$V/(\mathrm{m\cdot s^{-1}})$	$\Theta/(°)$	L/m	K_3	K_c	P	t_c	T_{c1}	T_{c2}	混凝土 σ_c	碎石基础 σ_c	压实土基 σ_c	R_T/m	$\beta/(°)$
300	0	0.2	3	1	0	1	5	6	67 Mpa	20 MPa	5 MPa	0.52	0

由以上公式和已知条件计算结果见表 2。

表 2 侵彻深度计算结果

半无限混凝土靶侵彻深度 X_{con}/m	混凝土靶板穿透速度 $V_{ex}/(\mathrm{m\cdot s^{-1}})$	综合侵彻深度 X/m
0.490 2	169.722	0.759 9

1.3.2 爆破效应计算

对机场跑道的毁伤效应最终体现在爆炸成坑和隆起破坏区域的大小上，而其影响因素很多，如战斗部结构、装药形状、侵彻体总重量、爆炸深度、介质材料性质等，而多层介质中的作用过程更加复杂，因此成熟的理论或经验公式几乎没有。首先做几项合理简化。

战斗部壳体对爆炸的影响：炸药部分能量消耗在壳体的变形和破坏以及赋予破片一定的初动能上，而破坏壳体所需要的能量很少，经计算也就是占炸药总能量的1%左右，可忽略不计；又因为是在介质内部爆炸，则弹壳所形成的破片动能也将参与成坑，也不必单独计算。

装药形状影响：本战斗部不具有较大的长径比，因此简化为球型集中装药。战斗部质心到头部顶端距离 $L_z=0.23$ m，则起爆位置即最小抵抗线 h 为

$$h=X-L_z=0.759\ 9-0.23=0.529\ 9\ \mathrm{m}$$

由贝庆汉的 Π 项理论建立的函数式，并以美国空军实验室的试验数据为依据给出了如下炸坑经验公式：

$$\begin{cases} R_1/h = 0.9683\,(w^{1/3}/h)^{0.8650}, H/h = 1.2290\,(w^{1/3}/h)^{0.6833} \\ W^{1/3}/h = 1.1083\,(w^{1/3}/h)^{0.8227}, R_2/h = 1.711\,(w^{1/3}/h)^{0.9831} \end{cases} \tag{3}$$

其中 w——装药 TNT 当量；

R_1——开坑半径；

H——开坑深度；

W——开坑体积；

R_2——破坏半径。

由式（3）和相关已知条件计算结果见表 3。

表 3 计算结果

w/kg	h/m	R_1/m	H/m	W/m³	R_2/m	$S=\Pi R_2^2$
1.832	0.530	1.058	1.154	1.598	2.064	13.39

2 数值仿真

AUTODYN－2D 是由美国世纪动力公司开发的非线性动力学软件，它的应用范围很广，与其他力学软件相比，它更适于解决几何非线性和材料非线性的大变形问题，如爆炸冲击、高速碰撞等。本软件只有平面对称和轴对称两种可选的建模方法，本文问题所包含均为轴对称模型，且没有涉及斜侵彻和攻角侵彻，因此采用 AUTODYN－2D 软件并建立轴对称模型是比较合适的。

2.1 战斗部侵彻过程仿真

2.1.1 跑道靶体仿真模型

跑道靶体按照实际结构建立三层有限元模型，即 0.4 m 混凝土层；0.6 m 碎石基础层；0.5 m 压实土基层；砂找平层忽略不计；为减少边界影响，靶体宽设 0.84 m；均采用 SPH 算法，光滑粒子尺寸为 Φ5 mm；靶体模型侧边界和底部边界用速度边条加以约束。模型如图 3 所示。

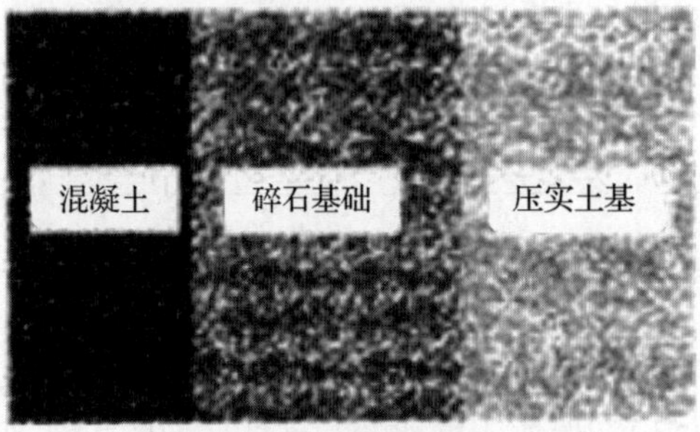

图 3 跑道靶体模型

2.1.2 战斗部模型

战斗部侵彻体模型选择 Lagrange 算法，内部为装药，侵彻体后部如动力舱、控制舱等相关结构作配重加在其尾端，并一体化处理。因壳体和装药头部是主要的受力部位，因此网格划分较密，并设观察点，模型如图 4 所示（对称映象）。

图 4 战斗部仿真模型

2.1.3　材料模型及参数设定

本计算中共包含5种材料模型，其基本参数由表4给出。因为采用带与不带损伤的混凝土模型侵彻深度计算结果相差不大，各材料均采用不带损伤模型，并采用几何应变侵蚀算法以真实的模拟混凝土的破坏情况。装药因算法问题未做冲击起爆模型处理，设置Vonmises模型计算侵彻冲击过程中装药中应力强度考察装药安全性，装药临界应力炮射法为250 MPa，模拟法为600 MPa。所设置的侵彻体质量与设计值一致。

表4　5种材料模型的基半参数

模型	材料	状态方程	强度模型	$\rho/(g \cdot cm^{-3})$	σ_s/MPa
侵彻壳体	高强度合金钢	Shock	Johnson – Cook	7.75	1 539
装药	8702装药	JWL（Explosive）	Vonmises	1.794	250
头层靶	混凝土	Porous	Mohr – Coulomb	2.41	67
中层靶	碎石	Linear	Mohr – Coulomb	2.3	20
后层靶	压实黏土	Linear	Vonmises	2.0	5

2.1.4　仿真结果

（1）因初始子弹距靶板10 mm间距，初始速度300 m/s，故子弹自0.033 3 ms时开始侵彻靶板；混凝土层板的穿透速度160 m/s。

（2）子弹在7.089 ms时达到最大侵深731 mm，此时子弹开始具有x轴负方向速度。

（3）子弹在侵彻约1.107 ms时出现x轴方向最大过载值1.486×1 049。

（4）战斗部壳体头部应力最大瞬时脉冲达到1 100 MPa，小于壳体的抗压强度1 539 MPa。

（5）装药头部应力最大瞬时脉冲达到230 MPa，小于装药的安全限250 MPa；仿真模型及结果如图5～图11所示。

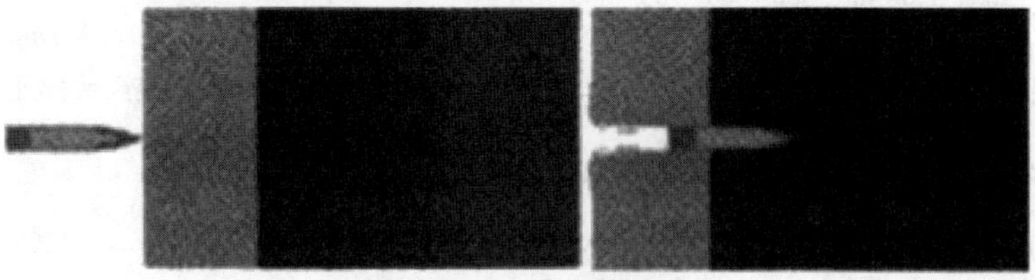

图5　侵彻前后弹靶模型

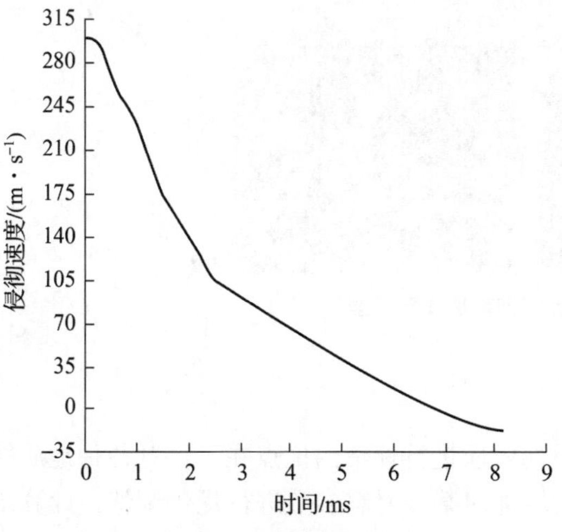

图6　侵彻速度曲线

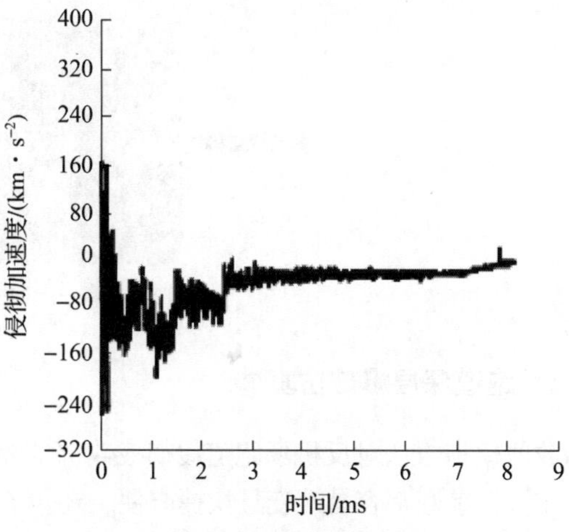

图7　弹体加速度曲线

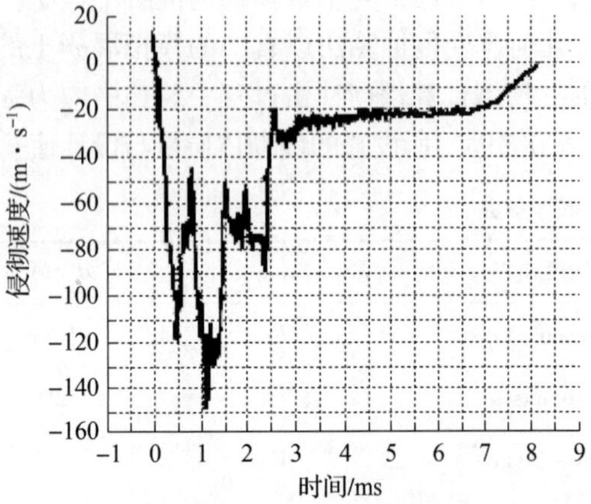

图 8　弹体加速度滤波厚曲线

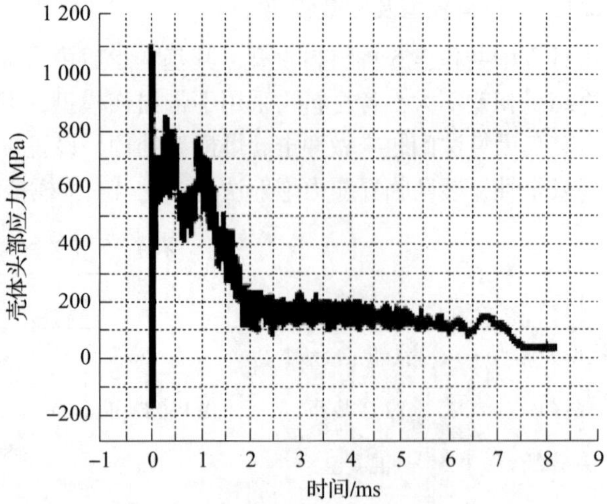

图 9　壳体头部压力曲线

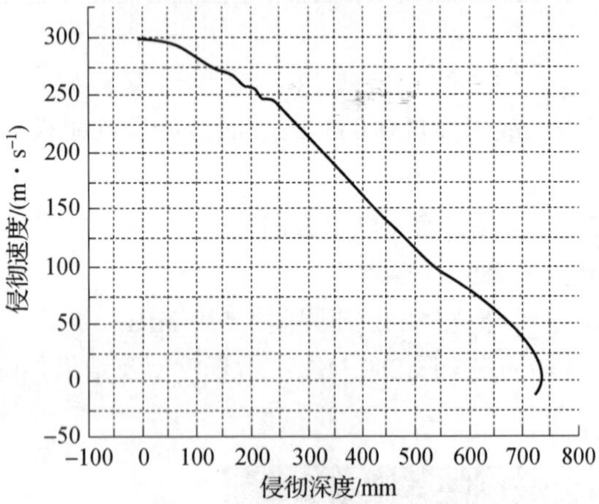

图 10　侵彻速度随侵深变化曲线

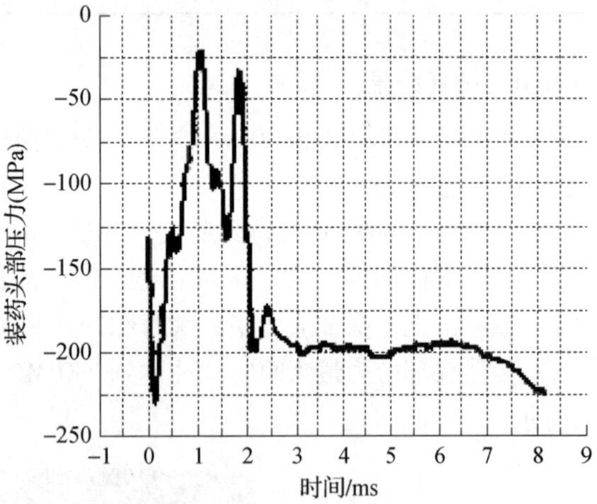

图 11　装药头部压力曲线

2.1.5　单层混凝土仿真结果

为了与该战斗部现有试验数据进行比较，本文对战斗部零攻角正侵彻单一混凝土介质的过程进行了仿真，靶板模型厚度 1 m，其他参数设置与上面多层靶中相同。结果如图 12 所示。2.7 ms 时侵彻极限深度 453 mm；背面崩落深度 225 mm。

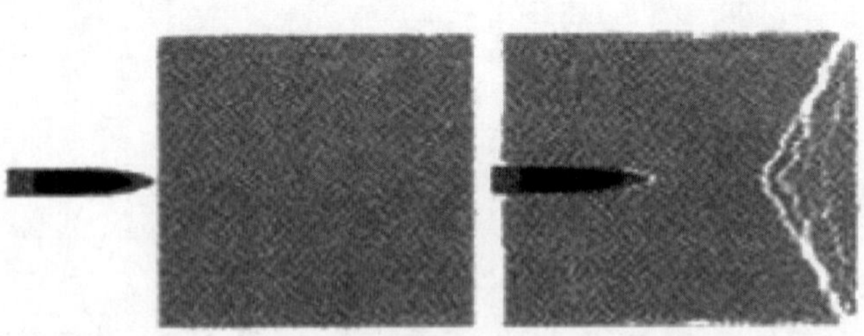

图 12　单一混凝土靶板侵彻前后弹靶模型

2.2　跑道深层爆破仿真

用数值模拟方法对反机场跑道侵彻战斗部的爆炸毁伤效应进行评估的优点在于：对各种地质结构，几何参数和材料参数都容易改变且快捷方便，还可有针对性地对某些材料参数进行优化调整，计算结果既能记录又能比较，而且直观。

2.2.1　弹靶模型及材料模型

为了真实地模拟混凝土层地毁伤效果，弹靶均采用SPH算法建模。为了降低战斗部建模的复杂程度优化算法从而得到相对准确的仿真结果，并考虑到壳体变形和破坏所消耗的炸药能量仅占总能量的1%左右，因此这里将战斗部做简化处理，即不考虑战斗部的结构和壳体，又因装药长径比较小，这里简化为球型装药。根据前面理论计算的结果，并考虑到减小边界影响，设靶板宽度8 m。靶体厚度按照实际结构即0.4 m混凝土层；0.6 m碎石基础层；0.5 m压实土基层；砂找 平层忽略不计。光滑粒子尺寸靶板为Φ10 mm、装药为Φ5 mm；靶体模型侧边界和底部边界用速度边条加以约束。高强度混凝土的上表面为自由面，靶板的两个侧面设为无反射压力界面，底部设为压力流出边界，以模拟侧面的无限大尺寸，这样可以在尽量满足计算精度的条件下减小靶板模型的几何尺寸。

跑道靶体材料模型的设定与表4相同，只是装药用当量TNT代替，计算得到的装药质量与设计当量值一致即1.832 kg，又TNT装药密度1.639/cm^3，得装药直径Φ130 mm。装药爆速6 990 m/s，爆热7.41 × 10^6 KJ/m^3，埋设深度采用理 论计算爆破中心位置近似取535 mm。模型如图13所示。

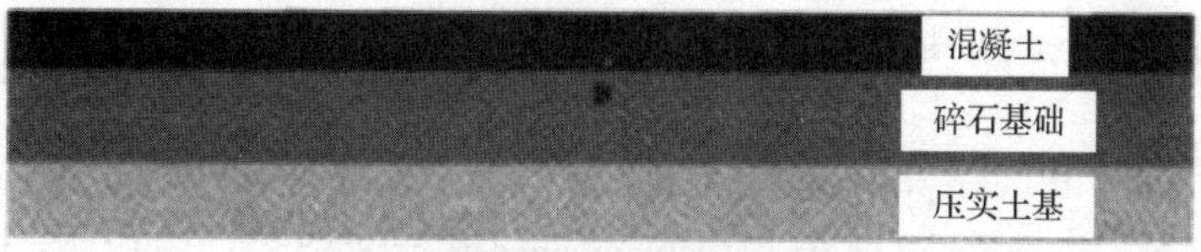

图13　弹靶初始状态

2.2.2　仿真结果

仿真共进行到36.4 ms，此时对混凝土层的破坏以基本定型，作用效果如图14所示。

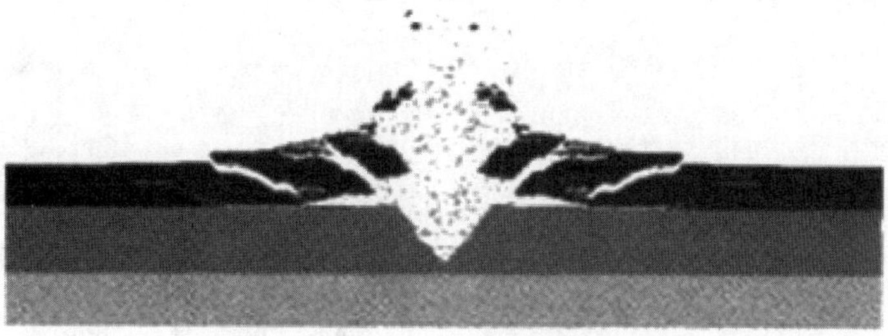

图14　爆炸36.4 ms时作用效果

仿真结果为：敞坑半径R_1 =1 110 mm，破坏半径R_2 =2 155 mm，破坏面积S =14.59 m^2，破坏坑深H = 980 mm。

3　结果对比分析

以上得到了战斗部侵彻爆破机场跑道多层靶板的经验公式计算结果与数值仿真结果，由于试验数据有限，这里只给出该战斗部以初始速度300 m/s零攻角正侵彻厚混凝土（各项参数不变）时的侵彻深度值，以对另外两种方法得到的结果做可信度的评价。其结果如表5所示。

表5

方法	X_{con}/m	V_{ex}/(m · s^{-1})	X(m)	R_1/m	R_2/m	S/m^2	H/m
经验公式	0.490 2	169.722	0.759 9	1.058 2	2.064 2	13.39	1.153 6
数值仿真	0.453	160	0.731	1.110	2.155	14.59	0.980
试验值	0.416 5	—	—	—	—	—	—
误差/%	15.03	5.73	3.80	-4.90	-4.40	-8.96	15.05

各组数据均显示该种结构的侵爆战斗部在以 300 m/s 的初速零攻角正侵彻机场跑道时，弹体强度满足要求不会产生明显变形，峰值过载不会使装药应力超过安全限而发生早炸。而且可以使爆破中心位于混凝土层面以下，从而使开坑半径超过 1 m，破坏面积超过 13 m^2，满足了战术指标要求。

通过表 5 数据可以看出无论是侵彻过程还是爆破效应，理论计算值与数值仿真结果相差都不是很大，其中对于单一混凝土介质的侵彻经验公式与试验值的误差也在可以接受的范围之内，三种方法相互印证，所得各项数据对于该侵爆战斗部的设计和进一步完善有着一定的指导和参考意义。理论计算与数值仿真的结果具有较高的可信度，也可为同类战斗部的方案设计提供参考。

参考文献

[1] 钟仁. 军用机场 [M]. 北京: 战士出版社, 1980: 33 – 37.

[2] 李晓军, 张殿臣. 常规武器破坏效应与工程防护技术 [M]. 洛阳: 总参工程兵科研三所, 2001: 51 – 140.

[3] 杨冬梅, 王晓鸣. 反机场弹药斜侵彻多层介质靶的三维数值仿真 [J]. 弹道学报, 2004, 16 (3): 1 – 3.

[4] 黄正祥, 张先锋, 陈惠武. 飞片起爆条件下聚能杆式侵彻体成型数值模拟 [J]. 系统仿真学报, 2005, 17 (8): 1 – 3.

HUANG Z X, ZHANG X F, CHEN H W. Simulation on the mechanism of jetting projectile charge under the condition of fragment detonation [J]. Journal of system simulation, 2005, 17 (8): 1 – 3.

[5] 雷宁利, 唐雪梅. 侵彻子母弹对机场跑道的封锁概率计算研究 [J]. 系统仿真学报, 2004, 16 (9): 1 – 3.

LEI N L, TANG X M. Research on blockage probability of intrusive submunition missile for airdrome runway [J]. Journal of system simulation, 2004, 16 (9): 1 – 3.

深层目标易损性与串联钻地战斗部动态侵彻分析①

冯顺山，董永香，王云峰②

（北京理工大学爆炸科学与技术国家重点实验室，北京 100081）

摘　要：本文针对目前常规钻地弹的作用效应，对深层目标的易损性进行了分析，并对常规钻地弹的易损性研究方法进行了探讨。通过串联钻地战斗部引战配合动态侵彻混凝土靶试验，分析了两级战斗部对混凝土靶的联合侵彻效应，从而得以实现两级串联战斗部对深层目标的有效打击。

关键词：深层目标；钻地弹；易损性；侵彻试验

引言

海湾战争和伊拉克战争后，各国更加重视钻地武器和防护结构的研究。尤其近几年随着钻地武器向着小型化、隐身化、智能化方向发展[1-4]，钻地弹对深层目标的易损性分析已成为国内外学者研究钻地武器和防护结构的热点，但也是大家公认的“硬骨头”，如何有效开展钻地武器对深层目标的易损性，成为首待解决的问题。本文结合国内外文献中钻地弹对深层目标的研究工作，分析了钻地弹对深层目标的作用效应，并在此基础上开展了深层目标易损性分析及其研究分析方法，最后结合研究工作，进行了两级串联钻地战斗部实现对深层目标有效打击的可行性动态侵彻试验分析。

1　钻地弹对深层目标的作用效应分析

1.1　毁伤元及作用效应

当钻地弹侵彻到地下一定深度时爆炸，与裸露爆炸相比将形成较为有效的威力场。钻地弹对目标的毁伤能力是通过毁伤元种类、数量及毁伤元的相对威力来度量的。毁伤元是战斗部爆炸后释放的对硬目标具有毁伤能力的毁伤元素，常规杀伤爆破战斗部释放的毁伤元主要包括侵彻冲击、载荷作用时间、冲量、能量、爆炸波以及破片[5]，此外还有炽热的烟尘、冲击波转换的应力波以及穿爆形成的联合作用效应，还包括由核战斗部产生的核辐射及热辐射，以及其他类战斗部形成的温压、耗氧、电磁干扰、燃料空气、化学战剂等其他类毁伤元。

1.2　对目标毁伤的影响因素分析

常规钻地战斗部对目标毁伤的有效性取决于弹目交汇条件、战斗部的侵彻能力、引信作用正确性和战斗部的爆炸威力等。当侵彻深度达到一定要求，爆炸威力较小时，对目标的毁伤效果并不理想，同样地侵彻能力达不到相应要求，即使爆炸威力相对较大，毁伤效果也不理想。因此，冲击爆炸传入地下的力学能比例同爆点位置关系非常密切，只有当弹目交汇位置正确、足够的侵彻能力与爆炸威力相匹配时，才可以发挥高效毁伤作用。装药几公斤、十几公斤的小当量战斗部，高精度命中钻入深层目标内部或打击超深层目标的通信系统、通风系统等薄弱部位时，同样可以发挥极好的作用，高侵彻能力、高命中精度可以弥补小型战斗部毁伤威力的相对不足，增加对深层硬目标作用的有效性。图 1 为常规钻地战斗部的毁伤作用框图。

---

① 原文发表于《战斗部与毁伤效率专业委员会第十届学术年会论文集》2007。

② 王云峰：工学博士，2002 年师从冯顺山教授，研究深层目标易损性及爆炸毁伤效应，现工作单位：吉林白城 31 基地。

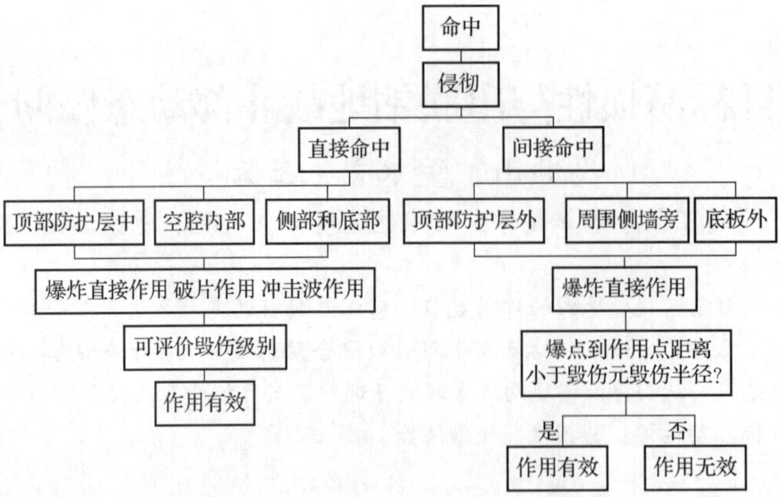

图1 常规钻地战斗部的毁伤作用框图

2 深层目标易损性分析

深层硬目标的易损性泛指目标对毁伤元作用的敏感性，包括结构易损性和功能易损性。常规钻地弹命中进入目标内部爆炸，是攻击地下坚硬目标的一种有效方法，可以较大程度地使目标功能彻底丧失，根据钻地武器落点的统计分析，当钻地弹对入口、通信与电子设备、电源及通风系统作用时，同样可以有效切断与外界的联系，内部系统功能失效，达到打击目的。

目前两种典型的常规钻地弹之一是通过弹丸较高的速度获得较大的侵彻深度，在深层爆炸对目标作用，主要通过发射和火箭助推方式来提高弹丸速度；另一种常规钻地弹是采用串联战斗部方式，前级通过高速侵彻体或射流开坑扩孔，后级动能弹随进侵彻爆炸，二者均可对深层硬目标进行有效打击，对深层目标的作用途径及目标的易损性[5-7]主要包括：

（1）近目标炸塌。当战斗部侵彻到深层目标顶部、侧壁附近爆炸后，爆源附近的介质会得到很大的速度随爆轰产物抛射出去，作用到防护目标的防护层上，使防护层被炸塌，抛出的介质及爆炸产物飞入目标内部，爆炸冲击波、战斗部破片及防护层碎片进入目标内部，毁伤内部人员及设备。

（2）局部拉伸破坏。当爆炸产生的能量不足以产生炸塌性的破坏机制时，爆炸形成的冲击波传播到主体防护层，防护层内层自由界面因拉伸波作用形成层裂，碎片抛掷到目标空腔内，毁伤内部人员和设备，破坏防护层结构及结构层内的管道、布线等。

（3）目标防护层中的爆炸毁伤作用。战斗部在主体防护层内爆炸，直接炸塌防护层，战斗部爆炸产物及防护层介质碎片直接毁伤工事内部人员和设备，毁伤效率较高。

（4）目标空腔内部的爆炸毁伤作用。当战斗部钻入目标空腔，高智能引信识别空腔后起爆，破片、冲击波对目标内人员和设备直接杀伤破坏，显然这是对目标最致命的毁伤，是爆炸对内部人员和设备的直接作用，毁伤效率最高。

（5）底板近处的爆炸毁伤作用。当战斗部终点弹道呈J形从远端侵入，钻入目标底板附近；另外战斗部穿过腔室，引信识别感知空腔，起爆时间延迟而钻入底板附近爆炸，此时对目标的毁伤效果近似于内爆毁伤。

（6）对目标保障系统的爆炸毁伤作用。破坏深层硬目标的口部、通风口、电源或通信系统等，可以造成整体目标功能的全部丧失或部分丧失。对一些深层、超深层目标，一般侵彻能力的战斗部很难对其主体结构构成威胁，但其口部的防护能力比起主体的防护能力来说相对脆弱。目标口部受到破坏，可以使目标功能部分或全部（或在一定时间内）丧失，如果战略导弹发射井口部受到攻击，可以使导弹发射井发射功能全部失效。

根据上述深层目标易损性特点，可以开展以下分析：①硬目标分析：包括目标分析、功能分析和结构分析。目标功能分析是研究目标所担负的战略战术任务；结构分析指分析目标各系统之间的相对关系及其与目标功能之间的关系。②毁伤元分析：分析战斗部产生毁伤元的原理，分析毁伤元的毁伤机理和毁伤能力，分析毁伤元的特征量及毁伤作用场。③毁伤元和目标交汇可能性分析：根据终点弹道的特点，找到具有共性意义的交汇位置典型分析。④目标易损概率分析：包括分析目标易损面积、目标纯易损面积、战斗部命中目标概率、命中目标局部后其局部毁伤的概率、局部毁伤后导致目标毁伤的概率、目标局部的权值等。⑤建立毁伤准则：分析目标防护层各部分在毁伤元的作用下的毁伤机理和毁伤模式，确定毁伤元的毁伤特征量与作用部位毁伤级别的关系等。⑥划分目标毁伤级别：根据目标结构毁伤状况或根据目标的功能毁伤来完成。⑦目标分系统毁伤评估：比如通风系统的毁伤评估、通信系统的毁伤评估、入口的毁伤评估、对生命保障系统的毁伤评估以及防护层或侧面墙壁等主体结构的毁伤评估等。⑧目标毁伤评估：根据目标主要分系统的毁伤情况确定与目标总体毁伤之间的关系，计算各个分系统的毁伤概率，或者根据目标的各个分系统的毁伤对目标总体功能的影响程度，确定目标的总体毁伤概率。深层目标易损性分析可以利用先进的实验测量手段进行少量野外实验或借助一些文献资料的数据基础上，开展大量的理论和数值研究，可以利用局部战争有用信息，利用改造废的旧矿坑、报废的地下工程进行等比或缩比试验[7]，获得相关数据。

3　串联钻地战斗部动态侵彻分析

侵彻能力是否达到相应要求，是实现钻地战斗部对目标有效打击的保障。下面将结合所研究的工作，在两级串联钻地战斗部引战配合动态侵彻综合试验基础上，分析两级串联钻地战斗部对深层目标作用实现有效打击的可行性。

图 2 为侵彻混凝土靶的串联战斗部原理结构示意图，前级为聚能装药，后级为随进战斗部，通过前级聚能侵彻体预侵彻形成孔道，后级侵爆战斗部随进侵彻，在预定侵深爆炸毁伤深层目标。目前，通过引战配合动态缩比试验和基于缩比试验的全尺度数值仿真，所设计的两级串联钻地战斗部可穿透大于 6 m 厚的钢筋混凝土（强度 30～40 MPa），或等同 6 m 厚的多层结构的钢筋混凝土目标。下面将主要分析串联战斗部对混凝土靶的动态侵彻性能。图 3 为试验布置图，图 4 为两级战斗部联合作用后毁伤混凝土靶板效果，由图 4 可知，两级战斗部在前级预开坑、后级随进作用下，对混凝土靶的侵彻性能大大增加[6]。

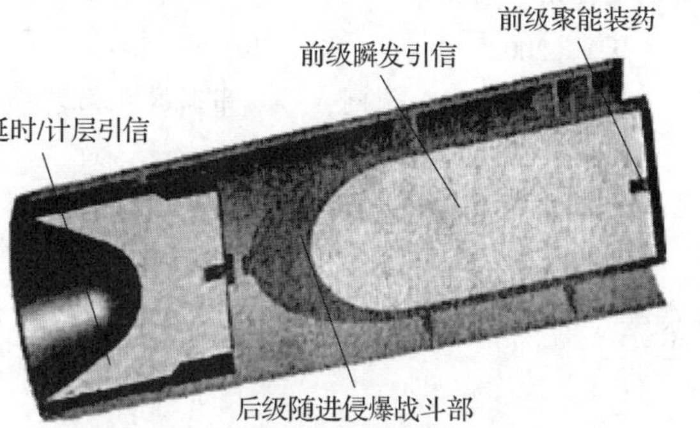

图 2　侵彻混凝土靶的串联战斗部原理结构示意图

图 3　试验布置图

图 4　两级联合作用后的混凝土靶板效果

通过串联战斗部动态侵彻混凝土靶试验，结果表明：

(1) 前级聚能战斗部的开坑和形成孔道，与后级侵爆战斗部在其基础上的继续侵彻，使得两级侵彻威力叠加，减小了混凝土靶的抗侵彻性能；

(2) 串联战斗部的两级在同一轴线上先后作用，由于前级的开孔效应使得后级抗弹道弯曲或跳弹能力显著加强，使其有效侵彻深度明显提高，由此可见串联战斗部在侵彻过程中，不跳弹，弹道不发生偏斜，对目标作用有效性好；

(3) 所设计的两级串联钻地战斗部可侵入深层目标有效作用区以实现进一步的有效打击。

4 结束语

深层目标易损性的分析需要选择合理可行、具有工程实践指导意义的科学方法，面对地下深层目标结构复杂、功能多样的系统以及非线性耦合动载荷，开展有效的目标易损性分析与高效能钻地弹研究既面临着挑战，也充满了机遇。

参考文献

[1] 石艳霞，郝丽萍．国外钻地武器的技术特点及发展趋势［J］，导弹与航天运载技术，2003（4）：50－53.

[2] 王涛，余文力，王少龙．美军钻地武器的现状及发展趋势［J］．飞航导弹，2004（8）：4－8.

[3] 林俊德．侵地武器及其气炮实验［J］．中国工程科学，2003，5（3）：25－33.

[4] 邓国强，周早生，郑全平．钻地弹爆炸聚集效应研究现状及展望［J］．解放军理工大学学报（自然科学版），2002，3（3）：45－49.

[5] 段建，杨黔龙，周刚，等．串联战斗部前级聚能装药和隔爆结构设计与实验研究［J］．高压物理学报，2006，20（2）：202－206.

[6] 李晓军，张殿臣，李清献，等．常规武器破坏效应与工程防护技术［R］．总参工程兵三所（内部），2001.

[7] 王云峰．深层目标易损性及爆炸毁伤效应分析［D］．北京：北京理工大学，2006.

不同硬度弹丸对中厚靶板作用的试验研究①

吴 广[1②]，冯顺山[1]，董永香[1]，赵宜新[2]

（1. 北京理工大学爆炸科学与技术国家重点实验室，北京 100081；
2. 安徽军工集团 9374 厂）

摘 要：弹体材料的热处理硬度对弹丸的侵彻性能有着重要的影响。通过三种不同硬度的30CrMnSiNi2A 弹丸在低速下对船用钢板的侵彻试验研究得到，弹材硬度在 HRC41－47 范围内，其弹头部的破碎程度随着其硬度的提高而降低，下限硬度弹丸其动能主要消耗在弹头部自身的破碎过程中，基本上不能在靶板上形成侵彻孔道，上限硬度弹丸动能主要消耗在侵彻过程中对靶板的扩孔过程中，并且弹体 的破碎也主要是在弹靶碰撞初期，当弹丸头部进入靶板后，基本不会再发生破坏，弹丸的侵彻能力也随着弹体硬度的提高而提高。本文研究结果可为弹靶作用分析和相关设计提供重要的参考价值。

关键词：侵彻；30CrMnSiNi2A；硬度；弹体破坏

0 引言

弹丸对靶板的侵彻问题一直以来是人们关注的重点，而影响弹丸侵彻过程的因素很多，其中弹靶材料性能对弹丸的侵彻能力有着重要的影响，不同的弹靶材料关系存在着不同穿甲现象。弹体的热处理硬度是衡量弹体材料性能的重要参数，不同硬度的弹丸其侵彻性能不同。文献［3－5］中作者研究了弹丸对不同硬度靶板的侵彻情况以及靶板的破坏模式。而硬度对弹丸对侵彻性能影响的研究较少。30CrMnSiNi2A 以其强度高、韧性好被广泛用来作为生产制造穿甲弹丸的材料。本文主要通过试验方法来研究不同硬度的 30CrMnSiNi2A 在弹丸低速情况下对靶板的侵彻性能。

1 试验材料和方法

1.1 试验材料

30CrMnSiNi2A 属于低合金超高强度钢，它是在高强度钢 30CrMnSiA 的基础上加入（1.4%~1.8%）的 Ni 而得到的。Ni 的加入，提高了钢的强度、韧性和塑性，并大大提高钢的淬透性。低合金超高强度钢原材料和制造费用相对较低，其性能较好，广泛用于各种导弹发动机壳体、飞机起落架以及穿甲弹壳体[6]。试验所用 30CrMnSiNi2A 材料化学成分如表 1 所示[7]。30CrMnSiNi2A 钢机械性能如表 2 所示[7]。

表 1 30CrMnSiNi2A 材料化学成分 单位：%

元素	C	Si	Mn	Cr	Ni	Cu	P	S
含量	0.26~0.32	0.90~1.20	1.0~1.3	0.90~1.20	1.40~1.80	≤0.20	≤0.035	0.03

① 原文发表于《兵工学报》2010（S1）。

② 吴广：工学博士，2006 年师从冯顺山教授，研究弹药对飞行甲板的半侵彻机理及技术，现工作单位：北理工重庆创新中心。

表 2　30CrMnSiNi2A 钢机械性能

材料	热处理毛坯	热处理	σ_b/MPa	$\sigma_{0.2}$/MPa	δ_5/%	ψ/%	a_k /(J·cm^{-2})
30CrMnSiNi2A	试样	250 ℃回火，或 900 ℃加热，(260 ±20)℃等温淬火	≥1 570		≥9	≥40	≥59
30CrMnSiNi2A	试样	890 ℃ × 30 min 油淬，290 ℃ ×3 h 回火空冷	$\frac{1\ 570 \sim 1\ 860}{1\ 730}$	$\frac{1\ 255 \sim 1\ 730}{1\ 630}$	$\frac{9 \sim 19}{12.3}$	$\frac{45 \sim 57.3}{50.5}$	$\frac{59 \sim 122}{96}$

1.2　试验方法

试验系统布置如图 1 所示，85 滑膛炮在距靶 6 m 处水平发射试验弹丸，撞击巧度斜靶板，靶前 1 m 处设置靶网测量弹丸着靶速度 v_0，在靶侧面 20 m 处设置高速运动分析系统，另一侧设置标志板，观测弹体着靶时和过靶后的弹道曲线及偏转角，同时测量弹丸着靶速度和过靶后的剩余速度。回收装置回收穿过靶板的弹体，观测弹体的变形和损伤情况。试验弹长径比为 2.5 ~ 3，头部为截锥形，弹体结构示意图如图 3 所示。

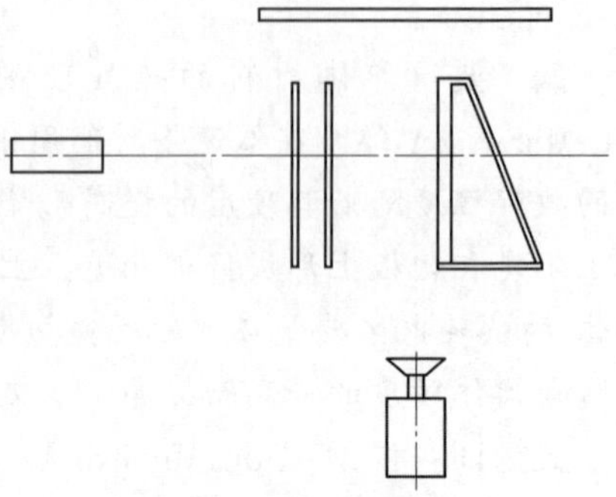

图 1　试验系统布置

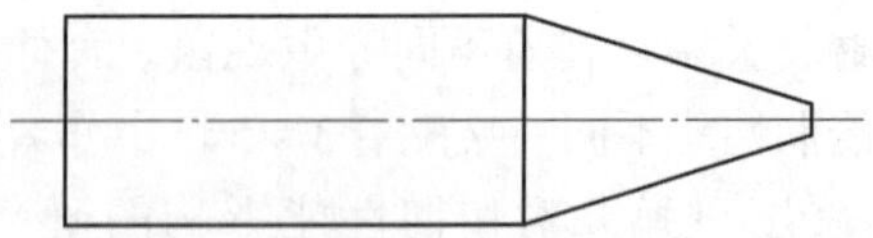

图 2　中硬度残余弹体断口图

图 3　弹体结构示意图

2　试验结果及分析

2.1　试验结果

弹体材料以（900 ±10)℃的温度进行淬火，通过不同的回火温度和保温时间来得到不同的弹体硬度；试验所用的弹丸的弹体硬度分别为：HRC41 ~ 42、HRC43 ~ 44、HRC46 ~ 47。由《钢铁硬度及强度换算表》可知[8]，三种硬度弹丸对应的平均抗拉强度分别为 1 322 MPa、1 405 MPa、1 547 MPa。

试验用靶板为船用钢板，靶板尺寸为 1 200 mm × 1 200 mm，厚度为 40 mm，其静态屈服强度为 550 ~ 600 MPa。通过调节发射药量来控制弹丸着靶速度。

试验共发射了 3 发试验弹，三种硬度弹丸各一发，速度均在 270 m/s 左右，试验结果如表 3 所示。由于弹丸着靶速度低于靶板的极限穿透速度，因此 3 发弹丸的靶厚剩余速度均为 0，试验中在靶前 1 m 处找到残余弹体，靶板整体变形很小。

表 3　试验结果

序号	弹体平均硬度（HRC）	着靶速度/(m·s^{-1})	弹靶作用效果	头部碎裂长度/mm
1	41.5（HRC -1）	272	未侵入，弹丸从靶上反弹	73
2	43.5（HRC -2）	278	半侵入，弹丸反弹出靶板	37
3	46.5（HRC -3）	268	靶板穿透，弹丸反弹出靶板	17

2.2 弹丸的侵彻性能分析

观察并测量残余弹体和弹坑情况，分析不同硬度弹丸的侵彻情况。3 发试验的弹体破坏情况与弹坑如图 4 ~ 图 6 所示。

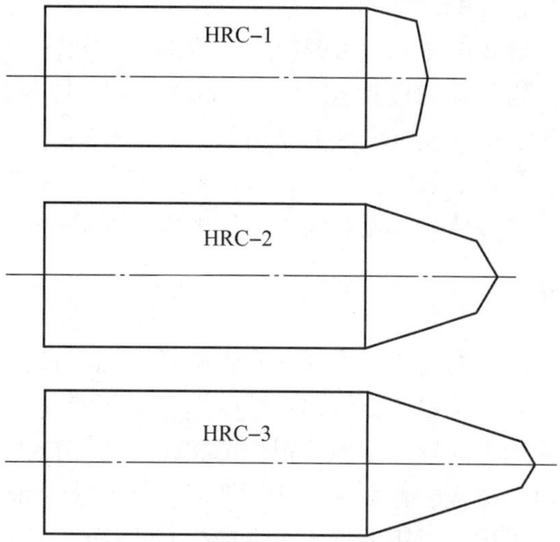

图 4 弹体破碎长度与弹材硬度关系

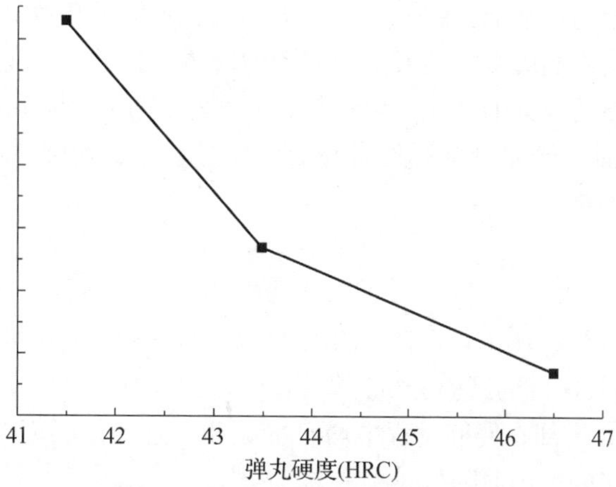

图 5 三种硬度弹丸弹体破坏示意图

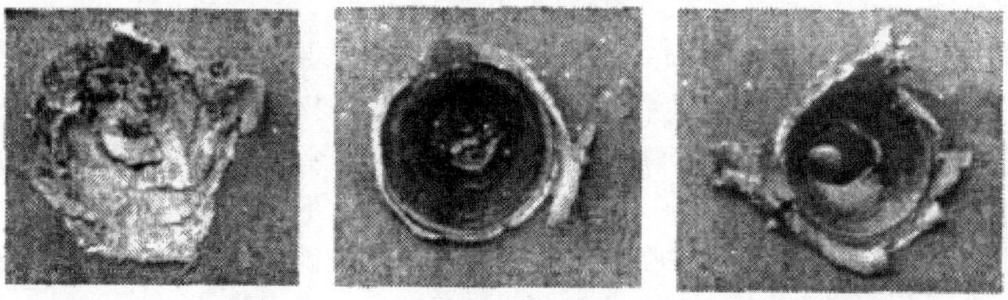

(a) (b) (c)

图 6 不同硬度弹丸侵彻的弹坑对比图

三种硬度的弹丸撞击靶板过程中，由于撞击初始阶段的高压将头部均产生破碎，由图 4 可知，弹体断面与水平面夹角为 45° ~ 50°，这主要是由于弹丸垂直撞击靶板过程中，弹体主要受到轴向的压缩应力，而由 Tresca 准则可知，材料屈服取决于最大剪应力，即当最大剪应力达到一定值时，材料开始屈服。而弹体头部受到单轴的压缩时，其最大剪应力方向与主应力方向成 45°夹角，并且断口部分区域发亮，这主要是由于弹丸在撞击过程中由剪应力引起的剪切破坏。

由图5和表3可知，三种不同硬度的弹丸撞击靶板后，低硬度弹丸头部破碎程度最大，高硬度弹丸破碎程度最小，中硬度弹丸居中。由图4可以看出，弹体头部破碎长度随着弹体硬度的提高而降低。

由于靶板材料具有良好的塑性，而且弹丸头部为圆锥形，靶板的主要破坏形式为延性扩孔为主，本试验为低速侵彻试验，因此弹靶材料的变形主要为塑性变形，弹坑形状、大小基本与侵入弹丸头部形状一致，这主要是由于在低速侵彻过程中，靶板材料将贴着弹丸的头部母线运动，并且在靶板正面形成翻起的唇边，从图6还可以看出，低硬度弹丸撞击靶板后形成的弹坑较浅，基本上没有形成侵彻孔道，而中硬度和高硬度弹丸均形成了一定长度的侵彻孔道，并且高硬度弹丸 侵彻时靶板已穿孔。由图6（b）未穿透靶板的弹坑可知，在弹坑底部并没有发现弹体头部损失的材料，同时弹坑底部形状与弹体头部的断口形状一致，因此可以推断，弹体头部的损坏是在着靶初期开坑阶段产生的，弹丸以破碎后形成的新的头部再进行侵彻，在侵彻隧道段弹体基本没有因侵蚀发生质量的损失。

2.3 侵彻过程的能量分析

在弹靶撞击瞬间，由于撞击产生的高压使得弹丸发生破坏，同时在靶板上形成开坑，之后弹丸将对靶板进行扩孔，在整个过程中伴随着靶板的整体弹塑性变形，由于弹丸的侵彻速度较低，弹丸不能完全穿过靶板，同时弹丸与靶板的作用时间较长，因此靶板的整体变形不能忽略。而当弹丸速度为0时，弹丸将不再对靶板发生侵彻。此时，靶板的弹性变形开始恢复，同时将会带动弹丸反方向运动最终脱离靶板，因此试验中在靶前发现残余弹体，而靶板整体将会产生自由振动，直到其动能完全消耗为止，此时靶板的整体变形为永久塑性变形。由以上分析可知，在侵彻过程中，弹丸的动能主要消耗以下几个方面：弹丸对靶板的扩孔过程中的扩孔能，弹丸头部的破坏能，弹丸的反弹动能，靶板整体的塑性变形能以及靶板的自由振动能，对比不同硬度下的残余弹体和弹坑图可以看出，低硬度弹丸的弹体破碎比高硬度弹丸消耗的能量大。而高硬度弹丸的动能主要消耗在对靶板的扩孔过程中。由图5可以看出，弹体硬度在HRC43以前，弹丸的动能主要消耗在弹体自身的破坏过程中，而弹体硬度在HRC43以后，弹丸的动能更多地用于在靶板上的扩孔。

3 结论

通过硬度分别为HRC41~42、HRC43~44、HRC46~47的三种不同硬度弹丸对40 mm厚、屈服强度为50~60 MPa的船用钢板的低速侵彻试验分析，得到以下结论：

（1）三种不同硬度的弹丸头部在侵彻过程中均产生破坏，且弹体断面与水平面夹角为45°~50°，弹丸头部的破碎程度随着其硬度的增加而降低。

（2）在本文试验条件下，弹丸对靶板的破坏形式主要表现为延性扩孔，并且弹坑形状与侵入靶板的弹丸头部形状一致。

（3）低速情况下，弹体硬度在HRC43以前，弹丸的动能主要消耗在弹体自身的破坏过程中，并且不能在靶板上形成侵彻隧道，而弹体硬度在HRC43以后，弹丸的动能更多地用于在靶板上的扩孔。

（4）低速侵彻范围下，在未穿透的弹坑底部没有弹体材料，弹体的破坏主要发生在弹靶撞击的初始阶段，而当由于撞击破坏形成新的头部进入靶板后，弹丸基本上不再发生破坏。

本文主要通过试验研究了不同硬度弹丸的侵彻性能，但影响弹丸侵彻性能的因素很多。为提高弹丸的侵彻性能，可优化弹丸结构或进一步提高弹丸材料性能。

参考文献

[1] 赵国志. 穿甲工程力学 [M]. 北京：兵器工业出版社，1992.
[2] 钱伟长. 穿甲力学 [M]. 北京：国防工业出版社，1984.

动能弹高速侵彻混凝土研究综述①

徐翔云[1,2]②，冯顺山[2]，高伟亮[1]，金栋梁[1]

（1. 总参工程兵科研三所，河南洛阳 471023；
2. 北京理工大学爆炸科学与技术国重点实验室，北京 100081）

摘　要：动能弹高速侵彻混凝土是目前相关领域研究的热点问题。当动能弹弹体超过极限速度时，弹体将出现弯曲、变形、断裂等情况，弹体的侵彻深度将显著降低，而弹体的质量损失明显增加。本文主要综述了动能弹侵彻混凝土时弹体极限速度、质量磨蚀方面的若干研究进展，包括试验研究、经验与理论模型分析，并对未来的研究方向提出一些建议。

关键词：高速侵彻；混凝土；质量损失；极限速度

1　引言

在未来战场上，高速钻地武器的使用对深埋地下坚固目标构成了严重威胁。开展弹体对混凝土类硬目标的高速侵彻效应研究对战斗部研制和防护结构设计都极具意义，当前该研究已成为相关领域共同关注的热点问题。

已有研究表明[1-5]，在常规速度范围（≤800 m/s）内对混凝土类目标侵彻弹体可视为刚体，随着侵彻速度的增加，侵彻机理逐渐由非变形体侵彻向半流体动力学侵彻转变，弹体不能单纯地作为刚体处理，随着速度的进一步提升，弹体的力学行为可按流体处理。弹体在半流体动力学侵彻阶段的侵彻机理比刚体侵彻阶段和流体侵彻阶段表现得尤为复杂，有许多的难点问题需要分析和解决，如上述不同侵彻阶段的临界速度、对应于最大侵彻深度的极限速度、弹体侵蚀和变形（破坏）的力学机理等。

本文主要综述了动能弹高速（弹体速度为800～1 500 m/s）侵彻混凝土时弹体的极限速度、质量磨蚀方面的若干研究进展，包括试验研究、经验与理论模型分析等，并对未来的研究方向提出一些建议。

2　试验研究进展

近年来，国内外学者针对高速动能弹侵彻混凝土类硬目标开展了大量研究。1996 年，Forrestal 等[1]利用4340 钢弹丸分别对砂浆和混凝土开展了动能高速侵彻试验，试验速度在1 200～1 700 m/s，获得不同撞击速度下弹体相应侵深，并完整记录弹体的质量磨蚀（图1）。试验发现，当弹体速度大于1 250 m/s，弹体会出现弯曲、断裂等现象；1998 年，Frew 等[2]利用 Aermet100Rc53 钢和 4340 钢弹体对混凝土开展类 似的高速试验，并对相同速度时不同弹体材料的质量损失率进行分析。以上两人共进行了 6 组试验数据，表 1 罗列出了相关弹、靶参数，其中工况 1～4 来自文献［1］，工况 5、6 来自文献［2］。初步分析表明，弹体质量损失与初始撞击动能存在线性关系，在刚体弹速度范围内，弹体质量损失可达 7%。2000 年，Jerome 等[3]对混凝土开展高速侵彻试验，侵彻速度范围为 442～1 500 m/s，积累了弹体高速正侵彻混凝土的基础数据。并在混凝土靶体中观察到了弹体侵彻路径偏移与头部磨蚀现象，如图 2 所示，验证了质量磨蚀使弹体发生轨迹偏离的可能。

---

①　原文发表于《第十二届学术交流会文集》2011 年第 13 期。

②　徐祥云：工学博士，2009 年师从冯顺山教授，研究高速动能弹侵彻混凝土/金属基混凝土效应与应用，现工作单位：军事科学研究院。

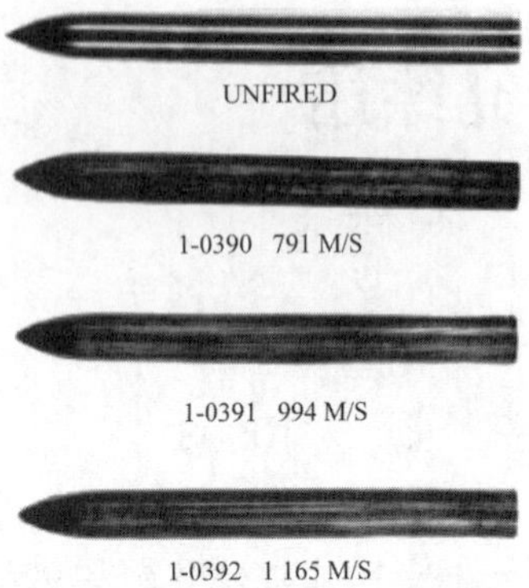

图 1 试验前后弹体对比图

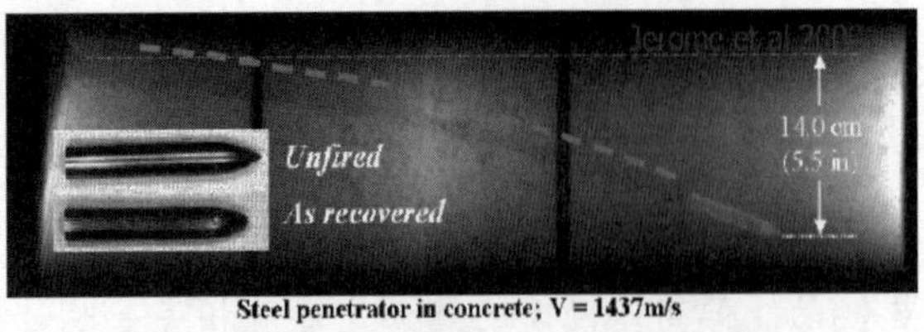

图 2 弹体头部侵蚀与侵彻路径偏移

表 1 文献 [1, 2] 中相关弹、靶参数

项目	工况 1	工况 2	工况 3	工况 4	工况 5	工况 6
靶体参数						
靶材	砂浆土	砂浆土	混凝土	混凝土	混凝土	混凝土
单轴抗压强度/MPa	13.5	21.6	62.8	51	58.4	58.4
密度/(kg · m^{-3})	2 000	2 000	2 300	2 300	2 320	2 320
弹体参数						
弹形	尖卵形空心弹			尖卵形实心弹		
头部 CRH	3.0/4.25	3.0/4.25	3.0	3.0	3.0	3.0
质量 M/g	64	64	478	1 600	478	1 620
弹径 D/mm	12.9	12.9	20.3	30.5	20.3	30.5
长径比 L/D	6.88	6.88	10	10	10	10
弹体材料及硬度 Rc	4 340 Rc39	4 340 Rc39	4 340 Rc45	4 340 Rc45	4 340Rc45 AerMet100 Rc53	4 340Rc45 AerMet100 Rc53

在国内，梁斌等[5]开展了先进钻地弹概念弹的高速侵彻缩比试验，最高撞击速度达 1 200 m/s。武海军等[6]开展了高速非正侵彻混凝土试验，撞击速度为 800 ~ 1 100 m/s。何翔等[4]将改进的次口径发射技术与常规火炮技术相结合，开展不同弹、靶组合下侵彻试验，侵彻速度为 800 ~ 1 500 m/s，分别考察了不同靶材（图 3、图 4）、不同弹体材料（图 5、图 6）对弹体侵彻深与相对质量损失的影响，得到弹体半流体转变时的极限侵彻深 P_m 和与之相对应的极限速度 V_m。发现弹体极限速度与极限侵深受弹体强度影响最大，靶体强度变化对它们的影响较小。而相对质量损失在半流体转变速度左右变化较大，刚性弹范围内，弹体质量损失在 2% ~ 6%，而超过半流体转变速度后，弹体质量损失陡增，可能达到 40%。何翔等[4]观察弹道轨迹时也发现，正侵彻弹体高速侵彻时弹道轨迹偏离原来的运动方向，与 Jerome 等[3]开展的相关试验中观察到的现象（图 2）一致。

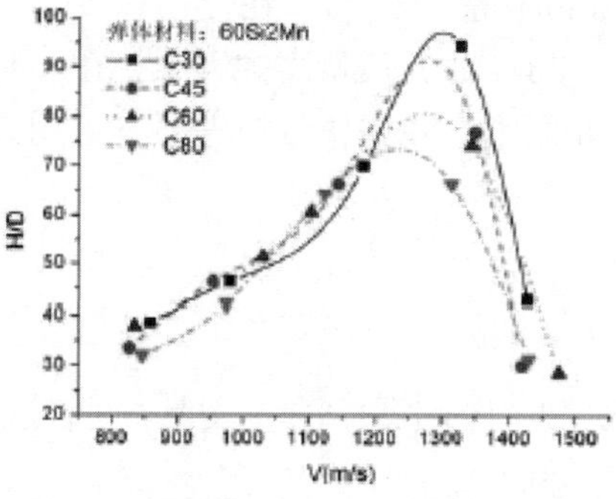

图 3 无量纲侵彻深度随速度的变化

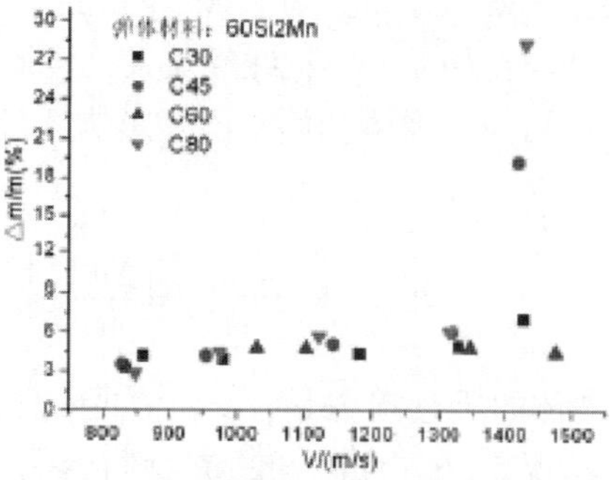

图 4 弹体质量损失率随速度的变化

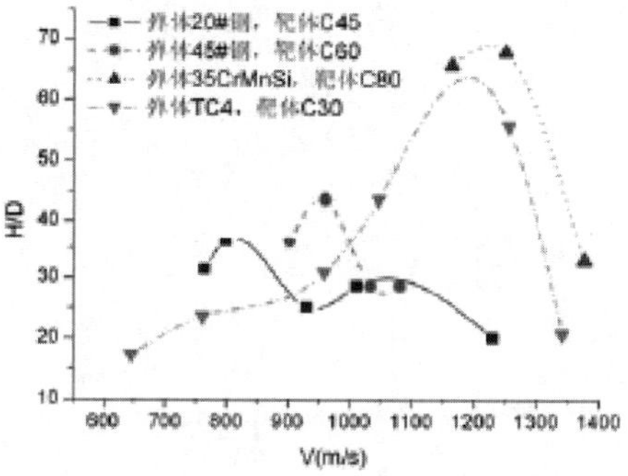

图 5 无量纲侵彻深度随速度的变化

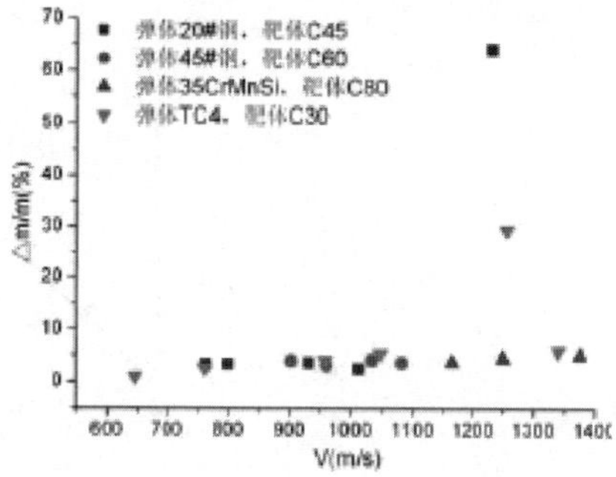

图 6 弹体质量损失率随速度的变化

3 理论研究进展

3.1 弹体质量损失

分析弹体的质量损失，必须正确理解弹体表面和靶体（如混凝土和岩石靶体）相互作用特性。不仅需要考虑不同的弹体材料，还需考虑不同的靶体材料，如混凝土、岩石和颗粒状材料（如砂）。混凝土中的骨料特性也可能令弹体的质量磨蚀机理更复杂。假定靶体力学性质已知（弹体表面局部压力可看作时间的函数，已知热传导系数等），相关理论研究可分为两部分：一是侵彻体表面动力学摩擦的理论和试验研究。主要着眼于摩擦系数的变化，或更一般地，涉及表面粗糙度的塑性动力学中的摩擦参数。有关这方面的工作在文献［7］中已比较完善，建立了单位表面积摩擦公式。二是极端条件下的磨损和侵蚀。尽管已有一些此领域的相关研究发表[8-9]，但在高或超高滑动速度（1 500 m/s 或更高速度）下，仍缺乏解决磨蚀热动力学的统一方法。在高速（1 200 m/s < 侵彻速度 < 1 700 m/s）侵彻时，弹体表面经历高温高压，在很短时间内发生了 磨蚀，高速磨蚀同时伴生若干物理现象，包括材料熔化、相变、金属屑氧化以及与靶板粒子间的化学反应，上述过程对弹体质量损失都有贡献。

由于问题的复杂性，Jones 等[10]假设弹体质量损失主要由头部材料熔化引起，忽略靶中骨料等切削引起的质量损失；摩擦功转化成热，全部作用于弹体，并由此建立了理论模型，提出了摩擦系数 μ。Davis 等[11]在此基础上进行了迭代计算，得到了剩余弹头轮廓函数，将它和剩余弹体头部照片对比，如图 7 所示，轮廓曲线和剩余弹体头部轮廓十分重合。

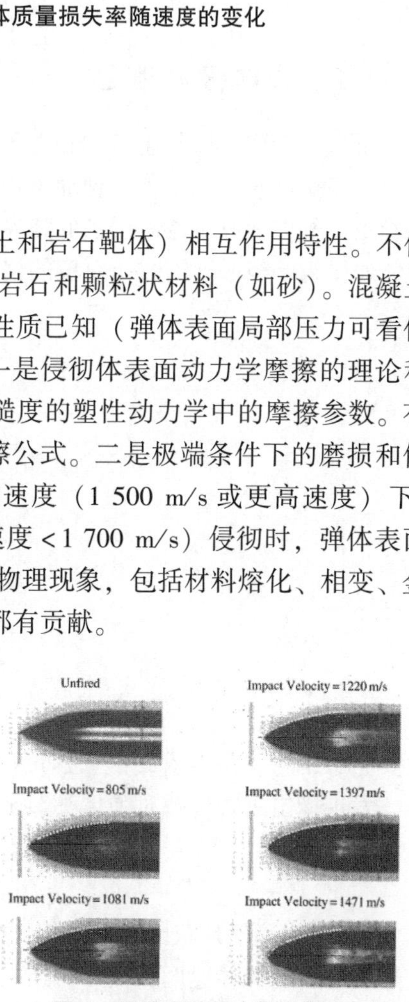

图 7 理论预测轮廓线和剩余弹头照片的重合情况图

在此基础上，陈小伟等[12]通过图形分析，明确了弹体质量损失主要发生于头部，并且侵彻后的弹头仍接近尖卵形，只是对应的曲径比 CRH 发生改变，可以通过弹头形状的变化预期弹体的质量损失。

另外，根据 Forrestal 等[1]和 Frew 等[2]的试验结果，Silling 等[13]指出，在侵彻速度 $V_i \leqslant 1$ km/s 时，弹体的质量磨蚀与初始动能存在线性关系；侵彻速度 $V_i > 1$ km/s 后，质量磨蚀则维持在一常数水平，如图 8 所示。由此可见，弹体初始动能是影响质量损失的主要因素。弹体质量损失率 γ 可用以下经验公式定量表述：

$$\gamma = \frac{\Delta m}{M} = \begin{cases} k \cdot V_i^2 & ,V_i \leqslant 1 \text{ km/s} \\ C & ,V_i > 1 \text{ km/s} \end{cases} \tag{1}$$

其中，Δm 是弹体的质量损失量，k 和 C 是常数，特别地，对于工况 1 ~ 4（文献［1］），$k_1 = 0.14$ 和 $C_1 = 0.07$，对工况 5 ~ 6（文献［2］），则有 $k_2 = 0.07$ 和 $C_2 = 0.035$。

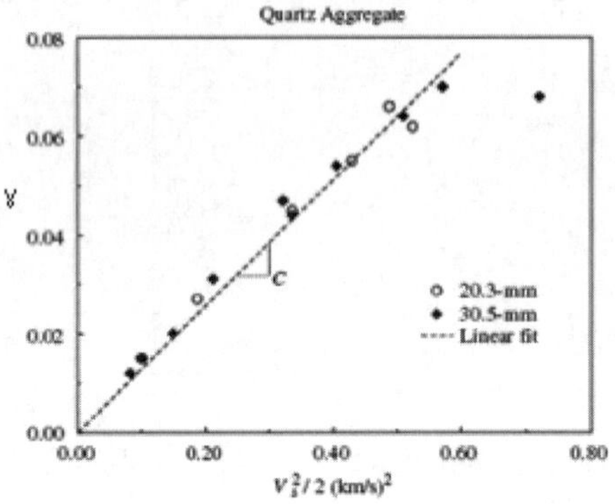

图 8　弹体的质量损失与初始撞击动能关系

3.2　弹体极限速度

早期关于弹体极限速度的相关研究主要集中在钢弹对金属靶的撞击上，Chen X W 等[14]、Forrestal 等[15]、Jones 等[10]在钢弹高速撞击铝靶方面都做了很多的研究，其中 Chen X W 等[14]利用撞击函数 I 分析确定了刚性弹和半流体侵彻的临界判据，如图 9 所示，分析了影响弹体极限速度的相关因素，并对当弹体速度大于极限速度时所对应的侵彻深度进行了分析。

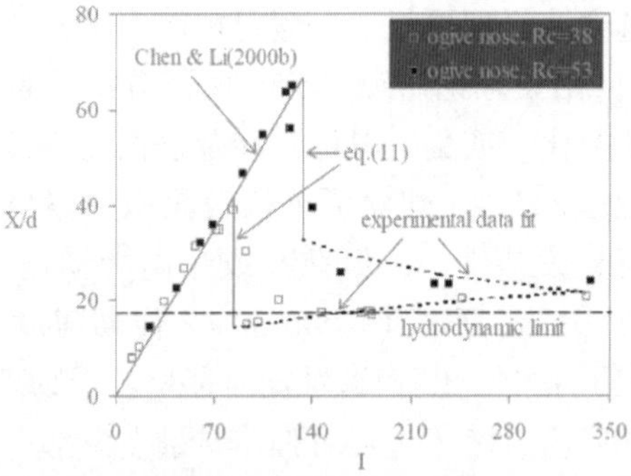

图 9　尖卵形弹侵彻铝靶的分析与实验

赵军等[16]提出了在考虑弹体头形磨蚀条件下极限侵深 P_m 和与之对应的极限速度 V_m 的估算公式。主要思想是将侵深表示成侵彻速度的单变量函数。拟合试验数据，将弹体头形函数与撞击初速度相联系，发现撞击过程中瞬时弹体头形函数与初始弹体头形函数存在线性关系，侵彻过程中的瞬时头形函数可以表示为

$$N^* = N_r^* - gV_i^* = N_i^* + g(V_s^2 - V^2) \tag{2}$$

式中，N_i^* 为弹头形状因子初始值；N_r^* 表示侵彻过程结束后的剩余弹体头形函数；V_s 与 V 分别表示撞击初速度与瞬时弹体速度；而 g 为经验常数，通过实验数据拟合得到。因此，可得到以撞击初速度为自变量的侵彻深度的显示表达式为

$$P=\frac{M_0}{2\pi d^2\rho_t g\chi}\left\{\left(8\rho_t g\left(1-\frac{1}{2}kV_s^2\right)+2\rho_t k(N_t^*+gV_s^2)\right)\left(\operatorname{arctan}h\frac{\rho_t(N_t^*+gV_s^2)}{\chi}-\operatorname{arctan}h\frac{\rho_t(N_t^*-gV_s^2)}{\chi}\right)\right.$$
$$\left.-k\chi\ln\left(1+\frac{\rho_t V_s^2 N_i^*}{Sf'_c}\right)\right\} \tag{3}$$

其中，M_0 为弹体初始质量，k 的定义见方程（1），ρ_t 为靶密度；Sf'_c 表示混凝土强度的影响，而 $\chi=\sqrt{4Sf'_c\rho_t g+(N_i^*+gV_s^2)}$。利用极限值原理可求得极限侵深及与之对应的极限速度 $dP/d(V_s^2)=0$，即赵军等[16]还发现，弹体的结构破坏一般发生在极限侵彻速度之前，如图 10 所示。

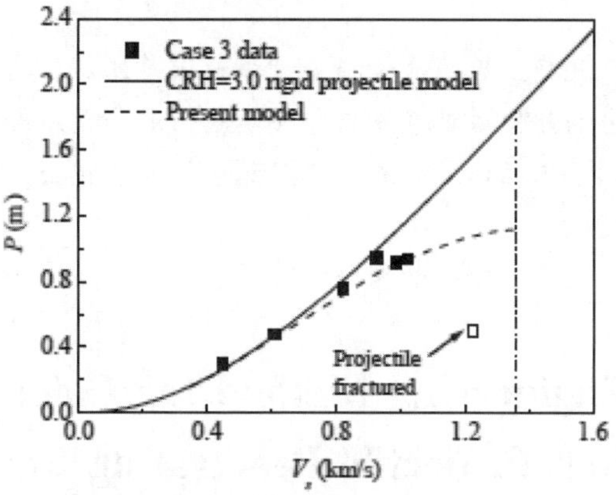

图 10　初始撞击速度与侵深关系

4　结论与展望

经过科研人员的多年努力，弹体高速侵彻混凝土研究已取得了一定的进展，相关的试验以及理论分析已初步成形，但是还存在着很大的不足：

（1）试验手段和方法还存在很大的不足，目前试验数据主要利用小口径火炮或空气炮作为发射手段，其发射能力非常有限，仅仅可以进行小尺寸的模型试验，对于更大尺寸和质量的弹体试验却无法进行，使得试验数据带有很大的局限性。同时目前还无法时时测量弹体在侵彻过程中的质量损失率、表面温度的变化、弹体的变形情况等，给理论分析带来了诸多困难。

（2）现有有关质量损失的理论分析，均以 Jones 等[8]提出的磨蚀模型为基础，即主要考虑弹体头部表面材料熔化对质量损失的贡献。摩擦功为弹体头部切向合力沿侵深的积分，且摩擦力的形式庞杂，具体哪种形式更行之有效尚无定论。缺乏研究动态摩擦的统一研究方法。隐含摩擦功转化成的热全部用于熔化弹体表面材料，且熔化的表面材料全部在磨蚀过程中损失的假设。而热量可能仅将弹体加热软化，不能达到熔点温度。因此，这种假设高估了摩擦功对质量损失的贡献。

针对以上不足，本文提出以下建议：

（1）发展试验技术，研发新的发射装置，提高现有发射设备的发射能力和次口径发射技术，使得可以进行不同尺寸的弹体高速试验，为理论研究提供充足的试验数据。利用多种试验手段，如 TEM，X 射线衍射和金相观察技术，分析剩余弹体，为更准确描述弹体侵彻过程，为研究侵彻机理提供基础。

（2）改善 Jones 等[8]提出的磨蚀模型中摩擦功的计算，统一动态摩擦研究方法。综合随机理论和绝热剪切理论等，提出切削引起质量损失的估算公式。融合切削和熔化现有两种主要质量磨蚀机制的影响，提出更合理，更全面的估算弹体质量磨蚀的计算公式。基于微观观察，提出描述质量磨蚀的细观模型，从更本质的角度解释质量磨蚀过程。

开展弹体高速侵彻混凝土研究对地下防护工程建设和武器战斗部设计都具有重要的现实意义，也是目前相关领域研究的热点问题，因此对高速侵彻混凝土的侵彻机理研究仍需科研工作者们的共同努力。

侵彻混凝土弹体质量损失的工程计算方法①

徐翔云[1,2]②，冯顺山[1]，何 翔[2]，金栋梁[2]，程守玉[2]

（1. 北京理工大学爆炸科学与技术国家重点实验室，北京 100081；
2. 总参工程兵科研三所，河南洛阳 471023）

摘 要：为了研究弹体侵彻混凝土后的质量损失，提出满足工程需求的计算方法，以侵彻混凝土靶后的弹体为研究对象，分析了影响弹体质量损失的相关参数。利用量纲分析的方法建立了弹体侵彻混凝土的相似模型和相似准则，回归了工程计算公式。研究结果表明，该公式具有良好的计算精度和较宽的适用范围。

关键词：侵彻；质量损失；混凝土靶；量纲分析

The Engineering Method for Calculating Mass Loss from Projectile Penetrating Concrete Target

Xu Xiangyun[1,2], Feng Shunshan[1], He Xiang[2], Jin Dongliang[2], Cheng Shouyu[2]

(1 State Key Laboratory of Explosion Science and Technology, Beijing Institute of Technology, Beijing 100081, China;
2 The Third Engineers Scientific Research Institute, Henan Luoyang 471023, China)

Abstract: In order to study the mass loss of projectile penetrating concrete target and work out the design formula that can meet engineering requirement, the relative parameters that influence projectile mass loss were analyzed. Through dimensional analysis, the reduced - scale model and similar criteria for high - velocity projectile penetrating concrete target were established for regressing concrete engineering computing formula. The test results indicate that the formula is of higher adaptability and accuracy.

Keywords: penetration; mass loss; concrete target; dimensional analysis

0 引言

钻地弹是攻击混凝土和岩石类加固与地下深埋防护目标的有效武器。随着侵彻速度增加，弹体质量损失亦不断增加，它可能导致弹道轨迹偏离，弹体破坏或失效。自 20 世纪 90 年代起，质量损失的相关研究已引起科研工作者们的注意，有关工作的详细回顾可参见文献［1］，文献［2－5］提供了某些实验数据。对钻地弹在侵彻过程中的受力机制并不是十分清楚，加之混凝土材料高应变率本构关系的复杂性，目前还无法得到弹体质量损失的解析解，为了满足工程的需要，文中利用量纲分析的方法，建立了无量纲方程，并利用试验结果回归了具体的计算公式。

1 弹体质量损失的影响参数

对于卵形弹高速侵彻混凝土靶体的过程，可以列出影响弹体质量损失的各相关物理力学参数如下：

---

① 原文发表于《弹箭与制导》2012 年第 1 期。

② 徐祥云：工学博士，2009 年师从冯顺山教授，研究高速动能弹侵彻混凝土/金属基混凝土效应与应用，现工作单位：军事科学研究院。

靶体参数：混凝土密度 ρ_t、混凝土的抗压强度 σ_t、混凝土的抗剪强度 σ_{ct}、混凝土的抗拉强度 $\sigma_{\tau t}$、靶板骨料莫氏硬度 H。

弹体参数：弹体长度 L、弹体直径 D、弹头长度 L_N、弹体质量 M、弹体综合密度 μ_p、弹体材料强度 σ_p、弹性模量 E_p。

发射参数：弹着角 α（初始时刻偏航角，射弹轴线与靶面法线夹角），攻角 θ（射弹质心速度方向与射弹轴线夹角），弹体速度 V。

设弹体侵彻完成时质量损失量为 ΔM，表征弹体质量损失率的量为 ξ，则它与各影响因素间的函数关系可写成如下的一般形式：

$$\xi = \frac{\Delta M}{M} = f(\sigma_t, \rho_t, \sigma_{ct}, \sigma_{\tau t}, H, L, D, L_N, M, \sigma_P, \rho_P, E_P, \theta, \alpha, V) \tag{1}$$

由于混凝土材料的抗压、抗拉、抗剪强度相互关联，故只保留 σ_t，对于材料的骨料莫氏硬度，可以作为单独量修正量出现，在这里先不予考虑，于是独立的靶板参数为：σ_t、ρ_t；弹体密度和弹体质量及弹体几何参数相关联，故可略去密度 ρ_p，弹体材料的杨氏模量近似于常数，可忽略 E_p 的影响，因此独立而有效的弹体参数为：L、D、L_N、M、ρ_p。对于正侵彻而言，α 和 θ 都为 0，所以发射参数只有 V。因此式（1）可改写为

$$\Delta M = f(L, D, L_N, M, \sigma_P, \sigma_t, \rho_t, V) \tag{2}$$

式中影响弹体质量损失的物理力学参数及其量纲如表 1 所示。

表 1　影响弹体质量损失的物理力学参数及其量纲

变量	基本量纲	变量	基本量纲
弹体直径 D	L	弹体抗拉强度 σ_p	$ML^{-1}T^{-2}$
弹体长度 L	L	靶板抗压强度 σ_t	$ML^{-1}T^{-2}$
弹头长度 L_N	L	靶体密度 ρ_t	ML^{-3}
弹体质量 M	M	弹体速度 V	LT^{-1}

根据 Π 定理，对式（2）进行无量纲化，可以得出弹体质量损失的质量表达式：

在基本假设下，土盘的径向运动可简化为一维的。由动量定理和质量守恒定律可得

$$\frac{\Delta M}{M} = f\left(\frac{L}{D}, \frac{L_N}{D}, \frac{M}{D^3\rho_t}, \frac{\rho_t V^2}{\sigma_t}, \frac{\sigma_p}{\sigma_t}\right) \tag{3}$$

2　计算公式的提出

2.1　质量损失公式的相似分析

用 C 表示相似比，带“′”者为模型参量，不带“′”者为原型参量，如果模型与原型的侵彻历程真实相似，则对应物理量成正比，见式（4）。

$$\begin{gathered} C_{M_t} = \frac{\Delta M}{\Delta M'};\ C_M = \frac{M}{M'};\ C_L = \frac{L}{L'};\ C_{L_N} = \frac{L_N}{L_N'};\ C_D = \frac{D}{D'}; \\ C_{\rho_t} = \frac{\rho_t}{\rho_t'};\ C_{\sigma_t} = \frac{\sigma_t}{\sigma_t'};\ C_{\sigma_p} = \frac{\sigma_p}{\sigma_p'};\ C_V = \frac{V}{V'} \end{gathered} \tag{4}$$

根据相似性要求，由式（3）、式（4）得到相似指标：

$$\frac{C_{M_1}}{C_M} = 1;\ \frac{C_L}{C_D} = 1;\ \frac{C_{L_N}}{C_D} = 1;\ \frac{C_M}{C_D^3 C_{\rho t}} = 1;\ \frac{C_{\rho t} {C_V}^2}{C_{\sigma t}} = 1;\ \frac{C_{\sigma P}}{C_{\sigma_t}} = 1 \tag{5}$$

靶板材料原型与模型相同，则有：$C_{\sigma_p}=1$；$C_{\rho_p}=1$；$\frac{C_M}{C_D^3 C_{\rho_t}}=1$，即要求：$C_D^3=C_M$，如果弹体材料原型与模型相同，此式自然满足$\frac{C_L}{C_D}=1$；$\frac{C_{L_D}}{C_D}=1$，在 $C_L=C_D=C_{L_N}$的条件下，即弹径、弹长和弹头长度按同样的比例缩小即可满足；由$\frac{C_{\rho_t}C_V^2}{C_{\sigma_t}}=1$ 得：$C_V=1$；由$\frac{C_{\Delta M}}{C_M}=1$，得：$C_{\Delta M}=C_M$。

由以上分析可知，如果靶板材料、弹体材料原型与模型相同，弹体几何参数是按同样的比例缩小，在弹体入射速度相同的情况下，模型试验的弹体质量损失模型与原型是相似的，且与弹径、弹长和弹头长度缩小的比例相符。

式（3）所包含的 5 个相似准则，它们各自具有明确的物理意义。$\frac{L}{D}$，$\frac{L_N}{D}$分别为射弹弹总长（L）、射弹 弹头长（L_N）与射弹直径（D）之间的比值，反映了弹体的几何形状；$\frac{M}{D_{\rho_t}^3}$为射弹特征密度$\left(\frac{M}{D^3}\right)$和混凝土密度（$\rho_t$）之比，$\frac{\rho_t V^2}{\sigma_t}$ 为射弹所受到的动压阻力（$\rho_t V^2$）与其所受到的静压阻力 σ_t 之比；$\frac{\sigma_p}{\sigma_t}$为射弹弹体强度（$\sigma_p$）和 混凝土强度（$\sigma_t$）之比。

2.2 侵彻深度公式的回归

将文献 E23 试验中的数据代入式（4）进行回归，得 到弹体质量损失的无量纲表达式如下：

$$\frac{\Delta M}{M}=0.405\left(\frac{L}{D}\right)^{-0.164}\varphi^{0.288}\left(\frac{M}{D^3\rho_t}\right)^{0.148}\left(\frac{\rho_t V^2}{\sigma_t}\right)^{0.733}\left(\frac{\sigma_p}{\sigma_t}\right)^{-0.317} \tag{6}$$

从量纲分析可知，弹体头部长径比$\frac{L_N}{D}$对弹体无量纲质量损失的影响是和其他因素不关联的，因此在数据拟合中，为了达到更高的精度，采用了 φ 代替$\frac{L_N}{D}$参与拟合，φ 为弹体头部曲率 CRH 的值，$\varphi=\left(\frac{L_N}{D}\right)^2+0.25$。

结合文献［1］中的数据，提出关于莫氏硬度的修正量 K，则有

$$\frac{\Delta M}{M}=0.405K\left(\frac{L}{D}\right)^{-0.164}\varphi^{0.288}\left(\frac{M}{D^3\rho_t}\right)^{0.148}\left(\frac{\rho_t V^2}{\sigma_t}\right)^{0.733}\left(\frac{\sigma_p}{\sigma_t}\right)^{-0.317} \tag{7}$$

其中：$K=1+(H-3)*0.203$，H 为混凝土骨料莫氏硬度值。

3 公式计算结果与试验结果的比较

为了验证式（7）的计算误差，将文献［4－5］中各次试验的实际参数代入无量纲公式（6），并将计算结果和实际测试数据进行了对比，图 1 为文献［4－5］中的部分工况实际测量值和本项目公式计算值的对比图，图中工况的相关参数如表 2 所示。图 1（a）～图 1（f）分别对应工况 1～6，从图中可以看出本公式具有良好的计算精度。

4 各无量纲量对弹体质量损失的影响

在式（7）中可以看出，影响弹体质量损失的无量纲量共有 6 个，去除单独修正的莫氏硬度 K，把包含射弹速度 V 的$\frac{\rho_t V}{\sigma_t}$作为自变量，则影响弹体质量损失 的无量纲量共有 4 个，图 2～图 5 分别为确定其他 3 个无量纲量的情况下，不同的无量纲量对影响弹体质量损失的影响。

表2　文献［4—5］中相关弹、靶参数

工况	靶体参数			弹体参数						
	靶材	单轴抗压强度/MPa	密度/(kg·m^{-3})	弹形	头部 CRH	质量 M/g	弹径 D/mm	长径比 L/D	弹体材料	弹体硬度 Rc
1	混凝土	68.8	2 300	尖卵形实心弹	3.0	478	20.3	10	*AerMet*100	Rc53
2	混凝土	51	2 300	尖卵形实心弹	3.0	1 620	30.5	10	*AerMet*100	Rc53
3	混凝土	62.8	2 300	尖卵形实心弹	3.0	478	20.3	10	4340	Rc45
4	混凝土	51	2 300	尖卵形实心弹	3.0	1 600	30.5	10	4340	Rc45
5	砂浆土	21.6	2 000	尖卵形空心弹	3.0	64	12.9	6.88	4340	Rc39
6	砂浆土	21.6	2 000	尖卵形空心弹	4.25	64	12.9	6.88	4340	Rc39

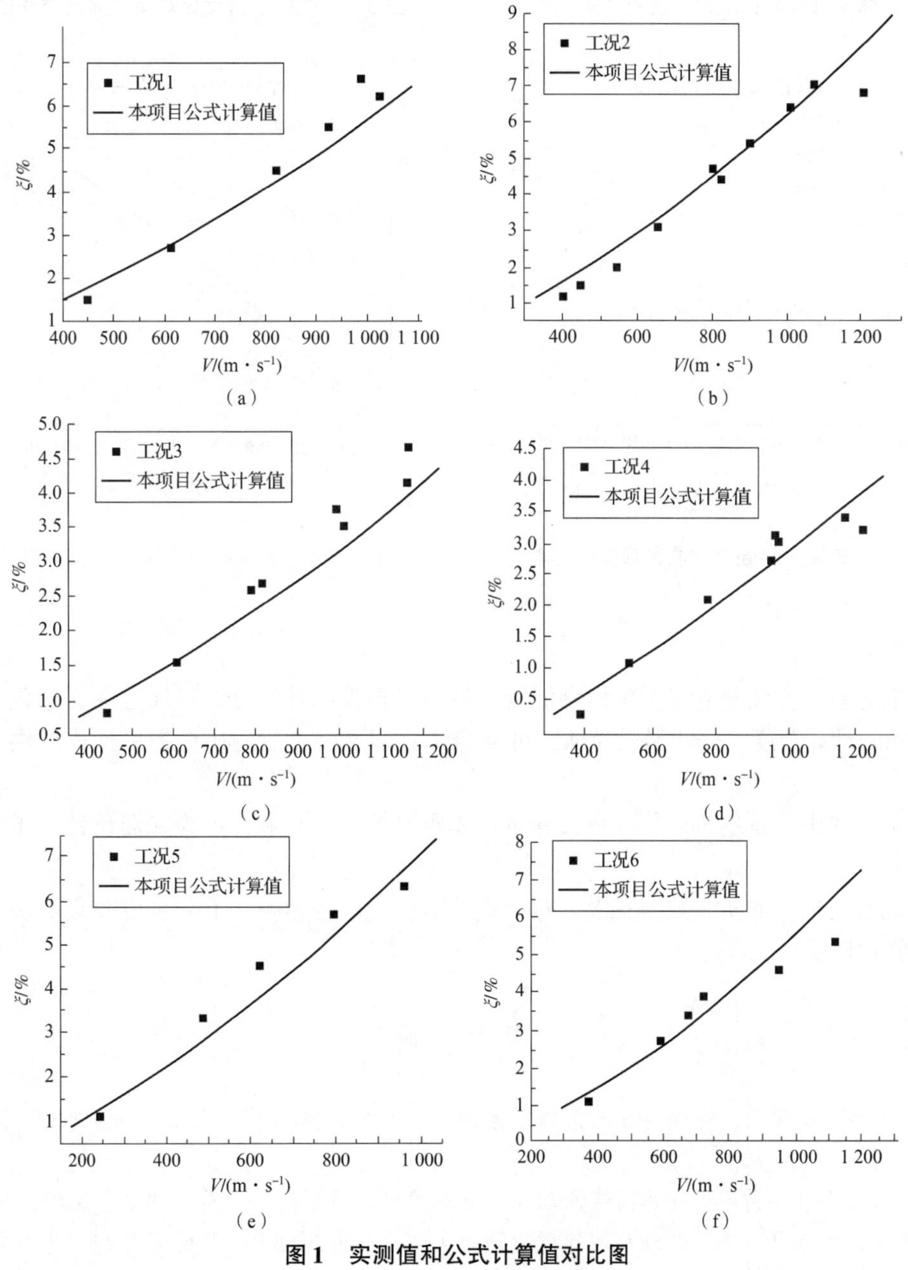

图1　实测值和公式计算值对比图

(a) 工况1；(b) 工况2；(c) 工况3；(d) 工况4；(e) 工况5；(f) 工况6

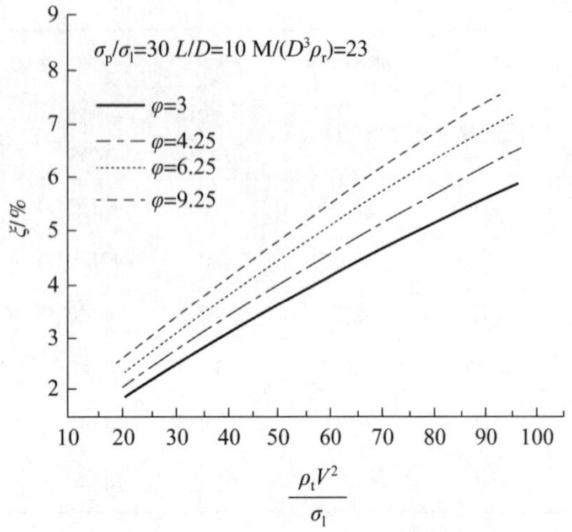

图 2　弹体头部长径比对弹体质量损失的影响

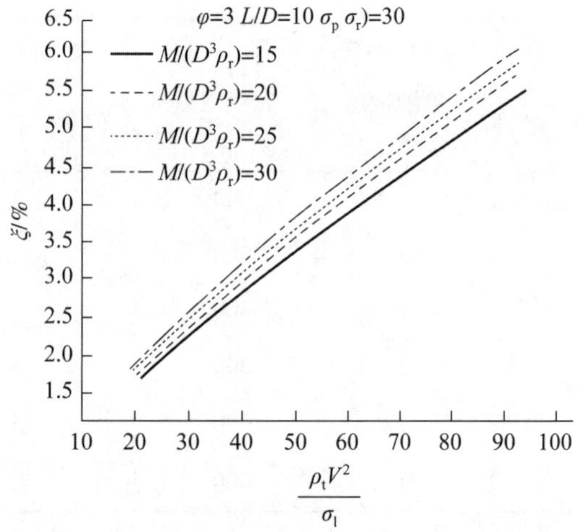

图 3　弹靶特征密度比对弹体质量损失的影响

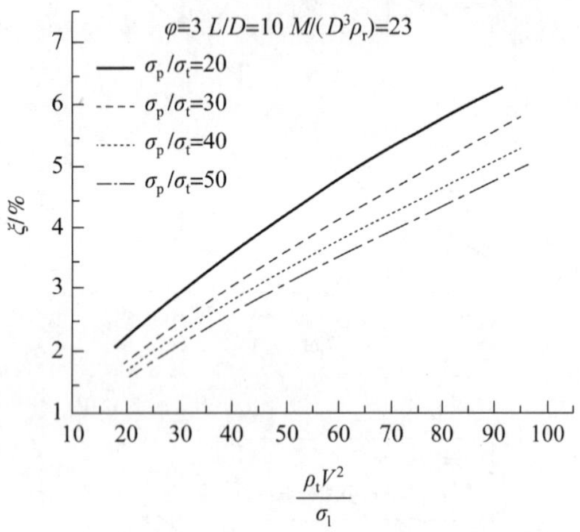

图 4　弹靶材料强度对质量损失的影响

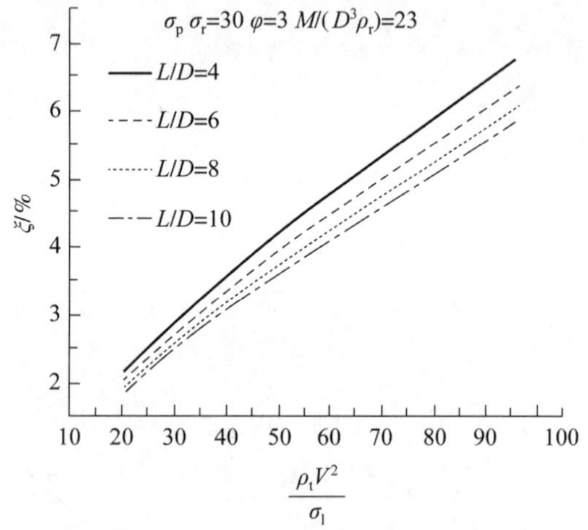

图 5　弹体几何形状对弹体质量损失的影响

5　结论

（1）用量纲分析的方法建立了弹体侵彻混凝土后质量损失的计算公式，通过和文献数据对比，证明了该公式有良好的计算精度。公式形式简洁，可为弹体侵彻混凝土后的质量损失计算提供了更加简单实用的方法。

（2）分析了公式中无量纲量的物理意义和对弹体质量损失的影响，可为武器设计和工程防护提供一定的参考。

（3）该公式参与拟合的实验数据有限，对公式的精度会产生影响，下一步应收集更多的实验数据参与拟合，提高公式精度。

参考文献

[1] 何丽灵，陈小伟，夏源明．侵彻混凝土弹体磨蚀的若干研究进展［J］．兵工学报，2010，31（7）：750－765.

[2] 何翔，徐翔云，孙桂娟，等．弹体高速侵彻混凝土的效应试验［J］．爆炸与冲击，2010，30（1）：1－6.

[3] 梁斌，陈小伟，姬永强，等．先进钻地弹概念弹的次口径高速深侵彻试验研究［J］．爆炸与冲击，2008，28（1）：1－9.

Numerical Simulation of the Influence of an Axially Asymmetric Charge on the Impact Initiation Capability of a Rod – Like Jet[①]

非轴对称装药对杆状射流冲击起爆能力影响的数值模拟研究

Li Yandong[②], Dong Yongxiang, Feng Shunshan

Abstract: With the aim to improve the impact initiation capability of a rod – like jet, this paper presents the influence of an axially asymmetric shaped charge on the jet studied by means of numerical simulations. According to Held's initiation criterion, the impact initiation capability of the jet is affected by the jet tip velocity and diameter. The detonation radius over the longitudinal axis, restricted by the charge radius over the same axis, affects the detonation wave in the charge, the force acting on the liner, and, therefore, the jet velocity and shape. Based on these laws, the structure of the axially asymmetric charge is optimized. Compared with axisymmetric jets, axially asymmetric rod – like jets possess a higher impact initiation capability.

Keywords: axially asymmetric; shaped charge; rod – like jet; impact initiation; numerical simulation

INTRODUCTION

Because of the higher damage capability demand of shaped charges, rod – like jets and axially asymmetric shaped charges came into being. The rod – like jet velocity value is intervenient between those of the explosively formed projectile (EFP) and the jet. There is a smooth transition between the jet and the slug, with a smaller tip and tail velocity contrast than in the jet. The rod – like jet is more powerful than the EFP and has a larger stand – off distance than the jet. The kinetic energy and damage area of the axially asymmetric shaped charge can be greater than those of the EFP and the jet, which allows achieving different tactical objectives, such as damaging armor, concrete, and I – beams [1 – 5].

Some scholars studied asymmetric shaped charges [6 – 8], but they did not study axially asymmetric shaped charges. To the authors' knowledge, there has not been any report about application of axially asymmetric shaped charges in impact initiation at present. This paper focuses on the formation mechanism of the axially asymmetric rod – like jet and then designs an axially asymmetric shaped charge that can form a rod – like jet with a high impact initiation capability.

MAIN FACTORS OF THE ROD – LIKE JET INITIATION CAPABILITY

With the purpose of studying the impact initiation phenomenon, some scholars put forward several initiation criteria. Held performed experiments on explosives initiated by the jet and derived the following criterion [9]:

$$v_j^2 d = \text{const}$$

(v_j and d are velocity and diameter of the jet tip). This paper adopts this criterion as the rod – like jet impact initiation criterion.

The axially asymmetric shaped charges studied in this work have the same liner structure and material, so the

---

① 原文发表于《Combustion Explosion and Shock Waves》2012。

② 李砚东：工学博士，2006 年师从冯顺山教授，研究拦截网成型装药对导弹战斗部作用效应，现工作单位：空军工程大学。

charge is the only factor responsible for the rodlike jet shape. Figure 1 illustrates the main parameters of the charge structure, where l_1 and l_2 are the charge radii over the longitudinal and transverse axes (in what follows, long and minor axis charge radii), r_1 and r_2 are the long and minor axis detonation radii, r_s is the radius of the liner bottom, and the charge height is maintained invariant. The parameters l_1 and r_1 are variable parameters, as is shown in Figure. 2, and they can influence the rod – like jet velocity and diameter. With the aim to present the structure of the axially asymmetric charge, we use the degree of axial asymmetry of the charge radius l_1/l_2 and the degree of axial asymmetry of the detonation radius r_1/r_2.

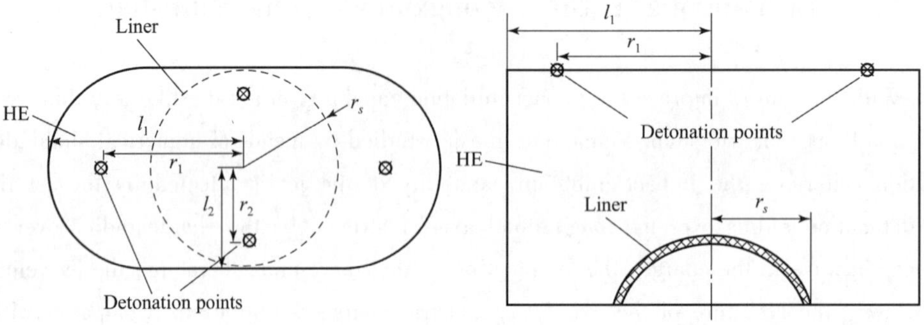

Figure. 1 Main parameters of the axially asymmetric charge structure.

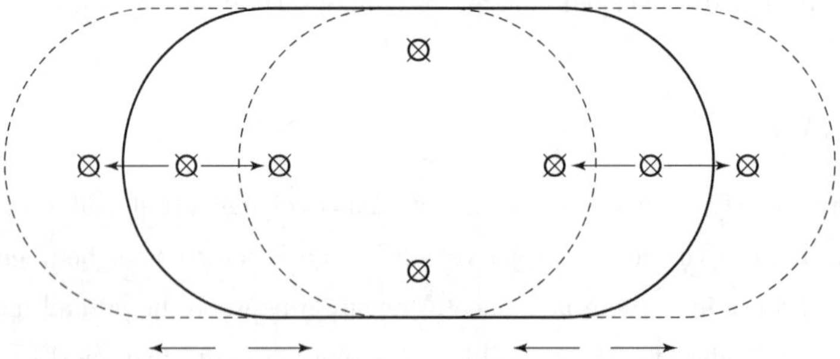

Figure. 2 Various forms of the axially asymmetric charge.

NUMERICAL SIMULATION OF THE AXIALLY ASYMMETRIC SHAPED CHARGE

Numerical Simulation Model

LS – DYNA software is adopted to simulate the shaped charge. The following explosives were studied: B explosive, red copper, and carbon steel [10]. For the B explosive, we used the HIGH EXPLOSIVE BURN constitutive model and the Jones – Wilkins – Lee (JWL) state model; for red carbon, we used the STEINBERG constitutive model and the Grüneisen state model; finally, for carbon steel, we used the PLASTIC KINETIC material model [10, 11].

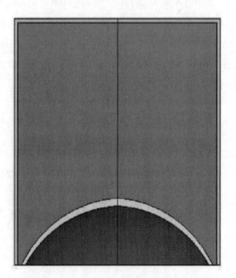

Figure. 3 Cross – sectional model of the axisymmetric shaped charge.

Some scholars carried on numerical simulations and experiments of certain shaped charge forms [11]. Imitating that shaped charge, we considered a numerical simulation model with the charge diameter $D_k = 52$ mm and four – point detonation at a distance $D_k/3$ from the center, as is shown in Figure. 3. Two shaped charges produce similar rod – like jets, as is shown in Figure. 4 (the calculation variants differ only in the values of some parameters). The rod – like jet tip velocity is 3 020 m/s at t = 50 μs. Compared with the experimental data (2 968.5 and 3 045 m/s) in the literature, the errors are less than 5%.

Figure. 4 Rod – like jet shapes at the time t = 50 μs.

Three configurations with the degree of axial asymmetry of the change $l_1/l_2 = 1.5$ (model 1), 2 (model 2), and 2.5 (model 3) were simulated (Figure. 5). The corresponding degrees of axial asymmetry of the initiation points are $r_1/r_2 = 1 - 1.5$, 1 – 2, and 1 – 2.5. The minor axis charge radius is $l_2 = D_k/2 = 26$ mm, and the minor axis detonation radius is $r_2 = D_k/3 = 17$ mm

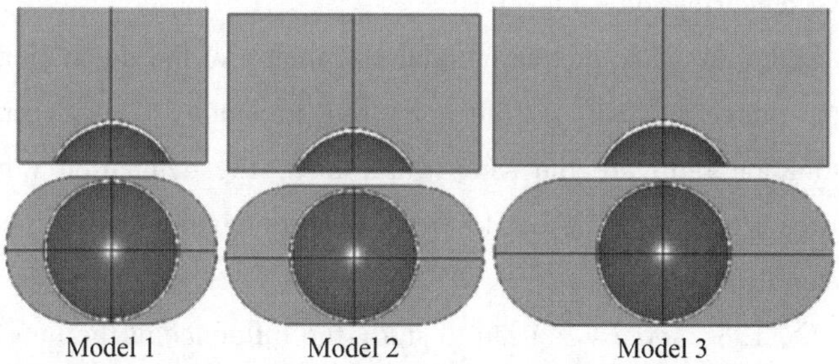

Figure. 5 Numerical models: the top row shows the axial cross sections of the charges, and the bottom row shows the charges from below.

Initial Velocity of the Axially Asymmetric Rod – Like Jet

The influence of the long axis charge radius l_1 on the detonation wave shape was studied at $r_1/r_2 = 1.5$ and $l_1/l_2 = 1.5$, 2, and 2.5. Because of the same long axis detonation radius r_1, the detonation wave shapes are similar, and the detonation wave peak pressures are all ≈42 GPa at t = 4 μs. Because of the similar detonation wave shape and pressure, the jet tip velocities are all ≈3400 m/s at t = 50 μs. The difference in the long axis charge radius l_1 only affects the detonation wave rim with the side rarefaction wave. Therefore, the long axis charge radius has little influence on the detonation wave shape and pressure, so it has little influence on the rod – like jet velocity.

The influence of r_1 on the detonation wave shape was studied in accordance with model 2 ($l_1/l_2 = 2$), and $r_1/r_2 = 1.2$, 1.8, and 2. Because of the difference in the long axis detonation radii r_1, the detonation wave shapes are different, and the detonation wave peak pressures at t = 4 μs are 41, 44, and 45 GPa for $r_1/r_2 = 1.2$, 1.8, and 2, respectively. As a consequence, the jet tip velocities are 3100, 3480, and 3400 m/s at t = 50 μs. The increase in the tip velocity does not follow the increase in the detonation wave pressure completely, because the velocity value is also affected by the detonation wave shape.

A larger long axis detonation radius r_1 makes the detonation wave interaction area bigger, and the peak pressure area is farther from the liner top, so the influence of the peak pressure on the tip velocity is smaller. An increase in r_1 also reduces the fixed long axis charge radius, so the side rarefaction wave affects the detonation wave rim more obviously. Thus, the value of r_1, affecting the detonation wave shape and pressure, has a pronounced effect on the rod - like jet velocity.

Figure 6 shows the behavior of the tip velocity as a function of the parameter r_1/r_2 at t = 50 μs. The tip velocity increases all the time. If r_1/r_2 is greater than 1. 8, the velocity increases slowly; if r_1/r_2 is smaller than 1. 8, the velocity increases faster.

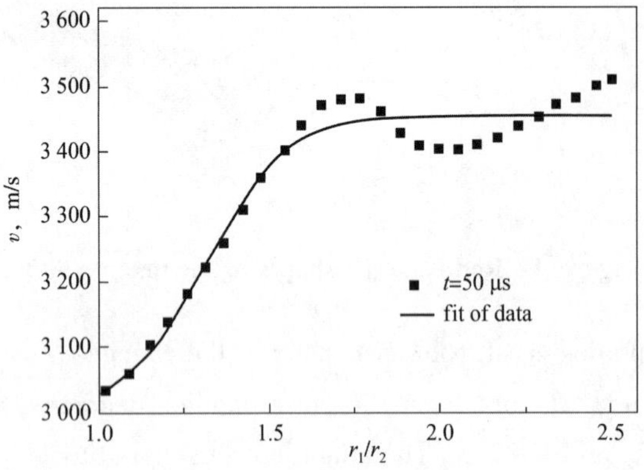

Figure. 6 Tip velocity versus r_1/r_2.

Shape of the Axially Asymmetric Rod - Like Jet

The simulation models adopt the same degree of axial asymmetry of the detonation radius $r_1/r_2 = 1.5$ to study the influence of the long axis charge radius l_1 on the rod - like jet shape. Though the long axis charge radii are different, the long axis detonation radii are the same, therefore, the detonation wave shapes are similar. The detonation wave peak pressures are all ≈42 GPa at t = 6 μs. The detonation wave shape and pressure are similar, so the rod - like jets have similar shapes.

Model 2 with $r_1/r_2 = 1.2$, 1. 8, and 2 was used to study the influence of the long axis detonation radius r1 on the rod - like jet shape. Because the long axis detonation radii r1 are different, the detonation wave shapes are also different. The wave peak pressures are, respectively, 41, 44, and 45 GPa at t = 6 μs. The detonation wave shape and pressure are different, so the rod - like jets have different shapes.

Figure. 7 gives the jet shape of model 2 with $r_1/r_2 = 1.2$ at t = 50 μs. Because of the force acting on the liner top, the axially asymmetric rod - like jet has a flattened jet tip and a wide jet tail. The liner bottom forms the rod - like jet fins. Section 4 is the transition between the liner top and bottom. An enhancing compression force in section 2 and section 3 would make the part velocity higher, form an inverse velocity gradient, produce jet mass piling - up, and this is the reason why there are bigger jet tips in sections 2 and 3.

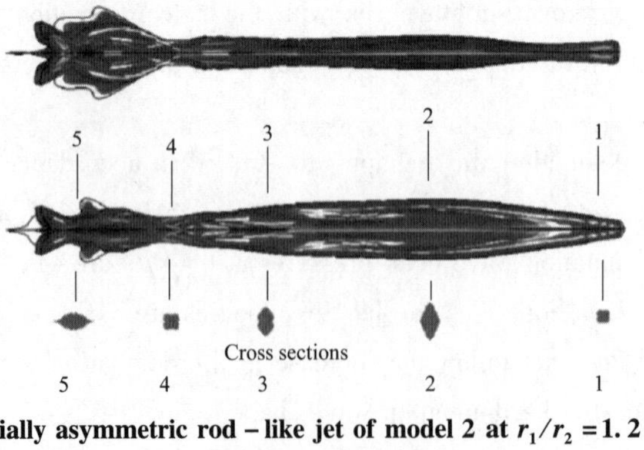

Figure. 7 Axially asymmetric rod - like jet of model 2 at $r_1/r_2 = 1.2$ (t = 50 μs).

We choose the half jet tip length r_{j1} and the half jet tip width r_{j2} of model 2 at t = 50 μs as the characteristic length and study the influence of r_1/r_2 on r_{j1}/r_s and r_{j2}/r_s (Figure. 8). Following the increase in r_1/r_2, the value of the ratio r_{j1}/r_s gets bigger all the time. If r_1/r_2 is smaller than 1.5, r_{j2}/r_s decreases, because the compression wave makes the jet tip flatter. If r_1/r_2 is greater than 1.5, r_{j2}/r_s increases, because the inverse velocity gradient makes the jet tip wider. If r_1/r_2 is equal to 1.8, r_{j2}/r_s is bigger than the original value. The empirical relations are

$$\frac{r_{j1}}{r_s}=0.45\frac{r_1}{r_2}-0.32, 1\leqslant\frac{r_1}{r_2}\leqslant 2 \tag{1}$$

$$\frac{r_{j2}}{r_s}=\begin{cases}0.41-0.24\dfrac{r_1}{r_2}, 1\leqslant\dfrac{r_1}{r_2}\leqslant 1.5\\[2ex] 0.97\dfrac{r_1}{r_2}-1.44, 1.5\leqslant\dfrac{r_1}{r_2}\leqslant 2\end{cases} \tag{2}$$

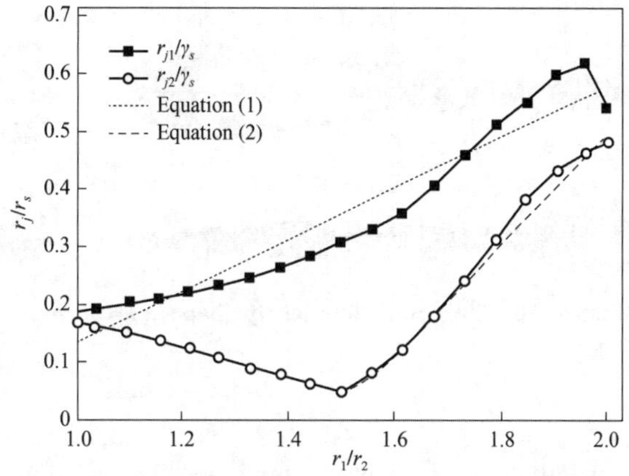

Figure. 8 Dependence of r_j/r_s on r_1/r_2 (t = 50 μs).

By means of numerical simulations, it appears that the long axis detonation radius r_1 affects the rod - like jet velocity and shape significantly. If $r_1/r_2=1.8$, the tip velocity increase becomes slow, and the tip size increases. Thus, we suggest that $l_1/l_2=2$ and $r_1/r_2=1.8$ should be chosen as the degrees of axial asymmetry of the charge radius and detonation radius.

IMPACT INITIATION CAPABILITY OF THE AXIALLY ASYMMETRIC ROD - LIKE JET

With the aim to illustrate the influence of the axially asymmetric charge on the impact initiation capability of the rod - like jet, the axially symmetric charge of an identical explosive is subjected to four - point detonation (the charge radius is 40 mm and the detonation radius is $D_k/3$).

Figure 9 shows a comparison of the axially asymmetric and symmetric rod - like jet formation processes. The axially asymmetric jet tip velocity is 3480 m/s, and the axisymmetric jet tip velocity is 3290 m/s, which is 5.78% smaller.

Figure 10 shows a comparison of two rod - like jet tip shapes at t = 50 μs. The axially asymmetric jet tip is obviously bigger than the axisymmetric jet tip. The axially asymmetric jet tip shape is close to an ellipse with the major semi - axis of 11.4 mm, minor semi - axis of 5.9 mm, and area of 211.2 mm^2. The ellipse area is equal to the area of a circle with a radius of 8.2 mm. The axisymmetric jet tip shape is close to a circle with a radius of 1.4 mm and an area of 6.2 mm^2.

The value of $v_j^2 d$ of the axially asymmetric jet is 198.61 mm^3/μs^2, and that of the axisymmetric jet is 30.31 mm^3/μs^2, which is smaller by a factor of 6.49. According to Held's impact initiation criterion [9], the axially asymmetric rod - like jet has a higher impact initiation capability.

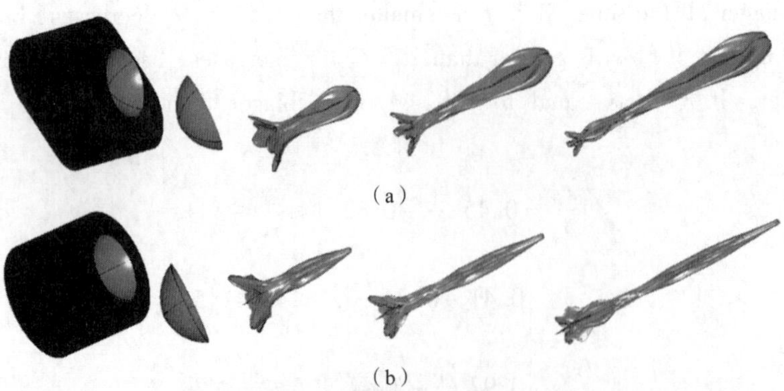

(a)

(b)

Figure. 9 Formation of the axially asymmetric (a) and axisymmetric (b) rod – like jets.

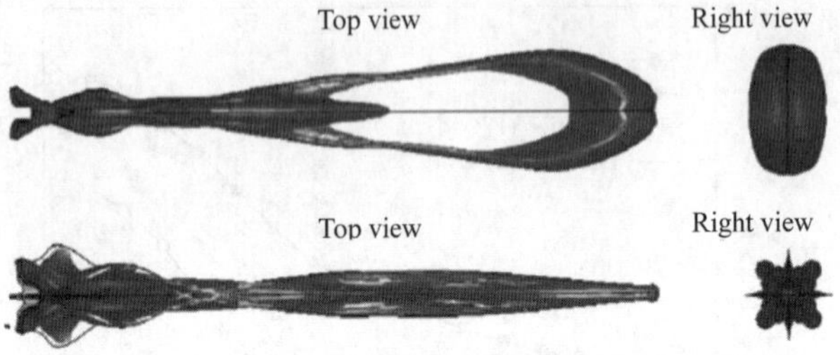

Figure. 10 Two rod – like jet tip shapes (t = 50 μs).

CONCLUSION

• According to Held's impact initiation criterion [9], the rod – like jet tip velocity and diameter are the main factors determining the jet impact initiation capability.

• By means of numerical simulations, it is demonstrated that the long axis detonation radius r_1 influences the detonation wave shape and pressure, the force acting on the liner, thus, determining the rod – like jet velocity and diameter. The long axis charge radius determines the range of variation of the long axis detonation radius r_1.

• Following the increase in the long axis detonation radius, the turning point of the rod – like jet tip velocity change rate is $r_1/r_2 = 1.8$, and the turning point of the rod – like jet tip width is $r_1/r_2 = 1.5$.

• Compared with the axisymmetric charge of an identical explosive, the axially asymmetric charge produces a rod – like jet with a higher impact initiation capability.

This paper is supported by the project of State Key Laboratory of Explosion Science and Technology (Beijing Institute. of Technology) (Grant No. QNKT10 – 06)

REFERENCE

[1] WALTERS W P, ZUKAS J A. Fundamentals of shaped charges. New York, John Wiley and Sons: 1989.

[2] BAKER E L, DANIELS A S, TURCI J P, et al. Selectable initiation shaped charges [C] //Proc. 20th Int. Symp. on Ballistics, 2002: 589 – 596.

[3] FONG R, NG W, WEIMANN K. Nonaxisymmetric waveshaped EFP warheads [C] //Proc. 20th Int. Symp. on Ballistics, 2002: 582 – 588.

[4] FONG R, KRAFT J, THOMPSON L, et al. 3D hydrocode analysis of novel asymmetrical warhead. ADA432103, 2005: 1 – 4.

[5] TAN G E B, LAM T K, THAM Y K. Planar cutting jets from shaped charges [C] //in Proc. 22th Int. Symp. on Ballistics, 2005: 507 – 513.

冲击作用下反应破片燃爆温度效应①

李顺平②，冯顺山，董永香，陈 赟
（北京理工大学爆炸科学与技术国家重点实验室，北京 100081）

摘 要：本文从一维冲击波和热力学以及物理化学角度出发，建立了反应破片冲击燃爆温度与破片反应材料组分配比的数学关系模型，对比分析了 PTFE/Al 和 PTFE/Ti 两种反应破片的变化规律，其冲击燃爆温度都先随配方金属粉含量的增加而增大，在金属粉含量一定值时有一个最大值，在该点后，都随其金属粉含量的增加而下降；PTFE/Ti 型反应破片的最大冲击燃爆温度要比 PTFE/Al 型反应破片高出 1 100 ℃左右。与相关试验结果进行对比，可以认为反应破片冲击燃爆温度越高，其燃爆效果越好。所建立的分析模型和研究方法可推广应用于其他类型反应破片的冲击反应特性等动态性能研究。

关键词：反应破片；冲击燃爆；温度效应；反应材料配比

Explosive Temperature Effect of Reactive Fragment Under Impact

Li Shunping, Feng Shunshan, Dong Yongxiang, Chen Yun
(State Key Laboratory of Explosion Science and Technology, BIT, Beijing 100081, China)

Abstract: Based on one – dimensional shock wave and thermodynamics and physical chemistry, the model of explosive temperature of reactive fragment and reaction material group assignment composition was established. The explosive temperature change of PTFE/ Al and PTFE/ Ti reactive fragment with the reaction material quality ratio was analyzed. There is a maximum value of explosive temperature, PTFE/ Ti reactive fragment's is 1 100 ℃ higher than PTFE/ Al' s, when certain content of the metal powder, before which the explosive temperature is increasing and after which the explosive temperature is decreasing. Compared with previous damage effect test of compound reactive fragment on the steel target, it can be supposed that the higher explosive temperature is, the better its blasting is. It is suggested that the established model and research method can be applied to dynamic properties research of other reactive fragment such as impact reaction characteristics and so on

Keywords: reactive fragment; shock blasting; temperature effect; reactive material composition

0 引言

目前已展开的新概念、新原理破片技术研究大多是将铝、钛、钨、锆等具有一定反应能力的金属与 PTFE、THV 等冲击反应非金属材料复合制成反应材料装进合金壳体内构成反应破片[1-2]，利用其与目标撞击时产生的高温高压引发反应金属与 PTFE 或 THV 发生反应，更有效的释放化学能和热能，进而产生更大的破坏能量。反应破片冲击燃爆机理的研究是控制反应破片适时释能以及优化反应破片配方设计的理论依据，而现阶段对反应破片在撞击靶板条件下的冲击燃爆机理研究较少，对 PTFE/Al 和 PTFE/Ti 型反应破片的冲击燃爆温度效应的对比研究更少。冲击燃爆温度是衡量反应破片发生燃爆反应能力的重要指标，也是反应破片热化学计算所必需的参数。从冲击波和热力学以及物理化学角度出发，给出了反应破片冲击燃爆温度的数学分析模型；然后，通过 Matlab 编程得出了反应破片冲击燃爆温度随反应破片质

---

① 原文发表于《弹箭与制导》2015 年第 2 期。

② 李顺平：工学博士，2009 年师从冯顺山教授，研究 PTFE/DU 材料冲击反应特性与作用后效，现工作单位：火箭军项目管理中心。

量配比的变化曲线，并比较了 PTFE/Al 和 PTFE/Ti 型反应破片（以下简称破片）冲击燃爆温度的变化规律，给出了 PTFE/Ti 型反应破片冲击燃爆温度变化规律的拟合公式，并结合前人复合反应破片对钢靶侵彻的试验研究对比分析了 PTFE/Al 和 PTFE/Ti 型反应破片发生燃爆反应的能力。

1 破片冲击燃爆温度数学模型

以复合结构型反应破片为研究对象，对破片冲击靶板作用过程参考李杰[3]研究可爆破片条件做以下几点假设：

（1）破片垂直侵彻靶板。

（2）只考虑在反应材料中传播冲击波的初始冲击波平面部分，忽略发散波，并不考虑侧向稀疏波的影响。

（3）破片撞击靶板后，在两者中产生的应力为准一维应力。反应破片与靶板撞击时，在反应破片和靶板接触部位产生冲击波，对于破片壳体有

$$P_f=\rho_f D_f u_f=\rho_f(a_f+b_f u_f)u_f \tag{1}$$

式中 P_f——破片壳体冲击波压力；

ρ_f——破片壳体密度；

D_f——破片壳体冲击波速度；

u_f——破片壳体质点速度；

a_f——压力为零时破片壳体中的声速；

b_f——Hugoniot 破片壳体经验参数。

对于靶板有

$$P_t=\rho_t D_t u_t=\rho_t(a_t+b_t u_t)u_t \tag{2}$$

式中 P_t——靶板冲击波压力；

ρ_t——靶板密度；

D_t——靶板冲击波速度；

u_t——靶板质点速度；

a_t——压力为零时靶板中的声速；

b_t——Hugoniot 靶板经验参数。

由连续条件 $P_f=P_t$，$u_f+u_t=v$（其中 v 为破片撞击靶板速度）并联合式（1）、式（2）可得

$$u_f=\frac{-A+\sqrt{A^2+4B(\rho_f b_f-\rho_t b_t)}}{2(\rho_f b_f-\rho_t b_t)} \tag{3}$$

式中，$A=\rho_f a_f+2\rho_t b_t+\rho_t a_t$，$B=\rho_t a_t+\rho_t b_t v^2$。

把式（3）代入式（1）可得 P_f。反应材料冲击波压力 P_e 是破片壳体中冲击波的透射波，即

$$P_e=P_f\cdot\frac{2}{1+n}\cdot\frac{R^2}{r^2},n=\frac{\rho_f C_f}{\rho_e C_e} \tag{4}$$

式中 R——壳体半径；

r——反应材料半径；

ρ_e——反应材料密度；

C_f 和 C_e——破片壳体和反应材料中声速。

利用 $P_e=\rho_e$（$a_e+b_e u_e$）u_e 反解出 u_e 得

$$u_e=\frac{-\rho_e a_e+\sqrt{(\rho_e a_e)^2+4\rho_e b_e P_e}}{2\rho_e b_e} \tag{5}$$

冲击波持续时间 τ 为

$$\tau = \frac{2r}{D_e} = \frac{2r}{a_e + b_e u_e} \tag{6}$$

式中　u_e——反应材料质点速度；

a_e——压力为零时反应材料中的声速；

b_e——Hugoniot 反应材料经验参数。

在冲击波波阵面上，假定热力学过程是绝热的，则冲击 Hugoniot 曲线如图 1 所示。

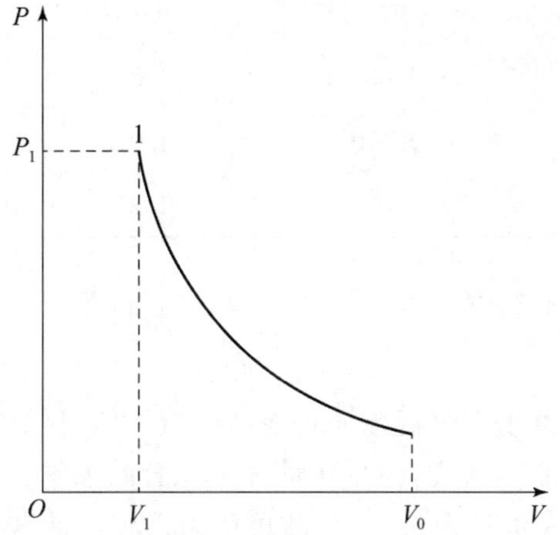

图 1　冲击 Hugoniot 曲线

并且 1 点处 P_1、V_1 满足下列关系[4]：

$$V_1 = V_0(1 - 1/b_e + a_e(a_e + b_e u_e)/b_e) \tag{7}$$

式中　V_0——破片初始体积；

V_1——破片冲击靶板完成时体积。

在冲击 Hugoniot 曲线上 1 点满足[4]：

$$T_1 = T_0 \exp\left[\frac{\gamma_e}{V_0}(V_0 - V_1)\right] + \frac{V_0 - V_1}{2C_{ve}}P_1 + \frac{\exp\left(-\frac{\gamma_e}{V_0}V_1\right)}{2C_{ve}}\int_{V_0}^{V} P_e \cdot \exp\left(\frac{\gamma_e}{V_0}V_1\right) \cdot \left[2 - \frac{\gamma_e}{V_0}(V_0 - V)\right] \mathrm{d}V \tag{8}$$

式中　T_1——破片冲击温度；

γ_e——反应材料 Gruneisen 常数；

C_{ve}——反应材料定容比热容。联合式（1）~式（8），并采用数值计算方法，可求解破片冲击温度。破片燃爆温度是指冲击引发破片发生反应后破片的温度变化。当反应破片得到的能量超过临界反应冲击能量后，反应破片开始反应，由盖斯定律可知，化学反应的反应热只与反应的始态和终态有关，而与具体反应进行的路径无关，如果一个反应可以分为几步进行，则各分步反应的反应热之和与该反应一步完成的反应热是相同的。所以反应破片释放的能量与反应过程无关，只与反应破片的初始状态和最终产物的状态有关。爆热 Q 会使反应区温度升高，假设爆热全部用于反应区温度升高，可得燃爆温度变化 ΔT_2 为

$$\Delta T_2 = Q/(C_P \cdot m) \tag{9}$$

式中，C_P 为反应材料定压比热容，并且 $C_P = C_v \cdot \gamma_0$。

由式（8）和式（9）联立可得破片冲击燃爆温度为

$$T = T_1 + \Delta T_2 \tag{10}$$

2　破片冲击燃爆温度随组分配比的变化规律

复合结构破片壳体材料采用钨材，反应材料采用烧结后的 PTFE/Al 和 PTFE/Ti，圆柱形破片壳体尺寸为 Φ10 mm×12 mm，反应材料尺寸为 Φ8 mm×10 mm；为计算方便靶板材料采用 1018 钢，厚度为

6 mm。

Al、Ti、PTFE、1018 钢和 W 的冲击参数和热力学参数如表 1 所示。

表 1　破片材料参数[4]

材料名称	密度/(g · cm^{-3})	a/(mm · μs^{-1})	b	γ_0	C_v/(J · g^{-1} · k)
Al - 2024	2. 785	5. 328	1. 338	2. 00	0. 838
Ti	4. 528	5. 22	0. 767	1. 09	0. 519
PTFE	2. 147	1. 682	1. 819	0. 59	1. 048
1018 钢	7. 850	3. 570	1. 920	1. 69	0. 439
W	19. 224	4. 029	1. 237	1. 54	0. 131

烧结后形成的反应破片的冲击参数和热力学参数可以通过以下计算方法得到[4]：

$$x = \sum m_i x_i \tag{11}$$

该模型通过 Matlab 编程实现冲击燃爆温度 T 随 PTFE 和 Al 以及 PTFE 和 Ti 质量配比的变化规律分析，基于 Matlab 的编程代码的核心实现式（8）中第 3 项积分的求解，文中采用四阶 Runge - Kutta 解法实现。计算过程中金属粉含量的递进量为 0. 1%，故可认为得到的曲线是连续的，计算得到的冲击燃爆温度 T 随 PTFE 和 Al 以及 PTFE 和 Ti 质量配比的变化曲线如图 2 和图 3 所示。为了明确表示冲击燃爆温度随材料配比（拟合公式中用 w 表示）的变化规律，PTFE/Ti 反应破片计算曲线进行了 6 次线性拟合，其拟合公式为

$$y = -1.821 \times 10^5 \cdot w^5 + 4.652 \times 10^5 \cdot w^4 - 3.963 \times 10^5 \cdot w^3 + 1.05 \times 10^5 \cdot w^2 + 8\,758 \cdot w + 79.49 \tag{12}$$

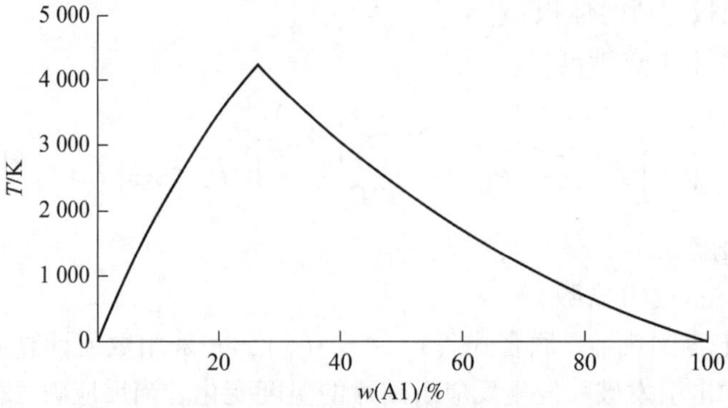

图 2　PTFE/Al 型反应破片冲击燃爆温度随铝粉含量的变化规律

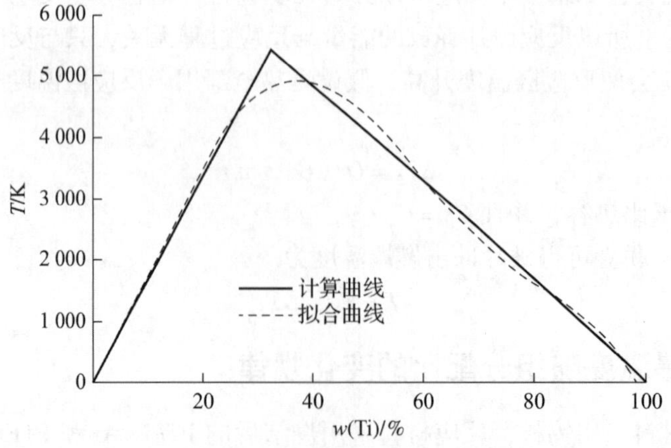

图 3　PTFE/Ti 型反应破片冲击燃爆温度随钛粉含量的变化规律

3　计算结果与试验结果的对比

由图2和图3可以看出，两种反应破片的冲击燃爆温度都先随其金属粉含量的增加而增大，在金属粉含量一定值时有一个最大值，在该点后，两种反应破片的冲击燃爆温度都随其金属粉含量的增加而下降；PTFE/Al型反应破片冲击燃爆温度在铝粉含量为26.5%时达到最大，PTFE/Ti型反应破片冲击燃爆温度在钛粉含量为32.4%时达到最大，而PTFE/Ti型反应破片的最大冲击燃爆温度要比PTFE/Al型反应破片高出1 100 ℃左右。

帅俊峰等[5]对PTFE/Ti型和PTFE/Al型反应破片对薄钢靶的侵彻毁伤效果进行了试验研究，利用12.7 mm口径弹道枪发射复合反应破片对6 mm厚A3钢板侵彻试验，采用高速摄影观察破片穿靶过程。试验结果显示PTFE/Ti型反应破片穿靶后出现的火光持续9 ms明显大于PTFE/Al型反应破片火光持续6 ms的时间，得出的结论为PTFE/Ti型反应破片爆炸效果好于PTFE/Al型反应破片效果。

由燃爆反应的短历时可以假定 $t=4.0$ ms时两种反应破片材料已完全反应，截取此时的试验图片以及最高冲击燃爆温度计算结果并列于表2中，由两种配方试验图片宏观对比以及试验结论可以认为，反应破片冲击燃爆温度越高，其燃爆效果越好。

表2　破片燃爆反应终点时刻的高速摄影图片及计算结果

破片类型	穿靶后4 ms时高速摄影图片	最高冲击燃爆温度/K
PTFE/Al型反应破片	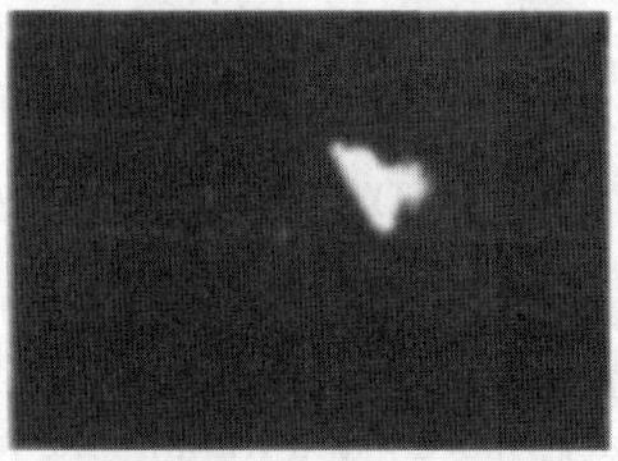	4 252
PTFE/Ti型反应破片		5 435

4　结论

基于一维冲击波和热力学以及物理化学理论建立的反应破片冲击燃爆温度与反应破片组分配比的数学关系模型，研究表明两种反应破片冲击燃爆温度都先随配方金属粉含量的增加而增大，在金属粉含量一定值时有一个最大值，在该点后，都随其金属粉含量的增加而下降；PTFE/Ti型反应破片的最大冲击燃爆温度要比PTFE/Al型反应破片高出1 100 ℃左右。与相关试验结果进行对比，可以认为反应破片冲击燃爆温度越高，其燃爆效果越好。研究结果可从冲击燃爆温度效应角度为高能反应破片的配方设计提供理论分析依据。

参 考 文 献

[1] FLIS W J. Reactive fragment warhead for enhanced neutral - lization of mortar, rocket&missile threats [EB/OL]. [2006 - 08 - 21]. http://www.detk.com.

[2] NIELSON D B. Reactive material enhanced munition compositions and projectiles containing same [P]. USA, US20050199323 Al, Sep. 15, 2005.

内置冲击反应材料弹丸壳体侵彻损伤效应研究[①]

冯顺山，李　伟[②]，周　彤，黄　岐，董永香
（北京理工大学爆炸科学与技术国家重点实验室，北京　100081）

摘要：基于冲击波理论、带侵蚀模型的空腔膨胀理论及圆柱壳体动力学模型建立了内置冲击反应材料弹体结构侵彻靶板过程模型和过靶后的壳体变形模型。通过 Matlab 编程得到反应材料段壳体变形的时间历程，并分析了冲击速度和锥角对弹头速度、加速度、侵蚀质量、等效锥角及侵透靶时壳体变形量的影响。研究成果为内置冲击反应材料弹丸侵彻靶板时的结构变形损伤研究提供理论分析方法，并为反应材料损伤和冲击反应条件研究提供了基础支撑与参考依据。

关键词：兵器科学与技术；冲击反应材料；圆柱壳体；膨胀变形

Research on Damage Effect of a Warhead Filled with Reactive Materials

Feng Shunshan, Li Wei, Zhou Tong, Huang Qi, Dong Yongxiang
(State Key Laboratory of Explosion Science and Technology, Beijing Institute of Technology, Beijing 100081, China)

Abstract: Based on theory of shock wave, theory of cavity expansion with erosion model and dynamic model of cylindrical shell, a model is established to describe the penetration of a warhead filled with reactive materials into the target and its shell deformation after penetrating through target. The time history of shell deformation is obtained by Matlab. And the effects of impact velocity and initial cone angle on war - head velocity, acceleration, erosion mass, equivalent cone angle and deformation of the middle shell are analyzed. The proposed model provides a theory method to study the structural deformation and damage of a warhead filled with reactive materials and a base support to study the damage and reactive conditions of reactive materials filled.

Keywords: ordnance science and technology; reactive material; cylindrical shell; expansion and de - formation

0　引言

金属破片作为常规弹药战斗部打击空中目标的毁伤元[1-3]，其毁伤机理单一（纯动能侵彻），制约了战斗部威力设计。冲击反应破片不仅具有动能侵彻功能，还可利用侵彻冲击效应诱发材料爆燃反应对目标实现动能和化学能联合作用，可使终端毁伤效能大幅度提升[4-7]。因此，冲击反应材料（以下称为 C 材料）在弹药战斗部设计中的应用受到很大关注。

战斗部是空中导弹/炸弹目标的要害舱段，是近程/超近程反导关注的主要对象。由于目标战斗部防御能力不断提高，其炸药装药低易损性水平也不断提高，这对近程反导穿甲型中小口径弹药的威力设计提出了挑战[8-10]。由于 C 材料自身强度和强度特征，单独设计对厚防护层目标侵彻的穿甲型弹丸有诸多困难[10-12]。本文提出一种内 C 材料穿甲型弹药结构，试图实现穿甲 + 穿后燃爆后效的双重作用，用于摧毁导弹目标战斗部。

---

① 原文发表于《兵工学报》2016.12.

② 李伟：工学博士，2012 年师从冯顺山教授，研究内置冲击反应材料弹丸对装药结构的侵爆机理，现工作单位：中国工程物理研究院。

根据 C 材料结构和强度特性，本文参考横向增强弹（PELE）的作用原理[13-14]，即利用内置装填物和外壳材料满足“软硬”结合的方式，在侵彻靶板的过程中，内置材料和壳体在轴向和径向强烈压缩并在侵透目标后卸载时剧烈膨胀致破裂，其产生径向膨胀的作用原理对本文问题的研究具有很重要的参考意义。本文根据 C 材料特点，提出一种内置冲击反应材料穿甲型弹丸结构（WSFCM）侵彻厚靶板。该弹丸具有较高长径比和侵彻比动能，壳体中段内置 C 材料，试图满足侵彻厚靶板且在过靶后充分诱发 C 材料爆燃，对目标造成高威力毁伤后效。通过分析冲击应力波及其在弹头内的传播和弹头在靶板中侵彻阻力及侵蚀情况，建立了中段壳体（指对应 C 材料的壳体）在弹头、弹尾、C 材料及靶板的综合受力作用下弹体结构侵蚀靶板的变形理论模型，从而得到弹头的速度、加速度、质量侵蚀和等效锥角的时间历程以及中段壳体过靶后的变形轮廓线。本文为内置 C 材料弹丸侵彻靶板的结构变形损伤研究提供理论分析方法，进而为内置材料的反应条件和其燃爆后效毁伤研究提供基础支撑和参考依据。

1　理论模型

对于内置冲击反应材料弹丸侵彻厚靶板，如图 1 所示，中段壳体作为 C 材料的外部接触边界，材料损伤与反应条件受制于壳体在侵彻过程中的受力变形情况。本文通过研究，表征弹丸出靶时壳体膨胀变形破坏的规律，为冲击反应材料损伤和冲击诱发爆燃条件的研究提供基础支撑。因此，本文重点研究弹丸侵彻过程中的中段壳体在侵彻过程中的变形和损伤问题，并将整个过程的理论模型分为三个阶段：第一阶段是头部侵彻冲击波传播模型；第二阶段是考虑侵蚀下的弹头阻力模型；第三阶段是中段壳体的受力变形模型。

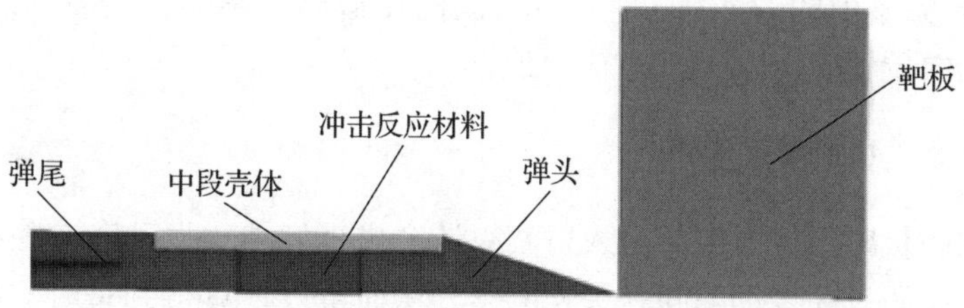

图 1　弹丸侵彻靶板示意图

1.1　头部侵彻冲击波传播模型

首先利用弹丸头部与靶板撞击面处的连续条件式（1）和弹头与靶板材料的线性 Hugoniot 关系式（2），并结合动量守恒得到撞击面处产生的冲击波压力 σ_0：

$$v_0 - u_h = u_t, \sigma_h = \sigma_t \tag{1}$$

$$U_h = c_h + S_h u_h, U_t = c_t + S_t u_t \tag{2}$$

式中　v_0——弹丸侵彻靶板的初速；

u_h——相对于移动参考坐标撞击处弹头粒子速度；

u_t——撞击处靶板粒子速度；

下标 h 和 t——弹头和靶板；

σ——撞击面处应力值，即冲击波压力 σ_0；

U——冲击波速度；

c——声速；

S——冲击波参数。

如图 1 所示，弹头在与靶板撞击过程中，冲击波由弹头部小截面向大截面传播。本文采用王礼立[15]提出的特征线方法计算冲击波压力 σ_0 在锥形弹头中的传播规律。

设截面积 $A(X)$ 是弹头轴坐标 X 的函数，则控制方程组为

$$
\frac{\partial \varepsilon}{\partial t} = \frac{\partial v}{\partial X}
$$
$$
\rho_0 A(X)\frac{\partial v}{\partial t} = \frac{\partial(\sigma A)}{\partial X} = A\frac{\partial \sigma}{\partial X} + \sigma\frac{\partial A}{\partial X}
$$
$$
\sigma = E\varepsilon \tag{3}
$$

式中　ε——应变；

v——粒子速度；

t——时间；

E——弹性模量。

从方程组（3）中消去 ε，并引入诸无量纲量

$$
\bar{X} = \frac{X}{L},\ \bar{t} = \frac{t}{t_L},\ \bar{\sigma} = \frac{\sigma}{\sigma_0},\ \bar{v} = \frac{v}{\dfrac{\sigma_0}{\rho_0 c_0}} \tag{4}
$$

式中　L——锥形弹头长度；

σ_0——入射波应力峰值；

$t_L = L/c_0$——特征时间。根据特征线理论可以确定两族特征线和特征相容关系：

$$
\begin{cases} \mathrm{d}\bar{X} \mp \mathrm{d}\bar{t} = 0 \\ \mathrm{d}\bar{\sigma} + \bar{\sigma}\,\mathrm{d}\ln A \mp \mathrm{d}\bar{v} = 0 \end{cases} \tag{5}
$$

$A(X)$通过特征相容关系中的 $\ln A(X)$ 项对应力波的传播产生影响，使波形发生畸变。通过式（5）可以得到在不同弹头轴向位置 X 处的冲击波应力值。

1.2　弹丸侵入靶板过程的受力模型

弹丸头部高速侵入靶板时不仅产生冲击入射波，还会出现侵蚀，而动态空腔膨胀理论适用于弹头为刚体的情况。因此，本文假设在侵彻瞬间弹头是刚体，从而可以计算得到每时刻的阻力，再计算该时刻弹头的侵蚀并得到新的弹头形状为下一时刻继续以刚体来计算阻力。通过 Matlab 编程计算就可以得到弹头阻力随时间的关系。

根据动态空腔膨胀理论[16]，靶板介质对弹丸的抗侵彻力可表示为

$$
F_{\mathrm{zd}} = \frac{\pi d^2}{4}(AY_{\mathrm{t}}N_1 + B\rho_{\mathrm{t}}v_0^2 N_2) \tag{6}
$$

式中　Y_{t} 和 ρ_{t}——靶板材料的屈服应力和密度；

d——弹丸圆柱部的直径；

A 和 B——无量纲参数；

N_1 和 N_2——无量纲弹头形状系数。

$$
A = \frac{2}{3}\left\{1 + \ln\left[\frac{E_{\mathrm{t}}}{3(1-v)Y_{\mathrm{t}}}\right]\right\}, B = 1.0 \sim 1.5 \tag{7}
$$

式中　E_{t}——靶板材料的杨氏模量。

本文锥形弹头函数为 $y = d/2 - x\tan\alpha$

$$
N_1 = 1 + \frac{\mu_{\mathrm{m}}}{\tan\alpha}, N_2 = \frac{\tan\alpha(\tan\alpha + \mu_{\mathrm{m}})}{1 + \tan^2\alpha} \tag{8}
$$

式中　μ_{m}——弹头材料与靶板的动摩擦系数；

α——弹头的半锥角。

中段壳体对弹头部和弹尾部的轴向应力为中段壳体的屈服强度 Y_{s}。将式（8）代入式（6）可得运动方程：

$$\frac{d^2z}{dt^2}m_0 = a^2\pi p_y + Y_s(b^2-a^2)\pi - \pi a^2 z^2 \cdot \left[AY_t\frac{\tan\alpha+\mu_m}{\tan\alpha} + B\rho\left(\frac{dz}{dt}\right)^2\frac{2\tan\alpha(\tan\alpha+\mu_m)}{1+\tan^2\alpha}\right] \tag{9}$$

式中，a 和 b 分别是中段壳体的内外径。求解上述偏微分方程，可以得到 $z(t)$，进而得到 $dz(t)/dt$，其中 p_y 由式（14）得到。

弹头部由于侵蚀使其形状随侵彻深度不断变化。为简化计算，把侵入靶板内弹头部分等效为锥形，如图 2 所示的双锥形；通过先计算该时刻弹头质量损失，再将进入靶内弹头被侵蚀剩余部分折算成等效锥角 α_e。

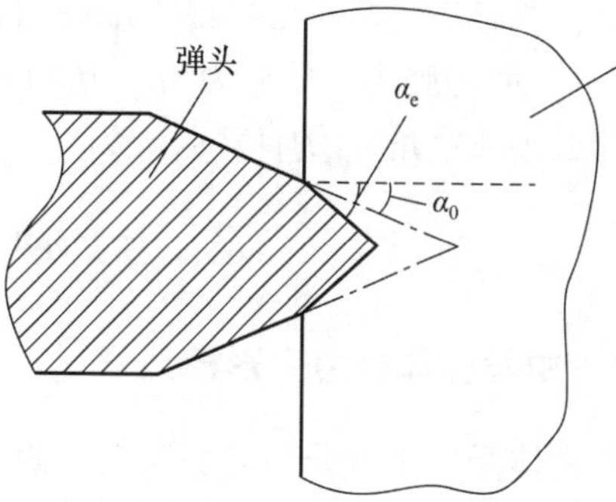

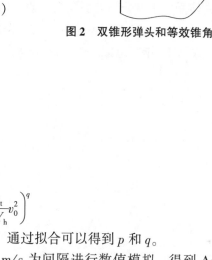

图 2　双锥形弹头和等效锥角

根据 Wen 等[17]提出的弹体侵蚀质量计算公式：

$$d\dot{m} = -2pq\left(\text{Moh}\,\frac{\rho_t}{2Y_h}\right)^q v_z^{2q-1}(\sigma_n\sin\theta+\mu_m\sigma_n\cos\theta)\,dA \tag{10}$$

式中　Moh——靶板的莫氏硬度；

ρ_t——靶板材料的密度；

Y_h——弹体材料的屈服强度；

v_z——给定时刻弹体轴线方向上的速度；

σ_n——弹体表面法向应力；

θ——弹体表面的切线与轴线的夹角；

μ_m——摩擦系数；

p 和 q——无量纲经验常数，其用于拟合的公式为

$$\Delta m/m = p\left(\text{Moh}\,\frac{\rho_t}{2Y_h}v_0^2\right)^q \tag{11}$$

式中，v_0 为弹丸初速，通过改变初速得到对应的质量侵蚀，通过拟合可以得到 p 和 q。

通过改变弹丸速度，从 1 000 m/s 到 1 400 m/s 每 50 m/s 为间隔进行数值模拟，得到 $\Delta m/m$ 与式（11）右边括号内乘积的关系，如图 3 所示。

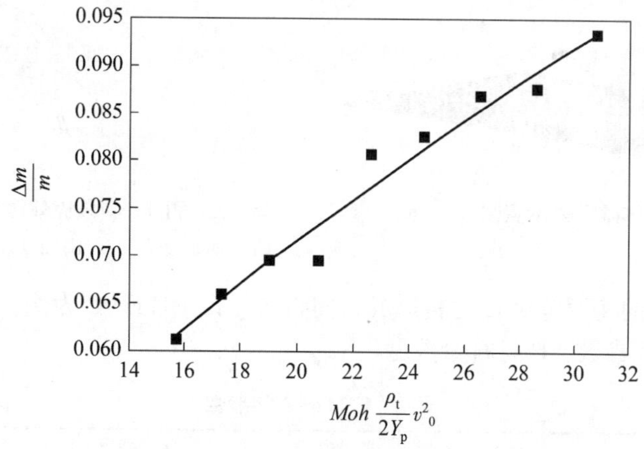

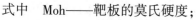

图 3　弹头侵蚀参数的拟合曲线和拟合数据点

用图 3 中数据点拟合式（11）得到曲线，并得到 p 和 q 的值为 0. 010 98 和 0. 626 03。

由于弹头部侵入靶板内部分始终等效为圆锥形，因此在每一个时刻，弹头部表面所有位置切向与轴向夹角 θ 是定值，等于锥角。因此，式（10）可以简化为

$$\dot{m} = -2pq\left(\text{Moh}\,\frac{\rho_t}{2Y_h}\right)^q v_z^{2q-1}(\sigma_n \sin\theta + \mu\sigma_n \cos\theta)A(t) \tag{12}$$

式中，$A(t)$为某一时刻侵入靶板内弹头的表面积。

综上所述，弹头受力模型的具体求解过程为：利用式（10）计算每个步长开始时弹头被侵蚀后的质量并折算成等效锥角。将等效锥角代入式（9）可得该步长的弹丸轴向加速度 a_z、速度 v_z、弹体表面法向应力 σ_n 及弹丸侵入长度 d_z，并将其代入式（12）可以得到该步长的弹体质量变化量以及下一步长的弹体质量，然后依次循环得到每个步长的弹丸轴向加速度 a_z、速度 v_z。设某一步长 i 到 $i+1$ 时，由式（10）得到此时的弹头的质量为 m_i，初始弹头质量为 m_0，初始弹头锥角为 α_0 时弹头部进入靶板的长度为 L_i，那么等效锥角 α_i 的计算公式为

$$\alpha_i = \arctan\left(\frac{\rho_h \pi L_i^3 \tan^3\alpha_0}{3m_i + \rho_h \pi L_i^3 \tan^2\alpha_0 - 3m_0}\right) \tag{13}$$

1.3 中段壳体受力变形模型

中段壳体受力主要分为 C 材料对中段壳体内表面的作用力 $P_{cr}(t,x)$、靶板对中段壳体外表面的作用力 P_{er}以及弹头尾对中段壳体的轴向挤压。为建立中段壳体的受力模型，选取一个微元体进行分析，如图4所示，微元体环向角度为 dθ。分析微元体的不同视图，如图5的1、2和3方向视图中段壳体微元体的受力。

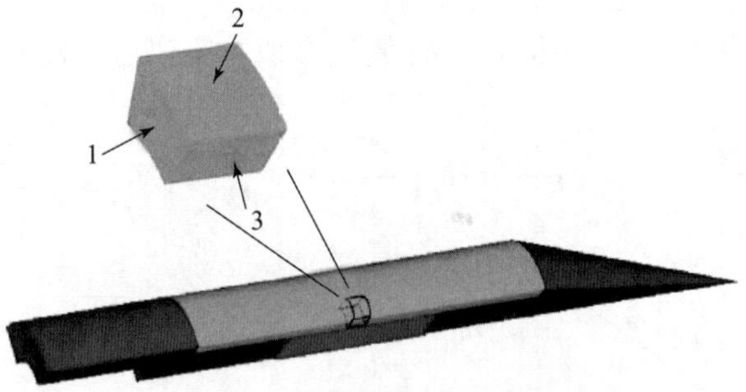

图4　中段壳体微元体示意图

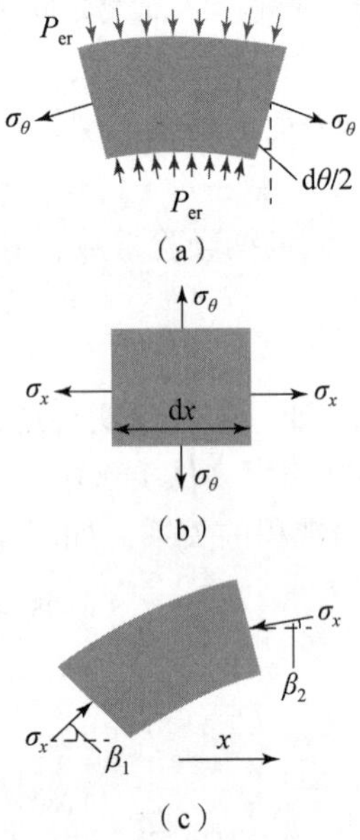

图5　中段壳体微元体受力分析图

(a) 1 方向受力图；(b) 2 方向受力图；(c) 3 方向受力图

此阶段弹头尾所产生的速度差对 C 材料形成轴向挤压。以 PTFE/Al 为例，采用 Raftenberg 等[11]得到的 Johnson - Cook 模型计算参数，C 材料参数见表 1。

表1　C 材料力学参数

ρ_{c0}/(kg · m^{-3})	G_c/MPa	A_c/MPa	B_c/MPa
2 270	666	8.044	250.6
C_c	n_c	T_{cm}/K	m_c
0.4	1.8	500	1.426

将 C 材料被挤压产生的轴向应变 $\varepsilon_y(t)$ 代入式（14）得到轴向压力 $p_y(t)$：

$$\sigma=(A_c+B_c\varepsilon^{n_c})\left(1+C_c\ln\frac{\dot{\varepsilon}}{\dot{\varepsilon}_0}\right)[1-(T_c^*)^{m_c}] \tag{14}$$

式中，$T_c^*=\dfrac{T-298}{T_{cm}-298}$。

由 1.1 节中式（5）可以得到弹头内锥形头部与圆柱形头部的结合处冲击波压力 P_{hz}。从该结合面到圆柱形头部与 C 材料的结合处，涉及冲击波的传播和衰减，设冲击波阵面到达距锥形头部和圆柱形头部结合处一定距离的圆柱形头部内 x_i 处的压力值为 p_{hi}，对应的时间步长 d_{ti} 为 t_i 到 t_{i+1}，那么，冲击波在 x_i 处的冲击波速度 U_{hi} 为

$$U_{hi}=\frac{\rho_h c_h+\sqrt{\rho_h^2c_h^2+4\rho_h S_h P_{hi}}}{2\rho_h} \tag{15}$$

式中 ρ_h——弹头材料的密度；

c_h 和 S_h——弹头材料的声速和冲击参数。那么，在 t_{i+1} 时刻，冲击波阵面的位置 x_{i+1} 为

$$x_{i+1}=x_i+U_{hi}\mathrm{d}t_i \tag{16}$$

将冲击波阵面位置 x_{i+1} 与弹头圆柱部的长度 L 对比，可以得到冲击波阵面到达 C 材料与弹头圆柱部界面的时刻。那么，冲击波阵面在 x_{i+1} 位置的压力为

$$p_{h(i+1)}=p_{hz}\exp(-\gamma_h x_{i+1}) \tag{17}$$

式中，γ_h 是冲击波在弹头材料中的衰减系数；x_{i+1} 的单位是 mm。以上的计算过程不断循环直到冲击波到达圆柱体弹头与 C 材料的结合面，计算得到 C 材料内初始受到的冲击波压力 p_c：

$$p_c=p_{ta}\exp(-\gamma_h L_1)2\Big/\left(1+\frac{\rho_h c_h}{\rho_c c_c}\right) \tag{18}$$

冲击波压力在 C 材料中传播的计算参考冲击波在弹头圆柱部传播的计算方法，即式（15）~式(17)按照时间步长循环计算。冲击波进入 C 材料后，从中段壳体进入 C 材料的稀疏波在冲击波进入该材料一定延时 t_y 后出现，即由稀疏波从外界穿过中段壳体进入 C 材料的时间：

$$t_y=(b-a)/c_s \tag{19}$$

式中，c_s 为中段壳体材料的声速。

综上所述，可以得到在 C 材料内部冲击波压力是时间和轴向位置的函数。由此得到对中段壳体的径向作用合力为

$$p_{cr}(t,x)=\frac{\upsilon_c[p_y(t)+p_c(t,x)]}{(1-\upsilon_c)+E_c/E_s(1-\upsilon_s)} \tag{20}$$

式中 下标 s 和 c——中段壳体和 C 材料；

υ——泊松比；

E——杨氏模量。

对于已侵入靶板内的中段壳体部分，认为径向作用力 p_{er} 为靶板的屈服强度 Y_t；对于尚未进入靶板内的部分，认为径向作用力为 0。

通过图 5 的受力分析，建立中段壳体微元在径向的受力方程：

$$\Delta m\frac{\mathrm{d}^2a}{\mathrm{d}t^2}=-F_e+F_i-F_\theta+F_x \tag{21}$$

式中，靶板对中段壳体在径向的受力 F_e［图 5（a）］为

$$F_e=p_{er}\mathrm{d}x2\pi b\frac{\mathrm{d}\theta}{2\pi}=p_{er}\mathrm{d}x\mathrm{d}\theta b \tag{22}$$

C 材料对中段壳体在径向的受力 F_i［图 5（a）］为

$$F_i=p_{cr}\mathrm{d}x2\pi a\frac{\mathrm{d}\theta}{2\pi}=p_{cr}\mathrm{d}x\mathrm{d}\theta a \tag{23}$$

中段壳体材料变形造成的环向受力在径向上的投影 F_θ ［图 5（b）］为

$$F_\theta = 2\sigma_\theta(b-a)\mathrm{d}x\sin\frac{\mathrm{d}\theta}{2} = \sigma_\theta(b-a)\mathrm{d}x\mathrm{d}\theta \tag{24}$$

中段壳体材料变形造成的轴向变形在径向上的受力投影 F_x ［图 5（c）］为

$$F_x = -\sigma_x\pi(b^2-a^2)\frac{\mathrm{d}\theta}{2\pi}(\sin\beta_2-\sin\beta_1) \tag{25}$$

式中，β_1 和 β_2 为中段壳体在变形时微元体轴向两端偏离轴向的角度。

将式（22）~式(25）代入式（21），可得到微元体的运动控制方程：

$$\frac{\mathrm{d}^2a}{\mathrm{d}t^2} = \frac{1}{\rho}\left[-p_{er}\frac{2\sqrt{a^2+b_0^2-a_0^2}}{b_0^2-a_0^2} + p_{cr}\frac{2a}{b_0^2-a_0^2} - \sigma_\theta\frac{a}{\sqrt{a^2+b_0^2-a_0^2}+a} + \frac{\sigma_x(\sin\beta_2-\sin\beta_1)}{\mathrm{d}x}\right] \tag{26}$$

下面采用二阶微分方程的四阶龙格－库塔法求解式（26）。式（26）中角度 β_1 和 β_2 是中段壳体在变形过程中形成的，可根据中段壳体的变形程度进行计算。

如图 6 所示，中段壳体的径向位移沿轴向先增大后减小，因此将计算方法分为两部分考虑：第 1 部分是径向位移增大的部分，如图 7（a）所示；第 2 部分是径向位移减小的部分，如图 7（b）所示。

图 6　弹丸过靶后中段壳体变形图

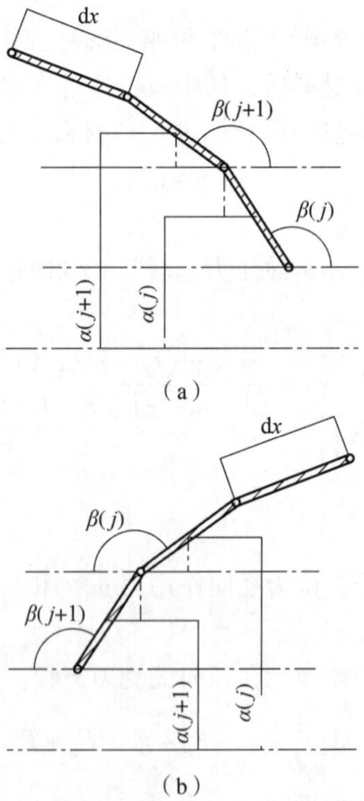

图 7　中段壳体轴向弯曲的角度计算示意图

（a）径向位移增大；（b）径向位移减小

第 1 部分，已知前一个微元段（j）的变形角度 $\beta(j)$，计算下一个微元段（$j+1$）的变形角度 $\beta(j+1)$，如图 7（a）所示，满足

$$\frac{a(j+1)-\{\sin[\pi-\beta(j)]0.5\mathrm{d}x+a(j)\}}{0.5\mathrm{d}x}=\sin[\pi-\beta(j+1)] \tag{27}$$

化简以后得到

$$\beta(j+1)=\pi-\arcsin\left\{\frac{a(j+1)-a(j)}{0.5\mathrm{d}x}-\sin[\beta(j)]\right\} \tag{28}$$

第2部分，已知前一个微元段（j）的变形角度$\beta(j)$，计算下一个微元段（$j+1$）的变形角度$\beta(j+1)$，如图7（b）所示，满足：

$$\frac{a(j)-\{\sin[\beta(j+1)]0.5\mathrm{d}x+a(j+1)\}}{0.5\mathrm{d}x}=\sin[\beta(j)] \tag{29}$$

化简以后得到

$$\beta(j+1)=\arcsin\left\{\frac{a(j)-a(j+1)}{0.5\mathrm{d}x}-\sin[\beta(j)]\right\} \tag{30}$$

2　计算实例

图1所示的弹丸弹头及弹尾采用钨合金材料，材料密度ρ_h为17 g/cm^3，动态屈服强度Y_h为1.5 GPa，声速C_h为4 029 m/s，冲击参数S_h为1，中段壳体采用45号钢，材料密度ρ_s为7.8 g/cm^3，动态屈服强度Y_s为0.48 GPa，声速c_s为3 570 m/s，杨氏模量E_s为200 GPa，泊松比υ_s为0.3。靶板材料采用A3钢，材料密度ρ_t为7.8 g/cm^3，动态屈服强度Y_t为0.48 GPa，杨氏模量E_t为200 GPa，靶板的莫氏硬度Moh为6。C材料参数如表1所示。这里主要分析在改变冲击速度、弹头锥角情况下弹头速度、加速度、等效锥角和质量在侵彻历程中的变化以及在指定轴向位置的中段壳体出靶时的变形情况。

2.1　中段壳体变形随时间的变化历程

首先分析在冲击速度为1 200 m/s、壳体厚度为2 mm、靶板厚度为30 mm、弹头锥角为7/26的条件下中段壳体出靶时变形随时间的变化历程。如图8所示，在64 μs时壳体变形基本为0，此后壳体中部径向膨胀随着时间的增加而增加。这是由于此刻中段壳体刚出靶板，此前受到靶板的径向约束而变形很小，只有壳体中部径向膨胀是由于在壳体的中部内填C材料挤压变形造成的。

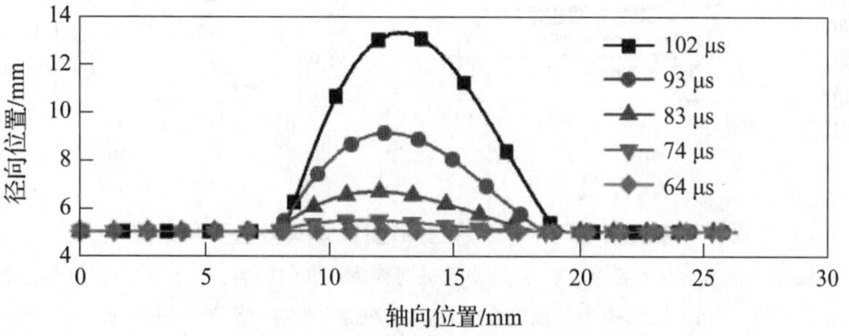

图8　不同时刻的中段壳体变形

2.2　不同冲击速度下中段壳体变形及头部参量变化

分析冲击速度为1 000～1 400 m/s、壳体厚度为2 mm、靶板厚度为30 mm、弹头锥角为7/26的条件下中段壳体的变形。为了便于比较，取中段壳体与弹头结合处离靶板背面80 mm处中段壳体变形情况对比。

如图9（a）所示，在靶板后相同位置，随冲击速度增大，中段壳体径向变形量相应增大，但增幅缩小。根据式可知，冲击速度越大，靶板对弹头阻力越大［由图9（c）前半段可知］，因此弹头与弹尾部速度差越大，致使内置C材料压缩程度越高，中段壳体内表面径向压力越大，则径向变形越大。由图9（b）和图9（c）可知，弹头的速度先减小后增大，而加速度先增大后减小，这是由于锥形弹头侵彻过程中与靶板的接触面积逐渐增大，受到的靶板阻力增大。在弹头穿透靶板后，接触面积减小，使得靶板阻力减小，且由于侵彻过程中弹头的阻力远大于弹尾的阻力，致使弹头出靶板后加速，因此图 9（c）中出现加速度从负值恢复到 0 再加速为正值的过程。通过对比不同速度下弹头速度和加速度曲线可知，冲击速度越大，弹头在靶板内速度下降越快且过靶之后恢复越快，在后半段趋向于恒定加速。由图 9（d）和图 9（e）可知，随着速度的增加，弹头质量侵蚀越快且出靶板后最终侵蚀量越大，弹头等效锥角增加越快且最终弹头越钝。

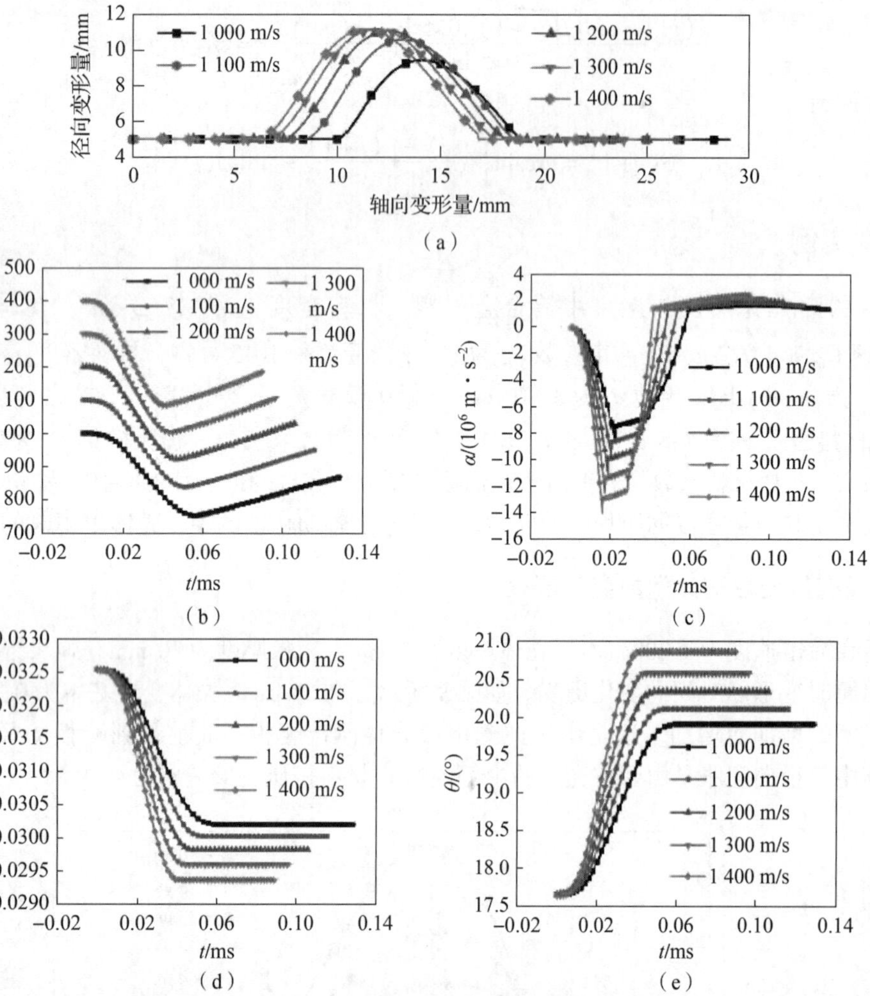

图 9　冲击速度对中段壳体变形量及弹头参量变化的影响

（a）冲击速度对中段壳体径向膨胀的影响；（b）冲击速度对弹头速度的影响；（c）冲击速度对弹头加速度的影响；（d）冲击速度对弹头质量的影响；（e）冲击速度对弹头等效锥角的影响

2.3　不同弹头锥角下中段壳体变形及头部参量变化

如图 10（a）所示，在靶板后相同位置，随着弹头锥角减小，中段壳体径向变形量先减小后增大，并在锥角弧度值为 7/24 时达到最小且越远离 7/24 则壳体径向变形增量越大。

由图 10（b）和图 10（c）可知，随着弹头锥角减小，弹头在靶板中的加速度减小及速度降幅减小。这是由于弹头锥角减小使在靶板中受到阻力的接触面减小。

如图 10（d）和图 10（e）所示，弹头锥角减小但弹头与中段壳体结合处的直径保持不变，即弹头长度增加，因此初始质量会增加。但是，从图 10（d）可以看出，弹头锥角越小，弹头最终被侵蚀掉的质量越小。由图 10（e）可知，随着弹头锥角减小，等效锥角在侵彻过程中的变化减小。

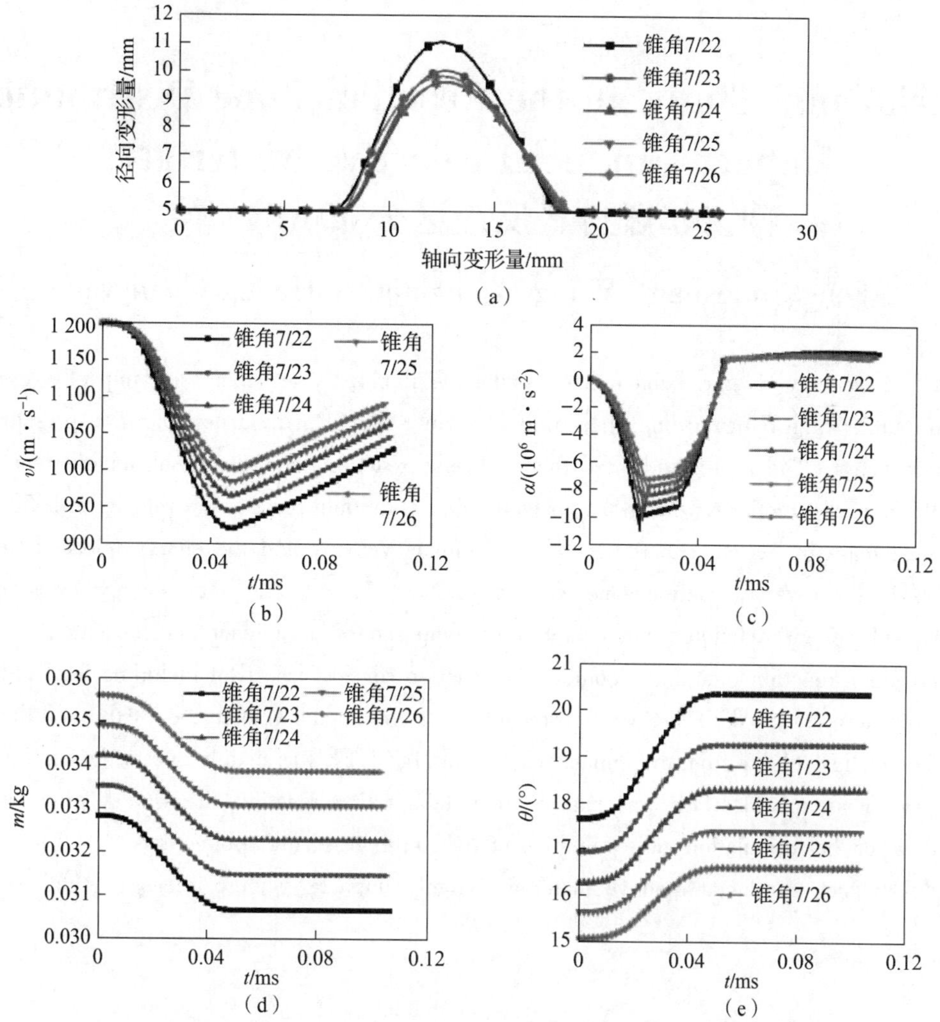

图 10 弹头锥角对中段壳体变形量及弹头参量变化影响

(a) 初始锥角对中段壳体径向膨胀的影响；(b) 初始锥角对弹头速度的影响；
(c) 初始锥角对弹头加速度的影响；(d) 初始锥角对弹头质量的影响；(e) 初始锥角对弹头等效锥角的影响

3 结论

本文重点对弹丸侵彻靶板时内置 C 材料的外部边界即中段壳体膨胀变形和碎裂进行了理论建模与分析。通过冲击波在锥形弹丸传播模型，并考虑头部侵蚀下的弹丸阻力模型和中段壳体受力运动模型，建立了弹丸中段壳体的损伤变形模型。对中段壳体提取微元体进行受力分析，得到了弹丸中段壳体结构在侵彻厚靶板过程中的径向变形控制方程［式（26）］和轴向弯曲计算模型［式（28）和式（30）］。本文考虑了在侵彻过程中弹丸头部被侵蚀的情况，并提出了弹头等效锥角模型［式（13）］用于修正弹头受靶板材料阻力模型。通过计算实例，首先得到了中段壳体变形的时间历程曲线，并得到随时间增加径向变形量增大。通过设定冲击速度梯度，得到了弹丸出靶时在靶后相同位置的中段壳体损伤变形随着速度增加而增加但增量逐渐减小的结论，还得到了不同速度下弹头速度、加速度、质量及等效锥角的时间历程。通过对比发现，冲击速度越大，受到的阻力越大且侵蚀和钝化情况越严重。最后，通过设置初始锥角梯度，在靶后相同位置的中段壳体变形随着锥角的减小而先减小后增大且锥角弧度值为 7/24 时达到极值，初始锥角越大，弹头受到的阻力越大且侵蚀和钝化情况越严重。本文研究为内置 C 材料弹丸侵彻靶板的结构损伤变形损伤研究提供一种理论分析方法，本文研究成果可为内置材料的反应条件及其燃爆后效毁伤研究提供基础支撑和参考依据。

Experimental Study on the Reaction Zone Distribution of Impact – Induced Reactive Materials[①]

冲击起爆材料反应区分布试验研究

Feng Shunshan, Wang Chenglong[②], Huang Guangyan

Abstract: The energy release behaviors of a metal – fluoropolymer composite impact – induced reactive material (IRM) under high dynamic impact loading were investigated using a new partition pressure test and the multipoint pressure test. The results indicated that a reverse reaction zone and a subsequent reaction zone were formed along the impact direction as the IRM impacted on an aluminum plate at the velocity of 1050 – 1450 ms^{-1}. The total energy release increased with the increase of impact velocity and the energy released from the impact reaction of IRM in the reverse reaction zone was only 20% – 30% of the total energy. Most energy release occurred in the subsequent reaction zone, which was composed of an impact decomposition reaction zone, a thermal decomposition reaction zone and a combustion reaction zone. Three IRM including Al/PTFE, Mg/PTFE, and Ti/PTFE were tested. The Ti/PTFE was most sensitive to the impact velocity, but exhibited the lowest energy release rate. The energy release from the impact reaction of Mg/PTFE was mainly due to the combustion reaction. The high energy release of Al/PTFE was mainly from deflagration reaction. This work provides experimental methods and data for the formulation and evaluation of IRM in engineering applications.

Keywords: Impact – induced reactive material; Highly impact; Energy release; Metal – fluoropolymer; Pressure test

1 Introduction

The impact – induced reactive material (IRM) is a class of functional materials that are inert under quasi – static and static loadings and trigger deflagration under high dynamic impact loadings [1]. A recent study indicated that the impact initiation of an IRM prepared by a special process could also be induced under quasi – static compression [2, 3]. IRM has extremely promising application prospects in the many industries, such as energetic materials, aeronautics and astronautics, security and defense and so on. It is usually prepared by compacting a fluoropolymer powder with a reactive metal power [4, 5]. The fluoropolymer component is subject to "explosive" decomposition under high – velocity impact, forming reactive fluorinated small molecules. The fluoride reactive molecules have strong oxidant activities and can rapidly react with the metal powder [6]. The heat energy produced by the high impact temperature and deflagration reaction can further heat and decompose the fluoropolymer, producing more fluoride reactive molecules to react with the metal (Figure 1). The deflagration reaction between the metal and fluoride reactive molecules and the impact decomposition of fluoropolymer are more vigorous than the thermal decomposition [7, 8].

---

① 原文发表于《Propellants, Explosives, Pyrotechnics》2017。

② 王成龙：工学博士，2013 年师从冯顺山教授，研究侵爆毁伤元冲击反应机理及作用后效，现工作单位：北京航天长征飞行器研究所（14 所）。

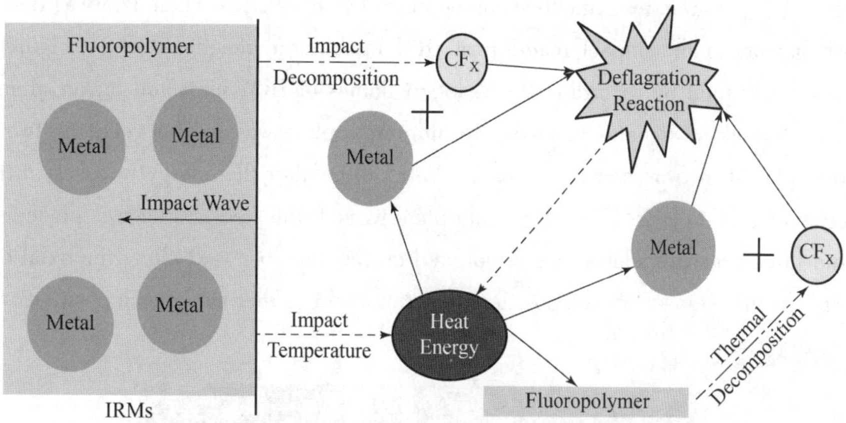

Figure 1 Block diagram of the reaction principle of IRM.

The impact – induced initiation and energy release behavior of IRM are a hot research topic. The pressure testing in a quasi – sealed test chamber, originally developed by Ames [9], is the most widely used method for the study of the impact – induced initiation and energy release behavior of IRM [10 – 13]. It is usually conducted in a special sealed cylindrical chamber with a thin aluminum plate attached on the front side. The IRM fragment perforates the aluminum plate, impacts on the steel anvil located a certain distance from the aluminum plate in the chamber and reacts violently [10, 12]. The high pressure produced inside the chamber is monitored using a pressure sensor on the chamber wall to determine the energy release behavior of the IRM. However, IRM can penetrate a relative thick target in the actual engineering applications, leading to more violent chemical re – actions. Therefore, in H. F. Wang, X. F. Zhang, and P. Luo research, the thin aluminum plates and ingots inside the test chamber can be replaced with thicker metal plates [11 – 13]. IRM is launched at a high impact velocity, penetrates the thick metal plate and triggers a series of severe chemical reactions, resulting in a dynamic reaction zone after passing through the plate. The heat released from the reactions then rapidly increases the pressure in the chamber.

Both methods described above are conducted in sealed test chambers with one pressure sensor on the chamber wall to measure the energy release efficiency. The first method assumes that the IRM completely reacted in the test chamber and the second method only monitors the energy release of IRM in the test chamber after penetrating the plate. According to the impact reaction principle and behavior of IRM, the impact of IRM on a metal plate forms two featured reaction zones, a reverse reaction zone and a subsequent reaction zone (Figure 2), which is composed of an impact decomposition reaction zone, a thermal decomposition reaction zone and a combustion reaction zone.

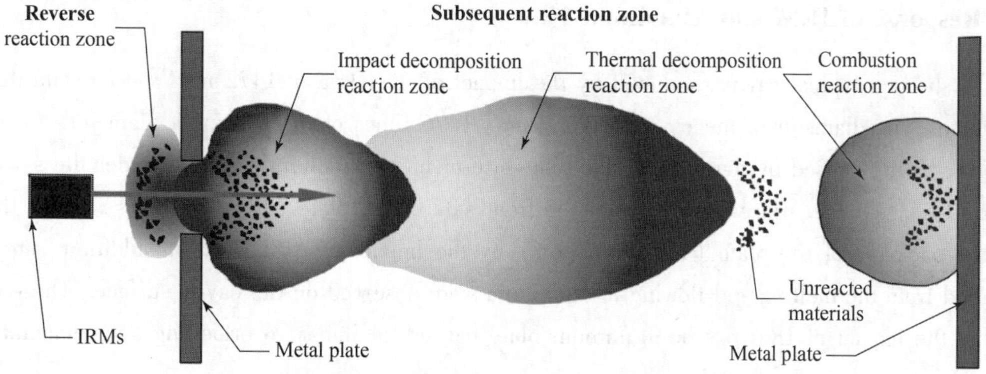

Figure 2 Impact reaction process of IRM.

In the present work, a tandem test chamber composed of two cylindrical chambers was designed to investigate the energy release behaviors of the impact reaction of IRM in different reaction zones (Figure 3). The reaction zone distribution was determined by recording the reaction flames of IRM impacted on an aluminum plate with a high – speed camera. A thick steel holder holding an aluminum plate was put between the two chambers and two pressure sensors were placed in two chambers, respectively, to monitor the energy releases in the reverse reaction zone and the subsequent reaction zone. The aluminum plate without the steel holder was placed on the front side of the test chamber and three pressure sensors were mounted on the chamber wall along the axial direction to monitor the energy releases in the impact decomposition reaction zone, thermal decomposition reaction zone and combustion reaction zone.

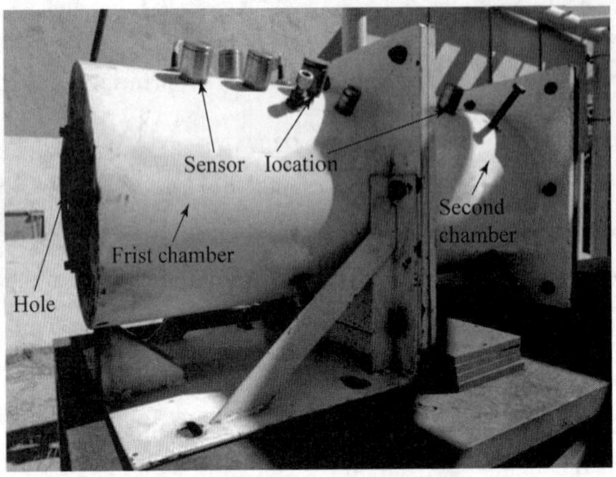

Figure 3 Tandem – test chamber.

2 Impact Reaction of IRM

2.1 Material Specimens

Al/PTFE (AP) was a typical formula used herein to study the impact reaction of IRM. The AP specimen was prepared by compressing a mixture of aluminum powder and PTFE powder into a 0.5 mm thick and 12318 mm cylindrical aluminum shell at 60 MPa with a dry powder mixing craft and a cold – compacting tool. Other specimens listed in Table 1 were prepared using the same procedure. P (s) and Al (s) were prepared with the pure PTFE powder and pure aluminum powder, respectively. Al (d) was a 12318 mm aluminum column. The material type, compositions, specimen mass, material mass, and theoretical energy of all test specimens are summarized in Table 1.

2.2 Response of IRM and Aluminum Plate

Figure 4 shows a typical cavity produced by the impact of Al/PTFE at 1472 ms^{-1} on a 6 mm thick LY – 12 aluminum plate. The diameter of the cavity was 22 mm, 1.76 times of the specimen diameter. In contrast, the diameter of the cavity caused by the impact of a same size of inert aluminum fragment under the same conditions was smaller [14]. A folded rim was formed on the front side of the plate (Figure 4 (b)) and a thin wall was formed on the backside of the plate (Figure 4 (c)) by the impact of Al/PTFE. In addition, numerous layer cracks resulted from the melting and flowing of aluminum were observed on the cavity surface. These observations indicated that the impact of IRM on the aluminum plate caused an impact overloading and temperature arising, evoking the chemical reaction of IRM.

Figure 4　A typical cavity caused by the impact of Al/PTFE on the aluminum plate

(a) Cavity in the aluminum plate; (b) front side of the cavity; and (c) backside of the cavity.

Table 1　Composition and properties of the IRMs for the impact test.

No.	Specimen	Material type	Mass ratio	Specimen mass [g]	Material mass [g]	Theoretical energy [Jg^{-1}]
1	AP	Al/PTFE	23.5wt−% Al+76.5wt−% PTFE	5.25	4.57	8 860
2	P(s)	PTFE powder	100% PTFE powder	4.39	4.98	–
3	Al(s)	Al powder	100% Al powder	4.71	4.12	–
4	Al(d)	Al column	100% Al solid	5.5	5.5	–
5	Mg/PTFE	Mg/PTFE	32.4wt−% Mg+67.6 wt−% PTFE	4.64	3.96	9 675
6	TP	Ti/PTFE	32.4wt−% Ti+67.6 wt−% PTFE	5.97	5.19	5 683

The impact pressure was determined by the one−dimensional shock wave theory. When a specimen impacts a stationary plate at the velocity of v_0, the shock wave propagates in the specimen and plate respectively [15].

For specimen:

$$P_f = \rho_f D_f u_f \tag{1}$$

For plate:

$$P_t = \rho_t D_t u_t \tag{2}$$

For shock wave velocity:

$$D_f = a_f + b_f u_f \text{ and } D_t = a_f + b_t u_t \tag{3}$$

By interfacial continuity

$$P_f = P_t \text{ and } v_0 - u_f = u_t \tag{4}$$

Where P is the shock wave pressure, 1 is the density, D is the shock wave velocity, u is the particle velocity, a and b are the Hugoniot coefficient, and subscript f and t represent the specimen and plate, respectively. By combining Equation (1), Equation (2), Equation (3), and Equation (4), the impact pressure P_f on specimen can be obtained.

According to the Hugoniot curve, a shock wave propagates through a specimen and rises its average temperature. Assuming that the thermodynamic process is adiabatic and the internal friction is negligible, the average impact temperature T_1 can be expressed as follows (Equation (5)) [15]:

$$T_1 = T_0 \exp\left[\frac{\gamma_0}{V_0}(V_0 - V_1)\right] + \frac{V_0 - V_1}{2C_v}P_1 + \frac{\exp\left(-\frac{\gamma_0}{V_0}V_1\right)}{2C_v}\int_{V_0}^{V_1} P\exp\left(\frac{\gamma_0}{V_0}V\right)\left[2 - \frac{\gamma_0}{V_0}(V_0 - V)\right]dV \tag{5}$$

Where g_0 is the Gruneisen coefficient, C_v is the heat capacity at constant volume, P_1 is the shock wave pressure, V_0 is the initial specific volume of the solid, and V_1 is the post - shock specific volume. As the IRM is a composite material of PTFE and metal powder, the state equation can be determined by mass averaging method as follows (Equation (6)):

$$x = \sum m_i x_i \tag{6}$$

The material parameters for calculation of impact pressure and average temperature are listed in Table 2.

Table 2 Material parameters for the calculation of impact pressure and average temperature.

No.	Material	1 [gcm^{-3}]	a [$mm\ \mu s^{-1}$]	b	γ	C_v[$Jg^{-1}K^{-1}$]
1	Al	2.96	5.35	1.34	2.0	0.89
2	Mg	1.74	4.49	1.24	1.6	1.02
3	Ti	4.54	5.22	0.76	1.1	0.52
4	PTFE	2.15	1.84	1.71	0.6	1.02

The calculation indicated that the impact of AP specimen on the aluminum plate at 1 450 ms^{-1} resulted in an impact pressure of 8.66 GPa an average impact temperature of 489.3 K. Under such extreme conditions, the high impact pressure induced the "explosive" decomposition of PTFE. PTFE can be rapidly thermal - decomposed at 693 K. Therefore, impact temperature caused the slow thermal decomposition of PTFE. Once PTFE was completely decomposed, the aluminum rapidly reacted with the decomposition products of PTFE.

2.3 Reaction of IRM Impacted on Aluminum Plate

The reaction process was recorded with a high - speed camera. Figure 5 shows the selected video frames of the Al/ PTFE fragment impacted at 1 446 ms^{-1} on a 6 mm thick LY - 12 plate. The AP specimen penetrated the plate in less than 28.6 ms and was fragmented into numerous tiny particles. A small portion of the material initiated the reaction in the front of the plate. Most of the material perforated the plate and formed a subsequent reaction zone. The flame of the reaction rapidly expanded within a few hundred microseconds after the material penetrated the aluminum plate (Figure 5, t = 0.3 ms). Meanwhile, the unreacted material flew away from the flame zone. The fireball reached the maximum in 1 ms and lasted several milliseconds (Figure 5, t = 1 ms). The unreacted materials impacted the platform to initiate the combustion reaction. The reaction process was completed in 4 ms.

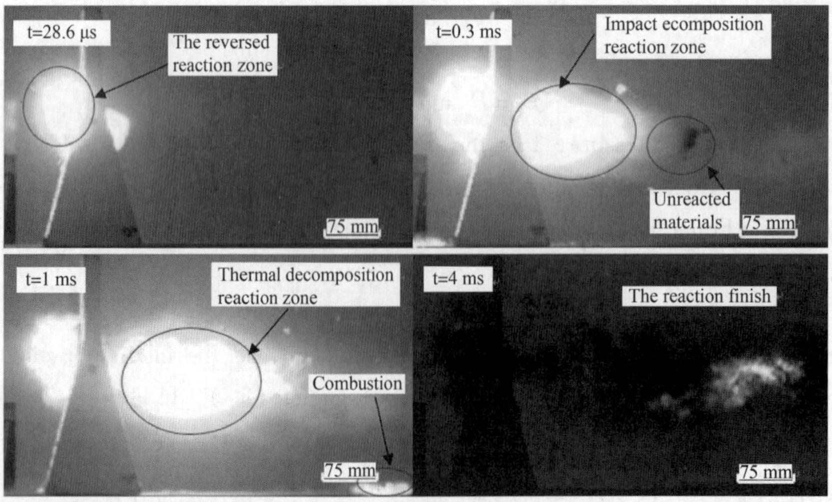

Figure 5 Selected high - speed camera video frames of the impact and reaction process of AP.

In the reverse reaction zone, the impact of AP specimen on the aluminum plate induced the impact decomposition and thermal decomposition of PTFE. The material was taken apart during the reaction. In the subsequent reaction zone, the AP particles exhibited a high residual velocity (about 1 000 ms^{-1}). The reaction zone outspread along the impact direction, passed the plate and formed an impact decomposition reaction zone where the impact decomposition products of PTFE reacted with aluminum. The thermal decomposition reaction zone was an area with stable flame where the thermal decomposition products of PTFE reacted with aluminum. The unreacted material flew away from the flame area at 1 000 ms^{-1} and subjected to combustion reaction as it impacted on metal plate at the bottom, forming a combustion reaction zone.

3 Energy Release in the Reverse Reaction Zone

The energy release in the reverse reaction zone is inevitable and undesired in engineering application. It should be reduced as much as possible in material design to allow more material passing through the plate into the subsequent reaction zone. In the presented work, a partition pressure test method was used for the first time to determine the energy release ratio in the reverse reaction zone and the subsequent reaction zone.

3.1 Partition Pressure Testing Method

The setup for the partition pressure testing was composed of two cylindrical chambers, a thick steel holder and a thick steel plate that was designed to isolate the reverse reaction zone and the subsequent reaction for the separated pressure measurement (Figure 6). The first and second cylindrical chambers had an inner diameter of 260 mm and were 300 mm and 280 mm long, respectively. The metal plate with a hole of 70 mm diameter was fixed on the front side of the first chamber to allow the projectiles passing through. The thick steel holder with a similar hole of 70 mm diameter was used to hold the aluminum plate and isolate the chambers. The thick steel plate on the back side of the test chamber was opened after each test to release pressure and clean up the reaction products.

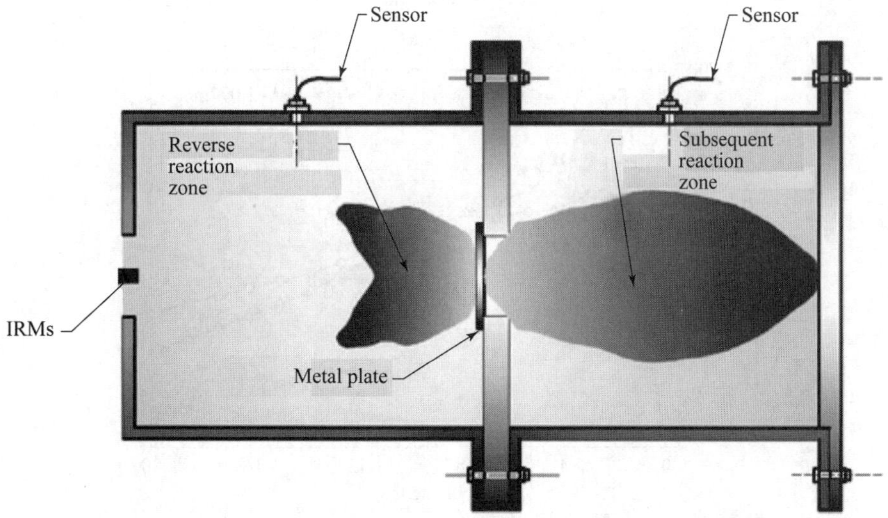

Figure 6 Test chamber setup for the partition pressure testing.

The specimen was shot by a 13.1 mm ballistic gun and impacted on a 6 mm thick LY – 12 aluminum plate fixed on the thick steel holder. Its velocity was measured with the high – speed camera. The parameters for the test are listed in Table 3.

Table 3 Specimen parameters.

No.	Specimen	Velocity [ms^{-1}]	Impact pressure [GPa]	Average impact temperature [K]
1	AP	1 050	5. 81	403. 7
2	AP	1250	7. 19	440. 0
3	AP	1 450	8. 66	483. 9

3. 2 Energy Release Distribution in the Reverse and Subsequent Reaction Zones

Figure 7 shows the smooth curves of the impact pressures in the reverse reaction zone (Figure 7 (a)) and the subsequent reaction zone (Figure 7 (b)) produced by the impact of AP at different velocities. It is clear that the peak pressures produced by the impacts of AP at all velocities in subsequent reaction zone (Figure 7 (b)) were much higher than those in the reverse reaction zone (Figure 7 (a)), indicating that most of the material passed though the aluminum plate and reacted to form the subsequent reaction zone.

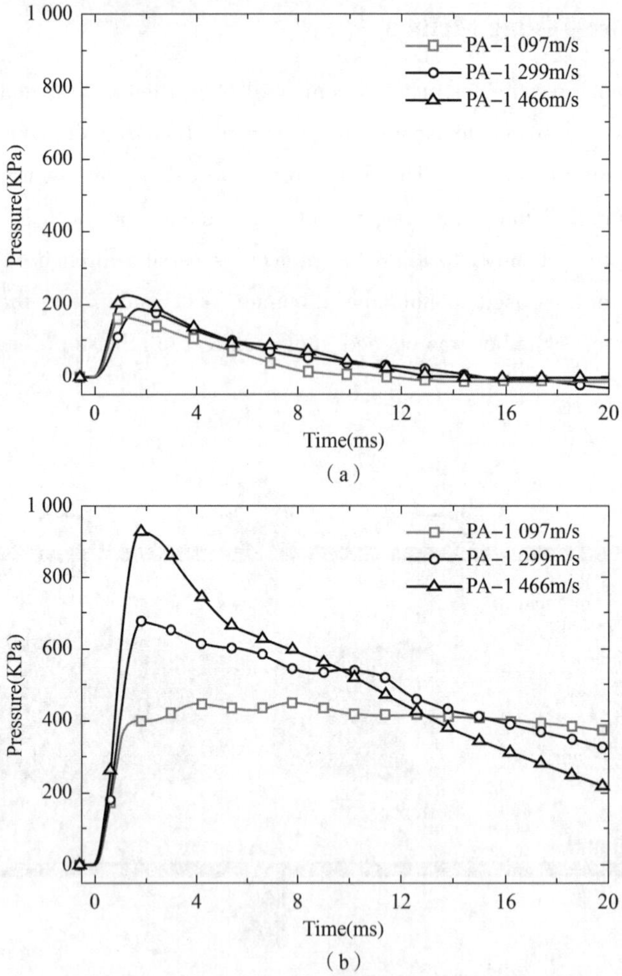

Figure 7 Smooth curves of the pressures produced by the impacts of AP at different velocities in the reverse reaction zone (a) and the subsequent reaction zone (b).

The peak pressures (DP) in the reverse reaction zone and the subsequent reaction zone were then used to calculate the reaction energy E by Equation (7):

$$E = \frac{V}{\gamma - 1} \Delta P \tag{7}$$

where V is the volume of the test chamber and g is the ratio between the specific gas heats in the first and second chambers.

Figure 8 shows the energy releases in the reverse reaction zone and the subsequent reaction zone produced by the AP impacted at 1 077 ms^{-1}, 1 266 ms^{-1}, and 1 466 ms^{-1}. The energy release of AP in the reverse reaction zone was only 20% – 30% of the total energy release. The rest energy release was measured in the subsequent reaction zone. The total energy releases from 40% to 70% with the increase of the impact velocity from 1 077 ms^{-1} to 1 466 ms^{-1}. It can be explained that more material penetrated the aluminum plate and reacted in the subsequent reaction zone at high impact velocities and released more energy. In addition, the average impact temperature also significantly affected the energy release in the react and increased with the increase of the impact velocity.

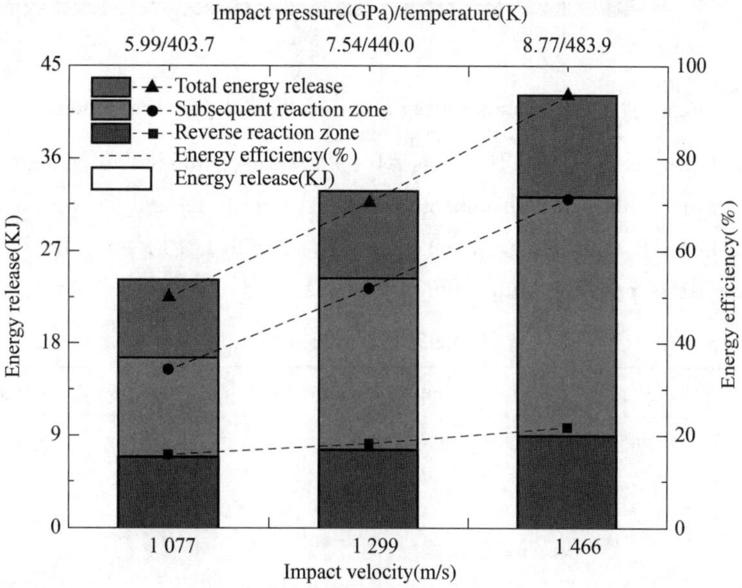

Figure 8　Energy distributions of the AP impacts at different velocities in the reverse and subsequent reaction zones.

4　Energy Release in the Subsequent Reaction Zone

4.1　Multipoint Pressure Testing Method

As discussed above, the subsequent reaction zone was composed of an impact decomposition reaction zone, a thermal decomposition reaction zone and a combustion reaction zone. Therefore, a multipoint pressure testing method was used to determine the distribution of these three reaction zones. The plate holder was removed to connect the two test chambers and the aluminum plate was fixed on the front side of the first chamber for the multipoint pressure test (Figure 9). The specimen impacted through the aluminum plate and formed three reaction zones in the tandem – test chamber. Three pressure sensors were mounted on the chamber wall along the impact direction to measure the pressures produced in each reaction zone.

The three pressure sensors were positioned based on the reaction flames recorded by the camera. The sensor #1 was located 75 mm away from the front plate to measure the pressure produced in the impact decomposition zone where the impact decomposition products of PTFE reacted with the metal. The sensor #2 was at 225 mm away from the aluminum plate to measure the pressure produced in the thermal decomposition reaction zone where the thermal decomposition products of PTFE reacted with the metal. The sensor 3# was mounted on the middle of the second chamber wall along the impact direction to measure the pressure produced in the combustion reaction zone where the unreacted metal powder impacted on the thick steel plate of the chamber backside and was combusted.

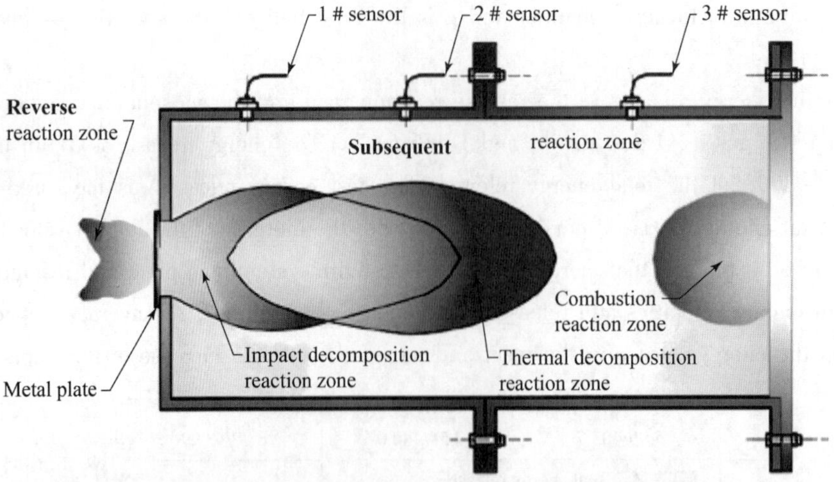

Figure 9 Test chamber setup of the multipoint pressure testing.

The test scheme is shown in Table 3. The tests #1 – #4 were conducted to compare the impact – induced reactions of IRM and inert materials at a high impact velocity. The tests 1# and #5 were performed to compare the impact – induced reactions of AP at different impact velocities. The tests #1, #6, and #7 were used to compare the impact reaction of the IRMs prepared with different formula.

Table 3 Test scheme.

No.	Specimen	Velocity [Ms^{-1}]	Impact pressure [GPa]	Impact temperature [K]	Comparison
1	AP	1 450	8.66	483.9	
2	A	1 450	–	–	Materials
3	P	1 450	–	–	
4	A	1 450	–	–	
5	AP	1 450	7.19	440.0	Velocity
6	MP	1 450	8.01	490.4	Formula
7	TP	1450	9.47	448.0	

4.2 Reaction Zone Distribution in Subsequent Reaction Zone

4.2.1 Reaction Zone Distribution of Reactive Materials and Inert Materials

Figure 10 shows the smooth curves of the pressures produced by impact of AP, A (s), P (s), and A (d) at 1450 ms^{-1} in the three reaction zones. As mentioned above, the impact induced reaction was completed in 4 – 10 ms. Therefore, the pressure curves were recorded for 20 ms.

As shown, the impact of the inert aluminum A (d) was unable to induce any reaction and thus its pressure curves in all reaction zones were flat. The impact of pure PTFE powder [P(s)] was able to induce the impact decomposition, but no deflagration reaction due to the lack of the active metal. Therefore, the impact decomposition and powder diffusing effect of P(s) only caused a slight pressure change. Al(s) was the aluminum powder pressed into an aluminum shell. The aluminum powder was combusted with oxygen and released energy, which was much milder than the deflagration reaction. Therefore, the pressure caused by the impact of AP was higher than that caused by Al(s) in the impact decomposition zone and thermal decomposition reaction zone. The pressure of produced by the impact of Al (s) in the combustion reaction zone was higher than that of AP due to more combustible component in Al (s).

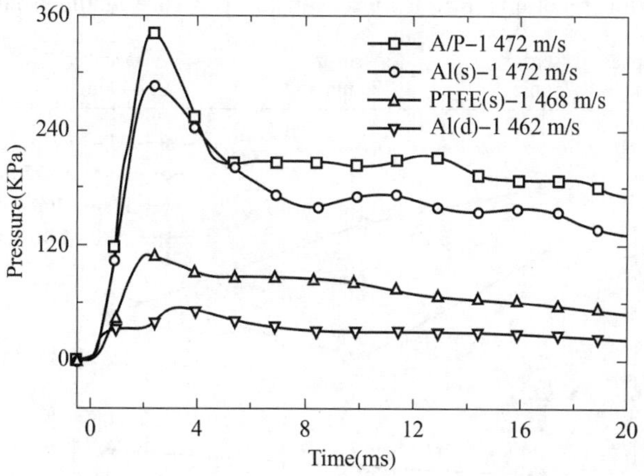

(a)

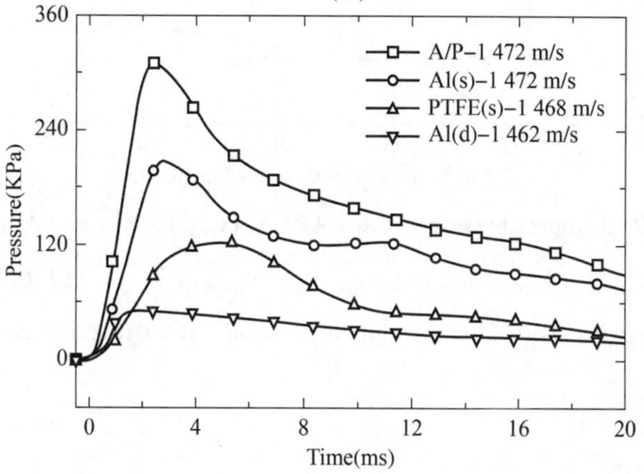

(b)

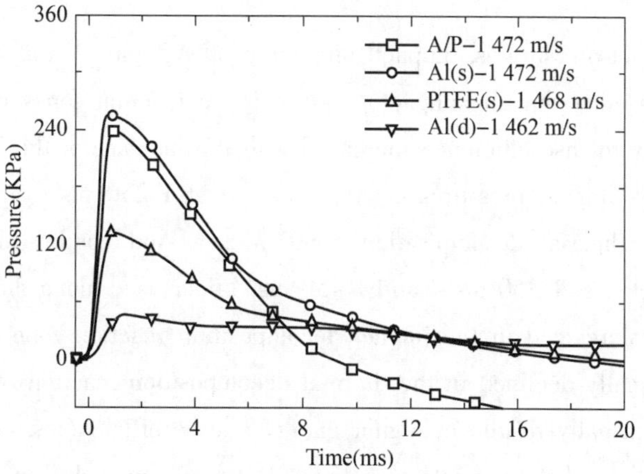

(c)

Figure 10　Smooth curves of the pressures produced by AP, A(s), and A(d) impacted at different velocities in the impact decomposition reaction zone (a), thermal decomposition reaction zone (b), and combustion reaction zone (c).

Figure 11 shows the peak impact pressures of AP, A(d), P(s), and A(s) recorded by the three sensors. The peak impact pressures along the impact direction of both A(d) and P(s) were evenly distributed and lower than those of IRM AP and combustible metal powder A(s). The peak pressure of AP decreased along the impact direction from sensor #1 to sensor #3, indicating that most of AP reacted in the impact decomposition reaction zone and thermal decomposition reaction zone and only a small amount of unreacted material was combusted in the combustion reaction zone. The aluminum powder reacted with oxygen after Al(s) penetrated the plate and significantly increased the pressure at the position of sensor #1. The pressure at the position of sensor #2 was very low because there were no reaction occurred. The unreacted aluminum powder impacted on the back plate of the chamber and induced combustion reaction, which increased the pressure at the position of sensor #3.

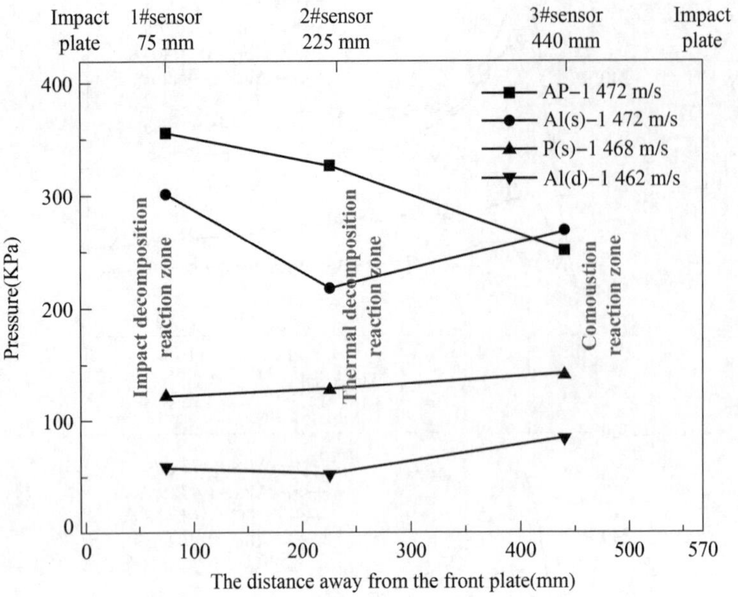

Figure 11 Peak impact pressures of the AP, A (d), P (s), and A (s) recorded

In all, the impact reaction of AP was more violent than those of A(s), A(d), and P(s) in the subsequent reaction zone and released energies mainly in the impact decomposition reaction zone and thermal decomposition reaction zone.

By the sensors #1, #2, and #3.

4.2.2 Reaction Zone Distributions of AP impacted at 1 250 ms^{-1} and 1 450 ms^{-1}

Figure 12 shows the impact pressure vs. impact time curves of AP impacted at 1 250 ms^{-1} and 1 450 ms^{-1} in the three reaction zones. It is clear that the impact pressure in all reaction zones increased with the increase of impact velocity and the energy release efficiency increased with the increase of the impact pressure.

Figure 13 shows the peak impact pressures of AP impacted at 1 250 ms^{-1} and 1 450 ms^{-1} in the impact decomposition reaction zone, thermal decomposition reaction zone and combustion reaction zone. Both peak impact pressures of AP impacted at 1 250 ms^{-1} and 1 450 ms^{-1} decreased along the impact direction from sensor #1 to sensor #3. Most material reacted in the impact decomposition reaction zone and the peak pressure of the impact at 1 250 ms^{-1} dramatically declined in the thermal decomposition reaction zone.

Higher impact pressures usually results in higher energy release efficiencies. The dotted lines in Figure 13 are the theoretical peak pressures of the impacts of AP at 1 250 ms^{-1} and 1 450 ms^{-1} at the positions of sensor #1, #2, and #3, assuming that the energy release was completely applied on the wall of the chambers. However, three reaction zones were formed in the subsequent reaction zone along the impact direction. The unreacted material flew away from the deflagration reaction zone, which reduced the total energy release. The deviation between the measured peak pressure and the theoretical peak pressure increased along the impact direction. These results indicated that the reaction zone division is consistent with the actual impact - induced reaction of AP and thus can be used to determine the reaction characteristics of IRM impacted on a metal plate.

4.2.3 Impact - induced Reaction Zone Distributions of Al/PTFE, Mg/PTFE, and Ti/PTFE

Al/PTFE (AP), Mg/PTFE (MP), and Ti/PTFE (TP) were tested at the impact velocity of 1 450 ms^{-1} to compare the impact induced reaction zone distributions of IRMs with different formula. MP is a pyrotechnic

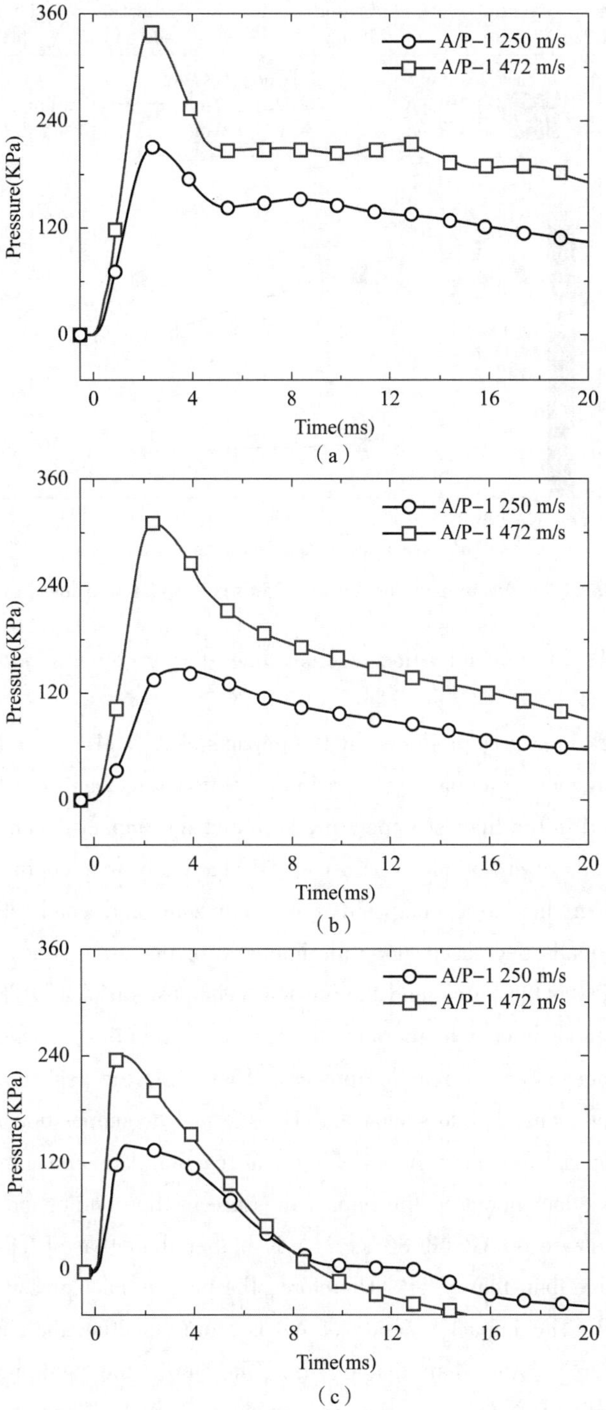

Figure 12 Impact pressure vs. time curves of AP impacted at different velocities in the impact decomposition reaction zone (a), thermal decomposition reaction zone (b), and combustion reaction zone (c).

composite widely used as the flare and solid rocket propellant. However, its impact reaction characteristics have been rarely reported. The theoretical energy release of TP is less than that of AP. However, its impact reaction is more sensitive and violent than that of AP and it can be used as an effective additive to enhance the deflagration reaction of IRMs.

Figure 13 shows the impact – induced pressure vs. time curves of AP, MP, and TP impacted at 1 450 ms^{-1} in the three reaction zones. The impact – induced pressures of AP and TP were higher than that of MP in the impact decomposition reaction zone. The highest pressure in the thermal decomposition reaction zone was

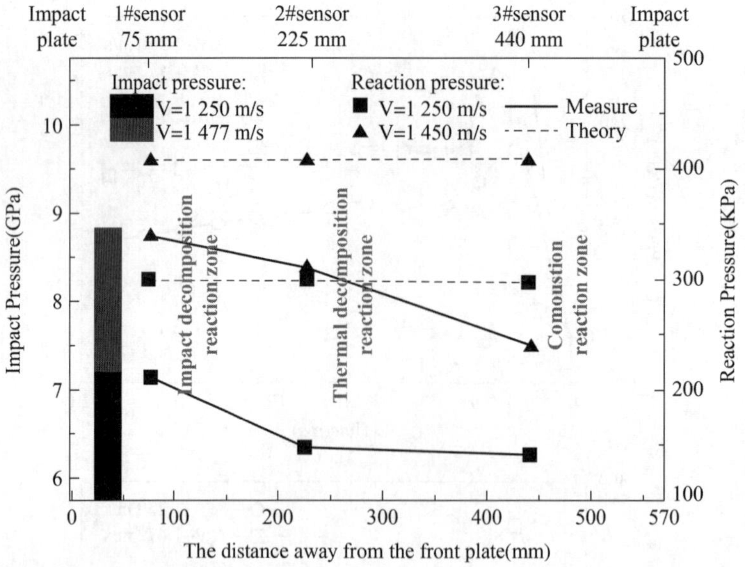

Figure 13 Peak pressures produced by the impacts of AP at 1 250 ms^{-1} and 1 450 ms^{-1} at different pressure test positions.

produced by the impact of AP. In the combustion reaction zone, the pressures produced by the impacts of AP and MP were higher than that of TP.

Figure 15 shows the peak pressures produced by the impacts of AP, MP, and TP at 1 450 ms^{-1} in the in the impact decomposition reaction zone, thermal decomposition reaction zone, and combustion reaction zone.

The impact of TP resulted in the highest impact pressure and the impact decomposition of its PTFE in impact decomposition reaction zone was most sufficient. Most of TP vigorously reacted in this reaction zone and only a small amount of TP reacted in thermal decomposition reaction zone and combustion reaction zone. The peak impact pressure of TP dramatically decreased along the impact direction from sensor #1 to sensor #2 and sensor #3. The impact pressure of MP was the lowest and the impact decomposition of its PTFE was insufficient. However, MP exhibited the highest average impact temperature under same conditions, which could improve the thermal decomposition of PTFE and accelerate the reaction process. Therefore, the peak impact pressure of MP increased along the impact direction from sensor #1 to sensor #2. The thermal decomposition reaction was slow, and thus a large amount of unreacted material flew away from deflagration reaction zone and impacted on the back plate of the chamber. The combustion reaction increased the impact pressure in the combustion reaction zone.

The theoretical energy release of AP (8.86 kJg^{-1}) is higher than that of TP (5.683 kJg^{-1}) although its impact reaction is less sensitive than that of TP. Therefore, the peak impact pressure of AP recorded by sensor #2 was higher than that of TP. The impact reaction of AP is more sensitive than MP and its theoretical energy release (8.86 kJg^{-1}) is slightly lower than that of MP. Therefore, the peak impact pressures of AP in the impaction decomposition and thermal decomposition zone were much higher than those of MP.

In summary, AP exhibited the best comprehensive impact performance. TP was most sensitive to the impact and could be used as an additive to improve the initial impact reaction. MP was least sensitive to the impact and was mainly subject to combustion reaction.

5 Conclusions

The impact – induced reaction zone distribution of IRM impacted on an aluminum plate at 1 050 – 1 450 ms^{-1} was determined using the partition pressure testing and multipoint pressure testing methods. Four reaction zones including the reverse reaction zone, impact decomposition reaction zone, thermal decomposition reaction zone and combustion reaction zone were formed during the impact of IRM on an aluminum plate and recorded

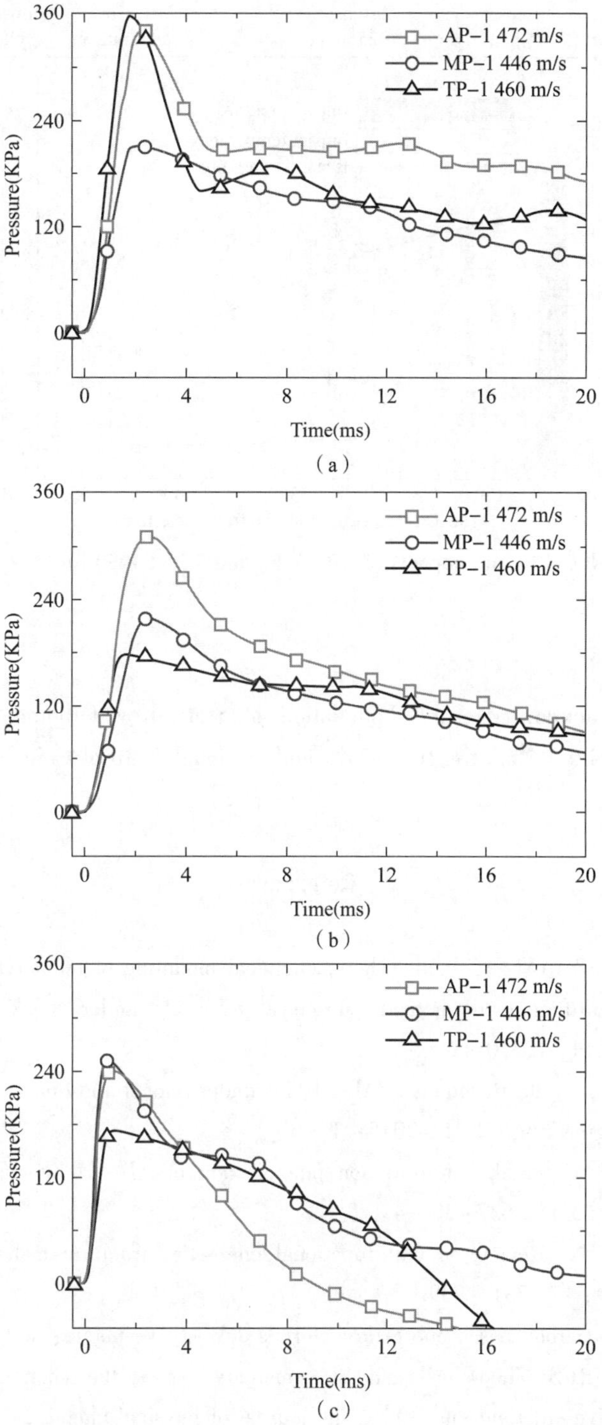

Figure 14 Impact - induced pressure vs. time curves of AP, MP, and TP impacted at 1 450 ms^{-1} in the impact decomposition reaction zone (a), thermal decomposition reaction zone (b), and combustion reaction zone (c).

using a highspeed camera. The total energy release of IRM increased with the increase of impact velocity. The energy release of IRM in the reverse reaction zone was only 20% - 30% of the total energy release and the rest mainly occurred in the subsequent reaction zone. The impact of Al/PTFE mainly caused a deflagration reaction in the impact decomposition reaction zone and thermal decomposition reaction zone with a high energy release. Ti/PTFE was more sensitive to the impacts than Al/PTFE but exhibited lower energy releases. The impact reaction of Mg/PTFE mainly occurred in the combustion reaction zone and it was less sensitive to the impact.

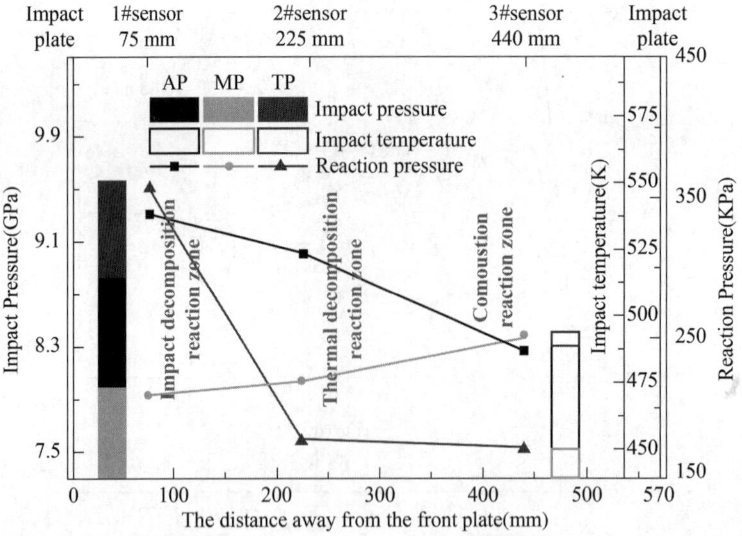

Figure 15 Peak pressures produced by the impacts of AP, MP, and TP at 1450 ms^{-1} at different pressure test positions.

Acknowledgements

The authors gratefully acknowledge the Foundation of State Key Laboratory of Explosion Science and Technology of China under Grant No. KFJJ13 - 1Z, and National Natural Science Foundation of China under Grant No. 11102023.

References

[1] ZHANG X F, SHI A S, ZHANG J, et al. Thermochemical modeling of temperature controlled shock - induced chemical reactions in multifunctional energetic structural materials under shock compression [J]. Journal of applied physics, 2012, 111, 12350112.

[2] FENG B, FANG X, LI Y, et al. Reactions of Al - PTFE under impact and quasi - static compression Advances in materials science and engineering, 2015, 2015: 1 - 6.

[3] FENG B, FANG X, LI Y, et al. An initiation phenomenon of Al - PTFE under quasi - static compression Chemical physics letters, 2015, 637: 38 - 41.

[4] ZHANG X F, ZHAO X N. Review on multifunctional energetic structural materials [J]. Chinese journal of energetic materials, 2009, 17: 731 - 739.

[5] KOCH E C. Metal - fluorocarbon based energetic materials [M]. Weinheim: Wiley - VCH, 2012: 36 - 65.

[6] LOSADA M, CHAUDHURI S. Theoretical study of elementary steps in the reactions between aluminum and Teflon fragments under combustive environments [J]. The journal of physical chemistry A, 2009, 113: 5933 - 5941.

[7] DOLGOBORODOV A Y, MAKHOV M N, KOLBANEV I V, et al. Detonation in an aluminum - Teflon mixture, JETP Letters, 2005, 81: 311 - 314.

[8] SULLIVAN K T, CERVANTES O, DENSMORE J M, et al. Quantifying dynamic processes in reactive materials: an extended burn tube test [J]. Propellants explosives, pyrotechnics, 2015, 40: 394 - 401.

[9] AMES R G. Vented chamber calorimetry for impact - initiated energetic materials [C] //43rd AIAA Aerospace Sciences Meeting and Exhibit, Reno, NV, USA, January 10 - 13, AIAA Paper 2005 - 279, 2005.

[10] AMES R G. Energy release characteristics of impact - initiated energetic materials [J]. Symposium H - multifunctional energetic Materials 2006, 896: 123 - 132.

反应破片对密实防护油箱的引燃效应研究①

王成龙②，黄广炎，冯顺山

（北京理工大学爆炸科学与技术国家重点试验室，北京　100081）

摘　要：反应破片具有冲击诱发剧烈化学反应的特性，可以对油箱实现显著引燃效应。通过弹道枪试验和理论分析计算，研究了单枚反应破片对密实防护油箱的引燃问题，获得了引燃机理和引燃判据。研究结果表明，油箱的引燃判据有闪燃和持续燃烧两种，对于密实防护油箱，单枚反应破片可引起油箱闪燃，通过增加油气量可以提高闪燃的规模和延长持续时间；持续燃烧的必要条件是燃油持续溢出与空气接触且燃油蒸发速率等于油气燃烧消耗速率，通过提高反应能量加热燃油或通过灯芯效应，可提高燃油蒸发速率从而实现燃油的持续燃烧。研究方法和所得结论可为反破片战斗部应用和引燃油箱威力评价提供参考。

关键词：反应破片；油箱；引燃；闪燃；持续燃烧；密实防护

Research on Ignition of Aircraft Fuel Tank by Reactive Fragment

Wang Chenglong[1,2], Huang Guangyan[2], Feng Shunshan[2]

（1. Beijing Institute of Space Long March Vehicle, Beijing 100076, China;

2. State Key Laboratory of Explosion Science and Technology, Beijing Institute of Technology, Beijing 100081, China）

Abstract: The reactive fragments will trigger deflagration under high dynamic impact loadings, which can ignite fuel tank. Impact ignition tests were performed on fuel tanks with well protection from single reactive fragments. The ignition mechanism and criterion are studied through the ballistic gun experiment and theoretical analysis. The ignition criteria include flashover and sustained burn. For a well - protected tank, the impact of single reactive fragment is able to induce flashover of fuel. The scale and duration of flashover can be improved by increasing the fuel gas. The necessary condition for sustained burn is that the fuel continues to spill out and the fuel evaporation rate is equal to consumption rate. The fuel continues to burn due to that the duration of heat source. The evaporation rate of fuel is improved by increasing the reaction energy or wick effect, thus achieving the sustained burn of fuel.

Keywords: reactive fragment; fuel tank; ignition; flashover; sustained burn; dense protection

引言

反应破片是一种新型高效毁伤元技术，它由高分子氟聚物与活性金属复合而成[1-2]，通过压制烧结工艺和结构设计可使其具备一定的机械强度[3-4]。如图 1 所示，当反应破片以一定速度 v 撞击目标时，在完成侵彻贯穿目标防护的同时，由于强冲击载荷发生破碎并诱发化学反应，在传靶后产生剧烈的燃爆效应[5]，其反应场温度可高达 3 300 K[6-7]，具备对靶后目标的引燃、引爆等后效威力。相对于惰性破片如钢破片或钨破片，借助破片对油箱壳体机械撞击侵彻产生的火花引燃燃油，反应破片对油箱的引燃效

---

① 原文发表于《兵工学报》2018. 12。

② 王成龙：工学博士，2013 年师从冯顺山教授，研究侵爆毁伤元冲击反应机理及作用后效，现工作单位：北京航天长征飞行器研究所（14 所）。

应更加显著[8]。

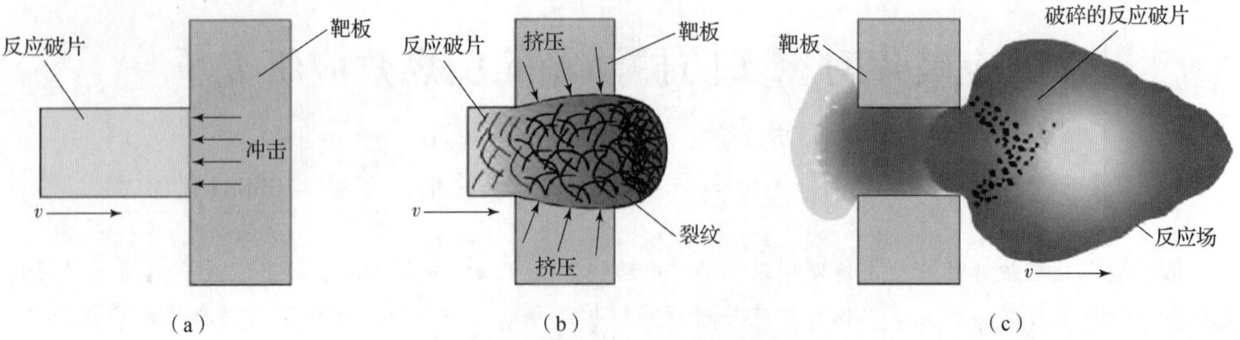

图 1　反应破片冲击反应作用示意图

(a) 冲击响应阶段；(b) 侵彻破碎阶段；(c) 穿靶后能量释放阶段

反应破片已经成为战斗部毁伤元领域的前沿技术和研究热点[6-7]，但是仍有许多问题尚未解决，其中对油箱引燃效应评估就是制约反应破片战斗部应用的关键问题之一。本文针对该关键问题，通过理论分析和试验研究相结合的方法，开展反应破片对密实油箱的引燃机理和闪燃/持续燃烧判据研究，以期为解决反应破片引燃油箱威力评价问题提供参考。

1　反应破片及油箱引燃机理

1.1　反应破片试件

本文研究所采用的反应破片试件为金属 - 聚四氟乙烯（PTFE）基反应破片且通过配方优化设计、克服原铝（Al）- PTFE 材料低密度缺点而得到的一种高密度、高能量反应破片，其参数如表 1 所示，成型密度大于 8.5 g/cm^3，同时单个反应破片含能不低于同尺寸的 Al - PTFE 材料（约为 37.82 kJ）。

表 1　反应破片试件参数

破片代号	时间尺寸/mm	试件质量/g	成型密度/(g·cm^{-3})	材料含能/kJ
YP - 1	φ11.8 × 17.09	16.35	8.75	67.26
YP - 2	φ11.8 × 17.70	16.46	8.51	67.72
YP - 3	φ11.8 × 17.23	16.47	8.74	67.76
YP - 4	φ11.8 × 16.96	16.35	8.82	67.26
YP - 5	φ11.8 × 16.94	16.20	8.75	66.68

1.2　燃油引燃机理

本文研究所采用的燃油类型为煤油，其主要性能参数参考值如表 2 所示[13-14]。煤油属于液体燃料，其蒸馏温度低于燃点，液体燃料的引燃和燃烧是首先蒸发、生成燃料蒸气，然后与空气混合，进而被引燃。

表 2　煤油主要性能参数及其参考值（化学式 $C_{9.71}H_{20.52}$）

参数	参考值
相似摩尔质量/(g·mol^{-1})	137.04
20 ℃下饱和蒸气压/mmHg	35

续表

参数	参考值
20 ℃下比热容/(kJ·kg^{-1}·℃$^{-1}$)	2.1
质量热值/(MJ·kg)	42.5
相对密度/(g·cm^{-3})	0.8
闪点/℃	>38
燃点/℃	>300
蒸发热/（kJ·kg^{-1}）	251

决定燃油是否能够被引燃并持续燃烧主要有以下4个条件：

条件1　燃油蒸发形成蒸汽。燃油燃烧首先由液态蒸发成油气态，燃油蒸发速率可以用式（1）表示：

$$u_c = (p_f/\rho_f)[M_f/(2\pi RT_f)]^{1/2} \tag{1}$$

式中：p_f——燃油饱和蒸汽压强；

ρ_f——燃油密度；

M_f——燃油摩尔质量；

R——气体常数；

T_f——燃油温度。其中 p_f 与 T_f 间的关系服从安瑞方程：

$$\lg p_f = a + \frac{b}{T_f + c} \tag{2}$$

a、b 和 c 为拟合常数。

条件2　蒸汽与空气混合，混合气体浓度在燃烧极限范围内。

燃油蒸发产生的蒸汽必须与空气混合，使油气与氧气充分接触，油气浓度过低或者氧气含量过低都不能支持燃烧。燃油在空间大小为 V 的环境内产生的混合气体浓度可用式（3）表示：

$$\eta = \frac{p_f V/(RT_{f,R}) + m_{f,q}/M_{f,R}}{p_0 V/(RT_{f,g}) + m_{f,q}/M_{f,e}} \tag{3}$$

式中：p_0——大气压力；

$T_{f,R}$——油气温度；

$m_{f,q}$——燃油汽化质量；

$M_{f,R}$——油气摩尔质量。

条件3　点火源温度大于燃油的着火温度。

当油气浓度在燃烧范围极限内、遇到明火且其火源温度 T_s 大于燃油的着火温度 T_i 时，油气就能发生燃烧。

条件4　燃油的燃烧速率等于其蒸发速率。

在满足上述条件后燃油被引燃，但是当燃油燃烧速率 u_b 大于其蒸发速率 u_c 时，新的蒸汽来不及补充被烧掉的蒸汽，油气层只能维持一瞬间燃烧，即闪燃；若要形成持续燃烧，则需要燃油燃烧速率 u_b 等于其蒸发速率 u_c，燃油的燃烧速率通过式（4）表示：

$$u_b = \frac{q_1 + q_2 + q_3 - C_f M_f (T_f - T_0)}{S\rho_f [Q_f - C_f (T_f - T_0)]} \tag{4}$$

式中：q_1——油箱壁向燃油的热流量；

q_2——燃烧的高温气体向燃油的热流量；

q_3——火焰和高温气体向燃油的辐射热流量；

C_f——燃油的比热容；

S——油池面积；

Q_f——燃油的蒸发热；

T_0——燃油初始温度。

1.3 油箱条件及引燃判据

油箱引燃情况受油箱结构、初始温度条件等影响很大。对于简易油箱，破片高速冲击产生的水锤效应会造成油箱结构损坏，导致出现多处漏油现象；对于密实防护油箱，单枚破片冲击仅会造成单一破孔，而油箱其他部位完好。另外，油箱初始温度越高，燃油蒸发形成油气的速度越快，从而更容易被引燃。

现有文献报道的油箱引燃试验方法和判定方法大都采用简易油箱设计，破片高速撞击会出现大量燃油泄漏，同时在很多工程实践引燃评价中，也用高速摄影观察到的闪燃现象判定燃油被引燃，反应破片对油箱的引燃机理、引燃判据和引燃效能评估还缺乏统一标准。为了使试验条件统一，同时考虑到目标防护能力日益增强的趋势，本文采用密实油箱设计，使反应破片在对油箱进行一次打击时油箱结构整体完好，并将油箱引燃判据定义为高速摄影观察到的闪燃现象和人员现场可以观察到的持续燃烧现象。

2 引燃密实防护油箱实验

2.1 密实防护油箱试验装置

密实防护油箱结构设计如图 2 所示，油箱材料均为 Q235 钢，壁厚 10 mm，内部尺寸为 280 mm × 200 mm，采用整体焊接方式。前面板和后面板中心均有开孔，可更换前靶板和后靶板，靶板材料为 Q235 钢，厚度 6 mm。

2.2 试验方法及引燃工况设计

反应破片由直径为 13.1 mm 的弹道枪发射，使用尼龙弹托。通过高速摄影观察反应破片的着靶速度以及油箱的引燃情况。试验方案布局如图 3 所示，燃油为煤油，油量为 10 L，液面高于弹道线 20 mm，试验当天室外温度为 -10 ℃，油箱前靶板侧空槽尺寸为 200 mm × 300 mm × 80 mm。

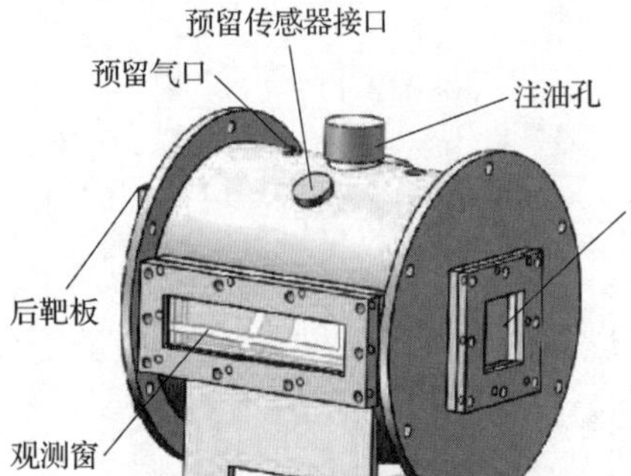

图 2 密实防护油箱结构设计

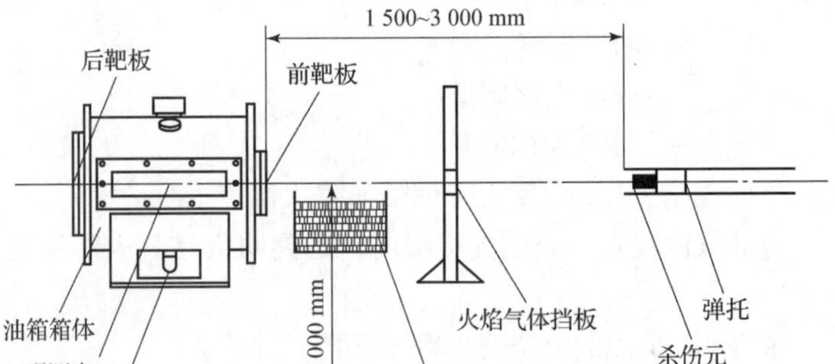

图 3 试验方案布局

为研究单枚反应破片对燃油的引燃机理和引燃判据，针对空槽采用以下试验工况进行设计，以模拟引燃条件：①空槽空置，用于接收溢出的燃油、回收试验流出的液态燃油；②空槽内倒入少量燃油，增加油箱外侧的油气量；③空槽放置少量浸油后的棉线，增加外部可燃物并加速燃油挥发。

3 3 种试验工况引燃结果

为了获得反应破片对 3 种不同试验工况的引燃结果，本文进行了多发试验，其中工况 1 和工况 2 均进行了 2 个破片速度的试验，以相互对照引燃现象。试验工况和试验现象如表 3 所示。

试验序号	破片速度 $v/(\mathrm{m \cdot s^{-1}})$	试验工况	试验现象
1	865	1	仅出现闪燃，未出现持续燃烧
2	997	2	仅出现闪燃，未出现持续燃烧
3	884	3	仅出现闪燃，未出现持续燃烧
4	973	4	仅出现闪燃，未出现持续燃烧
5	1 002	5	燃油持续燃烧

选取 3 个试验工况下速度接近的试验 2、试验 4 和试验 5 的试验结果高速摄影分时刻图，如图 4 所示。由图 4 可见，反应破片撞击油箱前靶板，一部分进入油箱内部，另一部分在靶前发生剧烈反应，3 发试验分别在 $t=1.00$ ms、1.75 ms 和 1.67 ms 时靶前火光达到最大。反应破片的内爆现象使油箱内的燃油大量汽化并从破孔处溢出，当 $t=25$ ms 左右时，反应破片在靶前的火光接近消失，此时可以在靶前观察到油雾现象，形成的油雾遇到油箱前侧的靶前反应火星发生燃烧，3 发试验分别在 $t=60$ ms、100 ms 和 75 ms 时油气燃烧火光尺寸达到最大。

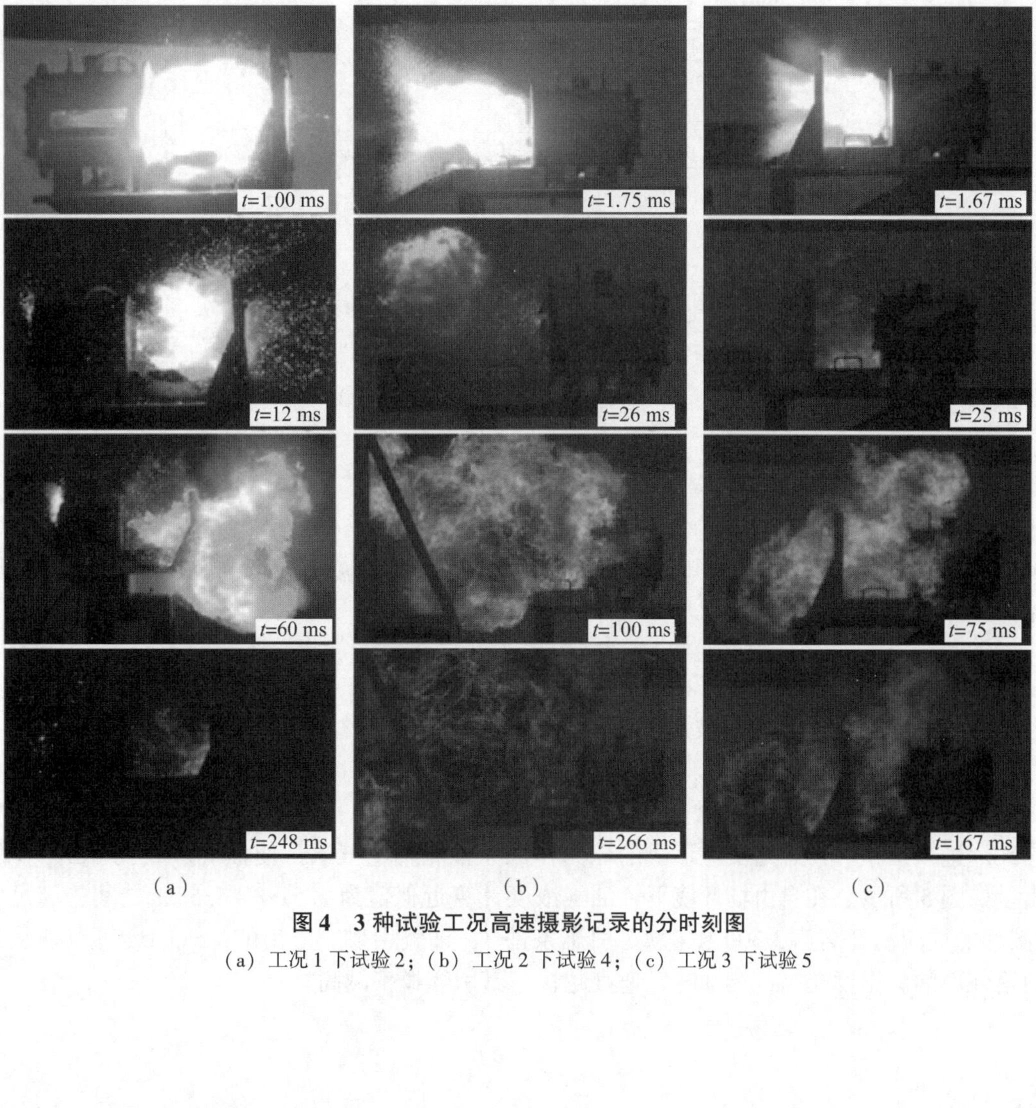

图 4 3 种试验工况高速摄影记录的分时刻图

(a) 工况 1 下试验 2；(b) 工况 2 下试验 4；(c) 工况 3 下试验 5

从工况 1 下试验 2 的高速摄影结果可知，随着反应破片的内爆效应消失，油箱内溢出的油气越来越少，火光也开始逐渐衰退，当 $t=248$ ms 时油气燃烧现象逐渐消失，此时未观测到液态燃油从破孔流出，试验现场空槽中有大量燃油，但是无持续燃烧现象。

从工况 2 下试验 4 的高速摄影结果可知，由于在油箱外部增加了油气量，油气闪燃火光面积最大，持续到 $t=266$ ms 火光仍未见衰退，如图 4 中红圈部分，靶前油槽中的燃油已经开始燃烧，并有液体燃油从油箱前靶板破孔处流出。但是在试验现场油箱前侧的油槽内尚有燃油并无持续引燃现象，表明燃油闪燃后熄火。

从工况 3 下试验 5 的高速摄影结果可知，在 $t=167$ ms 时刻油气燃烧火光已开始消退，试验现场人员观察到油槽内有燃烧的火焰，同时有燃油从破孔处不断流出，形成持续燃烧的状态。

4 油气溢出、闪燃和持续燃烧分析讨论

4.1 燃油溢出、引燃持续时间分析

对于上述 3 种试验工况，反应破片撞击油箱，首先对油箱结构造成破孔损伤，同时反应破片自身开始诱发燃爆反应，通过高速摄影对试验现象的观察可知，燃爆反应的持续时间在 30 ms 以内；反应破片撞击油箱以及内爆效应引起大量油气溢出，油气的溢出持续时间也近似为 30 ms，油气被靶前反应火源引燃，闪燃持续时间为 200 ms 甚至更长；液态的燃油在 200 ms 之后从油箱破孔处流出，若满足引燃条件 1 至条件 4 则持续燃烧直至燃油耗尽，否则闪燃熄火。

4.2 油气溢出量分析

燃油受反应破片的冲击作用和内爆效应汽化从靶板破孔处溢出，油气的溢出速率与油箱内压力 $p_{f,t}$ 有如下关系：

$$\frac{dm_{f,g}}{dt}=A\sqrt{\frac{\gamma m_{f,\#}p_{f,t}}{V_f}\left(\frac{2}{\gamma+1}\right)^{\frac{\gamma\gamma+1}{2(\gamma-1)}}} \tag{5}$$

式中：A——靶板破孔面积；

γ——多方气体指数；

$m_{f,g}$——箱体内油气总质量；

V_f——油箱内油气体积。

在油气溢出阶段，箱体内压力 $p_{f,t}$ 随时间 t 满足以下衰减关系：

$$p_{f,t}-p_0=\Delta p_{max}e^{-k(t-t_{max})} \tag{6}$$

式中：k——衰减系数；

t_{max}——压力最大持续时间；

Δp_{max}——反应破片冲击作用和内爆效应引起的压力上升，Δp_{max} 与反应破片冲击速度和反应破片含能呈正相关[15]：

$$\Delta p_{max}=\frac{\gamma-1}{V}Q\cdot\sigma(v) \tag{7}$$

Q 为反应破片含能，$\sigma(v)$ 为与破片速度 v 相关的能量释放率：

$$\sigma(v)=C_2-C_1\ln v \tag{8}$$

C_2、C_1 为通过试验拟合得到的常数。

反应破片进入油箱后，燃爆反应形成的压力作用于油箱内空气层，使油气从破孔溢出，5 发试验的油气溢出量如图 5 所示。在自由场环境下，油气浓度计算近似值为 4.51%~4.58% 之间，满足引燃条件 2；反应破片的反应温度高达 3 300 K，满足引燃条件 3，油气引燃。溢出的油气闪燃可以加热油箱外的燃油，油箱外燃油的温度越高，燃油挥发速度越快，其引燃概率越高。

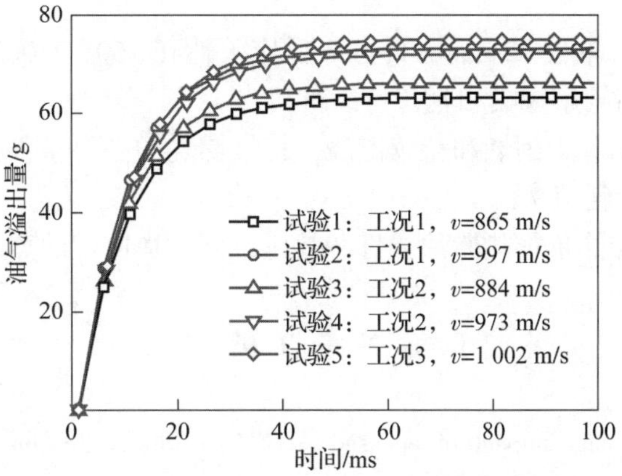

图 5　不同试验条件下的油气溢出量随时间的变化

4.3　燃油闪燃和持续燃烧分析

由引燃条件并通过对式（1）~式（4）进行解析计算，可得燃油的蒸发速率和燃烧速率随燃油温度的变化曲线如图 6 所示。由图 6 可见，随着燃油温度上升，燃油液面蒸发速率缓慢上升，同时需要更少的能量（燃油）消耗来加热燃油使其蒸发，因此燃油燃烧消耗速率快速下降。当燃油温度达到 343.25 K、油液面蒸发速率和燃烧消耗速率达到平衡时，燃油形成持续燃烧。假设棉线直径为 0.2 mm，其在油池局部形成毛细现象，增加了蒸发效应，从而在更低燃油温度下形成灯芯燃烧。

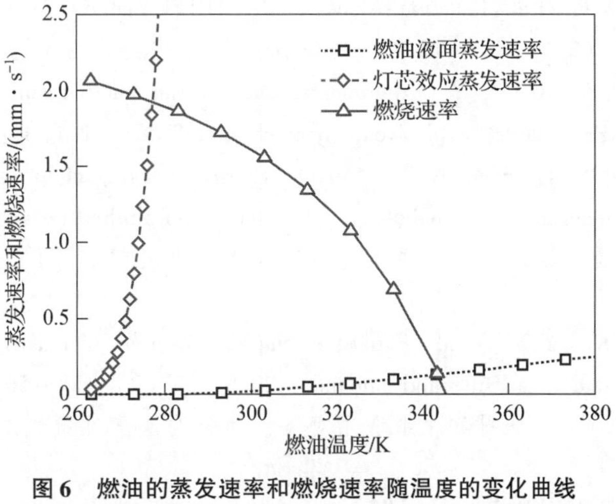

图 6　燃油的蒸发速率和燃烧速率随温度的变化曲线

结合计算分析结果对本文试验现象进行分析可知：试验工况 1 中，箱体外无燃油，溢出的油气燃烧结束后熄火；试验工况 2 中，箱体外燃油被闪燃的油气加热而加速蒸发，油气闪燃时间较试验工况 1 持续更长，但是试验工况 2 下油气的燃烧速率大于蒸发速率，因此仍然出现熄火；试验工况 3 中，棉线在油层表面形成灯芯效应，显著增加了蒸发速率，因此燃油得以持续燃烧。

上述理论分析结果和试验现象表明，反应破片对油箱的引燃试验判据为闪燃和持续燃烧两种现象，对于反应破片战斗部工程应用和后效引燃威力评估，应尽可能地使其出现持续燃烧毁伤现象而非闪燃现象。同时燃油持续燃烧可通过增加燃油温度或灯芯效应实现，前者可通过增加反应破片初始能量实现，后者可通过增加反应破片复合结构设计实现。

5　结论

本文通过试验研究和理论分析，阐明了单枚反应破片作用于密实防护油箱的引燃机理，获得了油箱闪燃和持续燃烧的引燃判据，可为反应破片战斗部应用设计和引燃油箱威力评价提供参考，所得主要结论如下：

（1）对于密实防护油箱，单枚 12 mm×18 mm 反应破片在 860~1 000 m/s 速度下仅能引起油箱闪燃，通过增加油气量可以提高闪燃规模和延长持续时间。

（2）油箱引燃的试验判据分为闪燃和持续燃烧，持续燃烧的必要条件是燃油溢出与空气接触，并且燃油蒸发速率等于油气燃烧消耗速率。

（3）通过提供更多的反应能量加热燃油或灯芯效应，可以加速燃油蒸发从而实现持续燃烧。

参考文献

[1] SORENSEN B. High - velocity impact of encased Al /PTFE projectiles on structural aluminum armor [J]. Procedia engineering, 2015, 103: 569 - 576.

[2] CUDZILO S, SZALA M, HUCZKO A, et al. Combustion reactions of poly (carbon monofluoride), (CF) n, with different reductants and characterization of the products [J]. Propellants, explosives, pyrotechnics, 2007, 32 (2): 149 - 154.

[3] 陈志优. 金属/聚四氟乙烯反应材料制备和动态力学特性 [D]. 北京: 北京理工大学, 2016.
CHEN Z Y. Preparation process and dynamic mechanical property of metal /PTFE reactive materials [D]. Beijing: Beijing Institute of Technology, 2016. (in Chinese)

[4] 任会兰, 李尉, 刘晓俊, 等. 钨颗粒增强铝/聚四氟乙烯材料的冲击反应特性 [J]. 兵工学报, 2016, 37 (5): 872 - 878.
REN H L, LI W, LIU X J, et al. Reaction behaviors of Al /PTFE materials enhanced by W particles [J]. Acta armamentarii, 2016, 37 (5): 872 - 878. (in Chinese)

[5] FENG S S, WANG C L, HUANG G Y. Experimental study on the reaction zone distribution of impact - induced reactive materials [J]. Propellants, explosives, pyrotechnics, 2017, 42 (8): 896 - 905.

[6] DENSMORE J M, BISS M M, HOMAN B E, et al. Thermal imaging of nickel - aluminum and aluminum - polytetrafluoroethylene impact initiated combustion [J]. Journal of applied physics, 2012, 112 (8): 084911.

[7] 李顺平, 冯顺山, 董永香, 等. 冲击作用下反应破片燃爆温度效应 [J]. 弹箭与制导学报, 2015, 35 (2): 54 - 56.
LI S P, FENG S S, DONG Y X, et al. Explosive temperature effect of reactive fragment under impact [J]. Journal of projectiles, rockets, missiles and guidance, 2015, 35 (2): 54 - 56. (in Chinese)

[8] 肖统超, 陈文, 王绍慧, 等. 不同破片杀伤元对飞机油箱的毁伤试验 [J]. 四川兵工学报, 2010, 31 (12): 32 - 34.
XIAO T C, CHEN W, WANG S H, et al. Experimental investigation on damage to aircraft fuel tank by different fragments [J]. Journal of Sichuan Ordnance, 2010, 31 (12): 32 - 34. (in Chinese)

[9] 汪秀明. 活性材料对燃油引燃增强效应研究 [D]. 北京: 北京理工大学, 2015.
WANG X M. Enhanced ignition behavior of reactive material fragment impacting fuel - filled tank [D]. Beijing: Beijing Institute of Technology, 2015. (in Chinese)

[10] 王海福, 郑元枫, 余庆波, 等. 活性破片引燃航空煤油实验研究 [J]. 兵工学报, 2012, 33 (9): 1148 - 1152.
WANG H F, ZHENG Y F, YU Q B, et al. Experimental research on igniting the aviation kerosene by reactive fragment [J]. Acta armamentarii, 2012, 33 (9): 1148 - 1152. (in Chinese)

[11] 许化珍, 李向东. 含能破片对柴油箱的引燃破坏效应 [J]. 弹箭与制导学报, 2012, 32 (2): 85 - 88.
XU H Z, LI X D. The igniting damage effect of energetic fragments on diesel oil box [J]. Journal of projectiles, rockets, missiles and guidance, 2012, 32 (2): 85 - 88. (in Chinese)

12.7 mm动能弹斜侵彻复合装甲的数值模拟研究①

王维占[1,2]②，赵太勇[2]，冯顺山[3]，杨宝良[4]，李小军[5]，陈智刚[2]

（1. 中北大学机电工程学院，山西太原 030051；
2. 中北大学地下目标毁伤技术国防重点学科实验室，山西太原 030051；
3. 北京理工大学机电工程学院，北京 100081；
4. 西安现代控制技术研究所，陕西西安 710065；
5. 军事科学研究院防化研究院，北京 102205）

摘 要：通过弹道枪实验对斜置角度为0°~60°的陶瓷复合装甲进行了弹道极限测试，分析了靶板斜置角度对穿燃弹的弹道极限和钢芯质量变化、破坏形态的影响。利用数值模拟的方法对上述实验结果进行验证计算，鉴于数值计算结果与实验结果较好的一致性，进一步研究了陶瓷复合靶板斜置角度对穿燃弹钢芯穿靶偏移角和等效Q235钢靶厚度的影响。结果表明，随陶瓷复合靶板斜置角度的增大：弹道极限近似指数型提高；在相同弹道极限速度下，穿燃弹对Q235钢靶板的极限穿深和对斜置陶瓷复合靶板的极限穿深的等效厚度的比也随之增大；同时，钢芯完整度逐渐降低，穿靶偏移角反向增大。

关键词：斜置；弹道极限；破坏；侵彻

Numerical simulation study on penetration of a 12.7 mm kinetic energy bullet into a composite armor

Wang Weizhan[1,2], Zhao Taiyong[2], Feng Shunshan[3], Yang Baoliang[4], Li Xiaojun[5], Chen Zhigang[2]

(1. College of Mechatronic Engineering, North University of China, Taiyuan 030051, Shanxi, China;
2. National Defense Key Discipline Laboratory of Underground Target Damage Technology, North University of China, Taiyuan030051, Shanxi, China;
3. School of Mechanical and Electrical Engineering, Beijing Institute of Technology, Beijing 100081, China;
4. Xi'an Institute of Modern Control Technology, Xi'an 710065, Shaanxi, China;
5. Institute of Defense Research, Academy of Military Science, Beijing 102205, China)

Abstract: Ballistic limit tests were carried out by using a ballistic gun system for the ceramic composite armors obliquely placed with the angles of 0° – 60°. The influences of the oblique angles were analyzed onthe ballistic limits, steel core mass change and damage forms of armor – piercing bullets. The numericalsimulations were performed to verify the above experimental results. Based on the fact that the calculated results were in agreement with the experimental ones, the influences of the oblique angles were further explored on the deflection angles of the bullet steel cores penetrating through the target plates, and the thicknesses of the equivalent Q235 steel target plates. Results show that with increasing the oblique angles of the ceramic composite targets: (I) the ballistic limit obeys an exponential increase law; (2) at the sameballistic limit, the ratio of the limit penetration depth of the Q235 steel target plate by the armor – piercing bullet to the equivalent thickness of the limit penetration depth of the obliquely – placed ceramic composite target by the armor – piercing bullet increases; (3) the integrity

① 原文发表于《爆炸与冲击》2019.12。

② 王维占：工学博士，2016年师从冯顺山教授，研究陶瓷复合枪弹及对陶瓷复合靶的侵彻效应，现工作单位：中北大学。

of the bullet steel core decreases gradually, its deflection angle increases reversely.

Keywords: oblique; ballistic limit; damage; penetration

随着高新技术在军事领域的广泛应用发展，反装甲武器与装甲防护技术的冲突愈加激烈，同时相互促进、共同发展、交替上升[1]。其中，陶瓷复合装甲的出现往往使现役轻武器弹药束手无策，因而对陶瓷复合装甲的毁伤效能研究成为轻武器弹药发展的重中之重，针对弹靶作用过程，已开展了大量的实验与理论研究。Rosenberg 等[2]通过开展正侵彻两种大块体氧化铝陶瓷靶实验，发现铜、钢和钨合金杆弹撞击 AD95 陶瓷靶的开始侵彻阈值速度分别为 1.15 km/s、0.99 km/s、0.66 km/s[3]。李继承等[4]、Anderson 等[5]通过长杆弹冲击陶瓷复合装甲的实验研究、理论分析与数值模拟等研究工作，较好地解释了金属弹与陶瓷复合装甲之间界面的击溃原理。Chi 等[6-7]利用数值模拟的方法得知提高约束预应力效应可明显提高陶瓷复合装甲的抗侵彻能力。李继承等[8]和 Li 等[9-10]以不同弹丸头部形状为实验变量，获知了在界面击溃条件下弹丸速度、长度、动量的演变规律，并给出界面击溃/侵彻转变速度和时间的理论表达式。谈梦婷等[11]利用数值模拟的方法研究了弹丸头部形状、陶瓷复合装甲的盖板厚度、陶瓷靶预应力效应对界面击溃效应的影响。汪建锋等[12]根据陶瓷复合靶板受力情况将金属弹侵彻陶瓷复合靶板的过程分为 3 个阶段：初始撞击阶段、烧蚀变形阶段、裂纹成型和断裂阶段。丁华东等[13-16]发现提高 Al203 基陶瓷的剪切模量可提高其抗侵彻性能。陈斌等[17]通过研究穿甲弹对陶瓷复合装甲的毁伤效应，发现弹着角是影响穿甲弹毁伤效能的重要因素。郭英男[18]通过 12.7 mm 制式穿甲弹冲击陶瓷复合装甲的实验研究与数值模拟，发现弹丸弹着点接近陶瓷靶边缘时，弹体侵彻姿态转变为斜侵彻。以上研究大多是对于动能弹正侵彻陶瓷/钢复合靶板的实验与数值模拟研究，而对于斜侵彻下 12.7 mm 穿燃弹对陶瓷/凯夫拉复合靶板的毁伤效能研究鲜有报道，因此开展此项研究对于较好地反映战场真实作战条件具有现实意义。本文拟开展 12.7 mm 穿燃弹斜侵彻陶瓷复合装甲实验，获取靶板在不同斜置角度下的弹道极限范围，利用LS-DYNA 软件对上述穿甲过程中制式穿燃弹的质量、弹道极限及破坏形态等参数进行较好的验证计算，并进一步对穿燃弹穿靶偏移角、陶瓷复合靶板等效 Q235 钢靶厚度进行预测。

1 实验准备

1.1 实验器材

实验在中北大学地下目标毁伤技术国防重点学科实验室的靶道内进行。主要进行 12.7 mm 穿燃弹对不同斜置角度下的陶瓷复合靶板的冲击实验。实验背面靶采用尺寸为 500 mm×500 mm×10 mm 的凯夫拉面板，表面靶采用尺寸为 50 mm×50 mm×8 mm 的 Al203 陶瓷面板，中间层采用尺寸为 500 mm×500 mm×2 mm 的 Q235 钢板，陶瓷面板、Q235 钢靶及凯夫拉背靶之间采用玻璃纤维层粘接和包覆。实验用器材如图 1 所示。

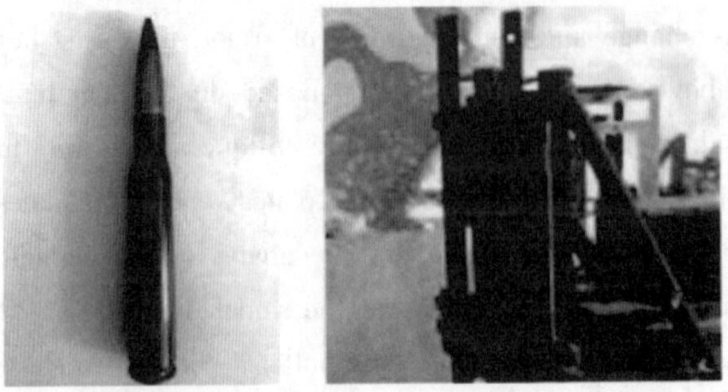

(a)　　　　(b)

图 1　实验用 12.7 mm 穿燃弹及陶瓷复合靶板

(a) 12.7 mm 穿燃弹；(b) 陶瓷复合靶板

1.2　弹道实验

进行了多发 12.7 mm 动能弹侵彻陶瓷复合靶板实验，采用 12.7 mm 口径实验弹道枪，通过调节火药装填量，控制发射速度在 521～1 213 m/s，测速设备采用中北大学自主开发的激光测速仪，测速误差为 ±1.7%，同时使用高速摄影设备对实验宏观现象进行记录。实验装置及场地布置见图 2。

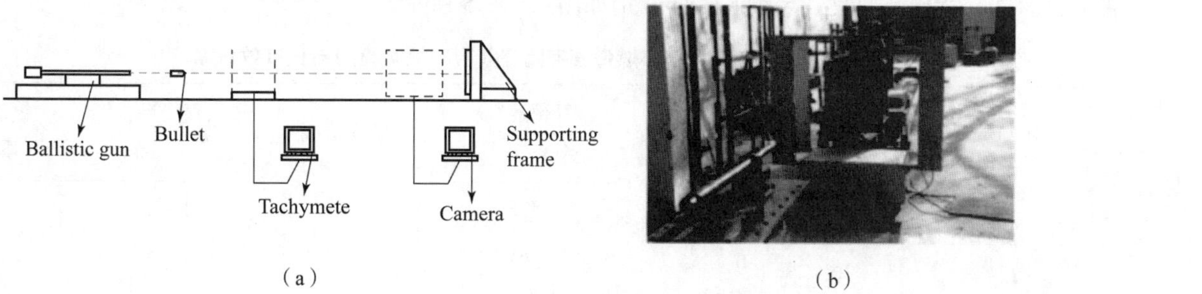

图 2　实验装置及场地布置

（a）实验装置；（b）场地布置

2　数值模拟

2.1　模型建立与参数选取

基于上述实验研究，数值模拟中 12.7 mm 制式穿燃弹的结构参考文献［19］，与本文实验用 12.7 mm 穿燃弹的质量、结构均一致。在陶瓷面板周向边界节点上施加应力流出边界条件，避免应力在边界上反射，影响计算结果。穿燃弹轴线方向与靶板法线方向的夹角 0 为靶板的斜置角度。利用 TrueGrid 软件建立 1/2 结构三维有限元模型，计算网格选用 Solid164 八节点六面体单元，并在 1/2 模型的对称面上设置对称约束条件。弹靶作用过程采用 Lagrange 算法，接触作用采用侵蚀接触算法[20]，有限元模型见图 3。数值计算中玻璃纤维层通过设置接触面为固连失效接触方式，陶瓷面板节点与 Q235 钢靶及凯夫拉背板单元间的法向失效力及剪切失效力分别取为 21 N 和 12 N[19]。

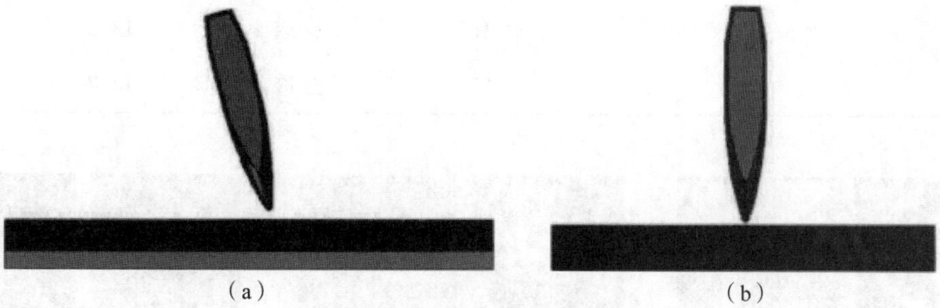

图 3　弹靶有限元模型

（a）Ceramic composite target；（b）Q235 steel target

2.2　材料参数

本文算例中，穿燃弹弹芯采用高碳钢材料，采用 * MAT_ADD_EROSION 裂纹控制附加失效模型，12.7 mm 穿燃弹及 Q235 钢靶材料模型选用 JOHNSON－COOK 材料模型和 GRUNEISEN 状态方程，AL203 陶瓷面板采用 MAT_JOHNSON_HOLMQUIST_CERAMICS 材料模型，陶瓷面板材料参数和金属材料参数见文献［20］，凯夫拉材料使用 COMPOSITE_DAMAGE 模型，材料参数见文献［1］。通过 LS－DYNA 软件对 12.7 mm 穿燃弹侵彻陶瓷复合装甲及 Q235 钢靶的过程进行数值模拟。

3 结果分析

3.1 符合计算分析

进行多发 12.7 mm 穿燃弹侵彻陶瓷复合靶板的实验，得到了部分有效实验数据，如表 1 所示，复合装甲表面靶、背靶及回收残余钢芯式样分别如图 4、图 5 所示。

表 1 12.7 mm 穿燃弹侵彻陶瓷复合靶板实验的部分有效数据

实验编号	靶板倾斜角度/(°)	着靶速度/($m \cdot s^{-1}$)	穿透情况	背靶穿孔尺寸/mm	钢芯剩余质量/g
1#	0	576	穿透	13.2	28.4
		593	穿透	14.7	30.1
		521	嵌入	5.9	26.8
2#	15	637	穿透	13.9	23.6
		593	嵌入	14.3	22.1
		579	嵌入	7.6	20.7
3#	30	693	穿透	13.1	17.6
		645	嵌入	7.1	13.9
		713	穿透	15.1	15.1
4#	45	789	嵌入	8.3	11.0
		765	未嵌入	—	15.6
		833	穿透	14.3	14.4
5#	60	1 086	嵌入	4.9	8.1
		1 213	穿透	13.3	6.0
		1 179	穿透	14.7	9.3

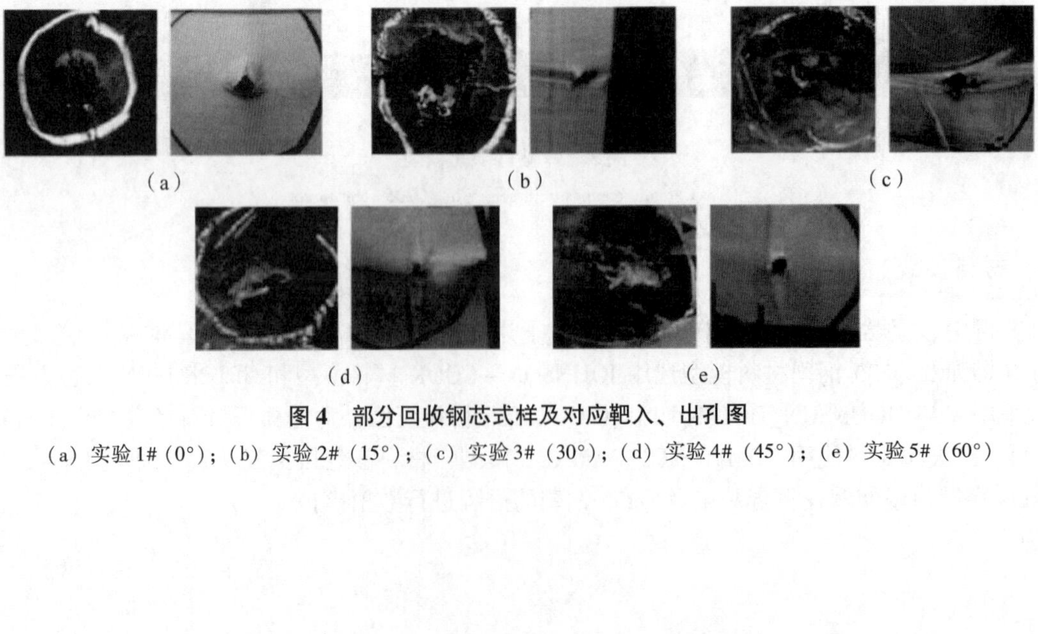

(a) (b) (c) (d) (e)

图 4 部分回收钢芯式样及对应靶入、出孔图

(a) 实验 1# (0°)；(b) 实验 2# (15°)；(c) 实验 3# (30°)；(d) 实验 4# (45°)；(e) 实验 5# (60°)

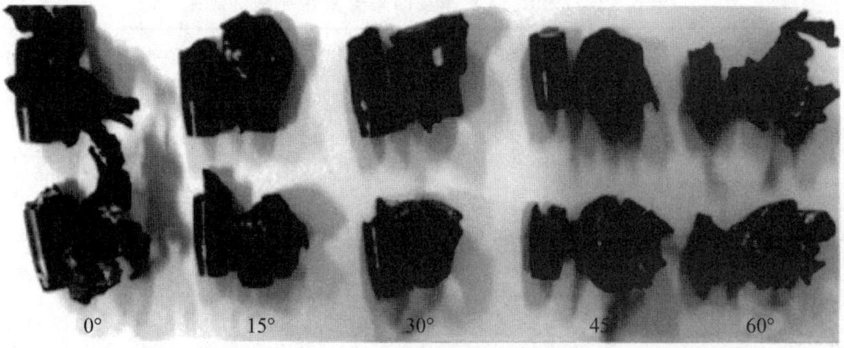

图 5　部分钢芯及弹头壳的破坏形态

由表 1 可知，陶瓷复合靶斜置角度在 0°~60°范围时，12.7 mm 制式穿燃弹对其斜侵彻的弹道极限小于 1 300 m/s，其中陶瓷复合靶板斜置角度为 0°、15°、30°、45°和 60°时，穿燃弹的弹道极限范围见表 2。在极限穿透的情况下，靶板正面纤维层撕裂，陶瓷破碎，背靶凯夫拉层呈瓣裂式穿孔破坏，见图 4。随着靶板斜置角度的增大，靶板背面鼓包越来越明显，且出现鼓包和穿孔偏移的现象，见图 4（c）~图 4(e)。穿燃弹钢芯以头部和圆柱部断裂破坏为主，在靶板存在斜置角度的情况下，钢芯头部呈现斜侧方断裂的现象，穿燃弹弹头壳发生倾斜翻卷断裂破坏，且愈发明显，如图 5 所示。

表 2　不同斜置角度下穿燃弹的弹道极限范围

靶板倾斜角度/（°）	弹道极限/($m \cdot s^{-1}$)	靶板倾斜角度/(°)	弹道极限/($m \cdot s^{-1}$)
0	521~576	45	789~833
15	579~637	60	1 086~1 179
30	645~693		

调整复合装甲的斜置角度依次为 0°、15°、30°、45°、60°，展开穿燃弹侵彻复合装甲过程的数值模拟，对钢芯剩余质量、弹道极限（嵌入靶板的最大速度与穿透靶板的最小速度的平均值（21））等性能参数进行分析。陶瓷复合靶及穿燃弹钢芯的破坏形态如图 6 所示。

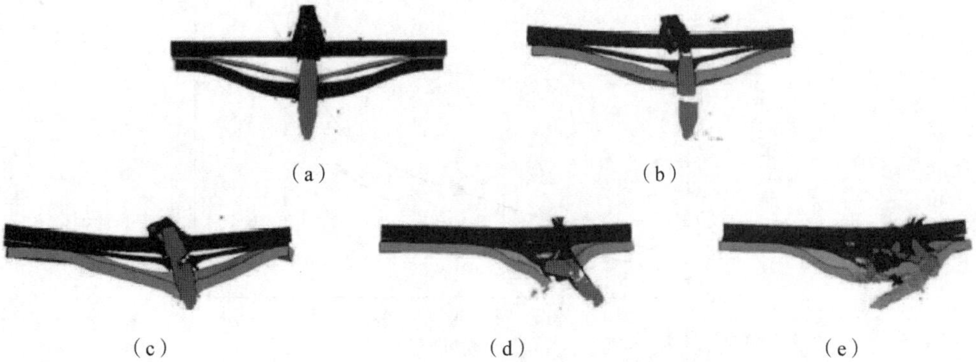

图 6　陶瓷复合靶及穿燃弹钢芯的破坏形态

(a) 0°；(b) 15°；(2) 30°；(d) 45°；(e) 60°

从图 6 可以看到，随着斜置角度的增大，穿燃弹钢芯发生不同程度的断裂破坏，钢芯圆弧头部断裂及质量侵蚀现象越来越明显。凯夫拉背靶出现不同程度的鼓包及穿孔现象，结合图 5 可知，这与实验现象基本一致。

由表 3 和图 7 可知，在弹道极限条件下，随着陶瓷复合靶板斜置角度的增大，穿燃弹钢芯的剩余质量逐渐减小。靶板斜置角度的增大，导致在碰撞点处弹丸头部所受应力随着弹丸冲击速度的升高而升高，因此钢芯所受应力过载逐渐增大，发生断裂侵蚀现象越来越严重，钢芯的剩余质量逐渐减小。

表3　12.7 mm 穿燃弹侵彻陶瓷复合靶板的结果

靶板倾斜角度/(°)	弹道极限/(m·s^{-1})	剩余质量/g	靶板倾斜角度/(°)	弹道极限/(m·s^{-1})	剩余质量/g
0	600	26.4	45	925	9.3
15	675	18.1	60	1 375	6.7
30	770	12.5			

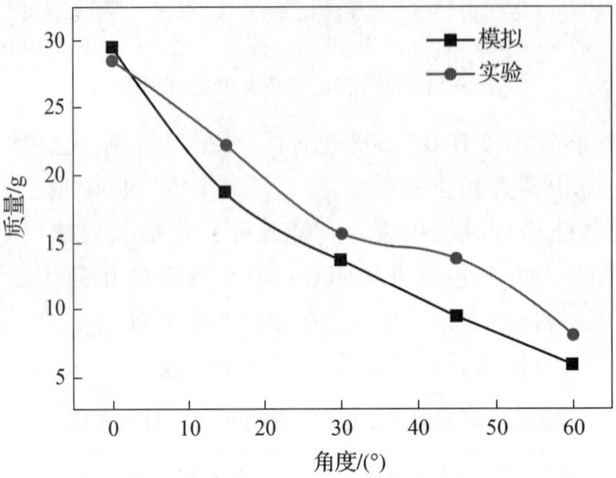

图7　钢芯剩余质量与靶板斜置角度的关系

从图8可以看出，随复合装甲斜置角度的增大，穿燃弹对复合靶板的弹道极限近似指数型升高。这是因为复合装甲斜置角度的增大，导致子弹侵彻靶板的等效厚度的增大，穿燃弹对靶板的冲击应力在垂直弹丸轴线轴方向发生分解，穿燃弹整体相对于质心所受偏转力矩增大，有发生跳弹的趋势，同时子弹用于垂直侵彻靶板的冲击应力减小。要保证子弹贯穿靶板，穿燃弹速度必然需随着靶板斜置角度的增大而提高。

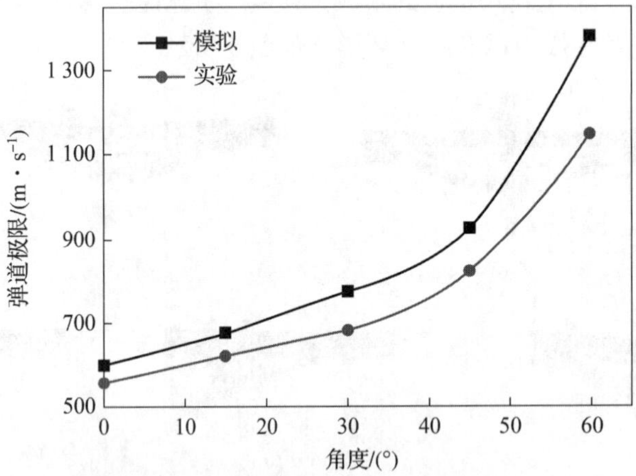

图8　弹道极限与靶板斜置角度的关系

对于以上分析，实验结果与数值计算结果虽然具有一定的偏差，但钢芯剩余质量、弹道极限的变化规律相同，说明本文选用的 LS－DYNA 软件及选取的材料参数具有一定的可靠性。

鉴于12.7 mm 穿燃弹对陶瓷复合装甲的穿甲过程中，穿燃弹钢芯的破坏过程对于弹丸结构设计具有重要意义，通过数值模拟研究了靶板斜置角度对穿燃弹钢芯穿甲过程中的破坏特性的影响。图9给出了靶板斜置角度在0°～60°范围内实验钢芯试样破坏形态和数值模拟得到的钢芯应力云图。由图9可知：随靶板斜置角度的增大，穿燃弹钢芯的穿靶偏移角逐渐近似反向线性增大；钢芯在侵彻过程中发生断裂，因为转动力矩的存在，其姿态发生不同程度的偏转。由0°～60°范围内数值模拟钢芯应力云图与实验钢芯断裂形态的对比可以发现，穿燃弹钢芯以圆弧头部断裂侵蚀和圆柱部断裂破坏为主，破坏形态较一致。在靶板存在斜置角度的情况下，钢芯头部呈现斜侧方断裂的现象，断裂截面位于迎弹面一侧，且断裂侵蚀区域与钢芯侵彻靶板过程中所受应力过载区域基本相同，数值模拟中钢芯受应力区与未受应力区分界截面与实验钢芯的迎弹面断裂方向基本一致。随靶板斜置角度的增大，迎弹面方向断裂截面角度增大。靶板斜置角度的增大导致穿燃弹的弹道极限提高，进而钢芯微元所受应力升高，应力峰值超过钢芯材料的弹性极限导致钢芯发生断裂破坏。斜置角度的增大改变了弹靶作用面，进而改变了钢芯微元的受力方向，其应力波主要作用区域位于背弹面一侧，随着斜置角度的增大，应力波作用区域增大，钢芯迎弹面一侧断裂侵蚀质量增大，剩余质量减小。由以上分析可知，高碳钢材质弹芯在斜侵彻陶瓷复合靶板时其主要破坏形式为背弹面脆性断裂，在弹道极限提升的同时，对于保证弹芯完整性具有消极作用，因此在设计弹丸弹芯时应考虑根据靶板目标特性及弹芯结构材质进行合理匹配设计。

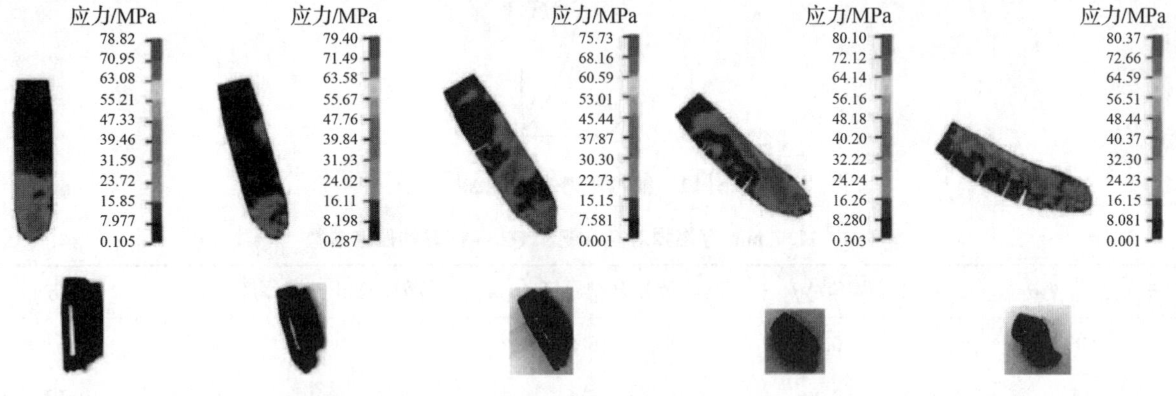

图 9　不同靶板斜置角度下实验钢芯试样破坏形态与数值模拟得到的钢芯应力云图

(a) 0°；(b) 15°；(c) 30°；(d) 45°；(e) 60°

3.2　等效威力分析

鉴于 3.1 节中陶瓷复合靶板斜置角度对 12.7 mm 穿燃弹剩余质量、弹道极限及破坏区域的影响规律的一致性，进一步研究靶板斜置角度对穿燃弹钢芯穿靶偏移角、等效 Q235 钢靶厚度的影响，等效斜侵彻复合靶厚度 H、等效 Q235 钢靶厚度 h、钢芯偏移角 $\Delta\theta$ 的示意图见图 10～图 12，表 4 为 12.7 mm 穿燃弹对陶瓷复合靶板的侵彻结果。

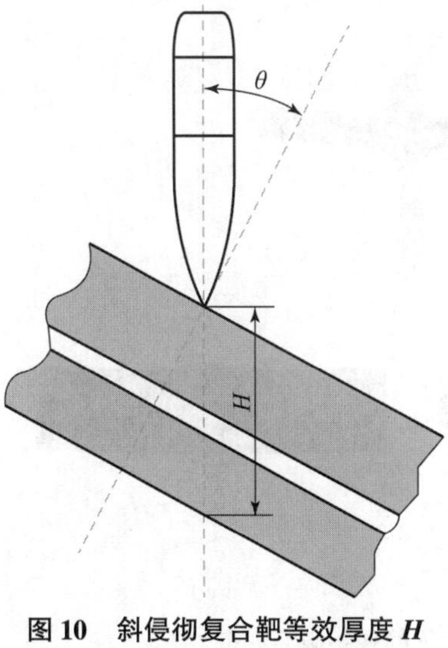

图 10　斜侵彻复合靶等效厚度 *H*

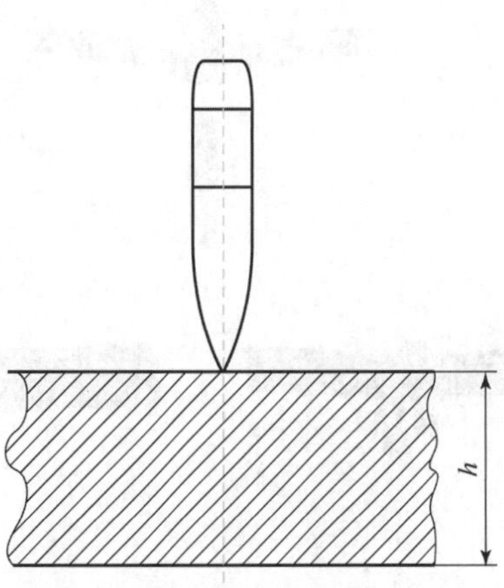

图 11　正侵彻等效 Q235 钢靶厚度 *h*

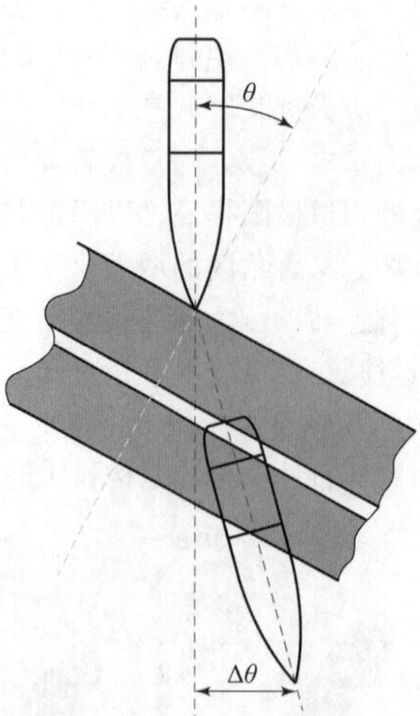

图 12　钢芯穿靶偏移角 Δθ

表 4　12.7 mm 穿燃弹对复合靶和 Q235 钢靶的侵彻参数

靶板倾斜角度 θ/(°)	钢芯偏移角 Δθ/ (°)	等效复合靶厚度 H/mm	等效 Q235 钢靶厚度 h/mm	H/h
0	-0.8	16.0	10	0.63
15	-2.8	16.6	12	0.72
30	-4.2	18.5	15	0.81
45	-7.6	22.6	19	0.83
60	-11.5	32.0	28	0.88

图 13 给出了靶板斜置角度为 0°～60°时，等效正侵彻 Q235 钢靶极限穿深条件下的弹靶破坏形态。由图 13 可以看出：随着 Q235 靶板厚度的增大，穿燃弹钢芯破碎程度增大，完整性降低；等效 Q235 钢靶厚度为 10 mm、12 mm 的条件下，穿靶后的钢芯完整性较好；当 Q235 钢靶厚为 15 mm 时，钢芯头部开始发生断裂；Q235 靶厚增大至 19 mm、28 mm 时，钢芯头部至圆柱部发生断裂破坏，钢芯呈现出整体断裂的趋势。

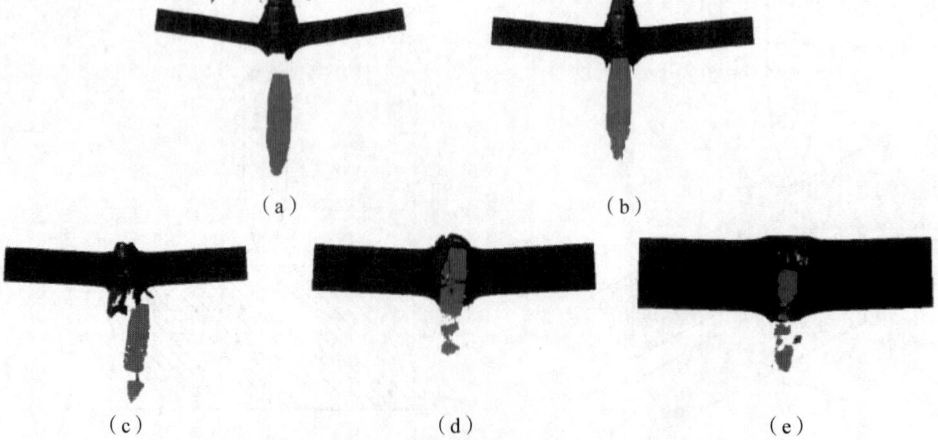

（a）　　（b）

（c）　　（d）　　（e）

图 13　等效正侵彻 Q235 钢靶极限穿深条件下的弹靶破坏形态

（a）10 mm；（b）12 mm；（c）15 mm；（d）19 mm；（e）28 mm

由图 14 可以看出，靶板斜置角度的增大导致弹丸钢芯向背弹面偏转，其穿靶偏转角度近似线性变化，负向增大。这是因为穿靶过程中，钢芯穿透陶瓷面板后，当侵彻 Q235 钢薄靶时，姿态进行反向调整，导致穿靶偏移角负向增大。

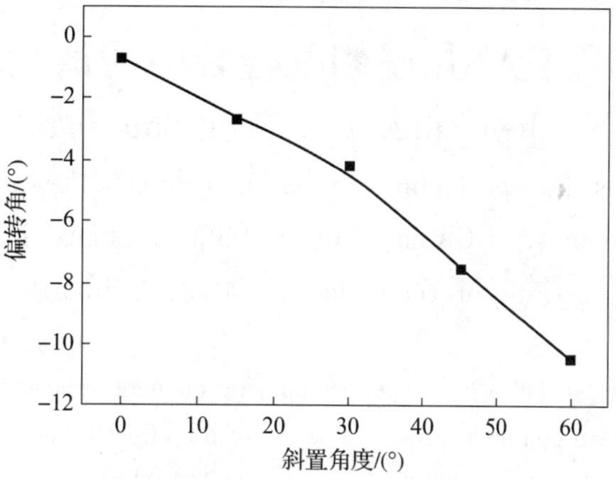

图 14　靶板斜置角度与钢芯穿靶偏移角的关系

图 15 为陶瓷复合靶及 Q235 钢靶等效厚度与靶板斜置角度的关系，结合图 10 ~ 图 12，可明显看出：随着靶板斜置角度的增大，陶瓷复合靶的等效厚度呈近似指数型增大；穿燃弹侵彻陶瓷复合靶和 Q235 钢靶时，在相同弹道极限情况下，靶板斜置角度的增大导致在弹丸轴线方向等效 Q235 钢靶厚度与陶瓷复合靶板的等效厚度均呈指数型增大，且等效 Q235 钢靶厚度的增大速率大于复合靶板的等效厚度的增大速率。可见，陶瓷复合靶板斜置角度的增大，有利于提高 Q235 钢靶与陶瓷复合靶板的等效厚度比（图 16），可有效提高其对 12.7 mm 穿燃弹的防御能力。

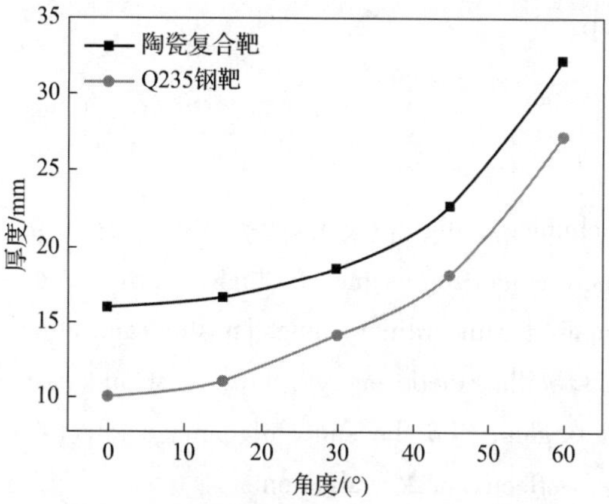

图 15　靶板等效厚度与靶板斜置角度的关系

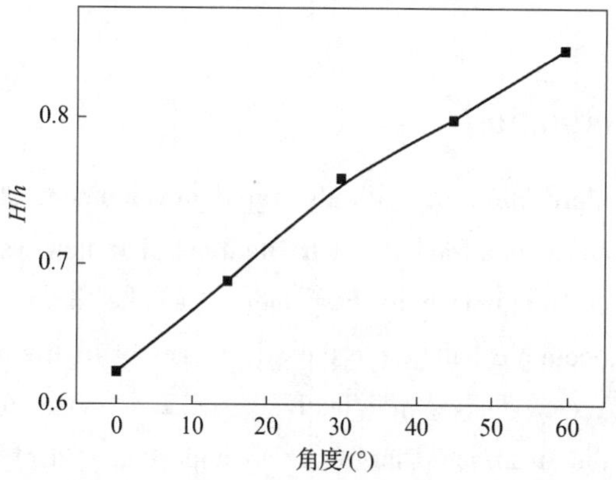

图 16　陶瓷复合靶板和 Q235 钢靶的等效厚度比与靶板斜置角度的关系

4　结论

（1）通过弹道枪实验测试了 12.7 mm 穿燃弹对实验选定的陶瓷复合靶板的弹道极限范围，发现随陶瓷复合靶板斜置角度的增大，弹道极限近似指数型提高，且穿燃弹钢芯背弹面斜侧方发生脆性断裂破坏，断裂倾角逐渐增大。

（2）基于数值模拟结果与实验结果的一致性，发现随靶板斜置角度的增大，穿燃弹钢芯穿靶偏移角反向增大，陶瓷复合靶板与 Q235 钢靶的等效厚度比也相应增大，同时由应力波引起的迎弹面应力响应区变大，弹芯剩余质量逐渐减小，完整度降低。

Experimental study on ceramic balls impact composite armor①

球形陶瓷弹丸侵彻复合装甲的试验研究

Wang Weizhan[1②], Chen Zhigang[1], Feng Shunshan[2], Zhao Taiyong[1]

(1 Underground target damage technology national defense key discipline laboratory of North University of China, Taiyuan 030051 Shanxi P. R. China;

2 Beijing Institute of Technology, 100081 Beijing P. R. China)

Abstract: Ceramic balls represent a new type of damaging element, and studies on their damaging power of composite armor are required for a comprehensive evaluation of the effectiveness of various types of weapons. The goal of this study was to determine the impact of ϕ7 mm toughened Al_2O_3 ceramic balls on a composite ceramic/metal armor. The influences of the ceramic panel and the thickness of the metal backing material on the destroying power of the ceramic balls were first determined. Based on the agreement between numerical simulation, experimental results, and calculation models of the target plate resistance, the response mechanism of the ceramic balls was further analyzed. The results indicate that for a back plate of Q235 steel, with an increasing thickness of the ceramic panel, the piercing speed limit of the ceramic balls gradually increases and the diameter of the out - going hole on the metal back decreases. Different conditions were tested to assess the effects on the piercing speed, the diameter of the out - going hole, the micro - element stress, and the integrity of the recovered ceramic bowl.

Keywords: Impact; Penetration; Infinitesimal; Interface collapse

Introduction

In modern warfare, with the rapid development of warhead technology, lightweight composite armor with strategic value has attracted much attention. The new warhead damage element represented by high - performance and low - cost ceramic balls has emerged as the times require. Compared with ordinary tungsten alloy and steel fragments, ceramic balls have the advantages of high strength, large specific kinetic energy, low - cost and easy production, etc. They can effectively strike at close range armored targets, and the study of damage effect of ceramic balls on armored targets is an important part of weapon system effectiveness evaluation.

In recent years, scholars at home and abroad have made some reports on the damage power and dynamic properties of ceramic damage elements. Nechitatio [1 - 2] and others have used numerical simulation methods to study the fracture of a ceramic rod impacting metal target, and compared the fracture morphology of the target under different impact velocities. Then Nechitatio carried out in the experimental study of penetrating concrete target with head - mounted polycrystalline diamond penetration projectile at different impact velocities based on the numerical simulation method. Takahashi [3] studied the fracture characteristics of ceramic balls under high pressure by means of an experimental study, theoretical analysis and numerical simulation. Matsuda [4 - 5] analyzed the fracture characteristics of ceramic balls under thermal shock, established the calculation theory of virtual crack model, and carried out the indentation contrast test on silicon nitride ceramic balls between air and

---

① 原文发表于《Defence Technology》2019。

② 王维占：工学博士，2016 年师从冯顺山教授，研究陶瓷复合枪弹及对陶瓷复合靶的侵彻效应，现工作单位：中北大学。

vacuum. Ma [6] used fractal theory to analyze the fracture characteristics of brittle spheres under high velocity impact. The theoretical analysis results are in good agreement with the experimental results. Wang [7 – 15] pointed out that ceramic balls have the advantages of high strength, high specific kinetic energy and high initial velocity, and have high penetrating efficiency and high aftereffect of close range targets (pine targets, aluminum plates, etc.). Chen [16 – 25] proposed replacing conventional metal warheads with non – metal ceramic warheads, and found that ceramic composite projectiles can effectively improve the damage power of composite armor.

At present, the damage mechanism and armor – piercing response characteristics of the damage elements of ceramic materials have been preliminaries understood through the research on the damage effectiveness of the damage elements of ceramic materials, but the damage law of light composite armor and the response law of the damage elements of ceramic materials are rarely studied. Based on the ballistic impact test, the influence of ceramic balls on the ultimate penetration velocity and impact breakage threshold of ceramic composite armor under different working conditions is studied in this paper. Based on the numerical simulation and impact pressure calculation theory, the interface breakdown effect of ceramic balls is predicted and analyzed, and the response characteristics of ceramic balls and composite armor are expounded.

1 Experiment Preparation

1.1 Laboratory Equipment

The experiment was conducted out at the target road of the National Defense Key Discipline Laboratory of Underground Target Damage Technology of the North University of China. The test composite back plate uses a Q235 steel target of 500 mm × 500 mm × 2 mm in size and a 2024 aluminum target. The ceramic panel includes Al_2O_3 ceramic panels with dimensions of 100 mm × 100 mm × 1 mm, 100 mm × 100 mm × 2 mm, and 100 mm × 100 mm × 3 mm. The middle layer is bonded with glass fibers. In view of the that the loading size of the 7.62 mm ballistic gun to the ceramic ball diameter is less than 7.2 mm and the loading speed is less than 1500 m/s, the penetration power (specific kinetic energy) of the ceramic ball is proportional to the radius R and the loading speed v, respectively, i.e. a 7.62 mm ballistic gun. There is a loading limit on the size and penetration power of the ceramic ball. Moreover, considering that the wavelength of the stress wave of the ceramic ball material is 5.2 mm, to explore the breaking law of the ceramic ball effectively it is necessary to ensure that the size of the ceramic ball is at least larger than the wavelength of the ceramic material. A tensile wave and a compression wave were utilized to analyze the breaking law of the ceramic ball. To penetrate the target plate comprehensively and effectively explore the breaking law of the ceramic ball, the fragment is made of a toughened Al_2O_3 ceramic bowl with a size of ϕ7 mm. The test equipment is shown in Figure 1.

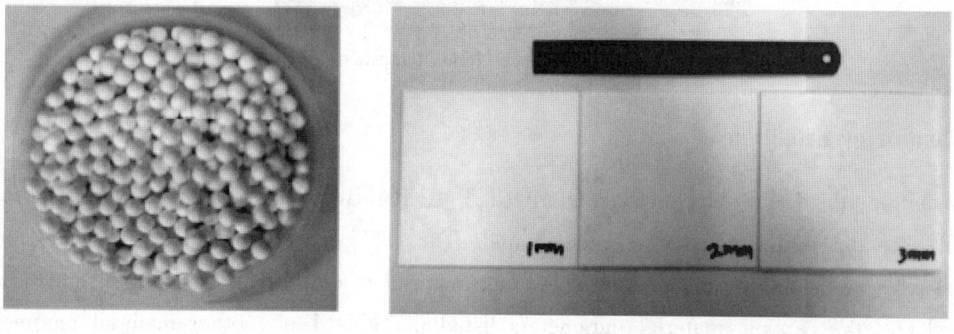

(a)　　　　　　(b)

Figure 1　Laboratory equipment

(a) ϕ7 mm Al_2O_3 ceramic ball; (b) Al_2O_3 ceramic panel

1.2 Ballistic Tests

In this study, several effective toughened Al_2O_3 ceramic ball impacting composite target panels was tested. A 7.62 mm caliber ballistic guns were used for testing, and the launch speed was adjusted by modifying the loading of the gunpowder. A laser speedometer is independently developed by the North University of China was used for the speed measurement, within an error of ± 0.3%. A schematic diagram of the (a) test arrangement and (b) site layout is shown in Figure 2.

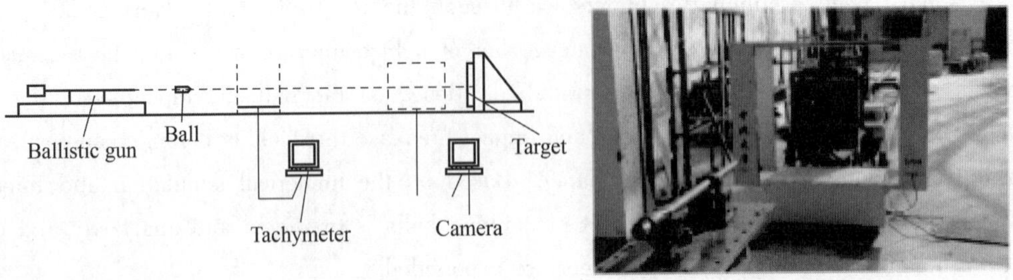

(a) (b)

Figure 2 Schematic diagram of (a) test arrangement and (b) site layout

(a) Test layout diagram; (b) Layout of the site

2 Numerical Simulation

2.1 Model Establishments

A finite element model was built using TRUEGRID software. To reduce the computation time, a three - dimensional finite element model was established based on a 1/2 structure, and symmetric constraints were set on the symmetry plane of the 1/2 model. The ceramic ball and ceramic panel were modeled using the SPH smooth particle algorithm, and the contact between the ball and panel was modeled using a particle - particle contact algorithm. The numerical simulation of the bonding layer utilized the solid joint between the ceramic panel/metal backplane interface. The normal failure force and shear failure force between the finite elements of the particle and the metal back plate were taken as 21 N and 12 N, respectively [25]. Figure 3 shows the finite element mesh model applied.

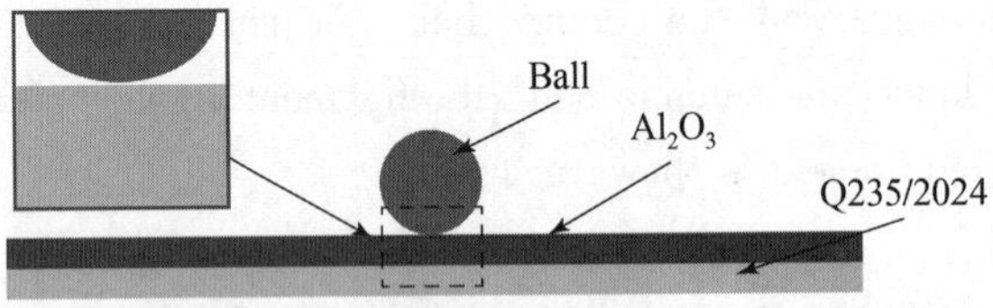

Figure 3 Finite element mesh model

2.2 Parameter Selections

In this example, the MAT_JOHNSON_HOLMQUIST_CERAMICS material model was used to describe the toughened Al_2O_3 ceramic ball and Al_2O_3 ceramic panel. The Q235 steel back target and 2024 aluminum back target material model were based on the Johnson - Cook material model and the Gruneisen equation of state. To toughen the Al_2O_3, the ceramic material parameters listed in Table 1 and other material parameters from the literature [16 - 24] were used in the dynamics software Autodyn to reproduce the process of a ceramic ball impacting a composite armor.

Table 1 Ceramic material parameters

Materials	G/GPa	$\rho/(g \cdot cm^{-3})$	A	B	M	N	K/GPa	HEL/GPa	SFMAX	T
Toughen Al_2O_3	268	3.73	1	0.28	0.6	0.64	201	3.3	1	0.36

3 Analysis of the Results

3.1 Analysis of the Response Characteristics of Ceramic Balls

When a ceramic ball impacts a composite target, different influences are imparted on the response characteristics of the ceramic bowl by the thicknesses of the metal back plate material and ceramic plate. Table 2 shows the test results of the impacting and crushing velocity thresholds for some ceramic balls, and Figure 4 shows the relationship between the impacting and crushing velocity thresholds of ceramic balls and the thicknesses of the ceramic plates. Figure 5 (a) – (c) present the change curve in the stress applied to the micro units of ceramic balls under two operating conditions.

Table 2 Impact test results of impact crushing speed of some ceramic balls

Serial number	Ceramic panel thickness/mm	Condition 1 $/(m \cdot s^{-1})$	Condition 1 broken or unbroken	Condition 2 $/(m \cdot s^{-1})$	Condition 2 broken or unbroken
		307	Unbroken	324	Unbroken
#1	1	289	Unbroken	338	Unbroken
		332	Broken	355	Broken
		299	Unbroken	301	Unbroken
#2	2	276	Unbroken	329	Broken
		312	Broken	320	Broken
		269	Unbroken	312	Broken
#3	3	259	Broken	271	Unbroken
		284	Broken	299	Broken

As shown in Figure 4, the impacting and breaking threshold of the ceramic balls decreased as the thickness of the ceramic panel increased, where as the impacting and crushing thresholds of ceramic balls on the ceramic plates of the same thickness were less under working condition 1 than those under working condition 2. This is because, compared with the Q235 steel back plate, the 2024 aluminum back plate can deform more readily under the smashing and plugging effect of the ceramic balls and ceramic cones [26]; hence, reducing the reaction of the composite ceramic ball to the target surface. At the same impact velocity, the force applied on the micro units of a ceramic ball was smaller than under condition 1 (see Figure 5 (a)), resulting in increased thresholds for the impact and crushing of the ceramic balls. Under the same impact velocity, the force applied on the micro units of ceramic balls under the two working conditions increased with an increase in the thickness of the ceramic panel (see Figure 5 (b)). The increase in the thickness of the ceramic panel increased the opportunity to form a ceramic cone. Under the same impact velocity, the force applied on the micro units of a ceramic ball continues to increase (see Figure 5 (c)), as does the extension rate of the fracture inside the ceramic ball, causing the ball to break more easily.

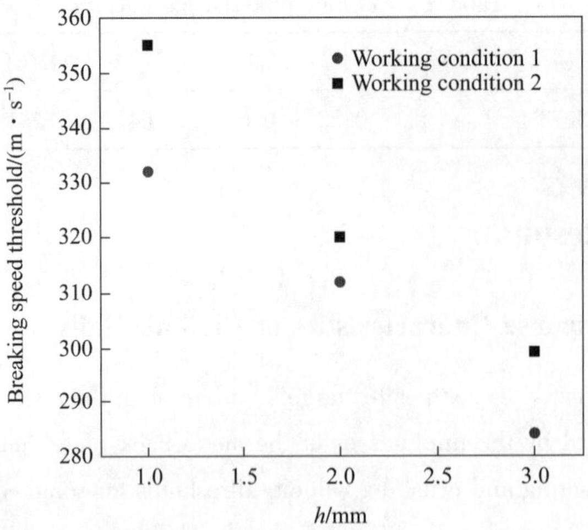

Figure 4　Relationship between ceramic plate thickness and impact fracture threshold

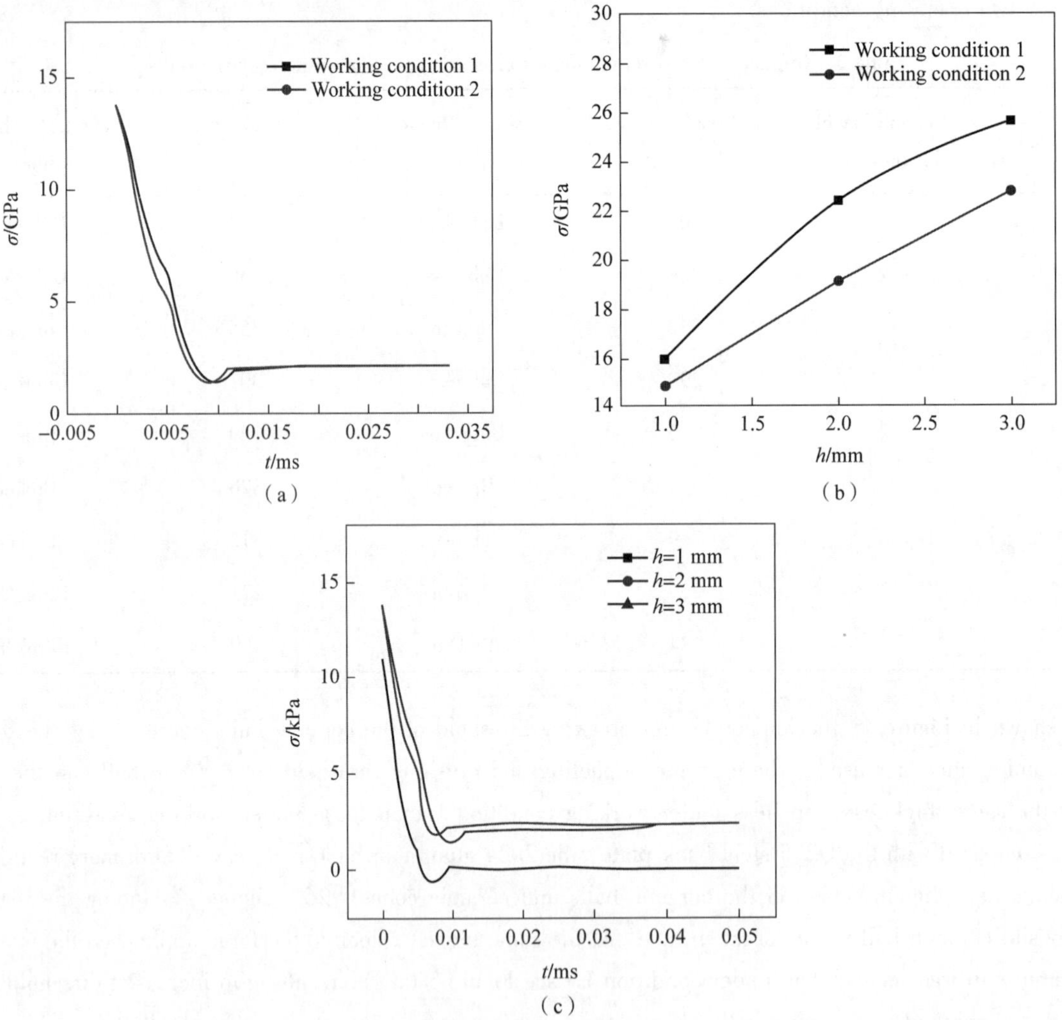

Figure 5　(a) Relationship between ceramic ball micro – element stress σ and t of $h = 3$ mm, (b) Relationship between ceramic ball micro – element stress σ and ceramic panel thickness (h) at $v = 700$ m/s, (c) Relationship between the stress σ and t of the ceramic ball micro – element under condition 2.

As a ceramic ball hits the target plate at a higher speed. A zone of extremely high pressure is formed at the impact point within a rather wide pressure range in which no impact phase changes are present. According to the momentum conservation law at the time of impact and the continuity condition at the interface, the pressure at the hitting point can be expressed by the following formula [27]:

$$P_p = \rho_p (a_p + b_p u_p) u_p \tag{1}$$

$$P_c = \rho_c (a_c + b_c u_c) u_c \tag{2}$$

According to Newton's third law

$$P_c = P_p \tag{3}$$

The real velocity on the impact interface is as follows:

$$v_p = u_p + u_c \tag{4}$$

Where P_P and u_P are the impact pressure and particle velocity acting on the projectile; P_c and u_c are the impact pressure and particle velocity acting on the ceramic panel; and a_P and b_P and a_c and b_c are the Hugoniot material parameters of the ceramic ball and the ceramic panel and RHA, respectively [16 – 23], where v_P is the velocity of the ceramic ball, and r_P is the density

Considering the heterogeneity of the structures, the backplate materials of the composite armor, and the concept of the composite target surface density, the unit body density in the composite target thickness direction is introduced, namely,

$$\rho_c = \rho_{cc} / h_c \tag{5}$$

where ρ_{cc} is the surface density of the composite target and h_c is the unit thickness of the composite target, the value of which is 1.

By using Eqs. (1) – (5), the impact pressure acting on the projectile when the ceramic ball impacts the ceramic composite target can be solved. Figure 6 shows the relationship between the composite target plate body density and the thickness of the ceramic panel h, and Figure 7 shows the relationship between the force applied on the micro units of the ceramic ball and the thickness of the ceramic panel under two working conditions for an impact velocity of 700 m/s. It can be seen from Figure 6 that the unit body density in the composite target thickness direction increases with an increase in the ceramic panel thickness. The theoretical calculation of the force applied on the micro units (see Figure 7) and the rules of the numerical simulation (see Figure 5 (b)) show good consistency.

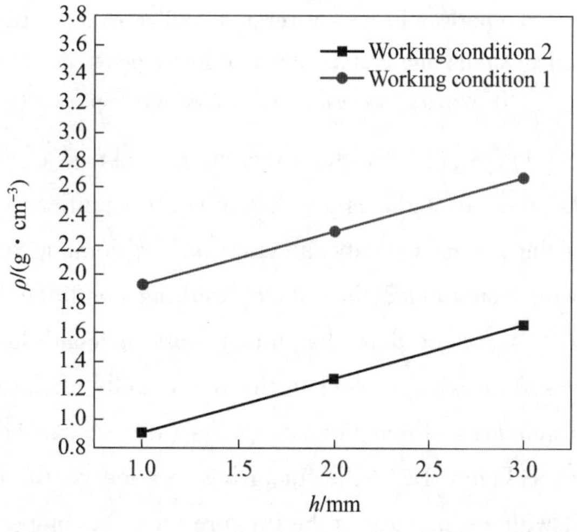

Figure 6 Relationship between unit density ρ_{cc} of composite target plate and thickness h of ceramic panel

Under the two working conditions, for the ceramic panels with different thicknesses, the impact energy distribution of a ceramic ball changes under the support of the elastoplastic back plate and the impact property changes. Figure 8 (a) and 8 (b) present the crushed pattern of the recovered ceramic ball under the condition of limiting piercing. Figure 8 (c) and 8 (d) display the numerically simulated damage cloud diagram of a ceramic

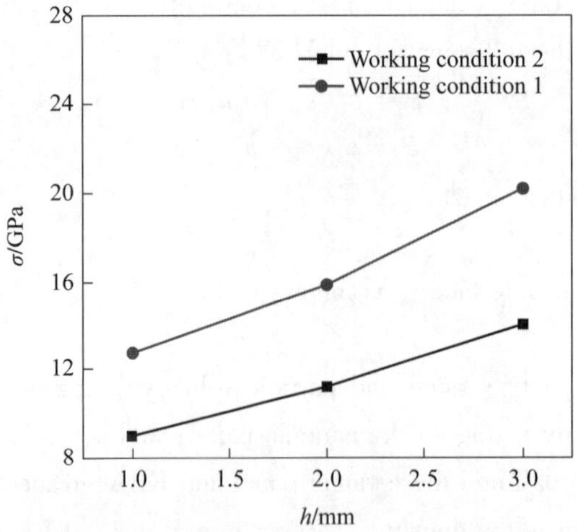

Figure 7　Relationship between ceramic ball micro – element stress σ and ceramic panel thickness h

ball for the same limit penetration velocity. Figure 9 shows the relationship between the thickness of the ceramic plate and the residual mass of the ceramic ball for a limited penetration velocity. Figure 10 presents the relationship between the force applied on the micro units of the ceramic bowl and the thickness of the ceramic panel at the limit penetration velocity.

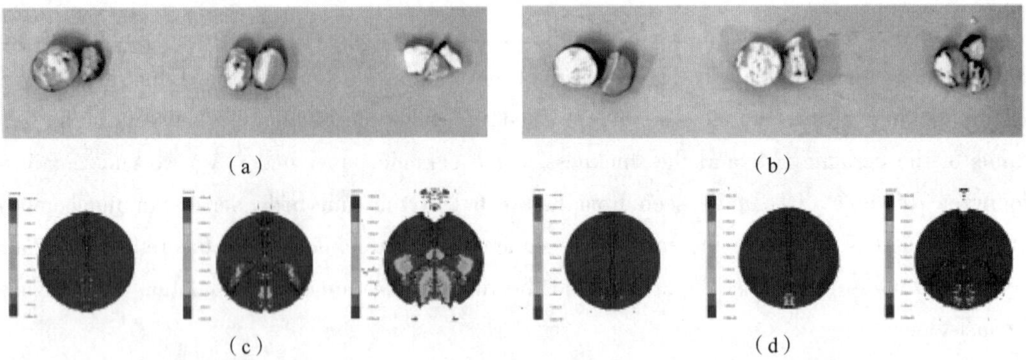

Figure 8　Comparison of partial ceramic ball recovery patterns and numerical simulation results under ultimate penetration speed

(a) Working condition 1; (b) Working condition 2; (c) Working condition 1; (d) Working condition 2

As can be seen from Figs. 8 (a) – (b), with an increased thickness of the ceramic panel, the quantity of recovered ceramic bowl fragments gradually reduces and the integrity significantly declines. The fracture cross – section passes through the axis of the impact velocity direction on the ceramic ball. Further, the integrity of the ceramic ball is better under working condition 2 than under working condition 1, because the limit penetration velocity under working condition 1 is greater than that under working condition 2 for the same ceramic panel thicknesses. The pressing and tensile stresses applied to the micro units of the ceramic balls is larger (see Fig 10), with obvious fractures and breakages. From Figs. 8 (c) – (d), it can be seen that, when increasing the limit of the penetration velocity (see Figure 12), the fracturing location on the ceramic ball changes, extending from an axial to a radial fracture, with an increase in the fracture area. A comparison of the ceramic ball fractures under the two types of working conditions revealed that the extent of crushing on the ceramic ball under working condition 1 is greater than that under working condition 2, for the same ceramic panel thicknesses. Moreover, its remaining mass is less than that under working condition 2 (see Figure 9). Comparisons of the numerical simulation results and the experimental phenomena reveal that the two are highly consistent.

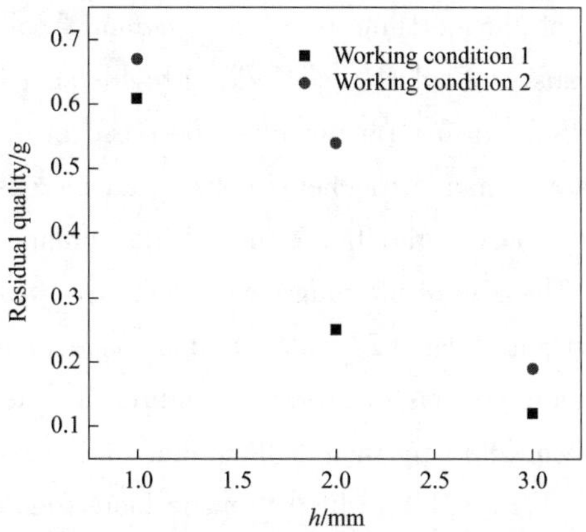

Figure 9 Relationship between ceramic plate thickness *h* and residual mass m of ceramic ball is under ballistic limit

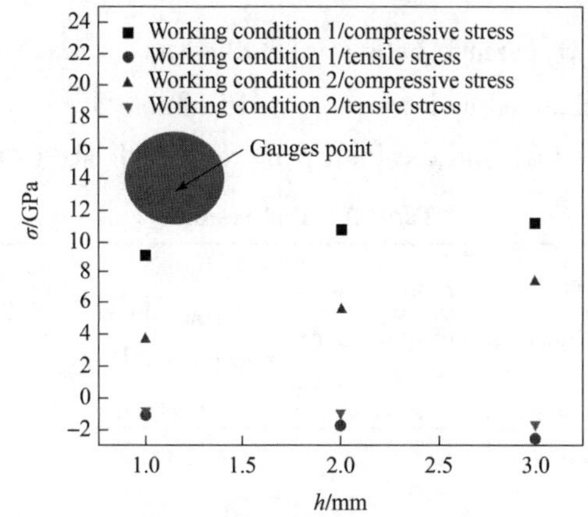

Figure 10 Relationship between ceramic panel thickness *h* and micro – element stress *σ*

Based on the good consistency demonstrated between the numerical simulation and the experimental results, a further numerical simulation of the fracture characteristics of a ceramic ball during a high – speed impact on the composite target plate was carried out. Figure 11 shows a cloud diagram of the smashing and damage occurring on the interface of ceramic ball under different impact velocities under working conditions 1 and 2.

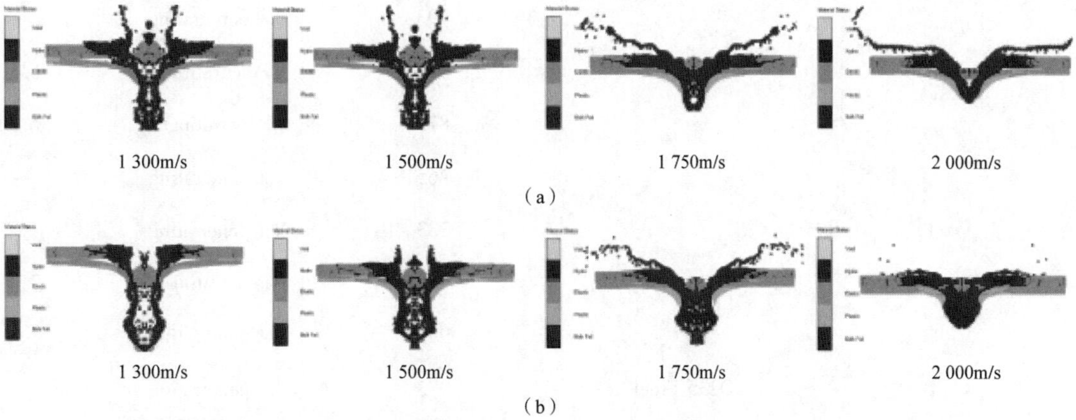

Figure 11 Damaged cloud image of ceramic ball interface under high – speed impact

(a) Working condition 1; (b) Working condition 2

As can be seen from Figs. 11 (a) and 11 (b), with an increase in the impact velocity, a more obvious breakage of the ceramic ball occurs, which remains on the side of the positive impact surface. The ability to penetrate the metal back plate became increasingly weaker, with changes in the damaged drums occurring along the direction of the valvular dehiscence, and perforation appearing on the metal back plate. The angle between splashing direction of the fragments of the ceramic panel and ceramic ball and the impact velocity direction increased, gradually changing into transverse splashing. Under a high - speed impact, there is an increased force applied on the ceramic bowl and the fracturing phenomenon becomes even more noticeable. In addition, the kinetic energy is distributed into smaller ceramic fragments causing an increased impact area on the target plate, with most of the kinetic energy consumed by the fracturing of the ceramic panel, the ceramic ball, and the deformation of the metal back plate. The overall piercing power declines, weakening the penetrating ability of the metal back plate. There is increased smashing [28, 29] and damage to the interface, without the ability to pierce the metal backboard. This is more obvious for working condition 1 than for working condition 2. Owing the higher backing plate strength under working condition 1 than that under working condition 2. This causes the ceramic ball to be more stressed (see Figure 10), which is made more fragile.

3.2 Analysis of Composite Armor Damage Characteristics

Multiple ϕ7 mm toughened Al_2O_3 ceramic balls were studied for an impact of a composite armor. Some limit penetration velocity test data were obtained and are presented in Table 3 below. Figure 12 shows the relationship between the ceramic panel thickness (h) and the limit penetration velocity (v).

Table 3 Test results

Serial number	Ceramic panel thickness/mm	Back target specification/mm	Target speed /($m \cdot s^{-1}$)	Whether the ceramic ball penetrates	Perforation diameter/mm
			412	Not penetrating	–
#1	1		470	Penetrating	7.72
			446	Not penetrating	=
			531	Not penetrating	–
#2	2	2024 Aluminum	547	Not penetrating	–
			593	Penetrating	7.23
			793	Not penetrating	–
#3	3		841	Penetrating	5.32
			875	Penetrating	6.71
			765	Not penetrating	–
#4	1		783	Not penetrating	–
			814	Penetrating	7.10
			951	Not penetrating	–
#5	2	Q235 Steel	973	Not penetrating	–
			1 028	Penetrating	4.93
			1 316	Not penetrating	–
#6	3		1 323	Not penetrating	–
			1 351	Penetrating	3.94

It can be seen from Figure 12 that under the two working conditions, with an increase in the thickness of the ceramic panel, the limit penetration velocity gradually increases. The limit penetration velocity and the increasing rate under condition 1 were greater than those under condition 2. Under the same conditions of the back plate, with an increase in thickness of the ceramic formed cone, more kinetic energy of the ceramic ball is consumed. When the speed of the ceramic ball is higher, the greater the impact pressure on the micro units [30] and the more likely that a crushing will occur (see Figs. 10 and 11). The increase in kinetic energy consumption is not conducive to the influence behavior. For the same ceramic panel thicknesses, when the ceramic balls impact the composite target plate at the constant speed, the ceramic cone forming sizes is almost the same [26], with the same area being affected on the metal backplane. Because the impact strength of a Q235 steel back plate is greater than that of a 2024 aluminum back plate, under the same kinetic energy conditions, the ceramic ball can more easily pierce the ceramic/aluminum composite target plate. This is identical with the experiment results and the published rules for a theoretical model calculation [31].

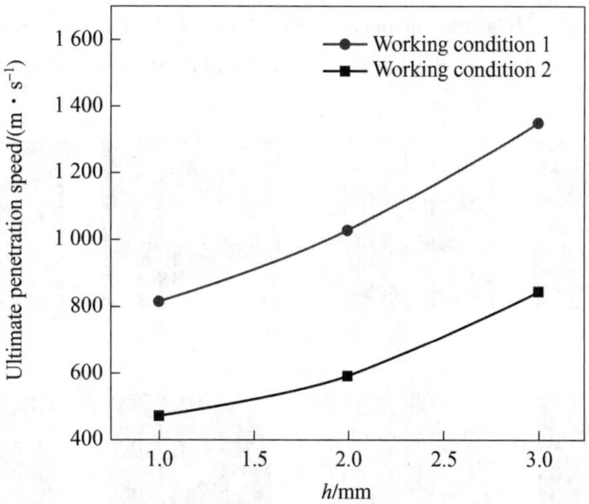

Figure 12　Relationship between ceramic panel thickness and h – limit penetration velocity *v*

The damaging characteristics of the ceramic composite targets under different impact velocities can differ to a certain extent. Understanding damaging characteristics of the ceramic composite targets will provide guidance for their design. Figure 13 shows the relationship between the thickness of the ceramic panel and the diameter of the penetrated hole on the back panel. Figs. 14 and 15 show the test results and numerical simulation results for a piercing of the composite ceramic target plate under the two working conditions.

Figure 14 shows that, under working condition 1, when the ceramic panel thickness is 1 mm, the damage on the metal back plate shows a punching and plugging pattern (see Figure 14 (a), where $h = 1$ mm). When the ceramic panel thickness is 2 or 3 mm, the penetrated hole on the Q235 steel metal back plate shows valvular dehiscence and perforation (see Figure 14 (a), where $h = 2$, $h = 3$ mm, and Figure 14 (b)), with identical numerical simulation results. With an increased thickness of the ceramic panel, the broken area around the edge of the penetrated hole on the ceramic panel decreases gradually, and the cracks extend circumferentially from the aperture. Under working condition 2, with an increased thickness of the ceramic panel, the convoluted pattern of the penetrated hole on the 2024 aluminum back plate gradually becomes obvious, transition from a punched and plugged penetrated hole pattern to a ductile penetrated hole pattern, with a good matching between the penetrated hole patterns and the numerical simulation results (see Figure 15).

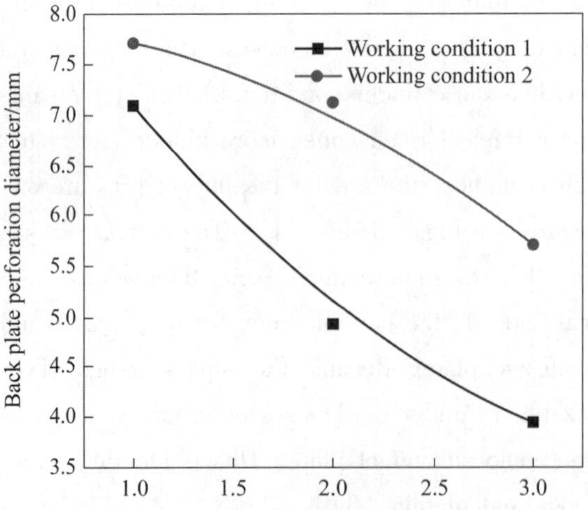

Figure 13　Relationship between the thickness *h* of the ceramic plate and diameter *r* of the back plate at the ultimate penetration speed

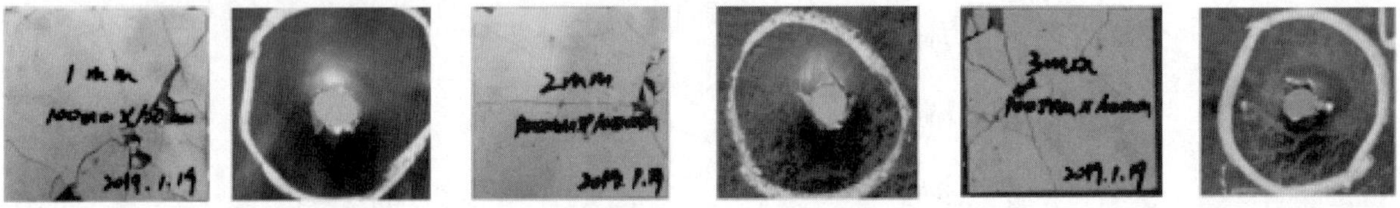

h=1 mm　　h=2 mm　　h=3 mm

(a)

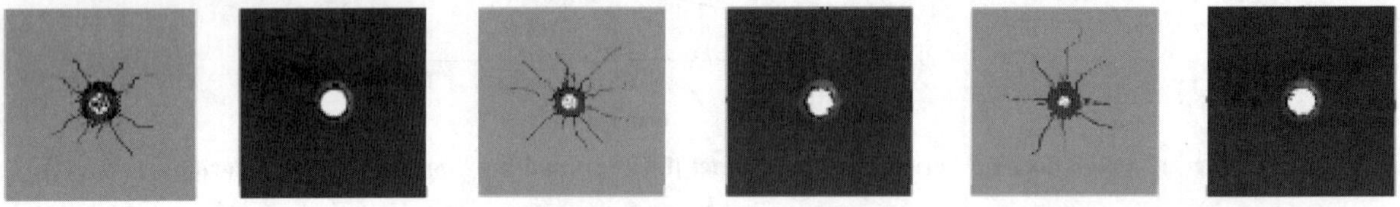

h=1 mm　　h=2 mm　　h=3 mm

(b)

Figure 14　Comparison of test results and numerical simulation results under working condition 1

(a) Test ceramic panel and e back target failure mode; (b) Numerical simulation of ceramic panel and metal back target failure modes

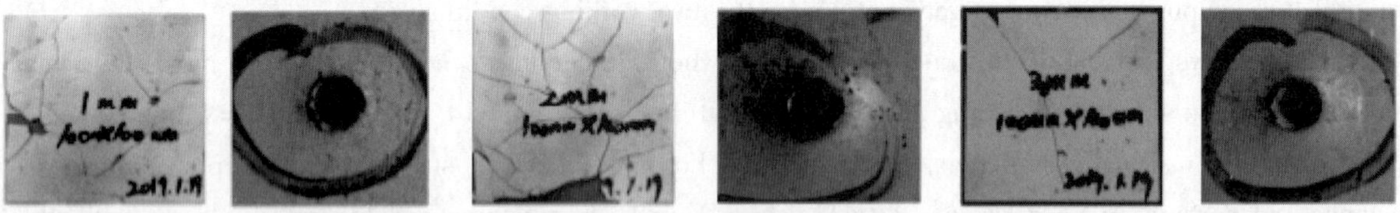

h=1 mm　　h=2 mm　　h=3 mm

(a)

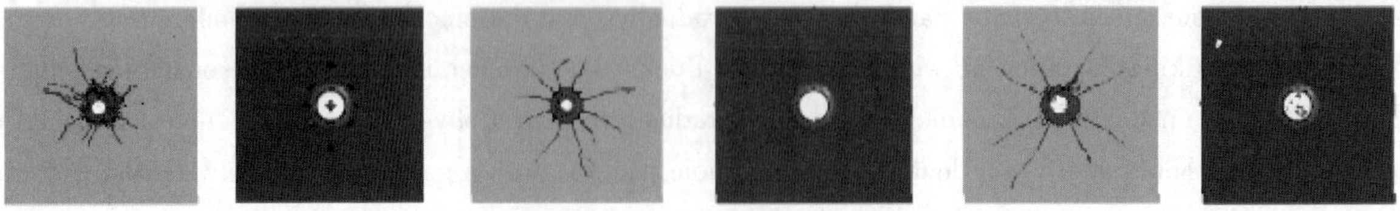

h=1 mm　　h=2 mm　　h=3 mm

(b)

Figure 15　Comparison of test and numerical simulation results under working condition 2

(a) Test ceramic panel and metal back target failure mode; (b) Numerical simulation of ceramic panel and metal back target failure modes

From the results presented in Figs. 14 and 15, it can also be seen that as the thickness of the ceramic plate increases, a decrease in the diameter of the penetrated hole gradually occurs, which is not obvious. This is because the ceramic plate is locally fractured and broken after being impacted, and the hole diameter formed by joining the ceramic fragments is incomplete. As the ceramic plate increases in thickness, the overall resistance of the ceramic plate to the shear strength increases, and the compaction effect of the ceramic bowl on the formed ceramic cone decreases until the ceramic ball can no longer pierce the ceramic panel. Hole piercing on the metal back plate mainly occurs by the ceramic core and some of the ceramic bowl fragments; hence, the penetrated hole diameter on the ceramic panel decreases. Considering the results of Figure 13, if the thicknesses of the ceramic panels are the same, the penetrated hole diameter on the back plate under working condition 1 was less than that under working condition 2, and the decreasing rate of the penetrated hole diameter was greater than that for working condition 2. This indicates that the support strength of the Q235 steel back plate is greater than that of the 2024 aluminum back plate, which causes the force applied to the micro units of a ceramic ball to greatly increase (see Figure 5 (a) -5(c)). The quantity of the ceramic fragments decreases during the impact process. But because their volumes are relatively small given the penetrated hole diameter on the metal back plate (see Figure 13).

4 Conclusion

Based on ballistic impact test results, a numerical simulation and a theoretical analysis were used to study the impact of ceramic balls on composite target plates, and the following conclusions were obtained:

1. Increasing the thickness of a ceramic panel and the strength of the metal back plate will lead to an increase in the stress applied to the micro units of a ceramic ball, and the fracturing of the ceramic ball will be more serious under the same impact velocity. For a ceramic panel 3 mm thick and a metal back plate of Q235 steel, the threshold values of the impact and crushing speed of the ceramic ball are at a minimum. With an increase in the impact speed, smashing and damage to the interface are more likely to occur, and there is a decreased chance that the ceramic ball will pierce the metal back plate.

2. When the ceramic panel thickness and strength of the metal back plate are increased, the limit penetration velocity of the ceramic ball increases and its integrity decreases. The hole - piercing mode on the ceramic ball changes from overall hole - piercing into hole piercing of the local fragments, which causes a change in the hole - piercing pattern on the metal back plate from a punched and plugged penetrated hole pattern to a valvular dehiscing/ductile penetrated hole pattern, with the penetrated hole diameter gradually decreasing in size.

3. Comparing and analyzing the numerical simulation results with the experiment and theoretical calculation results, good consistency was shown, indicating that the numerical simulation method adopted in this paper is reliable (to a certain extent).

References

[1] NECHITAILO N V, BATRA R C. Penetration/Perforation of aluminum, steel and tungsten plates by ceramic rods [J]. Int. J. Computers and Structures, 1998 (66): 571 -583.

[2] NECHITAILO N. Advanced high - speed ceramic projectiles against hard targets [J]. IEEE transactions, 2008 (45): 614 -619.

[3] TAKAHASHI M, OKABE N, ABE Y, et al. Fracture analysis for ceramic ball in backflow valve [J]. Fatigue & fracture of engineering mechanics, 2011 (4): 291 - 300.

[4] MATSUDA S, NAKADA T. Simple mechanics model and Hertzian ring crack initiation strength characteristics of silicon nitride ceramic ball subjected to thermal shock [J]. Engineering fracture mechanics, 2018 (197): 236 - 247.

[5] MA G, ZHOU W, ZHANG Y, et al. Fractal behavior and shape characteristics of fragments produced by the impact of quasi - brittle spheres [J]. Powder technology, 2018 (325): 498 - 509.

第 五 部 分
封锁类弹药技术

圆柱形弹体的入水弹道分析①

李　磊②，冯顺山
（北京理工大学爆炸科学与技术国家重点实验室，北京　100081）

摘　要：入水弹道的研究具有重要的科学意义，对弹体入水弹道进行数值仿真可以弥补理论分析和试验研究的不足。采用非线性动力学软件AUTODYN对一种无控的圆柱形弹体入水弹道进行了仿真计算，建立了4个计算模型，分析了弹体在以一定角度倾斜入水的条件下，弹体的质心位置、入水角度以及攻角对入水弹道的影响。通过数值仿真表明：入水弹道可以分为入水冲击弹道和下沉弹道两个阶段，弹体质心位置对这两个弹道阶段都有较大影响；入水角在45°时弹体出现侧表面贴水运动等。数值仿真结果可以对无动力水下弹药的入水弹道分析以及合理的结构形式设计提供参考。

关键词：弹药工程；圆柱形弹体；入水弹道；数值仿真

引言

入水弹道的研究具有重要的科学意义，空投鱼雷、深水炸弹以及其他水下灵巧弹药的入水弹道是考察其作战性能的重要指标之一，宇宙飞船和返回卫星的回收也涉及了入水问题。弹体以较高速度入水的过程是一个复杂的力学过程，涉及固体、液体和气体三种介质的运动，而且这三种介质的运动都具有强烈的不定常性。

1979年，英国学者Mackey给出了一种计算方法用以计算鱼雷带空泡航行时的水弹道[1]。颜开等[2]在Mackey的研究基础上，根据实验空泡的外形，重新拟合了空泡椭球模型中的经验常数，使计算结果更加接近试验结果。文献［3］给出了空投鱼雷有攻角运动时的曲线弹道：

$$\dot{x} = V\left[\cos\theta_0 - \sin\theta_0 \frac{a_2}{a_1}\alpha_0 \ln(1 + aV_1 t)\right]$$

$$\dot{y} = V\left[\sin\theta_0 - \cos\dot{\theta}_0 \frac{a_2}{a_1}\alpha_0 \ln(1 + a\ V_1 t)\right]$$

$$\theta(t) = \theta_0 + \frac{a_2}{a_1}\alpha_0 \ln(1 + aV_1 t)$$

其中，入水角θ_0，攻角α_0，入水初速度V_0。这个计算公式只能给出鱼雷在空泡名义消失时间内的弹道参数，而且是对实际情况做出假定而得到的结果。因此，虽然对入水弹道的理论求解取得了一定的成果，但是仍具有局限性。实验是研究入水弹道的重要方法，美国海军水下研究中心[4]进行了大量的入水试验，得到了一系列成果，但是用模拟试验方法研究入水问题，即使把入水物当作刚体，相似方案也极为复杂。物体的入水过程，可以看成入水物对水介质的侵彻过程，对问题进行数值仿真研究可以弥补解析求解和试验研究的不足。

对无控水下弹药进行入水弹道设计的一个关键环节是确保弹药在触底时保持一个好的姿态，但是在实战中严格限定入水条件和水域环境等外部因素是不现实的，因此如何改善弹药的构形以及水动力布局等设计，以保证其在一个比较宽松的外部条件下具有稳定的入水弹道就具有重要的意义。本文以某型弹药研制为背景，结合数值计算软件AUTODYN，研究无控的圆柱形弹体在以一定角度倾斜入水的条件下，

---

① 原文发表于《中北大学学报》2005.

② 李磊：工学博士，2003年师从冯顺山教授，研究港口水域封锁弹药系统总体技术，现工作单位：北京航天长征飞行器研究所（14所）。

弹体的质心位置、入水角度以及攻角对入水弹道的影响，结果可以为工程设计提供依据。

1 数值方法与计算模型

应用非线性动力学分析软件 AUTODYN 对弹体入水过程进行计算，弹体采用 Lagrange 单元法建模，水介质则采用 Euler 单元法建模，利用 Lagrange 和 Euler 耦合的算法模拟流体和结构的相互作用。弹体高速入水可能形成冲击波，AUTODYN 为能量和动量方程中的流体静压力 p 加上了一个额外项来解决计算中的人工黏性力问题，即

$$q=\rho\left[\left(C_Q d\left(\frac{\dot V}{V}\right)\right)-C_L c\left(\frac{\dot V}{V}\right)\right],\quad \frac{\dot V}{V}<0;$$

$$q=0,\quad \frac{\dot V}{V}>0$$

式中 C_Q 和 C_L——常数；

ρ——密度；

d——典型长度；

c——声速；

$\frac{\dot V}{V}$——体积变化。

图 1 为弹体入水示意图，弹体为圆柱形，长 1 m，半径 0.1 m，入水速度 100 m/s，弹体为实心体，质量 50 kg，本文把弹体分为体积相等的上、下两部分，建立了如表 1 所示的 4 种计算模型，在弹体轴线上等距高设置了 3 个观察点，1、3 两个观察点位于轴线两端；水池深度 20 m，宽度 20 m，水面为自由液面，水池四周和底部的边界条件设置为 transmit 无反射边界条件，即水介质不局限于水池尺寸。规定逆时针为角度正方向，单位为度；x、y 轴正方向如图 1 所示，单位为 m，坐标原点与观察点 1 重合。图 2 给出了计算模型的初始状态。表 2 给出了水介质和弹体的材料模型。

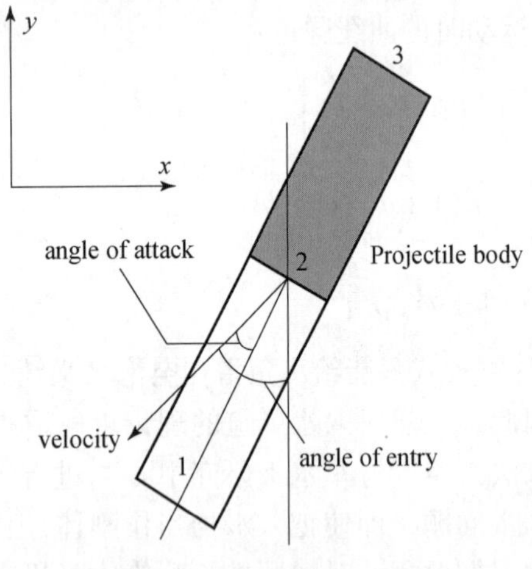

图 1 弹体入水示意图

图 2 有限元计算模型

表 1 弹体入水计算模型

模型	弹体密度/(g · cm^{-3})	入水角度/(°)	入水攻角/(°)
模型 1	均匀密度 1.6	-30	0
模型 2	2.2（下）1（上）	-30	0
模型 3	2.2（下）1（上）	-30	-10
模型 4	2.2（下）1（上）	-45	0

表 2 材料模型

	状态方程	强度模型	失效模型	体积模量/kPa	剪切模量/kPa	屈服应力/kPa
弹体	Linear	Von – mises	None	$2.0*10^8$	$9.0*10^7$	$2.0*10^5$
水	Linear	Hydro	Hydro	$2.0*10^6$	0	

2 计算结果及分析

2.1 弹体入水的数值仿真

2.1.1 均匀密度弹体的入水计算

在模型 1 的计算中，弹体密度是均匀的，弹体的质心位于形心处。当圆柱形弹体以一定角度入水时，开始只有头部与水接触，可以认为弹体做直线运动，如图 3 所示。在 8. 33 ms 左右的时刻弹体尾部与水接触，发生所谓的“尾击”，入水弹道发生变化：在 33. 3 ms 的时刻，弹体已经呈现竖直状态；之后，从 x、y 方向位移与时间曲线中可以看出，弹体在水中经历数次翻转，在 859 ms 左右的时刻，弹体上 3 个观察点的 x 方向速度趋近于 0，y 方向速度趋于一致，弹体以相对稳定的姿态在水中竖直下落。从纵向位移与时间的曲线中可以看出，弹体最先入水的一端在姿态稳定时已经朝上，这说明弹体质心与形心重合时，入水弹道是不稳定的。

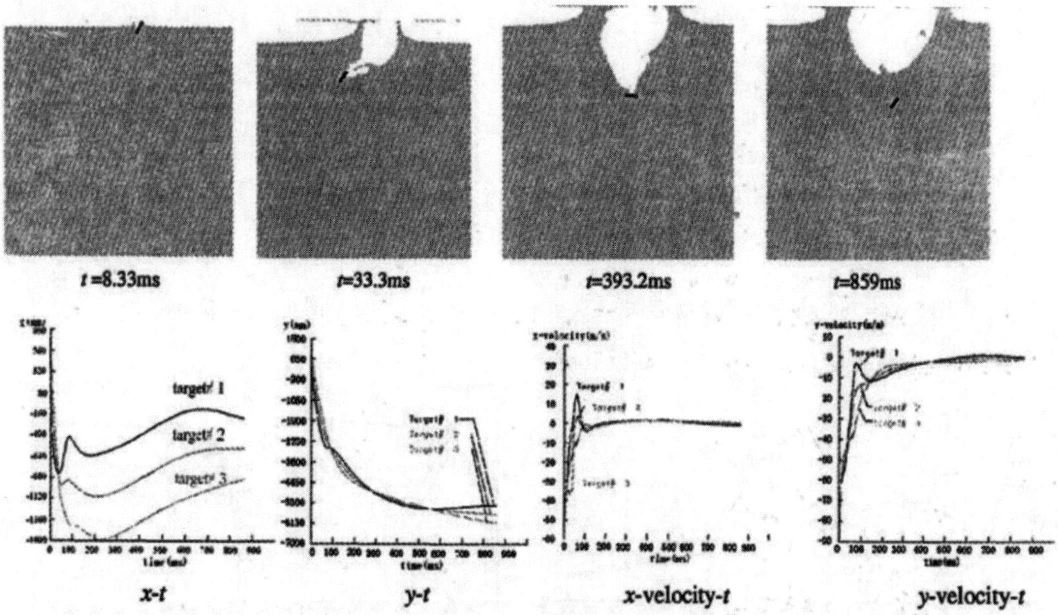

图 3 模型 1 入水状态图以及弹道参数随时间变化曲线

2.1.2 质心位置对入水弹道的影响

在模型 2 中，弹体被分为体积相等的两个部分，充填两种不同的密度，使弹体质心更靠近入水端。在发生“尾击”前弹道仍呈直线状，弹体尾部与水接触后，弹道发生变化。图 4 给出了入水弹道参数随时间变化曲线，从横向、纵向位移随时间变化曲线中可以看出，弹体入水后到姿态稳定的时间内，弹体在水中出现了来回摆动的现象。在 40 ms 左右时刻，弹体竖直；由于惯性作用，弹体逆时针翻转，在 131 ms 时弹体水平；之后，在力矩作用下，弹体出现顺时针旋转，在 181 ms 时再次呈水平状；弹体在水中姿态继续改变，直至达到力学平衡，表现为三个观察点的横向速度接近 0，纵向速度趋于一致。模型 2 入水后姿态稳定后状态见图 5。

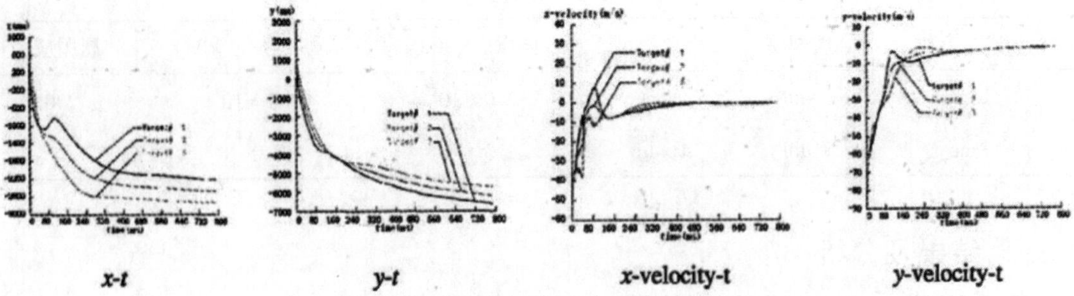

图 4　模型 2 入水弹道参数随时间变化曲线

2.1.3　有攻角时的入水弹道

模型 3 与模型 2 的差别在于弹体带有 -10°的攻角，弹体在入水初期并未出现“尾击”，因此入水弹道近似一条直线，如图 6 所示。从横向位移与时间关系曲线上可以看出，在 200 ms 左右时刻，弹体呈竖直状；之后，弹体在水中成逆时针翻转，在入水后 800 ms 左右时刻，3 个观察点纵向位移近似相等，弹体呈水平状，但是在这个时刻，3 个观察点的横向速度基本为 0。纵向速度趋于一致，说明弹体不再翻转，姿态基本稳定。由于弹体质心不在形心处，可以推断，弹体在 800 ms 时刻并不是皱后的稳定姿态，弹体将在力矩作用下顺时针翻转，使重心在形心以下，这将使得入水端仍保持朝下。

图 5　$t = 788$ ms 时模型 2 状态图

图 6　$t = 132$ ms

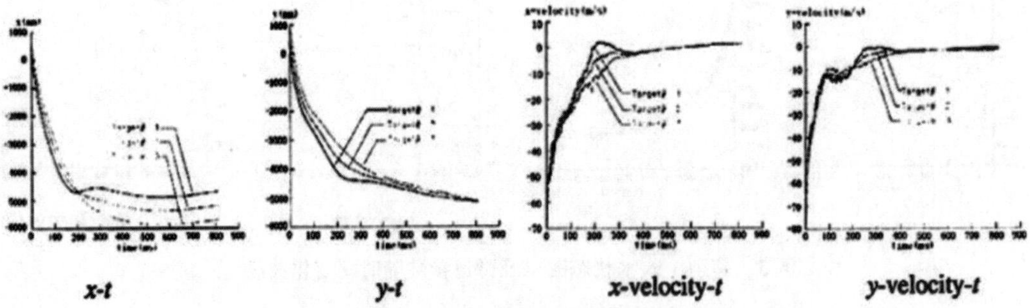

图 7　模型 3 入水弹道参数随时间变化曲线

2.1.4　入水角对入水弹道的影响

模型 4 的入水角为 -45°，从图 8 中可以看出，模型 4 出现了一个强烈弯曲的朝上运动，属于弹体侧表面贴水运动。在 804.5 ms 时刻，弹体上 3 个观察点的横向速度基本为 0，纵向速度基本一致，说明此时弹体的入水运动基本完成，但是由于弹体所受重力大于浮力，弹体将下沉，力矩将使弹体调整姿态，使重心位于形心以下。

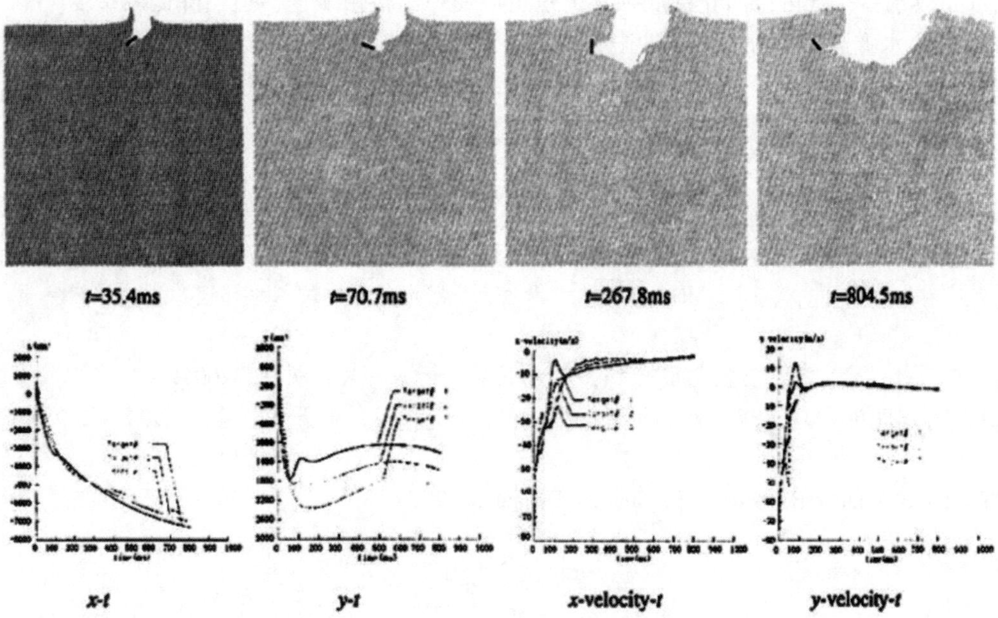

图 8　模型 4 入水状态图以及弹道参数随时间变化曲线

2.2　计算结果分析

不同的水下弹药对入水弹道有不同的要求，和鱼雷等有动力的水下兵器不同的是，无动力弹药入水后，不能通过控制系统调整姿态和改变弹道，只能通过弹药自身设计来改善水下性能。通过对以上 4 个模型的数值计算可以看出，圆柱形弹体入水弹道主要可以分为两个阶段：一个是入水冲击弹道；另一个是下沉弹道。弹体以 100 m/s 的速度入水以后，横向、纵向速度都急剧下降，到 200 ms 时刻，横向、纵向速度基本都下降到了 10 m/s 左右，在不到 1 s 的时间内，弹体的横向速度基本为 0，纵向速度下降到 3 m/s 以内，在这个阶段，弹药经历的激烈的入水冲击，弹道也剧烈变化，这个阶段弹道可以称为入水冲击弹道。入水冲击弹道结束以后，弹体的姿态不会再出现剧烈变化，只能在下降过程中，通过重心与浮心的偏差所造成的力矩调整姿态。入水冲击弹道结束时弹体的横向位移可以近似看成整个入水弹道的最大横向位移，表 3 近似地给出了弹体在入水冲击阶段的最大位移。

表 3　弹体的最大位移

	模型 1	模型 2	模型 3	模型 4
X 最大位移/m	−1.5	−2.5	−5.0	−7.5
Y 最大位移/m	−6.0	−6.0	−5.5	−2.0

3　结论

通过对圆柱形弹体入水弹道的数值仿真，可以得出以下几点结论：

（1）非线性动力学软件 AUTODYN 可以进行弹体入水弹道的数值仿真，应用数值计算软件对入水弹道的计算为入水弹道的研究提供了一个新的技术途径，可以弥补理论分析和试验研究的不足，为理论和试验研究提供参考。

（2）圆柱形弹体的质心位置对入水弹道和弹体的水下姿态产生影响，可以通过调整弹体质心和浮心位置来优化弹体水下运动品质。

（3）入水角在 45°时，弹体出现侧表面贴水运动，这种弹道对于大多数水下弹药来说是不理想的，但是对于浅水布放水雷，既可以防止撞底损坏，也可以使弹药很快减速，适当降低散布度。

（4）弹体的入水弹道可以分为入水冲击弹道和下沉弹道两个阶段，入水冲击弹道决定了入水弹道的最大横向位移和自由下沉的位置、初始姿态，弹体在下沉弹道阶段可以通过不平衡力矩调整姿态。

参考文献

[1] MACKEY A M. A mathematical model of water entry：AD A085774 [R]. 1979.

[2] 颜开，史液君，薛晓中，等. 用Mackey方法计算鱼雷带空泡航行时的入水弹道 [J]. 弹道学报，1998，10(2)：93－96.

[3] 杨世兴，李乃晋，徐宣志. 空投鱼雷技术 [M]. 昆明：云南科技出版社，2001.

[4] WAUGH J G，STUBSTAD G W. 水弹道学模拟 [M]. 陈九锡，张开荣，译. 北京：国防工业出版社，1979.

[5] AUTODYN Users Manual Revision 4.3，Century Dynamics，2003.

[6]《鱼雷力学》编著组. 鱼雷力学 [M]. 北京：国防工业出版社，1992.

一种可用于封锁弹药的地面运动目标智能识别方法①

徐德琛②，冯顺山，董永香

（北京理工大学爆炸科学与技术国家重点实验室，北京 100081）

摘 要：战场地面运动目标识别技术是智能雷弹、封锁子弹药设计的一个关键性技术，利用微加速度传感器进行地面运动目标识别的基本原理是用加速度传感器采集地面车辆目标运动时引起的地震波信号，并利用先进的信号处理方法进行目标的自动识别。这种基于微加速度传感器的探测系统具有重量轻、体积小、功耗低和可在多种恶劣环境下工作（比如用于战场探测识别）的优点。为了提取不同车辆目标行驶时引起的地震波的特征并对不同目标加以识别，本文研究了典型目标引起的地震波的特征。为了达到目标识别的目的，应用了人工神经网络（ANN）技术。为提高学习速度，避免误差曲线中的局部极值，采用了改进的反向递推（BP）算法。野外环境下的实验表明，用该算法所提取到的目标的地震波特征是正确的，ANN在解决运动车辆目标的识别问题上是有效的，微加速度传感器可以用于地面运动目标的识别问题。

关键词：微加速度传感器；地震波；人工神经网络（ANN）；目标识别

引言

MEMS目前已被应用于多个领域，包括环境检测、自动控制、服务业以及战场目标识别等。现在的MEMS器件包括加速度传感器、压力计、流速计等。从Roylance等开始对微加速度传感器的研究已有20多年的历史。[1]报道了1979年研制的第一个微加速度传感器，从那以后相继出现了很多关于加速度传感器的论文[2-6]。而目前的研究多集中在微加速度传感器的应用方面。

本文提出了微加速度传感器在运动车辆目标探测及识别上的一种应用，这一技术在一定程度上可推动智能雷弹、封锁子弹药技术的研究。在地面目标识别，尤其是地面运动车辆识别方面的研究已有不少方法，一般是利用图像、声信号、红外线和地震信号，其中地震信号的应用较少，原因是地震信号自身的复杂性，目标信号的特征提取难度很大。本研究之所以选用地震信号作为有用信号，是因为地震信号较之其他几种信号有其自身的特点和优势，如地震信号不受多普勒效应影响；受其他运动目标、天气等因素的影响方面也比其他几种信号弱；最重要的一点是，针对智能雷弹、封锁子弹药的应用，用于测量振动信号的传感器在价格、体积和鲁棒性方面有着不可忽视的优势。实验中对典型车辆目标的地震波信号进行了测量和分析。在目标的分类和识别的实现方面，基于多层前馈网络，设计了利用地震波信号进行目标识别的ANN结构。由于传统的BP算法不能满足速度的要求，在多层前馈网络中运用了改进的BP算法。实验结果证明，这种改进的BP算法在地震波信号的分类和识别中是有效的，而微加速度传感器可以用于运动车辆目标的识别。

1 地震信号探测系统的结构

实验中设计的地震信号探测系统由微加速度传感器、A/D转换器、存储器、可编程器件、自适应采集和存储控制器（ACCM）及电源等构成。微加速度传感器用的是AD公司的ADXL05，该器件具有价格

---

① 原文发表于《2005年弹药战斗部学术交流会论文集》。

② 徐德琛：工学硕士，2004年师从冯顺山教授，研究典型地面运动目标的地震波信号特性，现工作单位：北京昆天科微电子技术有限公司。

低、体积小、功耗低（电池供电）的特点，可用于静态和动态加速度的测量，4 K 的带宽对本实验而言是足够的。图 1 描述了地震波探测系统的结构。传感器的输出信号输入 A/D 转换器和 ACCM。其中 ACCM 用于控制 A/D 转换、压缩数据，并负责将代码写入存储器。一旦数据到达存储器，数据的分析工作便交由软件完成。

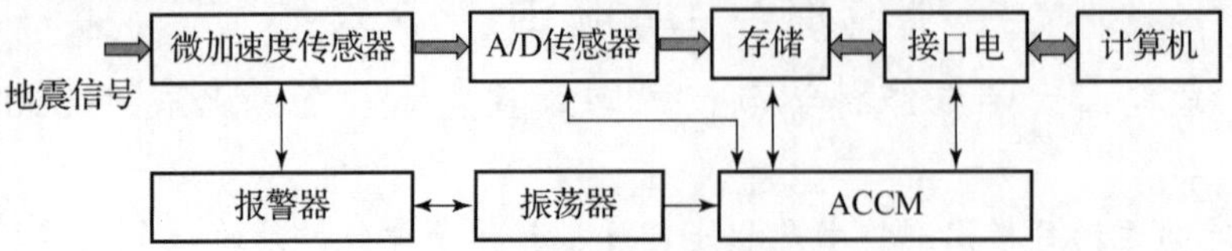

图 1　地震波探测系统的结构

2　车辆目标的地震信号特征获取实验

作为目标识别和分类的基础，在野外环境下测得不同目标引起的地震波的特征，图 2 描述了上述探测系统的一个应用实例，ADXL05 被固连在封锁子弹上，弹体吸收地震波，并将其传至加速度传感器，当车辆驶入加速度传感器的测量范围时，系统采集到地震信号。

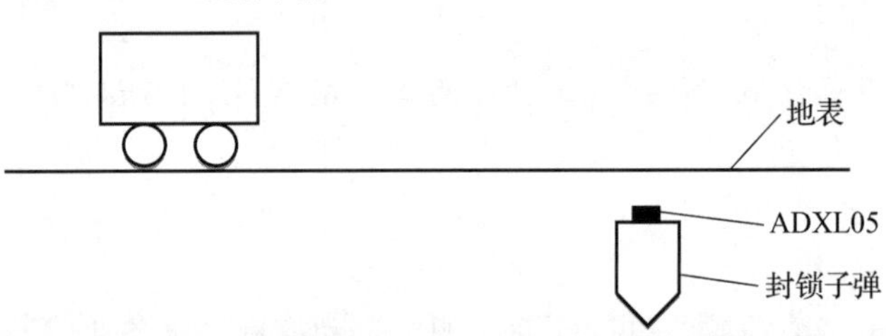

图 2　地震波探测系统在封锁子弹药中的应用

被测目标包括：普通柴油机车辆、重型柴油机车辆和汽油机车辆。实验在晴天下进行，气温 15 ~ 20 ℃，风力 2 ~ 3 级。系统测得地震信号后，将其转换为数字信号，并将该信号输入计算机进行时域和频域的分析。

3　运动目标引起的地震波信号的分析

3.1　运动目标引起地震信号的原理

运动目标引起地面震动，从而产生地震信号。目标不同的运动状态、不同的距离以及不同的路况引起不同的地震信号。车辆的引擎、驱动系统、车体是车辆行驶时引起地面震动的震源。而对于履带式车辆，除了上述震源外，车辆的履带还会周期性地引起地面震动，其频率随车辆行驶速度及履带结构的不同而有所不同。它们的关系大致是

$$F = \frac{S \times 17.6}{P} \tag{1}$$

式中　F(Hz)——上述周期性对应的频率；

S(miles/h)——车辆的运动速度；

P(in)——履带的尺寸。

3.2　目标引起的地震信号的传播

地震波可分为两类：体波和面波，面波占了地震波 70% 的能量，当震源位于地表时，传感器所测量到的主要是 Rayleigh 波。地震波在传播过程中的衰减与很多因素有关，比如地震波传播过程中的介质摩擦就是其中的一个因素。

3.3　典型车辆目标的地震信号特征

根据数据处理结果有如下结论：

（1）从图3～图5可见，微加速度传感器能有效地测量到中频和低频信号。

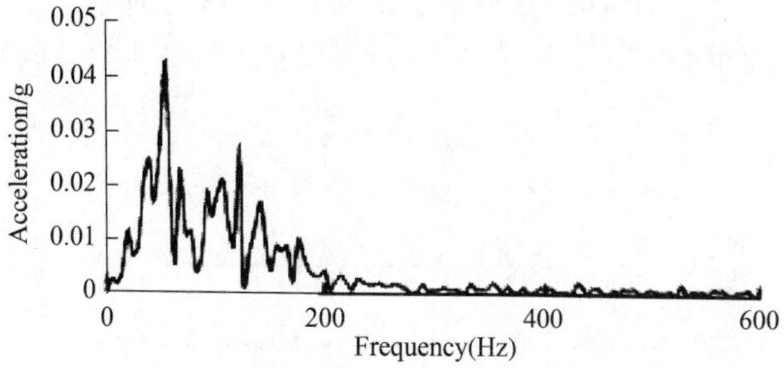

图3　普通柴油机车辆的频域特性曲线

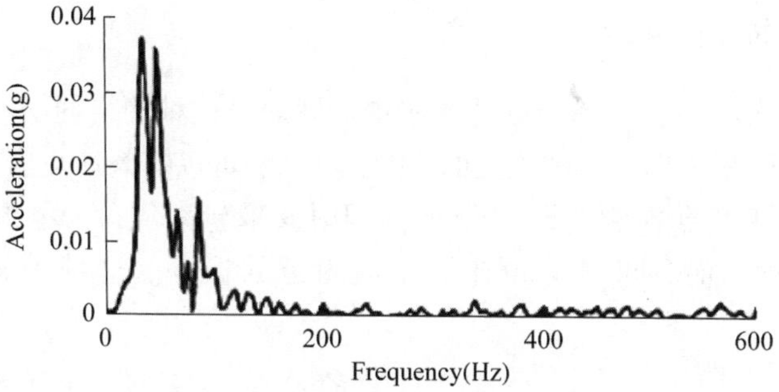

图4　汽油机车辆的频域特性曲线

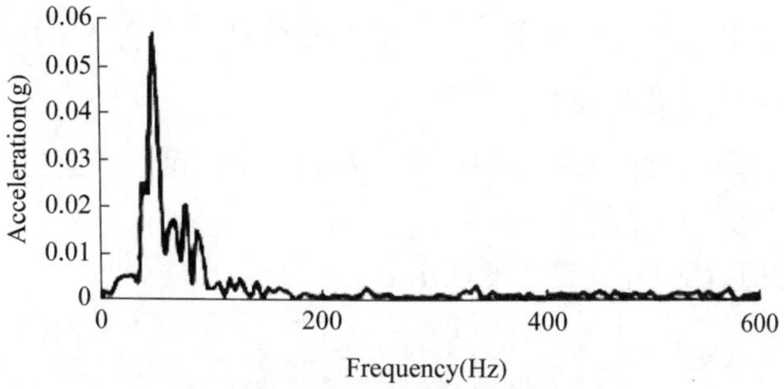

图5　重型柴油机车辆的频域特性曲线

（2）信号的能量集中在600 Hz以下，并且有明显的峰值。

（3）不同目标对应的信号在频谱和幅度上有很大的差别，所以利用地震信号的频谱和幅度进行目标的分类是可能的。

4　分类方法及结果

4.1　ANN目标分类结构

ANN运用网络对数据的响应来调整连接权系数，最终可以有效地逼近一个n元到m元的映射关系。目前进行的很多研究都运用了这一技术，在这里的地震波探测中也选用了ANN。

实验中设计的目标分类ANN的结构如图6所示。其中，n个输入表示地震信号，m个输出表示分类结果。在网络的输入和输出之间建立的n元到m元的映射关系是由网络的连接权系数决定的。R. Hecht－Nielson已经证明，三层感知器网络能以任意的精度逼近任意函数，所以这里就用三层网络来实现目标的分类。设在地震波信号分类 ANN 模型中有 n 个输入和 m 个输出，下式反映了前面所述的映射关系：

$$Y_k = g(W, X_k) \tag{2}$$

其中，

$$X_k = (x_{k1}, x_{k2}, \cdots x_{kn})^T \quad Y_k = (y_{k1}, y_{k2}, \cdots y_{kn})^T \quad \boldsymbol{W} = \{w_{ij}, i = 1, 2, \cdots, N_i, j = 1, 2, \cdots, N_j\}$$

$\boldsymbol{W}$ 为连接权系数矩阵，N_i 和 N_j 分别为第 i 层和第 j 层的神经元数。训练的过程实际上就是利用训练数据计算连接权系数的过程，一旦训练结束，这个 ANN 就可以快速地解决复杂问题了。

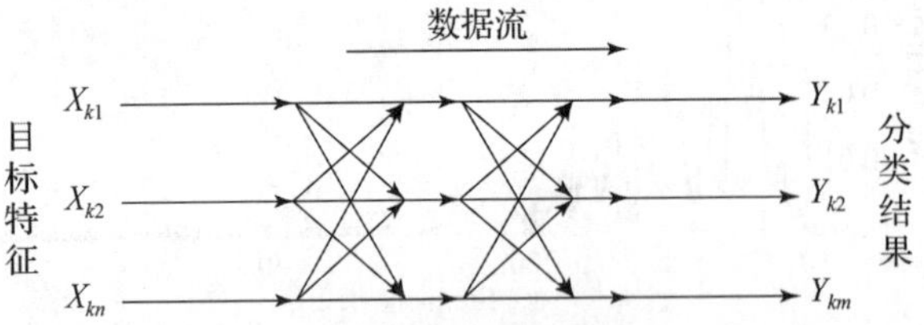

图 6　ANN 目标分类结构

4.2　ANN 目标分类技术的实现

BP 网络结构简单，工作稳定并且易于硬件的实现。然而，这种网络的学习速度不够快，并且误差曲线上存在局部极值[12,13]。这里提出了一种改进的 BP 学习算法可以有效地克服上述缺点。

（1）引入一种可动态调整的学习率 μ，刚开始学习时 μ 取较大值，为了提高学习速度，当网络的输出误差减小时，μ 取较小值，而且可以避免在极小值附近的振荡。这种动态调整的学习率由式（3）表示：

$$\mu = \beta \frac{\sum E(i)}{\sum E(i+1)} \tag{3}$$

其中 β 是一个介于 1.5 和 2.5 之间的常数，i 表示迭代次数，$\sum E(i)$ 表示第 i 次迭代的总误差，$\sum E(i+1)$ 表示第（$i+1$）次迭代的误差。

（2）用（-1，1）上的随机数作为初始的连接权系数，这样可以保证当权重变大时网络不至于达到饱和，而且可以避免网络收敛时的振荡。

（3）用带可变参数的 S 形函数作为输出变换函数：

$$f(x_{kj}) = \frac{1}{1 + e^{-\alpha x_{kj}}} \tag{4}$$

其中，α 是第 k 次迭代中的第 j 个神经元的参数。

传统 BP 算法的迭代公式为

$$w_{ij}(k+1) = w_{ij}(k) + \mu\delta_{kj}x_{kj} \tag{5}$$

其中，$w_{ij}(k)$ 为第 k 次迭代中，某层的第 i 个神经元和下一层的第 j 个神经元之间的连接权系数，δ_{kj} 为第 k 次迭代中，第 j 个神经元的输出，δ_{kj} 的输出层和中间层的表达式分别为式（6）、式（7）。

$$\delta_{kj} = (y'_{kj} - y_{kj}) f_j(x_{kj}) [1 - f_j(x_{kj})] \tag{6}$$

$$\delta_{kj} = f_j(x_{kj}) [1 - f_j(x_{kj})] \sum_l \delta_{kl} w_{lj} \tag{7}$$

其中，y'_{kj} 为期望输出，y_{kj} 为实际输出。当输出变换函数为式（4）时，式（5）~式(7）变为

$$w_{ij}(k+1) = w_{ij}(k) + \mu\delta_{kj}x_{kj} \tag{8}$$

$$\delta_{kj} = \alpha(y'_{kj} - y_{kj}) f_j(x_{kj}) [1 - f_j(x_{kj})] \tag{9}$$

$$\delta_{kj} = \alpha f_j(x_{kj}) [1 - f_j(x_{kj})] \sum_l \delta_{kl} w_{lj} \tag{10}$$

从式（8）~式(10）可以看出，通过参数 α 可以调整误差函数的曲率，加快网络的收敛速度。当误差较大时，α 取较小的值，这样使 S 形函数就相对平缓，饱和带加长。相反地，α 取较大的值。

建立了一个训练误差限。当训练误差达到这一误差限时，就停止对网络的训练，这也可以加快学习的速度。传统 BP 算法和改进的 BP 算法的误差曲线如图 7 所示。可见改进的 BP 算法明显地减少了训练时间，提高了学习速度。

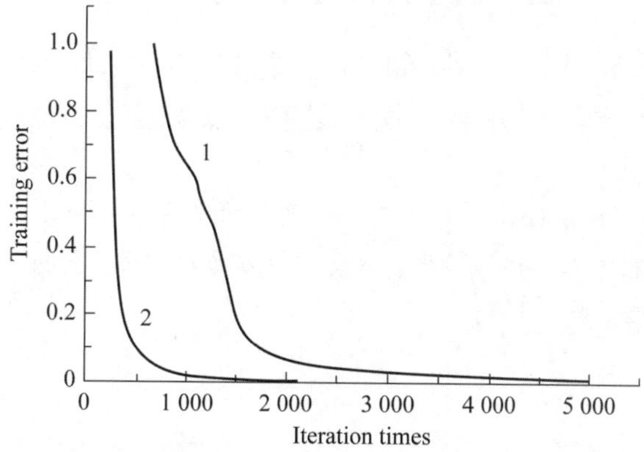

图 7　传统 BP 算法（1）和改进的 BP 算法（2）的误差曲线

4.3　分类结果及分析

运用改进的 BP 算法，输入实测的地震波信号数据得到的结果如表 1 所示。可见，对汽油机车辆、重型柴油机车辆和普通柴油机车辆的识别率分别为 96%、84% 和 82%，识别率依次降低。这一结果与它们的幅度和频谱特征是一致的。汽油机车辆对应的地震信号的幅度是最小的，而且其频谱与其他两种信号有较大的差别，汽油机目标的识别率高是合理的。相反，重型柴油机和普通柴油机车辆有相似的频谱，这与测量时距传感器的距离、环境、路况和车辆的速度有关，这导致了它们相近的识别率。虽然识别率还不能达到 100%，但微加速度传感器仍有较之其他系统的优势。特别是在战场环境下，微加速度传感器不但能够实现对目标的探测，而且在抗声传播的环境下有较之声传感器的优势。另外，微加速度传感器系统功耗小、体积小，适合于野外利用电池长时间供电的小型智能弹药系统。

表 1　基于改进的 BP 算法的目标识别结果

	柴油机车辆	汽油机车辆	重型柴油机车辆
采样总数	50	50	50
正确识别数	41	48	42
识别率/%	82	96	84

5　结论

实验中，用微加速度传感器系统成功地测量了运动车辆目标引起的地震信号，并将神经网络技术运用于地震信号的识别中，为了更有效地利用神经网络技术，对学习算法进行了改进。实验结果证明，微加速度传感器能够用于运动车辆的目标识别，将神经网络技术运用于目标的分类和识别中能够获得较高的识别率。鉴于该系统体积小、功耗低、识别率高的特点，可将其应用于智能雷弹、封锁子弹药等相关弹药的研究中。

可供进一步研究的一个方向是，发展一种新的特征提取算法，来获得鲁棒性强的特征向量，因为识别率之所以低于 100%，正是由于地震波和地况的复杂性；另外，由于环境的复杂性，至今还没有一种传感器能达到 100% 的识别率，所以一种可行的方法是，利用多传感器数据融合来获得更准确的信息，进一步提高识别率。

巡飞弹的发展与展望①

王　玥②，冯顺山，曹红松，王云峰
（北京理工大学爆炸科学与技术国家重点实验室，北京　100081）

摘　要：本文对巡飞弹做了概念描述，并介绍了巡飞弹的发展历程和目前各国的研究情况，介绍了巡飞弹的种类，通过对多弹种针对特定点目标的精度、效费比等方面的对比，分析了巡飞弹的优势及应用前景。

关键词：巡飞弹；无人机；灵巧弹药

引言

巡飞弹是指能够在攻击目标区上空一定范围内实现巡弋飞行，同时具备自主探测、识别、锁定和攻击预定目标的弹药。巡飞弹属于灵巧弹药的一种，是由攻击型无人机演变而来的。

从使用方式上分，攻击型无人机可分为两类，一类是可供多次使用的攻击机，这类多以大型无人机为主，机上可携带导弹、鱼雷、炸弹等武器，在攻击时启动机上武器摧毁目标。无人机可回收，多次使用，如美国的“捕食者”。另一类则是指一次性使用的攻击机，这类无人机多为中小型，机上装有寻的器及战斗部，由地面遥控或事先的航路规划飞向目标区上空进行待机巡逻飞行，一旦发现目标即由寻的器导引飞向目标并适时引爆战斗部将其摧毁。这类只能一次性使用的攻击型无人机实际上就是本文提到的巡飞弹的前身。

1　巡飞弹的发展概况

1.1　各国早期的巡飞弹及其特点

此阶段的巡飞弹在外形和结构上很大程度具有无人机的相关特征，如多采用活塞发动机、螺旋桨推进等，但能够携带战斗部并且不需要目标精确的位置信息即可对其实施攻击的特点则在功能上明确了它的范畴。

1.1.1　以色列的发展

以色列的 Happy 反辐射巡飞弹具有自动搜索能力，并具有全天候作战和同时攻击多目标的能力，主要用于攻击和摧毁敌方雷达目标。从地面发射时飞行速度为200 km/h，巡航速度为165 km/h，飞行高度3 000 m，续航时间3 ~5 h，战斗部质量15. 5 kg，射程超过100 km，攻击精度达到5 m。Happy 巡飞弹系统由一辆指挥车、三辆发射车和一个油机站组成。每辆发射车可携带18 个发射箱。该系统可多枚发射，同时攻击多个目标，发射后自动控制，无须精确的情报信息，特别适于作战初期使用，可有效地压制和破坏敌防空系统。

恶妇人也是以色列生产的反雷达型巡飞弹，以一台活塞发动机为动力，由螺旋桨推进。战斗部内装有近炸碰炸组合引信。翼展3. 35 m，弹长2. 055 m，起飞重量113. 4 kg，航速250 km/h，升限3 km，航程500 km，续航时间3 ~4 h。

1.1.2　德国的发展

德国为了满足其空军的要求，研制了 DAR 反辐射巡飞弹，该弹全部用复合材料制成，由玻璃纤维壳

---

① 原文发表于《弹药战斗部学术交流会论文集》2005（11）.

② 王玥：工学博士，2002 年师从冯顺山教授，研究巡飞子弹总体技术及对雷达目标毁伤，现工作单位：北京理工大学。

体组成一体化的弹翼和弹身。双冲程双缸汽油活塞发动机，螺旋桨推进，并由一台固体火箭助推器加速到巡航速度。起飞重量110 kg，战斗部13 kg，翼展2 m，弹长2.25 m，俯冲速度360 m/s，航速250 km/h，升限3 km，续航时间3 h。该弹采用被动反辐射导引头，全天候使用。DAR将装在集装箱式发射架中，每个集装箱发射架能装40枚。DAR反辐射巡飞弹系统由巡飞弹、发射架、地面控制站和动力系统组成。

德国的PAD/KDH反坦克型巡飞弹配备红外寻的器，活塞发动机，螺旋桨推进，起飞重量110 kg，折叠机翼，翼展2.26 m，弹长1.18 m，航速140～250 km/h，升限3 000 m。采用容器发射，一个标准的6.10 m的容器内可以装20枚PAD/KDH，此容器可用于储存、运输和发射巡飞弹。

1.1.3　美国的发展

美国的XBQM－106巡飞弹采用无线电指令遥控，活塞发动机螺旋桨推进，翼展3.63 m，弹长3.07 m，起飞重量106.6 kg，巡飞速度167 km/h，升限3.05 km，续航时间5 h。

此外美国还有“勇士”200，活塞发动机螺旋桨推进，翼展2.57 m，弹长2.12 m，航速225 km/h，升限3.96 km，航程大于805 km。

1.2　各国第二代巡飞弹的发展状况

相比早期的巡飞弹，第二代具有更强的自主探测与攻击能力，发射平台也变得灵活而多样化。目前主要有德国和美国进行了相关技术的研究。

1997年德陆军开始研制用于攻击装甲目标及重要固定目标如炮兵阵地、导弹发射场、机场及指挥中心的“台风”巡飞弹，作为炮兵的有力补充。该弹装有惯性导航及GPS系统，可使其自动导向预编程序的目标区域，经过与其他侦查器材的概略侦察后，“台风”可自动对目标进行探测，并按照起飞前的目标种类预编程序对目标进行分类，当所瞄准的目标与所存储的目标一致时，就自主实施攻击。“台风”采用毫米波雷达寻的，螺旋桨推进，起飞重量150 kg，航速200 km/h，升限2.5 km，续航4 h，能从2 500 m的空中攻击目标。

美国研制的低成本自主攻击系统（low－cost autonomous attack system，LOCAAS）可用SUU－64战术弹药布撒器进行布撒，也可以由飞机内部的武器发射架单个发射。或者可以通过多联火箭发射系统（MLRS）或陆军战术导弹系统（ATACMS）进行布撒。捕食者B和全球鹰等无人机也可以成为发射平台。LOCAAS被发射到预定区域上空后，能用激光探测和测距器发现、识别和锁定目标，并在恰当的时机和准确的地点起爆系统装载的多状态渗透性爆炸弹头。它在计算机存储器中存有多种军事目标，可将目标呈现为实时三维图像。自动目标识别处理器将这些图像用来探测和分类目标。低成本自主攻击系统还能穿透装甲，对机动发射架之类的目标实施软压制以及对雷达制导的防空导弹之类的目标进行战术作战。其携带的ATK多模战斗部，可以根据目标的坚硬程度来决定形成聚能射流、爆炸成型弹丸或是破片。LOCAAS内部有自毁装置，当不能发现目标或燃油耗尽及发生故障时便启动自毁装置。在中制导方面，LOCAAS采用GPS/INS导航系统。在飞行过程中飞行器之间可以通过数据链路进行通信。动力方面，LOCAAS采用0.13 kN TDI J45G涡轮喷气发动机。LOCAAS整枚弹的单价在33 000美元以下。LOCAAS尺寸、重量及性能参数如表1所示。

表1　LOCAAS尺寸、重量及性能参数表

弹长/宽/高 /m	翼展 /m	战斗部重 /kg	燃料重 /kg	最大起飞重量 /kg	巡航速度 /(km·h^{-1})	作用高度 /m	搜索区域 /km	射程 /km	续航时间 /min
0.91/0.25/0.18	1.18	7.7	0.45	45.4	370	230	86	185	30

美国在研的另一种巡飞攻击导弹（LAM），弹长1.27 m，弹径178 mm，弹重45.36 kg；巡航速度

200 m/s，最高盘旋速度 70 kg/min，飞行高度 200～225 m。可以在空中盘旋 45 min，搜索并攻击 100 km 距离内的软目标，或向其他精确制导武器及常规武器发送火力呼唤，也可执行战斗毁伤评估任务。在不进行盘旋时，射程可达 280 km。其特点是：采用具有目标自动识别能力的激光雷达导引头，可进行目标识别与定位并为战斗毁伤评估提供覆盖范围分别为直径 150 m 和 500 m 的图像；GPS 与惯性导航系统将保证对固定与运动目标的精确攻击，数字化传输系统将提供目标坐标和战斗毁伤评估数据，并能在导弹飞行中改变任务；具有灵活的杀伤能力，较小的破甲杀伤战斗部既可攻击轻型装甲车，也可攻击软目标；采用助推火箭和微型涡轮喷气发动机。

1.3 巡飞弹的种类

目前，根据发射方式的不同，巡飞弹主要分为两类：一类是单枚的形式，陆基或空基发射；另一类则是子弹形式，以战术导弹或布撒器作为运载工具。

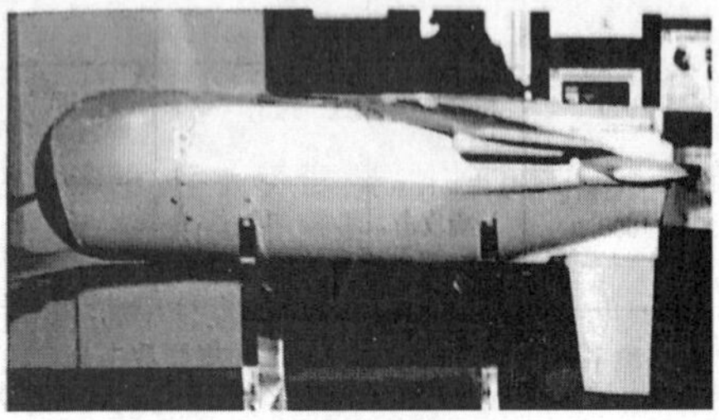

图 1 LOCAAS 样机

图 2 LAM 巡飞攻击导弹

在子弹药的研究方面，洛克希德－马丁公司在 LOCAAS 的基础上进一步将其改进成具有子弹药形式的垂直发射自动攻击系统（VLAAS），由 Mk15 战术弹药布撒器运送发射，以满足海军对水面快速巡逻艇及时间敏感目标如地面装甲部队、移动导弹发射架等目标实施精确打击的要求。VLAAS 发射后，在 15 000～20 000 英尺（1 英尺 = 0.304 8 米）高度将 4 枚子弹形式的 LOCAAS 释放，经过一段时间的滑翔后，LOCAAS 的发动机开始工作。每枚 LOCAAS 有大约 100 海里（1 海里 = 1.852 千米）的巡飞距离，四枚总共可以对 25 平方海里的区域实施搜索。弹与弹之间还具备通信能力以避免同时锁定相同的目标。

2 发展巡飞弹的必要性

2.1 巡飞弹攻击的目标

一般来说，分布相对集中的高价值点目标最能发挥巡飞弹的优势。如雷达阵地、导弹发射阵地、机场、火炮阵地、集群的坦克目标等。对于 50～100 km 较近距离的目标，可以以单发形式陆基或空基发射；对于较远目标，可以以子母弹的形式，用战术导弹或布撒器将若干枚巡飞子弹药运抵目标区域上空，待子弹被抛出后再进行巡飞，搜索预定目标。

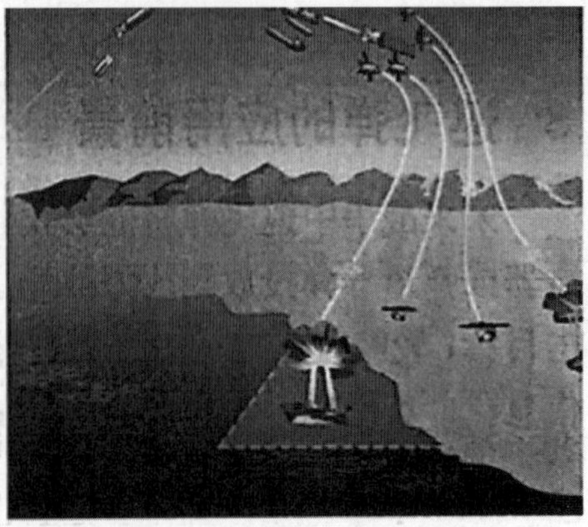

图 3 子母弹形式的 LOCAAS 巡飞弹

2.2 巡飞弹的优势

在射程方面，单枚发射的巡飞弹射程从几十千米到上百千米，而子母战斗部形式的巡飞弹射程则可以达到上千千米。对于这一范围的集群高价值点目标，常规火炮无能为力，而无控子母弹虽然射程上不存在问题，但精度却很差，这从后面的对比中也可以看到。在精度方面，无控子母弹属于覆盖式攻击，而巡飞弹属于精确攻击，后者与前者有本质的区别，精度提高了几十倍。我们以距离我方 300 km 的敌方地空导弹阵地为攻击目标，对三种类型的武器系统进行比较（其中巡飞弹考虑用子母战斗部的形式，与无控子母弹共用平台），结果见表 2。

表 2　巡飞子母弹与其他弹药的比较

弹药种类	射程 /km	命中精度 CEP/m	总价格 /万元	攻击方式	毁伤全部目标所需导弹数	毁伤全部目标的概率/%
无控子母弹	70～1 000（取决于母弹的射程）	70～250	1 800	覆盖式攻击	5	100
巡飞子母弹	70～1 000	6	2 000	精确攻击	1	96.9
巡航导弹	500～1 500	2～15	1 600	精确攻击	10	87.6

通常低空导弹阵地的一个火力单元由指挥控制车、雷达车、天线车、电源车和 6～8 辆发射车组成，共 10～12 个点目标分布在 500 m×500 m 区域内，计算时统一按 10 个目标考虑。无控子母弹母弹装弹量 60 枚；巡飞子母弹母弹装弹量 20 枚。

对于攻击点目标而言，巡飞子母弹具有无控子母弹无可比拟的优势。如果要攻击分布在同样的面积内的多个点目标，前者的攻击效率是后者的 5 倍。虽然巡航导弹具有与巡飞子弹相近的命中精度，但对于集群分布的而非单个的高价值点目标而言，一枚巡航导弹只能攻击其中一个目标，成本过高。

通过以上对比可以看出，对于集群分布的高价值点目标，巡飞子母弹相对于无控子母弹的效费比提高了约 4.5 倍，而相对于适合攻击单个点目标的巡航导弹，效费比提高了约 8 倍。在对特定目标的攻击方面，装备巡飞子母战斗部的弹道导弹具有很高的作战效费比和实用性能。

另外，巡飞弹还有一些以往其他武器所不具备的特点。如果延长巡飞弹药的巡飞时间，便可实现对目标区域上空的封锁和对目标区附近人员及设施的威慑。如果对巡飞弹进行改装，则还可实现实时的作战效能评估。

3　巡飞弹的应用前景

未来战争中巡飞弹将能够集侦察、探测、攻击、战场评估于一身。多枚巡飞弹以子弹形式由布撒器或弹道导弹投放到敌方战场上空，探测目标的同时实现弹与弹之间的通信，自主决定各自的攻击目标，较晚攻击目标的负责进行战场毁伤效能评估，并将信息通过数据链路及时传送到指挥中心。信息化，智能化，这将成为未来战争中灵巧弹药的发展方向。

参考文献

[1] MUNSON K, STREETLY M. Jane's unmanned aerial vehicles and targets [J]. Jane's defense weekly, 2004.

多模式封锁弹对机场跑道封锁效能的分析①

黄龙华②，冯顺山，王震宇

（北京理工大学爆炸科学与技术国家重点实验室，北京　100081）

摘　要：本文提出了使用多模式封锁弹对机场跑道进行封锁的概念，并与传统单纯对跑道进行硬毁伤的方法进行比较，利用数值模拟法近似求出子弹落点坐标，运用矩阵仿真法算出多模式封锁弹的封锁概率，还从封锁时间效率的角度分析了多模式封锁弹的优势。

关键词：子母弹；机场；矩阵仿真法；封锁效能

Study on the Interdiction Effectiveness of Multi－model Interdiction Submunition to Airport Runway

HUANG Longhua，FENG Shunshan，WANG Zhenyu

（State Key Laboratory of Explosion Science and Technology，

Beijing Institute of Technology，Beijing100081，China）

Abstract：In this paper，it is new idea that useing multi－model interdiction submunition to attack airport runway. The end points of interdiction submunition is simulated by numerical，and interdiction probability to one airport runway is calculated by matrix simulation method，and give the particular example. Finally，by comparing the interdiction time，the advantage of multi model interdiction submunition is analysed.

Keywords：submunition；airport；matrix simulation method；interdiction effectiveness.

1　引言

多模式封锁弹对机场跑道进行封锁，就是将两种或两种以上的功能的子母弹组合使用，联合对机场进行攻击，侵彻子母弹、反跑道动能弹等对跑道进行破坏，封锁目标子弹药撒布在跑道等机场上，采用多种引信起爆方式，能对滑行的飞机，修复跑道的人员和机械设备进行攻击。多种功能弹种的联合使用，实现了弹药组合“1＋1＞2”的效应，可有效地提高对机场跑道的封锁效能。为了更清楚地说明问题，文中以某型号的地地战役战术导弹（以下简称“导弹”）为载体，配备侵彻子母弹和封锁目标子母弹两种战斗部，多模式封锁弹就是指这两种子母弹的混合使用，并在考虑空间与时间封锁效率两个评价因素的基础上，运用矩阵仿真法对其封锁效能进行分析，可以更准确地反映出对机场跑道封锁的情况。

2　机场跑道封锁和多模式封锁弹

封锁机场跑道，就是跑道遭到破坏之后，飞机在一定时间内将无法利用这组跑道进行安全升降。封锁含有较强的时间概念。封锁机场跑道成功的准则应该是：机场跑道遭到破坏之后，产生一系列弹坑，在一定时间内，若在这组跑道上不存在供飞机起降的最小升降窗口，封锁就成功，否则，认为失败。最小升降窗口是指固定翼飞机在跑道上安全起降的最小完好矩形区域。不同机种的飞机需要有不同的最小

①　原文发表于《弹箭与制导学报》2006（7）.

②　黄龙华：工学博士，2003年师从冯顺山教授，研究弹药终端毁伤效能评估方法，现工作单位：华东交通大学。

升降窗口，一般为 800 m×20 m、400 m×20 m，现代作战飞机有向短距起降的方向发展的趋势，需要的最小升降窗口的尺寸将更小。

多模弹药封锁弹，将配备两种子母弹战斗部的导弹组合使用，联合对机场跑道攻击。一种是侵彻子母弹战斗部，对跑道进行破坏并形成弹坑；另一种是封锁目标子母弹战斗部，这是一种新概念弹药。封锁目标子母弹的杀伤目标不是跑道，而是飞机、人员和器材等目标。通过抛撒大量子弹药到机场跑道、机库、油库及其附近区域，采用随机延时起爆、感应起爆、声光起爆等多模式引信起爆方式，以预制、半预制破片对目标进行毁伤，图 1 为子弹破片的仿真示意图。该子弹药的破片杀伤范围可达 30 多米，形成一个较大幅员的封锁面积，如图 2、表 1 所示，可对其落点附近的飞机、人员和器材等目标构成极大的威胁。一旦有飞机试图突破封锁，在一定距离的一个或几个子弹药将按照预定起爆方式起爆，其破片对滑行的飞机给予有效杀伤。同时，也给对方探测、排除子弹药以及清理、修复跑道等工作造成障碍，能保持或延长多模式封锁弹对机场跑道形成的封锁状态。

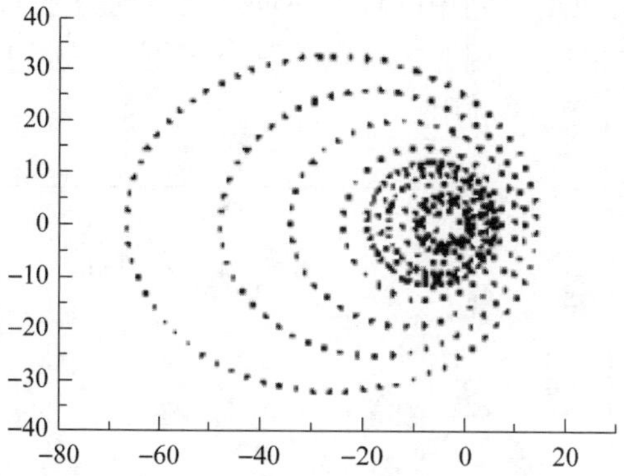

图 1　子弹破片的仿真示意图

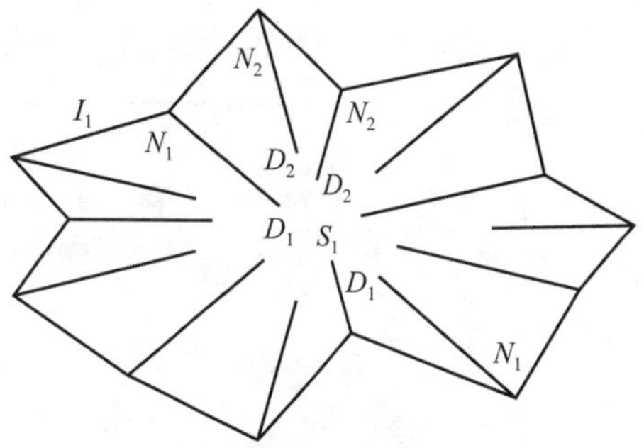

图 2　战斗部封锁幅员示意图

表 1　某战斗部的杀伤面积与封锁面积、封锁幅员

倾角/(°)	炸高/m					
	6			10		
	S_L	S_b	S_e	S_L	S_b	S_e
0	233.3	3 803.1	10 103	120.2	334.4	780.3
15	224.6	2 074.2	6 153.9	210.5	1 487.3	3 506.9
30	231.7	1 627.7	4 034.2	223.8	1 228.9	3 430.4

3　两种突击方式对机场跑道封锁效能的比较

当前，对机场跑道封锁的突击方式可分为以下两种：一种是单纯对机场跑道破坏。如使用侵彻子母弹、穿爆弹或反跑道动能炸弹等弹药在跑道上炸出一系列弹坑，使得跑道上不存在飞机所需的最小升降窗口，即认为封锁成功。另一种是多模式封锁弹对机场的攻击。即将多种功能的子母弹混合使用，不仅对跑道进行破坏，而且抛撒大量的封锁目标子弹药，采用多种起爆引信模式，对机场的飞机、人员和器材进行毁伤或构成威胁，使跑道也不存在最小升降窗口，并阻碍对方采取反封锁措施，即认为封锁成功。

为了更好地进行比较，先做几个假设：一是对同一个机场跑道进行封锁，并达到相同的封锁程度；二是两种子母弹的子弹数和子弹抛撒规律等情况相一致；三是封锁目标子弹没有起爆前，可将其视为“弹坑”来计算跑道上是否存在最小升降窗口。

一是所需兵力不同。机场跑道一般长 2 000 ~ 3 500 m，宽 40 ~ 80 m。国内武器系统在制导精度方面还存在不足，如导弹的射程为100～800 km，CEP却为150～200 m。子弹数为100～260枚，一枚侵彻子弹形成的弹坑为4.0 m×2.0 m，一枚封锁目标子弹的杀伤面积200 m^2左右，封锁目标子弹对飞机等目标构成的威胁明显更大[2]。经试验数据得出，一枚导弹在有效射程内正常抛撒子弹，约有20%的子弹能落入跑道上并造成破坏，假设落入跑道上的子弹数为80枚。如使用两枚导弹进行攻击，在第一种方式下，侵彻子弹在跑道上炸成160个弹坑；在第二种方式下，多模式封锁弹形成相同的“弹坑”分布，其中侵彻子弹炸成80个弹坑，撒布在跑道上的封锁目标子弹有80枚，可视为“弹坑”，如图3、图4所示。因此，对同样一条跑道进行封锁，第一种方式需要消耗更多的导弹。二是封锁的效果不同。第一种方式子母弹通常是采用触发或侵入跑道等方式控制起爆，侵彻子弹在跑道上形成炸成弹坑后，不再对飞机、人员等构成威胁。在没有安全威胁的情况下，对方利用现代工程机械能在1 h之内迅速修复被破坏的跑道。在第二种方式下，封锁目标子弹不仅能视为“弹坑”，而且因为采用了多种起爆模式的引信，能随机地对目标进行攻击，对飞机、人员等构成较大的安全威胁，给对方造成较大的心理压力，迟滞对方修复跑道的行动。因此，其封锁的效果更好。三是对机场再利用的程度不同。使相同数量的导弹进行攻击，第二种方式对跑道的破坏程度明显比第一种方式的小，因此，在第二种方式下，当我方夺取对方机场后，因跑道保持较好的完整性，更有利于我方对机场的再利用。

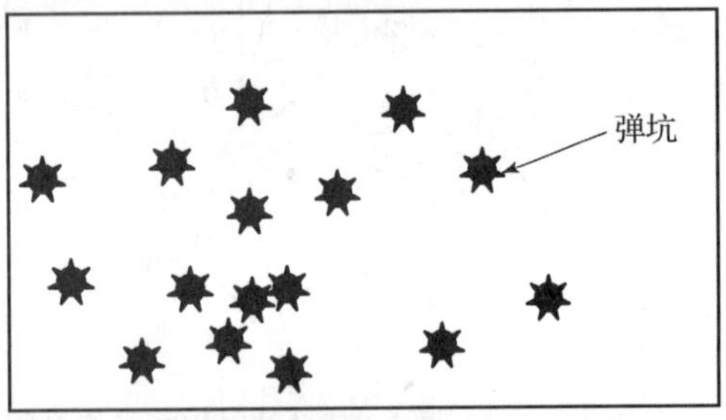

图3　单纯侵彻子母弹攻击跑道后示意图

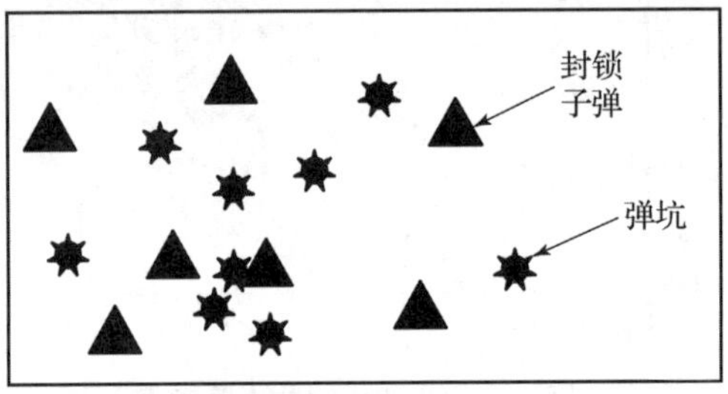

图4　多模式封锁弹攻击跑道后示意图

4　多模式封锁弹对机场跑道的封锁效能

多模式封锁弹对机场跑道的封锁效能应从两个角度来分析，一是从空间封锁概率[4]。将抛撒在跑道上的所有封锁目标子弹也视为“弹坑”，计算跑道上是否存在最小升降窗口，来判断封锁成功与否。二是从时间封锁效率。当多模式封锁弹对跑道的空间封锁能够满足预定的要求时，能将这种封锁状态延续多长时间就取决于子弹坑的数量、对方修复跑道的能力和排除封锁目标子弹的所需时间，从两个角度分析得出的结果更贴近实战情况。

4.1　从空间封锁概率的角度来分析

子弹随机散布和子弹毁伤面积的不规则性，不可避免地带来了子弹毁伤幅员的重叠。如何有效地处理好子弹毁伤幅员重叠问题是计算子母弹对跑道封锁概率的关键。有文献提出了像素—仿真法、矩阵仿真法[3,5]来解决这一问题，效果较好。矩阵仿真法的计算精度更高，并更适合于仿真计算机场等超大型目标。文中采用数值模拟法来模拟多模式封锁目标子弹落点，用矩阵仿真法计算其对机场跑道的封锁概率（图5）。

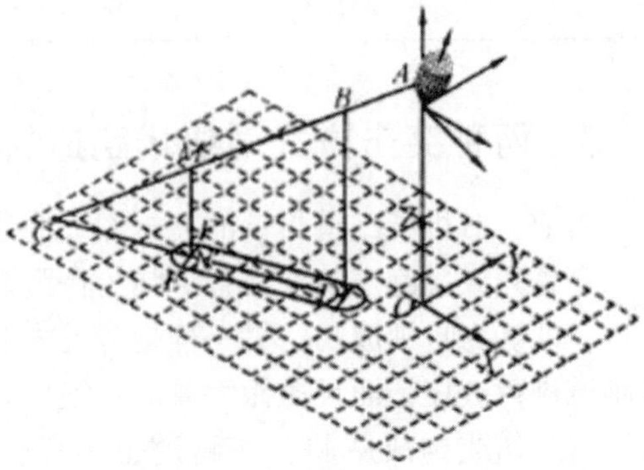

图5　矩阵仿真法原理图

4.1.1　数值模拟多模式封锁弹药落点的具体步骤

（1）建立地面坐标系。坐标系的原点在跑道几何中心 O。OX 轴沿跑道纵向，向右为正，OY 轴过原点垂直于 OX 轴，向上为正，OZ 轴与 OX、OY 轴构成右手直角坐标系。

(2) 模拟母弹抛撒点坐标。

$$\begin{cases} X = X_m + z_1 \cdot \text{CEP}/1.177\,4 \\ Z = Z_m + z_2 \cdot \text{CEP}/1.177\,4 \end{cases} \tag{1}$$

式中　X,Z——母弹抛撒点坐标；

X_m,Z_m——母弹抛撒点坐标的理论值；

CEP——导弹的圆概率偏差；

z_1、z_2——服从标准正态分布的随机数。

(3) 模拟抛撒圆环内子弹的落点坐标。假设子弹在抛撒圆环内服从均匀分布。

$$\begin{cases} r = r_P + (R_P - r_P) \cdot \eta_1 \\ \varphi = 2\pi \cdot \eta_2 \end{cases} \tag{2}$$

$$\begin{cases} x_{zd} = r \cdot \cos\phi \\ z_{zd} = r \cdot \sin\varphi \end{cases} \tag{3}$$

其中，抛撒圆环内外半径均为服从正态分布的随机变量；为（0，1）区间内均匀分布的随机数；x_{zd}、z_{zd}为子弹相对母弹抛撒点的相对坐标。由母弹抛撒点坐标和子弹相对母弹抛撒点位置可求出子弹在抛撒圆环内的落点坐标（x_t，z_t）为

$$\begin{cases} x_t = X + x_{zd} \\ z_t = Z + z_{zd} \end{cases} \tag{4}$$

4.1.2　矩阵仿真法计算多模式封锁弹对机场跑道的封锁概率[7-9]步骤

(1) 对机场跑道划分网格，同时建立一个大型矩阵，矩阵内的元素与网格数一一对应，每一个网格定义为一个矩阵元素，对矩阵元素进行初始化，赋值为0。

(2) 根据实际情况定义机场跑道上飞机的最小升降窗口。

(3) 将命中机场跑道的子弹的毁伤面积等效为面积相等的网格，将子弹毁伤面积覆盖的矩阵元素赋值为1。

(4) 用飞机的最小升降窗口来判定跑道上是否存在一块能供飞机正常升降的矩形区域，如果跑道上找不到这样的矩形区域，则该次模拟的封锁成功，否则，封锁失败。

用以上方法对机场跑道封锁情况模拟 N 次，将模拟中成功封锁的次数记为 N_c 次，则多模式封锁目标子弹对机场跑道的封锁概率 $P_f = N_c/N$。

4.2　从时间封锁效率的角度来分析

当多模式封锁弹实施攻击后，其维持封锁状态的时间取决于对方反封锁的能力，对方反封锁一般包括以下四部分的内容：对机场跑道损毁情况的判定；快速确定应急跑道修复方案；排弹分队在确定的修复区域及其周围一定距离内排除封锁目标子弹；跑道抢修分队迅速填补弹坑和修复跑道。这个反封锁过程所需要的时间 T_ω：

$$T_\omega = T_q + T_d + T_p + T_x \tag{5}$$

式中　T_q——机场跑道损毁情况判定需要的时间；

T_d——确定应急跑道修复方案需要的时间；

T_p——排弹分队在确定修复区域完成任务所需的作业时间；

T_x——跑道抢修分队在修复区域修复跑道所需的时间。

T_q、T_d、T_x 三个所需时间都比较好确定。其中 T_q、T_d 可取一个均值来表示。跑道抢修分队填补一个弹坑有标准时间，因此，当知道修复区域内有多少个弹坑，假设有 m 个，也就可以确定 $T_x = m \times T_k$。排除封锁目标子弹是影响整个过程的关键因素。因为封锁目标子弹采用了多模式引信控制起爆，且具有较大的杀伤威力，通过实际毁伤和心理威胁的双重作用，能有效延缓排弹的作业速度，甚至迫使这项工作

无法进行，从而达到封锁的目的。时间 T_p 的求取是一个比较复杂的过程，需要建设相应的数学模型来计算，文献［8］对此做了详细的分析。为了简化问题，假设排除一枚封锁目标子弹所需的时间为 T_f，可求出时间 T_p。最后，可得到 T_ω。

在实战中，当使用导弹对某一机场跑道进行锁时，通常要根据战场形势确定封锁机场的预期时间，记为 T_a。那么，多模式封锁弹对机场跑道的时间封锁效率 Q_f 可表示为

$$Q_f=\begin{cases}\dfrac{T_\omega}{T_a} & T_\omega<T_a\\ 1 & T_\omega\geqslant T_a\end{cases} \tag{6}$$

5 算例分析

计算的假设条件是：跑道为 2 500 m×50 m，最小升降窗口为 400 m×20 m，某型导弹的 CEP 为 100 m，配备侵彻子母弹和封锁子母弹两种战斗部，子弹数为 120 枚，子弹抛撒圆环内外半径的理论值分别是 30 m、60 m，侵彻子弹形成的弹坑为 4.0 m×2.0 m，封锁目标子弹采用随机延时引信，有效杀伤半径为 20 m，跑道损毁情况判定的时间为 10 min，确定应急跑道修复方案的时间为 5 min，排弹分队和跑道抢修分队各有 10 个小组，每个小组排除一枚封锁目标子弹所需的时间为 6 min，填补一个弹坑的标准时间为 2 min，确定修复区域内分布有 60 个弹坑或 30 个弹坑、30 枚封锁目标子弹。分别计算多模式封锁弹对机场跑道的封锁概率和两种突击方式对跑道的封锁效果，结果如表 2、表 3 所示。

表 2　多模式封锁弹对机场跑道的封锁概率

母弹数量/枚	3	4	5	6
封锁概率/%	34.27	83.54	94.36	98.64

表 3　不同突击方式对机场跑道的封锁效果

突击方式	损毁判定/min	确定方案/min	排除子弹/min	修复跑道/min	总时间/min
单一侵彻子弹	10	5	0	12	27
多模式封锁弹	10	5	18	6	39

通过计算可得出如下结论：

（1）使用多模式封锁弹对机场跑道进行封锁，不仅能够有效地提升封锁效果，提高弹药的费效比，而且能维持和延长封锁的状态，给对方的飞机、人员和器材造成较大的威胁。因此，利用多模式封锁弹的综合作战能力对机场跑道进行封锁将是未来的发展趋势。

（2）利用数值模拟法来模拟子弹的落点可以减少计算时间，矩阵仿真法能较好地解决子弹毁伤面积的重叠问题，而且巧妙地解决了像素—仿真法因受显示屏技术的限制不便于分析大型目标的难题，精度也有所提高，能较好地分析子母弹对机场跑道的封锁概率。但是，在判定跑道上是否存在最小升降窗口的过程仍比较复杂，还可对其进行算法优化，以提高运算效率。

（3）从空间和时间两个角度来分析多模式封锁弹对跑道封锁，能够得出更合理的结论，如果封锁目标子弹采用更多的毁伤机理、更灵活的引信起爆方式，增加子弹的智能化程度，增大排除子弹的难度，就能更加有效地提高其封锁效能。

参 考 文 献

［1］寇保华，杨涛，张晓今，等．末修子母弹对机场跑道封锁概率的计算［J］．弹道学报，2005（4）：22－26，49.

子弹协同弹道规划问题的效率分析[1]

王　玥[2]，冯顺山，曹红松，李国杰，王云峰

（北京理工大学爆炸科学与技术国家重点实验室，北京　100081）

摘　要：为了能使多枚子弹在巡飞过程中相互配合并对目标区域实现全面的探测，必须对其协同弹道进行规划。运用了遗传算法原理，结合具体问题编制了相应的算法程序，对多枚子弹的协同弹道进行了全局的优化设计，并通过对优化结果的分析对比提出了适用于此特定多目标优化问题的分时分目标的优化方法。

关键词：巡飞；协同弹道；遗传算法；全局优化

Efficiency Analysis on Coordinated Trajectory Planning of Submunition

Wang Yue，Feng Shunshan，Cao Hongsong，Li Guojie，Wang Yunfeng

（State Key Laboratory of Explosion Science and Technology，Being Institute of Technology，Being100081，China）

Abstract：In order to make the several submunition cooperated with each other and give the destination area a full detection，a coordinated trajectory planning is needed. We use genetic algorithm theory and associating with the practical problem，we program corresponding code and do overall optimization design to the several submunition coordinated trajectory and through analysis and comparison to the optimization result we put forward a time – sharing object optimization method which is suitable for the specific multi – object optimization.

Keywords：loitering；coordinated trajectory；genetic algorithm；overall optimization

1　引言

对于多目标优化问题，一般多个目标函数需要同时进行优化。但这样多个目标便形成了多维的搜索空间，使得搜索量急剧增大，算法的效率因此变得很低，经常难以在规定的种群大小和进化代数范围内找到满意的结果。文中采用了分时分目标的优化方法，在计算的初期只对关注程度较高的目标进行优化，经过一定世代数的进化，种群中已有一部分个体分布于满意解空间，在此基础上再对所有目标进行同时优化，能使算法在满意解附近进行搜索，在一定程度上减少了搜索的范围，提高了算法效率。

2　问题描述

母弹携带的多枚具有巡飞功能的子弹被抛撒到预定区域上空后，能够在一定范围内实现巡弋飞行，同时自主探测、识别、锁定和攻击预定点目标。事实上各子弹不能瞬时发现拟攻击的点目标，需要通过一段时间的搜索才会发现。为了能够达到较好的探测效果，各子弹之间需要和指挥弹进行必要的通信并由指挥弹下达指令，确定最优的协同巡飞弹道，从而实现子弹分区域搜索、最优分配攻击各个目标的任务。

子弹群被抛撒到预定区域后，指挥弹首先通过 GPS 获得自己的位置信息，并进一步根据抛撒点与目标区的距离、目标区面积及子弹个数等信息决策各枚子弹的巡飞路径。因此，必须预先规划好在不同的

---

① 原文发表于《弹箭与制导学报》2006（Mar）。

② 王玥：工学博士，2002 年师从冯顺山教授，研究巡飞子弹总体技术及对雷达目标毁伤，现工作单位：北京理工大学。

距离、面积及子弹个数的情况下多枚子弹的协同巡飞弹道，为指挥弹的决策提供依据。这也正是文中所要研究的主要问题。

如图1所示，假定目标区为一个以点 $B(x_B,\ y_B)$ 为圆心，半径为 R 的圆形区域 M，子弹群的抛撒点为 $A(x_A,\ y_A)$，则 A、B 点坐标皆为已知参数，A 与 M 的位置关系完全确定。子弹飞行时其高度不变，某一时刻在地面探测到的面积为半径为 r 的圆形区域；各枚子弹的图1子弹与目标区最大飞行距离 L 相同。

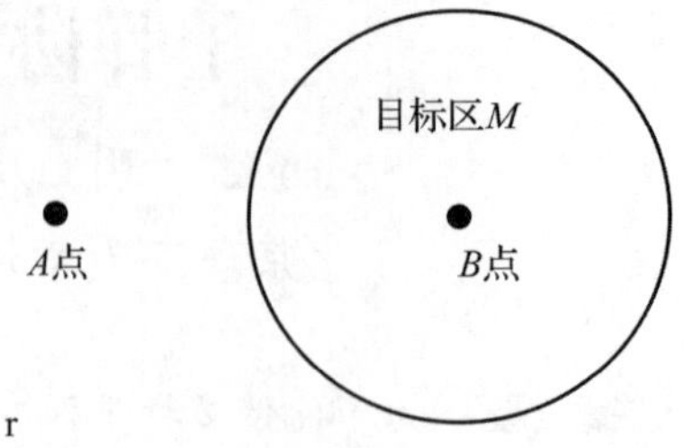

图1　子弹与目标区位置关系图

子弹协同弹道的规划要满足如下两个方面的条件：

（1）弹道的转弯半径大于子弹的最小转弯半径 R_{turn}。

（2）扫描范围优先覆盖区域 M，并且面积最大。

对于条件（1），因为会转化为对编码的限制，因此不必对其进行优化。而对于条件（2），其含义的算法解释为：

①各枚子弹飞行路径上的坐标点优先分布于 B 为圆心半径为 $R—r$ 的区域内；

②各子弹路径间重叠与交叉最少。

分析可知，这属于多目标优化问题，多个目标函数需要同时进行优化。由于目标之间的无法比较和矛盾现象，导致不一定存在在所有目标上都是最优的解[2]，因此，最终只要得到能满足要求的可行解即可。文中的算法在实际操作过程中，为了减少搜索空间，加快搜索速度，采用了分时分目标的优化方式，即在初始阶段，为了让解更多集中于期望的圆域内，只对①目标进行优化；而在执行一定的循环后，再对①、②目标共同实行优化。通过控制初始阶段单目标优化的世代数占总世代数的比例，能有效地加快解向期望值的收敛速度。

3　巡飞弹道规划方法

3.1　路径编码

由于路径的总长度 L 为定值，在满足精度的条件下，考虑将路径离散成若干段等长 L_{A} 的首尾相接的链路。这样可以减少计算时对这一参数的操作，能有效地降低计算量。将后一段链路相对于前一段转过的角度值 α 作为遗传操作的表现型进行编码。为了简化编、解码操作，便于实现交叉、变异，也考虑到搜索能力强等因素，文中选用二进制的编码方式。需要说明的是，当单段链路的长度固定后，约束条件 R_{turn} 就能转化为对角度值 α 的限制，从而缩小了可行解的搜索空间。

3.2　路径个体适应度函数的构造

路径个体的适应度函数采用罚函数的形式。对于不满足约束条件的部分路径施加一个适当的惩罚量值，以驱动其向着约束条件定义的空间进化。

适应度函数形式如下：

$$\text{Fitness} = k_1 \times \left(\frac{1}{2}\right)^{a \times D} + c + k_2 \times \operatorname{sech}(b \times R)$$

式中　k_1 和 k_2——两个约束罚函数的比例系数；

c——常数调整因子。

$R = [\overline{P_0B}\quad \overline{P_1B}\quad \cdots\quad \overline{P_nB}]$，表示路径上各点到目标区中心的距离的罚函数，$b$ 为其比例系数。

$D = \begin{pmatrix} D_{12} & D_{13} & \cdots & D_{1n} \\ & D_{23} & \cdots & D_{2n} \\ & & \ddots & \vdots \\ & & & D_{(n-1)n} \end{pmatrix}$，表示路径之间的距离的罚函数，$D_{(n-1)n}$ 表示第 $n-1$ 和第 n 个子弹路

径上各点之间的距离，a 为其比例系数。

3.3 路径的遗传操作

选择算子采用轮盘赌算法，其基本思想是先计算个体的相对适应度值。

$F_i = \dfrac{f_i}{\sum_{i=1}^{n} f_i}$，其中 f_i 是第 i 个个体的适应度值，然后生成［0，1］之间的随机数 r，如果 $\sum_{k=1}^{i-1} F_k < r < \sum_{k=1}^{i} F_k$ 则选择个体 i。在这种选择方式下个体的适应度值越大，被选中的机会越多，从而其基因结构被遗传到下一代的可能性越大[3]。另外，这种策略使得种群中的个体都有被选中的概率，保证了较小的个体有较大的适应度值。

交叉算子采用均匀式交叉，变异算子采用单点式变异。

4 仿真结果及分析

文中仅就两枚子弹及两个优化目标的情况进行了仿真计算，算法同样适用于多枚子弹及多目标优化的情况。计算时各参数的取值如下：$A(0,0)$，$B(800,0)$，$R=500$ m，$r=150$ m，$R_{\text{turn}}=300$ m，$L=2\ 010$ m，$L_A=67$ m。

采用多目标同时优化方法时两枚子弹协同弹道优化结果及效果如图 2 和图 3 所示。

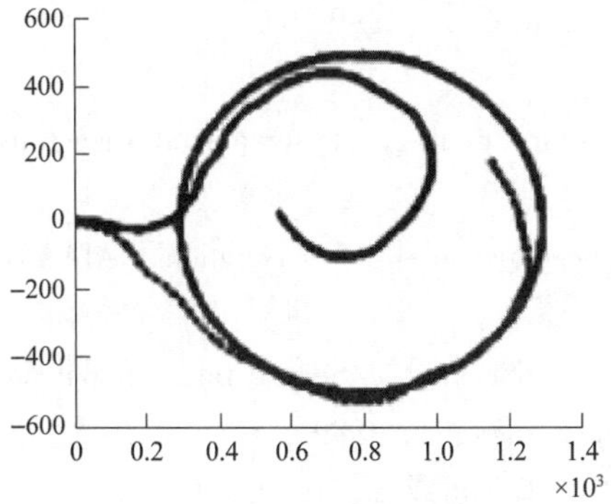

图 2 采用多目标同时优化方法时两枚子弹协同弹道优化结果

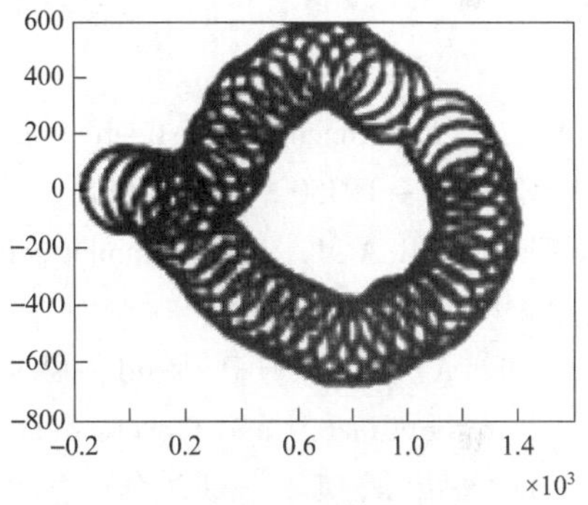

图 3 采用多目标同时优化方法时两枚子弹协同弹道优化效果

交叉概率 0.7，变异概率 0.05，种群数量 200，终止代数 150。

采用分时分目标优化方法时两枚子弹协同弹道优化结果及效果如图 4 和图 5 所示。

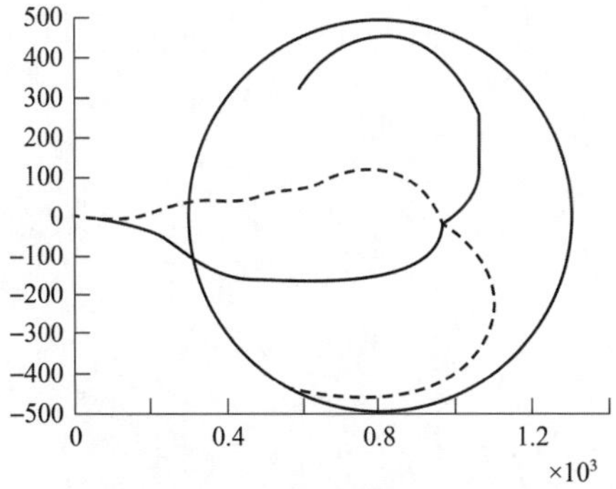

图 4 采用分时分目标同时优化方法时两枚子弹协同弹道优化结果

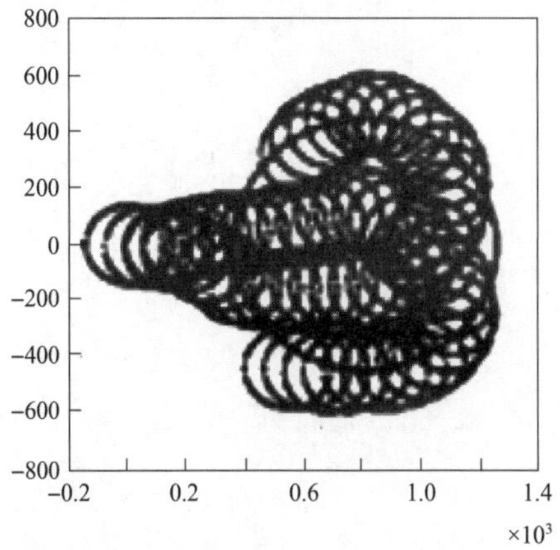

图 5 采用分时分目标同时优化方法时两枚子弹协同弹道优化效果

交叉概率0.7，变异概率0.05，种群数量200，目标优化终止代数40，①、②目标共同优化终止代数150。

图2和图4中，两枚子弹的起点为$A(0,0)$，实线和虚线分别代表计算所得的两条巡飞路径，圆形区域代表目标区。图3和图5中分别显示了优化的效果，被重叠的圆环所覆盖的部分代表能够扫描到的区域。经计算，在图3中，未扫描面积占目标区总面积的17.2%，而在图5中，未扫描面积仅占目标区总面积的1.8%。将分时分目标的优化方法应用于多目标的优化问题，在同样的进化代数下，搜索到的解明显优于多目标同时优化方法。由以上仿真结果及分析可知，分时分目标的优化方法显著地提高了算法效率，也有效地解决了子弹对目标区进行全面扫描的协同弹道规划问题。

5 结束语

文中提出了一种基于遗传算法在巡飞探测阶段多枚子弹协同弹道的规划方法，编制了相应的程序，实现了对指定全域最优搜索性探测，也为后续的子母战斗部作战效能的评估提供了依据。对于此类多目标优化问题，采用了分时分目标的优化方式，有效地加快了解向期望值的收敛速度。

参考文献

[1] 王小平，曹立明．遗传算法——理论、应用与软件实现［M］．西安：西安交通大学出版社，2002.

[2] 玄光男，程润伟．遗传算法与工程优化［M］．于歆杰，周根贵，译．北京：清华大学出版社，2004.

[3] 周明，孙树栋．遗传算法原理及其应用［M］．北京：国防工业出版社，1999.

[4]

[5] EMAMI S. Experimental investigation of inlet—combustor isolators for a dual - mode scramjet at a mach number of 4: NASA - 95 - TP3502 [R]. 1995.

[6] HAGENMAIER A M, et al. Scramjet solator modeling using reduced navier - stokes equations: AIAA97 - 3883 [R]. 1997.

[7] GODBOLE A, WYPYCH P. Pseudo - shock in supersonic injection feeder [C] //Second International Conference on CFD in the Minerals and Process Industried CSIRO, Melbourne, Australia, 1999.

[8] 赵鹤书，潘杰元，等．飞机进气道气动原理［M］．北京：国防工业出版社，1989.

[9] MAHONEY J J. Inlet for supersonic missile [M]. Washington, DC: AIAA Inc, 1990.

[10] CURRAN E T, MURTHV S N B. Scramiet propulsion [M]. Washington, DC: AIAA Inc, 2000.

[11] HEISER W H, PRAT D T. Hypersonic airbreathing propulsion [M]. Washington, DC: AIAA Inc, 1994.

封锁型子母弹对飞机从机场跑道强行起降的封锁效能分析①

黄龙华[1②]，冯顺山[1]，樊桂印[2]
（1. 北京理工大学爆炸科学与技术国家重点实验室，北京 100081；
2. 南昌陆军学院，南昌 330103）

摘 要：本文以飞机从跑道强行起降为研究对象。通过数值模拟子弹的随机落点和矩阵仿真法计算破片对目标毁伤等方法。分析了封锁型子母弹的封锁效能。计算结果表明，封锁型子母弹具有很强的封锁作战能力。同时也受到导弹射击精确度、子弹抛撒半径等因素影响。该计算方法对进一步研究封锁型子母弹对其他面目标的封锁效能评估具有重要的参考价值。

关键词：子母弹；子弹散布；飞机；封锁效能；数值模拟

Analysis on the Interdiction Effectiveness of Interdiction - target Submunition on Airplane Force to Fly or Land from Airport Runway

Huang Longhua[1], Feng Shunshan[1], Fan Guiyin[2]
(1. State Key Laboratory of Explosion Science and Technology, BIT, Beijing 100081, China;
2. Nanchang Army College, Nanchang 330103, China)

Abstract: Focused on the airplane force to fly or land from airport runway, the end points of interdiction submunition was simulated, and interdiction probability to one airplane was calculated by matrix simulation method, the particular examples were simulated, it confirmed that the interdiction - target submunition is feasible and valid for blanking off surface target. but it can be affected by other factors such as the CEP and the radius of submunition dispersion. The calculative result shows the analysis is feasible.

Keywords: submunition; submunition dispersion; airplane; interdiction effectiveness; numerical simulation

1 引言

封锁型子母弹是一种先进的新型弹药。与传统封锁弹药相比，封锁型子母弹具有子弹随机反抛空中起爆、杀伤威力大、难排除、封锁效率高等特点，它能够对机场、港口和交通枢纽等大型复杂面目标进行迅速、有效的封锁，同时，其对目标的封锁效能与母弹的射击精度、子弹的抛撒方式、子弹数量和引信模式等诸多因素有关。能否研制新型弹药对强行起降的飞机进行有效毁伤，实现封锁机场的目的就成为当前研究封锁机场作战的一个新热点。文中主要以军用固定翼飞机（以下简称“飞机”）从机场跑道强行起降的情况为研究对象，通过数值模拟和矩阵仿真法等方法分析封锁子母弹的封锁效能，并计算了各种影响因素带来的不同结果。

---

① 原文发表于《弹箭与制导学报》2008.

② 黄龙华：工学博士，2003 年师从冯顺山教授，研究弹药终端毁伤效能评估方法，现工作单位：华东交通大学。

2 封锁型子母弹对飞机强行起降进行封锁的过程

封锁型子母弹到达机场跑道上空后，在预定的解爆高度上以一定的抛撒半径将子弹抛出，子弹在下落的过程中，如遇到混凝土质的跑道时就散布在其上面；如遇到跑道周围的土壤时就能以近似垂直的角度侵入土壤中。当飞机试图从跑道强行起飞或着陆时，子弹可配用多模式引信，以随机延时、感应和触动等方式使子弹反抛至一定高度的空中起爆，以高速破片对经过其附近的飞机进行有效毁伤，如图 1、图 2 所示。

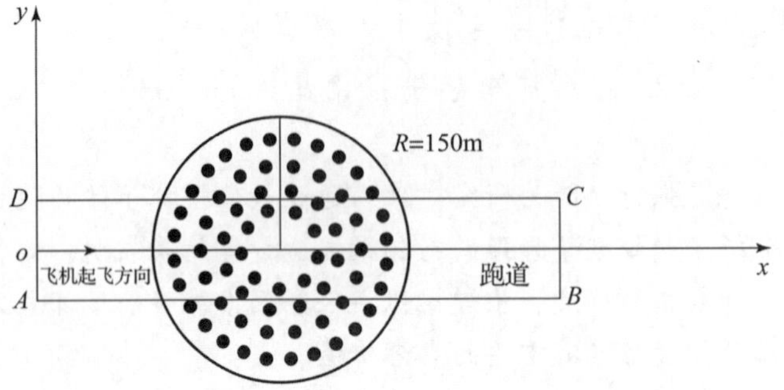

图 1　单枚封锁子母弹在圆散布模式下子弹落点示意图

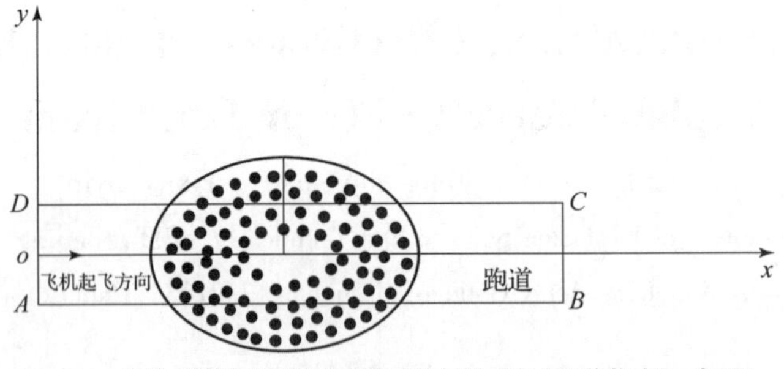

图 2　单枚封锁子母弹在椭圆散布模式下子弹落点示意图

由于子弹空中起爆，破片杀伤威力大，对几十米外的目标能构成极大的威胁，如英国 HB876 区域子地雷产生的破片，可击穿 20 m 处的装甲和 50 m 处的铝合金[1]，表 1 是某封锁型子弹的杀伤威力；子弹能侵入跑道周围的土壤中，具有隐蔽性强；多种模式引信起爆，让对方防不胜防，心理威慑效果明显。特别是将封锁型子弹与侵彻跑道的子弹结合使用，即在侵彻子弹对跑道进行"硬毁伤后"，将封锁型子弹散布到跑道周围，能更有效地对跑道进行封锁和对强行起降的飞机进行毁伤，实现"1 + 1 > 2"的效果。

表 1　某型封锁型子弹的杀伤面积与封锁面积、封锁幅员

倾角/(°)	炸高/m					
	6			10		
	S_L	S_b	S_e	S_L	S_b	S_e
0	230	3 503	10 050	110	300	760
15	210	2 100	6 000	200	1 300	3 100
30	220	1 500	4 000	221	1 200	3 300

注：表中 S_L 为杀伤面积，S_b 为封锁面积，S_e 为封锁幅员。

3　封锁型子母弹对飞机强行起降过程进行毁伤的分析

封锁型子弹对跑道上的飞机能构成有效毁伤的情形有三种：停放在跑道上、强行从跑道上起飞和着陆。文中主要研究后两种情形，首先来分析飞机强行起飞或着陆过程。强行起飞过程中，飞机在推力作用下滑跑加速，当速度增加到离地速度时，飞机离地。从在跑道上开始滑跑到飞机离地所经过的距离叫起飞滑跑距离。离地后，飞机转入小角度上升和继续增速，并收上起落架；然后平飞或小上升率上升增速，并收起襟翼；最后上升到安全高度（图3）。

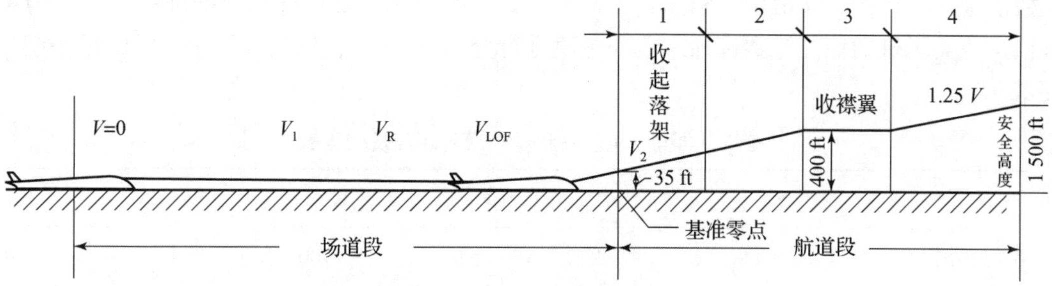

图3　飞机起飞剖面示意图

图中 $V=0$ 为起飞滑跑开始点速度；V_{LOF}表示到了起飞离地速度，$V_{LOF}=1.1\ V_S$；V_S 为在给定重力下，飞机定常平飞的最小理论速度，即该飞机的失速速度。

从图3中可以看出，当飞机试图强行起飞时，不仅在起飞滑跑距离上可能被子弹毁伤，而且离地后至开始收起落架的过程中，因其距地面高度仅有10.67 m，仍在封锁型子弹的有效杀伤范围内。

飞机强行着陆过程包括着陆进场和着陆两个阶段，飞机着陆是一种减速运动。着陆进场阶段开始于从放下起落架，结束于飞机到达跑道头；高度到达着陆安全高度15.24 m，速度不小于进场最小速度（>1.3 Vs）。飞机从跑道头下降，经过拉平（改平飞）接地，直到速度减小到滑行速度或停止滑跑，这一阶段叫着陆。飞机着陆所经过行程称为着陆距离，如图4所示。

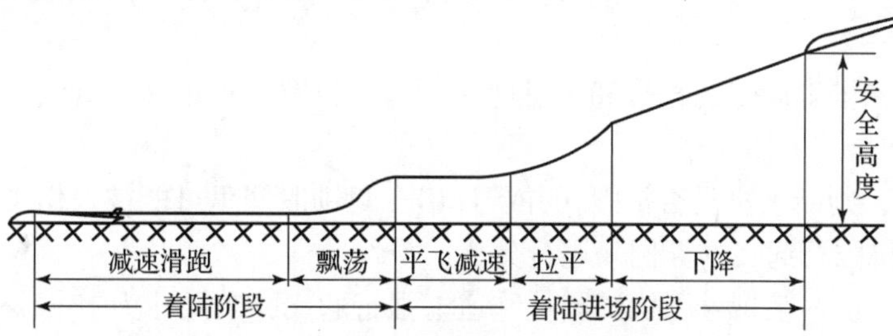

图4　飞机着陆剖面示意图

从图4中也可看出，飞机强行着陆过程中的两个阶段都处于封锁子弹的杀伤范围内。因此，封锁型子母弹非常适合用于杀伤试图强行起降的飞机。

飞机强行起飞和着陆的滑跑距离可通过以下公式求得

$$l_R=\frac{V_{LOF}^2}{2a}=\frac{V_{LOF}^2}{2g(P/W-f')} \tag{1}$$

$$l_{HR}=\frac{V_{JD}^2}{g(1/K+f)} \tag{2}$$

式中　l_R——起飞滑跑距离；

V_{LOF}——起飞离地速度；

a——飞机平均加速度；

g——重力加速度；

P——飞机推力；

W——飞机重力；

f'——起飞换算摩擦因数；

l_{HR}——着陆滑跑距离；

V_{JD}——飞机接地速度；

K——飞机接地时的升阻比；

f——跑道表面与机轮之间的滚动摩擦系数。

由于飞机的起飞和着陆的滑跑距离受飞机重量、跑道情况、机场自然环境等诸多因素影响，不同情况下，其数值是不一样的，实战中可根据具体情况按式（1），式（2）进行较准确的计算。为了简化计算，也可根据各类飞机的性能参数选择一组理想值进行分析（表2），取某型飞机所需的最小起降带尺寸为600 m×30 m。

表2　典型飞机起飞和着陆滑跑距离及轮距表

飞机型号	F-14	F-15	F-16	F-111	B-52G	幻影2000
起飞滑跑距离/m	214	274	533	900	3 050	460
着陆滑跑距离/m	488	869	808	900	3 050	646
主轮距/m	5	2.75	2.36	2.74	2.74	3.4
前主轮距/m	7.02	5.42	4	15.24	15.24	—

4　封锁型子母弹对飞机强行起降的封锁效能

封锁型子母弹对飞机强行起降的封锁效能，可通过以下步骤进行计算：

（1）运用数值模拟法对封锁型子弹的随机落点进行模拟仿真，具体方法可参见文献［2］；

（2）根据封锁型子母弹可能造成的封锁时间，计算飞机在此时间里可能从最小起降带强行起飞或着陆的架数；

（3）根据在某一时刻子弹起爆的枚数和飞机此时所处的位置，用矩阵仿真法[3]来计算其产生的破片对飞机的毁伤情况；

（4）根据毁伤准则，判断飞机是否被毁伤并统计出在封锁时间里可能被毁伤飞机的总数。

下面，通过一个飞机强行起飞的算例来进行分析。

计算的基本条件与假设：某型号飞机试图从跑道上强行起飞，其尺寸为10 m×1.4 m×2.1 m（长×宽×高），飞机机腹高度为2 m，比动能杀伤标准为800 J/cm^2，飞机起飞所需要的平均时间为23 s，飞机所需的最小起降带尺寸为600 m×30 m；封锁型子母弹的CEP=0、50、100、150时，抛撒半径为150 m，子弹数为100枚、150枚、200枚，子弹引信采用随机延时，子弹落点为圆散布；机场只有一条跑道，其尺寸为3 600 m×45 m。为了计算简化，可依据飞机所需的最小起降尺寸，取跑道的600 m×45 m区域进行研究，见图1。

（1）假设在$t_1=0$时刻，第一架飞机的质心正好位于跑道端点O处开始滑行准备强行起飞，其他飞机为排队等候起飞方式，即只有前一架飞机升空后，后一架飞机才正好来到O点准备滑跑强行起飞，每一架飞机滑跑起飞的时间均为23 s[4]；

（2）假设每枚母弹中子弹的随机延时期间为0～50 min，在此封锁时间内可能强行起飞的飞机架数约130架，但母弹能杀伤飞机的数量不能超过其携带子弹数；

（3）假设飞机只沿跑道Ox轴正方向以匀速滑行起飞，不考虑被毁伤的飞机对后续飞机起飞造成的障碍。

5 结论

通过上面的计算结果，如表3所示。可以知道封锁型子母弹的封锁效能受到以下因素的影响。

表3 封锁型子母弹封锁杀伤飞机的平均架数

母弹数量/枚	子弹数量/枚	CEP = 0 m	CEP = 50 m	CEP = 100 m	CEP = 150 m
1	100	19	13	10	6
	150	20	14	11	7
	200	22	15	12	7
2	100	39	27	21	13
	150	40	28	22	13
	200	43	30	24	14
3	100	54	38	29	18
	150	51	36	28	17
	200	56	39	31	18
4	100	67	47	37	22
	150	69	48	38	23
	200	70	49	39	23
5	100	82	57	45	27
	150	78	55	43	26
	200	83	58	46	27

（1）导弹的射击精度。导弹的射击精度（常用导弹的圆概率偏差CEP来表征）对封锁型子母弹的封锁效能影响较大。受杀伤范围限制，封锁型子弹只有尽可能多地散布在跑道上和跑道的附近，才能对飞机构成有效杀伤；如子弹远离跑道，将形成杀伤死角，则对飞机目标失去杀伤作用。只有射击精度较好，才能保证子弹群能较好地覆盖跑道，资料表明，当导弹的圆概率由200变为250、300时，其对机场跑道的封锁效率分别降低28%和66%。

（2）母弹瞄准点选择。母弹瞄准点的选择合理与否也将对封锁型子母弹的封锁效能造成较大的影响。选用爆破子母弹封锁某机场跑道，当要达到50%~75%的封锁概率时，需要子母弹2~8枚[5]。若母弹瞄准点选择合理的话，达到同样的封锁概率，则需要更少的子母弹，如需要2~5枚，可有效降低其弹药消耗量。文中的母弹瞄准点采用了沿跑道中轴线等距离分布，是为了计算的方便，并不是最优化的瞄准点选择，可进一步优化。

（3）子弹的数量。在导弹其他条件相同的情况下，一枚母弹能携带的子弹数量越大，散布在跑道周围的子弹数就越多。当采用随机延时引信时，在一定的封锁时间内，子弹数量增加，则其延时起爆间隔就更短，对强行起降的飞机就能造成更大的毁伤，从而提高了其封锁效能。因此，可设法提高单枚母弹携带子弹的数量，但是，应更注重通过提高导弹的命中精度来提高其封锁效能，因为后者的毁伤效果远比单纯增加子弹数量的毁伤效果明显，如文献［6］所说，单枚母弹中子弹的装填密度提高1%，其毁伤效能仅提高0.68%。

（4）子弹的抛撒模式。子弹的抛撒模式有两种：圆散布和椭圆散布。当子弹按照圆散布抛撒时，散布在跑道周周的子弹是一种等概率的情况，当子弹按椭圆散布抛撒时，散布在跑道周围的子弹将与火力运用方法有关，见图1、图2。文献［7］表明：在子弹圆散布的情况下，跑道的失效率是60%；在子弹椭圆散布的情况时，与子弹椭圆散布的长轴与跑道中轴线的夹角有关，当该夹角为0°、30°、60°、90°时，跑道的失效率分别是66%，63%，60%，57%。因此，子弹的抛撒模式也对子母弹的封锁效能构成一定的影响。尽管上述计算过程采用了一些假设条件，与复杂多变的实战环境仍有一定的差距，但该方法较好地反映出封锁型子母弹的封锁效能情况，为更科学合理地对其进行封锁效能的评估提供了思路，具有较好的参考价值。

参考文献

［1］午新民，王中华．国外机载武器战斗部手册［M］．北京：兵器工业出版社，2005.

［2］黄龙华，冯顺山，樊桂印．封锁型子母弹对机场跑道的封锁效能分析［J］．弹道学报，2007，19（3）：49－52.

［3］王震宇，冯顺山．矩阵仿真法在战斗部威力评估中的应用［J］．弹箭与制导学报，2005，25（4）：60－63.

［4］于颖贤．机场封锁子弹随机起爆问题研究［D］．北京：北京理工大学，2005.

［5］钱立新，卢永刚，杨云斌．反跑道动能战斗部威力评定方法研究［R］．中国国防科学技术报告，1999.

［6］李向东，董旭意．子弹数目对子母弹毁伤效能的影响研究［J］．弹道学报，2001，12（1）：36－38.

［7］李亚雄，刘新学，舒健生．两种子母弹抛撒模型对跑道失效率的影响分析［J］．弹箭与制导学报，2005，25（2）：48－50.

［8］李向东，郝传宏．机载布撒器对机场跑道封锁效率计算研究［J］．弹道学报，2004，15（1）：65－69.

［9］寇保华，杨涛，张晓会，等．末修子母弹对机场跑道封锁概率的计算［J］．弹道学报，2005，16（4）：22－26，49.

杀伤子母弹机场封锁效能仿真①

王震宇[1②]，冯顺山[2]

（1. 中北大学武器装备技术学院，太原 030051；

2. 北京理工大学爆炸科学与技术国家重点实验室，北京 100081）

摘 要：本文针对机场封锁作战中的间接封锁与直接封锁作战方式，讨论了利用破片弹道示踪法仿真子母杀伤封锁战斗部在子弹延时起爆条件下，单枚母弹对长度等于最小起降带长度的跑道区域内以一定规律起飞的作战飞机的封锁效能，其结果对封锁战斗部研制及机场封锁弹药战术应用策略具有指导作用。

关键词：机场封锁；子母弹；破片弹道示踪法；最小起降带；封锁效能

Airfield Blockade Effectiveness Simulation of Fragmentation Shrapnel

Wang Zhenyu[1], Feng Shunshan[2]

（1. School of Weapons and Armament Technology, North University of China, Taiyuan 030051, China;

2. State Key Laboratory of Explosion Science and Technology,

Beijing Institute of Technology, Beijing 100081, China）

Abstract: Focused on indirect blockade and direct blockade methods in airfield blockade warfare * the blockade effectiveness of submunitions from a single bomblet exploding at certain time intervals using fragment trajectory tracking method to defeat planes taking off from a minimum operation strip was studied. The simulation results may be referred to airfield warhead R&D and tactical application of these warheads.

Keywords: airfield blockade; shrapnel; fragment trajectory tracking method; minimum operation strip; blockade effectiveness

1 引言

机场是现代战争机器的重要载体，从海湾战争以及美国对阿富汗的战争可以发现，夺取制空权已经成为现代战争的必要前奏。战争初期对机场目标实施突然密集的攻击已经成为战争的一种基本模式，是夺取制空权的重要手段。

文中以子弹采用随机延时引信为例分析在机场跑道某最小起降带附近投放单枚母弹情况下，子弹随机延时起爆利用破片弹道示踪法仿真计算对于跑道区域上以一定规律起飞的作战飞机的封锁效能。

2 机场间接封锁与直接封锁

封锁飞机起降的作战方式可分为两种：一种是传统的破坏飞机跑道使飞机达不到最小起降带这一必要条件而达到封锁飞机起降目的间接封锁方式；另一种是采用某种杀伤元在机场跑道附近区域形成一定的时间、空间分布，直接杀伤威胁地面停放飞机以及起降过程中的飞机目标从而封锁飞机起降的直接封锁方式。

---

① 原文发表于《弹箭与制导学报》2008（S2）.

② 王震宇：工学博士，2003 年师从冯顺山教授，研究封锁作战与机场封锁子弹效能评估，现工作单位：兵器 5013 厂。

第一类弹药是传统的航空炸弹战斗部和地地导弹侵爆、穿爆战斗部等，如法国的迪兰达尔反跑道炸弹。第二类弹药是通过布撒能够形成大量杀伤元的区域封锁子弹药到飞机掩蔽库、机场跑道及其附近区域。这些弹药可采用随机延时起爆、感应起爆等多模起爆方式对其落点附近的飞机、器材和人员目标形成杀伤威胁，给对方探测，排除弹药及清理修复跑道等反封锁措施的展开造成障碍，如英国的 JP233 子母炸弹。

3 破片弹道示踪法

对于以预制及半预制破片为主的杀伤战斗部，在既定的战斗部结构、已知的弹目交汇起爆条件下，可通过解析方法或数值仿真方法（如 Hydrocode）得到战斗部起爆时形成的所有破片的离散初始参数，包括每个破片的质量、外形几何参数、在地面静止坐标系下的三维坐标值、破片速度和飞散方向。以这些初始条件为基础，根据空气中外弹道方法遍历跟踪每一个破片在与目标交汇前的整个空气弹道过程以及相关参数（速度、动能、比动能）变化情况，最后可得到整个破片空间威力分布作用场。这种以战斗部破片起爆初始参数为基础，遍历跟踪每一个破片弹道以得到破片空间威力分布作用场的方法称作破片弹道示踪法。

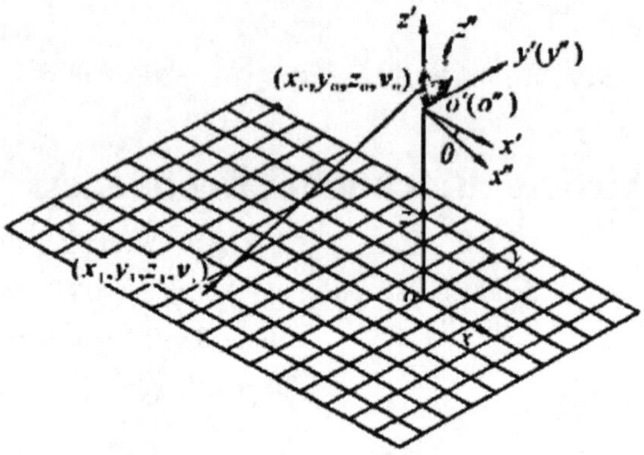

图1 破片弹道示踪法示意图

如图 1 所示，($O''x''y''z''$) 为弹体坐标系，($Oxyz$) 为地面坐标系，($O'x'y'z'$) 为与弹体坐标系原点重合且平行于地面坐标系的中间坐标系。假设战斗部某破片在零时刻（炸药作用结束）时，其在地面坐标系下的坐标为（x_0，y_0，z_0），其初始速度矢量为 v，不考虑重力对破片弹道的影响以及目标可能对破片的拦阻，经过一定时间 t 后破片经历了空气中弹道过程后到达地面，其在地面坐标系下的落点坐标为（x_1，y_1，z_1），落地时破片速度矢量为 v_1。

4 单枚杀伤子母弹对 600 m × 45 m 跑道区域封锁效能

4.1 条件与假设

（1）如图 2 所示，假设母弹瞄准点位于机场跑道长度方向轴线上，假定母弹 CEP = 0 m，假定母弹抛撒子弹的抛撒半径 R = 150 m，单发母弹携带子弹发数 n = 120，子弹在抛撒圆内服从均匀分布。假设机场只有单条跑道，尺寸为 3 000 m × 45 m，飞机最小起降带尺寸为 600 m × 20 m。

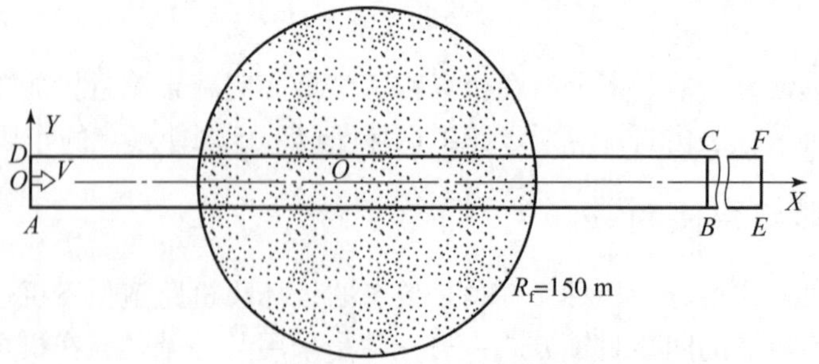

图2 单枚母弹子弹随机抛撒落点示意图

图中 $AD = EF = BC$ = 45 m，ABCD 为与最小起降带长度相等的 600 m × 45 m 跑道区域，假定飞机只从跑道一端即原点 O 处开始沿 OX 轴正向滑行起飞。OX 轴与跑道纵轴重合，母弹抛撒中心 Q 在 OXY 中坐标为 O_c(300，0)，V_q 为飞机起飞平均速度。

（2）采用蒙特卡罗（Monte－Carlo）随机仿真方法生成子弹随机落点与子弹起爆随机姿态子弹采用

延时起爆方式，根据封锁强度要求的不同可取相邻的两枚子弹间起爆间隔分别为 30 s、45 s、60 s，显然子弹起爆间隔越短封锁强度越高。

(3) 取第一发子弹起爆时间 $t_1=0$ 为计时起点，假设在 $t_1=0$ 时刻，第一架飞机形心正好位起飞所需平均时间为 25 s，前一架飞机起飞升空后，后一架飞机立即到达起飞位置开始滑跑起飞。

(4) 假设飞机沿 OX 轴正向以匀 V_q 滑行起飞，且滑跑时机身轴线与跑道纵轴重合。

(5) 不考虑被杀伤飞机可能给后续飞机起飞造成的障碍，即跑道上始终无其他障碍物。

(6) 假设飞机等效杀伤缩比尺寸为 $10\times1.4\times2.1$ $(L\times B\times H)$，机腹高度为 2 m，取 800 J/cm^2 的比动能杀伤标准，不考虑冲击波对目标的杀伤。

4.2　仿真结果与分析

根据以上条件以单枚母弹在不同子弹起爆延时下以一定速度顺序起飞飞机的杀伤效能进行 10 次仿真，结果如表 1 所示。

表 1　仿真 10 次得到的单枚母弹封锁杀伤飞机数目分布

延时	延时/s		
	$t_j=30$	$t_j=45$	$t_j=60$
1	22	19	20
2	19	16	20
3	25	18	18
4	21	22	25
5	20	17	23
6	22	18	27
7	20	23	24
8	21	21	19
9	17	22	20
10	25	23	19
avg	21.2	19.9	21.5

其中 avg 为 10 次仿真平均杀伤架次。由于单枚子弹起爆时，跑道区域上且只有一架飞机，单枚母弹携带子弹数即为可能杀伤的最大飞机数。定义母弹封锁效率为子弹杀伤飞机数目与子弹总发数的比值，则封锁效率分别为 17.7%、16.6%、17.9%。可从几个方面分析仿真结果。

首先由于飞机目标尺寸相对跑道长度尺寸很小，当取飞机目标长度 $L=10$ m 时，飞机最小起降带长度为 600 m 时，两者比值为 1.67%，因而飞机目标出现在子弹附近的概率也较低。此外单枚子弹封锁半径在 100 m 以内，假设在跑道区域纵向中心轴线与母弹抛撒圆边缘交点坐标（150，0），（450，0）处刚好各有一枚子弹，则以母弹抛撒圆圆心为圆心，半径 250 m 的范围为母弹的最大威力范围。在该段跑道两端距离端面各 50 m 的区域为该单枚母弹的杀伤死角，假设飞机目标位于跑道 OX 轴上任一点的概率相等，则有 $120\times100/600=20$ 架位于母弹威力范围之外。

即使在母弹最大威力半径 $R=250$ m 之内，直径为 500 m 的圆形区域内，由于子弹最大威力直径≤200 m，考虑最有利情况即某子弹起爆时起爆点都位于该跑道区域纵向轴线上，子弹威力直径为 200 m，则在母弹威力圆内沿该跑道区域轴线还有 $(500-200)/500=60\%$ 的区域在子弹威力范围之外。

其次母弹抛撒半径 150 m，会有一些子弹落点距离 OX 轴较远的位置，如图 3 中阴影区域，由于飞机目标宽度与子弹威力半径限制而成为对封锁仿真无效的子弹。据几何关系，阴影部分面积与抛撒圆面积之比为 21.91%，若子弹均匀分布，则有 26 枚子弹为无效子弹。

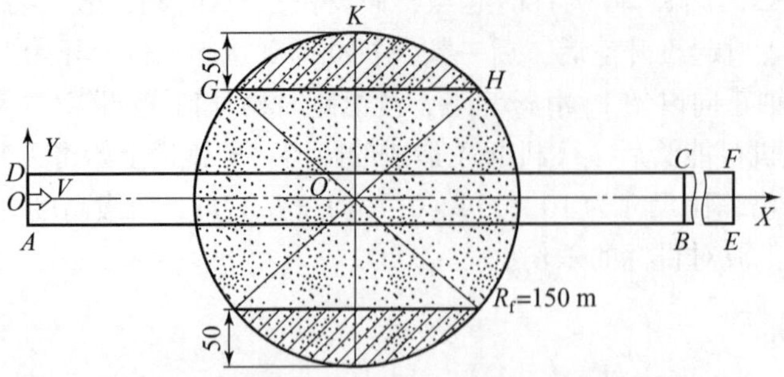

图 3　对于仿真封锁无效的子弹落点区域示意图

5　结论

从仿真结果可见，单枚母弹对 600 m×45 m 跑道区域的封锁效率约为 20%。文中单独讨论采用直接封锁作战方式下的封锁效能，实际应用时都是直接封锁弹药和间接封锁破坏跑道的侵爆弹混合使用，如果再考虑杀伤飞机残骸对后续飞机的起降形成的障碍，实战封锁效能会大大提高。

参考文献

[1] 方再根．计算机模拟和蒙特卡罗方法［M］．北京：北京工业学院出版社，1988.
[2] 李廷杰．导弹武器系统的效能及其分析［M］．北京：国防工业出版社，2000.
[3] 张廷良，陈立新．地地弹道式战术导弹效能分析［M］．北京：国防工业出版社，2001.

刚性弹丸低速正侵彻金属靶板分析模型——基于空穴膨胀理论①

吴　广②，冯顺山，董永香

（北京理工大学爆炸科学与技术国家重点实验室，北京　10081）

摘　要：本文通过空穴膨胀理论建立弹丸低速正侵彻金属靶板的分析模型，分析了侵彻深度小于弹丸头部长度情况下，不同头部形状弹丸的侵彻深度随初速的变化关系。并与文献中的试验结果进行对比，表明理论分析结果与试验结果吻合较好。

关键词：正侵彻：空穴膨胀理论；低速；金属靶板

1　引言

冲击问题一直以来是人们所关注的焦点[1,2]，Backman 和 Goldsmith[3]、Corbett 等[4]在文献中回顾了冲击问题 100 多年的研究成果，特别是 Goldsmith[5]全面回顾了非理想状态下冲击问题的研究成果。

目前，人们广泛采用空穴膨胀理论来分析弹丸对各种靶板材料的侵彻问题。Chen X W 和 Li Q M[6,7]通过空穴膨胀理论推导出多种头部形状的无变形弹丸深侵彻靶板的最大侵彻距离公式，并且提出了影响侵彻距离的两个函数，冲击函数 I 和头部形状函数 N，同时还得到多种头部形状的形状因子（包括卵形头部、锥形头部和截卵形头部以及钝头弹）；Jones S E 和 Rule W K[8]在忽略侵彻过程中弹丸于靶板之间的摩擦效应得到了无变形弹丸在侵彻过程中的轴向阻力与头部形状常数的关系，得到几种典型弹丸侵彻过程中最优头部函数关系式；后来 Jones S E 和 Rule W K[9]又提出了在考虑摩擦效应时，侵彻的最优头部形状函数关系式，并通过数值求解得到最优头部形状的弹丸与几种典型头部形状弹丸进行比较。Q. M. Li 等[10]考虑弹丸头部未完全进入靶板阶段，通过空穴膨胀理论建立了无变形弹丸侵彻中厚靶的理论分析模型，分析了抛物线形头部弹丸侵彻中厚靶的过程；覃悦等人在假设弹体在侵彻过程中的平均压力由靶板材料弹塑性变形引起静态阻力和由速度效应引起的动阻力的情况下研究了锥头弹丸撞击 FRP 层合板。

本文主要通过空穴膨胀理论来分析低速情况下弹丸侵彻深度小于弹丸头部长度时不同头部形状弹丸正侵彻金属靶板，得到侵彻深度随初速的变化关系，并对理论分析结果与已有试验结果进行比较和讨论。

2　空穴膨胀理论

在前人的大量工作下，目前空穴膨胀理论广泛应用于各种靶板材料的侵彻问题研究中。空穴径向压力 σ_r 表示为静态和动态分量之和，其表达式如下：

$$\sigma_r = AY + B\rho v^2 \tag{1}$$

式中　v——头部某点的空穴径向膨胀速度；

Y——靶板材料的屈服应力；

ρ——靶板材料的密度；

A、B——关于靶板材料的两个常数。

---

① 原文发表于《中国宇航学会无人飞行学会战斗部毁伤效率专业委员会第十一届学会交流论文集》2009。

② 吴广：工学博士，2006 年师从冯顺山教授，研究弹药对飞行甲板的半侵彻机理及技术，现工作单位：北理工重庆创新中心。

对于不同材料，参数 A 可表示为

$$A=\begin{cases}\dfrac{2}{3}\left\{1+\ln\left[\dfrac{E}{(1+\gamma)Y}\right]\right\} & \text{不可压缩材料}\\ \dfrac{2}{3}\left\{1+\ln\left[\dfrac{E}{3(1-\gamma)Y}\right]\right\} & \text{可压缩材料}\end{cases} \tag{2}$$

其中，γ 为泊松比，参数 B 的取值范围一般在 1 ~ 1.5。对于不可压缩弹塑性材料 $B=1.5$，然而对于可压缩的固体材料 B 值的计算较为复杂。铝合金材料取值为 $B=1.1$，混凝土材料 $B=1.0$，土介质材料一般取值为 $B=1.2$。

3　侵彻模型建立

任意弹丸头部结构以及弹体坐标系如图 1 所示。假设：①弹丸为刚性，在侵彻过程中不变形；②忽略靶板的整体变形情况；③不考虑靶板正面翻边和靶背边界对侵彻的影响。

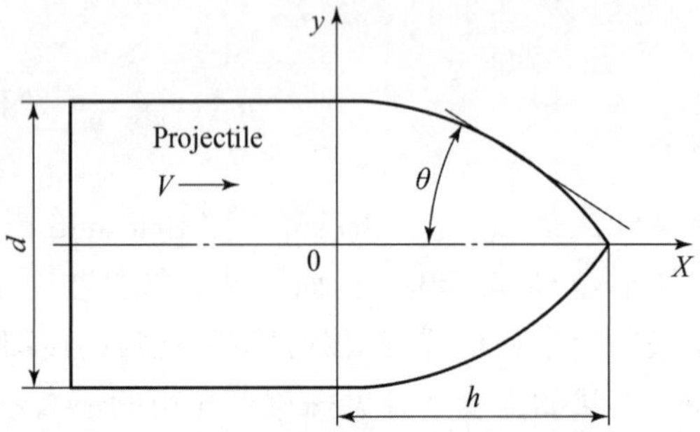

图 1　任意头部形状弹丸结构示意图

由于弹丸为刚体，其侵彻过程中的瞬时速度为 V，则弹靶界面处的质点速度（空穴径向膨胀速度）为

$$v=V\sin\theta \tag{3}$$

将式（3）代入式（1）得弹丸头部正向应力 σ_n 为

$$\sigma_n=\sigma_r=AY+B\rho\,(V\sin\theta)^2 \tag{4}$$

考虑弹丸侵彻过程中的摩擦效应，那么由于摩擦引起的弹丸头部切向应力 σ_t 为，由摩擦定律得

$$\sigma_t=\mu_m\sigma_n \tag{5}$$

式中，μ_m 为滑动摩擦系数。

将式（4）代入式（5）得

$$\sigma_t=\mu_m(AY+B\rho(V\sin\theta)^2) \tag{6}$$

那么正应力和切向应力引起的轴向分量为

$$\sigma_{nx}=\sigma_n\sin\theta=AY\sin\theta+B\rho V^2\sin^3\theta \tag{7}$$

$$\sigma_{tx}=\sigma_t\cos\theta=\mu_m AY\cos\theta+\mu_m B\rho V^2\sin^2\theta\cos\theta \tag{8}$$

所以弹丸侵彻到靶板中 x 深处时，弹丸所受到的靶板阻力为

$$F_x=\oint\!\!\int_{An}(\sigma_{nx}+\sigma_{tx})\,\mathrm{d}A \tag{9}$$

式中，A_n 为弹丸与靶板接触的那部分表面积。

将式（7）、式（8）代入式（9）得到弹丸侵彻过程中受到的轴向阻力为

$$\begin{aligned}F_x&=\oint\!\!\int_{A_n}(AY\sin\theta+B\rho V^2\sin^3\theta+\mu_m AY\cos\theta+\mu_m B\rho V^2\sin^2\theta\cos\theta)\,\mathrm{d}A\\&=\oint\!\!\int_{A_n}[AY(\sin\theta+\mu_m\cos\theta)+B\rho V^2\sin^2\theta(\sin\theta+\mu_m\cos\theta)]\,\mathrm{d}A\end{aligned} \tag{10}$$

低速情况下，当靶板厚度大于弹丸头部长度且弹丸侵彻深度小于弹丸头部长度时，此阶段弹丸的轴

向阻力随着弹靶接触面积的增加而增加，弹靶状态如图 2 所示。若弹丸头部母线方程为 $y=y(x)$，并且以 x 表示侵彻深度，那么

$$dA=\frac{2\pi y}{\cos\theta}dx \tag{11}$$

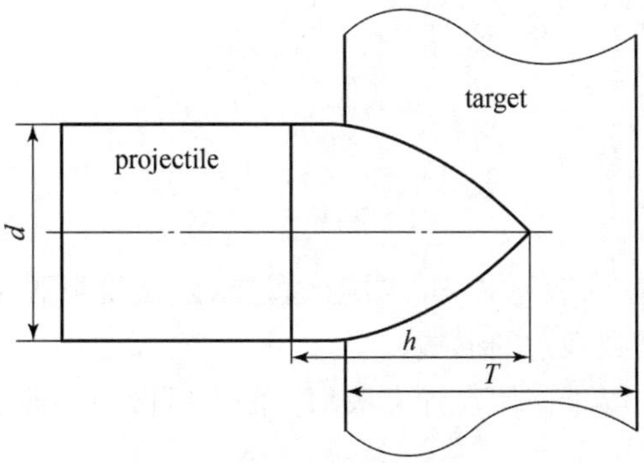

图 2　弹靶作用结构示意图

那么将式（11）代入式（10），得到弹丸的轴向阻力为

$$F_x=\int_{h-x}^{h}\left[AY(\sin\theta+\mu_m\cos\theta)+B\rho V^2\sin^2\theta(\sin\theta+\mu_m\cos\theta)\right]\frac{2\pi y}{\cos\theta}dx \tag{12}$$

将式（12）变换为下式

$$F_x=\pi y^2(A\overline{N_1}Y+B\overline{N_2}\rho V^2) \tag{13}$$

其中，$\overline{N_1}=1+\frac{2\mu_m}{y^2}\int_{h-x}^{h}y\,dx$，$\overline{N_2}=\bar{N}^*+\frac{2\mu_m}{y^2}\int_{h-x}^{h}\frac{yy'^2}{1+y'^2}dx$，$\bar{N}^*=-\frac{2}{y^2}\int_{h-x}^{h}\frac{yy'^3}{1+y'^2}dx$

3.1　锥形头部弹丸

锥形弹丸头部母线满足如下函数关系 $y=y(x)$

$$y=(-\tan\theta)x+\frac{d}{2}=-\frac{d}{2h}x+\frac{d}{2} \tag{14}$$

$$y'=-\tan\theta=-\frac{d}{2h} \tag{15}$$

那么锥形头部弹丸的头部阻力表示为

$$F_x=\pi AY\frac{d(d+2h\mu_m)}{4h^2}x^2+\pi B\rho\frac{d^3(d+2h\mu_m)}{4h^2(4h^2+d^2)}V^2x^2 \tag{16}$$

由牛顿第二定律可知

$$F_x=-M\frac{DV}{Dt}=-MV\frac{DV}{Dx} \tag{17}$$

令：$a_1=-\pi AY\frac{d(d+2h\mu_m)}{2Mh^2}$，$a_2=\pi B\rho\frac{d^3(d+2h\mu_m)}{2Mh^2(4h^2+d^2)}$

$$\frac{D(V^2)}{Dx}=a_1x^2+a_2V^2x^2 \tag{18}$$

因此，弹丸速度随着侵彻距离的变化关系式为

$$V^2=\left(V_0^2+\frac{a_1}{a_2}\right)e^{\frac{a_2}{3}x^3}-\frac{a_1}{a_2} \tag{19}$$

将式（22）代入式（14）得到弹丸轴向阻力与侵彻距离的变化关系式

$$F_x=-\frac{Ma_1}{2}x^2-\frac{Ma_2}{2}\left[\left(V_0^2+\frac{a_1}{a_2}\right)e^{\frac{a_2}{3}x^3}-\frac{a_1}{a_2}\right]x^2 \tag{20}$$

由式（22）可知，当速度为 0 时，可得到对应不同初速下的最大侵彻距离为

$$L_1 = \sqrt[3]{\frac{3}{a_2}\ln\left(\left(\frac{a_1}{a_2}\right)\Big/\left(V_0^2 + \left(\frac{a_1}{a_2}\right)\right)\right)} \tag{21}$$

3.2 截锥形头部弹丸

卵形弹丸其头部母线满足如下函数关系 $y = y(x)$

$$y = -\frac{d - d_1}{2h}x + \frac{d}{2} \tag{22}$$

$$y' = -\tan\theta = -\frac{d - d_1}{2h} \tag{23}$$

将以上两式代入$\overline{N_1}$、$\overline{N_2}$、$\overline{N}^*$的表达式中，并且通过求解截锥形弹丸的侵彻微分方程，弹丸侵彻过程中的状态参量不能求以初等函数表达的形式。

由弹丸侵彻过程中动能损失等于作用在弹丸头部上的力所做功，即

$$\Delta E = -F_x \Delta x \tag{24}$$

通过数值计算，Δx 为侵彻深度步长，第 i 步

$$\Delta E = \frac{1}{2}M(V_i^2 - V_{i-1}^2) = -\Delta x\left[\frac{F(x_{i-1}) + F(x_i)}{2}\right] \tag{25}$$

3.3 卵形头部弹丸

卵形弹丸其头部母线满足如下函数关系 $y = y(x)$，卵形头部母线方程的圆心坐标为（0，b）

$$y = \sqrt{r^2 - x^2} - b \tag{26}$$

$$y' = -\frac{x}{\sqrt{r^2 - x^2}} \tag{27}$$

同理，可通过数值计算求解弹丸侵彻过程中的状态参量。

3.4 抛物线形头部弹丸

抛物线形弹丸其头部母线满足如下函数关系 $y = y(x)$，抛物线形头部母线方程的顶点坐标为（h，0）

$$y = \sqrt{-2p(x - h)} \tag{28}$$

$$y' = -\sqrt{\frac{p}{-2(x - h)}} \tag{29}$$

同理，也可通过数值计算求解弹丸侵彻过程中的状态参量。

4 分析与讨论

通过数值计算了不同头部形状弹丸低速侵彻金属靶板的过程，参考文献中的弹丸结构与靶板参数。文献中截锥形弹丸和卵形头部弹丸结构如图 3 所示。

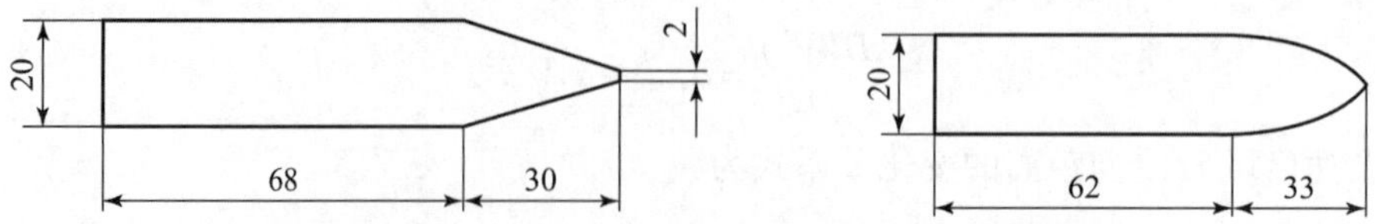

图 3　截锥形弹丸（左）和卵形头部弹丸（右，CRH = 3）

（1）截锥形弹丸头部母线方程为 $y = -0.3x + 0.01$，弹丸质量 197 g；

（2）卵形头部弹丸头部母线方程为 $y = \sqrt{0.06^2 - x^2} - 0.05$，弹丸质量 197 g；

（3）文献中弹丸的头部的头部母线为抛物线，其方程为 $y = 0.0707\sqrt{0.02 - x}$，弹丸质量 110 g；

（4）为比较不同头部弹丸，选取锥形头部弹丸，其头部母线方程为 $y = -\frac{1}{3}x + 0.01$，弹丸质量 197 g。

文献［11］中靶板材料 Weldox460E，密度 $\rho = 7.8\ \mathrm{g \cdot cm^{-3}}$，屈服强度 $Y = 449$ MPa，弹性模量 $E =$ 204 GPa，泊松比 $\gamma = 0.33$。上述四种弹丸低速侵彻 Weldox460E 靶板的侵彻深度与初速的关系曲线如图 4 所示。由图 4 可知，上述四种弹丸中，相同初速下抛物线头部弹丸侵彻深度最小，锥形弹丸侵彻深度最深，卵形头部弹丸和截锥形弹丸侵彻深度居中，但截锥形弹丸侵彻深度大于卵形头部弹丸。

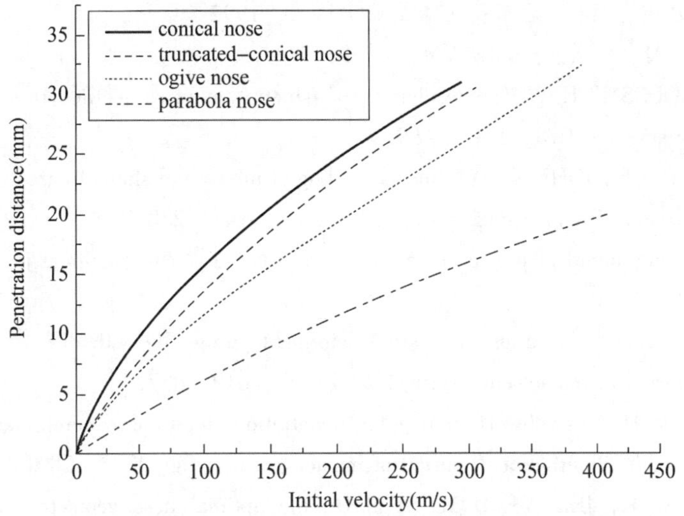

图 4　四种头部形状弹丸侵彻 Weldox460E 靶板的初速与侵彻深度关系曲线

文献［13］中，S. N. Dikshi 和 G. Sundararagjan 用上述抛物线头部弹丸进行试验，撞击 20 mm 厚屈服强度为 1 080 MP 靶板、40 mm 厚屈服强度为 920 MP 靶板、80 mm 厚屈服强度为 860 MP 靶板。图 5 为抛物线头部弹丸低速撞击三种靶板分析模型与试验结果对比图。由图可知，分析模型与试验结果符合较好。

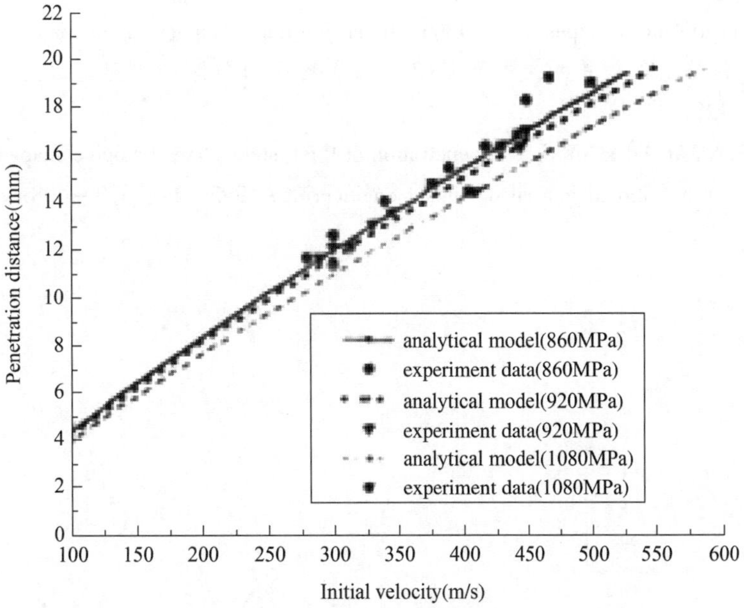

图 5　分析模型结果与试验结果对比

5 结论

本文通过空穴膨胀理论来分析低速情况下弹丸侵彻深度小于弹丸头部长度时不同头部形状弹丸正侵彻金属靶板的过程，得到侵彻深度随初速的变化关系，并且理论分析结果与已有试验结果符合较好。

参考文献

[1] 赵国志．穿甲工程力学［M］．北京：兵器工业出版社，1992.

[2] 钱伟长．穿甲力学［M］．北京：国防工业出版社，1984.

[3] BACKMAN E M, GOLDSMITH W. The mechanics of penetration of projectiles into targets［J］. International journal of engineering science 1978, 16: 1－99.

[4] CORBETT G G, REID S R, JOHNSON W. Impact loading of plates and shells by free－flying projectiles: a review［J］. International journal of impact engineering, 1996, 18: 141－230.

[5] GOLDSMITH W. Review: nonideal projectile impact on targets［J］. International journal of impact engineering, 1999, 22: 95－395.

[6] CHEN X W, LI Q M. Deep penetration of a non－deformable projectile with different geometrical characteristics［J］. International journal of impact engineering, 2002, 27: 619－637.

[7] LI Q M, CHEN X W. Dimensionless formulae for penetration depth of concrete target impacted by a non－deformable projectile［J］. International journal of impact engineering, 2003, 28 (1): 93－116.

[8] JONES S E, RULE W K, JEROME D M, et al. On the optimal nose geometry for a rigid penetrator［J］. Computational mechanics, 1998, 22: 413－417.

[9] JONES S E, RULE W K. On the optimal nose geometry for a rigid penetrator, including the effects of pressure dependent friction［J］. International journal of impact engineering, 2000, 24: 403－415.

[10] LI Q M, WENG H J, CHEN X W. A modified model for the penetration into moderately thick plates by a rigid, sharp－nosed projectile［J］. International journal of impact engineering, 2004, 30: 193－204.

[11] DEY S, BOVIK T, HOPPERSTAD O S, et al. The effect of target strength on the perforation of steel plates using three different projectile nose shapes［J］. International journal of impact engineering, 2004, 30: 1005－1038.

[12] 覃悦，文鹤鸣，何涛．锥头弹丸撞击下 FRP 层合板的侵彻与穿透的理论研究［J］．高压物理学报，2007, 21: 121－128.

[13] DIKSHIT S N, SUNDARARAJAN G. The penetration of thick steel plates by ogive shaped projectiles－experiment and analysis［J］. International journal of impact engineering, 1992, 12: 373－408.

靶板在弹丸低速冲击作用下的弹性恢复规律①

黄广炎②，冯顺山，吴　广
（北京理工大学爆炸科学与技术国家重点实验室，北京　100081）

摘　要：为了对不同靶板被尖头弹丸低速侵彻条件下的弹性恢复机理进行研究，同时考虑弹丸低速冲击靶板过程中靶板整体塑性变形及弹性变形恢复，更好地反映弹丸对靶板低速侵彻过程中的变形及材料失效问题，本文使用非线性动力学分析软件 ANSYS/LS－DYNA 计算分析了弹丸以相同速度冲击不同边长、厚度靶板时，弹丸反弹速度以及靶板回弹速度随着靶板厚度的变化情况。得到了靶板厚度、靶板边长和冲击速度对靶板的弹性恢复和弹丸反弹的影响规律，为低速穿甲弹对靶板侵彻的工程与实验设计研究提供参考。

关键词：固体力学；弹性恢复；弹道冲击；反弹

Elastic recovery in targets impacted by low velocity projectiles

Huang Guangyan，Feng Shunshan，Wu Guang
（State Key Laboratory of Explosion Science and Technology，Beijing Institute of Technology，Beijing 100081，China）

Abstract：By taking into account the whole plastic deformation and elastic deformation recovery of targets during the penetration of the rigid，sharp nose projectiles，the ANSYS/LS－DYNA code was used to calculate the rebound velocities of the projectiles and targets in the cases that the projectiles at the same velocities penetrated into the targets with different widths and thicknesses. Influences of the sizes of the targets and the impact velocities of the projectiles on the elastic recovery of the targets and the rebound of the projectiles were analyzed. The researched results are helpful for the engineering and experimental designs of the projectiles with low velocities penetrating into the targets.

Keywords：solid mechanics；elastic recovery；ballistic impact；rebound

弹丸对靶板的侵彻过程中，靶板将会发生局部和整体的弹塑性变形。在弹丸对靶板的侵彻过程研究中，往往忽略靶板整体的塑性变形，也不考虑靶板的弹性变形。当弹丸以较高的速度撞击靶板时，弹坑附近的变形区域仅扩展到2～3倍着靶弹体直径处，因此忽略靶板的整体弹塑性变形是可行的。而当弹丸以较低速度撞击靶板时，靶板变形区域会扩展到10～12倍弹径区域，靶板产生明显的永久挠屈变形。而在此过程中，弹丸的侵彻速度为零后靶板的弹性变形开始恢复，特别是当弹丸头部不能完全穿过靶板时，靶板的弹性变形恢复将会带动弹丸一起反向运动，最终将使弹丸反弹出靶板。因此在弹丸低速冲击靶板的过程中，不仅要考虑靶板整体的塑性变形，还要考虑靶板的弹性变形恢复过程。目前，研究弹丸低速冲击作用下靶板的弹性恢复以及弹丸的反弹的文章较少。J. Radin 等[1]研究了弹丸侵彻多层靶板过程中靶板的弹性振动现象，得到了双层靶板的振动频率。G. G. Corbett 等[2]、R. S. I. Corrant 等[3]分析了钝头弹撞击靶板过程中的能量吸收机理，并分为四部分：弹性变形能、塑性变形能、冲塞能、弹丸的镦粗耗能，通过实验得到靶板的弹性变形能先随着靶板厚度增加而降低后又增大。A. Neuberger 等[4]通过实验和数值方法研究了爆炸载荷作用下装甲钢板的回弹现象。本文中，主要通过有限元动力分析软件

---

① 原文发表于《爆炸与冲击》2011（1）。

② 黄广炎：工学博士，2004年师从冯顺山教授，研究封锁型弹药设计及效能分析方法，现工作单位：北京理工大学。

ANSYS/LS - DYNA 研究弹丸在低于弹道极限速度范围内冲击不同厚度和边长靶板的过程，分析靶板在弹道冲击作用下的弹性变形情况，以及弹丸在靶板回弹作用下的反弹情况。

1 问题描述

1.1 几何模型

假设弹体为刚性，采用圆柱形弹体锥形头部弹丸（图 1）。弹径 72 mm，弹长 205 mm，头部长度 115 mm，质量 4.3 kg；为了研究弹丸冲击不同靶板的弹性恢复和弹丸反弹情况，对 3 种厚度和 3 种边长的柱形靶板的侵彻进行数值模拟。

1.2 有限元计算模型

由于弹靶结构、冲击载荷、初边值条件都具有对称性，计算时采用 1/4 模型，采用 8 节点 3 维实体单元对模型进行单元网格剖分，采用单点积分和沙漏控制，以便更好地反映大变形和材料失效等非线性问题，并节省机时。典型弹、靶有限元计算模型见图 2。由于精度要求，在计算机硬件以及时间允许的情况下，尽可能将网格细化。为了提高运算速度，减少单元数目，保证计算精度，在对靶板进行有限元网格剖分时，在弹靶接触区域，网格剖分较密；距离弹、靶接触区域较远的地方，网格剖分较稀疏。弹、靶剖分后共有 106 424 个实体单元，弹体与靶板之间的接触界面采用面 - 面接触的侵蚀算法。在 z 和 z 对称面上施加对称约束，对侧面施加全约束。

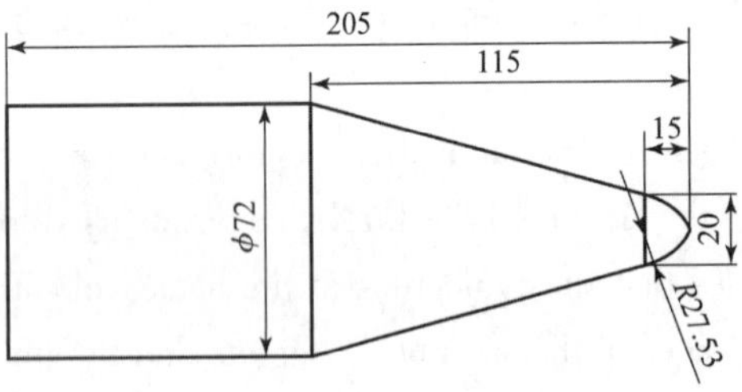

图 1　弹丸结构示意图

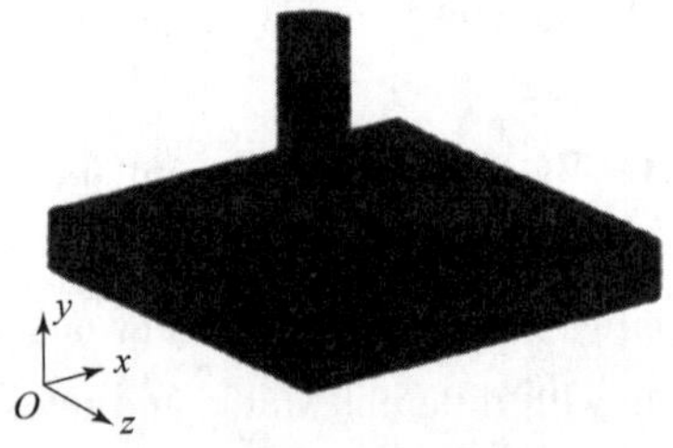

图 2　有限元计算模型

1.3 材料模型

1.3.1 弹丸的材料模型

假设弹丸在侵彻过程中不发生变形，弹丸材料选用刚性材料模型，密度 $\rho = 7.83\ \mathrm{g/cm^3}$，弹性模量 $E = 204$ GPa，泊松比 $\mu = 0.33$。

1.3.2 靶板的材料模型

利用 Johnson - Cook 强度模型描述靶板材料，具体形式为

$$\sigma = (A + B\tilde{\varepsilon}_p^n)(1 + C\ln\dot{\varepsilon}^*)[1 - (T^*)^m]$$

式中　A、B、n、C、m——材料参数；

$\tilde{\varepsilon}_p^n$——等效塑性应变；

$\dot{\varepsilon}^*$——等效应变率；

$\dot{\varepsilon}^* = \dot{\bar{\varepsilon}}_p/\dot{\varepsilon}_0$，$\dot{\varepsilon}_0$——参考塑性应变率，一般取值$\dot{\varepsilon}_0 = 1s^{-1}$；

$T^* = (T - T_r)/(T_m - T_r)$——相对温度；

T_r——参考室温；

T_m——熔化温度。

由于局部区域的塑性功产生的热没有足够的时间传导到周围材料中，通常认为高速撞击过程是绝热过程，温升

$$\Delta T = \frac{\chi}{\rho c_p}\int_0^{\varepsilon}\sigma(\varepsilon_p)\,d\varepsilon_p$$

式中　χ——塑性功转化成热的因子；

ρ——材料密度；

c_p——材料比定压热容。

韧性材料的破坏失效与所承受载荷状况密切相关，适当运用失效准则才能较好地描述弹靶撞击的破坏过程和特征。J-C 积累损伤失效模型考虑应力三轴度、应变率和温度效应，并通过累积损伤的概念考虑变形路径的影响，模型定义单元损伤

$$D = \sum \frac{\Delta\bar{\varepsilon}_p}{\varepsilon_f}$$

式中　D——损伤参数，在 0~1 之间变化，初始时 $D=0$，当 $D=1.0$ 时表示材料失效；

$\Delta\bar{\varepsilon}_p$——1 个时间步的塑性应变增量；

ε_f——当前时间步的应力状态、应变率和温度下的破坏应变，表达式为

$$\varepsilon_f = [D_1 + D_2\exp(D_3\sigma^*)(1 + D_4\ln\dot{\varepsilon}^*)(1 + D_5T^*)$$

式中　$D_1 \sim D_5$——材料参数；

$\sigma^* = p/\sigma_{\rm eq}$——应力状态参数，$p$——压力，$\sigma_{\rm eq}$——等效应力。

靶板为钢板，材料参数分别为：$\rho=7.85\ \mathrm{t/m^3}$，$E=204$ GPa，$\mu=0.33$，$c_p=452\ \mathrm{J/(kg\cdot K)}$，$\chi=0.9$，$T_r=298$ K，$T_m=1\ 763$ K，$\dot{\varepsilon}_0=10^{-5}\ \mathrm{s^{-1}}$，$A=601$ MPa，$B=356$ MPa，$n=0.586$，$C=0.022$，$m=1.05$，$D_1=0.636$，$D_2=-2.969$，$D_4=0.014$，$D_5=1.014$。

2　计算结果及分析

2.1　弹丸低速冲击靶板的反弹现象

利用上述材料模型参数，考虑靶板厚度 h、边长 b 以及冲击速度 ν_0，对刚性弹丸垂直侵彻不同状态靶板进行数值模拟，研究各种弹靶状态下靶板的弹性恢复和弹丸的反弹。图 3 为弹丸以 250 m/s 的速度侵彻边长 400 mm、厚 30 mm 靶板时的状态图；图 4 为不同时刻靶背变形图。由图 3 可以看出，1 300 μs 时靶板正面的位移明显小于 930 μs 时的位移；由图 4 可以看出，1 300 μs 时靶背各点的位移明显小于 700 μs 时的位移。图 5 为弹丸以 150 m/s 的速度侵彻靶板 1 200 μs 时状态图，可以明显看出，后期弹丸随着靶板的回弹一起做反向运动，并最终导致弹靶分开。

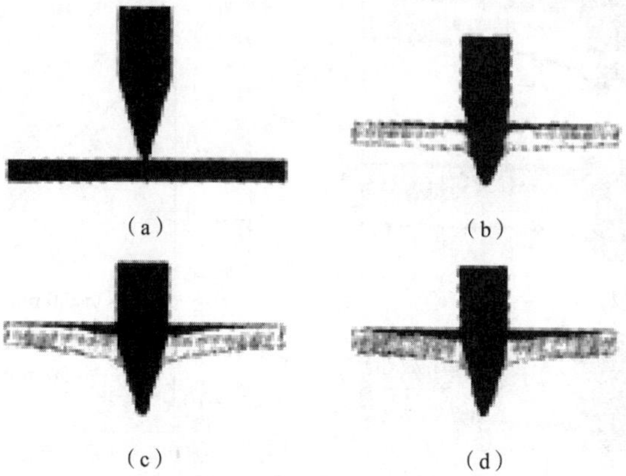

图 3　弹丸侵彻靶板状态

(a) $t=0$；(b) $t=400$ μs；(c) $t=930$ μs；(d) $t=1\ 300$ μs

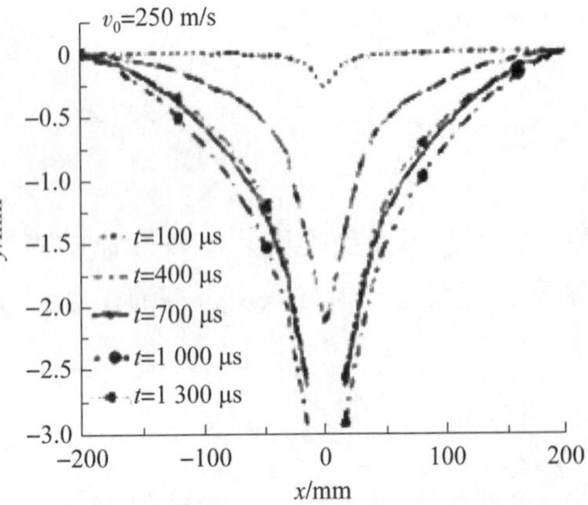

图 4　靶板在弹丸侵彻下的背面变形

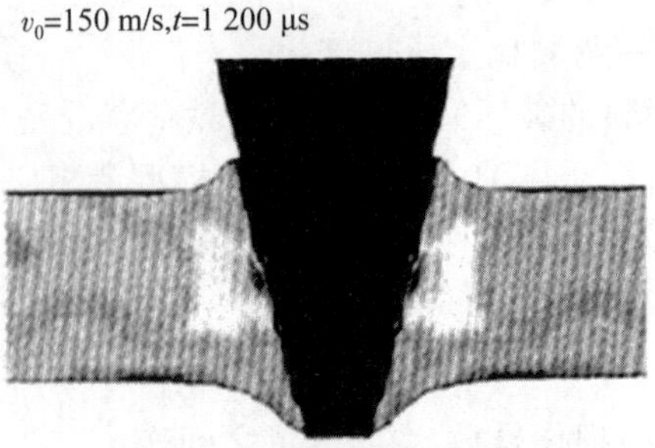

图 5　弹丸侵彻靶板状态

2.2　冲击不同厚度靶板

弹丸以 250 m/s 的速度冲击边长 400 mm、厚度分别为 30 mm、40 mm 和 50 mm 的靶板。图 6 为侵彻 3 种厚度靶板，弹丸速度降低至零时，靶板的变形情况。可以看出，厚 30 mm 靶板除了靶孔附近的局部变形外，靶板整体出现明显的碟形弯曲；厚 40 mm 靶板整体出现了微小挠曲变形；而厚 50 mm 靶板整体基本上没有出现挠曲变形。图 7 为距离靶板中心约 1 倍弹径（80 mm）处节点的位移历程曲线。从图中可以看出，侵彻厚 30 mm 靶板时，位移远大于侵彻其余 2 种厚度靶板时的位移，也就是说侵彻厚 30 mm 靶板时弹丸的很大一部分动能消耗于靶板整体变形。图 8、图 9 分别是弹丸以 250 m/s 速度冲击边 400 mm 的 3 种厚度靶板时，靶板整体平均速度历程曲线和弹丸速度历程曲线。由图 8 可以看出，由初始冲击引起的靶板整体负向运动过程中，厚 30 mm 靶板的最大负向运动速度最大，而厚 50 mm 靶板整体的最大负向运动速度和最大回弹速度都是 3 种厚度靶板中最小的，且厚 50 mm 靶板整体回弹速度大于由初始冲击引起的靶板最大负向运动速度，而由初始冲击引起的厚 30 mm 靶板的最大负向运动速度大于靶板的最大回弹速度。由图 9 可以看出，弹丸侵彻不同厚度靶板时，靶板越厚，弹丸的速度下降越快，且越早出现反弹。图 10、图 11 分别是弹丸以 250 m/s 速度冲击边长 400 mm 的 3 种厚度靶板时，弹丸最大反弹速度随靶板厚度变化关系曲线和靶板整体最大回弹速度随靶板厚度变化关系曲线。可以看出，相同初速下弹丸的最大反弹速度随着靶板厚度的增加而减小；靶板整体最大回弹速度则是随着厚度的增加先增大后降低。

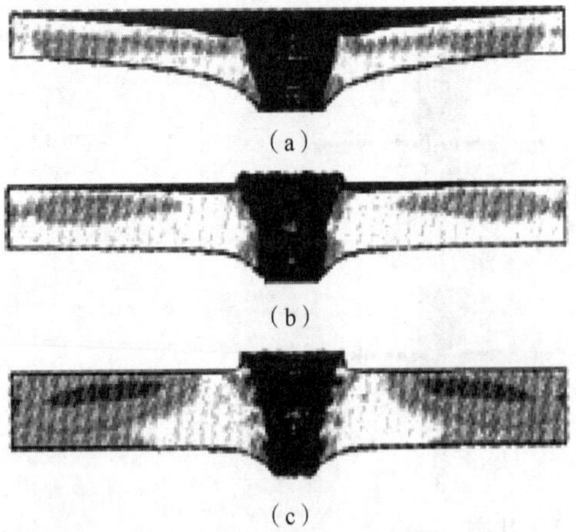

图 6　弹丸速度为零时靶板的变形情况

(a) $h=30$ mm；(b) $h=40$ mm；(c) $h=50$ mm

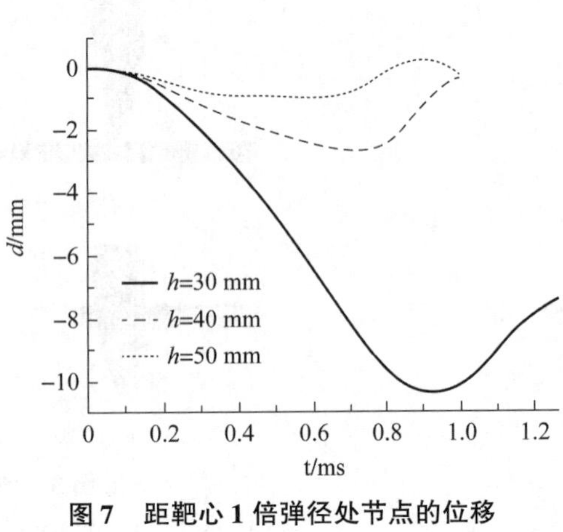

图 7　距靶心 1 倍弹径处节点的位移

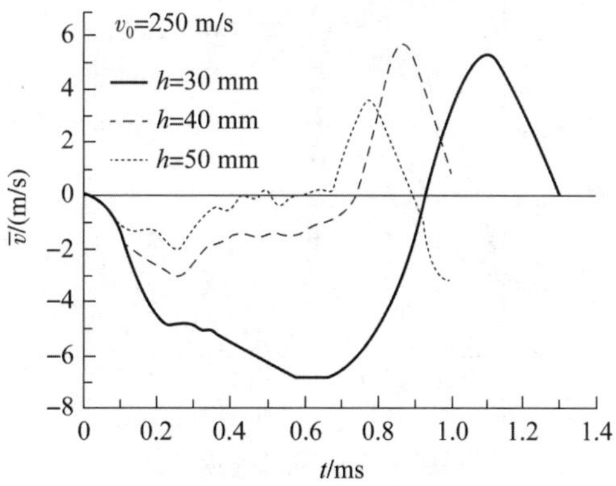

图 8　不同厚度靶板被弹丸侵彻时速度

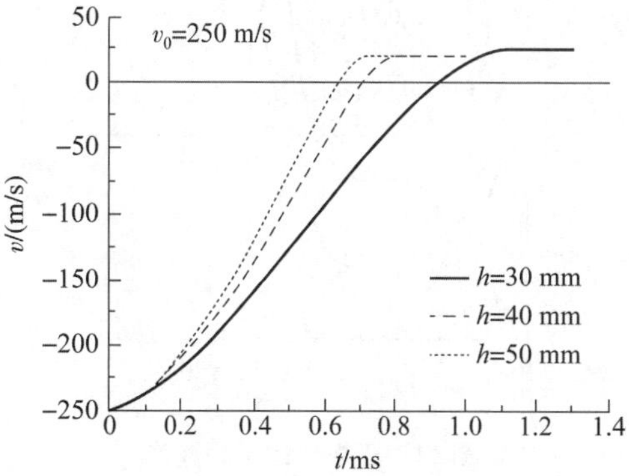

图 9　弹丸冲击靶板时的速度

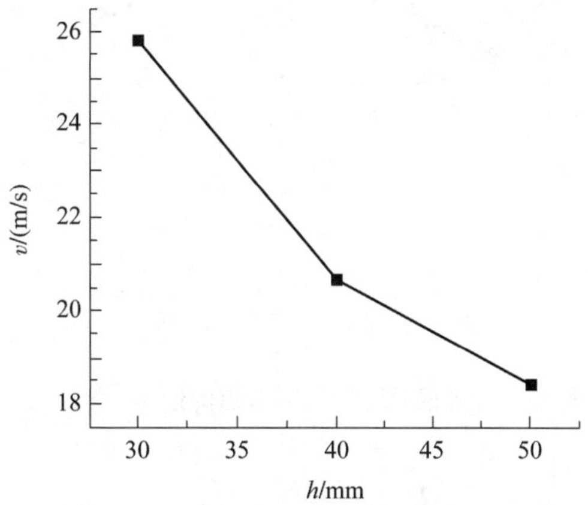

图 10　弹丸最大反弹速度随靶板厚度变化

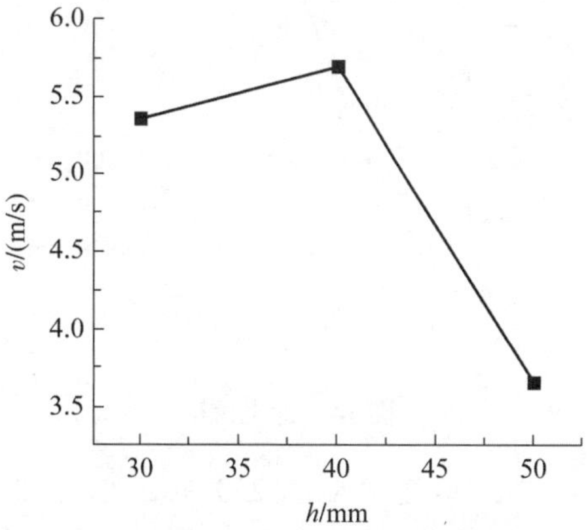

图 11　靶板整体最大回弹速度随靶板厚度变化

2.3　冲击不同边长靶板

弹丸以 250 m/s 的速度冲击厚度为 40 mm 的靶板，靶板边长分别为 400 mm、600 mm 和 800 mm。图 12 ~ 图 15 分别为弹丸冲击不同边长的靶板时，靶板整体平均速度、弹丸加速度、弹丸速度、弹丸位移历程曲线。由图 12 可以看出，靶板边长越大，由初始冲击引起的靶板最大负向运动速度越大。边长 400 mm 和边长 600 mm 靶板的回弹速度达到最大后马上开始减小，而边长 800 mm 靶板的回弹速度达到最大后则保持较长一段时间后才开始减小。由图 13 可以看出，靶板边长对侵彻过程中弹丸轴向加速度影响较小，但对弹丸反弹阶段的加速度影响较大。从弹靶碰撞接触到弹靶板分离，侵彻边长 400 mm 和 600 mm 靶板时，弹丸加速度出现了 2 次峰值，而侵彻边长 800 mm 靶板时出现了 3 次加速度峰值，在反弹阶段，弹丸的加速度峰值随着靶板边长的增加而增加，但侵彻边长 800 mm 靶板的峰值出现在第 3 次。同样由图 14 可以看出，靶板边长对弹丸速度下降影响不大，而对反弹阶段弹丸速度影响较大，靶板边长越小，弹丸越早出现反弹；侵彻边长 600 mm 和边长 800 mm 靶板时，由于靶板边长相对较大，振动周期长，靶板回弹过程中，弹丸受靶板驱动反向运动，且轴向无约束，弹靶分离，弹丸恒速运动，但回弹初期，靶板速度不断增加，当弹坑周围靶元速度等于弹速时，再次驱动弹丸运动，直到弹靶最终分离，如图 15 所示。因图 15 分界面处弹靶相邻节点的速度，图 16 弹丸位移此，弹丸的反弹速度曲线均出现 2 个平台；由图 16 可见，靶板边长对弹丸的最大位移影响很小。

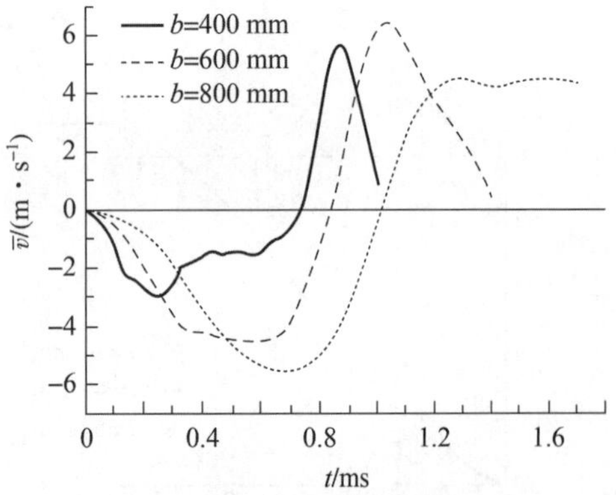

图 12　靶板整体平均速度

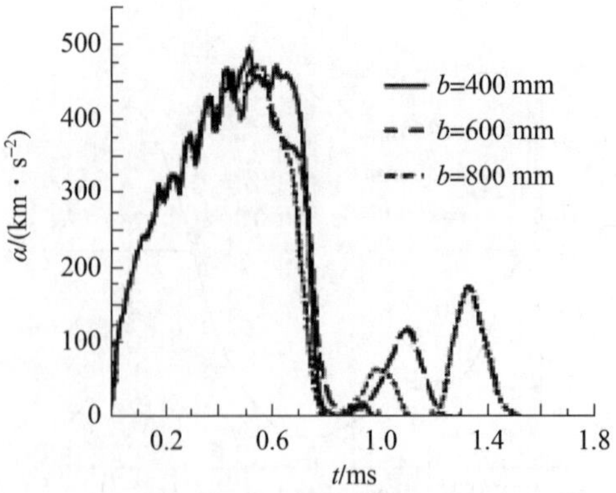

图 13　弹丸加速度

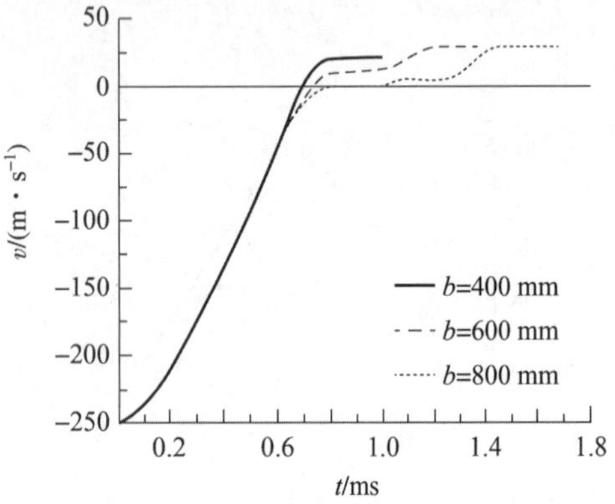

图 14　弹丸速度

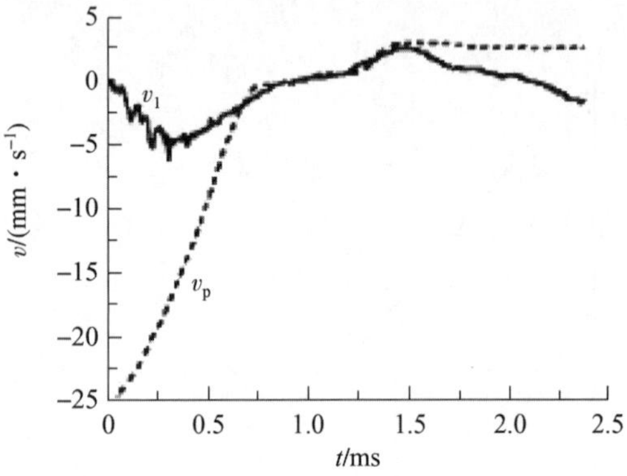

图 15　分界面处弹靶相邻节点的速度

图 17、图 18 是弹丸以 250 m/s 速度冲击不同边长、厚 40 mm 靶板时，弹丸最大反弹速度随靶板边长变化关系曲线和靶板整体最大回弹速度随靶板边长变化关系曲线。由图 17 可以看出，弹丸最大反弹速度随靶板边长的增加先增大后减小，弹丸侵彻边长 600 mm 靶板时最大反弹速度最大，侵彻边长 400 mm 靶板时最大反弹速度最小，侵彻边长 800 mm 靶板时最大反弹速度略小于侵彻边长 600 mm 靶板的最大反弹速度；由图 18 可以看出，弹丸侵彻边长 600 mm 靶板时靶板最大回弹速度最大，而侵彻边长 800 mm 靶板时最大回弹速度最小。

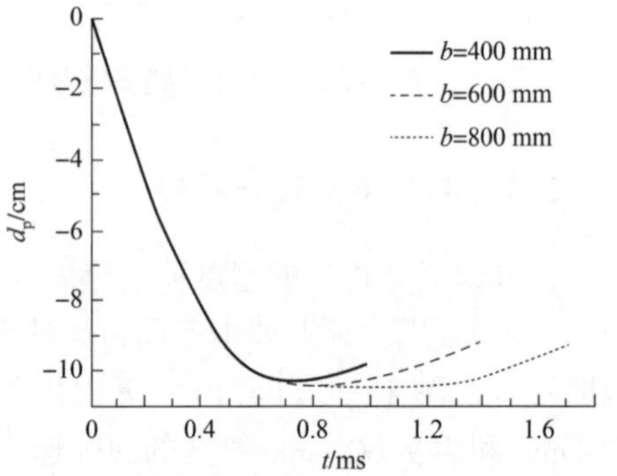

图 16　弹丸位移

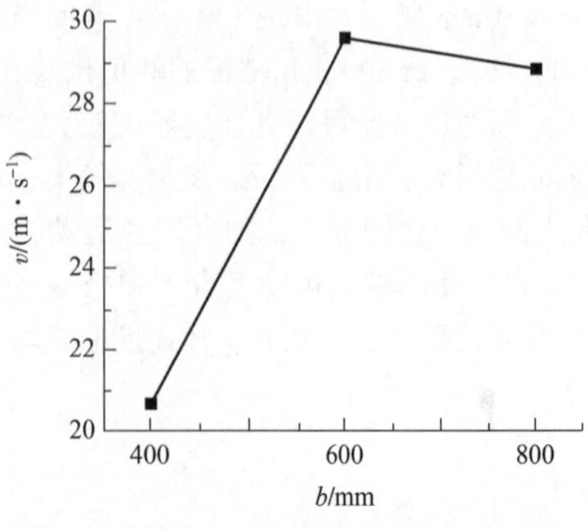

图 17　弹丸最大反弹速度随靶板边长变化

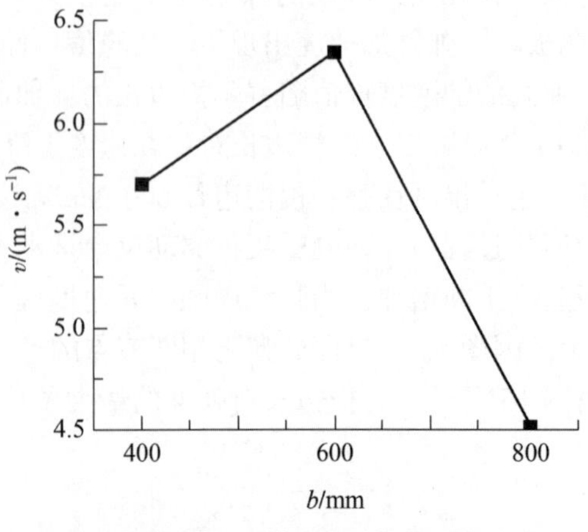

图 18　靶板最大回弹速度随靶板边长变化

2.4　不同速度冲击靶板

通过数值模拟分别研究了弹丸以 150 m/s、200 m/s、250 m/s、300 m/s 和 350 m/s 的速度冲击边长 400 mm、厚 40 mm 靶板的情况。图 19、图 20 分别是弹丸最大反弹速度随冲击速度变化关系曲线和靶板整体最大回弹速度随冲击速度变化关系曲线。图 19 中速度为 350 m/s 已经穿透靶板，没有反弹速，可以看出，低于弹道极限速度随冲击速度变化时，弹丸的反弹速度随着冲击速度的增加而增加；由图 20 可以看出，低于弹道极限速度时，靶板的最大回弹速度也是随着冲击速度的增加而增加的，弹丸穿透靶板的情况下，靶板的回弹速度降低，这主要是在冲击速度小于弹道极限时，弹丸与靶板的接触时间长，靶板整体产生较大的变形，而冲击速度大于弹道极限时，速度越大弹丸与靶板的接触时间越短，靶板仅在局部发生变形。

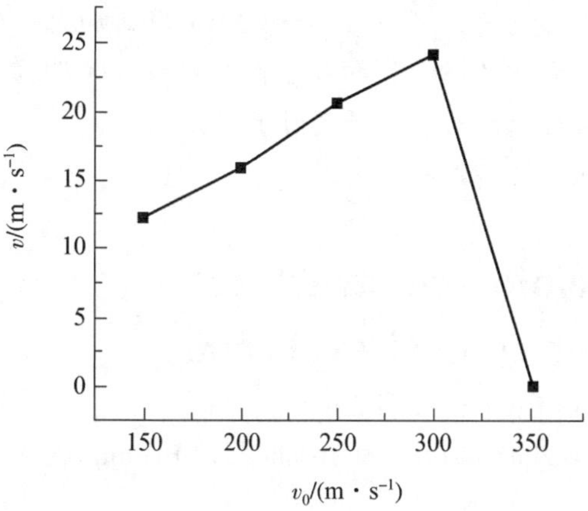

图 19　弹丸最大反弹速度随冲击速度变化

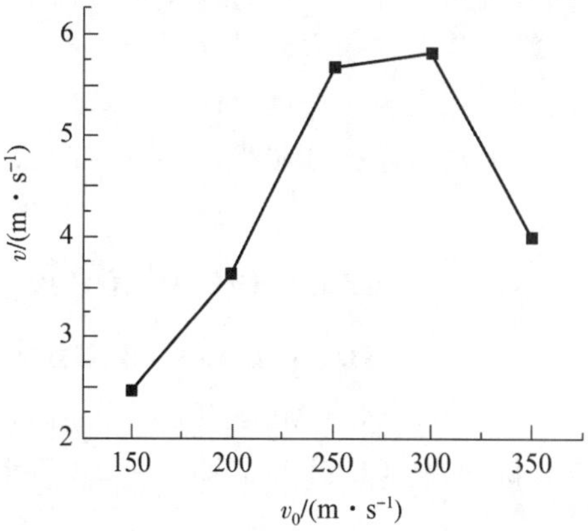

图 20　靶板最大回弹速度随冲击速度变化

3　结论与讨论

通过有限元分析软件 ANSYS/LS－DYNA 对弹丸以不同速度冲击不同厚度和边长的靶板进行了数值模拟，研究了弹丸在低于弹道极限速度范围内冲击不同靶板的过程，分析了靶板在弹道冲击作用下的弹性变形情况，以及弹丸在靶板回弹作用下的反弹情况，得到了：

（1）弹丸以相同速度冲击相同边长、不同厚度靶板时，弹丸的最大反弹速度随着靶板厚度的变化曲线以及靶板的最大回弹速度随着靶板厚度的变化规律；

（2）弹丸以相同速度冲击相同厚度、不同边长靶板时，弹丸的最大反弹速度和靶板的最大回弹速度随着靶板厚度的变化情况；

（3）弹丸以不同速度冲击相同厚度、相同边长靶板时，在小于弹道极限速度时，弹丸的最大反弹速度和靶板的最大回弹速度都随着冲击速度的变化情况。

以上关于靶板厚度、靶板边长和冲击速度对靶板的弹性恢复和弹丸的反弹的影响规律的研究，可为低速穿甲弹对靶板侵彻的工程与实验设计研究提供参考。另外，我们在低速弹道实验时，从高速摄影上也观察到了靶板回弹现象。且对弹丸的低速侵彻弹道影响很大。因为实验中并未对靶板的回弹特性参量进行重点测试，还没有实验现象的详细描述。

参 考 文 献

［1］RADIN J，GOLDSMITH W. Normal projectile penetration and perforation of layered targets［J］. International journal of impact engineering，1988，7：229－259.

不同加筋结构在水中接触爆炸下的破损规律[①]

赖　鸣[②]，冯顺山，黄广炎，边江楠
（北京理工大学爆炸科学与技术国家重点实验室，北京　100081）

摘　要：采用 LS－DYNA 对不同加筋强度和不同加筋位置的筋板结构的水中接触爆炸进行了数值模拟，定义描述不同加筋强度、位置的强度因子和距离因子，建立不同强度因子的 6 种加筋模型和不同距离因子的 3 种加筋模型。通过对水中接触爆炸下典型加筋结构破损过程分析和不同模型纵、横破口长度和破口面积的对比，得到了加筋板在水中接触爆炸下强度因子、距离因子对破 VI 形状和大小的影响规律。结果表明：增大纵横方向强度因子能有效抑制破口面积，增大横向强度因子对破口无明显影响；药量一定情况下，距离因子与破口面积成正比，破口形状与距离因子、强度因子相关。

关键词：爆炸力学；破损规律；LS－DYNA；加筋结构；破口

Damage of different reinforced structures subjected to underwater contact explosion

Lai Ming, Feng Shunshan, Huang Guangyan, Bian Jiangnan
(State Key Laboratory of Explosion Science and Technology, Beijing Institute of Technology, Beijing 100081, China)

Abstract: LS－DYNA was used to simulate the responses of different reinforced structures subjected to underwater contact explosion by considering different reinforcement strengths and locations. Strength and distance factors were defined to describe the strength and locations of stiffeners, respectively. Six models were proposed with different strengths of stiffeners and three models were developed with different locations of stiffeners. Typical damage processes of the reinforced structures were analyzed. The horizontal and vertical crack lengths of crevasse were compared as well as crevasse area. The effects of the strength and distance factors were found on the reinforced structures subjected to underwater contact explosion. The results show that it can be effective to restrain crevasse area when the strength factor rises in the vertical direction and it is not significant when the strength factor rises in the horizontal direction. The distance factor is in direct proportion to crevasse area and the crevasse shapes are related to both of the two factors defined.

Keywords: mechanics of explosion; damage; LS－DYNA; reinforced structure; crevasse

水中接触爆炸是水中武器对舰船常见毁伤方式。舰船筋板等复杂板壳结构在水中接触爆炸下出现的变形、破裂问题，不仅涉及了流体介质与板架结构的相互耦合作用，而且还需要考虑几何非线性与材料非线性及接触问题，同一般的动力学问题相比具有较高复杂性。解析方法只能在极度简化情况下简单描述[1-2]，对加筋板的研究常常利用实验与经验公式相结合的办法[3]。由于实验耗资巨大，利用数值模拟研究加筋板在爆炸冲击载荷作用下的动态响应，已成为研究爆炸冲击响应的重要辅助手段[4-5]。

不同结构舰船的加筋强度和分布位置不同，研究水中接触爆炸下不同强度、不同加筋位置对加筋板破坏作用的影响，不但对船舶防护能力的提高有理论意义，对水中战斗部设计也具有重要的参考价值。

---

① 原文发表于《爆炸与冲击》2012（6）.

② 赖鸣：工学博士，2007 年师从冯顺山教授，研究水中接触爆炸对船舶筋板结构的破坏效应，现工作单位：中国船舶重工集团公司第 714 研究所。

1　数值模型

为研究不同筋板结构在水中接触爆炸下的破坏及变形，按加强筋强度和分布位置不同建立了9种工况模型，见图1（a）~（c）。图1（a）~（c）所示分别为加筋板结构的俯视图、主视图和俯视图；图1（a）中面板大小为500 cm×500 cm，模型1~3所示为面板上厚度分别为1.1 cm、1.3 cm和1.5 cm的加强筋；图1（b）可以看出加筋的最大高度为107 cm，模型4~6分别表示加筋高度为最大加筋高度107 cm 25%（26.75 cm）、50%（53.50 cm）和75%（80.25 cm）；图1（c）中模型7~9分别表示加强筋所处位置在面板对称线一侧的距离为62.5 cm、125.0 cm和187.5 cm。

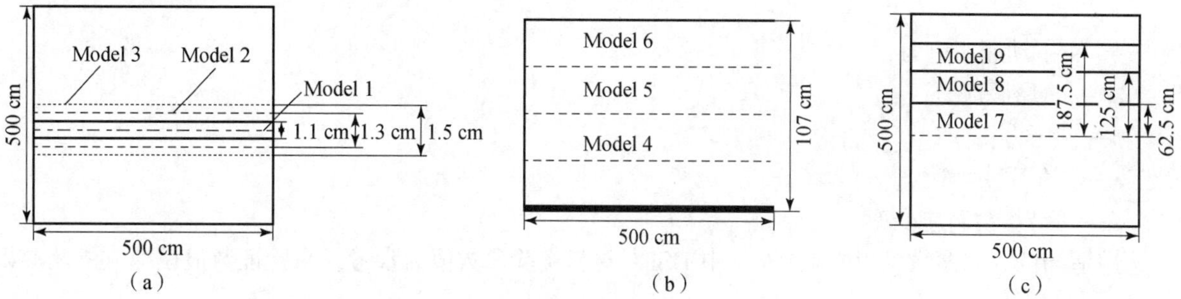

图1　数值模型结构示意图

（a）俯视图；（b）主视图；（c）

在LS-DYNA中，采用流固耦合算法进行数值计算。采用全六面体Lagrange单元建立加筋板结构模型，并增加靶板材料失效参数。利用侵蚀算法得到破裂图像。靶板位于水气交界处且与水中炸药接触。炸药、空气和水均采用多物质Euler单元描述。模型1~6建立1/4模型，模型7~9建立1/2模型，并在靶板固定边界施加全固支条件，流场外围和对称边界处分别施加非反射边界和对称边界条件。

2　材料模型及状态方程[6]

2.1　炸药的材料定义和状态方程

采用TNT材料模型定义炸药，并以JWL状态方程来描述炸药爆轰产生的压力

$$p(V,E)=A\left(1-\frac{\omega}{R_1V}\right)e^{-R_1V}+B\left(1-\frac{\omega}{R_2V}\right)e^{-R_2V}+\frac{\omega E}{V} \tag{1}$$

式中　p——压力，

A、B、R_1、R_2和ω——JWL状态方程的5个参数；

V——相对体积；

E——单位体积内能。

2.2　水和空气

对水和空气在冲击波下的压力、密度关系分别采用Mie-Grüneisen状态方程和多项式状态方程

$$p=\frac{\rho_0c^2\mu\left[1+\left(1-\frac{\gamma_0}{2}\right)\mu-\frac{a}{2}\mu^2\right]}{\left[1-(S_1-1)\mu-S_2\frac{\mu^2}{1+\mu}-S_3\frac{\mu^3}{(1+\mu)^2}\right]^2}+(\gamma_0+\alpha\mu)E \tag{2}$$

$$p=C_0+C_1\mu+C_2\mu^2+C_3\mu^3+(C_4+C_5\mu+C_6\mu^2)E \tag{3}$$

式中　P——压力；

E——单位体积的比内能；

ρ_0——介质初始密度，$\mu=1/(V-1)$；

c——介质中声速；

γ_0、S_1、S_2、S_3、C_0、C_1、C_2、C_3、C_4、C_5、C_6——常数，其中 $C_2=C_6=0$，

α——Grüneisen 系数修正项。

2.3 靶板的材料和本构关系

舰船结构材料为 HSLA 钢，本构关系采用双线性随动强化模型来描述。该模型不但结构形式简单，还能够有效描述材料的应力硬化效应，其动屈服应力与静屈服应力关系为

$$\sigma_{dy}=\sigma_y\left(1+\frac{\dot{\varepsilon}}{c}\right)^{\frac{1}{p}} \tag{4}$$

式中 σ_{dy}——动屈服应力；

σ_y——屈服应力；

p——压力；

c——介质中声速；

$\dot{\varepsilon}$——结构材料应变率。

选用适当的材料参数，以文献［7］中 HSLA 材料实验参数值为参考，来验证数值模拟方法及参数。

3 数值模拟方法的实验验证

按前述计算方法及材料参数进行水中接触爆炸数值模拟，并与文献［7］实验结果进行对比，相关结果如表 1 和图 2 所示。

表 1 数值模拟与实验结果对比

编号	变形/mm		破口/mm		δ/%	
	实验	数值模拟	实验	数值模拟	变形	破口
1			120.0	122.8		2.33
2	35.5	38.0	6.6	6.0	7.04	10.00

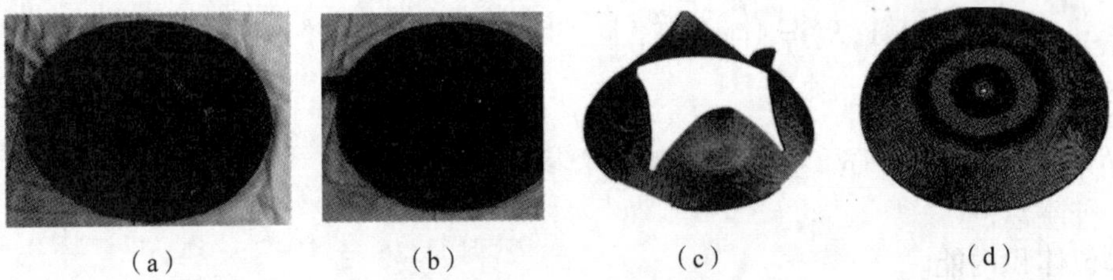

图 2 数值模拟与实验结果对比

(a) 实验 1；(b) 实验 2；(c) 实验 3；(d) 实验 4

炸药起爆位置在面板结构的几何中心处。实验 1 药量为 20 g，实验 2 药量为 10 g。实验 1 药量较大，产生较大破口，未对结构变形进行测量。从表 1 中看出，小药量时数值模拟的破口长度尺寸误差较大，达到 10%，原因是圆板结构网格密度不够细密，结果中仅破裂了 2 个网格，第 3 个网格并未达到消去条件，因而形成较大误差。由此可见，对于水中接触爆炸，在破口较大的条件下，数值模拟与实验结果误差在 5% 以内，在破口较小条件下误差也控制在 10% 以内，满足工程计算要求，可认为数值模拟的方法和参数是合理的。

4　数值模拟结果分析

4.1　加强筋强度对破口的影响

为研究相同药量下不同加筋强度对板架破坏效应的影响，设计了模型 1 ~6。6 种数值模型的结果在加强筋两侧均出现椭圆形破口且对称，加强筋本身在接触炸药的一侧受到冲击波侵蚀、破坏，形成类似拱形的破损带。模型 1 ~6 破口形状相似大小不同，其中模型 1 的数值模拟结果如图 3 所示。

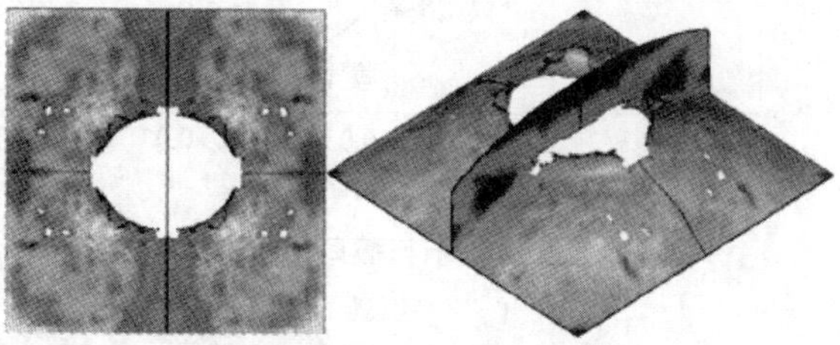

图 3　模型 1 的数值模拟结果

加强筋在横、纵 2 个方向的变化导致强度不同，将加强筋体积定义为描述强度的强度因子 v，从纵、横 2 个方向改变加强筋训值，其中模型 1 ~3 从横向改变，模型 4 ~6 从纵向改变。不同模型的数值模拟结果见表 2，其中，$L_{H,\max}$为横向最大长度，$L_{V,\max}$为纵向最大长度，S 为破口面积。

表 2　不同强度加筋模型数值模拟结果

模型	v/m^3	$L_{H,\max}/cm$	$L_{V,\max}/cm$	S/m^2
1	0. 058 9	235. 22	201. 32	3. 72
2	0. 069 6	236. 06	222. 46	4. 12
3	0. 080 1	260. 02	203. 36	4. 15
4	0. 014 7	277. 00	309. 18	6. 73
5	0. 029 4	326. 84	246. 84	6. 34
6	0. 044 1	236. 06	223. 60	4. 15

对应表 2 中数据，强度因子 v 与破口面积 S 变化趋势如图 4 所示。从图 4 中看出，破口面积总的趋势是随着加筋强度增大而减小，但破口面积最小时并不是加强筋强度最大，说明对于一定药量的接触爆炸，加强筋强度有一个最佳值，并不是强度越大越好。整个趋势单调下降阶段代表模型 4 ~6，当加强筋高度减少到原高度的 75% 以下时，破口面积对加筋强度很敏感；余下部分代表模型 1 ~3，当加强筋强度增大时，加强筋和板之间的应力集中更严重，造成更大的横向撕裂，导致破口面积增大，同时由于加筋强度整体变大，使破口面积增大量不多。当加强筋高度减少时（模型 6）与加强筋宽度增加时（模型 2）破口面积相近，说明调整加强筋高度比调整宽度更加有效（加强筋强度因子更小）。

4.2　加强筋位置对破口的影响

模型 7 ~9 模拟结果显示不对称加强筋造成不对称破口形状，破口裂纹和花瓣向无筋方向延伸、翻转，翘曲高度在有加强筋的一侧得到有效抑制。其中模型 7 数值模拟结果如图 5 所示。

定义加强筋距离炸药所处位置的距离 L 为距离因子，以 L 的变化来研究不同加筋位置对破口尺寸的影响。模型 7 ~9 距离因子 L 与破口面积 S 关系如图 6 所示。

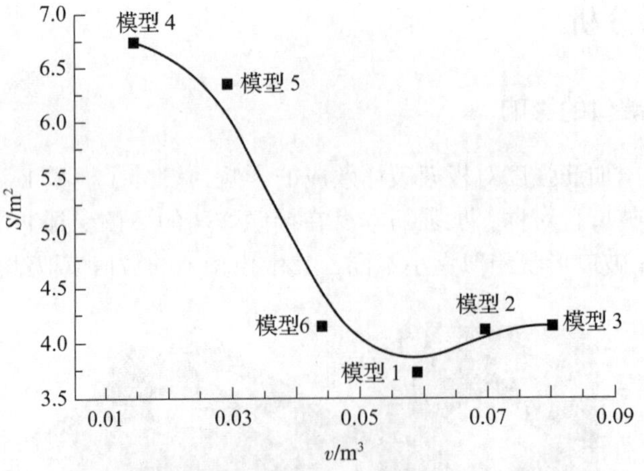

图 4　强度因子与破口面积曲线

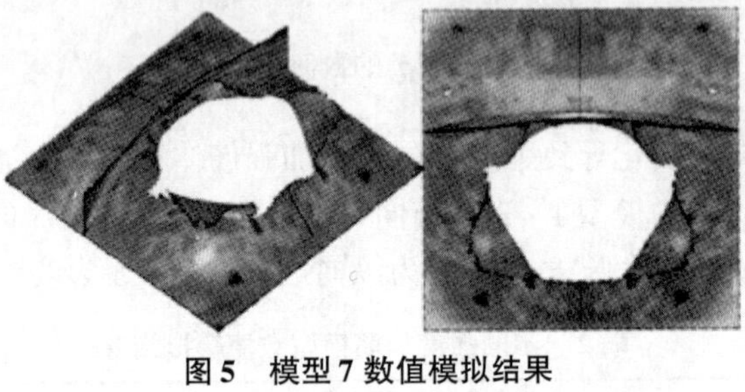

图 5　模型 7 数值模拟结果

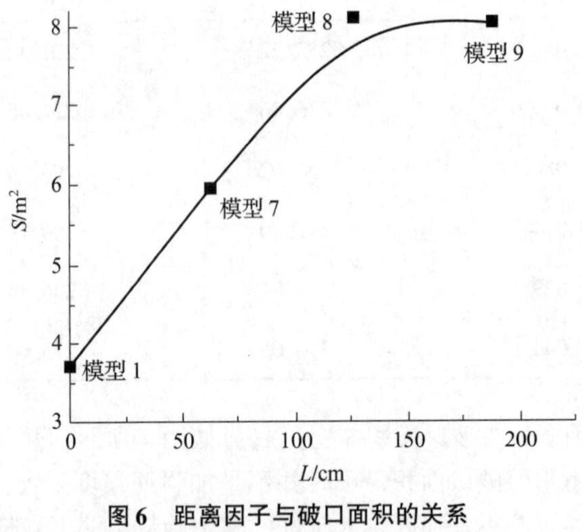

图 6　距离因子与破口面积的关系

从图 6 可以看出，距离因子与破口面积成正比。加强筋位于炸药正上方时（模型 1）破口面积最小，随着加强筋远离炸药，其阻滞破口能力越来越低。在炸药作用半径内变化加强筋布置位置能很大程度改变破口面积和形状。模型 7 比模型 1 的破口面积增大约 60%，模型 8 仅比模型 7 增加 36%。这说明加强筋布置位置离炸药的作用半径越远，影响能力越弱。在炸药作用半径以外设置加强筋对破口抑制基本没有作用。

4.3　筋板结构变形过程

模型 1 和模型 7 分别代表加强筋位于炸药正上方和加强筋在炸药一侧的 2 种结构。图 7 所示为模型 1 的变形过程。图 7（a）中为初始时筋板结构在冲击波作用下出现一个比装药半径略大的初始破口，同时，炸药正上方加强筋出现小范围破损；图 7（b）中为破口在冲击波作用下继续扩张，沿筋布置方向和垂直于筋的方向出现应力集中，表现出撕裂状态；图7（c）中撕裂口继续延伸，同时破口花瓣开始翻转；图7（d）所示为最终筋板结构状态。图8所示为模型7的变形过程。图8（a）中，面板结构出现初始破口，破口形状与筋板上的应力分布近似为圆形；图8（b）中破口扩张受筋板的限制；图8（c）中，明显表现出筋板对破口的阻滞作用，破口向未加筋的一侧延伸，直到如图8（d）中筋板动响应停止。

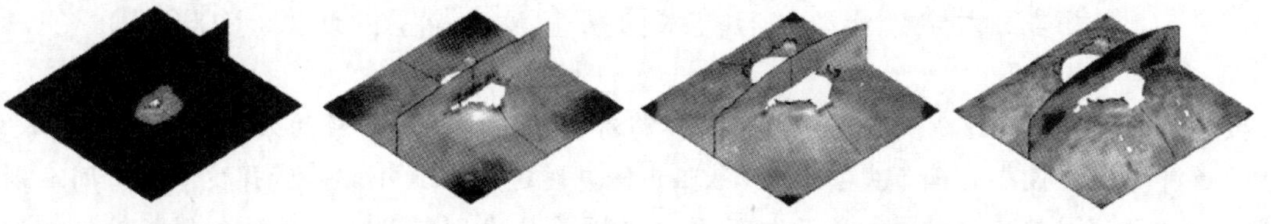

（a）　（b）　（c）　（d）

图7　模型1变形过程

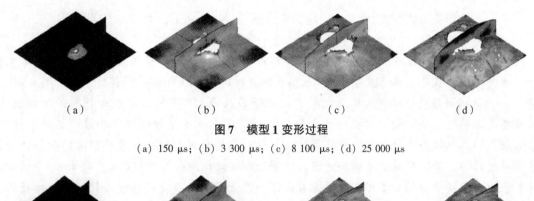

（a）150 μs；（b）3 300 μs；（c）8 100 μs；（d）25 000 μs

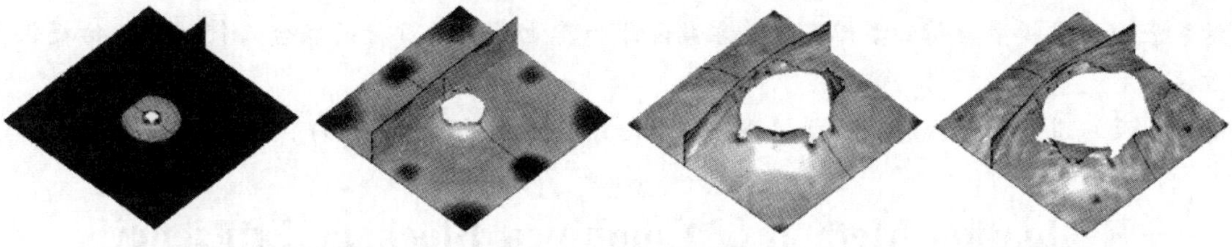

（a）　（b）　（c）　（d）

图8　模型7变形过程

（a）150 μs；（b）1 350 μs；（c）10 950 μs；（d）25 000 μs

从上述变形过程看出，加筋板的破坏过程主要分为三个阶段：①接触爆炸形成初始破口；②破口迅速扩大，形状与加强筋位置和强度相关，一般分为加强筋断裂与不断裂两类；③破口继续延伸，出现花瓣并翻转。同时整个板架出现整体大变形，直到爆炸能量耗尽。

5　结论

利用LS－DYNA有限元软件对水中接触爆炸进行数值模拟，能够清楚地显示对加筋板结构的破坏效应和加筋板结构的瞬态变形及破口扩张过程，一定药量接触爆炸时局部破坏先于整体破坏产生。用加强筋体积ν描述不同强度加强筋对破口大小的影响规律。对于一定药量的接触爆炸，破口面积随着ν的增大先减小后增大，并具有一个最佳值。增大加强筋高度能有效减小破口面积，高度每增高25%破口面积分别减少6%、63%和12%，高度在50%~75%变化时对破口面积影响最大。加强筋宽度增加时导致破口面积轻微增大。用加强筋离炸点的距离L描述不同位置加强筋对破口形状和大小的影响，加强筋距离炸药越近，破口面积越小，L每增大25%，破口面积分别增大60%、36%和1%，说明在一定药量下，L越小，对破口影响越大，当加强筋大于炸药作用半径时，对破口面积基本无影响。

参考文献

[1] 李磊．某弹药系统总体技术研究［D］．北京：北京理工大学，2008．

[2] 盖京波，王善，唐平．薄板在接触爆炸载荷作用下的破坏分析［J］．哈尔滨工程大学学报，2006，27（4）：523－525．

GAI J B，WANG S，TANG P. Damage of thin plate subjected to contact explosion loading［J］. Journal of Harbin Engineering University，2006，27（4）：523－525.

反跑道与区域封锁子母弹联合封锁效能的评估方法①

黄广炎②，邹　浩，王成龙，冯顺山
（北京理工大学爆炸科学与技术国家重点试验室，北京　100081）

摘　要：为了研究反跑道弹药与区域封锁型弹药对机场跑道联合封锁的效能评估方法，本文提出了基于判断跑道破坏的“最小起降带失去准则”、阻止修复跑道的“最小起降带弹坑修复时间准则”，以及跑道修复后的“区域封锁型子弹对起降飞机的再次威慑杀伤准则”3种准则联合判断封锁效能的方法，通过建立反跑道弹药与区域封锁型弹药的落点模型、判断反跑道子弹封锁概率的最小起降带计算方法，以及区域封锁型弹药对起降飞机封锁概率计算方法，得到了2种弹药联合封锁作用时的封锁效能评估模型，计算了单发母弹、不同反跑道子弹和区域封锁型子弹装配比例时对长600 m，宽45 m跑道带的封锁概率，结果表明，该模型为封锁型弹药封锁方案设计和其他高价值目标的联合封锁效能评估提供了实用的分析方法。

关键词：反跑道子弹；区域封锁型弹药；联合封锁；机场跑道；效能评估

Evaluation Method of Combined Blockage Efficiency of Anti－runway Warhead and Zone Blockage Warhead Carried by Dispenser

Huang Guangyan，Zou Hao，Wang Chenglong，Feng Shunshan
（State Key Laboratory of Explosion Science and Technology，Beijing Institute of Technology，Beijing 100081，China）

Abstract：To study the evaluation method of efficiency of combined anti－runway warhead and zone blockage warhead blockading airport runway，the method to judge the efficiency of blockading airport runway was presented based on three criteria，such as the criteria for judging the destruction of airport runway by checking whether the minimum flying area is still existing，the time criteria of preventing the repair of craters within the minimum flying area，and the killing criteria of zone blockage warhead to airplane after the minimum flying area on runway was repaired. The model of drop points of anti－runway warhead and zone blockage warhead was established，and the calculation method of minimum flying area to judge the blockage probability of anti－runway warhead to runway was proposed as well as the calculation method of killing probability of zone blockage warhead to flying airplane. The evaluation model of runway blockage efficiency under combined blockage by the two kinds of blockage warhead was obtained. Under the conditions of different assembly scale of anti－runway and zone blockage warhead carried by one dispenser，the probability of blockading runway with 600 m long and 45 m wide was calculated. The result shows that the evaluation model can offer an important reference value for the design of blockage method of submunition and the efficiency evaluation of other blockage effect.

Keywords：anti－runway ammunition；zone blockage warhead；combined blockage；airport runway；efficiency evaluation

---

①　原文发表于《弹道学报》2013（1）。

②　黄广炎：工学博士，2004年师从冯顺山教授，研究封锁型弹药设计及效能分析方法，现工作单位：北京理工大学。

多种类封锁型子母式弹药联合封锁作用是对机场跑道、交通枢纽、航母飞行甲板、大型电力设施等高价值目标低附带毁伤的重要发展方向，已成为终点弹道与毁伤领域的研究热点。新的联合封锁作用样式使得子弹群终点弹道更具复杂和多样性，引发了高价值目标封锁效能评估方法的重大变革。

以机场跑道封锁为例，目前国内外均已开展反跑道侵爆子弹药与区域封锁子弹药复合的多种类封锁型子母弹药的设计与使用工作。通过反跑道侵爆子弹药与区域封锁弹药的综合使用来同时达到损坏飞机起降跑道和阻止已损坏跑道修复的目的。其中反跑道侵爆子弹药一般装载侵爆战斗部，可侵入机场跑道一定深度爆炸，形成一定数量弹坑，使机场跑道暂时失去起降功能。而区域封锁型弹药一般为杀伤战斗部，抛撒在跑道的弹坑附近，阻止或延缓机场跑道的修复，也可有效杀伤在已强行修复跑道上执行起降任务的飞机，实现多重封锁效应叠加的目的。

针对单一类型封锁型子母弹对机场跑道的封锁效能分析，国内外学者均已开展了深入系统的研究，并得到了科学的封锁概率和打击效果评估计算方法。如反跑道侵爆子弹药方面，建立了基于最小起降带的封锁概率与打击效果评估计算方法[1-3]，在区域封锁型弹药方面，基于破片打击迹线方法进行了封锁概率计算模型的研究[4]。但针对多种类封锁型子母式弹药联合封锁作用的情况，国内外还较少展开其封锁效能评估的研究工作，尚缺乏科学的评估方法和数学分析模型，因此本文将以对机场跑道的封锁为例，研究区域封锁型子弹与反跑道子弹联合封锁作用情况下的封锁效能数学分析模型和评估方法。

1 联合封锁作用的封锁判断准则

1.1 机场跑道的联合封锁准则构成

根据区域封锁型子弹与侵爆子弹对机场跑道联合封锁的作用特点，可通过建立判断跑道破坏的“最小起降带失去准则”，阻止破损跑道修复的“最小起降带弹坑修复时间准则”，以及跑道修复后的区域封锁子弹破片杀伤元对强行起降飞机的“威慑杀伤准则”来综合分析2种封锁型弹药联合作用情况下机场跑道的空间封锁效应和时间封锁效应。

1.2 最小起降带失去的空间封锁准则

假设母弹平台的子弹抛撒样式为：以母弹落点为中心，抛撒区域内子弹随机分布，每一个击中机场跑道的反跑道子弹均能够在跑道上形成一个阻碍飞机滑行起降的弹坑，将各弹坑均等效为半径为 R_b 的标准圆形区域。衡量反跑道侵爆子弹群封锁机场跑道成功的“最小起降带失去准则”为：在机场跑道遭到破坏之后的一段时间内，若找不到任何一块可满足飞机起降滑跑所需的最小跑道长度和宽度的跑道面，则认为空间封锁成功，否则即为失败。现代作战飞机起飞最小滑行段长为300～450 m，再考虑飞机轮距宽度9～14 m，一般可选取最小起降带为300 m×10 m，真实评估中可根据实际飞机性能参数调整最小起降带标准。

1.3 阻止跑道最小起降带弹坑修复的时间封锁准则

维持机场跑道破损封锁状态（阻止最小起降带修复）的时间取决于机场快速修复跑道的能力，反封锁（2种弹药的联合封锁时间）所需时间为 t_w：

$$t_w = t_s + t_d + t_p + t_x \tag{1}$$

式中 t_s——机场跑道损毁数据处理所需时间；

t_d——确定跑道修复方案所需时间；

t_p——在确定修复区域排除影响修复的所有区域封锁型子弹所需时间，也即区域封锁子弹阻止跑道弹坑修复的时间；

t_x——修复待抢修区域中所有弹坑所需的时间。

式（1）中的 t_s，t_d 所需时间都比较好确定，可取一个均值来表示；t_x 可通过公式计算：

$$t_x = n_q t_q \tag{2}$$

式中　n_q——最小起降带区域内的弹坑数；

t_q——修复每个弹坑的时间。

为了简化问题，假设能对欲清理出的最小起降带起到杀伤威慑作用的区域封锁型子弹数为 n_w，排除其中第 h 枚区域封锁型子弹所需的时间为 $t_{p,h}$，则有

$$t_p = \sum_{h=1}^{n_w} t_{p,h} \quad h = 1,2,\cdots,n_w \tag{3}$$

通常要根据战场形势来确定封锁机场的预期时间，记为 t_a。那么，侵爆型子弹与区域封锁型子弹联合对机场跑道的时间封锁效率 Q_f 可表示为

$$Q_f = \begin{cases} t_w/t_a & t_w < t_a \\ 1 & t_w \geqslant t_a \end{cases} \tag{4}$$

1.4　区域封锁子弹对跑道的威慑杀伤准则

即使跑道上某条最小起降带被修复，附近的未排除区域封锁型子弹可对正在强行起飞的飞机进行再次威慑杀伤。区域封锁子弹一般为随机延时起爆，杀伤战斗部群产生的某枚破片 p_e 将具备时间 t_e、空间 S_e 以及杀伤威力参数 Q_e3 种杀伤因素。

$$\begin{cases} p_e = F(t_e, S_e, Q_e) \\ t_e = F(t_1, t_2) \\ S_e = F(x_1, y_1, z_1, x_2, y_2, z_2) \\ Q_e = F(\nu, m, \varphi(e), \xi(e)) \end{cases} \quad e = 1,2,\cdots,n_f \tag{5}$$

式中　t_e，S_e——破片 p_e 从战斗部爆炸驱动开始至速度衰减到不具备杀伤能力或着地所经历的时间段和空间位置；

Q_e——破片 p_e 的杀伤威力参数；

ν，m，$\varphi(e)$，$\xi(e)$——破片的初速、质量、形状及材料特性；

n_f——杀伤战斗部产生的破片总数。区域封锁型子弹对跑道的威慑杀伤概率可通过破片打击迹线对起降飞机的时间、空间交汇统计来计算[4]。

2　联合封锁效能评估模型

2.1　子母弹抛撒落点计算模型

假设 K 枚母弹抛撒瞄准点地面投影坐标分别为 $O_i(x_i, y_i)$，实际落点坐标为 $O'_i(x'_i, y'_i)$，$i=1,2,3,\cdots,K$ 令第 i 枚母弹的圆概率偏差为 C_i，则该母弹的实际落点坐标为

$$\begin{cases} x'_i = x_i + \dfrac{C_i}{\lambda}\cos(2\pi r_1)\sqrt{-2\ln r_2} \\ y'_i = y_i + \dfrac{C_i}{\lambda}\cos(2\pi r_1)\sqrt{-2\ln r_2} \end{cases} \tag{6}$$

式中　r_1，r_2——（0，1）区间上相互独立的均匀分布随机数；

λ——母弹的均方差与圆概率偏差之间的换算系数。

假设每枚母弹携带子弹数为 n，子弹抛撒方式为以 $O'_i(x'_i, y'_i)$ 为椭圆圆心，抛撒椭圆长短半轴为 a，b，抛撒椭圆死区的长短半轴为 c，d 的随机布撒方式。第 i 枚母弹中第 j 枚子弹落点地面投影坐标 $O_{z,xy}(x_{z,ij}, y_{z,ij})$，$j=1$，2，…，$n$，则

$$\begin{cases} x_{z,ij} = x'_i + r_3\cos(2\pi r_5) \\ y_{z,ij} = y'_i + r_4\cos(2\pi r_5) \end{cases} \tag{7}$$

式中　r_3，r_4，r_5——(c,a)，(b,d)，$(0,1)$区间上相互独立的均匀分布随机数。

因篇幅有限，这里不对子母弹抛撒点计算模型做更多阐述，详细的子母弹落点计算方法可参考笔者此前的研究工作[4]。

2.2　反跑道子弹对跑道封锁概率计算模型

利用蒙特卡罗方法的原理，在模拟母弹、子弹的随机落点的基础上，判断机场跑道上是否存在最小升降窗口，图1所示为有效反跑道侵爆子弹的落点区域范围，图中，①、②、③、④区域的宽为 r_z，⑤、⑥、⑦、⑧的半径为 r_z，只有落在上述区域或图中 W 区域内的子弹才能有效覆盖最小起降窗口，并可称之为有效子弹落点。分别用（w_{x1}，w_{y1}），（w_{x2}，w_{y1}），（w_{x2}，w_{y2}），（w_{x1}，w_{y2}）记录最小起降窗口移动过程中左下角起逆时针方向的4个端点的位置坐标。

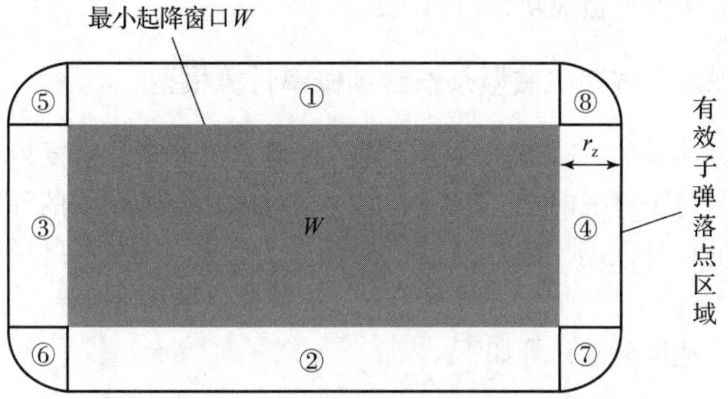

图1　有效反跑道侵爆子弹的落点区域范围

具体可使用窗口检验法，首先确定一个最小起降窗口 W（简称为窗口），其尺寸为 $W_x \times W_y$，如图2所示。

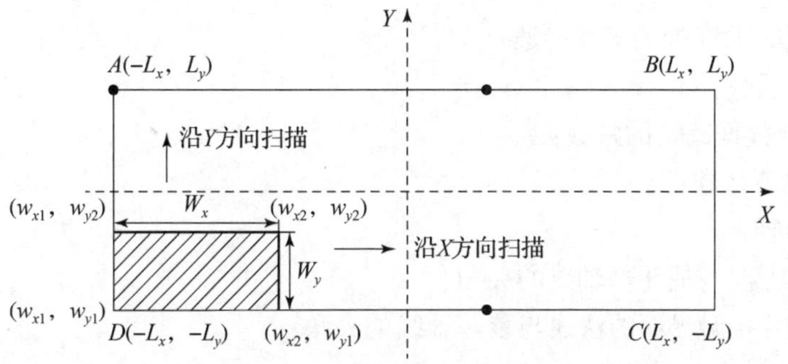

图2　最小起降窗口验证法示意图

在计算中，首先将最小起降窗口从跑道左下顶点 $D(-L_x, -L_y)$ 开始沿 X 方向扫描，判断是否有子弹落点 $E(X_k, X_k)$ 在窗口范围内。

（1）如果没有子弹落在上述范围内，则封锁失败，开始下一次的模拟；如果子弹落点 E 在①、⑤、⑧内，则窗口平移到 $w_{x1} = X_k + [r_z^2 - (Y_k - w_{y2})^2]^{1/2}$；如果落点在②、⑥、⑦内，则窗口平移到 $w_{x1} = X_k + [r_z^2 - (w_{y1} - Y_k)^2]^{1/2}$；如果落点在③、④或 W 内，则窗口平移到 $w_{x1} = X_k + r_z$。

（2）重复（1）进行判断、平移，直到 $w_{x2} \geqslant L_x$。

（3）窗口左下顶点移至（$-L_x$，$L_y + r_z + W_y$），重复（1）、（2）。

（4）窗口逐次上移 W_y，重复（1）、（2），直到 $w_{y2} \geqslant L_y$。

（5）将最小起降窗口从跑道左下顶点开始沿 Y 方向扫描，判断是否有子弹落点在图2所示的范围内。如果没有子弹落在上述范围内，则封锁失效，开始下一次的模拟；如果落点 E 在①、②或 W 内，则

窗口平移到 $w_{y1}=X_k+r_z$；如果落点在③、⑤或⑥内，则窗口平移到 $w_{y1}=X_k+[r_z^2-(w_{x1}-X_k)^2]^{1/2}$；如果落点在④、⑦或⑧内，则窗口平移到 $w_{y1}=X_k+[r_z^2-(X_k-w_{x2})^2]^{1/2}$。

(6) 重复 (1) 进行判断、平移，直到 $w_{y2}\geqslant L_y$。

(7) 窗口左下顶点移至 $(-L_x+W_x,\ -L_y+r_z+W_y)$，重复 (1)、(2)。

(8) 窗口逐次右移 W_x，重复 (1)、(2)，直到 $w_{x2}\geqslant L_x$。

对于一定长度和宽度的跑道目标，采用 Monte – Carlo 方法进行模拟统计试验，根据大数定理（伯努利定理），在相同条件下进行无限次独立试验时，跑道被成功封锁的事件出现的频率 f_b 将收敛于跑道被封锁的概率 P_b，因此可得到一定数量反跑道子弹攻击条件下机场跑道的封锁概率：

$$P_b\approx f_b=N_b/N \tag{8}$$

式中 N——对机场跑道进行模拟打靶试验的次数；

N_b——N 次模拟试验中跑道被成功封锁的次数。

2.3 区域封锁型子弹对起降飞机威慑杀伤封锁概率计算模型

区域封锁型子弹对修复跑道的再次封锁具体计算，可通过统计在已修复最小起降带强行起飞的飞机与杀伤战斗部群形成的破片打击迹线的交汇数据来实现。对破片打击迹线的计算，需建立战斗部破片杀伤作用场各威力特性参数的计算模型。破片对目标杀伤威力性能参数主要包括质量、初速、飞散方向、速度（能量）衰减等[5]。

破片初速、飞散方向、速度衰减计算如下：

$$\begin{cases} v_{0,f}=[1-A_1e^{-Bf/d(f)}][1-A_2e^{-C(L-f)/d(f)}]\times\sqrt{2E_g}\sqrt{\beta(f)/[1+0.5\beta(f)]} \\ \varphi_f=\gamma_f-\arcsin\left[\dfrac{v_{0,f}}{2v_u}-\dfrac{v'_{0,f}\tau}{2}-\dfrac{(v'_{0,f}\tau)^2}{5}\right] \\ v_{f,s}=v_{0,f}\,e^{-\alpha S}=v_{0,f}\mathrm{e}^{\frac{-C_D\rho\psi m^{-1/3}}{2}S} \end{cases} \tag{9}$$

式中 $v_{0,f}$——距离起爆端面 f 处的破片初速；

A_1，A_2——修正系数；

B，C——由实验数据确定的常数；

$\beta(f)$——爆炸载荷系数；

$d(f)$——装药直径；

φ_f——破片飞散方向与战斗部轴线的夹角；

γ_f——炸药、破片接触界面的法线与装药轴线的夹角；

$v'_{0,f}$——破片初速轴向梯度；

τ——破片加速常数；

v_u——爆轰波扫过炸药壳体交界面的速度；

$v_{f,s}$——破片的速度衰减，

C_D——常数；

ρ——空气密度；

S——破片迎风面积；

ψ——破片形状系数。

联立方程组 (9) 即可计算得到各子母弹战斗部所有破片打击迹线的空间与时间分布[6]。

在已知区域封锁型子弹打击条件及目标机场跑道、飞机起降等相关参数条件，以及破片打击迹线与起降飞机的交汇情况，即可得到飞机目标被有效威慑杀伤的统计数据。更为具体的基于破片打击迹线的区域封锁型弹药对飞机目标的封锁概率计算模型和方法可参考文献 [6]。

3　联合封锁作战效能评估计算

3.1　计算条件与假设

选取长 600 m、宽 45 m 的飞机跑道段进行分析。假设战场形势要求该发携带反跑道侵爆子弹与区域封锁型子弹的子母式弹药对机场该跑道段联合封锁的预期时间为 $t_a=90$ min。飞机起降所需的最小起降带区域为 300 m×10 m；母弹预定落点为该机场跑道正中央，圆概率偏差为 20 m，可装配子弹 200 枚，子弹抛撒方式为椭圆环内随机分布，抛撒椭圆长短半轴分别为 320 m、42.5 m，椭圆死区的长短半轴分别为 40 m、5 m。单发侵爆子弹毁伤半径为 2 m；假设修复 1 个跑道弹坑所需的标准时间 $t_q=240$ s，判定机场跑道损毁情况需要的时间 $t_s=300$ s，确定应急跑道修复方案需要的时间 $t_d=300$ s，排除 1 枚区域封锁型子弹的平均时间为 600 s。假设区域封锁型子弹群在 90 min 内全部随机延时起爆完成。这些子弹在跑道失去最小起降带时，主要对抢修跑道的人员和车辆产生威慑，阻止跑道修复；在最小起降带修复后，可对强行起降飞机进行威慑杀伤封锁。

3.2　联合封锁效能计算与分析

计算单发母弹在不同反跑道子弹和区域封锁型子弹装配比例情况下对 600 m×45 m 跑道带的封锁概率，包括反跑道子弹对机场跑道的空间封锁概率（失去最小起降带的概率）、欲修复最小起降带内的弹坑修复时间、区域封锁型子弹对最小起降带的跑道修复阻止时间，以及对在已修复最小起降带强行起飞飞机的威慑杀伤封锁能力。最终统计得到 2 种类型子弹在不同配比下对机场跑道的联合封锁时间及效能，如表 1 所示，表中，P_b 为侵爆子弹对机场跑道的空间封锁概率，t_x 为修复待抢修区中所有侵爆子弹弹坑所需的时间，t_p 为区域封锁子弹的威慑封锁时间，P_w 为区域封锁型子弹对强行起降飞机威慑杀伤概率，t_w 为 2 种类型子弹的联合封锁时间。

表 1　2 种类型子弹在单发母弹不同配比情况下对 600 m×45 m 机场跑道的联合封锁效能

母弹携带 2 种类型封锁子弹配比		反跑道侵爆子弹封锁效能		区域封锁子弹封锁效能		2 种类型子弹联合封锁效能	
反跑道	区域封锁	P_b/%	t_x/s	t_p/s	P_w/%	t_w/s	Q_f/%
0	200	—	—	8 460	71.6	8 460	100
30	170	12.5	552	6 975	65.6	7 527	100
50	150	30.1	1 080	6 030	56.4	7 110	100
80	120	46.9	1 704	5 040	38.6	6 744	100
100	100	54.9	2 208	4 140	28.3	6 348	100
120	80	64.8	2 688	3 195	25.8	5 883	100
150	50	76.3	3 216	2 025	16.6	5 241	97.1
170	30	85.3	3 720	1 035	12.6	4 755	88.1
200	0	92.3	4 512	—	—	4 512	83.1

对表 1 统计获得的数据分析，随着单发母弹携带反跑道侵彻子弹的增加，其对该段机场跑道的空间封锁概率和封锁时间均增加，当其全部携带反跑道侵彻子弹时，空间封锁概率可达到 92.3%，但封锁时间仍然不到 90 min(4 512 s)，达不到时间封锁期望；同样，区域封锁子弹的子弹可有效阻止跑道弹坑修复，且提高对强行起降飞机威慑杀伤概率，但其单独使用效果不明显。

2种弹药联合使用可明显提高对机场跑道的综合封锁效能，但装配比情况决定着不同的空间封锁概率、时间封锁效能和对强行起降飞机的杀伤威慑能力。其中反跑道子弹决定了空间封锁概率及弹坑修复时间，区域封锁型子弹则决定弹坑修复的难度及对强行起降飞机的杀伤威慑能力。如单发母弹携带反跑道子弹和区域封锁子弹分别为120枚和80枚时，其空间封锁概率为64.8%，封锁时间为5 883 s，超过时间封锁期望（108.9% >1），而且该封锁期间内对强行起降飞机的杀伤概率为25.8%，具备较好的联合封锁效果。在本计算案例中，反跑道子弹与区域封锁型子弹的装配比例在1.5∶1时，其对该跑道区域的联合封锁效果较理想。

由以上的计算数据可知，当选取2种类型子弹药单发母弹装填对机场跑道进行联合封锁时，根据不同类型子弹的技术指标、性能参数、目标机场特性和不同封锁作战的要求（封锁时间期望、空间封锁要求、威慑杀伤强度），可通过使用本效能评估方法和计算模型，获得评估数据，依此合理调整子弹配比，适应不同封锁作战需求，提高子弹效率，加大对机场跑道的综合封锁效能。

另外，虽然本计算案例所做的假设和计算条件均有一定的局限性，但所建立的综合效能评估模型，通过一定的修正，可灵活适应于其他高价值目标的综合封锁效能分析。

4 结论

以多类型封锁型子弹对机场跑道的联合封锁效能评估为对象，提出了区域封锁型子弹与反跑道子弹联合封锁作战情况下的机场跑道综合封锁效能分析方法，建立了空间封锁概率、时间封锁效能和对修复后的最小起降跑道上强行起飞飞机的威慑杀伤概率3种封锁综合的效能评估模型。所得到的联合封锁综合效能评估模型，可适应于不同技术指标、性能参数的封锁型子弹、典型目标特性和不同封锁作战要求的高价值目标联合封锁作战的综合封锁效能计算和评估，为联合封锁子弹配比和战术设计提供数据支撑。

参考文献

[1] 杨玉斌，张建伟，钱立新．反跑道集束战斗部毁伤概率研究［J］．计算机仿真，2003，20（8）：12－15.
YANG Y B，ZHANG J W，QIAN L X. The study of kill probability by anti－runway cluster warhead［J］. Computer simulation，2003，20（8）：12－15.

[2] 朱近，夏德深，戴奇燕．侵彻子母弹对跑道封锁概率与打击效果评估［J］．火力与指挥控制，2007，32（4）：106－115.
ZHU J，XIA D S，DAI Q Y. Research on blockage runway probability with intrusive submunition missile and evaluation of attacked effectiveness［J］. Fire control and command control，2007，32（4）：106－115.

[3] 黄龙华，冯顺山．封锁型子母弹对机场的封锁效能［J］．弹道学报，2007，19（3）：49－52.
HUANG L H，FENG S S. Interdiction effectiveness of interdiction submunition on airport［J］. Journal of ballistics，2007，19（3）：49－52.

[4] 王芳，黄广炎，冯顺山．一种基于破片打击迹线的机场封锁概率计算方法［J］．弹道学报，2010，22（1）：24－28.
WANG F，HUANG G Y，FENG S S. Calculation method of blockage probability on airport based on fragment shot－line［J］. Journal of ballistics，2010，22（1）：24－28.

[5] HUANG G Y，FENG S S. Research on blocked damage effect of fragment warhead to target on ground［C］//The 25th International Symposium Ballistics. Beijing：China Science and Technology Press，2010：918－215.

[6] 黄广炎，冯顺山．基于VC和MATLAB的战斗部破片对目标打击迹线计算方法［J］．爆炸与冲击，2010，11（3）：413－418.
HUANG G Y，FENG S S. A visual C（++）and Matlab－based computational method for shot－lines of warhead fragments to a target［J］. Explosion and shock waves，2010，11（3）：413－418.

灵巧航行体半实物仿真系统设计方法与应用①

边江楠[1②]，冯顺山[1]，邵志宇[1]，方　晶[1]，段相杰[2]
（1. 北京理工大学爆炸科学与技术国家重点实验室，北京　100081；
2. 湖北航天飞行器研究所，武汉　430040）

摘　要：本文以新型灵巧航行体研发为背景，针对其近水面运动易受波浪力干扰、舵面受力复杂等问题，通过建模与 仿真解耦设计、环境条件自动切换等方法设计了模拟灵巧航行体水中工作环境的半实物仿真系统，并在此基础上进 行了实时仿真计算能力试验、灵巧航行体半实物仿真试验研究和波浪干扰试验研究 结果表明，应用所建立的设计 方法构建的半实物仿真系统解决了灵巧航行体近水面运动姿态控制系统半实物仿真的需求，可为灵巧航行体的姿 态控制系统设计和实航试验提供重要参考依据。

关键词：半实物仿真；灵巧航行体；近水面；波浪力

Design and Application of Hardware – in – the – Loop Simulation System For Smart Underwater Vehicles

Bian Jiangnan[1]，Feng Shunshan[1]，Shao Zhiyu[1]，Fang Jing[1]，Duan Xiangjie[2]
（1. State Key Laboratory of Explosion Science and Technology，Beijing Institute of Technology，
Beijing 100081，China；2. Hubei Aerospace Flight Vehicle Institute，Wuhan，Hubei 430040，China）

Abstract：The new smart underwater vehicle which runs near free – surface of the water is subject to wave disturbance and its rudder moment is complex. To simulate the water environment，a hardware – in – the – loop simulation（HILS）system was designed using methods of modeling，simulating decoupling design and automatic switching environment conditions The real – time computation ability test of the simulation computer was implemented The HILS and wave disturbance tests of the underwater vehicle were done Test results show that the designed HILS system could meet the demands of the attitude control system HILS of the underwater vehicles. The approach presented in this research might be very useful to the design of the attitude control system and the undergoing shipping test of the underwater vehicles.

Key words：hardware – in – the – loop simulation；smart underwater vehicle；near free surface；wave force

引言

为了研究水中灵巧航行体系统性能，可对系统各部件或全系统进行数学仿真、半实物仿真试验或实航试验等[1]。数学仿真过程中一些硬件实物无法准确建模，影响仿真精度和准确性。实航试验容易受试验经费、时间和其他条件的限制，在水中灵巧航行体的研制过程中，其试验难度高，试验次数也有限。半实物仿真（hardware – in – the – loop simulation）技术可以克服这些缺点，将无法准确建立模型的实物，如弹载计算机、舵机装置、目标探测器等，接入仿真回路，通过模型和实物之间的切换，进一步检验数学仿真结果的准确性，校准数学模型；检验姿态控制系统的性能指标及其可靠性，优化姿态控制系统相

① 原文发表于《北京理工大学学报》2013（1）。
② 边江楠：工学博士，2010 年师从冯顺山教授，研究水域封锁弹药控制总体技术，现工作单位：空军研究院航空兵研究所。

关设计参数，并直接检验姿态系统实物各部分的功能，从而提高系统研制质量。

水中灵巧航行体因需满足高机动性、快速性和稳定性特点，以及在近水面灵巧运动过程中需克服海浪、近岸浪和自由表面等的影响，其数学模型（包括近水面动力学、运动学模型和运动控制模型等）一般是强耦合、高度非线性的[2]。同时，为了模拟水中灵巧航行体的水中受力环境，主要是舵面上的流体动力形成的负载和水深及波浪的影响，需要针对性地设计负载力矩仿真和水深及波浪压力仿真。

本文建立了灵巧航行体半实物仿真的模型和仿真方法，设计了仿真软硬件架构，进行了实时仿真计算能力试验、灵巧航行体半实物仿真试验和波浪干扰试验研究。

1　软件架构

灵巧航行体的姿态控制软件必须满足多任务需求，这些任务对应不同的姿态控制算法，姿态控制软件程序的结构框架如图 1 所示。

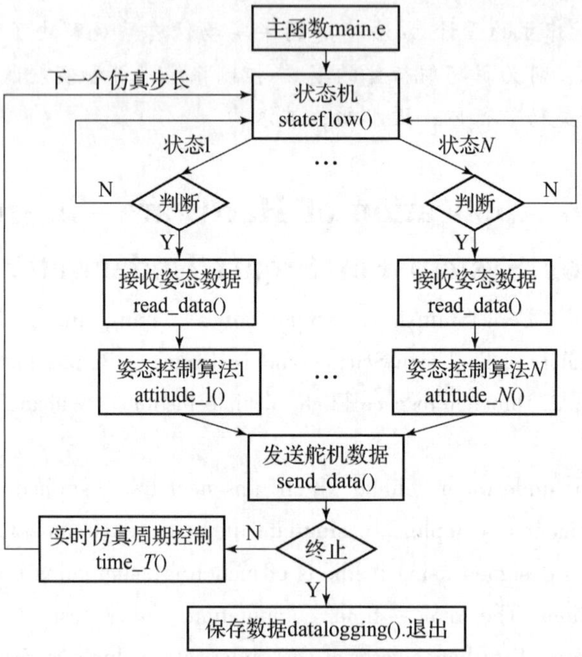

图 1　姿态控制软件架构

每次得知目前灵巧航行体的姿态数据后，首先通过状态机 stateflow( )来判断灵巧航行体下一步的运动状态，比如直航、下潜、爬升等，不同的运动状态对应不同的姿态控制算法 attitude_N( )。姿态控制算法解算出舵偏角指令后通过串口发送函数 send_data( )发送给仿真主机，准备下步长的运算。

2　建立模型

2.1　灵巧航行体数学模型

灵巧航行体 6 自由度运动由体坐标系和惯性坐标系定义。体坐标系 $Oxyz$，原点为灵巧航行体的浮心 O，x 轴指向灵巧航行体前进方向 y 轴垂直于 z 轴向下，z 轴与 x 轴和 y 轴构成右手直角坐标系；惯性坐标系 $O_Ex_Ey_Ez_E$ 选用北东地坐标系，原点位于零时刻的灵巧航行体浮心 O_E。

2.1.1　运动学模型

运动学方程可被表示成下列向量形式[3]

$$\dot{\eta}=J(\eta)v \tag{1}$$

式中

$$J(\eta)=\begin{bmatrix} R_b^n & 0_{3\times3} \\ 0_{3\times3} & T_\Theta(\Theta) \end{bmatrix} \tag{2}$$

且，$\eta\in \boldsymbol{R}^3\times \boldsymbol{S}^3$（三维环曲面，定义在区间［0，2π］内），$v\in \boldsymbol{R}^6$，欧拉角转动矩阵 $\boldsymbol{R}_b{}^n(\Theta)\in \boldsymbol{R}^{3\times3}$，其他符号含义参见文献［3］。

2.1.2　动力学模型

灵巧航行体6自由度动力学方程为[4]

$$\boldsymbol{A}\begin{bmatrix} \dot{u} \\ \dot{v} \\ \dot{w} \\ \dot{p} \\ \dot{q} \\ \dot{r} \end{bmatrix}=\begin{bmatrix} F_{sx}+F_{gx}+F_{tx}+F_{wx} \\ F_{sy}+F_{gy}+F_{ty}+F_{wy} \\ F_{sz}+F_{gz}+F_{tz}+F_{wz} \\ W_{sx}+W_{gx}+W_{tx}+W_{wx} \\ W_{sy}+W_{gy}+W_{ty}+W_{wy} \\ W_{sz}+W_{gz}+W_{tz}+W_{wz} \end{bmatrix} \tag{3}$$

式中：$\boldsymbol{A}$ 为质量矩阵，包括航行体质量 m、附加质量 λ_{ij}、转动惯量 J_i 和质心坐标（x_c，y_c，z_c）等，表达式

$$\boldsymbol{A}=\begin{bmatrix} m+\lambda_{11} & 0 & 0 & 0 & mz_c & -my_c \\ 0 & m+\lambda_{22} & 0 & -mz_c & 0 & -my_c+\lambda_{26} \\ 0 & 0 & m+\lambda_{33} & my_c & -mx_c+\lambda_{35} & 0 \\ 0 & -mz_c & my_c & J_x+\lambda_{44} & 0 & 0 \\ mz_c & 0 & -mx_c+\lambda_{35} & 0 & J_y+\lambda_{55} & 0 \\ -my_c & mx_c+\lambda_6 & 0 & 0 & 0 & J_z+\lambda_{66} \end{bmatrix}\lambda_{35} \tag{4}$$

将其按类型不同分为流体动力分量（下标用s表示）、重力和浮力分量（下标用g表示）和重力对浮心的作用力（下标用t表示）。用CFD仿真的方法得到理想情况下的流体动力参数，然后经如下推导给出波浪力干扰力和力矩因素。

2.1.3　波浪力和力矩干扰

波浪力和力矩干扰波浪的速度势函数为[5]

$$\phi(x,z,t)=\xi \tag{5}$$

$$\sigma_0=\frac{ig}{\omega}\exp[-kz_E-ik(x_E\cos\gamma+z_E\sin\gamma)] \tag{6}$$

式中　ξ_a——波幅；

g——重力加速度；

ω——波浪角频率；

k——波数，$k=\omega^2/\mathrm{g}$；

γ——波浪方向角；

$x_E y_E z_E$——灵巧航行体在北东地坐标系中的坐标，可根据相应的坐标转换矩阵转换到体坐标系中。

水中各点处的压力可以用拉格朗日积分求出，根据拉格朗日积分并考虑动坐标系，可以得到波浪产生的压力场为

$$p=-\rho Re\, i\omega\xi_a\exp(i\omega t)(\sigma_0+\sigma_1) \tag{7}$$

式中，σ_1 为灵巧航行体对波浪的衍射作用而产生的衍射流场。

将压力 p 在灵巧航行体的表面上积分，可以得到波浪扰动力 F_w 和扰动力矩 M_w

$$F_w=\iint_s p\cdot ndS;M_W=\iint_s p\cdot r\times n\mathrm{d}S \tag{8}$$

式中 S——灵巧航行体表面积；

$\boldsymbol{r}$——体坐标系到灵巧航行体表面任一点的矢径。

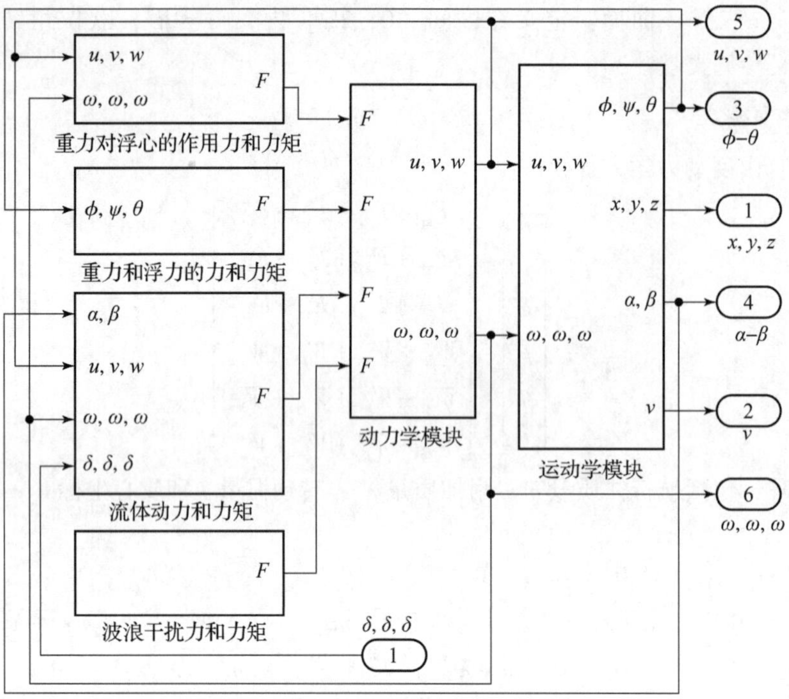

图 2　灵巧航行体数学模型

2.2　负载力矩模型

在水中运动的灵巧航行体，作用在舵面上的流负载，其对舵机输出轴是一种反作用力矩，包含惯性力矩、阻尼力矩、攻角铰链力矩和舵面铰链力矩，其中舵面铰链力矩 M_{h} 是主要因素，其是水中灵巧航行体速度和海水密度的函数[6]

$$M_h = \frac{1}{2} m_\delta \rho v^2 \delta S_\delta b_\delta \tag{9}$$

式中 m_s——舵铰链力矩系数；

ρ——海水密度；

v——水中灵巧航行体速度；

δ——舵偏角；

S_δ——舵面积；

b_δ——舵的平均弦长。

式（9）为经验公式，对于水中灵巧航行体的舵面铰链力矩的计算，尤其是近水面情况下较不准确。因此，本文采用 CFD 仿真方法建立了数值水洞，对典型舵偏角情况下的舵面铰链力矩进行了仿真，对式（9）计算的结果进行修正，达到了较好的效果。

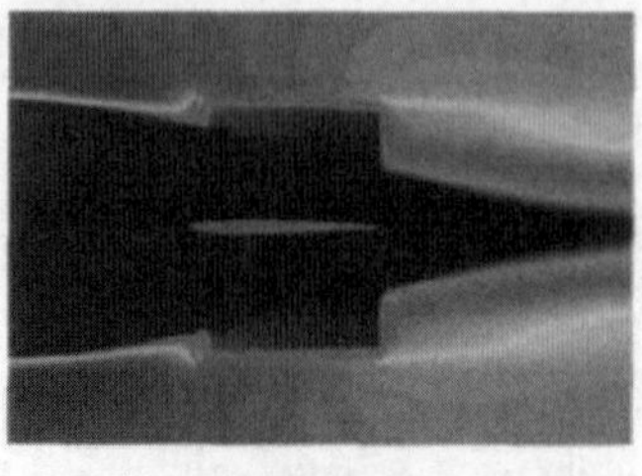

图 3　舵面受力分析示意图

2.3　波浪对水深干扰模型

水中灵巧航行体的运动深度范围较大，一般需经历水面至水下，再经水下至水面的往复运动，故水压仿真不但需要考虑在深水区压力的仿真问题，也需考虑在近水面附近时波浪力的影响。在水深大于 3 m 时，根据 $p=\rho gh$ 计算水下压力并经水压仿真器模拟；在离水面较近的位置（$h\leqslant 3$ m）时，需考虑波浪的影响，参考式（7）计算水压。

半实物仿真时，根据水中灵巧航行体不同工作水深等条件，根据深度条件自动切换不同的压力模拟算法，既能准确体现水中灵巧航行体的工作环境，也为深度控制提供参考。

3　半实物仿真系统设计

3.1　半实物仿真系统组成

为了验证上述半实物仿真系统设计方法的正确性，建立了如图 4 所示的半实物仿真系统。该半实物仿真系统主要由仿真设备（仿真计算机、三轴转台、目标模拟器、负载力矩模拟器、水压仿真器）、参试设备（弹载计算机、目标探测器、舵机装置等）、接口设备（以太网、反射内存网、串行接口等）和试验控制台（主控、监视及视景计算机等）组成。

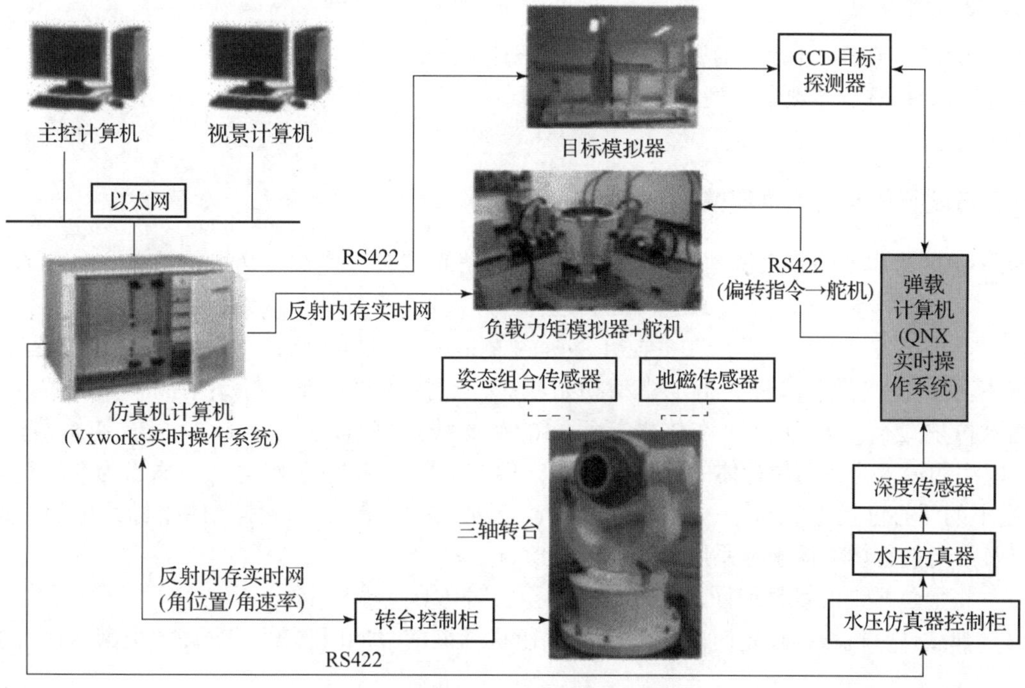

图 4　灵巧航行体半实物仿真系统

仿真计算机运行 VxWorks 实时操作系统，根据弹体动力学模型实时解算弹体姿态、速度、位置以及铰链力矩、弹体与目标间的相对位置等参数，然后将各参数通过接口设备（反射内存实时网等）分别输出给三轴转台、目标模拟器、负载力矩模拟器及水压仿真器等，分别模拟弹体角运动、弹目相对运动、舵机装置负载、弹体高度或深度等；另外将舵面偏角及弹体运动参数输出给视景计算机，显示弹体的三维运动；目标探测器及弹载计算机安装在三轴转台上，目标探测器跟踪目标模拟器的模拟目标，进行控制律解算，生成运动控制指令；弹载计算机接受运动控制指令，根据测得的转台姿态及模拟深度等信息，通过控制算法得到驱动指令，驱动舵机装置偏转，同时将舵轴偏转量送仿真计算机，准备进行下一轮解算。该过程迭代执行，直至完成全过程仿真。

3.2 基于 QNX 的弹载计算机系统

水中灵巧航行体的弹载计算机系统是一种高可靠、高性能、多任务、实时的嵌入式电子系统，它包含嵌入式的计算机硬件、监控软件和应用软件等。

弹载计算机硬件系统以低能耗 X86 处理器为核心进行研发，系统主要实现以下硬件功能：1 个千兆以太网接口，3 个 USB2.0 接口，5 个 RS232 串口，5 个 RS422 串口，1 个 VGA 接口，18 路 TTL 1/0 口，板载 GPS 定位模块，惯性测量模块，压力测量模块，并可实现 8 路温度采集。系统具有实时性、高可靠性以及可快速启动等特点。

4 灵巧航行体半实物仿真试验研究

4.1 仿真计算机实时计算能力试验

为检验仿真机的实时性能，在本仿真机硬件环境中进行了矩阵运算性能试验。

4.1.1 试验内容

矩阵运算：20 阶浮点矩阵相乘，循环 200 000 次后计算运行时间取均值，所有数据采用双精度浮点数（64 位）表示。

4.1.2 试验结果

经测试，编译器 GUB，DIAB 的实时计算用时分别为 156.3 μs、49.6 μs。

且仿真机对灵巧航行体数学模型解算周期为 109 μs，中断响应延迟小于 1 μs，完全能够满足 1 ms 的仿真周期需求。

4.2 灵巧航行体半实物仿真试验

根据图 4 中的半实物仿真系统结构连接各设备参与仿真试验，进行灵巧航行体半实物仿真，主要步骤如下。

步骤 1 全部设备上电启动后，在主控计算机的 Matlab/Simulink 软件中建立 2.1 节的灵巧航行体 6 自由度数学模型，设置仿真步长，给定模型初始状态，包括在北东地坐标系中的初始位置（x_{E0}，y_{E0}，z_{E0}），初始欧拉角（ϕ_0，θ_0，ψ_0），在体坐标系下的初始线速度（u_0，v_0，ω_0）以及初始角速度（ω_p，ω_q，ω_r）。模型的输入为灵巧航行体的俯仰舵角 δ_e，偏航舵角 δ_r 和差动舵角 δ_d，输出为灵巧航行体在北东地坐标系中的浮心位置（x_{CE}, y_{CE}, z_{CE}），欧拉角（ϕ，θ，ψ），灵巧航行体在体坐标系中的线速度（u，v，ω），以及灵巧航行体在体坐标系中角速度（ω_p，ω_q，ω_r）。

步骤 2 将上位机中的模型编译后经网络传输至实时仿真计算机中。

步骤 3 弹载计算机根据设定时间自动启动运行状态该信号同时传输至仿真主机激活仿真主机中的运动模型。

步骤 4 弹载计算机按照航线要求进行全局控制结束后发出停止信号仿真主机停止仿真其他设备进行断电等。

步骤 5 分析仿真数据并存档。

图 5 为试验过程中记录的下潜 - 爬升位移曲线，图 5 中 d_N，d_E 分别为北、东向的位移，半实物仿真试验结果表明通过本文的设计方法建立的半实物仿真系统可较为准确地仿真出灵巧航行体下潜 - 爬升位移情况且基本符合预订航线状况。

4.3 波浪干扰试验

灵巧航行体在航行时所受波浪干扰主要取决于波浪入射角、航行深度、波浪波幅等因素，图 6 给出了灵巧航行体在水下 3 m 和 15 m 深度航行的深度误差曲线，波浪的波幅为 0.6 m，周期为 5 s。可见，航行深度为 15 m 时其纵向运动误差很小基本不受波浪干扰，航行过程较为平稳。

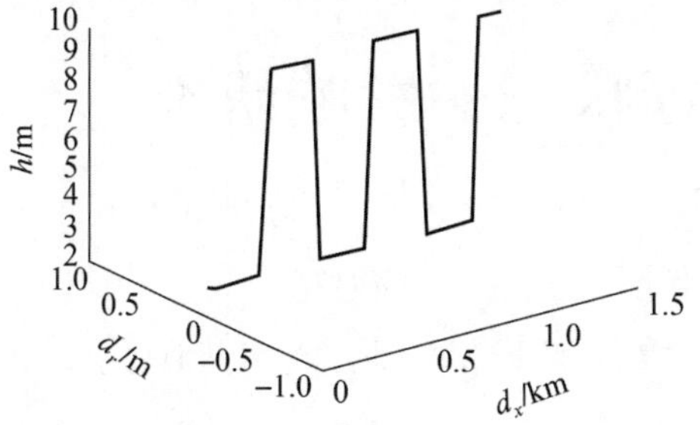

图 5　灵巧航行体下潜－爬升往复运动位移曲线

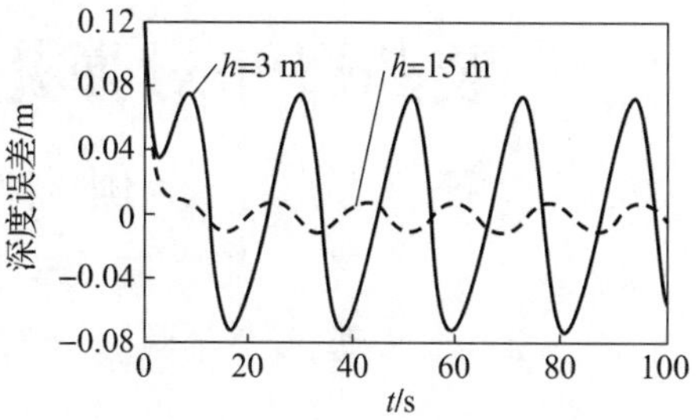

图 6　深度误差曲线

5　结论

针对近水面运动的细长状灵巧航行体易受波浪、近岸浪和自由表面的影响，其半实物仿真中负载力矩模拟和深度模拟情况复杂等问题。通过建模与仿真解耦设计、环境条件自动切换等方法建立了相应的模型和仿真方法，并针对多灵巧航行体近水面运动半实物仿真中环境模型使用问题给出了合理的具有鲁棒性的设计方法。

应用本方法建立了半实物仿真系统，并通过建立的半实物仿真系统进行了存在波浪力等干扰的灵巧航行体姿态控制系统半实物仿真试验，试验结果基本反映了实航状态下的灵巧航行体运动情况。

参考文献

[1] 单家元，孟秀云，丁艳．半实物仿真［M］．北京：国防工业出版社，2008：245－246.
SHAN J Y MENG X Y, DING Y. Semi－physical simulation［M］. Beijing: National Defense Industry Press, 2008: 245－246.

[2] FOSSEN T I. Nonlinear modeling and control of underwater vehicles［D］. Trondheim: Department of Engineering Cybernetics Norwegian University of Science and Technology, 1991.

[3] ROBERTS G N, SUTTON R. Advances in unmanned marine vehicles［M］. Stevenage: Institution of Engineering and Technology, 2006: 17－23.

[4] 石秀华，王小娟．水中兵器概论：鱼雷分册［M］．西安：西北工业大学出版社，2005：48－49.
Shi X H, WANG X J. Introduction to underwater weapons: volume of torpedo［M］. Xi'an: Northwestern Polytechnical University Press, 2005: 48－49.

[5] 严卫生．鱼雷航行力学［M］．西安：西北工业大学出版社，2005：57－63.
YAN W S. Torpedo sailing mechanics［M］. Xi'an: Northwestern Polytechnical University Press, 2005: 57－63.

[6] 康凤举，杨惠珍，高立娥．现代仿真技术与应用［M］．北京：国防工业出版社，2010：392－393.
KANG F J, YANG H Z, GAO L E. Modern simulation technology and application［M］. Beijing: National Defense Industry Press, 2010: 392－393.

装药爆炸对弹丸嵌立封锁状态的影响研究①

孙　凯[1②]，李国杰[1]，刘坚成[1]，王成龙[1]，冯顺山[2]

（1. 北京航天长征飞行器研究所，北京　100076；
2. 北京理工大学爆炸科学与技术国家重点实验室，北京　100081）

摘　要：本文利用 LS－DYNA 有限元分析软件，对已嵌立在舰船飞行甲板上的弹丸爆炸过程进行数值模拟，得到了弹丸嵌入深度、靶板厚度、装药位置以及装药量对弹丸嵌立状态的影响规律，并进行了典型工况验证分析。研究结果表明：在一定条件下，当弹丸初始嵌入牢固度较好并且靶板超过一定厚度时，装药爆炸后弹丸头部仍具有嵌立状态，而其剩余的嵌立高度与装药一靶面距离有直接联系，但装药量对弹丸嵌立状态影响较小。并通过典型工况试验，验证了装药爆炸后弹丸仍具有嵌立状态的可实现性，与仿真结果吻合。

关键词：装药爆炸；嵌立状态；数值模拟；封锁

Influence of Projectile Embedded Blocking State Under Charge Explosion

Sun Kai[1], Li Guojie[1], Liu Jiancheng[1], Wang Chenglong[1], Feng Shunshan[1]

(1. Beijing Institute of Space Long March Vehicle, Beijing 100076, China; 2. State Key Laboratory of Explosion Science and Technology, Beijing Institute of Technology, Beijing 100081, China)

Abstract: LS－DYNA finite element analysis software was used to simulate the projectile explosion process embedded on the flight deck of a ship, and the influence rules of projectile embedding depth, target plate thickness, charge position and charge amount on the projectile embedding state were obtained. The results show that under certain conditions, when the initial embedding firmness of the projectile is good and the target plate exceeds a certain thickness, the head of the projectile still has embedded state after the charge explosion, and the remaining embedded height is directly related to the distance between the charge and the target surface, but the amount of charge has little influence on the embedded state of the projectile. Through the typical working condition test, it is verified that the projectile still has the embedded state after charging explosion, and the simulation results are consistent.

Keywords: charge explosion; embedded state; numerical simulation; lockout

封锁舰船飞行甲板，阻碍舰载机起降，在一定时间内降低航母或大型舰船作战效能的毁伤模式已提出较多年。封锁型弹药根据封锁效能要求的不断提高也经历了 3 种概念阶段。第一阶段概念是弹丸利用某种原理使头部牢固嵌立在甲板上，中后部露出甲板形成“障碍元”，封锁舰载机起降。第二阶段概念是在第一阶段嵌立封锁的基础上增加炸药装药和延时起爆引信，形成“活性障碍元”防止弹丸被“暴力”拆除和增加对舰面人员、载机的毁伤。第三阶段概念是在第二阶段基础上，合理设计装药结构，使装药爆炸后，弹丸仍有部分凸出甲板，具有“惰性”封锁功能，进一步增加封锁时效性。

① 原文发表于《兵器装备工程学报》2020（7）。
② 孙凯：工学硕士，2013 年师从冯顺山教授，研究封锁型子弹战斗部威力，现工作单位：航天科技一院 14 所。

第一阶段重点解决弹丸头部如何牢固嵌立问题，国内较多学者也给出了解决方案，如李真等[1]提出了串联嵌入式子弹，采用前级聚能射流开孔，后级嵌入模式实现子弹对甲板的牢固嵌立；张险峰等[2]提出了整体式嵌入子弹药，实现了整体式子弹对甲板的嵌立与封锁；杭贵云等[3]提出设障子弹结构设计方案，并通过仿真进行了验证；董永香等[4-6]利用半侵彻作用机理解释并实现了弹丸稳定嵌立于靶板的问题。第二阶段主要涉及装药抗过载和引信延时问题，技术较为成熟[7-9]。而第三阶段中装药爆炸后弹丸是否仍具有嵌立封锁状态的工程问题鲜少有人研究。但伊文静等[10]通过仿真与试验结合的方式研究了弹丸爆炸驱动的问题，陈小伟等[11-13]对穿甲及金属空腔膨胀理论的研究可作为本文的研究基础，为解决爆炸弹丸爆炸后的嵌立封锁问题提供理论基础。

因此，本文基于弹丸侵彻甲板时的靶材流动效应[4-6,14-15]。以弹丸头部牢固嵌立在靶上为前提，研究了弹丸装药爆炸对嵌立状态的影响。通过仿真分析，分别研究了初始嵌入深度、靶板厚度、装药—靶面距离以及装药量4个参量对爆炸后连接状态的影响规律，并通过典型工况进行了试验验证。

1　作用过程和力学模型

弹丸装药爆炸在形成大量破片的同时会驱动嵌立头部产生沿弹轴方向的速度，冲击靶板。因此将作用过程分为两个阶段：第一阶段为嵌立头部在爆炸冲力与侵彻阻力作用下的二次侵彻过程，第二阶段为嵌立头部在靶板反弹力和凹槽内靶材约束力作用下的反弹过程，如图1所示。其中侵彻阻力与靶材约束力可由动态空穴膨胀理论[14-15]得出，空穴径向压力 σ_r 可近似表示为静态和动态分量之和，即

$$\sigma_r = A\sigma_y + B\rho v^2 \tag{1}$$

式中　σ_r——靶板材料的屈服应力；

ρ——靶板材料的密度；

A——静阻力系数，与空腔周围的材料性质有关；

日——动阻力系数，与侵彻头部形状有关；

B——嵌立头部速度。

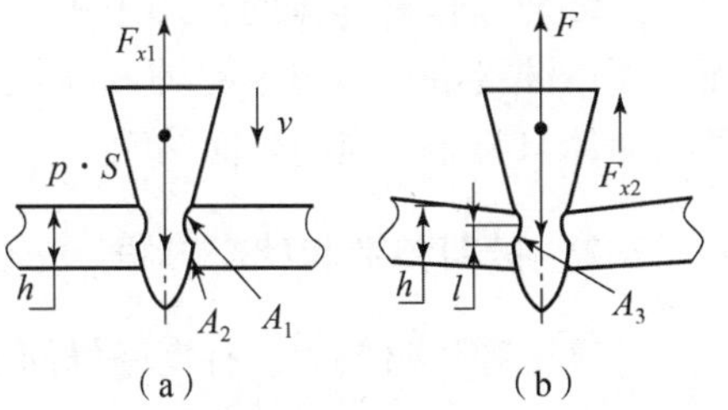

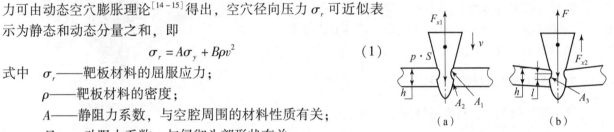

图1　弹丸二次侵彻阶段和反弹阶段示意图

(a) 2次侵彻阶段；(b) 反弹阶段

将 σ_r 转化为弹轴方向所受应力：

$$\sigma_x = \sigma_n \sin\theta + \sigma_t \cos\theta \tag{2}$$

式中　θ——嵌立头部某点切向与弹轴方向夹角；

σ_r——正应力 $\sigma_r = \sigma_n$；

σ_t——切应力 $\sigma_r = \mu_m \sigma_n$；

μ_m——滑动摩擦系数，由于侵彻过程中由靶材强度产生的阻力远大于摩擦阻力，因此后一项可忽略不计。所以嵌立头部受到的靶板阻力为

$$Fx = \iint_{An} \sigma_x \mathrm{d}A \tag{3}$$

式中，A 为弹丸与靶板接触面积。

对于爆炸驱动嵌立头部的问题，根据牛顿定理，二次侵彻阶段可表示为

$$\frac{\int_0^t p \cdot S\mathrm{d}t - \int_0^t F_{x1}\mathrm{d}t}{m} = \frac{\int_0^t p \cdot S\mathrm{d}t - \int_0^t \iint_{An} \sigma_x \mathrm{d}A\mathrm{d}t}{m} = v \tag{4}$$

式中　P——爆压；

S——爆轰产物作用面积，积分面积 $A = A_1 + A_2$；

m——嵌立头部质量；

v——嵌立头部所获得最大驱动速度。

对于反弹阶段，认为嵌立头部的动能均转化为靶板的弹性变性能，得

$$\Delta = \frac{1}{2}mv^2 - F_{x2} \cdot h = \frac{1}{2}mv^2 - \iint_{An} \sigma_x \mathrm{d}A \cdot h \tag{5}$$

$$p = A\left(1 - \frac{w}{R.V}\right)\exp(-R_1V) + B\left(1 - \frac{w}{R_2V}\right)\exp(-R_2V) + \frac{wE}{V}$$

式中，h 为反弹阻力面高度，积分面积 $A = A3$。

由此可得，当 $\Delta > 0$ 时，嵌立头部弹出靶板；当 $\Delta < 0$ 时，嵌立状态不变。

2 仿真模型及材料参数

2.1 仿真模型和计算方法

本文采用 LS—DYNA 有限元分析软件构建有限元模型。由于结构的对称性，计算时采用 1/4 三维模型，并在对称面施加对称约束。考虑到董永香等[4-6]已研究了弹丸侵彻靶板时的靶材流动能够使弹丸与靶板牢固连接，且爆炸时弹已稳定，故将真实弹丸进行简化和等效，将弹丸头部的特殊结构等效为一个弧形凹槽，靶材填充在凹槽内，将其他结构等效为一定厚度的弹材以密闭装药。计算模型如图 2 所示。

计算方法为 LAGRANGE 算法，采用 S0LID164 八节点单元模型。装药爆炸对弹体的作用采用共节点算法。弹靶之间添加面—面侵蚀接触，体现弹靶的相互作用，不考虑摩擦阻力。此外，弹体自身施加自动单面接触，防止爆炸后自身的穿透问题。

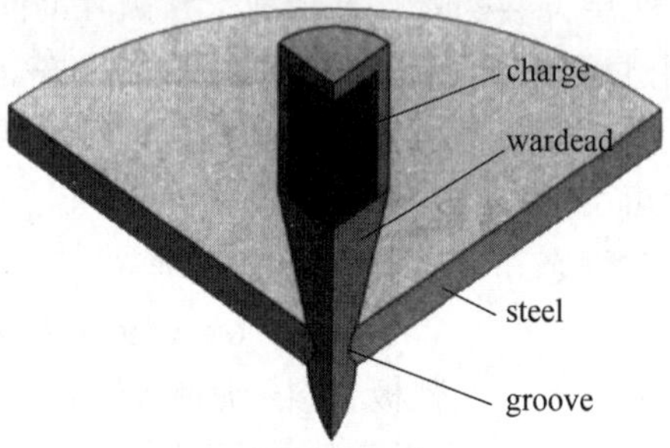

图 2 计算模型

2.2 材料参数和状态方程

计算中弹体材料为具有良好侵彻性能的 30CrMnSiNi2A 合金钢，靶板材料为 921A 舰船用钢。两种材料均由 Johnson—Cook 强度模型描述。Johnson—cook 强度模型能够较为准确的描述高应变率下金属材料的动态行为，该模型利用变量乘积关系分别描述应变、应变率和温度的影响，具体形式为

$$\sigma = (A + B\bar{\varepsilon}_p^n)(1 + C\ln\dot{\varepsilon}^*)[1 - (T^*)^m] \tag{6}$$

式中 A、B、n、C、m——材料参数；

$\bar{\varepsilon}$——等效塑性应变；

$\dot{\varepsilon}^*$——等效应变率；

T^*——相对温度。具体材料参数见表 1[16]。

表 1 具体材料参数

材料	$\rho/(\mathrm{g \cdot cm^{-3}})$	E/GPa	μ	A/MPa
30CrMnSiNi2A	7.85	210	0.39	1 500.5
921A	7.8	205	0.28	760
材料	B/MPa	C	n	m
30CrMnSiNi2A	1 045	0.019	0.57	1.03
921A	500	0.014	0.53	1.13

装药采用 8701 商能混合炸药，用 JWL 状态方程对兵进行描述。JWL 状态方程形式如下：

$$p=A\left(1-\frac{w}{R_1V}\right)\exp(-R_1V)+B\left(1-\frac{w}{R_2V}\right)\exp(-R_2V)+\frac{wE}{V} \tag{7}$$

式中　$V=v/v_0$，E——比热容力学能；

A、B、R、R、w——常数。具体参数见表 2[17]。

表 2　8701 炸药参数

材料	$\rho/(\mathrm{g\cdot cm^{-3}})$	$D/(\mathrm{m\cdot s^{-1}})$	PC－J/GPa	A/GPa
8701 炸药	1.72	8 315	29.5	854.5
材料	B/GPa	$R1$	$R2$	ω
8701 炸药	20.493	4.6	1.35	0.25

3　数值计算结果及分析

3.1　初始嵌入深度对嵌立状态的影响

结构参数采用无量纲化，定义初始嵌入深度系数 λ 为初始嵌立状态时靶材嵌入弹体凹槽轴向高度比例系数，在相同的装药结构和靶板厚度条件下，λ 取 0.25、0.5、0.75、1 时，分别代表靶材嵌入凹槽的 1/4、1/2、3/4 以及完全嵌入状态。计算后得到不同嵌入深度时弹靶相互作用情况以及弹头部速度随时间的变化曲线，结果如图 3 和图 4 所示。

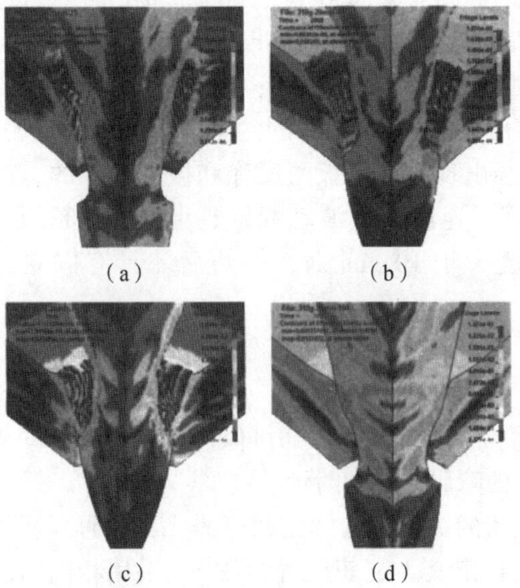

图 3　不同嵌入深度系数时弹靶作用情况

(a) $\lambda=0.25$；(b) $\lambda=0.5$；(c) $\lambda=0.5$；(d) $\lambda=1$

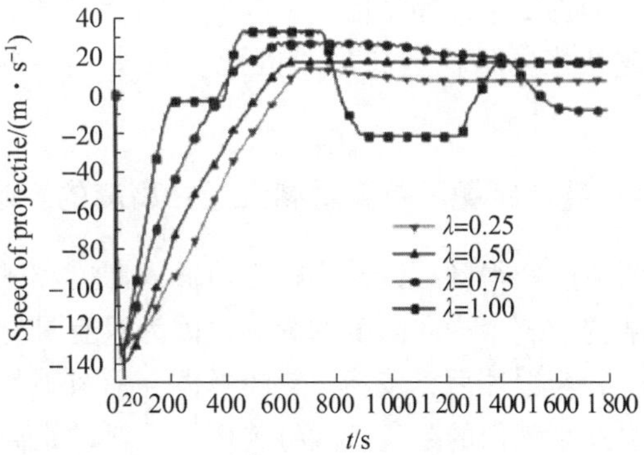

图 4　不同嵌入深度系数弹头部速度随时间变化曲线

从图 4 可以看出，弹头部受爆炸冲击后，均在 20 μs 左右达到最大速度，此后，由于嵌入深度不同，速度呈不同变化趋势。当 $\lambda=1$ 时，即初始嵌立状态时凹槽内完全嵌入靶材，弹头部向下冲击靶板，但受凹槽内靶材的约束，速度迅速减小；在 400 μs 时，由于受靶板的反弹力，速度转而向上；在 800 μs 时，弹头部继续受凹槽内靶材约束造成速度转而向下，此后由于靶板的振动，带动嵌立头部速度上下波动。当 $\lambda=0.75$ 时，弹头部速度变化趋势与类似，但由于凹槽内靶材未满，靶板对弹头部的约束较小，因此速度减小较慢，速度振动周期较长。从速度的正负波动可以看出，$\lambda=0.75$ 时，弹头部仍卡在靶板内，仍具有嵌立封锁功能。当 $\lambda=0.5$ 和 $\lambda=0.25$ 时，弹头部受到靶板向上反弹作用后，产生向上的速度，不再反向，说明弹体已弹出靶板侵孔，嵌立失效。

$\lambda=1$ 和 $\lambda=0.75$ 对应理论模型中 $\Delta<0$ 的情况，$\lambda=0.5$ 和 $\lambda=0.25$ 对应理论模型中 $\Delta>0$ 的情况。由此证明只有当 $\lambda=0.5$ 时，爆炸后弹丸仍嵌立于靶板上。此外，从图 3 可以看出，当 $\lambda=1$ 时，弹靶相互作用均会产生侵蚀损伤，造成嵌立松动；只有当 $\lambda=1$ 时，弹靶损伤最小。

3.2　靶板厚度对嵌立状态的影响

对舰船甲板进行目标特性分析可知，不同位置处的甲板厚度并不一致，一般介于 10～50 mm，因此采用相同装药结构和嵌立状态对不同靶厚条件下进行数值仿真，得到不同厚度靶板的变形情况如图 5 所示和弹头部速度随时间变化情况如图 6 所示。

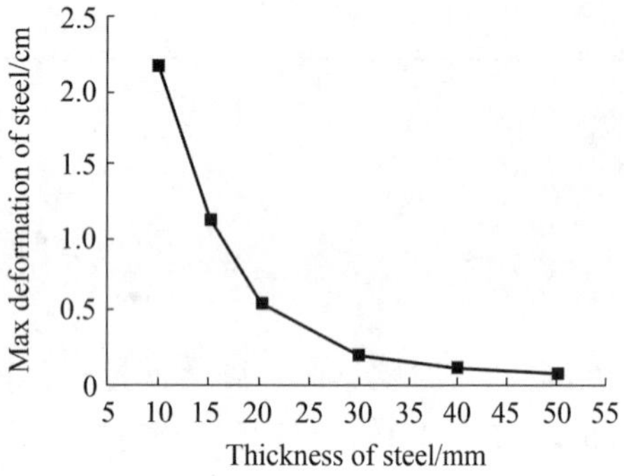

图 5　不同厚度靶板的变形曲线

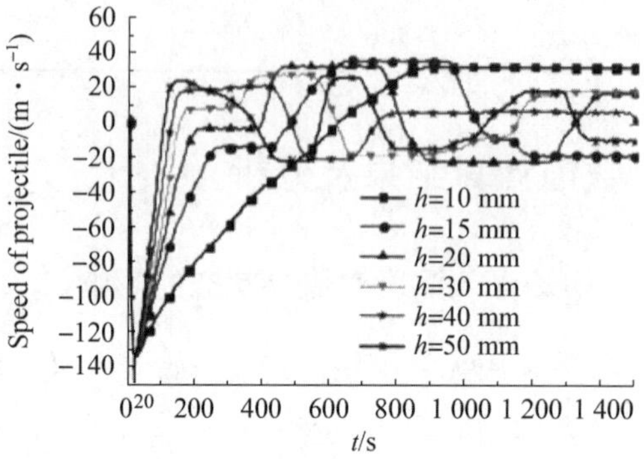

图 6　不同靶厚弹头部速度随时间变化曲线

结果显示，靶板越厚，弹头部冲击靶板导致靶板变形越小，尤其靶厚在 30～50 mm 时，靶板几乎不受爆炸冲击的影响，变形低于 0.5 cm；从不同厚度靶的弹头部速度曲线可以看出，弹头部受爆炸驱动后均在 20 μs 时达到的最大速度约 133 m/s，此后受靶板的反弹作用速度逐渐减小；10 mm 靶厚时，头部速度反弹为正向后并保持下去，说明弹头部受到靶板的反弹力大于凹槽内靶材的约束力，弹头飞离靶板，嵌立状态失效，对应 $\Delta>0$ 情况；其他靶厚工况下，头部均会出现由于靶板的反弹和振动导致的速度正负波动，对应 $\Delta<0$ 情况且靶板越厚，对弹头部的约束越强，速度正负波动幅度越小，周期越短，作用过程越快恢复到稳定状态。因此，只有当目标靶板厚度大于 10 mm 时，爆炸后才具有嵌立封锁状态。

3.3　装药距靶面距离对剩余嵌立高度的影响

前两节已经得到结论，当特定地嵌入深度和靶板厚度时，爆炸后弹丸头部可保持嵌入状态。本节研究装药结构对剩余嵌立高度的影响，因为只有具有一定高度的障碍才具有封锁功能。

由于装药爆炸后，装药位置的弹体会形成破片，只有残留的头部作为续留的障碍元，障碍高度取决于装药位置距靶面的距离，故将装药距靶板的距离与弹径之比定义为装药位置参数 α。采用相同的嵌入深度、靶板厚度和装药量，对不同装药位置进行数值研究。α 分别设置为 0.25、0.5、0.75 和 1，观察爆炸后弹头部剩余高度。爆炸结果如图 7 所示，障碍元损失高度/弹径与 α 的关系、障碍元损失高度/初始设定高度与 α 的关系如图 8 所示。

从图 8 可以看出，随着装药距靶面距离的增加，障碍元剩余高度与弹径的比值呈减小趋势，从 0.12 降到 0.09，但变化不大，说明爆炸对障碍元高度减小的量与装药的位置关系不大；而剩余高度与初始距离的比值由 0.53 降到了 0.11，说明装药距靶面距离越远，不仅剩余的高度越大，而且损失高度的比例也越小。根据以上计算结果，说明弹头部可由弹径和所需障碍元剩余高度设计装药距靶面的距离。

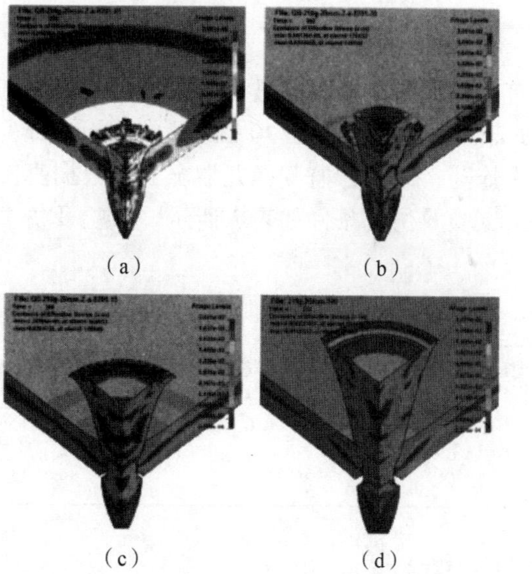

图 7　爆炸后弹丸头部的嵌立状态

(a) $\alpha=0.25$; (b) $\alpha=0.50$; (c) $\alpha=0.75$; (d) $\alpha=1.00$

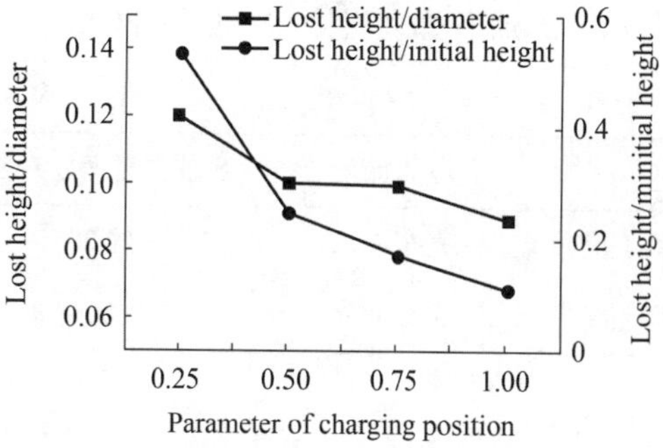

图 8　障碍元高度随装药距凹槽距离关系曲线

3.4　装药量对嵌立状态的影响

分别选取了 150 g、180 g、210 g 和 240 g 四种装药量时做数值模拟，并保持其他参量一致。结果显示不同装药量条件下，弹头部产生的驱动速度和弹头部冲击靶板造成的变形略有不同，但对嵌立状态影响不大。靶板最大变形与弹头部最大速度之间的关系如图 9 所示，弹头部速度变化情况如图 10 所示。

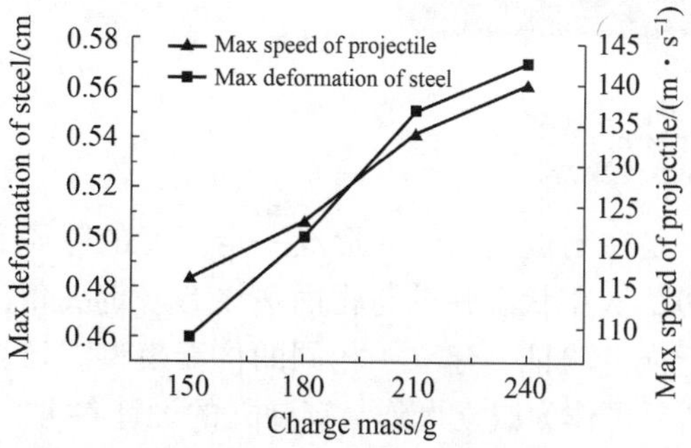

图 9　不同装药量下靶板最大变形与弹头部最大速度关系曲线

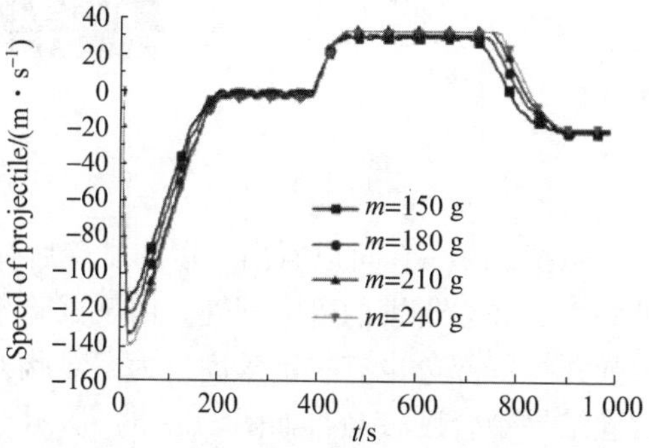

图 10　不同装药量弹头部速度随时间变化曲线

从结果来看，150 g 装药时，靶板的最大变形为 0.46 cm，此后，药量每增加 30 g，变形增加约 10%；但由 210 g 药量增加到 240 g 药量时，变形仅增加 4%，说明靶板变形与药量不具有严格正比关系。从弹头部获得的最大驱动速度角度显示，装药量越大，对弹头部的最大驱动速度越大；但是弹头部速度变化曲线基本一致，可得知弹头部受靶板反弹力产生正向速度后（600 ~ 800 μs 时间段），由于凹槽内靶材对弹头部的约束和靶板的振动导致速度负向变化（800 ~ 1 000 μs）。

说明，装药为 150 ~ 240 g 时，弹头部均不会弹出靶孔，爆炸后都具有良好的嵌立状态。

同时注意到，3.3 节装药位置的变化引起了弹丸头部质量 m 的变化，3.4 节装药量的变化引起了弹丸头部速度 v 的变化，但是弹丸的嵌立状态并没有因此改变，均保持了 $\Delta<0$，说明在反弹阶段 $\iint_{An}\sigma_x \mathrm{d}A \cdot h$ 占据主导作用，靶材的约束力较强。

4 试验

为验证装药爆炸后弹丸的嵌立状态及仿真研究的有效性，对典型工况进行了试验研究。根据董永香等[4-6]的研究结果，利用火炮将具有头部凹槽的弹丸加载到一定速度侵彻 30 mm 厚的钢板，弹丸成功嵌立在钢板上，测量弹丸嵌入状态，然后在弹丸后端装入炸药，起爆并监测弹丸情况，并根据真实的试验条件建立仿真分析模型，针对特定工况进行对比分析。试验及仿真条件如表 3 所示。试验及仿真结果如表 4 所示，爆炸后弹丸嵌立状态如图 11 所示。

表 3　试验与仿真条件

	嵌入深度	弹轴与靶面发现夹角	靶板厚度/mm	装药距靶面距离 α	装药量/g
试验工况	完全嵌入	约 20°	30	0. 92	150
仿真工况	完全嵌入	20°	30	0. 92	150

表 4　试验与仿真条件

	是否仍嵌立在靶上	损失的高度/弹径	损失的高度/初始高度
试验工况	是	0. 12	0. 13
仿真工况	是	0. 10	0. 11

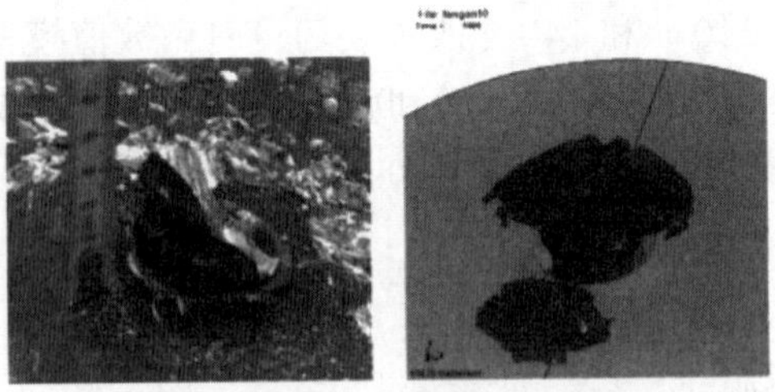

（a）　（b）

图 11　爆炸后弹丸嵌立状态

（a）试验结果；（b）仿真结果

由图 11（a）可以看出，爆炸试验后弹丸头部仍嵌立在靶板之内，并形成一定高度的障碍元，证明了爆炸后嵌立封锁的可实现性。由于试验条件限制，初始嵌立状态弹轴与靶面法线有一定夹角，因此导致弹丸续留部分也有一定角度，为了对此典型工况进行充分验证，建立了相同的仿真模型。仿真结果与试验结果吻合度较高，如图 11（b）所示，说明仿真结果可作为嵌立爆炸封锁弹丸的设计参考。

5 结论

（1）经典型工况试验证明，在一定条件下，弹丸爆炸后仍具有嵌立状态并形成具有一定高度的障碍元。

（2）试验结果与仿真结果吻合度较高，证明了仿真方法和材料参数的正确性，可作为嵌立爆炸封锁弹丸的仿真设计参考。

（3）只有初始嵌入深度和靶板厚度符合一定要求时，爆炸后弹丸才处于嵌立状态，而装药位置和装药量对嵌立状态影响较小，装药位置对障碍元剩余高度影响较大。

参 考 文 献

[1] 李真．串联嵌入弹的终点效应优化设计和数值模拟［D］．南京：南京理工大学，2008.

刚性尖头弹对甲板的半侵彻特性研究①

邓志飞②，白　洋，孙　凯，冯顺山
（北京理工大学爆炸科学与技术国家重点实验室，北京　100081）

摘　要：本文开展了刚性尖头弹对船用钢靶的半侵彻特性研究。首先，基于圆柱形空腔膨胀理论，建立了刚性尖头弹正侵彻钢靶的延性扩孔模型，并推导出半侵彻弹道极限速度公式。然后通过正交试验的极差分析法，获得影响半侵彻弹道极限由大到小的因素依次为：靶板厚度、弹头锥角、弹丸质量，其中后两个因素的影响程度基本相同。此外，对斜侵彻过程进行了简化，引入体积修正系数χ_1和平均径向应力修正系数χ_2，并通过仿真结果获得修正值，并加以验证。结合数值仿真，对理论计算结果的误差分析表明：本文理论公式可用于预测在靶板厚度为30～60 mm时半侵彻弹道极限，仿真结果与理论计算吻合较好。

关键词：半侵彻特性；刚性尖头弹；正交试验法

0　引言

研究穿甲弹的侵彻性能及影响因素具有较强的军事应用价值。蒋志刚等[1-5]研究了多种弹头垂直撞击塑性金属靶板的扩孔冲塞型和延性扩孔型穿孔模式，基于空腔膨胀理论提出最小穿透能量的两阶段工程模型。薛鸿艳[6]、邹运[7]和史忠鹏[8]分别进行了不同初速度、长细比、装甲倾角条件下侵彻装甲的数值研究。Rosenberg Z 和 Dekel E[9]对刚性尖头弹侵彻延性靶板的仿真研究表明，碟形破坏和扩孔型破坏的转折点在$H/D=1/3$左右（H为靶厚，D为弹径），对于中厚靶（$1/3<H/D<1.0$）无量纲侵彻阻力$\sigma_r/Y_t=2.0$；对于较厚靶板($H/D>1.0$)σ_r/Y_t随H/D单调递增。

大量文献对弹－靶侵彻特性和弹道极限进行研究，而本文着重于研究半侵彻特性。其中，半侵彻的定义如下：弹体低速侵彻靶板，当速度降为零且未能穿过靶板时，靶板对弹药产生反弹作用，使弹丸具有显著的向外运动趋势。通过一定的侵立方法可克服靶板的反弹作用，最终实现侵立在靶板上。本文主要研究以下两个问题：第一，推导得出正侵彻和斜侵彻的半侵彻弹道极限公式；第二，结合正交试验方法进行数值仿真，确定弹丸质量、靶板厚度和弹头锥角等因素对半侵彻弹道极限速度影响的主次关系。此外，本文利用仿真数据对也用于理论公式的计算结果进行了误差分析以验证其准确性。

1　问题描述

选择舰船甲板作为典型目标，美国舰艇用钢主要为HY/HSLA系列钢（主要为HY/HSLA－80/100）。根据材料性能和舰船甲板结构调研，选取921A钢作为靶板材料；靶板厚度t范围为30～50 mm。另外，弹体材料选用30CrMnSiNi2A，弹体质量m的取值范围为2.0～2.4 kg，简化弹形如图1所示，其中$h_1=74$ mm，$d=60$ mm，$2\beta'=24°$，$2\beta=28°\sim32°$。在上述所有的参数中，弹头锥角2β和弹体质量m决定了研究的弹体形状。

根据半侵彻机理，通常在弹头部设计特殊结构以克服靶板的反弹作用。为简化问题，本文忽略异型弹头的结构，研究刚性穿甲弹开始侵彻到速度为零的过程。

---

① 原文发表于《第十四届战斗部与毁伤技术学术交流会论文集》2015。

② 邓志飞：工学硕士，2014年师从冯顺山教授，研究弹丸对甲板半侵彻效应，现工作单位：美国 Virgnia polytechnic institute and state university。

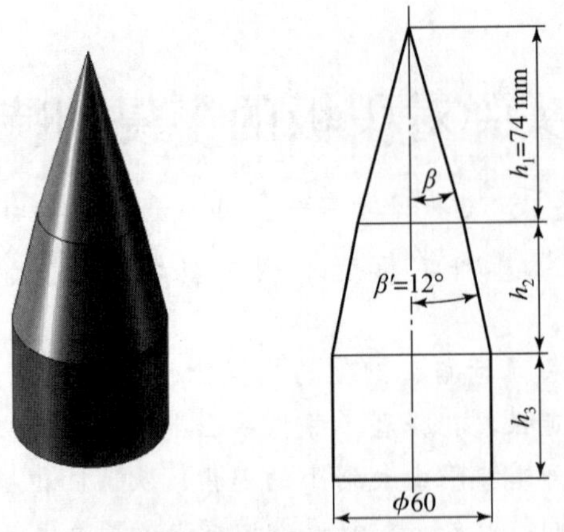

图 1　弹体简化示意图

定义无量纲侵深 $\lambda=t/L$，其中，t 为靶板厚度，L 为侵彻深度（图 2）。相应地，当 λ 取不同临界值时，存在不同临界侵彻速度，定义为半侵彻弹道极限 v_{50}。半侵彻的有效侵深 L 存在一定的取值范围，若 λ 取值太小，则弹体侵深过大，无法起到侵立封锁效果；若 λ 取值太大则弹体嵌立不稳。因此，在本文研究中取 λ 有效范围为 0.7～1.1，则半侵彻弹道极限 v_{50} 也存在一个取值范围。

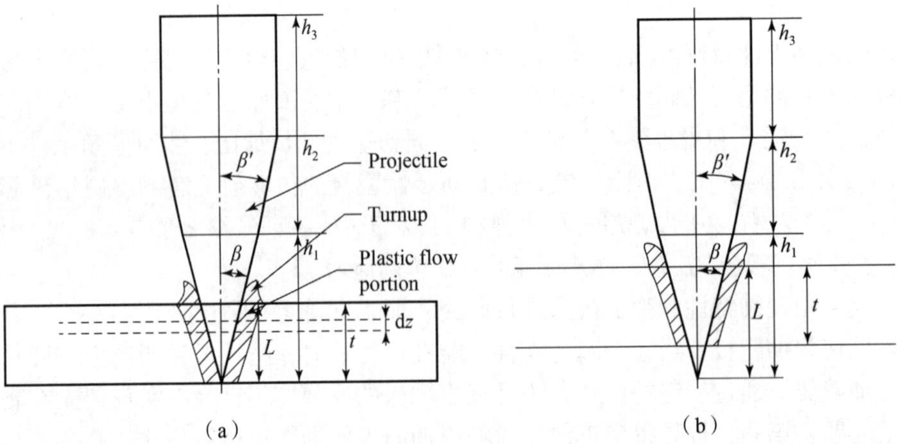

图 2　正侵彻临界侵深示意图

(a) $\lambda\geqslant1.0$；(b) $\lambda<1.0$

2　正侵彻特性研究

2.1　理论模型

本文的理论模型是基于以下几个假设：①弹体始终呈刚性，且靶板在侵彻过程中仅出现单一延性破坏；②忽略了靶板背面自由面的效应以及靶板材料的应变强化效应；③靶板材料视为理想弹塑性材料。

因此，本文研究的最大侵深为（$t=50$ mm，$\lambda=0.7$）$L_{\max}=71.43$ mm $<h_1$。设延性扩孔体积为 U，靶板薄层微元 dz 扩孔半径 r 从零增大到 R 过程中孔壁处径向力做功为

$$\mathrm{d}W=\int_0^R\sigma_r(2\pi r\mathrm{d}z)\,\mathrm{d}r=\pi R^2\sigma_r\mathrm{d}z \tag{1}$$

忽略摩擦力以及靶板的弹性变形，由功能原理，侵彻扩孔耗能近似等于弹体侵彻扩孔过程中孔壁处径向力所做的功。由式（2）得靶板延性扩孔耗能为

$$W = \int \mathrm{d}W = \int \pi R^2 \sigma_r \mathrm{d}z = \sigma_r U \tag{2}$$

在理想弹塑性靶板的假设下，由圆柱形空腔膨胀理论[10,11]，得径向应力表达式

$$\sigma_r = \kappa Y, \kappa = 0.5\left\{1 + \ln\left[\frac{2E}{(5-4\mu)Y}\right]\right\} \tag{3}$$

式中：κ——侧限系数（反映非塑性区约束作用）；

E——杨氏模量；

μ——泊松比；

Y——靶板材料单向压缩塑性流动应力。

由于不考虑靶板材料的应变强化效应，本文中 Y 用材料的屈服极限代替 σ_y。根据能量守恒，当弹丸初始动能等于靶板的延性扩孔耗能时，可计算出弹丸半侵彻的极限速度：

$$\frac{1}{2}mv_{50}^2 = \sigma_r U, v_{50} = \sqrt{2\sigma_r U/m} \tag{4}$$

（1）当 $\lambda \geqslant 1.0$ 时，弹头完全嵌在靶板内部。如图 2（a）所示，延性扩孔呈圆锥状，侵彻过程中的延性扩孔体积 U_1 为

$$U_1 = \int_0^L \pi(z\tan\beta)^2 \mathrm{d}z = (\pi/3)L^3\tan^2\beta,\ v_{50} = \sqrt{\frac{2\pi\sigma_r t^3\tan^2\beta}{3\lambda^3 m}} \tag{5}$$

（2）当 $\lambda < 1.0$ 时，弹头刺穿靶板。如图 2（b）所示，$t < L < h_1$，扩孔呈圆台状，此时设延性扩孔体积为 U_2，则

$$U_2 = \int_{L-t}^{L} \pi(z\tan\beta)^2 \mathrm{d}z = (\pi/3)[L^3 - (L-t)^3]\tan^2\beta, v_{50} = \sqrt{\frac{2\pi\sigma_r t^3[1-(1-\lambda)^3]\tan^2\beta}{3\lambda^3 m}} \tag{6}$$

2.2　数值仿真

本文采用 ANSYS/LS－DYNA 软件建立 3D 拉格朗日有限元算法对侵彻问题进行模拟。根据弹－靶结构和加载方式的对称性，垂直侵彻建立 1/4 模型，在对称界面上施加对称约束。弹－靶碰撞采用面－面接触侵蚀算法。弹靶均采用八节点六面体（solid164）实体单元。弹体网格尺寸约为 0.2 mm。靶板采用渐变网格，在弹－靶直接作用区域网格较为密实，并在侧面定义非反射边界，如图 3 所示。

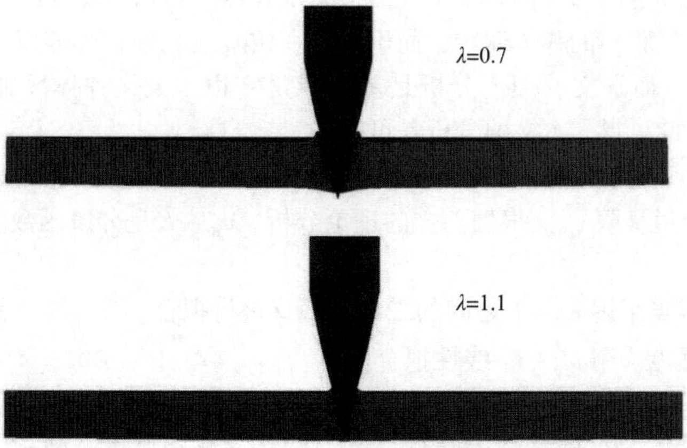

图 3　不同 λ 条件下的正侵彻临界侵深

刚性弹体的材料模型采用＊MAT_RIGID，而靶板采用 Johnson－Cook 本构模型，它适用于描述材料在大变形、高应变率和高温条件下的本构状态，表 1 给出了弹靶材料 30CrMnSiNi2A 和 921A[12,13] 的参数。此外，在计算靶板的径向应力 σ_r 时未考虑应变强化，即单向压缩塑性流动应力 Y 取屈服强度 $Y \approx \sigma_y$。

表 1　弹靶的材料参数

参数	ρ(kg·m^{-3})	σ_y/MPa	E/GPa	μ
30CrMnSiNi2A	7.80	—	205	0.28
921A	7.80	898.6	205	0.3

2.3　正交试验和极差分析

本文的指标参量为在 λ 取不同值时的半侵彻弹道极限 v_{50}；而变量因素包括 m（2.0～2.4 kg）、弹头锥角 2β（28°～32°）和靶板厚度 h（30～50 mm）。本试验为 3^3 型试验，采用拟水平法构造 L_9（3^4）试验正交表，空置第四列（表 2）。

表 2　正交试验的仿真拟合结果

试验	因素及水平			不同 λ 取值时的拟合结果 v_{50}/(m·s^{-1})				
	2β	m/kg	t/mm	0.7	0.8	0.9	1.0	1.1
1	28°	2.0	30	119.40	104.09	89.38	77.80	66.90
2	28°	2.2	40	171.73	146.66	127.55	112.26	97.10
3	28°	2.4	50	230.32	195.80	168.25	148.66	129.33
4	30°	2.0	40	192.47	165.43	142.68	124.71	108.35
5	30°	2.2	50	245.88	215.93	189.32	165.04	143.29
6	30°	2.4	30	116.81	99.83	85.59	70.82	61.14
7	32°	2.0	50	282.50	247.56	213.51	179.05	155.55
8	32°	2.2	30	121.64	105.90	91.33	77.10	66.57
9	32°	2.4	40	186.75	163.33	140.81	121.71	105.87

改变弹丸初速，以侵彻速度为零作为侵彻终点，记录弹体沿初始速度方向的位移，即为弹体侵深 L。不考虑弹体反弹和斜侵彻过程中的弹体偏转，而事实上在仿真过程中，弹体以一个较小的速度反弹（初速度的 0.37%～10.8%），而本文的理论分析是基于能量守恒出发，弹体反弹的能量耗散最大值仅为 1.17%，可忽略不计。由此可得，本文的分析是可靠的。

为保证精确度，每次试验的速度区间能够覆盖各临界状态的侵彻深度，并采用多项式拟合的方法获得 λ 取不同临界值时的弹道极限 v_{50}。根据前文的理论分析，$v_{50}-L$ 是分段函数关系，因此在拟合时进行分段拟合。

（1）对于 $L>t$，数据点采用 v_{50}^2-L 之间的二次多项式进行拟合；

（2）对于 $L\leqslant t$，数据点采用 $v_{50}^2-L^3$ 线性拟合。

拟合结果见表 2，图 4 则为每个因素的极差柱状图，其中波动越大表明该因素对弹道极限速度 v_{50} 影响越大。易得 $R_C>R_A\approx R_B$，即各因素对 v_{50} 影响主次顺序为：靶板厚度、弹头锥角和弹丸质量，其中后两个因素对 v_{50} 的影响程度相近。

2.4　正侵彻仿真结果与理论计算的误差分析

根据表 1 中的参数可获得理论计算的参数，$Y\approx\sigma_y$，由此可得 $\kappa=2.894$，$\sigma_r=2\,601$ MPa。理论计算的相对误差值列于表 3，从中可得如下结果：

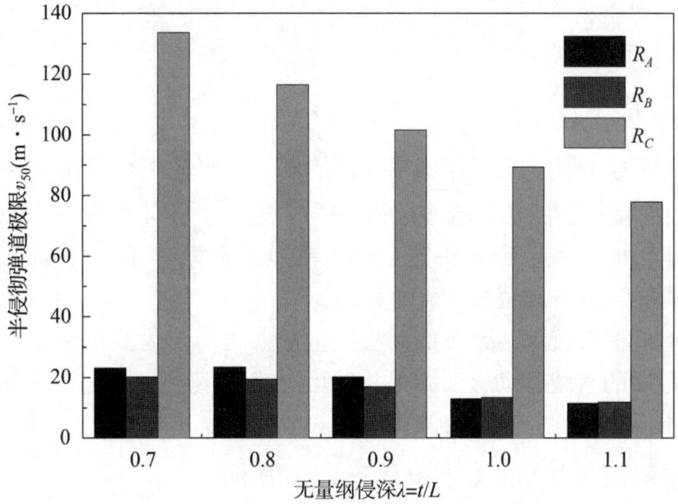

图 4　极差分析柱状图

表 3　正侵彻理论计算误差表

试验	t/mm	不同 λ 取值时的相对误差/%				
		0.7	0.8	0.9	1.0	1.1
1	30	-4.62	-9.59	-11.45	-13.10	-12.39
2	40	-2.66	-5.80	-8.91	-11.59	-11.40
3	50	-2.88	-5.59	-7.60	-10.66	-11.00
4	40	-2.10	-5.87	-8.21	-10.30	-10.50
5	50	2.11	-3.90	-7.82	-9.68	-9.83
6	30	-4.36	-7.51	-9.28	-6.34	-5.95
7	50	-0.25	-5.93	-8.27	-6.55	-6.76
8	30	2.66	-2.55	-4.97	-3.84	-3.46
9	40	-1.44	-6.86	-9.14	-10.20	-10.52

（1）误差存在正负。其中误差为正表示理论计算值大于仿真结果，误差为负表示理论计算结果小于仿真结果。这主要是由靶板材料的应变强化效应和靶板背面自由面对侵彻过程产生影响这两者共同作用产生的。应变强化效应使得靶板材料在经历屈服阶段后，又增强了抵抗变形的能力。而在理论计算中采用屈服强度计算径向应力，因此理论结果偏低。而另外，靶板背面自由边界对侵彻过程的影响为，当靶板金属塑性流动区扩展到靶板背面时，弹体所受的阻力下降，侧限系数 κ 下降，扩孔所需的能量比理论计算小，即理论计算结果偏高。

（2）当靶板厚度取值范围为 30~50 mm，靶板材料在高应变率的条件下，其应变强化效应占据主导地位，因此理论计算误差基本呈负值。

（3）对于 $\lambda<1.0$，靶板被贯穿后导致作用在弹体上的阻力骤减，抵消了部分应变强化效应带来的负误差，这一点可以用于解释仅有的正相对误差仅出现在 $\lambda=0.7$ 时。此外，靶板背面自由面效应随 λ 的递减而增大，进一步对负误差进行补偿。因此，相对误差的绝对值随 λ 的递减而减小。

（4）对于 $\lambda\geqslant1.0$，由于靶板背面未被贯穿，靶板背面自由面的效应相对较小，这可以用于解释误差值在 $\lambda=1.0$ 和 $\lambda=1.1$ 时的差别较小（小于0.71%）。

（5）根据以上分析，前文推导的理论计算公式的适用于靶板厚度 30～50 mm 的高强度船用钢 921A 钢，其绝对误差均小于 12.39%。

3 斜侵彻特性研究

斜侵彻过程中，弹体所受非轴对称力作用不经过质心，且弹－靶的接触面积不断变化。假设着靶角为 α，弹体在头部完全侵入靶体前的初始轨迹近似为螺旋线[14]，弹体存在一个偏转角，非轴对称的弹靶间作用力逐渐趋于与弹体轴向一致。当弹头完全侵入靶板，阻力 F 成为完全轴对称，偏转角 δ' 不再变化，弹体在后续的运动轨迹为沿着斜角的 $\delta'+\alpha$ 直线。

本文的弹头较长，靶板相对较薄，弹头部无法完全侵入靶体。为简化模型，不考虑弹体偏转，在侵彻过程中一直沿着斜角为 α 的直线运动，而 δ' 是相对可忽略的值。为简化计算并获得与正侵彻模型相似的理论公式，引入扩孔体积和平均径向力的修正系数 χ_1、χ_2。

设斜侵彻延性扩孔平均径向力为 $\bar{\sigma}_r$，扩孔体积 U'，则

$$\bar{\sigma}_r=\chi_1\sigma_r, U'=\chi_2 U \tag{7}$$

其中，χ_1 为斜侵彻与正侵彻的延性扩孔体积比，与 α，β 和 λ 相关；而 χ_2 与靶板材料的性能相关。χ_1 与弹头形状和无量纲侵深有关，通过椭圆锥（台）和圆锥（台）的体积关系计算可得表 1，而 χ_2 通过数值仿真试验结果获得。

3.1 修正系数 χ_1

斜侵彻的延性扩孔形状如图 5 所示（$\alpha=20°$），可表示为斜椭圆锥。在斜侵彻理论模型中，侵彻深度 L 是弹体沿其初始速度方向的位移。此外，在该延性扩孔与靶板上表面的交面为标准椭圆：$\dfrac{x^2}{a^2}+\dfrac{y^2}{b^2}=1$。

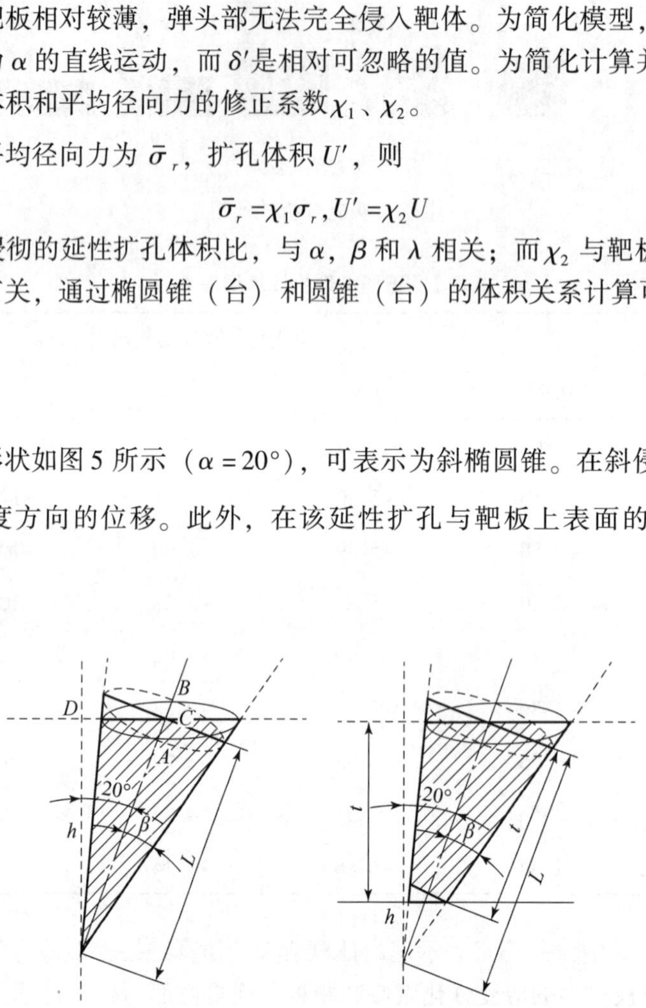

图 5 斜侵彻的临界侵深示意图

（a）$\lambda\geqslant1.0$；（b）$\lambda<1.0$

（1）当 $\lambda\geqslant1.0$，如图 5（a）所示，根据几何关系可得椭圆长轴如下方程：

$$2a=h[\tan(\alpha+\beta)-\tan(\alpha-\beta)]=(t\cos\alpha/\lambda)[\tan(\alpha+\beta)-\tan(\alpha-\beta)] \tag{8}$$

为获得短轴 b 的值，取底面椭圆如图 6 所示。直线 AB 是正侵彻和斜侵彻时的延性扩孔的交线，因此根据几何关系，设 A 点的坐标为（x_0，y_0），则

$$\begin{cases} x_0=-\overline{oc}=(t\cos\alpha/\lambda)[\tan\alpha-\tan(\alpha-\beta)]-a \\ y_0=r_0=L\tan\beta \end{cases} \tag{9}$$

$$b=\sqrt{a^2y_0^2/(a^2-x_0^2)}, U_1'=(\pi/3)abh=(\pi/3)abL\cos\alpha \tag{10}$$

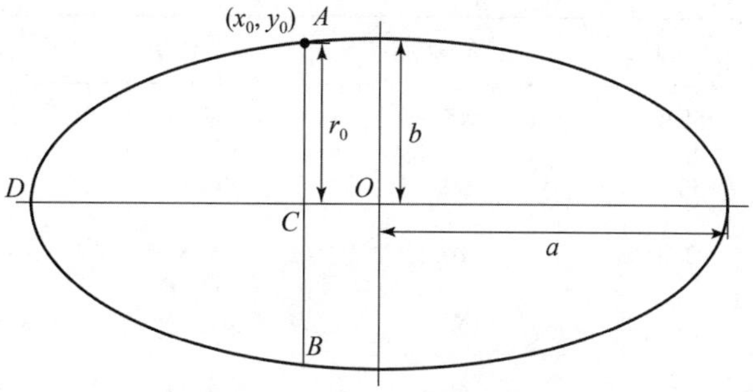

图6　斜侵彻的底面椭圆示意图

（2）当 $\lambda<1.0$，如图4（b）所示，与上述方程相似，延性扩孔体积可表示为

$$h'=L\cos\alpha-t=t(\cos\alpha/\lambda-1) \tag{11}$$

根据几何相似原理 $\frac{a'}{a}=\frac{b'}{b}=\frac{h'}{h}=\frac{\cos\alpha-\lambda}{\cos\alpha}$，有

$$U'_2=(\pi/3)(abh-a'b'h')=(\pi/3)abL\cos\alpha[1-(1-\lambda/\cos\alpha)^3] \tag{12}$$

根据上述分析，对于 $\alpha=20°$，$\beta=30°$时，对于 $\lambda\geqslant1.0$，$\chi_1\approx1.014$；而当 $\lambda=0.9/0.8/0.7$，$\chi_1\approx1.015/1.019/1.025$。

3.2　修正系数 χ_2

根据几何斜椭圆锥的截面相似性，易得在不同弹体截面处的径向受力相似，因此可以引入平均径向应力修正系数 χ_2，如图6所示。因此，当 α 保持不变时 χ_2，即 $\bar{\sigma}_r/\sigma_r$ 比值保持不变。在斜侵彻仿真过程中，为提高计算效率仅建立1/2模型。

根据式（4）和式（7）可得，若正/斜侵彻的半侵彻弹道极限分别用 v_{50}/v'_{50} 表示，则平均径向力修正系数 χ_2 为

$$\chi_1=(v'_{50}/v_{50})^2\chi_2^{-1} \tag{13}$$

由前文定义，χ_2 仅与靶板材料有关，但考虑到靶板应变强化和自由边界效应，在此正侵彻理论计算结果 v_{50}。因此，本文获得的修正系数 χ_2 是一个耦合修正系数。选取弹－靶系统：靶板厚度40 mm，弹头锥角30°，弹体质量2.3 kg。设理论计算与仿真结果分别用 v_t 和 v_s 表示，可得在 λ 取不同值时的平均径向力修正系数 χ_2 见表4。

表4　平均径向应力修正系数 χ_2

Results	$\lambda=0.7$	$\lambda=0.8$	$\lambda=0.9$	$\lambda=1.0$	$\lambda=1.1$
v_t	175.71	145.21	122.12	104.32	90.43
v_s	154.78	127.63	108.85	95.97	83.52
χ_2	0.757 0	0.758 1	0.782 7	0.834 6	0.841 2

根据表4，耦合修正系数 χ_2 有明显的分段，这与延性扩孔体积出现分段的现象一致；此外当 $\lambda\geqslant1.0$ 时，受靶板背面自由边界的影响较小，这也是修正系数比 $\lambda<1.0$ 时大的原因。当 $\lambda<1.0$ 时取 $\chi_2=0.766$；而当 $\lambda\geqslant1.0$ 时 $\chi_2=0.838$。

为检验该耦合参数的实用性，建立新的仿真模型 $t=20/30/50/60$ mm（$2\beta=30°$，$m=2.3$ kg，$\alpha=20°$）。采用上述修正系数 χ_1 和 χ_2 的理论计算结果和仿真数据列于表5。

表 5　斜侵彻修正系数验证

t/mm	Results	$\lambda=0.7$	$\lambda=0.8$	$\lambda=0.9$	$\lambda=1.0$	$\lambda=1.1$
	v_t (m·s^{-1})	55.05	45.36	38.07	34.00	29.47
20	v_s (m·s^{-1})	56.67	50.73	44.44	37.60	32.24
	Errors	−2.86%	−10.59%	−14.33%	−9.57%	−8.59%
	v_t (m·s^{-1})	101.12	83.33	69.94	62.46	54.14
30	v_s (m·s^{-1})	102.25	86.08	73.71	62.95	55.27
	Errors	−1.11%	−3.19%	−5.11%	−0.78%	−2.04%
	v_t (m·s^{-1})	217.58	179.29	150.49	134.40	116.49
50	v_s (m·s^{-1})	216.42	184.13	158.71	138.94	121.56
	Errors	0.54%	−2.63%	−5.18%	−3.27%	−4.17%
	v_t (m·s^{-1})	286.02	235.69	197.83	176.67	153.13
60	v_s (m·s^{-1})	284.63	235.04	197.49	170.21	148.94
	Errors	0.49%	0.28%	0.17%	3.80%	2.81%

由此可见，由于引入修正系数，斜侵彻理论计算精度得到提高。由于本文的耦合修正系数χ_2是从$t=40$ mm 的试验案例中推算出来的，更能体现在较厚靶板条件下各误差因素的耦合情况，因此更适用于靶板相对较厚的情况（30～60 mm），其最大相对误差的绝对值在 5.18%。

对于$t=20$ mm，尽管理论计算的最大绝对误差为 6.37，其最大相对误差的最大绝对值却达到了 14.33%。由于靶板相对较薄，靶板内部的材料塑性流动区很快膨胀到达背面的自由面，因此在薄靶的侵彻过程中自由面的影响较大；而且由于达到临界侵深的极限速度较小，在侵彻过程中靶板材料的应变率效应较小，因此材料应变强化效应对侵彻过程影响较小，与在较厚靶板条件下两个因素的影响程度不同。综上所述，本文获得的修正系数χ_2不适用于薄靶条件下。

4　结论

本文通过大量的数值仿真对刚性尖头弹半侵彻延性靶板进行研究。基于弹丸半侵彻时不反弹的假设，定义了半侵彻弹道极限，即当无量纲侵深$\lambda(\lambda=t/L)$取不同值时v_{50}相应地取不同临界值。

首先，根据空腔膨胀模型建立了正侵彻的简化模型，从能量守恒的角度推导出半侵彻弹道极限v_{50}的理论公式。结合正交试验法进行仿真研究，通过极差分析法得出了影响v_{50}的各因素主次关系。对比理论计算与仿真结果，其误差表明靶板的材料应变强化效应和背面自由面效应共同对侵彻过程产生影响。

其次，对于斜侵彻研究，本文引入延性扩孔体积修正系数χ_1和平均径向应力修正系数χ_2，并通过几何分析与仿真结果获得了具体取值。此外，通过四组斜侵彻仿真实验验证了修正系数用于预测斜侵彻v_{50}的可行性。此外，在靶板厚度为 30～60 mm 的条件下本文的理论计算与仿真结果吻合较好。

参考文献

[1] 蒋志刚，曾首义，周建平．金属薄靶板冲塞破坏最小穿透能量分析［J］．工程力学，2004，21（5）：203－208.

[2] 蒋志刚，曾首义，周建平．分析金属装甲弹道极限的两阶段模型［J］．工程力学，2005，22（4）：229－234.

弹丸变攻角侵彻间隔靶弹道极限研究①

黄　岐②，周　彤，白　洋，兰旭柯，冯顺山

（北京理工大学爆炸科学与技术国家重点实验室，北京　100081）

摘　要：攻角对弹丸侵彻能力有重要影响，而侵彻过程中攻角的改变对弹丸侵彻多层靶能力的影响更为显著。针对厚度相同的双层接触靶和间隔靶，通过理论分析和数值模拟，研究弹丸贯穿前靶后的攻角变化对侵彻后靶能力的影响。研究结果表明，贯穿前靶后弹丸攻角随着靶板间距的增大而增大，当攻角变化大于30°时，间隔靶的弹道极限开始高于等厚接触靶的弹道极限；弹丸变攻角侵彻双层靶过程的动能损耗机理不同于无攻角的正侵彻和斜侵彻，它包含弹靶之间非对称的相互作用；贯穿前靶后弹丸的攻角变化取决于弹轴角速度和靶板间距，而弹轴角速度取决于弹丸初速、靶板厚度、初始攻角、初始倾角等；弹丸侵彻能力越强，弹靶分离后弹轴角速度越大。

关键词：兵器科学与技术；弹道极限；变攻角；双层靶板；数值模拟

Ballistic Performance of Projectile Penetrating Large – spaced Multi – layer Plates at Variable Angle – of – attack

Huang Qi，Zhou Tong，Bai Yang，Lan Xuke，Feng Shunshan

（State Key Laboratory Explosion Science and Technology，Beijing Institute of Technology，Beijing 100081，China）

Abstract：Angle of attack has an important effect on penetration，and the change of angle – of – attack has more important effect on the penetration of projectiles into multi – layered targets. The ballistic performance of projectile penetrating into double – layer targets is simulated. Results show that the angle – of – attack of projectile increases with the increase in the space between target plates. When the angle – of – attack exceeds 30°，the ballistic limit of multi – layer spaced plates is higher than that of contact – target. Kinetic energy loss mechanism of variable angle – of – attack is different from normal penetration or oblique penetration without angle – of – attack，which contains the process of bullet side interact with target. Axial deflection angle depends on the angular rate and the space between target plates. The angular rate depends on initial velocity，thickness of target，initial yaw angle，etc. The axial angular rate is found to increase with the increase in penetration ability of projectile.

Keywords：ordnance science and technology；ballistic limit；variable angle – of – attack；double – layered target；numerical simulation

0　引言

穿甲和侵彻技术是武器弹药传统技术，但终点效应的新要求促进了技术的发展。随着防护技术的进步，多层装甲防御技术得到应用，因此动能体对多层靶的侵彻理论研究得到了关注。金属防护结构可分为单层靶和多层靶。多层靶按结构间隙可分为接触靶和间隔靶[1]。本文研究弹丸对双层金属靶的侵彻机理和效应，为弹药侵彻能力的提高或防护结构的设计提供技术支持和理论基础。

---

① 原文发表于《兵工学报》2016年37卷Suppl. 2。

② 黄岐：工学博士，2016年师从冯顺山教授，研究弹药对飞行甲板侵立机理及封锁效应，现工作单位：北京理工大学博士后。

研究表明，单层靶和多层靶的抗侵彻性能与靶板材料、厚度、分层数目、叠层顺序和叠层间隙相关[2]。尽管有不少研究人员对单层靶和多层靶进行了大量实验、数值计算和理论分析研究[3-5]，然而由于考察的撞击条件不同，得到的结论也有差异。目前对双层间隔靶的抗侵彻性能的研究较少，且很少考虑靶板间距对弹丸姿态和后续侵彻效应的影响。

弹丸在带攻角侵彻或斜侵彻非均质靶板时会受到不对称作用力，过靶后其姿态会受到影响。大靶距条件下弹丸的攻角和倾角会随距离变化。当靶板间距足够大时，弹轴角度的变化较为显著，并对弹丸侵彻后靶的状态造成较大影响。弹丸变攻角侵彻双层靶过程的动能损耗机理不同于无攻角的正侵彻和斜侵彻，它包含弹靶之间非对称的相互作用。本文基于国内外研究现状，考虑弹丸带攻角贯穿前靶后，研究其姿态变化对后靶侵彻能力的影响。

1　模型的建立

1.1　理论模型

杆状弹丸侵彻金属靶板分为两个阶段：第1阶段为冲塞形成阶段：在杆状弹丸作用下，受到冲击的靶材部分硬化并产生相对于其余部分的运动形成剪切带，剪切带不断进入热塑失稳状态，直至塞块完全形成[6]；第2阶段是弹靶侧向作用阶段，主要耗能机制为靶板的塑性变形吸能。

图1为杆状弹丸攻角侵彻的5个典型阶段[7]：①初始侵彻时弹丸姿态未改变，如图1（a）和图1（b）所示；②弹丸无攻角侵彻时失效，如图1（c）和图1（d）所示；③非对称作用力导致弹轴偏转，弹轴与靶板法线夹角增大，如图1（c）所示；④形成冲塞后角度继续增大，如图1（f）所示；⑤弹丸侧面与靶板作用，靶板呈花瓣形破坏，弹轴获得偏转角速度，如图1（h）所示。

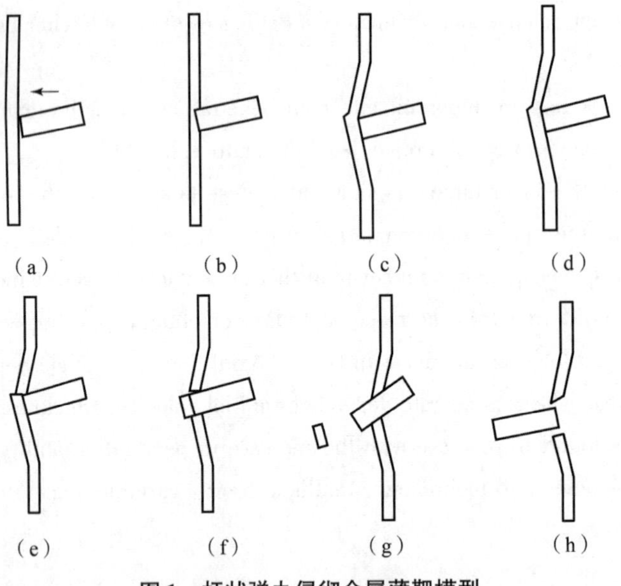

图1　杆状弹丸侵彻金属薄靶模型

（a）弹靶初始接触；（b）靶板初始变形；（c）弹丸开始偏转；（d）靶板拉伸失效；
（e）靶板破坏；（f）冲塞破坏；（g）塞块飞出/弹丸继续侵彻；（h）弹丸贯穿靶板

图2显示出了在准静态阶段后作用在弹靶接触面的正应力分量 σ_{rr}、σ_{xx} 与剪切应力 $\sigma_{r\theta}$，应力分量对弹丸质心产生力矩，剪切应力不会导致弹丸两侧不对称。径向应力 σ_{rr} 导致的力矩可表示为

$$M_{rr}=Sdb'(L-b')/\pi \tag{1}$$

式中　S——弹丸横截面积；

d——弹丸直径；

L——弹丸长度；

$b' = d\tan\alpha$——弹接触面最大的接触距离。

同理，轴向应力 σ_{xx} 导致的力矩可表示为

$$M_{rr} = Sd^2 b'/\pi \tag{2}$$

这两个力矩方向相反，而另一个力矩 $M_G = M_{rr} - M_{xx}$ 导致了弹丸的攻角加速度 $\ddot{\varphi}$ 用积分形式的动能定理分析弹丸贯穿靶板前后两种状态，有 $T_2 - T_1 = \sum W$，其中

$$T_1 = \frac{1}{2}mv_1^2, T_2 = \frac{1}{2}mv_2^2 + \frac{1}{2}J\dot{\varphi}^2$$

$$\sum W = \frac{1}{2}mgl(\cos\alpha + h + \cos\varphi) - W_1 \tag{3}$$

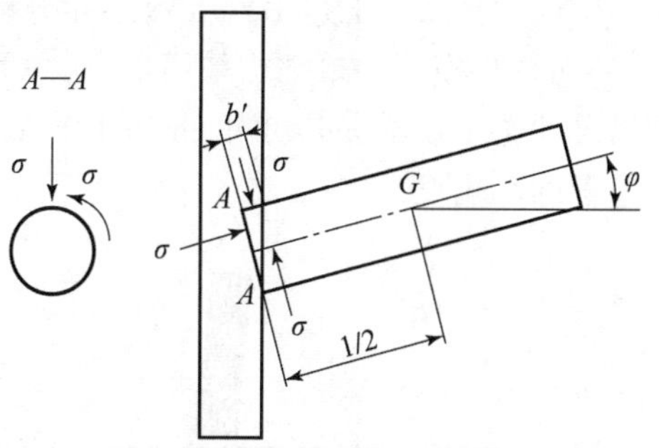

图 2　杆状弹丸嵌入靶板时的应力状态

式中　v_1——着靶速度；

v_2——剩余速度；

J——转动惯量；

α——初始攻角；

φ——弹靶分离后弹丸的攻角；

W_1——弹靶作用消耗的能量[7]。

从式（3）可知，弹丸贯穿靶板后获得的角速度与多重因素有关。下面仅考虑弹丸初速、靶板厚度、初始攻角对弹轴偏转角速度的影响。

1.2　数值模型

选用长径比为 5 的平头弹丸结构，弹丸质量为 13 g，材料选用 45 号钢，其具体结构尺寸如图 3 所示。多层靶的材料采用 A3 钢。

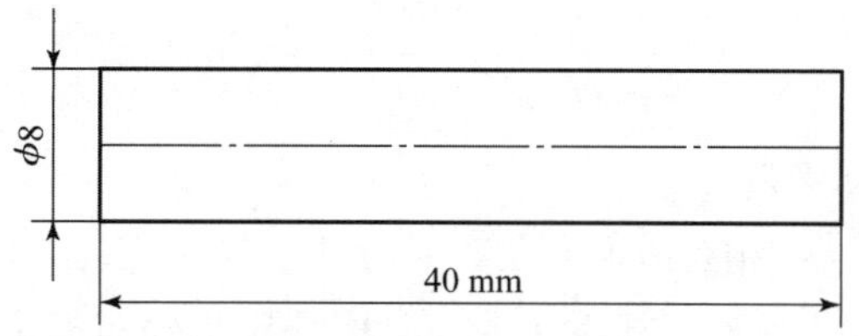

图 3　弹丸结构尺寸

图 4 为弹丸侵彻大间距多层靶的结构。定义靶板间距系数 $\lambda = s/L$ 其中 s 为靶板间距，为弹丸长度。$\lambda = 0$ 时为接触靶，侵彻时后靶延缓了靶板背面稀疏波对前靶背面的拉伸，且在一定程度上限制了前靶的弯曲变形；当 $\lambda > 0$ 时为间隔靶，侵彻时靶板间无直接的相互作用；当 $0 < \lambda < l$ 时，弹丸未完全贯穿前靶就已与后靶接触，因此弹丸在侵彻后靶时会受到已贯穿靶板的干扰；当 $\lambda \geqslant 1$ 时，弹丸侵彻任意层靶板均不会受到其他靶板的干扰。特别是，当 $\lambda \gg 1$ 时为大间距多层靶结构，弹丸在靶板间隔中飞行的姿态改变不容忽视。本文研究的靶板结构为大间隔多层靶，间隔系数为 $\lambda > 1$，因此弹丸在靶板间自由飞行时其姿态的改变是影响其侵彻下一层靶板能力的主要原因。

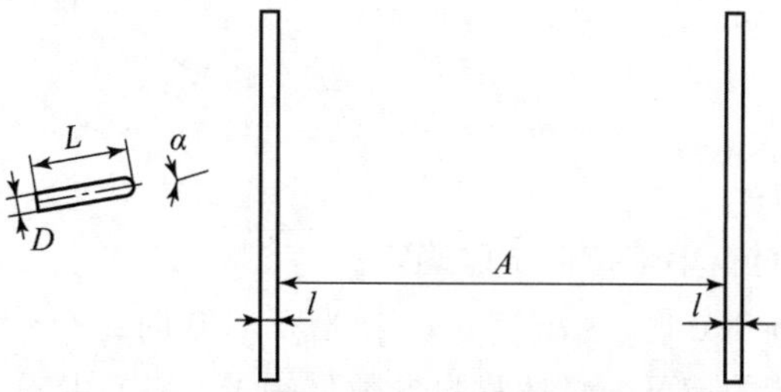

图 4　弹丸侵彻大间距多层靶的结构

数值仿真采用 LS－DYNA 有限元分析软件。根据弹丸攻角侵彻多层靶过程中载荷的对称性特点，建立 1/2 的 3D 模型。单靶均采用 8 节点 6 面体实体单元，在对称界面上施加对称约束。单靶接触区靶板的网格尺寸选用 0.4 mm×0.4 mm×0.4 mm。单靶碰撞采用面面接触侵蚀算法，同时忽略靶板间空气对侵彻过程的影响[8]。

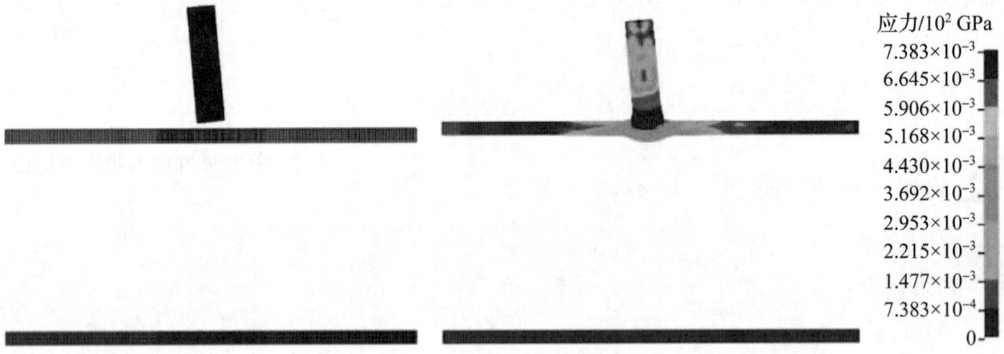

（a）　　　　　　　　　　　　（b）

图 5　弹丸侵彻多层靶有限元网格

（a）有限元网格；（b）应力云图

对靶板尺寸为 120 mm×120 mm 的方靶，单靶接触范围假设为靶板中心 30 mm×30 mm 的区域。接触区外的网格尺寸逐渐变粗，但网格厚度保持不变。固定靶板侧面边界，并定义成非反射边界。弹丸网格数量为 6 336，靶板网格数量为 141 120。单靶的材料模型均采用著名的 Johnson－Cook 方程，它适用于描述材料在大变形、高应变率和高温条件下的本构状态。其本构方程可以表示为

$$\sigma_{eq} = (A + B\varepsilon_{\mathrm{eq}}^{n})(1 + \varepsilon)^{C}(1 - T^{*m}) \tag{4}$$

式中　σ_{eq}——等效应力；

$\varepsilon_{\mathrm{eq}}$——等效塑性应变；

A、B、C、m、n——材料常数；

$\dot{\varepsilon}_{\mathrm{eq}}^{*} = \dot{\varepsilon}_{\mathrm{eq}}/\dot{\varepsilon}_{0}\varepsilon_{\mathrm{eq}} = \cdot \mathrm{eq}/2^{\circ}$——相对应变率；

同系温度 $T^{*} = (T - T_{\mathrm{r}})/(T_{\mathrm{m}} - T_{\mathrm{r}})$，其中 T、T_{m}、T_{r} 分别为绝对温度、材料的熔化温度和室温。绝热升温导致的温度增量可表示为

$$\Delta T = \int_{0}^{\varepsilon_{\mathrm{eq}}} \chi \frac{\sigma_{\mathrm{eq}} d\varepsilon_{\mathrm{eq}}}{\rho C_{\mathrm{p}}} \tag{5}$$

式中　ρ——材料密度；

C_{p}——比热；

χ——功热转化的 Taylor－Quinney 比例系数。

材料模型中使用的 Cockcroft－Latham 失效准则表达式如下：

$$W = \int_{0}^{\varepsilon_{\mathrm{eq}}} < \sigma_{1} > d\varepsilon_{\mathrm{eq}} \leqslant W_{\mathrm{cr}} \tag{6}$$

式中　W——单元体积的塑性功；

W_{cr}——W 的临界值，由材料单轴拉伸试验确定；

σ_{1}——最大主应力，当 $\sigma_{1} \geqslant 0$ 时，$\langle \sigma_{1} \rangle = \sigma_{1}$；当 $\sigma_{1} < 0$ 时，$\langle \sigma_{1} \rangle = 0$。Johnson－Cook 材料模型能较好地模拟钢靶受到冲击载荷的情况，具体材料参数见表 1。表 1 中 μ 为曲线截距，$D_{1} \sim D_{5}$ 为材料失效参数，η 为参考应变率系数。

表 1　弹靶材料参数[9]

参数	45 钢 506 钢	A3 钢	参数	45 钢	A3 钢
A/MPa	506	410	p_t/(kg · m^{-3})	7 800	7 800
B/MPa	320	20	E/GPa	200	200
C	0.064	0.100	D_1	0.1	0.3
μ	0.3	0.3	D_2	0.76	0.90
n	0.28	0.08	D_3	1.57	2.80
m	1.06	0.55	D_4	0.005	0
T_r/K	300	300	D_5	− 0.84	0
T_m/K	1 795	1 795	η	1	1

2　仿真结果及分析

2.1　弹轴偏转对剩余速度的影响

图 6 为弹丸在侵彻过程中弹轴偏转的时程曲线。图 6 中曲线可分为 4 个阶段：①水平段；②曲线上升段；③线性下降段；④线性上升段。水平段弹轴的偏转角度未发生改变。该段包含两个状态：弹靶接触前的自由飞行阶段和初始侵彻时无角度变化的阶段，对应图 1（a）和图 1（b）。曲线上升段并不是线性增长，而是近似 S 形的增长曲线。侵彻初期，由于弹轴倾斜，靶板对弹丸的不对称作用力导致弹轴倾斜程度加剧，即攻角逐渐增大。随着弹丸不断侵彻，弹体侧面径向力的作用逐渐明显，导致曲线上升段后期攻角的增幅放缓，该阶段对应图 1（e）和图 1（f）。线性下降段对应图 1（g）和图 1（h），弹丸剩余速度较高，速度和方向几乎未改变。此时已形成完整塞块，弹丸已侵彻至靶板背面，同时弹体侧面仍与靶板接触，使椭圆形通孔的长轴继续增大。弹体侧面的径向力作用愈发明显，并使弹轴朝反向偏转。线性上升段斜率的绝对值与线性下降段的相等，因为其物理意义为弹轴偏转的角速度。忽略空气阻力，弹丸在线性上升段内不受外力作用，可视为处于以恒定角速度偏转的自由飞行状态。

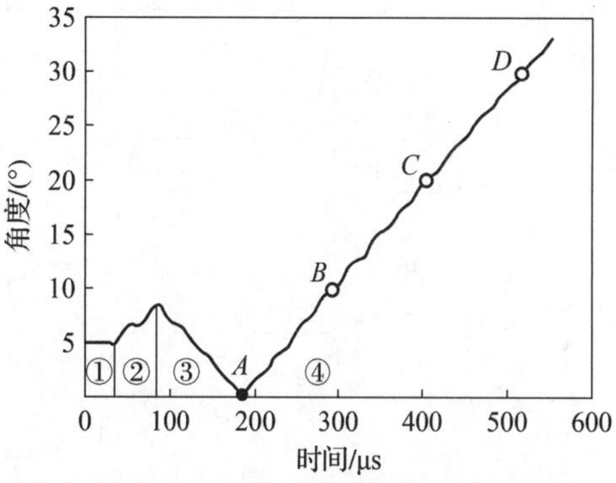

图 6　弹丸在侵彻过程中弹轴偏转的时程曲线

弹丸在 A 点（图 6）的弹轴偏转角度为 0°，即弹轴与靶板法线平行。假设此时弹丸恰好接触下一层靶板，则弹丸的倾角和攻角均为 0°。通过计算弹丸的行程，可得靶板的间距 s。B 点为弹轴与靶板呈 10°时的状态，C、D、E、F、G 点的含义和状态以此类推，具体情况见表 2。

表 2　弹丸过靶后自由飞行时的典型状态

参数	A	B	C	D	E	F	G
弹轴角度/(°)	0	10	20	30	40	50	60
侵彻用时/μs	186	295	404	518	632	746	860
弹丸行程/mm	53	88	122	158	193	228	260
间距系数	1.33	2.20	3.05	3.95	4.83	5.70	6.50

注：弹丸行程指弹轴呈特定角度时弹头离前靶背面的 Y 轴投影距离，在下文中用来定义靶板间距。

基于上述数值模拟，可对弹丸自由飞行阶段中弹轴特定角度进行标定，并将靶板的间距分别定为53 mm、88 mm、122 mm、158 mm、193 mm、228 mm、260 mm。弹丸以5°初始攻角贯穿首层靶板［图7 (a)］，忽略速度方向的微小改变，并以0°、10°、20°、30°、40°、50°、60°的攻角［图7（b）~图7（h）］侵彻后靶。从该组对比试验可观察不同靶距对变攻角弹丸侵彻能力的影响，具体表现为弹丸剩余速度的差异。

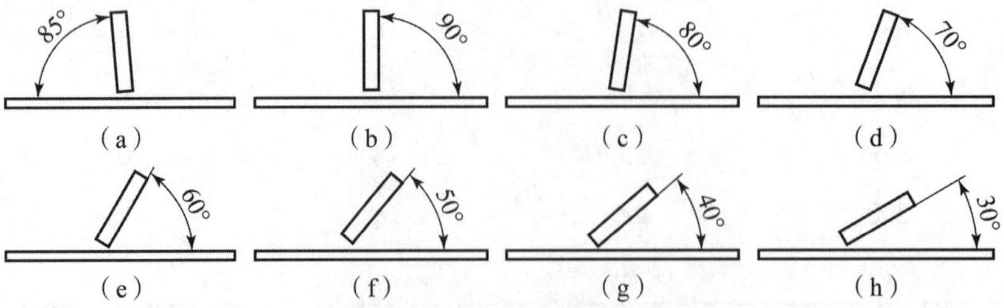

图7　弹丸以不同攻角侵彻后靶示意图

(a) $\alpha=-5°$；(b) $\varphi=0°$；(c) $\varphi=10°$；(d) $\varphi=20°$；(e) $\varphi=30°$；(f) $\varphi=40°$；(g) $\varphi=50°$；(h) $\varphi=60°$

以图7（b）为例，通过观察弹丸以0°攻角侵彻后靶的状态，分析侵彻多层靶过程的特点。弹丸贯穿两层靶板的过载情况类似，均分为两个阶段，在图8（b）中表现为两级波峰：首级波峰（A_1/B_1）由弹头与靶板接触形成，过载随着侵彻的深入而逐渐增加，形成冲塞后则开始减少；次级波峰（A_2/B_2）由弹攻角所致，随着侵深增加，倾斜的弹体侧面与靶板相互作用。次级过载与首级过载值相比较小，表现为弹体对靶板进行切割。与过载对应的速度值见图8（a），贯穿次层靶板的速度可明显分为两个衰减阶段（B_1/B_2），与两级波峰相对应。贯穿首层靶板时，由于弹轴偏转角度较小，弹丸侧面与靶板作用对速度的衰减影响甚微，因此在图8（a）中较难观察到速度分段。

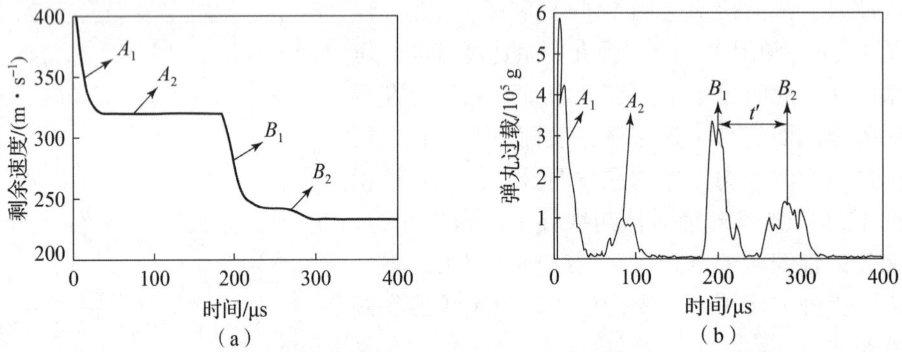

图8　弹丸以0°攻角侵彻后靶的状态

(a) 弹丸速度随时间的变化；(b) 弹丸过载随时间的变化

图9（a）为弹丸剩余速度随弹轴偏转角度的变化。弹丸均以400 m/s的初速贯穿前靶，并保持相同的剩余速度（323 m/s），飞行不同的距离，弹轴偏转不同角度，忽略速度方向的微小改变，最终弹丸以不同攻角接触后靶。当弹丸以0°攻角碰靶时，剩余速度最大，为231 m/s。随着攻角的增加，剩余速度逐渐减小甚至趋近于0 m/s。特别是，当弹丸侵彻次层靶板的攻角大于30°时，剩余速度已经小于弹丸侵彻接触靶的剩余速度。这说明弹轴的偏转增大了侵彻阻力，削弱了弹丸的侵彻能力。弹轴偏转角度超过90°后，虽然攻角绝对值开始减小，但由于弹丸飞行距离过长（入M10），不能忽略空气阻力的作用，因此弹丸剩余速度的变化较为复杂。

图9（b）显示了弹丸过载随弹轴偏转角度的变化。由于靶板间距不同，弹丸接触次层靶板时，过载出现的时间略有差别，并且第二次波峰的极值也随着弹丸攻角的增加而增大［图9（b）拟合线］。定义两个波峰间的时间差为t'［图8（b）］，显然t'随着攻角的增大而逐渐减少。弹丸侧面与靶板相互作用的程度是上述两种现象出现的主要原因。图10为弹丸以5°初始攻角贯穿首层靶板后以0°、10°、20°、

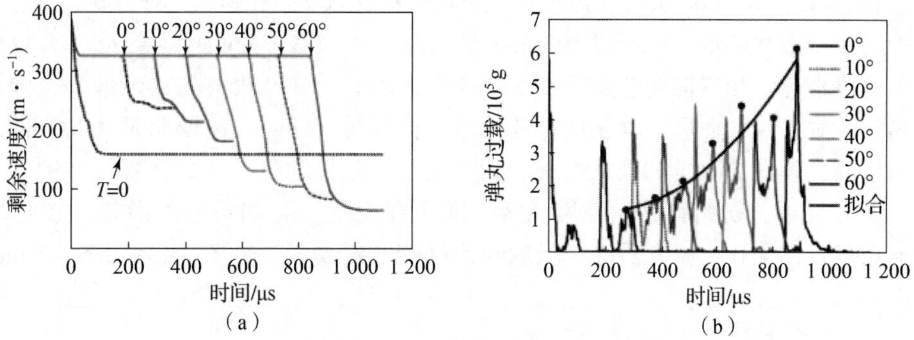

图 9　弹丸以不同攻角侵彻后靶的状态

（a）弹丸剩余速度随弹轴偏转角度的变化；（b）弹丸过载随弹轴偏转角度的变化

30°、40°、50°、60°的攻角侵彻后靶的状态。从靶板的破损程度也能观测出弹丸耗能的差异。当攻角在锐角范围内，弹丸对次层靶板的攻角越大，靶板通孔的横截面积也越大。

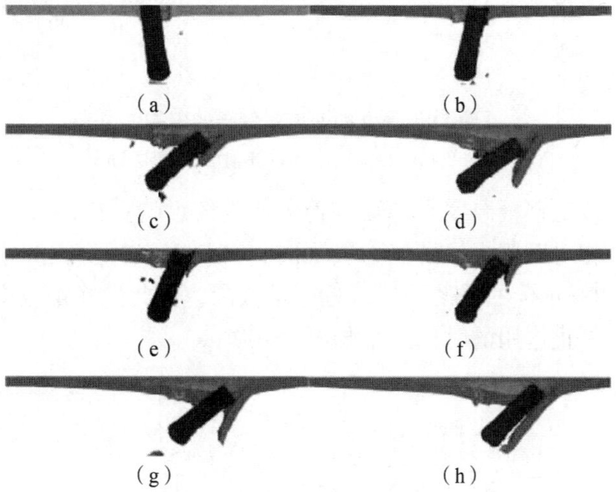

图 10　前靶与后靶穿孔情况

（a）首层靶；（b）0°；（c）10°；（d）20°；（e）30°；（f）40°；（g）50°；（h）60°

综上所述，弹轴偏转角度取决于弹轴角速度和靶板间距。当靶板的间距增大到一定程度，即不能忽略弹轴的偏转时，弹丸的侵彻能力明显下降，靶板抵御弹丸侵彻的能力大幅上升。与侵彻双层间隔靶相比，弹丸在侵彻多层间隔靶时，其侵彻能力下降的程度会更严重。

2.2　影响弹轴偏转角速度的因素

弹轴偏转程度与贯穿后获得的角速度有关。现利用上述材料模型，分别研究弹丸初速、靶板厚度初始攻角对弹轴偏转角速度的影响。影响因素的参数见表 3。弹丸以 5°攻角 >400 m/s 初速侵彻 4 mm 厚的靶板为侵彻条件，其参数作为对比试验的参考。

表 3　影响弹轴偏转角速度的主要因素

影响因素	参数				
弹丸初速/($m \cdot s^{-1}$)	200	300	400	500	600
靶板厚度/mm	2	3	4	5	6
初始攻角/(°)	1	3	5	7	9

图 11 显示出弹丸以 200 ~600 m/s 初速侵彻 4 mm 厚靶板的攻角和弹轴角速度的变化。图 11（a）中的曲线与图 6 中的曲线形状类似。由于以 200 m/s 低速接近靶板的弹道极限，因此弹丸攻角的变化规律不同于其他 4 种速度情况：当侵彻初速高于 1.5 倍弹道极限时，弹丸攻角先逐渐增加，弹丸侧面与靶板相互作用后攻角达到临界值，此后弹轴朝反方向偏转。特别是，弹丸初速较低时（200 m/s），其保持原有速度方向能力较弱，弹丸的攻角和倾角均有较大改变，因此弹轴角速度变化规律不同于其他曲线通过弹轴角速度表示上述规律：初始弹轴角速度为零，弹靶接触后，弹轴角速度沿初始攻角增大的方向增加，达到极大值后朝反向变化，弹丸贯穿靶板后角速度值不再改变，最终恒定在 0.5 ~2 rad/s 范围内。

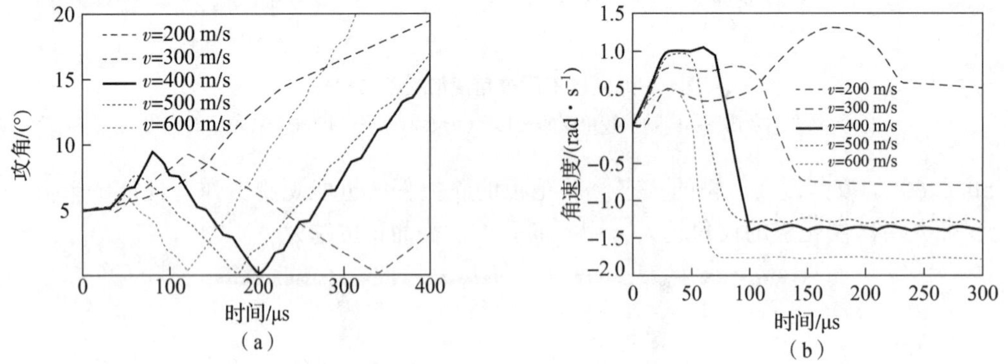

图 11　弹丸初速度对偏转的影响

（a）弹丸攻角时程曲线；（b）弹轴角速度时程曲线

与改变侵彻初速相比，改变靶板厚度对弹轴偏转角速度的影响更规律。随着靶板厚度的增加，角速度极大值在稳步增长。弹丸相对于靶板侵彻能力较强时，其自由飞行阶段的角速度约为 -1.5 rad/s，当靶板厚度增大到 6 mm 时，400 m/s 的侵彻初速接近弹道极限，因此最终角速度值较小。由图 12（a）可知，改变靶板厚度后，弹靶停止作用的时刻几乎相同（100 μs）。

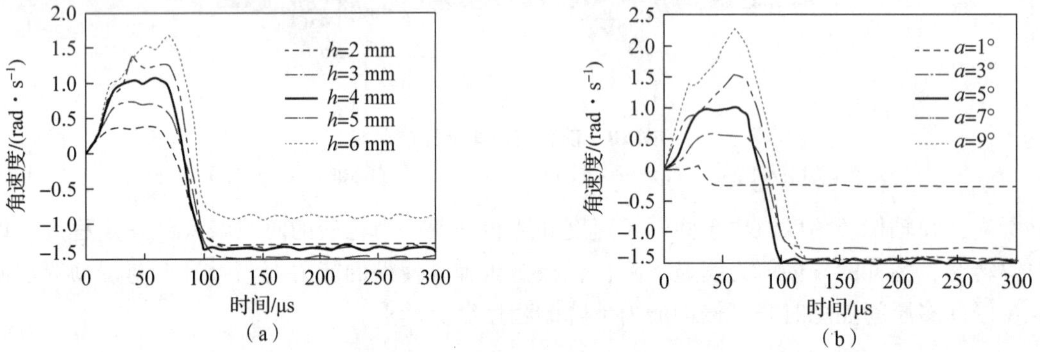

图 12　弹轴角速度时程曲线

（a）靶板厚度的影响；（b）初始攻角的影响

改变弹丸初始攻角不仅影响弹轴偏转角速度的极值，也影响弹靶停止作用的时刻。攻角为 1°时，弹丸侧面与靶板作用的程度较低，所受不对称作用力的时间较少，因此在弹靶分离前角速度已到达极值。随着初始攻角增大，弹丸侧面与靶板作用程度越发剧烈，弹轴角速度极值亦逐渐上升。角速度变化的趋势与上文相似。

3　结论

本文对弹丸变攻角侵彻双层金属靶的情况进行了理论分析和数值模拟，定义了靶板间距系数，对比弹丸以相同初始状态、不同攻角侵彻后靶的剩余速度，并研究了影响弹轴偏转角速度的弹丸初速、靶板厚度和初始攻角，得出以下结论。

（1）弹丸变攻角侵彻多层靶过程的动能损耗机理不同于无攻角的正侵彻和斜侵彻，它包含弹靶之间非对称的相互作用。

（2）贯穿前靶后弹丸的攻角随着靶板间距增大而增大，攻角变化大于30°时，间隔靶的弹道极限开始高于等厚接触靶的弹道极限。

（3）弹丸在足以贯穿前靶的情况下，弹轴角速度变化趋势相似；弹丸侵彻能力越强，弹靶分离后弹轴偏转的角速度越大。

参考文献

[1] IQBAL M A, GUPTA P K, DEORE V S, et al. Effect of target span and configuration on the ballistic limit [J]. International journal of impact engineering, 2012, 42: 11 -24.

[2] 邓云飞，张伟，曹宗胜. 间隙对A3钢薄板抗卵形头弹侵彻性能影响的实验研究 [J]. 振动与冲击，2013，32 (12): 95 -99.

DENG Y F, ZHANG W, CAO Z S, et al. Effect of gap on behavior of a double - layered A3 steel shield against penetration of ogival rigid projectiles [J]. Explosive and shock waves, 2013, 32 (12): 95 -99. (in Chinese)

[3] ALMOHANDES A, ABDEL - KADER M S, ELEICHE A M. Experimental in - vestigation of the ballistic resistance of steel - fiberglass reinforced polyester laminated plates [J]. Composites part B: engineering, 1996, 27 (5): 447 -458.

[4] ELEK P, JARAMAZ S, MICKOVI D. Modeling of perforation of plates and multi - layered metallic targets [J]. International journal of solids & structures, 2005, 42 (3/4): 1209 -1224.

[5] GOLDSMITH W. Non - ideal projectile impact on targets [J]. International journal of impact engineering, 1999, 22 (2): 95 - 395.

[6] 赵国志. 穿甲工程力学 [M]. 北京：兵器工业出版社，1989.

ZHAO G Z. Armour piercing engineering mechanics [M]. Beijing: Publishing House of Ordnance Industry, 1989. (in Chinese)

[7] GOLDSMITH W, TAM E, TOMER D. Yawing impact on thin plates by blunt projectiles [J]. International journal of impact engineering, 1995, 18 (3): 479 -498.

[8] 寇兴华，黄岐，冯顺山. 考虑弹丸侵蚀的多层间隙靶侵彻弹道研究 [C]//第八届全国强动载效应及防护学术会议. 北京：爆炸科学与技术国家重点实验室，2016.

KOU X H, HUANG Q, FENG S S. Investigation on penetration trajectory of multi - spaced targets considering projectile erosion [C]// The 8th National Symposium on Dynamic Load Effectiveness and Protection. Beijing: State Key Laboratory of Explosion Science and Technology, 2016. (in Chinese)

[9] 陈刚，陈小伟，陈忠富，等. A3钢钝头弹撞击45钢板破坏模式的数值分析 [J]. 爆炸与冲击，2007，27 (5): 390 -397.

CHEN G, CHEN X W, CHEN Z F, et al. Simulations of A3 steel blunt projectiles impacting 45 steel pates [J]. Explosive and shock waves, 2007, 27 (5): 390 -397. (in Chinese)

Planar Path – following Tracking control for An Autonomousunderwater Vehicle in the Horizontal Plane①

自主水下航行器的水平面轨迹跟踪控制

Nie Weibiao②, Feng Shunshan

State Key Laboratory of Explosion Science and Technology, Beijing Institute of Technology, Beijing 100081, P. R. China

Abstract: The combined problems of path – following tracking control for an under – actuated autonomous underwater vehicle (AUV) on the horizontal plane are addressed in this paper. Given three types of smooth and 2 – D desired trajectory including straight line path, curve path and circular path in the horizontal plane, the designed path – following control algorithm uses AUV kinematic and dynamic models with cross – track error method and line of sight to compute the desired orientation and velocities, at the same time it also forces the AUV to tracking the desired three types of path which are specified in advance. Numerical simulation is presented to validate the proposed control approach. Simulation results illustrate the good performance of the control system proposed with tracking errors converge to zero fast and smoothly in exponential form at last, which show that the tracking effect remains good and satisfactory that can provide an effective theoretical guidance and technical support for the control system design of the AUV in future.

Keywords: Autonomous underwater vehicle (AUV); Kinematic and dynamic model; Path – following; Tracking contro

1 Introduction

The sea covers approximately two – thirds of the earth surface area which contains a large amount of resources. Unfortunately, the extent and depth exploration of the ocean is still in a relatively superficial stage. The sea which is also one of the four human strategic development spaces has been the world's major power strategic competitive position for long. Nowadays the thalassocracy is paid more and more attention by more countries. The ocean not only has a great influence on the ecological environment, but also it has a profound impact on human society. With the increasing depletion of land resources, the ocean is more concerned to the country's major interests and it also has become the focus attention of the world. Entering the 21st century, the coastal countries accelerate the development of marine science and technology, adjust the military strategy and strengthen the maritime rights and interests struggle. The marine development is an important research area and also becoming one of the main battlefields of the world in the 21st century [1 –3].

Autonomous underwater vehicle (AUV) is a very effective tool in the human sea activities. The AUV which has no towing cable can be autonomous navigating in three – dimensional space in the water according to the requirements of the task of the pre mission with its own power. It has the characteristics of far range, intelligence, concealment, mobility and economy. It also has a wide range of applications in both civilian and military areas. Military aspects are such as theater reconnaissance, detection and eliminate mine, submarine warfare, maritime warning, blockade routes or ports, assault to the enemy ships or submarines, damage to enemy oil

① 原文发表于《Journals in Optics – Elsevier》2016。

② 聂为彪：工学博士，2012 年师从冯顺山教授，研究小型水下航行器运动建模与控制方法，现工作单位：中国舰船研究院。

facilities and communication network, underwater relay communication and so on; civil aspects are such as marine resources exploration and exploitation, monitoring of temporal and spatial changes of marine environment, submarine topography investigation and surveying and deep - sea technology, underwater facilities inspection, ocean rescue and salvage and so on [4 - 6].

AUV technology is a new field in the development of marine engineering technology which is of great military and economic value and also has been widely concerned. Therefore, many countries are competing to develop high technology underwater equipment. AUV's research began in 1950s and developed to a considerable level from 1970s to 1980s. At the end of the 1980s, with the development of microelectronics, computer, artificial intelligence, small navigation equipment, high precision control system, high performance software technology, dense energy and other high technology development and marine engineering and military needs, the AUV has become hotspot research of marine technology in the developed countries.

The United States is one of the earliest countries in the research and development of the AUV and has most product types and research units with the highest technology level. At present it has formed a series of mature products. Britain has mainly developed two kinds of underwater unmanned vehicle series, which is the unmanned underwater vehicle "Autosub" series for civil use and the "Tailsman" series for military use. ECA Company is the main R & D unit of France which has rich experience in the research and development of the AUV. In 1970s, the world famous PAP104 was developed by it. Germany has developed three types of unmanned underwater vehicles of MK "SeaOtter" which are marked as "SeaOtter" MKI, "SeaOtter" MKII and "SeaOtter" MKIID. The main R & D units are ATLAS MARIDAN and Germany's ATLAS group. Norway has mainly developed the "Hugin" series of unmanned underwater vehicles which "Hugin" 3 000 is mainly used for civilian areas and "Hugin" 1 000 is mainly used in the military field. The main R & D units are the Norway National Defense Research Institute and Konsberg Company. Japan Marine Science and technology research and development center and the University of Tokyo are the main R & D units of the AUV which is mainly used in the field of civil field in Japan. China started this research late and the technology is relatively backward. Based on domestic and foreign technology progress, high starting point for the development of AUV technology to promote the development of marine science and technology and improve the level of marine development and utilization and protection of maritime rights and interests has important significance [7 - 9].

Over the past few decades, rapid progress in the AUV is steadily affording some scientists advanced tools for ocean exploration, long range survey, underwater pipelines tracking, scientific sampling and mapping maintenance and construction, search and rescue military applications and so on [10 - 13]. Besides their numerous practical applications, the AUV presents a challenging control problem since most AUVs are under - actuated because they have more degrees of freedom to be controlled than the number of independent control inputs. In addition, the AUV's kinematic and dynamic models are highly nonlinear, time - varying and strong coupled, which making the control design a hard work. If classical control systems designed for fully or over - actuated systems are directly used on the under - actuated AUV system, the resulting performance is significantly deteriorated or the control objectives cannot be achieved. The control method used should get rid of the dependence on the precise mathematical model, so it has great important theoretical and practical significance to study this issue. Path - following tracking requires the design of control laws to force the AUV to track an inertial desired trajectory. The 6 - DOF AUV is decoupled into two reduced dynamical systems: a depth - pitch model that considers the motion in vertical plane and the plane - yaw model in the horizontal plane. The trajectory tracking problems of the AUV have been discussed in the last few years. The major solution methods of the trajectory tracking problems for the AUV are sliding mode control [14, 15], cascade system method [16, 17], backstepping technique [18, 19], robust control [20], switching control [21], Lyapunov's direct method

[22, 23] and so on. A dynamic formation model was proposed and several algorithms were developed for the complex underwater environment in Ref [24]. where virtual potential point based formation – keeping algorithm was employed by incorporating dynamic strategies which were decided by the current states of the formation. The design of an output feedback controller for an AUV to achieve the task of path – following in vertical plane was presented in [25], where a static output feedback controller is designed and implemented on the nonlinear plant based on the linearized model, but this method cannot guarantee the tracking error to a minimum. In Ref. [26], it provides an experimental implementation and verification of a hybrid controller for semi – autonomous and autonomous underwater vehicles in which the missions imply multiple task robot behavior, but the authors didn't consider the influences which were caused by the non – diagonal term, nonlinear damping and time – varying environmental disturbance. The path – following tracking control of the AUV in the horizontal plane is more complicated compared with that in the vertical plane, because the displacement of the AUV on the horizontal plane can not be measured directly. It leads to the path – following tracking carry out only through the measurement of yaw angle and yaw angle rate.

Motivated by the work above, a solution to the path – following tracking control problem for an AUV in 2 – D space was proposed. In this paper, we are especially interested in the combined problems of path – following control in 2 – D space for an AUV. The goal of path – following tracking control is forcing the AUV to tracking the desired path and making the tracking errors converge to zero fast and smoothly. In order to design a controller that can meet the requirements of low tracking error and high stability, a sliding mode controller with line of sight and cross – tracking method was designed for the horizontal path – following tracking control of the AUV based on the kinematic and dynamic models. We specify three types of path – following tracking in the horizontal plane including a straight line path, curve path and circular path respectively to verify the performance and effectiveness of the proposed controller. Simulation results show that the path – following tracking control remains very satisfactory and the results that demonstrate the performance of the developed sliding mode control design which uses line of sight guidance and cross – track errors of the AUV are also presented and discussed.

2 AUV Kinematic and Dynamic Model

In this section, the kinematics and dynamics equations of the motion of an AUV moving on the horizontal plane are described. In general, most AUVs are designed to have symmetric structures. We assume that the body – fixed frame is located at the center of gravity with neutral buoyancy. In order to analyze the motion of the AUV, the earth – fixed frame and body – fixed frame which is a moving coordinate frame fixed to the AUV are described as shown in Figure 1.

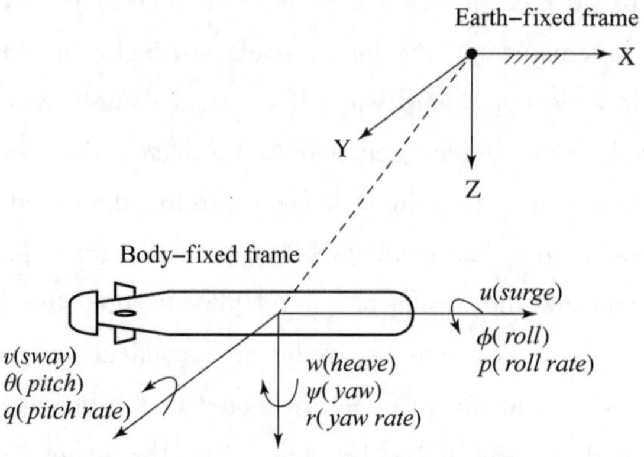

Figure 1 Earth – fixed frame & body – fixed frame for the AUV.

The kinematic equations of AUV moving on the horizontal plane can be described as follow:

$$\begin{bmatrix} \dot{x} \\ \dot{y} \\ \dot{\psi} \end{bmatrix} = \begin{bmatrix} \cos\psi & -\sin\psi & 0 \\ \sin\psi & \cos\psi & 0 \\ 0 & 0 & 1 \end{bmatrix} \begin{bmatrix} u \\ v \\ r \end{bmatrix} \tag{1}$$

Where x and y represent the earth – fixed coordinates of the AUV and the angle ψ describes the orientation of the AUV named yaw, while u and v are the surge and sway velocities which are defined in the body – fixed frame and r is the yaw velocity. The following is the dynamic model of the AUV:

$$\begin{cases} \dot{u} = \dfrac{m_{22}}{m_{11}} vr - \dfrac{X_u}{m_{11}} u - \dfrac{X_{u|u|}}{m_{11}} u|u| + \dfrac{F}{m_{11}} \\ \dot{v} = -\dfrac{m_{11}}{m_{22}} ur - \dfrac{Y_v}{m_{22}} v - \dfrac{Y_{u|u|}}{m_{22}} u|u| \\ \dot{r} = \dfrac{m_{11} - m_{22}}{m_{33}} uv - \dfrac{N_r}{m_{33}} r - \dfrac{N_{r|r|}}{m_{33}} r|r| + \dfrac{N}{m_{33}} \end{cases} \tag{2}$$

Where F denotes the control force along the surge motion of the AUV and N is the control torque producing angular motion. The constants m_{11} and m_{22} are the combined rigid – body and added mass terms and m_{33} is the combined rigid – body and added moment of inertia about the z_b axis. In addition, X_u, $X_{u|u|}$, Y_v, $Y_{v|v|}$, N_r, and $N_{r|r|}$ are the linear and quadratic drag terms coefficients. Using the inertial position and orientation and body – fixed states of the AUV, the tracking errors are defined as follows:

$$\begin{cases} x_e = x - x_d \\ y_e = y - y_d \\ \psi_e = \psi - \psi_d \\ u_e = u - u_d \\ v_e = v - v_d \\ r_e = r - r_d \end{cases} \tag{3}$$

It is obvious that the convergence of x_e, y_e, ψ_e, u_e, v_e and r_e to zero implies the convergence of $x - x_d$, $y - y_d$, $\psi - \psi_d$, $u - u_d$, $v - v_d$ and $r - r_d$ to zero. So we can define the path – following tracking error in the horizontal plane as:

$$d_e = \sqrt{x_e^2 + y_e^2} \tag{4}$$

We substitute u, v, r, x, y, ψ in Eq. (1) and Eq. (2) into Eq. (3), then we can derive the error kinematics as:

$$\begin{bmatrix} \dot{x}_e \\ \dot{y}_e \end{bmatrix} = \begin{bmatrix} \cos\psi & -\sin\psi \\ \sin\psi & \cos\psi \end{bmatrix} \begin{bmatrix} u_e \\ v_e \end{bmatrix} + \begin{bmatrix} \cos\psi - \cos\psi_d & -\sin\psi + \sin\psi_d \\ \sin\psi - \sin\psi_d & \cos\psi - \cos\psi_d \end{bmatrix} \begin{bmatrix} u_d \\ v_d \end{bmatrix} \tag{5}$$

The error dynamics is obtained as:

$$\begin{cases} \dot{u}_e = \dfrac{m_{22}}{m_{11}}(v_e r_e + v_e r_d + v_d r_e) - \dfrac{X_u}{m_{11}} u_e - \dot{u}_d + \dfrac{m_{22}}{m_{11}} v_d r_d - \dfrac{X_u}{m_{11}} u_d - \dfrac{X_{u|u|}}{m_{11}}(u_e + u_d)|u_e + u_d| + \dfrac{F}{m_{11}} \\ \dot{v}_e = -\dfrac{m_{11}}{m_{22}}(u_e r_e + u_e r_d + u_d r_e) - \dfrac{Y_v}{m_{22}} v_e - \dot{v}_d - \dfrac{m_{11}}{m_{22}} u_d r_d - \dfrac{Y_v}{m_{22}} v_d - \dfrac{Y_{v|v|}}{m_{22}}(v_e + v_d)|v_e + v_d| \\ \dot{r}_e = \dfrac{m_{11} - m_{22}}{m_{33}}(u_e v_e + u_e v_d + u_d v_e) - \dfrac{N_r}{m_{33}} r_e - \dot{r}_d + \dfrac{m_{11} - m_{22}}{m_{33}} u_d v_d - \dfrac{N_r}{m_{33}} r_d - \dfrac{N_{r|r|}}{m_{33}}(r_e + r_d)|r_e + r_d| + \dfrac{N}{m_{33}} \end{cases} \tag{6}$$

The magnitude of the velocity vector V_s of a reference path – following point S at time t is given as:

$$v_S = \sqrt{\dot{x}_d^2 + \dot{y}_d^2} \tag{7}$$

The direction of velocity V_s is defined as the angleα:

$$\alpha = a\tan\left(\frac{\dot{y}_d}{\dot{x}_d}\right) \tag{8}$$

The first derivatives of is given as:

$$\dot{\alpha} = \frac{\dot{x}_d\ddot{y}_d - \dot{y}_d\ddot{x}_d}{\dot{x}_d^2 + \dot{y}_d^2} \tag{9}$$

At the same time, the derivative of the velocity V_s is:

$$\dot{v}_s = \frac{\dot{x}_d\ddot{x}_d + \dot{y}_d\ddot{y}_d}{\sqrt{\dot{x}_d^2 + \dot{y}_d^2}} \tag{10}$$

The feasible path including the feasible orientation is completely known, the control is given by:

$$\begin{cases} F_d = m_{11}\dot{u}_d - m_{22}v_d r_d + X_u u_d + X_{u|u|}u_d^2 \\ N_d = m_{33}\dot{r}_d + (m_{22} - m_{11})u_d v_d + N_r r_d - N_{r|r|}r_d^2 \end{cases} \tag{11}$$

3 Line of Sight Guidance and Controllers of The AUV

Guidance is the action of the AUV that continuously computes the desired position, velocity and even acceleration of a vessel to be used by the control system [27]. Systems for the guidance consist of a waypoint generator with human input. One solution to design the system is to store the desired waypoints in the waypoint database and use them to generate a trajectory path for the AUV. Another solution to the systems can be linked to the waypoint guidance system as the case of weather routing, obstacle avoidance, collision and mission planning, etc [28]. The AUV guidance is most simply accomplished by a heading command to the vehicle's steering system to approach the line of sight between the present position of the vehicle and the waypoint to be reached. The concepts of path – following tracking and maneuvering control should be distinguished when designing waypoint guidance systems. These systems consists of a waypoint generator with human's commands. The command waypoints are stored in a waypoint database and used for generation of a trajectory or a path for the moving AUV to follow. Both trajectory and maneuvering control systems can be designed for this purpose. The position definition and nomenclature are as given by reference to Figure 2.

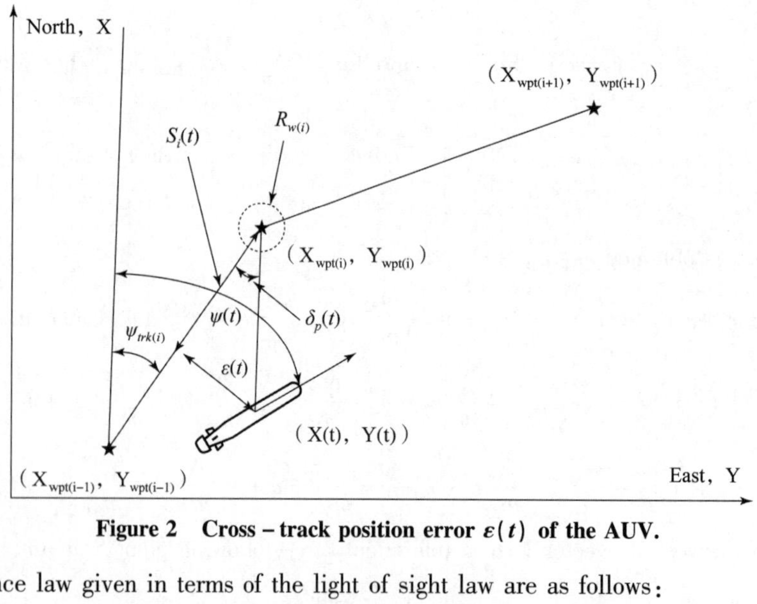

Figure 2 Cross – track position error $\varepsilon(t)$ of the AUV.

The basic guidance law given in terms of the light of sight law are as follows:

$$\begin{cases}\psi(t)los(\tan(\tilde{Y}(t)_{wpt(i)}/\tilde{X}(t)_{wpt(i)})_{com}\\ \tilde{Y}(t)_{wpt(i)}=(Y_{wpt(i)}-Y(t))\\ \tilde{X}(t)_{wpt(i)}=(X_{wpt(i)}-X(t))\end{cases} \tag{12}$$

The path - following tracking control of the AUV on horizontal plane means that the AUV cruises in accordance with a predetermined route within a limited region of the horizontal plane. The predetermined route is usually composed of a number of connections between different path points. The control algorithm can be designed based on measurable yaw angle and yaw angle rate. The sliding mode control algorithm containing cross - track error method and line of sight method is used in this work. The cross - track error method adjusts the AUV's position constantly according to its position error so as to get the desired position. The cross - track error $\varepsilon(t)$ is the vertical distance from the center of the AUV to the adjacent routes, and the control strategy of the cross - track error method is to keep $\varepsilon(t)$ at minimum. The track angle and the total track length between way - points i and $i-1$ are defined as:

$$\begin{cases}L_i=\sqrt{(X_{wpt(i)}-X_{wpt(i-1)})^2+(Y_{wpt(i)}-Y_{wpt(i-1)})^2}\\ \psi_{trk(i)}=a\tan2(Y_{wpt(i)}-Y_{wpt(i-1)},X_{wpt(i)}-X_{wpt(i-1)})\end{cases} \tag{13}$$

Where the ordered waypoints $(X_{wpt(i)},\ Y_{wpt(i)})$ *and* $(X_{wpt(i-1)},\ Y_{wpt(i-1)})$ are the current and previous path points, respectively.

The cross - track heading error $\tilde{\psi}(t)_{CET(i)}$ for the ith track segment is defined as

$$\tilde{\psi}(t)_{CET(i)}=\psi(t)-\psi_{trk(i)}+\beta(t) \tag{14}$$

Where $\beta(t)$ is the AUV's sideslip angle and the heading error $\tilde{\psi}(t)_{CET(i)}$ should normalized between $-180°$ and $180°$. The distance to the ith way - point projected to the track line $S_i(t)$ can be calculated as a percentage of track length using Eq. (15) as follow:

$$S_i(t)=(\tilde{X}(t)_{wpt(i)}(X_{wpt(i)}-X_{wpt(i-1)})+\tilde{Y}(t)_{wpt(i)}(Y_{wpt(i)}-Y_{wpt(i-1)}))/L_i \tag{15}$$

In order to make the AUV accurately track the predetermined route, the control system on the horizontal plane should be able to ensure $\varepsilon(t)$ 0. According to the current position point, current path point and previous path point, the cross - track error can be defined as:

$$\varepsilon(t)=S_i(t)\tan(\delta_p(t)) \tag{16}$$

Where $\delta_p(t)$ is the angle between the line of sight to the next waypoint and the current track line, which must be normalized between $-180°$ and $180°$

$$\delta_p(t)=a\tan2(Y_{wpt(i)}-Y_{wpt(i-1)},X_{wpt(i)}-X_{wpt(i-1)})-a\tan2(\tilde{Y}(t)_{wpt(i)}-\tilde{X}(t)_{wpt(i)}) \tag{17}$$

When it comes to the cross - track defined, a sliding surface can be as follows in terms of derivatives of the errors:

$$\begin{cases}\dot{\varepsilon}=u\sin(\tilde{\psi}(t)_{CET(i)})\\ \ddot{\varepsilon}=u(r(t)+\dot{\beta}(t))\cos(\tilde{\psi}(t)_{CET(i)})\\ \dddot{\varepsilon}=u(\dot{r}(t)+\ddot{\beta}(t))\cos(\tilde{\psi}(t)_{CET(i)})-u(r(t)+\dot{\beta}(t))^2\sin(\tilde{\psi}(t)_{CET(i)})\end{cases} \tag{18}$$

We assume that $\dot{\beta}(t)$ and $\ddot{\beta}(t)$ are negligible because they are small compared with the angular velocity of the yaw angle. The sliding surface of sliding mode controller based on cross - track error can be designed as a second - order polynomial form as:

$$\sigma(t) = \ddot{\varepsilon}(t) + k_1\dot{\varepsilon}(t) + k_2\varepsilon(t) \tag{19}$$

The derivation of Eq. (19) and the reaching condition of the reduction of the error can be expressed as:

$$\dot{\sigma}(t) = \dddot{\varepsilon}(t) + k_1\ddot{\varepsilon}(t) + k_2\dot{\varepsilon}(t) = -\eta\tan h(\sigma/\phi) \tag{20}$$

For getting the control input, the AUV's heading dynamics are simplified to neglect the side slip dynamics and can be expressed as:

$$\begin{cases}\dot{r}(t) = ar(t) + b\delta_r(t)\\ \dot{\psi}(t) = r(t)\end{cases} \tag{21}$$

Substituting Eq. (20) using Eq. (18) and Eq. (21) we can get the following expression:

$$\begin{aligned}\dot{\sigma}(t) = & u(ar(t) + b\delta_r)\cos(\tilde{\psi}(t)_{CTE(i)}) - ur(t)^2\sin(\tilde{\psi}(t)_{CTE(i)}) + \\ & k_1ur(t)\cos(\tilde{\psi}(t)_{CTE(i)}) + k_2u\sin(\tilde{\psi}(t)_{CTE(i)})\end{aligned} \tag{22}$$

Rewriting Eq. (19), the sliding surface $\sigma(t)$ can be expressed as:

$$\sigma(t) = ur(t)\cos(\tilde{\psi}(t)_{CTE(i)}) + k_1u\sin(\tilde{\psi}(t)_{CTE(i)}) + k_2\varepsilon(t) \tag{23}$$

By using Eq. (20), Eq. (21) and Eq. (22), the rudder input can then be expressed as:

$$\begin{aligned}\delta_r(t) = & (ub\cos^{-1}(-uar(t)\cos(\tilde{\psi}(t)_{CTE(i)}) + \\ & ur(t)^2\sin(\tilde{\psi}(t)_{CTE(i)}) - k_1ur(t)\cos(\tilde{\psi}(t)_{CTE(i)}) - \\ & k_2u\sin(\tilde{\psi}(t)_{CTE(i)}) - \eta\tanh(\sigma/\phi)\end{aligned} \tag{24}$$

The heading command can be determined as:

$$\psi(t)_{com(i)} = a\tan 2(Y_{wpt(i)} - Y(t), X_{wpt(i)} - X(t)) \tag{25}$$

And the line of sight error can be expressed as:

$$\tilde{\psi}(t)_{Los(i)} = \psi(t)_{com(i)} - \psi(t) \tag{26}$$

The slide mode function is selected as:

$$\begin{aligned}\sigma_1(t) &= \tilde{\phi}(t) + k_3r(t)\\ \dot{\sigma}_1(t) &= -\eta_1\tan h(\sigma_1(t)/\phi_1)\end{aligned} \tag{27}$$

The control law of the sliding mode controller based on the line of sight method is designed as:

$$\delta(t) = -(k_3b)^{-1}((k_3a-1)r(t) + \eta_1\tan h(\sigma_1(t)/\phi_1)) \tag{28}$$

The transition condition of activating the next path - point is in Eq. (29) and ρ is the cross - track error control distance.

$$\sqrt{(X_{wpt(i)} - X(t))^2 + (Y_{wpt(i)} - Y(t))^2} < \rho \tag{29}$$

4 Reference Paths and Simulation

4.1 Straight Line Path

In order to verify the control performance of the AUV path - following tracking, we consider an AUV that is assigned to converge to three types of 2 - D path - following trajectory specified by straight line path, curve path and circular path. 2 - D path - following is commonly used for guidance of surface vessels, however, when it comes to autonomous underwater vehicles, we assure that the path following control problem is limited to the horizontal plane's motions. Simulations are conducted to prove effectiveness of the controller while AUV follows a given straight line path. We consider a reference straight line path described as follows:

$$\begin{cases} x(t) = t \\ y(t) = t \end{cases} \tag{30}$$

Where $t \in [0:0.01:7]$, here we set the initially path tracking point is $(0, -2)$. Through simulation, we can find the desired and tracking trajectory curves of the AUV in Figure 3. The tracking error curve of the AUV is demonstrated in Figure 4. Integrated looking from these two figures, we can find that the tracking effect is good and the tracking error converges to zero smoothly at about 4.6 s in exponential form.

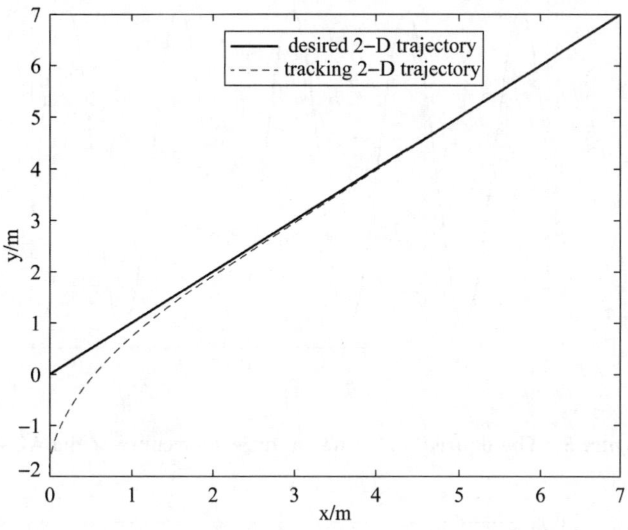

Figure 3　The desired and tracking trajectory curve of the AUV.

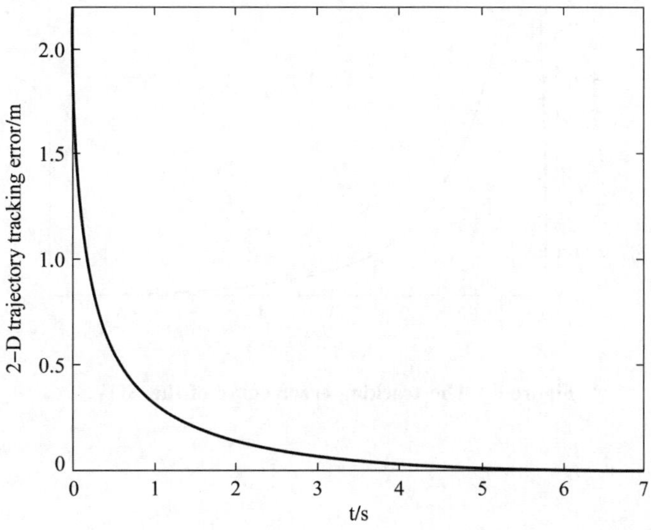

Figure 4　The tracking error curve of the AUV.

4.2　Curve Path

Simulations of curve path tracking are conducted for further verification. Desired curve path is expressed as:

$$\begin{cases} x = t \\ y = a * \bar{\omega} * \sin(\bar{\omega} * t) \end{cases} \tag{31}$$

Where $a = 1.3$, $\bar{\omega} = 1.3 * pi$, here we set the initially path tracking point is $(0, 0)$. Through simulation, we can find the desired and tracking trajectory curves of the AUV in Figure 5. The tracking error curve of the AUV is demonstrated in Figure 6. The two figures also show that the tracking error reach its maximum at about 0. 3 s, after that it convergences to zero smoothly at about 5. 6 s in exponential form. Overall, the effect of path tracking is also good and satisfactory.

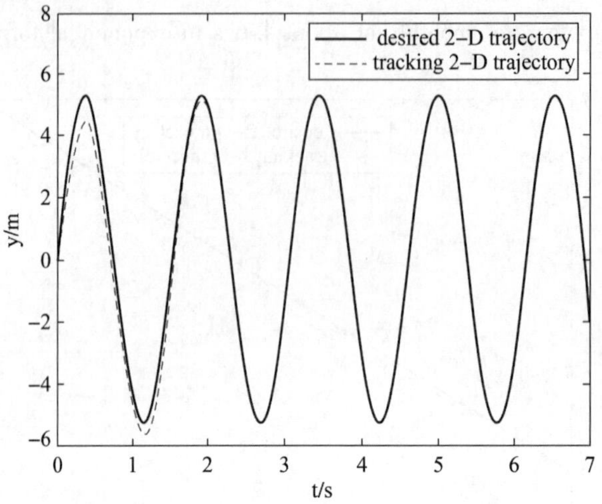

Figure 5 The desired and tracking trajectory curve of the AUV.

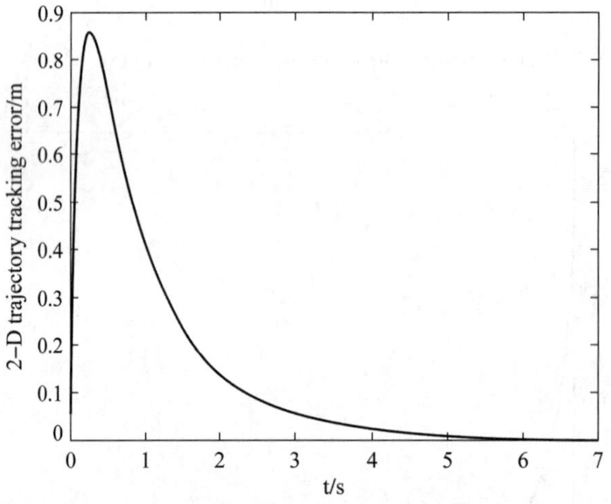

Figure 6 The tracking error curve of the AUV.

4. 3 Circular Path

Simulations of circular path tracking are also conducted for further verification. Desired circular path is expressed as:

$$\begin{cases} x = 2 * a * \bar{\omega} \sin(\bar{\omega} * t) \\ y = 2 * a * w\cos(\bar{\omega} * t) \end{cases} \tag{32}$$

Where $a = 1.3$ $\bar{\omega} = 1.3 * pi$ $t \in [0 : 0.01 : 7]$. Here we set the initially path tracking point is (0, 10). Through simulation, we can find the desired and tracking trajectory curves of the AUV in Figure 7 and the tracking error curve of the AUV is demonstrated in Figure 8. The two figures show good performance of the tracking effect and the error curve converges to zero at about 5 s fast and smoothly in exponential form.

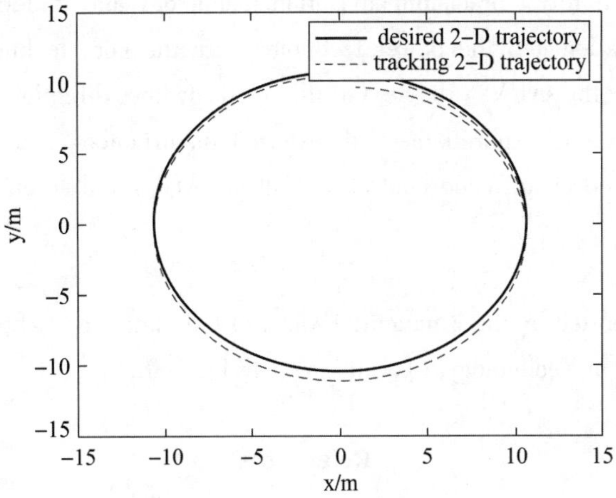

Figure 7　The desired and tracking trajectory curve of the AUV.

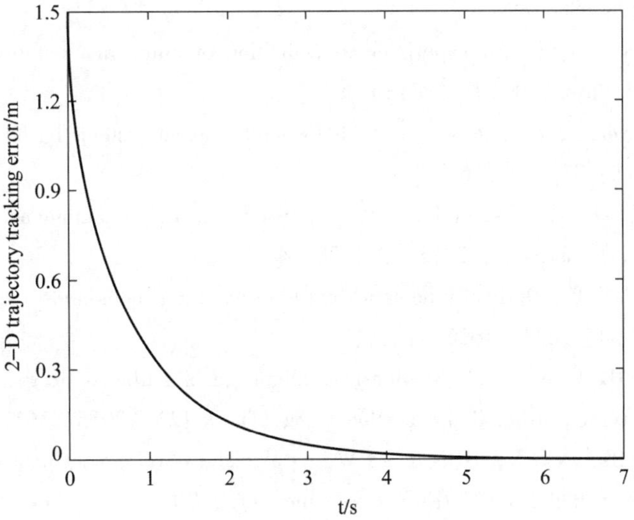

Figure 8　The tracking error curve of the AUV.

5　Conclusions and Future Work

In this paper, the combined problems of planar path – following tracking control for an under – actuated AUV in the horizontal plane were addressed. The kinematic and dynamic model of the AUV is established and the line of sight guidance of AUV is also designed. The horizontal plane path – following sliding mode controllers with cross – track error method and line of sight are designed to fulfill high – precision path – following tracking control, which ensures the robustness and control accuracy when the error is too large. Given three types of smooth and 2 – D desired trajectory including straight line path, curve path and circular path. The simulation results show that the controllers have good path – following tracking performance that the

AUV can tracking the predetermined three types of curves from the specified point and the errors of the tracking trajectory converge to zero fast and smoothly in exponential form at last. Results also show that the proposed controllers have good adaptive and robustness performance.

The proposed method can provide us a reference road of solving the control problems of waypoint tracking, trajectory planning, course – changing maneuvers, attitude control, parking, sea – keeping, etc. At present, there are many key technologies and problems needed to be studied in the development of the AUV which mainly include as follows: advanced propulsion technology with high endurance, advanced control technology with autonomous decision – making ability, precision navigation technology and underwater communication technique and so on. The control, navigation and the power technology are the core technology of the AUV. If these key technologies are all resolved, the AUV will be on the road in the direction of deeper, farther and more intelligent. Future work will include experiments with external disturbances, such as waves, winds and ocean currents, obstacle avoidance and coordinated control of multiple AUVs and so on.

Acknowledgments

This work was partly supported by the Financing Project of the State Key Laboratory of Explosion Science and Technology of Beijing Institute of Technology (Grant No. QNKT13 – 02)

References

[1] LOC M B, CHOI H S, SEO J M, et al. Development and control of a new AUV platform [J]. Int. J. Control Autom. Syst., 2014, 12: 886 – 894.

[2] WU Z, HU X, WU M, et al. An experimental evaluation of autonomous underwater vehicle localization on geomagnetic map, Appl. Phys. Lett. 103 (2013).

[3] LI Y M, GUAN X P. Nonlinear consensus protocols for multi – agent systems based on centre manifold reduction, Chin. Phys. B, 2009, 18: 3355 – 3366.

[4] BONIN – FONT F, OLIVER G, WIRTH S, et al. Visual sensing for autonomous underwater exploration and intervention tasks [J]. Ocean Eng., 2015, 93: 25 – 44.

[5] BIGGS J, HOLDERBAUM W. Optimal kinematic control of an autonomous underwater vehicle [J]. IEEE Trans. Autom. Control, 2009, 54: 1623 – 1626.

[6] WANG Q Y, LI Y B, DIAO M, et al. Moving base alignment of a fiber optic gyro inertial navigation system for autonomous underwater vehicle using Doppler velocity log, Optik 126 (2015) 3631 – 3637.

[7] TANGORRA J L, DAVIDSON S N, HUNTER I W, et al, The development of a biologically inspired propulsor for unmanned underwater vehicles [J]. IEEE J. Oceanic Eng., 2007, 32: 533 – 550.

[8] SMITH S M, AN A, KRONEN D, et al. The development of autonomous underwater vehicle based survey and sampling capabilities for coastal exploration, OCEANS '96 MTS/IEEE. Prospects for the 21 St Century. Conference Proceedings (1996) (pp. 35 suppl).

[9] KONDO H, SATO M, HOTTA T, et al. Development of a marine ecosystem and micro structure monitoring AUV for plankton environment, in: autonomous Underwater Vehicles (AUV), IEEE/OES 2014 (2014) 1 – 5.

[10] ZHU D, HUA X, SUN B. A neurodynamics control strategy for real – Time tracking control of autonomous underwater vehicles [J]. J. Navig., 2014, 67: 113 – 127.

[11] LI J H, LEE P M. Design of an adaptive nonlinear controller for depth control of an autonomous underwater vehicle, Ocean Eng., 2005, 32: 2165 – 2181.

[12] ZHANG Y, BELLINGHAM J G, GODIN M A, RYAN J P, Using an autonomous underwater vehicle to track the thermocline based on peak – gradient detection [J]. IEEE Journal of oceanic engineering, 2012, 37: 544 – 553.

[13] XU Y, MOHSENI K. Bioinspired hydrodynamic force feedforward for autonomous underwater vehicle control [J]. IEEE – Asme Trans. Mechatronics, 2014, 19: 1127 – 1137.

[14] SALGADO – JIMENEZ T, JOUVENCEL B, IEEE I. Using a high order sliding modes for diving control a torpedo autonomous underwater vehicle, 2003.

[15] ASHRAFIUON H, MUSKE K R, MCNINCH L C, et al. Sliding – mode tracking control of surface vessels [J]. IEEE Trans. Ind. Electr., 2008, 55: 4004 – 4012.

水面航行体对舰船目标的图像检测方法①

方　晶[1②]，冯顺山[1]，冯　源[2]

(1. 北京理工大学机电学院，北京　100081；2. 中国兵器科学研究院，北京　100089)

摘　要：针对水面航行体在近岸水域条件下对舰船目标进行实时光学检测时，易受到光照、相似颜色背景和海面波浪反射等干扰的问题，提出了基于改进视觉注意模型的舰船目标检测方法，采用小波变换方法提取图像的低频、高频特征，将任务水域图像从 RGB 颜色空间转化到 HSV 颜色空间来提取图像的色调、饱和度和明度特征，应用高斯金字塔、归一化算子等图像处理方法融合了各类特征。仿真结果表明，提出的舰船目标检测方法能够准确地实现复杂背景条件的舰船目标检测，具有良好的抗干扰能力。

关键词：水面航行体；视觉注意；小波变换；舰船目标检测

Image Algorithm of Ship Detection for Surface Vehicle

Fang Jing[1], Feng Shunshan[1], Feng Yuan[2]

(1. School of Mechatronical Engineering, Beijing Institute of Technology, Beijing 100081, China;
2. Science Research Institute of China North Industries Group Corp, Beijing 100089, China)

Abstract: In order to solve the problem that the surface vehicle detecting ship targets in the nearshore waters was vulnerable to light, similar color background and wave reflection, ship detection algorithm based on improved visual attention model was proposed. First, low frequency and high frequency features of images were extracted by using wavelet transform theory. Then, the hue, saturation and value features of images were also extracted by converting the images of task waters from RGB color space to HSV color space. Finally, various features of images were merged in the application of image processing method of Gaussian Pyramid and normalization operator. The simulation results show that the proposed ship detection method can accurately detect the ship targets under complicated backgrounds and has satisfactory anti – interference capability.

Key words: surface vehicle; visual attention; wavelet transform; ship detection

水面航行体是一种能够通过更换任务模块实现在海洋环境中自主航行并执行环境勘探、海洋搜索和军事侦察等任务的无人小型航行器。当前，中国海上安全形势十分严峻，相关邻国觊觎、侵占中国海洋国土，敌对势力干涉中国内政、阻挠统一进程。因此，水面航行体发展和应用为中国海洋权益的维护提供了重要保障[1]。

舰船目标检测方法的目的是运用数字图像处理技术在任务水域图像中实时检测舰船位置，为舰船目标识别、跟踪算法提供有效的目标区域，缩小搜索范围，降低算法的复杂度和运算量，以提高水面航行体对舰船目标侦查、打击能力。

针对此种情况，国内外学者对基于图像的舰船目标检测方法进行了研究。Yang G 等[2]研究了海面图像灰度分布，利用海面与目标的灰度差异实现对舰船目标的检测。王焜[3]改进了 Itti L 等[4]提出的视觉

① 原文发表于《北京理工大学学报》2017（37 卷 12 期）。

② 方晶：工学博士，2011 年师从冯顺山教授，研究小型水中弹药探测与控制系统设计方法，现工作单位：中国船舶工业系统工程研究院。

注意模型，将 RGB［红（red）、绿（green）、蓝（blue）］图像转换到 HSV（色调、饱和度、明度）颜色空间提取亮度和颜色特征，实现了舰船目标的初步检测。周伟等[5]利用多尺度相位谱重构生成舰船目标的显著图，应用最小距离分类器对 ROI 进行鉴别得到最终的检测结果。

在高度动态和不可预测的海洋环境执行任务时，由于山、岸上建筑物和海面波浪反射等因素的干扰，水面航行体采集的任务水域视频图像背景复杂，上述方法不能有效地提取舰船目标所在区域。因此，设计一种稳定的图像检测方法满足水面航行体对舰船目标高效检测的需求，具有重要的意义。

由于任务水域图像中舰船目标相比天空、海面和岸上建筑等自然背景更容易吸引人的视觉注意，因此本文提出了基于改进视觉注意模型的舰船目标检测方法，该方法采用小波变换方法提取任务水域图像的高频和低频特征，将图像从 RGB 空间转换到 HSV 空间用于提取图像色调、明度和亮度特征，改进 Itti 视觉注意模型，实现舰船目标的检测。

1 视觉注意模型

视觉注意模型是用于模拟人类视觉系统对信息的处理过程。根据处理信息的不同方式，视觉注意模型可分为自顶向下和自底向上两种[6]：自顶向下的视觉注意模型受意识支配、依赖于具体任务的先验知识，能够根据具体任务的需求，有意识地调整选择准则，将注意力集中于图像中任务对象；自底向上的视觉注意模型在缺乏先验知识引导情况下，视觉注意处理完全由外部视觉刺激信号驱动，并能够按照一定的优先级顺序有选择地对图像中的显著区域进行处理。

在视觉注意模型众多的研究成果中，最为经典的是由 Koch & Ullman 模型派生出的 Itti 模型[7]，其显著区域检测过程为：①提取输入图像颜色、亮度和方向特征，并建立各类特征的多分辨率金字塔表示；②将各类特征金字塔形式存在的多尺度分量图像经过一个模拟人眼感受野特性的“中央－周边”算子运算后，形成多尺度特征图；③将各类特征的多尺度特征图经过跨尺度组合和归一化形成相应的颜色、亮度和方向的分量显著图；④将 3 张不同特征的显著图线性融合为 1 张用来表征图像中各区域显著度的总显著图。

2 特征提取

2.1 基于小波变换的频率特征提取

小波变换理论的基本思想是用一组函数去表示或逼近一信号[8]。设 $\varphi(t)$ 为平方可积函数，即 $\varphi(t)\in L^2(\boldsymbol{R})$，$L^2(\boldsymbol{R})$ 为 $\boldsymbol{R}$ 上平方可积函数构成的函数空间。若 $\varphi(t)$ 的 Fourier 变换 $\psi(t)$ 满足条件：

$$\int_R \frac{|\psi(\omega)|^2}{\omega}\mathrm{d}\omega < \infty \tag{1}$$

则称 $\varphi(t)$ 为小波母函数。将 $\varphi(t)$ 进行伸缩和平移，设其伸缩因子（尺度因子）为 a，平移因子为 b，则平移伸缩后的函数为

$$\varphi_{a,b}(t)=|a|^{-\frac{1}{2}}\varphi\left(\frac{t-b}{a}\right) \tag{2}$$

式中，$\varphi_{a,b}(t)$ 称为参数 n，b 的连续小波基函数，并且 $a,b\in\boldsymbol{R},a\neq 0$。

对于任意的 $f(t)\in L^2(\boldsymbol{R})$，若 $\varphi(t)\in L^2(\boldsymbol{R})$，则函数 $f(t)$ 的连续小波变换为

$$W_f(a,b)=\int_{-\infty}^{+\infty}f(t)\,\overline{\varphi_{a,b}(t)}\,\mathrm{d}t=|a|^{-\frac{1}{2}}\int_{-\infty}^{+\infty}f(t)\varphi\left(\frac{t-b}{a}\right)\mathrm{d}t \tag{3}$$

式中，$\overline{\varphi_{a,b}(t)}$ 为 $\varphi_{a,b}(t)$ 的共轭函数。

水面航行体采集的任务水域图像为离散信号形式，需要对连续小波基函数进行离散化。在离散化时，通常对尺度按幂级数进行离散化，取 $a=a_0^m$（m 为整数，$a_0\neq 1$），$b=na_0^m b_0$，则离散小波基函数为

$$\varphi_{m,n}(t)=\frac{1}{\sqrt{a_0^m}}\varphi\left(\frac{t}{a_0^m}-b_0\right) \tag{4}$$

则离散小波变换可定义为

$$W_f(m,n) = \int_{-\infty}^{+\infty} f(t)\,\overline{\varphi_{m,n}(t)}\,\mathrm{d}t = \frac{1}{\sqrt{a_0^m}}\int_{-\infty}^{+\infty} f(t)\,\varphi\left(\frac{t}{a_0^m} - b_0\right)\mathrm{d}t \tag{5}$$

水面航行体采集的任务水域图像中舰船目标是出现在不同大小尺度上，因此本文利用小波变换的多分辨率特性对任务图像进行分解，从而获得图像的高频和低频特征。

二维图像的小波分解是通过对图像的行和列分别进行两次一维小波变换和下采样来实现。首先将任务水域图像转换为灰度图像，然后对灰度图像先“逐行”做小波变换，分解为低频 L 和高频 H 两个分量，在“逐列”做一维小波变换，分解为 LL_1、LH_1、HL_1 和 HH_1 4 个分量，其中 LL_1 包含图像的低频数据，LH_1、LH_1 和 HH_1 分别代表水平、垂直和对角线方向细节的高频数据。下一层小波分解是在前一层低频子图 LL_1 上进行的，重复可得多层小波分解子图。2 层小波分解如图 1 所示。

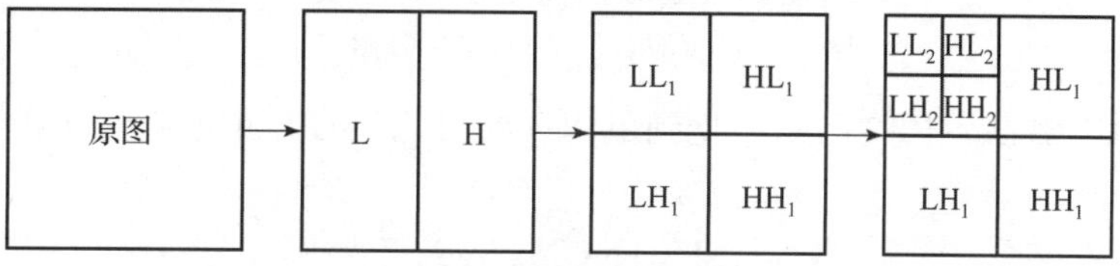

图 1　2 层小波分解

本文利用 Daubechies 小波对水面航行体采集的任务水域图像进行 3 层小波分解，选取一幅舰船目标在海天背景下航行的图像进行仿真，如图 2 所示。

图 2　3 层小波分解图

图 2 左上角 4 副图像为 3 层小波分解结果，将第 3 层小波分解求得的 LH_3、HI_3、HH_3 3 个方向的高频分量进行线性叠加、插值得到图像的高频特征，LL_3 进行插值得到原图的低频特征，如图 3 所示。

(a)　(d)　(c)

图 3　图像高频和低频特征

(a) 原图像；(b) 图像的低频特征；(c) 图像的高频特征

2.2　颜色、饱和度、明度和方向特征提取

由于颜色信息对于舰船目标发生形变、旋转有较强的适应性，因此舰船目标检测算法中，一般会选取颜色作为特征信息。由于 RGB 颜色空间的 3 种颜色分量均与光照强度有关，其对光照强度变化较为敏感。而 HSV 颜色空间色调 H 分量和 V 分量是相互独立的，并且更符合人类视觉系统观察颜色的方式。

因此，本文将 RGB 颜色空间转换到 HSV 颜色空间来提取任务水域图像的色调、饱和度和明度特征，用于改进 Itti 视觉注意模型。

Itti 模型计算的方向特征图，如图 4 所示。

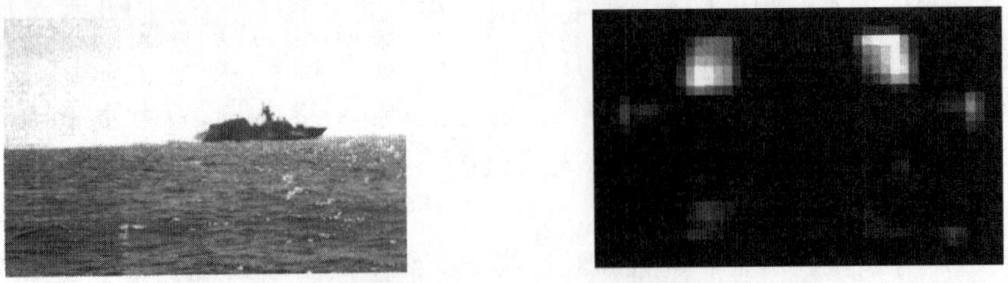

（a）　　　　（b）

图 4　Itti 模型计算的方向特征图

（a）原图像；（b）方向特征图

假设所有的 R，G，B 颜色值均已归一化到范围［0，1］，RGB 颜色空间到 HSV 颜色空间的转换转化公式为

$$V=\max(R,G,B)$$

$$S=\begin{cases}0, & V=0\\ \dfrac{V-\min(R,G,B)}{V}, & 其他\end{cases}$$

$$H=\begin{cases}\dfrac{G-B}{V-\min(R,G,B)}\times 60^\circ & (R=V)\\ \left(2+\dfrac{B-R}{V-\min(R,G,B)}\right)\times 60^\circ & (B=V)\\ \left(4+\dfrac{R-G}{V-\min(R,G,B)}\right)\times 60^\circ & (G=V)\end{cases} \tag{6}$$

在 Itti 视觉注意模型中，提取了图像的方向特征，由于受到海浪、天空和岸上建筑等因素的干扰，方向特征比较杂乱，因此，本文摒弃了方向这个特征。

3　特征融合

本文提取了任务水域图像的高频、低频、色调、饱和度和明度特征，需要对这些特征进行融合，以求得图像中舰船目标的所在区域。特征融合步骤如下。

步骤 1　构造任务水域视频图像的高频、低频、色调、饱和度和明度特征的 Gauss 金字塔，分别为 $P_L(\sigma)$、$P_H(\sigma)$、$H(\sigma)$、$S(\sigma)$和 $V(\sigma)$，其中 $\sigma\in(0,8)$表示 Guass 金字塔级数，共 8 级。

步骤 2　对高频、低频、色调、饱和度和明度特征的 Gauss 金字塔进行"中央 - 周边"运算[4]，求得各特征的多尺度特征图，以色调 Gauss 金字塔 $H(\sigma)$为例，计算公式为

$$H(c,s)=|H(c)\Theta H(s)| \tag{7}$$

式中　c——中心尺度；$c\in\{2,3,4\}$；

s——边缘区域的尺度，$s=c+\xi$；$\xi\in\{3,4\}$；

Θ——一个跨尺度的差减算子，将粗尺度图像差值到细尺度上再逐个像素做减法。

由于 c 和 s 之间有 3×2——6 种组合，因此每个特征均可以求得 6 张多尺度特征图。

步骤 3　对高频、低频、色调、饱和度和明度的多尺度特征图进行归一化处理，首先计算特征图的全局最大值 $M_{\max}$，将特征图归一化到［0，$M_{\max}$］区间内，计算特征图除 $M_{\max}$之外的所有局部最大值的平均值 $\bar{m}$，用 $(M_{\max}-\bar{m})^2$ 乘以特征图，最后对特征图进行插值、叠加运算，求得与原图像尺寸相同的各特征的分量显著图。

步骤 4　将高频、低频、色调、饱和度和明度特征的分量显著图进行叠加，并采用直方图阈值分割

方法，最后获得任务水域图像中舰船目标所在区域。

综上所述，海天背景、波光干扰条件下任务水域图像的舰船目标特征融合结果如图 5 所示。

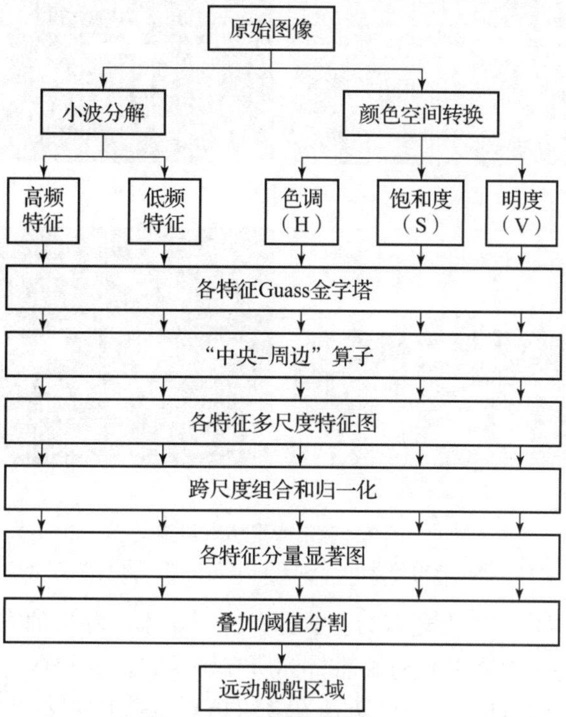

图 5　海天背景特征融合结果

从图 5 可以分析出，在海天背景条件下，虽然视频图像背景简单，但是由于受到波光、舰船目标航行尾流等因素的干扰，各个特征的显著图均受到不同程度的影响，本文通过归一化算子最大限度地保留了各特征图中的有效部分，并通过多特征融合消除了各类干扰的影响，获得了较为清晰的舰船目标区域。

4　仿真分析

本文基于 Opencv 图像库，在 VC ++ 环境中编写了仿真程序，实现了水面航行体对舰船目标的图像检测方法。检测方法流程如图 6 所示。

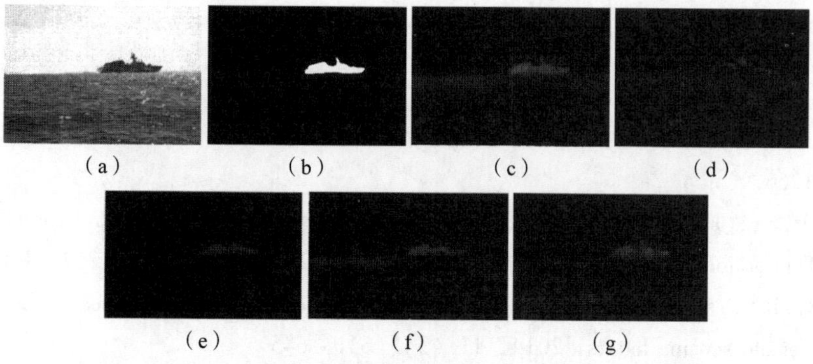

图 6　检测方法流程

（a）原图像；（b）舰船目标检测结果；（c）低频分量显著图；（d）高频分量显著图；（e）色调显著图；（f）饱和度显著图；（g）明度显著图

当舰船目标在山、港口等复杂背景条件下航行时，舰船目标信息易受到相似颜色背景的干扰，给舰船目标检测造成影响，本文选取了两幅以山、岸上建筑为背景的任务水域图像进行仿真，结果如图 7 所示。

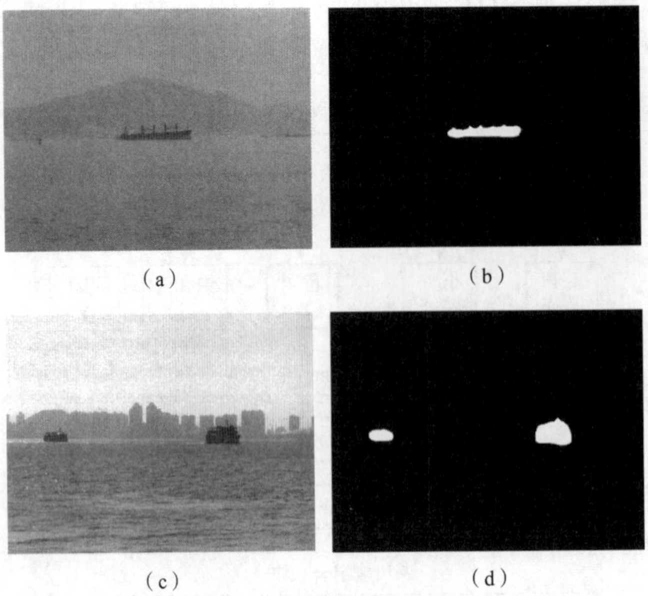

图7　复杂背景仿真结果

（a）原图像，例1；（b）本文提出的方法，例1；（c）原图像，例2；（d）本文提出的方法，例2

从图7可以分析出，在以山、岸上建筑为背景的条件下，本文提出的方法能够准确地实现对舰船目标的检测。而Itti方法虽然能够检测出运动舰船目标所在区域，但轮廓区域与目标实际大小有较大的偏差，同时由于受到岸上建筑、山的干扰，Itti方法在2组仿真结果中均发生了误检的情况。

5　结论

针对水面航行体在港口、码头和锚地等近岸水域条件下对舰船目标进行实时光学检测时，视频图像背景复杂，舰船目标的图像信息易受到岸上相似颜色建筑物和海面波浪反射等干扰的问题，本文应用颜色空间、小波变换等理论，提取了任务水域图像的低频、高频、色调、饱和度和明度特征，采用视觉注意算子将各类特征融合，有效地克服了波光、尾流和岸上建筑等干扰对舰船目标检测效果的影响。

仿真结果表明，本文提出的舰船目标检测方法能够准确地实现复杂背景条件下舰船目标检测，具有良好的可靠性和稳定性，有效地提高了水面航行体对舰船目标侦查、打击能力，为各武器平台对舰船目标检测提供了算法依据，具有重要的参考价值。

参考文献

[1] 方晶，邵志宇，冯顺山，等．小型航行体鲁棒动态逆姿态控制［J］．北京理工大学学报，2015，35（12）：1262－1266.

FANG J，SHAO Z Y，FENG S S，et a1. Robust dynamic inversion attitude control for miniature under－water vehicle［J］．Transactions of Beijing Institute of Technology，2015，35（12）：1262－1266.（in Chinese）

[2] YANG G，LI B，JI S，et al. Ship detection from optical satellite images based on sea surface analysis［J］．IEEE geoscience and remote sensing letters，2014，11（3）：641－645.

[3] 王焜．基于视频的运动船只识别与跟踪技术研究［D］．厦门：厦门大学，2009.

WANG K. Maritime objects recognition and tracking based on video［D］．Xiamen：Xiamen University，2009.（in Chinese）

[4] ITTI L，KOCH C，NIEBUR E. A model of saliency－based visual attention for rapid scene analysis［J］．IEEE transactions on pattern analysis and machine intelligence，1998，20（11）：1254－1259.（in Chinese）

水下主动攻击弹药的弹道特性分析[①]

李　磊[②]，冯顺山
（北京理工大学爆炸科学与技术国家重点实验室，北京　100081）

摘　要：弹道特性的分析与仿真对水下主动攻击弹药的研制具有重要的意义。本文从理论上分析了水下火箭航行体的弹道特性，结合海流和弹药质量变化等实际情况对水下弹道进行了数值仿真，得到了弹道参数随舵角和弹药初始发射速度的变化规律，海流和弹药质量、质心浮心距的变化对弹道的影响。弹道特性的研究结果可以为水下主动攻击弹药的总体设计提供参考和依据。

关键词：水下弹药；火箭航行体；弹道；数值仿真

引言

在封锁作战中，水雷战是一个重要手段，但是水雷一般体积大，比较笨重，用传统方法难于布放到诸如港口水域、重点航道等敌方重点保护的目标附近以实现封锁目的，因此一种能够应用航空布撒器或子母弹远程投放的、具有主动攻击能力的水下封锁弹药可以弥补水雷的不足，具有重要的作战需求背景。

本文主要对一种新型的水下主动攻击弹药的进行弹道特性分析。该弹药经历空中弹道、入水弹道和水下弹道后到达目标区域海底并实现自主固定，等待攻击时机。得到攻击指令后，弹药对目标进行主动攻击。本文主要从理论上分析了火箭助推水下主动攻击弹药的弹道特性，并结合实际情况对弹道进行了数值仿真，为新型的水下封锁弹药的总体性能、弹道规划以及控制系统设计等研究提供一些参考和依据。

1　水下航行体的运动数学模型[1,2]

1.1　作用在水下航行体上的力和力矩

作用在水下航行体上的力主要有重力 $G=mg$、浮力 $B=pgv$、推力 T 和流体动力等。可以近似地认为重力、浮力和推力的大小是常数。

在速度坐标系上取流体动力的分量为

$$\begin{cases} X = c_x \dfrac{1}{2}\rho v^2 s \\ Y = \dfrac{1}{2}c_y^{\alpha}\rho v^2 s\alpha + \dfrac{1}{2}c_y^{w_z}\rho sLvw_z + \dfrac{1}{2}c_y^{\delta_e}\rho v^2 s\,\delta_e + \lambda_{22}\dfrac{\mathrm{d}(v\alpha)}{\mathrm{d}t} - \lambda_{26}\dfrac{\mathrm{d}w_z}{\mathrm{d}t} \\ Z = \dfrac{1}{2}c_z^{\beta}\rho v^2 s\beta + \dfrac{1}{2}c_z^{w_y}\rho vsLw_y + \dfrac{1}{2}c_z^{\delta_r}\rho v^2 s\,\delta_r + \lambda_{33}\dfrac{\mathrm{d}(v\beta)}{\mathrm{d}t} + \lambda_{35}\dfrac{\mathrm{d}w_y}{\mathrm{d}t} \end{cases} \tag{1}$$

在雷体坐标系上取流体力矩分量为

---

① 原文发表于《战斗部与毁伤效率专业委员会第十届学术年会论文集》2007 年 10 月。

② 李磊：工学博士，2003 年师从冯顺山教授，研究港口水域封锁弹药系统总体技术，现工作单位：北京航天长征飞行器研究所（14 所）。

$$\begin{cases} M_{x1} = \frac{1}{2}m_x^{\beta}\rho v^2 sL\beta - \frac{1}{2}m_x^{w_y}\rho vs\ L^2 w_x - \frac{1}{2}m_x^{\delta_d}\rho v^2 sL\ \delta_d + M_{xp}\rho v^2 sL - \lambda_{44}\frac{\mathrm{d}w_x}{\mathrm{d}t} \\ M_{y1} = \frac{1}{2}m_y^{\beta}\rho v^2 sL\beta - \frac{1}{2}m_y^{w_y}\rho vs\ L^2 w_x - \frac{1}{2}m_y^{\delta_r}\rho v^2 sL\ \delta_r - \lambda_{35}\frac{\mathrm{d}(v\beta)}{\mathrm{d}t} - \lambda_{55}\frac{\mathrm{d}w_y}{\mathrm{d}t} \\ M_{z1} = \frac{1}{2}m_z^{\alpha}\rho v^2 sL\alpha - \frac{1}{2}m_z^{w_y}\rho vs\ L^2 w_z - \frac{1}{2}m_z^{\delta_e}\rho v^2 sL\ \delta_e + \lambda_{26}\frac{\mathrm{d}(v\alpha)}{\mathrm{d}t} - \lambda_{66}\frac{\mathrm{d}w_z}{\mathrm{d}t} \end{cases} \tag{2}$$

1.2 动力学方程

重心运动方程

$$\begin{bmatrix} m\frac{\mathrm{d}v}{\mathrm{d}t} \\ mv\frac{\mathrm{d}\Theta}{\mathrm{d}t} \\ -mv\cos\Theta\frac{\mathrm{d}\Psi}{\mathrm{d}t} \end{bmatrix} = \begin{bmatrix} P\cos\alpha\cos\beta - X - \Delta G\sin\Theta \\ P(\sin\alpha\cos\Phi + \cos\alpha\sin\beta\sin\Phi) + Y - Z\sin\Phi - \Delta G\cos\Theta \\ P(\sin\alpha\sin\Phi - \cos\alpha\sin\beta\cos\Phi) + Y\sin\Phi + Z\cos\Phi \end{bmatrix} \tag{3}$$

旋转运动方程

$$\begin{bmatrix} J_{x1}\frac{\mathrm{d}w_{x1}}{\mathrm{d}t} \\ J_{y1}\frac{\mathrm{d}w_{y1}}{\mathrm{d}t} \\ J_{z1}\frac{\mathrm{d}w_{z1}}{\mathrm{d}t} \end{bmatrix} = \begin{bmatrix} 0 & -w_{z1} & w_{y1} \\ w_{z1} & 0 & -w_{x1} \\ -w_{y1} & w_{x1} & 0 \end{bmatrix}\begin{bmatrix} J_{xp} & w_{x1} \\ J_{y1} & w_{y1} \\ J_{z1} & w_{z1} \end{bmatrix} = \begin{matrix} M_{x1} \\ M_{y1} - mg(x_G\cos\theta\sin\varphi) \\ M_{z1} - mg\cos\theta\cos\varphi \end{matrix} \tag{4}$$

1.3 运动学方程

重力运动的运动学方程

$$\begin{cases} v_{x0} = \frac{\mathrm{d}x_0}{\mathrm{d}t} = v\cos\Psi\cos\Theta \\ v_{y0} = \frac{\mathrm{d}y_0}{\mathrm{d}t} = v\sin\Theta \\ v_{z0} = \frac{\mathrm{d}z_0}{\mathrm{d}t} = -v\sin\Psi\cos\Theta \end{cases} \tag{5}$$

旋转的运动学方程

$$\begin{bmatrix} \frac{\mathrm{d}\psi}{\mathrm{d}t} \\ \frac{\mathrm{d}\theta}{\mathrm{d}t} \\ \frac{\mathrm{d}\varphi}{\mathrm{d}t} \end{bmatrix} = \begin{bmatrix} 0 & \sec\theta\cos\varphi & -\cos\theta\sin\varphi \\ 0 & \sin\varphi & \cos\varphi \\ 1 & -\tan\theta\cos\varphi & \tan\theta\sin\varphi \end{bmatrix}\begin{bmatrix} w_{x1} \\ w_{y1} \\ w_{z1} \end{bmatrix} \tag{6}$$

1.4 几何学方程组

$$\begin{cases} \sin\Theta = \sin\theta\cos\alpha\cos\beta - \cos\theta\cos\varphi\cos\beta\sin\alpha - \cos\theta\sin\varphi\sin\beta \\ \sin\Psi\cos\Theta = \cos\alpha\cos\beta\sin\psi\cos\theta + \sin\alpha\cos\beta\cos\varphi\sin\psi\sin\theta + \\ \quad \sin\alpha\cos\psi\sin\varphi\cos\beta - \sin\beta\cos\psi\cos\varphi + \sin\beta\sin\psi\sin\theta\sin\varphi \\ \sin\Phi\cos\Theta = \sin\theta\cos\alpha\sin\beta - \cos\theta\cos\varphi\sin\alpha\sin\beta + \cos\theta\sin\varphi\cos\beta \end{cases} \tag{7}$$

结合作用在水下航行体上的力和力矩以及动力学方程、运动学方程和几何学方程，可以得到一般航行体水下运动的数学模型。

2　一些实际问题在弹道建模中的考虑

以上所建立的水下航行体的运动模型是在没有考虑实际情况下的一种通用的、理想的模型。但是火箭助推水下弹药的实际工作环境是相当复杂的，如火箭发动机工作不断消耗燃料，导致弹药的质量不断发生变化，同时质心与浮心的距离也在不断变化。另外，海流的存在也必然对火箭航行体的水下弹道产生一定的影响。本文假设弹药为轴对称体，质心、浮心均在对称轴上，发动机推力始终沿轴线推动弹体运动，忽略了发动机存在的一些干扰问题。

2.1　质量变化方程

以火箭发动机提供动力的弹药在水下运动时的质量方程为

$$m = m_0 - \int_0^t m_c(t)\,\mathrm{d}t \tag{8}$$

式中　m_0——弹药的起始质量；

$m_c(t)$——单位时间内燃料消耗量。

近似认为固体火箭发动机推进剂是稳定燃烧，药柱截面积不变，则单位时间内减少的燃料质量为

$$m_c = A_b r \rho_b \tag{9}$$

式中　A_b——推进剂装药的燃烧面；

r——燃速；

ρ_b——推进剂密度。

弹体质量的变化，使得弹药质心浮心距也相应发生变化，其变化规律可以认为是

$$x_G(t) = x_{G0} - \frac{1}{2}rt \tag{10}$$

式中　x_{G0}——初始质心浮心距。

2.2　海流对道的影响[3]

海流是很复杂的一种自然现象，其大小和方向是空间与时间的函数，并且有很大的随机性。这里假设海流速度是常量，即

$$\vec{v}_{w0} = v_{wx0}\vec{i}_0 + v_{wy0}\vec{j}_0 + v_{wz0}\vec{k}_0 \tag{11}$$

式中，$\vec{v}_{w0}$是海流在地面坐标系中的速度，v_{wx0}，v_{wy0}，v_{wz0}是海流速度在地面坐标系中的 3 个分量，都是不随时间和空间变化的常量。

当弹药在海流中运动时，相当于在原流场上叠加海流流场，即

$$\vec{v}_{T0} = \vec{v}_{Tw0} + \vec{v}_{w0} \tag{12}$$

式中　$\vec{v}_{T0}$——弹药在地面坐标系中的速度；

$\vec{v}_{Tw0}$——弹药相对于海流的速度。

作用在弹药上的流体动力取决于弹药相对于流体的运动参量，因此在考虑到海流的情况下，其运动模型应该根据式（12）进行修正。

结合方程（1）~式(12）可以得到考虑了质量变化和海流影响的水下弹道方程组。

3　水下主动攻击弹药的仿真与分析[4,5]

3.1　弹道仿真的初始条件

在弹道仿真中忽略了随机扰动因素，仿真中弹药的尺寸外形为：弹体长度 1.5 m，弹体直径 0.3 m，

尾部直径 0.1 m，鳍的面积 0.018 75 m^2，舵的面积 0.002 625 m^2，弹体质量 100 kg。流体动力参数可以根国内外的经验公式计算得出[5]。

3.2 水下主动攻击弹药的理想弹道仿真

弹药的水下运动是空间运动，可以分解为纵向运动和侧向运动，分别由横舵和直舵控制。当弹药仅做纵向运动时，直舵不工作，由横舵控制弹道，与侧向运动有关的运动参数为零。假设弹药的工作环境是理想环境，即不存在水流扰动，忽略弹药的质量变化，弹药无横滚运动。

图 1 给出的水下主动攻击弹药纵向运动的弹道仿真情况。图 1（a）为弹药以 20 m/s 的初速垂直发射的弹道曲线，1、2、3 和 4 号曲线分别是横舵角为 0°、5°、10°和 15°时的弹道曲线。可以看出舵角越大，弹道越弯曲。如果水深 60 m，舵角不大于 15°时弹药垂直发射可以命中水面半径 10 m 以内的目标。图 1（b）、（c）和（d）分别给出了当弹药以 20 m/s 的初速垂直发射时，俯仰角、俯仰角速度和攻角随时间变化情况，曲线号的意义与图 1（a）相同。可以看出，当舵角为 0 时，弹药垂直向上运动，各种与回转有关的弹道参数保持为 0。舵角越大，弹道的回转半径越小，俯仰角减小越快，攻角越大，但攻角大小几乎不变，弹道曲线近似为圆弧。图 1（e）中的 1、2 和 3 号曲线分别代表弹药以 10 m/s、20 m/s 和 30 m/s 初速垂直发射时的弹道曲线，舵角都为 10°。图 1（f）为弹药初速 20 m/s，初始弹道倾角 60°时的情况，1、2、3 和 4 号曲线分别代表横舵角为 0°、5°、15°和 −15°时的弹道曲线。

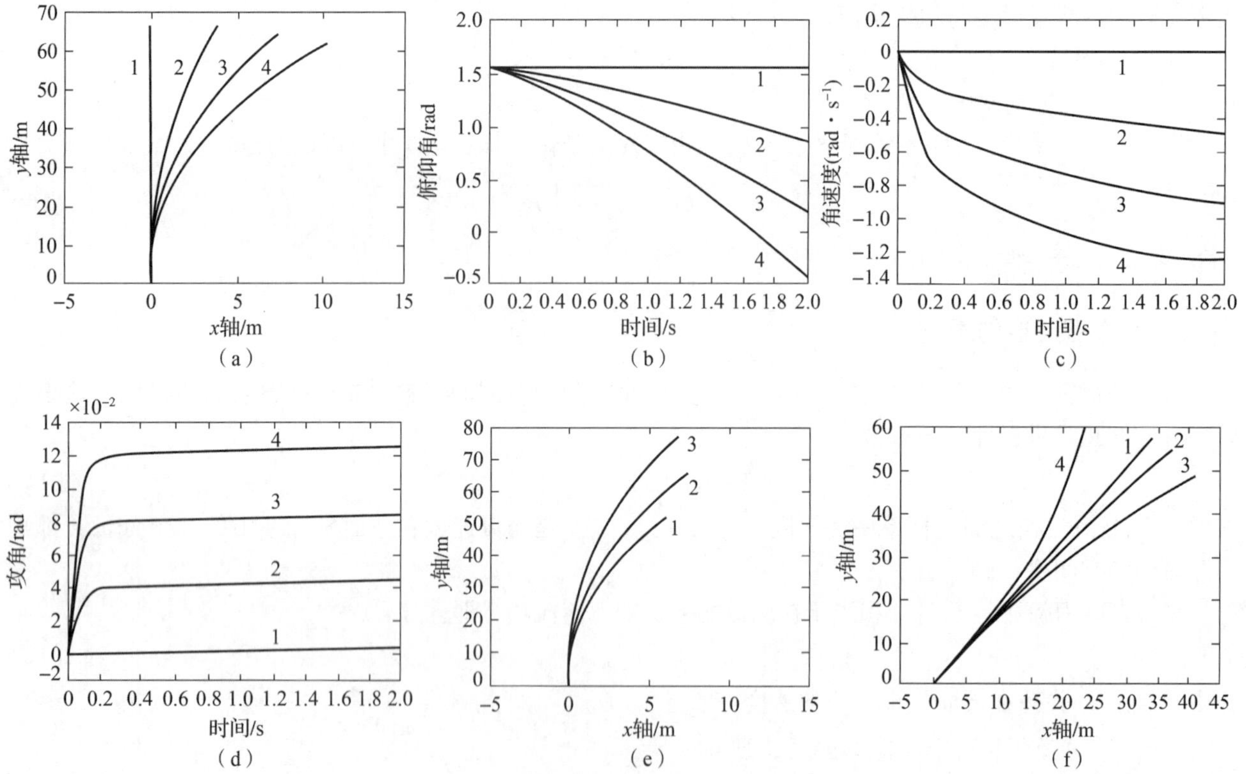

图 1　水下主动攻击弹药垂直平面内弹道仿真

（a）垂直平面内子弹弹道（不同舵角）；（b）俯仰角随时间变化曲线；（c）俯仰角速度随时间变化曲线；
（d）攻角随时间变化曲线；（e）垂直平面内子弹弹道（不同初始速度）；（f）垂直平面内子弹弹道（初始俯仰角为 60°）

当弹药仅做侧向运动时，横舵不工作，仅由直舵控制弹药的运动。图 2（a）给出了弹药初速 20 m/s、弹道倾角 60°保持不变的弹药运动情况，1、2 和 3 号曲线分别表示直舵角为 5°、10°和 15°时弹道曲线，1′、2′和 3′曲线是对应的 1、2 和 3 号曲线在水平面内的投影。图 2（b）给出的是弹道倾角为 60°不变时不同初速时的弹道曲线情况，1、2 和 3 号曲线分别对应初速 10 m/s、20 m/s 和 30 m/s。

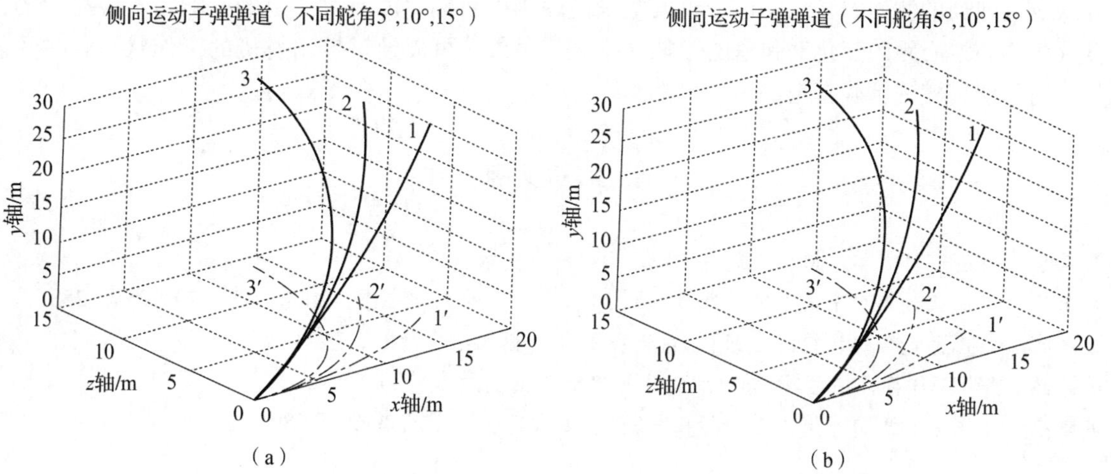

图 2 弹道倾角一定时水下主动攻击弹药的侧平面内弹道仿真

3.3 海流对弹道的影响

图 3 给出了海流对弹道的影响情况，假设海流速度大小为 0.5 m/s，方向为 x 轴正方向。1、2 和 4 号曲线分别代表无海流干扰下，发射初速为 20 m/s、横舵角为 0°、10°和 −10°时的弹道曲线，3、5 号曲线是考虑海流影响下舵角仍为 10°和 −10°的弹道曲线。当水深 50 m 时，图 3 海流对弹道的影响海流的存在导致弹药水面攻击范围平移 4 m 以上，如果海流速度大于 0.5 m/s 时影响将会更大。

3.4 变质量弹药的水下弹道

假设推进剂燃烧速度 20 mm/s，推进剂为圆柱形，等截面燃烧。考虑弹药质量和质心浮心距不断变化的情况，弹道仿真曲线见图 4。曲线 1 代表不考虑质量变化时的弹道曲线，曲线 2 考虑了质量和质心浮心距的变化，可以看出，质量变化对弹道的影响并不明显。

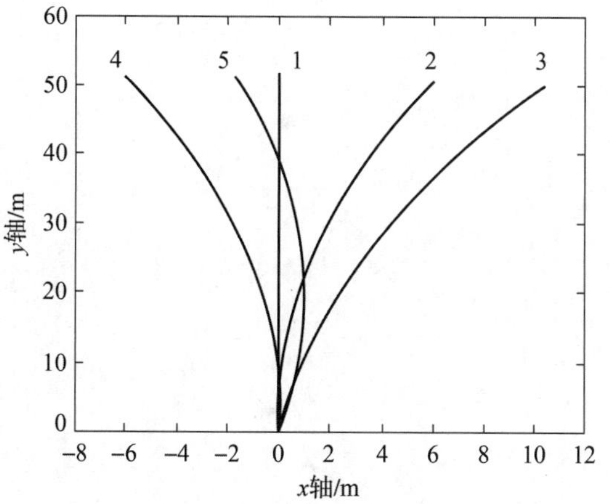

图 3 海流对弹道的影响

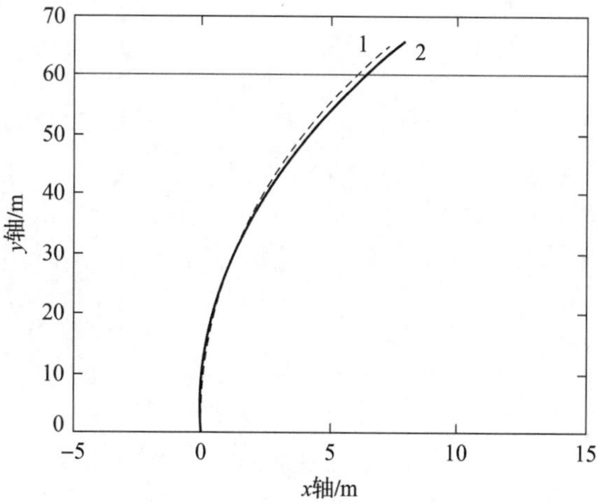

图 4 质量变化对弹道影响

4 结束语

本文从理论上分析了火箭助推水下主动攻击弹药的弹道特性，考虑了海流以及弹药质量和质心浮心距的变化对弹道的影响，把弹药的水下弹道分成纵平面内的弹道和弹药的侧运动弹道进行了数值仿真研究。通过仿真结果可以看出，弹药垂直发射时弹道回转半径随横舵角增大而减小，在相同横舵角相同水深发射时，初速越大，控制水面的范围越小。以一定弹道倾角发射的弹药到达水面时，其横向位移较大，可以通过弹药的侧向回转运动攻击较大范围内的水面目标。海流对弹道的影响是不容忽视的，但是弹药质量和质心浮心距的变化对弹道的影响并不大。另外，可以通过控制舵机的工作状态和时间来控制弹道曲线的形状，实现精确打击。

参考文献

[1] 严卫生. 鱼雷航行力学 [M]. 西安：西北工业大学出版社，2005.
[2] 宗瑞良. 火箭航行体水中运动的数学模型 [J]. 西北工业大学学报，2000，18 (2)：254－258.
[3] 张宇文. 鱼雷弹道与弹道设计 [M]. 西安：西北工业大学出版社，1999.
[4] 王沫然. MATLAB 与科学计算 [M]. 北京：电子工业出版社，2005.
[5] 黄景泉，张宇文. 鱼雷流体力学 [M]. 西安：西北工业大学出版社，1989.

地面运动目标的地震动特性研究①

刘　畅②，冯顺山，董永香，王　超
（北京理工大学爆炸科学与技术国家重点实验室，北京　100081）

摘　要：本文研究了人员、轮式车、限带式车辆等地面运动目标在运动时产生的地震动信号，利用动圈式地震动传感器组成的地面运动目标探测系统进行了外场测试，取得了运动目标在不同速度和距高条件下的震动信号，应用时域分析方法得出信号的时域特征，并应用 Welch 法进行了频域分析，研究所得的信号时域、频域特征向量可作为目标识别的依据。

关键词：地面目标；震动；时域特征向量；频域特征向量

引言

地面目标在地面上运动时会引起地震动信号，设置在一定距离远处的地震检波器可检测到此震动信号作为封锁类弹药识别和检测运动目标是否进入有效杀伤区域内的依据[1]。实质上该震动信号是运动目标对地面的激励信号经过与地面的耦合、地表介质质点的振动、地层介质质点间的耦合、地层介质与地震检波器的耦合而得到的，而且信号中不可避免地夹杂着大地自然振动信号[2]。假设地面介质是各向同性的半无限弹性均匀介质，震动信号在地面介质的传播过程中随传播距离越远而衰减越大，则对同一震动信号在不同距离处检测到的信号是有所变化的，反映在信号上就是时域幅值和频率组成不同。因此，在利用地震动传感器进行探测并识别目标的系统中，对于目标信号的特征分析应兼顾时域、频域特性，并进行时域、频域的特性研究。

1　典型信号的外场测试

对信号的目标识别，一般采用时域和频域分析，在时域上，运动车辆（轮式车和履带车）之间的区别并不显著，但人和运动车辆的时域波形之间有明显区别。不管是人员还是车辆的震动信号，在频域都能找到各自对应的特征。尤其是轮式车辆和履带式车辆之间，在时域特征不明显的情况下，在频域对其进行分析是十分必要的。因此，文中通过时域与频域两种方法对典型信号进行了分析。

测试中，以人、轻型车辆、卡车、坦克作为典型目标，轻型车辆、卡车代表轮式车，坦克代表履带车，将动圈式传感器分别布置在不同距高远处并通过地震动记录仪分别采集目标在不同距离和行驶速度的信号。实验地点的地质条件为：良好土质地面。气象条件为：晴朗、风约 4 级，地面干燥。

根据实验结果，这几种信号的频率主要分布都低于 150 Hz。

2　时域分析

人的信号波形幅度变化是随着脚步的落地变化的，而车辆由于履带和车轮与地面的连续接触，其运动产生的震动信号是连续的。由于重量和发动机等因素的影响，在车辆目标出现后，信号的幅值慢慢变大，然后变化缓慢，信号逐渐变小，最后目标消失，脚步的频数远低于车辆信号的主频。

2.1　过零分析算法原理

信号的过零数分析也简称为过零分析[3]。过零分析就是对确定时间段内的时域信号幅值与设定阈值

---

① 原文发表于第十届全国战斗部与毁伤技术学术交流会论文集（2007 年）。

② 刘畅：工学硕士，2005 年师从冯顺山教授，研究典型地面运动目标的地震动探测与识别特性，现工作单位：华为公司。

比较，计算得到信号幅度绝对值超过阈值的次数，在这个计算出的过零数基础上开展进一步的分析研究。

信号的过零分析与频谱分析有相似之处，但又有本质不同。对于一些简单波形，如正弦波、方波等，过零率和频率有直接的正比关系，进而和过零次数也有简单正比关系，对于复杂波形，过零率 N 和频谱关系不明显，但它和功率谱 $G(f)$ 有关：

$$N = 2\sqrt{\int_{f_L}^{f_H} f^2 G(f)\,\mathrm{d}f \Big/ \int_{f_L}^{f_H} G(f)\,\mathrm{d}f} \tag{1}$$

式中 f_H——上限截止频率；

f_L——下限截止频率。

式（1）反映出若信号频谱处在高频段，则信号过零率比较高。

2.2 两类目标的过零分析

图 1 为人员脚步信号，图 2 为车辆的地震动信号，从图的对比中可以看出，脚步作用时间短，信号可近似看作周期性的脉冲信号，并且间隔使间比较长，而车辆信号却近似连续的。分别取两类信号数秒时间内的信息，并按一定规则设定阈值[6]，计算两类目标的过零数。若两类信号的过零数差别足够大，就可以将两类目标区别开来。

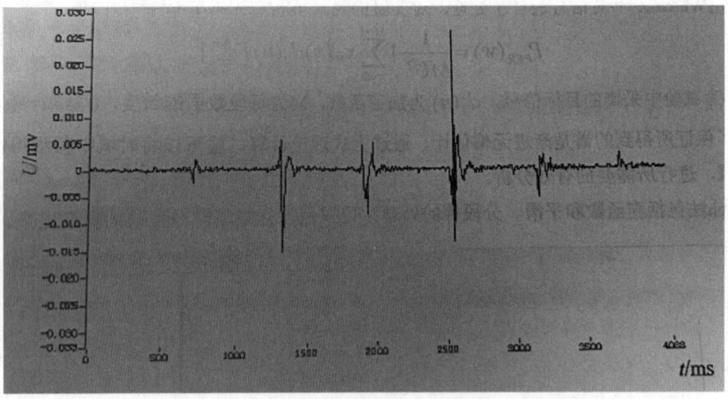

图 1 人员脚步信号

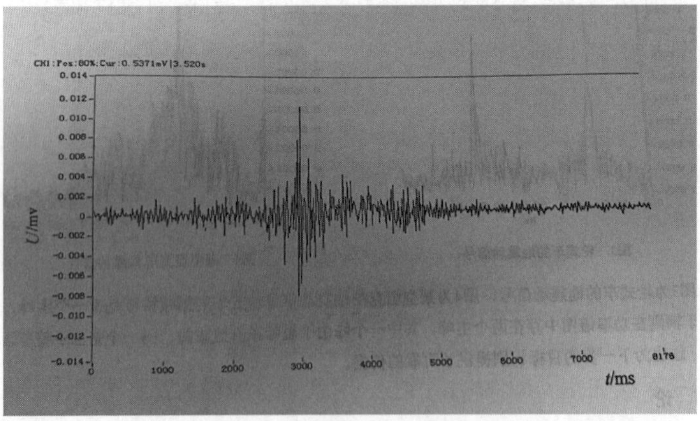

图 2 车辆的地震动信号

在已经确定目标出现的前提下，由于信号幅值与目标和传感器距离有很大关系，所以阈值不能选取固定值，只能选取某一浮动值。具体是利用信号当前时刻幅值和所采集范围内最大幅值的关系或者和前一时刻幅值差异不会太大的这种关系[7]，把当前阈值的设定和当前时刻最大值或以前时刻的幅值联系起来。后一种方法可以直接利用已得到的信号，速度较快，若进一步利用前两时刻所得信号，则可以使识别精度得到改善，并且提高抗噪声能力。

3 频域分析

本文利用 Welch 法[3,4]对信号进行了处理，对实验的轮式车和坦克做了功率谱分析：

$$P_{\mathrm{PER}}(w) = \frac{1}{MUL}\left|\sum_{n=0}^{M-1} x_N(n) d_2(n) \mathrm{e}^{-jwn}\right|^2 \tag{2}$$

其中，为试验中采集的目标信号，$d_2(n)$为加窗函数，M 为每段数据的长度，U 是归一化的因子，使用它是为了保证所得到的谱是渐进无偏估计。通过式（2）理论计算，就可以将时域中所测得的目标信号转换到频域，进行所需要的有效分析。

Welch 法包括窗函数和平滑、分段等的处理，可以使特征频率能很直观地观察出来。

图 3 为轮式车的地震动信号，图 4 为某型坦克的地震动信号，轮式车的频域特征是有一个主峰，而履带式车辆则在功率谱图中存在两个主峰：其中一个是由于履带拍打造成的，另一个是由于整车运动造成的。这也为下一步的目标识别提供了可靠的依据。

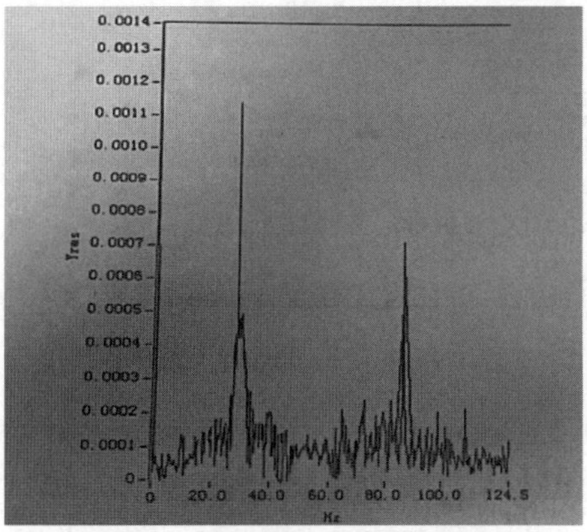

图 3 轮式车的地震动信号

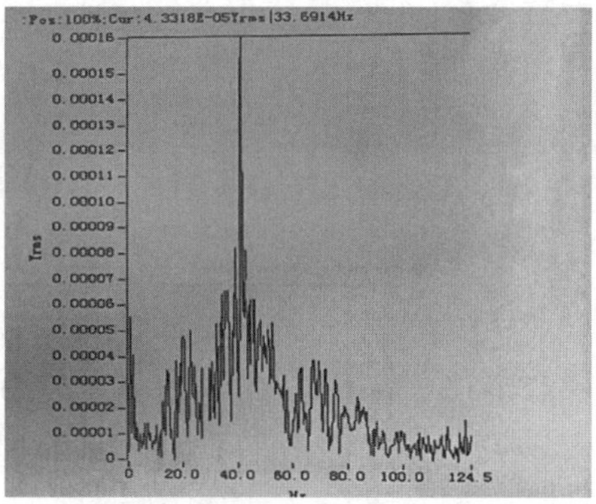

图 4 某型坦克的地震动信号

4 结论

通过上述实验所得到的数据以及在时域、频域的分析，可以得出以下结论：

（1）人员和车辆在时域上幅度差异显著且过零数方面也有很大差别，可为封锁类弹药识别运动状态的人员和车辆振动特征提供有效的依据。

（2）在地面运动目标探测系统中，目标引起的地震动信号跟目标距离传感器的临商和目标的速度均存在紧密的关系，但在对采集数据进行整理和分析后得到比我检定的信号特征。

（3）轮式车的频域特征是有一个主峰，而履带式车辆则在功率谱图中存在两个主峰：其中一个是由于履带拍打造成的，另一个是由于整车运动造成的。这也为下一步弹药对运动目标识别提供了可靠的依据。在本文研究工作的基础上，今后可针对考虑更多实际条件的时域、频域相结合的方法，提取不同目标的特征向量，利用最小错分样本数法则和人工神经网络方面的知识对各种典型目标进行识别。

参考文献

[1] 牛强. 漫谈地面传感器侦察 [J]. 现代军事，1999（12）：59.

一端全约束战斗部破片空间分布研究①

高月光②，钱海涛，张　博，边江楠，冯顺山
（北京理工大学爆炸科学与技术国家重点实验室，北京　100081）

摘　要：为研究一端全约束另一端带盖战斗部自然破片形成过程及空间分布规律，本文采用仿真的分析方法，设计了圆柱形战斗部装药结构，仿真结果表明一端全约束战斗部破片形成过程有自身特点，一端起爆条件下，爆轰波传播至全约束端面时会产生马赫杆，对未反应炸药作用，近似2.5倍正常爆轰波压力驱动壳体，进而获得更高的速度；圆柱壳体首先在与全约束端接触面处破裂，其次在起爆端的端盖处破裂，然后破裂延伸至整个圆柱部壳体；爆炸产物从破裂口开始泄出，驱动破片飞散，形成两个密集飞散区。通过战斗部静爆试验，验证了其破片空间分布与仿真结果的吻合性，可为有关战斗部的破片威力设计提供参考。

关键词：战斗部；破片；空间分布规律；仿真；静爆试验

Research on the spatial distribution of fragment of one end fully constrained warhead

Gao Yueguang，Qian Haitao，Zhang Bo，Bian Jiangnan，Feng Shunshan
(State Key laboratory of Explosion Science and Technology，Beijing Institute of Technology，
Beijing 100081，PR China)

Abstract：In order to study the process and spatial distribution of natural fragmentation in the warhead with end caps under one full constraint condition，this paper designs a cylindrical shell with two end caps which one of them is fully constrained using the simulation analysis. The result shows that fragmentation under one full constraint condition has its own characteristic. The Mach rod is generated when the detonation wave propagates to the fully restrained end face under the condition of one end detonation，working on unreactive explosives and causing the nearby fragment subjected to nearly 2.5 times the normal pressure to obtain a higher speed. The cylindrical casing first ruptures at the contact surface with the fully restrained end，and then at the end cover of the initiating end，and then the rupture extends to the whole cylindrical shell，The explosive products start to leak out from the rupture. driving fragments to fly，and forming two dense flying areas. The warhead static explosion test is carried out to verify the spatial distribution of the fragment of the simulated cylindrical shell，which can provide a reference for the fragment power design of the warhead.

Key words：Warhead；Fragment；Spatial distribution；Simulation；Static explosion test

引言

当评价破片对目标的杀伤效果时，通常研究的都是考虑具有一定速度和质量的单枚破片的毁伤效应，但更为合理的应该是考虑多枚破片和冲击波作用下目标结构的整体响应。在这种情况下，破片的空

① 原文发表于《第十六届全国战斗部与毁伤技术学术交流会论文集》（2019年）。

② 高月光：工学博士，2017年师从冯顺山教授，研究半侵彻弹药战斗部对甲板目标威力效应，现工作单位：在校生。

间分布规律的研究就显得尤为重要。

黄广炎[1]用 X 光捕捉到了两端开口的圆柱形壳体起爆后破片的空间分布情况。Arnold[2]用见证靶记录了两端开口的圆柱形壳体的破片分布情况，通过他的试验可以发现破片分布是不均匀的，中间一部分破片具有较大质量，其余两侧是较小的破片。Kong[3]采用仿真的方法研究了两端带端盖的圆柱形壳体产生的破片空间质量分布。

一端全约束战斗部的破片空间分布规律目前研究得不是很多，对其破片的形成过程及破片的空间分布规律也缺乏相应的研究。本文将通过仿真的方法建立一端全约束战斗部，并进行相应战斗部的静爆试验，来探究此类型战斗部破片的形成规律和空间分布。

1　数值模拟

数值模拟采用有限元软件 AUTODYN 进行仿真，该软件广泛应用于分析非线性动力学问题[4]。本文利用 SPH（光滑粒子流体动力学）算法模拟子弹在爆炸载荷作用下的破碎过程，由于圆柱壳体的对称性，仿真只需要模拟 1/4 模型。圆柱壳体从端盖中心起爆。

1.1　材料模型

合理选择材料的状态方程、本构关系，对数值模拟最后的结果起到很关键的作用，结合后期试验，仿真中炸药采用 B 炸药，材料状态模型采用 Jones - Wilkins - Lee（JWL）状态方程，壳体采用 30CrMnNi2A 材料，材料模型采用经典的 Johnson Cook（JC）模型[5]，将子弹一端设置为全约束，仿真模型如图 1 所示，同时在轴向方向的壳体上均匀设置 12 个观测点，如图 2 所示。

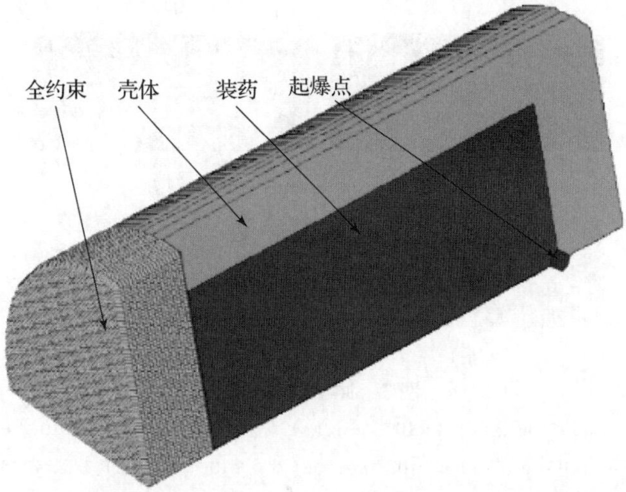

图 1　端部中心点起爆、一端为全约束的 1/4 仿真模型图

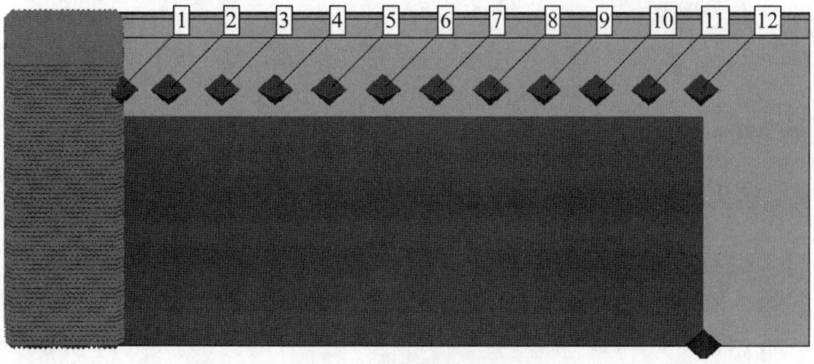

图 2　观测点位置

1.2 计算结果分析

1.2.1 冲击波的传播

当自由端中心起爆点起爆后，炸药开始快速反应并在炸药中产生冲击波，如图3（a）所示。开始的冲击波在炸药中为球形波，当冲击波传播到圆柱部壳体后发生反射，反射回来的冲击波与之前的球形波叠加，并继续沿装药传播，如图3（b）~（d）所示。当冲击波传播到全约束端盖处时，由于壳体波阻抗大于炸药波阻抗，其反射回来的为反射冲击波，与炸药中的冲击波不断叠加，如图3（e）~（g）所示，并且由于反射回来的冲击波也近似是以全约束端盖中心为圆心的球形波，两波叠加会在端盖附近炸药中产生马赫杆效应，如图4中的1号观测点所示，可以明显看到其压力在0.008 5 ms时达到峰值，且峰值远高于其他观测点，约为其他观测点的2.5倍。图5为仿真中观测到的全约束端附近的马赫杆现象，全约束端反射回的强冲击波继续在炸药中传播，但是其强度随着时间逐渐衰减，其对端盖和圆柱壳体的加速强度也逐渐减弱。

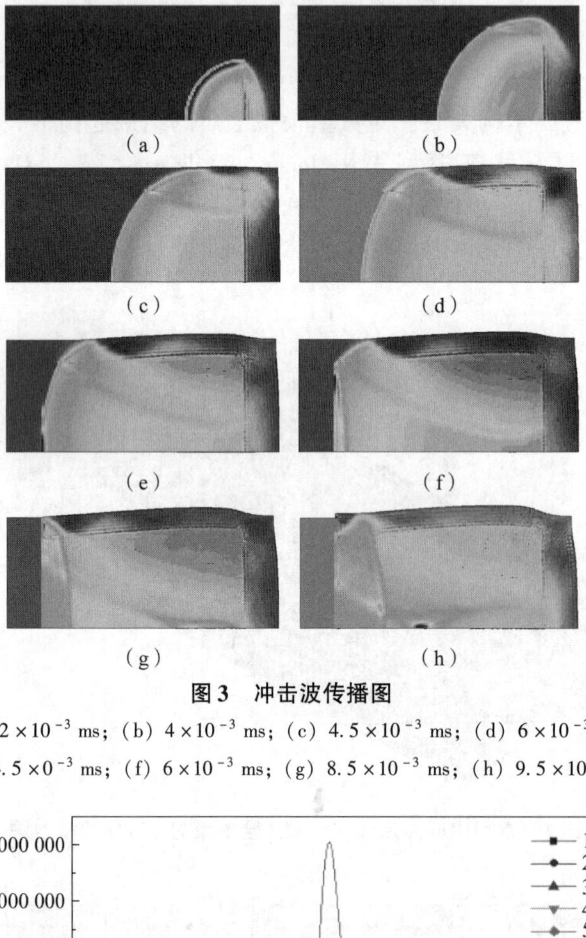

图3 冲击波传播图

（a）2×10^{-3} ms；（b）4×10^{-3} ms；（c）4.5×10^{-3} ms；（d）6×10^{-3} ms；（e）4.5×0^{-3} ms；（f）6×10^{-3} ms；（g）8.5×10^{-3} ms；（h）9.5×10^{-3} ms

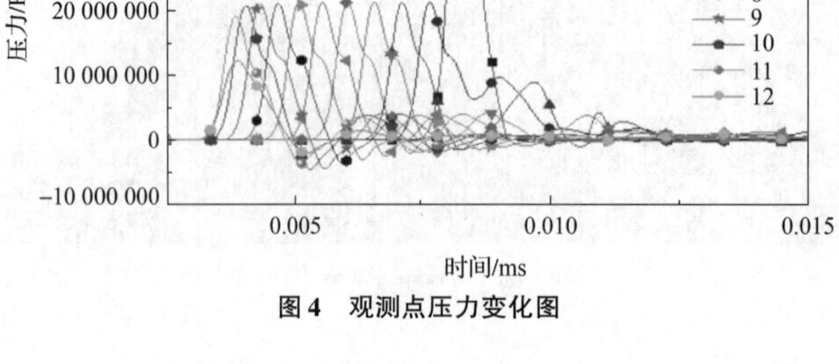

图4 观测点压力变化图

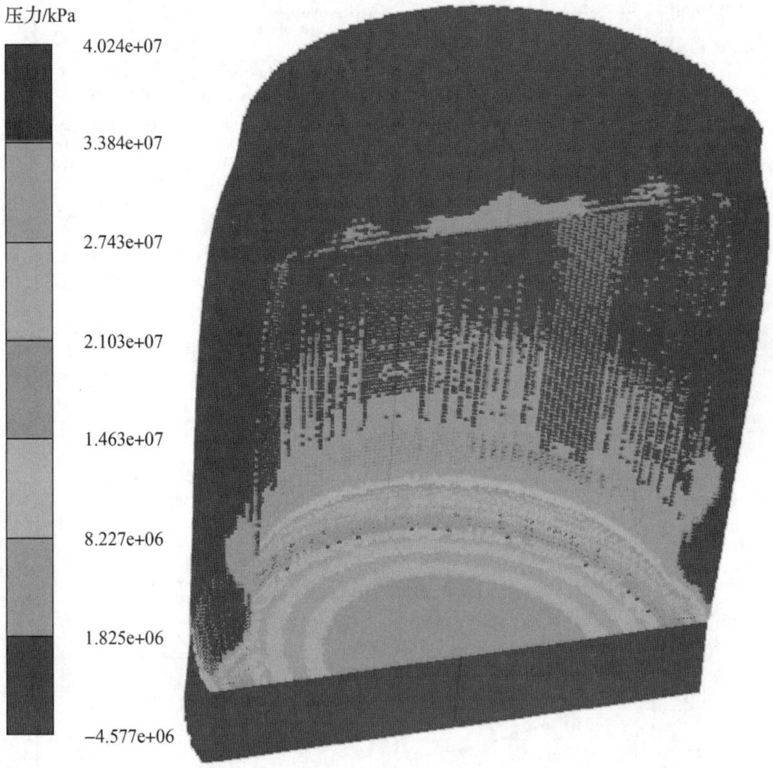

图 5　全约束端盖附近的马赫杆效应

1.2.2　壳体的膨胀与破碎过程

当装药引爆后，产生的强烈的冲击波会在壳体中产生强烈的压缩波并沿壳体快速传播，壳体首先在冲击波作用下开始膨胀加速，如图 6（a）所示，接着由于炸药产生的高温高压的爆炸产物开始膨胀做功，壳体进一步膨胀，当膨胀超过壳体材料的强度极限后，壳体就会发生破裂形成破片。由图 6 可知，由于圆柱壳体靠近全约束端获得更高的速度，以及开始的马赫杆加速效应，圆柱部壳体在靠近全约束端处率先发生破裂，随着爆炸产物的加速作用，壳体继续膨胀，并且在靠近端盖附近的圆柱壳体紧接着在 0.025 ms 左右观察到明显的破裂，如图 6（c）所示，然后端盖与圆柱部分离，在爆轰产物的作用下继续加速并产生破碎，如图 6（d）~（e）所示。端盖大体分为两部分：一部分在中间形成质量比较大的破片；另一部分是靠近圆柱壳体的直径比较的一圈发生破裂，形成一圈相对大一点的破片，在 0.1 ms 左右，破片速度已基本趋于稳定，爆炸产物作用在破片上的力与空气阻力基本平衡，由图 7 所示，并且由图 8 可知，圆柱部壳体产生破片在靠近全约束端的速度最大，证明了之前所述的马赫杆效应，靠近端盖部分由于与端盖分离后稀疏波传入，造成了爆轰产物压力下降，对破片的加速作用大幅度降低。

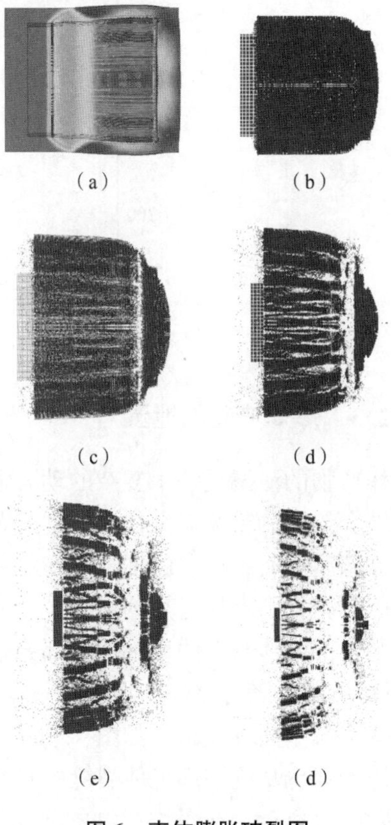

图 6　壳体膨胀破裂图

(a) 0.006 ms；(b) 0.02 ms；(c) 0.025 ms；(d) 0.04 ms；(e) 0.075 ms；(f) 0.12 ms

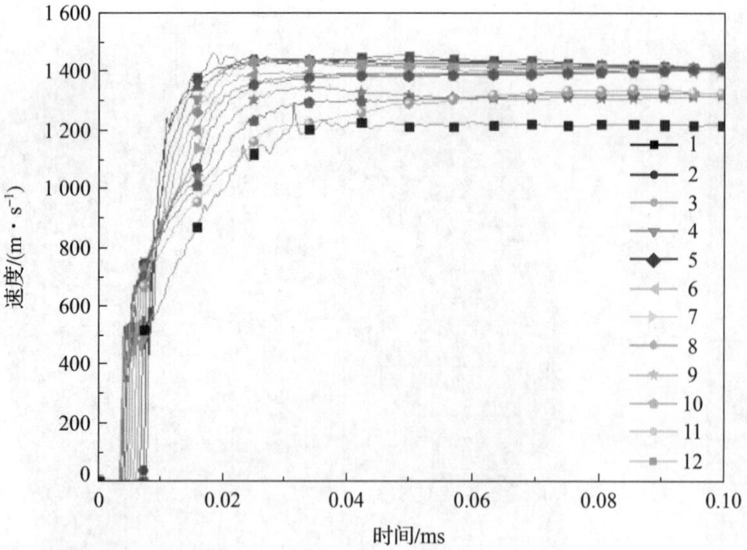

图7　观测点速度

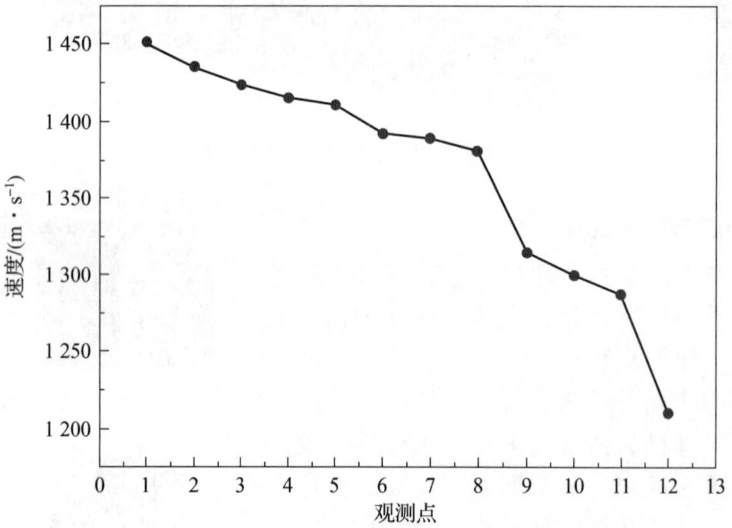

图8　圆柱壳体速度分布

1.2.3　爆炸产物的泄漏

爆炸产物的泄漏是一个复杂的动力学过程，高温高压的产物从失效的壳体缝隙中飞出。本节将详细讨论一端全约束的带端盖圆柱壳体中的爆炸产物的泄漏过程。

当炸药起爆后，产生的冲击波在炸药迅速传播，炸药发生爆轰，如图3（a）~（h）所示。当冲击波传过整个装药，并在接近全约束端产生强烈的马赫杆之后，爆炸产物开始膨胀运动，由于强烈的马赫杆作用，圆柱部壳体与全约束端连接部分率先失效，爆炸产物从失效部分产生的缝隙中开始往外飞出，如图9（a）所示。随着爆炸产物的进一步膨胀做功，圆柱部壳体与端盖连接处也开始失效，爆炸产物紧随其后，从缝隙中向外飞出，如图9（b）所示。随着壳体的进一步膨胀，圆柱部壳体也开始失效，产生裂缝，爆炸产物也从圆柱部壳体裂缝中开始向外飞出，如图9（c）所示。如图6（d）所示，端盖与圆柱部壳体发生分离后，在爆炸产物的作用下继续加速运动，但由于圆形端盖周围的稀疏波传入爆炸产物中，造成了位于端盖边缘处的爆炸产物的压力迅速下降，其对端盖的加速能力也相应下降，端盖中心部分和边缘部分开始产生加速度差和应力差，并且超过壳体材料的强度极限，端盖开始发生环形破裂，爆炸产物从裂缝中飞出，如图 9（d）所示。接下来环形破裂的端盖继续发生破碎，如图 6（c）、（d）所示，此时破片的加速度逐渐下降，在 0.075 ms 左右，速度基本趋于稳定，如图 7 所示。

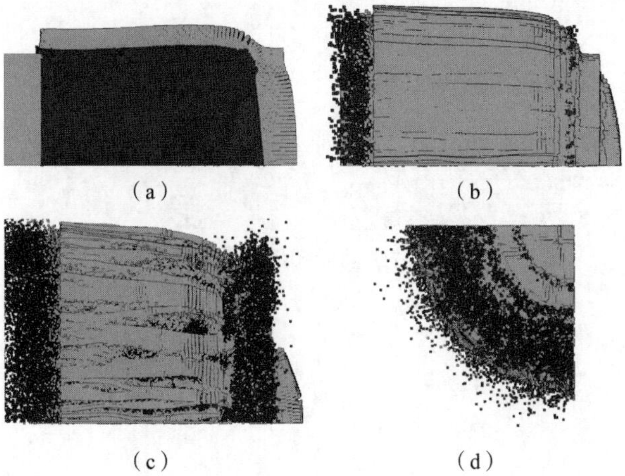

图 9　爆炸产物的泄漏过程

（a）0.015 ms；（b）0.022 ms；（c）0.038 ms；（d）0.045 ms

1.2.4　破片空间分布

仿真最后，破片的速度趋于稳定，破片的数量也基本确定，将仿真中形成的破片数据导入 Matlab 中，编写相应的程序，求出破片飞散角中破片数量与飞散角度的对应关系，以壳体赤道面为 0°位置，靠近全约束端角度为负值，靠近自由端角度为正值，结果如图 10 所示，由图可知，破片分布包含两个密集区，其中较多的一部分集中分布在 -20°～20°，-8°～-16°分布较多，因为圆柱部壳体的破片基本在这个范围内形成，且因为马赫杆的作用，会产生更多更小的破片，数量也会因此变多；另一部分的破片分布在 30°～70°，由图 7（c）-（d）可知，这一部分破片主要是来自端盖，且由统计图可知，破片在靠近端盖边缘处形成的破片数量较多，根据动压破坏理论，边缘处由于受到侧向稀疏波的影响，壳体内压缩应力区域会更快地减少，径向裂纹会更快地沿拉伸应力区域传播，裂纹会更快地从壳体外表面传到壳体内表面，因此会形成更多数量的破片，端盖中心处会形成质量较大的破片，但数量不多，这与仿真中观测到的一致，破片分布如图 11 所示。

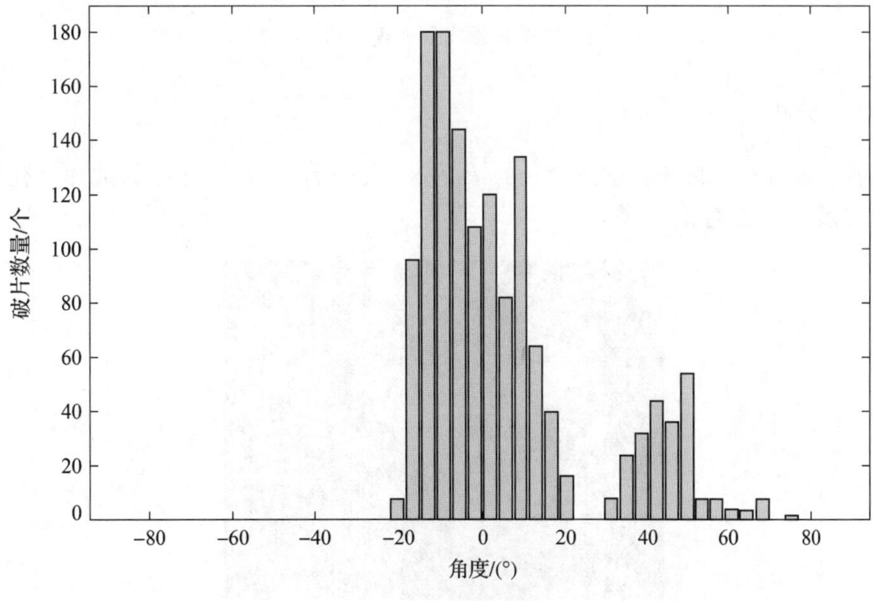

图 10　破片数量分布

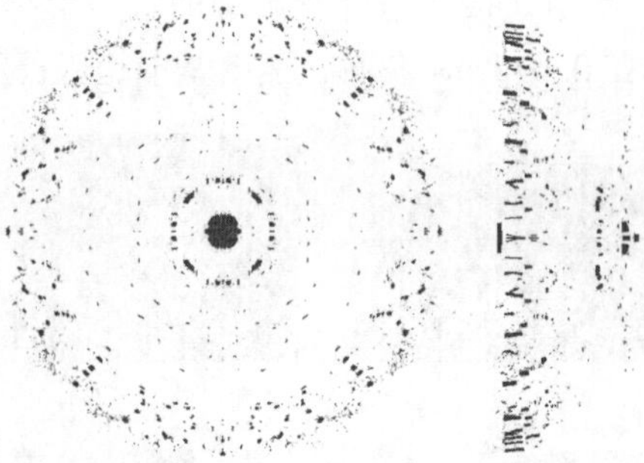

图 11　破片空间分布俯视图和侧视图

2　静爆试验

2.1　试验现场布局

将战斗部一端固定约束，在距战斗部中心 1.4 m 处布置飞散靶，采用 0.2 mm 厚铁皮，长 1 m，高 1 m，试验布置示意图如图 12 所示。

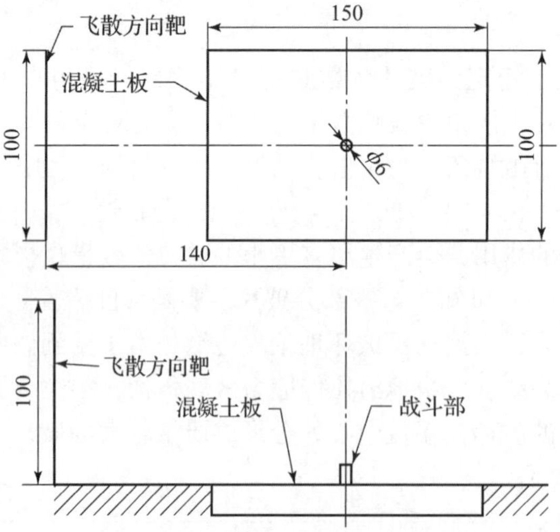

图 12　静爆试验布局俯视图和正视图

2.2　试验结果分析

试验进行 1 发，试验后飞散靶被破片穿孔的情况如图 13 所示，对飞散靶的破片穿孔情况进行统计，建立破片数量与飞散角度的关系。

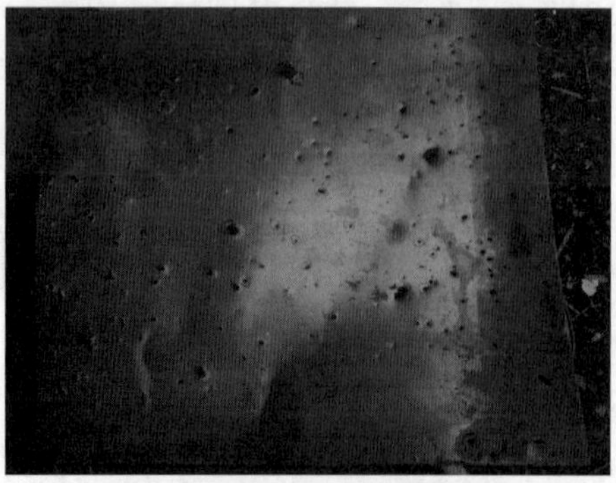

图 13　破片飞散靶穿孔情况

由于试验中测量并统计所有破片的飞散情况较为困难，为节约成本，试验假设破片在圆周方向呈均匀分布，因此仅需研究飞散靶内破片的分布情况，因为飞散靶板占战斗部圆周方向的1/9，将飞散靶上获得的每个角度的破片数量乘以9即可得到整个圆周方向上破片的分布情况。将试验中获得的破片数量关于角度的关系与仿真中获得的数据进行对比，如图14所示。

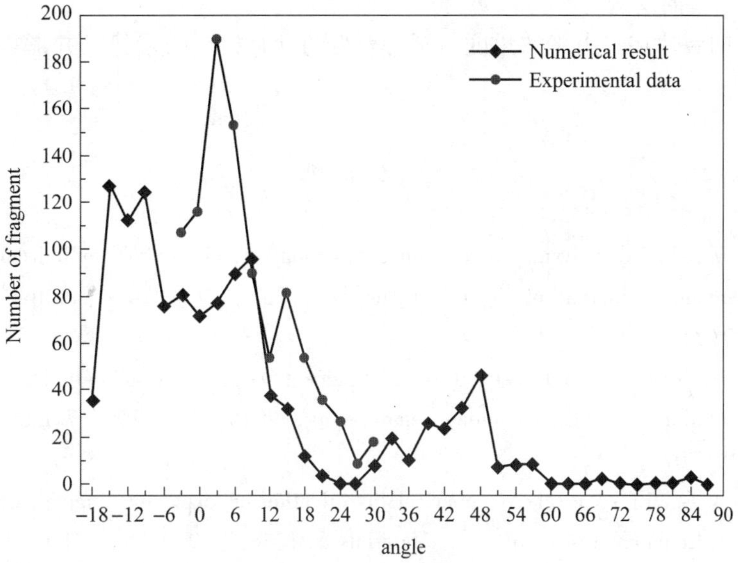

图14 试验和仿真数据对比

由图14可知，一端全约束的带端盖的圆柱形壳体，其破片分布首先区别于两端开口的圆柱形壳体的破片分布，两端开口的圆柱形壳体基本在中间会形成一条“密集带”分布[6,7]，两边稀疏分布一定数量的破片。一端为全约束的带端盖圆柱形壳体破片分布基本分为两个密集区，其中较多数量的破片是由圆柱形壳体产生的，另一区域破片是由端盖产生的，且端盖边缘处会产生数量相对较多的质量较小的破片，而端盖中心会形成数量较少而质量较大的破片，但由于试验中靶板为竖直的直板，无法捕捉到由端盖产生的破片，因此第二区域的破片无法测量，后续试验应该相应进行完善。

另外，仿真中破片在 −20°～0°分布有较多数量的破片，试验中破片在此角度范围内几乎没有发现破片，然而在0°～10°分布的破片数量远远高于相邻角度，其原因是仿真中并没有考虑到现实中破片会撞击到混凝土地面并发生反射的现象，因此试验中在 −20°～0°的破片由于反射作用而偏离了原先的飞散方向，增加了0°～10°破片的分布数量，因此后续仿真应考虑到实际破片会发生反射的情况，如添加混凝土地面用来模拟实际中破片撞击混凝土的反射。

仿真中在10°～30°破片数量分布与试验中获得的具有相似的趋势，试验中共测得破片数量为936枚（整个周向），仿真中圆柱部壳体产生975枚破片、端盖产生194枚破片，破片数量也较为接近，初步说明仿真和试验中破片的形成及其空间分布规律具有较好的吻合性。

3 结论

本文利用AUTODUN中的SPH（光滑粒子流体动力学）算法对一端全约束另一端带盖的战斗部进行了模拟仿真，模拟了战斗部中冲击波的传播、壳体的膨胀与破碎过程、爆炸产物的泄漏及破片的空间分布情况，并进行了战斗部的静爆试验验证仿真中破片的空间分布规律，经过仿真和试验的相关分析得到以下结论：

（1）一端全约束战斗部在端盖中心起爆后会在全约束端产生马赫杆效应，附近壳体的压强几乎为其他部分圆柱壳体的2.5倍；

（2）战斗部在端盖中心起爆后圆柱部壳体将先从与全约束段接触面处发生材料失效，其次是在起爆端的端盖处，最后延伸到整个圆柱部壳体，爆炸产物从破裂口开始泄出；

（3）战斗部爆炸形成破片分为两个密集区，一个是由圆柱部壳体产生的破片，分布在 -20° ~20°之间，实际战斗部 -20° ~0°的破片会反射到0° ~10°，另一个是由端盖产生的破片，分布在30° ~70°，且此部分破片由一圈环形小破片和中间数枚大破片组成；

（4）试验中战斗部由于破片会在地面反射，因此会改变自由分散的破片方向，飞散角会相应改变，仿真中未考虑反射作用，需要进一步深入研究；

（5）仿真中获得的破片在自由飞散角度与试验获得的具有相同的趋势，并形成数量接近的破片，初步验证仿真真实可行。

参考文献

[1] HUANG G Y, LI W, FENG S. Axial distribution of fragment velocities from cylindrical casing under explosive loading [J]. International journal of impact engineering, 2015, 76: 20 - 7. http: //dx. doi. org/10. 1016/j. ijimp eng. 2014. 08. 007.

[2] HUANG G, LI W, FENG S. Axial distribution of fragment velocities from cylindrical casing under explosive loading [J]. International journal of impact engineering, 2015, 76: 20 - 7. http://dx. doi. org/10. 1016/j. ijimp eng. 2014. 08. 007.

[3] KONG X H, WU W G, LI J, et al. A numerical investigation on explosive fragmentation of metal casing using Smoothed Particle Hydrodynamic method [J]. Materials & design, 2013, 51 (Oct.): 729 - 741.

[4] DYNAMICS C. Release 14. 0 documentation for ANSYS AUTODYN [Z] . Canonsburg, PA: ANSYS Inc, 2011.

[5] JOHNSON G R, COOK W H. A constitutive model and data for metals subjected to large strain, high strain rates and high temperatures [C]//The 7th International Symposium on Ballistic. Netherlands: Hague, 1983.

[6] GRISARO H Y, DANCYGIER A N. Spatial mass distribution of fragments striking a protective structure [J]. International journal of impact engineering, 2018, 112: 1 - 14.

[7] KRAPP R, PREDEBON W. Calculations of fragment velocities from naturally fragmenting munitions [R]. Report 2509. USA Ballistic Research Laboratories, Aberdeen Proving Ground, Maryland, USA, 1975.

Partial Penetration of Annular Grooved Projectiles Impacting Ductile Metal Targets①

环形槽弹丸对韧性金属靶的半侵彻效应

Huang Qi②, Feng Shunshan, Lan Xuke, Song Qing, Zhou Tong, Dong Yongxiang

State Key Laboratory of Explosion Science and Technology, Beijing Institute of Technology, Beijing, 100081, China

Abstract: Changing and optimizing the projectile nose shape is an important way to achieve specific ballistic performance. One special ballistic performance is the embedding effect, which can achieve a delayed high - explosive reaction on the target surface. This embedding effect includes a rebound phase that is significantly different from the traditional penetration process. To better study embedment behavior, this study proposed a novel nose shape called an annular grooved projectile and defined its interaction process with the ductile metal plate as partial penetration. Specifically, we conducted a series of low velocity - ballistic tests in which these steel projectiles were used to strike 16 - mm - thick target plates made with 2024 - O aluminum alloy. We observed the dynamic evolution characteristics of this aluminum alloy near the impact craters and analyzed these characteristics by corresponding cross - sectional views and numerical simulations. The results indicated that the penetration resistance had a brief decrease that was influenced by its groove structure, but then it increased significantly - that is, the fluctuation of penetration resistance was affected by the irregular nose shape. Moreover, we visualized the distribution of the material in the groove and its inflow process through the rheology lines in microscopic tests and the highlighted mesh lines in simulations. The combination of these phenomena revealed the embedment mechanism of the annular grooved projectile and optimized the design of the groove shape to achieve a more firm embedment performance. The embedment was achieved primarily by the target material filled in the groove structure. Therefore, preventing the shear failure that occurred on the filling material was key to achieving this embedding effect.

© 2021 China Ordnance Society. Publishing services by Elsevier B. V. on behalf of KeAi Communications Co. Ltd. This is an open access article under the CC BY - NC - ND license (http://creativecommons.org/licenses/by - nc - nd/4.0/).

Keywords: Partial penetration; Embedment behavior; Ballistic impact; Annular grooved projectile (AGP); Microscopic experiments

1 Introduction

Semi - armor - piercing high - explosive incendiary (SAPHEI) [1] is a new concept ammunition, one of which includes annular grooved projectiles (AGPs) and that has obtained increasing attention for multifunctions in its firm embedment behavior with delayed reactions. AGPs not only can penetrate multilayer structures [2] like other traditional projectiles but also can effectively achieve embedment in the target without rebounding. In addition to this performance, the delayed high - explosive reaction could be more controllable and reliable. Therefore, the embedment process and characteristics of this novel penetrator are considerably important

---

① 原文发表于《Defence Technology》17 (2021) 1115 - 1125。

② 黄岐：工学博士，2016 年师从冯顺山教授，研究弹药对飞行甲板侵立机理及封锁效应，现工作单位：北京理工大学博士后。

and require further research.

In this series of embedment - related studies, we defined the embedment process as "partial penetration" to distinguish it from "classical penetration". Considerable research [3 - 6] of the latter has focused primarily on material properties of targets [7 - 9], ballistic limits of projectiles [10 - 12], or structure optimizations [13, 14] focusing on the nose shape. Among these past studies, a famous phase diagram for projectile impact targets summarized by Backman and Goldsmith [15] categorized different experimental phenomena, including perforation, embedment, and ricochet under various initial conditions. Gupta et al. [16, 17] investigated the effect of projectile nose shape, impact velocity, and target thickness on the deformation behavior of aluminum plates. Iqbal et al. [18] analyzed the failure mode of several double - nosed projectiles penetrating into the thin aluminum plates. Liu et al. [19] designed a double - nosed projectile and calculated the projectile's penetration depth using an equivalent nose - shape coefficient model. Deng et al. [20] analyzed the penetration performance of axisymmetric U - shape - nosed grooved projectiles on aluminum targets. Kpenyigba et al. [21] studied the effect of projectile nose shape on the ballistic resistance of interstitial - free steel targets. Han et al. [22, 23] discussed the effect of grooves on the double - nosed projectile during penetration. As for the role of groove, it generally is believed that it can provide space for the inflowing target material during the penetration process. Most ballistic experiments and numerical simulations, however, have concentrated on the perforation, and few studies have involved the embedment behavior of projectiles, particularly the partial penetration [24]. As one of the ballistic modes between projectile and target, partial penetration shows the embedment, rather than intact perforation or rebound, of certain projectiles under certain impact conditions, including impact velocities, target material properties, or groove shapes. This process involves several physical phenomena and is more complex than classical impacts; thus, partial penetration is considered to be an innovative study that is derived from the classic penetration question.

The nose shape of projectile has a significant influence on the penetration performance; to date, however, few studies have been conducted on the embedding mechanism of AGPs. In this study, we experimentally and numerically analyzed the flowing feature of a target made by ductile metal material close to the grooved outline and the deceleration characteristics of AGPs during partial penetration. This study obtained a better understanding of the embedding mechanism of AGP and achieved a more firm and reliable embedding effect. Accordingly, we based the projectile used in this study on the application of a certain SAPHEI, and we simplified its shape to a projectile with one annular groove on the head. We studied partial penetration through the way by which an AGP struck a ductile metal target at several velocities below its ballistic limit. We employed gas gun apparatus in ballistic impact tests to precisely control the initial velocity range of AGPs. We performed corresponding numerical simulations and microscopic experiments to observe the dynamic evolution characteristics of the target material around the crater. Therefore, we were able to clearly analyze the inflow process and distribution of the target material in a certain groove. To distinguish the target material inside and outside the groove, we defined the part that flowed into the groove as filling material. The results indicated that the plastic deformation of the filling material principally contributed to the embedding ability of the AGP. Furthermore, the results also revealed the embedment mechanism of the AGP impacting the ductile metal targets and optimized the design of the groove shape to achieve a more firm embedment.

2 Experimental Study

We conducted all ballistic impact tests in a ballistic laboratory at Beijing Institute of Technology, and we used a gas gun apparatus to launch the AGPs (Figure 1). The detailed description of the experimental setup and the function of corresponding facilities is available in our previous work [25]. We used a Phantom 710 high - speed camera operating at 44 000 fps to measure the impact velocities of AGPs accurately. We recorded high -

resolution videos of the partial penetration process.

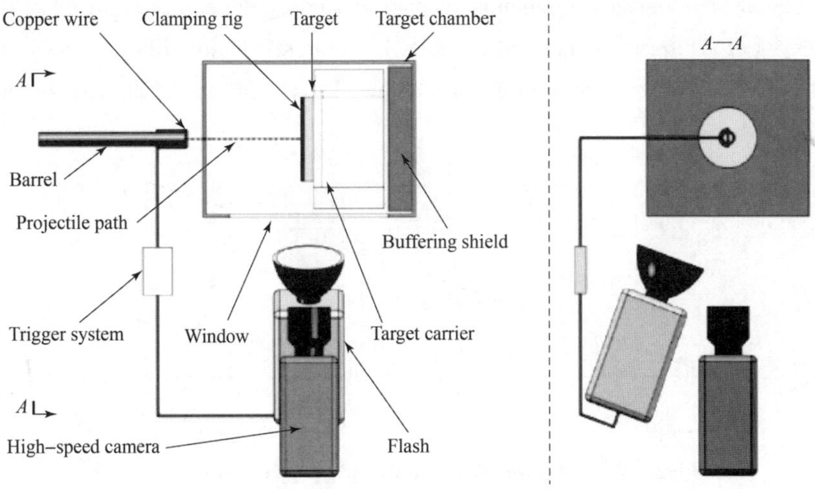

Figure 1 Gas gun apparatus.

We installed square targets on a C – shaped support plate by inserting four bolts in the target chamber. The support plate was welded to the bottom base structure to keep its stability during impact loading (Figure 2). Each shot was fired at the center of each target. The construction of a buffering shield located behind the target was used to stop and recover the possibly intact perforated projectile.

The dimensions of every target plates were 140 mm × 140 mm × 16 mm and all were made with 2024 – O aluminum alloy. The chemical compositions (in weight%) of this metal [26, 27] are shown in Table 1. To obtain ductile superiority instead of brittle fracture during partial penetration, the target plates underwent a heat treatment processes to achieve an annealed temper of O after production. Table 2 details the process. In the current tests, the aluminum alloy plates were impacted by AGPs shot at a variety of velocities. Cylindrical projectiles were 16 mm in diameter and had a nominal weight of 65 g (Figure 3). The nose length of each projectile was 1.5 times the diameter. The cylindrical cavity inside the projectile was designed for subsequentstretching tests and delayed explosion experiment, which we do not discuss in this study. The AGP material was AISI 1045 steel, which had an ultimate tensile strength of 625 MPa and was much higher than that of the target material. According to previous experiments [25], projectiles made with this relatively high – strength material experienced negligible inelastic deformation when penetrating 2024 – O aluminum alloy, and as such, they were considered to be rigid [28]. We used two – piece sabots made by nylon 66 to guarantee ideal acceleration conditions and reduce the yaw angle inside the firing barrel.

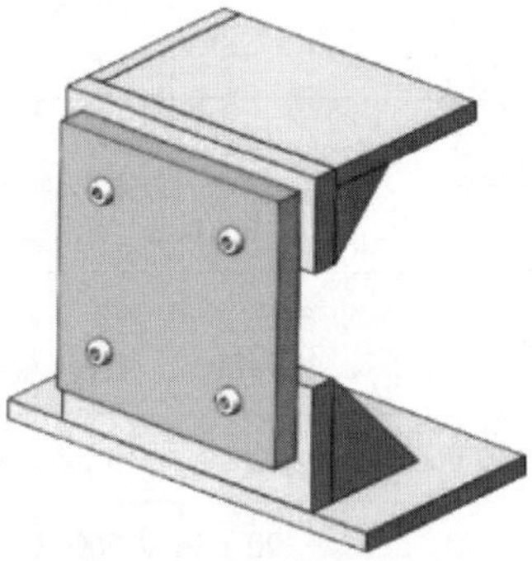

Figure 2 Target supports.

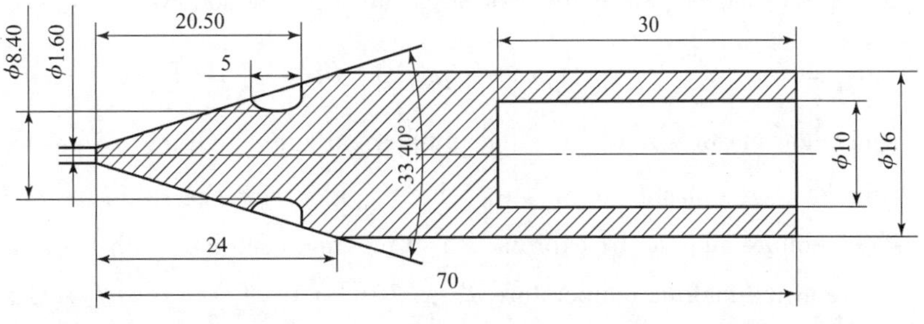

Figure 3 Dimensions of the AGP and corresponding conical projectile (dotted contour instead of the groove structure). All measures are in mm.

The annular groove structure originally was designed to provide sufficient room for the inflowing material. This structure was a possible obstacle during the rebound process caused by the elastic recovery of the target. In theory, AGP can contain grooves of arbitrarily shape and number. In contrast, to simplify the analysis, the AGP used in this study contained only one groove with a line – arc shape. Thus, the projectile was simplified into the three parts as shown in Figure 4.

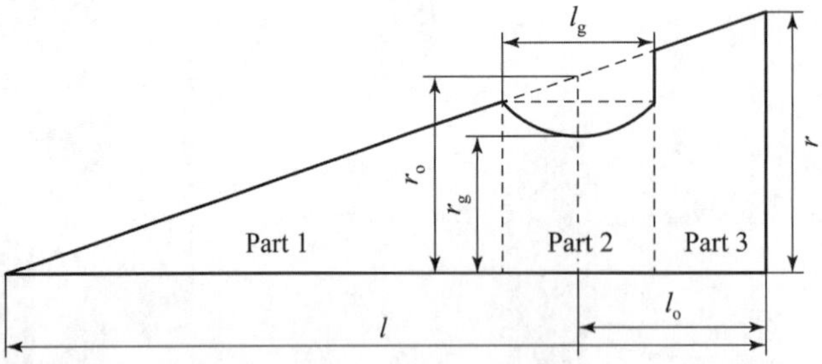

Figure 4　The simplified contour of an AGP nose shape.

The dimensions of a simplified annular groove structure can be characterized according to three main dimensionless parameters, namely, the groove relative position $L = l_0/l$, the relative width $W = l_g/l$, and the relative depth $H = (r_0 - r_g)/r$, where l_g, l_0, and r_g are the width, center position, and bottom diameter of the groove, respectively. In particular, we took the L, W, H in as 0.25, 0.2, and 0.225. Other shapes and sizes of grooved structures and their effects will be analyzed in further studies.

Table 1　Chemical components of 2024 aluminum alloy.

Al	Cu	Mg	Zn	Mn	Fe	Si	Cr	Ti
Balance	3.8 – 4.9	1.2 – 1.8	0.25	0.3 – 0.9	0.5	0.5	0.1	0.15

Table 2　Heat treatment processes of 2024 aluminum alloy to obtain the temper O [26]

Temper	Solutionizing	Cooling	Annealing/Artificial aging	Cooling
O	90 min at 560 ℃ (+5 ℃)	Water quench	24 h at 350 ℃ (+5 ℃)	Slow cooling

3　Finite Element Model

We modeled the target material according to a modified Johnson – Cook constitutive model. The quasi – static strain hardening was described by an extended Voce rule [29]. According to comprehensive research on the aluminum alloy behavior presented by Børvik [30 – 32], we modeled the constitutive behavior with the von Mises yield criterion, which we reasonably assumed to be isotropic, although the target material exhibited a little anisotropy [33]. In the current paper, the equivalent stress is written as follows:

$$\sigma_{eq} = \left(A + \sum_{i=1}^{2} Q_i\left(1 - \exp\left(-C_i\varepsilon_{eq}^{n}\right)\right)\right)\left(1 + \varepsilon_{eq}^{*}\right)^{C}\left(1 - T^{*m}\right) \tag{1}$$

where ε_{eq} is the equivalent plastic strain, A is the initial yield stress, Qi and Ci are hardening parameters, and $\varepsilon^{*}_{eq} = \varepsilon_{eq}/\varepsilon_0$ is a dimensionless plastic strain rate, where ε_0 is a reference strain rate defined by user [34]. $T*$ is the homologous temperature and can be expressed by a function combined with the absolute temperature T, the room temperature T_r, and the melting temperature T_m as $T* = (T - T_r)/(T_m - T_r)$. C and m are the model parameters related to the rate sensitivity and the thermal softening of the material, respectively [35]. The temperature change due to adiabatic heating is expressed as follows:

$$\Delta T = \int_0^{\varepsilon_{eq}} \chi \frac{\sigma_{eq} d\varepsilon_{eq}}{\rho C_p} \tag{2}$$

where ρ is the material density; χ is the Taylor – Quinney coefficient, which refers to the proportion of plastic work converted into heat; and C_p is the specific heat of target material. The Cockcroft and Latham (CL) fracture criterion [36] was used to model failure and can be expressed as follows:

$$W = \int_0^{\varepsilon_{eq}} \langle \sigma_1 \rangle \, d\varepsilon_{eq} \leqslant W_C \tag{3}$$

where σ_1 is the major principal stress, $\langle \sigma_1 \rangle = \sigma_1$ when $\sigma_1 \geqslant 0$, and $\langle \sigma_1 \rangle = 0$ when $\sigma_1 < 0$. Eq. (3) shows that failure will not occur when no tensile stresses are operating. W_c is the model constant about the value of W at failure. Note that, according to the material's anisotropic behavior and the uncertainty in the calibration of the CL criterion [26], W_c should not be considered as a material characteristic. To determine the constitutive relation and the failure criterion for 2024 – O aluminum alloy used in this study, we employed LS – DYNA, and the corresponding constants in the MJC model are given in Table 3.

Table 3 Main material constants for 2024 – O aluminum alloy target [25, 26, 37]

A/MPa	B/MPa	n	C	E/GPa	υ	ρ/(kg · m^{-3})
85	325	0.40	$8.3 \cdot 10^{-3}$	73.1	0.33	2780
C_p/(J · kg^{-1} K^{-1})	α/(K^{-1})	ε_0	χ	T_r/K	T_m/K	m
910	$2.3 \cdot 10^{-5}$	$5 \cdot 10^{-4}$	0.9	293	893	1.0

As mentioned, the hardened AGP made by high – strength steel experienced negligible deformation during ballistic tests. Furthermore, the dynamic evolution characteristics of ductile metal material during the partial penetration process is the major focal point in this study, so we assumed that the projectile was rigid in numerical simulations. The density, Young's modulus, and Poisson's ratio of the projectile material were considered as ρ = 7 850 kg/m^3, E = 203 GPa, and v = 0.29, respectively [25].

We performed all numerical simulations in this study using LS – DYNA 971, a widely used nonlinear finite element (FE) software program. Considering the axisymmetry in this ideal penetration simulation without yaw or obliquity, we modeled only half of the projectiles and plates. Moreover, we modeled the target plate as a smaller (50×50 mm^2) plate instead of a full – size plate (140×140 mm^2). Theoretically, two – dimensional (2D) simulations might have an acceptable deviation compared with three – dimensional (3D) simulations, which generally are used in classic penetration research. Previous studies [38], however, have indicated that 2D simulations exhibit a good agreement with realistic a phenomenon and could reduce the CPU running time significantly without overly affecting the results [26]. Therefore, we determined that this simplified method was reasonable and feasible in partial penetration studies.

Figure 5 shows the FE meshes that contained a refined mesh method used in the numerical simulations. The mesh size in the impact region (Zone I) was significantly smaller than that of other regions (Zone II). Earlier studies [39] have evaluated the mesh – size dependency and the effect of friction in this problem. The results have shown that an element size of $0.062\,5 \times 0.062\,5$ mm^2 is the ideal choice for the impact region when considering the compromise between calculation accuracy and CPU running time. Moreover, we ignored the effect of friction between the AGP and the target because changing the coefficient of friction from 0.5 to 0 resulted in a mere 4% reduction in depth of penetration (DOP). We applied an automatic – single – surface contact algorithm to describe the contact between the AGP and the target [40]. We assigned a target edge with an hourglass setting using the Flanagan – Belytschko stiffness [25] and fully clamped the target by limiting displacement, velocity, and acceleration in all directions.

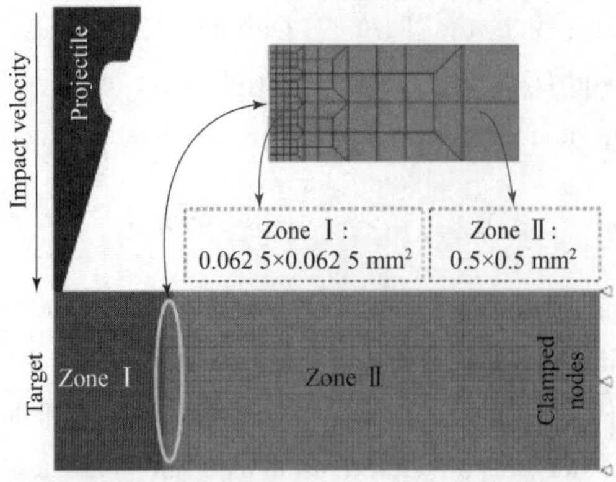

Figure 5 Fe meshes and corresponding refined mesh method of different regions used in simulations.

4 Results

The experimental data, including initial projectile mass m, impact velocity vi, rebound velocity v_r, initial kinetic energy E_i, and DOP, are listed in Table 4. In particular, we controlled the initial velocities of projectiles below the ballistic limit using the gas gun to prevent intact perforation.

Table 4 Partial penetration data of a 16 mm 2024 – O aluminum alloy plate impacted by AGPs

No.	Projectile type	m/g	$v_i/(\mathrm{m \cdot s^{-1}})$	$V_r/(\mathrm{m \cdot s^{-1}})$	$E_i/$	DOP/mm
1	Annular Grooved Projectiles (AGPs)	65. 15	91. 6	0	273. 32	15. 4
2		65. 01	109. 3	0	388. 32	19. 2
3		64. 98	114. 8	0	428. 19	20. 4
4		64. 98	123. 6	0	496. 35	21. 6
5		65. 07	132. 0	0	566. 89	22. 2
6		64. 67	140. 7	0	640. 12	23. 1
7		65. 18	147. 4	0	708. 08	23. 8
8		65. 00	157. 7	0	808. 25	25. 7
9		65. 02	185. 3	0	1 116. 27	32. 0

Typical high – speed video images of a projectile with 140. 7 m/s are shown in Figure 6. After a relatively long period (t > 3 ms) of observation, the projectile kept embedding in the target instead of rebounding. This phenomenon revealed that the AGP had an embedding ability at low – velocity impact. Photographs of three typical embedded AGPs are shown in Figure 7, and a significant difference in depth of penetration can be observed comparatively. As the initial speed increased, the global bending of the metal plate gradually became obvious when the impact velocity was higher than 140 m/s. Note that the projectile's rebound velocity in No. 1 was zero, but it was not firmly embedded because it fell off during manual shaking. Other AGPs (Nos. 2 – 7) also were shaken manually but did not fall from the target. This indicated that an unignorable cavity existed between the embedded AGPs and the target, which further explained that the annular groove structure was not filled completely by the ductile material. The remaining two AGPs (Nos. 8 – 9) were sufficiently firmly embedded and could not be shaken loose. This embedment firmness needs to be evaluated further and will be a topic of a future study.

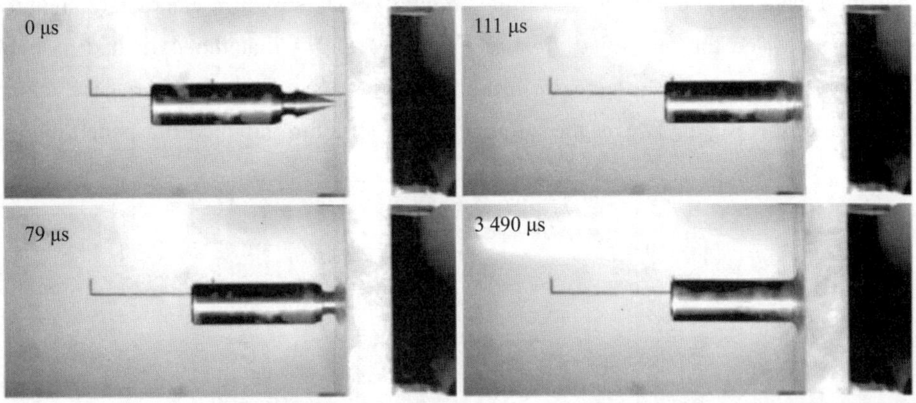

Figure 6　Typical illustrative images of a 16 mm 2024 – O aluminum alloy plate impacted by an AGP at 140. 7 m/s, captured by the high – speed camera.

Figure 7　Embedment situation of AGPs (Test No. 1/5/9); a significant difference in depth of penetration can be observed comparatively in the images from left to right.

Typical simulations of the partial penetration of 16 – mm – thick 2024 – O targets impacted by the AGPs at 185 m/s are shown in Figure 8. The impact velocity (185. 0 m/s) was the same in this test as that in test No. 9 (185. 3 m/s; given in Table 4) to facilitate comparison of results. The simulation captured some of the typical physical behaviors of high – ductile aluminum targets under low – velocity impact loading, such as petals and denting on the impact surface [41]. According to the brief observation of partial penetration, a visible cavity appeared between the annular groove and the target material in the early stages of penetration. With increasing DOP, the target material gradually filled the cavity when pressed by the upper surface (marked in Figure 12). This process also explains the reason why the von Mises stress of the target material inside the groove generally was higher than that of the surrounding area.

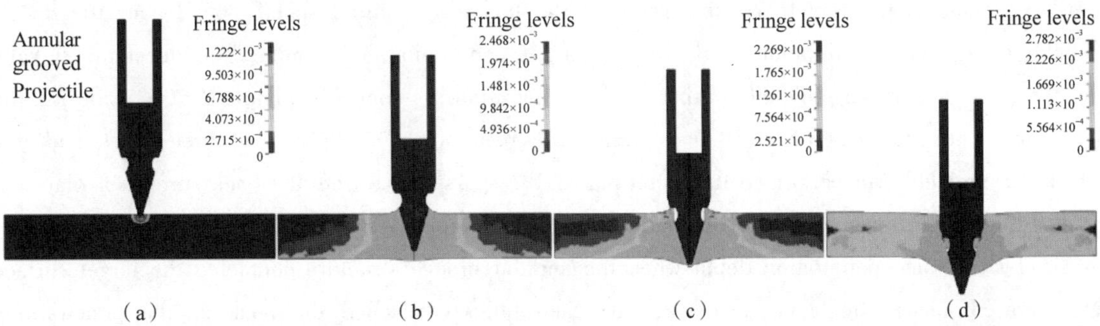

Figure 8　Penetration process of 16 mm 2024 – O aluminum alloy plate impacted by an AGP with an initial velocity of 185 m/s, where t refers to the time after impact.

(a) $t = 1$ μs; (b) $t = 74$ μs; (c) $t = 108$ μs; (d) $t = 260$ μs

The results from a large number of numerical simulations are represented by the dimensionless DOP versus impact velocity curves in Figure 9, and the experimental DOP data are given for comparison. Moreover, the horizontal dashed line represents the DOP at which the annular groove structure completely penetrated the plate. A short area under this line (see the magnified view in Figure 9) shows that the AGP had a slightly higher DOP because a cavity in the groove caused a decrease in the axial resistance. With the increasing impact velocity, the extrusion surface (marked in Figure 12) of the annular groove structure affected the penetration resistance.

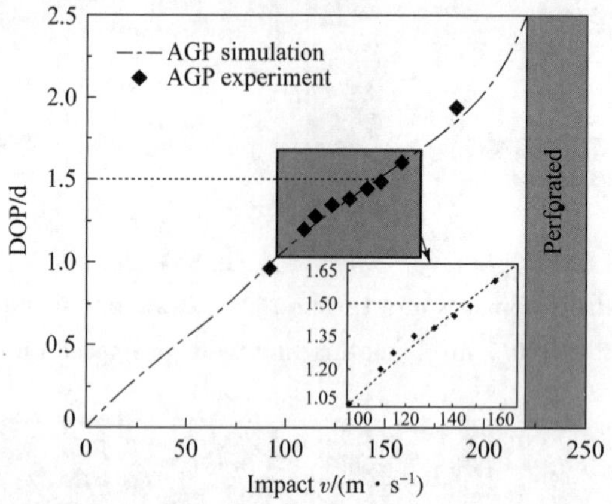

Figure 9 A comparison of DOP data between experimental and simulated results of a series of AGPs.

Generally, we achieved good agreement between the simulations and experiments, which indicated the feasibility and accuracy of the simulation model for partial penetration. The over – and under – deviations may have been caused by the thickness dependency and anisotropy of the material properties of 2024 – O aluminum alloy [9, 42]. These results, which were caused by a complex inflowing process of the ductile metal material, remain acceptable when considering the limitations of the constitutive relation and fracture criterion [43].

We selected a series of tests (No. 4/6/9 in Table 4) and conducted corresponding simulations to analyze the deceleration characteristics (deceleration against time and penetration depth curves is shown in Figure 10). Figure 10 clearly shows that the partial penetration could be divided into two distinct stages according to the positive and negative deceleration value: the invasion phase and the rebound phase. The time information in the impact tests shown in Figure 6 revealed that the rebound duration was much longer than the invasion process (i. e. , only the first 450 μs of data). Figure 10 (a) shows the invasion stage and one complete rebound phase. In general, the deceleration curves increased gradually in the early period, but an ephemeral descending phase occurred in which the annular groove structure barely contacted the target. The partial penetration situation at this moment is indicated by the inset in Figure 10. The deceleration value then rapidly reached its peak after the upper surface contacted the target. We observed that the higher the initial speed, the longer the deceleration lasted at the peak, but the duration of the entire partial penetration was similar. The deceleration value also reflected the resistance of the projectile during partial penetration. As shown in Figure 10 (b), during the early period of partial penetration, the resistance against the penetration depth in the three cases had a similar trend. After the projectile tip perforated the metal plate, the sparse wave from the back surface of plate caused a significant difference in resistance. The fluctuation of force maintained a good correlation with the shape of the warhead. Under the same penetration depth when the annular groove structure contacted the target surface, the resistance showed almost the same decrease, and then quickly reached the peak as the penetration depth increased.

Moreover, the deceleration curve became negative and fluctuated at the rebound phase of partial penetration. This was significantly different than the traditional penetration process. According to the analysis of the simulation and experiment results, the negative deceleration was caused primarily by the interaction between the filling material in the groove and the elastic recovery of the target during the rebound phase [39], which limited

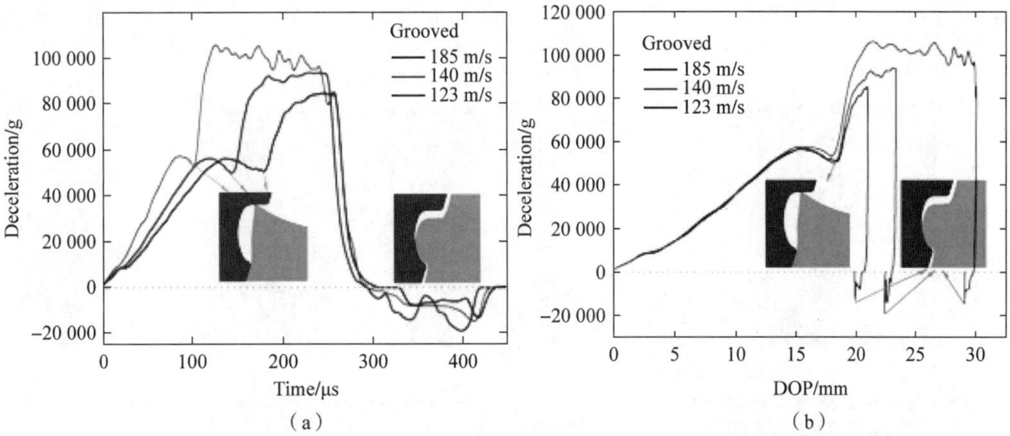

Figure 10　Deceleration history of AGPs at three different initial velocities.

(a) Deceleration against time; (b) Deceleration against penetration depth

the reverse movement of the projectile and thus achieved the firm embedment behavior. This situation also is shown in the insert in Figure 10. In addition, as the initial velocity increased, the negative peak did not show a visible difference. Because the rebound was caused by the elastic recovery of the ductile metal plate, we considered that during partial penetration, the AGP's reverse deceleration reached a maximum value and was related to the material properties of the metal plate. These curves showed the characteristics of the partial penetration process and further provided an intuitive explanation for the embedment mechanism of the AGP.

5　Discussion

We selected several cross - section images of AGP partial penetration tests to exhibit a partial penetration process. We arranged these images by ascending impact velocity (Figure 11). In general, the target material was in close contact with the projectile and eventually filled the annular groove structure almost completely. The gap in the groove explained why the AGP could be manually shaken but remained firmly embedded in the target.

5.1　Inflow process of the target material

The magnified views in Figure 12 show, in detail, the deformation of three 16 - mm - thick 2024 - O aluminum alloy plates after ballistic impact tests, including the evolutionary characteristics of the material and the changes in cavity volume. With increasing DOP, the target material was pressured by the extrusion surface (red line) so that the target material gradually filled from the top to the bottom along the contour of the groove. Because the embedding ability was caused primarily by the plastic deformation of the filling material [39], a relatively blunt extrusion surface on the groove was necessary for the target material to flow in. This phenomenon indicated that the groove shape was an important factor to achieve embedment behavior.

A comparison between the simulation and experiment images is shown in Figure 13. The bottom labels, Nos. 2 -9, correspond to the test numbers in Table 4. The simulation images show the main penetration process of AGP with a 185 m/s impact velocity. These images show that the AGPs did not have a subsequent rebound process, and the experimental images show the various final states of the target subjected to different impacts. The good agreement of the shape of the filling material between numerical and experimental results illustrated that the various experiments could approximately reflect the dynamic process of partial penetration. Note the irregular deformation of the filling material near the lower surface in experimental images. This deformation was caused by an extrusion during AGP reverse movement, but the simulations did not contain the subsequent rebound process. During this period, the filling material acted on the lower surface, and the plasticity deformation energy and friction work offset the rebound kinetic energy of the projectile. This phenomenon revealed the embedment mechanism of AGP.

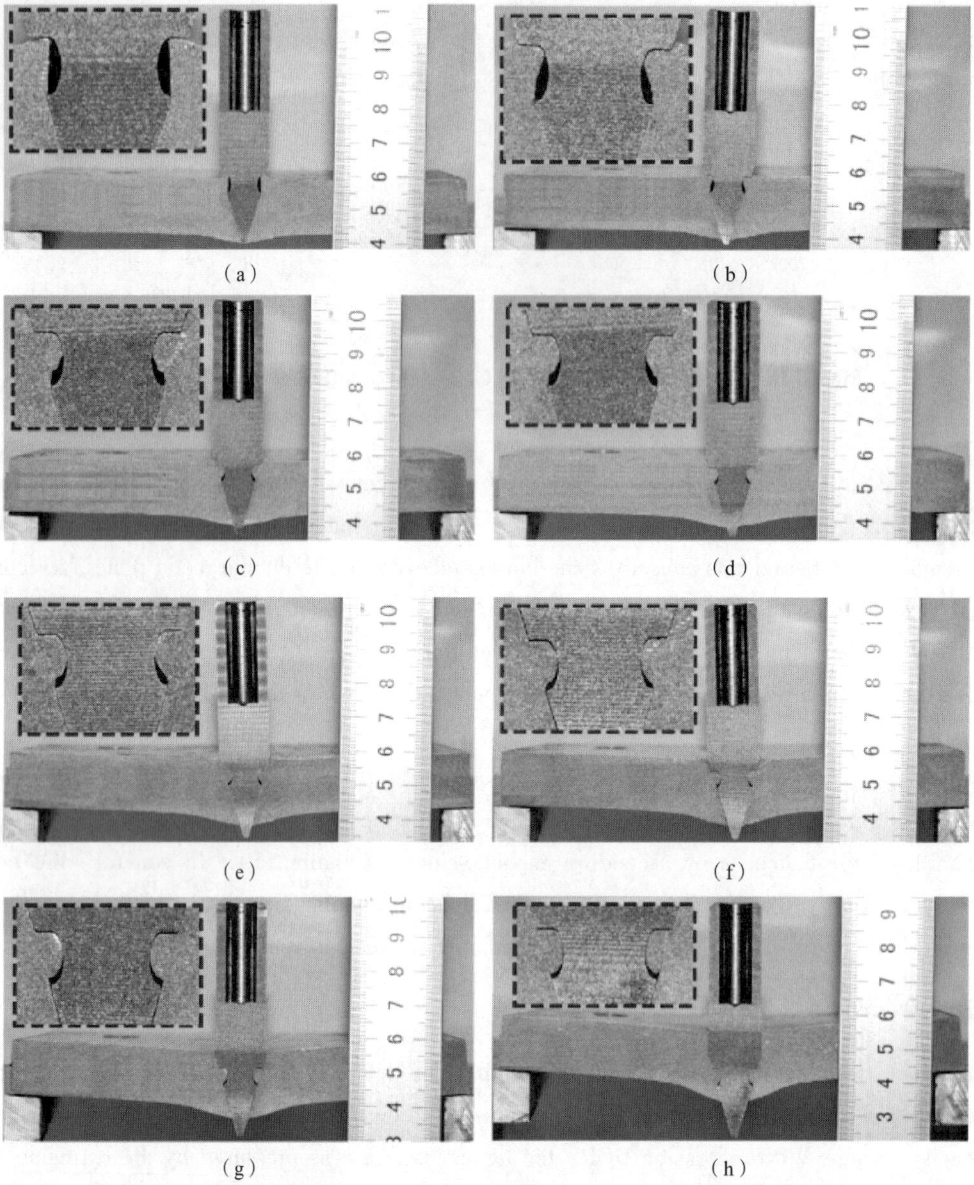

Figure 11　Deformation and crater profiles of a variety of 16 – mm – thick 2024 – O aluminum alloy plates.

(a) No. 2 109. 3 m/s; (b) No. 3 114. 8 m/s; (c) No. 4 123. 6 m/s; (d) No. 5 132. 0 m/s; (e) No. 6 140. 7 m/s; (f) No. 7 147. 4 m/s; (g) No. 8 157. 7 m/s; (h) No. 9 185. 3 m/s

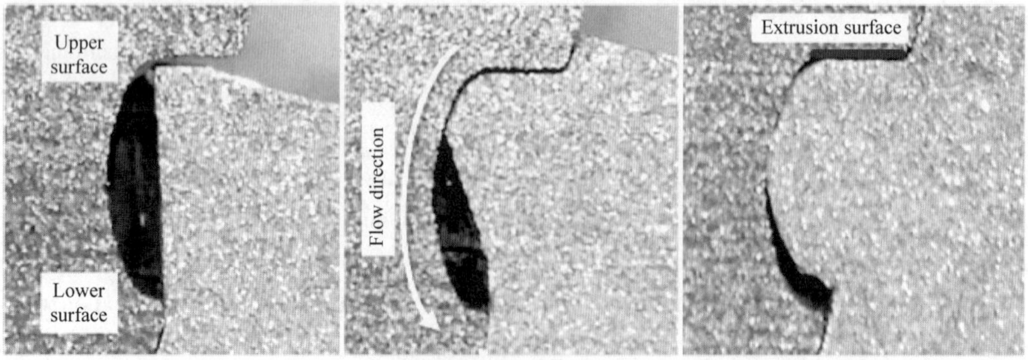

Figure 12　The target material flows from the upper surface to the lower surface.

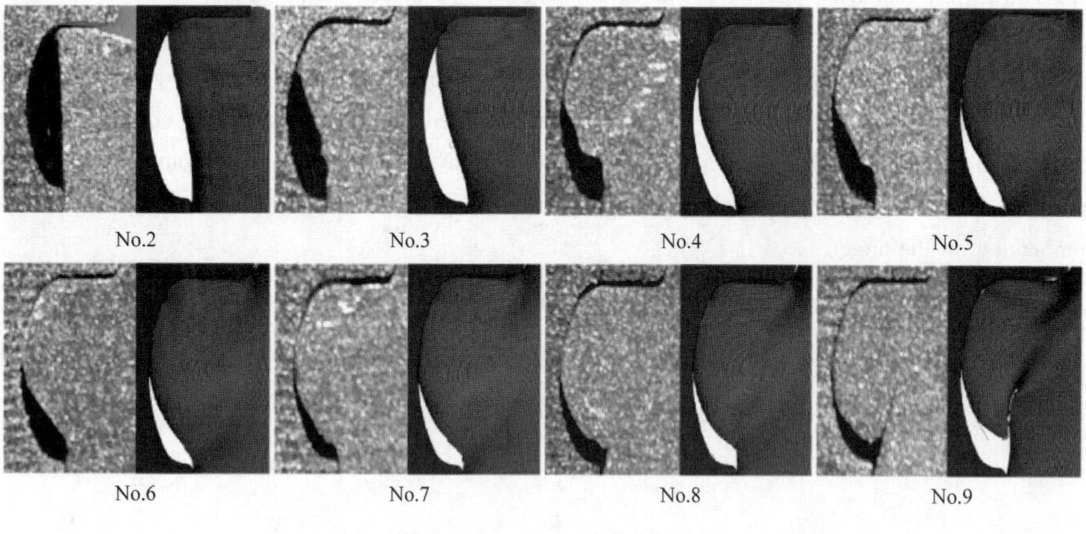

Figure 13 A comparison of the filling material in the groove between various experiments and one continuous simulation with an impact velocity v = 185 m/s.

Moreover, the experimental and numerical results in No. 9 both showed that small cracks occurred at the ends of the groove. These cracks represented that the shear fracture of the filling material started on a relatively high DOP. If the filling material was sheared completely and the projectile could not penetrate the plate, the entire bound resistance came only from friction. In this case, the embedded firmness decreased to a level similar to that of a conical projectile.

5.2 Inflow Volume of the Target Material

We extracted the contours of the cavity in the groove using Photoshop software (Figure 14). The cavity volume V_c can be calculated by integrating the contour shape. We took the moment when the target material contacted the upper surface of the groove as the initial time to record the inflow process, as shown by the No. 2 test. At this moment, the material was not affected by the groove and the cavity volume was defined as V_{c0}. With the continuous inflow of target material, the instantaneous cavity volume V_{ci} continued to decrease until it reached a minimum value. Therefore, the instantaneous volume of the inflow material V_{fi} could be calculated as $V_{fi} = V_{c0} - V_{ci}$. We defined the fill ratio as V_{ti}/V_{ci}, which could describe the real - time situation of the filling material. Moreover, those parameters (given in Table 5) exhibited the relationship between the inflow situation of target material and the DOP of the AGP.

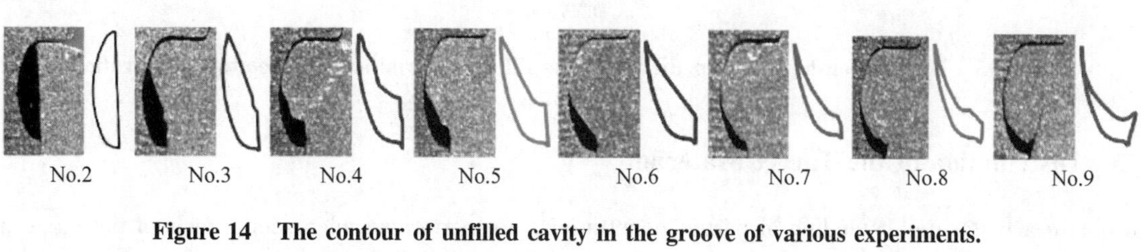

Figure 14 The contour of unfilled cavity in the groove of various experiments.

The experimental data shown in Table 5 can be fitted and are shown in Figure 15. The fitting curve showed that in the early stage of partial penetration, the volume of the filling material and the DOP of the AGP had an approximately linear relationship, which was related to the effect of the extrusion surface. As the DOP increased, the fill rate reached a critical value. This value generally was affected by several factors, including elastic modulus and failure of target material, volume and extrusion surface shape of groove structure, and impact velocity of the projectile. The fill rate could be defined as one of the indicators to quantify the embedment firmness and this will be discussed in detail in future research. When the kinetic energy of the AGP was not sufficient to increase its DOP, and the projectile moved in reverse because of the overall elastic recovery of the target. At this stage, the filling material acted on a lower surface of the groove and absorbed the remaining kinetic energy of the projectile through its plastic and elastic deformation. As a result, the groove was difficult to be completely filled with the target material after the experiments, which indicated that the projectile could be shaken manually but remain firmly embedded on the target.

Table 5 The relationship between the volume of filling material in the groove and depth of penetration.

No.	DOP/mm	$V_c/(mm^{-3})$	$V_f/(mm^{-3})$	Fill ratio %
2	19.17	122.76	0.00	0.00
3	20.35	69.72	53.04	43.21
4	21.57	46.71	76.05	61.95
5	22.20	37.16	85.61	69.74
6	23.14	33.50	89.26	72.71
7	23.84	18.16	104.60	85.21
8	25.73	23.16	99.60	81.13
9	31.97	22.00	100.76	82.08

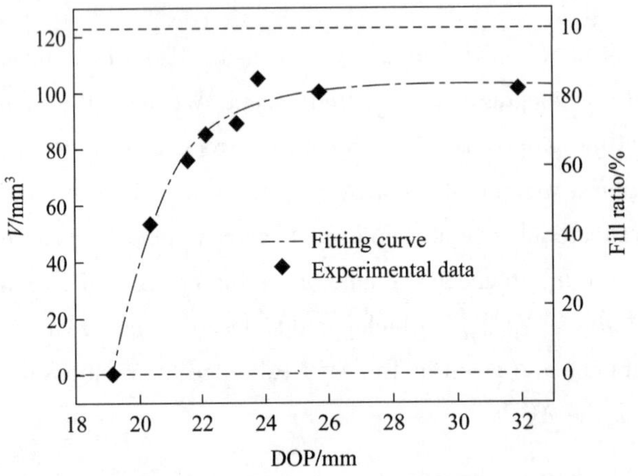

Figure 15 The relationship between the volume of filling material and the depth of penetration.

5.3 Distribution of the Target Material

To obtain a better understanding of partial penetration, we performed microscopic tests of the target material to analyze the dynamic evolution characteristics. We employed a chemical etchant called Keller reagent [25] in microscopic tests to obtain a clear observation. According to the different DOP, we selected three typical specimens of targets near the crater from tests No. 6, No. 8, and No. 9. The cross - sections of these three specimens, which were immersed in the Keller reagent, are shown in Figure 16. Corresponding microscopic images are shown in Figure 17.

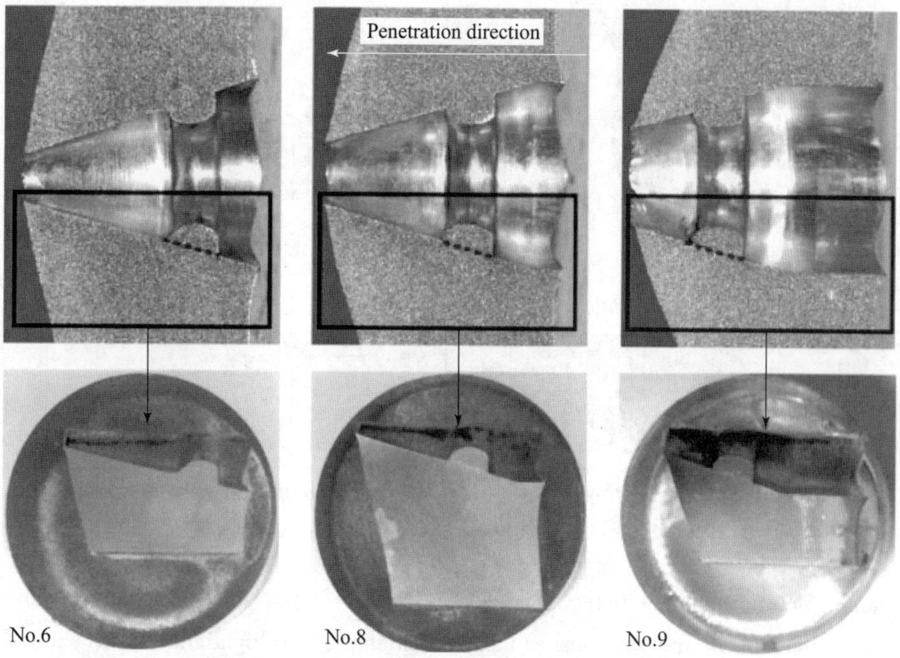

Figure 16　Cross－section specimen（bottom）of three 2024－O aluminum alloy plates（top）in tests No. 6，No. 8，and No. 9.

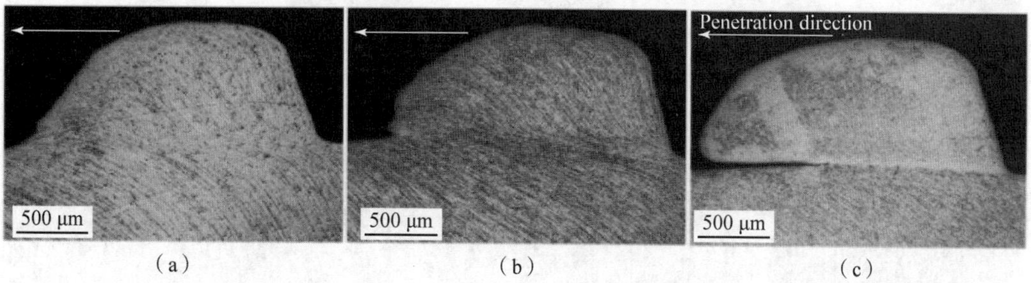

Figure 17　Grain microscopic images of three specimens in tests No. 6，No. 8，and No. 9.
(a) No. 6；(b) No. 8；(c) No. 9

The "rheology lines" of the target material can be observed clearly in Figure 17. These lines showed an "S－bent" shape around the interface inside and outside the groove and were more pronounced as the DOP increased. In tests No. 6 and No. 8，the lines were connected inside and outside the groove，whereas in test No. 9 they were not. Also，cracks appeared in the penetration direction and the interface was significantly brighter compared with the surrounding region. This phenomenon suggested that this part of the material underwent a large deformation of stretching and shearing，which caused a higher temperature on the interface during penetration. Furthermore，a large amount of heat was generated in the target material inside the groove on a relatively large DOP penetration，which resulted in a refinement of material grain and inconspicuous rheology lines.

The related simulation showed similar results and enabled us to analyze the dynamic processes of partial penetration. We marked six horizontal lines through the highlighted meshes in the target，which are shown in Figure 18. Because the groove width was 5 mm，the distance between the mesh lines and the target surface was set to 0.5，1，2，3，4，and 5 mm.

The bending，stretching，and shear deformation of the six highlighted lines，shown in Figure 19，further

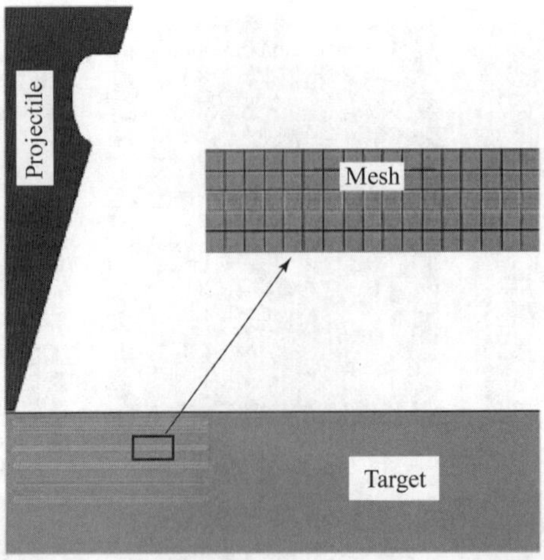

Figure 18　Highlighted meshes mark the six horizontal lines, which are 0.5, 1, 2, 3, 4, and 5 mm away from the target surface.

exhibited the dynamic evolution characteristics of the target material impacted by an AGP with an initial velocity of 185 m/s. Although we deleted the fracture meshes in the simulation, the highlighted lines showed good agreement with the rheological lines observed in the microscopic experiment. The highlighted lines illustrate the distribution of the material in the groove and provide an intuitive explanation for the partial penetration on the metal target.

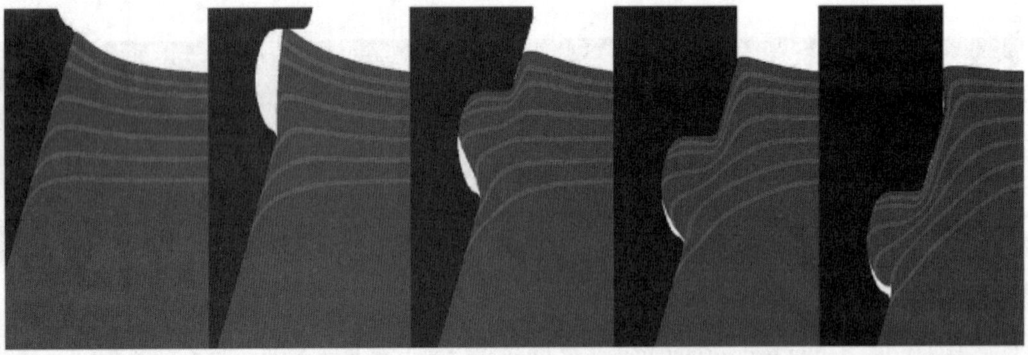

Figure 19　Dynamic evolution characteristics of target material of a 16 - mm - thick 2024 - O aluminum alloy plate impacted by an AGP with an impact velocity of 185 m/s, where t provides the time after impact.

Moreover, the material deformation reflected the fact that shear failure would happen at a relatively large DOP so that a relatively high - velocity impact should not be considered in partial penetration.

6　Conclusions

To examine the embedment behavior under low - velocity impact, we investigated the partial penetration of AGPs impacting medium - thickness 2024 - O aluminum alloy plates under corresponding ballistic velocities through a series of ballistic tests, numerical simulations, and corresponding microscopic experiments. The results showed how an annular groove structure affected deceleration and rebound behavior when the initial velocity of projectile was below its ballistic limit, which helped to achieve the embedment phenomenon. Overall, the following conclusions of partial penetration were drawn:

(1) During the partial penetration of AGPs impacting ductile metal targets, the target material gradually filled from the top to the bottom along the contour of the groove because of the compression of the grooved upper surface. This result indicated that the grooved shape was an important factor to achieve this embedment behavior.

(2) The penetration resistance during partial penetration briefly decreased when the grooved structure

contacted the target surface, and then it increased significantly to the maximum. This result indicated that the fluctuation of penetration resistance was affected by the special nose shape.

(3) The bottom filling material underwent plastic deformation when the AGP moved in reverse because of the overall elastic recovery of the target. For a certain embedded AGP, the peak value of deceleration during the rebound phase was related primarily to the target condition instead of the projectile's initial velocity.

(4) The distribution of the material in the groove and its inflow process could be visualized clearly through the rheology lines in microscopic tests and the highlighted mesh lines in the simulations. The results indicated that the embedment was achieved mainly by the target material filled in the groove structure; therefore, preventing shear failure from occurring on the filling material was key to achieving the embedment behavior.

The firmness of embedment could be characterized by the filling situation of the target material in the groove, which was closely related to many factors, including the material properties of the target, the shape of the groove, the impact velocity, and the DOP. Further numerical and experimental research should be carried out, however, to support these findings.

Declaration of Competing Interest

The authors declare that they have no conflict of interest.

Acknowledgments

The financial support of this research is from the National Natural Science Foundation of China (NSFC) [No. 11472053 and 11872121].

References

[1] Nammo ammunition handbook. second ed. 2014.

[2] LAN X, FENG S, HUANG Q, et al. A comparative study of blast resistance of cylindrical sandwich panels with aluminum foam and auxetic honeycomb cores [J]. Aero Sci Technol, 2019, 87: 37-47.

[3] BORVIK T, LANGSETH M, HOPPERSTAD O S, et al. Ballistic penetration of steel plates [J]. Int J Impact Eng, 1999, 22: 855-85.

[4] CORBETT G G, REID S R, JOHNSON W. Impact loading of plates and shells by free-flying projectiles: a review [J]. Int J Impact Eng, 1996, 18: 141-230.

[5] GOLDSMITH W. Non-ideal projectile impact on targets [J]. Int J Impact Eng, 1999, 22: 95-395.

[6] PANG C, HE Y, SHEN X, et al. Experimental investigation on penetration of grooved projectiles into concrete targets [J]. Acta armamentarii, 2015, 36: 46-52.

[7] BORVIK T, FORRESTAL M J, WARREN T L. Perforation of 5083-H116 aluminum armor plates with ogive-nose rods and 7.62 mm APM2 bullets [J]. Exp Mech, 2010, 50: 969-78.

[8] BORVIK T, CLAUSEN A H, HOPPERSTAD O S, et al. Perforation of AA5083-H116 aluminium plates with conical-nose steel projectiles-experimental study [J]. Int J Impact Eng, 2004, 30: 367-84.

[9] BORVIK T, HOPPERSTAD O S, PEDERSEN K O. Quasi-brittle fracture during structural impact of AA7075-T651 aluminium plates [J]. Int J Impact Eng, 2010, 37: 537-51.

[10] LI Q M, TONG D J. Perforation thickness and ballistic limit of concrete target subjected to rigid projectile impact [J]. J Eng Mech, 2003, 129: 1083-91.

[11] ZHOU D W, STRONGE W J. Ballistic limit for oblique impact of thin sandwich panels and spaced plates [J]. Int J Impact Eng, 2008, 35: 1339-54.

斜截头弹体入水的弹道特性①

邵志宇[1,2]，伍思宇[1,2]②，曹苗苗[1,2]，冯顺山[1,2]

（1. 北京理工大学机电学院，北京 100081；
2. 北京理工大学爆炸科学与技术国家重点实验室，北京 100081）

摘　要：斜截头弹体入水会发生姿态偏转和弹道弯曲。本文从理论上分析了弹丸在水中的偏转过程，得出了斜截头弹体姿态偏转方程。利用 AUTODYN 软件对斜截头弹体入水过程进行了数值模拟，基于模拟结果提出了描述弹体质心轨迹的经验公式，并通过试验验证了其准确性。设计了炮射垂直入水试验，利用高速摄像机记录了弹丸的撞击速度和运动过程，该试验验证了理论分析得出的姿态偏转方程，并验证了数值模拟和基于数值模拟得出的质心弹道经验公式的预报精度。

关键词：流体力学；斜截头弹体；入水；曲线弹道；姿态偏转

Water - entry trajectory of asymmetric projectile

Shao Zhiyu[1,2], Wu Siyu[1,2], Cao Miaomiao[1,2], Feng Shunshan[1,2]

(1. School of Mechatronical Engineering, Beijing Institute of Technology, Beijing 100081, China
2. State Key Laboratory of Explosion Science and Technology, Beijing Institute of Technology, Beijing 100081, China)

Abstract: The trajectory of the projectile with oblique nose is studied. The deflecting process of the projectile in water is theoretically analyzed. Finite element software AUTODYN is used to simulate the water entry of projectile with oblique nose, Based on the simulation results, an empirical formula is proposed to describe the trajectory of projectile mass center accurately, and its accuracy is verified by experiments. Verification tests were carried out, the impact velocity and motion process were recorded by high speed camera. The prediction accuracy of numerical simulation and empirical formula of centroid trajectory based on numerical simulation is verified

Keywords: fluid mechanics; oblique nose; water - entry; bend trajectory; attitude deflection

0　引言

随着智能水中弹药的发展，弹药开始起到侦查、识别、封锁等作用，抗冲击能力变弱。而弹药作用的有些水域深度较小，有些弹药为了提高突防能力，入水速度较快。传统的对称型头部的弹体高速入水时形成空泡，与水的接触面较小，减速较慢，在浅水域容易撞击水底。非对称型头部产生的弯曲弹道可以较好地满足入水减速的要求。

对不同形状的对称头部弹体入水的研究已经比较深入，如郭子涛等[1,3]研究了高速弹丸水平撞击水面的过程以及弹丸头部形状对速度衰减和相关力系数的影响。Mirzaei 等[4]讨论了高速弹丸入水的空化现象，建立了尾部撞击空泡壁时弹丸在空泡内运动的理论动力学模型。伍思宇等[11]则通过试验和数值模拟研究了对称弹体入水过程中的缓冲问题。魏卓慧[5]等理论分析了刚性截锥形弹体入水速度、弹体头部参数和沾湿因子对入水冲击载荷的影响。从冯・卡门[7]在 1929 年提出刚体在入水过程中作用在刚体上的

---

① 原文发表于《兵工学报》2022。

② 伍思宇：工学博士，2017 年师从冯顺山教授，研究空投弹药入水冲击防护及弹道特性，现工作单位：中国万宝工程有限公司。

冲击力的数学解后，产生了对入水问题的大量研究。而对于头部非对称的偏头弹的相关研究较少，Shams A 等[8]通过粒子图像测速实验得出了非对称楔形结构体入水的特性，并得出了水动力载荷的变化规律。王云等[9]通过高速入水试验研究了椭圆斜截头弹体入水弹道的特性，华扬等[10]也做了类似研究，其拍摄的弯曲弹道可为高速入水武器的研究提供参考。

目前，对于非对称弹体的理论研究还比较少，其形成的弯曲弹道并没有很好的理论模型来进行分析和预测。本文从理论出发，通过力学分析得出了斜截头弹体在入水过程中姿态偏转的数学模型；基于数值模拟提出了精准描述斜截头弹体入水弹道的经验公式；通过试验对姿态偏转数学模型、数值模拟和入水弹道经验公式进行了验证。

1 斜截头弹体垂直入水力学分析

图 1 为斜截头弹体的剖面图，头部斜度定义为 $\psi = c/d$。在入水初始阶段的受力情况如图 2 所示。由于所受到的力 F_1 不通过质心，其分力 F_3 会产生旋转力矩 M，F_2 则为阻力。由于高速入水会产生空泡，此时弹体与水的接触面积很小，因此受力也很小。图 2 中，弹体只受到头部的水动力的影响。水动力 F_1 由牛顿第二定律可得为

$$F_1(t) = \frac{1}{2}\rho_w A_0 C_d v_p{}^2(t) \tag{1}$$

式中 F_1——水动力；

ρ_w——液体密度；

A_0——沾湿横截面面积；

C_d——力系数；

v_p——侵水瞬时速度，是时间 t 的函数；

F_2 和 F_3——F_1 的分力；

F_2——水平阻力，它降低了弹丸的水平速度，而 F_3 则提供了旋转力矩，使弹丸旋转，从而失去稳定性：

$$F_3(t) = F_1(t)\cos\alpha(t) \tag{2}$$

式中，α 为 F_1 和 F_3 的夹角。

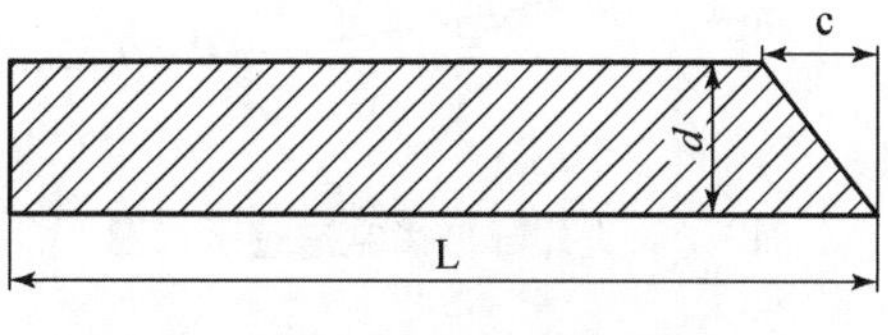

图 1 斜截头弹体的剖面图

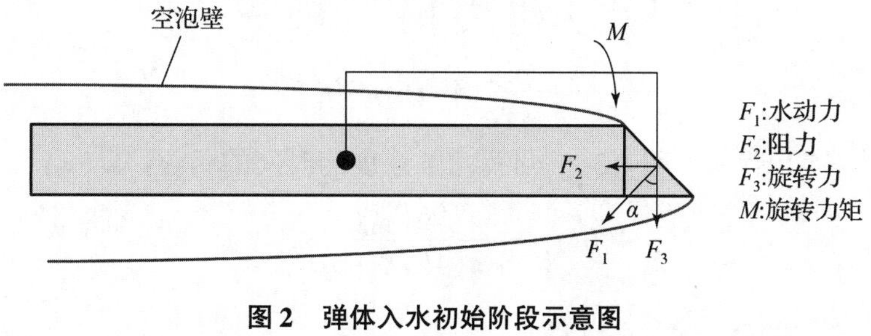

图 2 弹体入水初始阶段示意图

当弹丸旋转时，垂直于弹丸轴线的分量产生旋转力矩，导致旋转加速。角加速度与合扭矩之间的关系为

$$\frac{\mathrm{d}\omega}{\mathrm{d}t} = \beta = \frac{M}{I} \tag{3}$$

式中 ω——角速度；

β——角加速度；

M 和 I——合旋转力矩和转动惯量。

由于本文所研究的弹体长径比较大，非对称头部造成的质量损失相比整个弹体质量非常小，弹体可以视为圆柱体。因此，质心位于几何图形的中心，弹丸的姿态偏差可以看作是围绕质心旋转。在式（3）中，合力扭矩可以写成

$$M(t) = F_3(t) \times \frac{1}{2}l \tag{4}$$

式中，l 为弹丸长度。图 3 为转动惯量计算示意图，将弹丸视为绕 Z 轴旋转的刚体。因此，需要确定 Z 轴的转动惯量。转动惯量方程为

$$I = \iiint_V r^2 \mathrm{d}m = \iiint_V r^2 \rho \mathrm{d}V \tag{5}$$

式中 V——入水体体积；

r——物体到旋转轴的距离；

m——质量；

ρ——入水体密度。

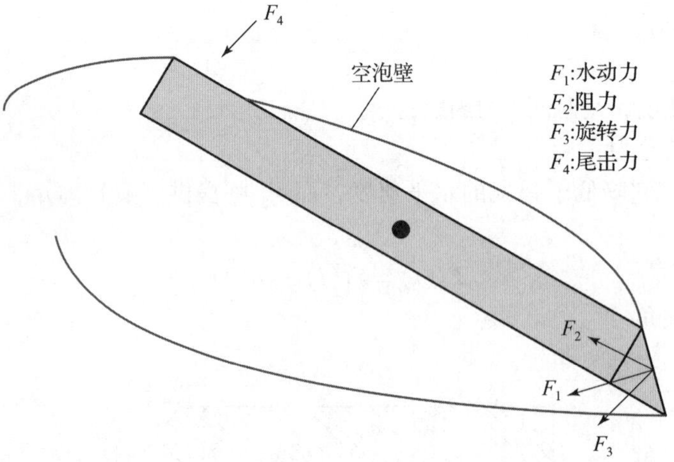

图 3　弹丸尾部碰撞空泡壁示意图

Z 轴的转动惯量为

$$I_z = \int_{-\frac{l}{2}}^{\frac{l}{2}} \mathrm{d}x \int_{-\frac{D}{2}}^{\frac{D}{2}} 2\rho(x^2 + y^2) \sqrt{\left(\frac{D}{2}\right)^2 - y^2} \mathrm{d}y \tag{6}$$

$$I_z = \frac{\rho l \pi}{192}(4l^2D^2 + 3D^4) \tag{7}$$

将方程（4）和式（7）代入方程（3），可得角加速度随时间的函数：

$$\beta(t) = \frac{96F_3(t)}{\rho\pi(4l^2D^2 + 3D^4)} \tag{8}$$

定义长径比为 $\delta = \dfrac{l}{D}$，方程（8）可以写成

$$\beta(t) = \frac{96F_3(t)}{\rho\pi D^4(4\delta^2 + 3)} = \frac{48\rho_w A_0 C_d v_p{}^2(t)\cos\alpha(t)}{\rho\pi D^4(4\delta^2 + 3)} \tag{9}$$

角速度表达式为

$$\omega(t) = \int\beta(t)\,\mathrm{d}t \tag{10}$$

以上计算分析了弹丸的姿态偏转，在弹丸姿态发生偏转后，弹丸失稳，尾部撞击空泡壁，导致产生尾流，如图3所示。随后弹丸受到 F_4 的作用，对弹体产生反向扭矩，减慢弹体角速度和水平速度，将 F_4 引入式（4）中可得

$$M(t) = [F_3(t) - F_4(t)] \times \frac{1}{2}l \tag{11}$$

将式（11）代入式（9）得

$$\beta(t) = \frac{96[F_3(t) - F_4(t)]}{\rho\pi D^4(4\delta^2 + 3)} \tag{12}$$

F_4 同时使弹丸向垂直方向移动，进而发生弹道偏转。

根据式（9）、式（10）和式（12）可知，偏转角速度与阻力系数 C_d 和头部斜度角 α 有关，而阻力系数 C_d 和方向角 α 是由头部形状、侵彻速度 v_p、弹体密度 ρ 和长径比 δ 决定的。从式（9）可以总结出：侵彻速度越大，偏转越快；密度、长径比越大，偏转越慢。

2　数值模拟

用 AUTODYN 的显式有限元软件进行数值模拟。采用二维数值模型进行入水计算，使用 SPH 算法对入水过程进行模拟，SPH（光滑粒子流体动力学）水域加上限制运动边界，模拟水箱以便于通过试验进行校核。采用拉格朗日网格法模拟弹体，研究重点是弹体头部形状对弹道和姿态的影响，因此弹体被设置为刚体，模型如图4所示。弹体质量为65～1 000 g，长径比为8～20，它们的初始速度为50～150 m/s，并且在模拟中考虑了8种斜度的头部（ψ = 0.2，0.3，0.4，0.5，0.8，1.0，1.4，2）。在控制变量的基础上研究以上了4种参数对弹体偏转的影响，由于控制变量时进行模拟的组别过多，表1仅列出了数值模拟的初始条件范围。忽略空气阻力和重力，将数据库中的材料参数用于弹体和水。

表1　数值模拟初始条件

质量/g	直径/mm	长度/mm	初速度/(m·s^{-1})	头部斜度（ψ）
65～1 000	20	160～400	50～150	0.2～2

图4和图5展示的是质量为65 g、长径比为8（直径20 mm，长度160 mm）、头部系数 ψ = 0.2 的弹丸以50 m/s入水的模型图和入水过程。入水初期发生偏转，出现空泡，随着侵彻深度的增加，偏转角增大，弹丸尾部撞击空泡壁，尾部的拍击使弹丸垂直移动。

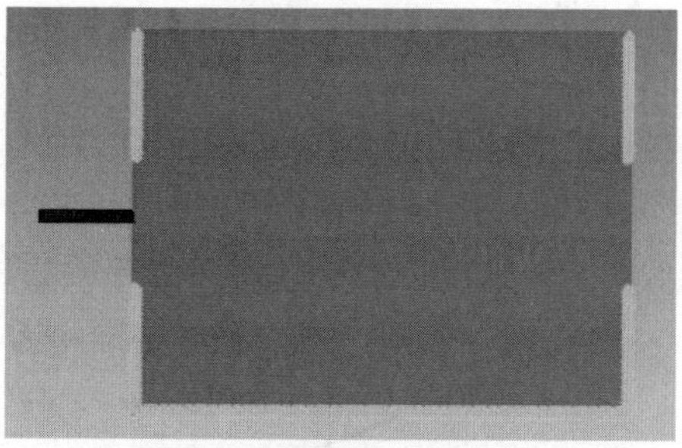

图4　数值模拟模型

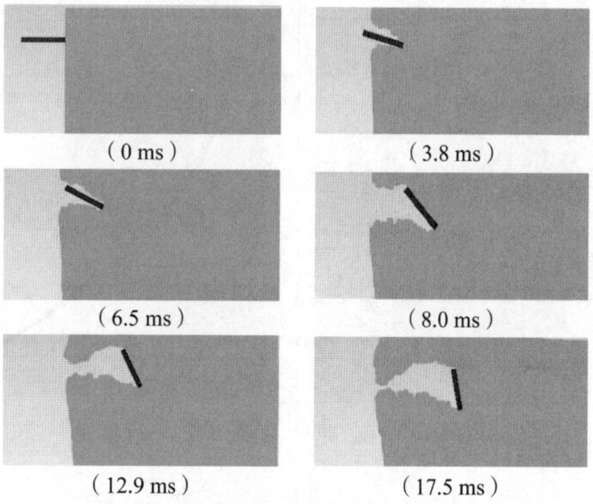

图5　数值模拟弹体入水过程

通过控制变量法进行多组模拟发现，弹体入水偏转符合式（8）~式（11），满足“侵彻速度越大，偏转越快；密度、长径比越大，偏转越慢”的结论，在后文的试验研究中将对该结论进行进一步的论证。但是头部斜度 ψ 对姿态偏转的影响却是有一定限度的。图6为不同斜度头部的弹丸入水姿态偏转结果，它们的质量均为65 g，长径比为8（直径20 mm，长度160 mm），入水初速度为50 m/s。从图中可以发现当 $0.2<\psi<0.5$ 时，姿态偏转速度随着头部斜度增大而增大，与方程相符，偏转90°所需要的时间明显缩短；当 $\psi>0.5$ 时，偏转90°所需要的时间差别不大，也就是说头部斜度在一定的范围内对弹道有影响，当达到某一值之后，对弹道的影响就变得很小了。

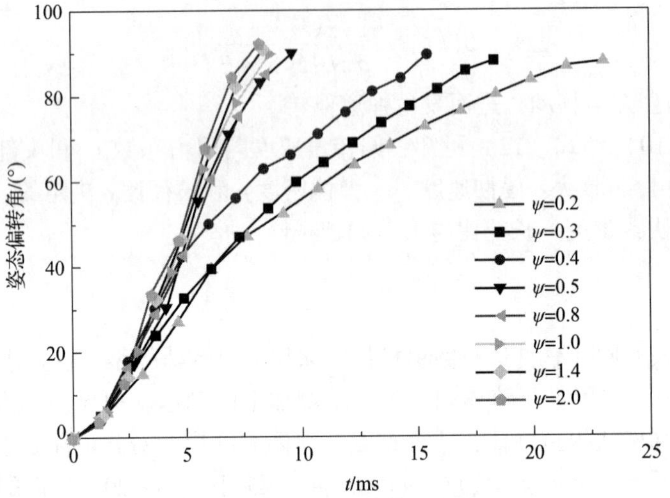

图6　不同斜度头部的姿态偏转

为探讨速度对弹道的影响，我们对比分析了相同弹体（质量为65 g，头部斜度为1，长径比为8）在不同速度下的质心弹道，并做了更高速度入水的数值模拟（低于声速），弹道对比如图7所示。由于在真实情况下竖直入水研究更有意义，本文中的研究均忽略了重力，因此得出的弹道图均等效为竖直入水的情况。经对比发现它们的弹道基本相同，只是通过弹道的时间有所不同。入水初速度虽然会影响弹体弹道偏转和姿态偏转的快慢，但对于弹道没有影响。本质上弹体的受力是由瞬时速度和阻力系数决定的，而阻力系数与弹体的沾湿面积和表面形状有关。虽然不同速度入水的弹体受力大小不同，但是水平和竖直方向上所产生的位移比是定值。由于不同速度下的相同弹丸弹道重合，因此可以推断不同速度下的弹体所经历的受力过程相同，只不过快慢不同，弹道由弹体的头部形状、长径比、质量决定。

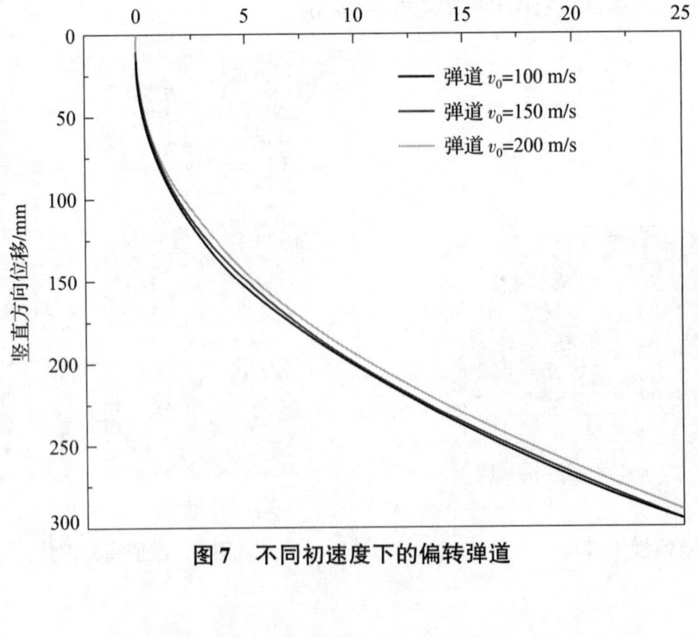

图7　不同初速度下的偏转弹道

3　质心弹道经验公式

斜截头弹体入水的各项参数有头部斜度 ψ、入水初速度 v_0、长径比 δ 和质量 m，对于姿态偏转来说，它们的变化规律满足式（9）和式（10）。本小节基于数值模拟提出斜截头弹竖直入水的质心轨迹方程。由于斜截头入水弹道为弯曲弹道，对这种弹道进行预测具有重要工程意义。由于头部斜度 $\psi>0.5$ 和入水速度（低于声速）对质心弹道影响非常小，对弹道影响最大的因素是弹丸质量 m 和长径比 δ。斜截头弹体质心轨迹具有良好的幂函数关系，考虑到实际情况中是竖直入水，据此建立一个新的方程来描述：①竖直入水弹道；②弹体质量；③头部斜度（$\psi>0.5$）的弹长径比之间的关系。方程的形式是

$$y=f(x,m,\delta) \tag{13}$$

式中　y——竖直深度；

x——水平位移；

m——弹体的质量；

δ——长径比（$\delta=l/d$）。

要拟合三元函数关系式，就需要大量数据，因此按照上述 SPH 仿真方法，改变弹体参数和入水条件进行了大量模拟，参数如表 2 所示。分别选用了 5 种长径比，每种长径比设置 5 种质量，弹体头部斜度均为 1，入水初速度为 100 m/s。

表 2　竖直入水质心轨迹模拟初始条件

头部斜度 ψ	1	入水初速度 v_0	100 m/s
长径比 δ	质量 m（g）	长径比 δ	质量 m/g
4	200	16	200
	400		400
	600		600
	800		800
	1 000		1 000
8	200	20	200
	400		400
	600		600
	800		800
	1 000		1 000
12	200		
	400		
	600		
	800		
	1 000		

图 8 为不同质量的弹体在不同长径比下的质心轨迹，根据图 9 中的曲线，质心轨迹满足幂函数，可表示为

$$y=ax^b \tag{14}$$

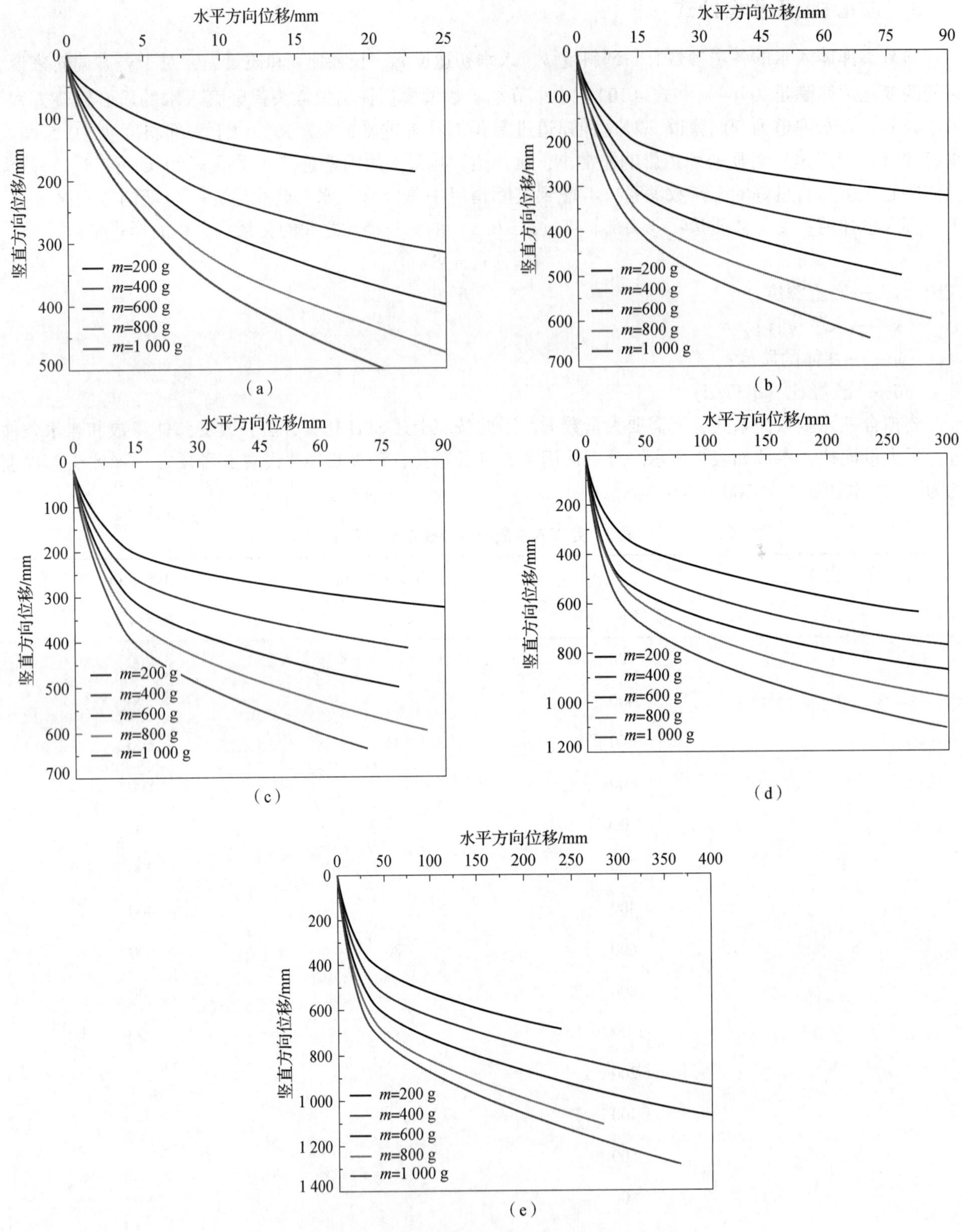

图 8　数值模拟出的不同状态下斜截弹体入水质心轨迹

(a) 长径比 $\delta=4$ 时；(b) 长径比 $\delta=8$ 时；(c) 长径比 $\delta=12$ 时；(d) 长径比 $\delta=16$ 时；(e) 长径比 $\delta=20$ 时

式中的参数 a、b 与质量和长径比有关。参数 a 和 b 根据图 9 的拟合值如表 3 所示，拟合优度 R^2 值接近 1，表示拟合准确度高。

表 3　不同条件下参数 a、b 的拟合值

长径比 δ	弹体质量/g	参数 a	参数 b	优度 R^2
4	200	47.95	0.446 63	0.993 27
	400	73.71	0.460 66	0.993 09
	600	91.64	0.462 64	0.993 46
	800	105.8	0.473 69	0.993 09
	1 000	118.62	0.478 47	0.994 05
8	200	70.61	0.347 57	0.995 55
	400	85.05	0.372 37	0.976 82
	600	101.83	0.375 09	0.979 73
	800	116.27	0.375 82	0.982 83
	1 000	128.3	0.386 83	0.985 47
12	200	87.74	0.322 18	0.983 5
	400	108.76	0.330 92	0.983 74
	600	124.61	0.339 02	0.979 12
	800	145.47	0.336 39	0.978 67
	1 000	153.51	0.345 26	0.977 54
16	200	88.84	0.354 43	0.980 19
	400	132.72	0.308 92	0.987 35
	600	135.9	0.326 14	0.984 63
	800	159.34	0.324 39	0.996 09
	1 000	180.61	0.324 06	0.982 95
20	200	91.76	0.373 04	0.980 99
	400	125.8	0.339 26	0.982 95
	600	155.8	0.324 22	0.996 09
	800	166.02	0.342 07	0.984 63
	1 000	181.32	0.335 16	0.982 73

图 9 表示的是参数 a 与质量 m 的关系，可以看出长径比 δ 一定时 a 与质量是线性关系，因此 a 可以表示为

$$a = km + t \tag{15}$$

k 和 t 为 a 的参数。

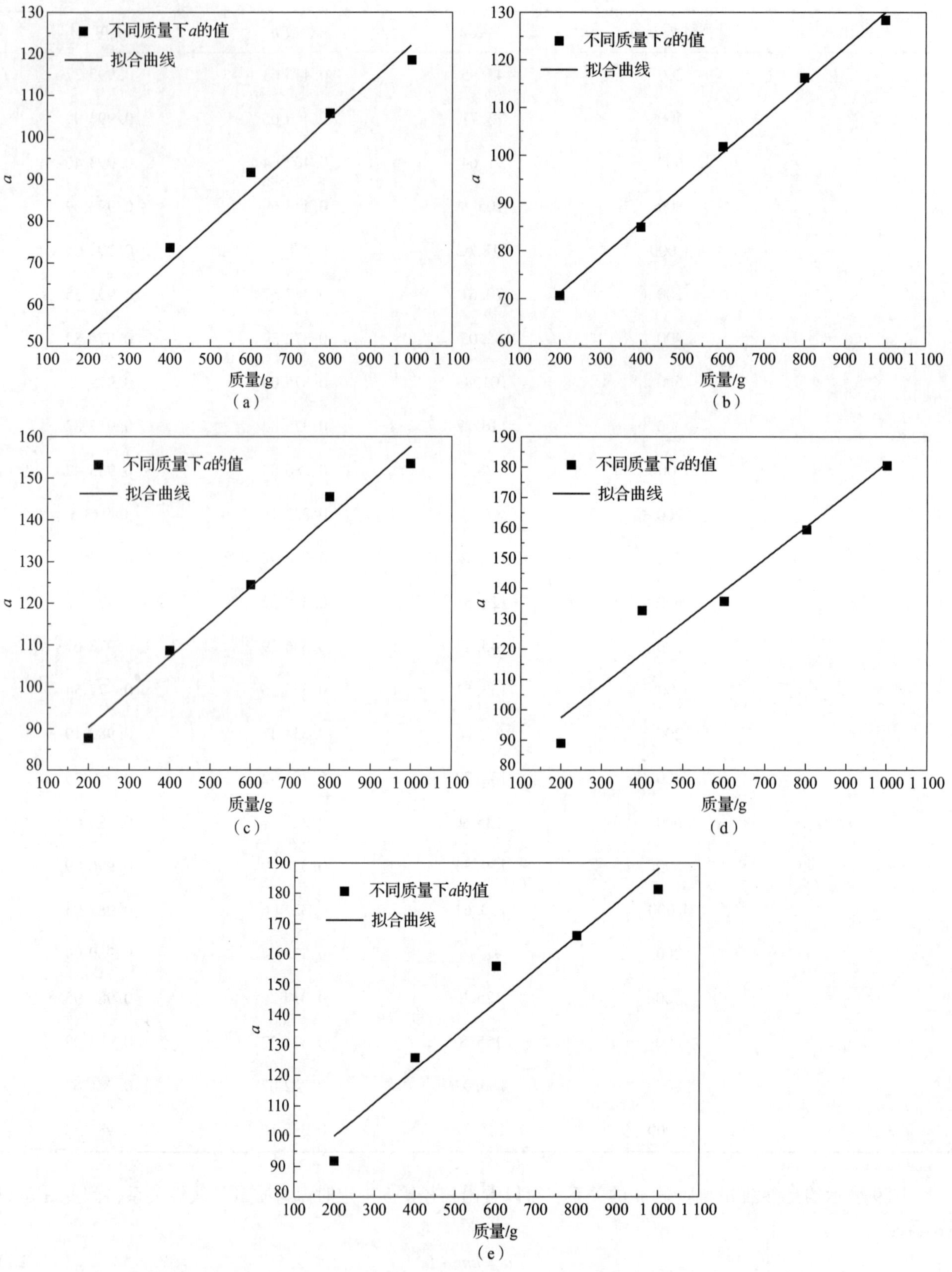

图 9　参数 *a* 在不同长径比条件下与质量的关系

(a) 长径比 $\delta=4$ 时；(b) 长径比 $\delta=8$ 时；(c) 长径比 $\delta=12$ 时；(d) 长径比 $\delta=16$ 时；(e) 长径比 $\delta=20$ 时

图 10 表示的是 b 与质量 m 的关系，可以看出长径比 δ 一定时 b 随质量的变化量非常小，可以视为一个常数，因此 b 可以表示为

$$b = c \tag{16}$$

式中，c 为 b 的参数，在质量一定时 c 为常数。

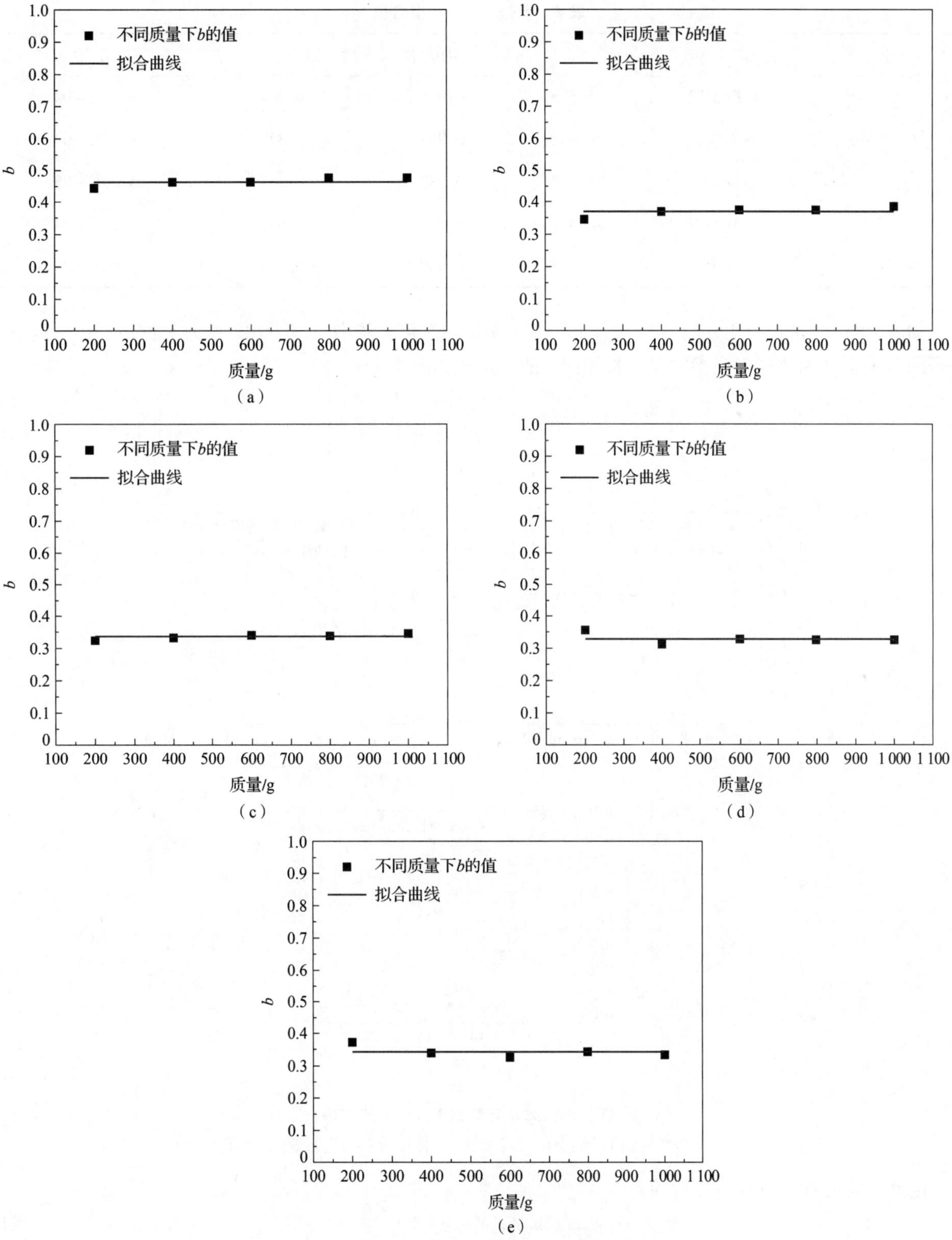

图 10　参数 b 在不同长径比条件下与质量的关系

（a）长径比 $\delta=4$ 时；（b）长径比 $\delta=8$ 时；（c）长径比 $\delta=12$ 时；（d）长径比 $\delta=16$ 时；（e）长径比 $\delta=20$ 时

将 a 和 b 代入式，质心轨迹方程可以写为

$$y=(km+t)x^{c} \tag{17}$$

参数 k、t、c 在不同长径比下的数值如表 4 所示，表中给出了线性拟合式（15）的拟合优度和常数拟合式（16）的方差。

表 4　参数 k、t、c 拟合值

长径比 δ	k	t	优度 R^2	c	方差
4	0. 086 72	35. 515	0. 978 20	0. 464	1. 24E－04
8	0. 073 3	56. 432	0. 997 11	0. 372	1. 68E－04
12	0. 084 13	73. 543	0. 983 19	0. 335	6. 09E－05
16	0. 105 08	76. 434	0. 937 93	0. 328	2. 19E－04
20	0. 109 67	78. 338	0. 947 60	0. 343	2. 66E－04

图 11 表示的是长径比 δ 与参数 k、t、c 之间的拟合关系，从中可以看出，k 为常数值，t、c 与 δ 符合幂函数关系，因此这 3 个参数关于长径比 δ 的表达式为

$$k=c_1 \tag{18}$$

$$t=-c_2c_3^{\delta}+c_4 \tag{19}$$

$$c=c_5c_6^{\delta}+c_7 \tag{20}$$

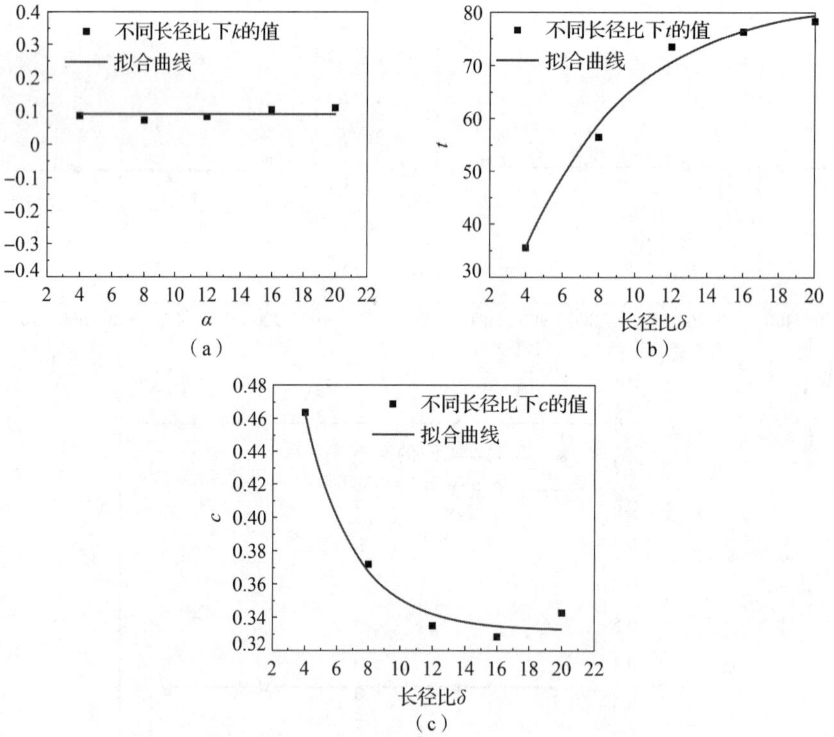

图 11　长径比 δ 与参数 k、t、c 的关系

（a）δ 与参数 k 的关系；（b）δ 与参数 t 的关系；（c）δ 与参数 c 的关系

因此，质心弹道方程可以写为

$$y=(c_1m-c_2c_3^{\delta}+c_4)x^{c_5c_6^{\delta}+c_7} \tag{21}$$

式中拟合的参数值由表 5 给出（表中给出了常数拟合的方差和曲线拟合的优度 R^2），其中 x、y 的单位为 mm，m 的单位为 g。

表 5　式（21）中的拟合参数值

c_1	方差	c_2	c_3	c_4	优度 R^2
0.092	1.85E-04	95.23	0.84	45.38	0.988 21
c_5	c_6	c_7	优度 R^2		
0.5	0.72	0.46	0.963 26		

4　试验研究

4.1　试验方案

图 12 为实验装置的原理图。水箱为 0.8 m×0.8 m×1.0 m 的聚碳酸酯复合材料制成的开放、透明、抗冲击立方体。其前后有两个孔，用两块 0.2 mm 厚的丙烯酸隔膜密封，确保不会影响入水过程。利用高速摄影机记录流体、弹体、弹道偏转的初始速度和运动过程。获得图像后，可以通过软件测量弹体速度、位移、偏转角等参数。试验用的缩比弹从一个标准火炮装置发射，瞄准水箱的隔膜。为了确保弹体偏转被高速摄像机记录，入水的角度被设置为尽可能垂直于水面，并防止弹体的旋转。另外，在发射管内填充尼龙弹托，减少发射气体对弹体的冲击，保证斜头弹体的入水姿态。此外，采用低燃速推进剂来降低发射气体对弹体姿态的影响。通过调整装药中推进剂的质量来调整弹体的加速度和初速度。针对弹体入水偏转的研究中采用直径为 20 mm 的试件，它们通过前孔水平射入水箱，每做完一次试验都要重新装水，并用新的隔膜密封水箱。

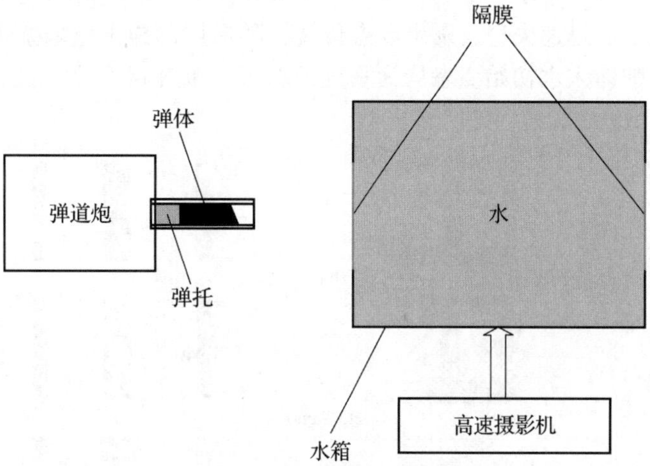

图 12　实验装置的原理图

试件截面形状与图 1 相同，表 6 为 10 种不同试验弹体的初始状态，包括质量、长径比、头部斜度和入水初速度，为了对式（9）进行验证，用控制变量法。

表 6　入水偏转试验诸元

编号	材料	质量/g	入水初速度/(m·s^{-1})	头部斜度（ψ）	长径比（δ）
1	铝合金	62.2	47	1	8
2	铝合金	60.1	47.7	0.5	8
3	铝合金	60.66	50.2	0.2	8
4	钛合金	230	31.2	0.2	8

续表

编号	材料	质量/g	入水初速度/$(m \cdot s^{-1})$	头部斜度（ψ）	长径比（δ）
5	钢	411.3	45.4	0.2	8
6	钢	238	48.1	0.2	5
7	铝合金	242	51.9	0.2	14
8	钛合金	230	52.0	0.2	8
9	钛合金	230	89.3	0.2	8
10	铝合金	315	106.0	0	8

（1）样本3、5、8选用了三种不同密度的金属材料，以保证它们除了质量以外，其他初始条件一样，可通过对比分析找出质量对弹体偏转的影响；

（2）样本6、7、8除了长径比不同，其他初始条件相同，可以对比分析出长径比对弹体偏转的影响；

（3）样本1、2、3除了头部斜度不同，其余条件相同，可以对比分析头部斜度对弹体偏转的影响；

（4）样本4、8、9除了入水初速度不同，其余条件相同，可以对比分析撞水速度对弹体偏转的影响；

（5）样本10为平头弹体，作为对照组。

4.2 试验结果分析

图13为平头弹和斜截头弹入水过程对比，可以看到平头弹体由于其对称性，能较长时间在水中保持稳定，形成空泡，因此沾湿面非常小，不利于减速。平头弹体最终以较高的速度从水箱的另一侧隔膜穿出。反观斜截头部弹体入水后迅速失稳，发生姿态偏转，随后尾部撞击空炮壁，沾湿面瞬间增大，有利于弹在水中的迅速减速。弹体入水初始姿态均为垂直于水面，如图14所示，定义弹体偏转后弹轴与竖直方向夹角为弹体的偏转角 α。

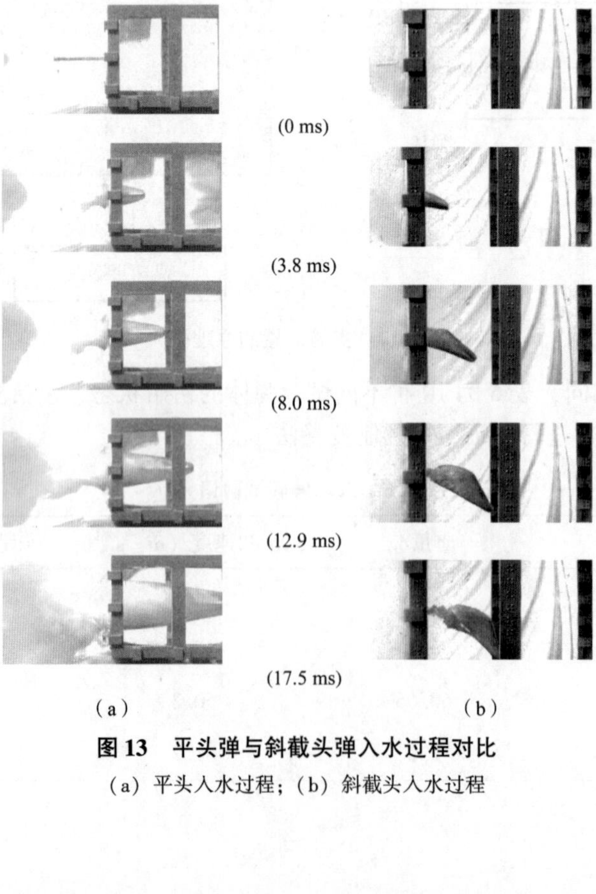

图13 平头弹与斜截头弹入水过程对比

（a）平头入水过程；（b）斜截头入水过程

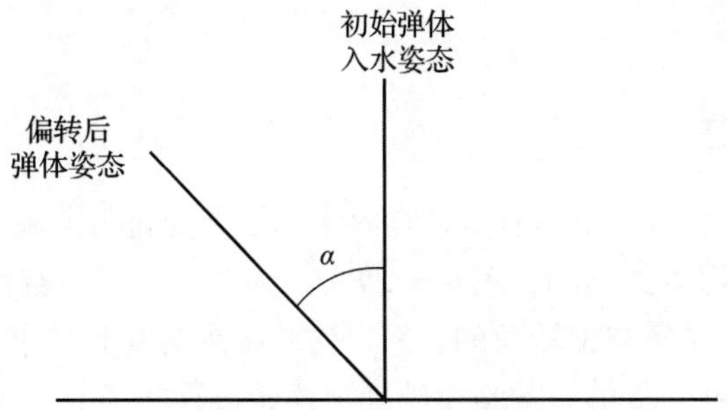

图 14　偏转角 α 示意图

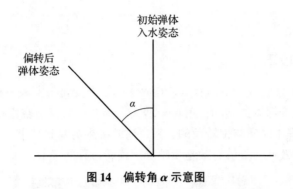

图 15 反映的是试验中通过高速摄像机测出的不同条件下弹体姿态随时间的偏转曲线，图 15（a）中头部斜度 $\psi=0.2$ 与 0.5 对姿态偏转影响很大，但是 $\psi=0.5$ 与 1 时对姿态偏转影响不大，符合数值模拟结果。初速度越大，姿态偏转越快，而弹体的长径比与质量越大会导致偏转越困难。综上所述，通过试验得出的结论与理论分析和数值模拟的结论吻合。

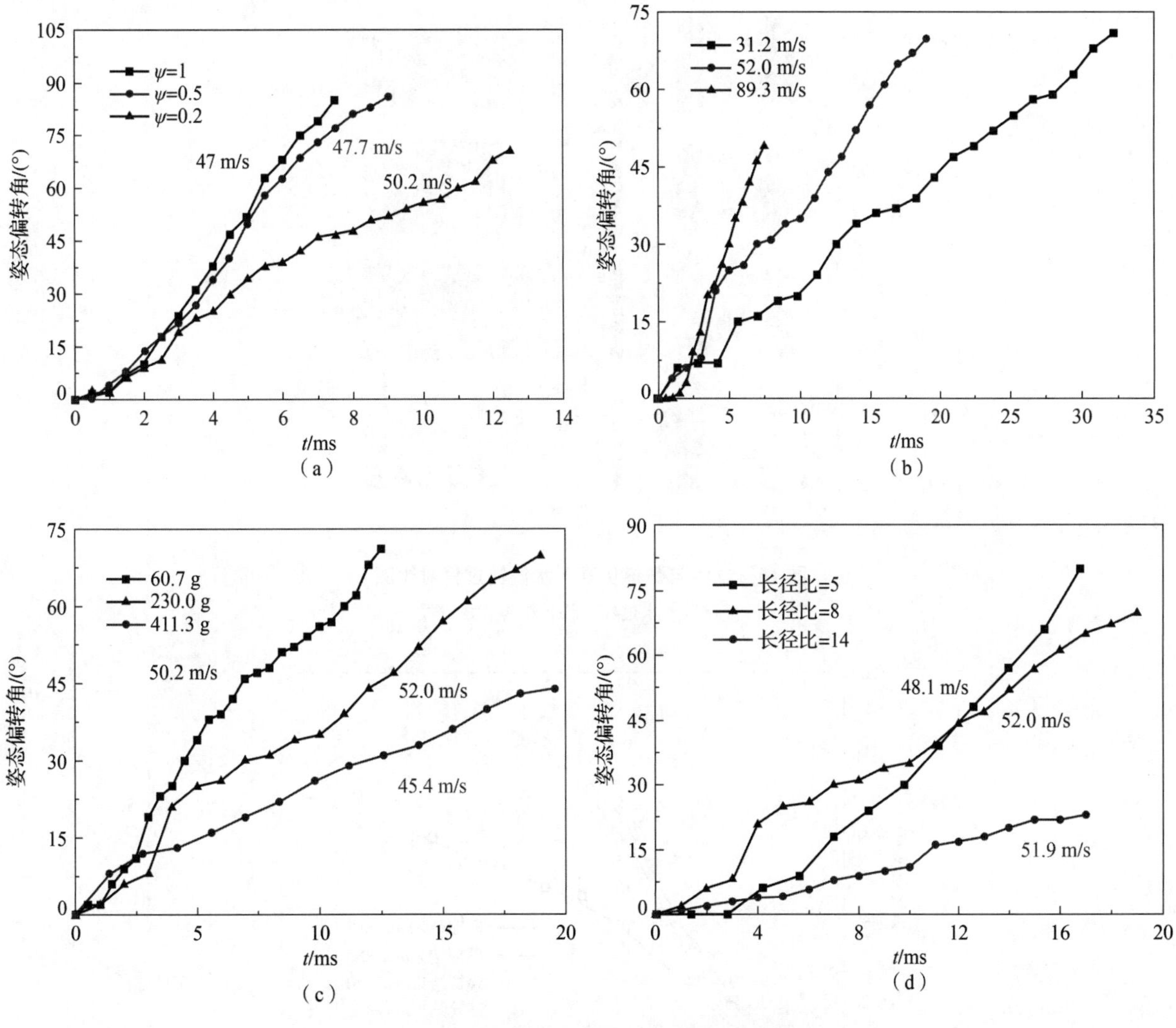

图 15　不同条件下弹体姿态随时间的偏转曲线

（a）不同头部斜度弹体姿态偏转曲线；（b）不同初速度弹体姿态偏转曲线；
（c）不同质量弹体姿态偏转曲线；（d）不同长径比弹体姿态偏转曲线

5　对比验证

5.1　数值模拟有效性验证

$\psi=0.2$ 弹丸的入水过程如图 16（b）所示。在入水的姿态变化与实验一致性较好。入水初期发生偏转，出现空化效应，随着侵彻深度的增加，偏转角增大，弹丸尾部撞击空泡壁。尾部的拍击使弹丸垂直移动。图 17 表示了入水过程中时间与偏转角的关系。仿真结果与实验结果一致性较好，误差在可接受范围内，证明了仿真方法的有效性，能够有效地反映弹道偏转过程的细节。

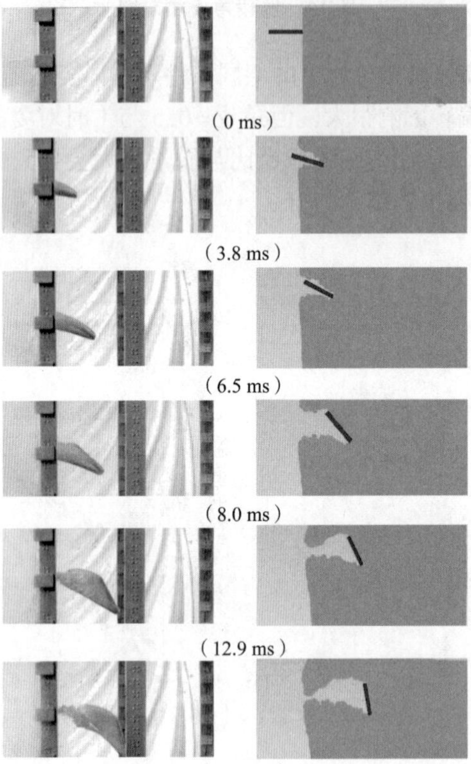

图 16　试验与数值仿真入水偏转过程对比图

(a) 试验入水过程；(b) 模拟入水过程

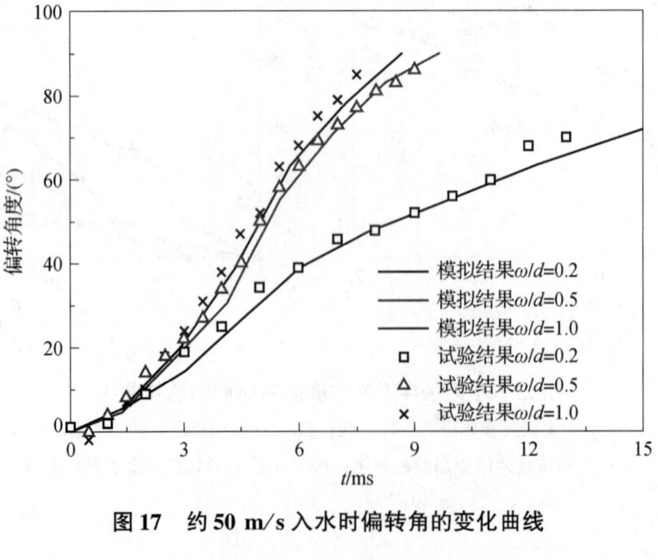

图 17　约 50 m/s 入水时偏转角的变化曲线

5.2　弹道经验公式有效性验证

通过拟合数值模拟得出了头部斜度 $\psi>0.5$ 的弹体入水的质心轨迹方程，试验部分中 1 号样本为 $\psi>0.5$ 的弹体。提取 1 号样本的质心轨迹，提取方法如图 18 所示。并将长径比 $\delta=8$、质量 $m=62.2$ g 代入经验公式，计算出质心轨迹方程。由于忽略了重力，因此水平入水的弹道与竖直入水的弹道是相同的。将两种方法得到的质心轨迹方程进行对比，如图 19 所示，其一致性较好，误差在可接受范围内，证明弹丸质心轨迹经验公式是有效的。

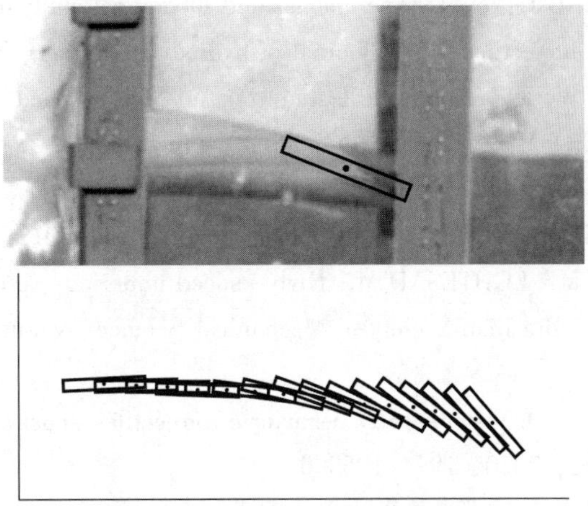

图 18　描边法提取弹体位置与姿态

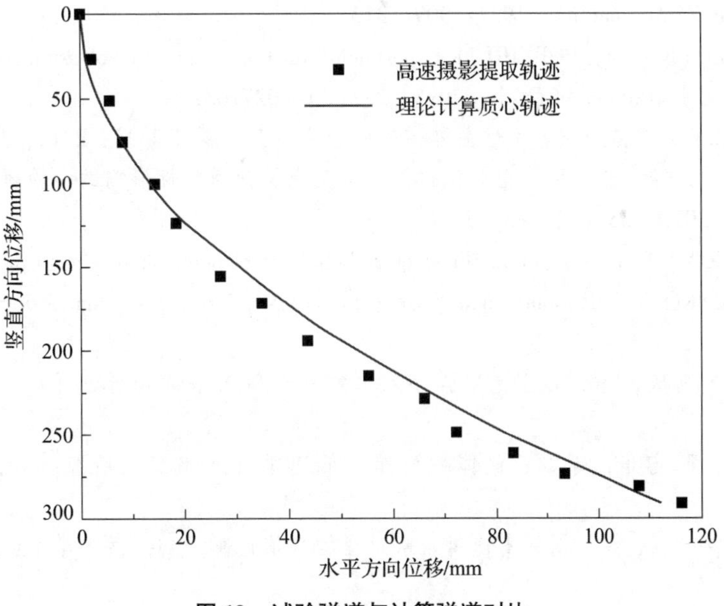

图 19　试验弹道与计算弹道对比

6　结论

斜截头弹体在入水后可以向预定的方向偏转，有助于弹体在竖直方向的减速，防止弹药以高速撞击水底导致失效。该研究对空投入水武器的设计具有重要意义。本研究可以得出下列主要结论：

（1）针对斜截头弹体，结合数值仿真和试验结果，得出当头部斜度 $\psi<0.5$ 时，弹道的偏转会随着 ψ 的增大而增大；当 $\psi>0.5$ 时，其对轨迹的影响较小。

（2）提高撞击速度引起水动力偏转力矩增大，进而提高了角加速度，因此姿态偏转会更快。

（3）质量和长径比越大，转动惯量越大，偏转角加速度越低，因此弹道的弯曲程度与质量和长径比成反比。

（4）数值模拟得到的入水过程和偏转角度与实验结果吻合较好。

（5）基于数值模拟，提出了一个描述斜截头弹体入水过程中质心轨迹的经验公式，并通过试验对其准确性进行了验证。

参考文献

[1] GUO Z T, ZHANG W, WANG C, et al. Experimental and theoretical study on the high - speed horizontal water entry behaviors of cylindrical projectiles [J]. Journal of hydrodynamics ser. B, 2012 (2): 217 - 225.

[2] GUO Z T, WEI Z, GANG W, et al. Numerical study on the high - speed water - entry of hemispherical and ogival projectiles [C]//ELERT M L, et al. Aip Conference Proceedings. AIP: Melville, NY, 2012: 64 - 67.

[3] GUO Z T, WEI Z, XIAO X, et al. An investigation into horizontal water entry behaviors of projectiles with different nose shapes [J]. International journal of impact engineering, 2012, 49: 43 - 60.

[4] MIRZAEI M, ALISHAHI M M, EGHTESAD M. High - speed underwater projectiles modeling: a new empirical approach [J]. Journal of the Brazilian Society of Mechanical Sciences & Engineering, 2015, 37 (2): 613 - 626.

[5] WU S, SHAO Z, FENG S, et al. Water - entry behavior of projectiles under the protection of polyurethane buffer head [J]. Ocean engineering, 2020, 197: 106890.

[6] 魏卓慧，王树山，马峰．刚性截锥形弹体入水冲击载荷［J］．兵工学报，2010（S1）：118 - 120.

[7] VON KARMAN T. (1929) The impact on seaplane floats during landing, No. 321 [R]. Washionton: National Advisory Committee for Aeronautics, 1927: 309 - 313.

[8] SHAMS A, JALALISENDI M, PORFIRI M. Experiments on the water entry of asymmetric wedges using particle image velocimetry [J]. Physics of fluids, 2015, 27 (2): 027103.

[9] 王云，袁绪龙，吕策．弹体高速入水弯曲弹道实验研究［J］．兵工学报，2014，35（12）：1998 - 2002.

[10] 华扬，施瑶，潘光，等．非对称头型对航行器入水空泡及弹道特性影响的实验研究［J］．水动力学研究与进展（A辑），2020，35（1）：65 - 71.

[11] WU S Y, SHAO Z Y, Feng S S, et al. Water entry behaviors of hemisphere - head projectiles with high velocity [C]//KHARCHENKO V. IOP Conference Series: materials science and engineering. IOP Publishing, 2019: 1 - 9.

[12] 侯宇，黄振贵，郭则庆，等．超空泡射弹小入水角高速斜入水试验研究［J］．兵工学报，2020（2）：126 - 135.

[13] 唐楚淳，黄振贵，陈志华，等．斜截体头型弹丸低速垂直入水实验研究［J］．兵工学报，2020，41（S1）：56 - 60.

[14] 孙士明，颜开，褚学森，等．射弹高速斜入水过程的数值仿真［J］．兵工学报，2020，41（S1）：124 - 129.

[15] WU S Y, SHAO Z Y, ZHANG B, et al. Experimental study on water - entry of body with buffer at high velocity [C]//FENG C G, WOODLEY C. Journal of Physics Conference Series. IOP Publishing, 2020: 102006.

第六部分
爆炸冲击作用及防护

密实介质中冲击波衰减特性的近似计算①

王海福②，冯顺山

（北京理工大学，北京　100081）

摘　要：本文对密实介质中冲击波峰值压力的衰减建立了一个近似的计算模型，并利用该模型分别对铝、钢和有机玻璃等介质中的冲击波衰减特性进行了计算。算例表明与实验结果吻合较好，从而在一定程度上验证了模型的有效性。

关键词：冲击波衰减；冲击波压力；冲击绝热线

AN APPROXIMATE THEORETICAL MODEL FOR THE ATTENUATION OF SHOCKIPRESSURE

Wang Haifu，Feng Shunshan

（Beijing Institute of Technology，Beijing，100081）

Abstract：An approximate theoretical model for the calculation of attenuation of shock pressure in solid materials is presented. Based on this model，the attenuation characteristics of shock pressure in aluminium，steel and PAMM are obtained. The calculated results show that they fit well with experimental data.

Key words：shock pressure，shock attenuation，shock Hugoniot

冲击波在密实介质中的传播和衰减，是一切从事与爆炸、防护等有关的研究人员所关心并急需解决的问题。长期以来，围绕衰减问题曾提出过多种计算模型[1-6]，尽管他们都从一定意义上描述了冲击波在密实介质中的衰减特性，但仍存在诸多不足。本文就一般爆炸冲击波峰值压力在密实介质中的衰减，尤其是在其传播中、前期的衰减特性，从介质温升效应出发，建立了一个近似的计算模型。算例表明，它与实验结果吻合较好。

1　模型和理论推导

密实介质在等温和绝热条件下的冲击压缩曲线如图1所示。图中 AH 代表冲击绝热线，AT_0 代表等温压缩线。鉴于能量守恒，冲击波后介质热能的增加可由图中斜线部分的面积来表示，即

$$\frac{1}{2}P(V_0 - V) - \int_V^{V_0} P_{\tau_0} \mathrm{d}V = \int_{T_0}^{T} C_V \mathrm{d}T \qquad (1.1)$$

另外，如果把 Grüneisen 状态方程[7]的参考点选在等温线上，那么在同一比容 V 处，冲击绝热线上任一状态点 $H(P,V)$ 与等温线上相对应的状态点 $T_0(P_{\tau_0},V)$ 间的关系为

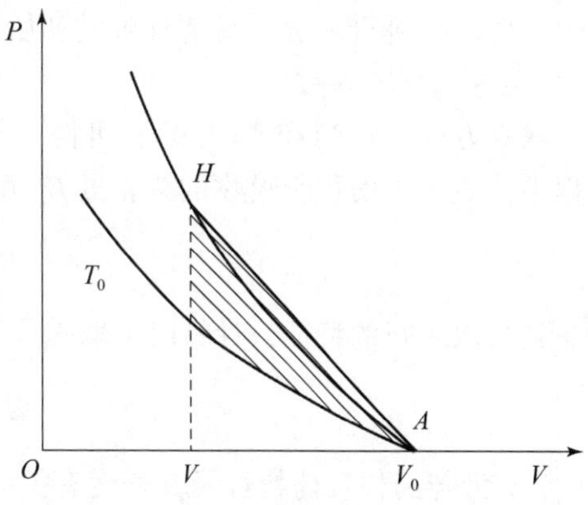

图1　密实介质的冲击压缩曲线

① 原文发表于《兵工学报》1996（1）。

② 王海福：工学博士，1993年师从冯顺山教授，研究多孔材料对爆炸载荷的弱化效应及其机理，现工作单位：北京理工大学。

$$P - P_{\tau_0} = \frac{\Gamma}{V}C_V(T - T_0) \tag{1.2}$$

微分式（1.1），并把式（1.2）代入，得

$$\frac{\mathrm{d}T}{\mathrm{d}V} + \frac{\Gamma}{V}(T - T_0) = \frac{1}{2C_V}\left[P + (V_0 - V)\frac{\mathrm{d}P}{\mathrm{d}V}\right] \tag{1.3}$$

式中　Cv，Γ——介质的等容比热和 Grüneisen 系数；

T——冲击波后介质的温度；

P，V——冲击 Hugoniot 压力和比容；

下标 0——相应参数的波前值。有关 P 和 V 之间的关系可通过冲击波速度与粒子速度间的线性关系和冲击波阵面守恒方程得到[7]

$$P = \frac{a^2(V_0 - V)}{[V_0 - b(V_0 - V)]^2} \tag{1.4}$$

式中，a、b 为与介质有关的常数。有关非常态下的系数 Γ 可通过式（1.5）求得[7]

$$\frac{\Gamma}{V} = \frac{\Gamma_0}{V_0} \tag{1.5}$$

C_V 可通过 Debey 公式求得[8]

$$C_V = \left(\frac{3R}{\mu}\right)D\left(\frac{\Theta}{T}\right) \tag{1.6}$$

式中　R——普适气体常数；

μ——介质的摩尔质量；

Θ——Debey 温度。

Debey 函数 $D(X)$ 为[8]

$$D(X) = \frac{3}{X^3}\int_0^X \frac{\exp(Z)Z^4 DZ}{[\exp(Z) - 1]^2} \tag{1.7}$$

式中，$X = \Theta/T$ 。联立方程（1.3）~式(1.7)，可求得冲击波后介质的 T 与 P 之间的关系

$$T = T(P) \tag{1.8}$$

另外，实验表明，密实介质中冲击波峰值压力随传播距离具有指数衰减的特征。鉴此，假定具有形式：

$$P(X) = P_m\exp(-\alpha X) \tag{1.9}$$

式中　P_m——炸药—介质分界面处的初始冲击波峰值压力；

α——衰减系数。

联立方程（1.8）和式(1.9)，可得 T 与传播距离 X 之间的关系：$T = T(X)$。这样，冲击波在其传播过程中介质单位面积所吸收的总能量 E_m 可表述为

$$E_m = \int_0^{X_0} C_V T(X)\rho_{m0} DX \tag{1.10}$$

考虑到指数衰减的特征，计算时可取 $X_0 = 5/\alpha$ 。冲击波传入介质单位面积的总能量 E_n 为

$$E_n = \int_{L/D}^{\tau} \frac{1}{\rho_{m0}D_b}P_{0X}^2(t)\,\mathrm{d}t \tag{1.11}$$

式中，τ 为与药柱装药特性及其长度有关的常数，计算时可取 $\tau = 3L/D$。

炸药—介质分界面处的压力 $P_{0X}(t)$ 为[8]

$$P_{0X}(t) = \frac{64\,P_H}{27}\left(\frac{L}{Dt}\right)^3\left\{1 + \frac{\beta}{1-\beta}\left(\frac{U_b}{D}\right)\frac{Dt}{L} - \frac{U_{bX}}{(1-\beta)D}\right\}^3 \tag{1.12}$$

$$\beta = \gamma/(0.241\,9 + 0.364\,3\gamma);U_b = U_{bX}[L/(Dt)]^3$$

$$\gamma = (\rho_{m0}D_{bX})/(\rho_0 D);D_b = a + bU_b;P_H = (1/4)\rho_0 D^2$$

式中　D——炸药的爆轰速度；

L——药柱的长度；

U_{bX}——炸药—介质分界面处质点的最大速度；

ρ_{m0}——介质的初始密度；

ρ_0——炸药的装药密度。

考虑到能量守恒关系：$E_m = E_n$，即可求得冲击波在密实介质中的衰减系数 α。

2　计算结果和分析

为了与文献［9］的实验结果进行比较，本文在计算中选用密度 $\rho_0 = 1.57\ \mathrm{g/cm^3}$，药柱长 $L = 50$ mm 的 TNT 装药。有关计算中所使用的介质参数[8]列于表 1。

表 1　介质参数

介质	$\rho_{m0}/(\mathrm{g \cdot cm^{-3}})$	$a/(\mathrm{km \cdot s^{-1}})$	b	Γ_0	Θ/K
Ly-12 铝	2.785	5.328	1.338	2.00	350
45 * 钢	7.850	3.574	1.920	1.69	170
有机玻璃	1.184	2.572	1.536	0.97	266

利用上述参数，本文分别对 3 种介质中的冲击波峰值压力随传播距离的衰减特性进行了计算，结果如图 2.1 所示。从图中可以看出，比较本文所得计算结果与文献［9］的实验结果，冲击波在中、前期的衰减特性基本吻合。但随着 X 的增大，两者的偏差变得较为显著，这主要是因为模型中没有考虑侧向稀疏波和自由面弹性卸载波轴向追赶塑性冲击波等的衰减作用。从图 2 中还可以看出，对于装药特性相同的炸药爆炸后，在钢材料中所形成的初始冲击波压力最高，但其衰减也最快；在有机玻璃中初始压力最低，衰减也最慢；在铝材料中的初始压力及其衰减均介于二者之间。这是因为密实介质中冲击波初始压力及其衰减特性与介质材料的冲击阻抗 $\rho_{m0}C_0$ 密切相关，冲击阻抗越大，其可压缩性就越小，要达到相同的压缩程度所需的冲击压力也就越高。

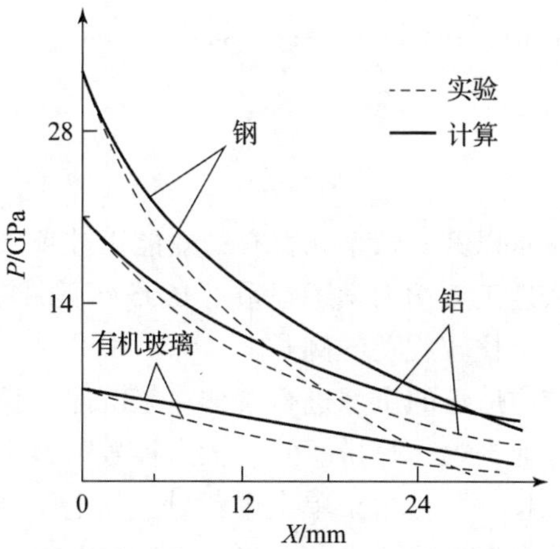

图 2　冲击波压力随传播距离的衰减曲线

3　结束语

本文给出了一种计算冲击波峰值压力在密实介质中衰减的近似方法。其优点在于不受作用载荷形式的限制，计算中仅以材料的冲击 Hugoniot 数据作为输入参数，因此计算结果较为可靠。算例表明，它与实验结果，尤其是冲击波峰值压力在中、前期的衰减特性吻合较好，从而在一定意义上验证了模型的有效性。

水下强爆破的近场地面运动观测与分析①

孙为国②，冯顺山

（北京理工大学，北京市 100081）

何政勤，丁志峰，叶太兰

（国家地震局地球物理所，北京市 100081）

摘 要： 通过对南通港主航道水下除障强爆破的观测，结合爆炸地震波理论，分析了爆破近场地面运动的特点，指出经验公式估算值与实际观测结果在评价地面运动对场地及地表建筑安全性评价的意义。

关键词： 水下强爆破；地震波；地面运动

Analysis and Measurement of Ground Motion at underwater Blasting of Port NanTon

Sun Weiguo，Fen Shushen

（Beijing Institute of Technology 100081）

He Zhengqin，Ding Zifen，Ye Tailan

（Institute of Geophysics SSB 100081）

Abstract：In this paper，based on analysis and measurement of ground motion at underwater blasting demolishing of port NanTon and analysis for characteristic of ground motion by seismic wave，discussed relation between empiric formula and field measurement data at evaluate of Safety of buildings in the ground motion.

Key words：underwater explosion；seismic wave；ground motion

1 前言

在我国现代化建设中，爆炸技术以其高效率和特有的效能正被越来越广泛地应用于各种科研、工业生产和某些特殊工程领域中，北京理工大学力学工程系于1995年元月在长江入海口南通港主航道进行的水下强爆破的工作可以说是其中一个比较成功的例子。

这次水下爆破是对沉于南通港主航道的沉船进行解体。该沉船，长多米，吨位大，加上船体已部分陷入江底淤泥中，因此，当时号称亚洲第一打捞船的"大力神号"也不能将其整体打捞上来。水下爆炸解体是一项技术要求很高的爆炸工程，而且由于单次用药量大加上距港口近，建筑物密集，因此，水下强爆破引起的近场强地面运动是否会对近场区内的人员与建筑构成威胁是一个必须认真对待的问题。

2 水下爆破近场振动观测分析

1995年1月，在南通港主航道内成功地对沉船进行了水下解体爆破。因用药量大而爆破位置离岸较

---

① 原文发表于《爆破》1997（3）。

② 孙为国：工学博士，1995年师从冯顺山教授，研究岩土介质中爆炸震源与地震波及其效应，现工作单位：中国地震局地球物理研究所。

近，最近距离不足测点位置如图 1 所示。为了确保陆上建筑物的安全和为今后类似工程爆破提供可参考的设计依据，我们对这次水下爆破进行了近场观测。

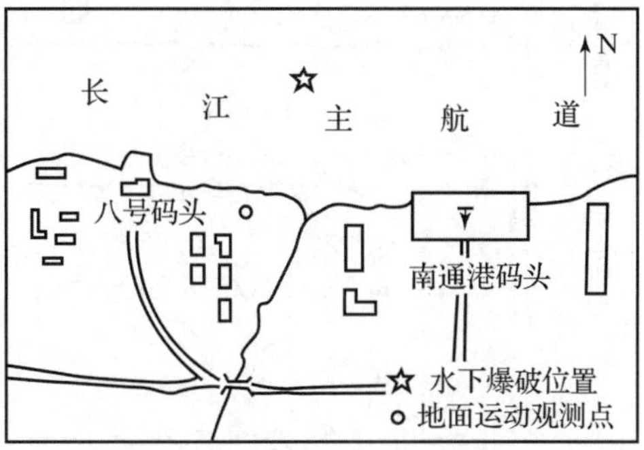

图 1　南通港水下爆破观测点位置示意图

2.1　观测仪器与测点选择

本次观测使用的是 DCS－302 型强震数字记录仪。该仪器具有动态范围大、频率特性稳定可靠等优点，是国际上广泛用于近场强地面运动观测的设备。拾震器是与该地震仪记录仪配套的三分向加速度传感器，记录时将传感器的方向对准爆破点（即保证 N 方向分量在径向，在地震动中通常对建筑物影响最大的是径向和垂直向分量，而切向次之）。采样频率为 100 个点秒，记录时间为 2 分钟。

为了保证记录资料的质量和可靠性，应当从三个方面来考虑选择测点，这就是合适的震中距。观测点介质应坚实，避免能量过多衰减，环境干扰背景小。根据这个原则观测点最后被选在距爆源 900 m 的一条坚实便道上。记录时将地面捣平后用 4 个长钉将拾震器牢固地固定在板结的地面，以保证与地面有很好的耦合关系。

2.2　数据处理方法

DCS－302 数字地震仪记录到的是地面运动加速度，为了得到速度和位移的时程曲线，我们对加速度进行积分计算，一次积分得到速度，二次积分得到位移。由于爆破地震的地面运动一般无法用解析函数表达，因此不可能通过求原函数的方法得到积分，而只能进行数值积分，积分求和表达式为

$$\int_a^b f(x)\,\mathrm{d}x = \lim_{n\to\infty}\sum_{k=1}^{n} f(X_k)\,\Delta X_k$$

若两采样点之间是直线，求积按矩形法计算。如假设采样点间为二次曲线，即为 Simpson 方法，其表达式为

$$\int_a^b f(x)\,\mathrm{d}x = (b-a)\left[\frac{1}{6}f(a) + \frac{4}{6}f(c) + \frac{1}{6}f(b)\right.$$

式中，$f(c)$为积分区间（a，b）中插值点的函数。

由于积分运算在频率域相当于除以频率的平方，这会使得原始记录的低频成分受到很大的扩充，甚至会淹没有用的信号，在数据处理中为消除背景干扰，我们在积分前在时间域采用四阶 Butterworth 带通滤波器对数据进行了滤波处理，通频带为 0.2～40 Hz。

为了消除因拾震器安装时未调整到完全水平位置，而使记录中出现并不表示地面运动的直流分量，以保证数值积分的稳定性，在数据处理中，首先进行直流分量扣除，具体做法是首先对整个记录长度范围内各点求和后除以记录的总点数作为该记录的直流偏移量，然后逐点减掉此值，经试算表明这种方法是方便有效的。

2.3　观测结果与分析

南通港水下爆破的原始观测资料与计算结果如图 2 和表 1 所示。

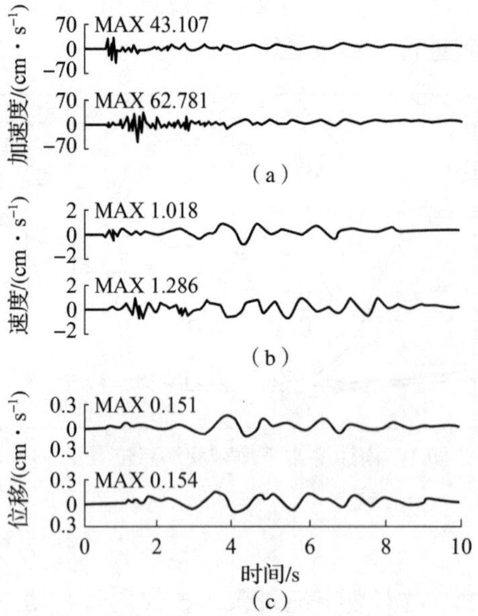

图2　南通港水下爆破近场加速度、速度和位移时程曲线

(a) 观测的加速度记录曲线；(b) 地面运动速度地程曲线；(c) 地面位移时程曲线

表1　水下爆破数值计算结果表

最大加速度峰值/($cm \cdot s^{-2}$)		最大地面运动速度值/($cm \cdot s^{-1}$)		最大地面运动位移值/mm	
水平径向	垂直向	水平径向	垂直向	水平径向	垂直向
43.107	62.781	1.018	1.286	1.51	1.54

对南通港水下爆破观测记录的频谱分析表明，有效信号的能量主要集中在1～20 Hz，垂直运动幅度大于水平径向幅度，峰值作用持续时间短，因此对普通构筑物不构成危害。

根据中华人民共和国交通部制定的《水运工程爆破技术规范》JTJ 286—90中有关建筑物安全振动速度的标准，南通港水下强爆破所产生的最大地面振动速度值，在安全距离内符合建筑物的安全要求，爆破设计是成功的，对比数据如表2所示。

表2　爆破数据对比

水运工程爆破技术规范			南通港水下爆破
序号	主要构、建筑物类型	安全振动速度/($cm \cdot s^{-1}$)	振动速度峰值/($cm \cdot s^{-1}$)
1	土窑洞、土坯房、毛石房屋	1	水平径向1.018
2	一般砖房、非抗震的大型砌块建筑物	2－3	
3	钢筋混凝土框架房屋	5	垂直向1.286

3　结束语

爆炸能正广泛地应用于工农业各项领域，它以特有的高效能、高效率受到工程技术界重视。但在一些特殊环境下的强爆破，如像南通港水下爆破，由于一次引爆药量较大，而在爆源近场区的建筑和人员活动很多，特别是构筑物与建筑类型混杂，爆破地震效应则成为一个很重要的安全问题。

由于影响爆破振动效应的因素很多，它不仅与爆炸能量、接收距离和对介质或建筑物作用时间有关，而且也会因介质物理特性以及浅部基岩起伏等诸多因素不同而有较大差别，有时仅依靠经验公式和近似条件下的参数估算安全距离是不够精确的，因此，充分利用较大规模的工程爆破，开展近场地面运动观测是很必要的。通过积累各种条件下的观测数据为研究爆破地震效应及各种场地条件下的抗震设计提供可靠的参考依据。而且在某些特殊情况下，强地面运动观测的结果还可能成为法律裁决的科学依据。

参考文献

[1] 孟吉复，惠鸿斌. 爆破测试技术 [M]. 北京：冶金工业出版社，1992.
[2] 亨利奇. 北京：爆炸动力学及其应用 [M]. 北京：北京科学出版社，1987.
[3] 张雪亮，黄树棠. 爆破地震效应 [M]. 北京：地震出版社，1981.
[4] 霍永基. 水下爆炸 [J]. 爆破，1986 (3)：1 - 7；(4)：5 - 13.
[5] 董承全，蔡金祥，高建光. 水下爆破引起结构物振动的监测与分析 [J]. 爆破，1987 (2)：47 - 52.
[6] 陈华腾. 水下爆破的几个问题 [J]. 爆破，1989 (4)：32 - 36.

多孔材料中冲击波衰减特性的实验研究①

王海福②，冯顺山
（北京理工大学机电工程系，北京　100081）

摘　要：本文通过脉冲网络法对不同初始孔隙度的多孔铁中冲击波速度的测量，结合多孔材料完全压实区冲击 Hugoniot 方程，就初始孔隙度对冲击波在多孔铁中衰减特性的影响进行了研究。

关键词：多孔材料；冲击绝热线；冲击波衰减；脉冲网络

An Experimental Research on Shock Attenuations in Porous Materials

Wang Haifu, Feng Shunshan
(Department of Engineering Safety, Beijing Institute of Technology, Beijing 100081)

Abstract: Shock velocities in porous iron samples having different initial porosities have been measured using the pulse generation network. By combining the experimental data with complete compaction regime Hugoniot equations of porous iron, effects of initial porosities upon the shock attenuations in porous iron are studied.

Key words: porous materials; Hugoniot; shock attenuation; pulse generation network

多孔材料较其相应的密实材料具有较大的初始比容，从而决定其具有不同于一般密实材料的冲击压缩特性。如当多孔材料被冲击压缩到与其相应的密实材料具有同一终态比容或压力时，多孔材料所需的冲击压力和能量要高得多，材料内部将产生更高的冲击温升[1-3]。多孔材料因具有这种独特的冲击压缩特性而在军事和民用工程防护领域得到了广泛的应用[4,5]。多孔材料因其内部存在大量的孔隙，当初始孔隙度较大（或初始密度较小）时，孔隙的大小将有可能达到或超过普通锰铜压阻传感器敏感部分的尺度，从而影响利用锰铜计敏感部分的压阻效应来直接测量多孔材料中冲击波压力的实验测试精度，鉴此，本文通过对多孔材料中冲击波速度的测量，结合其完全压实区冲击 Hugoniot 方程，就初始孔隙度对冲击波在多孔铁中衰减特性的影响进行了研究。

1　试验装置与测试原理

实验测试装置如图 1 所示。其测试原理为：当平面波发生器所产生的平面冲击波传至每两块相邻的圆柱形被测多孔材料试件（隔板）的分界面时，冲击波对设置其间的聚四氟乙烯绝缘层的击穿作用，使脉冲网络发生短路，从而导致电容放电，通过存储示波器获得匹配电阻两端的放电脉冲信号。当冲击波依次击穿相邻隔板间的绝缘层时，存储示波器就逐一记录下所有放电脉冲信号，即冲击波传至一定厚度隔板时所需的时间，从而可测得冲击波在多孔材料中传播时，距离与时间关系的实验数据，对实验数据进行适当的数学处理，即可获得冲击波速度与传播距离间的关系。

实验所采用的平面波发生器的直径为 50 mm；加载 TNT 药柱的密度、直径及高度分别为 1.57 g/cm、3.45 mm 和 50 mm；圆柱形被测多孔试件的直径为 50 mm；铜箔和聚四氟乙烯薄膜的厚度均为 0.1 mm；脉冲网络的放电脉冲宽度为 0.5 μs。

---

① 原文发表于《北京理工大学学报》1997（1）。

② 王海福：工学博士，1993 年师从冯顺山教授，研究多孔材料对爆炸载荷的弱化效应及其机理，现工作单位：北京理工大学。

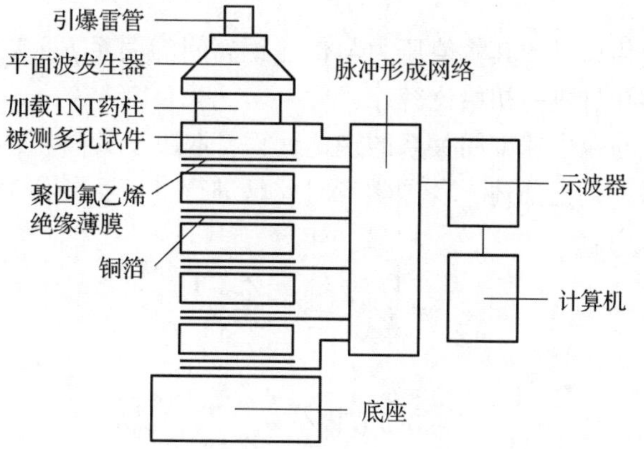

图 1　实验测试装置

2　实验测试结果

多孔铁中冲击波传播距离与时间关系的实验测试结果如图 2 所示。图中曲线 1 ~ 3 所对应的多孔铁初始密度 ρ_{00}分别为铁晶体理论密度 ρ_0 的 80%、70% 和 60%。对实验数据采用四阶多项式进行拟合，即

$$X = a_1 t + a_2 t^2 + a_3 t^2 + a_4 t^4 \tag{1}$$

微分式（1）得

$$D = a_1 + 2a_2 t + 3a_3 t^2 + 4a_4 t^3 \tag{2}$$

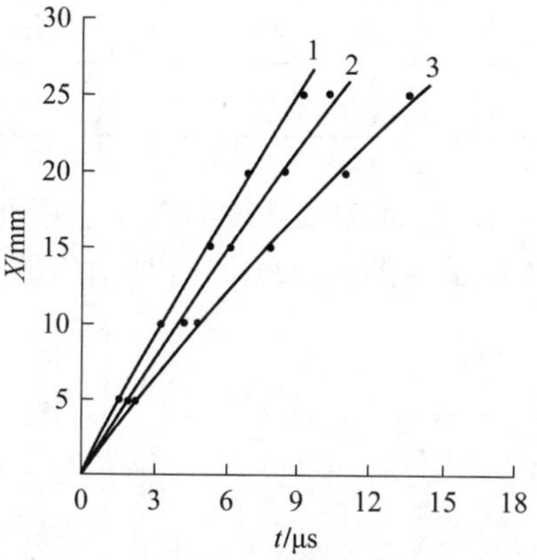

图 2　冲击波传播距离与时间关系的实验测试结果

联立式（1）和式（2），可得到冲击波速度与传播距离间的关系 $D(X)$。式中，X、t 分别为冲击波传播距离和时间；D 为冲击波速度；a_1、a_2、a_3 和 a_4 为拟合因素，列于表 1。

表 1　方程（1）中的拟合因素

ρ_{00}/ρ_0	a_1	a_2	a_3	a_4
0.6	2.547	-0.144	0.010 7	-0.000 30
0.7	3.042	-0.171	0.015 4	-0.000 51
0.8	3.665	-0.231	0.023 5	-0.000 94

3　初始孔隙度对冲击波衰减特性的影响

以上已得到多孔材料中冲击波速度与传播距离间的关系 $D(X)$，若能进一步求得多孔材料中冲击波速度与粒子速度间的关系 $D(u)$，则由冲击波阵面动量守恒方程

$$p = D(u)/V_{00} \tag{3}$$

可获得冲击波在多孔材料中传播时，其峰值压力与传播距离间的关系 p (X)。式中，p 为冲击波峰值压力；u 为粒子速度；V_{00} 为多孔材料的初始比容。

多孔材料中冲击波速度与粒子速度间关系的理论计算，取决于其高压冲击 Hugoniot 方程，有关多孔材料高压冲击 Hugoniot 关系的理论描述，各国学者对此已进行过广泛的研究[6-8]，为便于计算，本文选用如下形式的 Hugoniot 关系[8]：

$$V = \frac{V_0}{K}\left(1 - Z - \overline{\frac{Z_2 - 1}{S}}\right) \tag{4}$$

其中

$$K = \frac{1 + 0.5pTV_0/c_p}{1 + 0.5pTV_{00}/c_p}; Z = 1 + \frac{C_0^2 K}{2V_S P} \tag{5}$$

式中　T 和 c_p——相应密实材料的体膨胀系数及其等压比热；

C_0 和 S——相应密实材料中冲击速度与粒子速度间的关系常数；

p 和 v——多孔材料的 Hugoniot 压力和比容；

V_0——相应密实材料的初始比容。

将式（3）~式（5）与冲击波阵面质量守恒方程

$$D = V_{00}(D - u)/V \tag{6}$$

相联立，可求得多孔材料中的冲击波速度与粒子速度间的关系 $D(u)$。有关计算使用参数列于表 2[8,9]。

表 2　计算使用参数

$\rho_0/(g \cdot cm^{-3})$	$C_0/(km \cdot s^{-1})$	S	$C_p/(J \cdot kg^{-1} \cdot K^{-1})$	α/K^{-1}
7.85	3.57	1.92	724	3.51×10^{-5}

不同初始孔隙度（ρ_{00}/ρ_0）下多孔铁中冲击波速度与粒子速度间关系的计算结果如图 3 所示，初始孔隙度对冲击衰减特性的影响如图 4 所示。图中曲线 1 ~3 所对应的多孔铁初始密度分别为铁晶体理论密度的 80%、70% 和 60%。

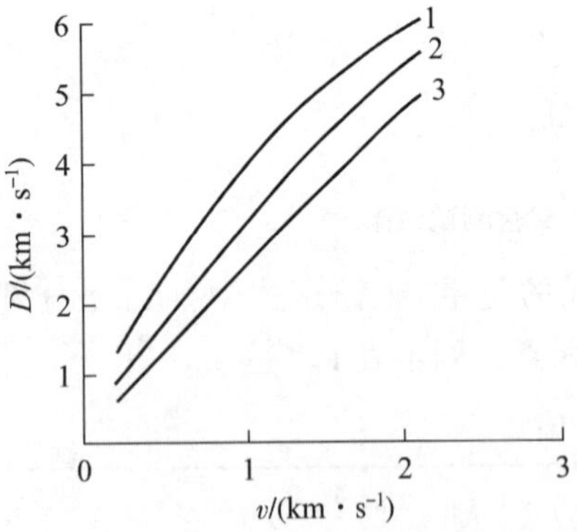

图 3　冲击波速度与粒子速度间关系的计算结果

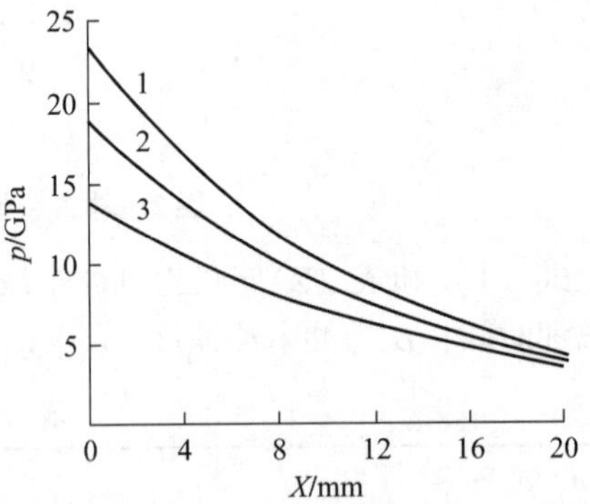

图 4　初始孔隙度对冲击波衰减特性的影响

从图 4 可以看出，当装药特性相同的 TNT 炸药与不同初始孔隙度（或初始密度）的多孔铁做接触爆炸作用时，在多孔铁中所形成的初始冲击波峰值压力随初始孔隙度的增大而显著下降，但初始孔隙度对冲击波峰值压力随传播距离衰减速率的影响并不十分显著。对应相同的传播距离，初始孔隙度较小的多孔材料中的冲击波压力相对较高。这表明，初始孔隙度对接触爆炸冲击波衰减特性的影响主要体现在两方面：一是对初始冲击波峰值压力的影响，二是对冲击波峰值压力随传播距离衰减的影响。鉴于本文实验结果，初始孔隙度对接触爆炸冲击波衰减特性的影响，主要反映在其对初始冲击波峰值压力的显著减弱效应，由传播过程所引起的冲击波衰减效应并不十分显著。

4 结束语

通过脉冲网络法对不同初始孔隙度的多孔铁中冲击波速度的测量，结合其完全压实区冲击 Hugoniot 方程，得到了初始孔隙度对冲击波在多孔铁中衰减特性的影响。结果表明，初始孔隙度对接触爆炸冲击波在多孔铁中衰减特性的影响，主要反映为其对初始冲击波峰值压力的显著减弱效应，而由传播过程所引起的冲击波衰减效应并不十分显著。另外，该实验方法对已知其相应密实材料中冲击波速度与粒子速度间关系常数 C_0 和 S 的多孔材料普遍适用，而且具有实验简便、可同时获得多个有效测点的优点。

参考文献

[1] HERRMANN W. Constitutive equation for the dynamic compaction of ductile porous materials [J]. Journal of applied physics, 1969, 40 (6): 2490 – 2499.

[2] CARROLL M M, Holt A C. Static and dynamic pore – collapse relations for ductile porous materials [J]. Journal of applied physics, 1972, 43 (4): 1626 – 1635.

[3] BOADE R R. Compression of porous copper by shock waves [J]. Journal of applied physics, 1968, 39 (12): 5693 – 5702.

[4] FRITZ J N, TAYLOR J W. An equation of state for adiprene foam and its application in producing low pressure – long time pulses [R]. LA – 3400 – M S, 1965.

[5] RICHARD A H, MANUEL J U. Shipping container for small samples of high explosives [R]. LA – 9107 – MS, 1981.

[6] MCQUEEN R G, MARSH S P, TAYLOR J W. High velocity impact phenomena [M]. Orlando: Academic Press, 1970.

[7] Ki – Hwan Oh, Per – Anders Persson. Equation of state for extrapolation of high – pressure shock Hugoniot data [J]. Journal of applied physics, 1989, 65 (10): 3852 – 3856.

[8] 李晓杰. 多孔材料冲击绝热线的近似计算 [J]. 高压物理学报, 1991, 5 (4): 301 – 306.

[9] 北京工业学院八系. 机电工程系爆炸及其作用: 上册 [M]. 北京: 国防工业出版社, 1979.

多孔铁中冲击波压力特性的研究①

王海福②，冯顺山
（北京理工大学机电工程系，北京 100081）

摘　要：本文的目的是研究孔隙度和粒度参数对爆炸载荷作用下多孔铁中冲击波压力特性的影响。采用锰铜压阻实验对多孔铁冲击波压力进行测量。在相同的爆炸载荷作用下，多孔铁中初始冲击波压力随孔隙度增大而下降，当孔隙度增大到0.35后，冲击波传播衰减效应随孔隙度增大而减弱；当粒度大于250目时，初始冲击波压力随粒度减小而显著下降；当粒度小于250目后，冲击波传播衰减效应随粒度减小呈逐渐减弱趋势。结论为：多孔铁中冲击波压力特性取决于孔隙度或粒度参数对初始冲击波压力以及冲击波传播衰减影响的综合效应。

关键词：多孔材料；冲击绝热线；冲击波衰减

Study on Shock Pressure Properties of Porous Iron

Wang Haifu, Feng Shunshan
(Department of Engineering Safety, Beijing Institute of Technology, Beijing 100081)

Abstract: Aim To study effects of porosities and particle sizes on shock pressures caused by explosion load in porous iron. Methods Using the strain compensated manganin gage to measure shock pressures in porous iron. Results Initial shock pressures in porous iron will decrease with increasing of porosity, and after porosity is larger than 0.35, shock propagating attenuations will become slow gradually with increasing of porosity. While particle size is larger than 250 mesh, initial shock pressures will decrease notably with decreasing of particle size, and after particle size is smaller than 250 mesh, shock propagating attenuations will be - come slow gradually with decreasing of particle size. Conclusion Shock pressure properties of porous iron are dependent on comprehensive effects of porosities or particle sizes on both initial shock pressures and shock propagating attenuations.

Key words: porous material; shock Hugoniot; shock attenuation

多孔材料内部存在大量孔隙，从而决定其不同于一般密实材料的冲击压缩特性。当多孔材料被冲击压缩到与其相应的密实材料具有相同的终态比容或压力时，所需冲击压力和能量较其相应的密实材料要高得多，材料内部将产生更高的冲击温升[1-5]。多孔材料因具有这种独特的冲击压缩性能而在军事和民用工程领域被广泛用作抗冲击防护材料。爆炸载荷作用下多孔材料中所形成的冲击波压力特性，是研究和评价多孔材料防护性能的前提与基础，对爆炸防护设计具有重要应用和参考价值。本文以多孔铁为研究对象，就爆炸载荷作用下孔隙度和粒度参数对其中冲击波压力特性的影响进行了较系统的实验研究。

1　实验测试系统

实验测试系统如图1所示，其中平面波发生器直径为50 mm，加载TNT药柱装药密度为1.58 g/cm^3，直径为45 mm，高度为50 mm，多孔铁试件厚度为3 mm，直径为50 mm。实验测试原理为：当平面波发

① 原文发表于《北京理工大学学报》1998（2）。
② 王海福：工学博士，1993年师从冯顺山教授，研究多孔材料对爆炸载荷的弱化效应及其机理，现工作单位：北京理工大学。

生器引爆加载 TNT 药柱后形成的平面冲击波，传至埋在两块相邻圆柱形多孔铁隔板间的锰铜压阻传感器时，锰铜计敏感部分电阻将随冲击波压力增大而增大，在恒流源恒定电流的供给下，由冲击波压力所引起锰铜计敏感部分电阻变化，可通过示波器所获电压 - 时间曲线体现。因此，通过锰铜压阻实验测出多孔铁中若干拉氏位置处的冲击波峰值压力，经数据处理后可得到孔隙度和粒度参数对冲击波压力影响特性。

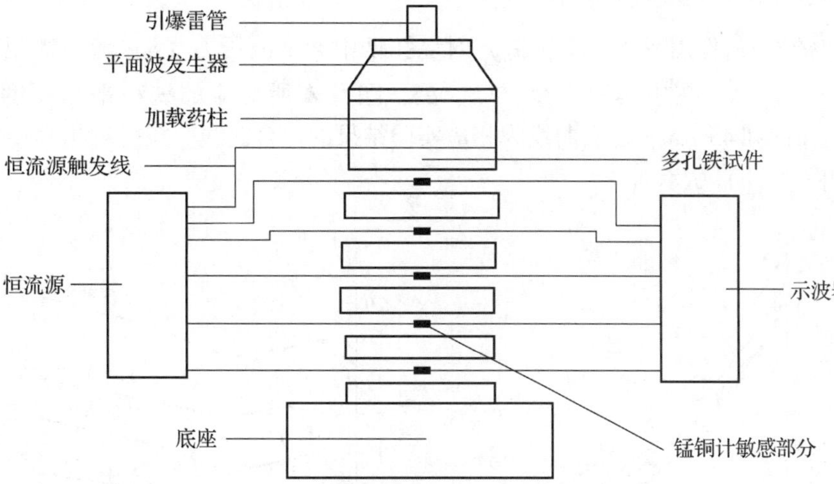

图 1　实验测试系统

2　孔隙度对冲击波压力特性影响实验结果

在强度相同的爆炸载荷作用下，孔隙度（$1-\rho_{00}/\rho_0$）对多孔铁中冲击压力（P）随传播距离（X）变化的影响特性如图 2（a）所示，图中实验点●、○、■、□及▲所对应的多孔铁初始密度（ρ_{00}）分别为铁晶体密度（ρ_0）85%、75%、65%、55% 和 45%（相应的孔隙度分别为 0.15、0.25、0.35、0.45、0.55），各实曲线为相应实验结果的拟合曲线。孔隙度对相同传播距离处冲击波压力的影响特性如图 2（b）所示。

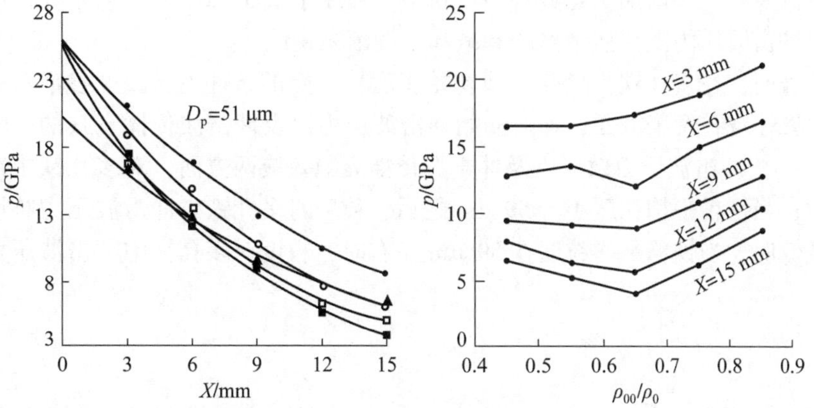

图 2　孔隙度对多孔铁中冲击波压力特性影响的实验结果

从图 2 可以清楚地看出，在装药特性相同的爆炸载荷作用下，当多孔铁初始密度从铁晶体密度的 85% 减小到 65%（或孔隙度从 0.15 增大到 0.35）时，在多孔铁中所形成的初始冲击波压力随孔隙度的增大（或初始密度的减小）而下降，但下降幅度并不显著，冲击波压力随传播距离的衰减呈逐渐加快趋势。当多孔铁初始密度从铁晶体密度的 65% 再减小到 45% 时，在多孔铁中所形成的初始冲击波压力随初始密度减小而显著下降，但冲击波压力随传播距离的衰减却呈逐渐减慢趋势。

上述实验结果表明，在铁粉平均粒度一定的条件下，孔隙度（或初始密度）对多孔铁中冲击波压力特性的影响主要体现在两个方面，即对初始冲击波压力以及冲击波传播衰减的影响。当孔隙度不是很大（初始密度小于铁晶体密度的65%）时，冲击波压力随传播距离衰减效应的决定作用较为显著；而当孔隙度较大时，初始冲击波压力效应的决定作用更为显著。当多孔铁初始密度在铁晶体密度的65%左右时，孔隙度对多孔铁中冲击波压力特性的综合影响效应最为显著。

3 粒度对冲击波压力特性影响实验结果

在强度相同的爆炸载荷作用下，粒度（D_p）对多孔铁中冲击波压力（p）随传播距离（X）变化的影响特性如图3（a）所示，图中实验点●、○、■、□及▲所对应的铁粉平均粒度分别为87 μm、71 μm、59 μm、51 μm和45 μm，各实曲线为相应实验结果的拟合曲线。粒度对相同传播距离处冲击波压力的影响特性如图3（b）所示。

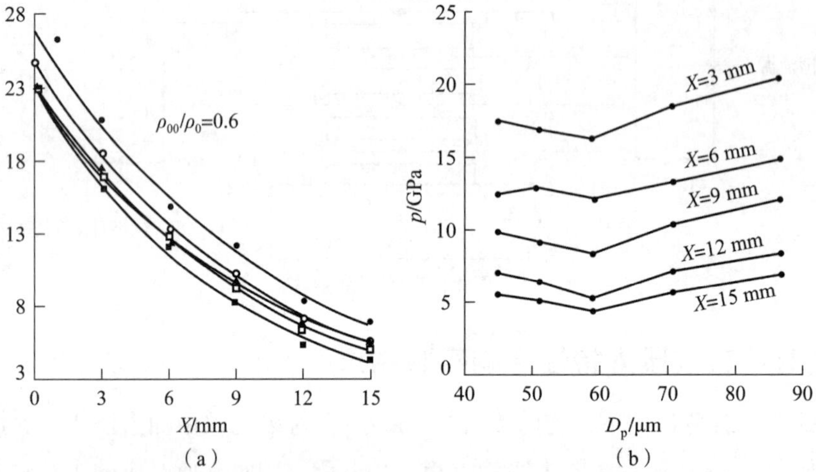

图3 粒度对多孔铁中冲击波压力特性影响的实验结果

从图3可以清楚地看出，在装药特性相同的爆炸载荷作用下，当铁粉平均粒度从87 μm（170目）减小到59 μm（250目）时，多孔铁中所形成的初始冲击波压力随铁粉平均粒度的减小而显著下降，冲击波压力随传播距离衰减呈逐渐加快趋势；当铁粉平均粒度小于59 μm后，初始冲击波压力基本不受粒度大小的影响，但冲击波压力随传播距离的衰减却呈逐渐减慢趋势。

上述实验结果表明，在多孔铁孔隙度（或初始密度）一定的条件下，粒度参数对多孔铁中冲击波压力特性的影响也主要体现在两个方面，即对初始冲击波压力以及冲击波传播衰减的影响。当铁粉平均粒度大于59 μm时，初始冲击波压力效应以及冲击波传播衰减效应两方面，对多孔铁中冲击波压力特性起着同等重要的作用；当铁粉平均粒度小于59 μm之后，粒度对多孔铁中冲击波压力特性的影响主要取决于冲击波传播衰减效应。当铁粉平均粒度在59 μm左右时，粒度对多孔铁中冲击波压力特性的综合影响效应最为显著。

4 结论

（1）孔隙度参数对多孔铁中冲击波压力特性的影响较粒度参数更为显著，无论是孔隙度还是粒度参数对多孔铁中冲击波压力特性的影响，均体现在初始冲击波压力以及冲击波传播衰减两方面。

（2）在强度相同的爆炸载荷作用下，多孔铁中初始冲击波压力随孔隙度增大而显著下降，当孔隙度小于0. 35时，冲击波传播衰减效应随孔隙度增大而加强；当孔隙度大于0. 35后，冲击波传播衰减效应随孔隙度增大呈逐渐减弱趋势。

（3）当铁粉平均粒度大于59 μm时，多孔铁中初始冲击波压力和冲击波传播衰减效应分别随粒度减小而下降和减弱；当铁粉平均粒度小于59 μm后，多孔铁中初始冲击波压力基本不受粒度的影响，但冲击波传播衰减效应随粒度减小呈逐渐减弱趋势。

节理岩体弹塑性动态有限元分析①

郭文章②，冯顺山，王海福

（北京理工大学机电工程系，北京　100081）

摘　要：目的研究节理岩体的弹塑性动态本构关系，并进行数值模拟．**方法**考虑动荷作用过程中，应力随时间变化这一应力历史，取单元为分离体，根据静力平衡原则，求出与之相应的节点集中力，再计算节点位移，得到节理岩体弹塑性动态有限元方程。**结果**给出了一种节理岩体在动荷作用下的有限元分析方法。**结论**通过动态有限元分析，有效地解决了节理岩体在爆炸载荷作用下的大塑性变形问题，并对爆破范围进行了数值模拟，从而向工程实用迈进了一步。

关键词：节理岩体；动态有限元；数值模拟

Analysis of the Elastic – Plastic Dynamic Finite Element on a Jointed Rock Mass

Guo Wenzhang，Feng Shunshan，Wang Haifu

（Department of Engineering Safety Beijing Institute of Technology Beijing 100081）

Abstract：Aim To study the elastic – plastic dynamical constitutive relations about a jointed rock mass under explosion load and its computer simulation. Methods Stress history is taken into account and stresses will follow changes in time during a period of explosion load. According to the principle of static force balance the corresponding nodal concentrated force is calculated and the nodal displacement is counted. The elastic – plastic dynamic finite element equations are thus obtained. Results. A finite element method is given for a jointed rock mass under explosion load. Conclusion The problem of large plastic deformation for jointed rock mass on blasting was efficiently resolved through dynamic finite element analysis and the range of damages by blasting simulated and this pushes forward the problem to engineering practice.

Key words：jointed rock mass；dynamic finite element；numerical simulation

Numerical method has widely been used by workers following popularization of the computer and accelerated developments in science and technology. Dynamic response of materials strides forward to engineering practice because of the appearance of dynamic finite element method. In the past years，Swoboda ［1］ and Li Ning ［2，3］ have set up finite element equations by the constraint joint element method and simulated the stress wave propagation through jointed rock mass but this method was based on the elastic theory and limited in this range i. e. by one – to – one correspondence for stress and strain. The nodal concentrated force and displacement was connected by the element stiffness matrix and then the element stiffness matrices were composed for the total stiffness matrix. Upon this the total line equations were obtained and finally solved. However，it will be plastically deformed even heavily plastically deformed under the explosion load as a con – sequence it cannot have one – to – one correspondence for stress and strain，so the stress history must be taken into account that is the stress changes

① 原文发表于《北京理工大学学报英文版》1999，8（2）。

② 郭文章：工学博士后，1998 年合作导师冯顺山，研究水中障碍的爆炸毁伤效应，现工作单位：北京城建勘测设计研究院。

should follow in time. Unlike the linear elastic case the solution of total stiffness matrix is found through the elementary stiffness matrix. Therefore the nodal concentrated force is computed by the principle of static force according to the elementary stress and the nodal displacement evaluated further forward. In this way a study of the dynamic response of the jointed rock mass under explosion load will approach the requirements in practice. Jointed rock mass blasting is a complex phenomenon controlled by many governing para – metric variables that generally cannot be included in a single analysis. The structured characteristic and the compose tensor of jointed rock mass are considered in building the constitutive relation. So it reflects the dynamic characteristics overall and systematically.

1 ANALYSIS OF DYNAMIC FINITE ELEMENT

1.1 Evaluation of the Nodal Concentrated Force

The nodal concentrated force is evaluated by the principle of static force with one element taken as a separate body and as a result the stress of the boundary becomes an external force which can be large or small corresponding to the orientation therefore the relevant nodal concentrated force can be obtained [4 5].

For the coplanar case as shown in Figure 1

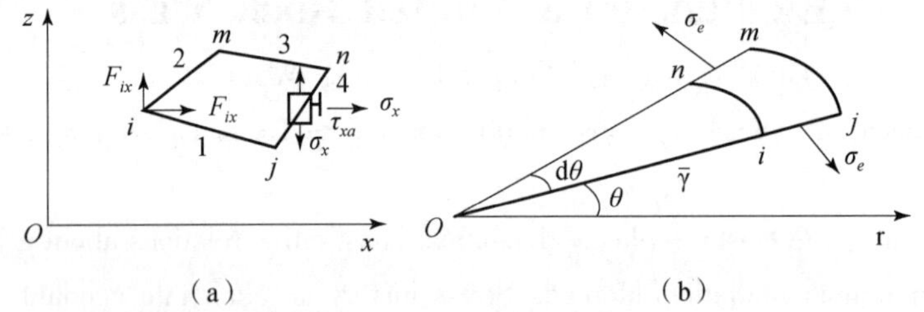

Figure 1 Evaluation for the nodal concentrated force

(a) coplanar case; (b) axisymmetric case

$$F_{ix1} = F_{jx1} = \frac{\delta}{2}[\sigma_x(Z_i - Z_j) + \tau_{zx}(X_j - X_i)] \tag{1}$$

$$F_{ix2} = F_{mx2} = \frac{\delta}{2}[\sigma_x(Z_i - Z_m) + \tau_{zx}(X_m - X_i)] \tag{2}$$

where F_{ix1} is the x direction concentrated force of nodal i in ij boundary with corresponding external force, others remaining the same, δ is the thickness, X, Z are coordinates of the node, so the total concentrated force of the node i in the x direction is given as

$$F_{ix} = F_{ix1} = F_{ix2} = \frac{\delta}{2}[\sigma_x(Z_m - Z_j) + \tau_{xz}(X_j - X_m)] \tag{3}$$

In a similar manner, it is given that

$$F_{iz} = \frac{\delta}{2}[\sigma_z(X_j - X_m) + \tau_{xz}(Z_m - Z_j)] \tag{4}$$

In a two dimensional axisymmetric case, the equations can be expressed similarly as follows:

$$F_{ir} = \pi \bar{r}[\sigma_r(Z_m - Z_j) + \tau_{r\theta}(R_j - R_m)] - \frac{2}{3}\pi A\theta, \tag{5}$$

$$F_{i\theta} = \pi \bar{r}[\tau_{r\theta}(Z_m - Z_j) \mid + \tau_{z\theta}(R_j - R_m)], \tag{6}$$

$$F_{iz} = \pi \bar{r}[\sigma_z(R_j - R_m) + \tau_{rz}(Z_m - Z_j)], \tag{7}$$

where A is an elementary area, $\bar{r}$ is the radius of the element center.

It can be obviously evaluated that the element contributes to other nodes of concentrated force of nodes j, m

and n using the same method. The nodal displacement of the next time step size will be computed after the nodal force is obtainted.

1.2 Computation of the Nodal Displacement

It has the nodal concentrated force, the nodal displacement will be evaluated by making use of Newton's second law according to the nodal concentrated force and the nodal mass.

Newton's second law according to the nodal concentrated force and the nodal mass. The displacements of node i with $u_{t+\Delta t}^{(i)}$, $u_{t+\Delta t}^{(i)}$ are respectively assumed in the x and y directions at the moment $t+\Delta t$, therefore it has

$$u_{t+\Delta_t}^{(i)} = u_t^{(i)} + \left(\dot{u}_t^{(i)} + \frac{F_x^{(i)} \Delta_t}{2m^{(i)}} \right) \Delta_t, \tag{8}$$

$$v_{t+\Delta_t}^{(i)} = v_t^{(i)} + \left(\dot{v}_t^{(i)} + \frac{F_y^{(i)} \Delta_t}{2m^{(i)}} \right) \Delta_t, \tag{9}$$

Where $F_x^{(i)}$, $F_y^{(i)}$, are respectively the component force at the node i in the x and y direction, $m^{(i)}$ is the mass of the node $\dot{i}$, $\dot{u}_t^{(i)}$, $\dot{v}_t^{(i)}$ are respectively the velocity components of the node i in the x and y direction.

1.3 The constitutive Relation of the Jointed Rock Mass

According to Ref. [5], we have

$$\Delta \bar{\sigma}_{ij} = C_{ijkl} \Delta \bar{\varepsilon}_{kl}, \tag{10}$$

Where $\Delta \bar{\sigma}_{ij}$ is the average stress, $\Delta \bar{\varepsilon}_{ij}$ is the average strain, and

$$C_{ijkl} = \frac{1}{V} \sum_{M=1}^{N} V^M C_{ijmr}^M H_{mrkl}^M,$$

$$V = \sum_{M=1}^{N} V^M$$

where N is the total number of elements; H_{mrkl}^M is the compose tensor; C_{ijkl}^M is a four matrix tensor of the M – th element; and

$$H_{mrkl}^M = \frac{\pi}{4} \rho \int_0^\infty \int_\Omega r^3 n_i n_j n_k n_l E(n,r) \, d\Omega dr,$$

ρ is the joint number of unit volume, r is the joint plane equivalent diameter, n is the joint plane perpendicular line vector, Ω is the angle field in space, $E(n, r)$ is the density function of probability distribution.

1.4 The Yield Criterion

Mohr – Coulomb criterion is adopted as the yield criterion about the jointed rock mass under explosion load when the joint produces shear move. Von Mises yield criterion is used when the jointed rock mass has a tension yield by tension stress wave.

The equation of elastic distortion energy is given as

$$U_d = \frac{1}{4G} S_{ij} S_{ij} = \frac{1}{2G} J_2 = \frac{1}{6G} \sigma_e^2,$$

where U_d is the elastic distortion energy, S_{ij} is the stress slanting tensor, $J_2 = 1/2 S_{ij} S_{ij}$ is the second unchanged amount of stress slanting measure, $\sigma_e = (3/2 S_{ij} S_{ij})^{1/2}$ is the efficient stress.

As it can easily be seen, all the above – mentioned parameters J_2 is in proportion to the elastic distortion energy (U_d), σ_e is in proportion to the square root of U_d. Therefore the judging criterion by J_2 or σ_e is the elastic distortion energy criterion. If σ_e is equal to the yield stress (σ_Y), of tensioned jointed rock mass when it is in tension by the blasting load, the jointed rock mass begins to yield.

2 ANALYSIS FOR EXAMPLES

Charge conditions: one hole 100 mm deep, 8 mm diameter, charge is 50 mm high, explosive used: RDX, priming charge: DDNP.

Jointed rock mass parameters, condensed force c = 0. 008 MPa, friction angle $\varphi = 40°$, modulus of elasticity $E = 3.1$ GPa, Poisson ratio $\upsilon = 0.25$, density $\rho = 2.08$ g/cm^3, dynamic yield strength $[\sigma_Y] = 28$ MPa, joint thickness 3mm, the distance between the joints 50 mm.

Numerical simulation is based on the DYNA - 2D program structure. The displacement field of different angles of jointed rock mass are shown in Figure 2 and Figure 3. The efficient stress fields at different angles are shown in Figure 4 and Figure 5.

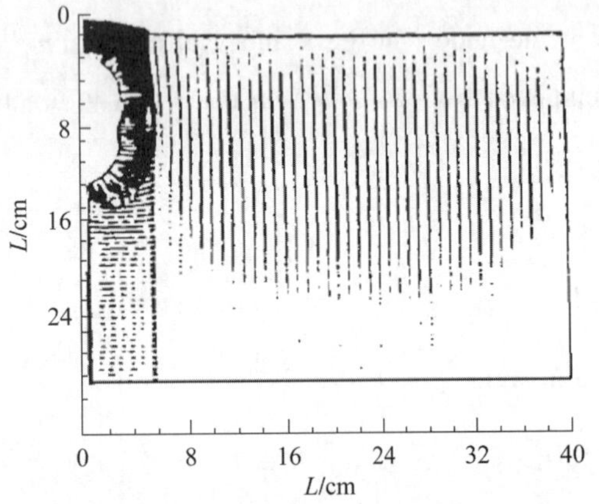

Figure 2 Displacement field of vertical joint

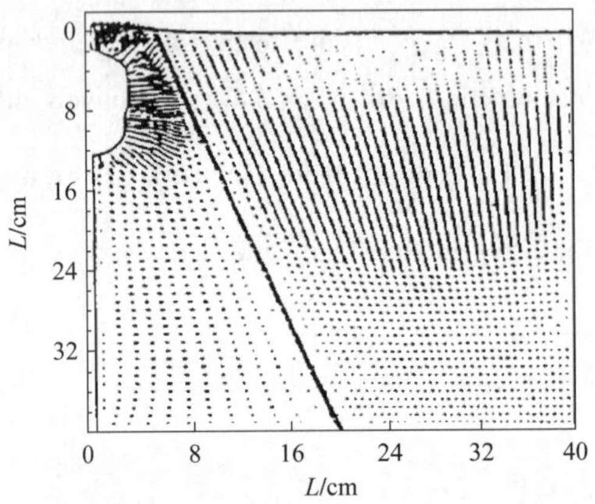

Figure 3 Displacement field of slope joint

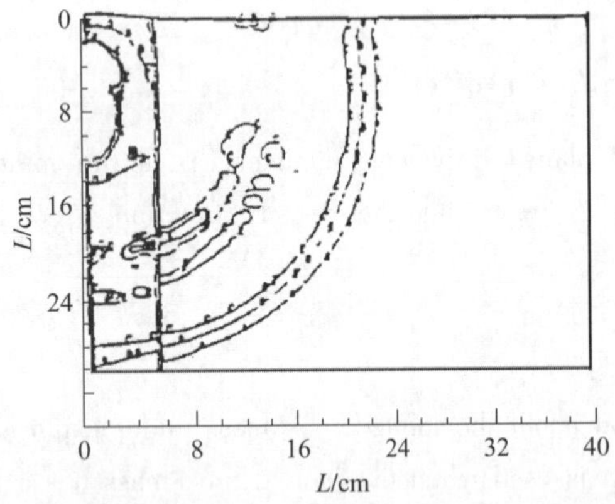

Figure 4 Efficient stress field of vertical joint

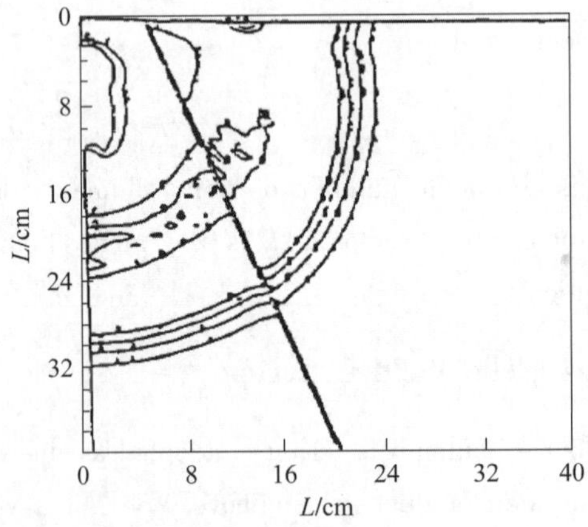

Figure 5 Efficient stress field of slope joint

It is obvious that the explosive solution is different when the difference angle included between the hole axis and the joint plane's vertical direction from the displacement field. The explosive gas flees early and effects shortly, so its effect is unsatisfactory because the angle between the hole axis and the joint plane's perpendicular line is 90°as given in Figure 2, in comparison with the case given in Figure 3. The *D* district is destroyed as shown in Figure 4 with a range of 9 cm and Figure 5 with a range of 12 cm. It is clear that the explosive effect is different obviously when the angle included between the hole axis and the joint plane's perpendicular line. The result shows a good agreement between test results and numerical simulation.

3　CONCLUSION

In this paper, the problem of large plastic deformation about a medium under dynamical load is solved effectively, and algorithm for setting up any stiffness matrices is avoided through evaluating the nodal concentrated force and further nodal displacement by the stress of element according to the principle of static force. By numerical simulation of jointed rock masses under explosion load, blasting design can be optimized, thus providing useful guidance to engineering practice.

References

[1] SWOBODA G, ZENZ G. Application of the 'decoupled finite element analysis'in tunneling [C]//RANDOLPH M F. Proceeding of the 6th ICONMIG. Austria: University of Innsbruck Press, 1988: 1465 - 1472.

[2] LI N. Wave propagation problems in the jointed rock mass [M]. Xi'an: Press of Northwest University of Technology, 1993.

[3] 李宁，雷晓燕，杜庆华．岩体节理在动荷作用下的有限元分析［J］．岩土工程学报，1994，16（1）：20－25.

[4] 恽寿榕，涂侯杰，梁德寿，等．爆炸力学计算方法［M］．北京：北京理工大学出版社，1995.

[5] 郭文章．节理岩体爆破过程数值模型及其实验研究［D］．北京：中国矿业大学，1997.

圆柱壳体结构在侧向爆炸冲击波载荷下的塑性破坏①

孙　韬[1②]，冯顺山[2]

(1. 北京北苑 89602 部队 23 分队，北京　100012；2. 北京理工大学，北京　100081)

摘　要： 在相关理论和假设的基础上，给出了自由状态下阶梯形质量分布金属圆柱壳体在侧向非对称脉冲载荷下的变形和运动方程，针对矩形脉冲载荷得出塑性响应解及临界断裂破坏判据，理论预测与实验结果吻合较好。

关键词： 圆柱壳体；冲击载荷；塑性变形；断裂破坏

PLASTIC DMAGE OF CYLINDRICAL SHELLS UNDER LATERAL IMPULSIVE LOADING

Sun Tao[1], Feng Shunshan[2]

(1. Research Institute of Artillery Equipment and Technology, Beijing, 100012

2. Department of Mechanical Engineering, Beijing Institute of Technology, Beijing, 100081)

Abstract: This paper uses a series of theories and hypo thesis for stepped cylindrical shells with free - free end subjected to lateral local distributed impulsive loading. Solutions for the deformation and velocity profile of the shell are obtained. Compared with experimental results, the theory predicted plastic response and the critical rupture impulse are performed w ell for the rectangular pressure pulse.

Key words: cylindrical shell; impulsive loading; plastic deformation; fracture failure

1　前言

自由壳体结构承受瞬时动态载荷下的变形和破坏分析，在空间方面具有重要的工程应用价值，如对飞行器在冲击载荷下的破坏准则研究或其防护设计问题等。瞬时动态载荷可以由外部物体的撞击、爆炸冲击波作用等产生，在作用于目标结构时，动载多非对称分布于结构的侧向或某一局部区域。图 1 所示两端自由敞口阶梯形质量分布圆柱壳体可近似为等效的飞行器结构或其基本组成单元之一[1]，其中两端 L_2 段为厚壳体部分，中间 $2L_1$ 段为薄壳体部分，两部分由同一种材料组成，脉冲载荷等效为 $P(x,\theta,t)=P(\theta,t)$。由于问题的复杂性和独特性，相关的理论分析和实验研究文献报道甚少。应用能量法可以大致给出壳体的冲击破坏判据，但其破坏系数近似为常数而与结构无关[2]。显然，应用理论和实验研究相结合的方法给出这种结构塑性破坏的近似分析解，建立破坏临界量与壳体结构相关参量之间的关系，是工程上更为关心的问题。

---

① 原文发表于《弹箭与制导学报》1999 (4)。

② 孙韬：工学博士，1995 年师从冯顺山教授，研究反舰导弹结构解体毁伤效应，现工作单位：陆军研究院炮兵防空兵研究所。

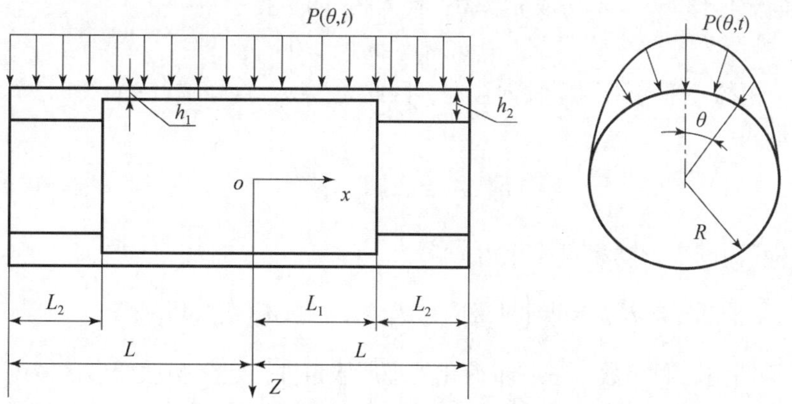

图1 阶梯形圆柱壳体几何形状及加载分布

2 基本方程

对图1所示圆柱壳，半径为 R，厚度 $h_2>h_1$。由结构对称性和加载特性，取一半进行研究，建立坐标系。压力载荷 $P(\theta,t)$ 作用在未变形壳体的侧向表面 S_0，载荷作用区域具有壳体半径的量阶或比半径小。压力分布分别对称于 $x=0$ 和 $\theta=0$。设材料为理想刚塑性的，流应力为 σ_0。由于质量分布不均匀，壳体承载后可能产生绕某一截面的弯曲，对应地产生沿轴向的位移。若总的转角及转动角速度分别为 ϕ 和 $\dot{\phi}$，总的轴向位移及其速度分别为 u 和 $\dot{u}$，假定 θ 和 u 远小于壳体 z 向和环向的最大位移 w 和 v，那么可以简单地取 $\phi=\dot{\phi}=0$，$u=\dot{u}=0$。这样，壳体的平衡可通过虚速度原理表示[3]

$$Q\dot{w}+\int_{S_0}T_i\dot{U}\mathrm{d}S_0+\int_{S_0}(-m\ddot{U})\dot{U}\mathrm{d}S_0=h\int_{S_0}\sigma_{ij}\,\dot{\varepsilon}_{ij}\mathrm{d}S_0 \tag{1}$$

式中 m——单位面积的质量；

σ_{ij}，$\dot{\varepsilon}_{ij}$——应力张量和应变率张量；

Q——自由端或表面作用的集中剪力；

T_i——表面应力矢量 [0，0，P]；

速度矢量为 $\dot{U}[0,\dot{v},\dot{w}]$，$\dot{v}$，$\dot{w}$——环向和 z 向的速度。壳体自由端边界条件为

$$Q=0,N=0,M=0 \tag{2}$$

N，M 分别为作用在端部的轴向拉力和弯矩。

设壳体承载后仅 L_1 段产生塑性运动，则式（1）右端单位体积内的变形功率

$$E_{\text{int}}=h_1\int_{S_0}\sigma_{ij}\dot{\varepsilon}_{ij}dS_0=\int_{S_0}(N_{\alpha\beta}\dot{\varepsilon}_{\alpha\beta}+M_{\alpha\beta}\dot{k}_{\alpha\beta})\,dS_0 \tag{3}$$

$\dot{\varepsilon}_{\alpha\beta}$、$\dot{k}_{\alpha\beta}$ 分别为中面的应变率张量和中面的曲率张量，$N_{\alpha\beta}$、$M_{\alpha\beta}$ 为壳体中的薄膜力张量和弯矩张量，它们必须满足一屈服条件

$$f(M_{\alpha\beta},N_{\alpha\beta})=0 \tag{4}$$

使得

$$\dot{k}_{\alpha\beta}=\dot{\lambda}\partial f/\partial M_{\alpha\beta},\dot{\varepsilon}_{\alpha\beta}=\dot{\lambda}\partial f/\partial N_{\alpha\beta} \tag{5}$$

式中 f——某一塑性势函数；

$\dot{\lambda}$——一比例常数。取无量纲参量

$$m_{\alpha\beta}=M_{\alpha\beta}/M_0,n_{\alpha\beta}=N_{\alpha\beta}/N_0 \tag{6}$$

其中，$M_0=\sigma_0h_1^2/4$，$N_0=\sigma_0h_1$ 分别为全塑性弯矩和全塑性轴向力。忽略壳中剪切变形，即令 $\dot{k}_{x\theta}=\dot{\varepsilon}_{x\theta}=0$ 之后，屈服条件（4）可以由 Huber－Mises 夹层塑性壳屈服条件表达[4]

$$n_{xx}^2-n_{xx}n_{\theta\theta}+n_{\theta\theta}^2+m_{xx}^2-m_{xx}m_{\theta\theta}+m_{\theta\theta}^2=1 \tag{7}$$

$$2n_{xx}m_{xx}+2n_{\theta\theta}m_{\theta\theta}-n_{xx}m_{\theta\theta}-n_{\theta\theta}m_{xx}=0 \tag{8}$$

考虑壳体在环向无延伸[2,5]，不妨取 $n_{\theta\theta}=0$，代入式（8）得

$$n_{xx}(2m_{xx}-m_{\theta\theta})=0 \tag{9}$$

变形的抵抗阻力中的一部分来源于轴向拉伸，因此 $n_{xx}\neq 0$，这样式（9）得 $m_{\theta\theta}=2m_{xx}$，代入式（7）得到屈服条件的新形式

$$n_{xx}^2+\frac{3}{4}m_{\theta\theta}^2=0 \tag{10}$$

同时式（3）单位体积内的变形功率可简化为

$$E_{\text{int}}=\int_{S_0}\left(N_{xx}\dot{\varepsilon}_{xx}+M_{\theta\theta}\dot{k}_{\theta\theta}+\frac{1}{2}M_{\theta\theta}\dot{k}_{xx}\right)\mathrm{d}S_0 \tag{11}$$

方程（10）为一个椭圆型屈服函数，进一步简化处理屈服条件，定义塑性变形区域内

$$|m_{\theta\theta}|=1,\ |n_{xx}|=1,\ |k_{xx}|\leqslant|\dot{k}_{\theta\theta}| \tag{12}$$

单位体积内的变形功率取其最大值

$$E_{\text{int}}=\int_{S_0}(|N_0\dot{\varepsilon}_{xx}|+1.5|M_0\dot{k}_{\theta\theta}|)\mathrm{d}S_0 \tag{13}$$

这样式（1）成为

$$\begin{aligned}&Q\dot{w}+2R\int_0^{\pi}\int_0^{L_1}P(\theta,t)\dot{w}(x,\theta,t)\mathrm{d}\theta\mathrm{d}x+2R\int_0^{\pi}\int_0^{L_1}[-m\ddot{w}(x,\theta,t)]\dot{w}(x,\theta,t)\mathrm{d}\theta\mathrm{d}x+\\&2R\int_0^{\pi}\int_0^{L_1}[-m\ddot{v}(x,\theta,t)]\dot{v}(x,\theta,t)\mathrm{d}\theta\mathrm{d}x=2R\int_0^{\pi}\int_0^{L_1}1.5|M_0\dot{k}_{\theta\theta}(x,\theta,t)|\mathrm{d}\theta\mathrm{d}x+\\&2R\int_0^{\pi}\int_0^{L_1}|N_0\dot{\varepsilon}_{xx}(x,\theta,t)|\mathrm{d}\theta\mathrm{d}x\end{aligned} \tag{14}$$

设刚体运动位移和塑性运动位移分别为 $W_0(x,\theta,t)$ 和 $W_1(x,\theta,t)$，即

$$w(x,\theta,t)=W_{01}(x,\theta,t)+W_{11}(x,\theta,t),\ 0\leqslant x\leqslant L_1 \tag{15a}$$

$$w(x,\theta,t)=W_{02}(x,\theta,t),\ L_1\leqslant x\leqslant L \tag{15b}$$

参照未变形构型，轴向应变率为

$$\dot{\varepsilon}_{xx}=\dot{u}'+W'_{11}\dot{W}'_{11} \tag{16}$$

这里$()'=\mathrm{d}()/\mathrm{d}x$。由前面假定 $\dot{u}=0$，式（16）简化为 $\dot{\varepsilon}_{xx}=W'_{11}\dot{W}_{11'}$。将截面上所有位置的塑性变形量均表示成 $\theta=0$ 处的函数，并引入等效参数[2,5]，将壳体在 $x=L_1$ 处分成两部分进行研究。不计 $x=L_1$ 处弯矩的影响，方程（14）等价于如下两个方程：

$$-Q\dot{W}_{02}(L_1,t)+\int_{L_1}^{L}(\overline{P}(t)-\overline{m}_{02}\ddot{W}_{02}(x,t))\dot{W}_{02}(x,t)\mathrm{d}x=0 \tag{17}$$

$$\begin{aligned}&Q\dot{W}_{01}(L_1,t)+(Q-\overline{N}_{11}W'_{11})\dot{W}_{11}\Big|_0^{L_1}+\int_0^{L_1}(P(t)-\overline{m}_{11}\ddot{W}_{11}(x,t)-\overline{m}_{01}\ddot{W}_{01}(x,t))\dot{W}_{01}(x,t)\mathrm{d}x\\&+\int_0^{L_1}(P(t)+\overline{N}_{11}W''_{11}(x,t)-\overline{m}_{11}(\ddot{W}_{01}(x,t)+\ddot{W}_{11}(x,t))-\overline{q}_{11})\dot{W}_{11}(x,t)\mathrm{d}x=0\end{aligned} \tag{18}$$

式中，Q 正负号不同表明在截面 $x=L_1$ 两端 Q 方向相反。对不同的变形截面形状，上述方程中的等效参数近似为

$$\overline{P}_{01}(t)=\overline{P}_{02}(t)=2R\int_0^{\pi}P(\theta,t)\mathrm{d}\theta \tag{19}$$

$$\overline{m}_{01}=2\pi Rm_1,\ \overline{m}_{02}=2\pi Rm_2,\ \overline{m}_{21}=\overline{m}_{11},\ \overline{m}_{31}=\overline{m}_{11} \tag{20}$$

$$\overline{P}_{11}(t)=2R\int_0^{\pi}P(\theta,t)\mathrm{d}\theta,\ \overline{q}_{11}=\frac{12M_0}{R},\ \overline{m}_{11}=\frac{2\pi R\,m_1}{3},\ \overline{N}_{11}=\frac{2\pi RN_0}{3},\text{平头形} \tag{21}$$

$$\overline{P}_{11}(t)=2RP(0,t)\theta_0,\ \overline{q}_{11}=\frac{12M_0}{R},\ \overline{m}_{11}=0.5Rm_1,\ \overline{N}_{11}=0.5RN_0,\text{凹陷形} \tag{22}$$

上述等效参量所取的值在 $0<W_{11}/R<0.5$ 时有效，m_1、m_2 分别为壳体 L_1 和 L_2 段单位面积的质量，

一般情况下，式（22）中的 $\theta_0=\pi/4$ 作为对这一变形的估算，此时 $\bar{P}_{11}(t)\approx 1.6RP(t)$。因此，在以下推导中不妨简单地取

$$\bar{P}(t)=\bar{P}_{11}(t)=\bar{P}_{01}(t)=\bar{P}_{02}(t) \tag{23}$$

由 L_2 段不变形的假定，则从式（17）得

$$Q=(\bar{P}(t)-\bar{m}_{02}\ddot{W}_{02})(L-L_1)=(\bar{P}(t)-\bar{m}_{02}\ddot{W}_{01}(L_1,t))L_2 \tag{24}$$

代入式（18）中第一项并做变换，近似取$\dfrac{\ddot{W}_{01}(L_1,t)}{\dot{W}_{01}(x,t)}\sim 1$，则

$$\begin{aligned}Q\dot{W}_{01}(L_1,t)&=\int_0^{L_1}(\bar{P}(t)-\bar{m}_{02}\ddot{W}_{01}(L_1,t))\frac{L_2}{L_1}\dot{W}_{01}(L_1,t)\,\mathrm{d}x\\&=\int_0^{L_1}(\bar{P}(t)-\bar{m}_{02}\ddot{W}_{01}(L_1,t))\frac{L_2}{L_1}\dot{W}_{01}(x,t)\,\mathrm{d}x\end{aligned} \tag{25}$$

由变分计算的一般方法，取 $\dot{W}_{01}\delta t=\delta W_{01}, W_{11}\delta t=\delta W_{11}$，则方程（18）化简为

$$\frac{L}{L_1}\bar{P}(t)-\left(\frac{L_2}{L_1}\bar{m}_{02}+\bar{m}_{01}\right)\ddot{W}_{01}(x,t)-\bar{m}_{11}\ddot{W}_{11}(x,t)=0 \tag{26}$$

$$\bar{P}(t)+\bar{N}_{11}W''_{11}-\bar{m}_{11}\ddot{W}_{11}-\bar{m}_{11}\ddot{W}_{11}-\bar{q}_{11}=0 \tag{27}$$

$$(Q-\bar{N}_{11}W''_{11})\delta W_{11}=0, x=0, L_1 \tag{28}$$

初始条件为

$$W_{01}(x,0)=\dot{W}_{01}(x,0)=W_{11}(x,0)=\dot{W}_{11}(x,0)=0 \tag{29}$$

由式（26）得刚体运动与塑性运动的关系为

$$\ddot{W}_{01}(x,t)=L\bar{P}(t)/(L_1\bar{m}_k)-\bar{m}_{11}\ddot{W}_{11}(x,t)\bar{m}_k \tag{30}$$

其中，$\bar{m}_k=(L_1\bar{m}_{01}+L_2\bar{m}_{02})/L_1$。将式（30）代入式（27）化简得塑性运动控制方程为

$$\bar{P}_m-\bar{q}_m+c_m^2W''_{11}-\ddot{W}_{11}=0$$

$$\bar{P}_m=\bar{P}\left(1-\frac{L\bar{m}_{11}}{L_1\bar{m}_k}\right)\Big/\left[\bar{m}_{11}\left(1-\frac{\bar{m}_{11}}{\bar{m}_k}\right)\right],\bar{q}_m=\bar{q}_{11}\Big/\left[\bar{m}_{11}\left(1-\frac{\bar{m}_{11}}{\bar{m}_k}\right)\right] \tag{31}$$

其中

$$c_m^2=\frac{\bar{N}}{\bar{m}_{11}}\Big/\left(1-\frac{\bar{m}_{11}}{\bar{m}_k}\right)=c^2\Big/\left(1-\frac{\bar{m}_{11}}{\bar{m}_k}\right),c^2=\frac{\bar{N}}{\bar{m}_{11}}=\frac{\sigma_0}{\rho}$$

方程（28）实际上为边界条件。显然，$x=0$ 处由对称性有 $Q=0$，及

$$W'_{11}(0,t)=0 \tag{32}$$

式（28）自然满足。$x=L_1$ 对应于如下两种运动形态。

运动形态 1：$x=L_1$ 处不发生塑性变形和转动，即

$$W_{11}(L_1,t)=0 \tag{33}$$

运动形态 2：L_2 段壳体相对于初始状态发生了刚体转动，$W_{11}(L_1,t)\neq 0$，则有

$$Q-\bar{N}_{11}W'_{11}(L_1,t)=0 \tag{34}$$

方程（31）的解与加载规律有关。在载荷作用下，运动形态 1 对应于壳体承载后塑性变形不大的状态；运动形态 2 对应于壳体塑性变形较大的情形，断裂破坏将在这种情形下发生。本文仅考虑运动形态 2 在矩形脉冲载荷下的解。

3 矩形脉冲载荷作用下的解

设方程（31）中

$$\bar{P}(t)=\begin{cases}\bar{P}, & 0\leqslant t\leqslant\tau \quad (35a)\\ 0, & t\geqslant\tau \quad (35b)\end{cases}$$

先对边界条件式（34）进行处理，由式（24）、式（30）、式（34）得

$$L_2\bar{P}(t)\left[1-\frac{L\bar{m}_{02}}{L_1\bar{m}_k}\right]+\frac{L_2\bar{m}_{11}\bar{m}_{02}}{\bar{m}_k}\ddot{W}_{11}(L_1,t)-\bar{N}_{11}W'_{11}(L_1,t)=0 \tag{36}$$

为求解方便，取如下无量纲量

$$\tilde{t}=tc/L_1,\ \tilde{w}=\frac{W_{11}\bar{N}_{11}}{L_1^2\bar{q}_{11}},\ \tilde{x}=x/L_1 \tag{37}$$

则运动控制方程形式转化为

$$\begin{aligned}&A_1+A_2\tilde{w}_{\tilde{x}\tilde{x}}-\tilde{w}_{\tilde{t}\tilde{t}}=0\\&\tilde{w}=\tilde{w}_{\tilde{t}}=0 \quad \tilde{t}=0\\&\tilde{w}_{\tilde{x}}=0 \quad \tilde{x}=0\\&B_1+B_2\tilde{w}_{\tilde{x}}-\tilde{w}_{\tilde{t}\tilde{t}}=0 \quad \tilde{x}=1\end{aligned} \tag{38}$$

其中

$$A_1=\left[\frac{\bar{p}}{\bar{q}_{11}}\left(1-\frac{L\bar{m}_{11}}{L_1\bar{m}_k}\right)-1\right]A_2,A_2=1\Big/\left(1-\frac{\bar{m}_{11}}{\bar{m}_k}\right) \tag{39}$$

$$B_1=\frac{L_1\bar{m}_k}{L_2\bar{m}_{02}}\left(1-\frac{\bar{m}_{11}}{\bar{m}_k}\right),B_2=\frac{\bar{P}(t)}{\bar{q}_{11}}\left(\frac{\bar{m}_k}{\bar{m}_{02}}-\frac{L}{L_1}\right)\left(1-\frac{\bar{m}_{11}}{\bar{m}_k}\right) \tag{40}$$

针对式（35），壳体的塑性响应将有两相。

对于第一相运动，$0\leqslant t\leqslant\tau,\bar{P}(t)=\bar{P}$。求解方程（38）得

$$\tilde{w}(\tilde{x},\tilde{t})=\psi_1(1-\tilde{x}^2)+\zeta_1\tilde{t}^2+\sum_{n=0}^{\infty}\hat{a}_n\cos\lambda_n\sqrt{A_2}\,\tilde{t} \tag{41}$$

λ_n 由如下特征方程确定：

$$tg\lambda_n=\lambda_n/B_1$$

及

$$\psi_1=\frac{A_1+B_2}{2(A_2-B_1)},\zeta_1=-\frac{A_1B_1+A_2B_2}{2(A_2-B_1)} \tag{42}$$

$$\hat{a}_0=-\frac{2}{3}\psi_1,\hat{a}_n=\frac{4\psi_1(B_1-1)\cos\lambda_n}{\lambda_n^2(B_1+\cos^2\lambda_n)} \tag{43}$$

对于第二相运动，$t\geqslant\tau$，此时运动方程变为

$$\begin{aligned}&A'_1+A_2\tilde{w}_{\tilde{x}\tilde{x}}-\tilde{w}_{\tilde{t}\tilde{t}}=0\\&\tilde{w}_{\tilde{x}}=0 \quad \tilde{x}=0\\&-B_1\tilde{w}_{\tilde{x}}+\tilde{w}_{\tilde{t}\tilde{t}}=0 \quad \tilde{x}=1\\&\tilde{w}=\tilde{w}(\tilde{x},\tilde{\tau}) \quad \tilde{t}=\tilde{\tau}\\&\tilde{w}_{\tilde{t}}=\tilde{w}_{\tilde{t}}(\tilde{x},\tilde{\tau}) \quad \tilde{t}=\tilde{\tau}\end{aligned} \tag{44}$$

其中，$A'_1=-A_2$，方程（44）的解为

$$\begin{aligned}\tilde{w}(\tilde{x},t)=&\ \Omega_0+\zeta_2\tilde{t}^2+\psi_2(1-\tilde{x}^2)+\\&\sum_{n=0}^{\infty}(\Omega_n\cos(\lambda_n\sqrt{A_2}\,\tilde{t})+\Lambda_n\sin(\lambda_n\sqrt{A_2}\,\tilde{t}))\cos\lambda_n\tilde{x}\end{aligned} \tag{45}$$

其中

$$\psi_2 = \frac{A'_1}{2(A_2 - B_1)}, \zeta_2 = -\frac{B_1}{2(A_2 - B_1)}, \Omega_0 = \hat{b}_0$$

$$\Omega_n = \hat{b}_n \cos\lambda_n \sqrt{A_2}\tilde{\tau} - (\tilde{c}/\lambda_n)\sin\lambda_n \sqrt{A_2}\tilde{\tau}$$

$$\Lambda_n = \hat{b}_n \sin\lambda_n \sqrt{A_2}\tilde{\tau} - (\tilde{c}/\lambda_n)\cos\lambda_n \sqrt{A_2}\tilde{\tau}$$

$$\hat{b}_0 = -\frac{2\psi_2}{3} + (\zeta_1 - \zeta_2)\tilde{\tau}^2 + \sum_1^{\infty} \frac{\tilde{a}_n}{\lambda_n}\cos\lambda_n \sqrt{A_2}\tilde{\tau}\sin\lambda_n$$

$$\hat{b}_n = \left[\left((\zeta_1 - \zeta_2)\tilde{\tau}^2 - \frac{2\psi_1}{3}\right)\frac{\sin\lambda_n}{\lambda_n} + 2(\psi_2 - \psi_1)\left(1 - \frac{1}{B}\right)\frac{\cos\lambda_n}{\lambda_n^2} + \frac{\hat{a}_n}{2}\left(1 + \frac{\cos^2\lambda_n}{B_1}\right)\cos\lambda_n \sqrt{A_2}\tilde{\tau}\right]/\Delta$$

$$\hat{c}_n = \left[2(\zeta_1 - \zeta_2)\tilde{\tau}\frac{\cos\lambda_n}{\lambda_n} - \frac{\hat{a}_n}{2}\left(\lambda_n + \frac{\sin 2\lambda_n}{2}\right)\sin\lambda_n \sqrt{A_2}\tilde{\tau}\right]/\Delta$$

$$\Delta = 0.5\left(1 + \frac{\cos^2\lambda_n}{B_1}\right) \tag{46}$$

简单地取式（45）中 $n=1$，化简有

$$\hat{w}(\tilde{x}, \tilde{t}) = \hat{b}_0 + \psi_2(1 + \tilde{x}^2) + \zeta_2\tilde{t}^2 + \left(\hat{b}_1\cos(\lambda_1 \sqrt{A_2}(\tilde{t} - \tilde{\tau})) + \frac{\hat{c}_1}{\lambda_1}\sin(\lambda_1 \sqrt{A_2}(\tilde{t} - \tilde{\tau}))\right)\cos\lambda_1\tilde{x} \tag{47}$$

比较得知，应变在 $\tilde{x}=1$ 处为最大，此处成为最危险截面。当此处最大应变达到材料的临界断裂应变值 ε_c 时，结构发生断裂破坏现象。因此，取临界断裂破坏判据为[5]

$$\frac{1}{2}\left(\frac{\partial W_{11}}{\partial x}\right)^2 = \varepsilon_c \tag{48}$$

将式（47）代入式（48）中得对应于壳体结构的临界破坏判据

$$\frac{L_1\bar{q}_{11}}{\bar{N}_{11}}\left(2\psi_2 + \sqrt{\hat{b}_1^2 + \frac{\hat{c}_1^2}{\lambda_1^2}}\lambda_1\sin\lambda_1\right) = \sqrt{2\varepsilon_c} \tag{49}$$

将相关项代入式（49），化简得如下形式：

$$\frac{4\bar{P}\tau\bar{m}_c\sin^2\lambda_1}{\bar{m}_{11}\lambda_1(B_1 + \cos^2\lambda_1)} = \sqrt{2\frac{\sigma_0\varepsilon_c}{\rho}} \tag{50}$$

其中

$$\bar{m} = \frac{\bar{m}_k}{\bar{m}_{01}}\left[\frac{\bar{m}_{01}}{\bar{m}_{02}} + \left(1 + \frac{L_1\bar{m}_{01} - L\bar{m}_{11}}{L_2\bar{m}_{02}}\right)\frac{\bar{m}_k}{\bar{m}_k - \bar{m}_{11}} - 1\right] - \frac{B_1(\bar{m}_{01} - \bar{m}_{11})}{2\bar{m}_{02}} \tag{51}$$

对应于不同的变形截面，得到壳体在 $x = L_1$ 处产生临界断裂破坏的比冲量为

$$I_c = \begin{cases} \dfrac{\sqrt{2}\pi\lambda_1(B_1 + \cos^2\lambda_1)}{12\bar{m}_c\sin^2\lambda_1}h_1\sqrt{\sigma_0\rho\varepsilon_c}, & \text{平头变形截面} \\ \dfrac{\sqrt{2}\lambda_1(B_1 + \cos^2\lambda_1)}{16\theta_0\bar{m}_c\sin^2\lambda_1}h_1\sqrt{\sigma_0\rho\varepsilon_c}, & \text{凹陷变形截面} \end{cases} \tag{52}$$

一般性地，可将上述临界破坏比冲量统一表示成

$$I_c = k_c h_1\sqrt{\sigma_0\rho\varepsilon_c} \tag{53}$$

式中，k_c 定义为临界破坏系数。对应不同的变形截面和结构参量，k_c 取对应值。

4　实验与计算

我们应用 LY11CZ 硬铝圆柱壳和 45 号钢圆柱壳进行了部分缩比试验。铝壳体材料力学性能经测定其

拉伸强度 $\sigma_b=370$ MPa，$\sigma_{0.5}=200$ MPa，$\varepsilon_c=0.12$。45 号钢壳体材料 $\sigma_s=360$ MPa，$\varepsilon_c=0.11$。壳体结构如图 1 所示，均由 3 个环壳加工组成。其中铝外壳总长度 57 mm，壁厚 1 mm，外径 40 mm，两端通过螺纹各内嵌有一长 8.5 mm、内径 28 mm、外径 38 mm 的厚壁环壳；钢壳结构有两种，外径 38 mm，壳体总长度 67 mm，壁厚分别为 0.9 mm 和 1.32 mm，$2L_1$ 和 h_2 均为 45 mm 和 7 mm。侧向爆炸冲击波载荷由 TNT/RDX（50/50）裸装药给出，装药质量及装药表面与壳体表面之间的距离见表 1。试验表明，变形较大时，壳体横截面均变形成凹陷形。由此，壳体产生临界断裂破坏的比冲量从式（52）得

$$I_c=\begin{cases}0.48h_1\sqrt{\sigma_0\rho\varepsilon_c}=126\ \mathrm{N\cdot s/m^2} & 铝壳(1.00\ \mathrm{mm})\\ 0.45h_1\sqrt{\sigma_0\rho\varepsilon_c}=226\ \mathrm{N\cdot s/m^2} & 钢壳(0.90\ \mathrm{mm})\\ 0.44h_1\sqrt{\sigma_0\rho\varepsilon_c}=322\ \mathrm{N\cdot s/m^2} & 钢壳(1.32\ \mathrm{mm})\end{cases}$$

临界破坏系数 k_c 对三种壳体分别为 0.48，0.45 和 0.44。

表 1　自由壳体冲击破坏试验结果

材料	装药质量 /kg	距离 /m	冲击波超压 /MPa	冲击波比冲量 /（Pa·s）	破坏现象
LY11CZ 铝	0.029 3	0.34	0.766	99	未变形
LY11CZ 铝	0.035 5	0.32	1.296	144	迎载面变形成凹陷形状，一端出现细微裂纹，长 7 mm
45 号钢（0.9 mm）	0.039 3	0.24	2.576	266	两端均撕裂开，迎载面中央凹陷严重并贴向另一面
45 号钢（1.3 mm）	0.044 3	0.26	2.390	256	未变形
45 号钢（1.3 mm）	0.045 3	0.23	3.198	334	迎载面变形成凹陷形状，一端宏观开裂，裂纹长 16 mm

表 1 列出了自由壳体在侧向爆炸冲击波作用下的试验结果，其中作用在壳体表面的压力场应当由反射冲击波和绕流波的相互作用场确定。载荷均等效成式（35）的矩形脉冲载荷，等效峰值载荷大约为反射冲击波峰值载荷和绕流波峰值载荷之差的 1/4[2,6]。对照计算结果可见，本文理论预测出的自由圆柱壳塑性变形临界压力值及壳体产生临界断裂破坏的比冲量与实验有相当的一致性。

5　结论

针对飞行器目标圆柱壳形结构体在侧向爆炸冲击波作用下的毁伤破坏问题，本文采用等效的自由阶梯形质量分布金属圆柱壳体进行近似的工程分析。在相关假设和理论分析的基础上，得出了壳体结构产生断裂破坏的临界比冲量与结构参量之间的函数关系。并应用小尺寸钢壳体结构和铝壳体结构进行了实验验证，计算与实验所得结果很接近。显然，壳形结构出现塑性变形和断裂破坏是结构最终产生解体破坏等现象的充分与必要条件。结构的临界开裂，一方面表明实际结构已失去其应起的某些功能或产生其他毁伤效应；另一方面，进一步的载荷有可能导致裂纹扩展到整个边界。本文的结果为相关空间飞行结构体在爆炸冲击波载荷下的破坏分析或工程设计奠定了基础，并具有一定的参考价值。

爆炸载荷下聚氨酯泡沫材料中冲击波压力特性[①]

王海福[②]，冯顺山

（北京理工大学爆炸灾害预防与控制国家重点实验室，北京 100081）

摘 要：本文采用间接测压法，对爆炸载荷作用下聚氨酯泡沫材料中的冲击波压力特性进行了研究。结果表明，在装药特性相同的爆炸载荷作用下，聚氨酯泡沫材料中的初始冲击波压力随初始孔隙度增大而显著下降；冲击波传播衰减效应受初始孔隙度影响同样显著，但不呈单一加快或减慢趋势；当初始孔隙度在0.25左右时，聚氨酯泡沫材料对爆炸载荷具有较强的抗冲击减压效能。

关键词：多孔材料；泡沫材料；冲击绝热线；冲击波衰减

PROPERTIES OF SHOCK PRESSURE CAUSED BY EXPLOSION LOADS IN POLYURETHANE FOAM

Wang Haifu, Feng Shunshan

(National Key Lab of Prevention Control of Explosion Disasters, BIT, Beijing 100081)

Abstract: Properties of shock pressure caused by explosion loads in polyurethane foam have been studied by means of an indirect experimental method. It is found that under the same charge and explosion condition, the initial shock pressure in polyurethane foam will decrease notably with increasing of the initial porosity. Effects of the initial porosity on propagating attenuation of the shock pressure are also notable, but its tendency is not completely corresponding to tendency of the initial porosity's changing.

Polyurethane foam with an initial porosity about 0.25 are more effective as protective buffer materials.

Key words: porous materials; cellular materials/foam; Hugoniot; shock attenuation

1 引言

聚氨酯泡沫材料在国防、军事领域广泛用于抗冲击防护材料，如复合装甲防护板、导弹头防护罩、大炮气密垫、飞机整流罩等。关于准静态加载条件下聚氨酯泡沫材料的压缩行为和能量吸收机理问题，人们已进行过多方面的研究，并取得了不少研究成果[1,2]，但对于爆炸载荷作用下冲击波压力传播衰减问题，研究则不多。爆炸载荷作用下聚氨酯泡沫材料中冲击波压力特性，是评价其抗冲击防护性能的前提与基础，对爆炸防护设计有重要的参考和应用价值。

采用传统锰铜压阻传感器对具有较大孔隙度的泡沫材料中的冲击波压力进行实验测试时，往往因孔隙与锰铜计敏感部分尺寸相当或更大，锰铜计无法获得较理想的压阻信号，从而影响测试精度。我们提出了一种间接测压实验方法，并就初始孔隙度对爆炸载荷作用下聚氨酯泡沫材料中冲击波压力特性的影响进行了研究。

2 实验系统与测试原理

实验测试系统如图1所示，其中平面波发生器直径为50 mm，加载TNT药柱密度、直径和高度分别

---

① 原文发表于《爆炸与冲击》1999（1）。

② 王海福：工学博士，1993年师从冯顺山教授，研究多孔材料对爆炸载荷的弱化效应及其机理，现工作单位：北京理工大学。

为1.58 g/cm^3、45 mm 和 50 mm，聚氨酯泡沫材料试件直径为 50 mm。鉴于标准材料须有准确冲击 Hugoniot 数据的要求，实验中选用直径为 50 mm 的 LY－12 铝做标准材料，锰铜计埋放在距分界面 5 mm 处。实验测试原理为，先测出 LY－12 铝对冲击波压力传播的衰减系数α；将不同初始密度和厚度的聚氨酯泡沫材料试件放置在 LY－12 铝板上，分别测出距分界面 5 mm 处 LY－12 铝中冲击波压力，并借助α值推算出入射到 LY－12 铝中的初始冲击波压力；最后，借助分界面阻抗匹配原理和聚氨酯泡沫材料高压 Hugoniot 方程，换算出聚氨酯泡沫材料中的冲击波压力。

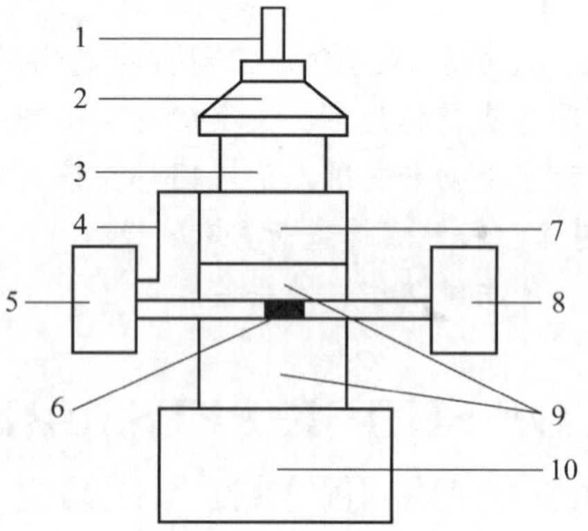

图 1　实验测试系统

1—Detonator；2—Plane wave generator；3—Explosive pad；4—Trigger line；
5—Constant current power；6—Manganin gage；7—Polyurethane foam；
8—Oscilloscope；9—LY－12 aluminium；10—Supporting pad

3　实验结果与冲击波压力换算

3.1　实验结果

研究表明[3]，冲击波峰值压力在密实介质中的传播行为呈指数衰减规律，衰减系数取决于介质的冲击 Hugoniot 参数，与冲击波强度基本无关。鉴此，实验中首先对冲击波在 LY－12 铝中的衰减系数进行了测试。表 1 给出了冲击压力与 LY－12 铝板厚度间的对应实验结果，表中 L 为锰铜计与加载面间的距离，p_L 为锰铜计所测冲击波峰值压力。

表 1　冲击波压力与 LY－12 铝板厚度间的对应实验结果

L/mm	4	6	8	10	12	14	16	18
p_L/GPa	17.8	15.1	13.9	11.8	11.3	9.8	8.4	7.9

对表 1 中实验数据用最小二乘法进行指数拟合，即 $p_L = p_0\exp(-\alpha L)$，得到 LY－12 铝中冲击波压力传播衰减系数为 $\alpha = 0.0583$。

在此基础上，对爆炸载荷作用下聚氨酯泡沫材料中的冲击波压力特性进行了实验测试，表 2 给出了聚氨酯泡沫材料初始孔隙度及厚度与冲击波压力间的对应实验结果。表中 H 为聚氨酯泡沫材料厚度，p_{LH} 为距聚氨酯泡沫材料－铝板分界面 5 处所测得的冲击波峰值压力，p_f 为根据 LY－12 铝中冲击波衰减规律外推而得到的入射到铝板中的初始冲击波峰值压力，ρ_{00}、ρ_0 分别为聚氨酯泡沫材料及相应密实材料的初始密度。

表 2　冲击波压力与聚氨酯泡沫材料初始孔隙度及厚度间的对应实验结果

ρ_{00}/ρ_0	0.85		0.75		0.65		0.55		0.45	
H/mm	p_{LH}/GPa	p_f/GPa	p_{LH}/GPa	p_f/GPa	p_{LH}/GPa	p_f/GPa	p_{LH}/GPa	p_f/GPa	p_{LH}/GPa	p_f/GPa
5	12.6	16.8	11.8	15.8	12.6	16.9	13.6	18.2	15.3	20.5
8	11.7	15.4	10.5	14.1	10.8	14.5	12.8	17.1	14.4	19.2
11	9.8	13.1	8.3	11.1	10.3	13.7	10.9	14.6	13.2	17.8
14	9.1	12.2	7.2	9.7	8.5	11.4	10.1	13.3	11.6	15.5
17	7.7	10.3	6.6	8.8	7.6	10.1	8.4	11.3	10.7	14.4
20	7.1	9.5	5.5	7.4	6.9	9.7	7.9	10.6	10.3	13.7

3.2　冲击波压力换算

聚氨酯泡沫材料中冲击波压力换算如图 2 所示，图中曲线Ⅰ和Ⅲ分别为聚氨酯泡沫材料中右传入射冲击波和左传反射冲击波 Hugoniot 曲线，且两者呈镜像对称关系[4]，曲线Ⅱ为 LY－12 铝中右传透射冲击波 Hugoniot 曲线。

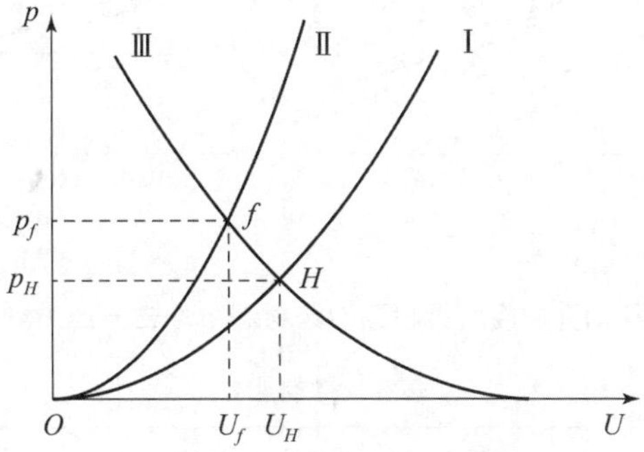

图 2　聚氨酯泡沫材料中冲击波压力换算

根据镜像对称原理，曲线Ⅲ、Ⅰ满足如下恒等函数关系：

$$p_{\mathrm{III}}(U_{\mathrm{III}}) = p_{\mathrm{I}}(2U_H - U_{\mathrm{III}}) \tag{1}$$

考虑冲击波阵面动量守恒以及分界面压力和速度连续条件，有

$$p_{\mathrm{I}}(U_{\mathrm{I}}) = \rho_{\mathrm{I}}(C_{\mathrm{I}} + S_{\mathrm{I}}U_{\mathrm{I}})U_{\mathrm{I}} \tag{2}$$

$$p_{\mathrm{II}}(U_{\mathrm{II}}) = \rho_{\mathrm{I}}[A + B(2U_H - U_{\mathrm{III}}) + C(2U_H - U_{\mathrm{III}})^2](2U_H - U_{\mathrm{III}}) \tag{3}$$

$$p_{\mathrm{III}}(U_f) = p_f = p_{\mathrm{II}}(U_f) \tag{4}$$

式中　U_{III}、U_{II}——曲线Ⅲ和Ⅱ上任一点处的粒子速度；

C_{II}、S_{II}——LY－12 铝中冲击波速度与粒子速度间的经验关系常数；

ρ_{I}、ρ_{II}——聚氨酯泡沫材料和 LY－12 铝的初始密度；

A、B 和 C——聚氨酯泡沫材料中冲击波速度与粒子速度间的拟合关系系数，并可通过联立以下 3 个方程而得到[5]

$$p_{H0} = \frac{C_{\mathrm{I}}^2(v_0 - v)}{[v_0 - S_{\mathrm{I}}(v_0 - v)]^2} \cdot \frac{1 - (0.5\Gamma_0/v_0)(v_0 - v)}{1 - (0.5\Gamma_0/v_0)(v_{00} - v)} \tag{5}$$

$$vD = v_{00}(D - U) \tag{6}$$

$$v_{00} = p_{H0}DU \tag{7}$$

式中　v_{00}、v_0——聚氨酯泡沫材料及其相应密实体的初始比容；

C_{I}、S_{I}——密实聚氨酯材料中冲击波速度与粒子速度间的经验关系常数；

Γ_0——常态 Grüneisen 系数；

p_{H0}——聚氨酯泡沫材料在终态比容 v 时的 Hugoniot 压力；

D、U——冲击波速度和粒子速度。

利用表 2 实验数据 p_f 和 U_f（对应于图 2 中 f 点），联立方程（1）~式（7），可求得图 2 中 H 点所对应的聚氨酯泡沫材料中的冲击波峰压 p_H 及粒子速度 U_H。

4 结果与分析

初始孔隙度对爆炸载荷作用下聚氨酯泡沫材料中冲击波压力特性的影响如图 3 所示。在图 3（a）中，实验点●、○、■、□及▲所对应的聚氨酯泡沫材料的初始孔隙度分别为 0.15、0.25、0.35、0.45 和 0.55（初始密度分别为其密实材料的 85%、75%、65%、55% 和 45%），各实曲线为相应实验结果的拟合曲线，p 为冲击波峰值压力，X 为传播距离，换算中所使用的材料参数列于表 3。

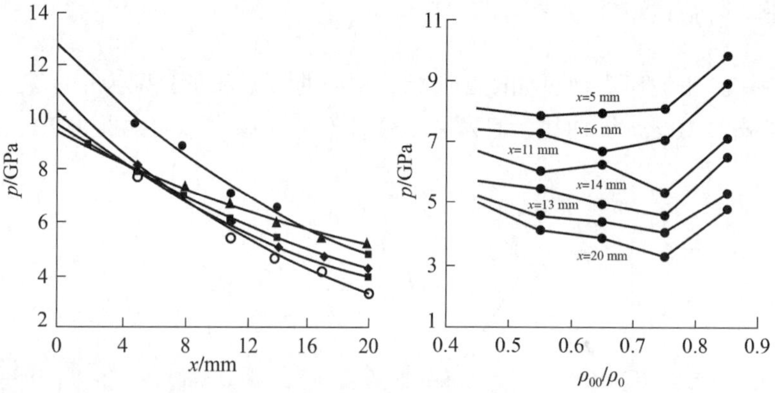

图 3 初始孔隙度对聚氨酯泡沫材料中冲击波压力影响特性

表 3 材料参数

$\rho_0/(\mathrm{g\cdot cm^{-3}})$	$\rho_{\mathrm{II}}/(\mathrm{g\cdot cm^{-3}})$	$C_{\mathrm{I}}/(\mathrm{km\cdot s^{-1}})$	$C_{\mathrm{II}}/(\mathrm{km\cdot s^{-1}})$	S_{I}	S_{II}	Γ_0
1.265	2.785	2.468	5.328	1.577	1.338	1.55

从图 3 可以看出，装药特性相同的炸药与聚氨酯泡沫材料做接触爆炸作用时，在聚氨酯泡沫材料中所形成的初始冲击波压力，随初始孔隙度增大（或初始密度减小）而显著下降。对于这一实验现象，可从材料声阻抗的角度得到很好的定性解释。理论计算表明[6]，随着初始孔隙度增大，多孔材料中声速 c_{00} 将显著下降。因而其声阻抗 $c_{00}\rho_{00}$ 也将随初始孔隙度增大而下降。因此，根据阻抗匹配原理，在装药特性相同的爆炸载荷作用下，多孔材料所形成的初始冲击波压力将随初始孔隙度增大而下降。

从图 3 还可以看出，当初始孔隙度在 0.15~0.25 范围内时，冲击波压力传播衰减效应随初始孔隙度增大而增强；当初始孔隙度在 0.25~0.55 范围内时，冲击波压力传播衰减效应随孔隙度增大呈逐渐减弱趋势。对于这一实验现象，下面从多孔材料在冲击载荷作用下的能量吸收机理角度对其做一定性分析。

多孔材料显著特征在于内部存在大量孔隙，在冲击波作用下材料首先被致密，以消除孔隙。致密过程可分为几个阶段[7,8]：首先孔壁发生弹性变形，部分冲击能量转变为弹性能，同时气隙被绝热压缩并吸收部分能量；继而孔壁发生塑性塌缩或脆性破碎，将部分冲击能量转变为塑性能，气隙绝热压缩过程基本结束；随后被逐渐压实直至接近密实材料。一旦多孔材料被完全致密，冲击波在其中的传播行为与相应密实材料基本相同。这表明，冲击波在多孔材料中的传播衰减效应，很大程度上取决于致密过程各阶段所吸收或耗散的能量，是初始孔隙度对多孔材料冲击压缩过程各阶段能量耗散效应的综合体现。由此可见，孔隙度参数所表征的只是多孔材料的宏观物理特性，其大小并不完全反映冲击压缩过程能量吸收或耗散的多少，即孔隙度大小与冲击波传播衰减效应强弱之间并不存在单一趋势的对应关系。根据本文实验研究结果，当初始孔隙度在 0.25 左右时，冲击波在聚氨酯泡沫材料中具有较强的传播衰减效应。

鉴于上述结果是通过间接实验所得到，其中无疑存在一定的误差，造成误差的主要因素包括两个方面：一是 LY－12 铝中冲击波衰减系数所引起的实验误差，可通过增加测点数量减小；二是多孔材料 Hugoniot 方程所引起的换算误差，鉴于高温高压下 Grüneisen 系数 Γ 的复杂力学行为，采用在 Grüneisen 假设 $v/\Gamma = v_0/\Gamma_0$ 基础上所导出的 McQueen 流体模型，进行冲击波压力换算无疑会造成一定的误差。在相同实验条件和换算方法下，上述两方面均属非随机性误差，虽对结果绝对值有一定影响，但不影响结果可比性。

5 结论

针对泡沫材料内部结构特征，提出了一种间接测压实验方法，并建立了间接测试参数外推计算关系，从而使根据间接实验结果获得泡沫材料中的冲击波参数成为可能，克服了利用锰铜计直接测量泡沫材料中冲击波压力的不足。

结果表明，在装药特性相同的爆炸载荷作用下，聚氨酯泡沫材料中所形成的初始冲击波压力随孔隙度增大而显著下降，冲击波传播衰减效应受孔隙度影响同样显著，但衰减速率不呈单一加快或减慢趋势；聚氨酯泡沫材料中的冲击波压力特性，主要取决于两个方面：一是初始孔隙度对初始冲击波压力的影响；二是初始孔隙度对冲击波传播衰减效应的影响，两方面综合影响效应决定聚氨酯泡沫材料中的冲击波压力特性；当初始孔隙度在 0.25 左右时，聚氨酯泡沫材料具有较强的抗冲击减压性能。

参考文献

[1] GIBSON L J, ASHBY M F, SCHAJER G S, et al. The mechanics of two dimensional cellular materials [J]. Proceedings of the Royal Society of London, 1982, 382 (1): 25－42.

[2] GIBSON L J, ASHBY M F. The mechanics of three dimensional cellular materials [J]. Proceedings of the Royal Society of London, 1982, 382 (1): 43－59.

[3] 张春生. 凝聚材料中矩形脉冲的衰减 [J]. 爆轰波与冲击波, 1986 (1): 43－53.

[4] DUVALL G E. High pressure physics and chemistry [M]. New York: Academic Press, 1963.

[5] MEQUEEN R G, MARSH S P, TAYLOR J W, et al. The equation of state of solids from shock wave studies [M] //KINSLOW R. High velocity impact phenomena. New York: Academic Press, 1970.

[6] 邵丙璜, 刘志跃, 王晓林. 粉末的爆炸烧结机制和加卸载状态方程 [M]//王礼立, 余同希, 李永池. 冲击动力学进展. 合肥: 中国科学技术大学出版社, 1992.

[7] HERRMANN W. Constitutive equation for the dynamic compaction of ductile porous materials [J]. Journal of applied physics, 1969, 40 (6): 2490－2499.

[8] CARROLL M M, HOLT A C. Static and dynamic pore－collapse relations for ductile porous materials [J]. Journal of applied physics, 1972, 43 (4): 1626－1630.

条形药包冲击波峰值超压工程计算模型①

周　睿②，冯顺山，吴　成
（北京理工大学，北京　100081）

摘　要：本文将条形药包分为起爆端、过渡区、中部和非起爆端几个部分，根据等效和相似律原理，结合条形药包冲击波压力场特性，对这几部分爆炸产生的冲击波的衰减特性进行了研究。对照实验结果，并引入修正系数，得到了各部分简化计算模型，可以推广应用到水中条形药包爆炸。本文提出的简化模型，可为实际工程中条形药包水中爆炸冲击波峰值超压的计算提供参考。

关键词：条形药包；冲击波；水下爆炸；峰值超压；计算模型

CALCU LAT ION MODEL FOR PEAK PRESSURES OF SHOCK WAVE FROM LINEAR CHARGES

Zhou Rui，Feng Shunshan，WU Cheng
（Beijing Institute of Technology，Beijing100081，China）

Abstract：A linear charge is divided into a few parts such as priming end，transition section，central section and non－priming end. On the basis of the principles of ex plosion equivalence and similitude and in the light of the characteristics of shock wave pressure from a linear charge，the attenuation of shock w aves from the explosion of these parts is studied. Compared with test results and introduced corrected factors，simplified calculation models for the explosion of these parts in the air are obtained and can be spread to underwater explosion of linear charges. The simplified models presented in this paper can be used as a reference for the calculation of peak shockwave pressure from the explosion of underwater linear charges in practical projects.

Key words：Linear charge；Shock wave；Underwater explosion；Peak shock wave pressure；Calculation model

1　引言

由于条形药包在工程实际中运用广泛，前人对条形药包冲击波峰值超压随距离衰减规律进行了大量研究[1]，但他们主要针对条形药包硐室爆破地震效应和爆破破岩，且得到的经验公式大多数只适用于岩土爆破。随着条形药包在水中爆炸修筑坝基、平整港口地基、移山填海和沉船爆破等工程应用的拓广，研究条形药包空气中特别是水中爆炸冲击波峰值超压衰减特性显得极其迫切而又重要。本文将条形药包分为起爆端、过渡区、中部和非起爆端几个部分，根据等效和相似律原理，结合条形药包冲击波压力场特性，对这几部分爆炸产生的冲击波的衰减特性进行研究；对照实验结果，得到各部分简化计算模型，并推广到水中爆炸，得到相应的简化计算模型。

2　理论模型的提出

由高速摄影光测试验[1]得到的条形药包爆炸冲击波波阵面发展图像可知，条形药包爆炸形成的冲击

① 原文发表于《工程爆破》2001 7（4）。

② 周睿：工学博士，1997 年师从冯顺山教授，研究条形装药水中爆炸对金属靶板作用效应及安全距离，现工作单位：兵器工业集团导航与控制所。

波波阵面是随着距离的变化而变化的。靠近药包时，波阵面是两端为半椭球形的圆柱面；较远时，波阵面近似为椭球面；趋于很远时，则接近于圆球面，如图 1 所示。条形药包爆炸冲击波峰值超压具有中部各位置相同，而端部明显偏小的特性。端部侧向和径向的冲击波峰值超压与中部的相比，差异较大，主要原因如下。

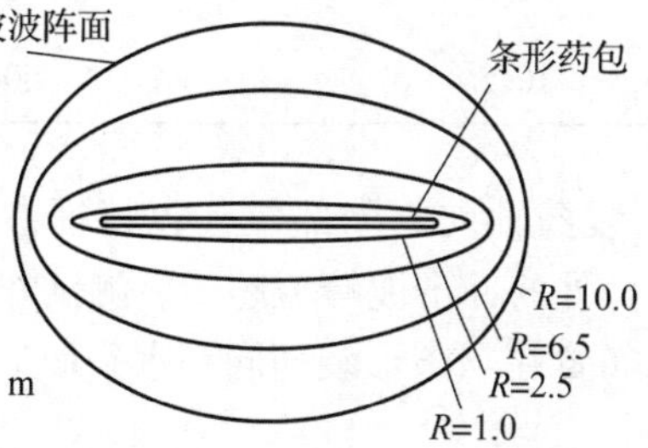

图 1　条形药包爆炸冲击波波阵面传播图

（1）端部的非稳定爆轰，是指起爆端。实验表明[2]，从炸药起爆开始到以特定爆速稳定传播，存在着一段不稳定的爆轰过程。如果条形药包直径满足大于临界传播直径、稳定爆轰的要求，且传入主装药中的冲击波速度低于主装药的特性爆速，其产生的爆压也低于稳定爆轰的爆压，则势必使端部效应明显。

（2）爆轰产物轴向扩散。药包端部冲击波波阵面的扩散具有明显的球形特点，且起爆端更明显于非起爆端。药包中间段（是药包长度的主体）的冲击波波阵面则具有明显的柱面扩散特点。正是这种扩散特征的不同，造成药包爆炸后端部与中间段侧向冲击波参数的差异（根据爆炸总能量不变原理，由于扩散面形状的不同，造成能量分布不均），从而导致端部效应明显。

（3）端部有效作用装药单元较之中部侧向少。本文结合爆炸相似律，提出计算模型并采用实验和前人已做的试验数据来验证模型的可行性。球形装药在介质（空气、水）中爆炸后，在距爆炸点一定距离处的峰值超压计算式可表示为[2]

$$\Delta p_m = f(\sqrt[3]{w}/R) \tag{1}$$

式中　w——球形装药 TNT 药重，kg；

R——距爆炸中心的距离，m。

根据以上分析，考虑到工程应用的方便，本文将采用球形装药，以在空气、水中爆炸冲击波超压计算公式的形式，对条形药包不同部分等效成为药量不同的球形装药，将起爆端、中部和非起爆端用不同的修正系数 $K(\lambda_x)$ 进行修正，而过渡区采用等效拟合系数得到条形药包冲击波超压计算模型，相应表达式如下：

$$\Delta p_{m条空} = 0.84[\sqrt[3]{K(\lambda_x)w}/R] + 2.7[\sqrt[3]{K(\lambda_x)w}/R]^2 + 7[\sqrt[3]{K(\lambda_x)w}/R]^3$$
$$\Delta p_{m条水} = 533K(\lambda_x)[\sqrt[3]{w}/R]^{1.13} \tag{2}$$

式中　w——与条形药包等径的球形装药 TNT 药重，kg；

R——装药轴心或端部与所求点的距离，如图 2 所示。

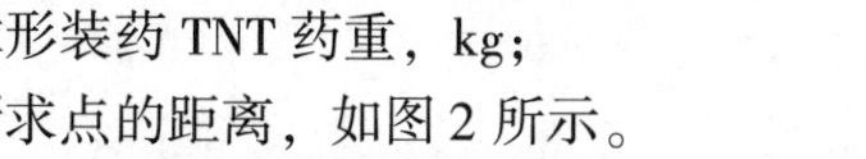

图 2　装药轴心或端部与所求点的距离

确定了系数 $K(\lambda_x)$，即可利用式（2）计算出条形药包空气或水中爆炸冲击波峰值超压。

3　实验及测试结果

实验的目的是测出条形药包不同装药半径、不同距离、不同位置产生的冲击波峰值超压大小。实验采用不同规格的 TNT 铸装的条形药包（密度为 1.6 g/cm³），其参数如表 1 所示。

表1　条形装药参数

装药型号	装药半径/mm	装药长度/mm	装药量/g
1#条形装药（大号）	14.0	600	591
2#条形装药（中号）	12.5	600	472
3#条形装药（小号）	10.5	600	330

采用一端起爆方式，测得条形装药轴向不同部位（以特征参数 λ_x 表示）、与爆炸中心的不同距离（R = 1.5、1.9、2.5、3.1、3.3 m）处的冲击波峰值超压。测试峰值超压时，为防止地面反射波的影响，条形装药高出地面 1.5 m，传感器布置在与条形装药同一水平面上，如图3所示。

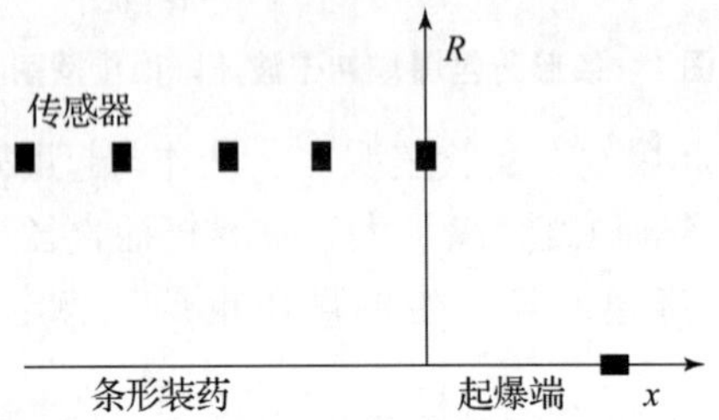

图3　实验场地药包和传感器的布置示意图

实验中测得的条形装药不同位置部分冲击波峰值超压见表2。

表2　1#、2#、3#条形装药空气中爆炸的冲击波峰值超压 Δp

条形装药	R/m	λ_x = 0	0.25	0.50	0.75	1.00
1#	1.5	0.40	0.75	0.75	0.75	0.60
	1.9	0.25	0.50	0.50	0.45	0.36
	2.5	0.18	0.30	0.30	0.30	0.25
	3.3	0.12	0.20	0.20	0.20	0.18
2#	1.5	0.36	0.70	0.70	0.70	0.56
	1.9	0.24	0.45	0.45	0.45	0.36
	2.5	0.16	0.28	0.28	0.28	0.24
	3.1	0.12	0.20	0.20	0.20	0.16
3#	1.5	0.25	0.50	0.50	0.50	0.40
	1.9	0.20	0.35	0.35	0.35	0.25
	2.5	0.12	0.20	0.20	0.20	0.18

4　理论计算与实验数据对比

根据表2、表3和表4可得到与条形药包轴心等距离处冲击波峰值超压沿轴线变化的规律以及条形药包冲击波峰值超压随距离衰减的规律，如图4、图5所示（图中1、2、3分别为1#、2#、3#条形装药）。基于以上分析、实验结果和式（2），由拟合法得到条形装药不同位置处的修正系数 $K(\lambda_x)$：起爆端、非起爆端和中部的 $K(\lambda_x)$ 分别为1、3.13和4.35。起爆端与中部之间以及非起爆端端部与中部之间的过渡区域的大小，一般被认为是条状药包直径的 3 倍[3]，此区域采用指数拟合，得到如下的 $K(\lambda_x)$ 表达式：

$$K(\lambda_x)=\begin{cases}4.36-3.36e^{-\lambda(x)}, & 0\leqslant\lambda(x)\leqslant6\\ 4.35-1.22e^{-[\lambda-\lambda(x)]}, & \lambda-6\leqslant\lambda(x)\leqslant\lambda\end{cases} \tag{3}$$

其中，λ——条形药包的长度与半径的比值。

表 3 条形药包水中爆炸冲击波峰值超压计算值与实测值的对比

组号	药包部位	药包重/kg	药长/m	药径/mm	测距/ m	计算超压/MPa	实测超压/MPa	误差/%
	起爆端	0.382	0.8	10	5	2.98	3.05	2.3
1①	中部	0.382	0.8	10	5	5.96	6.47	7.8
	非起爆端	0.382	0.8	10	5	4.47	3.91	12
	起爆端	22.68	15.2	17.2	0.61	31.20	35.00	13
2②	中部	22.68	15.2	17.2	0.61	46.80	52.70	13
	非起爆端	22.68	15.2	17.2	0.61	64.00	70.80	11

①为文献［4］、②为文献［5］中的数据。

表 4 条形药包水中爆炸冲击波峰值超压计算值与实测值的对比

组号	药包部位	药包重/kg	药长/m	药径/mm	测距/m	计算超压/MPa	实测超压/MPa	误差/%
	起爆端	0.382	0.8	10	5	298	3.05	23
1①	中部	0.382	0.8	10	5	5.96	6.47	7.8
	非起爆端	0.382	0.8	10	5	4.47	3.91	12
	起爆端	22.68	15.2	17.2	0.61	31.20	35.00	13
2②	中部	22.68	15.2	17.2	0.61	46.80	5 270	13
	非起爆端	22.68	15.2	17.2	0.61	64.00	70.80	11

①为文献［4］、②为文献［5］中的数据。

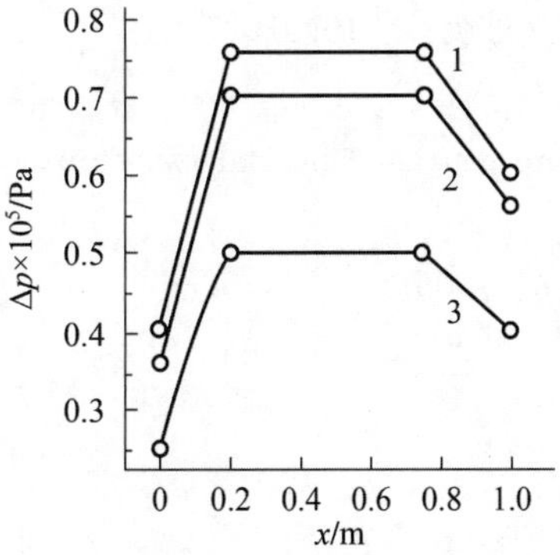

图 4 1.5 m 处条形药包不同位置的 Δp

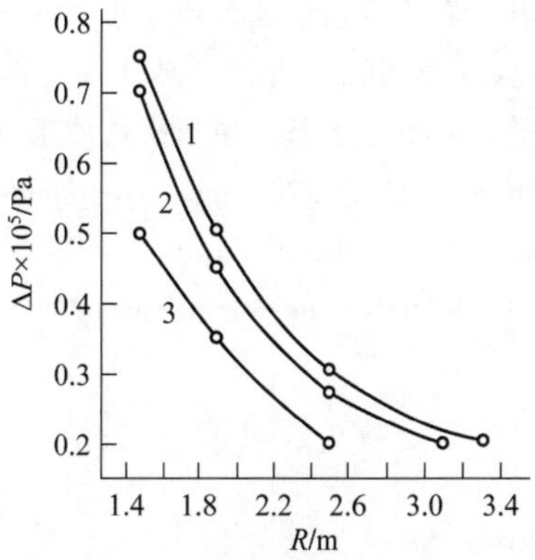

图 5 条形药包中部爆炸的 Δp 随距离的衰减

对比表明，计算结果与实验数据较为吻合，证明了本文所提出的条形药包空气中爆炸冲击波峰值超压计算模型是合理、可行的。

由有效装药理论也可以解释以上所得实验结果。对于起爆端，其有效装药可以看作长径比为 1 的装药。实验已经证明：长径比为1的条形药包完全可以等效为同径的球形药包，即装药侧向和径向冲击波峰值超压在等距离处近似相等[2]。这与拟合实验修正系数为1的情况相吻合。中部有效装药可被认为是起爆端的2倍，所得的实验数据与条形药包等径的球形药包冲击波峰值超压的2倍时的情况相符。对于非起爆端，考虑到条形药包爆炸爆轰波的前驱性，其有效装药量不同于起爆端，但可看成把中部截成两段，取走一部分后形成的端部，可等效为1.5倍起爆端的球形药包。以上分析是在实验数据的基础上进行的，具有较高的可信度。根据上述分析，可以将条形药包空气中爆炸的修正模型推广到条形药包水中的爆炸。装药在水中与空气中爆炸的超压计算式形式不同，相应的系数也不同，但同样可以由拟合法得到条形装药水中爆炸不同位置处的修正系数$K(\lambda_x)$：起爆端、非起爆端和中部的$K(\lambda_x)$分别为1、1.5和2；对于起爆端与中部之间以及非起爆端端部与中部之间的过渡区域，仍采用指数拟合得到$K(\lambda_x)$值，如式（4）所示。

$$K(\lambda_x)=\begin{cases}2-e^{-\lambda(x)}, & 0\leqslant\lambda(x)\leqslant 6\\ 2-0.5e^{-[\lambda-\lambda(x)]}, & \lambda-6\leqslant\lambda(x)\leqslant\lambda\end{cases} \tag{4}$$

将前人的实验数据代入上述公式进行对比检验，结果如表3所示。

由数据的对比可以看到，用修正模型计算与试验实测的超压值是基本吻合的。这说明，对于水介质中条形药包爆炸，采用相同的分析理论得到的计算模型进行冲击波峰压的计算，其结果是令人满意的。

5 结论

通过用有效装药量和爆炸相似律进行分析，可以得到条形药包在空气中和水中爆炸冲击波峰值超压计算的修正模型，计算值与实测值的对比表明，引入不同的修正系数计算条形药包在空气中和水中爆炸冲击波峰值超压的理论模型是合理的。用修正模型进行的计算证实：

（1）沿条形药包轴线不同部位，相同侧向距离处的超压值不同，侧向的冲击波峰值超压大于两端的超压，非起爆端的超压大于起爆端。

（2）条形药包起爆端端部超压等于相同距离处等径球形药包的超压，中部侧向和端部的超压大于相同距离处等径球形药包的超压，前两者分别约为后者的2倍和1.5倍。

（3）上述简化模型对于条形药包空气中、水中爆炸均适用，与实验测定的和前人已有的数据相比，误差较小，可用范围大。该模型具有普遍性，可用于实际工程计算。

参考文献

[1] 杨年华．条形药包爆破机理［R］．铁道部科学研究院，1994.

[2]《爆炸及其作用》编写组．爆炸及其作用［M］．北京：国防工业出版社，1979.

[3] 左宇军．条形药室爆破端部距离的确定［J］．爆破，1997，14（3）：12－15.

[4] MITSUI S，SASAKI N. Some experiments on the shock wave pressure generated by underwater explosions［J］. Journal of the Industrial Explosions Society，Japan，1972.

[5] COLE R H. Underwater explosion［M］. Mineola：Dover Publication，1965.

爆炸冲击波防护效应的评价方法研究①

冯顺山，王　芳②，胡浩江

（北京理工大学爆炸与安全国家重点实验室，北京　100081）

摘　要：本文以被防护对象在爆炸冲击波作用下的最低毁伤状态作为考虑防护设施在爆炸冲击波作用下防护能力的基础，基于等效靶板的试验方法及试验数据，建立了以冲击波超压—冲量为防护判据的爆炸冲击波防护效应的评价方法。利用该评价方法进行了某云爆装置的爆炸试验，对其威力进行了评价。结果表明了爆炸冲击波防护效应评价方法的可行性与实用性。

关键词：爆炸冲击波；防护效应；等防护曲线；等效靶板

0　引言

爆炸装置意外爆炸，可燃气体、云雾和粉尘意外爆炸都将产生严重的破坏作用，评价它们的TNT威力当量和毁伤能力是重要的，评价某种设施对它们的防护能力也同样重要。因此，对爆炸冲击波的防护效应提供一种科学、有效的评价方法具有重要的理论和工程应用价值。

本文从防护设施对爆炸冲击波的防护效应出发，以被防护对象在爆炸冲击波作用下的最低毁伤状态作为考虑防护设施在爆炸冲击波作用下防护能力的基础；同时，基于等效靶板的试验方法及试验数据，建立了以冲击波超压—冲量为防护判据的爆炸冲击波防护效应的评价方法。由某种装药对方形靶板的试验数据，结合本文的计算公式给出了相应的等防护曲线。利用该评价方法进行了某云爆装置的爆炸试验，对其威力进行了评价。结果表明了爆炸冲击波防护效应评价方法的可行性与实用性，可应用于今后爆炸冲击波防护效应的评价当中。

1　爆炸冲击波防护效应的评价方法

1.1　超压—冲量防护准则

毁伤与防护是一个问题的两个方面，作战时反映摧毁/杀伤目标，或目标生存/正常运行；在工业和日常生活中反映破坏、经济损失，或安全、减小损失。这两个方面互为对立面，且互相之间存在着有机联系，毁伤是基于目标防护能力的毁伤，防护是基于爆炸装置毁伤威力的防护。

超压—冲量准则既可以从毁伤的角度加以阐述，也可以从防护的角度给予说明。从毁伤角度讲，“超压—冲量”准则认为，要对给定目标产生某种程度（某一等级）的毁伤，ΔP必须不低于某一临界值，并且对目标持续作用的时间不小于某一临界值，只有具有这种条件的冲击波才能够对目标产生给定的毁伤效果；或者说，爆炸冲击波必须在某一临界时间内达到或超出某一最小临界比冲量，只有在临界时间内超出这个临界比冲量，目标才能产生某种等级的破坏。这实际上包含着要有一个最小超压，只有超出这个超压值，冲击波比冲量才能在临界时间内达到或超出最小临界比冲量。

从防护角度讲，超压—冲量准则认为，在爆炸冲击波作用下，要使给定被防护对象（或防护设施）不产生某种程度（某一等级）的损伤，ΔP必须不高于某一临界值，并且对被防护对象（或防护设施）持续作用的时间不大于某一临界值，只有具有这种条件的冲击波才不能够对被防护对象（或防护设施）

①　原文发表于《安全与环境学报》2004年第6期。

②　王芳：工学博士，1996年师从冯顺山教授，研究FAE战斗部威力的评价方法，现工作单位：北京理工大学。

产生给定的毁伤效果；或者说，爆炸冲击波不能在某一临界时间内达到或超出某一最大临界比冲量，如果在临界时间内没有超出这个临界比冲量，被防护对象（或防护设施）就不会产生某种等级的破坏。这实际上包含着一个最大超压值，只要不超出这个超压值，冲击波比冲量就不能在临界时间内达到或超出最大临界比冲量。

由此可知，爆炸冲击波能否在一段时间内对目标保持一定的压力作用决定着爆炸冲击波是否对目标具有一定的毁伤能力，相反也决定着在爆炸冲击波作用下防护设施是否具有一定的防护能力。大多数常规炸药产生的爆炸冲击波是一种具有较高幅值且持续微秒级至毫秒级时间的强间断压力波。它既具有一定的幅值大小，又具有时间的意义。爆炸冲击波中超压和比冲量是相互影响、相互关联的，如果在考虑问题时，忽略了任何一个，都不容易得到正确的结果。因此超压—冲量准则相比超压或冲量准则更具有考虑全面、评价准确和适用广泛的优点。

1.2 等防护曲线

在爆炸冲击波作用下，防护设施的防护原理如图 1 所示。

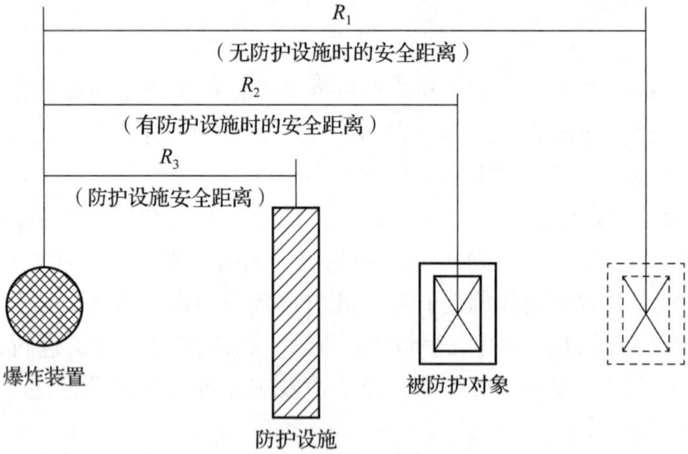

图 1　防护设施的防护原理

由图 1 可知，防护问题相对于爆炸对象（装置）的毁伤能力或损伤能力，一般是在弄清爆炸装置/爆炸物的毁伤/损伤能力以及防护方式基础上，确定防护准则并对防护设施进行设计的。

冲击波作用下的等防护曲线可基于等毁伤曲线进行研究。等毁伤曲线是在超压—比冲量坐标系中代表着对目标产生相同毁伤程度条件下的一系列超压—比冲量值的组合，在图上是一条将具有相同毁伤等级的点连起来的线。等毁伤/防护曲线示意图如图 2 所示。某一等级的等毁伤曲线一般划分为冲击加载区、准静态加载区和动态加载区三个区域。在冲击加载区，等毁伤曲线 A 接近于垂直渐近线 i_H，毁伤主要取决于比冲量。垂直渐近线对应的比冲量是目标产生此种毁伤等级的最小比冲量，此区域也是冲量准则适用的区域。而在准静态加载区，曲线趋近于水平渐近线 ΔP_{H}，毁伤主要取决于峰值压力。水平渐近线对应的压力值是目标产生此种毁伤等级的静压垮压力，也是等毁伤曲线上的最小压力。此区域也是超压准则适用的区域。动态加载区是连接冲击加载区和准静态加载区的过渡区域，也是等毁伤曲线研究中最主要关心的区域，毁伤由峰值压力和比冲量共同决定。

由图 2 可知，当冲击波的峰值超压和比冲量位于给定目标某一等级等毁伤曲线的右上部分时，就会对该目标造成一定程度的毁伤；而当它们位于等毁伤曲线的左下部分时，就不会对该目标造成相应的毁伤。

某一等级等防护曲线的特征及物理意义与等毁伤曲线相同，图 2 也表示了等防护曲线与等毁伤曲线之间的关系。通常曲线 A 表示目标处于最低毁伤状态下的等毁伤曲线，曲线 B 为被防护对象不发生损伤的等防护曲线，即许可承受的 $\Delta P—i$ 值，曲线 C 是防护设施所能承受的等防护曲线。等防护曲线 B、C

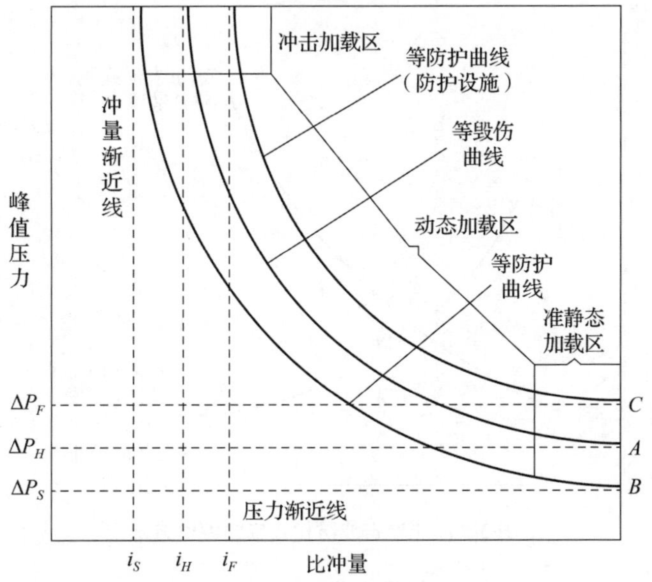

图 2　等毁伤/防护曲线示意图

形状与曲线 A 相近，但垂直渐近线与水平渐近线的值不同。被防护对象的等防护曲线在曲线 B 下面时，表明被防护对象更安全；防护设施的等防护曲线在曲线 C 上面时，表明其防护能力更强，被防护对象可更靠近爆炸装置，反之防护能力弱。

1.3　防护效应的等效靶板模型

通常认为，具有一般意义的被防护对象（或防护设施）是靶板，它经常被用来模拟坦克、装甲车的壳体、飞机的防护结构件和某些设施的防护结构等。由于应用具体真实的被防护对象（或防护设施）进行试验工程浩大，限制因素很多，且费用很高，爆炸冲击波防护效应的研究具有相当大的困难；而靶板结构简单，参数与试验条件易控制，且材料来源广泛，加工简单，采用等效靶板进行试验研究，相比真实被防护对象（或防护设施）来说是一种较科学有效且经济的做法。根据对某一被防护对象（或防护设施）的防护等级（程度）要求，总可以找到一种恰当的等效靶板。通过试验研究，还可将等效靶板与具体被防护对象（或防护设施）建立等效关系，从而得到各种被防护对象（或防护设施）在爆炸冲击波作用下防护效应的评价结果。

1.4　爆炸冲击波防护效应的评价方法

根据超压—冲量准则及其等毁伤/防护曲线，建立了一种基于靶板防护效应的爆炸冲击波防护效应评价方法。将不同爆炸装置的炸药装药量 w 与炸距 R 的关系曲线当作叠合线和等毁伤/防护曲线结合到一张图上来使用，如图 3 所示。

药量—炸距曲线与最低毁伤状态下的等毁伤曲线的交点，即为该爆炸装置产生最低毁伤状态时的临界炸距 R_H。只要炸距大于这个临界值，就对目标产生相应程度的毁伤，小于这个临界值，就不会产生相应程度的毁伤；药量—炸距曲线与某一等级下被防护对象的等防护曲线的交点 R_S，即为被防护对象某一防护等级下的安全距离；药量—炸距曲线与某一等级下防护设施的等防护曲线的交点 R_F，即为防护设施某一防护等级下的防护距离。一般情况下 $\frac{R_S}{R_H}\geqslant 1$，$\frac{R_F}{R_H}\leqslant 1$。

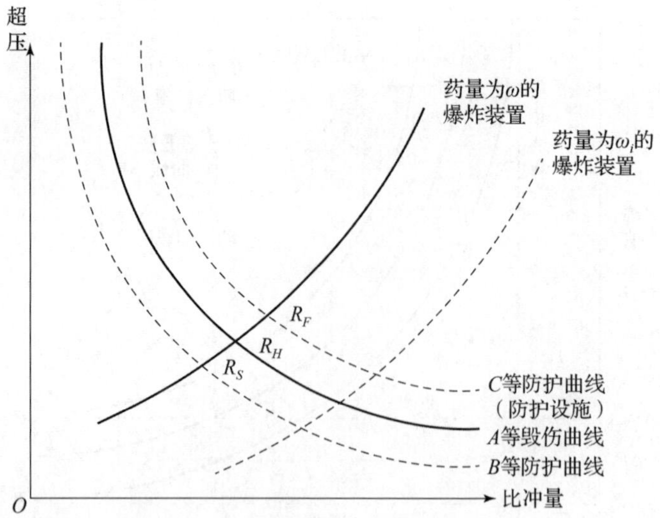

图 3　爆炸冲击波防护效应的评价方法

2　靶板等防护曲线的确定及敏感性与分辨率分析

2.1　基于靶板效应的等防护曲线的确定

本文主要对四边约束方形靶板在爆炸冲击波作用下发生塑性变形时的等防护曲线进行确定。文献［3］中已根据能量守恒和爆炸冲击波“等毁伤模型”得到了靶板在不同毁伤等级下等毁伤曲线的表达形式，本文在此基础上给出等防护曲线的表达形式。

由文献［3］可知，对于如图 4 所示的四边约束、厚度为 h、长宽尺寸分别为 $2a$ 和 $2b$ 的矩形板，不考虑弯曲效应，仅考虑膜力效应，由能量守恒可得到冲量加载区中在比冲量作用下矩形板的挠度解为

$$\begin{cases}\omega_{0i}=\dfrac{4abi_{0i}}{\pi h}\sqrt{(a^2+b^2\rho\sigma_Y)}\\[2ex] \omega_{0t}=\dfrac{4i_{0i}}{\pi h\sqrt{2\sigma_Y\rho}} \qquad \text{对于方形靶板}, a=b\end{cases} \tag{1}$$

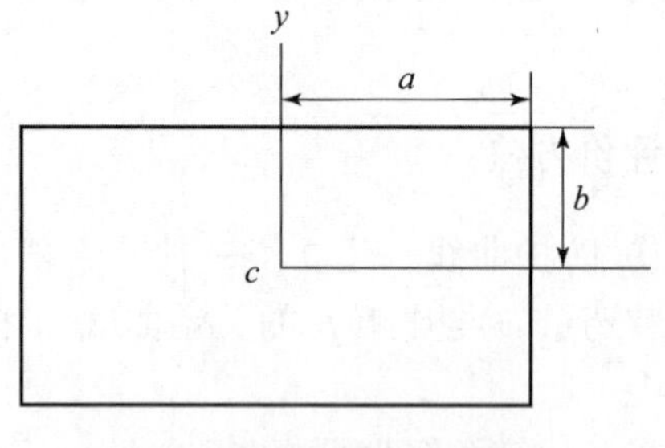

图 4　四边约束方形靶板

同理得到准静态加载区中在静压力载荷作用下矩形板的挠度解为

$$\begin{cases}\omega_{0i}=\dfrac{128}{\pi^4}\dfrac{a^2b^2}{a^2+b^2}\dfrac{P_{0i}}{\sigma_Y h}\\[2ex] \omega_{0t}=\dfrac{64a^2}{\pi^4}\dfrac{P_{0i}}{\sigma_Y h} \qquad \text{对于方形靶板}, a=b\end{cases} \tag{2}$$

式（1）和式（2）中，$i=H$ 时对应毁伤，$i=F$ 时对应防护。

由文献［3］可进一步得到适用于毁伤和防护的动态加载区表达式，即

$$
\begin{aligned}
(\Delta P - P_t) \cdot (i_+ - i_-^*) &= DN \\
(\Delta P - P_t) \cdot (i_+ - i_-^*) &= FN
\end{aligned}
\tag{3}
$$

式中 $i_t{}^*$——产生某种毁伤/防护等级的冲量加载区中的比冲量值 i_{0H} 或 i_{0F}；

$P_t{}^*$——产生相同毁伤/防护等级的准静态加载区中的静压力载荷 P_{0H} 或 P_{0F}；$i_t{}^*$ 和 $P_t{}^*$ 可由式（1）和式（2）得到，即

$$
i_t^* = \frac{\pi h}{4a}\sqrt{2\rho\sigma_Y}\omega_{0t}, P_t^* = \frac{\pi^4}{64a^2}\sigma_Y h\omega_{0i}
\tag{4}
$$

式中 DN——该种毁伤等级的准数；

FN——对应的防护等级的准数，$FN = k \cdot DN$。一般情况下，k 值取决于防护方式和防护等级，由等效靶试验方法确定。

2.2 靶板的敏感性与分辨率

在运用超压—冲量防护准则和等效靶板模型对爆炸冲击波作用下的防护效应进行评价时，还需要考虑靶板的敏感性与分辨率问题，只有选择具有很好的敏感性和分辨率的靶板进行理论及试验研究，才能够对爆炸冲击波的防护效应做出更为合理可靠的评价。

靶板的敏感性是指某种给定靶板在外界爆炸冲击波作用下发生变形破坏响应的程度大小。它包含如下两层意思。

（1）当爆炸冲击波载荷大小一定时，不同条件下的靶板所产生的变形破坏量是不相同的，变形破坏量大的靶板敏感性高；反之，敏感性低。

（2）当爆炸冲击波载荷发生变化时，不同条件下的靶板所产生的变形破坏的改变量也有所不同，变形破坏改变量大的靶板敏感性高；反之，敏感性低。

由上可知，靶板的敏感性大小是通过一定载荷下靶板的变形破坏量，以及载荷发生变化时靶板变形破坏的改变量体现出来的。对于前者，本文用靶板的敏感度来表示，对于后者，本文用靶板对载荷的分辨率来表示。

靶板的敏感度是指给定靶板在一定的爆炸冲击波作用下所产生的变形破坏量，可以用挠度与载荷之比表示。当载荷一定时，靶板的变形挠度越大，靶板的敏感度越高。

靶板对载荷的分辨率是指给定靶板相对于爆炸冲击波载荷的变化而产生的变形破坏的变化量，可以用靶板挠度变化量与载荷变化量之比表示。当载荷变化量一定时，靶板挠度变化量越大，靶板的分辨率越高。

靶板的材料、尺寸、形状、厚度、边界约束条件和载荷特性等因素决定着靶板的变形破坏量大小和变形破坏改变量的大小，是决定靶板敏感性大小的主要因素。

3 威力与防护等级确定的试验方法

3.1 等效靶板防护效应试验方法

本试验方法的特点是在不同药量下对四边约束方形靶板进行爆炸冲击波防护试验。试验实例之一是采用几种长径比为 1：1 的 TNT 裸装圆柱形药柱，靶板考虑敏感性和分辨率，采用不同厚度的 A3 钢和铝板等，边长为 50 mm。使用专用靶架，利用四周的螺钉及压板将靶板的四边夹紧。试验现场布置如图 5 所示。靶板在爆炸冲击波作用下沿载荷方向出现显著的塑性变形，严重情况下靶板四边会被拉出少许，并且螺钉孔处发生变形甚至断裂。典型靶板的变形破坏情况如图 6 所示。

图 5 试验现场布置

图 6 典型靶板的变形破坏情况

由试验结果可知，由于靶架约束条件的限制，不可能做到对靶板四边的绝对夹紧，在爆炸冲击波作用下，靶板四周被靶架压板压住的边宽均发生一定程度的向内移动现象，参与了靶板的变形，导致靶板的变形挠度相应增大。因此在靶板的变形挠度解中引入一个考虑靶板约束条件影响的修正系数，由此，式（3）也做相应修正，即

$$i_t^* = \frac{1}{\varphi'}\frac{\pi h}{4a}\sqrt{2\rho\sigma_Y}\omega_{0t}, P_t^* = \frac{1}{\varphi'}\frac{\pi^4}{64a^2}\sigma_Y h\omega_{0i} \tag{5}$$

根据本文试验数据，对于 A3 钢板，$\varphi' \approx 12.6$。

3.2 不同等级下的等防护曲线

为了合理准确地反映被防护对象或防护设施的防护情况，通常会根据它们的受损程度划分出多个防护等级，通过不同的防护等级反映出防护设施对被防护对象的防护能力。对爆炸冲击波载荷，防护设施通常表现为机械损伤，再由机械损伤导致防护设施的功能减弱或丧失。爆炸冲击波作用下防护设施的防护能力是指防护设施的抗破坏能力，在保证被防护对象处于安全条件下，防护设施的抗破坏能力越大，防护设施的防护等级越高。

可以用防护设施受载荷作用后出现的塑性变形挠度等 W_f 物理量定量地描述防护设施的受损，根据变形挠度的不同值，为防护设施划分出多个防护等级。本文根据靶板的最大变形挠度 W_{fm} 将靶板划分出 N 个防护等级。具体划分方法有两种：一种方法是每个防护等级对应一个挠度值，可用下式表示，即

$$W_{fj} = (j-1)\frac{W_{fm}}{N-1}, j = 1,2,3,\cdots,N$$

另一种方法是每个防护等级对应一个挠度区段，可用下式表示，即

$$(j-1)\frac{W_{fm}}{N} < W_{fj} \leqslant j\frac{W_{fm}}{N}, j = 1,2,3,\cdots,N$$

试验中将 1 mm 厚 A3 钢板划分成 4 个防护等级，每个等级所对应的挠度值分别为 0、39 mm、

77 mm、116 mm。根据试验数据，得到不同防护等级下的临界超压 P_t^*、临界比冲量 i_t^*，以及防护准数 FN，分别见表 1。

表 1　1 mm 厚 A3 钢板不同防护等级下的参数值

防护等级	$(\Delta P - P_t)\cdot(i_+ - i_-^*) = FN$		
	P_i^*/MPa	P_i^*/(Pa·s)	FN
一级	—	—	—
二级	0.028	23.3	53.4
三级	0.054	46	85
四级	0.082	69.3	226.3

1 mm 厚 A3 钢板各等级的等防护曲线如图 7 所示。

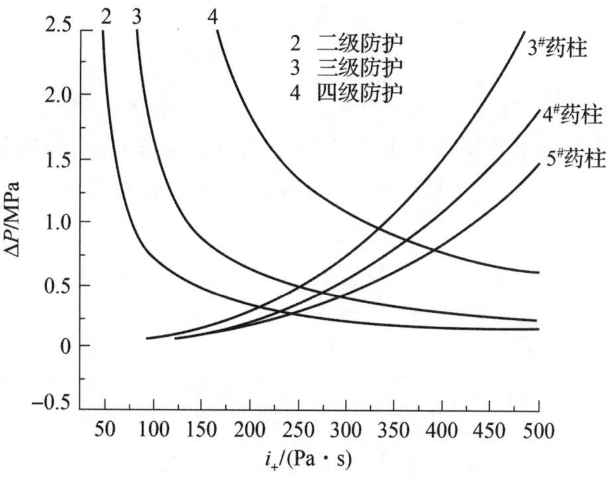

图 7　1 mm 厚 A3 钢板各等级的等防护曲线

在图 7 中还绘出了不同药量下的超压—比冲量曲线。对于某一种药量来说，它的超压—比冲量曲线与各等级等防护曲线的交点即为产生此种等级所需要的临界超压和比冲量，并可进一步得到对应的临界炸距。其中 3#药柱的超压—比冲量曲线与各等级等防护曲线交点处对应的临界超压、比冲量和临界炸距如表 2 所示。将临界炸距的计算值与相应的试验值进行比较，平均相对误差约为 7%。比较结果表明，计算值与试验值有着良好的一致性，该试验方法可用来确定不同药量对靶板作用后产生各等级防护时所需要的临界超压、比冲量和临界炸距。

表 2　3#药柱与各等级等毁伤/防护曲线的交点值

毁伤等级	临界超压/MPa	临界比冲量/(Pa·s)	临界炸距/m
二级	0.316	208.87	1.31
三级	0.474	248.75	1.1
四级	0.939	329.66	0.80

4　某云爆装置冲击波威力及防护效应评价[2]

利用本文提出的爆炸冲击波威力及防护效应评价方法，对某云爆装置冲击波威力效应进行了评价试验。在该云爆装置的静爆试验中，采用 0.5 mm 和 1.0 mm 厚 A3 钢板，并将 3 个固定靶板的专用靶架沿爆心径向的不同方位和不同距离放置于爆炸场中。通过对试验数据进行处理，得到不同毁伤等级下的临界炸距，也证明了该评价方法的实用性、直观性与可靠性。试验发现，对于1 mm厚的A3钢板而言，要在云爆装置爆炸场中产生较为恰当的变形量，至少需将靶架放置在4.5 m以内，4.5 m以外靶板基本上无变形，说明A3钢板在云爆装置爆炸场中对小值 $\Delta P—i$ 的敏感性不够。若要拓宽评价范围，就需要采用敏感性高的靶板，如更薄的A3钢板、薄铝板或其他敏感性更高的非金属材料靶板。

5 结论

本文从防护设施对爆炸冲击波的防护效应出发，以被防护对象在爆炸冲击波作用下的最低毁伤状态作为考虑防护设施在爆炸冲击波作用下防护能力的基础；同时，基于等效靶板的试验方法及试验数据，建立了以冲击波超压—冲量为防护判据的爆炸冲击波防护效应的评价方法。由某种装药对方形靶板的试验数据，结合本文的计算公式给出了相应的等防护曲线。利用该评价方法进行了某云爆装置的爆炸试验，对其威力进行了评价。结果表明了爆炸冲击波防护效应评价方法的可行性与实用性，可应用于今后爆炸冲击波防护效应的评价当中。

参考文献

[1] 贝克，等．爆炸危险性及其评估：上［M］．张国顺，等译．北京：群众出版社，1988.

[2] 王芳，冯顺山．某一次引爆火箭云爆弹防护威力评价试验报告［R］．内部资料，2000.

[3] 王芳，冯顺山，俞为民．“超压—冲量”毁伤准则及其等毁伤曲线研究［J］．弹箭与制导学报，2003(S2)：126－130.

地下目标防护层外介质中的爆炸作用分析[①]

王云峰[1,2②]，李　磊[1]，冯顺山[1]，陈　兴[2]

（1. 北京理工大学爆炸科学与技术国家重点实验室，北京　100081；
2. 中国白城兵器试验中心，吉林白城　137001）

摘　要：由于爆点到地下目标防护层的距离不同，爆炸可能以应力波的形式对目标混凝土防护层，也可能是爆炸产物直接加载，本文以现有的经验公式为基础，讨论了应力波与目标混凝土防护层的加载破坏问题，分析了不同装药量、不同爆炸位置对地下目标防护层可能产生的破坏程度，提出了所谓虚拟爆源原理，给出了计算混凝土中应力的一种方法，在深层目标毁伤研究方面具有一定的应用价值。

关键词：防护层；装药；爆炸作用；拉伸破坏；虚拟爆源

Blast Analysis of Outward Protection Layer of Underground Target

Wang Yunfeng[1,2], Li Lei[1], Feng Shunshan[1], Chen Xing[2]

(1. State Key Laboratory of Explosion Science and Technology,
Beijing Institute of Technology, Beijing 100081, China;
2. Baicheng Ordnance Test Center of China, Baicheng 137001, China. Correspondent:
WANG Yun - feng, E - mail: dandaoxue @ 163. com)

Abstract: Since the distance between explosion point and protection layer of underground target is different, explosion maybe load in the form of stress wave intensity, or it loads directly by explosion product to the protection layer of underground t n basis of current experiential formulas, relation between stress wave intensity and load/damage problem of protection layer of underground target was discussed. The effect of different blast amount and blast sit on protection layer of underground t was analyzed. Theory of suppositional blast site is put forward. A method to calculate stress in concrete is given. The result can be reference to research of deep buried target damage.

Key words: protection layer; charge; explosion effect; stretch damage; suppositional blast site

当装药量一定，爆点与地下目标混凝土防护层主体结构墙面距离的邻近，爆炸对防护层的加载增加，可能造成混凝土防护层的直接炸塌破坏或震塌破坏（应力波拉伸破坏）。文献［1］表明在接触爆炸时，由于正面成坑和背面震塌联合作用，防护层可能被贯穿，专门对一些震塌公式进行了较详细的对比研究，结合工程试验，验证了上述结论。如果混凝土的临界断裂应力σ_{cr}已知，也就相当于确定了防护层的破坏准则，但是，精确知道混凝土中应力变化的动态规律相当困难：①装药爆炸加载给防护层的应力波是球面波，对防护层的作用复杂，爆点距防护层多远才能把应力波视为平面波尚有待研究，通常处理成平面波与爆炸的实际情况存在差别；②目前还不能精确描述应力波在混凝土中的实际衰减规律[2]；③经验公式（或称工程表达式）都是在有限的样本量条件下获得的，可以很好地揭示研究对象当时当地的某些规律，从普遍意义上去类比还可能存在局限性。

爆炸冲击波转变为压应力波传入混凝土防护层内部，在自由面上反射为拉伸波，当在防护层介质中

---

① 原文发表于《东北大学学报》2006 年 8 月。

② 王云峰：工学博士，2002 年师从冯顺山教授，研究深层目标易损性及爆炸毁伤效应，现工作单位：吉林白城 31 基地。

合成的拉伸应力大于材料的动态抗拉强度时，防护层介质内的某个面将被拉断，形成所谓层裂，或曰剥落漏斗，这个过程是重复的，如果防护层内表面只有介质层崩落，而没有爆炸气体贯穿，称之为纯应力波拉伸破坏，介质剥落过程将剪断防护层内的布线或破坏其他适配管道等，同时具有能量的层裂介质碎块可能飞入目标空腔，毁伤墙体上或墙体附近的人员及设备。

1 介质交界面上的透射应力

根据文献［2］，假设目标防护层厚度为 4 m（30 MPa 混凝土），防护层外为天然黄土（6 m 厚），装药爆点在防护层外介质中，把传到土介质和混凝土界面（以下简称界面）的压缩波视为平面波，压缩波到达界面会发生透射和反射等复杂现象，σ_1、u_1、σ_r、u_r、σ_2、u_2 分别表示入射压缩波、反射的应力波、透射到混凝土内的应力波的应力和介质质点运动速度，应力传播工况示意图见图 1。

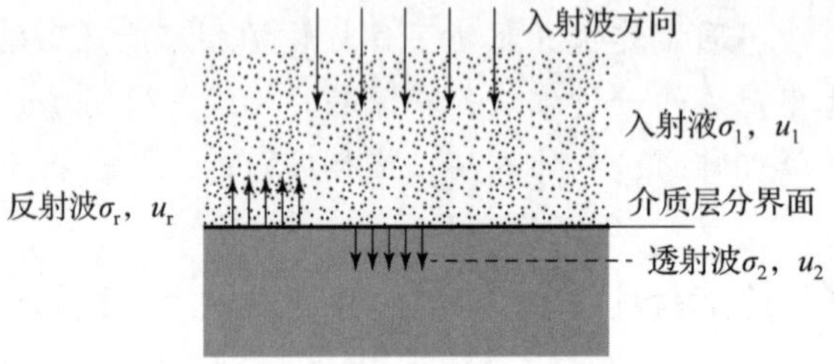

图 1 应力传播工况示意图

这里关心的是透射到混凝土中的应力 σ_2，天然黄土及混凝土的波阻抗分别为 $\rho_1 c_1$ 和 $\rho_2 c_2$，根据弹性理论，在分界面上应用应力和质点运动速度保持连续的条件，于是得到[4]

$$\sigma_2 = \frac{2\rho_2 c_2}{\rho_2 c_2 + \rho_1 c_1}\sigma_1 \tag{1}$$

式（1）即为透射到混凝土防护层中的应力 σ_2 计算公式，它是在弹性的假设条件下得出来的，实际上，波阻抗也会随爆炸过程演化，是一个动态量，不过从式（1）可以看出，透射应力 σ_2 显然介于 σ_1 和 $2\sigma_2$ 之间，动态的应力波阻抗参数一般很不易获得，工程上可用静态波阻抗参数估算。

研究不同介质交界面处的应力波的传递性质时，把入射波型简化为锯齿型[5]，见图 2。

应力波在自由面上的入射和反射见图 3。

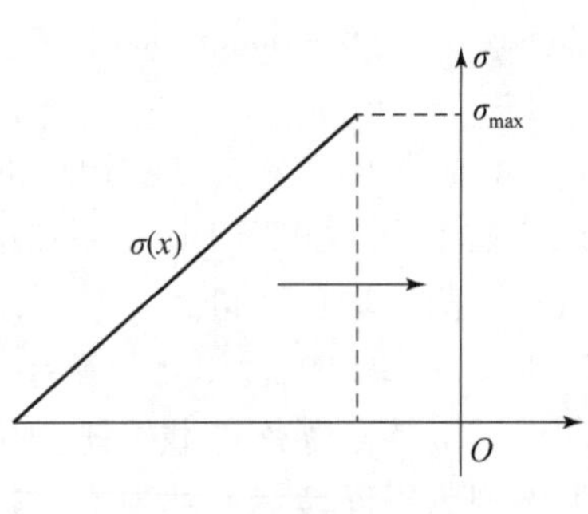

图 2 简化为锯齿型

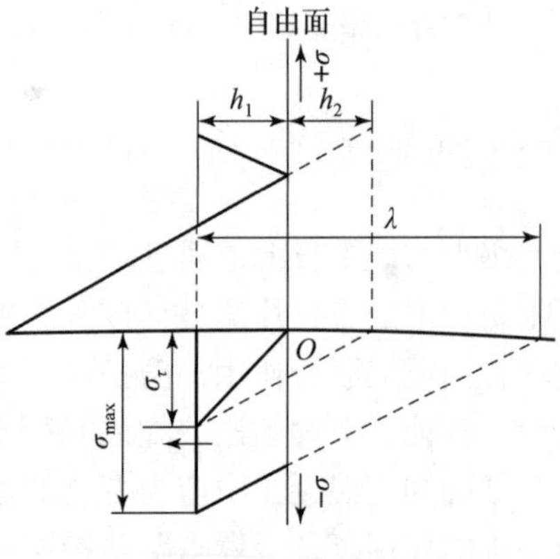

图 3 应力波在自由面上的入射和反射

波在自由面反射到距离 h 处，反射波和入射波合成动态拉应力的表达式为

$$\sigma(T) = \sigma_m \frac{2h_1}{\lambda_w}$$

式中，λ_w——应力波波长，当 $h_1 = \lambda_w/2$ 时，动态拉应力 $\sigma(T)$ 最大，即 $\sigma(T) = \sigma_m$，对于锯齿型波，介质瞬态断裂条件为 $\sigma(T) \geqslant \sigma_{cr}$，$\sigma_{cr}$ 为混凝土的临界断裂应力，经验的超压计算式为

$$\Delta P_m = \sigma_r = k\left(\frac{\sqrt[3]{\omega}}{r}\right)^{\alpha} \tag{2}$$

式（2）的应用条件：TNT 炸药，爆深 $h \geqslant 0.6\sqrt[3]{\omega}$，压应力为 0.098～4.5 MPa，装药量为 0.2～1 000 kg，可见式（2）的使用条件要受到很大的限制。计算时，对于天然黄土取 $\alpha = 2.8$，$k = 0.441\ 2$，装药量为 10～120 kg，爆点到交界面的距离为 3 m。当爆炸深度为 3 m、装药量为 120 kg 时对应的 $h \geqslant 0.6\sqrt[3]{\omega} = 2.96$ m，满足式（1）对爆点深度的要求，不同距离上的超压值计算结果见表 1。

表 1　不同装药量几个距离上的超压值

装药量/kg	距爆心处的超压/MPa		
	3.0 m	2.5 m	2.0 m
10	0.174 5	0.290 9	0.543 4
20	0.333 4	0.555 5	1.037 6
30	0.486 8	0.811 1	1.514 5
40	0.636 7	1.060 9	1.981 6
50	0.784 2	1.306 5	2.440 6
60	0.929 6	1.548 9	2.893 1
70	1.073 4	1.788 5	
80	1.125 9	2.025 9	
90	1.357 2	2.261 3	
100	1.497 5		
110	1.636 8		
120	1.775 3		

由表 1 可以看出，爆炸 TNT 量为 120 kg，距爆心 3 m 的混凝土界面处由黄土介质传到界面处透射前的超压约为 $\Delta P = 1.8$ Mpa，用式（1）估算。取 σ_2 约为 $1.9\sigma_1$，ΔP 取 σ_1 的值，得透射到混凝土方面的超压为 3.42 MPa。假设应力波在混凝土中的衰减略去不计，而且经自由面反射后，在 $\lambda_w/2$ 处动态拉应力叠加到最大为 3.42 MPa，混凝土的单轴抗拉强度和抗压强度的比值为 1/10 左右[6]，如果抗压强度为 30 MPa 的混凝土，则可得到静态抗拉强度约为 3.0 MPa，如果把破坏准则考虑为静态拉应力的 2 倍，那么可以得出结论：装 120 kg 药，在 4.0 m 厚混凝土上方 3.0 m 处的黄土中爆炸，混凝土防护层自由面不会产生拉应力破坏。考虑方法相同，在 2.5 m 距离，装 110 kg 和 120 kg 药爆炸后，可以对防护层造成拉伸破坏；同理在 2.0 m 距离，装 60 kg 以上药即可造成拉伸破坏，即如果爆点距混凝土防护层 2.0 m，战斗部装药至少要 60 kg。从表 1 还可以看出，超压对距离敏感，距爆心较近处时式（2）已经不能适用，梅尔尼科夫和鲁科夫对此[7]有过研究。

2　混凝土防护层中应力的另一种求解方法

装药在黄土中爆点 P 爆炸，假设爆点距自由面和交界面的距离 y 和 $d-y$ 都满足大于 $0.6\sqrt[3]{\omega}$ 的条件，见图 4。

交界面处透射前的冲击波超压的计算公式为

$$\Delta P_m = \sigma_r = k_t\left(\frac{\sqrt[3]{\omega_t}}{d-y}\right)^{\alpha}$$

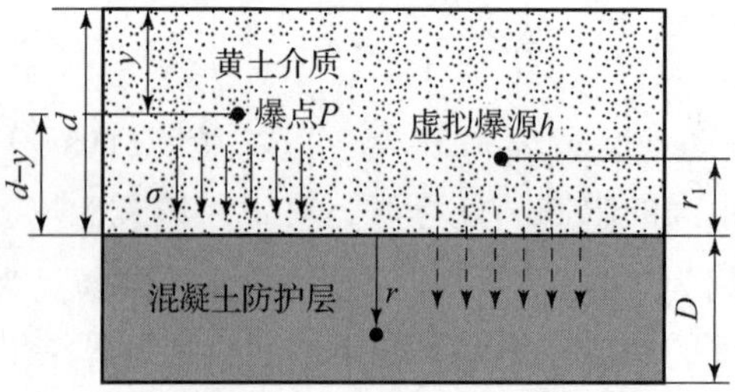

图4　虚拟爆源示意图

则透射到混凝土介质的应力为

$$\sigma_h = \frac{2\rho_h c_h}{\rho_h c_h + \rho_t c_t} k_t \left(\frac{\sqrt[3]{\omega_t}}{d-y} \right)^{\alpha_t}$$

式中，下脚标 t 代表黄土介质，h 代表混凝土介质。由于透射到界面上的超压值 σ_h 已知，把 σ_h 看成装药量 ω_t，在无限混凝土介质中距截面垂直距离为 r_1 处爆炸冲击波超压

$$\Delta P_h = \sigma_h = k_h \left(\frac{\sqrt[3]{\omega_t}}{r_1} \right)^{\alpha_h}$$

式中，k_h 和 α_h 为与混凝土有关的参数，由于混凝土为人工材料，很难做到所谓“无限介质”的尺度，可用与混凝土相当或类似的材料参数近似计算。

分析式（3）可知：“无限混凝土介质中的虚拟爆源 h”的位置由混凝土中爆炸的装药量 ω_t 决定，由应力相等条件解虚拟爆源位置 r_1：

$$k_t \left(\frac{\sqrt[3]{\omega_t}}{d-y} \right)^{\alpha_t} = k_h \left(\frac{\sqrt[3]{\omega_t}}{r_1} \right)^{\alpha_h}$$

解得

$$\frac{k_t}{k_h} = \frac{\dfrac{\left(\sqrt[3]{\omega_t}\right)^{\alpha_h}}{r_1^{\alpha_h}}}{\left(\dfrac{\sqrt[3]{\omega_t}}{d-y}\right)^{\alpha_t}} = \frac{\left(\sqrt[3]{\omega_t}\right)^{\alpha_h}}{r_1^{\alpha_h}} \frac{(d-y)^{\alpha_t}}{\left(\sqrt[3]{\omega_t}\right)^{\alpha_t}} = \left(\sqrt[3]{\omega_t}\right)^{(\alpha_h-\alpha_t)} \frac{(d-y)^{\alpha_t}}{r_1^{\alpha_h}}$$

$$r_1^{\alpha_h} = \frac{k_h}{k_t} \left(\sqrt[3]{\omega_t}\right)^{(\alpha_h-\alpha_t)} (d-y)^{\alpha_t}$$

$$r_1 = \sqrt[\alpha_h]{\frac{k_h}{k_t} \left(\sqrt[3]{\omega_t}\right)^{(\alpha_h-\alpha_t)} (d-y)^{\alpha_t}}$$

于是，虚拟爆源装药量 ω_t 在无限混凝土介质中的冲击波超压为

$$\Delta P_h = k_h \left(\frac{\sqrt[3]{\omega_t}}{r_1 + r} \right)^{\alpha_h}$$

式中，r 为混凝土交界面到板中的某个距离，则混凝土自由面上的超压（应力）为

$$\Delta P_r = k_h \left(\frac{\sqrt[3]{\omega_t}}{\sqrt[\alpha_h]{\dfrac{k_h}{k_t} \left(\sqrt[3]{\omega_t}\right)^{(\alpha_h-\alpha_t)} (d-y)^{\alpha_t}} + r} \right)^{\alpha_h}$$

当 $r = D$ 时，ΔP_D 为混凝土自由面上的超压，有了 ΔP_D，再由自由面反射叠加条件确定拉伸破坏情况。

如果试验能够确定混凝土的 k_h 和 α_h，混凝土中的应力分析就完全解决，k 和 a 可以通过试验方法或其他与混凝土近似的材料中得到。

3　结论

爆炸对防护层的作用是非常复杂的，理论解析分析十分困难，由于混凝土为人工材料，大体积工程试验一般很难，在岩石介质上试验虽然可以不受体积限制，但岩石的均匀程度、节理缝隙等特性与混凝土材料又存在很大差别，如何类比混凝土的试验结果还有待研究。本文是在经验公式的基础上，对防护层外介质层中爆炸破坏效应进行了分析，利用虚拟爆源的原理，提供了一种计算混凝土防护层中应力的方法，可能与工程实际存在差别，但对深层目标毁伤研究仍具有一定价值。

参考文献

[1] 郑全凭，钱七虎，周早生，等．钢筋混凝土震塌厚度计算公式对比研究［J］．工程力学，2003（6）：47－53.
[2] 王礼立．应力波基础［M］．北京：国防工业出版社，1985：180－190.
[3] Special feature adding new punch to cruise missiles specialized warheads can defeat hard and buried targets［J］. Jane's international defense，1998（1）：40－44.
[4] 北京工业学院八系．爆炸及其作用：下册［M］．北京：国防工业出版社，1979：302－327.
[5] 王儒策，赵国志．弹丸终点效应［M］．北京：北京理工大学出版社，1993：278－314.
[6] 过镇海．钢筋混凝土原理［M］．北京：清华大学出版社，2001：20－50.
[7] 亨利奇．爆炸动力学及其应用［M］．熊建国，译．北京：科学出版社，1987：183－216.

冲击波作用下目标毁伤等级评定的等效靶方法研究①

王　芳[1②]，冯顺山[1]，范晓明[2]
（1. 北京理工大学爆炸科学与技术国家重点实验室，北京　100081；
2. 国营第五一零三厂，南召　474675）

摘　要：为解决爆炸冲击波作用下目标毁伤等级的评定问题，提出了一种目标等效靶方法。通过在相同的冲击波作用下目标等效靶与目标毁伤程度之间的对应关系，快速地确定相应目标的毁伤等级。对冲击波作用下不同厚度的圆及方形 A3 钢板的变形情况进行了实验研究，结果表明形状对靶板的变形敏感性影响不大，而厚度对靶板的变形敏感性影响较大，0.3 mm 厚 A3 钢板的敏感性明显好于 1 mm 厚 A3 钢板。根据实验结果确定 0.3 mm 厚 A3 钢板为人员目标的等效靶，并在不同冲击波超压作用下目标等效靶与人员目标的毁伤等级之间建立了初步的对应关系。本文的研究成果可用于冲击波的威力效应评价、目标易损性分析，以及冲击波的防护效应评价。

关键词：爆炸冲击波；毁伤等级；目标等效靶；人员

1　引言

冲击波是云爆弹、杀爆火箭弹等爆破毁伤武器最主要的一种毁伤元。在冲击波的威力效应研究和冲击波作用下目标的易损性分析中，经常采用爆炸场的超压值或比冲量作为目标的毁伤准则和冲击波的威力参数，通过模拟试验、实物靶场试验或真实战场试验等方法，根据真实目标的具体毁伤情况进行目标易损性分析和冲击波威力评价。由于具体目标结构的复杂性，它们在爆炸冲击波作用下的响应情况及毁伤等级的确定也十分复杂，而且相关试验工程浩大，限制因素很多，费用极高。因此相比破片等其他毁伤元，冲击波的毁伤效应试验存在着许多的困难。

为此，本文提出了一种冲击波作用下目标毁伤等级评定的等效靶方法，通过在相同的爆炸冲击波作用下目标等效靶与目标毁伤等级之间的对应关系，可快速地确定相应目标的毁伤等级。根据上述方法对人员目标在爆炸冲击波作用下的毁伤情况进行了分析，通过实验研究，确定了人员目标等效靶的材料、尺寸与厚度，并在目标等效靶与人员目标的毁伤等级之间建立了对应关系。本文的研究成果不仅可应用于冲击波的威力评价和冲击波作用下目标的易损性分析，也可应用于冲击波的防护效应评价，具有重要的理论和工程应用价值。

2　冲击波作用下目标毁伤等级评定的等效靶方法

冲击波作用下目标毁伤等级评定的等效靶方法是指在相同的爆炸冲击波作用下，由一定约束条件下的、具有恰当敏感性的靶板的毁伤程度，去描述某一给定目标的毁伤程度，从而在靶板的毁伤程度与给定目标的相应毁伤程度之间建立对应关系，根据靶板的毁伤等级确定给定目标的毁伤等级。由于靶板材料来源广泛，加工简单，这种研究方法将大大减少试验经费，降低试验规模，相比真实目标的试验研究来说具有简单易行、投资少、有效性好的优势，是一种科学、低成本的研究方法。

在该等效靶方法中，选择具有恰当敏感性的靶板是进行等效的关键。靶板的敏感性是指给定约束条件下的靶板在爆炸冲击波作用下发生变形破坏响应程度的特性。敏感性的大小是通过给定载荷作用

---

①　原文发表于《第九届全国爆炸与安全技术学术会议论文集》。

②　王芳：工学博士，1996 年师从冯顺山教授，研究 FAE 战斗部威力的评价方法，现工作单位：北京理工大学。

下靶板的变形破坏量，以及载荷发生变化时靶板变形破坏的改变量表现出来的。当载荷大小一定时，变形破坏量大的靶板敏感性高；反之，敏感性低。当载荷发生变化时，变形破坏改变量大的靶板敏感性高；反之，敏感性低。具有恰当敏感性，是指在给定的载荷作用下，靶板能够产生适当的变形破坏量，而当载荷在一定范围内发生变化时，靶板的变形破坏量也能够产生适当的变化。

这里需要强调指出的是，靶板的敏感性并不是越高越好，而是需要根据爆炸冲击波对不同目标的具体作用情况来最终确定。爆炸冲击波所能有效毁伤的目标主要包括人员、军用车辆和建筑物等中软目标。这些目标所能承受的冲击波超压范围是十分不同的，人员所能承受的超压范围远远小于军用车辆和建筑物，这种情况下，对于人员目标，就需要选用敏感性较高的靶板；而对于军用车辆或建筑物来说，可以选择一些敏感性中等或稍低的靶板。影响靶板敏感性的因素有很多，包括靶板的材料、尺寸、厚度、形状、边界条件等。在选取靶板时，要进行综合考虑。下面在给定材料和尺寸情况下，对不同形状和厚度的 A3 钢板在爆炸冲击波作用下的变形情况及其敏感性进行了实验研究。

3 靶板在爆炸冲击波作用下的变形实验研究

3.1 实验条件

实验在爆炸洞中进行，靶板材料采用 A3 钢，厚度分别为 1 mm 和 0.3 mm。使用专用靶架，利用专用靶架上的 20 个特制螺栓和压板将靶板的周边夹紧固定。靶板有方形和圆形两种，靶板的边长，直径为 520 mm，靶板承受爆炸冲击波作用的边长/直径为 360 mm。装药为 TNT 圆柱形药柱，压装、药量分别为 270 g 和 50 g，采用端部垂直起爆方式起爆。实验现场布置如图 1 所示。

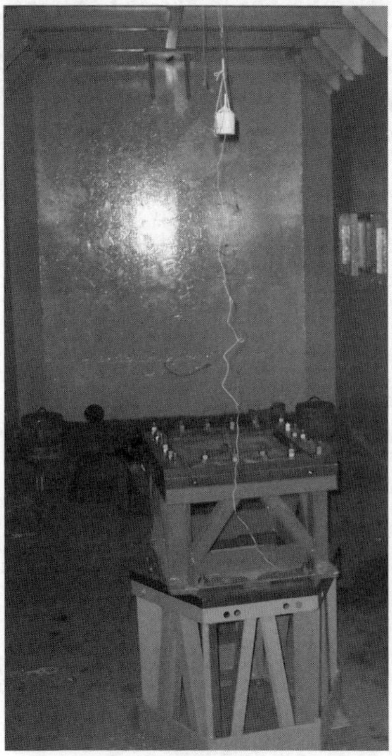

图 1 实验现场布置

3.2 实验结果及分析

在爆炸冲击波作用下，靶板产生了永久塑性变形，实验数据如表 1 所示，靶板变形情况如图 2 所示。

表 1　实验数据

序号	板材	板子形状	板厚/mm	药量/g	距离/m	超压/MPa	挠度/mm
1	A3	圆形	1	270	1.0	0.279 7	52
2	A3	圆形	1	270	1.2	0.186 9	40
3	A3	圆形	1	271	1.3	0.157 5	36
4	A3	方形	1	270	0.9	0.355 7	66.5
5	A3	方形	1	271	1.2	0.186 9	45
6	A3	方形	1	271	1.4	0.134 7	28
7	A3	圆形	0.3	50.1	0.9	0.102 6	25
8	A3	圆形	0.3	50.1	1.3	0.052	23.5
9	A3	圆形	0.3	50	1.8	0.029 4	19

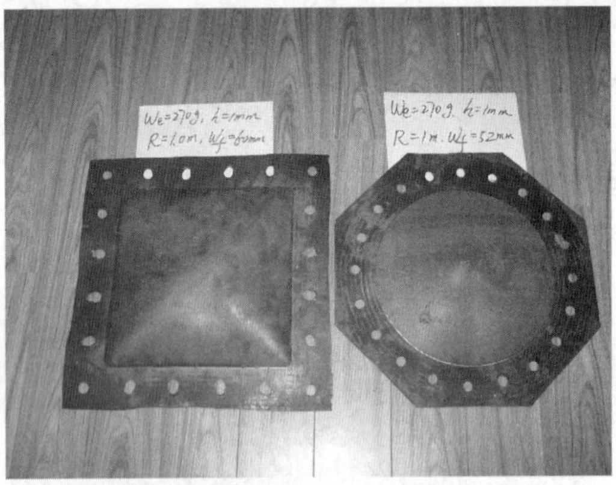

图 2　靶板变形情况

3.2.1　不同形状靶板的变形情况对比

图 3 为 270 g 药量不同距离条件下 1 mm 厚方形靶板和圆形靶板的变形情况对比。在相同药量和距离条件下，方形靶板与圆形靶板的变形挠度较为接近，但方形靶板的挠度一般稍大，而且在小距离条件下两者的变形情况差异较大，方形靶板即使产生严重的塑性变形，也不会出现剪切撕裂现象，只是螺栓孔

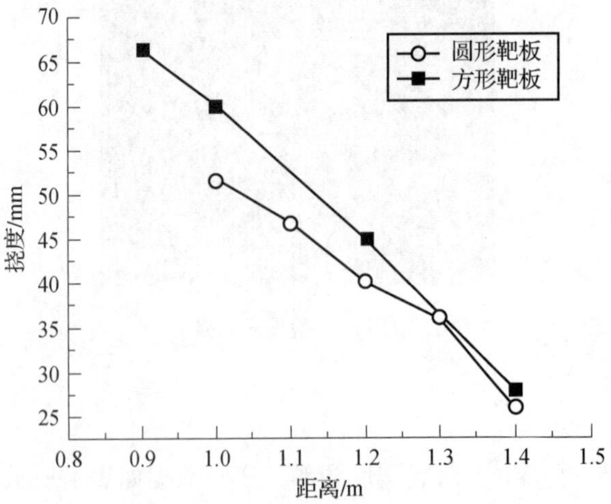

图 3　270 g 药量下圆形靶板和方形靶板变形情况对比

严重变形甚至拉断。而圆形靶板在螺栓孔基本发生变形的情况下已开始沿靶板边缘出现剪切撕裂。随着距离的增大，爆炸冲击波的超压不断减小，两者的挠度逐渐接近。经分析，造成这种现象的原因可能是方形靶板的受载面积比圆形靶板大，且方形靶板的边界约束要弱于圆形靶板。由实验结果可知，方形靶板的敏感性要稍好于圆形靶板，但形状对靶板敏感性的影响不是很大。

3.2.2　不同厚度靶板的变形情况对比

图4为不同厚度圆形靶板的挠度随超压变化的情况。在0.13～0.28 MPa范围内，1 mm厚A3钢板能够产生一定的变形挠度；随着超压的减小，靶板的变形迅速减小，在0.13 MPa左右，靶板的变形挠度已经很小，<0.13 MPa以后，根据实验数据推测，靶板将不会产生明显的变形。而0.3 mm厚A3钢板在<0.1 MPa的情况下，仍能够产生一定的塑性变形。由此可知，厚度对靶板的敏感性影响较大，0.3 mm厚A3钢板的敏感性要好于1 mm厚A3钢板。

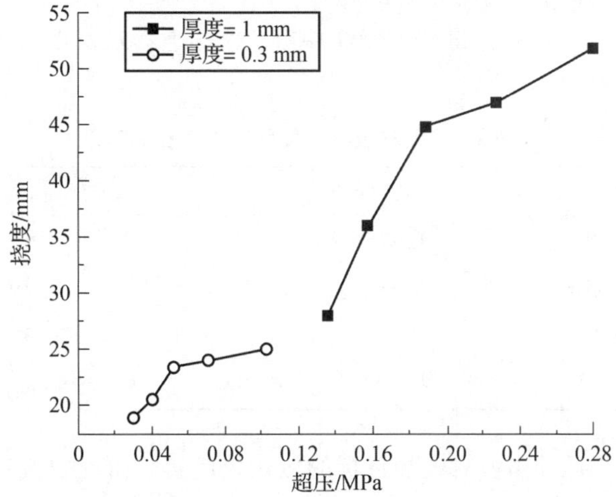

图4　不同厚度圆形靶板的挠度随超压变化的情况

4　人员目标等效靶的确定

4.1　人员目标在冲击波作用下的毁伤情况分析

人员受冲击波的作用情况可分为两种：直接冲击波作用和间接冲击波作用[1]。直接冲击波作用也即初始冲击波作用，可破坏人员的中枢神经系统，震击心脏，造成肺部出血，伤害呼吸及消化系统，震破耳膜等。这种因爆炸冲击波作用于人体所产生的损伤，称为爆炸冲击伤，这类损伤与冲击波波阵面的峰值超压直接有关，另外人员对冲击波到达后上升到峰值超压的速度和冲击波的持续时间也很敏感。间接冲击波作用可以分为次生作用、第三作用和其他作用。次生作用包括由爆炸装置本身或位于爆炸装置附近的物体所形成的抛掷物的撞击侵彻作用；第三作用是指在冲击波压力与气流相互作用下人员整个身体发生的位移，接着发生减速撞击；其他作用主要是指烟尘和热的伤害。

爆炸冲击波对人员的作用是直接作用和间接作用的综合结果，其中直接冲击波作用对人体的伤害最为严重。因此在考虑冲击波对人员的毁伤时，本文侧重于直接冲击波作用。通过研究发现，对于具体的身体部位来说，含有空气的肺组织比其他要害器官更容易受到初始冲击波的伤害。因为肺是含有大量气囊或气泡的组织，与周围邻近组织相比，密度较低，所以对冲击波伤害很敏感。本文在有关肺损伤实验研究成果的基础上[2]，以肺的损伤程度作为人员毁伤等级的主要划分依据，对文献［3］中不同冲击波超压作用下人员的损伤程度进行了修正和补充，得到了如表2所示的人员在空气冲击波超压作用下的毁伤等级。

表 2 人员在空气冲击波超压作用下的毁伤等级

毁伤等级	超压值/MPa	伤害程度	伤害情况
一级	0.02～0.03	轻微	轻微挫伤，肺部小灶性出血
二级	0.03～0.05	中等	听觉、气管损伤；肺大泡形成，肺体指数明显增加
三级	0.05～0.1	严重	内脏受到严重挫伤，肺体指数显著增加，可能造成伤亡
四级	>0.1	死亡	大部分人死亡

4.2 等效靶与人员目标等效关系建立

由表 2 可知，人员能够承受的冲击波超压很低，>0.1 MPa 就可以造成大部分人员死亡。因此在选择人员目标等效靶时，应该选择敏感性非常好的靶板，由实验结果看，0.3 mm 厚 A3 钢板比较合适。将 0.3 mm 厚 A3 钢板的实验结果外推后，得到了爆炸冲击波作用下等效靶板与人员目标毁伤等级之间一个初步的等效关系，如表 3 所示。

表 3 等效靶板与人员目标的等效关系

毁伤等级	超压值/MPa	人员伤害程度	等效靶板的变形程度 W_f
一级	0.02～0.03	轻微	<19 mm
二级	0.03～0.05	中等	19～23 mm
三级	0.05～0.1	严重	23～25 mm
四级	>0.1	死亡	>25 mm，出现撕裂破坏

根据这个等效关系，在今后冲击波对人员目标的毁伤效应实验研究中，就可以用等效靶替代人员目标，对等效靶进行不同距离下的毁伤试验，根据等效靶的毁伤程度，快速地给出人员目标的毁伤程度。由表 3 可知，0.3 mm 厚 A3 钢板在 0.1 MPa 以下各阶段的变形挠度差距不是太大，作为人员目标的等效靶来说，还不是最为理想的，还需要做进一步的研究工作。

5 结论

（1）提出了一种冲击波作用下目标毁伤等级评定的等效靶方法，通过在相同的爆炸冲击波作用下目标等效靶与目标毁伤等级之间的对应关系，可快速地确定相应目标的毁伤等级。

（2）在靶板材料和尺寸给定的情况下，对爆炸冲击波作用下不同形状及厚度 A3 钢板的变形情况及敏感性进行了实验研究，结果表明形状对靶板的变形敏感性影响不大，而厚度对靶板的变形敏感性有很大影响，0.3 mm 厚 A3 钢板的敏感性明显好于 1 mm 厚 A3 钢板。

（3）对人员目标在爆炸冲击波作用下的毁伤情况进行了分析，根据实验结果，确定 0.3 mm 厚 A3 钢板为人员目标的等效靶，并在等效靶与人员目标的毁伤等级之间建立了一个初步的对应关系。

（4）本文的研究成果不仅可应用于冲击波的威力评价和冲击波作用下目标的易损性分析，也可应用于冲击波的防护效应评价，具有重要的理论和工程应用价值。

参 考 文 献

[1] 贝克，威斯汀，等. 爆炸危险性及其评估：下 [M]. 张国顺，文以民，等译. 北京：群众出版社，1988：517－531.

[2] 安波，李兵仓. 某型榴弹静爆对生物杀伤效应的实验研究 [J]. 创伤外科杂志，1999，13（1）：161－163.

[3] 北京工业学院编写组. 爆炸及其作用（F）[M]. 北京：国防工业出版社，1988：286－287.

弹药安全性能评价模型研究①

董三强[1,2]②，冯顺山[1]，金　俊[3]

（1. 北京理工大学爆炸科学与技术国家重点实验室，北京　100081；
2. 第二炮兵工程学院核工程系，陕西　西安　710025；
3. 装甲兵工程学院机电工程系，北京　100072）

摘　要：安全性弹药是弹药未来发展的主要方向。本文依据弹药相对其寿命过程中可能遭受的主要环境刺激因素的安全水平提出了弹药的安全判据，引入安全度参数，建立了弹药安全性能评价的理论模型。模型中，依据弹药的安全度值划分了弹药的安全等级。基于该评价模型计算了改 Mk－82 钝感常规炸弹的安全性能，分析了未爆弹药（UXO）自失效设计对弹药整体安全性能的影响。结果显示，终点未爆自失效设计能明显提高弹药的安全水平。此模型较好地反映了弹药固有的安全性能，可以为弹药安全性设计及安全性能评价提供理论指导。

关键词：系统评估与可行性分析；安全工程；弹药；安全评价；自失效

Researches on Safety Assessment Model of Ammunitions

Dong Sanqiang[1,2], Feng Shunshan[1], Jin Jun[3]

(1. State Key Laboratory of Explosion Science and Technology, Beijing Institute of Technology, Beijing 100081, China;
2. Nuclear Engineering Department, The Second Artillery Engineering Institute, Xi'an 710025, Shaanxi, China;
3. Department of Mechanical Engineering, Academy of Armored forces Engineering, Beijing 100072, China)

Abstract: The main development trend of future ammunitions will be safety. The ammunition safety criteria was proposed on the basis of the effect of main environmental stimulus on the ammunitions safety. A general model for ammunition safety assessment was developed and a parameter of safety degree was introduced to rank the safety of the assessed ammunitions. With the model, the safety of a modified insensitive conventional bomb Mk－82 and the effect of terminal self－failing design for unexploded ordnance (UXO) were analyzed. The results show that the terminal sell－failing design can observably improve the safety level of the ammunition. The model in this paper can reflect the safety performance of ammunitions and provides a reference for safety design and evaluation of ammunitions.

Keywords: analysis on system assessment and feasibility; safety engineering; ammunition; safety evaluation; self－failing

弹药安全一直是世界各国普遍关注的问题。纵观弹药的发展历史，由于安全性设计方面的缺陷，以及管理和使用方面的不足，弹药爆炸事故频繁发生，造成重大的人员伤亡和财产损失。文献［1］提出安全性弹药的概念，要求在弹药中引入安全性设计来保证弹药在其寿命周期中各个环节的安全性。安全性弹药概念的提出顺应了社会的发展和人类文明的进步对弹药安全的新的要求，代表了弹药的未来发展方向。

---

①　原文发表于《兵工学报》2011（4）。

②　董三强：工学博士，2003 年师从冯顺山教授，研究弹药安全理论及相关应用，现工作单位：第二炮兵工程学院。

弹药安全评价是安全性弹药安全设计中的必要环节。目前弹药或危险性物质安全评估模型综合考虑人员管理、弹药以及环境等多方面的因素，分析比较简单，甚至是基于对诸多重要安全因素的模糊评价，不能准确反映弹药固有的安全能力[2-6]。本文依据弹药相对其寿命过程可能遭受到的主要环境刺激因素的安全水平提出弹药的安全判据，建立弹药安全性能评价模型。基于该模型对某钝感弹药的安全性能进行评价。

1 弹药的安全判据

安全与危险总是相互依存的。弹药一般都含有各种各样的含能材料，危险性是其本身所固有的特性，且弹药终归是为了完成某项作战任务，在执行完使命之前会一直存在于其寿命过程的某个环节中，因此弹药的安全具有相对性，与弹药固有的危险性因素（如引信、爆炸序列以及主装药等）对环境刺激因素的感度有关。通过定义弹药相对于其寿命过程可能遭受到的主要环境刺激因素的安全水平来评价弹药的安全性能。

设弹药在其寿命过程中可能遇到 n 类环境刺激因素，以集合的形式表示为

$$U=\{X_1,X_2,\cdots,X_i\cdots,X_n\} \tag{1}$$

式中，$X_i(i=1,2,\cdots,n)$ 表示第 i 个环境刺激因素。

每一类环境刺激因素可以用一组特性参数来表征

$$X_i=\{x_{i1},x_{i2},\cdots,x_{ij},\cdots,x_{im_i}\},i=1,2,\cdots,n \tag{2}$$

式中：$x_{ij}(j=1,2,\cdots,m_i)$ 表示刺激因素 X_i 的第 j 个特性参数。如破片 =｛破片速度，破片质量，破片形状因子｝，冲击波 =｛冲击波压力，冲击波持续时间｝。

对于任意一类环境刺激因素 X_i，以其某一水平（其特性参数分别取值）作为弹药相对于该刺激因素的安全标准，称为临界安全水平，记为 X_{e-i}。

规定：当弹药在所有低于该水平的环境刺激因素 X_i 的作用下均不发生危险响应时，是相对安全的，否则是相对不安全的。

分别规定集 U 中所有环境刺激因素的临界水平，得到弹药的临界安全水平

$$U_e=\{X_{e-1},X_{e-2},\cdots,X_{e-i}\cdots,X_{e-n}\} \tag{3}$$

式中，$X_{e-i}(i=1,2,\cdots,n)$ 为第 i 类环境刺激因素的临界安全水平。

对于给定弹药，以 U^* 表示其在集 U 中所有刺激因素作用下的最高安全水平：

$$U^*=\{X_1^*,X_2^*,\cdots,X_i^*\cdots,X_n^*\} \tag{4}$$

式中：$X_i^*(i=1,2,\cdots,n)$ 表示该弹药在第 i 类环境刺激因素的作用下的最高安全水平。

依据安全性弹药的安全理念，弹药的相对安全的判据可表示为，对于任意 $X_i^*\subset U^*$ 满足

$$X_i^*\geqslant X_{e-i},i=1,2,\cdots,n \tag{5}$$

则称式（5）中满足 $U^*=U_e$ 的弹药为临界安全弹药。

如果存在 $X_i^*\subset U^*$，使

$$X_i^*<X_{e-i},i=1,2,\cdots,n \tag{6}$$

则称弹药为低安全弹药或不安全弹药。

2 弹药安全性能评价模型的建立

2.1 基本约定

以变量 a 表示弹药的安全性能，即安全度值，并提出以下约定：

（1）a 在（0，1）间取值。

（2）以 $a=0.5$ 表示弹药的临界安全水平；$a\geqslant 0.5$ 则认为弹药是相对安全的，且 a 越大，弹药的相对安全性能越好；$a<0.5$ 则认为弹药为低安全或不安全，且 a 越小，弹药越不安全。

2.2　弹药安全性评价模型

仍以集合 U 表示弹药在其寿命过程中所主要遭受的环境刺激因素，以集合 U_e 作为弹药的安全判据，以集合 U^* 表示给定弹药的安全水平。

由于不同环境刺激因素之间以及同一环境刺激因素的不同特性参数之间一般不具有可比性，为了便于比较和计算，引入转换关系转换为统一的安全度形式：

$$a_{ij}^* = \frac{1}{2} \pm \frac{1}{\pi}\arctan\left(k_{ij}\left(\frac{x_{ij}^*}{x_{e-ij}} - 1\right)\right) \tag{7}$$

式中　a_{ij}^*——第 i 类环境刺激因素中第 j 个特性参数的安全度表示形式；

x_{ij}^* 和 x_{e-ij}——U^* 和 U_e 中第 i 类环境刺激因素中第 j 个特性参数的取值；

k_{ij}——转换因子。

式（7）中，当表示弹药在环境刺激因素X_i作用下的安全性能的特性参数 x_{ij}^* 的值随着弹药的安全性能的提高而增大时，取“+”；反之取“-”。

经过转换，可以得到反映弹药相对于刺激因素X_i的实际安全水平的安全度集合

$$A_i^* = \{a_{i1}^*, a_{i2}^*, \cdots, a_{im_j}^*\}, i = 1, 2, \cdots, n \tag{8}$$

记弹药相对于环境刺激因素 X_i 的安全度为 a_i^*，通过下面的方法评价弹药相对于环境刺激因素 X_i 的安全性。

如果任意 $a_{ij}^* \in A_i^*$，满足 $a_{i1}^* \geqslant 0.5$，则称弹药在环境刺激因素 X_i的作用下是相对安全的，可以通过加权平均或算术平均的方法计算 a_i^*。本文以算术平均计算

$$a_i^* = \frac{1}{m_i}\sum_{j=1}^{m_i} a_{ij}^* \tag{9}$$

如果存在 $a_{ij}^* \in A_i^*$，使 $a_{ij}^* < 0.5$，则称弹药在环境刺激因素 X_i的作用下是低安全或不安全的，不能通过加权平均或算术平均的方法来计算 a_i^*，否则可能丢失弹药最重要的安全信息。本文以最大安全原则处理此类情况：

$$a_i^* = \min(a_{ij}^*), j = 1, 2, \cdots, m_i \tag{10}$$

由式（9）和式（10）可以确定弹药相对于不同环境刺激因素的安全度集合

$$A^* = \{a_1^*, a_2^*, \cdots, a_n^*\} \tag{11}$$

以集合 W 表示环境刺激因素 U 的权重集：

$$W = \{w_1, w_2, \cdots, w_n\} \tag{12}$$

式中，$w_i(i=1,2,\cdots,n)$为集合 U 中元素 X_i 的权重值，表示不同环境刺激因素对弹药整体安全性能的影响程度，且有

$$\sum_{i=1}^{n} w_i = 1, w_i \geqslant 0$$

通过下面的方法计算弹药的安全度。

如果任意 $a_i^* \in A^*$，满足 $a_i^* \geqslant 0.5$，则称弹药为相对安全弹药，通过加权平均的方法计算弹药的安全度值

$$a = \sum_{i=1}^{n} w_i a_i^* \tag{13}$$

如果存在 $a_i^* \in A^*$，使 $a_i^* < 0.5$，则称该弹药为低安全或不安全弹药。不能通过加权平均的方法计算弹药的安全度，否则可能丢失弹药最重要的安全信息。本文采用式（14）计算弹药的安全度：

$$a = \frac{1}{2}\sum_{i=1}^{n-m} w_i + \sum_{j=1}^{m} \hat{w}_j \hat{a}_j^* \tag{14}$$

式中　m——使弹药未能达到安全标准的环境刺激因素的个数；

$\hat{a}_j^*$ 和 $\hat{w}_j(j=1,2,\cdots,n)$——弹药相对于某类环境刺激未能达到安全标准时的相应安全度值和权重值。

2.3 弹药的安全等级划分

依据式（13）和式（14）计算得到的弹药安全度值，可以将弹药划分为如表 1 所示的 5 个安全等级和 3 个不安全等级。

表 1 弹药安全等级划分

弹药分类	安全等级	a
安全性弹药 ($0.5 \leqslant a < 1$)	理想安全弹药	0.91～1.0
	高安全弹药	0.81～0.9
	安全弹药	0.71～0.8
	比较安全弹药	0.61～0.7
	一般安全弹药	0.51～0.6
不安全弹药 $a < 0.5$	低安全弹药	0.41－0.49
	不安全弹药	0.31－0.40
	非常不安全弹药	<0.30

依据弹药的安全度值而划分的弹药安全等级是一个相对安全的概念，参考标准是弹药的安全判据。

首先，集合 U 中环境刺激因素的选择要建立在对弹药的安全风险进行充分分析的基础之上，确保影响弹药安全的主要环境刺激因素不被遗漏。其次，集合 U 中环境刺激因素的临界水平的确定需要切实反映环境刺激因素的现实威胁水平，同时还要考虑当前弹药安全设计的技术水平，既不能定得太高，而限制弹药的有效生产及使用，又不能定得太低，降低弹药的安全等级而失去安全性弹药的意义。

3 算例

基于本文建立的弹药安全性能评价模型计算了美国空军研制的改 Mk－82 钝感常规炸弹的安全性能。改 Mk－82 钝感炸弹是美国空军 Wight 实验室于 20 世纪 90 年代在 Mk－82 的基础上通过应用钝感主装药及改装引信而开发的[7]。冷战期间，受到只具有 HD1.1 资格的弹药的 Q/D 的严格限制，美国空军对欧洲战区的弹药配备及运输能力被大大削减。因此，美国空军启动了该项研究，旨在通过钝感改进，提高弹药的安全等级，使其整弹达到 HD1.2 的标准，未装引信时达到 HDl.6 的标准。

模型中分析了弹药相对于撞击、冲击波、子弹冲击、慢速烤燃和快速烤燃 5 种不同环境刺激因素的安全性能。弹药的安全判据中，特性参数的设定及相关测试方法如表 2 所示，特性参数的临界取值参考 HDl.6 的安全标准给出[8-9]，计算过程及最终取值如表 3 所示。

表 2 改 Mk－82 弹药的安全判据

X_i	安全指标 X_{e-i}	指标说明	参考测试方法[8-9]
撞击作用 X_1	v_e/ms^{-1}，p_{max}/kPa	射弹临界速度，最大超压	UN Test7c（i）
冲击波作用 X_2	δ_e/mm	临界 PMMA 间隙	UN Test7b
子弹冲击 X_3	E_e/kJ	子弹临界动能	UN Test7j
慢速烤燃 X_4	η_e，$T/℃$	安全系数，响应温度	UN Test7h
快速烤燃 X_5	η_e，t_d/min	安全系数，反应延滞时间	UN Test7g

表3　改 Mk－82 安全评价结果（$a=0.65$）

X_i	W	A^*	A_i^*	X_{e-i}	X_i^*
撞击作用X_1	0.4	0.62	0.50，0.74	≥333，≤27	333，17
冲击波作用 X_2	0.1	0.72	0.72	≤70	41
子弹冲击 X_3	0.1	0.50	0.50	≥16.23	16，23
慢速烤燃 X_4	0.1	0.76	0.98，0.53	≥0.5，≥120	1，132
快速烤燃 X_5	0.3	0.69	0.74，0.64	≥0.5，≥15	1，22

根据计算结果，该钝感弹药的安全度值为 0.65，属于比较安全弹药。考虑到当前较为突出的弹药终点未爆安全问题，结合在相关研究中提出的自失效弹药的概念[1]，进一步分析了弹药终点未爆时引入自失效设计对改 Mk－82 安全性能的影响。

以弹药终点未爆时的自失效性能作为评价弹药安全性能的因素之一，引入前面所建立的弹药安全性能评价模型，以弹药自失效模式（即引信自失效、爆炸序列自失效和主装药自失效）和自失效时间（即弹药按照预定自失效模式失去规定效能的时间 t^0）作为评价弹药终点未爆安全性能的特性参数。取弹药终点未爆时规定的自失效时间（τ_{d-e}）为 5 d，计算了不同自失效模式下弹药终点未爆时的安全性能，如图 1（a）所示，以及考虑了弹药终点未爆自失效设计的改 Mk－82 钝感常规炸弹的整体安全性能，如图 1（b）所示。

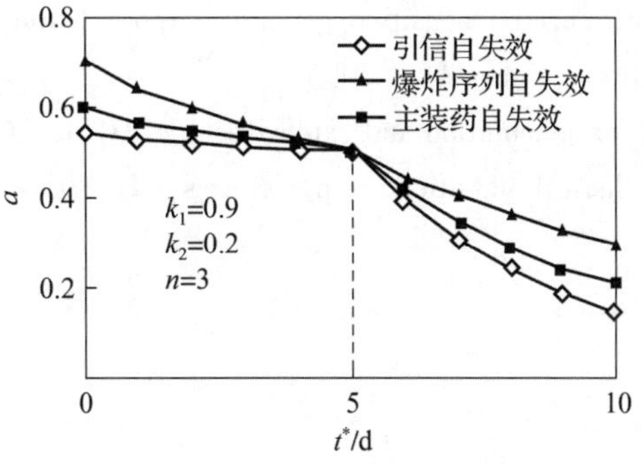

（a）

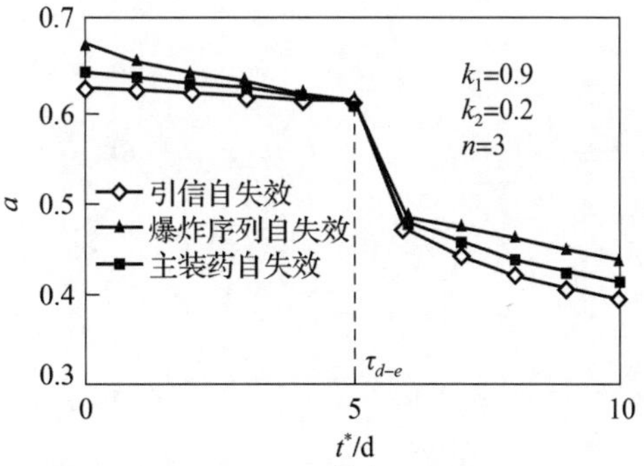

（b）

图 1　弹药终点未爆自失效设计对弹药安全性能的影响

（a）弹药终点安全性能；（b）改 Mk－82 钝感常规炸弹整体安全性能

结果显示，当弹药终点未爆时，引信自失效、主装药自失效和爆炸序列自失效 3 种不同的自失效模式下的弹药终点未爆安全度值最高分别为 0.546（一般安全弹药），0.598（一般安全弹药）和 0.701（比较安全弹药）。考虑终点自失效设计时，改 Mk－82 钝感常规炸弹在 3 种不同自失效模式下的安全度值分别达到 0.687（比较安全弹药）、0.695（比较安全弹药）和 0.711（安全弹药）。由此得出结论，引入弹药终点未爆自失效设计可以显著提高弹药的综合安全水平，且爆炸序列自失效设计对弹药的安全性能的贡献最大，引信自失效设计对弹药的整体安全性能影响最小。

4 结论

（1）根据弹药相对其寿命过程中可能遭受的主要环境刺激因素的安全水平提出了弹药的安全判据，引入安全度参数，建立了弹药安全性能评价的理论模型。依据弹药的安全度值划分了弹药的安全等级。

（2）基于该评价模型计算了改 Mk－82 钝感常规炸弹的安全性能，分析了弹药终点未爆时自失效设计对弹药整体安全性能的影响。结果显示，终点未爆自失效设计能明显提高弹药的安全水平。此模型较好地反映了弹药固有的安全性能，可以为弹药安全性设计及安全性能评价提供理论指导。

参考文献

[1] 董三强．弹药安全理论及相关应用研究［D］．北京：北京理工大学，2008.

[2] 宣兆龙．野战弹药环境安全的灰色模糊综合评判［J］．装备环境工程，2006，3（1）：56－59.

[3] 刘小春，黄嵩．基于模糊数学的民爆器材库房安全综合评价模型及应用［J］．中国安全科学学报，2007，17（7）：56－59.

[4] 宣兆龙，易建政，吴建华．灰色模糊理论在弹药安全管理系统评价中的应用［J］．军械工程学院学报，2005，17（2）：4－6.

[5] 刘铁民，张兴凯，刘功智．安全评价方法应用指南［M］．北京：化学工业出版社，2005：227－293.

[6] 罗云，樊运晓，马晓春．风险分析与安全评价［M］．北京：化学工业出版社，2004：176－178.

[7] CORLEY J D，STEWART A C. Fuzed insensitive general purpose bomb containing AFX－645［J］．Fuzed insensitive general purpose bomb containing Afx，1995.

[8] DOD Contractors's safety manual for ammunition and explosives：DOD4145.26－M［S］Washington：DOD，1997.

[9] DOD Ammunition and explosive hazard classification procedures：TB 700－2［S］．Washington：DOD，1998.

Dynamic Response of Spherical Sandwich Shells with Metallic Foam Core under External Air Blast Loading – Numerical Simulation①

球面金属泡沫夹芯板在外爆载荷作用下的动态响应

Li Wei[a]②, Huang Guangyan[a], Bai Yang[a], Dong Yongxiang[a], Feng Shunshan[a,]

[a] State Key Laboratory of Explosion Science and Technology,

Beijing Institute of Technology, Beijing, 100081, P. R. China

Abstract: The dynamic response of spherical sandwich shells with aluminum face sheets and aluminum foam core under external air blast loadings were investigated numerically by employing the LS – DYNA. To calibrate the numerical model, the experiments of cylindrical sandwich shells were modeled. And the numerical results have a good agreement with the experiment data. The calibrated numerical model was used to simulate the dynamic response of spherical sandwich shells subjected to the external air blast loadings. It is found that the spherical sandwich shells have a better performance than that of the cylindrical sandwich shells in resisting the blast loadings. The structural dynamic response process has been divided into three specific stages and the deformation modes have been classified and discussed systematically. According to parametric studies, it is concluded that with the decrease of radius of curvature, increase of the thickness of foam core and face sheets and decrease of blast intensity, the blast – resistance is increasing obviously; keeping the thickness summation of front and back face sheet almost constant, a big thickness of front face sheet will improve the blast – resistance performance. These simulations findings can guide well the theoretical study and optimal design of spherical sandwich structures subjected to external blast loading.

Key words: Sandwich shell; Metallic foam; Blast resistance; Numerical simulation

1 Introduction

Sandwich panels are made up of two stiff, strong shins separated by a lightweight core. The separation of the shins by the core increases the moment of inertia of the panel with little increase in weight, producing an efficient structure for resisting bending and buckling loads [1]. Additionally, good energy absorption can be achieved by employment of sandwich components: energy is dissipated by bending, stretching and fracture of the skins and by localized crushing of the core [2]. The cores, commonly, are made of balsa – wood, foamed polymers, glue – bonded aluminum or Nomex (paper) honeycombs. Using mental foams as cores can overcome the deficiencies which existed in the common cores – cannot be used much above room temperature, and their properties are moisture – dependent [3]. Therefore, sandwich structures with metal faces sheets and metal foam cores show advantages over other sandwich constructures under intensive impulse loading such as blast and impact [4]. The studies of the structural response of these sandwich structures with metal face sheets and metal foam core have

---

① 原文发表于《Composite Structure》2014 (sep. – oct.)

② 李伟：工学博士，2012 年师从冯顺山教授，研究内置冲击反应材料弹丸对装药结构的侵爆机理，现工作单位：中国工程物理研究院。

been conducted experimentally [5, 6], theoretically [7 – 11] and numerically [12, 13].

The curved panels are much stronger and stiffer than other structural forms and they generally have better performance under various loadings because they can support the external loads effectively by virtue of their spatical curvature. Combination of the advantages for both the shell and sandwich structures is of great importance for its applications [14, 15]. Some analytical models were developed to predict dynamic response of cylindrical sandwich shells including the buckling problem [16, 17], nonlinear sandwich shell theory [18], effect of blast loading [19, 20], and energy absorption [21]. Shen et al. [15] investigated the curved sandwich panels under blast loading experimentally, which perform better than equivalent solid counterpart and a flat sandwich plate. And a new deformation mode (global wrinkling) has been observed in the experiment. Xie et al. [22] conducted experiments on deformation and failure of clamped shallow sandwich arches with foam core subjected to projectile impact. And it is found that the traveling plastic hinge seems a dominant factor for significant difference of deformation modes in dynamic and quasi – static loading. Jing et al. [23 – 27] examined the effects of geometrical configurations and impulse loading on the deformation/failure of cylindrical sandwich shells with aluminum foam cores subjected to air blasting and projectile impact loading, experimentally and numerically. And it is found that the deformation modes are sensitive to the blast loading intensity and geometric configuration. In addition, the shear failure for the foam core and interfacial failure between foam core and face sheets were observed in the experiments under projectile impact loading, which was also observed in the numerical investigation under projectile impact loading.

Unlike the stress in cylinder shells, the stress of in spherical shells is distributed homogeneously because of central symmetry of their geometries and loadings. Under the same wall thickness, spherical shells show excellent bearing capacity [28]. At present, spherical shells are mainly applied in the storage of all kinds of gas, as the one end of capsule shell structures such as submarines, fuel tanks and coal mine refuge chambers, as the nuclear reactor containments designed to resist any external or internal blast load and as the roof coverings and domes to resist external pressure. And the dynamic response of spherical shells has been investigated extensively [29 – 31].

According to the aforementioned topics, spherical shell combined with sandwich structures may be have a better performance of resistance to the blast and impact loading than cylindrical sandwich shells. Li et al. [28] conducted the numerical investigate on the dynamic response of metallic spherical sandwich shells with graded aluminum foam cores under internal blast loading and obtained a better resistance to internal blast loading.

However, the investigations on the dynamic responses of spherical sandwich shells subjected to external blast loading have not yet reported. Therefore, the present research is concerned with the blast resistance of spherical sandwich shells with an aluminum foam core and aluminum face sheets under external blast loading, representing an extension of the previous works on cylindrical such panels. The commercial software Ls – Dyna 971 was applied to investigate the dynamic responses of metallic spherical sandwich shells under external air blast loading. To calibrate the accuracy of the numerical model, the numerical results of a series of cylindrical sandwich shells were compared with the experimental results recorded by Shen et al. [15]. The calibrated models were then employed to perform the simulations of the spherical sandwich shells. And the numerical results demonstrate that the spherical sandwich shells outperforms the cylindrical sandwich shells in resisting blast loading. The structural response processes and Deformation modes of the spherical sandwich shells have been discussed in this study. In addition, parametric studies were carried out to investigate the effects of shell geometrical configurations and blast loading on the capacities of resisting blast and absorbing energy. Details are presented in the following sections.

2 Numerical model calibrations

The numerical model was made by using commercial software Ls – Dyna 971 [32]. A nonlinear, explicit finite element numerical method was used in this software. Reliable numerical predictions of structural response to blast load have been proven [23, 24, 26]. So the numerical simulation of the responses of the spherical sandwich shells with metallic foam cores under blast loading was carried out by this software.

To validate the accuracy and reliability of the numerical model in this software, a cylindrical sandwich shells with aluminum foam cores tested by Jianhu Shen [15] was used to calibrate the model. A total of 33 explosion tests were conducted in the experiments. They were divided into three groups for parametric studies with respect to different charges, face sheets and core thicknesses. Two radius of shells (300 mm and 600 mm), three face - sheet thicknesses (0.5 mm, 1.0 mm and 1.6 mm) and three thickness of aluminum foam core (10 mm, 20 mm and 30 mm) were utilized. There were 16 bolts between matching cylindrical solid steel frames for clamping the each of the cylindrical sandwich shells and the frames were bolted onto the front face of the pendulum as shown in Figure 1 [15]. The impulse on shell and clamping, deflection history at the center point on the back face sheet and strain history at four points on the back face sheet were measured by the pendulum, laser displacement transducer and strain gauges, respectively [15]. Selected test datas were used to calibrate the numerical model made in Ls - Dyna in this study.

Figure. 1 A photograph of the experiment settings

2.1 Element, Mesh, Boundary Conditions and Contact Modeling

Since the sandwich shells and clamped frames is symmetric about $x-z$ and $x-y$ planes, only a quarter of the shells and frames was modeled, as shown in Figure 2. The numerical models have been built by using commercial software ANSYS and Ls - Prepost. After performing a mesh convergence analysis, a feature mesh size of 2mm was determined to be optimal for both the shell and solid elements, which balanced the numerical stability requirement, the accuracy of the FEA results and the computational efficiency. The sandwich shell sheets were modeled by Belyschko - Tasy four - node shell elements [33], while foam core and clamping frames were modeled by eight - node solid elements.

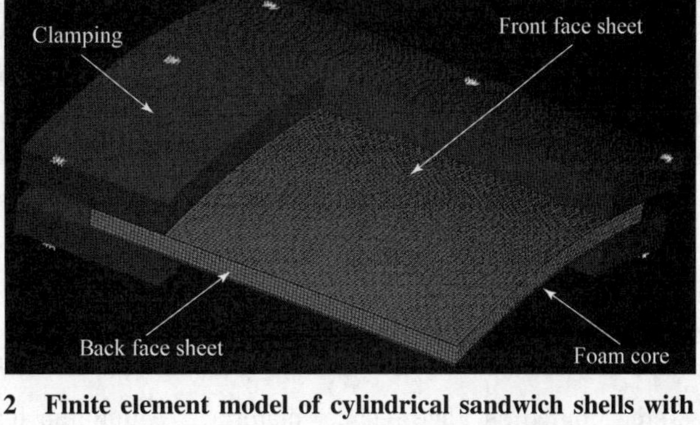

Figure 2 Finite element model of cylindrical sandwich shells with B. C. 1

In this numerical simulation, two approaches of boundary conditions were considered about for the bolts used in the experiments to clamp the sandwich shells. The first approach (B. C. 1) is to more closely represent the testing conditions, so the nodal constraints were used at positions of bolts on the top face of front clamping frame and bottom face of back clamping frame as showed in Figure 2. The second approach (B. C. 2) is the equivalent of the first approach. The surface constraint was used on the peripheral of top face of front clamping frame, as showed in Figure 3, instead of nodal constraints at positions of bolts.

The * CONTACT AUTOMATIC SURFACE TO SURFACE TIEBREAK model was adopted to account for the bond connection between the face - sheets and the foam core. The tiebreak failure criterion has normal and shear components [32]:

$$\left(\frac{|\sigma_n|}{NFLS}\right)^2 + \left(\frac{|\sigma_s|}{SFLS}\right)^2 \geqslant 1 \tag{1}$$

where σ_n is the normal stress; σ_s is the shear stress; NFLS and SFLS are the normal and shear failure stress, respectively, and listed in the Table 1.

To prevent the penetration between face sheets and core, the contact between them was defined by using * CONTACT AUTOMATIC SINGLE SURFACE with a contact static and a dynamic friction value of 0.28 and 0.20, respectively and it is a single surface contact, that is, the contact is defined wholly by the slave side and considers shell thickness. In addition, the contacts in all parts and the part self - contact are also considered in this contact type, which is suitable to simulate the composite structures with shells. The * CONTACT INTERIOR was adopted to the foam core to prevent self - penetration and is used for defining the interior for foam hexahedral and tetrahedral elements. Frequently, when foam core materials are compressed under high pressure, the solid elements used to discretize these materials may invert leading to negative volumes. In order to keep these elements from inverting, it is possible to consider interior contacts within the foam between layers of interior surfaces made up of the faces of the solids elements. The interior penalty contact factor K is determined by the formula [32]:

$$K = \frac{SLSFAC \cdot PSF \cdot Volume^{\frac{2}{3}} \cdot E}{Min.\ Thickness} \tag{2}$$

where SLSFAC is the factor for sliding interface penalty and set as 1.0. *Volume* is the volume of the brick element, E is a constitute modulus and *Min. Thickness* is approximately the thickness of the solid element through its thinnest dimension.

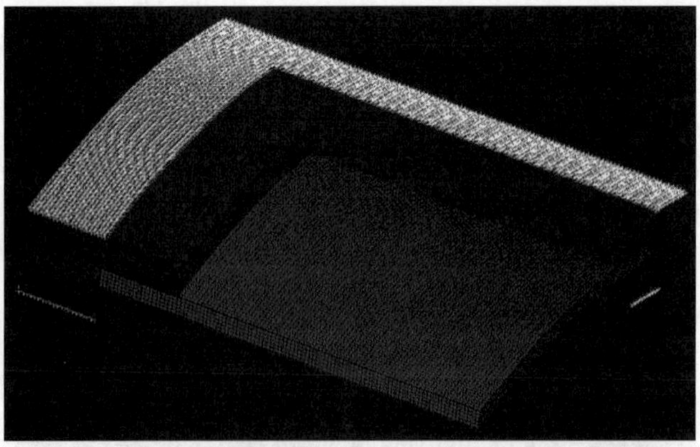

Figure 3 the finite element model of the cylindrical sandwich shells with B. C. 2

2.2 Material Model

According to the experiments of Jianhu shen [15], Al - 5005 H34, ALPORAS aluminum foam and Epoxy adhesive (FORTIS AD825) were adopted as Face - sheet material, Aluminum foam core and bond connection material in the simulation, respectively. And Steel 4340 was adopted as the steel clamping frames material.

The elastic – plastic material model 3 of the software Ls – Dyna named ＊MAT PLASTIC KINEMATIC was used to represent the mechanical behavior of Al – 5005 H34. It is good at modeling the bi – linear elastic – plastic constitutive relationship with two types of hardening formulation (isotropic or kinematic). Since Al – 5005 H34 has low strain rate sensitivity, the effect of strain rate was neglected in the simulation. The input parameters defined in this material model are based on the quasi – static material testing [15]. The material model 63 named ＊MAT CRUSHALBE FOAM is suitable for modeling the crushable foam with optional damping and tension cutoff. Tension was treated as elastic – perfectly – plastic at the tension cut – off value. The standard stress – strain curve for Aluminum foam was cited from the experiments of Jianhu shen [15]. The material model 15 named ＊MAT JOHNSON COOK is sometimes used for problems where the strain rates vary over a large range and adopted to represent the properties of Steel 4340 in this paper. The keyword ＊EOS GRUNEISEN was adopted as the equation – of – state of Steel 4340 and the pressure for compressed materials as [32]:

$$p = \frac{\rho_0 C^2 \mu \left[1 + \left(1 - \frac{\gamma_0}{2}\right)\mu - \frac{a}{2}\mu^2\right]}{\left[1 - (S_1 - 1)\mu - S_2 \frac{\mu^2}{\mu + 1} - S_3 \frac{\mu^3}{(\mu + 1)^2}\right]} + (\gamma_0 + a\mu)E \tag{3}$$

where C is the intercept of the $v_s - v_p$ curve; S_1, S_2 and S_3 are the coefficients of the slope of the $v_s - v_p$ curve; γ_0 is the Gruneisen gamma; a is the first order volume correction to γ_0; and $\mu = \rho/\rho_0 - 1$.

The main material types and mechanical properties of sandwich shells and clamping frame are shown in Table. 1

Table 1 Input data in the numerical simulation of sandwich shells

Material	Part name	LS – DYNAmaterial type, material property (unit = cm, g, us, Mbar)					
		＊MAT_CRUSHABLE_FOAM(＊MAT_63)					
Aluminium foam (ALPORAS[R]) Al + 1.5% Ca + 1.5% Ti	Core	Density	Young's modulus	Poisson ratio	Load curve ID	Tensile curve cutoff	Rate sensitivity via damping coefficient
		0.23	1.1E – 2	0.33	From[15]	1.6E – 5	0.1
		＊MAT_PLASTIC_KINEMATIC(＊MAT_3)					
Al – 5005 H34	Face sheet	Density	Young's modulus	Poisson ratio	Yield strength	Tangent modulus	Hardening parameter
		2.76	0.69	0.33	1.05E – 3	0.27	0.5
		＊CONTACT_AUTOMATIC_SURFACE_TO_SURFACE_TIEBREAK					
Epoxy adhesive: FORTIS AD825	Bonding	OPTION	Normal failure stress	Shear failure stress			
		6	12E – 5	2.47E – 5			

OPTION = 6 is a special number for use with solids and thick shells only.

2.3 Blast Load Modeling

In this study, blast loading was generated by the CONWEP (Conventional Weapons Effects Program) empirical model implemented as the ＊LOAD_BLAST[32] in Ls – Dyna. The CONWEP algorithms developed by

Kingery and Bulmash [34] account for angle of incidence by combining the reflected pressure (normal - incidence) value and the incident pressure (side - on - incidence) value. The pressure can be calculated by the following equation:

$$P_{load} = P_{reflected} \cdot \cos^2\theta + P_{incident} \cdot (1 + \cos^2\theta - 2\cos\theta) \tag{4}$$

where θ is the angle of incidence; $P_{incident}$ is the incident pressure; $P_{reflected}$ is the reflected pressure.

The keyword * LOAD_BLAST can provide the blast pressure on the certain surfaces predefined by the keyword * LOAD_SEGMENT_SET. The parameters required for * LOAD_BLAST include equivalent mass of TNT explosive, the spatial coordinates of the detonation point, and the type of blast. The blast loading was applied to the 5 cases of cylindrical sandwich shells listed in Table2 by a 40 g cylindrical TNT charge in the experiments [15], while the shape of charge used in * LOAD_BLAST is spherical. In order to get the equivalent mass of spherical TNT charge for CONWEP, the blast loading pressure applied to cylindrical sandwich shells by the 40 g cylindrical TNT charge detonated at the standoff distance 250 mm was simulated by using commercial software Autodyn [35]. The pressure time history result is shown in Figure 4 and the peak pressure is 1. 54 Mpa. In order to make the peak pressure of the spherical TNT charge exerted on the front face sheet same as that of the 40 g cylindrical TNT charge, the equivalent mass of spherical TNT charge is calculated by the empirical Equation (5) [36] and is equal to 50. 16 g with the standoff distance 250 mm and the peak pressure 1. 54 Mpa.

$$\Delta p = \frac{6.1938}{Z} - \frac{0.3262}{Z^2} + \frac{2.1324}{Z^3} \quad 0.3 \leqslant Z \leqslant 1 \tag{5}$$

Where Δp is the peak overpressure at the shock wave front, (kp/cm^2); Z is the scaled distance, ($m/kg^{1/3}$).

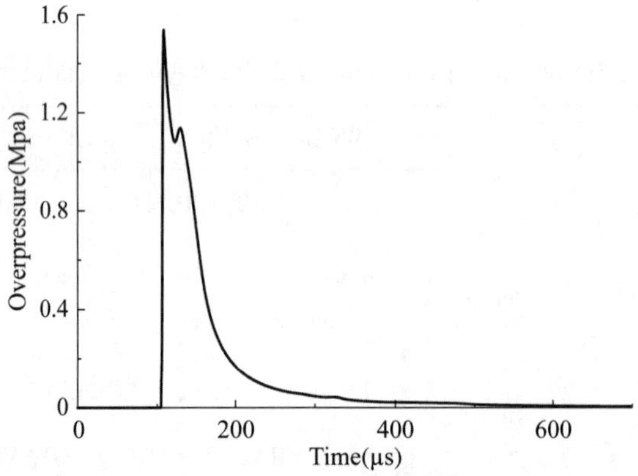

Figure 4　the pressure time history of the 40 g cylindrical TNT charge detonated at the standoff distance 250 mm in experiment

2.4　Result and Discussions

Figure 5 compares the experimental recorded with numerical calculated displacement time history of center point of back face sheet for a typical case NO. C2 as listed in Table 2. The peak displacements of experimental [15] and numerical results are in close agreement. The main difference occurs in the free - vibration phase after the action of blast load. It can be explained as: the big difference between experimental and numerical results after peak displacement may be caused by boundary condition. In the experiment [15], the sandwich shells is fixed on a four - cable ballistic pendulum which can swing in order to measure the impulse of blast. But in the numerical simulation, the sandwich shells is fixed on the clamping frames which cannot swing or move. So in the experiment, a part of energy exerted on the sandwich shells transforms into the kinetic energy of the ballistic pendulum, the kinetic energy of sandwich shells decrease sharply, which results in the phase that the experimental displacement after the peak is more or less constant with only a rather slight vibration superimposed. But in the numerical simulation, this part of energy transforms into the kinetic energy of the sandwich shells because the clamping frame cannot move, the sandwich shells will rebound and vibrate fiercely. However, the initial deformation phase up to the peak displacement may not be influenced by the boundary condition mentioned above. The reason can be explain as follow: the blast acting time on the sandwich shells is very short. For experiment [15], before peak displacement, the stress wave propagating in the sandwich shells may not reach the ballistic pendulum or the ballistic pendulum have a little kinetic energy; for numerical simulation, the reflected stress wave from the fully - constrained surface of clamping frame may not reach the sandwich shell or just a little. So both of initial deformation phases in experiment and numerical simulation only experience the blast wave and have the same phases. It can be concluded that the numerical simulation can predict accurately before peak displacement or at it. Since the primary concern in design is the peak response, the numerical model can give a reliable results before and at the peak displacement of cylindrical sandwich shells under blast loads. In addition, there is no difference between the displacement time history of NO. C2 with B. C. 1 and that of NO. C2 with B. C. 2, while the computational efficiency of NO. C2 with B. C. 2 is higher than that of NO. C2 with B. C. 1. Therefore, the B. C. 2 was employed in the subsequent models.

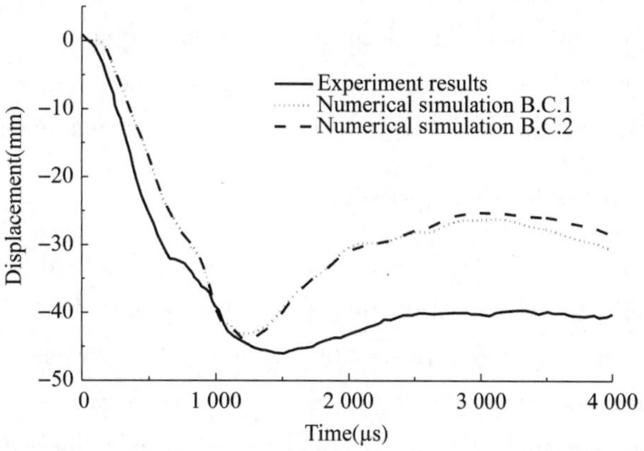

Figure 5 Comparison of displacement time histories between experiment and simulation

In order to validate the agreement between the numerical results and experimental recorded, extensively, some cases were selected from the experiments (Table 2) [15] to compare the experimental recorded with numerical calculated peak displacement of the center point on the back face sheet. And they have the same charge mass (a 50.16 g equivalent mass of spherical TNT) and standoff distance 250 mm. As listed in Table 2, the numerical results of the peak displacements agree well with the reported experiment data, indicating the numerical model gives reliable predictions of cylindrical sandwich shells under blast loads. Figure 6 shows the numerical displacement history of the Event C1 ~ C5.

Table 2 Experimental and numerical results of peak displacement of the centre point

Cases(NO.)	R (mm)	Tc (mm)	Ti (mm)	To (mm)	Experiment data δ_{max} (mm)	Numerical simulation δ_{max} (mm)	Numerical simulation Error(%)
R3Tc2Ti10To10(C1)	300	20	1.0	1.0	23.8	22.34	-6.54
R6Tc1Ti05To16(C2)	600	10	0.5	1.6	42.9	43.08	0.42
R6Tc2Ti10To10(C3)	600	20	1.0	1.0	35.4	36.05	1.84
R6Tc2Ti05To16(C4)	600	20	0.5	1.6	35.6	36.82	3.43
R6Tc2Ti16To05(C5)	600	20	1.6	0.5	36.8	37.90	2.99

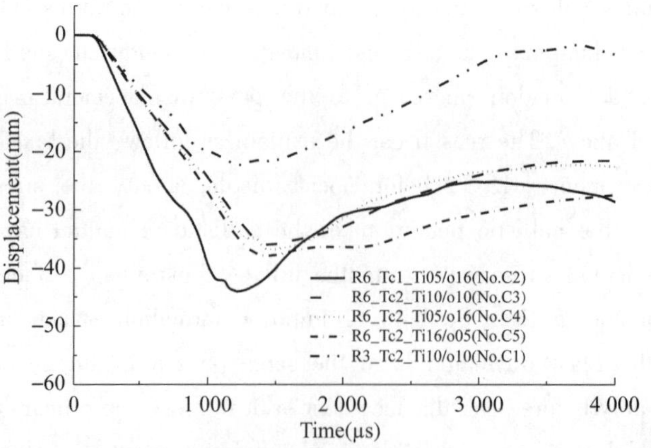

Figure 6 The displacement time history of the Event C1 ~ C5

3 Numerical Simulation

The calibrated numerical model above was used to perform a series of simulation of dynamic responses of spherical sandwich shells under external blast loads. The response quantities such as peak displacement, permanent displacement and internal energy (or plastic energy) absorption were compared to examine the effectiveness of different spherical sandwich shells designs on blast – resistance capacities.

3.1 Spherical Sandwich Shells Configurations

In order to compare the blast – resistance capacities between cylindrical and spherical sandwich shells, keeping the same radius of shell, the weight of all part (front face sheet, back face sheet and foam core) and the loading area of front face sheet as the corresponding cylindrical sandwich shells, a total of 5 equivalent spherical sandwich shells were considered in the study, as listed in Table 3. The equivalent thickness of front and back face sheet, the equivalent arc angle (Figure 7) and the equivalent thickness of foam core were listed in Table 3. The same method of name definition as cylindrical sandwich shells was adopted by spherical sandwich shells, while the number was characterized by S1 instead of C1, for instance. The schematic diagram and finite element model were shown in Figure 7.

Table 3 The equivalent spherical sandwich shells and the comparing of peak displacement between spherical and cylindrical sandwich shells

Case(No.)	Equivalent spherical sandwich shells						cylindrical sandwich		Reduction (%)
	R (mm)	Tc (mm)	Ti (mm)	To (mm)	α (°)	δ_{max} (mm)	No.	δ_{max} (mm)	
R3Tc2Ti10To10(S1)	300	20	0.982	1.018	40.274	**4.67**	C1	**23.8**	−80.38
R6Tc1Ti05To16(S2)	600	10	0.496	1.616	19.508	**28.93**	C2	**42.9**	−32.56
R6Tc2Ti10To10(S3)	600	20	0.982	1.018	19.508	**16.75**	C3	**35.4**	−52.68
R6Tc2Ti05To16(S4)	600	20	0.492	1.630	19.508	**15.78**	C4	**35.6**	−55.67
R6Tc2Ti16To05(S5)	600	20	1.572	0.508	19.508	**11.55**	C5	**33.8**	−65.83

3.2 Modeling

The 50.16 g equivalent spherical TNT was detonated at a distance of 250 mm above the internal layer as

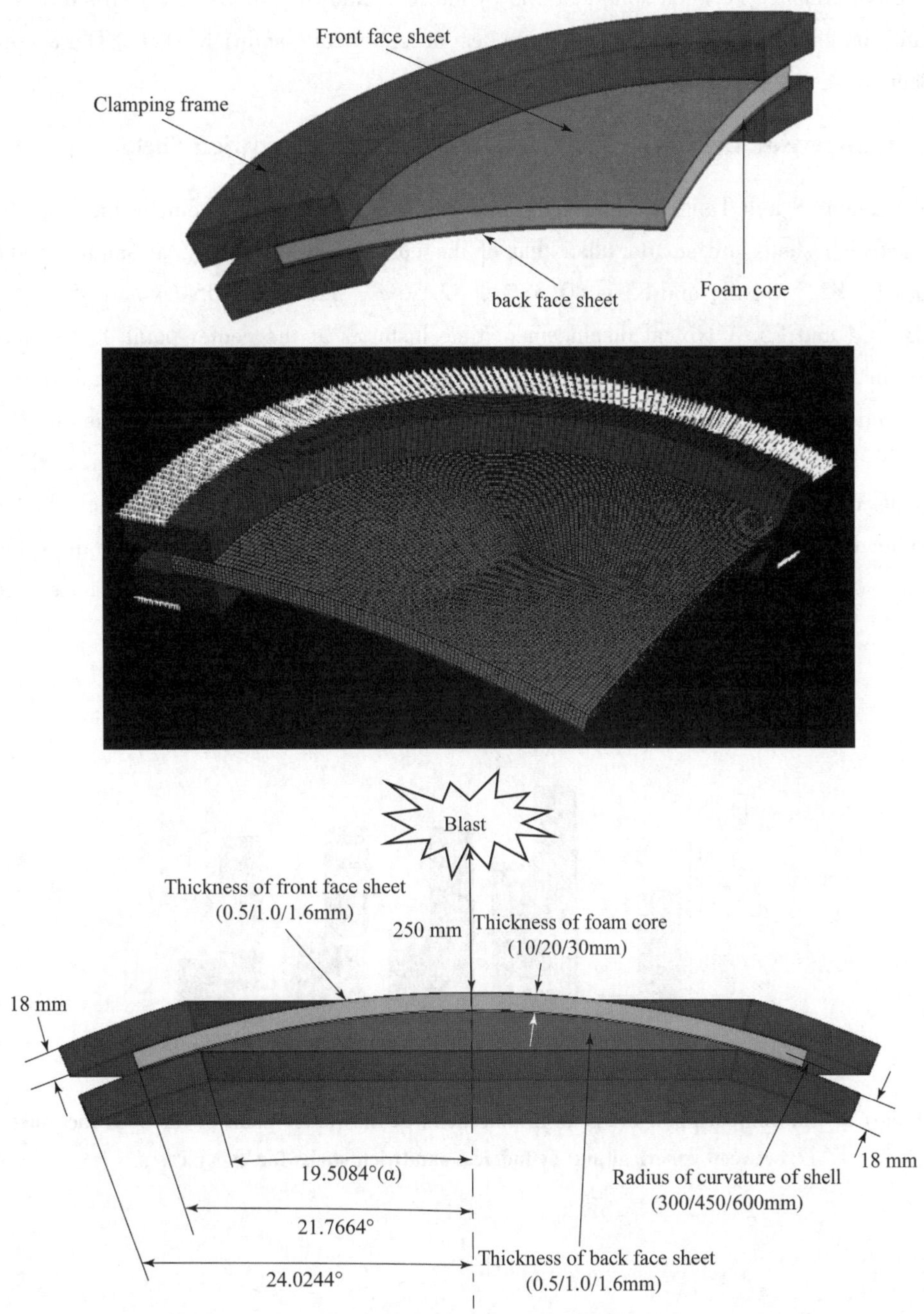

Figure 7　The schematic diagram and finite element model

shown in Figure 7. The scaled distance is 0. 68 m/kg$^{1/3}$. Blast load was modeled by using the keyword ＊ LOAD_BLAST in Ls－Dyna. Front and back face sheet were modeled with shell element of 2 mm size. Foam core and clamping frame were modeled with solid element of 2 mm size. The material properties used in the calculations are the same as those used in the calibration study. The connections between front/back face sheet and foam core were modeled by the same model as the calibration simulation. The contact models used in this simulation are same as the calibration simulation and the boundary condition B. C. 2 were used in this simulation.

4　Simulation Results and Discussion

The simulation results presented and discussed in this section include three aspects：（1） the comparison of blast – resistance performance between spherical and cylindrical sandwich shells; (2) structural response process of spherical sandwich shells; (3) deformation modes of spherical sandwich shells. These were individually described in Sections 4. 1 – 4. 3.

4. 1 The Comparison between Spherical and Cylindrical Sandwich Shells

As shown in Figure 8 and Table 3, the peak displacements at the center point of the back face sheet of all five spherical sandwich shells are smaller than that of the corresponding cylindrical sandwich shells, the peak displacement of S1, S2, S3, S4 and S5 is 80. 38%, 32. 56%, 52. 68%, 55. 67 and 65. 83% less than that of C1, C2, C3, C4 and C5. A typical displacement time histories at the center point of the back face sheet of C3/S3 are given in Figure 9 and the permanent displacements at the center point of the back face sheet of all five spherical and cylindrical sandwich shells are listed in Table 4. The permanent displacements of S1, S2, S3, S4 and S5 are 123. 99%, 70. 95%, 90. 64%, 93. 24 and 90. 97% less than that of C1, C2, C3, C4 and C5. These are caused by the center symmetric geometrical configuration of spherical sandwich shells under a center symmetric blast loading, resulting in a center symmetric stress field which is generated in spherical sandwich shells rather than cylindrical sandwich shells and provides a better performance of blast – resistance.

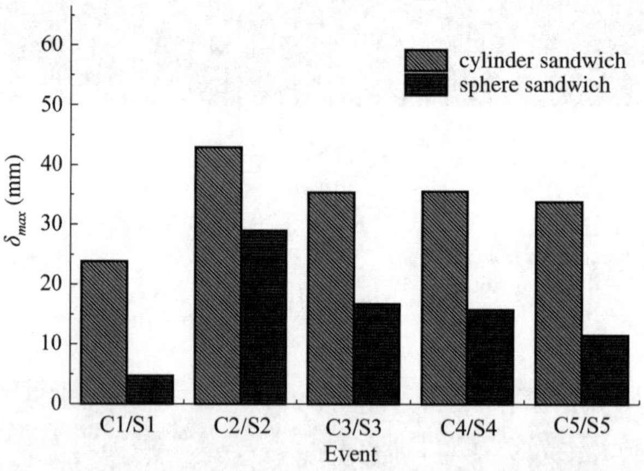

Figure 8 the comparing of the peak displacement at the center point of the back face sheet between spherical and cylindrical sandwich shells for every cases

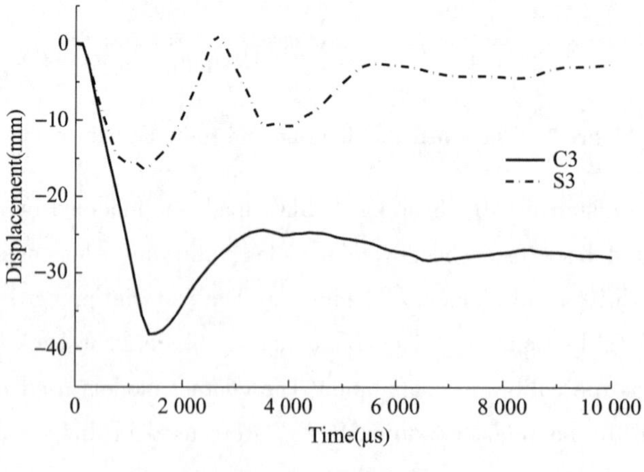

Figure 9 comparing of the displacement time history at the center point of the back face sheet between C3 and S3.

Figure 10 shows the internal energy absorption time histories of total sandwich shells (Ei − t), that of front face sheet (Ei − 1), that of foam core (Ei − 2) and that of back face sheet (Ei − 3) of C3/S3. It is clearly shown that every internal energy curve tends to a stable value after a peak value. And the stable value is the plastic energy dissipation by sandwich shells and the difference between the peak and stable value is the elastic energy which has been released completely when the dynamic response process of sandwich shells ends.

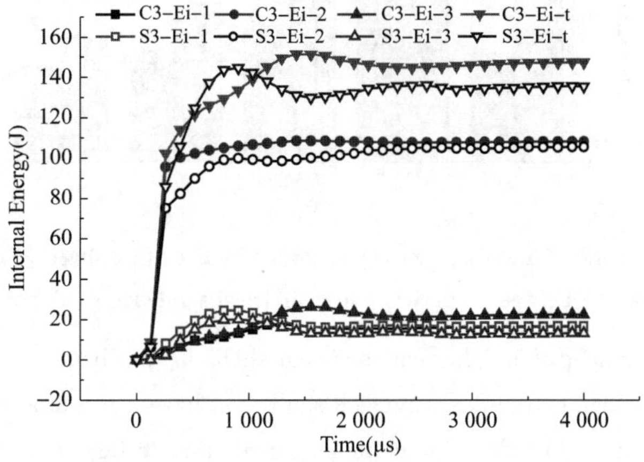

Figure 10 The comparing of internal energy absorption time history between C3 and S3

As shown in Figure 11 and Table 4, the ratio of the plastic energy absorbed by back face sheet (Ei − 3) to the total plastic energy absorbed by sandwich shells (Ei) of the most spherical sandwich shells except S2 are smaller than that of the corresponding cylindrical sandwich shells, this is, the ratio of S1, S3, S4 and S5 is 10.30%, 5.56%, 3.21% and 2.03% less than that of C1, C3, C4 and C5, indicating a better performance of the foam core and front face sheet of spherical sandwich shells for absorbing energy and preventing the blast energy passing into the back face sheet. However, Figure 11 and Table 4 shows the ratio of S2 is 8.46% more than that of C2, which implies that less blast energy is absorbed by the back face sheet of cylindrical sandwich shells than that of spherical sandwich shells. This may be because the thickness of foam core of C2/S2 is only 10 mm (that of others is 20 mm), leading to a complete crushing of foam core at the central area and a different deformation mode (Figure 13 (a)) occurring subsequently.

Table 4 The comparing of permanent displacement and plastic energy between spherical and cylindrical sandwich shells

No.	Equivalent spherical sandwich shells					No.	cylindrical sandwich				
	Plastic energy (J)				$\delta_{permanent}$ (mm)		Plastic energy (J)				$\delta_{permanent}$ (mm)
	Ei − t	Ei − 1	Ei − 2	Ei − 3			Ei − t	Ei − 1	Ei − 2	Ei − 3	
S1	252.61	9.36	224.2	19.05	2.07(+)	C1	113.91	15.08	78.51	20.32	8.63
S2	278.05	14.36	192.4	71.29	8.93	C2	316.72	13.30	249.0	54.42	30.74
S3	135.42	16.73	105.5	13.19	2.61	C3	147.25	16.53	108.2	22.52	27.89
S4	228.92	11.51	199.6	17.81	1.53	C4	269.32	29.61	210.1	29.61	22.63
S5	94.24	20.43	67.59	6.22	2.48	C5	88.37	19.41	61.33	7.63	27.46

(+) behind the number represents the permanent deformation above the original position of configuration

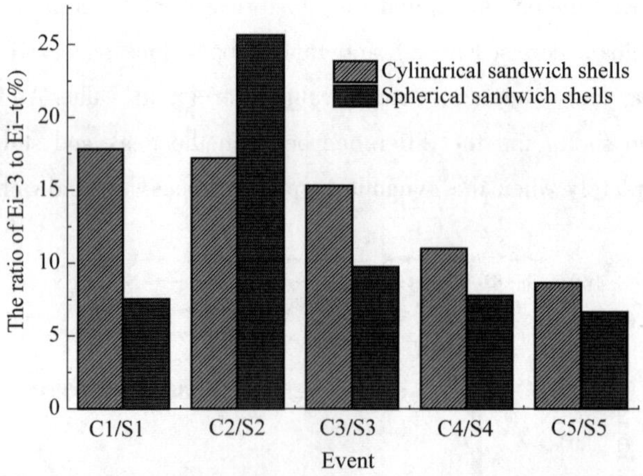

Figure 11 the comparing of the ratio of the plastic energy absorbed by back face sheet (Ei－3) to the total internal energy absorbed by sandwich shells (Ei－t) between spherical and cylindrical sandwich shells for every cases

Therefore, it can be concluded that spherical sandwich shells have a better performance of blast－resistance for protecting internal contents than cylindrical sandwich shells and have a potential to be applied in the one end of capsule shell structures such as submarines, fuel tanks, coal mine refuge chambers and the nuclear reactor containments to resist external blast load.

4.2 Structural Response Process of Spherical Sandwich Shells

In order to investigate the structural response of spherical sandwich shells, the process is divided into three specific stages, that is, First stage－compression at central region of foam core; Second stage－sharply localized descent at central region of shells with global deformation; Third stage－ obviously localized rebound at central region of shells with slight global rebound.

4.2.1 First Stage (100－280 μs)

Shock wave of blast begin to interact with the front face sheet of sandwich shell at this stage. And the central area of foam core is compressed sharply due to the large deformation of front face sheet as shown in Figure 12. According to the material properties of aluminum foam, the energy of shock wave is sharply absorbed by crushing compression of foam core until the pressure exerted by front face sheet is less than the plateaus pressure of aluminum foam. The front face sheet and foam core in the loading area are compressed progressively until to the external clamped boundaries (Figure 12) from t = 100 μs to t = 180 μs. The back face sheet has almost no deformation in this stage.

4.2.2 Second Stage (280－1 250 μs)

The central point of back face sheet begins to move in this stage. From t = 280 μs to t = 1 000 μs, all of the central region of the front face sheet, back face sheet and foam core descend sharply. And the bonding connection between them is still working. While the front face sheet begins to be separated from the foam core at t = 1 000 μs and descends sequentially. In all second stage, the bonding connection between back face sheet and foam core is working well. The global deformation of sandwich shell begins at 280 μs. It can be concluded that the tensile stress occurs between front face sheet and foam core in this stage due to the deformation of foam core larger than that of front face sheet. The second stage ends at t = 1 250 μs and the back face sheet reaches its peak displacement at this moment.

4.2.3 Thirdly stage (1 250－2 500 μs)

In this stage, duo to the clamped frame boundary, the weakening of shock wave and the increasing

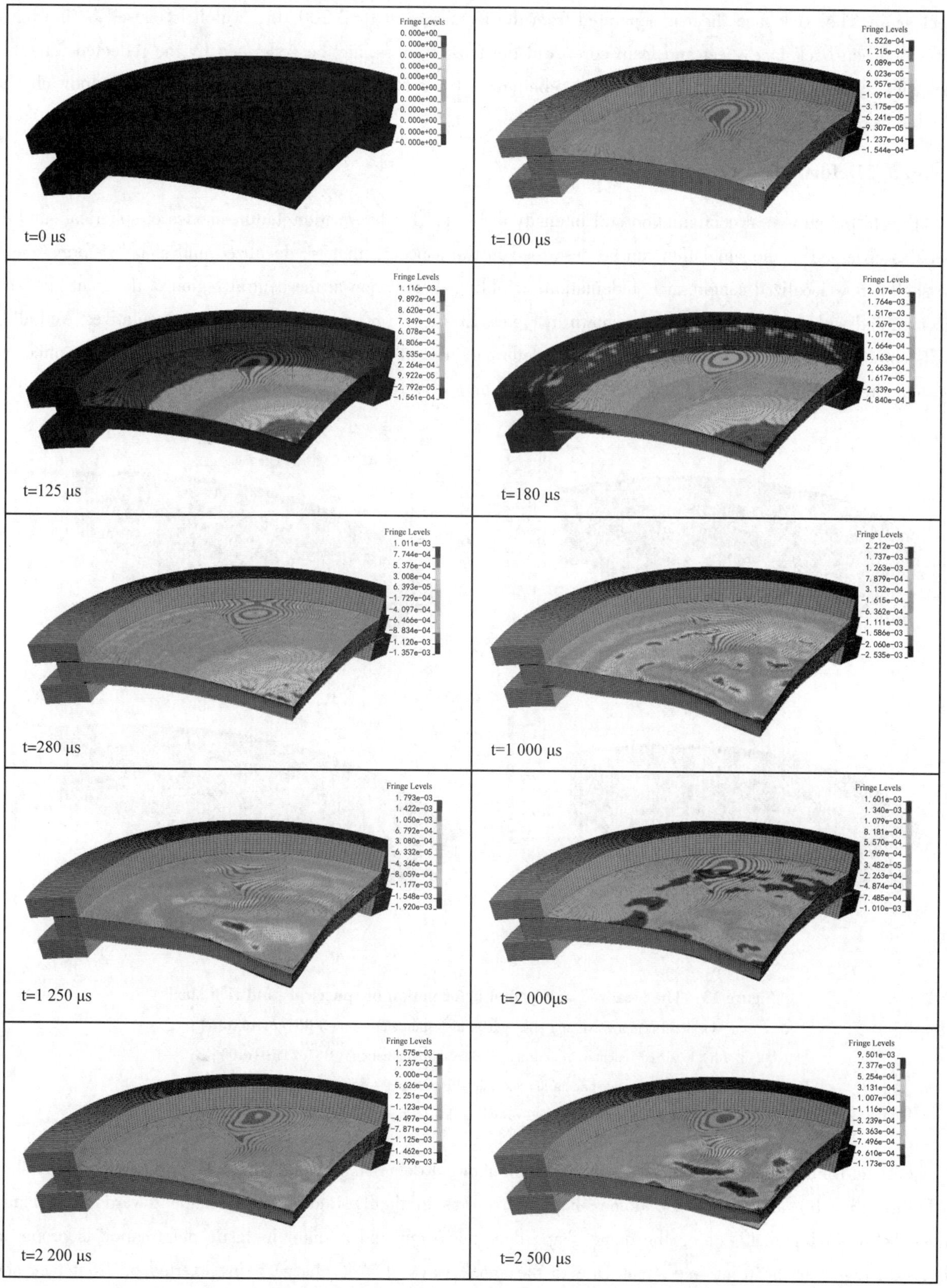

Figure 12 the structural response process of spherical sandwich shells

deformation energy of shells, the back face sheet and foam core begins to rebound at t = 1 250 μs. There is no significant rebound of the front face sheet. Therefore, at t = 2 000 μs, the foam core contacts with the front face sheet again. The back face sheet is separated from the foam core at t = 2 200 μs, which is caused by the tensile stress between back face sheet and foam core. And the tensile stress may be generated by the reflected wave from back face sheet. The sandwich shell ends its rebound at t = 2 500 μs. After t = 2 500 μs, the sandwich shell vibrates freely until the rest.

4.3 Deformation modes

Due to the structural configuration and intensity of blast, the deformation/failure modes of spherical sandwich shells as observed in the simulation can be described in two aspects, that is, localized and global deformation. All the shells show localized compressive/indentation/wrinkling deformation at the central region of the front face sheet accompanied with core compression, as shown in Figure 13 (a), (b) and (d) for the representatives. And all the shells show global inelastic/oscillating/wrinkling deformation at all region of the front face sheet accompanied with core compression, as shown in Figure 13 (a), (c), and (d) for the representatives.

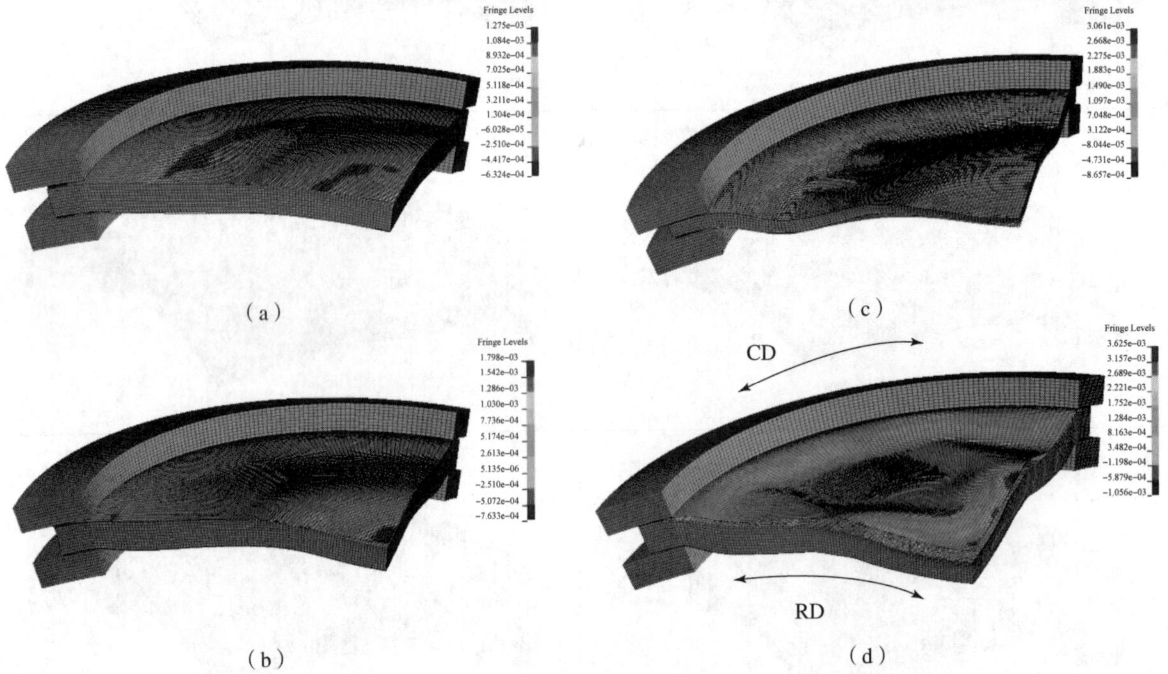

Figure 13 The localized and global deformation of spherical sandwich shells

(a) localized compressive and sight global inelastic deformation (R6Tc2Ti16To16);

(b) localized indentation and global inelastic deformation (R6Tc2Ti10To10);

(c) localized indentation and global oscillating (R6Tc1Ti05To16);

(d) localized and global wrinkling deformation (R6Tc2Ti05To05)

For the front face sheet, the global deformation shows a dome shape, starting out from the center. As shown in Figure 13 (b) and (c), the dome shape transforms gradually into a circle shape towards the clamped edges. While in Figure 13 (a), the dome shape does not occur and a slight inelastic deformation is propagated across all region of the front face sheet. Due to the shock wave of TNT charge being exerted on front face sheet from the center to the clamped edges, gradually, the foam core shows a gradually compressive deformation from the center to the clamped edges for all cases. Figure 13 (a), (b) and (d) show three types of localized deformation patterns. The thickness of the face sheet seems to be a dominant factor for these phenomenas. For the shells with the thickest face sheet, there is a slight compressive deformation as shown in Figure 13 (a). With decreasing the thickness of face sheet, the localized deformation changes from slight compressive to indentation (dome) and wrinkling deformation. A possible reason is the compressive instability of the thinner face sheet and foam core. Figure 13 (a), (c) and (d) show three types of global deformation patterns. The thickness of the foam core is the dominant factor for global oscillating deformation and the thickness of front face sheet is the dominant factor for global wrinkling deformation. According to the numerical investigation in section 5.4 (case R6Tc3Ti05To16), the intensity of blast is also a key factor for deformation modes. And for 50 g TNT, the global and localized deformation modes for 50 g, 75 g and 100 g TNT are similar to Figure 13 (a), (b) and (d), respectively. The plastic hinges can be observed in Figure 13 (b), (c) and (d), obviously. But the number and direction of plastic hinges of them are not same as each other. These differences can be explained by the phenomenon that the plastic hinges as seen in Figure 13 (b) and (c) are generated by the difference of localized displacement of the total sandwich shells from the center to the clamped edges and a good bonding is between face sheets and foam core, while the plastic hinges in Figure 13 (d) is created by squeezing the front face sheet in the direction of the circumferential direction (CD) and the radial direction (RD) and a wide range of separation is between front face sheet and foam core. In conclusion, the geometric configuration and blast intensity are the dominant factors for deformation modes of spherical sandwich shells.

5 Effect of Parameter

According to the results and discussions presented above, parametric studies are carried out to examine the effects of shell configurations (i. e. radius of curvature, thickness of foam core and face sheet combinations) on the structural response subjected to blast loading with TNT charges varying from 50 g to 100 g under constant standoff distance of 250 mm. In order to keep consistency with the comparisons mentioned in section 3, the following spherical sandwich shells investigated in this paper are also equivalent to the corresponding cylindrical sandwich shells. The equivalent thickness of face sheets and arc angle were listed in following tables.

5.1 Effect of Radius of Curvature

Three sandwich shells with spherical surfaces having $R = 300$ mm, 450 mm, and 600 mm, respectively, are compared in this section and they have the same foam core thickness 20 mm. The corresponding thickness of front/back face sheet and arc angle (Table. 5) are adjust to make the weights of every part in the shells and the loading area of front face sheet approximately the same. As shown in Figure 14, the peak displacements at the center point of the front face sheet, back face sheet and the back surface of foam core are compared, with increasing the radius of curvature, the peaking displacements of face sheets and foam core increase gradually. Using *R*3.0 ($R = 300$ mm) as benchmark, the peak displacements of back face sheet in the other two shells increase by 50.75% and 258.67%, respectively, and the permanent displacements increase by 14.15% and 219.32%. Duo to the foam core compression and no tension between foam core and back face sheet before the thirdly stage, the peak displacements of foam core is same as that of back face sheet and less than that of front face sheet. The displacement time histories at the center point of the back face sheet are given in Figure 15, with increasing the radius of curvature, the wavelength and amplitude of vibration increase gradually. The permanent displacement of *R*3.0 is a positive value, while that of *R*4.5 and *R*6.0 are negative values, therefore, compared the configuration of shells before blast loading, the permanent deflection of *R*3.0 is a protuberance and that of *R*4.5 and *R*6.0 are potholes.

Table. 5 the peak displacement, permanent displacement and Internal Energy with different radius of curvature

Case (No.)	R (mm)	Tc (mm)	Ti (mm)	To (mm)	α (°)	Displacement (mm) δ_{max}	Displacement (mm) $\delta_{permanent}$	Internal Energy (J) Ei-1	Internal Energy (J) Ei-2	Internal Energy (J) Ei-3
R3Tc2Ti10To10 (R3.0)	300	20	0.966	1.036	40.274	4.67	2.07(+)	9.36	224.2	19.05
R4.5Tc2Ti10To10 (R4.5)	450	20	0.974	1.027	27.726	7.04	0.86	15.13	124.2	14.68
R6Tc2Ti10To10 (R6.0)	600	20	0.982	1.018	19.508	16.75	2.47	16.73	105.5	13.19

(+) behind the number represents the permanent deformation above the original position of configuration

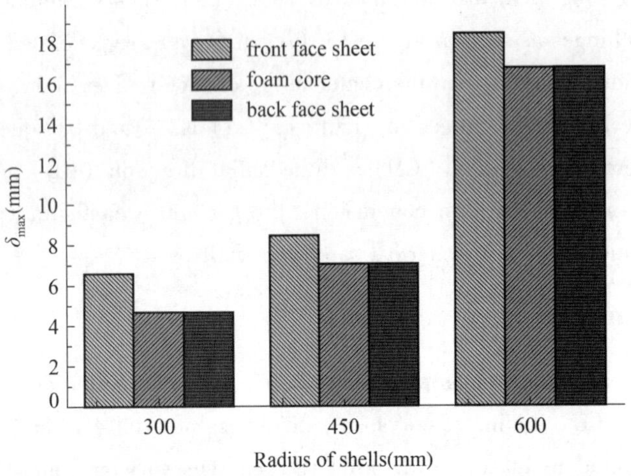

Figure 14 peak displacement of face sheets and foam core

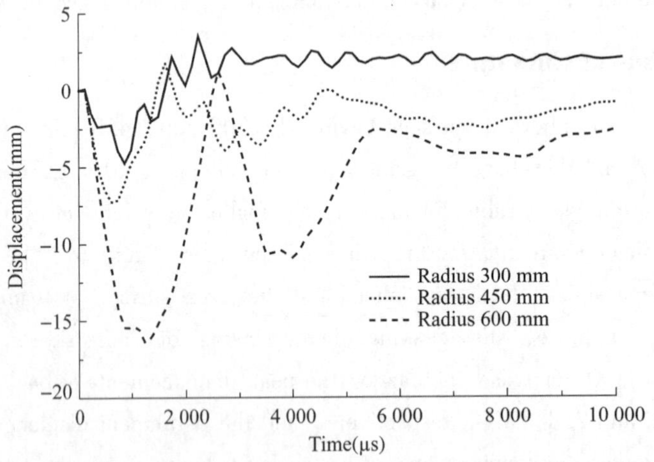

Figure 15 Time histories of center point displacement of back face sheet

Figure 16 indicates less plastic energy is absorbed by foam core (Ei-2) and back face sheet (Ei-3) while more plastic energy was absorbed by front face sheet (Ei-1) with the increase of radius of curvature. Therefore, the conclusion can be drawn that small radius will improve the blast-resistance performance of the spherical sandwich shells by reducing the peak and permanent displacement, changing the permanent deflection mode and increasing the energy absorption capacity.

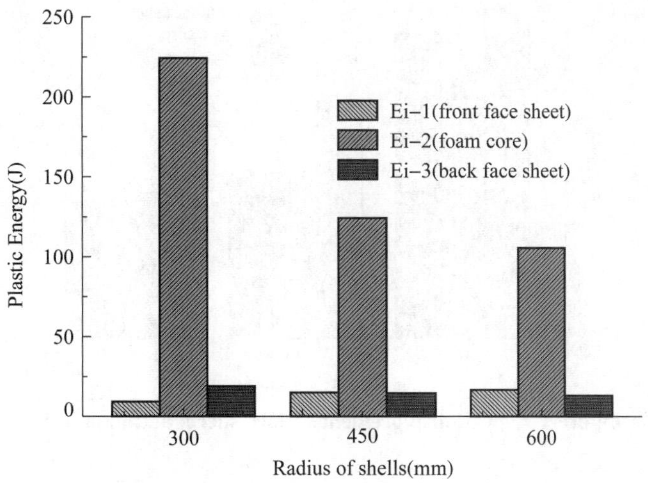

Figure 16 Plastic energy absorptions of face sheets and foam core

5.2 Effect of Thickness of Foam Core

Three sandwich shells with thickness of foam core Tc = 10mm, 20 mm, and 30 mm, are compared in this section respectively and they have the same radius of curvature 600 mm and arc angle 19.508°. The corresponding thickness of front/back face sheet and foam core (Table 6) are adjust to make the weights of every part in the shells and the loading area of front face sheet approximately the same. As shown in Figure 17, the peak displacements at the center point of the front face sheet, back face sheet and the back surface of foam core are compared, with increasing the thickness of foam core, the peaking displacements of face sheets and foam core decrease gradually. Using Tc10 (Tc = 10 mm) as benchmark, the peak displacements of back face sheet in the other two shells decrease by 45.45% and 54.93%, respectively, and the permanent displacements decrease by 48.26% and 56.66%, which can be concluded that no significant decrease of the peak and permanent displacement will occur if the thickness of foam core is more than 30 mm. The peak displacements of foam core of Tc = 20 and 30 mm are same as that of back face sheet while the peak displacement of foam core of Tc = 10 mm is less than that of back face sheet, which indicates that a separation between foam core and back face sheet occurs during the second phase when the thickness of foam core is 10 mm. It is clearly shown in Figure 13 (c) that the wave deflection causes the separation between foam core and back face sheet in the second stage of the structural response progress. The displacement time histories at the center point of the back face sheet are given in Figure 18, the wavelength and amplitude of vibration of Tc = 20 and 30 mm are similar while that of Tc = 1 0mm is much more than others. With increasing the thickness of foam core, The permanent displacements of back face sheet decrease gradually and the times tending towards stability increase gradually.

Table 6 peak displacement, permanent displacement and internal energy with different thickness of foam core

Case (No.)	R (mm)	Tc (mm)	Ti (mm)	To (mm)	α (°)	Displacement (mm)		Internal Energy (J)		
						δ_{max}	$\delta_{permanent}$	Ei - 1	Ei - 2	Ei - 3
R6Tc1Ti05To16 (Tc10)	600	10	0.496	1.616	19.508	28.93	8.93	14.36	192.4	71.29
R6Tc2Ti05To16 (Tc20)	600	20	0.492	1.630	19.508	15.78	4.62	11.71	188.7	24.94
R6Tc3Ti05To16 (Tc30)	600	30	0.488	1.644	19.508	13.04	3.87	11.51	199.6	17.81

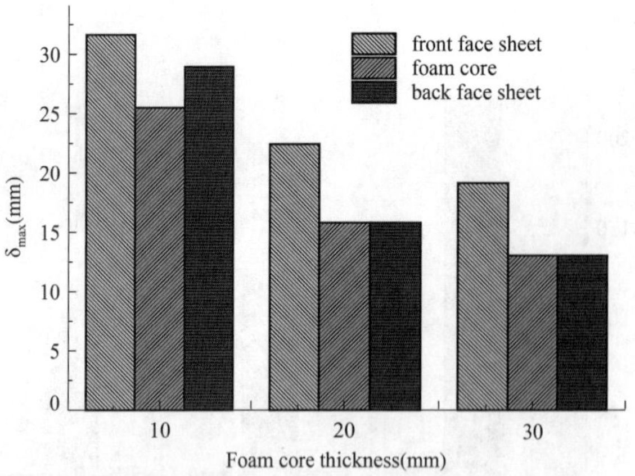

Figure 17 peak displacement of face sheets and foam core

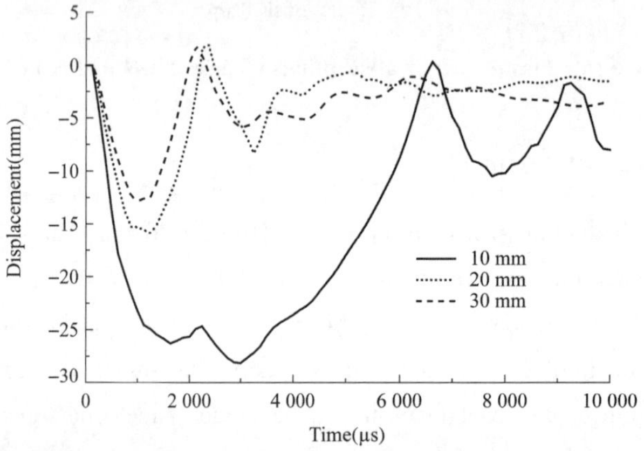

Figure 18 Time histories of center point displacement of back face sheet

Figure 19 shows that less plastic energy is absorbed by front (Ei − 1) and back (Ei − 3) face sheet with increasing the thickness of foam core, while the plastic energy absorbed by foam core (Ei − 2) decreases from 192.4J to 188.7J, then increases from 188.7J to 199.7 with increasing the thickness of foam core. It should be noted that, due to the wave deflection, Ei − 2 of Tc10 includes not only the energy absorbed by foam crushing, but also the energy absorbed by severe wave deflection. That is why Ei − 2 of Tc10 is more than that of Tc20. Therefore, it is concluded that big thickness of foam core can improve the performance on reducing the peak displacement, permanent displacement and the energy absorbed by back face sheet while increasing the energy absorbed by foam core, but the energy absorption capacity of whole shells is unimproved.

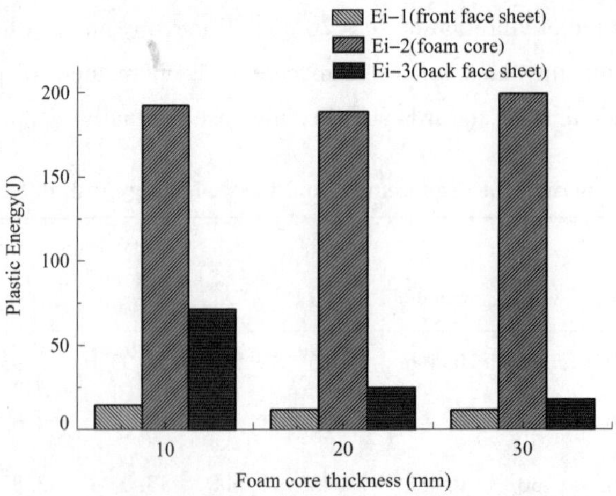

Figure 19 Plastic energy absorptions of face sheets and foam core

5.3 Effect of Face Sheets Combinations

Seven sandwich shells with different thickness combinations of front and back face sheet are compared in this section and they have the same radius of curvature 600 mm, thickness of foam core 20 mm and arc angle 19.508°. The corresponding thickness of front/back face sheet (Table 7) are adjust to make the weights of every part in the shells and the loading area of front face sheet approximately the same. As shown in Figure 20, the peak displacements at the center point of the front face sheet, back face sheet and the back surface of foam core are compared, with increasing the thickness of back face sheet, the peak displacements of face sheets and foam core of Ti/o05/05, Ti/o05/10 and Ti/o05/16 decrease gradually; with increasing the thickness of front face sheet, the peak displacements of front/back face sheet and foam core of Ti/o05/05, Ti/10/05 and Ti/o16/05 decrease gradually; with increasing both thickness of front and back face sheet and keeping them same values, the peak displacements of Ti/o05/05, Ti/o10/10 and Ti/o16/16 decrease gradually. Using Ti/o05/05 (Ti/To = 0.492/0.508 mm) as benchmark, the peak displacements of back face sheet of Ti/o05/10, Ti/o05/16, Ti/o10/05, Ti/o16/05, Ti/o10/10 and Ti/o16/16 decrease by 54.56%, 74.39%, 58.00%, 81.26%, 72.82% and 90.30% and the permanent displacements decrease by 94.02%, 97.22%, 76.33%, 95.49%, 95.25% and 98.49%. The displacement time histories at the center point of the back face sheet are given in Figure 21, with increasing the thickness of front face sheet, the thickness of back face sheet or both thickness of front and back face sheet and keeping them same values, the permanent displacements of back face sheet all decrease gradually.

Table 7 peak displacement, permanent displacement and internal energy with different face sheets combinations

Case (No.)	R (mm)	Tc (mm)	Ti (mm)	To (mm)	α (°)	Displacement (mm) δ_{max}	Displacement (mm) $\delta_{permanent}$	Internal Energy (J) Ei-1	Internal Energy (J) Ei-2	Internal Energy (J) Ei-3
R6Tc2Ti05To05 (Ti/o05/05)	600	20	0.492	0.508	19.508	61.62	54.96	22.50	186.3	17.98
R6Tc2Ti05To10 (Ti/o05/10)	600	20	0.492	1.018	19.508	28.00	3.82	16.83	185.5	20.86
R6Tc2Ti05To16 (Ti/o05/16)	600	20	0.492	1.630	19.508	15.78	1.53	11.71	188.7	24.94
R6Tc2Ti10To05 (Ti/o10/05)	600	20	0.982	0.508	19.508	25.88	13.01	25.77	102.4	9.37
R6Tc2Ti16To05 (Ti/o16/05)	600	20	1.572	0.508	19.508	11.55	2.48	20.43	67.59	6.22
R6Tc2Ti10To10 (Ti/o10/10)	600	20	0.982	1.018	19.508	16.75	2.61	16.73	105.5	13.19
R6Tc2Ti16To16 (Ti/o16/16)	600	20	1.572	1.630	19.508	5.98	0.83	12.13	72.14	10.42

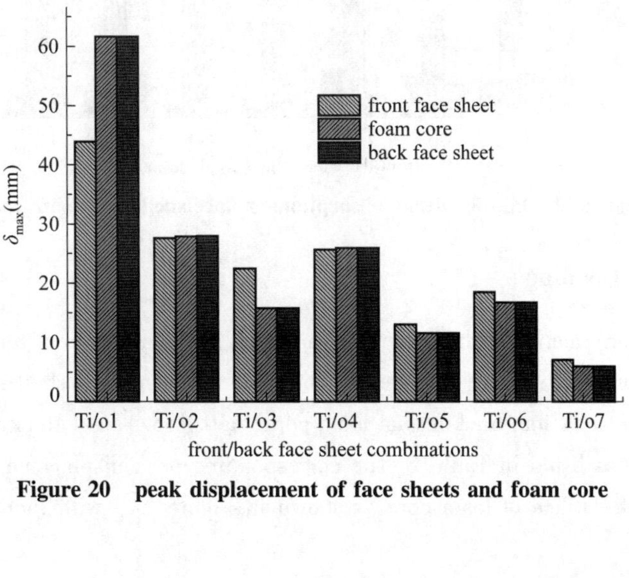

Figure 20 peak displacement of face sheets and foam core

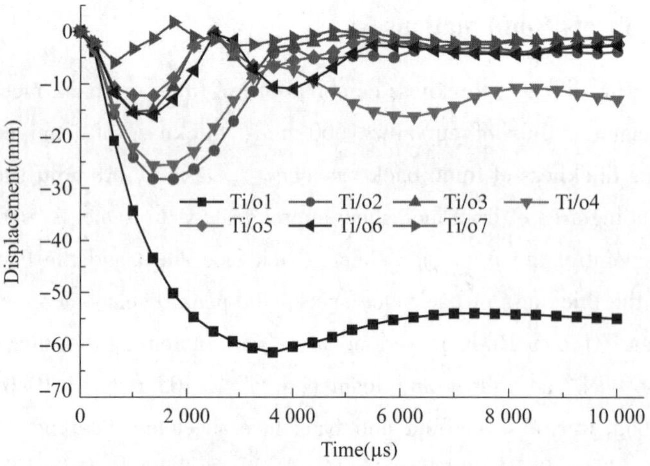

Figure 21 time histories of center point displacement of back face sheet

Figure 22 shows less plastic energy absorbed by front face sheet, more that by back face sheet and no big change in that by foam core with the increase the thickness of back face sheet (Ti/o05/05, Ti/o05/10, Ti/o05/16). With increasing the thickness of front face sheet (Ti/o05/05, Ti/o10/05, Ti/o16/05), the plastic energy absorbed by foam core and back face sheet decrease gradually. With increasing both thickness of front and back face sheet and keeping them same values (Ti/o05/05, Ti/o10/10, Ti/o16/16), the plastic energy absorbed by front face sheet, back face sheet and foam core decrease gradually. It is concluded that higher thickness of front and back face sheet can improve the performance on reducing the peak displacement and the permanent displacement, effectively. Keeping the thickness summation of front and back face sheet almost constant (Ti/o05/16, Ti/o10/10 and Ti/o 16/05), the case Ti/o16/05 has the best performance in resisting blast loading, which indicates that a big thickness of front face sheet with a small thickness of back face sheet will improve the blast - resistance performance.

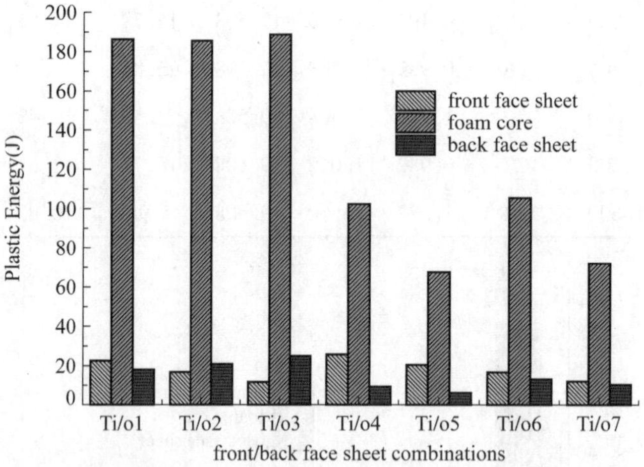

Figure 22 Plastic energy absorptions of face shells and foam core

5.4 Effect of Blast Loading

In order to study the performance of the spherical sandwich shells at different levels of blast intensity, three TNT weight of 50 g, 75 g and 100 g were considered. The spherical sandwich shell studied in this section has a 600 mm radius of curvature, a 30 mm thickness of foam core, a 0.488/1.644 thickness of front/back face sheet and a 19.508 of arc angle, as listed in Table. 8. The corresponding peak displacements at the center point of the front/back face and the back surface of foam core as shown in Figure 23, with increasing the TNT weight, the peak displacements of the front/back face sheet and the back surface of foam core increase gradually. Using B100 (TNT weight = 100 g) as benchmark, the peak displacements of back face sheet decrease by 39. 13% and 68. 60%, and the permanent displacements decrease by 48. 26% and 56. 66%, with the decrease of the TNT weight from 100 g to 75 g and 50 g. The displacement time histories at the center point of the back face sheet were given in Figure 24, with increasing the TNT weight, the permanent displacements of back face sheet increase gradually and slightly, and the times tending towards stability increase gradually. Figure 25 shows that more plastic energy is absorbed by front face sheet, back face sheet and foam core with increasing the TNT weight.

Table 8　peak displacement, permanent displacement and internal energy

No.	TNT equivalency	Scaled distance	Displacement (mm)		Internal Energy (J)		
	m_c (g)	m/kg$^{1/3}$	δ_{max}	$\delta_{permanent}$	Ei - 1	Ei - 2	Ei - 3
B50	50	0. 68	13. 04	3. 58 (-)	11. 51	199. 6	17. 81
B75	75	0. 59	25. 28	4. 09 (-)	21. 96	410. 8	40. 19
B100	100	0. 54	41. 53	4. 60 (-)	30. 53	672. 1	70. 36

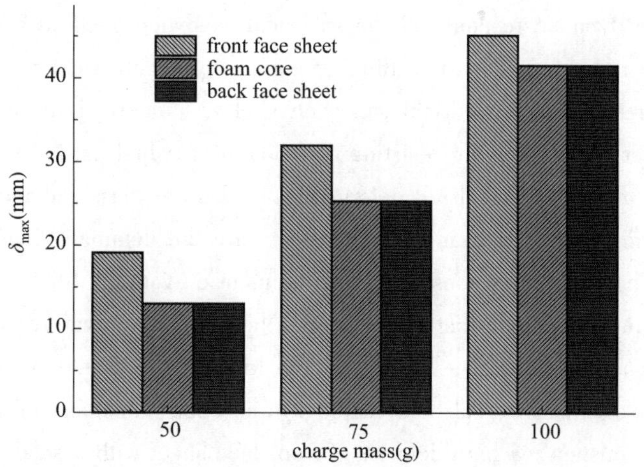

Figure 23　peak displacements of face sheets and foam core

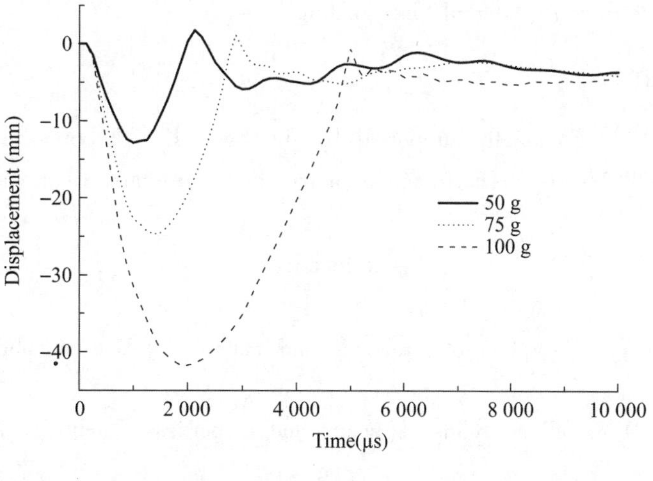

Figure 24　Time histories of center point displacement of back face sheet

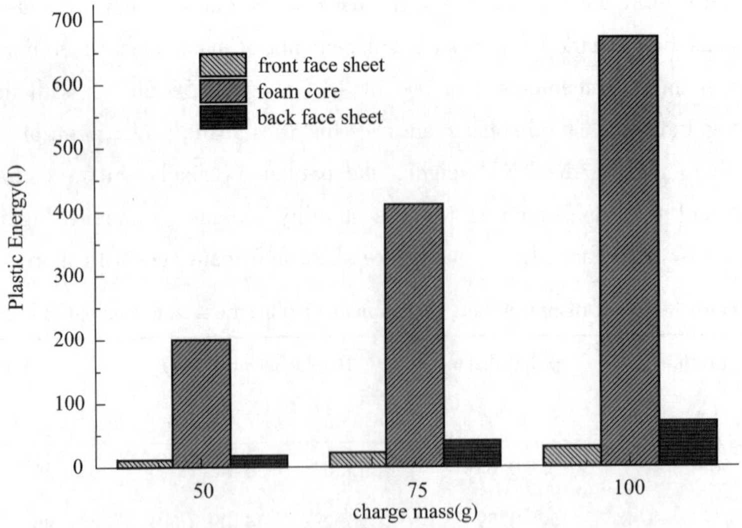

Figure 25 Internal energy absorptions of face sheets and foam core

6 Conclusions

This study presents the dynamic response of the spherical sandwich shell with an aluminum foam core and aluminum face sheets under the external blast loading by employing a numerical model validated. Compared with the experiment data of corresponding cylindrical sandwich shells, numerical results demonstrate that spherical sandwich shells have a better performance in resisting blast loading, which is due to the central symmetry of the spherical configuration. The global and localized deformation modes are classified into three types, respectively, which indicate that geometric configuration and blast intensity are the dominant factors for deformation modes of spherical sandwich shells. The parameters considered are radius of curvature, thickness of foam core, face sheets combinations and blast loading. It is found that the smaller is the radius of curvature, the thicker is the foam core, the thicker are both the front and back face sheet and the less intense is the blast loading, the better is the spherical sandwich shell in resisting external blast loading. In addition, keeping the thickness summation of front and back face sheet almost constant, a big thickness of front face sheet with a small thickness of back face sheet will improve the blast – resistance performance. Therefore, the optimized spherical sandwich shell can be used in the one end of capsule shell structures such as submarines, fuel tanks, coal mine refuge chambers and the nuclear reactor containments to resist external blast loading.

Acknowledgments

The reported research is financially supported by the State Key Laboratory of Explosion Science and Technology Funding (ZDKT13 – 01). The financial contributions are gratefully acknowledged.

References

[1] LU G, YU T X. Energy absorption of structures and materials [M]. Cambridge: Woodhead Publishing Ltd., 2003.

[2] GIBSON L, ASHBY M F. Cellular solids: structure and properties: Cambridge [M]. Cambridge University press, 1997.

[3] ASHBY M F, EVANS A G, FLECK N A, et al. Metal foams: a design guide [M]. oxford: Butterworth – Heinemann, 2000.

复合柔性结构防爆试验方法与判据研究①

刘春美[1②]，黄广炎[2]，由　军[1]，冯顺山[2]

(1. 公安部第一研究所，北京　100048；

2. 北京理工大学爆炸科学与技术国家重点实验室，北京　100081)

摘　要：为快速妥善处置随机发现的爆炸物恐怖威胁，本文研究一种由高强高弹复合材料和混合液组成的复合柔性结构。对复合柔性结构的防爆效果提出基于5级人员伤害/防护等级的防爆判据和等效靶试验方法。建立复合柔性结构抗爆作用的等效试验平台，进行抗爆减爆效应对比试验，得到了人员防护等级与等效靶挠度之间的对应关系。获得的人员一级/二级防护等级下的抗爆判据可用于防爆装备设计与开发。

关键词：兵器科学与技术；复合柔性结构；反爆炸恐怖；等效靶；防护判据

Research on Test Methods and Criteria of Anti - explosion of a Composite Flexible Structure

Liu Chunmei[1], Huang Guangyan[2], You Jun[1], Feng Shunshan[2]

(1. First Research Institute, Ministry of Public Security, Beijing 100048, China; 2. State Key Laboratory of Explosion Science and Technology, Beijing Institute of Technology, Beijing 100081, China)

Abstract: A composite flexible structure composed of high strength and high elastic composite material and mixed liquor is studied for the quick and proper disposal of explosives. An explosion proof criteria and the equivalent target test methods are proposed, which are based on 5 levels of personnel injury /pro - tection grade. An equivalent test platform of composite flexible structure is established. The contrast tests of explosion in the air and anti - explosion attenuation effect are carried out. The corresponding relationship between the level of protection and the maximum deflection of the equivalent target is obtained. The e - quivalent target criteria of the first level and second level of personnel protection under anti - explosion could be used to guide the design and development of explosion - proof equipment.

Key words: ordnance science and technology; complex flexible structure; anti - explosive terror; equiva - lent target; protection criteria

0　引言

爆炸恐怖因其具有行动隐蔽、实施简单、破坏规模大等特点，易造成持久、广泛的社会大众恐怖心理，已成为当今恐怖分子最常用、最普遍的活动形式。爆炸恐怖袭击、涉爆刑事犯罪已成为国内外社会公共安全的主要威胁之一。因此，及时有效反击爆炸恐怖袭击、处置各类爆炸物是公共安全部门所面临的最严峻挑战之一[1-4]。

由于爆炸恐怖活动的随机性和不确定性，即爆炸物触发方式未知、爆炸位置随机、爆炸当量难以准

---

① 原文发表于《兵工学报》2016 (S1)。

② 刘春美：工学博士后，2007年合作导师冯顺山，研究等待式拦截幕形成及对飞行器作用机理与应用，现工作单位：公安部第一研究所。

确判断，尤其在一些人员聚集区发现爆炸物时，如何将爆炸物采用轻便防爆器材隔离、转移并快速安全妥善处置（失效、销毁），是当前处置爆炸物所面临的重大难题之一。因此，亟须一种新型的便捷式、可随时组装、快捷实用的防爆炸抗冲击能伤害的吸能材料技术及装备。本文提出一种高强度高弹性固液复合的柔性防爆方法，建立防护人员的等效模型，对人员在爆炸冲击波作用下伤害/防护程度采用等效靶板变形来描述，以建立复合柔性结构防爆试验方法和等效判据。

1 复合柔性结构模型和防爆作用判据

1.1 复合柔性结构模型

复合柔性结构由复合材料和混合液介质采用一定方式组合而成，以实现将爆炸能高效耗散并安全转化成无害动能，如图 1 所示。复合材料由高强纤维材料和高弹性材料复合制成，混合液介质由高分子 MAP 胶体材料与水混合形成高水速凝胶体材料[3-7]。

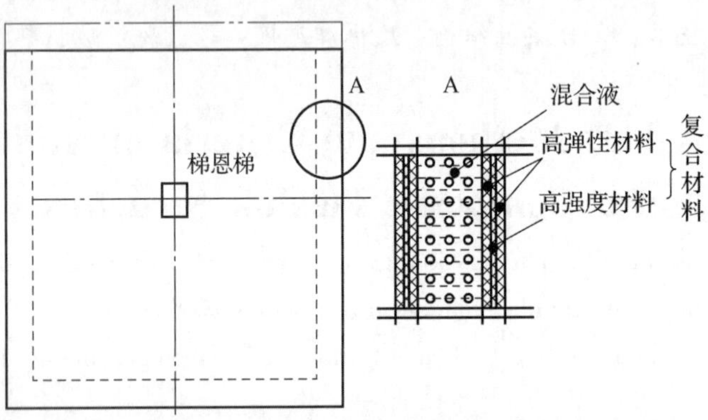

图 1 复合柔性结构组成示意图

在爆炸冲击波作用下，高强高弹性复合材料先发生变形，高弹性材料发生拉伸变形，高强度材料发生扭曲直至撕裂破坏，消耗应变能。冲击波在复合材料与混合液界面发生透反射，反射波压缩混合液。液体可压缩性小，对爆炸冲击波传递效率高。为了确保冲击波衰减效果的理想性，采用“气泡帷幕”方法进一步弱化冲击波传播，即混合液为在水中加入 MAP 基料形成的高分子胶体材料，各高分子之间相互枝联，并且大分子之间伴随大量气泡形成气泡幕帐。气泡帷幕对水中冲击波衰减能力与气幕中气泡含量和气幕带宽度成正比关系[8]。因此，防爆炸冲击波作用过程中复合柔性结构总消耗能量有：①复合材料发生变形、拉伸撕裂破坏的应变能；②混合液变形、多次反射消耗的热能；③传递给混合液使其解体、扩散、云雾化的热能和机械能、液体飞散运动带走的动能。

为降低试验成本，本文以防 3 kg 梯恩梯（TNT）当量爆炸物为原型，建立 1∶30 复合柔性结构防爆缩比模型。

1.2 防爆作用判据

在反爆炸恐怖中，对人员的防护是关键。根据文献［9］中人员受伤程度，将人员伤害等级划分为 5 级，如表 1 所示。

表 1 冲击波超压对人员的损伤阈值

人员伤害等级	冲击波超压值/MPa	伤害程度	损伤特征
一级	≤0.02	轻微伤	人员安全，轻微挫伤
二级	0.021～0.03	轻伤	轻挫伤，肺部小灶性出血

续表

人员伤害等级	冲击波超压值/MPa	伤害程度	损伤特征
三级	0.03～0.05	中等	听觉气管损伤、肺大泡形成，肺体指数明显增加
四级	0.05～0.1	严重	内脏受到严重挫伤，肺体指数显著增加，可能造成死亡
五级	>0.1	死亡	极严重，大部分死亡

本文所研究的复合结构防爆炸作用对人员伤害着重以一级轻微伤和二级轻伤为防护判据，即防爆后某点处冲击波超压值应满足不大于0.03 MPa的条件于$\omega=3$ kg TNT当量爆炸物，在距离爆源$r=3$ m处，其空爆时冲击波超压理论峰值为$\Delta p_+=0.147$ MPa。人若位于此处，将会受到五级伤害，大部分死亡。在采用复合柔性结构装置防爆后，同位置处的冲击波超压峰值能降到0.03 MPa以下，发生轻微伤/轻伤。

2　复合柔性结构防爆性能评价方法

2.1　等效靶评价方法

实际爆炸试验中，对人员在爆炸冲击波作用下伤害程度采用等效靶靶板变形来描述。相反，采用复合柔性结构抗爆以保护人员不受到伤害或减轻伤害，也可以用等效靶靶板变形来描述复合结构抗爆性能，可根据复合结构不同防爆等级建立对人员的防护等级。

在相同的爆炸冲击波载荷作用下，用有一定约束条件的、具有恰当敏感性的靶板的变形程度描述对人员造成伤害/防护的程度，即通过靶板的变形情况描述对保护人员的伤害/防护效应，并建立二者之间相应的对应关系。根据对靶板的破坏/防护等级建立对防护对象的伤害/保护等级，我们把这种靶板称为等效靶[10-11]。

等效靶靶板材料可选钢板或铝板，形状可方可圆，厚度可选。本文采用圆形2A12铝板作为等效靶板，采用周边压紧固支方式。这种方式试验成本低、有效性好、通用性强。

2.2　试验方法

试验中，把爆炸冲击波的作用转化为宏观可测量的等效靶靶板大挠度变形，通过测量等效靶在防爆前后的挠度变化来给出复合柔性结构防爆作用衰减冲击波的程度。建立人员损伤/保护与等效靶变形之间的对应关系，同时也进行爆炸参数测试作为辅助手段，得到不同位置处超压峰值，以对应1.2节中对人员的防护判据。总体试验方案及测试方案布置如图2所示。

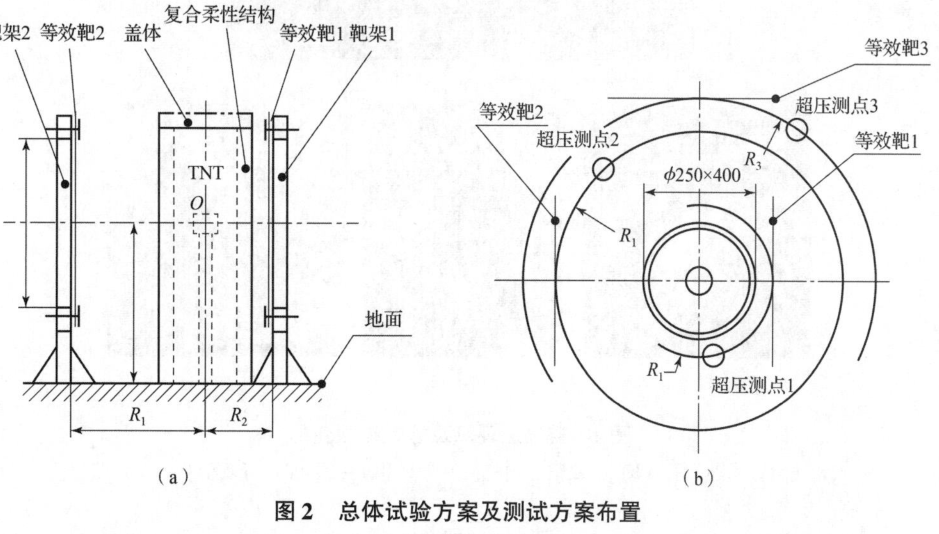

图2　总体试验方案及测试方案布置

(a) 主视图；(b) 俯视图

试验时，在炸药周边不同距离处放置等效靶，爆炸冲击波作用到等效靶上，通过记录冲击波超压值与靶板变形，得出人员受伤/防护程度与靶板最大挠度的关系。

3 防爆试验研究

3.1 防爆对比试验

3.1.1 空爆试验

以炸药在空气中爆炸作为对比组，试验现场及测试系统如图3所示。

图3 爆炸试验现场及测试系统布置

空爆条件：小药量的长径比为1的柱状B炸药距地面一定高度 h（满足 $h/\omega^{1/3} \geqslant 0.35$ 条件），用8号电雷管起爆。等效靶距爆源分别为 R_1、R_2、R_3 3个传感器距爆源的位置同等效靶，高度与爆源等高。空爆后现场图如图4（a）所示，各位置处的等效靶变形如图4（b）~图4（d）所示。试验数据见表2。

图4 空爆后现场及等效靶板变形

(a) 空爆后现场图；(b) 等效靶1-1；(c) 等效靶1-2；(d) 等效靶1-3（反凸）

表 2　等效靶板试验变形挠度记录表

工况	超压峰值 /MPa	等效靶板编号	靶板厚度 /mm	靶板距爆源位置/m	靶板挠度 /mm	备注
空爆	2.68	1-1	0.75	0.201	撕裂	等效靶中间大面积破裂掉约 4/5
	0.72	1-2	0.75	0.418	27.80	
	未测	1-3	0.75	0.836	15.58	
圆柱型（基本液）	2.31	4-1	0.75	0.201	撕裂	靶板变形前凸整个等效靶发生撕裂破坏
	0.06	4-2	0.75	0.418	20.20	
	未测	4-3	0.40	0.836	8.94	
圆柱型（混合液）	1.43	5-1	0.75	0.201	撕裂	沿固支边缘撕裂约占 2/3 周长，内部完整
	0.09	5-2	0.75	0.418	11.50	
	未测	5-3	0.40	0.836	5.98	

3.1.2　防爆试验

在圆柱型复合柔性结构防爆试验中，炸药尺寸与空爆条件相同，置于复合结构内居中悬挂，防爆结构厚度为 50 mm，内部充注两种液体介质，分别为基本液和混合液。等效靶和 3 个传感器距爆源的布置位置同空爆试验条件如图 5（a）所示。

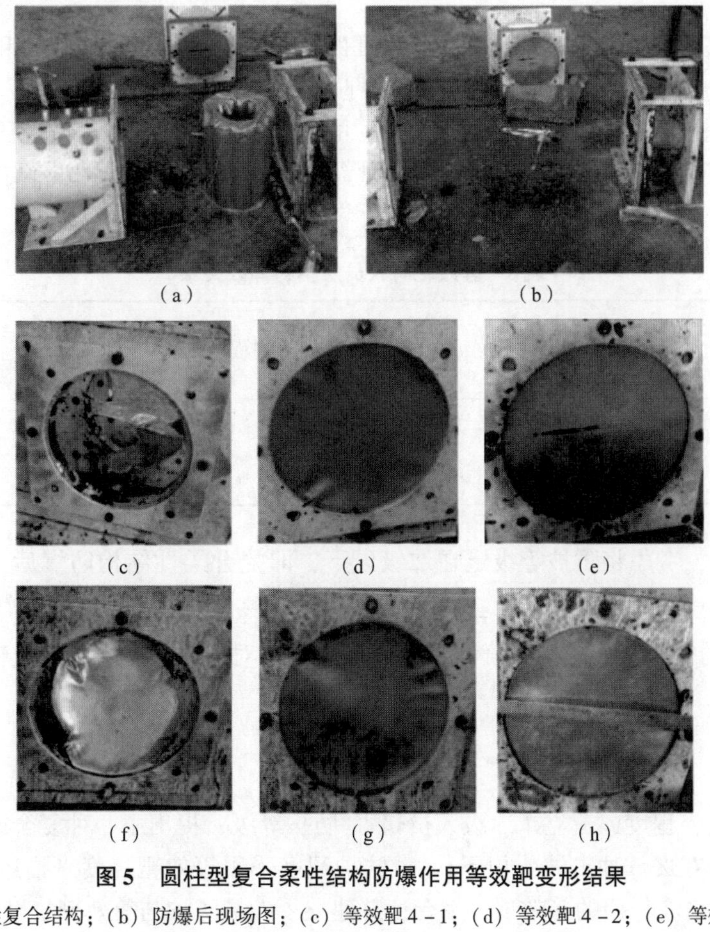

图 5　圆柱型复合柔性结构防爆作用等效靶变形结果

（a）柔性复合结构；（b）防爆后现场图；（c）等效靶 4-1；（d）等效靶 4-2；（e）等效靶 4-3；（f）等效靶 5-1；（g）等效靶 5-2；（h）等效靶 5-3

复合柔性结构防爆后现场图如图5（b）所示。典型等效靶变形结果如图5（c）~（h）所示。图5（c）~（e）为充注基本液防爆后等效靶变形结果，图5（f）~（h）为充注混合液防爆后效靶变形结果。

3.1.3 试验结果与分析

从图4、图5及表2可以看出，在同等小药量炸药爆炸冲击波作用下，空爆情况下等效靶的变形明显大于同位置处带有复合柔性结构组成防爆装置的等效靶的变形。

（1）在空爆情况下，在距爆源最近距离（$r_1=0.201$ m处，比例距离 $r=r/\sqrt[3]{\omega}=0.519$）处的等效靶大面积破裂，破坏最为严重，此处超压峰值为2.68 MPa。而在有复合柔性防爆结构的其他组试验中，破坏也很严重，沿靶板边缘大面积撕裂破坏，但圆柱形防爆装置在该处被破坏的程度明显小。

（2）在空爆情况下，距爆源 $r_2=0.418$ m（比例距离 $r=r/\sqrt[3]{\omega}=1.08$）处的等效靶挠度为 $f_{ae2}=27.80$ mm，其他2组试验最大挠度值均小于该值，分别为20.20 mm和11.50 mm，降幅达27%和59%，超压峰值从空爆时的0.72 MPa分别降为0.06 MPa和0.09 MPa。

（3）在第3个位置处，等效靶的挠度从空爆时的15.58 mm分别降为8.94 mm和5.98 mm，降幅分别达40%和62%。

从以上3组典型试验数据分析可知，两种复合柔性结构充注不同液体制成的防爆装置的对比试验（小当量缩比试验）可以得出：①复合柔性结构组成的防爆装置具有很强的减爆能力，可以大幅消减爆炸冲击波对人员造成的伤害；②复合柔性结构内充注混合液的防爆吸能效果优于充注基本液；③等效靶变形挠度比超压峰值能更好地反映防爆结构消减冲击波的能力，能更好地表达出超压和时间共同作用的效果，因此用挠度值作为防护判据合理可行。

3.2 复合柔性结构防爆效应的等效靶评价判据

（1）防护评价方法的选择。根据表述冲击波特征的超压峰值、正压作用时间和比冲量3个参数，可在前人研究成果和试验基础上选择超压准则进行人员防护效果评价。

（2）等效靶评价准则的确定。对应我们重点关注的人员防护等级、冲击波超压峰值、人员受伤害程度，初步建立与等效靶变形挠度的等效关系，并通过大量仿真模拟试验进一步确定各值之间的联系，如表3所示。

表3 等效靶与人员防护的等效关系

人员防护等级	冲击波超压值/MPa	人员伤害/防护程度	位于 $r=r/\sqrt[3]{\omega}=2.08$ 处等效靶板的最大挠度 f_0/mm
一级	<0.02	轻微伤	<5
二级	0.02~0.03	轻伤	5~9

为了保护人员安全，考虑将防护等级定在二级以下，即复合柔性结构防爆后人员受轻伤或轻微伤，甚至无伤，位于比例距离 $r=r/\sqrt[3]{\omega}=3/\sqrt[3]{3}=2.08$ 处，即TNT当量为3 kg，满足超压峰值小于0.03 MPa，距爆源距离 $r>3$ m处的等效靶板的挠度值在5~9 mm范围内对应二级防护；等效靶板挠度小于5 mm对应一级防护。

4 结论

（1）根据人员受伤害程度，建立了5级人员伤害/防护等级，提出了一种复合柔性结构防爆判据。

（2）提出了对人员在爆炸冲击波作用下伤害/防护程度采用等效靶挠度来描述的试验方法，建立了复合柔性结构抗爆炸冲击波作用的等效试验平台，并建立了人员防护与等效靶挠度之间的对应关系。

（3）圆柱型的复合柔性结构采用混合液胶体材料防爆效果优于充注基本液（水）的结构。获得了人员一级/二级防护等级抗爆判据，可用于防爆装备设计与开发。

参考文献

[1] 钱七虎. 反爆炸恐怖安全对策［M］. 北京：科学出版社，2005.

[2] YANG G. The terrorist attack cause and prevention in China' international contractor project based on fuzzy fault tree analysis［C］// 2006 International Conference on Construction &Real Estate Management. Orlando，US：University of Florida，2006：853-856

[3] 刘春美，黄广炎，由军，等. 复合柔性结构防爆试验方法与判据研究［R］. 北京：公安部第一研究所，2011.

[4] 贝克，威斯汀. 爆炸危险性及其评估［M］. 张国顺，文以民，刘定吉，译. 北京：群众出版社，1988.

[5] 石光明，刘春美，冯顺山，等. 高水速凝MAP材料的动态力学性能研究［J］. 兵工学报，2015，36（增刊1）：220-227.

[6] 石光明，刘春美，冯顺山，等. 高弹性体材料动态力学性能试验研究［J］. 北京理工大学学报，2005，35（增刊2）：182-184.

[7] 顾伯洪，孙宝忠. 纺织结构复合材料冲击动力学［M］. 北京：科学出版社，2012.

[8] 亨利奇. 爆炸动力学及其应用［M］. 熊建国，译. 北京：科学出版社，1987.

[9] 隋树元，王树山. 终点效应学［M］. 北京：国防工业出版社，2000.

[10] 冯顺山，王芳. 爆炸冲击波防护效应的评价方法研究［J］. 安全与环境学报，2004，4（增刊）：173-176.

[11] 王芳，冯顺山，范晓明. 冲击波作用下目标毁伤等级评定的等效靶方法研究［C］//第九届全国爆炸与安全技术学术会. 沈阳：中国兵工学会爆炸与安全技术专业委员会，2006.

Blast Response of Continues – Density Graded Cellular Material Based on the 3D Voronoi Model①
基于三维 Voronoi 模型的连续密度梯度胞状材料爆炸响应

Feng Shunshan, Lan Xuke②, Huang Qi, Bai Yang, Zhou Tong

State Key Laboratory of Explosion Science and Technology, Beijing Institute of Technology, Beijing 100081, China

Abstract: One – dimensional blast response of continuous – density graded cellular rods are investigated theoretically and numerically. Analytical model based on the rigid – plastic hardening (R – PH) model is proposed to predict the blast response of density – graded cellular rods. Finite element (FE) analysis is performed using a new model based on the 3D Voronoi technique. The FE results agree well with the analytical predictions. The blast response and energy absorption of cellular rods with the same mass but different density distributions are examined under different blast loading. As a blast resistance structure, cellular materials with high energy absorption and low impulse transmit is attractive. However, high energy absorption and low impulse transmit cannot be achieved at the same time by changing the density distribution. The energy absorption capacity increases with the initial blast pressure and characteristic time of the exponentially decaying blast loading. By contract, when the blast loading exceeds the resistance capacity of cellular materials, the transmitted stress will be enhanced which is detrimental to the structure being protected.

Key words: Gradient; Blast response; Cellular; 3D Voronoi model

1 Introduction

Cellular materials, such as honeycomb, foam, corrugated plate and metal hollow sphere, have been widely used in aerospace and defense industries as an energy absorption device to mitigate shock and impact by progressive local crushing of its micro – structure under dynamic loading [1]. Cellular materials are attached to the protect structure as sacrificial layers for blast and impact, the cladding is expected to attenuate the load on the structure behind it [2, 3].

Over the latest decade, the dynamic responses of cellular materials with uniform density have been well studied experimentally [4 – 11]. Based on these tests, two main phenomena were observed as follows: (a) when the impact velocity exceeds a critical value, the cellular material will be separated into crushed and uncrushed regions obviously by a propagating discontinuity; (b) the stress of proximal end is higher than the quasi – static plateau stress duo to the dynamic enhancement.

To analysis these phenomena, a one – dimensional shock – wave model with rigid – perfectly – plastic – locking (R – PP – L) constitutive relation was firstly developed by Reid and Peng based on the dynamic response of wood [4]. Tan et al. extend this shock – wave model to metal foam and honeycomb [5]. To depict the deformation behind the shock front at different velocities more accurately, various constitutive relations were developed considering the plastic hardening for different cellular materials. An elastic – perfectly plastic – rigid (E – PP – R) idealization and an elastic – plastic – rigid (E – P – R) idealization were employed by Lopatinikov

① 原文发表于2018 国际防务技术大会（2018）。

② 兰旭柯：工学博士，2016 年师从冯顺山教授，研究爆炸抛撒子弹药抗冲击与防护方法，现工作单位：北京理工大学博士后。

et al. to consider the effect of elastic [12－14]. Meanwhile Zheng et al. proposed a linearly hardening plastic－locking model (R－LHP－L) as a supplement for the R－PP－L model to capture the transition mode model [15]. Moreover, a dynamic rigid－plastic hardening (R－PH) idealization was proposed by Zheng et al [16]. This stress－strain idealization can well characterize the dynamic compression behavior of cellular materials under high velocity impact [17].

Blast responses of cellular material with uniform density were well investigated, and provide a useful guide for designing protective structures against blast loading [3, 18]. According to previous study, layered sacrificial cladding were observed as highly effective for energy absorption, with predictable behavior under blast loading. However experimental evidence has highlighted the fact that the presence of this "protective" layer can result in an enhanced loading of the structure, the influence of using cellular material as a protective layer remains debated [19－21]. To explain the increase in energy and impulse transferred to pendulum when foam panel exist, Hansen proposed an analytical solution based on shock－wave theory [20]. A systematic study was carried out, but the propagation mechanism of shock wave was beyond the scope of their study. Ma and Ye investigated the deformation of foam subjected to blast loading by proposing an analytical Load－Cladding－Structure (LCS) model [22, 23]. This study focused on the deformation behavior of the protected structure only, the crushing process of the foam was not considered. Aleyassin et al. focused on the attenuation/enhancement boundary based on the R－P－P－L model, a new method of accounting for fluid－structure interaction is derived [24]. However, the using of R－P－P－L model may overestimate the energy absorption ability of cellular material. Karagiozova et al. developed an analytical model to reveal the characteristic features of foam compaction [25]. The model is accurate in predicting the front of the propagation shock but failed to capture the impulse transmitted to the protected structure. Although the blast response of cellular materials has been studied experimentally, annalistically and numerically, the mechanism of cellular materials subject to blast loading needs further investigation.

In attempt to find an optimal design, investigations on the blast response of graded cellular materials have been done in recent years. Liu studied the blast resistance of sandwich－walled hollow cylinder with graded aluminum foam core [26]. It was found that the introduce of graded foam core can increase the energy absorption efficiency. MA and Ye derived the energy absorption capacity of double－layer foam cladding under blast loading based the R－P－P－L model [27]. However, the process of compaction wave propagation was lake of investigation due to the complex behavior of shock－wave propagation caused by different gradient distribution. With the development of manufacture, cellular material with continuous－density variation can be made [28]. Although some analytical and numerical study have been carried about the impact response of cellular material with continuous－density variation [29－33], seldom results of blast response can be found. And the numerical model used in these studies are limited to 2D Voronoi structures, they were not accurate enough compared with the 3D Voronoi structure.

The present study focused on the blast response and energy absorption of cellular materials with continuous－density gradient. To clarify the effect of density distribution an analytical model is proposed. Meanwhile, a new numerical model with continuous－density gradient is constructed based on the 3D Voronoi technology. The theoretical predictions are verified by numerical results, and the analytical predictions are compared with numerical simulations.

2 Analytical Analysis

2.1 Problem Formulation

Consider a density－graded cellular rod sandwiched by two rigid plates, as shown in Figure 1. Modeling of the cellular rod under blast loading. The front plate at the proximal end is regarded as a rigid mass, and the plate at the distal end is regarded as a rigid wall. The compressive blast loading p (t) acts directly on the front plate. According to Fleck et al [34]. when the distance between the explosive source and the sacrificial layer is much larger than the size of the protected structure, the spherical wave generated by the blast loading can be approximated as a plane wave which decays exponentially as follows:

$$p(t) = p_0 e^{-t/\tau} \tag{1}$$

Where p_0 is the initial peak pressure of blast loading, and τ is the characteristic time. The Impulse of the blast loading is $p_0\tau$.

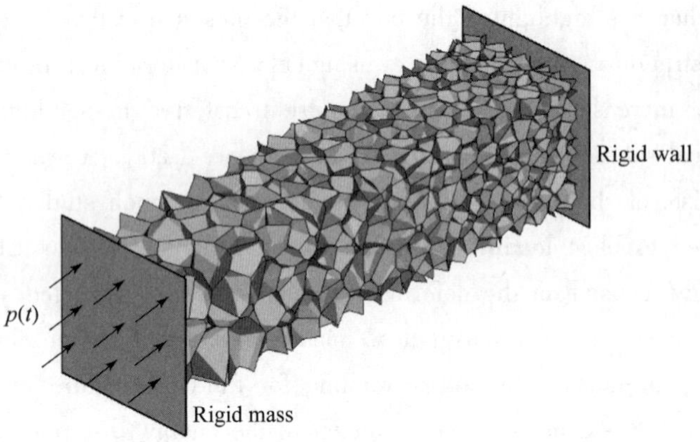

Figure 1 Modeling of the cellular rod under blast loading.

The liner density gradient of cellular rod is defined as $g = \rho_{distal} - \rho_{proximal}$, where ρ_{distal} and $\rho_{proximal}$ are the relative density of distal end and proximal end.

2.2 Cellular Material Model

The rate - independent R - PP - L model has been widely used to characterize the nominal stress - strain relation of the cellular material, as shown in Figure 2. This model is simply enough with only two parameters, and it can characterize the shock wave in cellular material well. However, the R - PP - L model failed to consider the plastic hardening, which exist in the real stress - strain curves of cellular material. In our study, a rigid - plastic hardening (R - PH) model is adopted to model the cellular material, written as:

$$\sigma = \sigma_0(\rho) + C(\rho)\varepsilon/(1-\varepsilon) \tag{2}$$

Where σ_0 is the initial crushing stress and C is an empirical fitting parameter, which characterizes the strain hardening behavior. The two material parameters are related to the relative density ρ of cellular material. Quasi - static compression tests are performed in section 3 to obtain the two parameters.

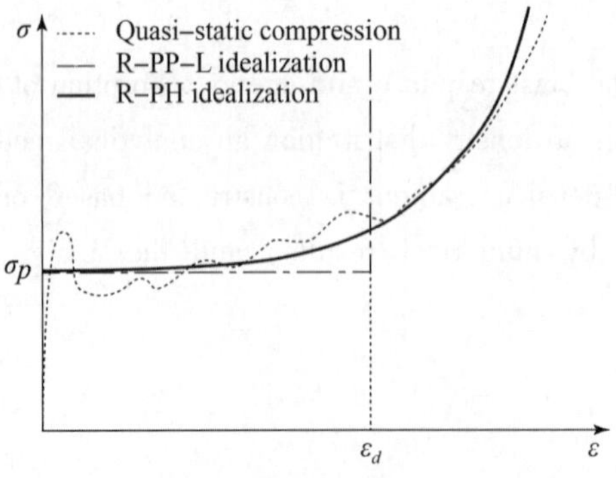

Figure 2 R - PP - L and R - PH idealization of stress - strain curve of a cellular material.

2.3 Analytical Model for Blast Response of Graded Cellular Rods

As shown in Figure 3, the compressive shock front travels along the cellular rod with a Lagrange coordinate: $\Phi(t)$. According to R – PH model, the speed of elastic wave is infinite while the shock front travels at a finite speed given by $\dot{\Phi}(t)$. It means that the shock front never catches the elastic wave. When the shock front travels across one position, the particle speed, strain and stress jumps from $\{0,0,\sigma_0(\rho(\Phi))\}$ to $\{v(t),\varepsilon_b(t),\sigma_b(t)\}$. where $\sigma_0(\rho(\Phi))$ is the initial crushing stress. According to the continuum – based stress wave theory, the conservation relation of mass and momentum across the shock front are given by

$$v(t)-0=\dot{\Phi}(t)(\varepsilon_b(t)-0) \tag{3}$$

$$\sigma_b(t)-\sigma_0(\rho)=\rho_s\rho\dot{\Phi}(t)(v(t)-0) \tag{4}$$

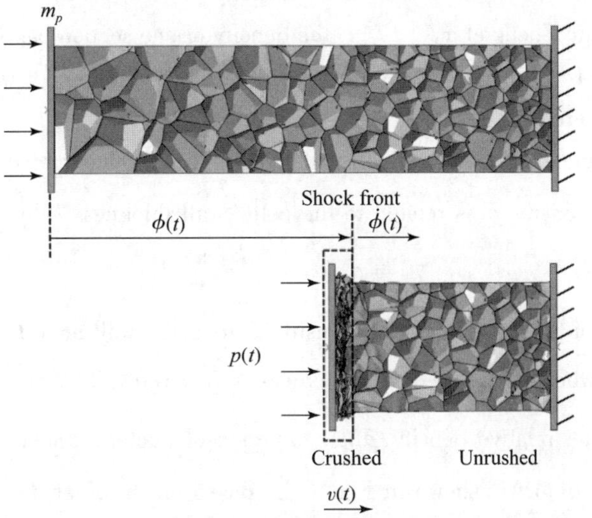

Figure 3 Diagrams for the anti – blast analysis of graded cellular rod.

Where ρ_s is the density of base material. Combining Eq. (3) and Eq. (4) gives the stress behind the shock front:

$$\sigma_b(t)=\sigma_0(\rho)+\rho_s\rho v^2(t)/\varepsilon_b(t) \tag{5}$$

Combining Eq. (2) and Eq. (5) gives the strain behind the shock front:

$$\varepsilon_b(t)=\frac{v(t)}{v(t)+c(\rho)} \tag{6}$$

Where $c(\rho)=\sqrt{\frac{C(\rho)}{\rho_s\rho}}$.

Mass conservation law gives the mass of crushed part as

$$m_s=\rho_s\int_0^{\Phi(t)}\rho(\xi)d\xi \tag{7}$$

Assumption that the crushed part moves with the front – plate at the same speed. The momentum conservation law for the front – plate and compacted part gives that:

$$\left(m+\rho_s\int_0^{\Phi(t)}\rho(\xi)d\xi\right)\frac{dv(t)}{dt}=p(t)-\sigma_b(t) \tag{8}$$

Combining Eq. (3), (5), (6) and (8) gives the governing equations:

$$\begin{cases}\dfrac{d\Phi(t)}{dt}=v(t)+c(\rho)\\[2ex]\dfrac{dv(t)}{dt}=\dfrac{p(t)-\sigma_0(\rho)-\rho_s\rho v(t)(c(\rho)+v(t))}{m+\rho_s\displaystyle\int_0^{\Phi(t)}\rho(\xi)d\xi}\end{cases} \tag{9}$$

With the initial condition, $\Phi(0)=0$ and $v(0)=0$, the governing equations can be solved numerically by using the Runge – Kutta method.

3 Cell – Based Finite Element Modeling

3.1 3D Graded Voronoi Structure

The 2D random Voronoi structure has been widely used to simulate cellular materials with uniform and graded density. The 3D random Voronoi structure can better simulate the microstructure of cellular materials [17, 35]. However cellular rods with continues – density gradient, generated by using 3D random Voronoi technology cannot be found in the literature. In order to generate 3D Voronoi structure with continues – density gradient, the cellular rod ($20\times20\times80\ mm^3$) was cut into 8 sections along the long axis, and each section has a volume of $20\times20\times10\ mm^3$. Referring to Zheng et al. [17], the density of one section can be changed by verify the cell – wall thickness or the cell size, in this study the later method is adopted which can better simulate the microstructure of cellular materials.

Cellular model with N nuclei and uniform cell – wall thickness is constructed in the volume of V_{foam} ($20\times20\times10\ mm^3$). The relative density $\bar{\rho}$ is related to the cell – wall thickness h by

$$\bar{\rho}\equiv\rho_0/\rho_s=\frac{h}{V_{foam}}\sum A_j \tag{10}$$

Where ρ_0 is the density of the foam, ρ_s is the density of the cell – wall material, A_j is the area of j – th cell – wall surface. In geometry, when the cell – wall thickness h is given (0.05mm in this study), $\sum A_j$ can be controlled by N. Therefor, the relative density $\bar{\rho}$ is a function of nuclei number N. Date fitting from $N=50$ to $N=600$ gives the relation of $\bar{\rho}$ to N, shown in Figure 4. Based on the liner density distribution, the nuclei of each section can be given and cellular rod with liner density gradient can be generate, as shown in Figure 5.

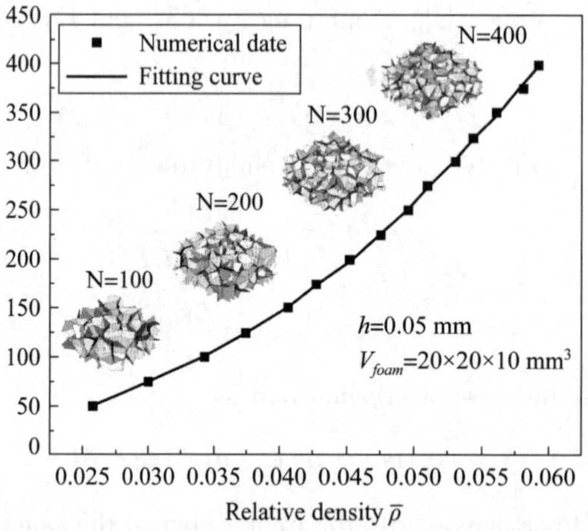

Figure 4 Relationship between relative density and number of nuclei.

3.2 Numerical Model

Numerical simulations are performed by the ABAQUS/Explicit software. The cellular rod is sandwiched by two rigid plates. The pressure of blast is load on the rigid plate at the proximal end, while the rigid plate at the distal end is fixed. The cell – wall material is assumed to be elastic – perfectly plastic with density ρ_s = 2 700 kg/m^3, Young's modulus $E_s=69$ Gpa, Poisson ratio $\nu_s=0.3$ and yield stress $\sigma_{ys}=170$ Mpa. The cell – wall is meshed by using ABAQUS shell elements S3R and S4R, as shown in Figure 5, the element size is set to

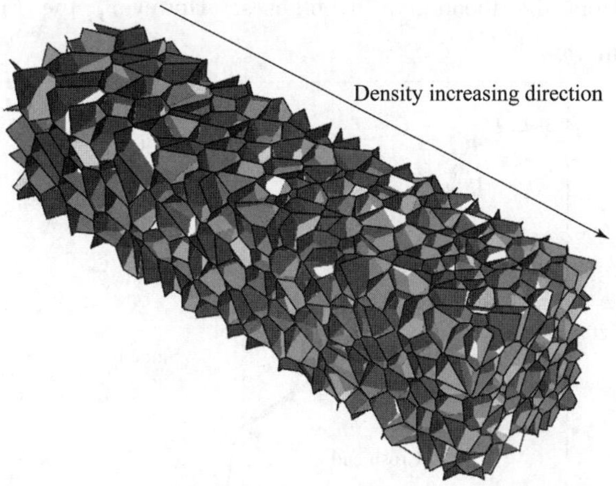

Figure 5 3D Voronoi structure with continuous density gradient.

be 0. 3 mm. To save computational time, shell elements of type S3R with sharp angle are eliminated. General contact is applied to all possible contacts during crushing with a friction coefficient of 0. 02.

A numerical compression test at low constant velocity is also performed to obtain a quasi – static nominal stress – strain relation. The quasi – static cellular specimens with different densities (by varying the cell size) are constructed in a volume of $20 \times 20 \times 30$ mm^3. The specimens are sandwiched between two rigid walls. One wall is fixed and the other wall travels at a constant velocity of 1m/s (the corresponding nominal strain rate is 33. 33/s) compressing the cellular specimen. The quasi – static nominal stress – strain relations of cellular material models is fitted by using the R – HP idealization, referring to Zheng [17] for detail. Repeating the compression tests on samples with different density gives the initial crushing stress and the hardening parameter in terms of relative density as follows:

$$\begin{cases} \sigma_0(\rho) = 2.52\sigma_{ys}\rho^2 \\ C(\rho) = 0.131\sigma_{ys}\rho^2 \end{cases} \tag{11}$$

4 Comparison between the oretical and Numerical Results

FE simulations for cellular materials subject to blast loading with an initial peak pressure 16. 9 MPa and characteristic time $\tau = 0.05$ ms are carried out and the results are presented in comparison with the results from the theoretical models. Figure 7 shows the velocity history of the front – plate under blast loading. The velocity history can be obviously divided into three main stages. In stage Ⅰ, the high blast pressure acts on the front – plate and drives the front – plate to the maximum velocity in a very short time. In stage Ⅱ, the high velocity front – plate compacts the cellular rods and the velocity decreases gradually till to the crush end. In stage Ⅲ, the front – plate and the fully crushed cellular rod impact on the protected structure which leads to a sharp velocity decrease. The inertia stress of cellular material rebounds the front – plate to achieve a negative velocity. The velocity predicted by theoretical analysis is compared with the FE results, and good agreement is achieved. The reaction stress on the blast end and fixed end are compared as depicted in Figure 8. There exists a perturbation in FE results, because the finite element model reflects the meso – structure of cellular

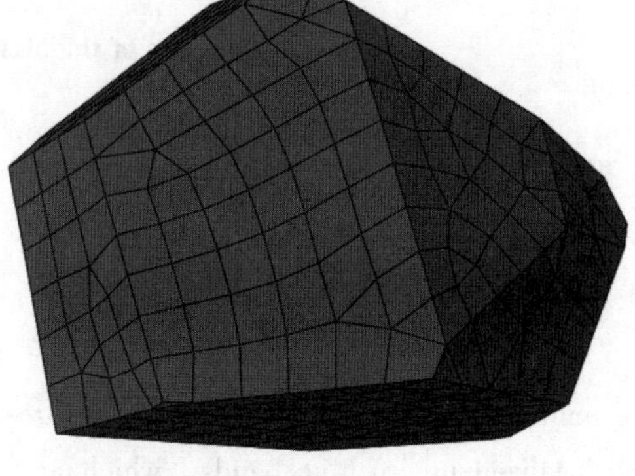

Figure 6 A cell of Voronoi structure with shell elements meshing.

materials which is different from the theoretical hypotheses. However, the FE results agree well with the theoretical analysis in the mean value.

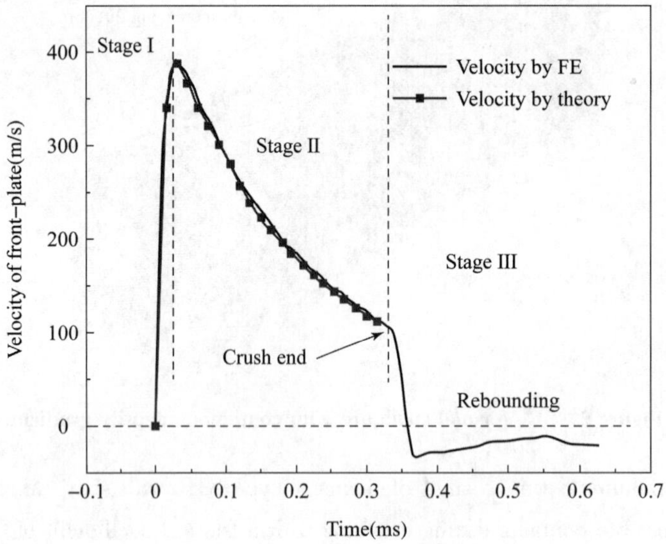

Figure 7　Velocity of front – plate under blast loading.

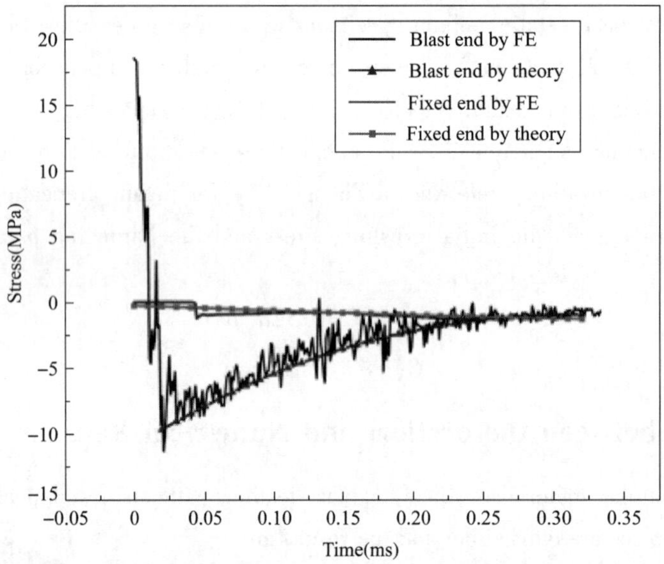

Figure 8　Stresses of the blast end and fixed end versus time of the cellular rod.

5　Results and Discussion

5.1　Deformation Mode

The macroscopic deformation of cellular materials is complicated under blast loading, and the related studies are limited, but it can be classified into three modes. In mode Ⅰ, the deformation initiates at the weakest cells and localized in crushing bands, which are randomly distributed at a low impact velocity. In mode Ⅱ, the crushing bands concentrate near the impact end because the inertia effect becomes crucial at a moderate impact velocity. In mode Ⅲ, the crushing bands highly localized at the impact end and progressive cell crushing is observed to propagate like a shock wave at a high impact velocity. Zheng cataloged the deformation into the Quasi – static mode, Transitional mode and Dynamic model. Figure 9 demonstrate the deformation process of cellular bar under initial blast peak pressure of 16. 9 MPa. As discussed in section 4, the motion of front – plate can be divided into three stages. In stage Ⅰ, the blast load drives the front - plate to the maximum speed in a very short time. According to Zheng et al. [15], the critical velocity for the occurrence of the impact - induced shock wave in cellular can be given as

$$v \geqslant v_{cr} = \sqrt{\frac{(\sigma_d - \sigma_0)\varepsilon_b}{\rho}} \tag{12}$$

Model Ⅲ occurs when the impact exceeds the critical velocity. Therefore, the cellular material always appears in model Ⅲ in stage Ⅱ. As the impact velocity decreases, the deformation model changes to model Ⅱ and finally to model Ⅰ.

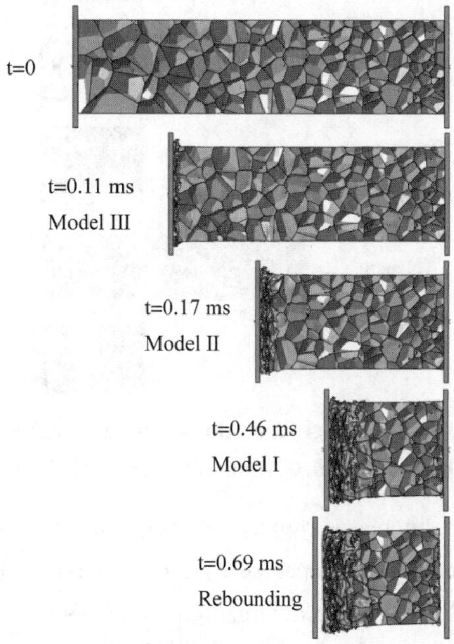

Figure 9 Deformation process of the graded cellular material under blast loading. (Initial peak pressure is 16.9 MPa and characteristic time is 0.05 ms)

As shown in Figure 10, difference deformation patterns are illustrated with initial blast peak pressure of 8.45 and 16.9 MPa. When the initial last peak is 8.45 MPa the cellular bars undergoes three deformation models and the deformation distributes randomly in the cellular bar. When the initial blast peak is 16.9 MPa, the crushing highly localized at the blast end and propagate to the fixed end like a shock wave. Only Model Ⅲ was observed, because the velocity of the front - plate exceeds the critical velocity though the deformation process.

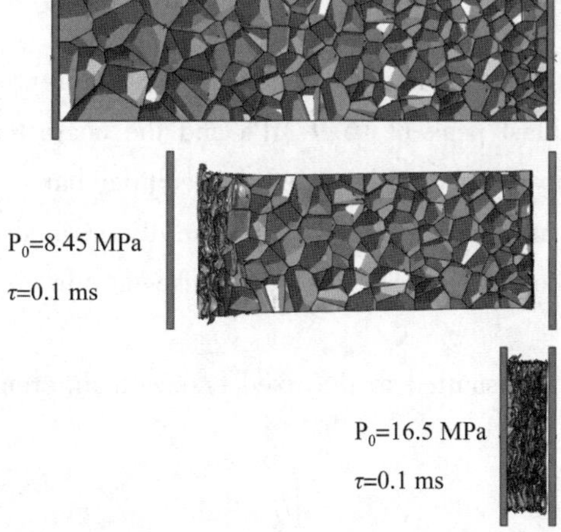

Figure 10 Deformation patterns of gradient cellular material under blast loadings with different initial peak pressure. (the gradient is +0.03)

5.2 Gradient Effects

The deformation patterns of cellular bars with different density gradients are presented here (Figure 11) to give an insight into the effects of gradient distribution on deformation mechanisms.

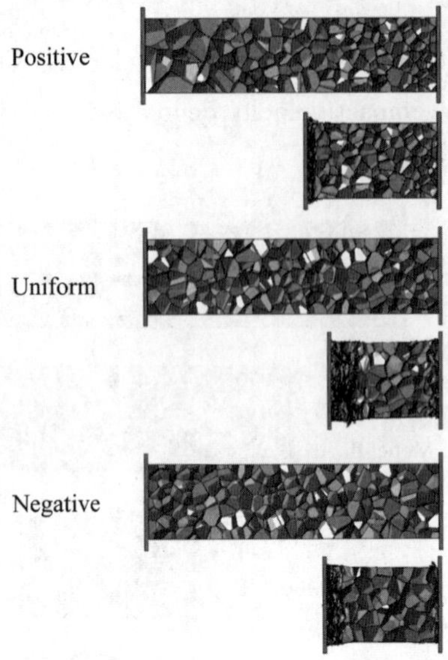

Figure 11 Deformation patterns of graded cellular rod with different density distributions (Initial peak pressure is 16.9 MPa and characteristic time is 0.1 ms)

Because the velocity of front plate increases rapidly in a very short period, the deformation process is similar to the phenomena in which cellular material is impacted by a rigid mass as presented in Ref [36]. The negative and the uniform density distribution have similar deformation mechanisms, the blast end crushes first, and the fixed end compacts slightly at a later period. When the gradient is positive, the compaction initiates from the blast end and propagates to the fixed end.

When evaluating the protective capability, the energy absorption capability and the impulse transferred to the fixed end are two important parameters. The cellular structure with high energy absorption ability and low impulse transmission is a good choice to mitigate shock and impact. In comparison with the plastic deformation energy of cellular material, the elastic deformation energy is negligible. Therefore, the energy absorption can be obtained by traversing the compaction wave from the coordinate to the maximum crushed distance as follows:

$$E = \int_0^{\Phi(t)} \frac{1}{2}[\sigma_b(\xi) + \sigma(\xi)]\varepsilon_b(\xi)d\xi \tag{13}$$

Where ξ is dummy variable for displacement. Cellular bars with different density gradient are fully compacted under the blast loading with initial blast peak of 16.9 MPa and the characteristic time $\tau = 0.1$ ms, Figure 12 depicts the energy absorption capacity. It should be noted that cellular bar is fully crushed when the shock front reaches the interface of cellular bar and the fixed end, and the fully crushed cellar bar along with the front plate will impact the fixed end as a rigid body. The present study focuses on the crushing process before the fully crush is reached.

Figure 13 shows the impulse I transmitted to the fixed end with different gradient distributions under blast loading, which is obtained by

$$I = \int f_{fixed} dt \tag{14}$$

where f_{fixed} is the force transmitted.

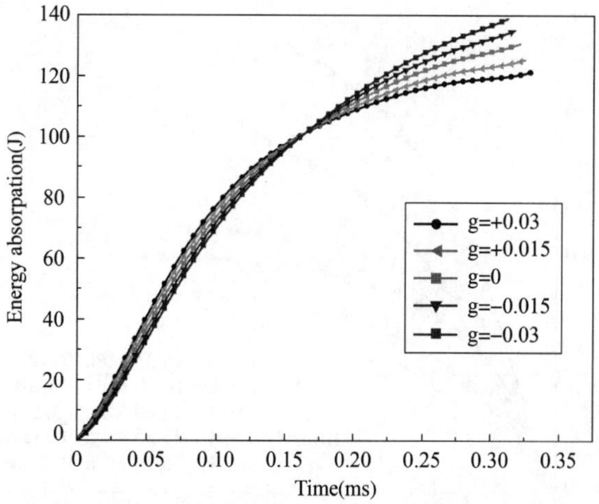

Figure 12　Energy absorption capacity of cellular rods with different gradients (Initial peak pressure is 16. 9 MPa and characteristic time is 0. 1 ms)

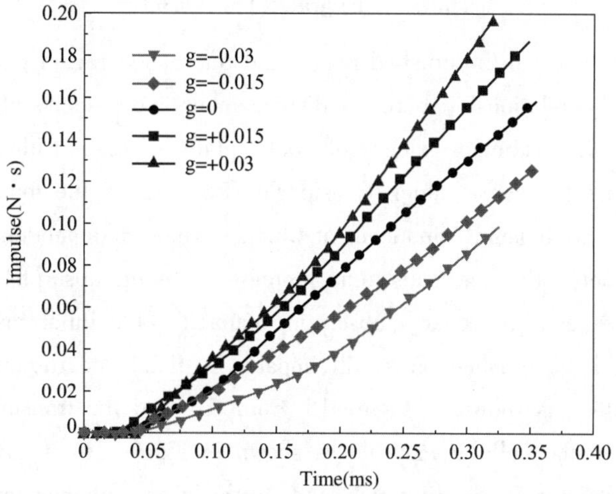

Figure 13　Transmitted impulse of cellular rods with different gradients under blast loading (Initial peak pressure is 16. 9 MPa and characteristic time is 0. 1 ms).

Combine Figure 12 and Figure 13, the cellular rods with positive density gradient shows the highest energy absorption and impulse transferred, whereas the cellular rods with the negative density gradient display the lowest energy absorption and impulse transferred. When the cellular materials are designed to mitigate shock and blast, the one with high energy absorption and low transferred impulse is the best choice. However, by improving density gradient variation, high energy absorption and low transferred impulse cannot be achieved at the same time, which are two conflicting objectives for blast resistance capacity of cellular materials.

5. 3　Blast Loading

The effect of blast loading on cellular materials are investigated. Two groups of blast loading with different initial peak pressure p_0 and characteristic time τ are introduced here, each group has the same impulse (group 1: $p_0\tau = 0.845$ MPa · ms, group2: $p_0\tau = 1.69$ MPa · ms). Figure 14 demonstrates the energy absorption history of cellular rods under different blast loading. The figure illustrates that, in each group higher initial peak pressure results in higher energy absorption rate and capacity. According to R – PH model the energy absorbed per unit volume at the shock front can be obtained by

$$\Delta U = \frac{1}{2}(\sigma_b + \sigma_0)\varepsilon_b = \frac{\sigma_0 v}{c + v} + \frac{1}{2}\rho_s \rho v^2 \tag{15}$$

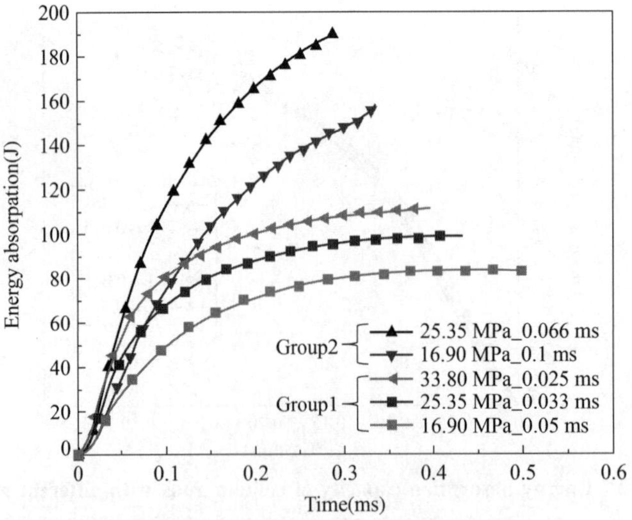

Figure 14 Energy absorption history of the cellular rods under different blast loadings. (the gradient is +0.03)

The difference in velocity between the crushed region and uncrushed region plays a dominant role in energy absorption. Figure 15 shows the relationship between displacement and velocity of front - plate under different blast loading. In each group, the maximum velocity of front - plate increases with increasing blast loading while the crushed displacements share the same value. Considering Eq. (13), the total energy absorption increases with the crushed distance and the dynamic enhancement which is velocity dependent. Therefore, the increase of initial peak pressure and characteristic time leads to an increase in energy absorption capacity. It should be note that, when the blast energy exceeds the energy absorption capacity of cellular material, full crushing will be reached. The front - plate and the crushed parts will impact the fixed end directly, which leads to a rapidly velocity decease of front - plate (as shown in Figure 15 group 2) and the transmitted pressure becomes much larger than the plateau stress of the cellular material as shown in Figure 16. Li and Meng also introduced this phenomenon and indicated that intensive loading may leads to the stress enhancement in cellular material [21]. Therefore, the energy absorption capacity increases with the blast loading, but when the full crush is reached the transmitted pressure will be enhanced.

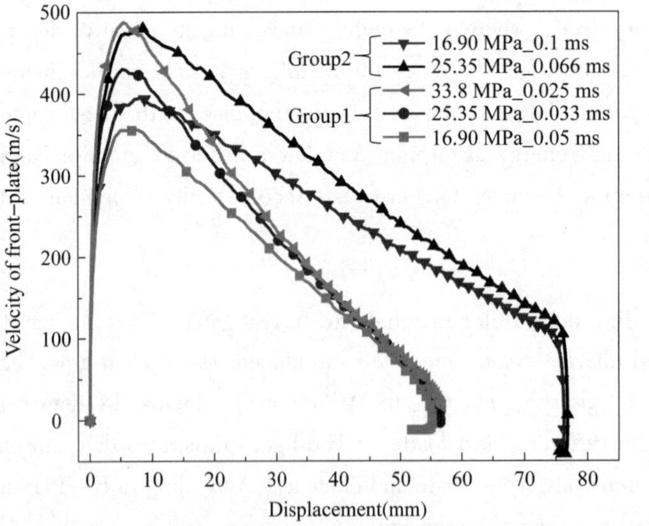

Figure 15 Relationship between displacement and velocity of the front plate under different blast loadings (the gradient is +0.03)

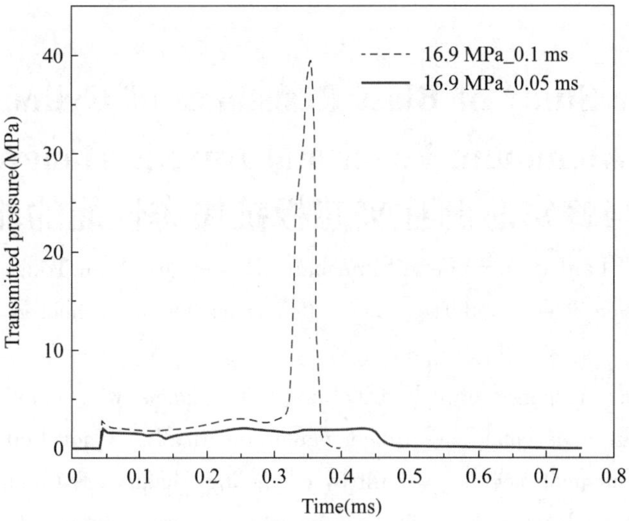

Figure 16　Transmitted pressure at the fixed end of cellular rod with different blast loadings. (the gradient is +0.03)

6　Conclusion

The blast responses of density - graded cellular materials are investigated theoretically and numerically. The theoretical model is developed based on the R - PH model. FE models with continues - density gradient are constructed through a novel method based on the 3D Voronoi technique. Numerical simulations are carried out by the ABAQUS/Explicit software and the theoretical predictions are verified by the numerical results. The cellular rod appears different deformation models as a result of different blast loading. When the initial blast peak pressure is high, the shock model propagates throughout the compression process and the cellular material is fully crushed. A partial crush will happen when the blast peak is not sufficient high. The blast response and energy absorption capacity of cellular rod with density gradient are investigated to clarify the effects of the density gradient distribution. The deformation first begins at the blast end and propagates to the fixed end, then the weakest part crushes subsequently. The deformation propagates through the weakest part and finally reaches the fixed end. When cellular material is used as a blast protective device, the ability to absorb energy while controlling the loading transmitted to the protected structure makes it attractive. However, a positive density gradient achieves the highest energy absorption and transmits a relatively high impulse to the protected structure, while a negative one shows a relatively low impulse transmission but attains the lowest energy absorption. Therefore, the introducing of density distribution cannot solve the contradiction between energy absorption and impulse transmitted to the protected structure. The effect of blast loadings with different initial peak pressure and characteristic time is also examined. The energy absorption capacity increase with the initial peak pressure and the characteristic time because of dynamic enhancement and crushed distance increased. When the blast loading exceeds the resistance capacity of cellular material, a fully crush will happen, and the transmitted stress will be enhanced which is a negative factor in engineering applications.

References

[1] ASHBY M F, EVANS A, FLECK N A, et al. Metal foams: a design guide [J]. Materials & design, 2002, 23 (1): 119.

A Comparative Study of Blast Resistance of Cylindrical Sandwich Panels with Aluminum Foam and Auxetic Honeycomb Cores①

泡沫铝与蜂窝芯圆柱夹芯板抗爆炸性能的比较研究

Lan Xuke②, Feng Shunshan, Huang Qi, Zhou Tong

State Key Laboratory of Explosion Science and Technology, Beijing Institute of Technology, Beijing 100081, P. R. China

Abstract: The dynamic response of cylindrical sandwich panels with aluminum foam core, hexagonal honeycomb core, and auxetic honeycomb core are compared numerically. A novel curved auxetic honeycomb core is designed, and the finite element models are built by employing Abaqus – Explicit. To calibrate the numerical models, the experiments of sandwich panels with honeycomb core and aluminum foam cores are modeled. And the numerical results have a good agreement with the experiment date. The calibrated numerical models are used to simulate the dynamic response of cylindrical panels subject to external blast loadings. It is found that the cylindrical panels with auxetic honeycomb cores have a better performance than that with aluminum foam cores and hexagonal honeycomb cores in resisting blast loadings. A material concentration effect was observed in the auxetic honeycomb core due to the negative Poisson's ratio (NPR) effect. According to parameter studies, it is concluded that with the increase of curvature and face sheet thickness the blast – resistance of panels with both auxetic honeycomb core, hexagonal honeycomb core, and foam cores increased obviously, especially the panels with auxetic honeycomb cores. For the panels with auxetic honeycomb cores, increasing the back face sheet thickness can improve the blast – resistance performance more efficiently than increasing the thickness of front face sheet, which is opposite for the panels with foam cores and hexagonal honeycomb cores. Auxetic cores with a smaller unit cell aspect ratio and a smaller unit cell length ratio has a larger Poisson's ratio, and achieves better blast resistance performance. These simulation findings can guide well the theoretical study and optimal design of cylindrical sandwich structures subject to external blast loading.

Key words: auxetic composite; blast resistance; finite – element simulation; negative Poisson's ratio; energy absorption.

1 Introduction

Cellular materials, such as honeycomb, foam, corrugated plate and metal hollow sphere, have been widely used in aerospace and defense industries as an energy absorption device to mitigate shock and impact by progressive local crushing of its micro – structure under dynamic loading [1]. Sandwich structures with metallic cellular fillers have shown a good energy absorption capacity while light in weight. Therefore, they have been employed as energy absorbers in a wide range of applications involving extreme loading conditions such as blast and impact [2 – 6].

In practice, the structures of many aerospace vehicles, marine vessels and civil infrastructures are curved [7]. The curved panels are much stronger and stiffer than other structural forms and they generally have better performance under various loadings because they can support the external loads effectively by virtue of their spatial

① 原文发表于 *Aerospace science and technology* 2019。

② 兰旭柯：工学博士，2016 年师从冯顺山教授，研究爆炸抛撒子弹药抗冲击与防护方法，现工作单位：北京理工大学博士后。

curvature [8]. Many researchers combine the advantages of curved panel with sandwich structures and cellular materials, and the curved sandwich panels with cellular material core have shown better blast and impact resistance than conventional flat panels [9 – 11].

Different from the traditional cellular materials used in the above studies, auxetic materials characterized by negative Poisson's ratio (NPR), present a counterintuitive behaviour as they contract laterally under compression and expand when stretched [12]. With such mechanical behaviors, they show enhanced performances in fracture toughness, indentation resistance, shear modulus and vibration absorption [13 – 15]. Therefore, in recent years, auxetic materials and structures have been investigated for protective purpose. Imbalzano et al [16]. compared the blast resistance of auxetic and honeycomb sandwich panels. Qi et al [17]. studied the impact and close – in blast response of auxetic honeycomb – cored sandwich panels experimentally and numerically. Yang et al [18]. presented a comparative study of ballistic resistance of sandwich panels with aluminum foam and auxetic honeycomb cores. In their study, auxetic panels demonstrate interesting crushing behaviour, effectively adapting to the dynamic loading by progressively drawing material into the locally loaded zone to thereby enhance the impact resistance. In addition, in their study, the back – face sheets of the sandwich structure have a high strength, which can support the auxetic cores to deform and shrink sufficiently and dissipate the imparted energy more efficiently. Similarly, the curved back face sheets may also have the same effect because they can support the external loads effectively by virtue of their spatial curvature.

However, few studies about blast response of auxetic cylindrical panels can be found. Recently, Duc et al [19]. developed an analytical approach to study nonlinear dynamic response and vibration of sandwich composite cylindrical panels with auxetic honeycomb core layer. Duc et al. [20] also analytically investigated the dynamic response of curved shells with auxetic honeycomb core subjected to blast loads. To further reveal the blast response of curved auxetic panels especially the deformation model and energy absorption capacity, more investigation should be carried out.

In order to combine the advantages of the curved sandwich structures with the advantages of the auxetic structures. In this paper, a novel auxetic cylindrical sandwich structure was designed. To reveal the role of the negative Poison's ratio characteristics in the blast response of sandwich structure. Finite element models of sandwich composite cylindrical panels with auxetic honeycomb core and other two representative type cores without negative Poison's ratio characteristics: metal foam core and hexagonal honeycomb core were established. The deformation models and energy absorption capacity were observed and compared through numerical simulation. The effect of curvature, face sheet thickness, and unit cell parameters were studied.

2 Problem Description

2.1 Geometry Description

The baseline dimensions of cylindrical sandwich panel considered in this work is shown in Figure 1. The panel consists of two cylindrical metallic face sheets and a cellular core. The panel has the following geometries: radius of curvature R, side length $L = 320$ mm, arc angle θ, inner and outer face thickness T_i, T_o.

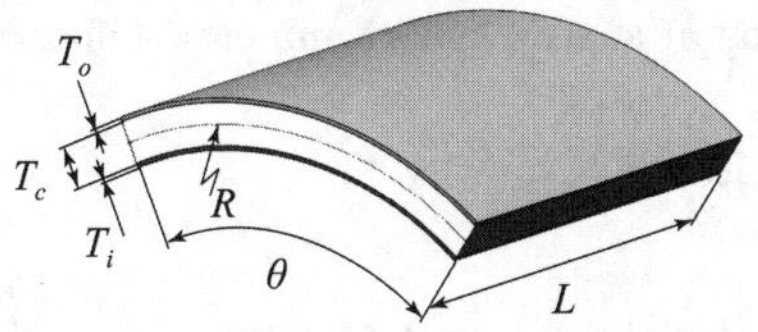

Figure 1 Baseline dimensions of cylindrical sandwich panel

Due to the symmetry of the sturctures, only quarter section models of the auxetic and foam curved panels with loading and boundary conditions are shown schematically in Figure 2. The panel was assumed to be fully clamped peripherally and subjected to a surface detonated blast load of 25 g TNT equivalent at a sand off distance (S) of 100 mm from the outer face along the centerline of the panel.

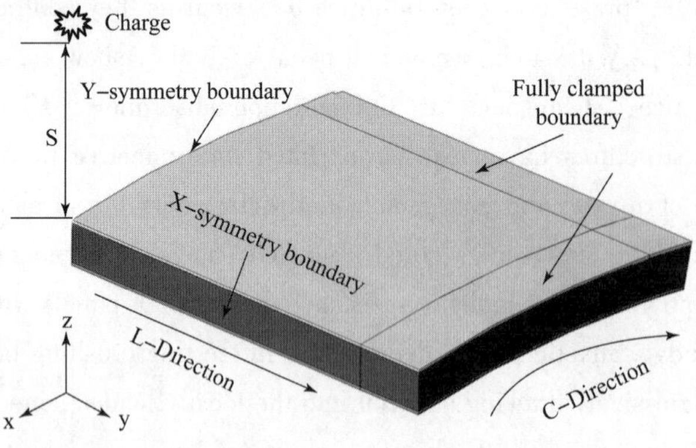

Figure 2 Loading and boundary conditions and cell description for 1/4 model of sandwich panels.

2.2 Curved Auxetic and Hexagonal Honeycomb Core Design

Three different cores: foam core, auxetic honeycomb core and hexagonal honeycomb core were designed. Unit cell of auxetic honeycomb as illustrated in Figure 3 (a) is characterised by four independent geometrical parameters: arch length L_1, arch length L_2, top and bottom side length H ($H = 2h$), and wall thickness t. These parameters can be adjusted by θ_1, θ_2 and R_1, R_2, H. In this study H is set to be 3 mm, the aspect ratio is $R_A = L_1/H = 1$, the length ratio is $R_L = L_2/L_1 = 1/3$. In this study the relative density of core layer is set to be 0.148. Similar to auxetic honeycomb unit cell, hexagonal honeycomb unit cell can be controlled by θ_1, θ_2 and R_1, R_2, R_3. The hexagonal honeycomob core is designed to have the same cell numbers with hexagonal honeycomb core along C-Directation and L-Directation. To ensure that the foam core, the auxetic honeycomb core and hexagonal honeycomb core have the same relative density, the shell thickness of auxetic honeycomb core varies from 0.126 mm to 0.128 mm and the shell thickness of auxetic honeycomb core varies from 0.183 mm to 0.186 mm. The unit cell is illustrated in increased curvature, the sides of actual unit cell is closed to a stright line. To construct the core structure, the unit are rotating arrayed in the C-Directation and then stretched in the L-Directation.

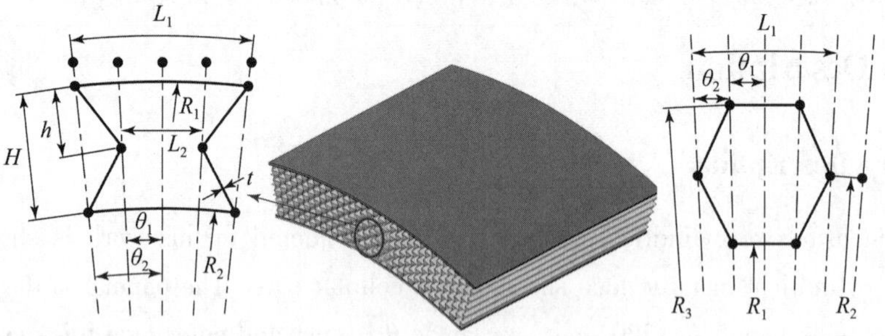

Figure 3 Unit cell description of auxetic honeycomb core (illustrated in increased curvature).

3 Finite Element Modeling

3.1 Geometry, Boundary Conditions, and Contact Modeling

All numerical simulations were carried out with the FE code ABAQUS/Explicit. Only a quarter of the panel was modeled to shorten the simulation time due to the symmetric nature of the problem. Corresponding constrains were defined on the two symmetric planes, while the other edges were fully clamped, as shown in Figure 2. The face sheets, auxetic honeycomb core and hexagonal honeycomb core of the sandwich panel were meshed by using ABAQUS shell element S4R (the 4 – node doubly curved general – purpose, reduced integration, hourglass controlled, finite membrane strain shell element). The element size was set to be 0. 5 mm. The foam core was meshed by solid element C3D8R and the element size was set to be 1 mm. Mesh sensitivity studies revealed that further refinement does not significantly improve the accuracy of the calculations but sacrifice longer of computational time. The face sheets were assumed to be perfectly bounded with the top and bottom surfaces of the core layer. And the normal contact of shell elements was defined using hard contact formulation, while the tangential behaviour was described with a penalty one with a friction coefficient of 0. 3.

3. 2 Material Properties and Modeling

3. 2. 1 Material Models of Face Sheets and Auxetic Honeycomb Core

In the research of Zhu [4], the face sheet was made from Al – 2024 aluminum alloy, while the honeycomb core was made from Al – 3104 aluminum. To validate the FE model of panel with auxetic honeycomb core in section 3. 4, the same materials were also used in the FE simulation. Aluminum materials is modeled as bilinear strain – hardening constitutive model without strain rate effect, and its mechanical properties are listed in Table 1.

Table 1 Material parameters [4]

Material	Density (Kg/m^3)	Young's modulus (Gpa)	Yield stress (Mpa)	Poisson's ratio	Tangent modulus (Gpa)
Al – 2024	2 680	72	75.8	0. 33	0. 737
Al – 3104 – H19	2 720	69	262	0. 34	0. 69

3. 2. 2 Material Model of Aluminum Foam

The plastic performance of the aluminum foam core is molded by using * CRUSHABLE FOAM and CRUSHABLE FOAM HARDENING options in the ABAQUS package. Deshpande and Fleck [21] model was used to describe the plastic crushable behaviour of the aluminum foam core in Abaqus. The yield criterion is defined as

$$\Phi = \hat{\sigma} - \sigma_y \leqslant 0 \tag{1}$$

where σ_y is the yield stress and equivalent stress $\hat{\sigma}$ is given as:

$$\hat{\sigma}^2 = \frac{1}{[1 + (\alpha/3)^2]}[\sigma_e^{\ 2} + \alpha^2 \sigma_m^2] \tag{2}$$

where σ_e is the von Mises effective stress and σ_m is the mean stress. Parameter α that controls the shape of the yield surface is a function of the plastic Poisson's ratio ν_p given as:

$$\alpha^2 = \frac{9(1 - 2\nu_p)}{2(1 + \nu_p)} \tag{3}$$

For the aluminum foam ν_p is equal to zero and $\alpha = \sqrt{9/2}$.

The yield stress can be expressed as:

$$\sigma_y = \sigma_p + \gamma \frac{\hat{\varepsilon}}{\varepsilon_D} + \alpha_2 \ln\left[\frac{1}{1 - (\hat{\varepsilon}/\varepsilon_D)^{\beta}}\right] \tag{4}$$

where $\hat{\varepsilon}$ is equivalent strain, σ_p, E_P, α_2, γ, β and ε_D are material parameters and can be related to the foam density as

$$\begin{cases} \left(\sigma_p, \alpha_2, \gamma, \dfrac{1}{\beta}, E_P\right) = C_0 + C_1 \left(\dfrac{\rho_f}{\rho_{f0}}\right)^k \\ \varepsilon_D = -\ln\left(\dfrac{\rho_f}{\rho_{f0}}\right) \end{cases} \tag{5}$$

where ρ_f is the foam density and ρ_{f0} is the density of the base material. C_0, C_1 and k are the constants as listed in Table 2.

Table 2 Material constants for aluminum foam [11, 22]

	σ_p (Mpa)	α (Mpa)	$1/\beta$	γ (Mpa)	E_p (Mpa)
C_0 (Mpa)	0	0	0. 22	0	0
C_1 (Mpa)	590	140	320	42	0. 33E6
k	2. 21	0. 45	4. 66	1. 42	2. 45

3. 3 Blast Load Modeling

In this study, blast loading was generated by The COMWEP (conventional weapons effects program) empirical model. The CONWEP algorithms developed by Kingery and Bulmash [23] account for angle of incidence by combing the reflected pressure (normal – incidence) value and the incident pressure (side – on – incidence) value. The pressure can be calculated by the following equation:

$$P_{load} = P_{reflected} \cdot \cos^2\theta + P_{incident} \cdot (1 + \cos^2\theta - 2\cos\theta) \tag{6}$$

Where θ is the angle of incidence; $P_{incident}$ is the incident pressure; $P_{reflected}$ is the reflected pressure. CONWEP model available in Abaqus/Explicit has two parameters: the equivalent mass of TNT and the standoff distance. The experiment used cylinder TNT, while the shape of charge used in CONWEP model is spherical. In order to get the equivalent mass of spherical TNT charge for CONWEP, Li [10] developed a numerical method. The blast loading pressure applied to the structure by cylindrical TNT charge detonated at a standoff distance was simulated by using commercial software Autodyn [24]. In order to make the peak pressure of the spherical TNT charge exerted on the front face sheet same as that of the cylindrical TNT charge, the equivalent mass of spherical TNT charge is calculated by empirical Eqs. (7) [25].

$$\Delta p = \frac{6.1938}{Z} + \frac{0.3262}{Z^2}\frac{2.1324}{Z^3} \tag{7}$$

Where Δp is the peak overpressure at shock wave front, (kp/cm^2); Z is the scaled distance, (m/kg$^{1/3}$) and $0.3 \leqslant Z \leqslant 1$. In addition, Dharmasena [26] has experimentally and numerically proved that the center detonated cylindrical charge with length to diameter aspect ratios close to 1 (the dimensions of the charge used in the experiment are as follows: height = 15 mm and diameter = 15 mm), the CONWEP model provides a reasonable estimate of the blast wave profile. Following these methods, the validation of FE model can be performed in next section. The typical pressure – time history of CONWEP loading of 25 g TNT equivalent at a sandoff distance (S) of 100 mm from the outer face along the centerline of the panel is shown in Figure 4.

3. 4 Validation of FE Model

3. 4. 1 Validation of Panel with Honeycomb Core

As no experiment about the dynamic response of this novel curved auxetic honeycomb been conducted in the existing literatures, the numerical approach was validated by a series of experiments about the response of square honeycomb panels under blast loading [4]. In this study the specimen ACG – 1/4 – TK – 5 was chosen and

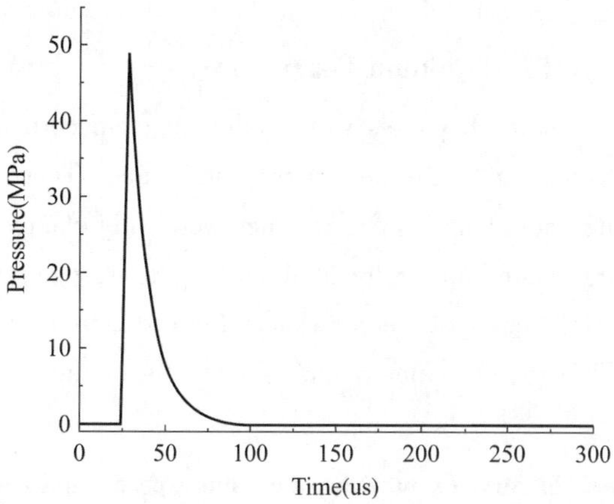

Figure 4 The typical CONWEP pressure time history of the 25 g spherical TNT charge detonated at the standoff distance 100 mm.

simulated in Abaqus/Explicit by using the same boundary conditions, materials and blast loading in section 3. 1, section 3. 2 and section 3. 3.

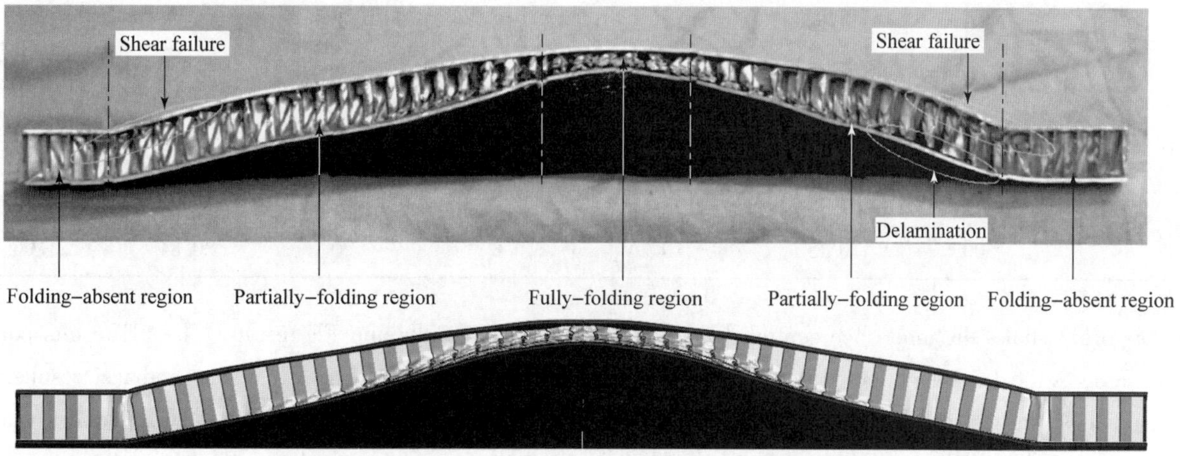

Figure 5 A comparison of the deformation and failure patterns of the honeycomb sandwich structure of experiment results and numerical prediction.

The results are shown in Figure 5, according to the deformation of the core layer, the specimen can be divided into three regions: fully - folding region, partly folding region and folding - absent region. It is obvious that the deformation/failure patterns of the honeycomb sandwich structures of the numerical predictions and experimental results are nearly the same. Furthermore, a comparison between the experimentally and numerically predicted back plate deflections was also made, as shown in Figure 6. It is clear that, the numerical predictions are close to the experimental results and the error is within an acceptable range. These satisfactory correlations indicate that, the cylindrical auxetic honeycomb model which used the same material models, boundary conditions, meshing, contact method and blast load with the calibrated

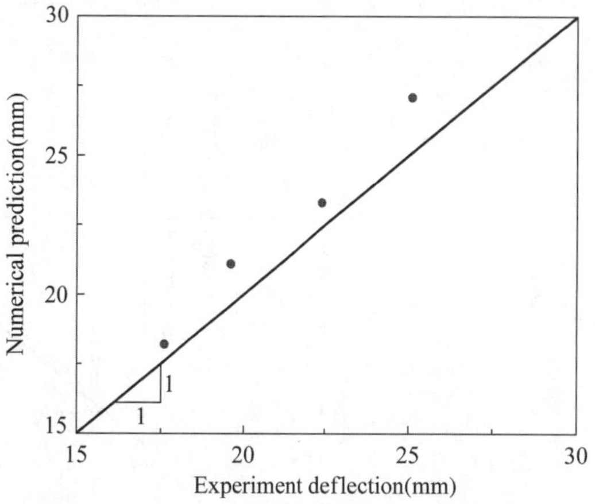

Figure 6 Comparison between the experimental and numerical predicted maximum central deflection of honeycomb panel.

flat hexagonal honeycomb panel can be trusted.

3.4.2 Validation of Panel with Aluminum Foam Core

The blast responses of curved sandwich panels with aluminum foam core tested by Jing et al. [11] were simulated and the results were compared with the experimental date. These panels consist of two identical aluminum face-sheets and an aluminum foam core, and they were fully clamped with an exposed area of $s \times l = 250 \times 250$ mm (s and l are the arc length and the longitudinal length, respectively.). Blast loading was applied to the specimens by detonating a TNT charge at a constant stand of distance (S) of 100 mm. The specifications of the investigated panels and the FE and the experimental test results of inner face central deflections are listed in Table 3.

Table 3 Specifications of curved sandwich panels and experimental and numerical results.

Number of specimens	Radius (mm)	Foam relative density	Core thickness (mm)	Face-sheet thickness (mm)	Mass of TNT charge (g)	Inner face central deflection (mm)	
						FE	Test
1	250	0.15	10	0.8	20	18.03	17.55
2	250	0.15	10	0.8	30	29.76	31.99
3	500	0.15	10	1.0	20	12.14	12.53
4	250	0.15	10	0.5	20	26.22	24.90
5	250	0.15	10	1.0	20	16.21	14.22
6	500	0.15	10	0.8	20	20.84	21.06

Figure 7 shows the inner face central deflection of experiment results and FE results. The FE results points gather around the line of perfect match, indicating good agreement between the FE results and test results. A comparison of the tested and the simulated deformation of specimen 1 in Table 3 is illustrated in Figure 8. A good agreement is achieved between the simulated deformation and the experimental observation, and the numerical methods are valid.

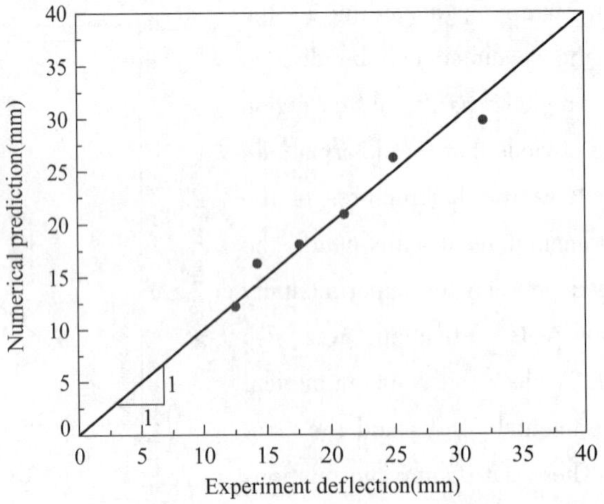

Figure 7 Comparison between the experimental and numerical predicted maximum central deflection of foam panel.

Figure 8　A comparison of the deformation and failure patterns of the honeycomb sandwich structure of experiment results and numerical prediction.

4　Results and Discussion

4.1　Deformation Mode Comparison

The numerical results are listed in Table 4, specimen number R6_Ti1_To1_F stands for the foam sandwich panel with radius of curvature of 600 mm, back face sheet (inner face) thickness $T_i = 1$ mm and front face sheet (outer face) thickness $T_o = 1$ mm. While R6_Ti1_To1_A and R6_Ti1_To1_H are the specimen with the same above parameters but has an auxetic honeycomb core and hexagonal honeycomb core respectively. And the To015

Table 4　Numerical results of maximum central deflection

	Specimen number	R (mm)	Ti (mm)	To (mm)	Maximum deflection of AU panel	Maximum deflection of HEX panel	Maximum deflection of foam panel
	R8_Ti1_To1_A/H/F	800	1	1	-15.43	-21.41	-26.19
	R7_Ti1_To1_A/H/F	700	1	1	-14.71	-19.13	-24.57
Group 1	R6_Ti1_To1_A/H/F	600	1	1	-13.67	-17.55	-22.80
	R5_Ti1_To1_A/H/F	500	1	1	-10.83	-14.09	-19.26
	R4_Ti1_To1_A/H/F	400	1	1	-7.95	-12.68	-17.78
	R6_Ti1_To1_A/H/F	600	1	1	-13.67	-17.55	-22.80
	R6_Ti1_To015_A/H/F	600	1	1.5	-11.64	-15.30	-20.52
Group 2	R6_Ti1_To2_A/H/F	600	1	2	-9.61	-12.74	-18.75
	R6_Ti1_To025_A/H/F	600	1	2.5	-8.11	-10.51	-16.93
	R6_Ti1_To3_A/H/F	600	1	3	-6.91	-8.08	-15.41
	R6_Ti1_To1_A/H/F	600	1	1	-13.67	-17.55	-22.80
	R6_Ti015_To1_A/H/F	600	1.5	1	-11.2	-16.29	-21.67
Group 3	R6_Ti2_To1_A/H/F	600	2	1	-8.56	-14.73	-20.46
	R6_Ti025_To1_A/H/F	600	2.5	1	-6.73	-13.15	-19.4
	R6_Ti3_To1_A/H/F	600	3	1	-4.45	-11.04	-18.2

stands for front face sheet thickness = 1. 5 mm. The deformation modes of the three sandwich panels are illustrated in Figure 9. The deflection histories at the central point of the back face sheets are shown in Figure 10. The deformation process can be roughly divided into three stages: loading stage, compression stage and rebounding stage. At loading stage, the blast load accelerates the front face sheet in a very short time, and the core layer shows almost no compression deflection. At compression stage, the front face sheet gradually compresses the core layer until the central point deformation of the back face sheet is maximized. At rebounding stage, the sandwich panels rebound and continues to vibrates for a period of time.

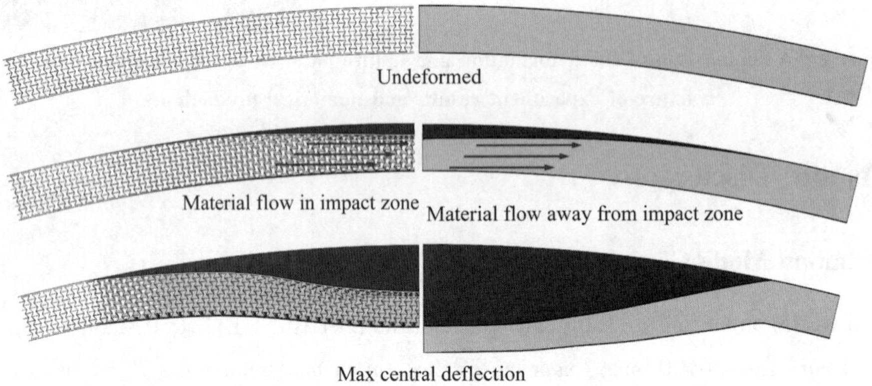

(a)

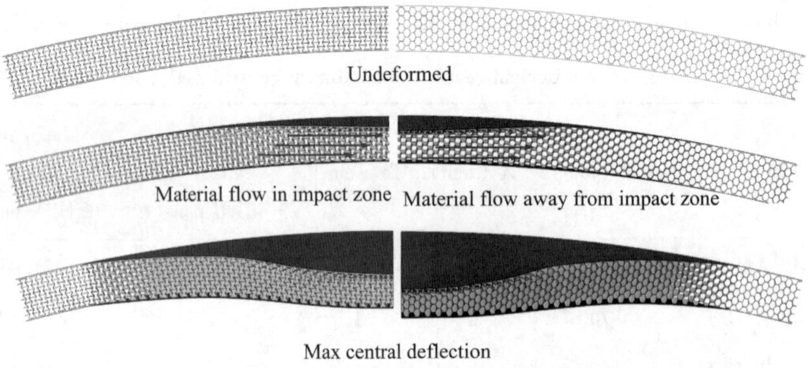

(b)

Figure 9 Comparison of deformation modes of panels with auxetic core, foam (a) core and honeycomb core (b) (specimen number R6_Ti1_To1_A and R6_Ti1_To1_F and R6_Ti1_To1_H) subjected to a surface detonated blast load of 25 g TNT equivalent at a sandoff distance (S) of 100 mm from the outer face along the centerline of the panel.

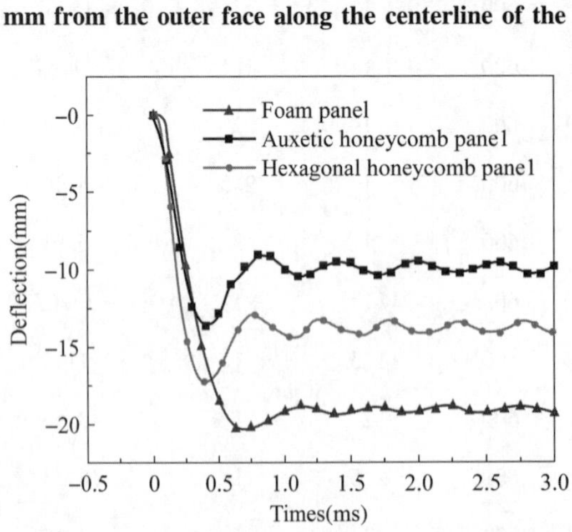

Figure 10 Deflection history at the central point of the back face sheet (specimen number R6_Ti1_To1_F, R6_Ti1_To1_A, and R6_Ti1_To1_H).

The global deformation of both the three panels shows a dome shape, starting out from the centeral. The main difference of the three panels is the deformation mode of the core. Auxetic honeycomb core shows a material concentration at compression stage. The core material displacement vectors are shown in Figure 11. When the face sheet compresses the auxetic honeycomb core, material flows in the impact region due to the negative Poisson's ratio (NPR) effect, while the material of hexagonal and foam core flows away from the impact region. The compression deformation of the foam core is concentrated in the central area, while the material concentration effect allows the core layer to have a larger deformation range, thereby distributing the blast energy more evenly.

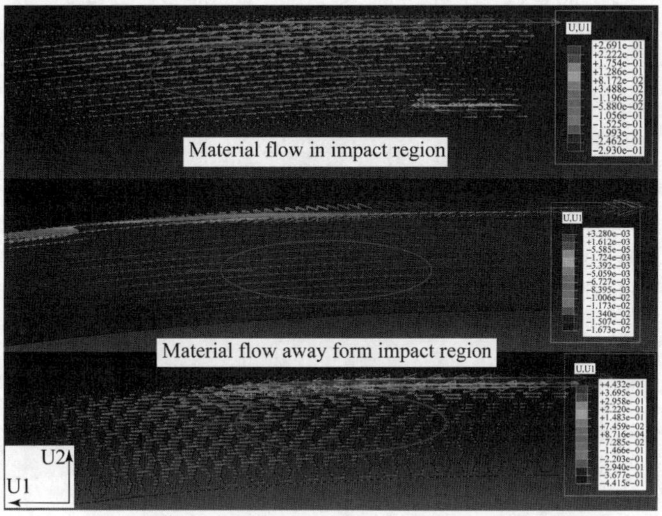

Figure 11 Comparison of displacement vectors of two sandwich core materials at compression stage.

4.2 Effect of Curvature of the Sandwich Panels

There are five different values of curvature in the present study, the effect of curvature of the sandwich panels on blast resistance is analyzed. Two major effects of curvature are identified. Firstly, it changed the reflective angle of the reflected blast wave, and result in a reduction of the impulse acting on the front face of sandwich panels which is experimentally verified [9].

Secondly, previous studies about foam panels concluded that the existence of curvature changes the deformation regimes in terms of bending and stretching regime, and the curved foam panels with an extended range for bending dominant deformation may outperform their corresponding curved solid over a wider range under blast loading. As shown in Figure 12, the increase of curvature result in a decrease of maximum center deflection. The auxetic honeycomb panels are more sensitive to changes in curvature. For auxetic honeycomb panels, the existence of curvature not only affects the deformation regimes, but also affects the material concentration effect. Figure 13 compared the deformation pattern of auxetic and foam panels with different curvatures. With the increase of curvature, the deformation pattern of both the foam panel and auxetic panel changes from bending domain to center compression domain. As the increasing curvature, the bending strength of the panel increases, the back face sheet and the part of the core layer adjacent to the back face sheet can support the core layer to compress. As a result, the sample R4_Ti1_To1_A shows a material concentration effect. As illustrated in Figure 14, the energy absorption capacity of foam panel decreases with the increase of curvature, but the auxetic panel shows less energy absorption reduction.

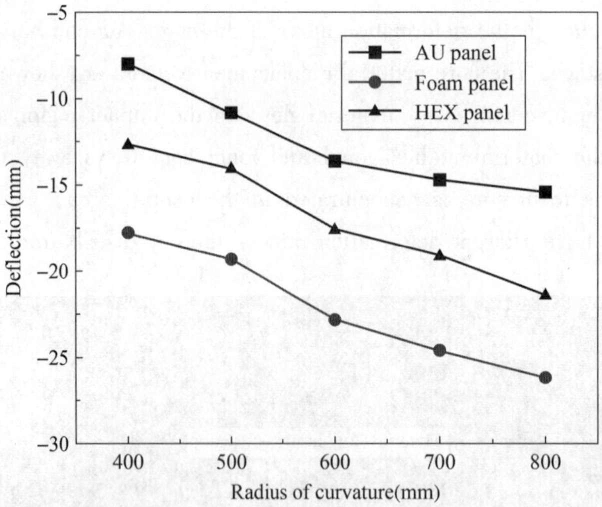

Figure 12 Effect of curvature on the maximum center deflection of panels with auxetic honeycomb core, hexagonal honeycomb core, and foam core.

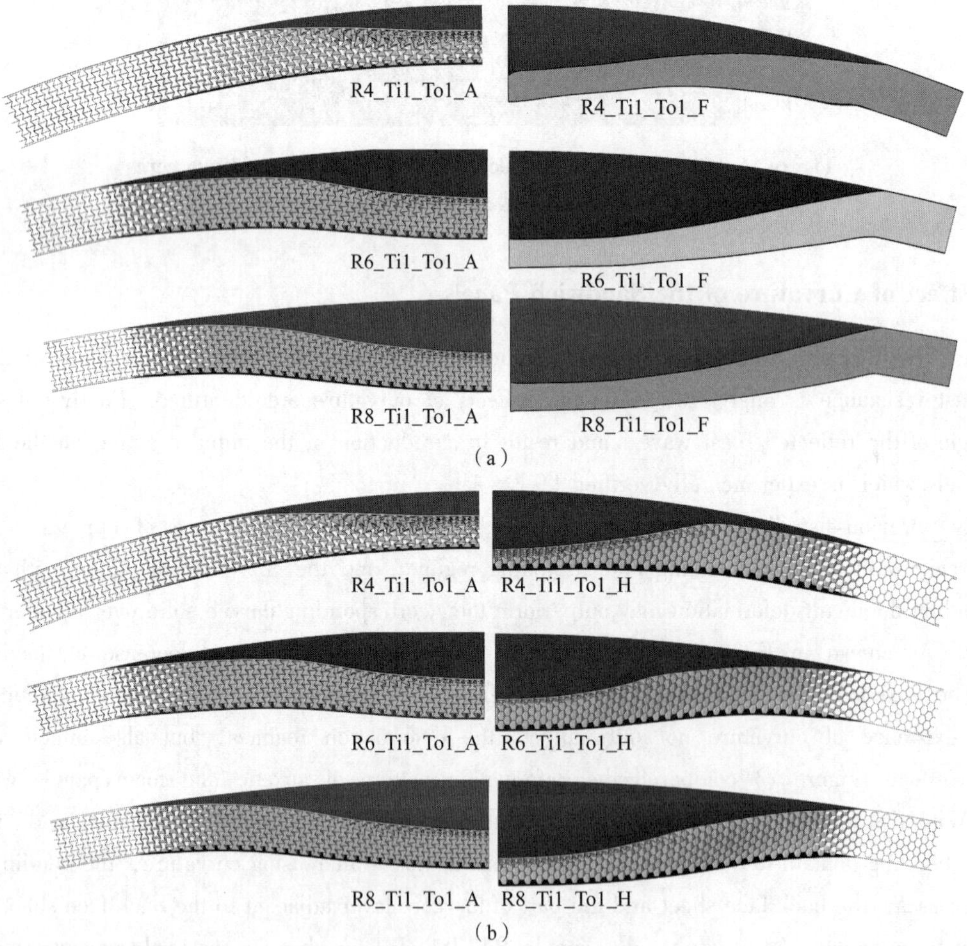

Figure 13 Comparison of effect of curvature on the deformation patterns on the maximum center deflection of panels with auxetic honeycomb core, hexagonal honeycomb core, and foam core.

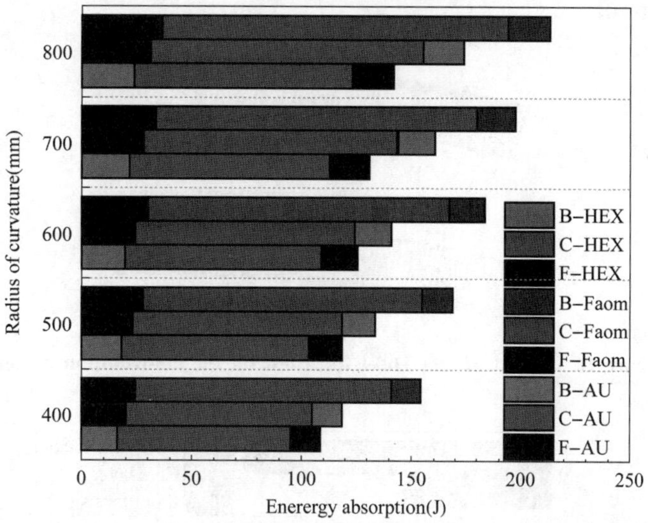

Figure 14 Comparison of the effect of curvature on the energy absorption of auxetic honeycomb, hexagonal and foam panels (B, C, F represent the back face sheet, the core, and the front face sheet, respectively).

4.3 Effect of Face Sheet Thickness

Five different values of the face sheet thickness are considered, the results are listed in Table 4 and the maximum center point deflection are compared in Figure 15. It can be observed that the central deflection of the both the curved sandwich panels decreased dramatically with the increase of the thickness of face sheet. The increase in the thickness of the front and back face sheet has a difference effects on the three sandwich structures. Foam panels and hexagonal honeycomb panels are more sensitive to the increasing of front face sheet thickness, while auxetic honeycomb panels are more sensitive to the increasing of back face sheet thickness. The ticker front face sheet can distribute the blast load more evenly and the compression deformation range of the core layer is enlarged, which is advantageous for the aluminum foam core to absorb energy more efficiently while decrease the maximum center deflection. A thicker back face sheet can provide support for the deformation of the auxetic honeycomb core during the deformation process, allowing more material to flow sufficiently towards the compression center to deform and absorb energy, as illustrated in Figure 16. The energy absorption ratio of each part of the sandwich structure can also reflect this phenomenon, as shown in Figure 17.

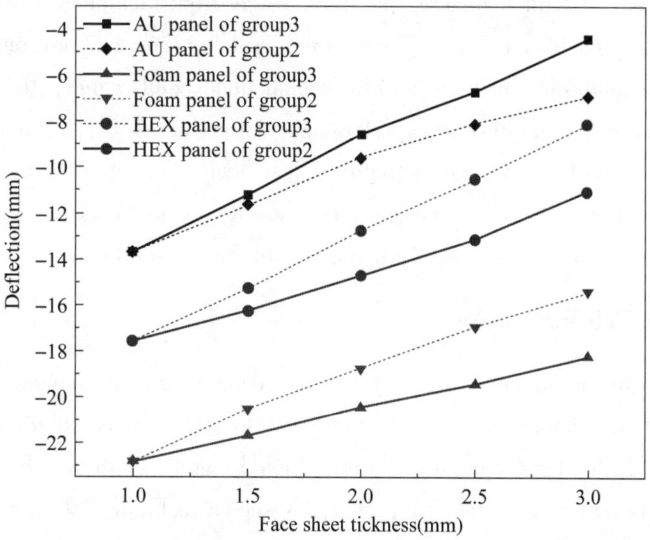

Figure 15 Comparison of the effect of face sheet thickness on the maximum center deflection.

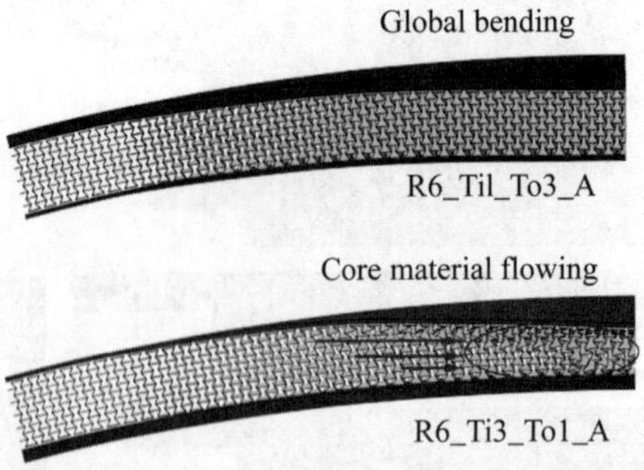

Figure 16 Effect of face sheet thickness on the deformation pattern.

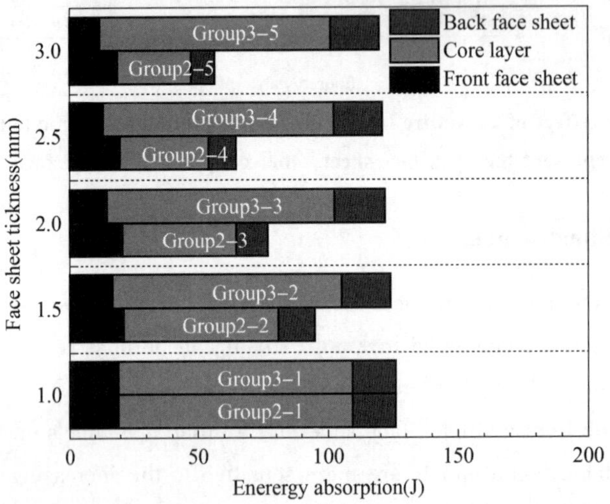

Figure 17 Comparison of energy absorption of auxetic honeycomb panel with different face sheet thickness

As the face sheet thickness increase, the maximum center deformation decrease, and the energy absorbed by global bending decrease, so the energy absorbed by group 2 (increased front face sheet thickness) is greatly reduced. However, due to the material concentration effect, the energy absorbed by group 3 (increased back face sheet thickness) mainly comes from the core compression deformation and lateral flow deformation of the core layer, rather than the global bending deformation, so the absorbed energy is almost constant.

As illustrated in Figure 18, with increasing the thickness of front face sheet or back face sheet, the energy absorbed by front face sheet, back face sheet, foam core, and hexagonal honeycomb core decrease gradually. Combing with Figure 16 for analysis, for foam and hexagonal honeycomb panel, the conclusion is obtained that higher thickness of face sheet can improve the performance on reducing the maximum central deflection while losing the ability to absorb. As a blast resistant structure, it is designed to reduce the maximum deformation while absorbing more energy. However, these two targets are contradictory in the sandwich with metal foam core and hexagonal honeycomb core but can be realized by using auxetic honeycomb core.

4.4 Effect of Unit Cell Parameters

To further analysis of the influence of the auxetic unit geometry on the dynamic response. The aspect ratio varies from 0.5 to 2, and the length ratio varies from 0.3 to 0.7. To calculate the effective Poisson's ratio (EPR) which is the ratio of transverse engering strain and axial enginging strain, a unixal load is applied on the top of the structure while its bottom is simply supported, as shown in Figure 19 (a). The stresses, strain, and EPRs are then obtained and shown in Figure 19 (b).

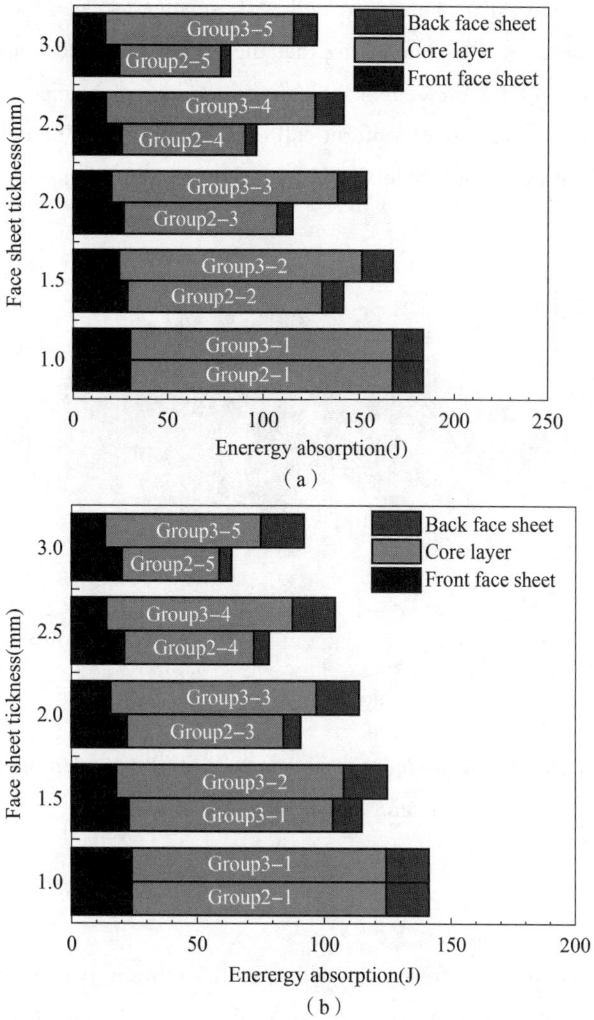

Figure 18　Comparison of energy absorption of foam panels (a), and hexagonal honeycomb panels (b) with different face sheet thickness.

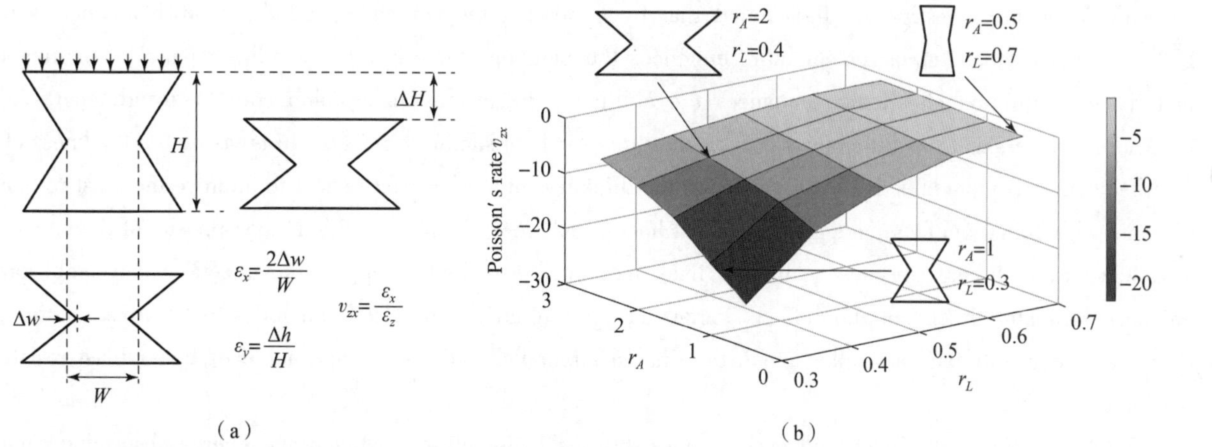

Figure 19　Calculation procedure for effective Poisson's ratio (EPR) of the auxetic unit cell (a) and the dependence of EPR on aspect ratio and length ratio of unit cell (b)

Figure 20 shows the maximum central deflection of sample R6_Ti1_To1_A with different auxetic core unit cell and same relative density subjected to a surface - detonated blast load of 25 g TNT equivalent at a sandoff distance (S) of 100 mm. It is obviously that, unit cell with smaller aspect ratio and length ratio shows better performance in resisting blast loading. It is interesting that the aspect ratio and length ratio have the similar effect on effective ratio and maximum central deflection. It can be explained geometrically, the core with a smaller aspect ratio has more cell numbers and a more uniform cell distribution, and the deformation of the auxetic core is more continuous, therefore more cells contribute to the material concentration effect.

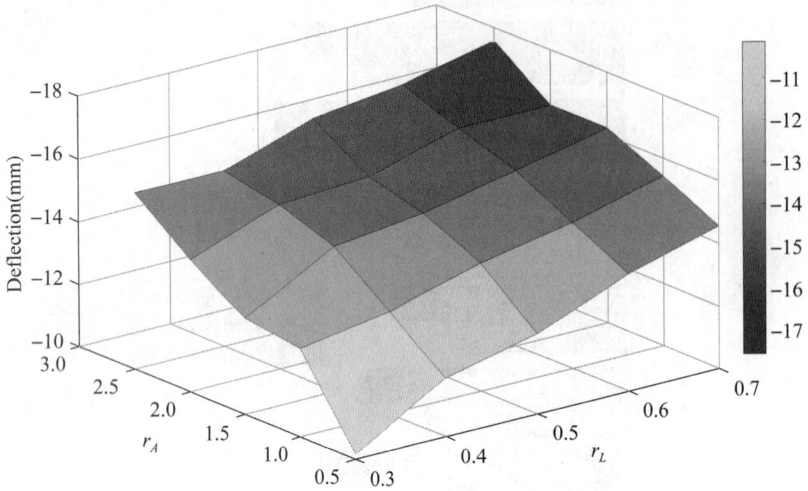

Figure 20 The dependence of maximum central deflection on aspect ratio and length ratio of unit cell.

5 Conclusions

This study presents the dynamic responses of cylindrical sandwich panels with auxetic honeycomb core, hexagonal honeycomb core and aluminum foam core under the external blast loading by numerical method. A novel cylindrical auxetic honeycomb core is designed, and the deformation mode is compared with the traditional aluminum foam core. A material concentration effect is observed, which is caused by the negative Poisson's ratio characteristics of the auxetic core and enhanced by the existence of curvature. The effects of curvature and face sheet thickness are investigated. It is found that by increasing the curvature, all the sandwich panels show a better blast resistance. A larger curvature enhances the bending stiffness of the sandwich panel, resulting in a material concentration effects that enhances the ability of the auxetic honeycomb core to absorb energy. The thickness of face sheet has different effects on the three kinds of sandwich panels. In creasing the thickness of the front face sheet is more effective than increasing the thickness of the back face sheet to improve the blast resistance of panel with aluminum foam core and hexagonal honeycomb core. However, this is the opposite of the panel with auxetic honeycomb core, because a thicker back face sheet can better support the auxetic honeycomb core to enhance the material concentration effect. Parameter study of auxetic unit cell concludes that, auxetic cores with a smaller aspect ratio and a smaller length ratio has a larger Poisson's ratio, and achieves better blast resistance performance.

Sandwich panel with auxetic honeycomb core shows a better blast resistance and energy absorption capacity than panel with aluminum foam core and hexagonal honeycomb core. Therefore, the auxetic honeycomb core can be used in one end of capsule shell structures such as the shell of aerospace vehicles, submarines, fuel tanks, coal mine refuge chambers and other curved structures to achieve a better performance than the commonly used cellular cores: aluminum foam core and hexagonal honeycomb core to resist external blast loading.

References

[1] ASHBY M F, EVANS A, FLECK N A, et al. Metal foams: a design guide [J]. Materials & design, 2002, 23 (1): 119.

[2] RADFORD D D, MCSHANE G J, DESHPANDE V S, et al. The response of clamped sandwich plates with metallic foam cores to simulated blast loading [J]. International journal of solids and structures, 2006, 43 (7-8): 2243-2259.

[3] YAHAYA M A, RUAN D, LU G, et al. Response of aluminium honeycomb sandwich panels subjected to foam projectile impact - An experimental study [J]. International journal of impact engineering, 2015, 75: 100-109.

[4] ZHU F, ZHAO L, LU G, et al. Deformation and failure of blast-loaded metallic sandwich panels—Experimental investigations [J]. International journal of impact engineering, 2008, 35 (8): 937-951.

[5] LI S, LI X, WANG Z, et al. Finite element analysis of sandwich panels with stepwise graded aluminum honeycomb cores under blast loading [J]. Composites part A: applied science and manufacturing, 2016, 80: 1-12.

[6] ALAVI NIA A, MOKARI S, ZAKIZADEH M, et al. Experimental and numerical investigations of the effect of cellular wired core on the ballistic resistance of sandwich structures [J]. Aerospace science and technology, 2017, 70: 445-452.

[7] ZENKERT D. Introduction to Sandwich Construction.

[8] FARSHAD M. Design and analysis of shell structures. Solid mechanics & its applications, vol. 16, 1992.

[9] SHEN J, LU G, WANG Z, et al. Experiments on curved sandwich panels under blast loading [J]. International journal of impact engineering, 2010, 37 (9): 960-970.

[10] LI W, HUANG G, BAI Y, et al. Dynamic response of spherical sandwich shells with metallic foam core under external air blast loading - Numerical simulation [J]. Composite structures, 2014, 116: 612-625.

[11] JING L, WANG Z, ZHAO L. Dynamic response of cylindrical sandwich shells with metallic foam cores under blast loading—Numerical simulations [J]. Composite structures, 2013, 99: 213-223.

[12] LAKES R. Foam structures with a negative poisson's ratio [J]. Science, 1987, 235 (4792): 1038-1040.

[13] EVANS K E, ALDERSON A. Auxetic materials: functional materials and structures from lateral thinking! [J]. Advanced materials, vol. 12, no. 9, pp. 617-+, May 3 2000.

[14] BEZAZI A, SCARPA F. Mechanical behaviour of conventional and negative Poisson's ratio thermoplastic polyurethane foams under compressive cyclic loading [J]. International journal of fatigue, 2007, 29 (5): 922-930.

[15] DUVAL E, DESCHAMPS T, SAVIOT L. Poisson ratio and excess low-frequency vibrational states in glasses [J]. Journal of chemical physics, vol. 139, no. 6, Aug 14 2013, Art. no. 064506.

[16] IMBALZANO G, LINFORTH S, NGO T D, et al. Blast resistance of auxetic and honeycomb sandwich panels: Comparisons and parametric designs [J]. Composite structures, 2018, 183: 242-261.

[17] QI C, REMENNIKOV A, PEI L Z, et al. Impact and close-in blast response of auxetic honeycomb-cored sandwich panels: Experimental tests and numerical simulations [J]. Composite structures, 2017, 180: 161-178.

[18] YANG S, QI C, WANG D, et al. A Comparative Study of Ballistic Resistance of sandwich panels with aluminum foam and auxetic honeycomb cores [J]. Advances in mechanical engineering, 2015, 5: 589216.

Optimal Design of a Novel Cylindrical Sandwich Panel with Double Arrow Auxetic Core under Air Blast Loading①

具有双箭型负泊松比芯圆柱防爆夹层板优化设计

Xuke Lan②, Qi Huang, Tong Zhou, Shanshan Feng

Abstract: The increasing threat of explosions on battle field and terrorist action requires the development of more effective blast resistance materials and structures. It is important to balance the blast resistance performance and light weighting property of anti – blast structures. Curved structure can support the external loads effectively by virtue of their spatial curvature, many curved sandwich panels with different cores were designed. And in review of the excellent energy absorption property of auxetic structure, applying auxetic structure as core in curved sandwich has the potential to improve the protection performance. In this study, a novel cylindrical sandwich panel with double arrow auxetic (DAA) core was designed and the numerical model was built by ABAQUS. Due to the complexity of the structure, systematic parameter study and optimal design are required. Two cases of optimal design were considered, case1 is focused on reducing the deflection and mass of the structure, while case 2 is focused on reducing the deflection and increase the energy absorption per unit mass. Parameter study and optimal design were conducted based on Latin Hypercube Sampling (LHD) method, artificial neural networks (ANN) metamodel and the nondominated sorting genetic algorithm (NSGA – Ⅱ). The Pareto front was obtained and the cylindrical DAA structure performed much better than its equal solid panel in both blast resistance and energy absorption capacity.

Key words: Auxetic structure; Blast response; Finite element analysis (FEA); Optimal design

1 Introduction

Nowadays, the increasing threat of explosions on battle field and terrorist action requires the development of more effective blast resistance materials and structures. Metallic sandwich structures with a cellular core such as honeycomb, foam and corrugated plate have attracted much attention as they have the capability to mitigate shock and impact by progressive local crushing of its micro – structure while still light in weight [1 –4].

Different from the cellular material used in the sandwich structures, Negative Poisson's ratio materials (as known as auxetic materials) show an interesting property as they contract laterally under compression and expand when stretched [5]. With such mechanical behaviors, they show enhanced performances in fracture toughness, indentation resistance, shear modulus and vibration absorption [6 – 8]. Therefore, in recent years, many 2D and 3D auxetic structures has been designed for impact and blast resistance. The most common used structure is re – entrant auxetic honeycomb, and double – V auxetic structures which are easy to manufacture. Imbalzano et al. compared the blast resistance of 2D re – entrant auxetic and honeycomb sandwich panels and numerically studied the blast response of 3D re – entrant auxetic composite panels [9, 10]. Wang et al. designed a novel sandwich panel with double – V auxetic structure core, and numerically studied the blast response [11]. Qi et al. experimentally and numerically studied the impact and close – in blast response of auxetic honeycomb – cored

① 原文发表于《Defence Technology》2020。

② 兰旭柯：工学博士，2016 年师从冯顺山教授，研究爆炸抛撒子弹药抗冲击与防护方法，现工作单位：北京理工大学博士后。

sandwich panels [12]. Jin et al. focused on graded effect of auxetic honeycomb cores on the dynamic response of sandwich structures under blast loading [13].

All these structures deigned in above studies are based on flat plate. In practice, many facilities and engineering structures are curved, such as high - speed rails, aerospace vehicles, marine vessels and civil infrastructures [14]. The curved structures can support the external loads effectively by virtue of their spatial curvature [15]. Therefore, researchers combined the advantages of curved panels with cellular core sandwich structures to design blast resistance structures [16 - 18]. And the curved sandwich panels with cellular material core perform better than traditional flat panels in blast resistance. However, few researches about blast response of curved auxetic panels can be found. Duc et al. analytically studied the nonlinear dynamic response and vibration of sandwich composite cylindrical and spherical panels with auxetic honeycomb core layer [19, 20]. In our previous study, a cylindrical sandwich structure with 2D re - entrant honeycomb cores were designed, and the negative Poisson's ratio (NPR) effect occurs only in one direction [21].

To design a cylindrical sandwich core with two direction NPR effect, a novel cylindrical 3D double arrow auxetic (DAA) structure is developed in this study. Since the curved auxetic structure is complex and highly nonlinear, no direct relations between and design variables and objective function can be found. Therefore, metamodel was employed to solve the objective function when the design variables were given. Several works in seeking the optimal design of blast resistance structures by using metamodels have been reported [22 - 24]. Qi has conducted an optimal design of curved metallic sandwich panel using artificial neural network metamodel [25]. Wang has employed Gaussian process metamodel to optimal the blast response of a double - V auxetic flat panel [11]. However, up to the present, no optimal design works about curved sandwich panel with auxetic cores can be found.

In this paper, a novel cylindrical sandwich panel with three - dimensional DAA structure core was proposed for air blast resistance purpose, and the numerical model was built by ABAQUS. Two cases of optimal design were considered, parameter study and optimal design were conducted based on Latin Hypercube Sampling (LHD) method, artificial neural networks (ANN) metamodel and the nondominated sorting genetic algorithm (NSGA - Ⅱ). Finally, the Pareto front was obtained which can guide the design of the anti - blast structures.

2 Cylindrical Sandwich Panel with DAA Structure Core

Based on the three - dimensional DAA structure, a novel cylindrical sandwich is proposed in this paper. As illustrated in Figure 1, a DAA unit is constructed by eight cylindrical rods, three layers of units are rigidly connected to contribute the whole DAA core. The DAA core is sandwiched by two face sheets to form a cylindrical panel. The panel has a radius of curvature R at the center line of the core, the length and width are both set to be L = W = 320 mm, the height of the core layer is H = 15 mm, and the thickness of the face sheet is T_f = 0.8 mm. In this study the DAA core is weld to the top and bottom face sheets.

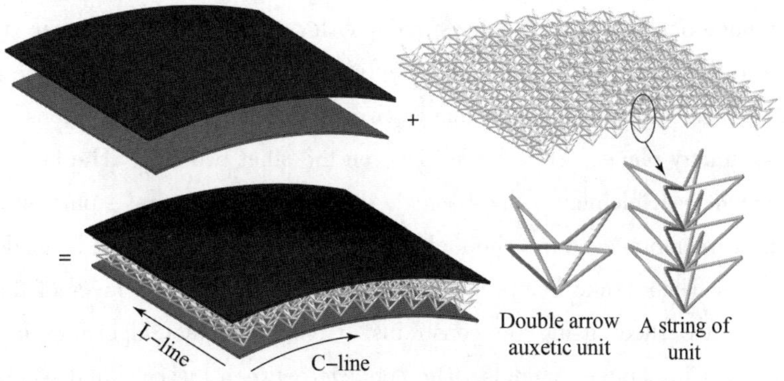

Figure 1　Schematic design of a novel cylindrical sandwich panel with DAA structure core.

Before parametric optimization, in order to select the appropriate parameters as variables, it is essential to understand the relations between geometric parameters. There are three layers of the DAA core, while the number of units along L – line and C – line are N_L and N_C respectively. Geometrically, a three – dimensional DAA unit is composed by two two – dimensional DAA units at L and C planes. As show in Figure 2, due to the symmetry, the cylindrical DAA structure has five parameters: three radius of curvature R_1, R_2, R_3, and θ at L – plane, one parameter: width $w = L/2n_L$ at C – plane, and the diameter d of the cylindrical rod. The height ratio k is defined as $k = (R_3 - R_1)/(R_2 - R_1)$. Each layer has the same height ratio k, thus the height of each layer can be expressed as $h = H/(2k+1)$. The volume of the a DAA core at layer i can be expressed as $V_{i-layer} = V_{C-plane} + V_{L-plane}$. Where $V_{C-plane}$ and $V_{L-plane}$ are volume of rods at C – plane and L – plane, which can be calculated by the cosine theorem as follows:

$$\begin{cases} V_{C-plane} = \pi \dfrac{d^2}{2}\left(\sqrt{R_1^2 + R_3^2 - 2R_1R_3\cos\dfrac{L}{n_cR}} + \sqrt{R_2^2 + R_3^2 - 2R_2R_3\cos\dfrac{L}{n_cR}}\right) \\ V_{L-plane} = \pi \dfrac{d^2}{2}\left(\sqrt{h^2 + w^2} + \sqrt{(1-k)^2h^2 + w^2}\right) \end{cases} \tag{1}$$

Thus, when the five parameters: R, k, d, N_L and N_c are given, R_1, R_2, R_3 can be calculated in each core layer according to Figure 2, and the mass of the DAA core can be obtained as:

$$M = \rho_{base}\sum_{i=1}^{3} n_c n_l V_{i-layer} \tag{2}$$

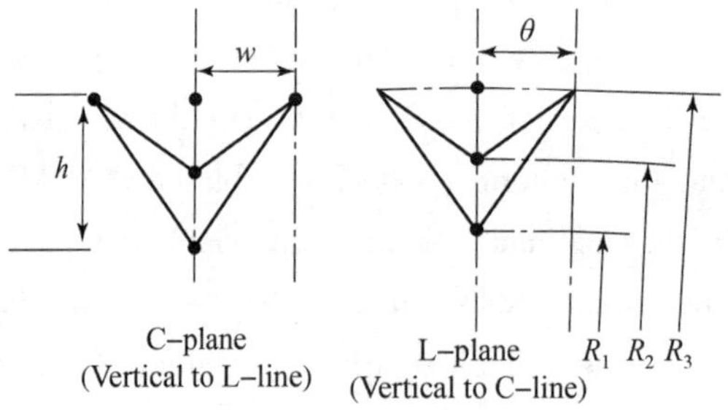

Figure 2 Geometric parameters of a DAA unit along L – line and C – line

Where ρ_{base} is density of base material. Therefore, in this study R, k, d, N_L and N_C are selected as the optimized parameters in section 4.

3 Numerical Modeling

3.1 Symmetrical Model for Cylindrical DAA Panel

In this paper, the air blast response of sandwich panel with DAA core was researched by employing ABAQUS/Explicit as a numerical solver, which is reliable for complicated nonlinear dynamic problems. As shown in Figure 3, due to the symmetry of the model, only a symmetrical quadrant of the cylindrical DAA panel is modelled in order to improve computational efficiency. Symmetrical boundary conditions are imposed on the C – symmetry plane an L – symmetry plane, while being fixed on the other two side. The face sheets are modelled by using ABAQUS shell element S4R element (the 4 – node doubly curved general – purpose, reduced integration, hourglass controlled, finite membrane strain shell element), and the DAA core is modelled by B32 element (three – node second – order Timoshenko beam elements). The top and bottom layer of the core are weld to the front face sheet and back face sheet using tie constrains. General contact algorithm in ABAQUS provide an efficient way to define contact in complex models. The contacts between the core and face sheets, as well as the self – contact of the DAA core are considered by using general contact.

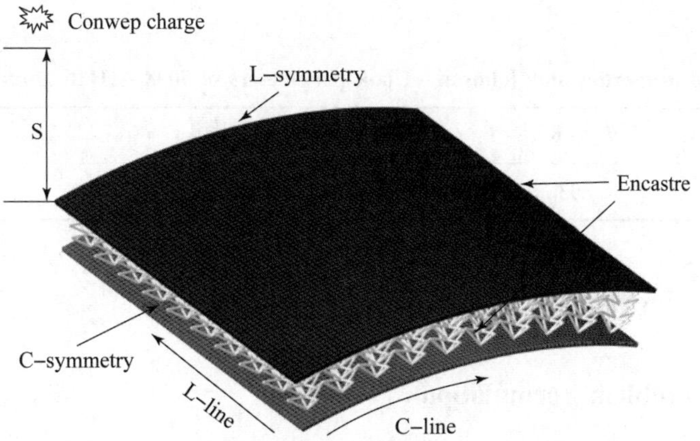

Figure 3 Boundary conditions and Conwep blast loading in FE model of a quarter cylindrical sandwich panel with DAA core.

Batch calculation in the optimization requires the model to have a high computational efficiency. Therefore, a mesh sensitive study and contact convergence study was performed, which indicate that model with B32 element mesh size of 0.5 mm and frictionless general contact can balance the calculation accuracy with the calculation efficiency. Therefore, all contacts are considered frictionless and element size of B32 is set as 0.5 mm while S4R is set as 1 mm. The numbers of elements and nodes varied largely depends on the geometric parameters. A typical model includes 50482 nodes and 56416 elements ($R = 460$, $k = 1.77$, $N_C = 24$, $N_L = 19$).

3.2 Blast Modeling

The air blast loading was generated by employing the COMWEP (conventional weapons effects program) empirical model. The CONWEP algorithms [26] developed by Kingery and Bulmash account for angle of incidence by combing the reflected pressure (normal – incidence) value and the incident pressure (side – on – incidence) value. The pressure can be calculated by the following equation:

$$P_{load} = P_{reflected} \cdot \cos^2\theta + P_{incident} \cdot (1 + \cos^2\theta - 2\cos\theta) \tag{3}$$

Where θ is the angle of incidence; $P_{incident}$ is the incident pressure; $P_{reflected}$ is the reflected pressure. CONWEP model available in Abaqus/Explicit has two parameters: the equivalent mass of TNT and the standoff distance. In this study, 30 g TNT was used as the explosive with a 100 mm standoff distance away from the center of the sandwich panel.

3.3 Material Properties and Modeling

The whole sandwich panel, including core and face sheets was designed to manufactured from aluminum alloy 5038 – H116. This aluminum alloy shows high energy absorption – to – weight ratio due to its low density, high ductility and anti – corrosive properties in comparison with steel alloys. In view of the plastic deformation and strain rate dependence, the Johnson – Cook plasticity, which includes strain hardening, strain rate and temperature effects was employed as a constitutive model in the simulations. According to this model, the flow stress σ_y is given by

$$\sigma_y = (A + B(\varepsilon_p^{eq})^n)(1 + c\ln\dot{\varepsilon}^*)(1 - (T^*)^m), \tag{4}$$

where $\dot{\varepsilon}^* = \dot{\varepsilon}_p^{eq}$, $T^* = (T - T_{room})/(T_{melt} - T_{room})$, where ε_p^{eq} and $\dot{\varepsilon}_p^{eq}$ are equivalent plastic strain and equivalent plastic strain rate, respectively. T, T_{room}, and T_{melt} are the material temperature, the room temperature and the melting temperature of material, respectively. A, B, n, c, $\dot{\varepsilon}_0$ and m are Johnson – Cook parameters determined by experimental curves fitting. The parameters from a series of Split Hopkinson Pressure Bar (SHPB) tests at high strain rate and Instron tensile tests at medium – to – low strain rate from literature are given in Table 1.

Table 1 Material properties and Johnson – Cook parameters of 5038 – H116 aluminum alloy [10].

ρ (kg/m^3)	E (Gpa)	v	T_m (K)	T_g (K)	A (MPa)	B (MPa)	n	C	$\dot{\varepsilon}_0$ (s^{-1})	m
2 750	70	0.3	893	293	215	280	0.404	0.0085	0.001	0.859

4 Optimization Process

4.1 Optimization Problem Formulation

When deign an auxetic structure, parameter optimization is an essential problem. To simplify the calculation, only the independent parameters of DAA core were optimized while the thickness of the face sheets was set to constant. Five independent parameters of the DAA core were selected as the design variables: R, k, d, N_L and N_C. The face sheets thickness $T_f = 0.8$ mm and core thickness $H = 15$ mm in all cases. The blast resistance performance of a structure can be evaluated by serval indices. First of all, the maximum displacement ($MaxD$) of the inner face sheet through the blast process needs to be reduced. In the meantime, the structure is expected to absorb more plastic energy to reduce the kinetic energy. Meanwhile, the mass of the anti – blast structure should be considered because of the requirement of lightweight in mechanical applications. In this study, the optimization problem of cylindrical sandwich panel with DAA core can be mathematically descripted as:

$$\text{Case1}: Min\{MaxD, m\} \tag{5}$$

$$\text{Case2}: Min\{MaxD, -SEA\} \tag{6}$$

$$s.\ t. \begin{cases} N_C \in \{10, 11, \cdots, 25\} \\ N_L \in \{10, 11, \cdots, 25\} \\ 300\ \text{mm} \leqslant R \leqslant 1\ 000\ \text{mm} \\ 1.2 \leqslant k \leqslant 1.8 \\ 0.35\ \text{mm} \leqslant d \leqslant 0.7\ \text{mm} \end{cases} \tag{7}$$

Where $SEA = E/m$ is the specific energy absorption. Case 1 focus on minimizing of both deflection and mass, while case 2 focus on minimizing the deflection and maximizing the specific energy absorption. Only the values within the above reasonable ranges were studied to ensure that the structure is meaningful in practice. A negative Poisson's ratio effect may not be observed if the parameter is too much deviation from the range, such as N_C, $N_L \leqslant 5$, $k \leqslant 1$. While large N_C, N_L, k, d may result in high density and reduce the calculation efficiency.

4.2 Artificial Neural Network (ANN) Metamodels

In this paper, if the optimization program algorithm was directly implemented on the numerical model, a large amount of performance evaluations was required to formulate objective and constraint functions. Therefore, metamodels was applied instead of expensive FE analysis to improve optimization efficiency. In this study, the artificial neural networks (ANN) metamodels was used to approximate the relationship between the blast resistance performance and design variables.

Similar to the biological neural system, ANN are formed of a diagram of simple processing elements called neurons that works together to solve problems. ANN has been used widely in optimization problems and has been proven effective and efficient in the design of composite structures. Typical feed forward ANNS has three layers of neurons as shown in Figure 4. Neurons at the input and the output layers represent the input variables ($x_1 \sim x_n$)

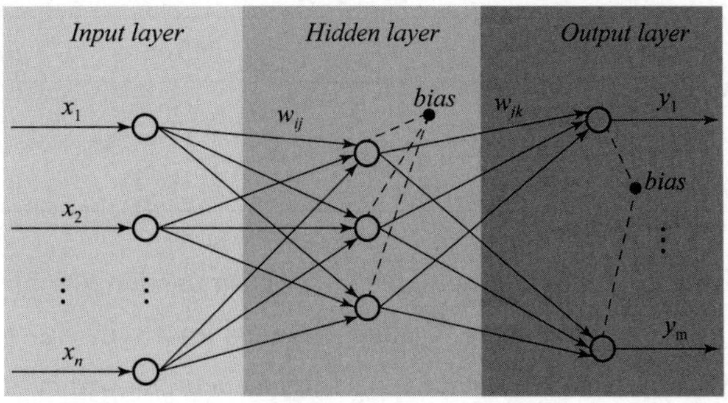

Figure 4　Schematic diagram of the topology of the three – layer feed forward neural network.

and the response of a system ($y_1 \sim y_m$), respectively. And the layer between the input and output layers is a hidden layer, which composed of nonlinear activation functions. w_{ij} and w_{jk} are the connection weight between the input layer and the hidden layer, the hidden layer and the output layer, respectively. The ANNS can be written in general form as

$$O_k = \sum_{j=1}^{l} w_{jk} f\left(\sum_{i=1}^{n} w_{ij} x_i + a_j\right) + b_k \quad j = 1,2,\cdots,l \quad k = 1,2,\cdots,m \tag{8}$$

where f is activation functions, a_j and b_k are constant called 'bias' neurons, n, l, m are the numbers of neurons of input layer, hidden layer, and output layer, respectively. In this study, the following function $y = 1/(1 + e^{-x})$ was employed as the activation function, which is included in the MATLAB Network Toolbox. The input layer has 5 nodes, the output has 2 nodes, and the hidden has about 15 nodes.

Before an ANN can be used, it needs to be trained by using design variables and the corresponding FE results as input and output data. Regardless of the type of applied metamodel, appropriate Design of Experiment (DoE) algorithm should be employed to guarantee the properly distribution of the samples in the design space. DoE aims at representing accurate relations between objective function and design variables with as few samples as possible. Latin Hypercube Sampling (LHD) method was used in this study to generate 200 DoE points filling in a uniform manner in the design space by maximally avoiding the design points of each other.

The quality of the ANN performance was established based on the specified values of errors: maximum absolute error MAX, maximum absolute percentage error MAPE, and R – square calculated by (9) – (11), respectively.

$$MAX = \max |y^i - \hat{y}^i|, \tag{9}$$

$$MAPE = \max\left(\frac{|y^i - \hat{y}^i|}{y^i}\right), \tag{10}$$

$$R^2 = 1 - \frac{\sum_{i=1}^{m} (y^i - \hat{y}^i)^2}{\sum_{i=1}^{m} (y^i - \bar{y}^i)^2}, i = 1,2,\cdots,M, \tag{11}$$

where y^i is the FE results, $\bar{y}^i$ is the mean value of y^i, $\hat{y}^i$ is the ANN metamodel approximation, and M = 200 is the number of DoE points. The values of the three quality indicate are given in Table 2.

Table 2 Error analysis of ANN metamodels

ANN metamodel	MAX	MAPE	R^2
MaxD	2. 653 8 mm	6. 620 5	0. 979 5
SEA	0. 231 4 J/g	7. 912 7	0. 954 3

The purpose of this study is to find the solutions to a constrained multiobjective problem. Usually, the solutions are a group of best trade - off designs, called "Pareto front" in the feasible domain where all the constrains are satisfied. In this study, the nondominated sorting genetic algorithm (NSGA - Ⅱ) were used. This algorithm has been successfully used to solve multiobjective engineering optimization problems . NSGA - Ⅱ has two effective sorting principles: the elitist nondominated sorting and crowing distance sorting, the details can be found in references [27, 28].

4. 3 The Interaction between MATLAB and ABAQUS

In order to realize parametric optimization and simplify the process of modeling, MATLAB platform was used in conjunction with the ABAQUS platform. As illustrated in Figure 5, on the MATLAB platform, 200 sets of parameters were generated by DoE parameter generator when the range of design parameters were given. DAA geometric modeling program calculated nodes coordinates as well as the connection relationship of nodes to form elements. Then, all the information of nodes and elements were written into . *inp* files through numerical modeling program. On the ABAQUS platform, the . *inp* files were solved by ABAQUS/Explicit solver and the results were written into . *odb* files. In order to carry batch post - processing in ABAQUS/CAE, . *py* files were generated through python compiling program by using the parameters from DoE. Finally, all the FE results were put back into neural network optimization program along with DoE parameters to get the results in form of Pareto front.

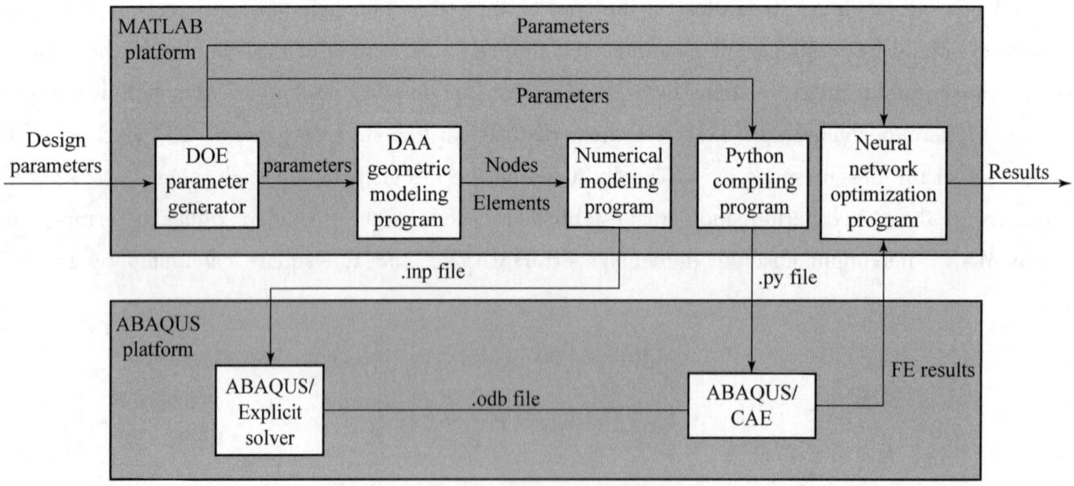

Figure 5 The flow chart for modeling, solution and optimization of the cylindrical DAA sandwich panel in MATLAB and ABAQUS.

5 Results and Discussion

Two ANN metamodels for case 1 and case 2 were built and trained by 200 DoE points and FE results. And the NSGA - Ⅱ were employed to solve the optimization problem of case 1 and case 2. The 200 DoE points and FE results were used as the first generation and the genetic algorithm iterated for 50 generations. All the 10 000 points were converged in the final results.

5.1 Case 1 Study

5.1.1 Pareto Front of Case 1

Figure 6 shows all the 10 000 optimization samples, represented by blue points. The Pareto Front can be obtained, which is not the precise front and was only a demonstration used black lines. A strong confliction of the maximum enteral deflection and mass of the DAA panel can observed from the Pareto Front. Along the pareto Front line, the only way to reduce MaxD is to increase the mass of the panel. On the purpose of validate the accuracy of the metamodel, three special points close to the Pareto Front were selected as illustrated in the red circle in Figure 6. The design of the three points were numerically simulated. The design parameters and the predicted results and numerical simulation results were listed in Table 3. It implied that the prediction of ANN metamodel was close to the FE results. To make a comparison, the blast response of an equivalent solid cylindrical panel with the same curvature, length, width, and mass of points 2 were simulated by using the same blast loading, material, boundary conditions as mentioned in section3. The deformation process of P2 and the equivalent solid cylindrical panel were shown in Figure 7. And in Figure 6 the panel was represented by a red square which is away from the Pareto Front. The results indicate that the DAA sandwich structure can massively improve the air blast resistance compared with solid panel.

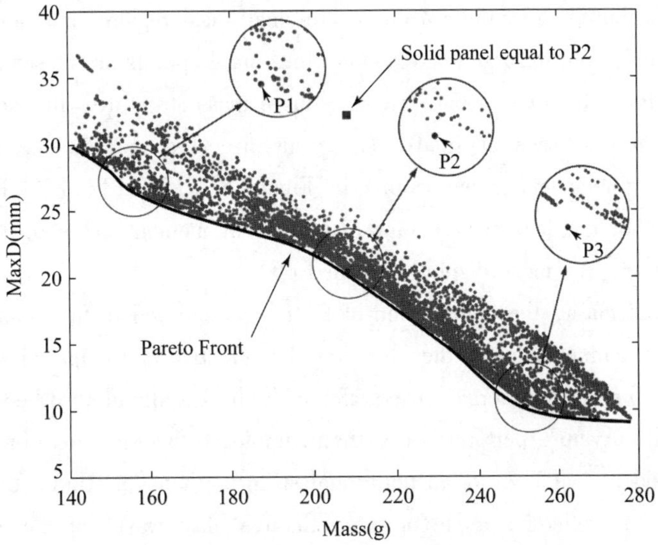

Figure 6 Optimization design results of case1 study.

Table 3 Optimum design parameters for case1.

Points	R (mm)	K (mm)	NC	NL	D (mm)	Mass (g)	MaxD (mm)	
							Predicted	FE
P1	998.57	1.25	12	18	0.35	155.79	26.65	28.37
P2	335.85	1.21	11	15	0.54	208.22	20.44	21.15
P3	309.36	1.22	11	18	0.70	251.38	10.24	10.88

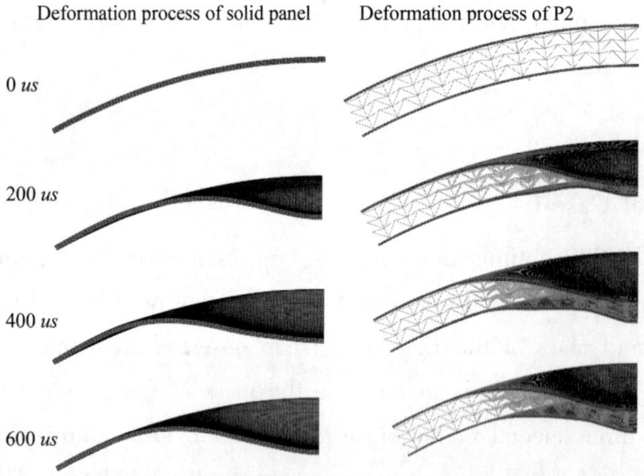

Figure 7　Comparison of deformation process of solid panel and P2

5.1.2　Parameter Study of Case 1

The influence of design parameters was studied by using scatter plots of all optimization samples points. MaxD vs Mass scatter figures with respect to R, k, N_L, N_C and D are shown in Figure 8 respectively. The value of a design parameter is represented by the color of sample points, and can be find in the color bar at each subplot. As shown in Figure 8 (a), the scatter plot can be divided into two regions: a large mass region and a small mass region by two curves. Sample points with small R are distributed along the blue curve, while sample points with large R are distributed along the red curve. In small mass region, red sample points are closer to the Pareto Front than blue points, and in large mass region blue sample points are closer to the Pareto Front than red points. This phenomenon indicates that in the case of a large mass structure, the performance increases as the curvature increases, and in the case of small mass structure, the performance decrease as the curvature increases. To explain this phenomena, four points in Figure 8 (a): P1, P3, P4, P5 were selected and simulated, the deformation of the maximum center displacement moment are shown in Figure 9 (To make a clearer display of deformation, the mesh display is turned off).

P1 and P4 have a small mass, the deformation of P4 is concentrated in the central area, although it has a large curvature the structure cannot disperse the blast impulse effectively. Compared with P4, P1 shows a global bending and compression, the impact energy can be effectively dispersed and absorbed. Therefore, in small mass region, samples with small curvature perform better than samples with large curvature. P3 and P5 have a large mass, by virtue of the curvature, P3 shows an additional stiffness than P5. Under the support of the core layer and the back face sheet, the reflective angle of the reflective blast wave can be changed, and results in a reduction of the blast impulse acting on the front face of sandwich panels. In addition, previous study [21] concluded that, the existence of the curvature can support the negative Poisson's ratio effect to pull material into the central deformation zone to absorb impact energy and enhance the strength of the structure. Therefore, in large mass region, samples with large curvature perform better than samples with small curvature.

As shown in Figure 8 (b), as the value of k decrease, the sample points are closer to the Pareto Front. Previous study of 2D double arrow auxetic structure provided that 2D double arrow auxetic structure with a smaller k has a higher yield strength [29]. And the 3D double arrow auxetic structure is composed of two 2D double arrow auxetic structure vertically combined, so its yield strength has a similar variation with the 2D double arrow auxetic structure. Therefore, blue sample points show a greater yield strength and closer to the Pareto Front than red sample points. N_L, N_C and D show a similar effect on the scatter plot that red points are more appeared in the right large mass region which means a large value of N_L, N_C and D can improve the blast resistance performance.

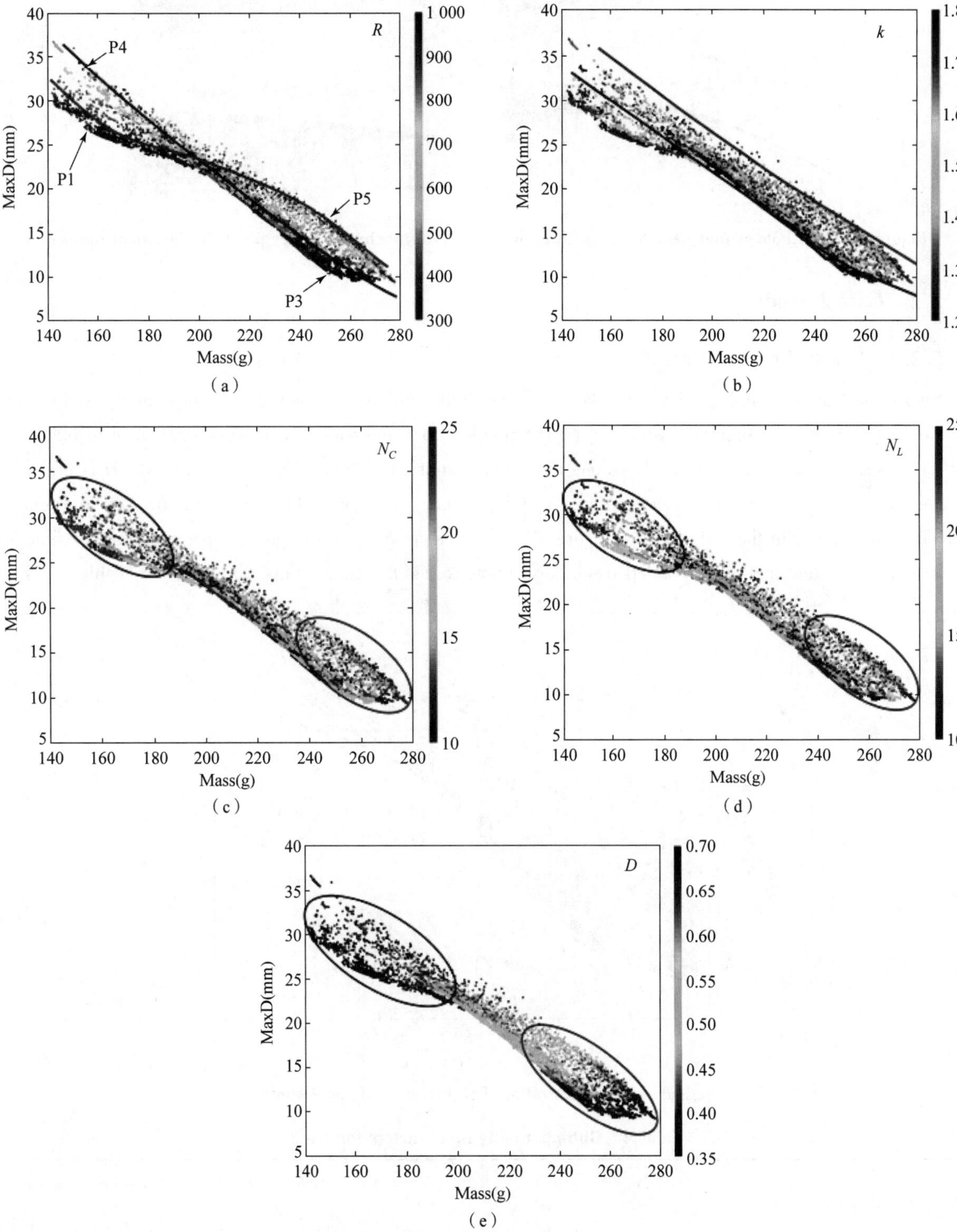

Figure 8　Scatter plots of the influence of five design parameters: *R*, *k*, N_L, N_C and *D*.

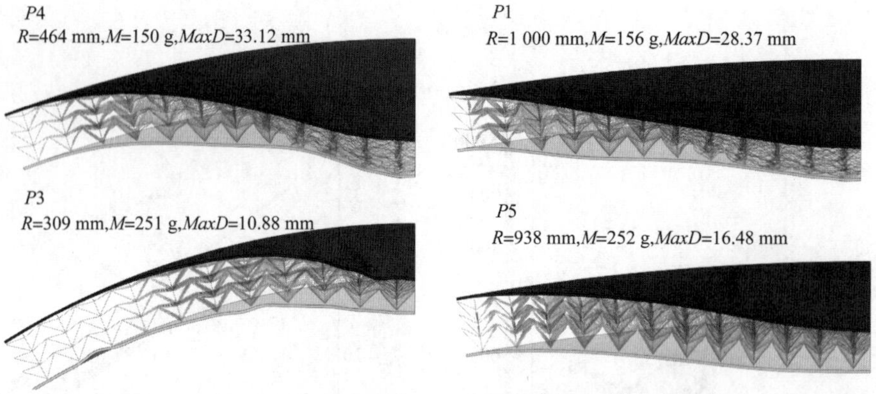

Figure 9 Snapshots of four sample points in Figure 8 (a) of maximum center point displacement moment.

5.2 Case 2 Study

5.2.1 Pareto Front of Case 2

Similar with case 1 study, Figure 10 shows all the 10000 optimization samples, represented by blue points. The Pareto Front can be obtained, which is not the precise front and was only a demonstration used black lines. MaxD is positively correlated with SEA. Along the Pareto Front, to increase SEA, a larger MaxD is needed. To validate the accuracy of the metamodel, three special points (P6, P7, P8) close to the Pareto Front were selected as illustrated in the red circle in Figure 10. The design of the three points were numerically simulated. The design parameters and the predicted results and numerical simulation results were listed in Table 4.

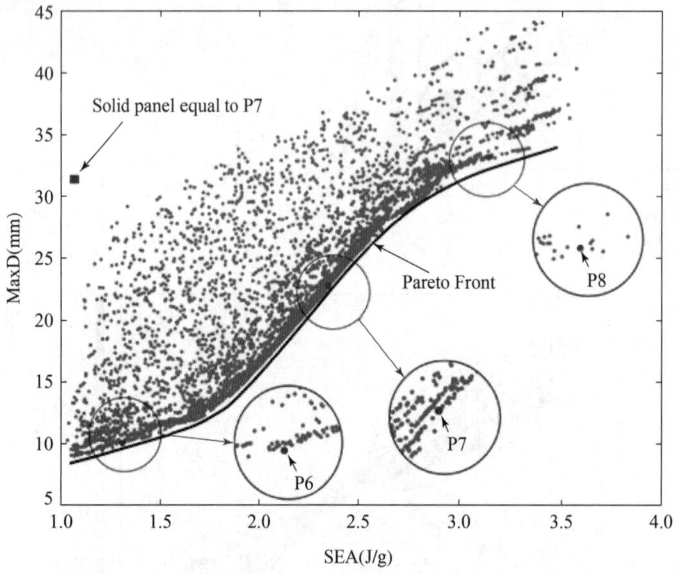

Figure 10 Optimization design results of case2 study

Table 4 Optimum design parameters for case2.

Points	R (mm)	K (mm)	NC	NL	D (mm)	SEA (J/g)		MaxD (mm)	
						Predicted	FE	Predicted	FE
P6	643.72	1.60	23	22	0.7	1.31	1.39	10.04	9.87
P7	995.29	1.63	12	13	0.59	2.34	2.40	22.79	24.50
P8	992.51	1.57	10	13	0.37	3.14	3.09	32.18	31.03

It implied that the prediction of ANN metamodel was close to the FE results. To make a comparison, the blast response of an equivalent solid cylindrical panel with the same curvature, length, width, and mass of points 6 was simulated by using the same blast loading, material, boundary conditions as mentioned in section3. The result is shown in Figure 10 represented as a red square, which is away from the Pareto Front. It indicate that the DAA sandwich structure can massively improve the SEA by core crushing while keeping a small MaxD compared with solid panel.

5.2.2 Parameter Study of Case 2

As the same with case1 study, the influence of design parameters was studied by using scatter plots of all optimization samples points. MaxD vs SEA scatter figures with respect to R, k, N_L, N_C and D were shown in Figure 11 respectively. The value of a design parameter is represented by the color of sample points and can be find in the color bar at each subplot. In Figure 11 (a), sample points with small R are distributed along the blue curve, while sample points with large R are distributed along the red curve. And most part of the Pareto Front consist of red points, which indicate that panels with small curvature perform better than panel with large curvature. As explained in case 1 study, deformation of panel with large curvature was concentrated in the center region of the panel, while deformation of panel with small curvature appeared in almost the whole panel, therefore panel with small curvature has more effective structural mass contributed to the energy absorption which results in a large SEA compared with panel with large curvature.

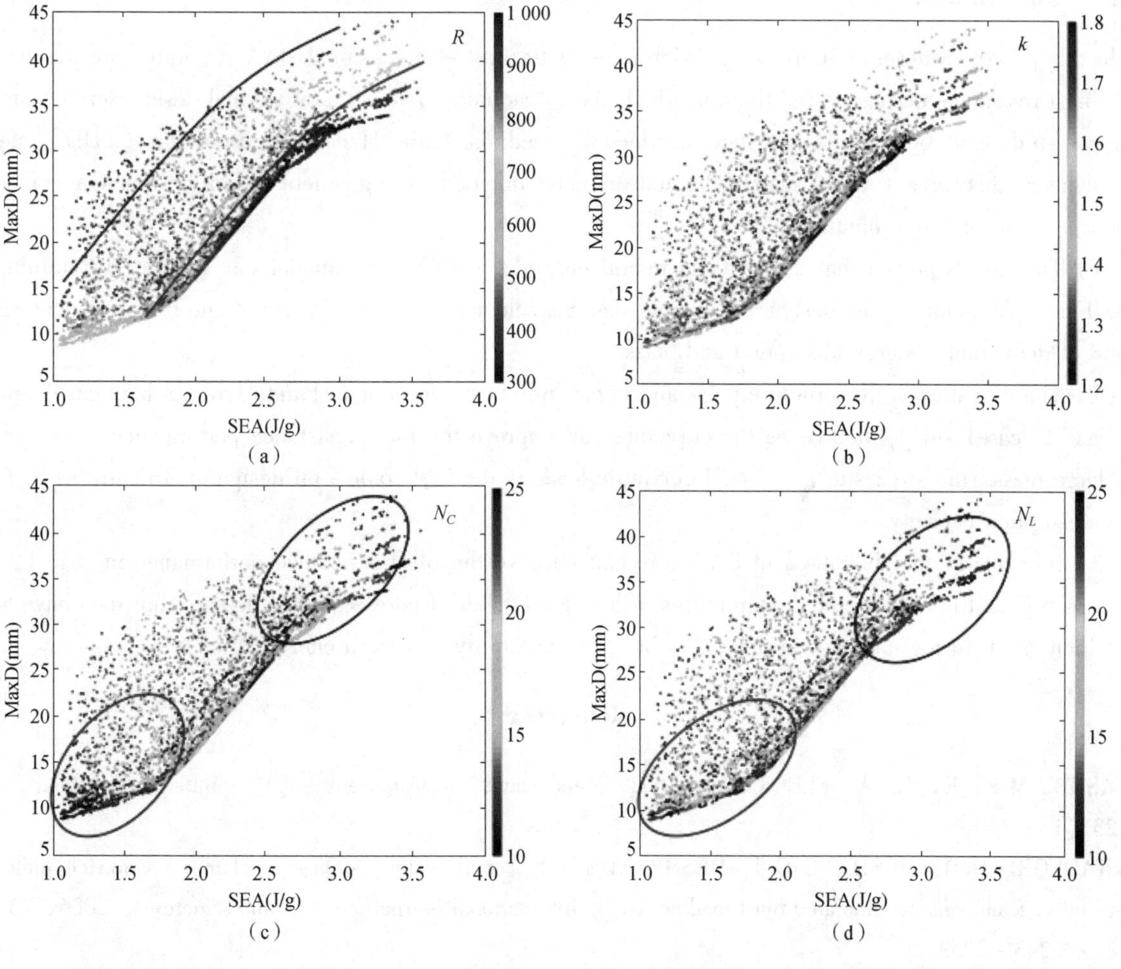

Figure 11 Scatter plots of the influence of five design parameters: R, k, N_L, N_C and D on case2

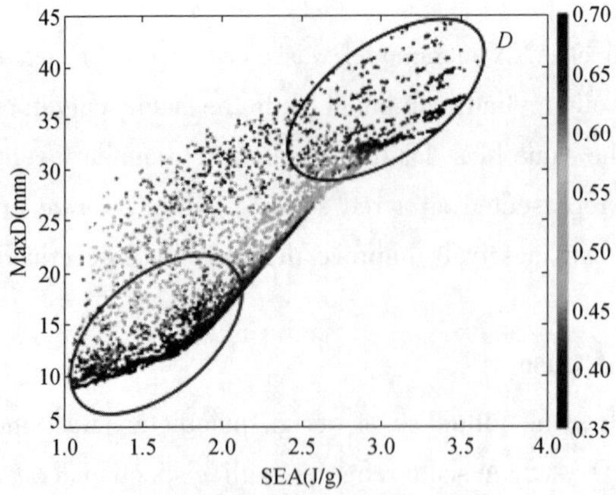

(e)

Figure 11 Scatter plots of the influence of five design parameters: *R*, *k*, N_L, N_C *and D* on case2（续）

The magnitude of the k value is almost evenly distributed in the scatter plot, that means the height ratio of DAA core unit shows no significant effect in the optimization design results. N_L, N_C and D shows similar effects on the scatter plot that red points are more appeared in the region close to the origin of the coordinate while blue points are opposite. A large value of N_L, N_C and D result in a large value of mass, which is the dominate factor of MaxD and SEA.

6 Conclusion

In this paper, a novel cylindrical sandwich panel with three - dimensional DAA structure core was designed for air blast resistance purpose, and the numerical model was built. Two cases of optimal design were considered, parameter study and optimal design were conducted based on Latin Hypercube Sampling (LHD) method, artificial neural networks (ANN) metamodel and the nondominated sorting genetic algorithm (NSGA - Ⅱ). The following conclusions were obtained:

(1) The results proved that the artificial neural networks (ANN) metamodel was accurate by training with 200 DoE sample points. The design parameters can be selected from the Pareto Front based on the specific demand of deflection, energy absorption and mass.

(2) The curvature of the structure plays and import role in the forming of Pareto Front in both case1 and case 2 studies. In case1 study, increasing the curvature can improve the blast resistance performance if the structure has a large mass. In case 2 study, a small curvature leads to the best trade - off design of structure with efficient energy absorption capacity.

(3) Increase the height ratio k of DAA core can improve the blast resistance performance in case 1, while show no significant effect on case 2. Structures with large number of core cells and thicker core rods have a large mass which result in the decrease of deflection but lose the ability to absorb energy efficiently.

References

[1] ASHBY M F, EVANS A, FLECK N A, et al. Metal foams: a design guide [J]. Materials & design, 2002, 23 (1): 119.

[2] RADFORD D D, MCSHANE G J, DESHPANDE V S, et al. The response of clamped sandwich plates with metallic foam cores to simulated blast loading [J]. International journal of solids and structures, 2006, 43 (7 - 8): 2243 - 2259.

[3] LAN X K, FENG S S, HUANG Q, et al. Blast response of continuous - density graded cellular material based on the 3D Voronoi model [J]. Defence technology, 2018, 14 (5): 433 - 440.

Water - entry Behavior of Projectiles under the Protection of Polyurethane Buffer Head①

基于聚氨酯泡沫头部防护结构的弹药入水行为

Wu Siyu, Shao Zhiyu②, Feng Shunshan, Zhou Tong

State Key Laboratory of Explosion Science and Technology, Beijing Institute of Technology, Beijing 100081, P. R. China

Abstract: The water entry behaviors of projectiles with cylindrical buffer head were studied experimentally and theoretically, focusing on projectile dynamics and impact reduction. Based on the elastic, perfectly plastic, strain rate dependent, compacting model, the equation of motion was developed for the projectiles to describe their motion before compaction of the buffer. The validity of the equation was proven both experimentally and numerically. The peak deceleration of the projectiles was investigated by numerical simulations in LS - DYNA using Lagrange - Euler coupling, and an empirical formula was proposed to accurately describe the impact reduction based on the experimental and numerical results. When the impact velocity and buffer dimension were held constant, an optimum buffer density existed for the reduction of impact by the buffer head. At a constant buffer density, the impact was larger when the initial velocity increased or when the buffer had a larger cross section compared with its length. Good agreements were observed between the analytical and experimental results.

Key words: water entry; cylindrical buffer; dynamics; impact

1 Introduction

Despite extensive studies, there is still no satisfactory theoretical solution to date for the water entry of projectiles because it is a very complex process. The initial developments of water entry theory can be traced back to 1929, when von Karman [1] proposed several mathematical solutions to describe the impact forces acting upon rigid bodies during water entry. In 1932, Wagner [2] further extended the earlier work of von Karman by mathematically analyzing the vertical water entry impact of two - dimensional wedge at a small deadrise angle. Sedov [3] later applied Wagner's theoretical solution to larger deadrise angles, and Yu [4] considered three - dimensional effects by applying correction factors. After World War II, the research interest in the water entry problem increasingly turned to underwater ordnance and was extended to the area of hydroballistics, and the impact and dynamic peak stress during water entry process received significant attention. Shiffman and Spencer [5 - 7] investigated the vertical impact of spheres and cones upon water entry. They emphasized added mass as an important concept in their early work [5, 6], and later extended Hillman's solution [8] for a 120° cone to give an improved approximate solution applicable to any cone angle [7]. Their conclusions were in good agreement with the experimental works of Watanabe [9, 10] as well as with Hillman's solution [8]. In the mean time, typical works on the water entry of projectiles were also undertaken by May, Schnitzer, Abelson, Waugh, and others [11 - 15]. These pioneering researchers have largely built a framework for the problem of water entry impact, although May pointed out in 1970 that more work on water entry was still needed to eliminate guesswork [12].

As theory developed in the field of water entry, research was also carried out on reducing entry impact to

---

① 原文发表于《Ocean engineering》2020。

② 伍思宇：工学博士，2017 年师从冯顺山教授，研究空投弹药入水冲击防护及弹道特性，现工作单位：中国万宝工程有限公司。

make sure that the projectile itself would stay intact upon water entry and remain functional to maximize damage only until reaching the intended target. In 1931, by taking into account the elasticity of float gear, Pabst [16] introduced his assumptions into the mechanical model of von Karman [1] by assuming the elastic function of the water entry structure as a linear spring. In order to mitigate the peak stress of water entry, Hinckley and Yang [17] analyzed the mitigation of shock and the absorption of energy under impact - loading conditions with crushable materials, and concluded that rigid polyurethane (PU) foam was an effective material to mitigate shock. Baldwin [18] tested various nose shapes of projectiles, especially their cone body, and derived a set of equations to calculate the total resistance during the vertical water entry of a projectile with an arch nose. These investigations constitute the theoretical basis of reducing water entry impact.

However, the water entry impact and its mitigation are still not well understood even in modern times. Zhang *et al.* [19] developed a theoretical model to simulate the impact buckling of the coupled fluid - structure system, and performed experiments in which an elastic - plastic column (its upper end attached to a large mass and its lower end to a flat plate) perpendicularly impacted water from various drop heights. In 2003, Park *et al.* [20] developed a numerical method to compute the impact forces and the ricochet behaviors of high - speed water entry bodies. Guo *et al.* [21 - 23] studied high - speed projectiles striking water horizontally and the effects of nose shapes on velocity attenuations and drag coefficient. Recent water entry studies took deformable materials into account as well. For example, Tate *et al.* [24] compared deformable elastomeric spheres with their rigid counterpart to describe the water entry characteristics of deformable spheres in terms of material properties and impact conditions. Nose shape and cushion materials are considered in many water entry studies, and it is thus of interest to find a method to describe the complex situation of water entry for projectiles with cushioning material.

In this work, extensive experiments with accelerometer and digital high - speed photography as well as numerical simulations were performed for the water entry of projectiles bearing full size cylindrical buffer heads made of rigid polyurethane (PU) foam. The dynamics of the projectiles during water entry were studied. The impact loads of the projectiles with buffer head were observed, and an empirical maximum deceleration model was proposed with consideration of the density of the rigid PU and the initial velocity.

2 Experimental Investigation

2.1 Device Setup

Since pool test was not suitable for inspecting high velocity water entry, we established an alternative experimental setup assembled as shown in Figure 1. The projectile equipped with an acceleration sensor was shot into the water tank. The initial speed and motion process were recorded by laser beam and high - speed camera, and the deceleration data were obtained by oscilloscope. A constant current power was used to supply power to the acceleration sensor.

The projectile was fired off from a standard artillery device aiming at the diaphragm of the water tank. A low burning rate propellant was employed to make sure that the buffer head would not be destroyed by large acceleration. The acceleration and initial velocity of the projectiles were adjusted by tuning the mass of the propellant in the cartridge.

A Phantom digital high - speed camera was deployed to record the fluid, the projectile, and the deformation of the buffer head upon the entry. The selected frame rate was 10 000 per second, *i. e.*, one frame was taken every 100 μs. These settings were chosen based on early numerical simulation results by LS - DYNA and AUTODYN. The transparent, impact - resistant water tank was made of polycarbonate composite material, and there was no cover on the tank. Two holes were cut in the front and the back pieces of the polycarbonate and were

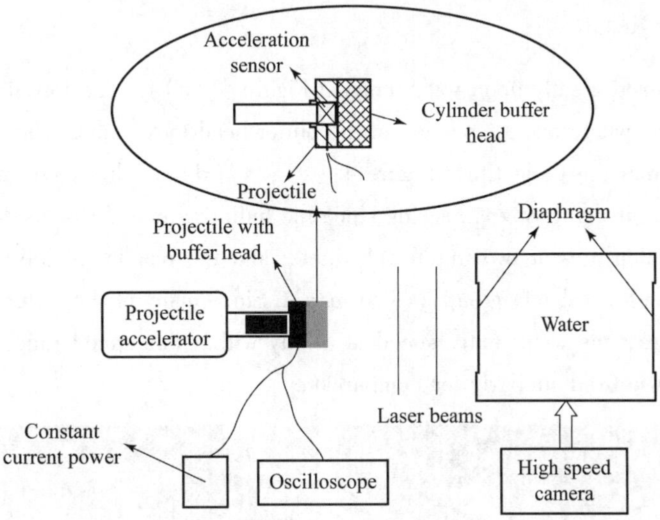

Figure. 1 Schematic of the experimental system

sealed by two pieces of diaphragms analogously to Guo Z, Zhang W, et al. [23] The diaphragms were made of 0. 2 mm thick acrylic sheets. When a projectile was fired at 71. 3 m/s to penetrate the diaphragms, the velocity and attitude of the projectiles showed no difference before and after the penetrate, proving that the 0. 2 mm thick acrylic sheets were ideal to seal the water tank and would not affect the water entry process. The projectile was fired into the front hole to penetrate the fluid horizontally and it either left the tank from the back hole by itself or was manually unloaded. Before each run, water was replenished and the tank was sealed with fresh diaphragms.

Projectiles with buffer heads of different density were fired into the water tank. All projectiles and buffer heads were of the same size. The projectiles were made of 2A12 aluminum (150 mm in diameter, 3 kg mass), and the buffer heads were made of rigid PU (also 150 mm in diameter). Inside each projectile was installed an accelerometer, which was attached to a flexible cable for power supply and collection of signal waveforms. Table 1 shows the parameters of the accelerometer. A tail rod was connected to the projectile with a thread to facilitate acceleration. Three PU densities (0. 12, 0. 18, and 0. 25 g/cm^3) of the buffer heads were tested (Figure 2). Figure 3 shows the assembled projectile.

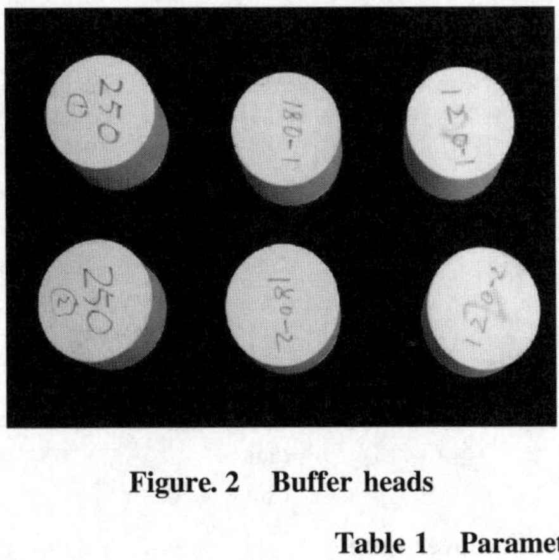

Figure. 2 Buffer heads

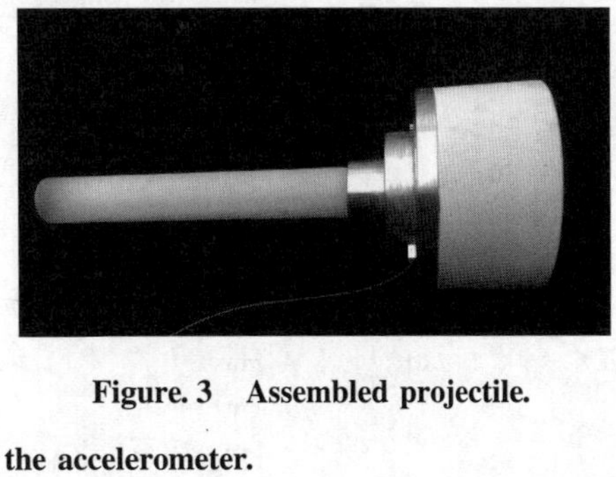

Figure. 3 Assembled projectile.

Table 1 Parameters of the accelerometer.

Sensitivity (20 ± 5 ℃)	Frequency range	Maximum acceleration	Thread	Nominal voltage	Rated current	Output mode
0. 003 mV/($m \cdot s^2$)	5 – 10 000 Hz	5×10^5 m/s^2	M5	5 V (DC)	2 – 10 mA	L5

2.2 Experimental Result

The buffer head deformed greatly upon water entry (Figure 4 (a)). The initial velocity and buffer density jointly determined when and where the compaction of the buffer head took place. The compacted buffer head then crushed and broke apart from the projectile (Figure 4 (c), 4 (d)). This automatic separation of the buffer head was a desirable feature in terms of engineering since the buffering effect was needed only in the initial stage. We tested the three buffer densities at two initial velocities, and the results were divided into the high velocity group (~95 m/s) and the low velocity group (~70 m/s). Since many of the water entry devices are equipped with air deceleration devices, the water entry speed is mostly within this speed range or even lower. Projectiles without buffer head were also fired similarly for comparison.

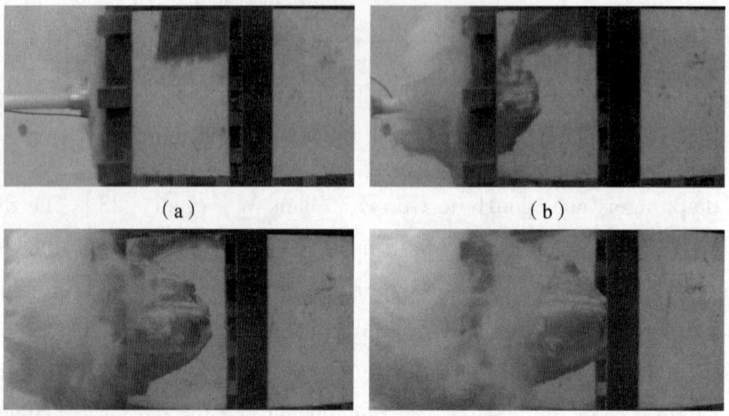

(a) (b) (c) (d)

Figure. 4 The water entry of projectile with 0.25 g/cm^3 PU buffer head.

(a) 0.6 ms; (b) 6 ms; (c) 13 ms; (d) 20.8 ms

Figure 5 demonstrates the velocity attenuation over time of the projectiles with buffer head. Camera images were collected only for the first two milliseconds because water entry was quick. For the projectiles with buffer head of 0.12 g/cm^3 and 0.18 g/cm^3 PU, deceleration became more significant when the projectile shot farther into the tank. However, this situation was not readily observed for the projectile with 0.25 g/cm^3 PU buffer head, probably because the dense PU compacted too early and the buffering effect was poorer.

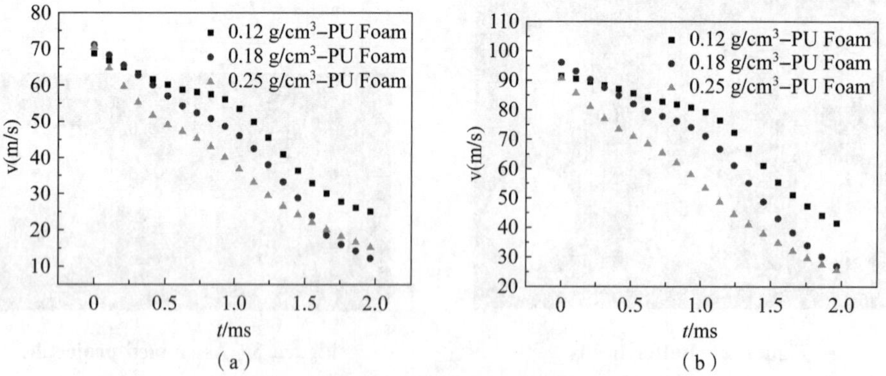

Figure. 5 Velocity attenuation over time.

(a) Projectiles with water entry at ~70 m/s; (b) Projectiles with water entry at ~95 m/s.

Several sensors were used on the projectiles to measure deceleration. Figure 6 draws a typical deceleration curve. Although the response frequency of the sensors was not high enough to measure real time deceleration in the very quick water entry process, the recorded maximum values sufficed for studying the buffering effect.

Because the sensor had a limited range of measurement, the maximum deceleration of the projectile with rigid head could not be experimentally obtained but could only be calculated by numerical simulations. Table 2 lists the data we obtained in experiments.

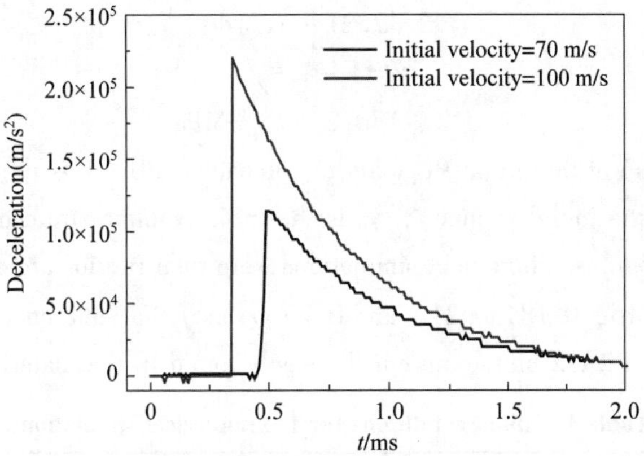

Figure. 6 Typical experimental deceleration curves (0.25 g/cm³ PU buffer head).

Table 2 Test results of maximum deceleration.

Buffer density (g/cm³)	Initial velocity (m/s)	Peak deceleration (G)
0.12	69.8	9 600
	91.7	48 700
0.18	71	3 400
	96	12 700
0.25	70.4	11 400
	90.7	22 000

3 Numerical Simulations

Numerical simulation was carried out to obtain more details and data during water entry. The water entry process was more clearly observed by considering cylindrical polyurethane foam as the buffer similarly as in the experiments.

3.1 Numerical Simulation Setup

Numerical simulations were conducted using the explicit finite element code of LS – DYNA, thus considering the fluid – solid interactions between the projectile and water. The coupled Lagrange – Euler technique was adopted to simulate the interaction problems, with the Lagrange mesh for the projectile and the Euler mesh for water. All projectiles had a same mass of 3 kg, their initial velocities ranged in 70 – 130 m/s, and five polyurethane densities were considered in the simulations. Table 3 lists the initial conditions of the numerical simulations. A 3D one – quarter symmetric numerical model was employed for the water entry. Figure 7 presents the numerical models and the mesh construction, in which the mesh size was determined from a convergence study. From the material database of LS – DYNA, * MAT_CLOSED_CELL_FOAM was selected as the material model for polyurethane foam, hence giving [25]

$$\delta_y = A + B(1 + C\lambda) \tag{1}$$

where δ_y is yield stress, A, B, and C are constants, and γ is a function of the volumetric strain, all of which can be obtained as follows:

$$A = 23.7 \times \phi^{1.676}\ \text{MPa} \tag{2}$$

$$B = 19.2 \times \phi^{1.645}\ \text{MPa} \tag{3}$$

$$C = 2.21 - 31.1\phi \tag{4}$$

$$\gamma = V - 1 + \gamma_0 \tag{5}$$

$$E = 3\,130.2 \times \phi^{2.20}\ \text{MPa} \tag{6}$$

where ϕ is the density ratio of the initial PU foam to compacted PU, V is the volumetric strain (*i. e.*, the ratio of the existing volume to the initial volume), γ_0 is the initial volume strain of the foam material (normally $\gamma_0 = 0$), and E is Young's modulus. Numerical simulations were then run for projectiles equipped with PU buffer of different density (0.12, 0.15, 0.18, 0.21, and 0.25 g/cm^3) that entered water at varying initial velocity. The material model of water and 2A12 aluminum can be easily found in the material database.

Table 3 Initial conditions for the numerical simulations.

M (kg)	D (mm)	L_0 (mm)	v_0	P_{PU} (g/cm^3)
3	150	75	70 - 130	0.12 - 0.25

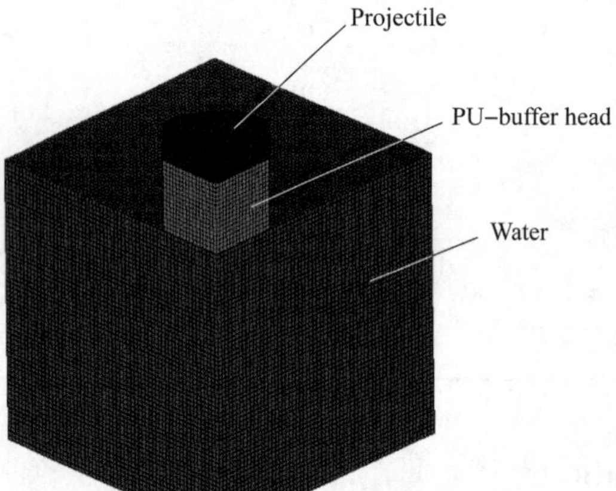

Figure. 7 Modeling and meshing of the numerical simulations.

3.2 Simulation Results

The numerical results were compared with the experimental results in the earlier section to verify simulation effectiveness (Figures 8 and 9). Because the response frequency of sensors was not high enough, the voltage could not be timely reset after the projectile passed the maximum point, which caused the discrepancy in Figure 8 (b) and 9 (b). Nevertheless, since the focus of this paper is peak acceleration, this discrepancy was inconsequential as the maximum value of acceleration obtained by the sensor showed good agreement with simulation. The velocity attenuation curve also showed good agreement. Therefore the numerical models were well validated and deemed to adequately capture the water entry behaviors.

Figure 10 shows the stress cloud map of a typical water entry process from the numerical simulation and illustrates both the deformation of the buffer head and the stress - time history of whole projectile body during water entry. High stress occurred upon contact between the buffer head and water, at which time the buffer head started to deform, giving stress waves that propagated backward. Thanks to the yield stress of the buffer, the projectile body did not experience significant stress. The buffer head then deformed further as the projectile was further extruded until the buffer fully compacted, and after the compaction of the buffer, peak stress and peak deceleration were observed upon strong impact.

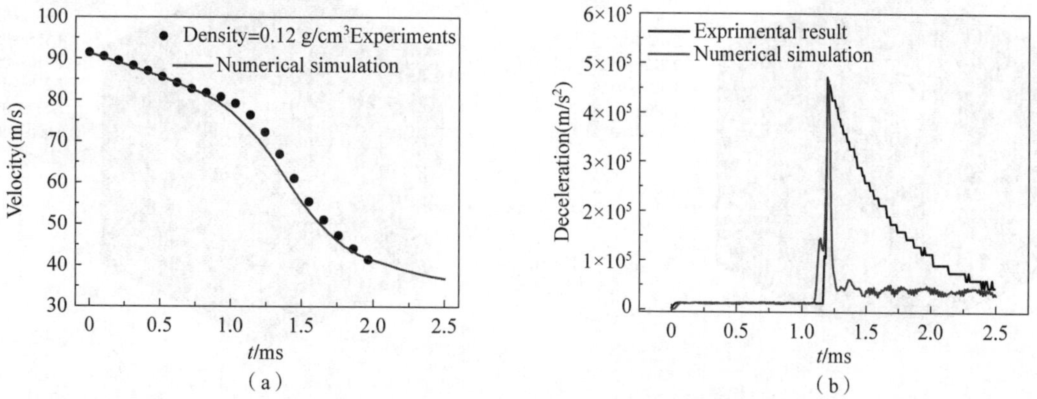

Figure. 8　Comparison between experiments and simulations for projectiles buffered with 0. 12 g/cm³ PU.

(a) Velocity attenuation with time; (b) Penetration time – deceleration history

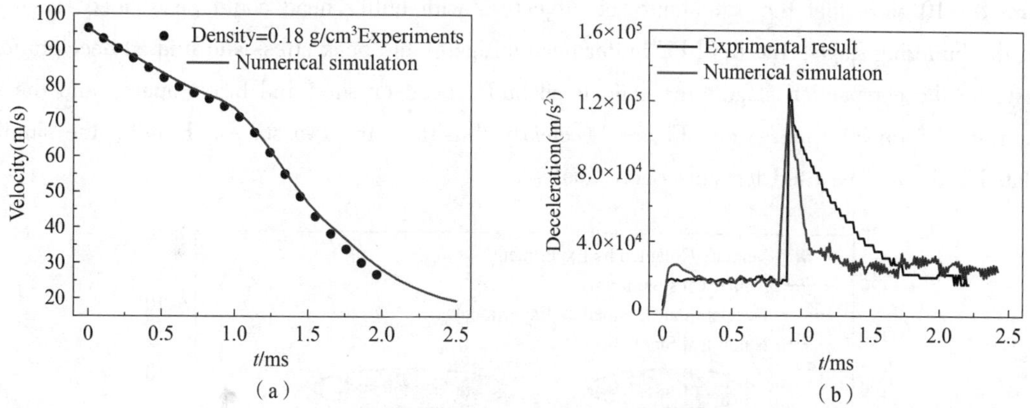

Figure. 9　Comparison between experiments and simulations for projectiles buffered with 0. 18 g/cm³ PU.

(a) Velocity attenuation with time; (b) Penetration time – deceleration history

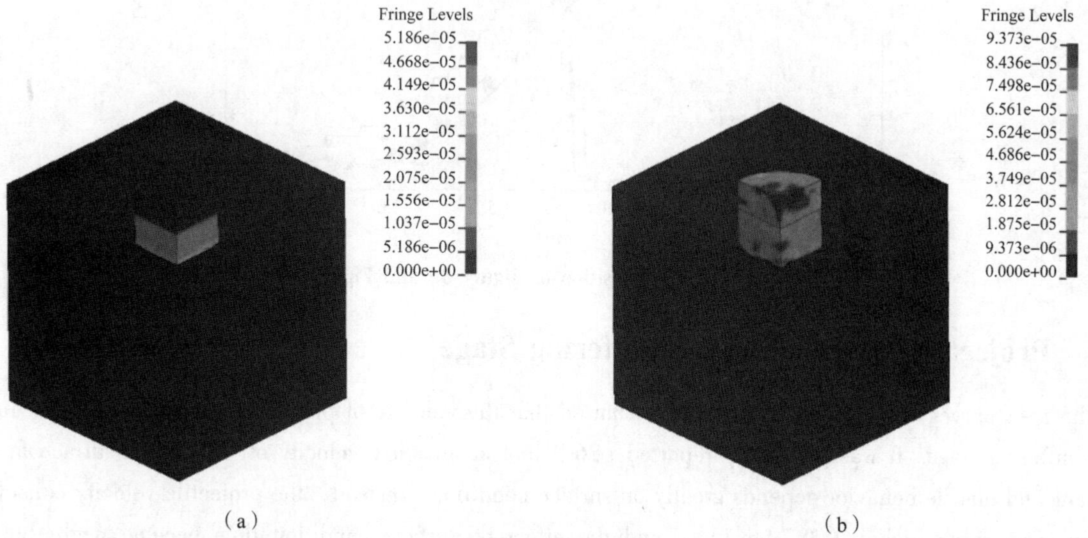

Figure. 10　Stress cloud map of 0. 18 g/cm³ PU buffer at v_0 = 100 m/s.

(a) 0. 08 ms; (b) 0. 22 ms

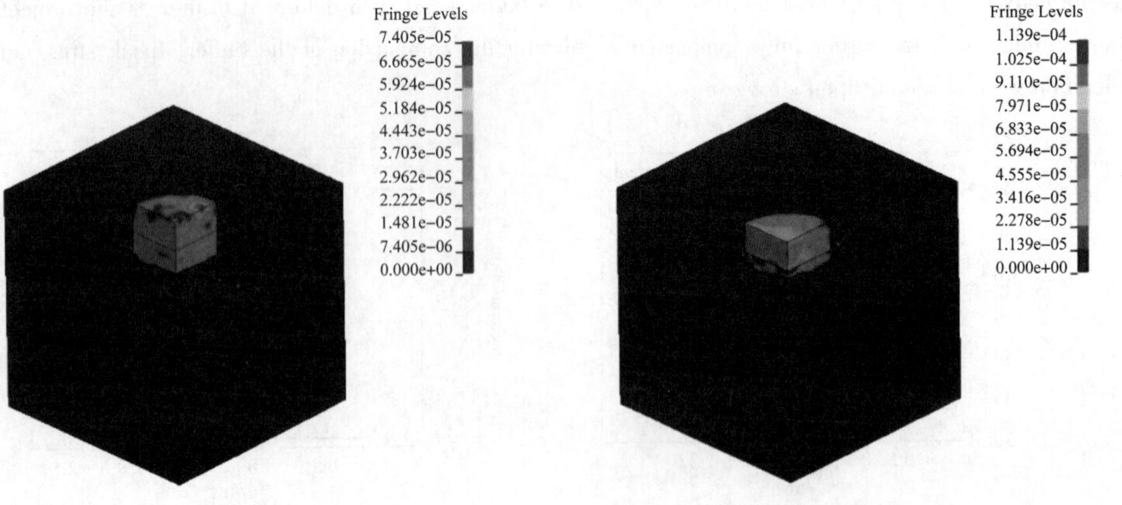

(c) (d)

Figure. 10 Stress cloud map of 0.18 g/cm³ PU buffer at v_0 = 100 m/s.

(c) 0.53 ms; (d) 1.10 ms

Figures 8 – 10 show that the water entry of projectiles with buffer head could be divided roughly into two stages. In the buffering stage, the rigid PU buffer head mitigated the peak stress and limited deceleration to a low level. Next, in the compaction stage, the compacted buffer head crushed and broke apart, and the stress and deceleration of the projectile increased. Figure 11 clearly illustrates the two stages. Hence, the simulation was well validated and could be used for subsequent analyses.

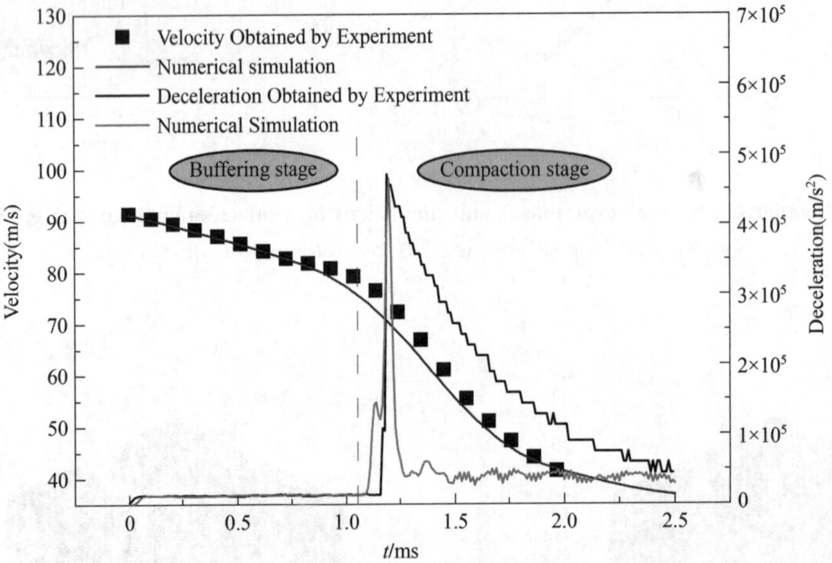

Figure. 11 Superposition of Figure 8a and Figure 8b.

4 Projectile Dynamics in the Buffering Stage

The $v-t$ curves of the buffered projectiles showed that the velocity of projectiles decreased almost uniformly in the buffering stage. It was previously reported [26] that at an entry velocity of < 8 m/s, surface forces are important and missile behavior depends greatly on surface condition. Instead, the projectile velocity considered in this work was much higher (≥70 m/s), and the effect of surface condition thus became negligible in the subsequent analyses. Consider an object with an initial velocity of v_0 penetrating into water along a linear trajectory in the $+x$ direction (Figure 12). The motion equation of the object can be described by Newton's second law while ignoring gravity and buoyancy:

$$\begin{cases} M\dfrac{d^2x_1}{dt^2} = F \\ \dfrac{d}{dt}\left(m\dfrac{dx_2}{dt}\right) = F \\ F = \sigma_T A = F(\varepsilon_T, \dot{\varepsilon}_T) \end{cases} \tag{7}$$

where x_1 and x_2 denote the front face of the projectile and the buffer respectively, M is the projectile mass, F is the drag force, m is the added mass, σ_T is the instantaneous stress, A is the stress area of the buffer (which can be assumed to be a constant), and ε_T and $\hat{\varepsilon}_T$ represent the strain and the strain rate, respectively.

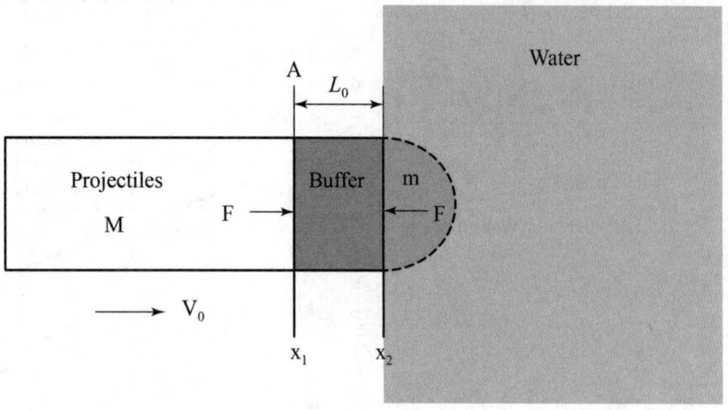

Figure. 12 Horizontal penetration of water by projectiles with cylindrical buffer.

Equation (1) can be integrated to give velocity as a function of time:

$$\begin{cases} \left(\dfrac{dx_1}{dt}\right)_T = -v_0 + \displaystyle\int_0^T \dfrac{F}{M}dt = -(v_1)_T \\ \left(\dfrac{dx_2}{dt}\right)_T = \displaystyle\int_0^T \dfrac{F}{m}dt = (v_2)_T \end{cases} \tag{8}$$

According to the elastic, perfectly plastic, strain rate dependent, compacting model [17], σ_T can be considered as a function of the strain rate $\dot{\varepsilon}_T$ before buffer compaction:

$$\sigma_T = \sigma_{yT} = (795 + 28.1\ln\dot{\varepsilon}_T)\rho^{1.64}\ \text{Pa} \tag{9}$$

where σ_{yT} is the instantaneous yield stress and ρ is the initial density of the rigid polyurethane foam (in kg/m^3). The strain rate in Equation (9) is an instantaneous value. Because the buffering stage is very short, the strain rate could not be easily measured with high precision but was instead assumed as a constant over the entire buffering stage of water entry. Therefore:

$$\dot{\varepsilon}_T = \bar{\dot{\varepsilon}} - \frac{v_0}{L_0} \tag{10}$$

where $\bar{\dot{\varepsilon}}$ is the average strain rate and L_0 is the length of the buffer. The motion equation can thus be written as

$$\begin{cases} (v_1)_T = v_0 - \dfrac{AT\rho^{1.64}}{M}(795 + 28.1\ln\bar{\dot{\varepsilon}}) \\ (v_2)_T = \dfrac{AT\rho^{1.64}}{m}(795 + 28.1\ln\bar{\dot{\varepsilon}}) \\ \bar{\dot{\varepsilon}} = \dfrac{v_0}{L_0} \end{cases} \tag{11}$$

The $(v_2)_T$ values were not obtained from the experiments since the front face of the buffer could not be readily observed. Figures 13 shows the measured, simulated, and calculated $(v_1)_T$ values for the projectiles with

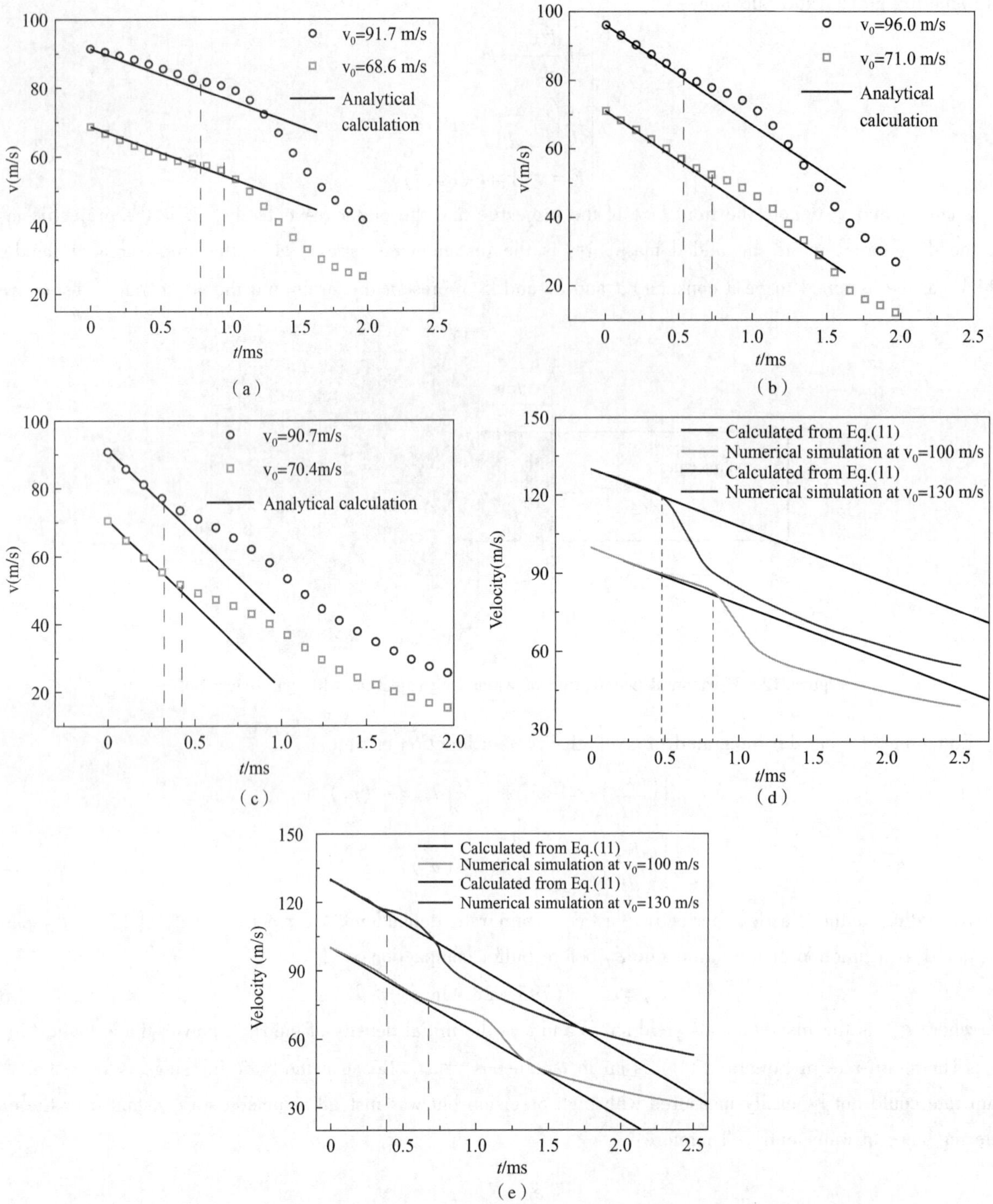

Figure. 13 Comparison between experimental/numerical results and model calculations from Equation (11).

(a) $\rho = 0.12$ g/cm^3; (b) $\rho = 0.18$ g/cm^3; (c) $\rho = 0.25$ g/cm^3; (d) $\rho = 0.15$ g/cm^3; (e) $\rho = 0.21$ g/cm^3

five different buffers. The red dotted – line is a rough dividing line between the buffering stage and the compaction stage. The model calculation gave $(v_1)_T$ values that were in good agreement with both experiments and simulations of the buffering stage. The buffering stage terminated when:

(1) the buffer head had become compacted,

(2) the extrusion stress fell below the yield stress $(\sigma_T < \sigma_y)$, or

(3) the buffer material broke apart before compaction took place.

5 Impact of Water Entry

5.1 Cushioning Effect of the Polyurethane Foam

Abuffer - free object is unlike a buffered object in that peak deceleration occurs at the beginning when it enters water (Figures 14 and 15). Since the peak deceleration is a great force on the object, it is imperative to effectively reduce the peak deceleration to ensure the structural integrity of the projectile upon water entry. Table 4 lists the maximum deceleration and reduction rate for each tested and simulated sample, which proved that the buffer head could effectively reduce the impact from water entry.

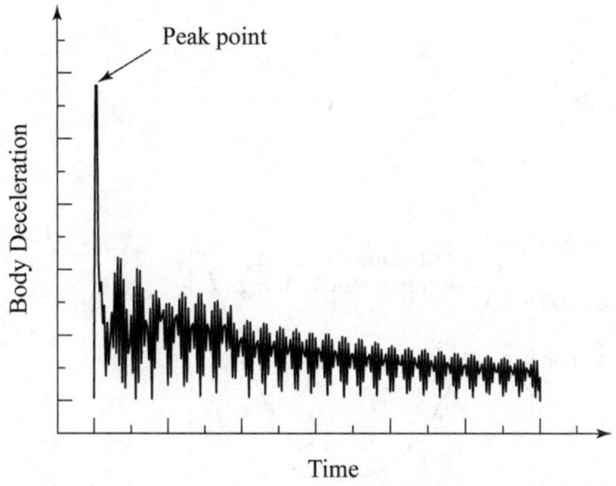

Figure. 14 Typical deceleration curve of the water entry of a rigid body.

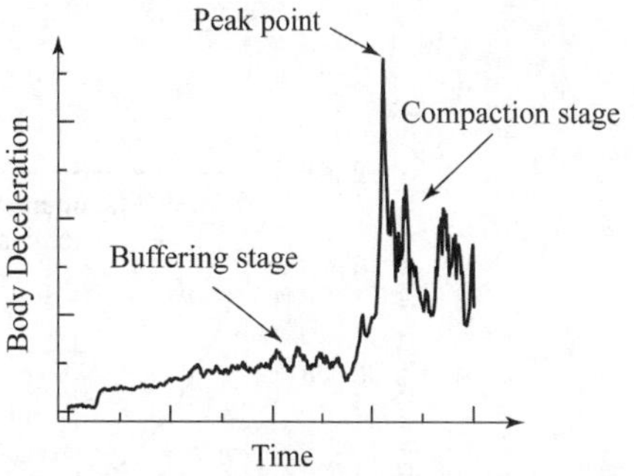

Figure. 15 Typical deceleration curve of projectile with cylindrical buffer head.

Table 4 Maximum deceleration and reduction rate.

Buffer density (g/cm^3)	Initial velocity (m/s)	Maximum deceleration (G)	Deceleration reduction
Buffer - free (Simulation results)	70.0	92 600	n/a
	98.4	161 300	n/a
0.12 (Experimental results)	69.8	9 600	89.6%
	91.7	48 700	69.8%
0.15 (Simulation results)	70	5 030	94.5%
	100	21 100	86.9%
0.18 (Experimental results)	71	3 400	96.3%
	96	12 700	83.9%
0.21 (Simulation results)	70	3 490	96.2%
	100	16 300	89.9%
0.25 (Experimental results)	70.4	11 400	87.7%
	90.7	22 000	86.4%

5.2 Empirical Formula of Maximum Deceleration

For projectiles equipped with cylindrical buffer head, peak deceleration is related to the density, diameter, and thickness of the cylindrical buffer, as well as the initial velocity of the projectile. The diameter and thickness of the cylindrical buffer were constants in the current experiment. Figure 16 shows that a good quadratic relationship could be found between peak deceleration and buffer density for the buffered projectiles fired at different initial velocity. This result prompted us to develop a new equation to describe the relations of (1) peak deceleration, (2) buffer density, and (3) initial velocity at a particular size of cylindrical buffer. By introducing the dimensionless parameters λ_1, λ_2, and λ_3, the new equation can be proposed as

$$\begin{cases} a_{\max} = a(\lambda_1, \lambda_2) \\ \lambda_1 = \dfrac{\rho_{PU}}{\rho_0} \\ \lambda_2 = \dfrac{v_0}{v_w} \\ \lambda_3 = \dfrac{D}{L_0} \end{cases} \tag{12}$$

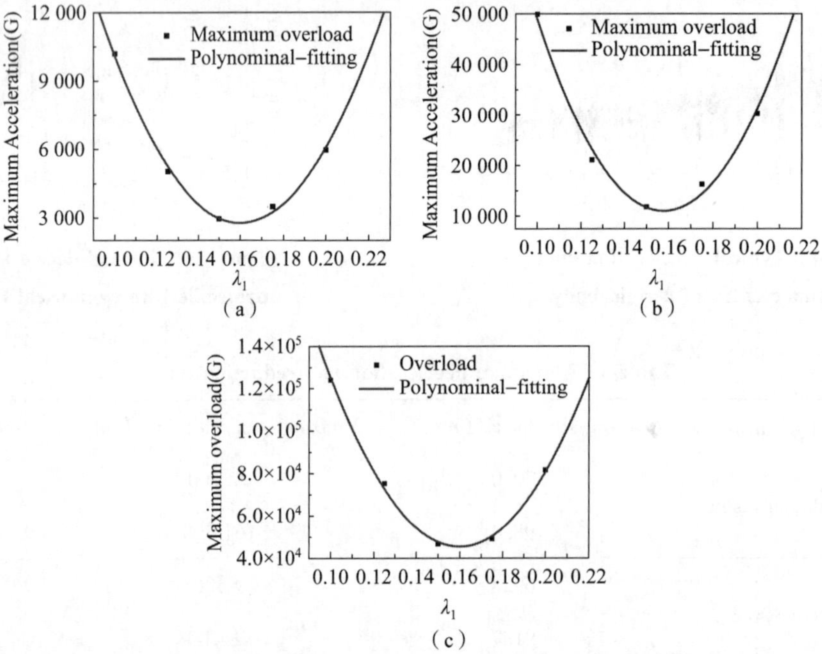

Figure. 16 Relationship between peak deceleration and average buffer density λ_1.

(a) $v_0 = 70$ m/s; (b) $v_0 = 100$ m/s; (c) $v_0 = 130$ m/s

where ρ_{PU} is the initial density of the rigid polyurethane foam, ρ_0 is the density of compacted PU, v_0 is the initial velocity, v_w is sound velocity in water ($v_w = 1\ 500$ m/s), and D and L_0 are the diameter and thickness of the buffer, respectively.

Table 5 Best fitted values of *a*, *b*, and *c* for different initial velocity.

Initial velocity (m/s)	λ_2	a	b	C
0	0	0	0	0
70	0.047	2.04×10^6	6.53×10^5	5.50×10^4
100	0.067	1.14×10^7	3.58×10^6	2.94×10^5
130	0.087	2.21×10^7	7.07×10^6	6.12×10^5

According to the fitting curves shown in Figure 16, at each initial velocity, the peak deceleration could be expressed as:

$$a_{max} = a(\lambda_1)^2 + b\lambda_1 + c \tag{13}$$

The parameters a, b, and c are associated with the initial velocity. Zero initial velocity would have meant no deceleration, and the curves thus passed through the origin. The values of a, b, and c were obtained by fitting the data points in Figure 16 and are listed in Table 5. Figure 17 shows the relationship between the parameters and λ_2, from which the parameters a, b, and c could be determined:

$$a = a_1\left(\frac{v_0}{v_w}\right)^{a_2} \tag{14}$$

$$b = -b_1\left(\frac{v_0}{v_w}\right)^{b_2} \tag{15}$$

$$c = c_1\left(\frac{v_0}{v_w}\right)^{c_2} \tag{16}$$

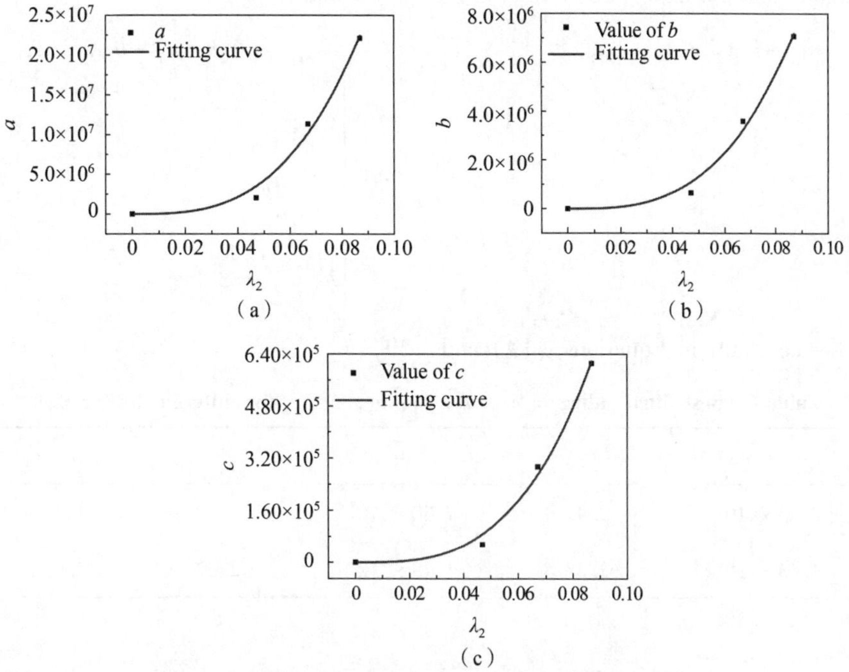

Figure. 17　Relationship between λ_2 and a, b, and c.

Hence, the peak deceleration for different initial velocity could be expressed as follows

$$\begin{cases} a_{max} = a_1\lambda_2^{a2} \cdot \lambda_1^2 - b_1\lambda_2^{b2} \cdot \lambda_1 + c_1\lambda_2^{c2} \\ \lambda_1 = \dfrac{\rho_{PU}}{\rho_0} \\ \lambda_2 = \dfrac{v_0}{v_w} \\ \lambda_3 = \dfrac{D}{L_0} = 2 \end{cases} \tag{17}$$

Table 6 lists the constants in Equation (17). The unit of a_{max} is G. The maximum acceleration of a projectile of any mass upon its water entry can be calculated according to Newton's second law as follows:

$$a_{max'} = \frac{ma_{max}}{M} \tag{18}$$

In this work, m = 3 kg. M is the mass of any other projectile, $a_{max'}$ is the corresponding maximum acceleration.

Table 6 Best fitted values of the constants in Equation (17).

a_1	a_2	b_1	b_2	c_1	c_2
3.56×10^{10}	3.02	1.23×10^{10}	3.05	1.46×10^9	3.18

5.3 Extension of the Empirical Formula

Section 4.2.1 only considered the case where diameter-to-thickness ratio of the PU buffer was $\lambda_3 = 2$ and proved that $\lambda_3 = 2$ could fit Equation (17). In order to verify the accuracy of the empirical model, $\lambda_3 = 1$ and $\lambda_3 = 3$ were considered additionally, and numerical simulations at different initial velocity and buffer density were carried out. Figures 18 and 19 illustrate the water entry for $\lambda_3 = 1$ and $\lambda_3 = 3$ respectively, assuming 0.12 g/cm^3 buffer head density and 70 m/s initial velocity. The deformation of the buffer head was basically the same as in $\lambda_3 = 2$. Both could well fit in the form of Equation (17) with a different set of parameters. The corresponding maximum deceleration could be expressed as follows for $\lambda_3 = 1$ and $\lambda_3 = 3$:

$$\begin{cases} a_{\max} = a_1 \lambda_2^{a_2} \cdot \lambda_1^2 - b_1 \lambda_2^{b_2} \cdot \lambda_1 + c_1 \lambda_2^{c_2} \\ \lambda_1 = \dfrac{\rho_{PU}}{\rho_0} \\ \lambda_2 = \dfrac{v_0}{v_w} \\ \lambda_3 = \dfrac{D}{L_0} = 1 \end{cases} \quad (19) \text{ and } \begin{cases} a_{\max} = a_1 \lambda_2^{a_2} \cdot \lambda_1^2 - b_1 \lambda_2^{b_2} \cdot \lambda_1 + c_1 \lambda_2^{c_2} \\ \lambda_1 = \dfrac{\rho_{PU}}{\rho_0} \\ \lambda_2 = \dfrac{v_0}{v_w} \\ \lambda_3 = \dfrac{D}{L_0} = 3 \end{cases} \quad (20)$$

Table 7 lists the constants in Equations (19) and (20).

Table 7 Best fitted values of a_1, a_2, b_1, b_2, c_1, c_2 for different buffer size.

λ_3	a_1	a_2	b_1	b_2	c_1	c_2
1	8.25×10^8	2.41	7.30×10^8	2.79	1.49×10^8	3.07
3	4.65×10^9	2.25	1.79×10^9	2.32	2.95×10^8	2.54

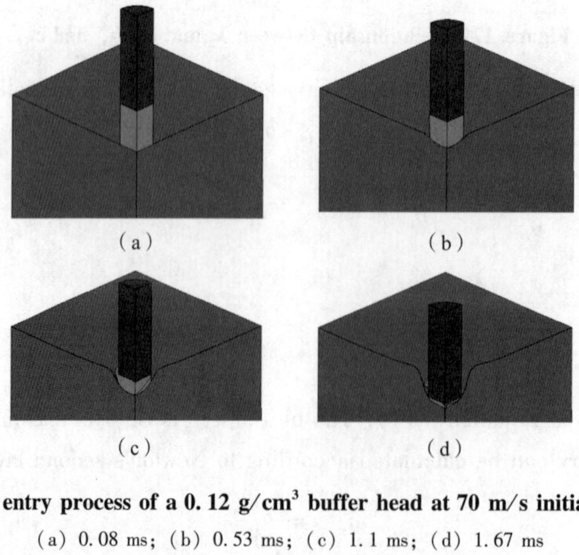

Figure. 18 The water entry process of a 0.12 g/cm^3 buffer head at 70 m/s initial velocity when $\lambda_3 = 1$.

(a) 0.08 ms; (b) 0.53 ms; (c) 1.1 ms; (d) 1.67 ms

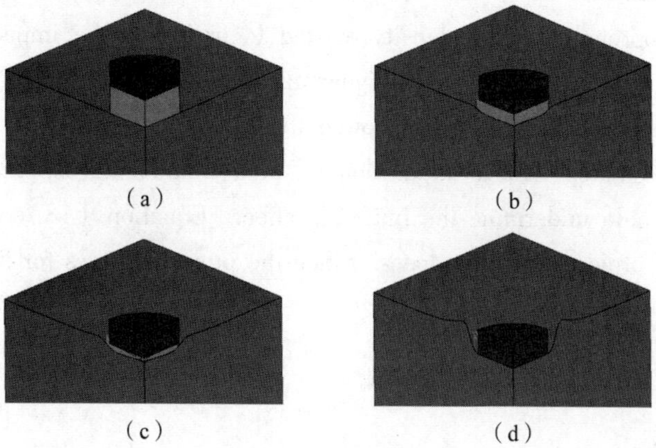

Figure. 19 The water entry process of a 0.12 g/cm^3 buffer head at 70 m/s initial velocity when λ_3 = 3.

(a) 0.08 ms; (b) 0.53 ms; (c) 0.79 ms; (d) 1.36 ms

6 Discussion

This paper presents a combined theoretical and experimental investigation onthe water entry behaviors of projectiles with cylindrical buffer head. Results showed that the buffer head made of polyurethane foam could greatly reduce impact, and the water entry process could be divided into two stages, *i. e.*, the buffering stage and the compaction stage. The projectile dynamics in the buffering stage was examined both experimentally and theoretically. The results showed that the deceleration of the projectiles was quite low in the first stage and depended on the mechanical properties of materials. Combining the elastic, perfectly plastic, strain rate dependent, compacting model, the equation of motion could be described as

$$\begin{cases} (v_1)_T = v_0 - \dfrac{AT\rho^{1.64}}{M}(795 + 28.1\ln\bar{\dot{\varepsilon}}) \\ (v_2)_T = \dfrac{AT\rho^{1.64}}{m}(795 + 28.1\ln\bar{\dot{\varepsilon}}) \\ \bar{\dot{\varepsilon}} = \dfrac{v_0}{L_0} \end{cases} \tag{21}$$

However, it should be noted that $\bar{\dot{\varepsilon}}$ in reality dose not remain constant during water entry, and its precise mathematical expression should be further studied. Nevertheless, good agreements were found between the experimentally observed velocity – time history and that predicted by the analytical cavity model (Figure 13). A number of major features, such as penetration distance, projectiles velocity, acceleration in the buffering stage with different kinds of buffer head, *etc.*, could also be well captured by the analytical model.

To accurately describe the reduction of impact, the maximum deceleration for different initial velocities and different buffer densities were further examined in numerical simulations. Analyses showed that the peak deceleration was reached after the buffer head had compacted, and Figure 16 shows a typical curve. The maximum deceleration could be described as

$$\begin{cases} a_{max} = a_1\lambda_2^{a_2} \cdot \lambda_1^2 - b_1\lambda_2^{b_2} \cdot \lambda_1 + c_1\lambda_2^{c_2} \\ \lambda_1 = \dfrac{\rho_{PU}}{\rho_0} \\ \lambda_2 = \dfrac{v_0}{v_w} \\ \lambda_3 = \dfrac{D}{L_0} \end{cases} \tag{22}$$

The parameters varied with the size of the buffer head (λ_3), whose typical values are shown in Table 6. The results showed that an optimal material density existed for reducing the impact under different condition. Projectiles with higher initial velocity suffered from higher impact during water entry. In addition, λ_3 also notably affected the impact, since numerical simulations showed that peak deceleration increased significantly when λ_3 was higher. It could be speculated that a larger contact area would increase resistance, although excessive thinning of the buffer head would undermine the buffering effect. Equation (18) could be applied to determine the maximum deceleration of projectiles of any mass. Since the projectile mass for fitting was 3 kg in the current tests, Equation (18) could be written as

$$a_{max'} = \frac{3a_{max}}{M} \tag{23}$$

Note that the unit of a_{max} is G.

7 Conclusions

In this paper, the water entry behavior of projectiles with cylindrical buffer head was studied experimentally and numerically. The current work, within its scope, allows the following main conclusions to be drawn:

(1) An effective numerical simulation method was developed for the water entry of projectiles with buffer heads. The velocity – time history and maximum deceleration of the projectile's water entry obtained from the simulation were in good agreement with experimental observations.

(2) An analytic kinematics equation was developed for projectiles before the compaction of buffer head. The predictions of the equation were in good agreement with experimental observations and numerical simulations.

(3) An empirical formula was proposed to accurately describe the maximum deceleration during the water entry process based on the experimental and numerical results.

(4) An optimal buffer density for reducing impact existed under each condition. In both the experiments and the simulations, the impact became stronger when the initial velocity was high and when the buffer's diameter – to – thickness ratio was large.

Acknowledgements

The authors thank Qian Haitao and Zhang Bo for their kind help in experiments and Dr. Zhu W. and Dr. An X. for the guidance on submission. All of them are members of the State Key Laboratory of Explosion Science and Technology, Beijing Institute of Technology, P. R. China.

References

[1] VON KARMAN T. The impact of seaplane floats during landing. NACA TN 321. Washington, USA: National Advisory Committee for Aeronautics, October1929. pp. 2 – 8.

[2] Wagner H. Trans. phenomena associated with impacts and sliding on liquid surfaces [J]. Math Mech, 1932, 12 (4): 193 – 215.

[3] SEDOV L. The impact of a solid body floating on the surface of an incompressible fluid [R]. Report 187, CAHI, Moscow, 1934.

[4] YU Y. Virtual masses of rectangular plates and parallelepipeds in water [J]. Journal of applied physics, 1945, 16 (11): 724 – 29.

[5] SHIFFMAN M, SPENCER D C. The force of impact on a sphere striking a water surface (approximation by the flow about a lens). AMP Report 42. 1R AMG – NYU no. 15, Applied Mathematics Panel, February 1945.

[6] SHIFFMAN M, SPENCER D C. The force of impact on a sphere striking a water surface (second approximation). Report 42. 2R AMG – NYU no. 133, Applied Mathematics Panel, July 1945.

第 七 部 分
其他弹药战斗部技术

火箭发射起始扰动仿真研究①

孔　炜②，朱春梅，冯顺山

摘　要：将图论引入多体动力学系统描述，采用 Roberson \ Wittenburg 方法建立了火箭发射系统的多体动力学模型，在此基础上编制了火箭发射动力学仿真软件，对某火箭武器发射过程起始扰动进行了仿真计算。

关键词：火箭；起始扰动；数值仿真

1　引言

火箭武器具有火力突然、猛烈、机动能力强、快速反应性能好等优点，它主要利用其强大的面射火力，实施战役战术的突袭，毁坏敌纵深目标，是现代战争中不可缺少的主要压制支援兵器。然而，火箭武器射击密集度低、落点散布大是个致命缺点，成为其继续发展和战术上更广泛应用的主要障碍。

提高火箭武器射击密集度、减小落点散布是火箭武器面临的主要技术难题。火箭发射过程中产生的起始扰动是引起火箭落点散布的重要因素之一。发射过程中由于存在火箭发射装置振动、火箭本身缺陷(推力偏心、动不平衡、质量偏心等)、火箭燃气流冲击以及弹管相互作用等扰动源，火箭将产生偏离理想运动轴线的微小角运动。这种微小角运动将导致火箭的不良发射，产生火箭起始扰动。火箭武器射击密集度低下的主要原因之一就是未能深入研究引起火箭起始扰动的各种因素及其作用机理。进行火箭发射过程的动力学仿真研究，从而更深入地认识火箭起始扰动的生成机理，并设法利用或减小起始扰动，将会提高火箭密集度、减小射弹散布，对发展我国的火箭武器、提高其作战性能，具有极其重要的意义。

2　火箭发射动力学模型

2.1　火箭武器系统结构的物理模型

火箭武器系统是一个复杂的动力学系统，不仅发射装置是由许多弹性元件以多种联结方式组合而成，而且火箭在定向管内的运动也是一个在发射装置振动情况下弹管相互作用的过程。

罗伯逊（Roberson）和威腾伯格（Wittenburg）首先提出了利用图论方法的有向图来表示多体系统的结构联系。Roberson \ Wittenburg 方法具有物理概念明确，数学公式规范且易推导的特点，便于实现计算机程序化。

按照模型能够反映实际系统结构的主要动力特性的原则，对火箭武器系统进行模型简化如下所述。一个典型的火箭武器系统动力学模型的结构如图 1 所示。

（1）火箭发射装置系统由运载体 B_1、回转体 B_2、俯仰体 B_3 和定向管 B_n（$n=4$，…，N_{tube}）等组成，这些子结构均简化为刚体 v 各子结构本身的弹性由它们之间连接铰处的弹性考虑。

（2）各子结构之间以弹性连接和刚性连接两种方式结合，形成火箭发射装置整体。其中，运载体相对地面具有 6 个自由度，记为在运载体和地面之间存在虚铰 O_1；回转体相对于运载体只具有绕回转轴方向的弹性连接，其余方向均和运载体刚性连接，相当于在回转体和运载体之间存在一个单自由度转动铰

① 原文发表于《火炮发射与控制学报》1998（7）。

② 孔炜：工学博士，1996 年师从冯顺山教授，研究超近程反导火箭发射起始扰动，现工作单位：中央电视台。

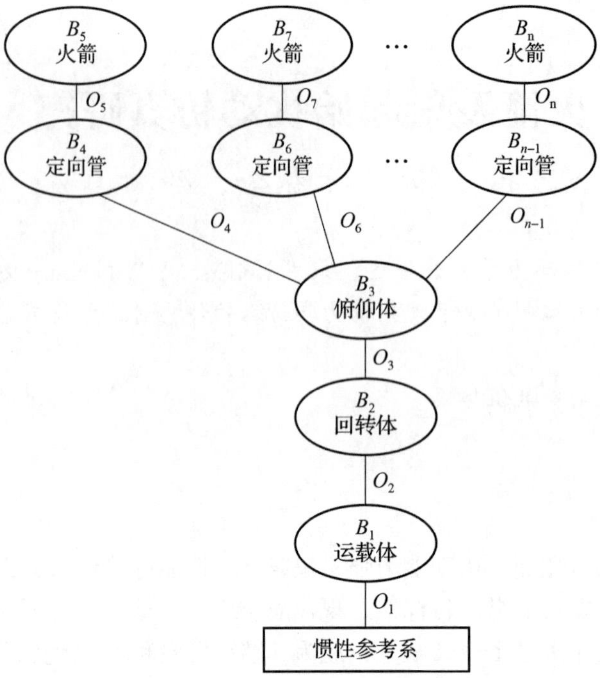

图 1　典型的火箭武器系统动力学模型的结构

一圆柱铰 O_2；俯仰体相对于回转体通过耳轴相联，发射过程中绕耳轴作俯仰转动，其余方向与回转体刚性连接，视为在俯仰体和回转体之间存在一个单自由度转动铰—圆柱铰 O_3；定向管相对于俯仰体绕刚性支承点具有俯仰和偏航两个方向上的弹性连接，其余方向和俯仰体刚性连接，简化为在定向管和俯仰体之间存在二自由度转动铰—万向铰 O_t($t=4$，…，N_{tube})。

（3）各子结构之间的弹性连接为无质量的等效弹簧、等效阻尼器并联而成，且阻尼为黏性阻尼；

（4）各发火箭的外形相同。火箭定心部与定向管之间存在间隙，火箭存在的某些缺陷（如火箭发动机推力偏心、质量偏心、动不平衡等）相对于弹体的方位不随时间变化。

（5）发射过程中，火箭具有相对定向管的移动和转动，其弹/管界面根据不同类型的火箭及管内不同运动时期而采用不同的铰。尾翼式火箭在约束期的弹/管界面用万向铰及螺旋铰，半约束期的弹/管界面用万向铵。涡轮式火箭在约束期的弹/管界面用万向铰及滑移柱铰，半约束期的弹/管界面用球铰。火箭飞离定向管时，和相应定向管之间的联系解除；

（6）各发火箭发动机推力大小遵从同一的推力－时间曲线，不考虑火箭发动机点火过程的影响。

2.2　火箭武器系统结构的数学描述

采用 Roberson\Wittenburg 方法利用有向图描绘火箭武器系统内各个刚体之间的联系状况，作为系统的结构。

典型的火箭武器系统动力学模型的结构有向图如图 2 所示。有向图中的每个顶点代表一个刚体，记作 $B_i(i=1,2,\cdots,n)$，下角标 i 代表该刚体在系统中的刚体编号。铰点用有向图中联结顶点的有向弧来表示，记作 0_j（$j=1$，2，…，n），下角标 j 代表该铰在系统中的铰编号。有向弧的方向确定两个相关联刚体的相对运动的参照关系。有向弧（铰）与所联系的两个顶点（刚体）之间的关系称为关联。如图 2 中铰 O_6 与 B_6、B_3 相关联。B_i 到 B_j 的一列弧组成了 B_i 到 B_j 路。

若系统中任意两个刚体之间的路有且仅有一条，即构成了所谓的树系统。对于树系统中刚体数与铰数相同（不包括零刚体）。为了方便描述树系统的结构，引入以下的术语和规则标号方法：

（1）零刚体 B_0。零刚体是已知运动规律的系统外刚体。例如，对于地面火箭武器系统，零刚体为地面惯性坐标系，其速度、加速度均可强制为零。对于舰载、机载火箭武器系统，可将舰艇、飞机作为零

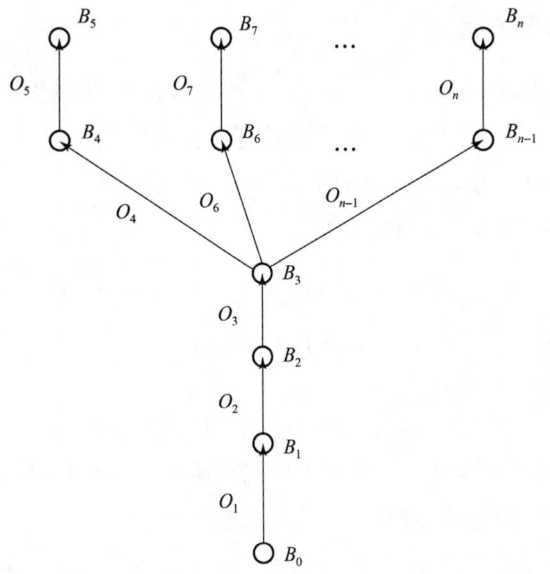

图 2　典型的火箭武器系统动力学模型的结构有向图

刚体处理，其运动规律取决于舰艇、飞机的运动时间历程。

（2）内侧刚体、外侧刚体。若系统中刚体 B_i 在零刚体 B_0 至刚体 B_j 的路上，则刚体 B_j 为刚体 B_i 的外侧刚体，刚体 B_i 为刚体 B_j 的内侧刚体。

（3）内接铰、内接刚体。将刚体 B_j 与其内侧刚体 B_i 相联结起来的铰为刚体 B_j 的内接铰，刚体 B_i 为刚体 B_j 的内接刚体。

（4）全关联矩阵 $\bar{\boldsymbol{S}}$。系统内各刚体以及零刚体与铰的关联状况可以用全关联矩阵 $\bar{\boldsymbol{S}}$ 来描述，矩阵的行号与列号分别与刚体和铰的标号相对应，其第 i 行第 j 列元素定义为

$\bar{S}_{i,j}=1$　O_j 铰与 B_i 刚体关联且以 B_i 为起点；

$\bar{S}_{i,j}=-1$　O_j 铰与 B_i 刚体关联且以 B_i 为终点；

$\bar{S}_{i,j}=-1$　O_j 铰与 B_i 无刚体关联。

$$(i=0,1,\cdots,n \qquad j=1,2,\cdots,n) \tag{1}$$

零刚体关联矩阵 $\boldsymbol{S}_0$：$\boldsymbol{S}_0$ 为全关联矩阵 $\bar{\boldsymbol{S}}$ 的第 1 行。

$$\bar{S}_{i,j}=\begin{cases}1(j=1)\\0(j=2,3,\cdots,n)\end{cases} \tag{2}$$

关联矩阵 $\boldsymbol{S}$：若 $\bar{S}$ 内不包括零刚体和 1 号铰，则得到关联矩阵 S。

（5）通路矩阵 $\boldsymbol{T}$。通路矩阵 $\boldsymbol{T}$ 用来描述系统内各刚体与零刚体之间的通路状况。与关联矩阵 $\boldsymbol{S}$ 相反，通路矩阵 $\boldsymbol{T}$ 行号对应于铰号，列号对应于刚体号，其第 i 行第 j 列元素定义为

$T_i=1$　　O_i 铰属于 B_0 到 B_i 的路且指向 B_0；

$T_1=-1$　　O_i 铰属于 B_0 到 B_i 的路且背离 B_0；

$T_i=1$　　O_i 铰不属于 B_0 到 B_i 的路。

$$(i,j=1,2,\cdots,n) \tag{3}$$

关联矩阵 $\boldsymbol{S}$ 和通路矩阵 $\boldsymbol{T}$ 均能精确地描述树系统的结构。$\boldsymbol{S}$、$\boldsymbol{T}$、$\boldsymbol{S}_0$ 之间有以下关系：

$$(\boldsymbol{S}_0\boldsymbol{T})^{\boldsymbol{T}}=\boldsymbol{T}^{\boldsymbol{T}}\boldsymbol{S}_0^{\boldsymbol{T}}=-1$$
$$\boldsymbol{TS}=\boldsymbol{ST}=\boldsymbol{E} \tag{4}$$

式中，$\boldsymbol{E}$ 为 n 阶单位矩阵。

2.3 火箭发射过程动力学方程

设火箭武器系统由 n 个刚体 $B_i(i=1,2,\cdots,n)$ 组成。刚体 B_i 的质量为 m_i，中心惯量张量为 J_i，质心 C_1，相对惯性基基点的矢径为 r_1，刚体 B_i 绕心转动角速度为 ω_i，作用于刚体 B_i 的外力的主矢及相对质心的主矩分别为 $\boldsymbol{F}_1^g$、$\boldsymbol{M}_1^G$，系统内力完成的虚功率总和为 δP。

则由若丹（Jourdain）形式的动力学普遍方程，有

$$\sum_{t=1}^{n}[\delta\dot{r}_i\cdot(m_i\ddot{r}_i-\boldsymbol{F}_1^g)+\delta\cdot\omega_1\cdot(J_1\cdot\omega_1+\varepsilon_i-\boldsymbol{M}_1^g)]-\delta P=0$$

$$(i=1,2,\cdots,n) \tag{5}$$

式中，ε_i 为转动时惯性的影响，定义为

$$\varepsilon_i=\omega_i\times(J_i\cdot\omega_i)(i=1,2,\cdots,n) \tag{6}$$

设 F_j^a、M_i^a 表示 O_i 铰联结的内接刚体 $B_{i(j)}$，作用于外接刚体 B_i 的铰内作用力的主矢及相对 O_j 点的主矩，F_k^e 表示力元作用力，则刚体之间的铰和力元的内力的总虚功率 δP 为

$$\delta P=\sum_{j=1}^{n}(\delta V_J\cdot F_J^a+\delta\Omega_J\cdot M_J^a)+\sum_{k=1}^{m}\delta V_J^e\cdot F_k^e \tag{7}$$

由于火箭武器系统内各刚体之间存在着铰的运动学约束，故动力学方程中的各个变分都不是独立变分，必须将其化作广义速度的变分 $\delta\dot{q}$表示，由上节火箭多体系统的运动学分析可得到这些变分表达式

$$\begin{aligned}\delta\omega^{\mathrm{T}}&=\delta\dot{q}^{\mathrm{T}}\beta^{\mathrm{T}}\\ \delta\dot{r}^{\mathrm{T}}&=\delta\dot{q}^{\mathrm{T}}\alpha^{\mathrm{T}}\\ \delta\Omega^{\mathrm{T}}&=\delta\dot{q}^{\mathrm{T}}p^{\mathrm{T}}\\ \delta V^{\mathrm{T}}&=\delta\dot{q}k^{\mathrm{T}}\\ \delta V^{eT}&=\delta\dot{q}^{\mathrm{T}}\gamma^{\mathrm{T}}\end{aligned} \tag{8}$$

式中，$\gamma=-(\alpha^{\mathrm{T}}S^e+\beta^{\mathrm{T}}\times C^e)$。

将上述表达式代入动力学普遍方程，经整理后得到

$$\delta\dot{q}^{\mathrm{T}}(\boldsymbol{A}\ddot{q}-B)=0 \tag{9}$$

式中，n 阶标量方阵 $\boldsymbol{A}$ 及标量列阵 $\boldsymbol{B}$ 为

$$\boldsymbol{A}=\alpha^{\mathrm{T}}\cdot m\alpha+\beta^{\mathrm{T}}\cdot J\cdot\beta$$

$$\boldsymbol{B}=\alpha^{\mathrm{T}}\cdot(F^g-mu)+\beta^{\mathrm{T}}\cdot(M^g-J\cdot\sigma-\varepsilon)\cdot M^u+\gamma\cdot F^e+(k+hH)F^a$$

由于各广义坐标均为独立变量，故有

$$A\ddot{q}=B$$

上式即为火箭武器系统动力学微分方程组。

对特定的火箭武器系统，其系统结构、各刚体的惯量参数、铰的位置和约束性质以及铰的作用力的变化规律均是确定的，动力学方程中的系数矩阵 $\boldsymbol{A}$ 就完全取决于广义坐标 q，矩阵取决于广义坐标 q 及其一次导数 $\dot{q}$。由于 $\boldsymbol{A}$ 为实对称矩阵，必然有逆阵存在，故方程组 $\boldsymbol{A}\ddot{q}=B$ 必存在解，可以利用各种方法对其进行数值积分。

由上述火箭武器系统动力学微分方程组可以求得发射过程中火箭/发射装置各部分的运动参数，发射装置残余振动及火箭在定向管内的各种运动状态。

3 火箭发射起始扰动仿真结果

3.1 火箭发射起始扰动数值仿真软件

根据本文建立的火箭发射过程动力学模型和推导的动力学方程，采用面向对象的 C ++ 计算机语言编制了计算机数值仿真的发射动力学仿真软件。源程序共有 227 个函数、19 种结构变量、4 种类变量。整个软件的源代码长度约为 12 000 行。有 DOS 和 Windows 两种版本。由于在软件设计过程中注意了结构

设计准则，并采用了面向对象的 C++ 语言编程，不但使源程序有较好的可读性和维护性能，而且大大缩小了源代码的规模。

3.2　54 mm 火箭发射起始扰动数值仿真

进行 54 mm 单管火箭武器系统发射过程的起始扰动仿真的目的主要是通过将数值仿真结果与该火箭起始扰动实弹射击试验数据对比，来检验本文建立的火箭起始扰动的发射动力学模型的合理性和推导的动力学方程的正确性。54 mm 单管火箭武器系统动力学模型结构图如图 3 所示。模型包括 4 个刚体和 4 个嵌，8~12 个自由度。图 4 和图 5 为通过计算机数值仿真得到的 54 mm 火箭武器系统发射过程的起始扰动角位移及角速度随时间的变化过程。

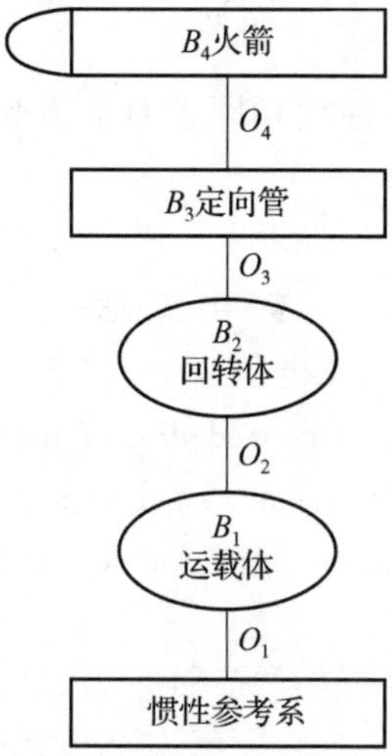

图 3　54 mm 单管火箭武器系统动力学模型的结构图

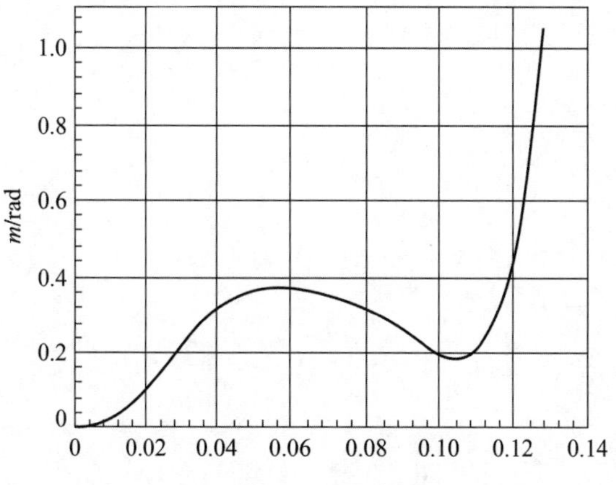

图 4　54 mm 火箭扰动角位移时间历程

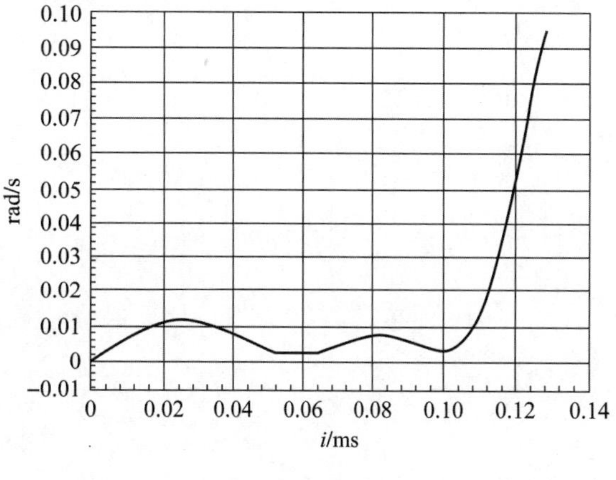

图 5　54 mm 火箭扰动角速度时间历程

54 mm 火箭的起始扰动试验实测结果与数值仿真结果如表 1 所示。

表 1　实测结果与数值仿真结果对比

	实测值	数值仿真值
$E_{\varphi 0}$/mrad	0.732 6	0.762 1
$E_{\varphi 0}$/mrad	0.566 2	0.963 5
$E_{\varphi 0}$/(rad · s^{-1})	0.040 2	0.032 0
$E_{\varphi 0}$/(rad · s^{-1})	0.044 0	0.042 1

由表 1 可见，试验实测起始扰动值与数值仿真结果基本一致。可知本文所建立的模型是合理的，能够真实地反映出火箭武器系统的发射过程中火箭的运动情况，数值仿真得到的结果是可信的。本文所建立的火箭发射动力学模型可以用来进行火箭发射起始扰动的研究。

4 结束语

(1) 根据多体动力学原理进行火箭起始扰动仿真研究，采用由 Roberson 和 Wittenburg 提出的将图论应用于多体系统描述的 R/W 方法，在解决火箭武器系统发射动力学问题中的常见的变自由度、结构时变问题时，R/W 方法有其独具的优越性。

(2) 针对火箭武器的结构特点，在模型中考虑了尾翼式火箭和涡轮式火箭发射时的弹/管相互约束界面，导出了其动力学相容方程。

(3) 研制的仿真软件可以用于计算各种扰动因素对火箭起始扰动的影响，具有微分方程自动生成功能。

参考文献

[1] ROBERSON R E. Computer - oriented dynamics modelling of spacecraft [R]. IAF paper77 - A11, 1977.

[2] ROBERSON R E. Dynamics equations for connected rigid bodies [J]. Journal of spacecraft Rockets, 1977.

[3] ZHAO W X, KONG W, WANG Z J. The measurement of thrust misalignment in rocket motors by using load identification technique. ISTM/95, 1995.

[4] 孔炜. 火箭推力偏心对发射起始扰动的影响研究 [D]. 太原：华北工学院，1996.

胶凝燃料的粘弹性与爆炸分散试验研究①

贵大勇[1②]，冯顺山[1]，曾学明[2]

(1. 北京理工大学机电工程学院，北京 100081；2. 防化研究院第五研究所，北京 102205)

摘　要：**目的**：研究胶凝燃料的粘弹性对分散性的影响。**方法**：分别测定胶凝燃料的黏弹性和爆炸分散效果，考查不同粘弹性对燃料分散性的影响规律，并建立函数关系。**结果**：得到了胶凝燃料粘弹性等物化性质对其分散性的影响规律。**结论**：胶凝燃料的黏性和弹性对其分散性均有影响，但弹性的影响更明显，所得分散性的函数表达式可用以预测胶凝燃料的分散特性。

关键词：胶凝燃料；黏弹性；爆炸分散

Experimental Research on the Viscoelasticities and Explosive Dispersion of Gelled Fuels

Gui Dayong[1], Feng Shunshan[1], Zeng Xueming[2]

(1 School of Mechano－Electronics Engineering, Beijing Institute of Technology, Beijing 100081;
2 Research Institute of Chemical Defense, Beijing 102205)

Abstract: Aim To study the influence of viscoelasticities of gelled fuels on the dispersion. **Methods** The viscoelasticities and the explosive dispersion effects of gelled fuels were measured respectively. The experimental data were investigated to obtain the relationship between the properties and the dispersion effects of gelled fuels. **Results** The influence relations of the viscoelasticities with the dispersion effects of gelled fuels were obtained. **Conclusion** The dispersion effects of gelled fuels are related to both the viscosity and the elasticity, in which the influences of elasticity is more significant. The functional relations of the dispersion with the viscoelasticities can be used to predict the dispersion proper ties of gelled fuel.

Key words: gelled fuel; viscoelasticities; explosive dispersion

中心爆炸分散机构可实现药剂的抛撒，其分散程度对药剂的燃烧或爆炸效果有着直接的影响。而装置中药剂的分散除了与其结构，如壳体、抛撒药等有关外，还与被抛撒药（药剂）本身的性质（如黏度、弹性等）有着直接的关系。在通常研究中主要从爆炸动力学、流体力学的角度出发，研究装置结构、抛撒炸药等对被抛撒药分散的影响[1-3]。美国 J. E. Matta 等对聚甲基丙烯酸甲酯的稀溶液进行过高速气流下的分散实验，考察黏弹性对其分散液滴大小的影响[4]。有关在爆炸分散方式下被抛撒药的性质对分散性的影响，国内外研究较少。本课题对不同配比的胶凝燃料的黏弹性进行了测定，将近年来发展起来的图像分析法用于分散效果的测试评价，研究胶凝燃料的黏弹性对其爆炸分散的影响规律。

1 实　验

1.1 燃料制备

胶凝燃料由油料与胶凝剂混合调制而成，胶凝剂为德国产聚异丁烯（PIB）（黏均相对分子质量

① 原文发表于《北京理工大学学报》1999 (S1)。

② 贵大勇：工学博士，1997 年师从冯顺山教授，研究固液燃料混合 FAE 战斗部设计，现工作单位：深圳大学。

47×10^5)；作为溶剂的油料为国内抚顺生产的2#喷气燃料煤油。制备4种不同配比的试样，以获得不同黏弹性的样品，代号及配比列于表1中。

表1 胶凝燃料代号及其配比

胶凝代号	F2	F3	F4	F6
配比①f/%	2	3	4	6

注：①试样配比指100 mL煤油中所添加的PIB的质量。

1.2 黏弹性测定

试样零切黏度的测定采用落球法进行，已知小球、试样密度和小球直径，根据斯托克斯公式，可计算出试样的零切黏度 Z_0。试样的表观黏度采用英国Dimos公司生产的管式黏度计测试，通过试验测量并计算得到剪切应力 f_w、剪切速率 V_w 等参数，试样表观黏度 Z_a 即可由公式 $Z_a=f_w/V_w$ 计算得到。并测试试样的挤出胀大比值 B 以表征其弹性。有资料报道，流体的相对松弛时间可以较好地同其分散粒子尺寸联系起来[4]。而相对松弛时间则可以通过挤出胀大曲线很容易得到。本课题以配比为2%的挤出胀大曲线为基准，各试样的相对松弛时间 $\theta/\theta_{2\%}$[4]可由各试样挤出胀大曲线平行移动到参考流体的挤出胀大比曲线所移动的量计算得到。

1.3 分散试验及其测试

用1 000 mL玻璃瓶装填740~800 g胶凝燃料作为爆炸装置，固定装置结构、装药结构、抛撒药条件，在野外进行不同黏弹性燃料的静爆分散试验。采用摄像机及VIPAS图像分析处理系统测量爆炸分散而落下的燃料液块大小和分布，得到液块有效平均面积 $\bar{S}$、有效分散个数 N、有效平均质量 $\bar{m}$ 和有效分散率 P，用于表征胶凝燃料的分散性[5]。这里有效的概念系指质量在0.2~10.0 g范围内的液块。

2 结果与讨论

2.1 胶凝燃料的黏弹性

表2所示为测得的试样密度 d 和零切黏度 Z_0。随着聚异丁烯配比的增加，试样密度略有提高，其表征样品黏性的零切黏度值提高明显。

表2 试样密度和零切黏度

试样	F2	F3	F4	F6
d/(g·cm^{-3})	0.778 7	0.779 9	0.781 0	0.783 4
Z_0/(Pa·s)	3.775 3	34.42	101.8	986.3

采用管式黏度计（tube viscometer）测试的表观黏度 Z_a 及挤出胀大比B随剪切速率的变化，经数据处理，结果如表3所示。

表3 不同剪切速率下各试样的表观黏度与挤出胀大比

V_w/s^{-1}	F2		F3		F4		F6	
	Z_a/(Pa·s)	B	Z_a/(Pa·s)	B	Z_a/(Pa·s)	B	Z_a/(Pa·s)	B
4	1.280	0.841	5.350	0.915	11.700	1.035	32.800	1.132

续表

V_w/s^{-1}	F2		F3		F4		F6	
	$Z_a/(Pa \cdot s)$	B	$Z_a/(Pa \cdot s)$	B	$Z_a/(Pa \cdot s)$	B	$Z_a/(Pa \cdot s)$	B
10	0.845	0.898	3.000	0.958	6.350	1.083	15.700	1.190
30	0.410	0.952	1.360	1.048	2.740	1.158	7.000	1.265
60	0.255	1.000	0.812	1.120	1.630	1.226	3.930	1.341
100	0.187	1.040	0.541	1.182	1.090	1.295	2.500	1.418
300	0.091	1.202	0.250	1.372	0.468	1.535	1.020	1.680

取各试样 B 值较大时的 $\theta/\theta_{2\%}$ 值作为相对松弛时间加以比较，结果如表 4 所示。

表 4　试样相对松弛时间

试样	F2	F3	F4	F6
$\theta/\theta_{2\%}$	1.0	1.5	2.8	4.5

2. 2　胶凝燃料的分散性及其与黏弹性的关系

各试样的分散性结果列于表 5 中。

表 5　各试样的分散性结果

试样	$\bar{S}/cm^2$	N	$P/\%$	$\bar{m}/g$
F2	19.786 3	903	82.59	0.749 5
F3	17.853 0	549	71.00	1.045 8
F4	16.613 0	391	56.12	1.157 8
F6	16.754 2	225	52.73	1.799 4

表 2 ~ 表 5 中所列数据表明油块分散个数、单位面积的质量以及有效分散率随着试样不同即黏弹性不同而有明显变化。其中零切粘度和表观黏度的对数与表示分散性的 N，P 和 $\bar{m}$ 基本上呈线性对应关系但在很宽的粘度范围内，分散性变化不大，表明了黏性行为虽对分散性有一定的影响，但影响不明显。从相对松驰时间 $\theta/\theta_{2\%}$ 对分散性影响关系来看，相对松弛时间的较小变化能引起分散个数、分散率以及液块平均质量较大的变化。液块体积或整体尺寸与液块平均质量呈正比关系，因此上述数据也说明了弹性较之粘性对分散液块的大小影响更大。

通过最小二乘法并编程分别求得 $\bar{m}$、N 和 P 对配比 f、零切黏度 Z_0、表观黏度 Z_a、相对松弛时间 $\theta/\theta_{2\%}$、密度 d 的函数关系。

$$\bar{m}=0.768\,335\,3f+9.542\,149\times10^{-4}Z_0-2.985\,862Z_a-0.053\,984\,4\theta/\theta_{2\%}+0.389\,444\,1d$$

$$N=-952.905\,6f-0.432\,576\,9Z_0+3\,908.941Z_a-23.420\,68\theta/\theta_{2\%}+1\,958.708d$$

$$P=-55.912\,02f-5.934\,612\times10^2Z_0+381.810\,3Z_a-29.545\,88\theta/\theta_{2\%}+171.474\,5d$$

求解时采用的 Z_a 是指 V_w 为 300 s^{-1}时的表观黏度值。

以上 3 个关系式对（f，Z_0，Z_a，$\theta/\theta_2\%$，d）5 个自变量的关系在给定数两端范围内是准确的，但同时起作用时为佳，实际上各自变量对 $\bar{m}$、N、P 的影响不是独立的，而是相互关联和制约的。求得的这些多元函数关系式比用一个自变量表示的关系式更能全面地反映出胶凝燃料的物化性质对其分散性的影响。

3 结论

通过试验研究，分别得到了胶凝燃料零切黏度、表观黏度、相对松弛时间（弹性）等物化性质对其爆炸分散性的影响规律。随着黏弹性的增加，分散个数和有效分散率下降，而液块有效平均质量逐渐增加。相比之下，黏性虽然对胶凝燃料的分散性有一定的影响，但不明显，而弹性对分散液块大小的影响更显著。

采用最小二乘法求得胶凝燃料分散性对黏弹性等参数的函数表达式为

$$\bar{m} = 0.768\,335\,3f + 9.542\,149 \times 10^{-4} Z_0 - 2.985\,862 Z_a - 0.053\,984\,4\theta/\theta_{2\%} + 0.389\,444\,1d$$

$$N = -952.905\,6f - 0.432\,576\,9Z_0 + 3\,908.941Z_a - 23.420\,68\theta/\theta_{2\%} + 1\,958.708d$$

$$P = -55.912\,02f - 5.934\,612 \times 10^2 Z_0 + 381.810\,3Z_a - 29.545\,88\theta/\theta_{2\%} + 171.474\,5d$$

由此可通过测定试样的黏弹性等参数和预测胶凝燃料的分散特性。

参考文献

[1] POPPOFF I G, THUMAN W C. Research studies on the dissemination of solid and liquid agents: AD - 827272 [R]. 1967.

[2] 薛社生，刘家骢，彭金华．液体燃料爆炸抛撒过程分析［J］．南京理工大学学报，1998，22（1）：34 - 37.

[3] GIDASPOW D, SYAMLAL M. The explosive dissemination of particulate pyrotechnic and explosive powders [C]//Proceedings of 13th International Pyrotechnics Seminar. Chicao, USA, 1988: 347 - 372.

[4] MATTA J E. Viscoelastic breakup [C] //Symposium on Chemical Operations, U. S. Army Chemical School Fort Mcclellan, Alabama, 1982: 116 - 118.

[5] 贵大勇，曾学明，肖凯涛．油基燃烧剂分散性测试研究［J］．火工品，1997，4：5 - 8.

片状火药作用下物体分离运动规律研究①

冯顺山，贾光辉②

（北京理工大学，北京　100081）

摘　要：文中对火药气体作用下，无限长圆管中圆柱体的双向相对运动过程进行分析，运用内弹道学理论建立了该问题的物理模型和数学模型，并对数学模型进行求解，得到了等截面燃烧的片状火药在燃烧阶段和膨胀阶段的物体运动速度、药室压力等解析关系式。

关键词：火药；圆柱体；分离运动；物理模型

A Study of Objects Separating Move Law Under Slice Powder

Feng Shunshan, Jia Guanghui

(Bei jing institute of Technology, Beijing 100081, China)

Abstract: In this paper, two cylinders' separating move process is analysed under the slice propellant gas, the physics and math modes are set up, a math solution is get about cylinders' velocity and powder chamber' pressure in the burning period and free expanding period.

Key words: powder; cylinder p separating move; physics model

1　问题描述

在内径为 D、截面积为 S 的刚性两端无限长的等截面绝热圆管中，有两个外径为 D 质量各为 M_1、M_2 的刚性圆柱体，$t=0$ 时刻，二圆柱体距离为 L_0，与圆管内壁密闭光滑接触，构成容积为 V_0 的初始药室，内部装填质量为 ω 的火药，在 $t=0$ 时刻，火药药粒同时燃烧，产生的高压火药气体推动二物体相对运动 L_1、L_2，在火药气体作用下的运动距离，$t=tk$ 时刻火药燃烧完毕之后二圆柱体在高压火药气体作用下，各自继续向相反方向运动，见图 1。

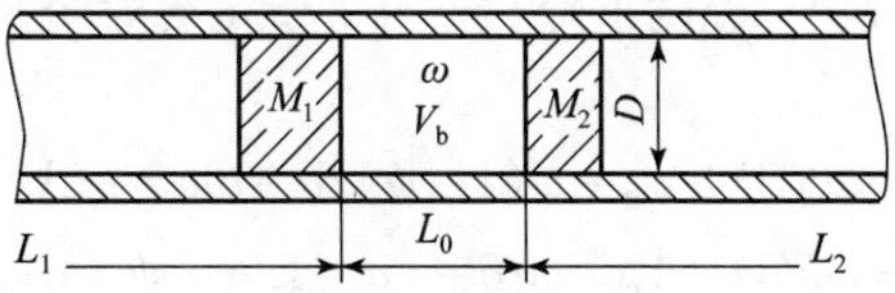

图 1　圆管及物体模型

2　物理模型

2.1　燃气生成率

按照内弹道学理论，火药燃烧遵从几何燃烧定律，所有药粒都将按照平行层燃烧，并始终保持相同的几何形状和尺寸。

为抓住该问题的物理本质，文中选择较为简单的片状火药进行分析。由于片状火药厚度尺寸（$2e_1$）

① 原文发表于《弹箭与制导学报》2003 年第 1 期。

② 贾光辉：工学博士，2000 年师从冯顺山教授，研究反航母分离式战斗部技术，现工作单位：北京航空航天大学。

远小于长、宽尺寸（长为$2a_1$，宽为$2b_1$），时间t，火药燃去的肉厚为2ε，其厚度尺寸为$2e=2(e_1-\varepsilon)$，宽为$2b=2(b_1-\varepsilon)$，考虑到$\delta<e_1\ll b_1<a_1$，火药燃烧的质量百分比ψ为

$$\psi=(2e_1\cdot 2a_1\cdot 2b_1-2e\cdot 2a\cdot 2b)/(2e_1\cdot 2a_1\cdot 2b_1)=1-e/e_1 \tag{1}$$

$t=0$时，e的初值为e_1；$t=tk$，火药燃烧完毕，$e=0$，所以$de/dt=-u<0$，u为火药的线性燃烧速度，ψ的导数为

$$\mathrm{d}\psi/\mathrm{d}t=-\mathrm{d}e/e_1\mathrm{d}t=u/e_1 \tag{2}$$

火药的燃烧速度与压力呈正比例关系：

$$u=-\mathrm{d}e/\mathrm{d}t=u_1p/p_a \tag{3}$$

积分式（3）可得到：$\int_{e_1}^{0}-\mathrm{d}e=e_1=(u_1/p_a)\int_{0}^{t_k}p\mathrm{d}t=u_1I_k/p_a$，即$I_k=(e_1p_a)/u_1$，$I_k$为确定火药的压力全冲量，为常量。

2.2 火药气体状态方程

火药气体燃烧后，生成的燃气分子所占的比例较大，分子间的斥力不可忽视，所以火药气体不再遵从理想气体的状态方程，而是范德瓦尔状态方程，但是由于火药燃烧时温度达到很高，所以分子之间引力作用不明显，参照内弹道学，采用如下所示的状态方程：

$$P(v-a)=n\,R_0T=RT \tag{4}$$

2.2.1 定容状态方程

定容状态下，火药不对外界做功，式（4）中T为火药气体的定容燃烧温度T_v，可写为

$$P(v-a)=f \tag{5}$$

式中，$f=RT_v$为火药力。考虑质量为ω火药的燃烧进程，则瞬时火药的燃烧质量为$\psi\omega$，记药室容积为V_0，火药密度为δ，当考虑火药燃烧的质量百分数ψ时，引入火药装填密度$\Delta=\omega/\delta$，火药余容α，式（5）可变为

$$\begin{aligned}P&=\omega f\psi/[V_0-\omega(1-\psi)/\delta-\alpha\omega\psi]\\&=\omega f\psi/[V_0-\omega/\delta-(\alpha-1/\delta)\omega\psi]\\&=\Delta f\psi/[1-\Delta/\delta-(\alpha-1/\delta)\Delta\psi]\end{aligned} \tag{6}$$

2.2.2 增容状态方程

在增容状态下，火药对外界做功，温度下降，不再等于定容燃烧温度，记某时刻，药室容积比原容积增加了$\mathrm{d}V$，则状态方程为

$$\begin{aligned}P&=\omega RT\psi/[V_0-\omega(1-\psi)/\delta-\alpha\omega\psi+\mathrm{d}V]\\&=\omega RT\psi/[V_0-\omega/\delta-(\alpha-1/\delta)\omega\psi+\mathrm{d}V]\\&=\Delta RT\psi/[1-\Delta/\delta-(\alpha-1/\delta)\Delta\psi+\mathrm{d}V/V_0]\end{aligned} \tag{7}$$

问题中，火药药室增大量为$S(L_1+L_2)$，可适用增容状态方程，可得到如下关系：

$$\begin{aligned}P&=\Delta RT\psi/[1-\Delta/\delta-(\alpha-1/\delta)\Delta\psi+\Delta V/V_0]\\&=\Delta RT\psi/[L\psi/L_0+(L_1+L_2)/L_0\\&=[(\omega/SL_0)\cdot RT\psi]/[L_4/L_0+(L_1+L_2)/L_0]\\&=\omega RT\psi/[S(L\psi+L_1+L_2)]\end{aligned} \tag{8}$$

式中，

$$L\psi=L_0[1-\Delta/\delta-(\alpha-1/\delta)\Delta\psi \tag{9}$$

2.3 动力学及运动学方程

假设各瞬时，药室中压力为均匀分布，则动力学方程为

$$PS = M_1 \mathrm{d}V_1/\mathrm{d}t = M_2 \mathrm{d}V_2/\mathrm{d}t \tag{10}$$

由于 $\mathrm{d}V_1/\mathrm{d}t = \mathrm{d}V_1 \mathrm{d}L_1/\mathrm{d}L_1 \mathrm{d}t = V_1 \mathrm{d}V_1/\mathrm{d}L_1$，式（10）也可写成

$$PS\mathrm{d}L_1 = M_1 \mathrm{d}V_1, PS\mathrm{d}L_2 = M_2 \mathrm{d}V_2 \tag{11}$$

由式（10）可得到：$M_1 a_1 = M_2 a_2$，积分并引入符号 M_{12}，得到

$$M_1 V_1 = M_2 V_2 = M_{12}$$

$$V_1 + V_2 = (1/M_1 + 1/M_2) M_{12} \tag{12}$$

引入符号 E_{12}，得到

$$M_1^2 V_1^2/2 = M_2^2 V_2^2/2 = E_{12}$$

$$M_1 V_1^2/2 + M_2 V_2^2/2 = (1/M_1 + 1/M_2) E_{12} \tag{13}$$

积分引入符号 L_{12}，可得到

$$M_1 L_1 = M_2 L_2 = L_{12}$$

$$L_1 + L_2 = (1/M_1 + 1/M_2) L_{12} \tag{14}$$

2.4　能量守恒方程

在本问题中，火药能量 E 全部转化物体的动能。

质量为 ω 的火药，可释放的能量 E 为

$$\begin{aligned} E &= \omega C_V (T_V - T) = \omega R (T_V - T)/(k-1) \\ &= \omega R (T_V - T)/\theta = \omega f/\theta - \omega RT/\theta \end{aligned} \tag{15}$$

考虑火药的燃烧百分比 ψ 时，燃烧质量为 $\psi\omega$ 的火药，可释放能量为

$$E = \omega\psi f/\theta - \omega\psi RT/\theta \tag{16}$$

二物体的动能之和见式（13），所以能量守恒方程可写为

$$\psi\omega(f/\theta - RT/\theta) = M_1 V_1^2/2 + M_2 V_2^2/2 = (1/M_1 + 1/M_2) E_{12} \tag{17}$$

3　数学模型

3.1　火药燃烧阶段

在火药燃烧阶段，由式（2）得

$$\mathrm{d}\psi/\mathrm{d}t = -\mathrm{d}e/e_1 \mathrm{d}t = u/e_1$$

由式（3）得

$$u = -\mathrm{d}e/\mathrm{d}t = u_1 p/p_a$$

变容状态方程式（8）：

$$P = \omega RT\psi/[S(L_\psi + L_1 + L_2)]$$

能量守恒方程式（17）：

$$\psi\omega(f - RT) = \theta(M_1 V_1^2 + M_2 V_2^2)/2 = (1/M_1 + 1/M_2)\theta E_{12}$$

由动力学及运动方程（10）或式（11）构成了该问题火药燃烧阶段的数学模型。

从式（8）与式（17）消去 T，可得到内弹道基本方程：

$$PS(L_\psi + L_1 + L_2) = \psi\omega f - \theta(M_1 V_1^2 + M_2 V_2^2)/2 \tag{18}$$

对上述模型，选择 ψ 为自变量，可以解出 V_1、V_2、L_1、L_2、P 关系式，导出该阶段的解析解：

由式（2）、式（3）可得

$$\mathrm{d}\psi/\mathrm{d}t = -\mathrm{d}e/e_1 \mathrm{d}t = u/e_1 = (u_1/e_1)(p/p_a) \tag{19}$$

结合式（10）$\mathrm{d}V_1/\mathrm{d}t = PS/M_1$：

消去 P，可得到：$M_1 V_1 = (e_1 P_a S)\psi/u_1$，同理得到 $M_2 V_2 = (e_1 P_a S)\psi/u_1$，于是

$$M_1 V_1 = M_2 V_2 = (e_1 P_a S)\psi/u_1 \tag{20}$$

考虑到 $I_k=(e_1P_a)/u_1$，式（20）可改写为：$M_1V_1=M_2V_2=I_kS\psi$，可见 V_1、$V_2\propto\psi$。

$T=0$ 时，$\psi=0$，$V_1=V_2=0$；$t=tk$ 时，$\psi=1$，$M_1V_{1k}=M_2V_{2k}=I_kS$。

由内弹道基本方程（18）以及式（11）的另一种法：

$$PS\mathrm{d}(L_1+L_2)=\mathrm{d}(M_1V_1^2+M_2V_2^2)/2 \tag{21}$$

将式（20）中的速度表达式代入（21），可得到

$$\mathrm{d}(L_1+L_2)/(L_1+L_2+L_\psi)$$

$$=(1/\theta)\frac{\mathrm{d}[(1/M_1+1/M_2)I_kS^2\theta\psi^2/2]}{\omega f\psi-(1/M_1+1/M_2)I_kS^2\theta\psi^2/2}=-(2/\theta)\frac{\mathrm{d}[\omega f-(1/M_1+1/M_2)I_kS^2\theta\psi^2/2]}{\omega f-(1/M_1+1/M_2)I_kS^2\theta\psi^2/2}$$

L_ψ 值介于 $L_0(1-\Delta/\delta-(\alpha-1/\delta)\Delta\cdot 0)$ 和 L_0 $(1-\Delta/\delta-$ $(\alpha-1/\delta)$ $\Delta\psi)$ 之间，其中值为 $L_{\bar\psi}=L_0(1-\Delta/\delta-(\alpha-1/\delta)\Delta\psi/2)$，由于 L_ψ 变化不大，对上式左边积分时，采用中值定理，积分并考虑初值条件得到［从 $L_0(1-\Delta/\delta)$ 到 $L_0(1-\alpha\Delta)$，一般 $\delta=1.6$，$0.5\sim1$］：

$$L_1+L_2=L_\psi[1-(1/M_1+1/M_2)(I_kS)^2\theta\psi^2/2\omega f]^{2/\theta}-L_{\bar\psi} \tag{22}$$

$t=0$ 时，$\psi=0$，$L_1+L_2=0$；$t=tk$ 时，$\psi=1$，记：$L_1+L_2=L_{1k}+L_{2k}$：

$$L_{1k}+L_{2k}=L_{\bar\psi}[1-(1/M_1+1/M_2)(I_kS)^2\theta\psi^2/2\omega f]^{2/\theta}-L_{\bar\psi}$$

将速度表达式和位移表达式代入内弹道基本方程（18），则得到压力表达式：

$$P=\frac{[\omega f-(1/M_1+1/M_2)(I_kS)^2\theta\psi^2/2]}{S(L_1+L_2+L_\psi)} \tag{23}$$

3.2 火药膨胀阶段

火药膨胀阶段，$t>tk$，火药已经燃烧完毕，数学模型由式（18）和式（11）构成：

$$PS(L_\psi+L_1+L_2)=\psi f-\theta(M_1V_1^2+M_2V_2^2)/2 \tag{24}$$

式中，$L_\psi=L_0(1-\alpha\Delta)$。

由式（21）和式（24）可得

$$\frac{\mathrm{d}(L_1+L_2)}{L_1+L_2+L_\psi}=\frac{[\mathrm{d}(M_1V_1^2+M_2V_2^2)/2]}{\omega f-\theta(M_1V_1^2+M_2V_2^2)/2}$$

积分并考虑 $t=tk$ 时的速度和位移条件有

$$\omega f-\theta(M_1V_1^2+M_2V_2^2)/2=[\omega f-\theta(1/M_1+1/M_2)(I_kS)^2/2]\times[(L_1+L_2+L_\psi)/L_{1k}+L_{2k}+L_\psi]^{-\theta} \tag{25}$$

代入内弹道基本方程，有压力关系式：

$$PS=(L_1+L_2+L_\psi)=[\omega f-\theta(1/M_1+1/M_2)(I_kS)^2/2]\times[(L_1+L_2+L_\psi)/L_{1k}+L_{2k}+L_\psi]^{-\theta} \tag{26}$$

4 结论

（1）对于文中所描述的问题，可知二物体的动量相等，动能之比与质量之比成反比，位移之比与质量之比成反比。

（2）二物体的动量与火药压力全冲量成正比，与药室截面积成正比。

（3）片状火药作用下，二物体的动量与火药燃烧百分比成正比；另文中研究表明：球状火药作用下，二物体的动量与火药燃烧百分比组合$[1-(1-\Psi)^{1/3}]$成正比。

（4）该系统二物体的极限动量为

$$\sqrt{(2\omega f/\theta)/(1/M_1+1/M_2)}$$

（5）最大压力点必然发生在 $0<\psi<1$ 范围内，此处 $\mathrm{d}p/\mathrm{d}\psi=0$ 适当选择参量，可使 $P_{\max}$ 位置得到优化。

参考文献

[1] 张柏生，李云娥．火炮与火箭内弹道学［M］．北京：北京理工大学出版社，1996.

灵巧弹药技术的研究进展①

冯顺山，张旭荣②，刘春美
（北京理工大学机电工程学院，北京　100081）

摘　要：灵巧弹药技术是未来导弹武器精确毁伤目标的主要手段，本文对灵巧弹药概念的多种定义方法进行了比较，并根据武器弹药的灵巧程度，采用坐标轴的表示方法对灵巧弹药进行了新的定义。在综合国内外灵巧弹药技术研究进展的基础上，分析了灵巧弹药技术在导弹武器中应用的研究重点和发展方向，指出具有宽视场、高分辨力的光学成像末制导子弹药是导弹武器未来发展的方向和重点。

关键字：灵巧弹药；精确毁伤；发展方向

The Development of Smart Munitions Technology

FENG Shun－shan，ZHANG Xu－rong，LIU Chun－mei
（School of Mechatronic Engineering，Beijing Institute of Technology，Beijing 100081，China）

Abstract：Smart munitions technology is an important manner to damage the target accurately using missile weapon. This article compared the definition of Smart munitions，and defined smart munitions using axes according to the degree of intelligent of ammunitions. It summarized the advances in researching on the smart munitions of technology. It also analyzed research focus and developing direction of smart munitions technology applied for missile. It points out that developing direction of missile is infrared imaging terminal guidance with the wide angle optical system.

Key words：smart munitions；damage accurately；develop direction

1　灵巧弹药的基本概念

在一些国外文献中，20 世纪 70 年代初已有“灵巧炸弹”一词，主要是指美国的激光制导炸弹。20 世纪 80 年代开始，特别是 1985 年以后，逐渐出现讨论灵巧弹药概念的论文，对“灵巧炸弹”一词也逐步有了较详细的解释，扩大了它的内涵。在 1990 年版的英汉辞海中对“灵巧弹药”一词解释为“一种制导（如激光制导）以击中目标的炸弹”。《世界兵器博览词典》中有“灵巧智能弹药”一词，它解释为“灵巧弹药是一种现代电子技术和子母弹技术相结合，以提高毁伤面积和命中精度的炮弹”。1992 年出版的《当代军事大全》在“精确制导武器”词条中，把灵巧弹药列为精确制导武器的一类，并对灵巧弹药的发展过程做了介绍[1]。从上述部分有关灵巧弹药的资料中不难看出，灵巧弹药的概念已引起关注，它的定义与内涵正逐步深入、逐步科学。具有代表性的定义有以下三种：①灵巧弹药是指自主地搜索、发现、识别和攻击目标的射弹、战斗部和子弹药，其实质是要求自主寻的，包括寻的末制导弹药、传感器引爆弹药、精确制导弹药，不包括人工照射式末制导弹药；②灵巧弹药是指传感器引爆弹药和末端制导弹药，包括人工照射式末制导弹药，不包括精确制导弹药；③具有制导系统能使其轨迹指向目标的弹药，其实质是只要弹上有寻的功能，不管其他条件，包括人工照射式末制导弹药，自动寻的式末制导弹药，传感器引爆弹药和精确制导弹药。

---

①　原文发表于《中北大学学报（自然科学版）》2006 第 27 卷增刊。

②　张旭荣：工学博士，2004 年师从冯顺山教授，研究对地攻击子弹药探测及识别技术，现工作单位：火箭军研究院导弹工程研究所。

上述三种定义的共同点是都在弹上有制导系统，可以导向目标，对目标进行攻击。但由于定义的内涵不同，所包括的弹种就不同。

为了对灵巧弹药给出更为科学的定义，必须对灵巧弹药的出现背景有所认识。灵巧弹药的出现，主要目的是在复杂多变的战场环境下，对各种目标提高命中率，特别是在弹箭远射程、航弹远程投放的条件下，提高命中率，使普通武器由射击面目标转向射击点目标。灵巧弹药是相对普通无控弹药而言的。无控弹药射击之后，只靠自身的惯性飞行，对目标无任何反应，从这种意义上讲，灵巧弹药是普通弹药的发展与改进，是高新技术运用于普通弹药的结果。因此，精确制导弹药、智能弹药，都应该是对原来的普通弹药而言，在类似的发射方式、战场环境下，使其具有寻的功能和精确制导功能。它们应该是由普通弹药发展起来的、在战术范围内使用的、采用高新技术的新型弹药。

基于上述考虑，本文采用坐标轴的方式对灵巧弹药进行重新定义，如图 1 所示。

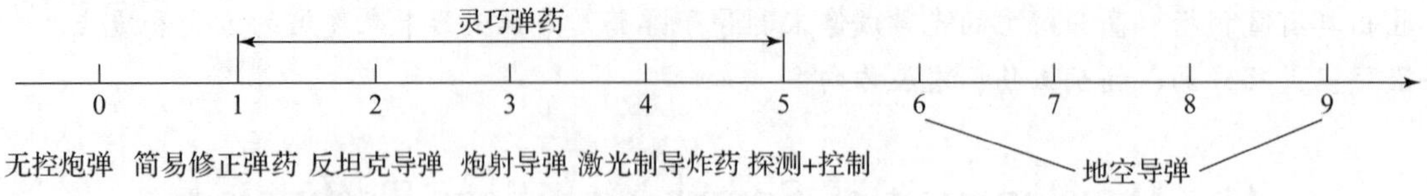

图1　灵巧弹药定义示意图

2　灵巧弹药的发展现状

2.1　国外灵巧弹药的发展现状

欧美各国与苏联自 20 世纪 70 年代以来，纷纷研制并装备了各种类型的灵巧弹药，早期的灵巧弹药以美国的“铜斑蛇”和苏联的“红土地”为代表。这两种灵巧弹药均采用了激光半主动末制导技术。目前已有除美俄之外的英国、法国、德国、瑞典和以色列等十几个国家进行了灵巧弹药的研制。

1990 年，瑞典当局开始在 155 mm Bonus 弹上装配两发“灵巧”子弹药战斗部的特种子母弹进行全面研制该子母弹采用底部排气结构，最大射程可达 25 km。飞抵目标区上方约 1 000 m 处时，该弹可释放出两发传感器引爆的子弹药。这种“灵巧”弹药和子弹药面临的一个问题是，在西欧和中欧，冬天的云层是很低的，如果导引头直到低于云层之后才开始搜索，那么其作用距离将受到严重限制。另外，Bonus 子弹药只是在飞抵地面上空约 150 m 时才能开始对目标进行扫描，此时，传感器的“观察”角与降落线倾斜约 30°，可提供一个螺旋形的扫描图形，一旦探测到目标，弹头就在最佳距离处起爆，以使自锻破片对目标的上表面进行攻击。

1993 年 10 月，法国陆军和瑞典国防当局开始联手研制能够满足瑞典和法国需要的 Bonus OBG 弹。该弹与美国 M864 弹丸相似，采用了底部排气结构。Bonus 子母弹由 39 倍口径的炮管发射，射程超过 28 km；由未来 52 倍口径的火炮发射时，其射程可增加到约 35 km。

英国空间公司研制的 Merlin“灵巧”破击炮弹，为了使之具有全天候的作战能力，选用了 94 GHz 的毫米波雷达导引头，导引头对活动目标第一次光栅扫描范围为 300 m × 300 m；如果没有探测到目标，导引头就开始第二次扫描，搜索 100 m × 100 m 范围内的固定目标。瑞典 FFV 公司和萨布公司开始研制 120 mm Strix 制导炮弹，采用红外成像制导。从理论上讲，这种导引头易受烟雾和现代热伪装的损害，但 Strix 研究小组却认为，破击炮弹的几乎垂直下落方式可以克服这些不足。同所有红外制导的“灵巧”炮弹一样，Strix 也必须到达云层下面后才能扫描目标。目标扫描在大约 700 m 高度和 750 m 斜距时开始，导引头可对 140 m × 120 m 的椭圆形区域进行扫描。在天气恶劣的情况下，由于云层高度为 100 m，扫描区域降至 15 ~ 20 m。随着 GPS 修正的小型惯导系统（GPS/INS）制导组件的到来，美国空军和海军的空投武器的精度已大大提高。最好的例子是联合直接攻击弹药（JDAM），这是一种标准的 900 kg 级炸弹，装有波音公司的尾部组件，内有可控制炸弹弹翼的 GPS/INS 装置，使炸弹在各种气候条件下精度均能达到 10 ~ 13 m。JDAM 首次实战被用于科索沃空战中。美国还在执行另一些计划，以便把 GPS/INS 组件引进到美国陆军的榴弹炮和海军舰炮炮弹中[2]。

20 世纪 70 年代末，美国空军和陆军开始研制空投和炮射小型末制导子弹药，可直接攻击行进中的装甲目标。目前，第一代灵巧末制导弹药已在美国空军和陆军服役，现正在研制更有效的末制导弹药。

2.2　美国几种典型的灵巧弹药

1. 杀伤坦克的杀手

美国空军的传感器引爆武器（SFW）编号为 GBU－97，它是世界上第一种可供作战用的装备末制导子弹药的空投集束炸弹，1996 年进入全速率生产。这种传感器引爆武器的质量为 450 kg，每个 SAW 有 10 枚可从圆柱形战术布撒器布撒的 BLU－108、B 子弹药，每枚子弹药又含 4 枚 Skeet 反装甲战斗部，所以一枚 SFW 可以投放 40 枚 Skeet，这是美国空军的战术飞机飞越目标一次就能杀伤多辆装甲车，也使杀伤一个目标阵列所需飞机出动的架次和武器数量有所减少。1997 年 2 月，F－16 战斗机首次使用 GUB－97SFW，每架飞机可外挂 4 枚。Skeet 的直径为 12.5 cm，边下降，弹上红外传感器边对地扫描，一旦探测到坦克发动机，Skeet 就向下发射爆炸成形贯穿头攻击目标易损的顶部。每枚 Skeet 可搜索 2 700 m^2 区域扫描，40 枚 Skeet 总共可搜索的区域达 60 700 m^2。SFW 是一种真正的全天候武器，云、雾、雨和雪都不能影响其性能，其红外传感器能有效地“看透”烟雾。

1996 年起，美国空军开始对 GBU－97 进行改进，可使其对地扫描区域提高 1 倍，达 121 400 m^2，同时增加了主动红外传感器的性能，而在这样的环境中，被动红外传感器仍可单独地探测目标。改进型 SFW 在 Skeet 的爆炸成形贯穿头的铜药型罩的外圈增加了 16 个弹丸，以扩大战斗部的破坏面积，使其能有效地攻击较软的目标。在 1999 年春季进行的改进型 SFW 的首次飞行试验中，其所破坏的目标数和覆盖的面积均是基型的两倍，超出预先的期望。

2. 风力修正弹药布撒器和联合防区外武器

GBU－97SFW 的最佳投放条件是：飞机飞行速度为 250～650 km 的低空水平飞行或小角度俯冲，但是这种使用条件使载机易遭敌防空系统的攻击。为了能从更远的距离外进行精确的中高空投放，美国空军计划为 SFW 安装一种新型尾部组件，称为风力修正的弹药布撒器（WCMD）。WCMD 包括惯导装置、风力估计和补偿软件以及将取代 GBU－97 上固定尾翼的可动尾翼。一旦弹药被投放后这种尾部组件对风力、弯曲弹翼或其他发射瞬变值进行修正。WCMD 将使 SFW 制这种战术弹药布撒器飞到距计划布撒点的投放高度大于 12. 2 km，并将控制这种战术弹药布撒器飞到距计划布撒点 26 m 的范围内。

3. 敏感的破坏装甲炮弹（SADARM）

1995 年，美国陆军的 155 mm 榴弹炮发射的敏感和破坏装甲炮弹（SADARM）投产，它是世界上第一种炮射灵巧弹药。这种反装甲灵巧弹药可攻击固定不动的自行式火炮和火箭发射架，也可有效地攻击坦克和步兵战车。SADARM 使美陆军用几发炮弹就能压制敌方火炮，并减少自己暴露在敌火力中。每发 SADARM 有两枚直径为 145 mm 的子弹药，其抛射高度为 1 000 m。子弹药靠降落伞下降，弹上配主动毫米波雷达、被动毫米波辐射计和单色红外传感器三种传感器，以便能全天候探测目标。三种传感器可对地搜索直径为 150 m 的一个圆形区域。计算机可同时对不同传感器的数据进行相关，从而可以对战场环境、恶劣的天气和妨碍某传感器效能的干扰进行补偿，并能根据目标的特定的尺寸、形状和红外特征可靠地探测真正的目标。与 SFW 相似，SADARM 子弹药向下朝目标顶部发射爆炸成形贯穿头。如果子弹药在下降过程中未发现目标，它会在造成附带算上距地面某一高度自毁，以消除己方的威胁。

1997 年，美国陆军开始对 SADARM 进行改进。改进型 SADARM 对地面的搜索区域是基型的 3 倍，原因是子弹药搜索目标的高度从 130 m 提高到 165 m，悬挂角从 55% 增到 80%，而所需的炮弹数量降低 30%。

4. 智能反装甲（BAT）子弹药

1985 年，美国陆军开始研制更具技术挑战的末制导子弹药——BAT（智能反装甲），这是一种微型导弹。子弹药从火箭或导弹中布撒出来，在下降过程中可对一个很大区域内行进的坦克进行搜索，然后水平滑翔对其攻击，BAT 是一种无动力子弹药，弹长为 0.9 m，弹径为 0.14 m，计划中的陆军战术导弹系统（ATACMC）Block Ⅱ和Ⅱ A型可以在恶劣天气昼夜向敌前线后方纵深布撒BAT子弹药，这些子弹药按预编程以不同的飞行路线飞行以便攻击不同的目标。BAT挂在降落伞下，音响传感器向下探测以捕获行进中的装甲车。一旦BAT子弹药确定了行进中的装甲队的位置及其运动方向，它就变成了滑翔子弹。它切断主降落伞，并“向上旋转”前端的导引头。音响传感器提示红外引导头应探测何处，旋即红外导引头开始捕获目标。然后BAT子弹药切断二级降落伞并寻的目标，从导弹发射到与目标交战的全过程仅需几分钟，BAT的传感器对地的搜索范围大，因而可补偿投放误差和目标的运动。

5. 低成本自主攻击系统（LOCAAS）

美国空军1998年开始研制一种新型空面弹药，称为低成本自主攻击系统（CLOCAAS）。它是末制导弹药下一阶段的典型产品，体积小、可待机巡逻、自动化程度高，并装备可自行决定攻击什么目标的弹载计算机。LOCAAS由微型涡喷发动机推进，弹长为0. 775 m，质量为38 kg，升阻比高，机动性好，射程为185 km，使载机具有足够的防区外距离。LOCAAS将在229 m的高空以200 km的速度绕目标区巡逻飞行，由激光雷达导引头探测并截获目标。一个战术弹药布撒器可装4个LOCAAS，一架F-16战斗机可外挂4个布撒器，一架F-22战斗机或联合攻击战斗部可内装16个LOCAAS。

LOCAAS通过发射激光脉冲和接收其反射回波来确定目标的距离，并对目标的尺寸和几何形状进行高分辨率三维测量，然后测量值与预先存储目标数据相匹配。LOCAAS根据预编程的目标优先顺序表自行决定攻击哪个目标，并可对早先探测的目标再搜索。

3 灵巧弹药的发展趋势

综合分析高新技术弹药的发展趋势可以发现，在弹药的威力、射程和精度三大指标中，精度问题越来越突出和重要。末制导子弹药是灵巧弹药发展的必然趋势。与战术导弹相比，它只在末段制导，因而结构简单、价格便宜；又比无控炮弹精度高、首发命中率高，可对付静止和运动的点目标，因此受到世界各国的重视，得到了较快的发展。随着红外/毫米波成像技术和微型数字信号处理机技术的进步，末制导子弹药的发展在国外产生了两种趋势。俄罗斯专注于发展激光半主动末制导炮弹，但该末制导方式具有明显的缺点：在末制导阶段，激光指示器必须连续不断地照射目标，迫使地面的侦察人员或无人驾驶机长时间暴露在敌方阵地前沿，易受敌方的反击和干扰。因此，美国开始重视发展采用成像被动寻的方式的“发射后不管”的末制导炮弹。成像被动寻的末制导炮弹虽隐蔽性好，具有“发射后不管”的能力，但成本较高、技术难度大；激光半主动末制导炮弹命中概率高，成本低、技术成熟。因此，应综合权衡使用要求和技术因素，也可考虑同时装备两种类型的末制导炮弹，以应对不同的任务要求和作战环境。

4 成像末制导子弹药技术在导弹武器中的应用前景

成像末制导子弹药具有自主末制导能力，可以提高导弹武器的综合作战能力，尤其是对点目标的精确打击能力，其难点和重点是在高动态、高过载等环境中，完成对地面复杂背景下的小幅员或点目标的自动探测识别和精确打击。成像末制导子弹药的关键技术包括：导引头技术及环境适应性，导引头目标探测与识别技术，子弹制导控制技术，高过载高速度抛撒状态下的子弹药减速减旋稳态飞行技术。

综合分析表明：成像末制导子弹药技术的应用发展应侧重两个方向：一是追求武器的低成本和高命中概率，满足同时打击多目标的作战要求；二是尽可能加大探测器件的视场宽度。因此，带有智能探测识别功能的宽视场、高分辨力光学成像末制导子弹药是导弹武器未来发展的方向和重点。

参 考 文 献

[1] 郭锡福．关于灵巧弹药的探讨［J］．弹箭技术，1996（2）：57-60.
[2] 周军．现代空中打击带要灵巧炸弹［J］．飞航导弹，1999（11）：57-60.

地磁陀螺组合弹药姿态探测技术研究①

曹红松[1,2]②，冯顺山[1]，赵捍东[2]，金　俊[3]

（1. 北京理工大学爆炸科学与技术国家重点实验室，北京　100081；
2. 中北大学机电工程学院，太原　030051；3. 装甲兵工程学院，北京　100072）

摘　要：在对飞行弹体控制时，需要实时获取弹体的姿态角。文中提出了一种采用地磁传感器和硅微陀螺构建低成本姿态探测系统的方案，给出了利用地磁矢量和陀螺探测结果进行弹体姿态解算的数学模型。仿真结果显示，该方案在一定条件下可以达到较高的精度。

关键词：地磁；陀螺；姿态；探测；弹药

Researching Ammunition Attitude Detect Technique Combination of Geomagnetism and Gyro

CAO Hong - song[1,2], FENG Shun - shan[1], ZHAO Han - dong[2], JIN Jun[3]

(1 State Key Laboratory of Explosive Science and Technology, Beijing Institute of Technology, Beijing100081, China; 2 Institute of Mechatronic Engineering, North University of China, Taiyuan 030051, China) 3 The Academy of Armored Forces Engineering of PLA. Beijing 100072. China)

Abstract: Controlling a flying projectile, the attitude angles of the projectile body need to be obtained in real - time. Therefore, a low cost attitude detect system scheme is put forward. The system consists of geomagnetic sensor and mi - cro silicon gyro. A mathematic model is established, and the attitude resolve using geomagnetic vector and gyro detect result. The simulation result indicates that the system can achieve proper precision in some condition.

Key words: geomagnetism; gyro; attitude; detect; ammunition

1　引言

目前，具有多目标探测、识别、选择、精确攻击并进行战场侦测的智能型弹药是灵巧弹药的发展方向。在此类智能型弹药系统中，常需要确定弹体的飞行姿态，为控制系统提供姿态及方位。由于磁强计具有测量范围宽、高稳定性、无漂移、抗干扰、体积小、成本低等许多优点，全固态的微机械陀螺具有可靠性高、动态范围宽、抗冲击振动、体积小、重量轻、成本低、短时漂移小等优点。利用磁强计与微机械陀螺来组成低成本的弹药捷联姿态探测器，其全固态特性能很好地适合弹药的弹上环境，微小型器件又能满足安装空间的需要，所以是一种实用的低成本姿态探测方法。文中主要研究地磁陀螺捷联姿态探测系统的构建及其姿态解算算法。

2　系统构建

地磁陀螺组合探姿系统主要由地磁传感器、陀螺、制导处理器构成，系统原理框图如图 1 所示。

---

① 原文发表于《弹箭与制导学报》2006 年第 3 期。

② 曹红松：工学博士后，2004 年合作导师冯顺山，研究战场毁伤效果电视侦察子弹系统总体技术，现工作单位：中北大学。

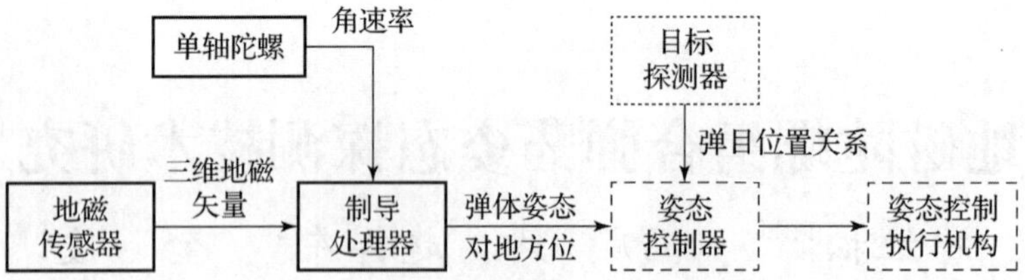

图1　地磁陀螺组合探姿系统原理框图

将单轴速率陀螺及三维地磁传感器捷联安装在弹体上，地磁传感器的敏感轴均对准弹体坐标系的三轴方向，陀螺敏感轴对应弹体纵轴。弹体飞行过程中，陀螺测定弹体滚转运动角速度、地磁传感器敏感地磁场在弹轴上的投影。由此可以解算弹体坐标系的三维姿态角。

制导处理器对陀螺及地磁传感器输出的信号进行采集及处理，并完成姿态角的解算[1]。制导处理器可利用 DSP 来实现，主要由 DSP 主芯片、A/D 转换芯片、Flash 存储芯片、通信转换接口等组成。

3　组合姿态解算模型

在弹体姿态确定系统中，姿态敏感器的测量通常不是太多就是太少，姿态确定问题常作为优化问题提出，让姿态的解在一定假设条件下达到某种最优。姿态确定算法可以分为单点算法和多点算法，单点算法要求有足够的测量，单个地磁矢量不足以确定姿态，因此单点算法要求至少两个矢量测量。多点算法是利用测量序列来确定姿态，需要利用姿态角速度或等价的参数将不同时刻的测量联系起来，如 KF（卡尔曼滤波算法）算法，能够获得较单点算法更高的精度[2]。但其也存在局限性，如需要精确的测量模型先验知识，这是难以精确得到的。这里地磁陀螺组合探姿系统中磁强计和陀螺能提供对两个物理量（地磁矢量、滚转角速率）的测量，所以可直接采用单点算法，单点算法计算复杂度低，易于满足实时性要求且误差不累积的特点，文中主要研究地磁陀螺组合姿态解算的单点算法。

3.1　姿态解算数学模型

记地理坐标系为 $OXYZ$ 是全系统的参考坐标系。定义当地磁场矢量在发射坐标系（坐标原点为炮位）下投影为 $\boldsymbol{H}$：记 H_x、H_y 和 H_z 为地磁矢量 $\boldsymbol{H}$ 在地理坐标系三轴上的分量，记 $OX'''Y'''Z'''$ 为弹体坐标系，磁测探头对准弹体系 $OX'''Y'''Z'''$ 三轴安装，则 H_x'''、H_y''' 和 H_z''' 为地磁矢量 $\boldsymbol{H}$ 在弹体系三轴上的投影。

地磁矢量在地理坐标系和弹体系各轴上的投影，可由两坐标系各轴间的方向余弦来确定，此方向余弦也体现载体坐标系相对地面坐标系的姿态。显然 H_x、H_y、H_z 与 H_x'''、H_y'''、H_z''' 间的转换关系可由式（1）表示：

$$\begin{bmatrix} H_X \\ H_Y \\ H_Z \end{bmatrix} = \begin{bmatrix} \cos\Psi & -\sin\Psi & 0 \\ \sin\Psi & \cos\Psi & 0 \\ 0 & 0 & 1 \end{bmatrix} \begin{bmatrix} 1 & 0 & 0 \\ 0 & \cos\gamma & -\sin\gamma \\ 0 & \sin\gamma & \cos\gamma \end{bmatrix} \begin{bmatrix} \cos\theta & 0 & \sin\theta \\ 0 & 1 & 0 \\ -\sin\theta & 0 & \cos\theta \end{bmatrix} \begin{bmatrix} H_X''' \\ H_Y''' \\ H_Z''' \end{bmatrix} \tag{1}$$

其中，Ψ 定义为偏航角，θ 定义为俯仰角，γ 定义为滚转角。

记 Q 为地面系到弹体系的坐标变换四元数：

$$Q = [q_0\ q_1 i\ q_2 j\ q_3 k]$$

则地磁量 $\boldsymbol{H}$ 在弹体系三轴上的投影为

$$H_T = QH \tag{2}$$

由式（2）不能同时解算 Ψ、γ、θ，从数学形式上来看，它是一个相关矩阵，即三分量磁强计不能提供 3 个独立的测量方程。因此需要预知 Ψ、θ、γ 三角中至少一角，才能计算另两角[3]，该系统由滚转陀螺提供滚转姿态角 γ。

由式（2）做三角变换可得

$$\begin{bmatrix}1&0&0\\0&\cos\gamma&-\sin\gamma\\0&\sin\gamma&\cos\gamma\end{bmatrix}\begin{bmatrix}\cos\Psi&-\sin\Psi&0\\\sin\Psi&\cos\Psi&0\\0&0&1\end{bmatrix}\begin{bmatrix}H_X\\H_Y\\H_Z\end{bmatrix}=\begin{bmatrix}\cos\theta&0&\sin\theta\\0&1&0\\-\sin\theta&0&\cos\theta\end{bmatrix}\begin{bmatrix}H_X'''\\H_Y'''\\H_Z'''\end{bmatrix}\tag{3}$$

即

$$H_X\cos\Psi+H_Y\sin\Psi=H_X'''\cos\theta+H_Z'''\sin\theta\tag{4}$$

$$(-H_X\cos\Psi+H_Y\sin\Psi)\cos\gamma+H_Z\sin\gamma=H_Y'''\tag{5}$$

$$(-H_X\sin\Psi-H_Y\cos\Psi)\sin\gamma+H_Z\cos\gamma=-H_X'''\sin\theta+H_Z'''\cos\theta\tag{6}$$

$$\pm\sqrt{(-H_X\sin\Psi+H_Y\cos\Psi)^2+H_Z^2}\cos\left(\gamma-\arctan\frac{H_Z}{-H_X\sin\Psi+H_Y\cos\Psi}\right)=H_Y'''\tag{7}$$

$$\mp\sqrt{(-H_X\sin\Psi+H_Y\cos\Psi)^2+H_Z^2}\sin\left(\gamma-\arctan\frac{H_Z}{-H_X\sin\Psi+H_Y\cos\Psi}\right)=\\-H_X'''\sin\theta+H_Z'''\cos\theta\tag{8}$$

由式（3）~式（8）有

$$\gamma=\arctan\frac{H_X'''\sin\theta-H_Z'''\cos\theta}{H_Y'''}+\arctan\frac{H_Z'''}{-H_X'''\sin\Psi+H_Y'''\cos\Psi}\tag{9}$$

由式（4）、式（6），并经三角变换得

$$\pm\sqrt{H_{X'''}^2+H_{Z'''}^2}\cos(\theta-\arctan\frac{H_{Z'''}}{H_{X'''}})=H_X\cos\Psi+H_Y\sin\Psi\tag{10}$$

$$\mp\sqrt{H_{X'''}^2+H_{Z'''}^2}\sin(\theta-\arctan\frac{H_{Z'''}}{H_{X'''}})=(H_X\sin\Psi-H_Y\cos\Psi)\sin\gamma+H_Z\cos\gamma\tag{11}$$

可得

$$\theta=\arctan\frac{-(H_X\sin\Psi-H_Y\cos\Psi)\gamma+H_Z\cos\gamma}{H_X\cos\Psi+H_Y\sin\Psi}+\arctan\frac{H_{Z'''}}{H_{X'''}}\tag{12}$$

若已知 H_X、H_Y，可写出

$$\Psi=\arctan\frac{-H_{X'''}\sin\theta\sin\gamma-H_{Y'''}\cos\gamma+H_{Z'''}\cos\theta\sin\gamma}{H_{X'''}\cos\theta+H_{z'''}\sin\theta}+\arctan\frac{H_Y}{H_X}\tag{13}$$

由式（12）、式（13）联立求解并经三角变换得

$$\Psi=\arcsin\frac{-H_Z\sin\gamma-H_{Y'''}}{\pm\sqrt{H_X^2+H_Y^2}\cos\gamma}+\arctan\frac{H_Y}{H_X}\tag{14}$$

θ 也可以进一步写为

$$\Psi=\arcsin\frac{H_Z-H_{Y'''}\sin\gamma}{\mp\sqrt{H_{X'''}^2+H_{Y'''}^2}\cos\gamma}+\arctan\frac{H_{Z'''}}{H_{X'''}}\tag{15}$$

至此，已经将 Ψ、θ 都表示为磁强计信号 H_x'''、H_y'''、H_z'''，地磁量 HX、HY、HZ 以及滚转角 γ 的算式。这样可以避免滚转角在特殊值时所带来的解算问题，式中的地磁场参数可在发射前进行装定。

3.2　姿态解算方法

直接利用式（14）、式（15）进行姿态角解算具有不完善性，主要体现为所求角的非单值性。其原因有两种：一是推导过程中的增根，二反三角函数的双值性。另外，有的计算公式具有局限性，表现为载体处于某些姿态时公式无解。如当 $\gamma=90°$ 或 $\gamma=270°$ 时，模型无解。

可以直接选用算式（12）、式（13）作为计算数学模型，其描述的是 $\gamma(n)$、$\Psi(n)$ 与 $\theta(n)$ 时刻 tn 的下标与姿态角自变量对应之间的关系。但不能由 $\gamma(n+1)$、$\Psi(n)$ 解得 $\Psi(n+1)$。可以采用一个近似方法：由 tn 时刻的姿态角 $\Psi(n)$、$\theta(n)$ 的进一步估计结果 $\hat{\Psi}(n+1)$、$\hat{\theta}(n+1)$，并结合由 tn 时刻的微机械陀螺测量值计算的 γ（$n+1$），由式（12）、式（13）计算新的 $\tilde{\Psi}(n+1)$、$\tilde{\theta}(n+1)$，由加权系数作用

$\hat{\Psi}(n+1)$、$\hat{\theta}(n+1)$、$\tilde{\Psi}(n+1)$、$\tilde{\theta}(n+1)$ 得到 $tn+1$ 时刻的真欧拉姿态角。另一个方法是将式（12）、式（13）改为递推形式，如式（16）、式（17）所示：

$$\theta(n)=\arctan\frac{-H_{X'''}(n)\sin\theta(n-1)\sin\gamma(n)-H_{Y'''}(n)\cos\gamma(n)+H_{Z'''}(n)\cos\theta(n-1)\sin\gamma(n)}{H_{X'''}(n)\cos\theta(n-1)+H_{Z'''}(n)\sin\theta(n-1)}+\arctan\frac{H_{Y'''}}{H_{X'''}} \tag{16}$$

$$\Psi(n)=\arctan\frac{-H_{X'''}(n)\sin\theta(n-1)\sin\gamma(n)-H_{Y'''}(n)\cos\gamma(n)+H_{Z'''}(n)\cos\theta(n-1)\sin\gamma(n)}{H_{X'''}(n)\cos\theta(n-1)+H_{Z'''}(n)\sin\theta(n-1)}+\arctan\frac{H_{Y}}{H_{X}} \tag{17}$$

4 仿真分析

4.1 仿真方法

为了验证模型的正确性和方法的可行性，考察算法精度，对姿态算法进行仿真计算。仿真中用白噪声叠加到标准滚转角速率时间序列上生成仿真的微机械陀螺输出；将白噪声序列叠加到地磁矢量在弹体系上的投影 $\{H_T\}$ 上作为磁强计的输出。不考虑传感器误差处理，按如下步骤进行仿真。

（1）以一种弹箭为实例，由六自由度刚体外弹道仿真程序给出弹体系下的角速率序列 $\{\omega_B\}$，姿态角序列 $\{\gamma、\Psi、\theta\}$，姿态四元数序列 $\{Q\}$，将以上结果视为真实弹道参数，作为仿真的原始数据；

（2）滚转角速率加上白噪声（给定方差的随机数序列）作为滚转陀螺的输出 $\{\tilde{\dot{\gamma}}\}$；

（3）积分滚转角速率得到滚转姿态角 $\{\gamma_{cal}\}$；

（4）将发射系下给定地磁矢量 H 投影到弹体系上生成 $\{H_T\}$，加上白噪声作为三分量磁强计输出 $\{\tilde{H}_T\}$；

（5）由 H、$\{H_B\}$、$\{\gamma_{cal}\}$ 计算其余姿态二角 $\{\Psi_{cal}、\theta_{cal}\}$，给出姿态角计算误差 $\{\gamma_{error}、\Psi_{error}、\theta_{error}\}$。

4.2 仿真结果及分析

仿真输入：某地区的地磁场参数 H 为：$H_X=297\ 12$ nt（地理北向）、$H_Y=-223\ 5$ nt（地理东向）、$H_Z=449\ 98$ nt（指向天顶）。利用刚体弹道解算再叠加白噪声后输出滚转陀螺的角速率再进行积分后得到的滚转角及误差如图 2 所示，积分滚转角的误差在 0.51°以内，滚转角积分误差曲线与滚转角速率曲线有相同的趋势，最大误差发生在主动段终点。

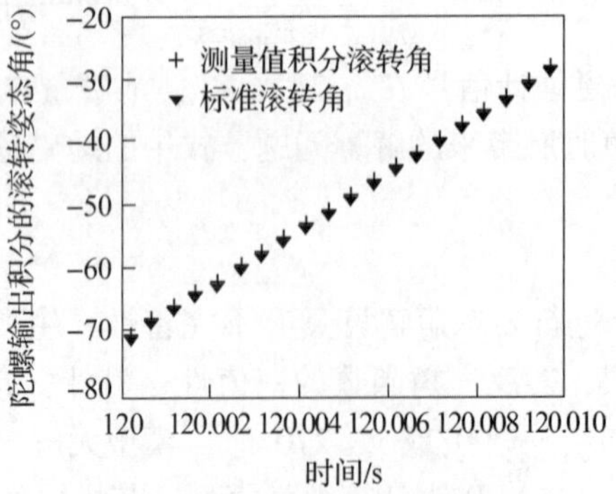

图 2　积分得到的滚转角及误差

地磁陀螺姿态探测组件其余姿态二角解算误差如图 3 所示。

从图 3 中可以看出，当解算不出现精度失控时，姿态计算结果具有较小的误差，且误差不随时间累积。但正当载体处于某些特殊角时，解算将出现较大的误差，这可以利用先验知识进行判断而剔除。

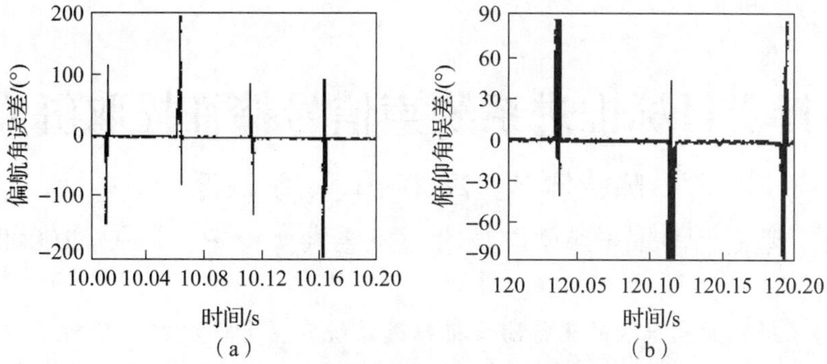

图 3　俯仰偏航姿态角解算误差（曲线片段）

5　结束语

通过上面的仿真可以看出，该方法在解算不发散的情况下，具有较高的精度，能应用于非高速滚转的弹体飞行姿态的探测。但由于制造工艺和环境等各种因素的影响，系统存在着各种误差，对于系统误差可以进行补偿。对于随机误差，可以采用适当的方法减弱。另外姿态探测元件本身也存在着各种误差，如硅微陀螺的初始温漂等，必须在使用中进行补偿。

参 考 文 献

[1] 曹红松．地磁/弹道探测技术研究［D］．太原：华北工学院，2003.

[2] LAI S T. Attitude determination of a spinning and tumbling rocket using data from two orthogonal magnetometers［R］. AFGL－TR－91－0180，AD A102908，1981.

[3] 张学孚，陆怡良．磁通门技术［M］．北京：国防工业出版社，1995.

飞行器目标推进系统声信号特征提取研究[①]

陈　赟[②]，冯顺山，冯　源

（北京理工大学爆炸科学与技术国家重点实验室，北京　100081）

摘　要： 本文为了提取低空来袭的飞行器目标推进系统声信号的特征，分析了典型飞行器目标推进系统辐射噪声的信号特性，并建立了其推进系统辐射噪声的数学模型及被动声预警系统的接收信号模型。进而提出了基于 Radon 变换的 LOFAR 图线谱特征提取算法，以获取飞行器的推进器转速特征。通过对实测飞行器噪声数据的分析，验证了信号模型算法的合理性。

关键词： 近程防卫武器；推进系统噪声；低频频率分析记录；线谱特征提取

research on extraction of acoustic characteristics of aircraft targets propulsion system

Chen Yun, Feng Shunshan, Feng Yuan

(State Key Laboratory of Explosion Science and Technology, Beijing Institute of Technology, Beijing 100081, China)

Abstract: For the short - range defensive weapons, extracting the acoustic characteristic of attacking aircraft at low altitude is an effective way to detect and recognize the target. Firstly, the signal characteristics of radiation noise of the typical vehicle propulsion system are analyzed, thus the mathematical model of propulsion system noise and the receiving signal model of passive acoustic early - warning system are given. Furthermore, the Radon transformation based LOFAR spectrogram extraction algorithm is proposed to obtain the rotation speed characteristics of the aircraft's propulsion system. Finally, the feasibility of the proposed algorithm is validated by analyzing the result of the measured aircraft noise.

Keywords: short - range defense system; propulsion system noise; low frequency analysis and recording (LOFAR) spectrogram; line spectrum feature extraction

近程防卫武器布设在重要的军用设施周围，用于防御突防的低空或超低空来袭的固定翼飞机、巡航导弹等攻击型飞行器目标[1]。由于某些重要军用设施常建在山区，低空威胁目标常常淹没在丘陵等背景杂波中，使得雷达等传统探测设备很难将其区分出来并及时报警。被动声探测系统可根据飞行目标推进系统机械部件的往复运动和旋转运动所辐射的噪声特征来探测和识别目标类型，且不易受到天气和地形等因素的影响，是一种隐蔽性好、性能可靠的近程预警方式。因此，近程防卫武器可以通过提取低空来袭的飞行器目标声信号特征，获取有效的目标特性信息，进而采取有针对性的拦截及毁伤措施。金业壮等[2]对飞行目标推进系统的辐射噪声进行了一定研究，主要面对推进系统故障诊断方面的应用，而非针对探测与分类识别飞行器的目的。因此本文详细分析了飞行器辐射的声信号特征，并建立了其辐射噪声信号的数学模型及被动预警系统的接收信号模型，进而提出了基于 Radon 变换的 LOFAR 图线谱特征提取算法，以获取目标推进器的转速特征。

---

① 原文发表于《北京理工大学学报》2010 年第 30 期。

② 陈赟：工学博士，2005 年师从冯顺山教授，研究超近程反导方法与威力效应，现工作单位：中国兵器装备研究院。

1　飞行器目标辐射噪声特性

1.1　飞行器辐射噪声

飞行器飞行过程中辐射的噪声主要为空气动力噪声和推进系统噪声。推进噪声是主要噪声源。推进系统噪声来自飞行器的发动机，主要包括风扇/压气机噪声、燃烧噪声、涡轮噪声和喷流噪声[3]。其中风扇/压气机噪声信号具有显著规律，且携带了发动机的特征信息，是被动声探测系统的主要信号源。当发动机涵道比较高（大于3）时，风扇/压气机噪声占绝对优势。然而，所需要防御的无人机、巡航导弹、战斗机等低空来袭飞行器多采用低涵道比涡扇发动机，以使它们的飞行速度能够达到跨声速或超声速。涡扇发动机的辐射噪声中风扇/压气机噪声和喷流噪声均占较大比重。图1给出了高、低涵道比涡扇发动机的噪声源分布[3]。

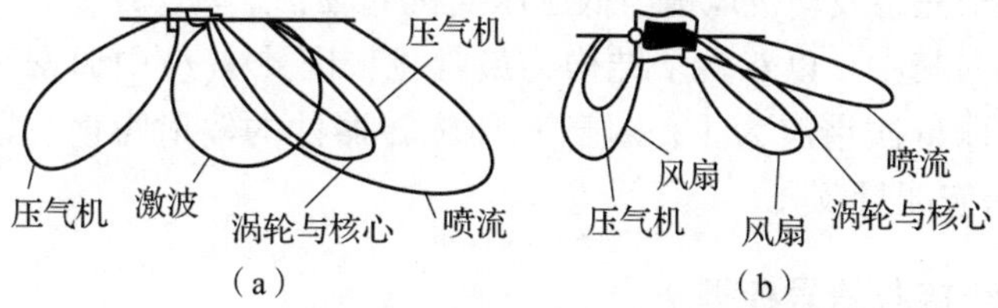

图1　高、低涵道比涡扇发动机的噪声源分布

（a）低涵道比；（b）高涵道比

1.2　涡扇发动机辐射噪声信号模型

风扇/压气机噪声由于声源辐射频率特性的不同，进一步分为宽频噪声和离散噪声，离散噪声占主导地位。离散噪声的产生机理是转子和静子间的相互干涉在叶片上产生周期性变化的非稳态气动力[4]，非稳态气动力是高效声辐射体，因此这种周期性变化导致叶片不断辐射呈谐波关系的离散信号。研究表明[2]：叶尖来流速度为亚声速的风扇/压气机，离散噪声主要是叶片通过频率下的多重单音噪声；而叶尖速度为超声速的风扇/压气机，还含有转子叶片前缘激波引起的激波分量，通常出现在轴流通过频率及其倍频处。以风扇为例，第 n 级风扇的叶片通过频率 f_{B_n} 与轴流通过频率 f_1 的关系为

$$f_{B_n} = B_n \times f_{(1)} \tag{1}$$

式中　$f_1 = n_1/60$，n_1——风扇的物理转速（r/min）；

B_n——第 n 级风扇叶片数（$n=1, 2, \cdots, N$），N 为风扇的级数。

分别对风扇、低压压气机、高压压气机的噪声信号予以建模，风扇的辐射噪声信号 $s_F(t)$ 可表示为

$$s_F(t) = \sum_{n-1}^{N}\sum_{k=1}^{\infty} A_{nk}\sin(2\pi k B_n f_1 t + \varphi_{nk}) \tag{2}$$

式中　k 为谐波次数；A_{nk} 和 φ_{nk} 分别为第 n 级风扇的 k 阶谐波噪声振幅系数和初始相位。

低压压气机辐射的声信号 $s_L(t)$ 为

$$s_L(t) = \sum_{p=1}^{p}\sum_{k=1}^{\infty} C_{pk}\sin(2\pi k E_p f_2 t + \varphi_{pk}) \tag{3}$$

式中　$f_2 = n_2/60$ 为低压压气机轴流通过频率，n_2 为其物理转速，r/min；E_p 为第 p 级低压压气机叶片数（$p=1, 2, \cdots, P$），P 为其级数；C_{pk} 和 φ_{p_k} 分别为第 p 级低压压气机的 k 阶谐波噪声振幅系数和初始相位。

高压压气机辐射的声信号 $s(t)$ 为

$$s_H(t) = \sum_{q=1}^{Q}\sum_{k=1}^{\infty} D_{qk}\sin(2\pi k F_q f_s t + \theta_{qk}) \tag{4}$$

式中，$f_3 = n_3/60$ 为高压压气机轴流通过频率，n_3 为其物理转速，r/min；F_q 为第 q 级高压压气机叶片数

($q=1, 2, \cdots, Q$)，Q 为其级数；D_{qk} 和 θ_{qk} 分别为第 q 级高压压气机的 k 阶谐波噪声振幅系数和初始相位。

当叶尖来流速度为超音速时，推进系统还存在的激波噪声信号 $s_C(t)$ 为

$$s_C(t) = \sum_{m=1}^{3} \sum_{k=1}^{\infty} C_{mk} \sin(2\pi k f_m t + \rho_{mk}) \tag{5}$$

式中，f_m 为各旋转器件的轴频，$m=1\sim3$；G_{mk} 和 ρ_{mk} 分别为第 m 个旋转器件的 k 阶谐波的噪声振幅系数和初始相位。

根据式（1）~式(5)，风扇/压气机总的辐射噪声信号 $s(t)$ 可表示为

$$s(t) = s_F(t) + s_L(t) + s_H(t) + s_C(t) \tag{6}$$

风扇/压气机所产生的宽频噪声能量微弱，一般可以忽略。对式（6）需要说明的是：①涡扇发动机所采用的结构有单转子、双转子和三转子三种，单转子涡扇发动机的风扇、低压压气机、高压压气机三者同轴，$f_1=f_2=f_3$；双转子涡扇发动机风扇与低压压气机同轴，$f_1=f_2<f_3$；三转子涡扇发动机三组旋转器件均不同轴，$f_1<f_2<f_3$。其中，以双转子结构为最常见，因此 $s(t)$ 中通常存在多组呈谐波关系的线谱。②通常发动机噪声谱以低次谐波为主，对于三种旋转器件的辐射噪声，通常叶频分量具有较大能量，激波噪声的 1~4 次谐波能量较大。

1.3 被动声探测系统接收信号模型

被动声探测系统接收到的信号是风扇/压气机噪声、喷流噪声、空气动力噪声、环境噪声的和，因此根据式（6）以及 1.1 节中的分析，被动声探测系统的接收信号 $r(t)$ 可以表示为

$$r(t) = s(t) + n_d(t) + n_B(t) + n(t) \tag{7}$$

式中，$n_B(t)$ 为宽带的空气动力噪声，服从高斯分布；$n(t)$ 为环境的高斯白噪声；$n_d(t)$ 为喷流噪声，可以表示成为一个有限带宽高斯噪声[5]，由油门位置参数决定能量。发动机辐射噪声频率主要集中在 1 kHz 以内，而喷流噪声在 3 kHz 以内都有作用[4]。

飞行器目标的推进系统旋转器件产生的离散噪声的成组谐波关系，适合用于判断飞行器发动机转速特征。所以从环境噪声、宽带的喷流噪声以及空气动力噪声中提取离散噪声的线谱分量，是预警系统需要解决的问题，因此作者提出了基于 Radon 变换的 LOFAR 图线谱特征提取的方法。

2 基于 Radon 变换的 LOFAR 图线谱特征提取算法

2.1 算法流程

对于低空目标预警系统而言，在探测与识别低空来袭的飞行器目标时，首先要提取混杂了环境噪声和宽带连续谱噪声的推进系统辐射噪声信号中的线谱特征，再通过线谱的位置及其分布的特性，推导发动机风扇、低压压气机、高压压气机的转速参数及判断目标类别，线谱特征提取算法的流程如图 2 所示。

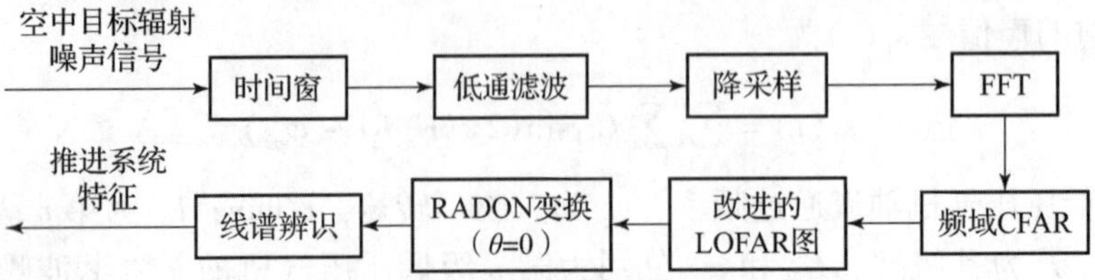

图 2　线谱特征提取算法的流程

2.2 基于 CFAR 处理的 LOFAR 图

提取线谱成分是被动声探测与识别目标的重要手段。直接利用单次输入提取线谱特征，在低信噪比时，易造成对目标的误判和漏判。为了提高在低信噪比下线谱提取能力，应利用多次数据、推迟决策进

行低频频率分析记录（low frequency analysis and recording，LOFAR）处理[6]。该方法广泛应用于被动水声探测系统中，但针对空中目标的辐射声特性探测及识别的相关研究较少。因此本文根据飞行器目标的辐射噪声信号特性，改进了 LOFAR 谱图的获取方法，使之符合飞行器辐射噪声线谱特征提取的应用需要。

LOFAR 谱图以瀑布图的形式将多次接收的数据予以显示，以 x 轴表示信号频率，y 轴表示时间，亮度表示信号幅度。在被动声呐信号处理中，通常采用基于排序截断平均算法（OTA）获取船舶螺旋桨辐射噪声的 LOFAR 谱图。但空中飞行目标推进器辐射噪声不具有舰船辐射噪声在频谱上每倍频程衰减 6 dB 的规律，再加上环境噪声的影响，在其传统 LOFAR 谱图上的线谱成分时明时暗，识别困难。加之飞行器旋转器件辐射声信号所需要分析的频谱范围比舰船辐射噪声的范围大得多，故 OTA 运算量巨大，难以满足实时报警的要求，需要针对目标特性改进 LOFAR 谱图的获取方法。因此对单次输入数据进行频域恒虚警处理（constant false alarm rate，CFAR），生成改进的 LOFAR 谱图。

恒虚警处理是目标检测中常用的处理手段。频域恒虚警处理不仅可以在变化的噪声背景下将淹没的单频或多重单频信号区分出来，且由于不需要进行排序的过程，因此计算量相对较小。所以，采用频域恒虚警率处理的方法，可在保持预期的虚警概率不变的条件下，将线谱成分更有效地提取出来。由于环境噪声和连续谱噪声均为高斯白噪声，而傅里叶变换为线性变换，因此它们两者之和的频谱幅度分布可由瑞利分布描述为[2]

$$p(y) = \frac{y}{\sigma_m^2}\exp\left(-\frac{y^2}{2\sigma_m^2}\right) \tag{8}$$

式中，σ_m 与无噪声干扰时均值 M_a 的关系为

$$M_a = \sqrt{\frac{\pi}{2}}\sigma_m \tag{9}$$

因此频域 CFAR 检测器可描述为

$$\begin{cases} \dfrac{X_p(u)}{\sigma_m} > \eta_{\mathrm{Th}}, \text{有线谱} \\ \dfrac{X_p(u)}{\sigma_m} \leqslant \eta_{\mathrm{Th}}, \text{无线谱} \end{cases} \tag{10}$$

式中，$X_p(u)$为接收信号的频谱；η_{Th}为 CFAR 检测器阈值：

$$\eta_{\mathrm{Th}} = \sqrt{2\ln(1/P_{\mathrm{fa}})} \tag{11}$$

式中，P_{fa}为虚警概率，通常可取 $10^{-6} \sim 10^{-4}$量级。对单次接收信号的频谱进行 CFAR 处理后，保留大于阈值的点，其余频点置 0。再以瀑布图的形式将多次 CFAR 处理的结果形成改进的 LOFAR 谱图，作为线谱提取部分的输入。

2.3　Radon 变换

在图像处理中，Radon 变换是常见的直线提取方法之一，也常用于时频图像的处理。Radon 变换对直线的提取算法简单，运算量较小，可提供直线的位置信息，尽管 Radon 变换无法提供直线的起点、终点信息，但这些信息对本文的应用是无关紧要的。因此，Radon 变换适合于在 LOFAR 谱图中快速提取谱线，并获得谱线对应的数值。

Radon 变换的原理是将原始图像经过线积分的形式变换到另一对参数域内，即

$$g(\rho,\theta) = \int_{-\infty}^{\infty}\int_{-\infty}^{\infty} f(x,y)\delta(\rho - x\cos\theta - y\sin\theta)\,\mathrm{d}x\mathrm{d}y \tag{12}$$

式中，ρ 为坐标原点到该直线的距离；θ 为垂线与 x 轴的夹角；$\delta(\cdot)$为冲激函数，表示沿着线谱的积分。由于改进的 LOFAR 图中的线谱是垂直于频率轴的，因此仅需要做 $\theta = 0$ 的 Radon 变换，形成积分值在 ρ 轴的分布平面图，其中的峰值代表谱线出现的位置，其在横轴上的投影可以换算得到线谱的频率值

f。换算公式为

$$f=\frac{1}{2}F_R+\rho\Delta F \tag{13}$$

式中 F_R——LOFAR 图所分析的频率范围；

Δf——频域分辨率。

2.4 谱线辨识准则

由于飞行器目标声信号产生的离散噪声在频域表现为成组的离散线谱，所以保留 Radon 变换后输出的成组谱线，剔除无成组关系的谱线，再根据式（1）可确定各转子的轴频，获得各转子的转速。

3 实验结果及分析

为验证本文中提出的算法的有效性，对 NATO－RSC－10 标准噪声库中实测的 F－16 战斗机在飞行过程中辐射的声信号进行分析。该战斗机采用 F100－PW－220 低涵道比涡扇发动机，涵道比为 0.6。发动机的主要旋转部件是风扇和高压压气机，包括 3 级风扇和 10 级高压压气机，为双转子结构。声信号的采样频率为 19.98 kHz，数据长度 90 s，在 1 kHz 以内的谱如图 3 所示。

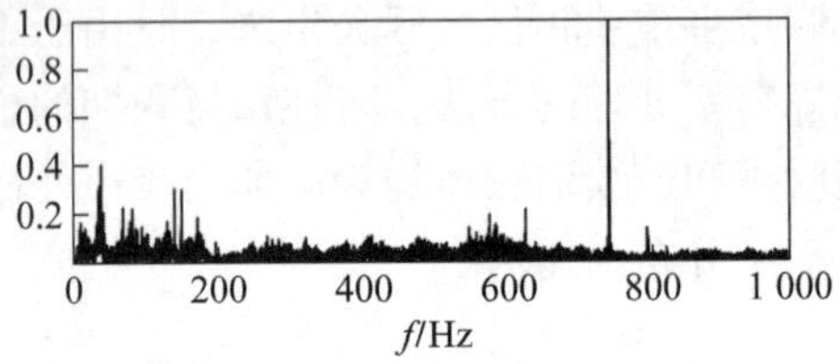

图 3 F－16 战斗机辐射噪声的频谱

从图 3 中可看到，飞行器辐射的噪声有多根明显线谱，是由推进系统旋转器件辐射产生的，还具有谱值较高的连续谱噪声。主要是由喷流噪声和环境噪声产生的。与第 1 节中描述的信号模型相符。

以 10 s 为单次接收信号的长度，分别取两次数据时刻 1 与时刻 2 的频域分析结果，如图 4 和图 5 所示。从图中可以看到，线谱出现的位置有较大的差异，因此仅根据单帧信号提取谱线是不准确的。

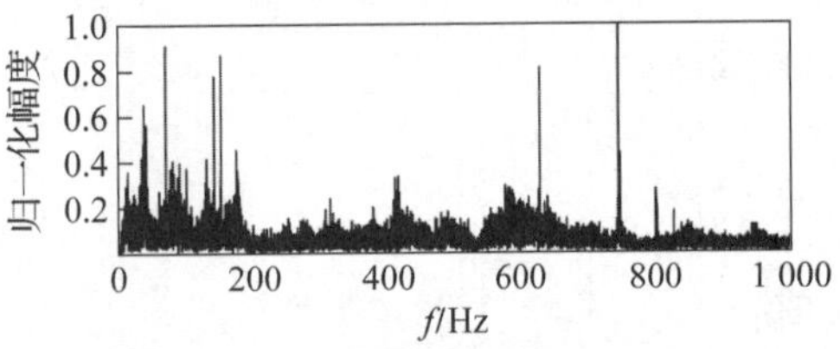

图 4 时刻 1 的频谱图

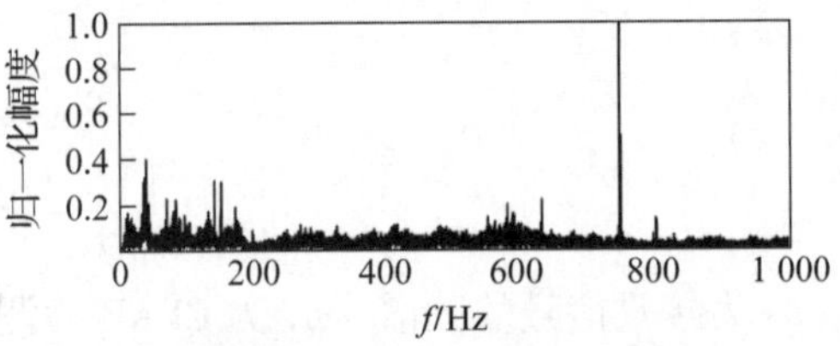

图 5 时刻 2 的频谱图

取每 10 s 数据为一帧，步进 1 s，在滤波、FFT 后进行频域 CFAR 处理，虚警概率取 10^{-6} 量级。由于线谱主要集中在 1 kHz 以内，因此改进的 LOFAR 谱图如图 6 所示，图中的亮线表示始终存在的涡扇发动机旋转频率及其谐波分量，而亮点仅是偶尔出现的噪声。

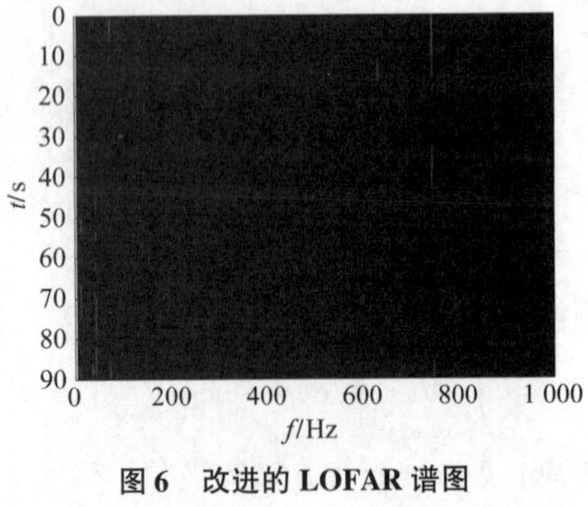

图 6 改进的 LOFAR 谱图

对 CFAR 处理后的谱图进行 $\theta=0$ 的 Radon 变换，结果如图 7 所示。相比图 3 而言，Radon 变换的输出不再有宽带连续谱的影响，谱线更加清晰且易于辨认。图 7 中 6 根主要的谱线分别为：39 Hz 的线谱对应风扇和低压压气机的轴频；155(Hz)≈39×4(Hz)对应轴频的 4 阶谐频；630(Hz)≈39×16(Hz)；749(Hz)≈39×19(Hz)；630 Hz 约是某级 16 个叶片转子的叶频；749 Hz 约是某级 19 个叶片转子的叶频；71.7 Hz 为高压压气机的轴频，143(Hz)≈71.7×2(Hz)对应该轴频的 2 阶谐频。所提取获得的这些谱线说明声信号具有两组成谐波关系的线谱结构，从而可推断该涡扇发动机具有两级转子结构。并且，通过线谱对应的频率，可以获得两级转子的转速 f 风扇转速为 2 340 r/min，高压压气机转速为 4 302 r/min。算法分析的结果与已知发动机情况是相符合的，从而说明谱线提取算法是可行的。

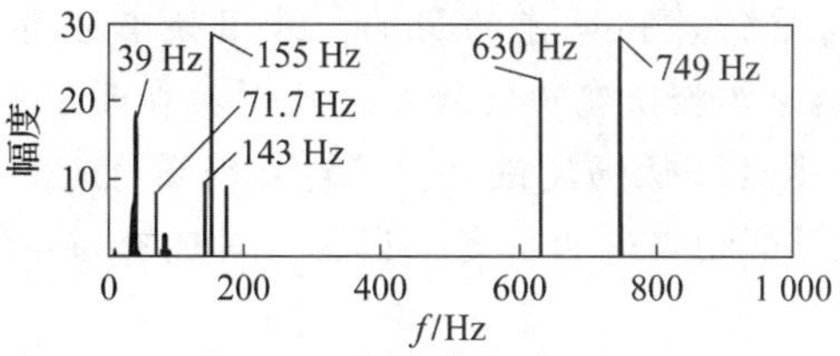

图 7　改进 LOFAR 谱图的 Radon 变换

4　结论

本文针对飞行器目标辐射噪声的特点，阐述了低涵道比涡扇发动机辐射噪声的信号特征，其中由推进系统旋转器件产生的声信号是由多组成谐波关系的离散谱分量构成，从而给出了被动声探测系统接收信号的模型。根据该信号模型，运用频域 CFAR 改进了 LOFAR 谱图的获取方法，并进而提出了基于 Radon 变换的改进 LOFAR 图线谱特征提取算法及其实现步骤。该算法可从环境噪声和宽带连续谱噪声中有效提取多组呈谐波关系的线谱成分，为进一步推导获得涡扇发动机转速的特征提供了依据。最后，通过对实测飞行器噪声数据的对比分析，结果验证了信号模型和算法的合理性。因此，作者的研究成果可为近程防卫武器的被动声预警系统探测识别低空来袭飞行器目标的提供一种信号处理解决方案。

参考文献

[1] 冯顺山. 巡航导弹研究报告 [R]. 北京：北京理工大学机电学院，2005.
FENG S S. Research on cruise missiles [R]. Beijing：School of Mechatronical Engineering，Beijing Institute of Technology，2005.

[2] 金业壮，盛元生. 某型涡轮喷气发动机噪声特性分析 [J]. 沈阳航空工业学院学报，1994，3 (25)：1－10.
JIN Y Z，SHENG Y S. An analysis of noise characteristics of a turbojet engine [J]. Journal of Shen－yang Institute of Aeronautical Engineering，1994，3 (25)：1－10.

[3] 唐狄毅，李文兰，乔渭阳. 飞机噪声基础 [M]. 西安：西北工业大学出版社，1995.
TANG D Y，LI W L，QIAO W Y. Noise of aircrafts [M]. Xi'an：Northwestern Polytechnical University Press，1995.

[4] CUMPSTY N A. Engine noise in aircraft engine Components [R]. [S. l.]：AIAA Press，1985.

[5] SHUNICHI A，ANDRZEJ C. Adapt blind signal processing－neural network approaches [J]. Proceedings of the IEEE，1998，86 (10)：2026－2048.

[6] 陈敬军，陆佶人. 被动声纳线谱检测技术综述 [J]. 声学技术，2004，23 (3)：57－60.
CHEN J J，LU J R. A review of techniques for detection of line－spectrum in passive sonar [J]. Technical acoustics，2004，23 (3)：57－60.

装药缺陷对熔铸炸药爆速影响的实验研究①

王 宇②，芮久后，冯顺山
（北京理工大学爆炸科学与技术国家重点实验室，北京 100081）

摘 要：为研究装药缺陷对熔铸炸药爆速的影响，采用缺陷分类的研究方法对含有不同缺陷的RDX/TNT熔铸混合炸药进行了爆速实验研究。实验结果表明存在于不同组分中的缺陷会对爆速产生不同程度的影响，TNT中的缺陷使爆速降低幅度最大，若对炸药爆速有较高要求则应重点去除TNT内缺陷，并且推断缺陷尺寸大小可能是影响爆速的决定性因素。对现有 $\omega-\Gamma$ 爆速计算公式进行改进使其能够体现装药缺陷对爆速的影响。

关键词：缺陷；组分密度；爆速；$\omega-\Gamma$ 公式

Experimental Research of the Charge Defects' Influence on Detonation Velocity of Melting – CastExplosive

WANG Yu，RUI Jiu – hou，FENG Shun – shan
（State Key Laboratory of Explosion Science and Technology，Beijing Institute of Technology，Beijing 100081，China）

Abstract：With the purpose of researching the defects' influence on detonation velocity of melting – cast explosive，the detonation velocities of RDX/TNT melting – cast explosives with different kind of defects incharge were studied using experimental method. It is found that the defects indifferent components lead to different influences on detonation velocity，and the defects in TNT decrease the detonation velocity to the great extent. So the defects removal in TNT should be emphasized to ensure high performance of explosive. By inference，the key factor is the size of defects is the key factor to weaken the performance of explosive. Considering the experimental results，the $\omega-\Gamma$ equation could be modified and the new equation would materialize the relationship between component density and detonation velocity.

Keywords：defects；component density；detonation velocity；$\omega-\Gamma$equation

以RDX和TNT为主要组分的熔铸混合炸药在当今世界各国有着极其广泛的应用，其装药技术的难点之一是避免产生气泡、裂纹等各种缺陷。而这些缺陷会对装药的能量、感度等性能产生明显影响。国内外多位学者针对这一问题进行了研究，研究结果显示缺陷增多会使装药的感度升高[1-4]，与之相应的热点起爆理论也已被广泛认可。另外，Borne[5]认为存在于不同组分中的缺陷会对装药的整体性能产生不同程度的影响，因此提出了缺陷分类的研究方法，以实验手段对比固相晶体内部、固相晶体表面、黏结组分内部三种缺陷与装药冲击波感度的关系，研究结果显示固相晶体内部的缺陷使冲击波感度升高幅度最大。Teipel[6]也在其著作中表达了相同观点. 这一研究成果的意义在于让研究者认清哪种类型的缺陷是影响装药冲击波感度的决定性因素，因此若对装药的低易损性有较高要求，则应该重点去除固相晶体内的缺陷。

缺陷同时也使装药的能量降低，但是在这方面并未受到足够的重视，鲜有缺陷分类对装药爆速影响

① 原文发表于《北京理工大学学报》2011年第3期。

② 王宇：工学博士，2007年师从冯顺山教授，研究基于球形硝基胍的钝感熔铸装药研究，现工作单位：航天科技第四研究院41所火工事业部。

的研究，即何种类型的缺陷对装药爆速的影响更大目前尚不清楚。因此本文在 L. Borne 研究成果的基础上，针对 RDX/TNT 熔铸装药中不同类型的缺陷与爆速的关系进行了实验研究，目的是要认清何种缺陷使爆速降低幅度最多，即若对装药能量有较高结果对现有要求，应重点去除哪类缺陷 ω－Γ 爆速计算公式。最后[7]，进行了改进结合本文的实验，使其可以区分组分密度变化与爆速的关系。

1　装药缺陷与组分密度

1.1　缺陷分类

混合炸药中的各种缺陷由于成因不同，会分布于不同的装药组分之中，RDX/TNT 熔铸装药主要由 RDX 颗粒和 TNT 两种成分组成，因此可以将其中的缺陷分成如下两类。

（1）RDX 内的缺陷。制备 RDX 晶体过程中混入晶体内部的空气或溶剂，在晶体内部以气泡或液泡的形式存在，如图 1（a）所示。

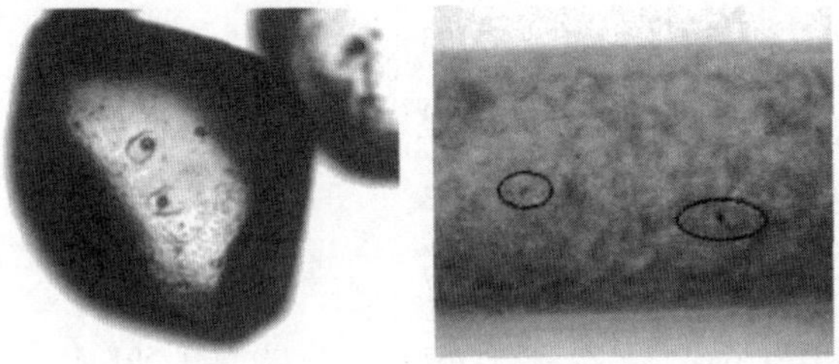

（a）　　（b）

图 1　装药缺陷的分类

（a）RDX 内缺陷；（b）TNT 内缺陷

（2）TNT 内的缺陷。装药过程中混入 TNT 中的空气形成的气泡，或冷却过快产生的缩孔、裂纹，如图 1（b）所示，为一个 TNT 药柱试样表面可见的气孔。

这种组分分类法的优点是便于进行实验研究，只要通过合理的工艺控制，就可以在不同组分中引入所需的缺陷并进行对比实验。

1.2　组分密度

由于缺陷将空气等低密度物质引入，必然会导致装药密度降低，然而按组分分类的两种缺陷实际上是降低各自所在组分的密度，使其无法达到理论密度值，因此针对缺陷进行分类研究，可以等价于针对组分密度的研究。

2　试样的制备及爆速测试

通过对 RDX 晶体品质和熔铸装药工艺的控制，制备出不同工艺质量的 RDX/TNT 装药试样，通过对比组分密度的差异区分不同的缺陷分布状况。

2.1　RDX 晶体试样的准备

本文中所采用的普通 RDX 晶体含有较多缺陷，平均密度为 1.767 g/cm^3，为理论密度的 97.3%。采用重结晶技术提高晶体纯度，减少晶体内缺陷，制备出更高密度的 RDX 晶体，平均密度达 1.793 g/cm^3，为理论密度的 98.7%。

2.2　RDX/TNT 熔铸装药试样的制备

选择不同晶体密度的 RDX 和控制熔铸装药工艺，制备 A、B 两组质量配比为 RDX/TNT－60/40 的熔铸装药试样，为减小随机误差的干扰，每组试样以完全相同的工艺制备 10 发，如表 1 所示。

表 1　RDX/TNT 熔铸装药试样

试样	孔隙率		组分密度		$\rho/(g\cdot cm^{-3})$
	$k_{RDX}/\%$	$K_{TNT}/\%$	$P_{RDX}/(g\cdot cm^{-3})$	$P_{TNT}/(g\cdot cm^{-3})$	
A-1	0.71	2.75	1.793	1.550	1.687
A-2	0.72	2.63	1.793	1.554	1.689
A-3	0.72	2.52	1.793	1.558	1.691
A-4	0.71	2.86	1.793	1.545	1.685
A-5	0.71	2.75	1.793	1.550	1.687
A-6	0.72	2.69	1.793	1.552	1.688
A-7	0.72	2.58	1.793	1.556	1.690
A-8	0.72	2.46	1.793	1.560	1.692
A-9	0.71	2.86	1.793	1.545	1.685
A-10	0.71	2.75	1.793	1.550	1.687
B-1	1.55	1.74	1.767	1.586	1.690
B-2	1.55	1.63	1.767	1.591	1.692
B-3	1.55	1.80	1.767	1.584	1.689
B-4	1.55	1.57	1.767	1.593	1.693
B-5	1.55	1.69	1.767	1.589	1.691
B-6	1.55	1.80	1.767	1.584	1.689
B-7	1.55	1.92	1.767	1.580	1.687
B-8	1.55	1.69	1.767	1.589	1.691
B-9	1.55	1.86	1.767	1.582	1.688
B-10	1.55	1.80	1.767	1.584	1.689

表 1 中孔隙率 k 为装药中孔隙体积占装药总体积的百分比，第 n 个组分的孔隙率为

$$k_n = \frac{w_n \rho}{\rho_n \rho'_n}(\rho'_n - \rho_n) \tag{1}$$

式中　w_n——组分质量分数；

ρ_n——组分实际密度；

ρ'_n——组分理论密度；

ρ——药柱密度。

应用式（1）进行计算时，RDX 的组分密度以及装药密度依据实验测定，TNT 的组分密度为

$$\rho_{TNT} = \frac{w_{TNT}\,\rho_{RDX}\,\rho}{\rho_{RDX} - \rho(1 - w_{TNT})} \tag{2}$$

式中　w_{TNT}——TNT 质量分数；

ρ_{RDX}——RDX 组分密度；

ρ——装药密度。

从表 1 可以看出，A、B 两组试样具有相同的成分和质量配比，药柱密度也接近，但组分密度存在

明显差异，通过以下对各试样进行爆速测试，观察组分密度的差异对爆速的影响。

2.3 爆速测试

爆速测试的原理是当爆轰波到达或接近电探极敏感部位所在剖面时，电探极开关状态突变，并输出记时信号[8]。将各试样均分4段进行切割，然后将各切断面打磨光滑平整，测量各小段药柱长度 Δx_i，在切断面处安装双丝式电探极，如图2所示。最后将各切断试样重新对接、固定、装配成如图3所示的爆速测试样品，并采用 TektronixTDS5000B 型数字示波器记录记时信号 Δt_i。

图2 双丝式电探极的安装

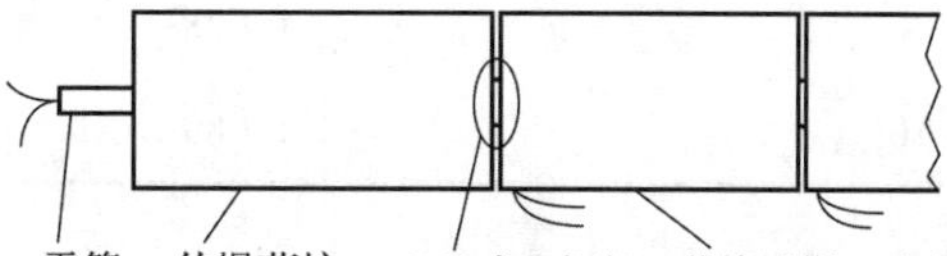

图3 爆速测试样品装配图

3 实验结果与分析

3.1 爆速测试结果与分析

根据实验结果采用最小二乘法处理4段爆速测试数据（Δx_i，Δt_i），爆速的线性回归公式为

$$v_D = \frac{n\sum_{i=1}^{n}\Delta x_i \Delta t_i - \sum_{i=1}^{n}\Delta x_i \sum_{i=1}^{n}\Delta t_i}{n\sum_{i=1}^{n}\Delta t_i^2 - \left(\sum_{i=1}^{n}\Delta t_i\right)^2} \tag{3}$$

计算所得各试样的爆速测试值如表2所示。

表2 试样爆速测试值

试样	$\rho/(\mathrm{g\cdot cm^{-3}})$	v_D(测试)/$(\mathrm{m\cdot s^{-1}})$
A-1	1.687	7 634
A-2	1.689	7 650
A-3	1.691	7 675
A-4	1.685	7 635
A-5	1.687	7 626
A-6	1.688	7 657
A-7	1.690	7 655
A-8	1.692	7 678
A-9	1.685	7 672
A-10	1.687	7 662
B-1	1.690	7 766

续表

试样	$\rho/(g\cdot cm^{-3})$	v_D(测试)/$(m\cdot s^{-1})$
B-2	1.692	7 772
B-3	1.689	7 778
B-4	1.693	7 785
B-5	1.691	7 789
B-6	1.689	7 791
B-7	1.687	7 767
B-8	1.691	7 782
B-9	1.688	7 768
B-10	1.689	7 775

表2中A、B两组试样的平均爆速分别为7 655 m/s、777 6 m/s，平均密度分别为1.688 $g\cdot cm^{-3}$、1.690 $g\cdot cm^{-3}$，爆速均值相差121 m/s。由于2组试样成分、配比完全相同，平均密度接近，所以理论上爆速差异不应该如此明显，这说明两组试样中组分密度的不同对爆速产生了影响。由于A组试样中的缺陷更多地集中于TNT中，且其平均爆速较低，说明存在于TNT中的缺陷对爆速的负面影响相对较大。产生这种现象的原因可能是存在于TNT中的缺陷通常在尺寸上更大，某些较大的气泡甚至可以达到毫米级别，而RDX内缺陷基本上为微米级别，因此缺陷尺寸可能是对爆速产生影响的决定性因素之一，即缺陷尺寸越大使爆速降低幅度越大。所以若对装药爆速有较高要求，应重点去除TNT内存在的缺陷。

3.2 对ω-Γ爆速计算公式的改进

吴雄于1985年提出的$\omega-\Gamma$爆速计算公式[7]，具有较高的计算精度，近年来得到了广泛的应用。将其计算值与本文实验测试值进行对比，误差最大仅为3.12%，足以满足工程应用的精度要求。此公式的具体形式为

$$\begin{gathered} v_D = 33.1Q^{0.5} + 243.25l\rho_0 \\ Q = \sum_{i=1}^{n} w_i Q_i, l = \sum_{i=1}^{n} w_i l_i \end{gathered} \tag{4}$$

式中 v_D——爆速；

ρ_0——装药密度；

w_i——各组分的质量分数；

Q_i——各组分的爆热；

l_i——各组分的位能因子。

式（4）只体现出了装药整体密度与爆速的关系，无法区分各组分密度的变化，即缺陷存在于炸药不同组分对爆速的影响无法区分。

结合本文中的实验研究，得出改进的$\omega-\Gamma$爆速计算公式为

$$\begin{gathered} v_D = 33.1Q^{0.5} + 243.25l/m \\ m = \sum_{i=1}^{n} k_i(\rho_i^0 - \rho_i)Q_i + \sum_{i=1}^{n} w_i/\rho_i \end{gathered} \tag{5}$$

式中 ρ_{i0}——各组的理论密度；

ρ_i——组分密度；

k_i——各组分密度的权值。

改进的 $\omega-\Gamma$ 爆速计算公式中体现了炸药各组分密度变化对爆速所产生的影响，式中组分密度的权值 k_i 体现了各组分密度与爆速的相关性，影响权值大小的重要因素之一为缺陷的尺寸，其他因素目前还不清楚，所以目前尚无炸药各成分权值的准确数值，还需继续进行更加深入的研究。

4　结论

基于缺陷分类的研究方法，以实验手段研究了 RDX/TNT 熔铸装药中存在的缺陷对爆速的影响，实验结果显示集中于 TNT 中的缺陷使爆速降低幅度更大，因此若对爆速有较高的要求，应更加注重 TNT 中缺陷的去除。最后，结合实验结果，针对现有的 $\omega-\Gamma$ 爆速计算公式进行改进，使其可以区分组分密度变化与爆速的关系。

参考文献

[1] JOHANSEN H，KRISTIANSEN J D，GJERSOE R，et al. RDX and HMX with reduced sensitivity towards shock initiation - RS RDX and RS - HMX [J]. Propellants，explosives，pyrotechnics，2008，33 (1)：20 - 24.

[2] 花成，黄明，黄辉，等. RDX/HMX 炸药晶体内部缺陷表征与冲击波感度研究 [J]. 含能材料，2010，18 (2)：152 - 157.

[3] 黄亨建，董海山，舒远杰，等. HMX 中晶体缺陷的获得及其对热感度和热安定性的影响 [J]. 含能材料，2003，11 (3)：123 - 126.

[4] STOLTZ C A，MANSON B P，HOOPER J. Neutron scattering study of internal void structure in RDX [J]. Journal of applied physics，2010，107 (5)：1 - 6.

[5] BORNE L，MORY J，SCHLESSER F. Reduced sensitivity RDX (RSRDX) in pressed formulations：respective effects of intra - granular pores，extra - granular pores and pore sizes [J]. Propellants，explosives，pyrotechnics，2008，33 (1)：37 - 43.

[6] TEIPEL U. Energetic materials：particle processing and characterization [M]. Weinheim：WILEY - VCH Verlag，2005：524 - 534.

[7] 孙业斌，惠君明，曹欣茂. 军用混合炸药 [M]. 北京：兵器工业出版社，1995：85 - 87.
SUN Y B，HUI J M，CAO X M. Mixed explosives for military application [M]. Beijing：Ordnance Industry Publishing House，1995：85 - 87.

[8] 黄正平. 爆炸与冲击电测技术 [M]. 北京：国防工业出版社，2006：71 - 94.
HUANG Z P. Explosion and shock measuring technique [M]. Beijing：National Defense Industry Press，2006：71 - 94.

重结晶法制备球形化 RDX[①]

赵　雪[②]，芮久后，冯顺山

（北京理工大学爆炸科学与技术国家重点实验室，北京　100081）

摘　要：为解决以 RDX 为基的熔铸炸药固态含量低的问题，本文采用环已酮作为溶剂重结晶的方法，得到了球形化 RDX 晶体。分析了重结晶工艺条件中结晶温度、搅拌速率、杂质等参数变化对于 RDX 晶体形状的影响，制备了球形化 RDX 晶体。测试了球形化 RDX 的晶体形状、流散性、机械感度和以其为组分的 PBX 冲击波感度。结果表明，采用环已酮重结晶法制备的球形化 RDX 晶体形状规则、表面光滑、棱角少、流散性好，撞击感度、摩擦感度比普通 RDX 略有降低，以球形化 RDX 为基的 PBX 相比以普通 RDX 为基的 PBX 冲击波感度降低约 25%。

关键词：RDX；球形化；重结晶

Recrystallization Method for Preparation of Spherical RDX

Zhao Xue，Rui Jiuhou，Feng Shunshan

（State Key Laboratory of Explosion Science and Technology，Beijing Institute of Technology，Beijing 100081，China）

Abstract：To improve the solid content of casting explosive based on RDX，the spherical RDX was prepared by recrystallization using Cyclohexane as a solvent. The influence of crystallization temperature，stirring speed and impurity on RDX crystal shape was analyzed. The crystal shape，free – flowing property，impact sensitivity and friction sensitivity of both spherical RDX and ordinary RDX were tested. Spherical RDX and ordinary RDX were used to prepare PBX with the same casting formula and their shock sensitivity was tested，respectively. The results show that the crystal shape of spherical RDX is regular，the surface is sleek and lack of edge angle. Spherical RDX has better free – flowing property than ordinary RDX. The impact sensitivity and friction sensitivity of spherical RDX are lower than ordinary RDX. PBX shock sensitivity based on spherical RDX is 25% lower than based on ordinary RDX.

Keywords：RDX；spherical；recrystallization

RDX 是一种爆炸性能良好的高能单质炸药，但晶体形状不规则、流散性差，导致熔铸炸药固态含量低，制约了其在混合炸药上的应用。为了拓展 RDX 的应用范围，发挥其本身爆轰性能高的优势，迫切需要改善 RDX 晶体不规则的形状。球形化 RDX 晶体具有形状规则、表面光滑、晶体致密、流散性好等特点，能够明显提高以 RDX 为基的混合炸药的爆轰性能。美国已有专利介绍制备球形 RDX 的方法[1]，但其使用的原料 RDX 并非普通 RDX，具体制备方法并未做介绍。国内关于球形化 RDX 的研究比较少，与国外存在一定的差距。国内外改变炸药晶形的方法很多[2-3]，包括物理研磨法、化学重结晶法、超临界法、超声波法等，由于重结晶法具有产物纯度高、能耗低、工艺简单和易于规模化生产等优点，本实验采用重结晶方法制备球形化 RDX，分析重结晶工艺中影响晶体形状的因素，制备出的球形化 RDX 晶体规则透明、流散性好、表面光滑、棱角少，机械感度和普通 RDX 相比略有下降，以其作组分的高聚物黏结炸药相比以普通 RDX 作组分的高聚物黏结炸药冲击波感度有所降低。

---

① 原文发表于《北京理工大学学报》2011 年第 1 期。

② 赵雪：工学博士，2007 年师从冯顺山教授，研究 NQ 球形化制备方法及其在装药中应用研究，现工作单位：北京理工大学。

1 实验方法与流程

本实验采用溶剂重结晶方法制备球形化 RDX，重结晶工艺简单、能耗低、污染小、易于规模化生产，实验采用的溶剂价格低廉且易得，环己酮的饱和溶液能够重复利用，制备出的球形化 RDX 晶体形状规则、透明、表面光滑、流散性好。球形化 RDX 工艺流程如图 1 所示。称适量 RDX 和环己酮混合，加热溶液，直到 RDX 已基本溶解于环己酮中，调节溶解后的溶液酸碱性，使 RDX 重结晶，过滤，烘干，后处理得到球形化 RDX 晶体。

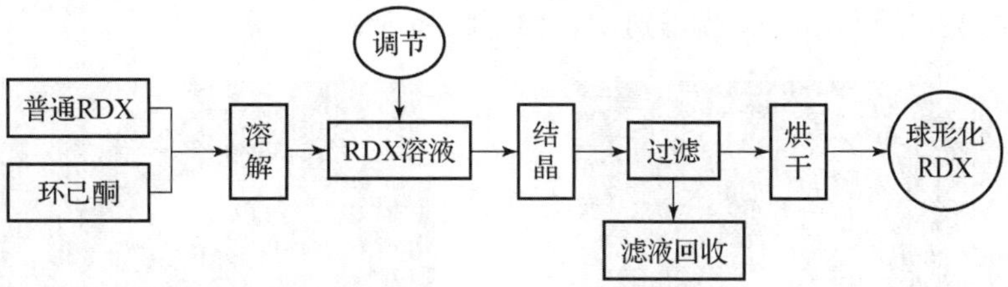

图 1 球形化 RDX 工艺流程

2 实验结果与分析

2.1 工艺条件对 RDX 晶体球形化的影响

2.1.1 结晶温度的影响

RDX 的晶体形状随着结晶温度的升高会成为规则的类球形。结晶过程是一个复相化学反应过程，温度降低，反应常数变小，晶体生长速度下降。温度是影响晶体生长速度的因素之一，在其他条件都相同的情况下，晶体生长的速率随温度的升高而加快，从而影响到粒子的扩散速度和生长界面。晶体生长实验表明，导致界面不稳定的原因之一是太低的温度梯度和温度的波动，温度适当升高有利于晶体生长界面的稳定，更有利于制备球形化晶体。

2.1.2 搅拌速率的影响

随着搅拌速率的增加，RDX 晶体的形状逐渐规则且成为类球形。搅拌速率对 RDX 晶核的形成和生长以及晶体的形态有重要影响。晶体生长的过程中，搅拌速率低，晶核扩散慢可能使细小晶体形成聚晶，从而使产品溶剂含量偏高，搅拌速率的增加有利于结晶体系中 RDX 晶体稀释剂和溶剂的分散速度，使这些细小晶体难以形成聚晶，有利于提高 RDX 晶体质量；搅拌的加强有利于将局部区域稀释结晶热量尽快转移，以免局部温度过高，有利于降低和控制结晶温度；搅拌的增强有助于消除晶体各晶面上过饱和度的差别，使晶体生长易处于平衡状态，使产品晶形趋于球形，且不易产生缺陷或包藏母液[4]。RDX 晶体在环己酮溶液中搅拌速率的提高，使得 RDX 晶体表面光滑且无棱角。在环己酮的溶液中高速率的搅拌能够使重结晶得到的 RDX 晶体棱角溶解，表面变得光滑，晶体形状更接近球形。搅拌速度的增加有利于 RDX 晶体棱角的溶解，并能缩短反应时间。

2.1.3 杂质的影响

RDX 中的杂质很多，杂质对于 RDX 形状的影响不容忽视，有的杂质对晶体生长极为敏感，杂质原子进入晶体后，使晶体内产生的结构缺陷不仅直接影响到晶体的物理性能，而且会使晶体在生长过程中改变形态。杂质的存在对晶体生长速率有明显的影响，杂质较多对晶体生长起抑制作用，也有的可加快晶体生长，溶液中的杂质常常导致晶体生长形态的改变。由于结晶作用的专一性，生长中的晶体对外来杂质有排斥作用，但晶体的各向异性，杂质在晶体的不同晶面上经常发生选择吸附，这种吸附作用常使某些晶面的生长受到阻碍，因而改变了晶体各晶面的相对生长速率，而促成了晶体形态的改变。RDX 中杂质对于形状的定量影响仍需进一步探索研究。

2.2 性能测试结果

2.2.1 晶体形状

通过光学显微镜和 SEM（扫描电镜）观察球形化 RDX 晶体内部缺陷及形状。为了识别普通 RDX 和球形化 RDX 的晶体形状，分别采用了适当倍数的扫描电镜照片。图 2 为 220 倍扫描电镜下的普通 RDX 晶体照片，可以看出普通 RDX 晶体形状不规则、棱角多、表面凹凸不平。图 3 为 30 倍球形化 RDX 的扫描电镜照片，球形化 RDX 晶体的表面光滑，棱角少，形状规则，为类球形。图 4 为球形化 RDX 的光学显微镜照片，可以看出，球形化 RDX 光滑透明，内部缺陷少。

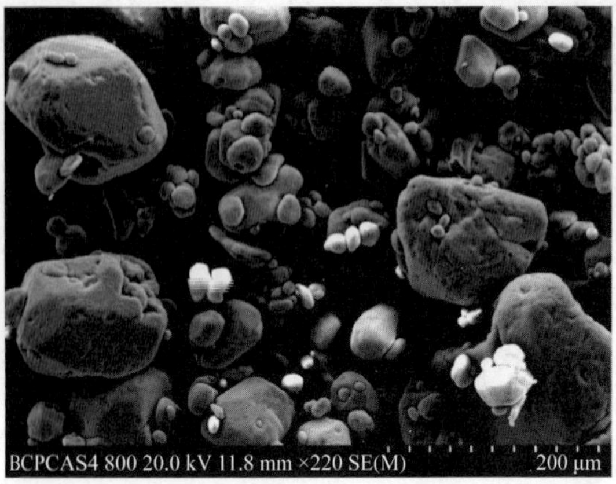

图 2　普通 RDX 的 SEM 照片

图 3　球形化 RDX 的 SEM 照片

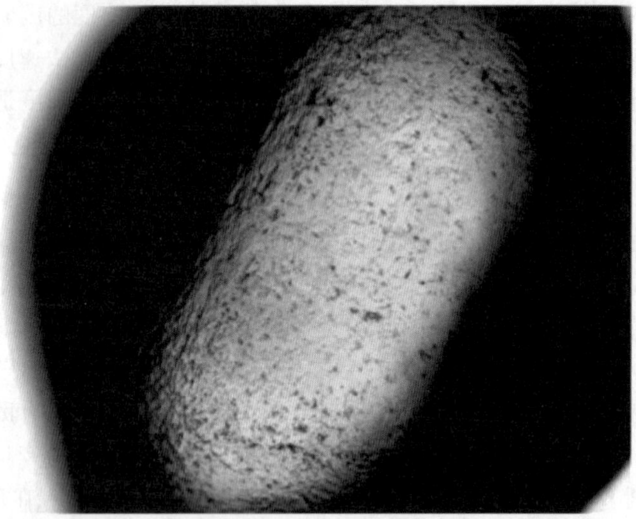

图 4　球形化 RDX 的光学显微镜照片

2.2.2 流散性

将球形化处理前后的 RDX 晶体用水配制成 60% 固体含量的浆状炸药进行流散性实验，结果表明，由球形化 RDX 配制的浆状炸药搅拌阻力小，说明球形化 RDX 的流散性较好。

2.2.3 感度

（1）机械感度。分别采用 GJB772A－97 炸药试验方法 601.2 特性落高法和 602.1 爆炸概率法测试了粒径 100～300 μm 普通 RDX 和球形化 RDX 的撞击感度和摩擦感度。普通 RDX 和球形 RDX 的特性落高 H_{50}分别为 42.1 cm 和 45.2 cm，摩擦爆炸概率分别为 38% 和 35%；球形化 RDX 的机械感度略低于普通 RDX。根据热点学说理论，球形化 RDX 晶体致密，内部的缺陷少，有可能形成的热点数量少，机械感度自然比普通 RDX 低。

（2）冲击波感度。将球形化 RDX 和普通 RDX 分别与铝粉、高氯酸铵、端羟基聚丁二烯等机械搅拌混匀，浇铸成型固化制备成高聚物黏结炸药（PBX），比较两组相同配方和工艺的样品的冲击波感度，如表 1 所示。

表 1　两组配方的 PBX 冲击波感度

样品	L_{50}/cm
普通 RDX 制备的 PBX	3.51
球形化 RDX 制备的 PBX	2.62

球形化 RDX 制备的 PBX 的隔板厚度小于普通 RDX 的隔板厚度，降低了 1.89 cm，降低幅度约 25%。RDX 安全性能的提高大大促进了 PBX 安全性能的提高，使 PBX 具有良好的生存能力，冲击波感度明显降低。

3　结论

采用溶剂（环己酮）重结晶的方法制备球形化 RDX。讨论了工艺条件中温度、搅拌速率、杂质对 RDX 晶体形状的影响。制备的球形化 RDX 晶体形状规则、表面光滑、棱角少、流散性好，相同粒径范围的球形化 RDX 机械感度及 PBX 冲击波感度低于普通 RDX。球形化 RDX 的晶体性能比普通 RDX 有明显提高，实验证明重结晶方法是制备球形化 RDX 的有效方法。球形化 RDX 的制备为高性能混合炸药、发射药和固体推进剂的制备奠定基础。

参考文献

[1] LAVERTU R R, CHARLESBOURG A G. Process for spheroidization of RDX crystals: USA, 4065529 [P]. 1997-12-07.

[2] 芮久后，王泽山，刘玉海，等．超细黑索金制备新方法 [J]．南京理工大学学报，1996，20 (5)：385-388.

RUI J H, WANG Z S, LIU Y H, et al. A newmethod for preparation of ultrafine RDX crystals [J]. Journal of Nanjing University of Science and Technology, 1996, 20 (5): 385-388.

[3] 王元元，刘玉存，王建华，等．降感 RDX 的制备及晶形控制 [J]．火炸药学报，2009，32 (2)：44-47.

WANG Y Y, LIU Y C, WANG J H, et al. Preparation and crystal control of desen sitized-RDX [J]. Chinese journal of explosives and propellants, 2009, 32 (2): 44-47.

[4] 陈国光，董素荣．弹药制造工艺学 [M]．北京：北京理工大学出版社，2004：25-30.

CHEN G G, DONG S R. The preparation technology of ammunition [M]. Beijing: Beijing Institute of Technology Press, 2004: 25-30.

最大熵原理在钝感熔铸炸药配方设计中的应用①

冯顺山，王　宇②，芮久后

（北京理工大学爆炸科学与技术国家重点实验室，北京　100081）

摘　要：本文针对含 HMX 或 RDX 的熔铸炸药降低冲击感度、提高使用安全性的需要，以最大熵为评价准则提出了钝感熔铸炸药的配方设计方法，并以此方法获得了能量和安全性俱佳的钝感熔铸炸药配方。选取 NQ/HMX/TNT - 36. 1/33. 9/30. 0 和 NQ/RDX/TNT - 34. 5/25. 5/40. 0 配方制备熔铸装药试样并进行爆速和枪击感度测试，结果显示两试样分别相对 Octol 70/30 和 Cyclotol 60/40 爆速仅降低2% ~4%，但枪击感度大幅度降低，说明两试样在能量损失较小的情况下安全性大幅度提高，使能量和安全性俱佳，证明了本配方设计方法的可行性。

关键词：最大熵原理；钝感熔铸炸药；能量；感度

Application Research of Maximum Entropy Theory in Insensitive Melting - Cast Explosive Formulation

Feng Shunshan，Wang Yu，Rui Jiuhou

（State Key Laboratory of Explosion Science and Technology，Beijing Institute of Technology，Bering 100081，China）

Abstract：To meet the needs of reducing the impact sensitivity and improving the safety of melting - cast explosives containing HMX or RDX，created a formulation design method of insensitive explosive based on the rule of maximum entropy. With this method，insensitive explosive formulations that are powerful and safe could be achieved. Choose the explosive formulations of NQ/HMX/TNT - 36. 1/33. 9/30. 0 and NQ/RDX/TNT - 34. 5/25. 5/40. 0 to make two experimental samples，and tested these samples' detonation velocity and rifle shooting sensitivity. Experimental results show that，the detonation velocity only reduces 2% to 4% but it is much more insensitive in rifle shooting experiment，compared with Octol 70/30 and Cyclotol60/40. It means that these two samples become much safer with a little energy reduce，that proves the formulation design method which created in this article is feasible.

Key words：maximum entropy theory；insensitive melting - cast explosive；energy；sensitivity

以 HMX 或 RDX 为主要固相组分的熔铸混合炸药因能量较高、装药工艺性能较好等优点被广泛应用于榴弹、航弹、导弹等各类弹药中。但是随着弹药战斗部设计中对装药钝感化关注度的提高，此类炸药安全性的不足也暴露得越来越明显。以应用最广泛的 B 炸药为例[1]，美国就曾发生过多次关于 B 炸药的安全事故，尤其是装填 B 炸药的榴弹曾频繁发生膛炸事故，致使美国曾一度放弃使用 B 炸药装填高初速榴弹。所以提高此类炸药的使用安全性，研发钝感的熔铸炸药已成为当前的迫切需求。

针对这一需求，多种钝感单质炸药被合成出来，其中 TATB、NQ 和 NTO 因安全性能突出而受到较多关注。现有的研究成果表明[2-4]，这 3 种钝感单质炸药同样可以作为钝感剂在混合炸药中使用，并且钝感效果良好。因此本文中参考最大熵原理在信息论中的应用[5]，建立钝感混合炸药的配方设计方法。该

---

① 原文发表于《北京理工大学学报》2011 年第 31 卷第 11 期。

② 王宇：工学博士，2007 年师从冯顺山教授，研究基于球形硝基胍的钝感熔铸装药研究，现工作单位：航天科技第四研究院 41 所火工事业部。

方法的特点是以 TATB、NQ、NTO 作为钝感剂，以 HMX、RDX 作为高能组分，以最大熵为评价准则进行配方设计，获得能量和安全性俱佳的钝感混合炸药配方。

1　钝感混合炸药配方设计方法的建立

1.1　最大熵原理在配方设计中的应用

以混合炸药作为一个系统进行分析，原理如图 1 所示，其性能是各成分性能的集合体现，将性能抽象成信息，因此各成分可以认为是携带着信息的单元，系统总的信息量即为各成分携带信息的总和。依据信息论中的最大熵原理“熵达最大时对应着这个系统所包含的信息量最大”[5]，可以将最大熵作为准则进行混合炸药配方设计，以增加系统信息量为手段优化钝感混合炸药配方。

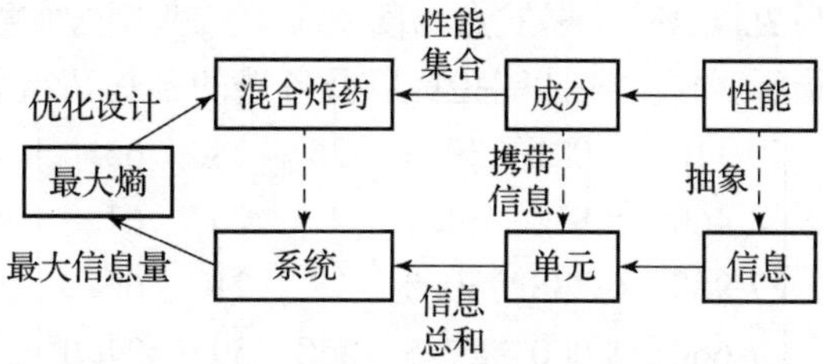

图 1　最大熵原理在混合炸药中的应用原理

依照此原理，以 RDX 或 HMX 为主要固相组分的熔铸混合炸药在安全性能上的不足即对应着此系统部分信息的缺失，通过合理的配方设计将安全性能提高，即可填补缺失的信息使系统的信息量增大。此时以熵作为准则，熵达最大时系统信息量最大，即得安全性能优化的混合炸药配方。

1.2　混合炸药配方设计过程

采用统计手段获得混合炸药各组分能量和感度的相关性能指标并建立性能矩阵，假设共有 m 个备选组分和各组分的 n 个性能指标，则 m 行 n 列的性能矩阵为

$$\boldsymbol{A} = \begin{bmatrix} x_{11} & x_{12} & \cdots & x_{1n} \\ x_{21} & x_{22} & \cdots & x_{2n} \\ \cdots & \cdots & \cdots & \cdots \\ x_{m1} & x_{m2} & \cdots & x_{mn} \end{bmatrix}$$

由于性能矩阵 $\boldsymbol{A}$ 中各指标的量纲不同而无法直接用于统计计算，所以必须先将数据进行量纲一化处理。采用极差变换法以 x_{imax} 和 $x_{j\text{max}}$ 分别代表性能矩阵 $\boldsymbol{A}$ 中每列大于和小于所有性能指标的边界数值，并通过式（1）计算量纲一化性能指标 $\overline{x_{ij}}$。

$$\overline{x_{ij}} = \frac{x_{j\text{max}} - x_{ij}}{x_{j\text{max}} - x_{j\text{min}}}，\text{或}\ x_{ij} = \frac{x_{ij} - x_{j\text{max}}}{x_{j\text{max}} - x_{j\text{min}}} \tag{1}$$

式（1）中前者用于处理结果随数值增大而渐优的指标，后者用于处理结果随数值减小而渐优的指标，计算所得的 $\overline{x_{ij}}$ 组成量纲一化性能矩阵

$$\boldsymbol{A} = \begin{bmatrix} \overline{x_{11}} & \overline{x_{12}} & \cdots & \overline{x_{1n}} \\ \overline{x_{21}} & \overline{x_{22}} & \cdots & \overline{x_{2n}} \\ \cdots & \cdots & \cdots & \cdots \\ \overline{x_{m1}} & \overline{x_{m2}} & \cdots & \overline{x_{mn}} \end{bmatrix}$$

针对矩阵 $\boldsymbol{B}$ 中各量纲一化数据，令

$$f_{ij} = x_{ij} / \sum_{i=1}^{m} x_{ij} \tag{2}$$

可得任意为组分 i 和 k 相混合后的熵 S 为

$$s = -t\left(w_i \sum_{j=1}^{m} f_{ij}\ln f_{ij} + w_k \sum_{j=1}^{m} f_{kj}\ln f_{kj}\right) \tag{3}$$

式中　$t = 1/\ln n$；

w_i 和w_k——两组分的质量分数，且$w_i + w_k = 100\%$。

2　混合炸药配方设计实例

2.1　组分的选取及熵值计算

通过文献［6－7］收集到各组分的性能指标分别为代表能量的爆速 D(m/s)、爆热 Q(kJ/kg)、爆压 P(GPa) 和代表感度的特性落高 H_{50}(cm)、爆发点温度 t(℃)、冲击波感度 G_{50}(mm)，以上数据除 G_{50}外均为结果随数值增大而渐优的指标，以这些性能指标以及备选的各炸药组分建立性能矩阵为

$$\begin{bmatrix} 9\,110 & 6\,021 & 39.3 & 26 & 280 & 65.0 \\ 8\,460 & 5\,812 & 34.7 & 22 & 257 & 61.8 \\ 7\,980 & 6\,238 & 30.0 & 12 & 210 & 66.0 \\ 7\,660 & 5\,000 & 27.5 & 350 & 397 & 21.9 \\ 7\,650 & 3\,724 & 26.8 & 350 & 230 & 5.0 \\ 7\,955 & 4\,370 & 27.8 & 146 & 260 & 21.2 \\ 6\,940 & 4\,402 & 19.1 & 154 & 330 & 46.4 \end{bmatrix}$$

矩阵 $\boldsymbol{A}$ 中列数据从左到右依次分别为 D、Q、P、H_{50}、t、G_{50}的值；行数据从上到下依次分别对应组分 HMX、RDX、PETN、TATB、NQ、NTO、TNT。

采用极差变换法针对性能矩阵 A 拟定的边界值为

$$x_{j\max} = [10\,000 \quad 7\,000 \quad 40 \quad 400 \quad 400 \quad 70]$$

$$x_{j\min} = [6\,000 \quad 3\,000 \quad 10 \quad 10 \quad 200 \quad 4]$$

通过式（1）的计算得到的量纲一化矩阵，并经过式（2）的转换即可以将 TATB、NQ、NTO 作为钝感剂，以 HMX、RDX 作为高能组分设计混合炸药配方，依照式（3）得出各混合炸药配方的熵值。

2.2　混合炸药配方设计结果及分析

2.2.1　HMX 基混合炸药配方

分别以 TATB、NQ、NTO 作为 HMX 的钝感剂得到的熵值－钝感剂含量曲线如图 2 所示。图 2 中 3 种钝感单质炸药与 HMX 混合后熵值均有提高，说明它们都对 HMX 发挥了降感作用，且降感效果从强到弱依次为 TATB、NQ、NTO。另外各曲线均存在极值点，各极值点的数据如表 1 所示。配方达到曲线极值点后钝感剂的降感效果已不再突出而使能量的降低开始凸显，所以能量损失在极值以后开始逐渐加剧导致熵值下降。

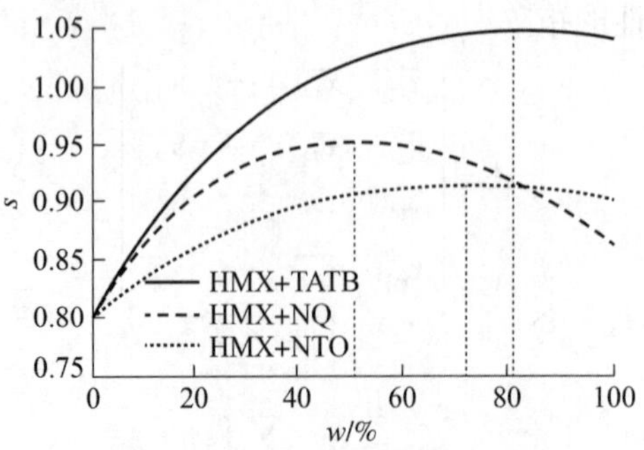

图 2　HMX 基混合炸药

图2中各曲线在极值点位置对应着系统包含最大的信息量，因此以极值点为基准进行配方设计可获得能量和安全性加和最大的配方，配方设计如表1所示。

表1　HMX基混合炸药极值点及配方设计

钝感剂	熵增幅度/%	极值点质量分数/%	配方设计
TATB	31.7	82.3	57.6 : 12.4 : 30.0
NQ	19.5	51.6	36.1 : 33.9 : 30.0
NTQ	14.9	72.8	51.0 : 19.0 : 30.0

注：配方设计组分为钝感剂/HMX/TNT的质量比，固含量为70%。

2.2.2　RDX基混合炸药配方

分别以TATB、NQ、NTO作为RDX的钝感剂，得到各配方的熵值－钝感剂含量曲线如图3所示。

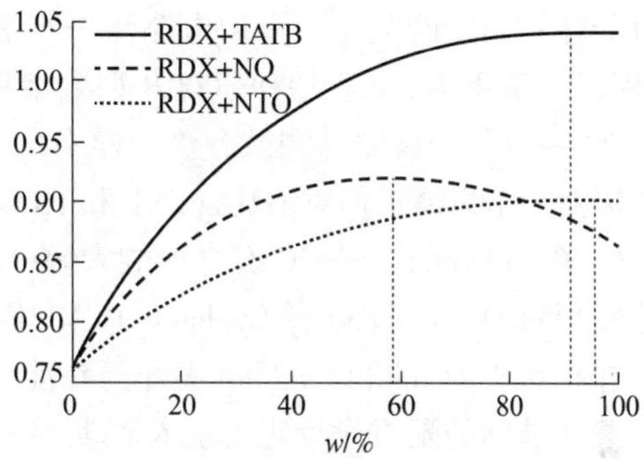

图3　RDX基混合炸药

图3显示3种钝感单质炸药对RDX仍会发挥降感作用且降感效果从强到弱依次为TATB、NQ、NTO，其中NQ在极值点以后的能量损失仍然最为严重。不同之处在于TATB和NTO曲线在极值点以后无明显下降，这是由于TATB和NTO的能量与RDX的差距较小所致。

表2所列为图3中各曲线极值点位置相关数据及混合炸药的配方设计。其中含TATB和NTO的两个配方中，RDX含量较少，因此RDX只能起到敏化剂的作用。

表2　RDX基混合炸药极值点及配方设计

钝感剂	熵增幅度/%	极值点质量分数/%	配方设计
TATB	37.6	91.2	54.7 : 5.3 : 40.0
NQ	21.4	57.4	34.5 : 25.5 : 40.0
NTQ	19.3	95.4	57.2 : 2.8 : 40.0

注：配方设计组分为钝感剂/RDX/TNT的质量比，固含量为60%。

3　实验验证

3.1　配方选取与试样制备

针对本文提出的以最大熵为准则的钝感炸药配方设计方法进行实验验证，分别选取表1中的配方NQ/HMX/TNT－36.1/33.9/30.0制备试样1和表2中的配方NQ/RDX/TNT－34.5/25.5/40.0制备试样2。

普通的NQ由于晶形呈长针状导致装填系数过低而无法用于装药，所以采用重结晶技术将NQ进行晶形改良，制备出的 NQ 晶体长径比更小，装填系数得到大幅度提高达到可以用于装药的水平。

采用 TNT 基熔铸炸药制备工艺，制备出的药柱试样如图 4（a）所示，两试样的密度值如表 3 所示。

表 3　测试结果与现有炸药数据对比

炸药	密度/($g \cdot cm^{-3}$)	爆速测试值/($m \cdot s^{-1}$)	枪击测试反应概率/%
试样 1	1.71	8 060	0
Octol 70/30[6]	1.8	8 377	70
试样 2	1.65	7 690	0
Cyclotol 60/40[6]	1.68	7 900	85

3.2　试样性能测试及结果分析

以爆速和枪击感度作为评价能量和感度的判据。采用电测法进行爆速测试[8]，将试样切割、打磨并按图 4（b）所示安装双丝式电探极，采用 Tektronix TDS5000B 型数字示波器记录爆轰波间隔时间的记时信号；枪击感度测试遵照 GJB 772A—97[9] 中规定的测试方法，样本量为 10 发。两项测试的结果见表 3。

试样 1 和试样 2 分别相当于以 NQ 作为钝感剂对 HMX/TNT 和 RDX/TNT 类炸药进行降感，两试样对枪击刺激无反应，相对于 Octol 70/30 和 Cyclotol 60/40 安全性能大幅度提升。说明 NQ 发挥的降感效果非常明显；同时两试样的爆速仅分别相对 Octol 70/30 和 Cyclotol 60/40 降低 2%~4%，说明 NQ 的加入并未造成过多的能量损失。因此两试样相对于 Octol 和 Cyclotol 类炸药在能量上仅略微降低，但安全性能大幅度提高。与本文中提出的以最大熵为准则的配方设计理论相符，实验验证了此方法的可行性。同时以两试样配方的炸药装填各种弹药战斗部可有效提高武器弹药的耐冲击、防殉爆能力，用于装填高初速炮弹可有效降低发生膛炸的概率，且毁伤能力仍然较高，因此采用这两种配方制备的炸药具有取代 Octol 或 Cyclotol 类炸药的潜力。

4　结论

（1）采用最大熵原理提出的钝感炸药配方设计方法有助于解决熔铸炸药钝感化的需要，获得能量和安全性俱佳的钝感炸药配方，该方法经过理论分析和实验验证，证明可行，具有在钝感炸药配方设计中的应用前景。

（2）选取 NQ/HMX/TNT－36.1/33.9/30.0 和 NQ/RDX/TNT－34.5/25.5/40.0 配方制备试样并进行性能测试，结果两试样分别相对于 Octol 70/30 和 Cyclotol 60/40 爆速仅有 2%~4% 的降低，但枪击感度大幅度提高。说明试样在能量损失较小的情况下安全性大幅度改善，使其能量和安全性俱佳。

参 考 文 献

[1] 董海山．钝感弹药的由来及重要意义［J］．含能材料，2006，14（5）：321－322.
DONG H S. The source and importance of insensitive ammunition［J］. Chinese journal of energetic materials，2006，14（5）：321－322.

[2] JOSEPH M D，JANGID S K，SATTPUTE R S. et a1. Studies on advanced RDX/TATB based low vulnerable sheet explosives with HTPB binder［J］. Propellants，explosives，pyrotechnics，2009，34：326－330.

[3] POWALA D，ORZECHOWSKI A. Insensitive high explosives nitroguanidine（NQ）［J］. Przemysl chemiczny，2005，84（8）：582－585.

[4] SMITH M W. CLIFF M D. NTO－based explosive formulations：a technology review：DSTO－TR－0796［R］.［S. 1.］：DSTO，1999.

基于磁强计和MEMS陀螺的弹箭姿态探测系统①

杜　烨[1,2]②，冯顺山[1]，苑大威[3]，方　晶[1]，郑　奕[4]

（1. 北京理工大学爆炸科学与技术国家重点实验室，北京　100081；
2. 中北大学弹箭模拟仿真研究中心，山西太原　030051；
3. 中国兵器工业第二〇八研究所，北京　102202；
4. 上海空间电源研究所，上海　200245）

摘　要：为解决弹道修正弹箭中捷联式姿态测量系统误差随时间不断积累的问题，设计了一种由二轴磁强计和MEMS陀螺构建的低成本弹体姿态磁－惯性测量系统，利用磁强计测量的地磁信息修正MEMS陀螺解算的姿态角误差。在此基础上，提出了将两轴地磁信号解算滚转角融入陀螺解算的姿态优化算法，研制的原理样机在二轴转台上进行了测试。有限的试验表明：在一定条件下，该测量系统可有效抑制陀螺漂移引起的姿态误差，能可靠地用于弹道修正弹箭的姿态测量。

关键词：MEMS陀螺；磁强计；旋转矢量；姿态探测

An Attitude Determination System Using Magnetometers and MEMS Gyro

Du Ye[1,2], Feng Shunshan[1], Yuan Dawei [3], Fang Jing[1], Zheng Yi [4]

(1. State Key Laboratory of Explosive Science and Technology, Beijing Institute of Technology, Beijing 100081, China;
2. Research Center of Projectiles and Rockets Simulation, North University of China, Taiyuan, Shanxi 030051, China;
3. No. 208Research Institute of China Ordnance Industries, Beijing 102202, China;
4. Shanghai Institute of Space Power Sources, Shanghai 200245, China)

Abstract: To solve a problem that systematic errors accumulated with time in the strapdown attitude determination system for the trajectory correction , a low - cost magnetic - inertial attitude measurement system composed of two - axis magnetometers and MEMS gyroscope was proposed. The geomagnetic information were used to correct attitude error from MEMS gyro and geomagnetic. Geomagnetic signals were devoted to the rotation vector in calculating attitude. A principle prototype was designed and fabricated, and it was tested on two - axis turntable. The determination system can effectively suppress attitude error caused by the gyro drift. This paper proves that the attitude determination system can be reliably used for trajectory correction on some condition.

Key words: MEMS gyro; magnetometer; rotation vector; attitude determination

近年来，使用磁强计、MEMS陀螺等惯性器件进行姿态探测成为研究捷联式低成本、全固态姿态探测系统的热点之一，磁强计具有无漂移、功耗低、测量范围宽、稳定性高、体积小，MEMS陀螺具有动态范围宽、可靠性高、体积小、重量轻等优点，其全固态特性能很好地适合弹药的弹上环境[1]。

2001年，Marins[2]在扩展卡尔曼滤波器的基础上，研究使用MARG传感器（三轴磁强计、三轴速率陀螺和三轴加速度计的捷联组合）姿态探测的算法，并提出了基于四元数扩展的卡尔曼滤波[3]。2005

---

① 原文发表于《北京理工大学学报》2014年第34卷第12期。

② 杜烨：工学博士，2011年师从冯顺山教授，研究基于脉冲发动机二维修正火箭弹控制技术，现工作单位：中北大学。

年，黄旭等[4]结合测量重力场和地磁场研究出一种 EKF 算法，使用高斯－牛顿迭代算法求解非线性方程组获得测量四元数。2006 年，曹红松等[1]提出了一种单轴陀螺和三轴磁强计组合的全姿态测量方案。2008 年，鲍亚琪等[3]用三轴磁强计和三轴陀螺的测量，然后用解析算式获得了全姿态。

本文采用两轴地磁、三轴 MEMS 陀螺进行解算，将两轴地磁信号解算滚转角融入陀螺解算的旋转矢量法中，并做成样机进行了转台试验。

1 姿态探测方案设计

该姿态探测系统主要由二轴磁强计、MEMS 陀螺、数据处理器、姿态控制器构成。安装时，将陀螺敏感轴对准弹体坐标系的三轴，用于解算 3 个姿态角，磁强计探头对准弹体系中 y_1 轴和 z_1 轴安装，用于测定地磁在两轴上的投影，如图 1 所示。

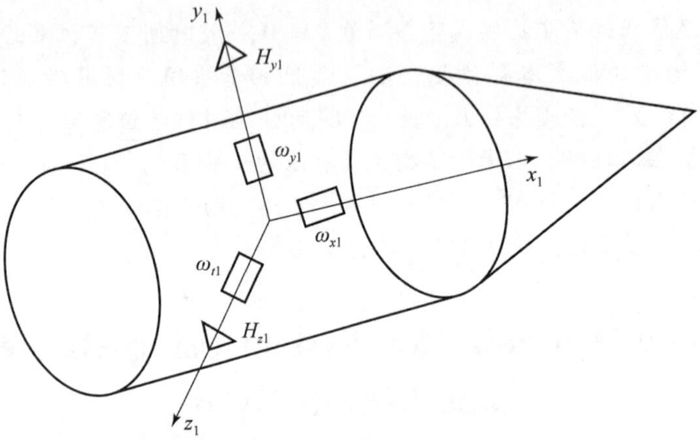

图 1　姿态探测系统捷联示意图

具体算法如图 2 所示，等效旋转矢量法积分陀螺信息解算出 3 个姿态角，二轴磁强计地磁信号解算出滚转角，与等效旋转矢量法解算的滚转角进行比较，当误差大于一定阈值时，令滚转姿态角归 0，重新计算消除前面的累积误差。一旦进入地磁盲区，地磁停止工作，全用陀螺信号解算，在非盲区，地磁重新工作，保证姿态解算的可靠性。

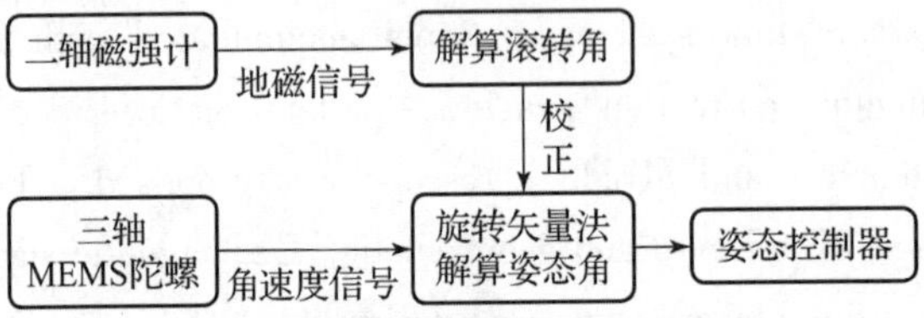

图 2　姿态算法原理

2 陀螺信息算法

假设 q 为弹体坐标系相对于地面坐标系的旋转四元数，ω 为弹体坐标系相对于地面坐标系的角速度矢量，其在弹体坐标系的投影为 ω_{x1}，ω_{y1}，ω_{z1}，则姿态运动学方程可用四元数表示为[5]

$$\frac{\mathrm{d}\boldsymbol{q}}{\mathrm{d}t}=\frac{1}{2}\boldsymbol{\omega}\boldsymbol{q} \tag{1}$$

即

$$\begin{bmatrix}\dot{\boldsymbol{q}}_0\\ \dot{\boldsymbol{q}}_1\\ \dot{\boldsymbol{q}}_2\\ \dot{\boldsymbol{q}}_3\end{bmatrix}=2\begin{bmatrix}0 & -\omega_{x1} & -\omega_{y1} & -\omega_{z1}\\ \omega_{x1} & 0 & \omega_{z1} & -\omega_{y1}\\ \omega_{y1} & -\omega_{z1} & 0 & \omega_{x1}\\ \omega_{z1} & \omega_{y1} & -\omega_{x1} & 0\end{bmatrix}\begin{bmatrix}\boldsymbol{q}_0\\ \boldsymbol{q}_1\\ \boldsymbol{q}_2\\ \boldsymbol{q}_3\end{bmatrix} \tag{2}$$

初始时刻，只要 $\boldsymbol{q}_0^{\mathrm{T}}\boldsymbol{q}_0=1$ 在理论上成立，在以后的时刻 $\boldsymbol{q}_0^{\mathrm{T}}\boldsymbol{q}_0=1$ 则恒成立，不管 $\boldsymbol{\omega}$ 为何值，都不能改变 $\boldsymbol{q}$ 的模，而且在实际计算中，这种归一化约束是能保持的[6]。

旋转矢量可以表示为 $\boldsymbol{\Phi}(h)=[\Phi_x \quad \Phi_y \quad \Phi_z]^{\mathrm{T}}$，用来确定弹体在 h 时刻的运动姿态，即以 $\boldsymbol{\Phi}(h)$ 为轴，转动角度大小等于旋转矢量 $\boldsymbol{\Phi}(h)$ 的幅值。当弹体姿态运动可用如下的近似旋转矢量微分方程表示为[7]

$$\boldsymbol{\Phi}=\boldsymbol{\omega}+\frac{1}{2}\boldsymbol{\Phi}\times\boldsymbol{\omega}+\frac{1}{12}\boldsymbol{\Phi}\times(\boldsymbol{\Phi}\times\boldsymbol{\omega}) \tag{3}$$

式中 $\boldsymbol{\Phi}$——等效旋转矢量；

$\boldsymbol{\omega}$——弹体角速度矢量。

在一个姿态更新周期中，旋转矢量微分方程中角速度可以表示为 $\omega=a+bh+ch^2$，即可得出旋转矢量的等效三子样法计算公式：

$$\boldsymbol{\Phi}(h)=\Delta\theta_1+\Delta\theta_2+\Delta\theta_3+0.45(\Delta\theta_1\Delta\theta_3)+0.675\Delta\theta_2(\Delta\theta_3-\Delta\theta_1) \tag{4}$$

三子样算法可采用如下角增量提取公式[7-8]：

$$\begin{cases}\Delta\theta_1=5\omega(t)+8\omega\left(t+\dfrac{h}{3}\right)-\dfrac{\omega\left(t+\dfrac{2h}{3}\right)h}{36}\\ \Delta\theta_2=\dfrac{\left[-\omega(t)+8\omega\left(t+\dfrac{h}{3}\right)+\omega\left(t+\dfrac{2h}{3}\right)\right]h}{36}\\ \Delta\theta_3=\dfrac{\left[-\omega\left(t+\dfrac{h}{3}\right)+8\omega\left(t+\dfrac{2h}{3}\right)+\omega(t+h)\right]h}{36}\\ \Delta\theta_4=\dfrac{\left[\omega(t)+3\omega\left(t+\dfrac{h}{3}\right)+3\omega\left(t+\dfrac{2h}{3}\right)+\omega(t+h)\right]h}{36}\end{cases} \tag{5}$$

姿态更新可以表达为

$$Q(t+1)=Q(t)\otimes q(h) \tag{6}$$

其中，$q(h)=\cos\dfrac{\varphi}{2}+\dfrac{\boldsymbol{\Phi}}{\varphi}\sin\dfrac{\varphi}{2}$，为了区分姿态四元数 $\boldsymbol{Q}(t)$，定义 $\boldsymbol{q}(h)$ 为姿态更新周期内的姿态变化四元数，$\boldsymbol{\Phi}$ 为等效旋转矢量，$\varphi=|\boldsymbol{\Phi}|$。

所以等效旋转四元数为

$$\boldsymbol{q}(h)=\left[\cos\Delta\theta/2 \quad \frac{\Delta\theta_x\sin\Delta\theta/2}{\Delta\theta} \quad \frac{\Delta\theta_y\sin\Delta\theta/2}{\Delta\theta} \quad \frac{\Delta\theta_z\sin\Delta\theta/2}{\Delta\theta}\right]^{\mathrm{T}} \tag{7}$$

式中，$\Delta\theta_x$，$\Delta\theta_y$，$\Delta\theta_z$ 为陀螺角增量。

$$\Delta\theta=\sqrt{\Delta\theta_x^2+\Delta\theta_y^2+\Delta\theta_z^2}$$

解出姿态更新四元数后，根据方向余弦矩阵与姿态变换矩阵各元素对应关系[9]，则姿态角公式为

$$\begin{cases}\theta=\tan^{-1}\dfrac{2(q_2q_3-q_1q_3)}{q_0^2-q_1^2+q_2^2-q_3^2}\\ \psi=\sin^{-1}2(q_0q_3-q_1q_2)\\ \gamma=\tan^{-1}\dfrac{2(q_1q_3-q_0q_2)}{q_0^2-q_1^2+q_2^2-q_3^2}\end{cases} \tag{8}$$

3 地磁信息算法

定义炮口为坐标原点，地磁矢量 $\boldsymbol{H}$ 在地面坐标系的投影为 H_x、H_y、H_z，经坐标变换投影到弹体坐标系为

$$\begin{bmatrix} H_{x1} \\ H_{y1} \\ H_{z1} \end{bmatrix} = \boldsymbol{C}_2^b \boldsymbol{C}_1^2 \boldsymbol{C}_n^1 \begin{bmatrix} H_x \\ H_y \\ H_z \end{bmatrix} \tag{9}$$

式中，

$$\boldsymbol{C}_2^b(\gamma) = \begin{bmatrix} 1 & 0 & 0 \\ 0 & \cos\gamma & \sin\gamma \\ 0 & -\sin\gamma & \cos\gamma \end{bmatrix}$$

$$\boldsymbol{C}_2^b(\gamma) = \begin{bmatrix} \cos\psi & 0 & -\sin\psi \\ 0 & 1 & 0 \\ -\sin\psi & 0 & \cos\psi \end{bmatrix}$$

$$\boldsymbol{C}_2^b(\gamma) = \begin{bmatrix} \cos\theta & \sin\theta & 0 \\ -\sin\theta & \cos\theta & 0 \\ 0 & 0 & 1 \end{bmatrix}$$

文献［1］提出的递推算法，是利用任意一个姿态角来估计全姿态，避开了仅对地磁场一个矢量测量值无法实现单点定姿的问题。磁强计可测得弹体系 y_1、z_1 两个方向的地磁分量 H_{y1}、H_{z1}，如图 3 所示。

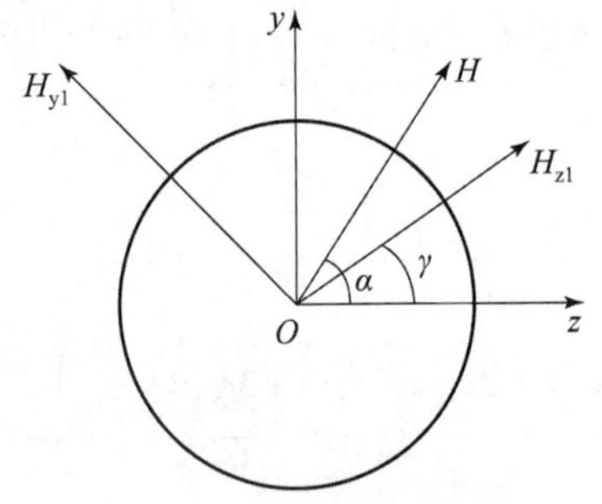

图 3　弹体截面示意图

由图 3 可以得出

$$\gamma = \alpha - \tan^{-1}\frac{H_{y1}}{H_{z1}} \tag{10}$$

式中，α 为“磁基准角”，当弹道弯曲比较缓慢时，通过查表法线性插值获取任意时刻“磁基准角”。

把用地磁信号解算的各时刻滚转角序列和陀螺信号解算的各时刻俯仰角和偏航角序列重新组合（图 4），作为下一步解算的初始四元数，重新进行三子样法姿态解算，解算精度可得到提高[6]。

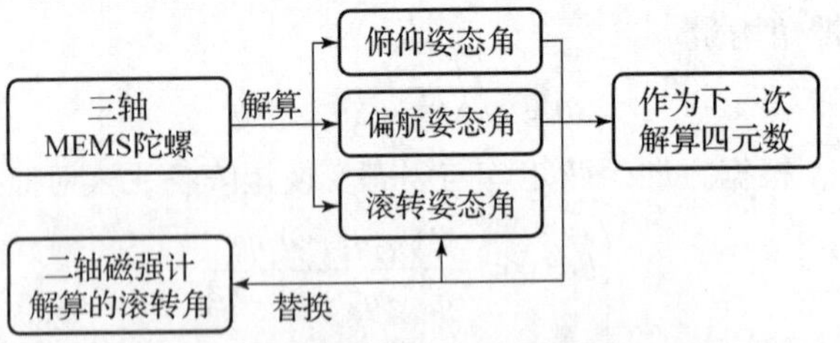

图 4　地磁信号解算角与陀螺信号解算角序列重新组合

同时通过测量地磁信号随弹体滚转时的变化曲线，判断弹体的滚转圈数，然后通过发送归零信号的方式使三子样解算一圈后归 0，下一圈开始后又重新开始解算，将积分误差减小在单圈内，修正由三子样姿态带来的积分累积误差。

然而试验中转台上的铁磁物质对磁强计的干扰非常大因为干扰的作用，因此，需要进行误差补偿，工程中，将椭圆圆型化的误差补偿方法称为椭圆拟合法。即在弹体绕滚转轴运动一周内，记录 y、z 轴的最大值和最小值，记为 H_{ymax}、H_{ymin} 和 H_{zmax}、H_{ymax}。计算椭圆 y、z 轴的伸缩因子和平移因子。

伸缩因子：

$$S_y = \max(1, (H_{z\max} - H_{z\min})/(H_{y\max} - H_{y\min})) \tag{11}$$

$$S_z = \max(1, (H_{y\max} - H_{y\min})/(H_{z\max} - H_{z\min})) \tag{12}$$

平移因子：

$$f_y = [0.5(H_{y\max} - H_{y\min}) - H_{y\max}]S_y \tag{13}$$

$$f_z = [0.5(H_{z\max} - H_{z\min}) - H_{z\max}]S_z \tag{14}$$

将磁强计原始测量值（H_{y1}，H_{z1}）通过平移和伸缩变换得到磁传感器的校正值（H'_{y1}，H'_{z1}）。

$$H'_{y1} = H_{y1}S_y + f_y \tag{15}$$

$$H'_{z1} = H_{z1}S_z + f_z \tag{16}$$

应用双轴转台系统，固定实验样机于转台工作台面，设置不同转速应用地磁信号解算滚转角，设定转台转动时间，记录转台停止位置对比有无经过经椭圆误差补偿情况下的滚转角解算误差。采集结果如表 1 所示。

表 1　有无经过经椭圆误差补偿情况下的滚转角解算误差

转速/[(°)·s^{-1}]	时间/s	未经补偿误差/(°)	补偿后误差/(°)
360	10	2.8	0.4
720	10	3.5	0.6
1 080	10	3.3	0.7
1 440	10	3.8	0.5

绘制两种情况下误差曲线对比结果，如图 5 所示。

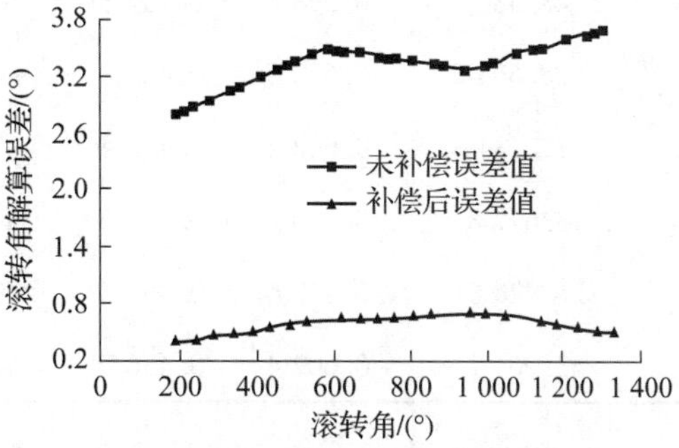

图 5　两种情况下误差值对比

4　转台试验

常规弹箭在飞行过程中，弹体的温度会存在较大变化，在不考虑安装隔热材料的情况下，MEMS 陀螺的温漂很大，所以在应用过程中要对 MEMS 陀螺进行零偏、温漂、刻度因子及非线性误差进行补偿，并对地磁信号进行校正，进而才能进行姿态解算精度实验。

将磁强计、MEMS 陀螺样机固定在转台工作面，陀螺仪 x 轴与转台滚转轴重合，y 轴与转台俯仰轴重合，z 轴指向遵循右手螺旋定则。当样机分别绕 3 个轴运动时，采集各轴的角速度输出值和温度值，写入文中的姿态解算程序，使用 TMS320F28335 型数字信号处理器实时解算并输出姿态角。

当俯仰轴速度为 1.5°/s、滚转轴速度为 360°/s 时，融合地磁信号各轴地磁投影变化情况如图 6 所示。

其中，y 轴、z 轴地磁信号经误差校准后滚转角解算结果如图 7 所示。

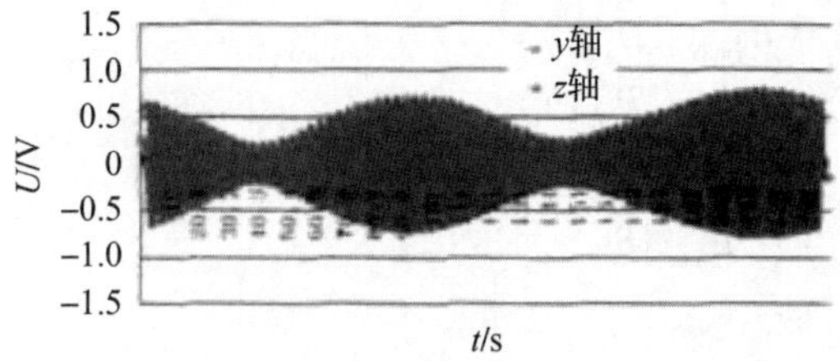

图6 地磁矢量在两轴投影变化

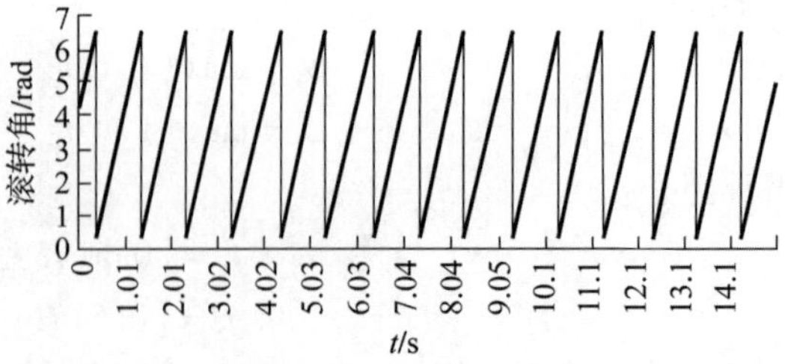

图7 滚转角解算结果

试验中给转台设置不同的转速及控制转动时间，转动完毕后，将转台停止位置 θ_{0x}、θ_{0y}、θ_{0z} 和样机的姿态角输出值 θ_{1x}、θ_{1y}、θ_{1z} 相减，即得到该系统的累计误差，如表2所示。

表2 姿态角绝对误差数据对比

转台速度/[(°)·s⁻¹]		时间/s	陀螺解算姿态绝对误差/(°)			融入地磁解算姿态角绝对误差/(°)		
x	y		x	y	z	x	y	z
360	−20	4.5	−10.805 7	0.591 5	0.401 2	−6.437 8	0.462 1	0.295 8
	−10	9.0	−5.851 7	0.584 3	0.395 6	−4.430 2	0.406 2	0.311 7
	−5	18.0	−2.564 0	0.602 1	0.327 2	−1.367 0	0.348 4	0.283 4
	5	18.0	2.586 8	0.602 1	0.328 8	1.356 3	0.397 4	0.202 3
	10	9.0	5.724 0	0.560 3	0.403 5	3.962 8	0.386 5	0.303 1
	20	4.5	10.590 6	0.585 4	0.407 1	6.455 7	0.435 7	0.363 2
1080	−20	4.5	−48.480 3	0.598 1	0.457 3	−3.705 6	0.588 3	0.407 8
	−10	9.0	−28.603 2	0.633 5	0.469 0	−17.653 3	0.425 4	0.386 1
	−5	18.0	−12.234 3	0.602 4	0.452 4	−18.274 9	0.523 6	0.386 2
	5	18.0	13.677 8	0.616 3	0.450 6	8.255 6	0.447 5	0.274 3
	10	9.0	28.226 5	0.599 6	0.466 8	19.452 9	0.416 5	0.327 8
	20	4.5	53.764 2	0.602 4	0.456 0	42.419 4	0.401 3	0.285 2

结果表明，融入地磁信号后，姿态角解算精度有了明显提高，尤其还能进行动态特性校正，当误差大于一定阈值时，重新计算来消除前面的累积误差。

5 结论

本文从实际出发，设计了一种磁－惯性微型姿态测量系统，该系统样机使用二轴磁强计、三轴角速率陀螺两个惯性器件输出的信息进行姿态解算，不仅解决了 MEMS 陀螺输出的信息解算存在漂移累积的缺点及仅使用磁强计探测盲区的缺陷，而且具有体积小、成本低、解算精度高、性能可靠的优点，在弹道修正弹箭姿态测量中具有广阔的应用前景。

参考文献

[1] 曹红松，冯顺山，赵捍东，等．地磁陀螺组合弹药姿态探测技术研究［J］．弹箭与制导学报，2006，26（3）：142－145.

双层球缺罩形成复合杆式射流的初步研究①

付建平[1②]，冯顺山[2]，陈智刚[1]，兰宇鹏[3]，张均法[3]，赵太勇[1]

（1. 中北大学地下目标毁伤技术国防重点学科实验室，山西太原 030051；

2. 北京理工大学爆炸科学与技术国家重点实验室，北京 100081；

3. 山东特种工业集团，山东淄博 255201）

摘　要：为进一步提高杆式射流在大炸高下的稳定性，本文设计了铜/铝双层球缺罩装药结构，运用脉冲 X 光照相技术获得了射流形态。结果表明：采用铜/铝双层球缺罩可以形成复合杆式射流，其形态与单铜罩形成的杆式射流基本类似，但是在压垮形成射流时，铜铝两种材料在初始界面发生了相互融合。同时，铜/铝双层球缺罩的射流整体速度略有提高，速度梯度却有所减小，最大侵彻深度相比单铜罩射流可以提高 16.7%。

关键词：爆炸力学；聚能装药；双层药型罩；射流；侵彻

Preliminary Study of Composite Jetting Penetrator Charge with Double Layer Spherical Segment Liner

Fu Jianping[1], Feng Shunshan[2], Chen Zhigang[1], Lan Yupeng[3], Zhang Junfa[3], Zhao Taiyong[1]

（1. National defense Key Laboratory of Underground Damage Technology. North University of China, Taiyuan 030051, China;

2. State Key Laboratory of Explosion Science and Technology, Beijing Institute of Technology, Beijing 100081, China;

3. Shandong Special Industry Group, Zibo 255201, China）

Abstract: In order to further improve the stability of rod – like jet at long stand – off distance, composite jetting penetrator charge (CJPC) with Cu – Al double – layer spherical segment liner was designed. Meanwhile jet shape was obtained by using flash X – ray photographic technology. The results show that the Cu – A1 double – layer spherical segment liner can form composite rod – like jet, and the shape of rod – like jet is substantially similar to single – layer copper liner. When the double – layer liner forms composite rod – like jet, copper and aluminum intermingle with each other at the initial interface. It is concluded that, the velocities of rod – like jet formed by the CJPC are increased overall while the jet velocity gradients is de – creased. The maximum penetration depth is increased by 16.7% than single – layer copper liner.

Key words: explosion mechanics; shaped charge; double – layer liner; jet; penetration

0　引言

随着复合装甲、屏蔽装甲、爆炸反应装甲等的使用，现代主战坦克的正面防护能力大幅度提高。因此，中小口径的攻顶式反坦克导弹成为发展趋势之一，其优点是可以直接打击坦克相对薄弱的顶装甲，

① 原文发表于《中北大学学报》2016 年第 37 卷第 1 期。

② 付建平：工学博士，2013 年师从冯顺山教授，研究基于新型毁伤元战斗部设计方法与威力效应研究，现工作单位：中北大学。

但缺点是战斗部的布置空间有限和对制导系统的要求较高。

攻顶式反坦克导弹战斗部一般为爆炸成型弹丸（EFP）战斗部，但其最大破甲深度约为装药直径（CD）的0.7~1.0倍[1]。杆式射流[2]作为一种侵彻性能介于EFP和聚能射流之间的聚能侵彻体，能够进一步提高攻顶式反坦克导弹的威力，高品质射流是其侵彻性能的关键。近年来，国内外大量学者针对杆式射流的形成及侵彻能力进行了研究[3-9]，结果表明大锥角和球缺罩聚能装药均能形成杆式射流，并证实了杆式侵彻体在15 CD炸高条件下对钢靶具有较好的侵彻性能。Fu Jianping[10]等对大炸高下杆式射流的侵彻性能进行了研究，在20 CD炸高下杆式射流的最大破甲深度可以达到1.5 CD以上。

然而关于杆式射流的大炸高稳定控制技术方面的研究相对较少，本文在前人工作基础上，设计了铜/铝双层球缺罩装药结构，并运用脉冲X光照相技术获得了射流形态，所形成的复合杆式射流在大炸高下稳定性比单层铜球缺罩杆式射流有所提高。拟为杆式射流战斗部的设计和工程应用提供参考。

1　铜/铝双层球缺罩的设计分析

根据小锥角药型罩射流形成原理，射流是由药型罩的少量内层金属形成，杵体则由药型罩的大量外层（接触炸药）金属形成的。在此基础上，设计了铜/铝双层球缺罩（这里规定，按照双层罩内层到外层的顺序，依次按材料进行命名），它的内层药型罩采用塑性好、密度大，易于形成侵彻射流的紫铜材料，而外层药型罩则采用密度、熔点、气化点以及阻抗均较低的铝材料[11]。通常，小锥角铜/铝双层罩的内层铜罩厚度约占总厚度的1/4。而文中内层铜球缺罩厚度约占总厚度的2/3，其目的是保证有效杆式射流部分（大于1.0 km/s）完全由内层铜罩形成。

2　试验设置

本文试验研究的对象是两种球缺形药型罩：单层铜药型罩和铜/铝双层药型罩，结构如图1和图2所示。两种药型罩对应的试件代号分别为S_1和S_2，装药直径为55 mm，装药高度为53.2 mm，药型罩底部外径为51.5 mm，药型罩高度17.5 mm，起爆方式均为中心点起爆。药型罩由紫铜板和纯铝板冲压而成，药柱采用8701炸药压制而成，靶板材料为45#钢。

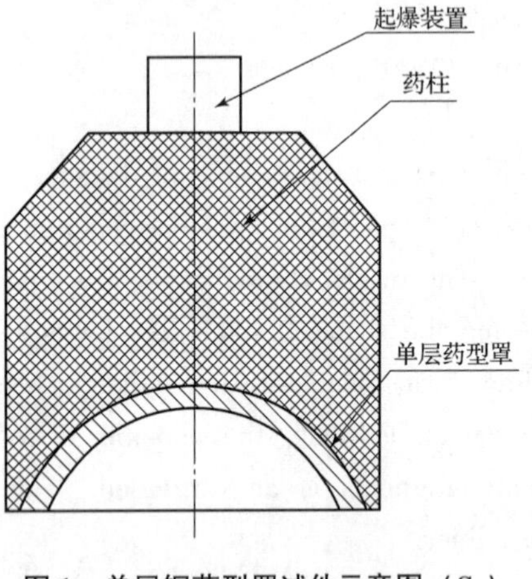

图1　单层铜药型罩试件示意图（S_1）

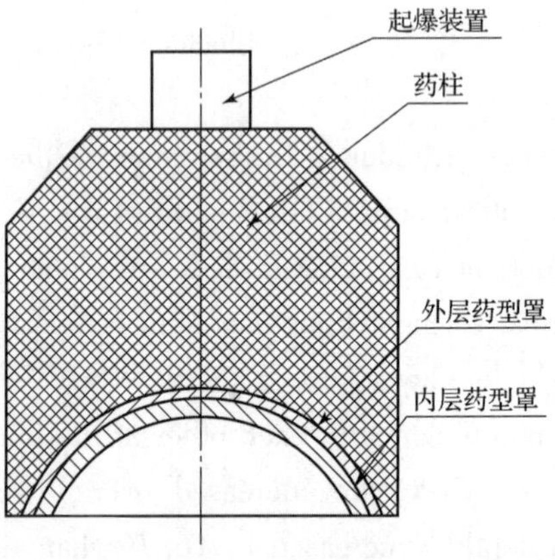

图2　铜/铝双层药型罩试件示意图（S_2）

试验的一些基本参数如表1所示。

表1　试验的一些基本参数

代号	内层罩参数		外层罩参数	
	材料	壁厚/mm	材料	壁厚/mm
S_1	紫铜	2.5	/	/
S_2	紫铜	1.5	纯铝	1.0

用脉冲X光照相法观察射流的形成过程，用钢靶测量射流的侵彻能力，X光照相试验布置如图3所示。采用了450 kV的X光设备，由于受到当时的试验条件限制（平行式支架被打坏），使用的是交叉式摄影的方法，两台闪光X射线机位于同一高度的两个不同方位，交汇成45°角。在该试验中，曝光延迟时间的设定非常重要，它是能否获得射流图像信息的关键。根据数值模拟得到射流头部速度在3 000 m/s左右，故在拍摄双层罩射流形成过程时，设置X射线管的曝光时刻为45 μs和120 μs，起爆前先在底片上拍摄静止照，静止照中的铜标杆长度为100 mm。将所得X光底片通过扫描仪数字化后，参照图上标杆像进行判读，由此可得到杆式射流的各项参数。

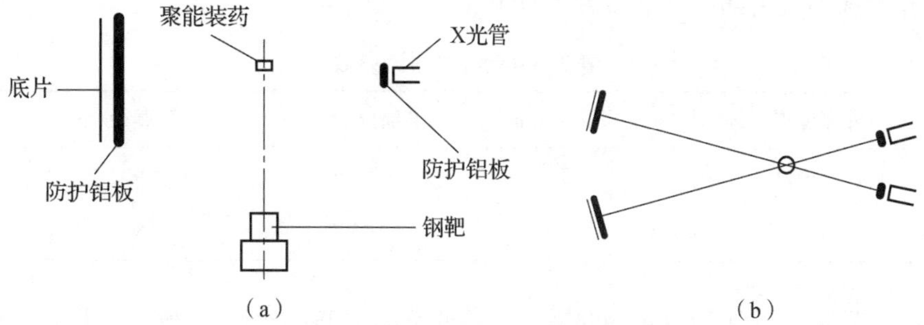

图3　脉冲X光照相试验布置

（a）前视图；（b）俯视图

3　结果与分析

3.1　复合杆式射流形成分析

文献［12］已对球缺形的单层铜罩形成杆式射流的过程进行了理论分析和X光照相试验。因此，文中仅对铜/铝双层球缺罩的射流形成过程进行X光照相分析，试验得到的射流形态如图4所示，图中的标尺是根据长度为100 mm铜标杆所获得的。

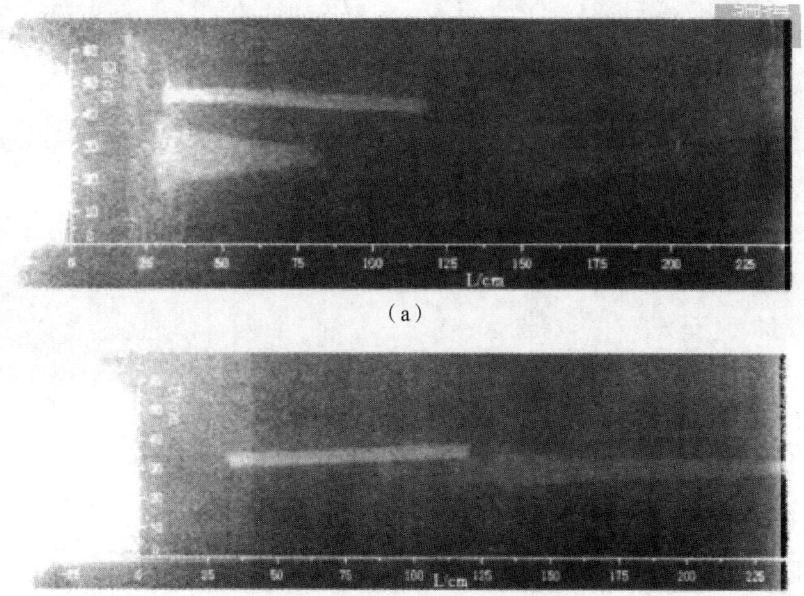

图4　铜/铝双层球缺罩射流形成过程

（a）t = 45.65 μs；（b）t = 120.6 μs

图4（a）揭示了铜/铝双层球缺罩装药在射流形成初期时的形貌，复合杆式射流在达到1.5倍口径位置处已经基本形成，除药型罩口部材料因受稀疏波影响形成少许飞边外，形态非常完整，没有明显的

杵体和射流之分，与文献［12］中常规单铜罩形成的杆式射流形态基本类似。射流的直径从头至尾逐渐增加，呈圆锥形。同时也发现，从 X 光照片中很难将铜材料与铝材料区分开来，分析其主要原因可能有：①由于铝材料的密度较小，并且和防护铝板材质相同，X 射线不能将铝材质清晰地在底片上拍摄出来；②由于铜铝两种材料在界面上部分融合；③球缺罩在翻转时铝被铜包覆在内部。

从图 4（b）来看，形成的杆式射流具有良好的延展性，射流在 120.6 μs 时刻已经得到了充分的拉伸，形态均匀细长。由于射流速度比预计的稍快，有少部分的射流头部正好飞出了第 2 幅底片边界，所以难以确定射流头部的速度以及各段的速度分布。因此本文仅对射流的头尾直径、初始长度等参数进行分析，测量基准参考药型罩底部，结果如表 2 所示。

表 2　复合杆式射流参数

实拍时刻/μs	头部位置/mm	头部直径/mm	尾部位置/mm	尾部直径/mm	长度/mm
45.65	80.1	9.5	25.6	26.2	54.5
120.6	/	/	106.7	15.9	>126

根据表 2 测得的数据可以计算出射流的尾部平均速度为 1 034 m/s，尽管头部位置难以判断，但是可以根据装药高度、爆速、药型罩高度和第一幅照片的时刻来估算出射流头部平均速度为 2 504 m/s。

3.2　两种结构形成的杆式射流断裂分析

为了比较两种药型罩结构形成的杆式射流在延伸至 6～10 CD 距离时的稳定性，增加曝光延迟时间后继续进行闪光 X 照相试验。

从图 5 的射流断裂图像可以发现，由于速度梯度的存在，铜/铝双层罩和单铜罩形成的杆式射流在拉伸过程中均出现了不同程度的断裂，与小锥角聚能射流断裂后图像有相似之处，但是各段长度、直径都要比小锥角聚能射流大得多，个数也要少得多。

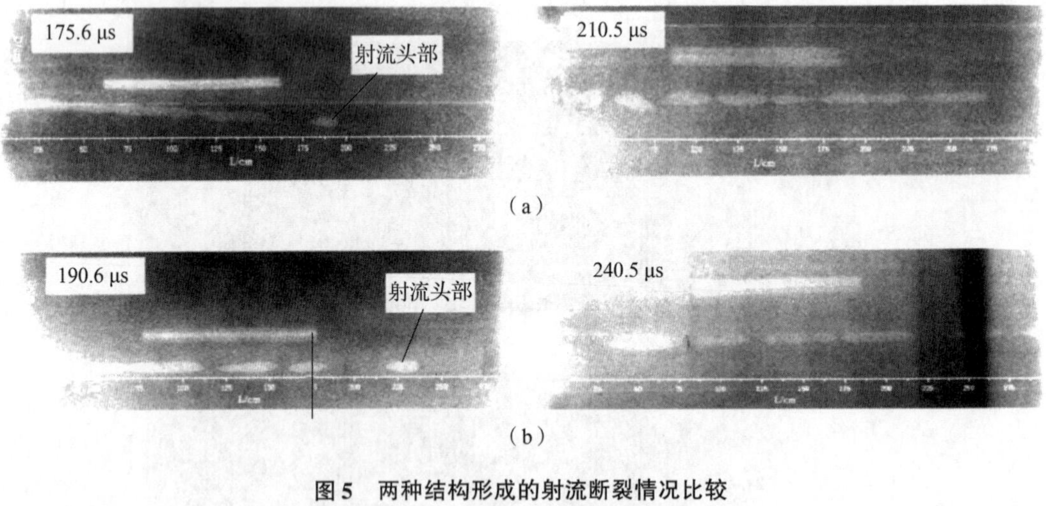

（a）

（b）

图 5　两种结构形成的射流断裂情况比较

（a）铜/铝双层球缺；（b）单层铜球缺罩装药

比较两种结构形成的射流达到相同位置的时刻可知，单铜罩射流要比铜/铝双层罩射流晚 15 μs 左右，说明铜/铝双层罩射流的平均速度要略高于单铜罩射流。

比较两种结构形成的射流的断裂情况可知，在飞行至相同位置时，两种杆式射流均断裂为 8 段较大的“枣核形”颗粒。但是铜/铝双层球缺罩形成的射流主体部分（断裂射流的前 4～5 段）速度梯度相对较小，射流颗粒在继续运动过程中更稳定一些。综合以上分析，预计铜/铝双层罩射流具有更好的侵彻效果，这在后面的侵彻试验中也得到了验证。

3.3　侵彻结果分析

由于炸高较大，断裂后的射流可能跑偏而无法打到靶板上，布置靶板时分为左、右两个部分，两部分钢靶紧贴在一起，左边靶板为一块100 mm厚的45#钢靶，右边靶板为两块60 mm厚的45#钢靶叠加而成。本次试验射流侵彻靶板后典型照片如图6所示。

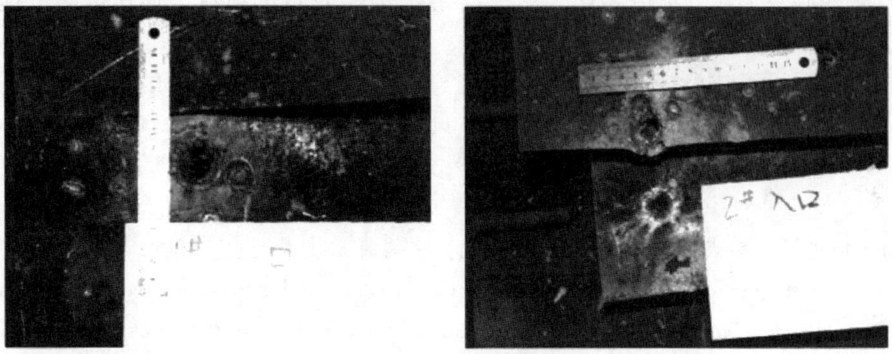

图6　射流侵彻靶板效果图

(a) S_1；(b) S_2

图6（b）中的靶板正面呈银白色，侵彻孔道呈黄色，与铜射流侵彻后截然不同，分析其原因为：①铜/铝双层罩在压垮形成复合杆式射流时，铜铝两种材料在初始界面发生了相互融合，②由于复合杆式射流从头部至尾部的材料依次为“铜、铜铝混合、铝”，在射流侵彻靶板时也按这样的先后顺序喷向侵彻孔洞。侵彻参数见表3，表中炸高是药型罩底部至靶板的距离。

表3　侵彻参数比较

试件号	炸高/mm	入口尺寸/mm	出口尺寸/mm	最大侵彻深度		备注
				深度/mm	深度/CD	
S_1	1 105	28	/	90	1.64	靶板入口处有少许浅坑，开坑比较集中
S_2	1 105	24	18	105	1.91	穿透第一块60 mm厚钢靶，第二块60 mm厚钢靶侵彻深度为45 mm

从表3中可看出，在20 CD的大炸高情况下，与单层铜罩相比，铜/铝双层罩形成的复合杆式射流侵彻孔径虽然有所减小，但是最大侵彻深度却提高了16.7%。针对单层铜罩，之前已经做了50多发试验，平均深度为80 mm。因此虽然本次试验只安排了2发铜/铝双层罩，数量偏少，但是从最大侵彻深度来看，双层球缺罩可以在一定程度上提高侵彻深度。

3.4　铜/铝双层球缺罩的初步分析

文中设计的双层球缺罩，内层铜罩厚度约占总厚度的0.6倍，保证了有效杆式射流部分（大于1.0 km/s）完全由内层铜罩形成。铜/铝双层球缺罩射流形成过程与单铜罩基本相同，但最大的特点是：铜、铝两种材料没有发生分离，而是在药型罩初始分界面处相互融合，形成了复合杆式射流。

由于复合杆式射流中铝射流密度低，且速度也不高，因此在侵彻钢靶时起决定性作用的是铜射流。与单铜罩相比，在增加外层铝罩后，药型罩微元速度增大，尤其是药型罩底部微元（单铜罩时形成射流的低速段）得以加速，铜射流的尾部速度明显提高，同时射流尾部速度的提高幅度大于射流头部，因而射流的头尾速度差有所减小。在20 CD的大炸高情况下，复合杆式射流侵彻深度有所提高的主要原因是铜射流的速度梯度减小，射流断裂情况有一定改善。

4 结论

通过对两种球缺形药型罩形成杆式射流及侵彻靶板的试验研究，可以得出以下结论：

（1）采用铜/铝双层球缺罩可以形成复合杆式射流，其形态与单层铜球缺罩形成的杆式射流基本类似，但是在压垮形成复合杆式射流时，铜铝两种材料在初始界面发生了相互融合。

（2）与单层铜球缺罩形成的射流相比，铜/铝双层球缺罩形成的射流整体速度得以提高，速度梯度却有所减小，最大侵彻深度提高了16.7%。

（3）外层铝药型罩可以减小扰动对铜罩的影响，射流品质好，大炸高下的稳定性和侵彻性能好。

参考文献

[1] 林加剑. EFP成型及其终点效应研究[D]. 合肥：中国科学技术大学，2009.

[2] WHELAN A J, FURNISSS D R, TOWNSLEY R G. Experimental and simulated (analytical & numerical) elliptical - form shaped charges [C]//20th International Symposium on Ballistics. Florida: National Defense Industrial Association, 2002: 446 - 454.

[3] MATTSSON K, SORENSEN N, OUVE R, et a1. Development of the K - charge, a short L/D shaped charge [C]//REI - NECKE W G. 18th International Symposium on Ballistics. San Antonio, Texas, Terminal Ballistics Vulnerability Wound Ballistics, 1999: 528 - 534.

[4] 谭多望，孙承纬，赵继波. 大锥角聚能射流实验研究[J]. 高压物理学报，2003，17（3）：204 - 208.
TAN D W, SUN C W, ZHAO J B. Experimental investigation of shaped charge with large cone angle [J]. Chinese journal of high pressure physics, 2003, 17 (3): 204 - 208.

[5] 黄正祥. 聚能杆式侵彻体成型机理研究[D]. 南京：南京理工大学，2005.

[6] 吴晗玲，段卓平，汪永庆. 杆式射流形成的数值模拟研究[J]. 爆炸与冲击，2006，26（4）：328 - 332.
WU H L, DUAN Z P, WANG Y Q. Simulation investigation of rod - like jets [J]. Explosion and Shock Waves, 2006, 26 (4): 328 - 332.

[7] 李伟兵，王晓鸣，李文彬，等. 装药长径比对聚能杆式侵彻体成型的影响[J]. 弹道学报，2011，23（4）：61 - 65.
LI W B, WANG X M, LI W B, et a1. Effect of length - diameter ratio of charge on jetting projectile charge forming [J]. Journal of ballistics, 2011, 23 (4): 61 - 65.

[8] 李成兵，沈兆武，裴明敬. 高速杆式弹丸初步研究[J]. 含能材料，2007，15（3）：248 - 252.
LI C B, SHEN Z W, PEI M J. Preliminary study of high velocity rod - shaped projectile [J]. Chinese journal of energetic materials, 2007, 15 (3): 248 - 252.

[9] 廖海平，黄正祥，张先锋. 亚半球罩聚能装药的杆式射流特性研究[J]. 弹箭与制导学报，2008，28（6）：91 - 94.
LIAO H P, HUANG Z X, ZHANG X F. Rod - like jet properties of semispherical liner shaped charge [J]. Journal of projectiles, rockets, missiles and guidance, 2008, 28 (6): 91 - 94.

[10] FU J P, CHEN Z G, HOU X C, et a1. Simulation and experimental investigation of jetting technology [J]. Defence technology, 2013, 9 (2): 117 - 124.

[11] 臧涛成，胡焕性，邵琦. 破甲弹复合罩性能研究[J]. 火炸药学报，1998（4）：45 - 48，59.
ZANG T C, HU H X, SHAO Q. The performance study of shaped charge liner [J]. Chinese journal of explosives & propellants, 1998 (4): 45 - 48, 59.

[12] 黄正祥，张先锋，陈惠武. 起爆方式对聚能杆式侵彻体成型的影响[J]. 兵工学报，2004，25（3）：289 - 291.
HUANG Z X, ZHANG X F, CHEN H W. Influence of modes of detonation on the mechanism of jetting projectile charge [J]. ACTA Armamentar II, 2004, 25 (3): 289 - 291.

冯顺山教授从教46年教学、科研和学科建设重要记事

一、主要教学和学科建设

主讲“理论力学”（本科，助教） 1977—1978

主讲“高等数学”（本科，助教） 1977—1978

主讲“爆炸及其作用”（本科，实验教学部分） 1983—1988

主讲“火箭战斗部设计原理”（本科，实验教学部分） 1983—1989

指导毕业论文/毕业设计（本科生） 1983—2019

主讲“战斗部实验技术”（本科） 1984—1986

主持“弹药战斗部科技进展”（硕/博士，学位必修课） 1993—2000

主持“毁伤与弹药工程前沿科技”（硕/博士，学位必修课） 2001—2020

主持“战斗部与终点效应”实验室建设 1983—1989

主持“爆炸科学与技术国家重点实验室”创建、验收、运行和发展 1988—1999

受聘北京理工大学首席教授，主持“火炮、自动武器与弹药工程”二级学科建设发展 1998—

作为执行负责人，实施“211”一期“爆炸与安全”、二期“毁伤与安全”学科平台、三期“灵巧与毁伤学科平台”建设项目，评为优 1996—2008

主持“毁伤与弹药工程”国家重点二级学科申报、建设和发展，全国评估排名第一，支撑“兵器科学与技术”评为全国重点一级学科（A+） 2001—

“弹药工程与爆炸技术”本科专业评为国防特色学科 2002—

主持国防科工委“北理工战斗部基础条件建设项目”，建成东花园“爆炸与毁伤实验基地”，成为教学科研主体实验基地 2006—2010

主持“新型/新概念战斗部”国防创新团队申报、建设和发展，首批国防创新团队 2008—

自主率先调查发现怀来“小山口村”适合建火工实验基地，并论证适用性和环境安全性，获得国防科工局批准建设 2008—

主持北京理工大学机电学院无人飞航工程系建设及学科可持续发展 2010—

作为学科责任教授，“弹药工程与爆炸技术”本科专业获批教育部首批“卓越工程师”教育培养计划 2010—

作为主任委员，将“战斗部与毁伤效率专业委员会”更名为“战斗部与毁伤技术专业委员会”，拓展专业领域，优化学术交流平台 2011—

作为学科责任教授，“弹药工程与爆炸技术”本科专业获批教育部首批“国家一流本科专业建设点” 2019—

作为学科责任教授，主持“弹药工程与爆炸技术”本科专业通过“中国工程教育专业认证”（全国武器类学科首个通过认证） 2020—

作为责任教授，无人飞航工程系授予首批国防科技工业先进集体 2021—

主持国家安全领域“973”重大基础项目“先进战斗部毁伤机理与终点效应”项目立项论证报告编写和立项答辩，获批 2000

主笔“弹药战斗部技术的发展与创新”报告，向总装建议设立“高效毁伤科技专项” 2000

参与编写“十五期间切实加强安全科技工作的建议”（报吴邦国副总理，获批示）	2000
主笔编写“非核高效毁伤战斗部技术”“非核终端毁伤技术”发展论证报告，向国防科工委和军方建言献策	2004
参与我国“弹箭技术”/“枪炮弹箭技术”“八五”~“十二五”预研指南编制	1991—2016
参与二炮“战斗部技术”“十五”“十一五”预先研究计划指南编制，撰写多份预研项目建议报告	1991—2016

二、主要科研工作

研究课题（项目负责人或技术总师）

战斗部破片参数精确测量脉冲X光摄影系统研制	设计定型	1983.3—1985
××鱼雷聚爆战斗部研制		1986—1987
××定向杀伤战斗部研制	设计定型	1988—1989
“长征二号”捆绑火箭爆炸分离螺母研制	设计定型	1988—1990.5
空间目标毁伤概念研究		1989—1990
高威力战斗部技术研究		1991—1995
超近程反导弹药技术研究		1996—2000
碳纤维子母战斗部技术		1999—2001
先进常规战斗部毁伤机理与终点效应（973项目）	课题负责人	2001—2005
反航母侵爆燃战斗部技术		2001—2004
DF－××碳纤维子弹关键技术综合集成演示验证		2001—2002
串联钻地战斗部技术		2001—2005
“1022工程” ××导电纤维子弹研制	设计定型	2002—2005
防核爆高抗力活门快速关闭系统研制	设计定型	2002—2007
子母战斗部抛撒技术		2003—2007
战场毁伤实时侦察评估弹研制		2003—2005
侵彻反抛毁伤技术集成演示验证		2004—2005
DF－××导电纤维子弹产品批产	批产负责人	2005—2008
港口水域智能封锁弹药技术		2006—2015
封锁航母飞行甲板侵立封子弹药技术		2007—2008
无人机载纤维丝弹研制和装备生产（“920”国家奥运安保应急攻关项目）	设计定型	2007—2007
新型/新概念战斗部技术（国防创新团队基金）	团队带头人	2007—2008
等待式毁伤飞行器方法、机理及应用技术研究		2009—2012
××航空火箭导电纤维弹导电纤维丝团研制	设计定型	2011—2019
贫铀复合燃爆杀伤元技术（探索一代重大项目）		2011—2014
智能定向杀伤战斗部技术		2011—2015
反海上目标金属风暴子弹药技术		2014—2015
DF－××侵立封子弹研制	设计定型	2017—2020
高效毁伤战斗部技术（国防创新团队基金项目）		2020—2021
BT011－××导电纤维子弹产品延寿项目		2018—2023
反巨型深钻地目标弹药关键技术研究		2016—2017
大型沉船解体清障水下爆破工程（6条沉船）	工程验收	1994—2000

三、主要社会兼职

科技学术兼职

中国兵工学会弹药学会第二～第五届副主任委员	1990—2021
红宇机械厂、北理工八系“战斗部教学科研联合体”主任委员	1991—1997
中国劳动保护学会理事	1993—2001
中国科协第四届、第五届“中国青年科技奖”学科组评委	1994—1998
“爆炸科学与技术国家重点实验室”学术委员会委员	1994—2017
中国宇航学会第三届、第四届理事	1995—2001
中国科学技术协会第五届委员会委员	1996—2001
国防科技奖/国家科技奖兵器专业评审组成员	1996—2003
“弹道学报”编委会委员	1999.4—2005
“弹箭与制导学报”第三届编委会委员	1999.5—2005
中国兵工学会理事	1999—2019
国家科技奖“公安武警专业”评审组成员	2000—2003
国防科工委“安全标准化技术”委员会委员	2001.3—2005
国防科工委“导弹与舰炮标准化技术”委员会委员	2001—2005
中国兵工学会“爆炸与安全技术专委会”、第二～第四届主任委员	2002.2—2017.7
中国宇航学会无人飞行器分会“战斗部与毁伤技术专委会”第三～第六届主任委员	2003—2022
中北大学兼职教授、博导	2004.7—2021
北京理工大学“机械与运载学部”第一届、第二届委员会委员	2009.7—2017.6
“公安部警用装备重点实验室”学术委员会委员	2011—2015
《含能材料》第七届编委会编委	2014—2018
“爆炸科学与技术国家重点实验室”学术委员会顾问	2017—2022

科技专业组/专家组兼职

兵器工业总公司“弹箭技术”“八五”“九五”预研专家组副组长	1992—2001
劳动部安全生产专家，“爆炸专业组”副组长	1995—2000
总装备部“枪炮弹箭技术”专业组专家	1996.9—2010
总装备部陆装科订部“枪炮弹箭技术”专家组顾问（2011年5月）	2011—2015
“国家安全生产”专家，“爆炸组”副组长、“建筑消防组”副组长	2001—2010
海军装备部预研专家组“导弹与舰炮技术”“舰炮技术专业组”专家	2001—2011
公安部“灭火救援专家组”专家	2003.3—2010

四、所获主要奖励

国家级科技奖励

带壳战斗部脉冲X光摄影系统	国家科技进步三等奖	第一获奖人	1985
××导电纤维弹	国家技术发明二等奖	第一获奖人	2008
先进常规战斗部毁伤机理与终点效应	国家科技进步二等奖	第二获奖人	2008

省部级主要科技奖励

HY-××导弹战斗部研制	航天部重大科技成果三等奖	完成人之一	1984
××鱼雷聚能战斗部可行性论证	军队科技进步三等奖	第三获奖人	1987

××定向杀伤破片式战斗部	中国兵器工业总公司科技进步三等奖	第一获奖人	1991
子母战斗部爆炸抛撒原理研究	航天工业总公司科技进步二等奖	第四获奖人	1996
北京市普通高等学校教学成果	北京市教学成果一等奖	第二获奖人	1997
地地导弹碳纤维子母弹头技术	国防科技进步二等奖	第三获奖人	2004
高抗力防爆波活门研制	军队科技进步二等奖	第四获奖人	2006
××导电纤维弹	国防技术发明一等奖	第一获奖人	2007
先进常规战斗部毁伤机理与终点效应	国防科技进步一等奖	第二获奖人	2007
幕式拦截飞行器弹药技术	国防技术发明二等奖	第一获奖人	2009
无附带损伤的无人飞行器拦截技术	教育部技术发明一等奖	第一获奖人	2011
中国兵工学会科技进步奖	一等奖	个人奖	2013
贫铀复合燃爆杀伤元技术	军队科技进步二等奖	第二获奖人	2015
北京理工大学高效毁伤战斗部科技创新团队	国防科技创新团队奖	第一获奖人	2018
柔性复合防爆技术	公安部科技一等奖	第二获奖人	2020
反大型水面目标新技术	国防技术发明二等奖	第一获奖人	2022

个人荣誉称号

第二届中国青年科技奖	中国科学技术协会	1990
首届全国高校先进科技工作者	国家教委，国家科委	1990
首届国家有突出贡献的硕士学位获得者	国务院学位委员会，国家教委	1990
国务院政府特殊津贴	国务院	1992
国家有突出贡献中青年专家	国家人事部	1994
“××三型新型弹头”研制荣立个人二等功	国防科工委	2005
国防“511 人才”学术技术带头人	国防科工委	2006
国防科技创新团队带头人	国防科工委	2007
中国发明协会第七届发明创业奖	中国发明协会	2012
北京市优秀教师	北京市教育工作委员会等	2013
庆祝中华人民共和国成立 70 周年纪念章	中共中央国务院、中央军委	2019
光荣在党 50 年纪念章	中共中央	2021

集体荣誉奖

任“机电工程系”（八系）系主任期间，该系被评为“北京理工大学队伍建设一等奖”	北京理工大学	1996
任“机电工程系”系主任期间，该系被评为“学科建设优秀集体一等奖”	北京理工大学	1998
“科技奥运先进集体”（城市空中安保项目）（团队负责人，技术总师）	科技部、第 29 届奥林匹克运动会科学技术委员会、奥运科技（2008）行动计划领导小组	2008
“十五”“十一五”“十三五”科技工作先进团队（团队带头人）	北京理工大学	2006—2021
任学科责任教授期间，无人飞航工程系被授予“首批国防科技工业先进集体”	人力资源社会保障部国防科工局	2021

五、指导的研究生

北京理工大学博士研究生：

1993年入学　王海福
1994年入学　彭昆雅
1995年入学　孙为国、孙　韬、李卫星
1996年入学　王　芳、孔　炜
1997年入学　周　睿、贵大勇（副导师：刘吉平教授）
1999年入学　徐立新、吴　成、郝向阳（副导师：刘吉平教授）
2000年入学　贾光辉、黎春林、梁永直
2001年入学　罗　健、蒋健伟、张国伟、周　玲
2002年入学　王云峰、李国杰、王　玥
2003年入学　王震宇、王　刚、董三强、李　磊、黄龙华
2004年入学　黄广炎、刘春美、张旭荣
2005年入学　赵太勇、陈　赟
2006年入学　吴　广、完颜振海、李砚东、孙　昊
2007年入学　赖　鸣、赵　雪（副导师：芮久后教授）、王　宇（副导师：芮久后教授）
2009年入学　张晓东、张会锁、徐翔云、周　彤、李顺平
2010年入学　边江楠
2011年入学　方　晶、赵宇峰、张广华、杜　烨（副导师：曹红松教授）
2012年入学　李　伟、聂为彪、白　洋
2013年入学　王成龙（副导师：黄广炎教授）、成　曦（副导师：黄广炎教授）
2015年入学　陈超南
2016年入学　兰旭柯、黄　岐
2017年入学　伍思宇（副导师：邵志宇副教授）
2018年入学　高月光
2019年入学　张　博、肖　翔

中北大学博士研究生（兼职博导）：

2004年入学　李学林（副导师：陈智刚教授）
2005年入学　李守苍（副导师：陈智刚教授）
2012年入学　薛　震（副导师：陈智刚教授）
2013年入学　沈冠军（副导师：曹红松教授）、付建平（副导师：陈智刚教授）
2015年入学　李中明（副导师：陈智刚教授）
2016年入学　王维占（副导师：陈智刚教授）
2017年入学　刘鹏飞（副导师：曹红松教授）、任　凯（副导师：陈智刚教授）

北京理工大学硕士研究生：

1988年入学　宗国庆
1991年入学　武锐军
1992年入学　李卫星
1993年入学　高怀孝、李　滔
1994年入学　赵忠和
1995年入学　李　斌
1996年入学　徐立新

1998 年入学	杨　卓
1999 年入学	张彦存
2002 年入学	冯高鹏、于颖贤、高　磊
2003 年入学	杨　彬
2004 年入学	崔晓刚、金　俊、吴俊斌、徐德琛
2005 年入学	刘　晶、刘　畅
2006 年入学	杨　宁、景建清、田传勇（5013 厂联合指导）、刘建东（5013 厂联合指导）
2007 年入学	周　元、龙　骁、赵　黎
2008 年入学	张首鹏
2009 年入学	王振兴
2011 年入学	闵冬冬
2012 年入学	杨正有、齐任超
2013 年入学	孙　凯、吴　丹
2014 年入学	寇兴华、邓志飞
2016 年入学	钱海涛
2019 年入学	刘云辉
2020 年入学	胡文龙

博士后研究人员：

1998 年进站	郭文章
2000 年进站	何玉彬
2004 年进站	曹红松
2007 年进站	刘春美（王富耻教授联合指导）
2008 年进站	李淑华（王富耻教授联合指导）
2012 年进站	石光明
2020 年进站	黄　岐（钱新民教授联合指导）

致 谢

本文集选编冯顺山教授及其指导的研究生撰写的学术论文共计123篇及相关的科研等简况照片，内容涵盖了弹药战斗部科技和非核终端毁伤方面的发展战略、新型/新概念弹药战斗部研发思路，以及目标毁伤及杀爆战斗部技术、短路软毁伤与导电纤维弹技术、侵彻作用效应、封锁类弹药技术、爆炸冲击作用及防护技术、其他弹药和装药技术等七个学术研究方面，从一个侧面反映我国毁伤科技发展和冯顺山教授科技研究的历程。

感谢无人飞航工程系“毁伤理论与技术”学科组王杏花助理对原始档案资料的收集和信息统计，感谢学科组黄广炎教授、董永香教授、王芳副教授、邵志宇副教授、周彤副教授、王涛博士、王超实验师、黄岐博士后和学科组研究生们在收集“发表学术论文”等信息和初稿的校阅等方面的帮助，以及张博、高月光、肖翔三位博士生和刘云辉、胡文龙两位硕士生对论文整理和分类所做的工作。

文集出版过程中，北京理工大学无人飞航系系主任、博士生导师王海福杰出教授从文集的组织规划到具体筹稿编辑都给出具体的指导，并对局部文字进行修订；还得到蒋建伟、姜春兰、周霖、芮久后、门建兵、马天宝和余庆波等教授，以及多位老师的热情支持；北京理工大学出版社国珊为高质量出版这部学术文集给予了大力帮助，在此一并表示深切谢意。

部分论文的参考文献从略，详细信息可由论文的原始刊物查得，对参考文献的原作者再次一并致谢。

《冯顺山教授学术文集》编辑组

王海福、黄广炎、蒋建伟、王芳、赵雪

2023年元月

附录：冯顺山教授学习、工作、科研和人才培养有关照片

冯顺山教授兵团工作和上大学期间部分照片

（1969 年 5 月—1976 年 12 月）

1969 年 5 月，“知青上山下乡”之前在上海全家合影留念

1969 年 5 月—1971 年 8 月，黑龙江生产建设兵团
六师六十团三连（武装连队），农工、司号员、通信员、
卫生员、售货员，1971 年 2 月加入中国共产党

1971 年 8 月，黑龙江生产建设兵团
六师（建三江垦区）司令部警卫班班长

1973 年 9 月，在北京工业学院火箭战斗部专业 81731 班上大学，任班长、党分总支委员、民兵营长

1976 年在上海国防研究所长达 11 个月的本科毕业论文研究，参与 HY－××导弹战斗部威力性能研究，完成毕业设计研究报告（全班唯一）。照片是在浙江某靶场进行战斗部冲击波超压试验

1976 年 12 月大学毕业，全班与院领导贾克院长、吕育新系主任等领导和老师们合影留念

冯顺山教授在北京理工大学工作期间部分照片

（1976 年 12 月—2022 年 9 月退休）

1976 年 12 月大学毕业留校任教（八系 81 教研室），
1977 年 3 月担任 81761 班师生合编党支部书记，
并作为助教辅导高等数学和理论力学

1982 年 12 月在职硕士研究生毕业，获得
炮弹、火箭弹、导弹战斗部专业工学硕士
学位，继续在 81 教研室工作

1983 年任“战斗部与终点效应实验室”主任，大力推动实验室建设和发展，在国内研制成功带壳战斗部脉冲 X 光摄影及防护技术和多套新型试验手段，服务于教学科研和国防企业或研究院所，推动战斗部技术进步，荣获校“科研先进实验室”

1984 年，战斗部与终点效应实验室（81 实验室）全体成员合影

1985 年，弹药战斗部工程专业（81 专业）部分教师合影

1987 年破格晋升副教授，1991 年元月破格晋升教授，1993 年国务院学位委员会批准为博士生导师

1988 年任系主任助理，1989 年任力学工程系（8 系）副系主任，1993 年任系主任，
1994 年任机电工程学院院长，照片中为历届系主任

左起：冯顺山（第五任）、马宝华（第三任）、丁儆（首任）、吕育新（第二任）、徐更光（第四任）

1988 年起负责国家重点实验室申报、可研、立项、建设和发展工作，1996 年通过国家验收。
1995—1999 年任爆炸科学与技术国家重点实验室主任

1996 年，北京理工大学获得“国家有突出贡献的中青年专家”人员合影留念

2002 年 7 月 16 日，作为技术总师创新研发的新型导电纤维弹药演示验证试验圆满成功，
与参试的校领导焦文俊书记和匡镜明校长在试验靶场合影留念

2007 年，匡镜明校长为首批三个国防科技创新团队授牌，冯顺山教授为“新型/新概念战斗部技术”创新团队带头人

2009 年 1 月，导电纤维弹研究成果获得
国家技术发明二等奖

左起：第三获奖人王芳、第二获奖人何玉彬、
第一获奖人冯顺山、第六获奖人董永香

2006 年，冯顺山教授主持“战斗部基础条件建设”项目立项和建设，并设计了“东花园爆炸与毁伤实验基地”（2010 年 8 月正式开工）建设方案，现成为教学科研主体实验室，为学科持续发展发挥重要作用

1998 年起，冯顺山任学科首席教授/责任教授，守正出奇，面向现代化，凝炼学科方向，
2010 年成立无人飞航工程系，毁伤与弹药工程学科内涵拓展，学科发展取得长足进步

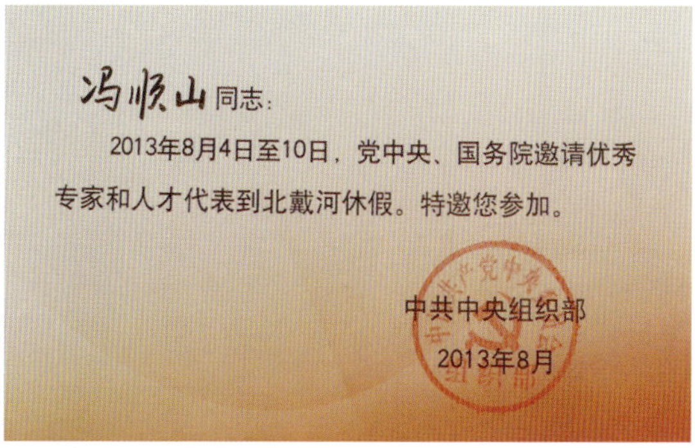
冯顺山同志：

2013年8月4日至10日，党中央、国务院邀请优秀专家和人才代表到北戴河休假。特邀您参加。

中共中央组织部

2013年8月

2013 年 8 月，中组部和国防科工委安排专家在北戴河休假，
中共中央政治局常委刘云山同志接见全体专家

国防创新团队取得标志性的研究成果，2018 年获国防科技创新团队奖（第一获奖人），
校长张军院士、陈鹏万院长等人员与获奖人员冯顺山、黄风雷、王海福、董永香合影留念

2021 年，无人飞航工程系被授予首批国防科技工业先进集体

冯顺山教授部分科研成果简介

（1976 年 12 月—2022 年 9 月）

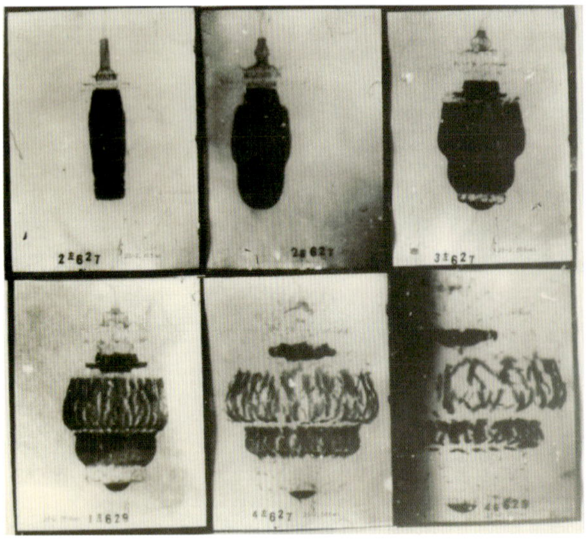
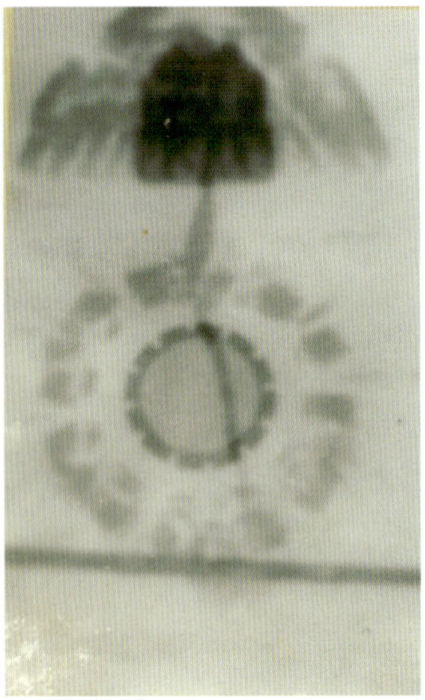

1983—1985 年，创新研发的带壳战斗部脉冲 **X** 光摄影系统，解决了战斗部破片威力参数轴向/径向分布精准测量问题，指导工程设计显著提升了威力性能，填补了国内空白，达到国际先进水平，**1985** 年获国家科技进步三等奖，第一获奖人

1986—1987 年，作为技术负责人完成鱼 ×× 聚爆战斗部研制，
可高效毁伤潜艇外壳和耐压层，**1987** 年获得海军科技进步三等奖

1988—1989 年，创新研制的偏心引爆定向杀伤战斗部技术，应用于某对地导弹战斗部，国内首型。
1991 年获兵器工业总公司科技进步三等奖，第一获奖人

1993—1996 年，反跑道子母战斗部爆炸抛撒技术研究，爆抛后子弹药性能及空中散布正常。
1996 年获航天工业总公司科技进步二等奖

2002—2004 年，作为技术负责人，创新研发反航母侵爆燃战斗部技术，
2004 年获中国兵器装备集团公司科技进步二等奖，第二获奖人

2003—2005 年，作为项目负责人，创新研发聚能侵彻两级串联反深层目标战斗部技术，
缩比样机通过侵靶演示验证，为有关型号研制打下技术基础

2003—2006 年，作为项目负责人，创新研发战场毁伤效果实时侦察弹技术，由卫星实时传输侦测武器毁伤效果图像信息，**1∶1** 样机通过动态演示验证，为有关工程研制提供支撑

2004—2006 年，作为项目负责人，创新研发高效封锁机场跑道的侵抛封锁子弹药技术（演示验证项目），完成炮射演示验证研制任务

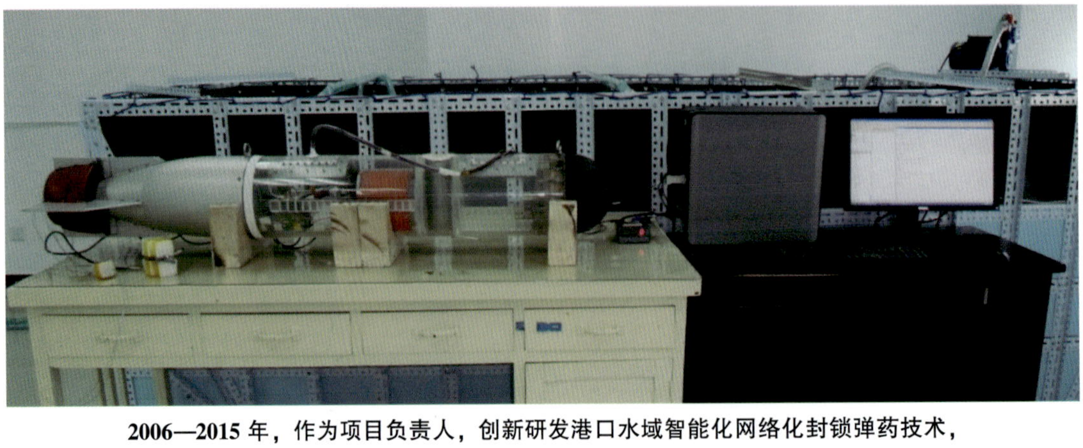

2006—2015 年，作为项目负责人，创新研发港口水域智能化网络化封锁弹药技术，完成技术样机研制任务，为有关应用研究打下技术基础

1999—2002 年，作为项目负责人和技术总师，完成碳纤维子母战斗部技术预研任务和碳纤维子弹（导电纤维弹）演示验证研究任务。2002 年 6 月，刘石泉总师陪同总装预研局辛毅局长检查北理工研制工作完成情况

2002 年 7 月 16 日，中央军委张万年副主席、总装备部部长曹刚川、二炮杨国梁司令等领导在东花园试验场听取冯顺山教授汇报导电纤维弹的研制情况，并观看动态短路毁伤电网靶标的演示试验，圆满成功。立项型号研制并完成批产装备部队，填补我军软毁伤手段的空白，研究成果 2007 年获国防技术发明一等奖，2008 年获国家技术发明二等奖，第一获奖人

校领导、科技处和机电工程学院领导参加试验并祝贺试验圆满成功

（左起：姜毅、赵家玉、俞为民、唐水源、赵长禄、冯顺山、徐云、崔占忠、杨树兴）

2011—2018 年，软毁伤元技术创新发展，导电纤维弹技术推广应用于直升机载陆航火箭导电纤维弹（总体 203 所），通过设计定型，装备部队，形成特殊战力，研究成果 2022 年获国防科技进步奖

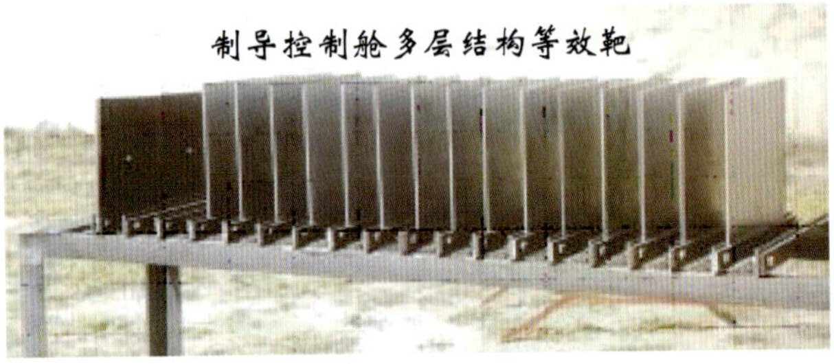

2001—2005 年，作为“973”项目立项负责人和毁伤巡航导弹机理及终点效应课题负责人，并完成常规战斗部毁伤机理与终点效应项目研究。研究成果 2007 年获国防科技进步一等奖，2008 年获国家科技进步二等奖，第二获奖人

2007—2008 年，负责承担国家“**920**”奥运安保专项，原创软拦截软毁伤“低慢小”无人机目标技术

孙建国副总参谋长检查无人飞行器拦截装备及性能

公安部副部长刘京和孙建国副总参谋长等专家领导，验收真实环境条件下作战系统拦截“低慢小”飞行器目标的作战效果，获得圆满成功。软拦截装备交付有关部队使用，在北京设置多个拦截阵地，圆满完成 **2008** 年北京奥运会城市空中安保任务，获得科技部的表彰。幕式拦截弹药技术 **2009** 年获得国防技术发明二等奖，无附带损伤无人飞行器拦截技术 **2011** 年获得教育部技术发明一等奖，均为第一获奖人

2011—2014 年，作为技术负责人，承担总装探索一代重大项目贫铀复合燃爆杀伤元技术，成功原创一种新型高威力侵爆杀伤元，拓展研究应用到有关战斗部型号研制。研究成果获中核集团公司科技进步二等奖，2015 年获军队科技进步二等奖，第二获奖人

火箭军装备部副部长何玉彬考查侵立试验情况

火箭军、航天科技一院和科工四院
有关人员考查侵立效应情况

2007—2020 年，作为项目负责人和技术总师，针对大型水面舰艇甲板封锁难题，原创反大型水面目标新技术（QLF 弹药技术）理念、方法和技术，完成型号研制任务，产品批产装备 ×× 反舰导弹，填补终端“打得巧”手段空白，可形成“区域拒止”的新质战力。2022 年获国防技术发明二等奖，第一获奖人

冯顺山教授其他科技创新成果

（1990—2022 年 9 月）

感谢信

北京理工大学：

1989—1990 年，作为技术负责人，针对“长征二号”捆绑火箭需求，创新爆炸分离螺母技术，北理工实验室生产的产品用于火箭多次发射，实现助推火箭与中心火箭的可靠分离。推广应用于后续的“天舟、神舟、天宫、问天”等捆绑火箭发射。**1990** 年获得北京理工大学科技进步一等奖，第一获奖人

2003—2011 年，作为技术负责人，针对核爆打击条件下地下重要军事工事在口部防护难题，原创非电实时探测、信号非电安全传输、强力驱动活门提前关闭等技术，性能指标达到国际领先水平，××型产品装备某地下重要工事，填补空白。**2006** 年获军队科技进步二等奖

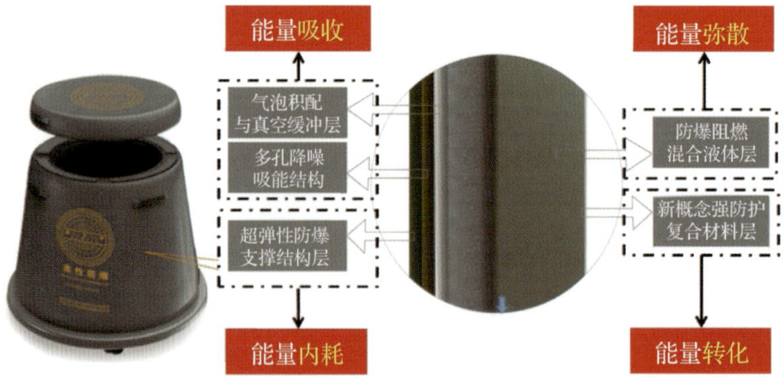

2010—2019 年，作为技术指导者，针对反恐防爆重大需求，创新研发柔性复合防爆技术，形成特色显著的系列化产品，应用于机场、车站、港口等重要场所。2020 年获公安部科技一等奖，第二获奖人

冯顺山教授是国内首批爆破高级工程师，1994—2000 年，应对交通部门解决大型沉船阻碍航道的清障难题，创新水下爆破沉船技术，突破了水下爆破清障、爆破解体、爆炸开舱和爆破安全等关键技术，成功应用于工程。在万山群岛、川山群岛、渤海湾、北海湾和长江等水域清除了 6 条大型沉船，工程通过港监和甲方的验收

冯顺山教授学术交流部分照片

（1990—2022 年 9 月）

1992 年赴瑞典国参加“国际弹道学术交流会”

1993 年在瑞典与有关公司洽谈科技合作

1995 年赴美国参加“冲击波物理国际会议”
与美国国家实验室科技人员洽谈合作事宜

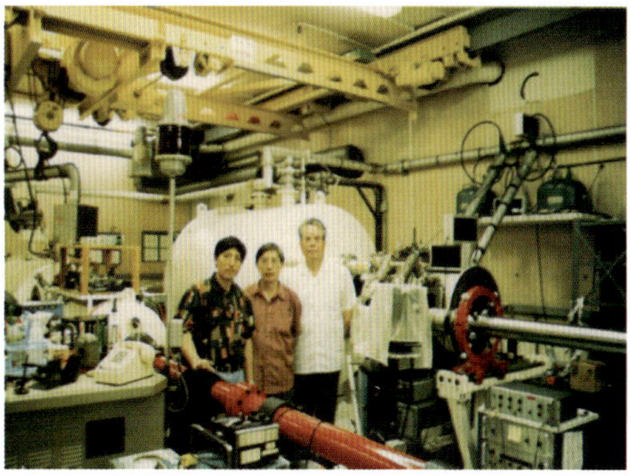

1995 年访问美国加州理工大学

1996 年作为中国青年科学家代表团成员赴美国
学术交流，访问美国耶鲁大学等学术机构

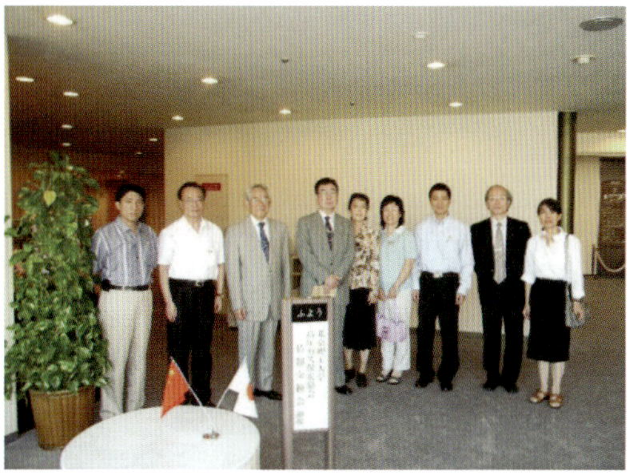

2006 年赴日本安全技术研究所学术
交流和洽谈科技合作

2006 年赴澳大利亚学术访问

2009 年赴越南洽谈高新科技合作

1991 年俄罗斯包曼大学弹药专业
教授在本学科讲学和学术交流

2009 年西班牙马德里大学教授参观
实验室并开展学术交流

2013 年冯顺山教授与英国剑桥大学两位
教授在本专业实验室研究试验情况

2014 年吉林大学黄大年教授来冯顺山团队
开展水下武器弹药前沿技术学术交流

冯顺山教授作为“战斗部与毁伤技术”（2003—2022 年）和“爆炸与安全技术”（2002—2017 年）专业委员会主任委员，主持学术会议并多次作大会学术报告

冯顺山教授与所指导的部分研究生

（1996 年～2022 年 9 月退休）

2000 年，郭文章博士后出站（第一个博士后）

2005 年，何玉彬博士后出站

2007 年，曹红松博士后出站
2009 年，刘春美博士后出站（联合指导）

2012 年，李淑华博士后出站（联合指导）

2015 年，石光明博士后出站（左一）

2022 年，黄岐博士后出站（联合指导）

1996 年，王海福通过答辩获得博士学位（第一个博士生）

1998 年，彭昆雅通过答辩获得博士学位

1998 年，孙韬通过答辩获得博士学位

1998 年，孙为国通过答辩获得博士学位

1999 年，孔炜通过答辩获得博士学位

2001 年，周睿通过答辩获得博士学位

2001 年，贵大勇通过答辩获得博士学位

2001 年，王芳通过答辩获得博士学位

2003 年，贾光辉通过答辩获得博士学位

2004 年，蒋建伟通过答辩获得博士学位

2004 年，吴成通过答辩获得博士学位

2005 年，周玲通过答辩获得博士学位

2005 年，梁永直通过答辩获得博士学位

2006 年，王震宇通过答辩获得博士学位

2006 年，王云峰通过答辩获得博士学位

2007 年，刘春美、李国杰、黄龙华通过答辩获得博士学位

2008 年，张旭荣通过答辩获得博士学位

2008 年，董三强通过答辩获得博士学位

2010 年，吴广通过答辩获得博士学位

2010 年，陈赟和王刚通过答辩获得博士学位

2011 年，与部分博士毕业生合影

（左起：李守苍博士、张国伟博士、徐立新博士、冯顺山教授、罗健博士、黎春林博士、贵大勇博士）

2011 年，与部分博士毕业生合影

（左起：李磊博士、王玥博士、冯顺山教授、王云峰博士、黄广炎博士）

2011 年，王宇通过答辩获得博士学位

2011 年，赵雪通过答辩获得博士学位

2011 年，完颜振海通过答辩获得博士学位

2011 年，李砚东、赖鸣通过答辩获得博士学位

2013 年，张会锁、孙昊通过答辩获得博士学位

2015 年，边江楠通过答辩获得博士学位

2015 年，张晓东通过答辩获得博士学位

2015 年，李顺平、周彤通过答辩获得博士学位

2015 年，徐翔云通过答辩获得博士学位

2015 年，杜烨通过答辩获得博士学位

2017 年，张广华、李伟通过答辩获得博士学位

2017 年，方晶通过答辩获得博士学位

2017 年，聂为彪通过答辩获得博士学位

2018 年，白洋通过答辩获得博士学位

2018 年，王成龙通过答辩获得博士学位

2019 年，成曦通过答辩获得博士学位

2019 年，赵宇峰通过答辩获得博士学位

2020 年，兰旭柯、黄岐、陈超南通过答辩获得博士学位

2022 年，伍思宇通过线上答辩获得博士学位

2005 年，冯高鹏、于颖贤、高磊通过答辩获得硕士学位
2006 年，杨彬通过答辩获得硕士学位

2006 年，崔晓刚通过答辩获得硕士学位

2006 年，吴俊斌通过答辩获得硕士学位

2007 年，刘晶、刘畅通过答辩获得硕士学位

2008 年，杨宁、景建清通过答辩获得硕士学位

2009 年，龙骁、赵黎、周元通过答辩获得硕士学位

2016 年，孙凯通过答辩获得硕士学位

2016 年，吴丹通过答辩获得硕士学位

2022 年，刘云辉通过答辩获得硕士学位

2010 年，中北大学李学林通过答辩获得博士学位

2021 年，中北大学刘鹏飞、任凯通过答辩获得博士学位

2007 年 7 月，师生参观军博武器装备展

左起第1排：张广华、宋卿、李伟、聂为彪、闵冬冬、方晶、徐晓松、曹琦、邹浩、赵宇峰、万九亮；第2排：王海福、孙韬、彭昆雅、徐云、冯顺山、李娉、何玉彬、吴成、周睿、王振兴；第3排：贾光辉、罗健、李学林、李守苍、徐立新、张国伟、贵大勇、黎春林、王云峰；第4排：梁永直、冯高鹏、于颖贤、高磊、李磊、刘春美、董永香、曹红松、杜烨、王杏花、张会锁、王玥、边江楠、李砚东、完颜振海、伍鹰；第5排：杨宁、方东洋、邵志宇、赖鸣、孙昊；第6排：黄广炎、吴俊斌、金俊、李顺平、赵雪、陈赟、冯源、王超、王芳、王宇、徐翔云、段相杰、肖李兴

2011 年 10 月，部分毕业及在校研究生（本学科组）聚会合影

2019 年 2 月，喜得孙女小双双（未来研究生），晋升爷爷

2021 年 12 月，冯顺山教授和部分在京硕士 / 博士毕业生聚会合影
（一排左起：蒋建伟、徐云、冯顺山、王海福、彭昆雅；二排左起：董永香、王芳、孙韬、周睿、李磊；
三排左起：黄广炎、余庆波、徐翔云、李国杰、王玥、刘海峰、冯源、陈赟、崔晓刚、边江楠）

2022 年 6 月，伍思宇获得博士学位、刘云辉获得硕士学位，和实验室四位在读生合影
（左起：胡文龙硕士生，张博博士生，刘云辉硕士，伍思宇博士，高月光博士生，肖翔博士生）